Applied Science

Applied Science

Technology

Editor

Donald R. Franceschetti, Ph.D.

The University of Memphis

SALEM PRESS

A Division of EBSCO Information Services, Inc.

Ipswich, Massachusetts

GREY HOUSE PUBLISHING

∞ The paper used in these volumes conforms to the American National Standard for Permanence of Paper for Printed Library Materials, Z39.48-1992 (R1997).

Library of Congress Cataloging-in-Publication Data

Publisher's Cataloging-In-Publication Data
(Prepared by The Donohue Group, Inc.)

Applied science. Technology / editor, Donald R. Franceschetti.
-- [1st ed.].

p. : ill. ; cm.

Comprises articles on technology extracted from the five-volume reference set Applied Science. Salem Press, 2012.
Includes bibliographical references and index.
ISBN: 978-1-61925-242-4

1. Technology. 2. Science. I. Franceschetti, Donald R., 1947- II. Title: Technology

T45 .A672 2013
600

PRINTED IN THE UNITED STATES OF AMERICA

CONTENTS

PUBLISHER'S NOTE

Applied Science: Technology is a comprehensive volume that examines the many ways in which technology affects daily life. Drawn from the technology articles in Salem's five-volume *Applied Science* reference set, this single volume is designed to help those specifically planning for a career in technology understand the interconnectedness of the different and varied branches of this field. Toward that end, essays look beyond basic principles to examine a wide range of topics, including industrial and business applications, historical and social contexts, and the impact a particular field of science or technology will have on future jobs and careers. A career-oriented feature details core jobs within the field, with a focus on fundamental and recommended coursework.

Applied Science: Technology is specifically designed for a high-school audience and is edited to align with secondary or high-school curriculum standards. The content is readily accessible, as well, to patrons of both academic and university libraries. Librarians and general readers alike will also turn to this reference work for both basic information and current developments, from nanotechnology to biofuels and synthetic fuels, presented in accessible language with copious reference aids. Pedagogical tools and elements, including a bibliographical directory of scientists, a timeline of major scientific milestones, and a glossary of key terms and concepts, round out this unprecedented and unique resource.

Scope of Coverage

Comprising 99 lengthy and alphabetically arranged essays on a broad range of technology subjects, this excellent reference work addresses technology fields as long-established as anthropology and glassmaking, and as cutting-edge as cryogenics and telemedicine. You will find entries in areas as diverse as drug testing, jet engine technology, photography, and plant breeding, as well as a multitude of entries that fall under the emerging fields of alternative fuel and wireless technologies. *Applied Science: Technology* also includes charts and diagrams, as well as "Fascinating Facts" related to each applied-science field.

Essay Length and Format

Essays in this volume range in length from 3,000 to 4,000 words. All the entries begin with ready-reference top matter, including an indication of the relevant discipline or field of study and a summary statement in which the writer explains why the topic is important to the study of the applied sciences. A selection of key terms and concepts within that major field or technology is presented, with basic principles examined. Essays then place the subject field in historical or technological perspective and examine its development and implication. Also discussed are applications and applicable products and the field's impact on industry. Cross-references direct readers to related topics, and further reading suggestions accompany all articles.

Special Features

Several features continue to distinguish this reference series from other works on applied science. The front matter includes a table detailing common units of measure, with equivalent units of measure provided for the user's convenience. Additional features are provided to assist in the retrieval of information. An index illustrates the breadth of the reference work's coverage, directing the reader to appearances of important names, terms, and topics throughout the text.

The back matter includes several appendixes and indexes, including a biographical dictionary of scientists, a compendium of the people most influential in shaping the discoveries and technologies of applied science. A time line provides a chronology of all major scientific events across significant fields and technologies, from agriculture to computers to engineering to medical to space sciences. Additionally, a complete glossary covers all technical and other specialized terms used throughout the set, while a general bibliography offers a comprehensive list of works on applied science for students seeking out more information on a particular field or subject.

Acknowledgments

Many hands went into the creation of this work. Special mention must be made of editor Donald R.

Franceschetti, who played a principal role in shaping this volume. Thanks are also due to the many academicians and professionals who worked to communicate their expert understanding of the applied sciences to the general reader; a list of these individuals and their affiliations appears at the beginning of the volume. The contributions of all are gratefully acknowledged.

EDITOR'S INTRODUCTION: TECHNOLOGY

This volume presents the technology articles from the five-volume reference set *Applied Science,* for the benefit of students considering careers in technology, their teachers, and their counselors. Other volumes cover science and medicine, and engineering and mathematics.

Something should be said at the outset about the relationship between science and technology. Terence Kealey states in his book *The Economic Laws of Scientific Research* (1996) that technology is the activity of manipulating nature, while science is the activity of learning about nature. For much of history, the two realms were relatively separate, technologies being developed by artisans, craftsmen and farmers and the sciences by natural philosophers. Leonard Mlodinow, in *Feynman's Rainbow* (2003) makes a distinction between the Greek way of approaching science and the Babylonian way. The Greeks, the great mathematicians of the ancient world, greatly admired pure thought. The Babylonians, technologists at heart, didn't care much about fine theoretical points but made important practical discoveries. The two realms began coming together at the time of the Industrial Revolution. In England the process was tied to the emergence of the Royal Institution, founded in 1799, which made it possible for the general public to learn about technical matters in their spare time and without the training in classical languages and social standing expected at the universities of Oxford and Cambridge. At about the same time, these august institutions were joined by the "red brick" institutions, which emphasized the training of students for industrial leadership. In the United States, the venerable Ivy League schools, including Harvard, Yale, and Princeton, were joined by the newly established state universities. Some of these were designated as land-grant colleges under the Morrill Act (1862) and were set up to advance agriculture and the mechanical arts. The Morrill Act provided that each state in the Union designate a parcel of land, income from the sale of which would be used to support public colleges and provide for agricultural experimental stations where new ideas in agriculture could be tested.

Educational practice is sometimes slow to recognize changing patterns in society. Colleges in the United States still award bachelor's, master's and doctoral degrees that have their roots in the Middle Ages. In fact the Bachelor of Arts, Master of Arts, and the Doctor of Philosophy degrees are still awarded by the majority of American universities, although the coursework required and the major subjects offered differ greatly from those of a century ago. The bachelor's degree is still awarded to individuals who have completed a four-year course of study, though the emphasis may be on computer science or psycholinguistics instead of the liberal arts. Many schools now award the Bachelor of Science degree to graduates with majors in the sciences (although some, particularly in the Ivy League, award the traditional B.A. degree to all graduates). Additional study is required for the master's degree, and a period of intensive research and the publication of a dissertation is required for the Doctor of Philosophy degree.

Despite the antiquity of the bachelor's, master's and doctoral designations, what the degrees actually meant was, for a time, quite flexible in early-twentieth-century America. Eventually a measure of quality control was achieved, with colleges being chartered by state governments and subject to review by regional accrediting agencies and, in some fields, by additional specialized agencies.

To meet the needs of rapidly growing economy, a number of credentials below that of the bachelor's degree had to be introduced. In modern usage a technologist is someone who has completed a program of study (usually four years in length) and been awarded a bachelor's degree in some field of technology. Technology programs may be distinguished from engineering programs by their mathematics requirement. Programs leading to a bachelor's degree in, say, electrical engineering usually include at least three semesters of calculus and one in differential equations before the Bachelor of Science in Electrical Engineering (B.S.E.E.) is awarded. In contrast, programs leading to a bachelor's degree in electrical engineering technology would require a single introductory course in calculus. In the United States, both types of programs receive accreditation from the American Board for Engineering and Technology (ABET), which is responsible for maintaining standards of instruction in those fields.

Many workers in technology areas do not, in fact,

have bachelor's degrees. These workers are generally referred to as technicians rather than technologists. A two-year program in technology at a community college or technical institute could lead to an Associate of Applied Science (A.A.S.) degree. There is also the possibility of various certificates attesting to technical competence, especially in rapidly changing computer-related fields. Further routes toward technical training are provided by the U.S. Armed Forces, and many a technically-trained soldier or sailor moves from a few years of enlistment to a good job in industry. Many large school systems include vocational-technical schools where high school- and college-age students can study such subjects as air-conditioning technology or computer programming.

TO THE STUDENT

Your decision to enter a technology-based field is one that, with persistence, will lead to a comfortable income and, more importantly, a variety of interesting work assignments. While in junior high and high school, be sure to practice writing and organizing presentations. If possible do not interrupt your study of mathematics, as mathematics courses build on each other. Resist the temptation to take an easy semester. Start or join a technology club at your school. Talk with people who are working in jobs you think you might enjoy. Visit nearby sites that employ technicians or technologists. Learn as much as you can before you commit to a particular career track.

As a technology-minded student, you are likely to be affected by two recent trends. One is the recent growth in on-line education. Although the widespread availability of computers and the Internet make it possible for students almost anywhere to study almost anything, there is still much to be said for the pre-professional socialization that goes on in college. For science majors, needing to keep to a schedule of assignments, working collaboratively with other students, developing presentation skills, and finding information by research are all established aspects of college life. The second trend is the increasing importance of the adult student and the need for continuing education. While there are still individuals who work at the same job for 30 or more years, it is likely that you will change jobs several times over the course of your career, and you can count on needing periodic retraining.

A BRIEF SURVEY OF TECHNOLOGIES

A clear distinction between science and technology is not possible. As scientific knowledge developed, it was occasionally put into the service of technology, particularly in the nineteenth- and twentieth century. In modern times, advances in scientific knowledge have often stimulated entirely new technologies. When American inventor Thomas Alva Edison sought a long-lasting filament for the first electric light, he proceeded mainly by trial and error. His notebooks report numerous attempts with different filament materials. Edison, however, did proceed with a background knowledge of fundamental science. He understood that most materials, when heated in air, would react with the oxygen to form an oxide, and therefore he focused on filaments carrying current in either a vacuum or an inert atmosphere. In 1883 Edison stumbled on the "Edison effect"—the fact that electric current would flow from a heated metal filament across an evacuated space to a positively charged conductor. Edison recorded the effect in his notebooks, but seeing no application for it, he did not pursue it. In 1897 the English physicist Joseph John Thomson discovered the electron, and it soon became apparent that the Edison effect was simply the emission of electrons from a hot metal surface. By 1904 English physicist John Ambrose Fleming had invented the vacuum-tube diode, and in 1906 the American inventor Lee de Forest showed how incorporating a third electrode would allow a very small voltage to control a much larger current, the effect that made radio and television possible.

At the close of World War II, there was considerable debate as to the proper relationship between science and government in peacetime. Some conservatives wanted to impose strong military control over scientific research. In 1945 Vannevar Bush, who had been director of the wartime Office of Scientific Research and Development, published the report "Science: The Endless Frontier," which painted a very rosy picture of the gains to be derived by public investment in basic scientific research. Among other things, Bush emphasized the role that government could play in supporting research in universities, research to fill in the middle ground between purely academic research and research directed toward immediate objectives. Further, universities had a natural role to play in training the next generation of scientists and engineers. Bush's arguments led to establishment of the National Science Foundation and to expanded research funding within and by the National Institutes of Health in the United States. This trend was accelerated by the Soviet Union's launch of Sputnik 1, the first Earth satellite, in 1957, an event that shook the American public's faith in the inevitable superiority of American technology. In the twenty-first century, science and technology are supported by many sources, public and private, and the pace of development is perhaps even greater than at any time in the past.

The Preliterate World

According to archaeologists and anthropologists, the development of written language occurred relatively recently in human history, perhaps about 3000 B.C.E. Many of the basic components of technology date to those preliterate times, when humans struggled to secure the basic necessities—food, clothing, and shelter—against a background of growing population and changing climate. When food collection was limited to hunting and gathering, knowledge of the seasons and animal behavior was important for survival. The development of primitive stone tools and weapons greatly facilitated both hunting and obtaining meat from animal carcasses, as well as the preservation of the hides for clothing and shelter. Sometime in the middle Stone Age, humans obtained control over fire, making it possible to soften food by cooking and to separate some metals from their ores. Control over fire also made it possible to harden earthenware pottery and keep predatory animals away at night.

With gradually improved living conditions, human fertility and longevity both increased, as did competition for necessities of life. Spoken language, music, magical thinking, and myth developed as a means of coordinating activity. Warfare, along with more peaceful approaches, was adopted as a means of settling disputes, while society was reorganized to ensure access to the necessities of life, including protection from military attack.

The Ancient World

With the invention of written language, it became possible to enlarge and coordinate human activity on an unprecedented scale. Several new areas of technology and engineering were needed. Cities were established so that skilled workers could be freed from direct involvement in food production. Logistics and management became functions of the scribal class, the members of which could read and write. Libraries were built and manuscripts collected. The beginnings of mathematics may be seen in building, surveying, and wealth tabulation. Engineers built roads so that a ruler could oversee his enlarged domain and troops could move rapidly to where they were needed. Taxes were imposed to support the central government, and accounting methods were introduced. Aqueducts were needed to bring fresh water to the cities.

Astronomy, the Calendar, and Longitude

One might think that astronomy could serve as a paradigmatic example of fundamental science, but astronomy—and later space science—provides an example of a scientific activity often pursued for practical benefit. Efforts to fix the calendar, in particular, provide an interesting illustration of the interaction among pure astronomy, practicality and religion.

The calendar provided a scheme of dates for planting and harvesting. Ancient stone monuments in Europe and the Americas may have functioned in part as astronomical observatories, to keep track of the solstices and equinoxes. The calendar also provided a means for keeping religious feasts, such as Easter, in synchrony with celestial events, such as the vernal equinox. Roman emperor Julius Caesar introduced a calendar based on a year of 365 days. This calendar, called the Julian, would endure for more than 1,500 years. Eventually, however, it was realized that the Julian year was about six hours shorter than Earth's orbital period and so was out of sync with astronomical events. In 1583 Pope Gregory XIII introduced the system of leap-year days. The Gregorian calendar is out of sync with Earth's orbit by only one day in 3,300 years.

Modern astronomy begins with the work of Nicolaus Copernicus, Galileo Galilei, and Sir Isaac Newton. Copernicus was the first to advance the heliocentric model of the solar system, suggesting that it would be simpler to view Earth as a planet orbiting a stationary sun. Copernicus, however, was a churchman, and fear of opposition from church leaders (who were theologically invested in an Earth-centered hierarchy of the universe) led him to postpone publication of his ideas until the year of his death. Galileo was the first to use a technological advance, the invention of the telescope, to record numerous observations that called into question the geocentric model of the known universe. Among these was the discovery of four moons that orbit the planet Jupiter.

At the same time, advances in shipbuilding and navigation served to bring the problem of longitude to prominence. Out of sight of land, a sailing ship could easily determine its latitude on a clear night by noting the elevation of the pole star above the horizon. To determine longitude, however, required an accurate measurement of time, which was difficult to do onboard a moving ship at sea. Galileo was quick to propose that occultations by Jupiter of its moons could be used as a universal clock. The longitude problem also drew the attention of the great Isaac Newton. Longitude would eventually be solved by John Harrison's invention of a chronometer that could be used on ship, and the deciphering of it also fostered an improved understanding of celestial mechanics. John Harrison was a carpenter, an early technologist.

Since the launching of the first artificial satellites, astronomy—or, rather, space and planetary science—has assumed an even greater role in applied science. The safety of astronauts working in space requires understanding the dynamics of solar flares. A deeper understanding of the solar atmosphere and its dynamics could also have important consequences for long-range weather prediction.

The Scientific Revolution

The Renaissance and the Protestant Reformation marked something of a rebirth of scientific thinking. This "scientific revolution" would not have been possible without Gutenberg's printing press and the technology of printing with movable type. With wealthy patrons, natural philosophers felt secure in challenging the authority of Aristotle. Galileo published arguments in favor of the Copernican solar system. In the *Novum Organum* (1620; "New Instrument"), Sir Francis Bacon formalized the inductive method, by which generalizations could be made from observations, which then could be tested by further observation or experiment. In England in 1660, with the nominal support of the British Crown, the Royal Society was formed to serve as a forum for the exchange of scientific ideas and the support and publication of research results. Need for larger-scale studies brought craftsmen into the sciences, culminating in the recognition of the professional scientist. Earlier Bacon had proposed that the government undertake the support of scientific investigation for the common good. Bacon himself tried his hand at frozen-food technology. While on a coach trip, he conceived the idea that low temperatures could preserve meat. He stopped the coach, purchased a chicken from a farmer's wife, and stuffed it with snow. Unfortunately he contracted pneumonia while doing this experiment and died forthwith.

The Industrial Revolution followed on the heels of the scientific revolution in England. Key to the Industrial Revolution was the technology of the steam engine, the first portable source of motive power that was not dependent on human or animal muscle. The modern form of the steam engine owes much to James Watt, a self-taught technologist. The steam engine powered factories, ships, and, later, locomotives. In the case of the steam engine, technological advance preceded the development of the pertinent science—thermodynamics and the present-day understanding of heat as a random molecular form of energy.

It is not possible, of course, to do justice to the full scale of applied science and technology in this short space. In the remainder of this introduction, consideration will be given to only a few representative fields, highlighting the evolution of each area and its interconnectedness with fundamental and applied science as a whole.

Chemical Technology

In 1792 the Scottish inventor William Murdock discovered a way to produce illuminating gas by the destructive distillation of coal, producing a cleaner and more dependable source of light than previously was available and bringing about the gaslight era. The production of illuminating gas, however, left behind a nasty residue called coal tar. A search was launched to find an application for this major industrial waste. An early use, the waterproofing of cloth, was discovered by the Scottish chemist Charles Macintosh, resulting in the raincoat that now carries his name. In 1856 English chemist William Henry Perkin discovered the first of the coal-tar dyes, mauve. The color mauve, a deep lavender-lilac purple, had previously been obtained from plant sources and had become something of a fashion fad in Paris by 1857. The fad spread to London in 1858, when Queen Victoria chose to wear a mauve velvet dress to her daughter's wedding. The demand for mauve outstripped the supply of vegetable sources. The discovery of several other dyes followed.

The possibility of dyeing living tissue was rapidly seized on and applied to the tissues of the human body and the microorganisms that afflict it. German bacteriologist Paul Ehrlich proposed that the selective adsorption of dyes could serve as the basis for a chemically based therapy to kill infectious disease-bearing organisms.

Optical Technology, the Microscope and Microbiology

The use of lenses as an aid to vision may date to China in 500 B.C.E. Marco Polo, in his journeys more than 1,700 years later, reported seeing many Chinese wearing eyeglasses. In 1665 English physicist (and curator of experiments for the Royal Society) Robert Hooke published his book *Micrographia* ("Tiny Handwriting"), which included many illustrations of living tissue. Antonie van Leeuwenhoek was influenced by Hooke, and he reported many observations of microbial life to the Royal Society. The simple microscopes of Hooke and van Leeuwenhoek suffered from many forms of aberration or distortion. Subsequent investigators introduced combinations of lenses to reduce the aberrations, and good compound microscopes became available for the study of microscopic life around 1830.

While van Leeuwenhoek had reported the existence of microorganisms, the notion that they might be responsible for disease or agricultural problems met considerable resistance. Louis Pasteur, an accomplished physical chemist, became best known as the father of microbiology. Pasteur was drawn into applied research by problems arising in the fermentation industry. In 1857 he announced that fermentation was the result of microbial action. He also showed that the souring of milk resulted from microorganisms, leading to the development of "pasteurization"—heating to a certain temperature for a specific amount of time— as a technique for preserving milk. As a sequel to his work on fermentation, Pasteur brought into question the commonly held idea that living organisms could generate spontaneously. Through carefully designed experiments, he demonstrated that broth could be maintained sterile indefinitely, even when exposed to the air, provided that bacteria-carrying dust was excluded.

Pasteur's further research included investigating the diseases that plagued the French silk industry. He also developed a means of vaccinating sheep against infection by Bacillus anthracis and a vaccine to protect chickens against cholera. Pasteur's most impressive achievement may have been the development of a treatment effective against the rabies virus for people bitten by rabid dogs or wolves.

Pasteur's scientific achievement illustrates the close interplay of fundamental and applied advances that occurs in many scientific fields. Political scientist Donald Stokes has termed this arena of application-driven scientific research "Pasteur's Quadrant," to distinguish it from purely curiosity-driven research (as in modern particle physics); advance by trial and error (for example, Edison's early work on the electric light); and the simple cataloging of properties and behaviors (as in classical botany and zoology). The study of applied science is a detailed examination of Pasteur's Quadrant.

Electromagnetic Technology

The history of electromagnetic devices provides an excellent example of the complex interplay of fundamental and applied science. The phenomena of static electricity and natural magnetism were described by Thales of Miletus in ancient times, but they remained curiosities through much of history. The magnetic compass was developed by Chinese explorers in about 1100 b.c.e., and the nature of Earth's magnetic field was explored by William Gilbert (physician to Queen Elizabeth I) around 1600. By the late eighteenth century, a number of devices for producing and storing static electricity were being used in popular demonstrations, and the lightning rod, invented by Benjamin Franklin, greatly reduced the damage due to lightning strikes on tall buildings. In 1800 Italian physicist Alessandro Volta developed the first electrical battery. Equipped with a source of continuous electric current, scientists made electrical and electromagnetic discoveries, practical and fundamental, at a breakneck pace.

The voltaic pile, or battery, was employed by British scientist Sir Humphry Davy to isolate a number of chemical elements for the first time. In 1820 Danish physicist Hans Christian Ørsted discovered that any current-carrying wire is surrounded by an electric field. In 1831 English physicist Michael Faraday discovered that a changing magnetic field would induce an electric current in a loop of wire, thus paving the way for the electric generator and the transformer. In Albany, New York, schoolteacher Joseph Henry set his students the challenge of building the strongest possible electromagnet. Henry would move on to become professor of natural philosophy at Princeton University, where he invented a primitive telegraph.

The basic laws of electromagnetism were summarized in 1865 by Scottish physicist James Clerk Maxwell in a set of four differential equations that yielded a number of practical results almost immediately. These equations described the behavior of electric and magnetic fields in different media, including in empty space. In a vacuum it was possible to find wavelike solutions that appeared to move in time at the speed of light, which was immediately realized to be a form of electromagnetic radiation. Further, it turned out that visible light covered only a small frequency range. Applied scientists soon discovered how to transmit messages by radio waves, electromagnetic waves of much lower frequency.

The Computer

One of the most clearly useful of modern artifacts, the digital electronic computer, as it has come to be known, has a lineage that includes the most abstract of mathematics, the automated loom, the vacuum tubes of the early twentieth century, and the modern sciences of semiconductor physics and photochemistry. Although computing devices such as the abacus and slide rule themselves have long histories, the programmable digital computer has advanced computational power by many orders of magnitude. The basic logic of the computer and the computer program, however, arose from a mathematical logician's attempt to answer a problem arising in the foundations of mathematics.

From the time of the ancient Greeks to the end of the nineteenth century, mathematicians had assumed that their subject was essentially a study of the real world, the part amenable to purely deductive reasoning. This included the structure of space and the basic rules of counting, which led to the rules of arithmetic and algebra. With the discovery of non-Euclidean geometries and the paradoxes of set theory, mathematicians felt the need for a closer study of the foundations of mathematics, to make sure that the objects that might exist only in their minds could be studied and talked about without risking inconsistency.

David Hilbert, a professor of mathematics at the University of Göttingen, was the recognized leader of German mathematics. At a mathematics conference in 1928, Hilbert identified three questions about the foundations of mathematics that he hoped would be resolved in short order. The third of these was the so-called decidability problem: Was there was a foolproof procedure to determine whether a mathematical statement was true or false? Essentially, if one had the statement in symbolic form, was there a procedure for manipulating the symbols in such a way that one could determine whether the statement was true in a finite number of steps?

British mathematician Alan Turing presented an analysis of the problem by showing that any sort of mathematical symbol manipulation was in essence a computation and thus a manipulation of symbols not unlike the addition or multiplication one learns in elementary school. Any such symbolic manipulation could be emulated by an abstract machine that worked with a finite set of symbols that would store a simple set of instructions and process a one-dimensional array of symbols, replacing it with a second array of symbols. Turing showed that there was no solution in general to Hilbert's decision problem, but in the process he also showed how to construct a machine (now called a Turing machine) that could execute any possible calculation. The machine would operate on a string of symbols recorded on a tape and would output the result of the same calculation on the same tape. Further, Turing showed the existence of machines that could read instructions given in symbolic form and then perform any desired computation on a one-dimensional array of numbers that followed. The universal Turing machine was a programmable digital computer. The instructions could be read from a one-dimensional tape, a magnetically stored memory, or a card punched with holes, as was used for mechanized weaving of fabric.

The earliest electronic computers were developed at the time of World War II and involved numerous vacuum tubes. Since vacuum tubes are based on thermionic emission—the Edison effect mentioned above—they produced immense amounts of heat and involved the possibility that the heating element in one of the tubes might well burn out during the computation. In fact, it was standard procedure to run a program, one that required proper function of all the vacuum tubes, both before and after the program of interest. If the results of the first and last computations did not vary, one could assume that no tubes had burned out in the meantime.

World War II ended in 1945. In addition to the critical role of computing machines in the design of the first atomic bombs, computational science had played an important role in predicting the behavior of targets. The capabilities of computing machines would grow rapidly following the invention of the transistor by John Bardeen, Walter Brattain and William Shockley in 1947. In this case, fundamental science led to tremendous advances in applied science.

The story of semiconductor science is worth telling. Silicon is unusual in displaying an increase in electrical conductivity as the temperature is raised. In general, when one finds an interesting property of a material, one tries to purify and refine the material. Purified silicon, however, lost most of its conductivity. On further investigation, it was found that tiny concentrations of impurities could vastly change both the amount of electrical conductivity and the mechanism by which it occurs. Because the useful properties of semiconductors depend critically on the impurities, or "dirt," in the material, solid-state (and other) physicists sometimes refer to the field as "dirt physics." Adding a small amount of phosphorus to pure silicon resulted in n-type conductivity, the type due to electrons moving in response to an electric field. Adding an impurity such as boron produced p-type conductivity, in which electron vacancies (in chemical bonds) moved through the material. Creating a p-type region next to an n-type produced a junction that let current flow in one direction and not the other, just as in a vacuum-tube diode. Placing a p-type region between two n-types produced the equivalent of Lee de Forest's diode—a transistor. The transistor, however, did not require a heater and could be miniaturized.

The 1960s saw the production of integrated circuits—many transistors and other circuit elements on a single silicon wafer, or chip. Currently hundreds of thousands of circuit elements are available on a single chip, and anyone who buys a laptop computer will command more computational power than any government could control in 1950.

Donald R. Franceschetti

Further Reading

Bell Telephone Laboratories. A History of Engineering and Science in the Bell System: Electronics Technology, 1925–1975. Ed. by M. D. Fagen. 7 vols. New York: Bell Laboratories, 1975–1985. Provides detailed information on the development of the transistor and the integrated circuit.

Bodanis, David. Electric Universe: How Electricity Switched On the Modern World. New York: Three Rivers Press, 2005. Popular exposition of the applications of electronics and electromagnetism from the time of Joseph Henry to the microprocessor age.

Burke, James. Connections. Boston: Little Brown, 1978; reprint, New York: Simon & Schuster, 2007. Describes linkages among inventions throughout history.

Cobb, Cathy, and Harold Goldwhite. Creations of Fire: Chemistry's Lively History from Alchemy to the Atomic Age. New York: Plenum Press, 1995; reprint, Cambridge, Mass.: Perseus, 2001. History of pure and applied chemistry from the beginning through the late twentieth century.

Garfield, Simon. Mauve: How One Man Invented a Color That Changed the World. New York: W. W. Norton, 2000. Focuses on how the single and partly accidental discovery of coal tar dyes led to several new areas of chemical industry.

Kealey, Terence. The Economic Laws of Scientific Research. New York: St. Martin's Press, 1996. Makes the case that government funding of scientific research is relatively inefficient and emphasizes the role of private investment and hobbyist scientists. Mlodinow, Leonard, Feynman's Rainbow: A Search for Beauty in Physics and in Life. New York: Warner Books, 2003; reprint, New York: Vintage Books, 2011.

Schlager, Neil, ed. Science and Its Times: Understanding the Social Significance of Scientific Discovery. 8 vols. Detroit: Gale Group, 2000–2001. Massive reference work on the impact of scientific and technological developments from the earliest times to the present.

Sobel, Dava. Longitude. New York: Walker & Company, 2007. Story of the competition among scientists and inventors to develop a reliable means of determining longitude at sea.

Stokes, Donald E. Pasteur's Quadrant: Basic Science and Technological Innovation. Washington, D.C.:

Brookings Institution Press, 1997. Presents an extended argument that many fundamental scientific discoveries originate in application-driven research and that the distinction between pure and applied science is not, of itself, very useful.

CONTRIBUTORS

Jeongmin Ahn
Syracuse University

Raymond D. Benge
Tarrant County College

Lakhdar Boukerrou
Florida Atlantic University

Michael A. Buratovich
Spring Arbor University

Byron D. Cannon
University of Utah

Edward N. Clarke
Worcester Polytechnic Institute

Christopher Dean
Curtin University of Technology, Perth, Australia

Joseph Dewey
University of Pittsburgh

Ronald J. Ferrara
Middle Tennessee State University, Murfreesboro

Glenda Griffin
Newton Gresham Library, Sam Houston State University

Gina Hagler
Washington, D.C.

Wendy C. Hamblet
North Carolina Agricultural and Technical State University

Howard V. Hendrix
California State University, Fresno

Robert M. Hordon
Rutgers University

Ngaio Hotte
Vancouver, British Columbia

Micah L. Issitt
Philadelphia, Pennsylvania

Jerome A. Jackson
Florida Gulf Coast University

Bruce E. Johansen
University of Nebraska at Omaha

Marylane Wade Koch
Loewenberg School of Nursing, University of Memphis

Narayanan M. Komerath
Georgia Institute of Technology

Jeanne L. Kuhler
Benedictine University

Dawn A. Laney
Atlanta, Georgia

M. Lee
Independent Scholar

Donald W. Lovejoy
Palm Beach Atlantic University

Sergei A. Markov
Austin Peay State University

Julia M. Meyers
Duquesne University

Randall L. Milstein
Oregon State University

M. Mustoe
Eastern Oregon University

Terrence R. Nathan
University of California, Davis

David Olle
Eastshire Communications

Robert J. Paradowski
Rochester Institute of Technology

Ellen E. Anderson Penno
Western Laser Eye Associates

George R. Plitnik
Frostburg State University

Michael L. Qualls
Fort Valley State University

Corie Ralston
Lawrence Berkeley National Laboratory

Steven J. Ramold
Eastern Michigan University

Richard M. J. Renneboog
Independent Scholar

Barbara J. Rich
Bend, Oregon

James L. Robinson
University of Illinois

Charles W. Rogers
Southwestern Oklahoma State University

Carol A. Rolf
Rivier College

Lars Rose
Author, The Nature of Matter

Charles Rosenberg
Milwaukee, Wisconsin

Julia A. Rosenthal
Chicago, Illinois

Joseph R. Rudolph
Towson University

Elizabeth D. Schafer
Loachapoka, Alaska

Martha A. Sherwood
Kent Anderson Law Office

Ruth Waddell Smith
Michigan State University

Martin V. Stewart
Middle Tennessee State University

Robert E. Stoffels
St. Petersburg, Florida

Bethany Thivierge
Technicality Resources

Christine Watts
University of Sydney

George M. Whitson III
University of Texas at Tyler

Edwin G. Wiggins
Webb Institute, Glen Cove, New York

Thomas A. Wikle
Oklahoma State University

Bradley R. A. Wilson
University of Cincinnat

Barbara Woldin
American Medical Writers Association

Robin L. Wulffson
Faculty, American College of Obstetrics and Gynecology

Susan M. Zneimer
U.S. Labs, Irvine, California

COMMON UNITS OF MEASURE

Common prefixes for metric units—which may apply in more cases than shown below—include *giga-* (1 billion times the unit), *mega-* (one million times), *kilo-* (1,000 times), *hecto-* (100 times), *deka-* (10 times), *deci-* (0.1 times, or one tenth), *centi-* (0.01, or one hundredth), *milli-* (0.001, or one thousandth), and *micro-* (0.0001, or one millionth).

Unit	***Quantity***	***Symbol***	***Equivalents***
Acre	Area	ac	43,560 square feet 4,840 square yards 0.405 hectare
Ampere	Electric current	A *or* amp	1.00016502722949 international ampere 0.1 biot *or* abampere
Angstrom	Length	Å	0.1 nanometer 0.0000001 millimeter 0.000000004 inch
Astronomical unit	Length	AU	92,955,807 miles 149,597,871 kilometers (mean Earth-Sun distance)
Barn	Area	b	10^{-28} meters squared (approx. cross-sectional area of 1 uranium nucleus)
Barrel (dry, for most produce)	Volume/capacity	bbl	7,056 cubic inches; 105 dry quarts; 3.281 bushels, struck measure
Barrel (liquid)	Volume/capacity	bbl	31 to 42 gallons
British thermal unit	Energy	Btu	1055.05585262 joule
Bushel (U.S., heaped)	Volume/capacity	bsh *or* bu	2,747.715 cubic inches 1.278 bushels, struck measure
Bushel (U.S., struck measure)	Volume/capacity	bsh *or* bu	2,150.42 cubic inches 35.238 liters
Candela	Luminous intensity	cd	1.09 hefner candle
Celsius	Temperature	C	1° centigrade
Centigram	Mass/weight	cg	0.15 grain
Centimeter	Length	cm	0.3937 inch
Centimeter, cubic	Volume/capacity	cm^3	0.061 cubic inch
Centimeter, square	Area	cm^2	0.155 square inch
Coulomb	Electric charge	C	1 ampere second
Cup	Volume/capacity	C	250 milliliters 8 fluid ounces 0.5 liquid pint

Unit	*Quantity*	*Symbol*	*Equivalents*
Deciliter	Volume/capacity	dl	0.21 pint
Decimeter	Length	dm	3.937 inches
Decimeter, cubic	Volume/capacity	dm^3	61.024 cubic inches
Decimeter, square	Area	dm^2	15.5 square inches
Dekaliter	Volume/capacity	dal	2.642 gallons 1.135 pecks
Dekameter	Length	dam	32.808 feet
Dram	Mass/weight	dr *or* dr avdp	0.0625 ounce 27.344 grains 1.772 grams
Electron volt	Energy	eV	$1.5185847232839 \times 10^{-22}$ Btus $1.6021917 \times 10^{-19}$ joules
Fermi	Length	fm	1 femtometer 1.0×10^{-15} meters
Foot	Length	ft *or* '	12 inches 0.3048 meter 30.48 centimeters
Foot, square	Area	ft^2	929.030 square centimeters
Foot, cubic	Volume/capacity	ft^3	0.028 cubic meter 0.0370 cubic yard 1,728 cubic inches
Gallon (British Imperial)	Volume/capacity	gal	277.42 cubic inches 1.201 U.S. gallons 4.546 liters 160 British fluid ounces
Gallon (U.S.)	Volume/capacity	gal	231 cubic inches 3.785 liters 0.833 British gallon 128 U.S. fluid ounces
Giga-electron volt	Energy	GeV	$1.6021917 \times 10^{-10}$ joule
Gigahertz	Frequency	GHz	—
Gill	Volume/capacity	gi	7.219 cubic inches 4 fluid ounces 0.118 liter
Grain	Mass/weight	gr	0.037 dram 0.002083 ounce 0.0648 gram
Gram	Mass/weight	g	15.432 grains 0.035 avoirdupois ounce

Unit	*Quantity*	*Symbol*	*Equivalents*
Hectare	Area	ha	2.471 acres
Hectoliter	Volume/capacity	hl	26.418 gallons 2.838 bushels
Hertz	Frequency	Hz	$1.08782775707767 \times 10^{-10}$ cesium atom frequency
Hour	Time	h	60 minutes 3,600 seconds
Inch	Length	in *or* "	2.54 centimeters
Inch, cubic	Volume/capacity	in^3	0.554 fluid ounce 4.433 fluid drams 16.387 cubic centimeters
Inch, square	Area	in^2	6.4516 square centimeters
Joule	Energy	J	$6.2414503832469 \times 10^{18}$ electron volt
Joule per kelvin	Heat capacity	J/K	$7.24311216248908 \times 10^{22}$ Boltzmann constant
Joule per second	Power	J/s	1 watt
Kelvin	Temperature	K	-272.15° Celsius
Kilo-electron volt	Energy	keV	$1.5185847232839 \times 10^{-19}$ joule
Kilogram	Mass/weight	kg	2.205 pounds
Kilogram per cubic meter	Mass/weight density	kg/m^3	$5.78036672001339 \times 10^{-4}$ ounces per cubic inch
Kilohertz	Frequency	kHz	—
Kiloliter	Volume/capacity	kl	—
Kilometer	Length	km	0.621 mile
Kilometer, square	Area	km^2	0.386 square mile 247.105 acres
Light-year (distance traveled by light in one Earth year)	Length/distance	lt-yr	5,878,499,814,275.88 miles 9.46×10^{12} kilometers
Liter	Volume/capacity	L	1.057 liquid quarts 0.908 dry quart 61.024 cubic inches
Mega-electron volt	Energy	MeV	—
Megahertz	Frequency	MHz	—
Meter	Length	m	39.37 inches
Meter, cubic	Volume/capacity	m^3	1.308 cubic yards

Unit	*Quantity*	*Symbol*	*Equivalents*
Meter per second	Velocity	m/s	2.24 miles per hour 3.60 kilometers per hour
Meter per second per second	Acceleration	m/s^2	12,960.00 kilometers per hour per hour 8,052.97 miles per hour per hour
Meter, square	Area	m^2	1.196 square yards 10.764 square feet
Metric. *See* unit name			
Microgram	Mass/weight	mcg *or* µg	0.000001 gram
Microliter	Volume/capacity	µl	0.00027 fluid ounce
Micrometer	Length	µm	0.001 millimeter 0.00003937 inch
Mile (nautical international)	Length	mi	1.852 kilometers 1.151 statute miles 0.999 U.S. nautical miles
Mile (statute or land)	Length	mi	5,280 feet 1.609 kilometers
Mile, square	Area	mi^2	258.999 hectares
Milligram	Mass/weight	mg	0.015 grain
Milliliter	Volume/capacity	ml	0.271 fluid dram 16.231 minims 0.061 cubic inch
Millimeter	Length	mm	0.03937 inch
Millimeter, square	Area	mm^2	0.002 square inch
Minute	Time	m	60 seconds
Mole	Amount of substance	mol	6.02×10^{23} atoms or molecules of a given substance
Nanometer	Length	nm	1,000,000 fermis 10 angstroms 0.001 micrometer 0.00000003937 inch
Newton	Force	N	0.224808943099711 pound force 0.101971621297793 kilogram force 100,000 dynes
Newton meter	Torque	N·m	0.7375621 foot-pound
Ounce (avoirdupois)	Mass/weight	oz	28.350 grams 437.5 grains 0.911 troy or apothecaries' ounce

Unit	Quantity	Symbol	Equivalents
Ounce (troy)	Mass/weight	oz	31.103 grams 480 grains 1.097 avoirdupois ounces
Ounce (U.S., fluid or liquid)	Mass/weight	oz	1.805 cubic inch 29.574 milliliters 1.041 British fluid ounces
Parsec	Length	pc	30,856,775,876,793 kilometers 19,173,511,615,163 miles
Peck	Volume/capacity	pk	8.810 liters
Pint (dry)	Volume/capacity	pt	33.600 cubic inches 0.551 liter
Pint (liquid)	Volume/capacity	pt	28.875 cubic inches 0.473 liter
Pound (avoirdupois)	Mass/weight	lb	7,000 grains 1.215 troy or apothecaries' pounds 453.59237 grams
Pound (troy)	Mass/weight	lb	5,760 grains 0.823 avoirdupois pound 373.242 grams
Quart (British)	Volume/capacity	qt	69.354 cubic inches 1.032 U.S. dry quarts 1.201 U.S. liquid quarts
Quart (U.S., dry)	Volume/capacity	qt	67.201 cubic inches 1.101 liters 0.969 British quart
Quart (U.S., liquid)	Volume/capacity	qt	57.75 cubic inches 0.946 liter 0.833 British quart
Rod	Length	rd	5.029 meters 5.50 yards
Rod, square	Area	rd^2	25.293 square meters 30.25 square yards 0.00625 acre
Second	Time	s *or* sec	$^1/_{60}$ minute $^1/_{3600}$ hour
Tablespoon	Volume/capacity	T *or* tb	3 teaspoons 4 fluid drams
Teaspoon	Volume/capacity	t *or* tsp	0.33 tablespoon 1.33 fluid drams

Unit	*Quantity*	*Symbol*	*Equivalents*
Ton (gross or long)	Mass/weight	t	2,240 pounds 1.12 net tons 1.016 metric tons
Ton (metric)	Mass/weight	t	1,000 kilograms 2,204.62 pounds 0.984 gross ton 1.102 net tons
Ton (net or short)	Mass/weight	t	2,000 pounds 0.893 gross ton 0.907 metric ton
Volt	Electric potential	V	1 joule per coulomb
Watt	Power	W	1 joule per second 0.001 kilowatt $2.84345136093995 \times 10^{-4}$ ton of refrigeration
Yard	Length	yd	0.9144 meter
Yard, cubic	Volume/capacity	yd^3	0.765 cubic meter
Yard, square	Area	yd^2	0.836 square meter

Applied Science

Technology

AIR-QUALITY MONITORING

FIELDS OF STUDY

Meteorology; electronics; engineering; environmental planning; environmental studies; statistics; physics; chemistry; mathematics.

SUMMARY

Air-quality monitoring involves the systematic sampling of ambient air, analysis of samples for pollutants injurious to human health or ecosystem function, and integration of the data to inform public policy decision making. Air-quality monitoring in the United States is governed by the federal Clean Air Act of 1963 and subsequent amendments and by individual state implementation plans. Data from air monitoring are used to propose and track remediation strategies that have been credited with dramatically lowering levels of some pollutants in recent years.

KEY TERMS AND CONCEPTS

- **Aerosol:** Suspension of solid particles or very fine droplets of liquid, one of the major components of smog.
- **Air Pollutants:** Foreign or natural substances occurring in the atmosphere that may result in adverse effects to humans, animals, vegetation, or materials.
- **Air Quality Index (AQI):** Numerical index used for reporting severity of air pollution levels to the public. The AQI incorporates five criteria pollutants — ozone, particulate matter, carbon monoxide, sulfur dioxide, and nitrogen dioxide —into a single index, which is used for weather reporting and public advisories.
- **Carbon Monoxide (CO):** Colorless, odorless gas resulting from the incomplete combustion of hydrocarbon fuels. CO interferes with the blood's ability to carry oxygen to the body's tissues and results in numerous adverse health effects. More than 80 percent of the CO emitted in urban areas is contributed by motor vehicles.
- **Continuous Emission Monitoring (CEM):** Used for determining compliance of stationary sources with their emission limitations on a continuous basis by installing a system to operate continuously inside of a smokestack or other emission source.
- **Criteria Air Pollutant:** Air pollutant for which acceptable levels of exposure can be determined and for which an ambient air-quality standard has been set. Examples include: ozone, carbon monoxide, nitrogen dioxide, sulfur dioxide, PM10, and PM2.5.
- **Ozone:** Reactive toxic chemical gas consisting of three oxygen atoms, a product of the photochemical process involving the Sun's energy and ozone precursors, such as hydrocarbons and oxides of nitrogen. Ozone in the troposphere causes numerous adverse health effects and is a major component of smog.
- **PM 10 and PM 2.5:** Particulate matter with maximum diameters of 10 and 2.5 micrometers (μm), respectively.
- **Regional Haze:** Haze produced by a multitude of sources and activities that emit fine particles and their precursors across a broad geographic area. National regulations require states to develop plans to reduce the regional haze that impairs visibility in national parks and wilderness areas.
- **Volatile Organic Compounds (VOCs):** Carbon-containing compounds that evaporate into the air. VOCs contribute to the formation of smog and may themselves be toxic.

Definition and Basic Principles

Air-quality monitoring aims to track atmospheric levels of chemical compounds and particulate matter that are injurious to human health or cause some form of environmental degradation. The scope varies, from single rooms to the entire globe. There has been a tendency since the 1990's to shift focus away from local exterior air pollution to air quality inside buildings, both industrial and residential, to regional patterns including air-quality issues that cross international boundaries, and to worldwide trends including global warming and depletion of the ozone layer.

Data from air-quality monitoring are used to alert the public to hazardous conditions, track remediation efforts, and suggest legislative approaches to environmental policy. Agencies conducting local and regional exterior monitoring include national and state environmental protection agencies, the National Oceanic and Atmospheric Administration (NOAA, popularly known as the Weather Bureau), and the National Aeronautics and Space Administration (NASA). Industries conduct their own interior monitoring with input from state agencies and the Occupational Safety and Health Administration (OSHA). There is an increasing market for inexpensive devices homeowners can install themselves to detect household health hazards.

The Clean Air Act of 1963 and amendments of 1970 and 1990 mandate monitoring of six criteria air pollutants—carbon monoxide, nitrogen dioxide, sulfur dioxide, ozone, particulate matter, and lead—with the aim of reducing release into the environment and minimizing human exposure. With the exception of particulate matter, atmospheric concentrations of these pollutants have decreased dramatically since the late 1980's.

A key concept in implementing monitoring programs is that of probable real-time exposure, with monitoring protocols matched to real human experience. Increasingly compact and automated equipment that can be left permanently at a site and collect data at intervals over a period of weeks and months has been a real boon to establishing realistic tolerance levels.

Background and History

The adverse effects of air pollution from burning coal were first noticed in England in the Middle Ages and were the reason for the ban on coal in London in 1306. By the mid-seventeenth century, when John Evelyn wrote *Fumifugium or, the Inconvenience of the Aer and the Smoake of London Dissipated* (1661) the problem of air pollution in London had become acute, and various measures, including banning certain industries from the city, were proposed to combat it. Quantifying or even identifying the chemicals responsible exceeded contemporary scientific knowledge, but English statistician John Graunt, correlating deaths from lung disease recorded in the London bills of mortality with the incidence of "great stinking fogs," obtained objective evidence of health risks, and scientist Robert Boyle proposed using the rate of fading of strips of dyed cloth to monitor air quality.

Chemical methods for detecting nitrogen and sulfur oxides existed in the mid-eighteenth century. By the late nineteenth century, scientists in Great Britain were undertaking chemical analyses of rainwater and measuring rates of soot deposition with the aim of encouraging industries either to adopt cleaner technologies or to move to less populated areas. Early examples of environmental legislation based on scientific evidence are the Alkali Acts, the first of which was enacted in 1862, requiring industry to mitigate dramatic environmental and human health effects by reducing hydrogen chloride (HCl) emissions by 95 percent.

Until the passage of the first national Clean Air Act in 1963, monitoring and abatement of air pollution in the United States was mainly a local matter, and areas in which the residents were financially dependent on a single polluting industry were reluctant to take any steps toward abatement. Consequently, although air quality in most large metropolitan areas improved in the first half of the twentieth century, grave health hazards remained in some industries. This discrepancy was highlighted by the 1948 tragedy in Donora, Pennsylvania, when twenty people died and almost 6,000 (about half the population) became acutely ill during a prolonged temperature inversion that trapped effluents from a zinc smelter, including highly toxic fluorides. Although autopsies and blood tests revealed a pattern of chronic and acute fluoride poisoning, no specific legislation regulating the industry followed the investigation. Blood tests and autopsies have also been used to track incidences of lead and mercury poisoning, some of it from atmospheric pollution.

How It Works

Structure of Air-Monitoring Programs. In the United States, state environmental protection agencies, overseen by the federal government, conduct the largest share of air-quality monitoring out of doors, while indoor monitoring is usually the responsibility of the owner or operator.

The Clean Air Act and its amendments require each state to file a State Implementation Plan (SIP) for air-pollution monitoring, prevention, and remediation. States are responsible for most of the costs of implementing regulations, which may be more stringent than federal guidelines but cannot fall below them. Detailed regulations specify which pollutants must be monitored, frequency and procedures for monitoring, and acceptable equipment. The regulations change constantly in response to developing technology, shifting patterns of pollution, and political considerations. There is a tendency to respond rapidly and excessively to new threats while grandfathering in older programs, such as those aimed at asbestos and lead, which address hazards that are far less acute than they once were.

The United States Environmental Protection Agency (EPA) collects and analyzes data from state monitoring stations to track national and long-term trends, make recommendations for expanded programs, and ensure compliance. Some pollutants can be tracked using remote sensing from satellites, a function performed by NASA. The World Health Organization (WHO), a branch of the United Nations, integrates data from national programs and conducts monitoring of its own. Air pollution is an international problem, and the lack of controls in developing nations spills over into the entire biosphere.

Workplace air monitoring falls under the auspices of state occupational safety and health administrations. In addition to protecting workers against the by-products of manufacturing processes, monitoring also identifies allergens, ventilation problems, and secondhand tobacco smoke.

Monitoring Methods and Instrumentation. Methods of monitoring are specific to the pollutant. A generalized monitoring device consists of an air pump capable of collecting samples of defined volume, a means of concentrating and fixing the pollutants of interest, and either an internal sensor that registers and records the level of the pollutant or a removable collector.

An ozone monitor is an example of a continuous emission sampler based on absorption spectroscopy. A drop in beam intensity is proportional to ozone concentration in the chamber. Absorption spectrometers exist for sulfur and nitrogen oxides. This type of technology is portable and relatively inexpensive to run and can be used under field conditions, for example monitoring in-use emissions of motor vehicles. Absorption spectroscopy is also used in satellite remote sensing and has been adapted to remote sensing devices deployed on the ground to measure vehicular emissions.

There are a number of methods for measuring particulate matter, the simplest of which, found in some home smoke detectors, involves a photoelectric cell sensitive to the amount that a light beam is obscured. More sophisticated mass monitors measure scattering of a laser light beam, with the degree of scattering proportional to particle size and density. Forcing air through a filter traps particles for further analysis. X-ray fluorescence, in which a sample is bombarded with X rays and emitted light is measured, will detect lead, mercury, and cadmium at very low concentrations. Asbestos fibers present will turn up either by X-ray fluorescence or visual inspection of filters. Pollen and mold spores, important as allergens, are detected by visual inspection. A drawback of filters is cost and the skilled labor required to process them.

A total hydrocarbon analyzer used by the auto industry uses the flame ionization detection principle to identify specific hydrocarbons in auto exhaust. Volatile organic compounds (VOCs) present a challenge because total concentration is low outside of enclosed spaces and certain industrial sites, and rapid efficient methodology is not available for distinguishing between different classes of organic compounds.

Applications and Products

Reporting and Predicting Air Quality. Media report air-quality indices along with other weather data, and the general public has become accustomed to using this information to plan activities such as outdoor recreation. Projections are also used to schedule unavoidable industrial-emissions release to coincide with favorable weather patterns and minimize public inconvenience. Predicting air quality is an evolving and inexact science involving predicting and tracking weather patterns but also integrating myriad human activities.

Vehicular Emissions. Federal law mandates that urban areas with unhealthy levels of vehicular pollution require testing of automobile emissions. Laws concerning testing vary considerably from state to state and jurisdictions within states. Additionally, new vehicles manufactured or sold in the United States must undergo factory testing, both of the model and of the individual units, to ensure the vehicle meets federal standards.

Typical vehicle-inspection protocol requires motorists to bring the vehicle to a garage where automatic equipment samples exhaust and analyzes it for CO, aggregate hydrocarbons, nitrogen and sulfur oxides, and particulate matter. Some state departments of motor vehicles operate their own inspection stations, while others license private garages. There are a number of compact units on the market that provide the required information with little operator input.

With improving air quality and a decreasing proportion of older cars that lack pollution-control equipment, some jurisdictions are withdrawing from vehicle testing.

Public-health effects of air-pollution monitoring and mitigation on public health deserve mentioning. Adverse effects on human health were the principal rationale for instituting laws curbing air pollution, and the decline in certain health problems associated with atmospheric pollution since those laws were enacted is testimony to their effectiveness.

Rates of lung cancer and chronic obstructive pulmonary disease (COPD) have declined, and ages of onset have increased substantially since the mid-twentieth century. Although the bulk of this is due to declining tobacco use, some of it is related to pollution control.

Impact on Industry

Impact on industry is twofold, including the effects, both positive and negative, of complying with emissions regulations, and the creation and marketing of new products to meet the needs of monitoring.

Costs of installing pollution-mitigation equipment and monitoring emissions are nontrivial, and the benefits to a community of cleaning up an industry do not necessarily compensate the manufacturer's bottom line. Earlier fears that stringent environmental regulations would force plants to close or relocate elsewhere have to some extent been realized, although the relative contributions of environmental legislation, labor costs, tax structure, and other government policies to the exodus of heavy industry from the United States are debated. When existing facilities are exempted from regulations but new facilities must comply, it discourages growth and investment. Few people doubt that the benefits of clean air outweigh the costs.

Air-pollution monitoring and abatement is a major industry in its own right, employing people in state and federal environmental agencies, university- and industry-based research laboratories, and in the manufacture, sale, and servicing of pollution-control devices and monitoring equipment. Because devices and their deployment must comply with constantly shifting state and federal regulations, this is an industry in which the manufacturing end for industrial and research equipment remains based in the United States.

The consumer market for pollution-detection devices and services is growing rapidly, particularly for interior air. Smoke detectors have become a standard household fixture and CO monitors are recommended for fuel-oil- and gas-heated homes. Testing for formaldehyde, mold, and in some localities radon is often part of building inspections during real estate sales.

Although the field is not expanding at the rate it once was, identification of additional areas of concern, the rapid expansion of industry in other parts of the world, and the continuing need for trained technicians make this one of the better employment prospects for would-be research scientists.

Careers and Course Work

Air-pollution monitoring is a field with solid career prospects, in government and private industry. The majority of openings call for field technicians who supervise monitoring facilities, conduct a varying amount of chemical and physical analyses (now mostly automated), and collect and analyze data. For this type of position a bachelor's degree in a field that includes substantial grounding in chemistry, mathematics, and data management, and training on the types of equipment used in monitoring, is usually required. A number of state colleges offer undergraduate degrees in environmental engineering that provide a solid background. Online programs offered by for-profit institutions lack the rigor and hands-on experience necessary for this demanding occupation.

For research positions in government laboratories and educational institutions, an advanced degree in meteorology, environmental science, or environmental engineering is generally required.

This is a rapidly evolving field for which knowledge of the latest techniques and regulations is essential. Degree programs that offer a solid internship program, integrating students into working government or industrial laboratories, offer a tremendous advantage in a job market where actual experience is essential.

The development, sale, and servicing of monitoring equipment offers other employment options. While course work can provide the general level of knowledge necessary to sell, adjust, and repair sophisticated automated electronic equipment, such a career objective will also require extensive on-the-job training. Some manufacturers offer factory training for service people.

Social Context and Future Prospects

The regulations and remediation efforts that monitoring informs and supports have clearly had a positive impact on the health and well-being of Americans in the nearly half-century since the passage of the Clean Air Act of 1963. Heavy-metal exposure from atmospheric sources has dropped dramatically. Mandatory pollution-control devices on new passenger vehicles have curbed emissions in states where annual vehicle emission tests are not even required. Older diesel trucks, farm vehicles, and stationary engines remain a concern in the United States.

Air pollution remains a significant problem in the developing world, especially in China, where the rapid growth of coal-fired industries has created conditions in urban areas reminiscent of Europe in the nineteenth century. Addressing these problems is a matter of international concern, because air pollution is no respecter of national boundaries.

Low-end environmental monitoring devices for the consumer market are a growth industry. With respect to genuine hazards such as smoke and carbon monoxide, this is a positive development, but there is concern that overzealous salespeople and environmental-consulting firms will exaggerate risks and push for costly solutions in order to enhance their bottom line, as occurred with asbestos abatement in the 1970's and 1980's.

Although often criticized, an integrated approach to environmental policy that includes "cap and trade"—allowing industries to use credits for exceeding standards in one area to offset lagging performance in other areas, or to sell these credits to other industries so long as an industry-wide target is met—helps ease the nontrivial burdens of complying with constantly evolving environmental standards.

Much of the information-gathering and tracking

Fascinating Facts About Air-Quality Monitoring

- In 1987, the Environmental Protection Agency (EPA) ranked indoor air pollution from cigarette smoke, poorly functioning heating systems, and formaldehyde from building materials as fourth in thirteen environmental health problems analyzed.
- Ninety percent of Californians were exposed to unhealthful air conditions at some point in 2008.
- The earliest known complaint about noxious air pollution is one registered in Nottingham, England, in 1257 by Queen Eleanor of Provence.
- Medieval and early modern physicians were quick to associate the smell of sulfur in coal smoke as a health hazard because of the prevailing belief that disease was caused by miasmas, or vaporous exhalations from swamps and rotting material including sewage and garbage.
- In 1955 the city of Los Angeles instituted the first systematic air-monitoring program to track levels of the city's notorious smog, caused mainly by motor vehicles. The first monitors used the rate of degradation of thin rubber strips to determine when smog levels were unhealthful and vulnerable individuals should stay indoors.
- Tetraethyl lead, a gasoline additive and significant pollutant, was removed from gasoline not because of human health issues but because it poisoned catalytic converters, needed for fuel efficiency.
- Haze from coal-fired electric generating plants is adversely affecting tourism in Great Smoky Mountains National Park and other Southeastern vacation destinations noted for splendid vistas.
- Ironically, children who engage in vigorous exercise out of doors are vulnerable to respiratory damage from smog, making real-time monitoring of sulfur, nitrogen oxides, and ozone and prompt reporting to the public an important public-health measure.

system developed to address emissions of criteria pollutants is being integrated into the effort to slow global warming due to carbon dioxide (CO_2) emissions from fossil fuel burning. While elevated CO_2 does not directly affect human health, and is actually beneficial to plants, the overall projected effects of global warming are sufficiently dire that efforts to reduce CO_2 emissions deserve a high priority in environmental planning.

Martha A. Sherwood, Ph.D.

FURTHER READING

Brimblecombe, Peter. *The Big Smoke: A History of Air Pollution in London Since Medieval Times.* 1987. New York: Routledge, 2011. A readable and fact-filled account of the history of air pollution.

Collin, Robert W. *The Environmental Protection Agency: Cleaning Up America's Act.* Westport, Conn.: Greenwood Press, 2006. Detailed history of environmental regulation, aimed at the general college-educated reader.

Committee on Air Quality Management in the United States, Board on Environmental Studies and Toxicology, Board on Atmospheric Sciences and Climate, Division on Earth and Life Studies. *Air Quality Management in the United States.* Washington, D.C.: National Academies Press, 2004. A reference for public agencies and policymakers, with much specific information.

Magoc, Chris J. *Environmental Issues in American History: A Reference Guide with Primary Documents.* Westport, Conn.: Greenwood Press, 2006. The chapters on tetraethyl lead and the Donora disaster deal with air pollution.

World Health Organization. Text edited by David Breuer. *Monitoring Ambient Air Quality for Health Impact Assessment.* Copenhagen: Author, 1999. Good coverage of efforts outside of the United States.

WEB SITES

Air and Waste Management Association
http://www.awma.org/Public

Ambient Monitoring Technology Information Center
http://www.epa.gov/ttn/amtic

American Academy of Environmental Engineers
http://www.aaee.net

National Association of Environmental Professionals
http://www.naep.org

United States Environmental Protection Agency

AIR TRAFFIC CONTROL

FIELDS OF STUDY

Meteorology; aviation laws and regulations; instrument flying; theory of flight; aviation; crew resource management; navigation; radar fundamentals; communication; telecommunication.

SUMMARY

Air traffic control is responsible for keeping aircraft safely separated when moving, both on the ground and in the air. It relies heavily on technology and a highly trained professional workforce. The equipment utilized in air traffic control is continually improving as technology evolves. The latest iteration of air traffic control technology involves the replacement of ground-based radar systems with satellite-based systems.

KEY TERMS AND CONCEPTS

- **Airspace:** Air within which aircraft operate; it may be controlled or uncontrolled.
- **Automatic Terminal Information System (ATIS):** A recording giving the pilot the latest weather and airport information.
- **Clearance:** Approval issued to an individual aircraft by air traffic control for a specific activity or route.
- **Controlled Airspace:** Airspace subject to positive control by air traffic controllers.
- **Encroachment:** Violation of controlled airspace or runway by an aircraft.
- **Instrument Flight Rules:** Flight rules that require aircraft to be under positive air traffic control.
- **NextGen:** Acronym for Next Generation Air Transportation System. The latest technology in development for air traffic control, it involves satellite tracking and global positioning systems.
- **Radar:** Acronym for radio detecting and ranging; measures range, bearing, speed, and other information concerning individual aircraft.
- **Transponder:** Onboard electronic equipment that returns a radar signal to air traffic control depicting aircraft identification, altitude, speed, and course.

Definition and Basic Principles

Air traffic control is the means by which separation of aircraft in flight and on the ground is maintained. This service is provided by ground-based personnel utilizing electronic systems and two-way communication. Present-day air traffic control relies primarily on radar. Radar allows air traffic controllers to identify aircraft and to determine altitude, speed, and course. This, in turn, provides the controllers the information required to maintain separation and guide aircraft to their destinations. Air traffic control is divided into three distinct entities: air traffic control towers (ATCTs); terminal radar approach control (TRACON); and air route traffic control centers (ARTCCs). Each has a distinct function, but all activities are coordinated among the sections. Flight service stations, an advisory service, are also a part of the air traffic control network.

Historically, air traffic control has been a function of government. In the United States, that responsibility falls to the Federal Aviation Administration (FAA). A number of countries, including Canada, have been experimenting with privatizing air traffic control by contracting with independent companies. In some countries, such as Brazil, air traffic control is completely under military control. In any case, air traffic controllers must coordinate the operation of thousands of aircraft every day. In the United States alone, there are as many as 5,000 aircraft operating at any moment. This can result in as many as 50,000 operations each day. Air traffic controllers have the responsibility to ensure the safety and efficiency of each operation.

Background and History

In the early days of aviation, the airplane was considered a novelty that served no useful purpose. As the number of aircraft and the performance increased, accidents became more frequent. By 1925 the U.S Postal Service was attempting to establish commercial passenger service by contracting with private companies to fly mail. As this service expanded, the need for some type of air traffic control became apparent.

Until the early 1930's, aircraft operated under the "see and be seen" method of collision avoidance. As the number, size, and performance of aircraft increased, this approach proved inadequate. The earliest form of air traffic control consisted of an individual positioned at the airport with red and green flags clearing aircraft to take off or land. This system was impractical for use at night or in bad weather. The Cleveland airport was the first to establish a modern type of air traffic control system that included two-way radio communication.

In 1934, the Bureau of Air Commerce was assigned the responsibility of controlling air traffic on the newly established airways. In 1937, the Department of Commerce took over the air traffic control function. Following a number of midair collisions, the Civil Aeronautics Authority (CAA) was created. Twenty-three airway traffic control centers were established. Following World War II, the use of radar was implemented. Improved radar systems remain the primary method of controlling traffic to date, although satellite systems are being introduced.

How It Works

The entire air traffic control system relies on radar, two-way radio communication, electronic navigation aids, and highly trained professional personnel.

Air Route Traffic Control Centers (ARTCCs). The air traffic control system is divided into twenty-two ARTCCs that manage traffic within specific geographical areas. The ARTCCs are responsible for all traffic other than that controlled by the terminal radar approach control (TRACON) and the control tower facilities. Primarily utilizing constant radar surveillance, the ARTCCs provide separation for aircraft operating in controlled airspace under instrument flight rules.

ARTCCs control traffic traveling between airports. When an aircraft departs the geographical area of responsibility of one ARTCC it is handed off to the next ARTCC along its flight route. The next ARTCC will control the aircraft until it is handed off to the next ARTCC or a TRACON facility.

Terminal Radar Approach Control. The TRACON controller accepts the aircraft from the ARTCC as it approaches within 30 to 50 miles of its destination. Again, depending primarily on radar, the TRACON controller issues instructions to the pilot, sequences aircraft for approach or departure, and provides separation between aircraft within the controlled airspace. The TRACON controller will hand off the traffic either to an ARTCC if the aircraft is departing or to a local controller at an air traffic control tower (ATCT) if it is landing.

Air Traffic Control Towers. The ATCT will clear the aircraft to land or take off. The tower controller, using primarily visual contact with the aircraft, will also provide all the necessary information to the pilot, such as the runway in use, the current altimeter setting, winds, ceiling and visibility, and the aircraft's position number in the landing or takeoff sequence.

Ground Control. The ATCT provides additional services while the aircraft is on the ground. The tower typically hands off a landed aircraft to a ground controller. The ground controller accepts responsibility for the aircraft as soon as it clears the active runway. Even though the aircraft is on the ground, two-way communication is still required. This is important since an aircraft on the ground could taxi onto or across an active runway and cause an accident. At larger airports ground-surveillance radar is used to direct the aircraft in taxiing.

Clearance Delivery. The last function of the ATCT is that of clearance delivery. When an aircraft is preparing to begin a flight under an instrument flight plan, the pilot must receive clearance prior to moving the aircraft. Clearance delivery will inform the pilot of the approved route, altitude, radio frequencies, and departure procedures prior to handing the pilot over to ground control for taxi instructions.

Flight Service Stations. The final component of the air traffic control structure is flight service. Flight service is an advisory-only function whereby the pilot can receive a weather briefing and file a flight plan. Flight service is also available to aircraft in flight for updated weather forecasts or to file a flight plan. Flight service can also lend assistance to aircraft in distress or pilots who become lost.

Coordination of Services. A typical scenario for a flight involves contacting flight service for a weather briefing and filing a flight plan. This would be followed by monitoring the airport terminal information frequency for automated weather and airport conditions. The pilot then contacts clearance delivery and receives the approved flight plan. After confirming the details of the flight plan, the pilot contacts ground control for taxi clearance. Ground control turns the aircraft over to the control tower for takeoff clearance. After takeoff the aircraft is handed off to departure control. As the aircraft approaches the limits of the area covered by departure control it is turned over to the ARTCC, which controls the aircraft until it approaches its destination. The aircraft is then handed over to approach control, followed by the control tower, and finally ground control. Throughout this entire procedure the aircraft would be under positive air traffic control and typically would be under radar surveillance.

Applications and Products

Modern air transportation could not exist without a highly developed and efficient system of air traffic control. The constantly developing technology allows controllers to guide the ever-increasing number of aircraft flying at any given time. There are differences, but virtually every country has an air traffic control system that is compatible with others. While some developing countries do not have the level of sophistication and may lack full radar coverage, the systems are standardized under international agreements.

Radar. With the advent of jet transport aircraft in the 1950's, an improved system of controlling these larger, faster aircraft was needed. Radar had been developed during World War II to identify approaching enemy aircraft so that fighters could be launched and directed to intercept the approaching aircraft. This system was effective from a military perspective, but the requirements for air traffic control and aircraft separation were quite different from the military application.

Air Route Surveillance Radar (ARSR). In 1956, ARSR, the first radar system specifically designed for tracking and separating civilian aircraft, was developed. Twenty-three of these units were ordered for the ARTCCs. Simultaneously, the first air traffic control computer system was installed at the Indianapolis International Airport. During the following twenty years, computer-generated display systems were incorporated into the ARSR systems. This display depicted the aircraft identification, altitude, and airspeed. It also computed the aircraft flight path and indicated possible conflicts. The aircraft signal flashed as it approached the limits of the controller's airspace so it could be safely handed off to the next controller. Because of the differing requirements of the ARTCC and TRACON, two entirely different systems were designed. The system used by ARTCCs is called radar data processing, while the system used in approach control and the tower is called automated radar terminal system (ARTS).

En Route Automation Modernization (ERAM). A total modernization of the air traffic control system is presently under way. The FAA is committed to increasing the capacity and efficiency of the entire air traffic control system. The target date for complete implementation of this system is 2018. As a bridge between the present system and the future next generation air transportation system (NextGen), the FAA has introduced interim measures. These include new, upgraded displays and new computer systems that allow a better interface between ARTCCs and TRACONs. To bridge the changeover to NextGen, a system known as en route automation modernization (ERAM) has been developed and is operational. ERAM incorporates a modern computer language and doubles the capacity of the air traffic control system. It is also designed to integrate satellite-based navigation and communication. Along with the implementation of ERAM, both navigation and communications capabilities are being upgraded to utilize satellite-based data.

Next Generation Air Transportation System (NextGen). NextGen is an integrated new system and procedure that will replace the existing ground-based radar systems with a satellite-based system and use digitally transmitted information in place of air traffic control. Weather information will be embedded in transmitted information to assist the pilot with decision making. NextGen not only will increase safety and efficiency but also will eliminate most of the problems that exist with ground-based systems, because much of the present technology has limited capacity and is antiquated and difficult to maintain.

NextGen is projected to reduce aircraft delays by 21 percent, reduce fuel consumption by 1.4 billion gallons, and reduce CO_2 emissions by 14 million tons. The FAA estimates that the total economic benefit will exceed $22 billion.

One integral component of the new system is the aircraft-based equipment automatic dependent surveillance-broadcast (ADS-B). This system is far more accurate than ground-based radar, and it provides the pilot with terrain maps, traffic information and location, weather, and critical flight information.

The satellite-based components of NextGen are expected to provide aircraft with the capability to fly shorter, more direct, and thus more efficient routes. It will also increase the capacity of existing runways to handle traffic and reduce delays as well as aircraft noise. At Dallas/Fort Worth International Airport alone the FAA projects a 45 percent reduction in departure delays and an increase of ten additional departures per hour per runway when the system is in place. This is extremely important given the finite capacity of runways to handle traffic, particularly in poor weather, as well as the virtual impossibility of constructing new commercial airports.

Yet another component of NextGen is system-wide information management (SWIM). SWIM is a streamlined network over which NextGen information will be exchanged. SWIM will provide secure information-management architecture for sharing national airspace data utilizing off-the-shelf hardware and software.

Impact on Industry

While the majority of the research and development is initiated and funded by national governments, the actual design and production of air traffic control equipment is in the hands of private manufacturers. Government agencies around the world are working together to address the issues confronting what has become a global air traffic control network.

A number of international corporations have entered the competition for air traffic control hardware and software. Companies such as the Harris Corporation, an international communications equipment manufacturer, produce a complete air traffic management system. Japan Digital Communications provides in-flight communications access including high-speed data communications. MacDonald Dettwiler and Associates provides air traffic control software. Raytheon Canada Limited manufactures solid-state air traffic control radars as well as integrated air traffic control systems. All of these companies contribute to a compatible, worldwide air traffic control network.

Careers and Course Work

While the demand for qualified air traffic controllers is increasing, there are a limited number of potential employers. In the United States, except for a few private companies staffing nonfederal control towers, the FAA is the sole employer. In some countries, the military handles all air traffic control, while some countries have contracted with private companies to run the air traffic control system. In all cases, air traffic controllers must be highly competent in areas such as communication, decision making, planning, and weather analysis.

Fascinating Facts About Air Traffic Control

- The first air traffic controller was Archie W. League, who sat at the end of the runway at Lambert Airfield in St. Louis. He controlled traffic with two flags, a red one for "hold" and a checkered one for "go," while sitting under an umbrella.
- In August, 1981, PATCO, the Professional Air Traffic Controllers Organization, called for a strike after unsuccessful contract negotiations. More than 12,000 strikers were terminated by President Ronald Reagan for participating in the strike, which violated the 1955 law banning government workers from striking. Reagan gave the striking air traffic controllers 48 hours to get back to work, citing the strike a "threat to national safety." The terminated controllers were then banned by Reagan from ever being hired by the Federal Aviation Administration again.
- In August, 1993, President Bill Clinton overturned Reagan's hiring ban, and about 850 controllers who went on strike in 1981 were eventually rehired.
- Stress is a big component of an air traffic controller's job, which may be one of the reasons controllers are required to retire earlier than most other federal employees. Controllers age fifty and over are eligible for retirement, provided they have completed twenty years of active service; and all controllers who have completed twenty-five years of active service can retire regardless of age.

The major portion of an air traffic controller's training typically takes place at the FAA Academy in Oklahoma City. In the United States, a number of colleges and universities are approved to provide initial training for potential controllers. This program is called the air traffic-collegiate training initiative (AT-CTI) and utilizes an approved curriculum to prepare candidates for the FAA Academy. The curricula typically include courses in theory of flight, aviation laws and regulations, aviation weather, navigation, instrument flight fundamentals, and basic air traffic control. Many also include topics in crew resource management and air traffic control computer simulations. AT-CTI candidates are required to complete at least a bachelor's degree to be eligible for employment. Historically, the FAA has hired former military controllers and even some candidates with no background or training in aviation. This process is changing, and more emphasis is being placed on the AT-CTI programs to meet the needs of training new controllers.

Social Context and Future Prospects

Given the globalization of industry and transportation, an efficient worldwide air traffic control network is critical. Every year, the number of air travelers and air freight increases. More than 2 billion passengers fly annually. Many countries, such as China, are on the cutting edge of developing high-technology air traffic control networks similar to NextGen. While the annual growth rate of air traffic in the United States is projected to be 4.5 percent, China's commercial aviation segment is projected to grow at a rate of 7.9 percent, and India's is projected at 10 percent. By 2020, 25,000 new commercial aircraft will enter service with 31 percent destined for the fast-growing Asia Pacific market. As more aircraft fill the skies the technology required to manage these aircraft successfully becomes more complex. This situation will result in increased emphasis on air traffic control and the associated technological development.

Ronald J. Ferrara

Further Reading

Gleim, Irvin N., and Garrett W. Gleim. *FAR/AIM: Federal Aviation Regulations and the Aeronautical Information Manual.* Gainesville, Fla.: Gleim Publications, 2011. Contains the FAA regulations concerning instrument flight operations and has an extensive section detailing air traffic control and air traffic control procedures.

Illman, Paul E. *The Pilot's Air Traffic Control Handbook.* 3d ed. New York: McGraw-Hill, 1999. Although somewhat dated, this work remains an excellent overview of air traffic control from a pilot's perspective.

Montgomery, Jeff, ed. *Aerospace: The Journey of Flight.* Maxwell Air Force Base, Ala.: Civil Air Patrol National Headquarters, 2008. This text is a basic overview of the aerospace industry and includes valuable sections on careers and a well-written section on air traffic control.

Nolan, Michael S. *Fundamentals of Air Traffic Control.* 5th ed. Clifton Park, N.Y.: Delmar Cengage Learning, 2010. The most comprehensive work in print about the air traffic control system, employment requirements, and equipment in use.

Preston, Edmund, ed. *FAA Historical Chronology: Civil Aviation and the Federal Government, 1926-1996.* Washington, D.C.: Department of Transportation, 1998. A very interesting chronological snapshot of the FAA, from the inception of federal regulations to 1996.

Robbins, Billy D. *Air Cops: A Personal History of Air Traffic Control.* Bloomington, Ind.: iUniverse, 2006. This is the memoir of an air traffic controller who experienced the transition to computerized systems.

Web Sites

Federal Aviation Administration
http://www.faa.gov

Federal Aviation Administration Academy
http://www.faa.gov/about/office_org/headquarters_offices/arc/programs/academy

U.S. Centennial of Flight Commission
http://www.centennialofflight.gov/index.cfm

See also: Maps and Mapping; Vertical Takeoff and Landing Aircraft.

ANTHROPOLOGY

FIELDS OF STUDY

Biological anthropology; cultural anthropology; linguistic anthropology; archaeological anthropology; biology; archaeology; linguistics; psychology; sociology; social psychology; forensics; epistemology; evolutionary biology; political science; religion; ethnomusicology; history; geography; geology; genetics; anatomy; economics; public health.

SUMMARY

Anthropology is the scientific study of all aspects of the human species, across the whole geographic and temporal span of human existence. It covers languages, customs, family structures, social behavior, politics, morality, health, and biology. As such, possible studies range from comparative analyses of human and chimpanzee family units to historical examinations of how languages change and to neurobiological studies of the roots of altruism in the brain. At its core, anthropology is concerned with what makes human beings who they are and why.

KEY TERMS AND CONCEPTS

- **Assimilation:** Assumption of the social behaviors and beliefs of a different group to the point where one's own cultural traits are eliminated.
- **Consanguinity:** Being related by blood or descended from a common ancestor.
- **Cosmology:** Set of beliefs about the origin of the universe.
- **Cultural Transmission:** Process by which patterns of thought and behavior are passed down from generation to generation; also known as enculturation and socialization.
- **Descent Group:** People who share an ancestor; relationships may be defined through members' mothers (matrilineal descent), their fathers (patrilineal descent), or some other means.
- **Ethnocentrism:** Using the practices of one's own culture to appraise the practices of another culture; the opposite of cultural relativism.
- **Ethnography:** Study of a particular culture, made through observing its members.
- **Ethnology:** Study of cultures in order to note their differences and similarities.
- **Etic Approach:** Method in which a culture is studied from outside in order to create objectivity; the opposite of an emic approach.
- **Kinesics:** Study of body movements and gestures used to communicate.
- **Modal Personality:** Cohesive set of personality traits, common to the majority of the people in a culture.
- **Morphology:** Study of morphemes, or meaningful units of sound within a given language.
- **Syncretism:** Tendency for different cultures, religions, belief systems, or varieties of a word within a language to converge and become one.

Definition and Basic Principles

Anthropology is concerned with the origin, evolution, behavior, beliefs, culture, and physical features of humankind. Rather than focusing on a single facet of human existence, anthropology is characterized by an inclusive approach that treats each feature of culture and society as interrelated parts of a whole. An anthropological study of psychology, for instance, might place it in the context of language and culture; a study of medicine might include an analysis of politics and human adaptation. This approach is known as holism. In addition, no matter how large or small the question an anthropologist asks—a research topic might be as specific as the types of wedding ornamentation used in a particular culture or as broad as how language evolved in different societies over millennia—its underlying aim will be to gain a deeper understanding of the species as a whole. Therefore, cross-cultural comparisons, or the practice of analyzing the similarities and differences between distinct human communities, is an essential component of anthropological research.

A few other basic principles distinguish anthropology as a science. One is an emphasis on immersion fieldwork, in which the scientist enters into the community being studied in order to observe it from the inside rather than the outside. Similarly, anthropology—in this way greatly influenced by feminism and postmodernism—breaks down the notion of scientists as objective observers, acknowledging that their mental models of the world are bounded by their own cultures. In other words, scientists' particular social backgrounds, experiences, and worldviews are inseparable from the perspectives they bring to ethnography. Anthropologists strive to overcome this limitation by consciously applying the principle of cultural relativism, judging each society by its own internal system of ethical and social guidelines. For example, an anthropological study of the ancient Chinese practice of foot binding (tightly tying cloth around female babies' feet in order to keep them small as the girl grows) would refrain from making a moral judgment about the tradition and instead would focus on describing its origins and the rationale for it within Chinese society.

Background and History

The word "anthropology," meaning "study of man," predates the development of anthropology as a modern scientific field. It was first used in the sixteenth century to describe a philosophical or theological examination of the soul. The term was later used by nineteenth-century German scientist Johann Friedrich Blumenbach in a sense closer to its modern meaning, to denote the study of both the physical and the psychological aspects of humankind. In the late nineteenth-century, the science of anthropology was dominated by an ethnocentric approach that peered at global cultures through the lens of strict Victorian mores and that largely labeled these cultures as primitive curiosities. Early anthropologists believed that cultural differences could be traced to genetic variances in the human species that resulted in moral and mental disparities.

German American anthropologist Franz Boas is given credit for formulating the principle of cultural relativism at the beginning of the twentieth century. Boas insisted that culture was not passed on from generation to generation through the genes but instead consisted of learned behaviors acquired over time through immersion in a group and eventual habituation to the behavior of others. He rejected the idea that culture developed toward a destination and that any single culture was inherently superior to or more advanced than another. Although anthropology has changed in the years since Boas first laid down his ideas, these principles continue to inform the field as a whole.

Over the course of the field's development, the focus has shifted away from broad descriptions of an entire culture or way of life and toward narrower examinations of specific features of community life. Thus, a cultural anthropologist may specialize in studies of specific topics such as music, marriage practices, or taboos.

How It Works

As in other fields of science, empirical research—research based on observed and verifiable data—forms a cornerstone of anthropology. The classic type of anthropological study is fieldwork, in which the scientist travels to and lives among a group of people with the intent of documenting their cultural practices. Other tools of archaeological investigation, common to the four main subfields of anthropology, include surveys, interviews, archival research, recordings, and statistical analysis.

Cultural Anthropology. Cultural anthropology is the most widely practiced and well-known subfield of anthropology. It is the study of human thought, knowledge, and practices—any behavioral trait, in other words, that is passed on not through the genes but through language, art, and ritual. The most important tool of the cultural anthropologist is the ethnographic study. Researchers directly observe members of a group as they go about their daily lives. They interview subjects, record oral histories, and make detailed reports of all that they see and hear. The goal of these observations is to make sense, in a deep way, of the reasoning behind the cultural practices of a given society. For example, it can be difficult for Western minds to apprehend the facial tattooing rituals that exist in many native cultures, including that of the Maori of New Zealand. Rather than seeing these practices as masochistic or stigmatizing, cultural anthropologists seek to uncover the principles that motivate them. In the case of the Maori, tattooing serves as a status symbol—the more complex one's facial tattoos, the higher one's rank—and as a sign of affiliation between group members.

Archaeological Anthropology. Archaeological anthropology is the study of human cultures through the material artifacts they produce, including such objects as mechanical devices, toys, writings, paintings, pottery, religious icons, buildings, and funerary items. Archaeological anthropologists collect, categorize, and describe these artifacts, then use them to piece together plausible theories about the belief systems and traditional practices of the societies from which they came. Anthropologists specializing in societies that have long passed out of existence, such as ancient Greek, Roman, Indian, Egyptian, and Mayan cultures, set up stations known as digs to unearth buried fragments and objects. Those anthropologists interested in the behavior of prehistoric hunter-gatherer societies or even the protohuman ancestors of the human species focus their attention on clues from fossilized skeletons and primitive tools. However, later human cultures provide equally valid subjects for archaeological-anthropological study. Archival maps, photographs, and other historical documents, for instance, can help scientists build up a clearer impression of the habits and ways of life practiced in communities such as those of nineteenth-century American coal miners or rural fishermen in colonial Malaya (now Malaysia).

Linguistic Anthropology. Linguistic anthropologists are concerned with the origin, development, and structure of the roughly 6,000 human languages that exist in the world, as well as with how language shapes and is shaped by culture. They study technical aspects of language such as phonetics, phonology, grammar, syntax, vocabulary, and semiotics. More broadly, researchers investigate how language is used to define social groups and transmit meaning from generation to generation, how the grammatical forms of a community's language affect common patterns of reasoning and thought, and how natural and artificial phenomena are represented or symbolized in language. For example, the vocabulary of some Chinese dialects includes an elaborate set of words describing specific family relations—terms that distinguish older siblings from younger ones, paternal from maternal relatives, and relatives by blood or marriage. The existence of such words in the language reflects an intense focus in Chinese culture on questions of kinship links and rankings.

Other fruitful areas of study include the relationships among languages within and across linguistic families and the question of how the pronunciation and usage of words evolved over time. Computational linguists, for example, may use software to conduct complex statistical analyses of thousands of instances of a word to reconstruct how different speakers meant and interpreted it. Linguistic anthropologists approach such issues through analyses of written texts and audio and video recordings and seek out language that occurs naturally in social discourse rather than language used in the context of formal interviews.

Biological Anthropology. Biological anthropology seeks to understand humankind by studying its physical anatomy. It examines the origin and evolution of the human brain, body, and nervous system; the physical diversity of individuals and groups within the human species; the place of humankind in relation to other animal species, both living and extinct, within the natural world; and the physiological bases of psychological processes and behaviors. There are various subfields within biological anthropology. Researchers who focus on human evolution conduct comparative analyses of human and nonhuman primates, making use of fossil evidence, DNA studies, and field research of animals in the wild. Medical anthropologists study issues of epidemiology (how diseases and health-related behaviors spread across human populations) and ethnomedicine (how different human societies think about, explain, and treat disease and health).

Applications and Products

Medicine and Public Health. The work of medical anthropologists leads to many useful applications in the world of public health. For example, an anthropologist who spent nearly two years immersed in the daily lives of families with children with the genetic disease cystic fibrosis produced a detailed set of practical clinical guidelines for physicians working with such families. The guidelines help doctors communicate effectively with children and their parents about cystic fibrosis symptoms and treatment. Another common medical application of anthropological research is the development of public health campaigns targeted to the specific worldviews and cultural traditions of a population. For example, anthropologists have worked with community health workers in Malaysia to educate locals about steps they can take to prevent the spread of the deadly virus responsible for dengue fever. Through door-to-door surveys and interviews, the anthropologists determined the Malaysian public's common misconceptions about dengue fever and helped create a more accessible list of recommendations for local inhabitants.

Crime Investigation. Forensic anthropologists, who apply the tools of anthropological study to skeletal remains such as bones and teeth, are indispensable members of any crime investigation team.

Since the 1930's, for instance, physical anthropologists employed at the Smithsonian Institution in Washington, D.C., have served as consultants to the Federal Bureau of Investigation (FBI) on a large number of criminal cases. The anthropologists help with tasks such as recovering physical evidence from the scene of the crime, determining victims' age and gender, and estimating the time and mechanisms of their deaths. One of the most crucial anthropological tools involved in crime investigation is facial reproduction. In facial reproduction, a forensic anthropologist uses computer simulation to reconstruct a picture of a deceased person's facial features that have been degraded by trauma or decomposition. This is done using scientific knowledge about the relationships between the shape of the hard tissue of the skull and the soft tissues that make up the face.

Military Applications. Anthropologists began assisting in military operation in the second half of the twentieth century. In the United States, for instance, cultural anthropologists working in the Human Terrain Team—a program under the umbrella of the U.S. Army Training and Doctrine Command—travel with soldiers to the field of warfare, interviewing local inhabitants and collecting data about the culture, sentiments, and needs of the people who live in battle zones in Afghanistan and Iraq. Their expertise helps the military improve security, target its aid and reconstruction efforts more effectively, work better with local governments and organizations, and conduct better counterterrorism programs. For example, army officials hope to use knowledge gained from the cultural advice of field anthropologists to persuade local tribal leaders to work with the Afghan police. Such assistance, however, comes at a cost. In 2008, two anthropologists lost their lives while working with the military, and the American Anthropological Association has expressed serious ethical concerns about the use of anthropology as a tool to aid the army in determining specific targets for military operations.

Social Problems. Anthropological studies are used by governments and various organizations across the globe to analyze and address a variety of social problems, such as malnutrition, poverty, teenage parenthood, unemployment, and drug abuse. For example, anthropological studies of population trends in countries such as Swaziland have shown that the number of children a woman has is inversely related to the amount of education she receives. Formal education raises women's awareness of birth control, opens doors to new jobs, and reduces their willingness to participate in practices such as arranged marriages and polygyny (two or more wives at the same time). As a result, one simple initiative applied in many developing nations with overpopulation problems is to increase the educational opportunities available to girls and young women.

Similarly, anthropological studies have been conducted of microcredit banks such as India's Grameen Bank. Microcredit agencies lend very small amounts of money to individuals in deep poverty, without requiring collateral, to help fund income-generating activities and housing. Anthropological research reveals that, despite its good intentions, microcredit can lead to the serious problem of debt cycling (paying off previous loans with new ones) among people who are already in financial difficulty. Anthropological research can thus be essential in evaluating the effectiveness of social programs that have already been put into place and in making recommendations for the future.

Impact on Industry

Government and University Research. Governments around the world use anthropological studies to help them evaluate the needs of their populations and plan public policy. They also rely on anthropologists to help them preserve information about national history. In the United States, the National Council for the Humanities and the National Science Foundation are the two major government agencies that provide funding to support anthropological research. In 2009, for example, the National Science Foundation approved funding for studies on the use of racial imagery depicting Asian Americans in advertising, how culture and national policy affect the safety of birthing, and how community television programs affect citizen participation in local, state, and national government.

Product Development. Anthropologists frequently work for companies that produce appliances, electronics, cosmetics, packaged foods, and a host of other consumer products. Their job is to use tools such as questionnaires and in-home observations to survey customers' needs and frustrations. They identify areas of potential for product development,

Fascinating Facts About Anthropology

- Ann Dunham, the mother of U.S. president Barack Obama, was a trained anthropologist whose doctoral dissertation was a study of the agricultural blacksmiths of Indonesia.
- Anthropological studies have documented a few cases of cannibalism (the eating of human flesh) in tribal societies such as the ancient Aztecs, usually as a ritual associated with war or funerals.
- Many early cultures subscribed to the belief, known as animism, that all natural objects or forces—such as rivers, mountains, and winds—are animated by a divine spirit.
- The language of the Plains Indian tribe known as the Assiniboine, with only fifty native speakers left, would be lost forever if anthropologists were not working to preserve it through recordings of stories, songs, and oral histories.
- Retail anthropologists are tracking the growth of self-service kiosks selling everything from food to newspapers to portable MP3 players. They say more and more consumers are drawn to these devices because they prefer not to have to interact with someone at a checkout.
- In 2008, the National Science Foundation awarded $100,000 to a group of anthropologists at the University of California, Irvine, to conduct an ethnographic study of the online role-playing game World of Warcraft.
- According to linguistic anthropologists, the claim that Eskimos have a hundred (or more) words for "snow" is a myth. The most snow words any scholar can come up with for an Inuit or Aleut language is two dozen, no more than the number of snow-related words in English—which include such terms as "frost," "hail," "ice," "slush," and "sleet."

suggest how products can be made more usable, and evaluate the impact and effectiveness of items that are already on the market. Anthropologists are a huge asset to companies wishing to ensure that the telephone, video game, or photocopier they have spent millions of dollars researching and developing is not one of the three-quarters of all new products that fail. Among the companies that have used the services of anthropologists are Xerox, Intel, Nokia, IBM, and Motorola. To fill this need, a new breed of consulting company has emerged that focuses on conducting anthropological studies of the retail market for clients. The Brazilian-based Anthropos Consulting group, which specializes in corporate anthropology, has developed products for pet food companies, airlines, hotels, and banks.

Marketing and Retail. Retail anthropologists, who study people's shopping habits, help clients such as grocery stores, shopping malls, mail-order companies, and Internet businesses increase the number and quality of the sales they make. Anthropologists study, for instance, the walking patterns and behavior of customers in stores, making observations such as the fact that mirrors or other shiny surfaces cause people to slow down, or that narrow aisles may deter a customer from spending time looking at products. They watch how people read their mail, noting which kinds of envelopes they tend to open and which they throw away. They also conduct statistical analyses of large amounts of purchasing data, looking for trends—such as whether women buy more shoes at a certain time of year or whether shoppers tend to purchase certain products one after the other—that can be used by marketers.

Careers and Course Work

A typical anthropology degree at the undergraduate level will involve a comprehensive series of courses in anthropological theories and methodologies, classes on broad topics such as human evolution and migration, and specific course work pertaining to a particular subfield. For example, a student interested in linguistic anthropology might enroll in classes on the structure of language, the neurobiology of language, and the sociological impact of language. No matter what subspecialty he or she chooses, an anthropology student should also gain a strong grounding in research methodologies, statistics, computational analysis, interview skills, and formal writing. Work experience or internships are helpful not only for gaining practical experience but also for making contacts in the field who can serve as professional mentors and advisers later in the student's career. Although such opportunities certainly exist at archaeological digs for those interested in archaeological anthropology, many other types of placements are also appropriate. For instance, a student might conduct an ethnographic study of breast-feeding among minority mothers at a community health clinic or study the history of the

English language in the dictionary department of an academic publishing house.

Once a student has graduated, his or her training in anthropology serves as an excellent preparation for virtually any career in which the understanding of the principles behind human behavior would be an asset. Possible career paths include archaeology, museum curation, linguistics, forensic science, marketing, human resources, social work, and consulting of all kinds. Anthropologists' ability to place problems within the context of human communities makes them valuable interpreters between organizations (corporations or governments) and the individuals they serve. For example, professionals with anthropology backgrounds have been hired to design more user-friendly software, create public awareness campaigns around pressing health issues, and interview families to discover the ways in which they use greeting cards to communicate with one another. Also, government agencies frequently employ anthropologists as policy researchers, program evaluators, planners, and analysts of all kinds.

Social Context and Future Prospects

In an ever-changing world, anthropology is a means of preserving the social, artistic, linguistic, and cultural heritage of different human communities for curious future generations. The significance of anthropological research, however, is far from merely intellectual. Increasingly, it is as common to find anthropologists engaged in solving practical problems for corporations, hospitals, or government agencies as it is to find them writing academic research papers or studying remote tribes. For example, the American Anthropological Association has a membership of about 10,000 professional anthropologists; the number of members who subscribe to its mailing list for applied anthropologists—scientists employed outside of the academic setting—totals more than 6,000. Also, contemporary anthropologists are more likely than ever before to turn their attention toward the beliefs and practices of their own cultures rather than conducting studies far from home. One example is British anthropologist Kate Fox's 2004 study of the thought patterns, rules, behaviors, and cultural rituals of modern British society.

M. Lee, B.A., M.A.

Further Reading

Brenneis, Donald. "A Partial View of Contemporary Anthropology." *American Anthropologist* 106, no. 3 (2004): 580-588. Addresses issues and practices within the field and makes proposals for sustaining the value of anthropological studies in the future. Refers to specific anthropological sites in the United States and abroad.

Brown, Peter, and Ron Barrett, eds. *Understanding and Applying Medical Anthropology.* 2d ed. Boston: McGraw-Hill, 2010. Each chapter includes an introduction to its author, thought questions, and citations.

Haviland, William A., et al. *Anthropology: The Human Challenge.* Belmont, Calif.: Thomson Wadsworth, 2008. A comprehensive textbook of cultural anthropology. Each chapter has a glossary, reflection questions, and sidebars such as "Anthropologists of Note" and "Biocultural Connections."

Metcalf, Peter. *Anthropology: The Basics.* New York: Routledge, 2005. A brief introduction to the field of cultural anthropology. Includes photographs and text boxes highlighting important concepts and case studies.

Park, Michael Alan. *Biological Anthropology.* 6th ed. Boston: McGraw-Hill, 2010. This discussion of biological anthropology contains chapter summaries, charts, drawings, a glossary, a pronunciation guide, and suggested readings.

Stephens, W. Richard, and Elliot M. Fratkin. *Careers in Anthropology.* Boston: Allyn and Bacon, 2003. A practical guide to careers in anthropology and related fields, organized as a series of real-life case studies describing the paths that sixteen different professionals followed after graduation.

Web Sites

American Anthropological Association
What Is Anthropology?
http://www.aaanet.org/about/WhatisAnthropology.cfm

National Science Foundation
Cultural Anthropology Program Overview
http://www.nsf.gov/sbe/bcs/anthro/cult_overview.jsp

See also: Forensic Science

APIOLOGY

FIELDS OF STUDY

Apiculture; melittology; entomology; agriculture; agricultural geography; biology; zoology; biochemistry; chemistry; physical science; statistics; botany; insect and pest management; food production; animal science; ecology.

SUMMARY

Apiology is the scientific study of the honeybee. It is a subdiscipline of melittology, which is the study of all bees and is a branch of entomology. Apiologists study the evolution of the honeybee and answer questions surrounding its biology, particularly its social behavior and the ecological role the insect plays in its habitat. Other areas of study include the reproduction cycle of the honeybee, its proficiency to gather nectar, and its production of honey. A critical topic for the apiologist is diseases of the honeybee. Given the migratory nature of commercial beekeeping, the spread of bee diseases and their treatment have become a major challenge for beekeepers.

KEY TERMS AND CONCEPTS

- **Apis Mellifera:** Genus and species name for the honeybee.
- **Bee Behavior:** Actions a bee takes in its instinctual inclination to survive.
- **Beehive:** Container that can house a colony of bees.
- **Brood:** Bees not yet developed out of the egg, larval, or pupal state.
- **Colony Collapse Disorder:** Collapse of a colony after a seemingly healthy colony of worker bees falls into a sudden, fatal decline.
- **Comb:** Wax material produced by bees and formed into hexagonal cells to store brood, nectar, honey, and pollen.
- **Drone:** Male bee.
- **Honey:** Product of the hive produced by the bees through a process of the fermentation of nectar.
- **Parthenogenesis:** State of being born of virgin female from an unfertilized egg.
- **Pollination:** Process of plant fertilization in which bees carry pollen from plant to plant.
- **Queen Bee:** Fertile female bee.
- **Swarm:** Process whereby part of a colony of honeybees leaves the hive in the company of a queen to propagate a new colony.
- **Worker Bee:** Sterile female bee.

Definition and Basic Principles

The terms "apiology" and "apiculture" (another word for apiology) are derived from the Latin word for bee, *apis*. The science to which these terms refer focuses specifically on the honeybee. Although beekeepers are required to manage and maintain their colonies using up-to-date methods and, in effect, are practicing apiculture, in contrast, it is the outcome of the research of apiologists and those scientifically investigating the well-being of the honeybee that sets the standard for good practice by beekeepers. Managers of commercial bee operations that might contain thousands of colonies are highly dependent on good science to help them stay productive as well as competitive.

Background and History

Apiary research and development emerged from a basic understanding of the behavior of bees. The outcome of this research has had considerable practical impact.

The standard removable-frame beehive—a design that was patented in 1852—was developed by observing bee behavior. Patent holder Lorenzo Langstroth, a beekeeper and Congregationalist preacher living in Ohio, noticed that bees maintained a prescribed space before they sealed over the frames of the hive. His response to this bee space was to develop a structure to function as the hive body that could be completely taken apart for harvesting honey and inspecting the bees. The outcome of this observation radically changed the way bees were housed and maintained.

The science of apiology has greatly contributed to understanding the genetic character and reproductive behaviors of bees. By the 1800's, it was generally accepted that the drone mated with the queen. The queen had the potential of laying eggs that became worker bees or drones. The Austrian monk Gregor Mendel, known for his genetic experiments with pea plants, also explored heredity by keeping and breeding bees. However, his bees were housed in the traditional stationary bee house, and he had no control over which drones bred with the queen.

In the early 1900's, apiology researchers had perfected the technique of instrumental insemination of the queen bee. This method made it possible to control the breeding fprocess and has allowed for dramatic advances in the crossing of bees. The practical outcome of this science for the beekeeper is a well-tempered and productive bee.

Apiology also has provided insight into the diseases and enemies of the honeybee. Extensive research is conducted on mites and their control within the colony. Out of the study of bacterial problems such as foulbrood (a disease that attacks honeybee larvae), new practices in apiary management have emerged that can control these devastating honeybee enemies. Likewise, apiology provides an understanding of the symbiotic relationships that exist between bees and other insects as well as other animals.

How It Works

Beekeeping. The process of keeping bees requires an understanding of the biology of bees and the instinctual behaviors they exhibit within their colony. The colony's ability for honey production, fecundity (potential for reproduction), and the tendency to swarm, are also factors that must be managed by the beekeeper to maintain the viability of the colony. The beekeeper must become aware of the environmental conditions that are conducive to pollination and nectar gathering. This requires an understanding of the blooming cycles of various flora within a region as well as weather conditions best suited for flying times by the bees. The outcome of the proper application of this knowledge will be expressed in the form of stronger and more productive colonies.

Bee Diseases. Diseases within the hive take a variety of forms. Some of these are found exclusively in the brood, thereby weakening the colony by lowering its overall population. In contrast, some of these diseases affect the adult bee, destroying the working population. The impact of these maladies can be substantial. Some diseases are the result of bacteria, such as American and European foulbrood; a virus, such as sacbrood; or a parasite such as *Nosema apis.* Mites such as *Varroa jacobsoni* are considered to be the number one killer of honeybees globally. These huge ectoparasites live on the outside of the bodies of the bees and the brood. They survive by attaching themselves to the insect and drawing out its hemolymph, a fluid similar to blood. During the early 1900's, the acarine mite (*Acarapis woodii*), also known as the tracheal mite, was responsible for killing all the black honeybees in England. These mites live in the trachea of the bee, cutting off its airway and choking it. Understanding the biology of this mite is the result of extensive research in apiology. Another pest that can wreak havoc on a hive is the wax moth (*Galleria mellonella*), which destroys the comb and the brood. Because of the nature of commercial migratory beekeeping, these diseases and pests can spread rapidly as colonies from one apiary come into contact with colonies from other apiaries.

Research into the diseases impacting honeybees is crucial to the survival and management of the honeybee. Apiologists identify these diseases by noting symptoms, tracking their spread geographically, and finding methods for their control. Integrated pest management systems using both cultural mechanical controls (such as hive modifications) and genetic controls (such as breeding bees that are resistant to a particular kind of disease) have all emerged from apiologists' research.

Bee Genetics. The domestication of the honeybee has been an ongoing process in societies as far back as those in ancient Egypt. In the twenty-first century, the crossbreeding of honeybees is controlled through scientific methods such as the instrumental insemination of the queen. This requires surgical-grade laboratory instruments, which hold and prepare the queen and the selection of drone bees that contribute semen for the insemination. This technique of breeding ensures that the traits desired in these bees will be secured. Techniques such as these have allowed for the quick development of new hybrid strains of the honeybee. Benedictine monk Karl Kehrle, known as Brother Adam, of Buckfast Abbey in England, is best known for his contribution to honeybee breeding. By selecting queens with particular traits and behaviors from around the world, he created what is now known as the Buckfast bee. The Buckfast bee was bred for good temper, honey production, and resistance to the acarine mite. Apiology also has genetically produced queens that can combat varroa mites. For example, varroa-sensitive hygiene (VSH) queens have been developed with

Fascinating Facts About Apiology

- Honeybees provide about 80 percent of crop pollination. Some of the crops they pollinate include apples, pears, cherries, plums, almonds, avocados, cantaloupes, blueberries, cranberries, cucumbers, watermelons, sunflowers, and pumpkins.
- The first successful instrumental insemination of a queen bee was performed in 1926.
- Some studies suggest that the economic contribution through pollination by honeybees is greater than $14 billion annually.
- Apiologists studying the foraging habits of honeybees have found that a bee makes as many as ten flower stops per minute collecting pollen and nectar from as many as six hundred flowers before returning to its colony.
- Africanized bees made their way north through South and Central America following an accidental release in Brazil in 1957. The bees entered North America in the early 1990's and have come to reside in Texas, California, Louisiana, Arkansas, New Mexico, Mississippi, Alabama, and Florida. The massive number of stings these bees inflict can be fatal to both animals and humans, which is why they are also called killer bees.
- Varroa mites are the number one killer of honeybees worldwide and are one of the largest ectoparasites (parasites that live on the outside of the host). If this mite's size was considered on a human scale, it would as big as a basketball.
- The acarine mite, also known as the tracheal mite, resides within the trachea of the bee. These parasites lay their eggs in the airway of the bee, where they mature. Only mated mites leave the trachea of the host bee to find another bee in which to reside. Eventually, all the mites will leave a host bee after the infested bee dies. These mites must reside within a host and will die if they cannot find another.

the ability to detect the presence of varroa within a colony. Upon detection, they begin to clean the brood chambers of larvae that contain the mite.

APPLICATIONS AND PRODUCTS

Honey. Honey production is the primary commercial output of the honeybee colony. About 200 million pounds of honey are produced yearly within the United States. Consumption in the United States is twice that amount. Thus, honey is imported from sources around the world. The floral sources for nectar, the substance from which bees produce honey, is variable and determines the type and color of honey produced. Perhaps as many as three hundred varieties of honey can be produced in the United States alone, all having their own unique taste and color. Lighter-grade honeys and honey still in the comb are consumed directly and are usually sold at relatively high prices. Darker-grade honey may be used to sweeten baked goods. Popular floral sources for honey include clover, apple blossom, orange blossom, and huckleberry. Honey is sold in dehydrated form, with fruit flavors added to it, in comb form with capped and uncapped cells filled with honey, and as either organic or nonorganic regular or creamed honey. Aside from its use for humans as the world's oldest sweetener, honey is used in the production of paints, adhesives, cosmetics, and medicinal products.

Pollination. A major source of value for the beekeeper and the grower, and hence the public at large, is the immense contribution of pollination that bees make to agriculture. Honeybees are used to pollinate the blossoms of many vegetables and fruits. The distribution of bees to locations timed to coincide with the seasonal arrival of blossoms is known as migratory beekeeping and constitutes a large and important industry. Some research suggests that nearly one-third of a human's diet is composed of products pollinated by insects, and about 80 percent of this pollination is accomplished by honeybees.

Medicine from the Hive. Substances coming from the beehive as well as the sting of the bee itself have been considered medicinal products. "Apitherapy" is a collective term referring to the application of bee venom and other products from the hive for medical purposes. Although modern apitherapy takes a scientific approach, it has a rich history in folk medicine. It began as a healing art practiced by the early Greeks, Egyptians, and Chinese. In the 1800's, the idea of bee-venom therapy was introduced to the United States. Bee venom, taken either directly from being stung or by injection with a syringe, is used in the treatment of arthritis, rheumatism, tendinitis, and multiple sclerosis; it has also been used to treat the pain from gout, shingles, and other maladies. It is believed that bee-venom therapy can increase circulation of the blood. Apitherapy is also used to decrease inflammation of tissues and to stimulate immune responses.

Royal jelly is the substance made by worker bees to feed potential queen bees within the hive. This substance can also be used to treat open wounds and as an energy tonic. The extraction of royal jelly is a labor-intensive process. Propolis is a bee-produced substance reddish to brown in color with the consistency of sticky clay. The bees utilize propolis to bridge over areas within the hive that exceed bee space so as to seal in the hive. This product has antifungal and antibacterial properties, so it can be used medicinally.

Bees collect plant pollen as a nutritional source for the colony. For some people, pollen is an allergen. But some studies suggest that human consumption of raw pollen from particular locales, as well as the raw honey that contains these pollen spores, can build immunity to pollen allergies from the same locales.

Beeswax. The worker bees of the hive secrete wax through glands on the sides of their bodies. In the colony, this wax is used to make comb, which is the base for building up hexagonally shaped cells. After the advent of the removable-comb by Langstroth, frames within the hive contained a foundation made from beeswax with a hexagonal imprint. From this paper-thin wax sheet, the bees produce the honeycomb that is used in rearing brood and storing food for the colony.

Beeswax finds its way into many common substances and products. Beeswax is well known in its use in candle production. In art, melted beeswax is used as a base for batik designs on fabric and in producing encaustic paints. Additionally, beeswax can be carved into figures that are used as models in the production of metal casting.

Many cosmetics, as well as some foods, use beeswax. Beeswax is also used as a seal in tree grafting.

Impact on Industry

Migratory Beekeeping Industry. The process of commercial pollination by bees translates into millions of bees doing labor that would otherwise have to be done by hand or by spraying pollen with an airplane. Both of these processes are costly. Bees—by virtue of both their efficiency, in visiting each flower in an orchard or field, and the fact that they do not need to be paid for their time—translate into millions of dollars saved in the production of food for human and animal consumption. Pollination by honeybees is limited only when growers misjudge blossom times, do not set sufficient numbers of colonies in the field, or confront weather that does not cooperate with the bees' flying time.

A symbiotic and economic relationship exists with honeybee pollinators and the plants they pollinate. Bees pollinating plants derive food and nectar for their colonies to survive for the winter. This can translate into an economic savings to the beekeeper. If plants or trees to be pollinated are profuse with pollen and nectar, the bees can store this food and feed themselves over the winter, requiring limited supplemental feeding from the beekeeper. Additionally, many migratory beekeepers are in business for pollination only and not honey production. This savings is critical to the migratory beekeeper's economic survival.

Perhaps the most attractive aspect of this industry is that it is among the few agricultural practices that have, for centuries, maintained the ecological principles of sustainability and have caused limited environmental degradation by avoiding mechanical means of pollination.

Research. Both public and private organizations spearhead the research efforts in apiology. At the national level, the U.S. Department of Agriculture maintains research stations in Tucson, Arizona; Baton Rouge, Louisiana; Logan, Utah; Beltsville, Maryland; and Weslaco, Texas. A number of universities have conducted ongoing research into areas such as colony collapse syndrome, Africanized bees or killer bees (an aggressive strain that began infiltrating the southern United States during the early 1990's), bee diseases, bee pests, and bee genetics, as well as apiary economics. At the state and local levels, multiple beekeeping associations in all fifty states, as well as the District of Columbia and Puerto Rico, assist in disseminating information and in some cases collecting and providing data to scientists conducting research.

The International Bee Research Association in the United Kingdom was established in 1949 and publishes a series of research journals, books, and pamphlets for the world beekeeping community. It also sponsors international conferences dealing with beekeeping issues. In the United States, publishing outlets for scientific research include the *American Bee Journal*, established in 1861 and published by Dadant and Sons, and Bee Culture, now online but originally established in 1873 and published by A. I. Root and Sons.

Honeybee Genome Project. Funded by the National Human Genome Research Institute and the U.S. Department of Agriculture, the effort to sequence the genes of *Apis mellifera* began in 2003 and was completed three years later. Interest in tracing the genes of the honeybee was based on its importance in agricultural production and its social structure. However, the project may also provide insight into human health conditions, such as immunity to disease and allergies.

Colony Collapse Disorder. Sometimes referred to as CCD, colony collapse disorder has made international news and describes the demise of millions of honeybee colonies. Given the economic benefit that bees provide in agricultural production, research into this problem is crucial. Although researchers think that a virus might be the cause of this bee malady, scientists are quick to point out that since the mid-twentieth century the varroa mite has been the most damaging pest to the bee industry worldwide. Many researchers suggest that varroa or a combination of this mite and some other factor might be the cause of CCD. Fighting the varroa-mite outbreak has required both integrated pest management systems and treatment with pesticides.

Careers and Course Work

Pursuing a career as an apiologist or a bee researcher requires a formal education that includes at least a bachelor of science (B.S.) degree. A master of science (M.S.), and in some cases a Ph.D., is required in most research and teaching posts at the university level. In all cases, a curriculum that develops the student's ability to establish systematic scientific studies, utilize a variety of computer programs that model environmental and biological conditions, and apply statistics is foundational for a career in apiology. In the United States, only a handful of colleges offer degrees in the field of apiology or apiculture. A degree in biology or entomology might be the key to employment. Ideally, such academic training would include apiary management or some related field of agriculture. Practical experience is also important and in some cases might pave the way to a related college program or even a career.

Researchers are found in both the private and public sectors. Chemical companies dealing with apiary pesticides and other chemicals seek trained individuals for laboratory and field research. At the federal level, the U.S. Department of Agriculture employs research scientists in their apiary units around the country. Additionally, some states have active apiary programs within their own departments of agriculture. At the state level, the trained apiologist may find duties both in a laboratory and in the field, inspecting hives.

Depending on the size of the apiary, migratory beekeepers individually maintain the seasonal requirements of their apiary operation as well as own and operate semi-truck transportation for the movement of their bees. In some cases, transportation is leased out to companies that are specialists in handling loads of bees. Forklifts and tractors are used to move palletized beehives from flatbed trucks to positions within the field of pollination and then back again to flatbed semitrailers for hauling to the next pollination or nectar-gathering site. A number of skills are employed to maintain the apiary. These might include building and repairing bee boxes and frames, "pulling of supers" (boxes loaded with honey), operating honey-harvesting equipment, and working in the bee yard requeening (introducing a new queen) and checking hives for disease.

Social Context and Future Prospects

Although mason bees, such as the orchard bee (*Osmia lignaria*), have garnered some attention as an alternative pollinator, as a honey producer and pollinator the honeybee still reigns supreme. The impact of honeybee pests such as varroa has already exposed the public to the importance of the honeybee in the production of food and to the potential of rising food costs if the demise of the honeybee continues. Further research into the interaction of bees with pesticides and herbicides as well as other environmental agents will continue to be critical research topics. Bee management systems will continue to be updated to accommodate environmental changes. Such changes are already being observed in some areas where reestablishing new colonies on a yearly basis is common practice. This requires the raising of more bees as well as new queens. Techniques such as instrumental insemination, which speeds the process of genetic modifications outside simple breeding programs, will also continue to develop as demand for bees with resistance to pests and diseases increases. Out of economic necessity, research will continue to develop around the honeybee. Its integration into the economic fabric of agriculture is far too great to ignore.

M. Mustoe, Ph.D.

Further Reading

Bradbear, Nicola. *Beekeeping and Sustainable Livelihoods: FAO Diversification Booklet No. 1.* Rome, Italy: Food and Agriculture Organization of the United Nations, 2004. This short treatise discusses apiary management as a sustainable agricultural enterprise.

Connor, Lawrence John, ed. *Asian Apiculture: Proceedings of the First International Conference on the Asian Honey Bees and Bee Mites.* Cheshire, Conn.: Wicwas Press, 1993. Overview of the impact of Asian mites such as varroa.

_______. *Queen Rearing Essentials.* Kalamazoo, Mich.: Wicwas Press, 2009. A discussion of the process of grafting and selecting queen bees to improve production.

Cook, Albert John. *Manual of the Apiary.* 3d ed. Chicago: T. G. Newman & Son, 1878. This text is an interesting view on apiary science from the past, much of which still has applications.

Graham, Joe M., ed. *The Hive and the Honey Bee: A New Book on Beekeeping Which Continues the Tradition of "Langstroth on the Hive and the Honeybee."* Rev. ed. Hamilton, Ill.: Dadant, 1992. Offers history and background on modern foundations of apiculture.

Root, A. I., and E. R. Root. *The ABC and XYZ of Bee Culture: A Cyclopedia of Everything Pertaining to the Care of the Honey-Bee: Bees, Hives, Honey, Implements, Honey Plants, Etc.* 41st ed. Medina, Ohio: A. I. Root, 2007. As the rest of the subtitle of this venerable source states, this volume contains "facts gleaned from the experience of thousands of beekeepers and afterward verified in our apiary." A key resource for apiologists.

Seeley, Thomas D. *Honeybee Democracy.* Princeton, N.J.: Princeton University Press, 2010. An entomologist surveys the collective behavior of honeybees.

Spivak, Marla, and Gary S. Reuter. *Honey Bee Diseases and Pests.* St. Paul: University of Minnesota, 2010. This book is an overview of bee diseases and treatment methods.

Web Sites

Bee Culture
http://www.beeculture.com

International Bee Research Association
http://www.ibra.org.uk

National Human Genome Research Institute
Honey Bee Genome Sequencing
http://www.genome.gov/11008252

U.S. Department of Agriculture
http://www.usda.gov

See also: Plant Breeding and Propagation; Wildlife Conservation.

AUTOMATED PROCESSES AND SERVOMECHANISMS

FIELDS OF STUDY

Electronics; hydraulics and pneumatics; mechanical engineering; computer programming; machining and manufacturing; millwright; quality assurance and quality control; avionics; aeronautics.

SUMMARY

An automated process is a series of sequential steps to be carried out automatically. Servomechanisms are systems, devices, and subassemblies that control the mechanical actions of robots by the use of feedback information from the overall system in operation.

KEY TERMS AND CONCEPTS

- **CNC (Computer Numeric Control):** An operating method in which a series of programmed logic steps in a computer controls the repetitive mechanical function of a machine.
- **Control Loop:** The sequence of steps and devices in a process that regulates a particular aspect of the overall process.
- **Feedback:** Information from the output of a process that is fed back as an input for the purpose of automatically regulating the operation.
- **Proximity Sensor:** An electromagnetic device that senses the presence or absence of a component through its effect on a magnetic field in the sensing unit.
- **Weld Cell:** An automated fabrication center in which programmed robotic welding units perform a specified series of welds on successive sets of components.

Definition and Basic Principles

An automated process is any set of tasks that has been combined to be carried out in a sequential order automatically and on command. The tasks are not necessarily physical in nature, although this is the most common circumstance. The execution of the instructions in a computer program represents an automated process, as does the repeated execution of a series of specific welds in a robotic weld cell. The two are often inextricably linked, as the control of the physical process has been given to such digital devices as programmable logic controllers (PLCs) and computers in modern facilities.

Physical regulation and monitoring of mechanical devices such as industrial robots is normally achieved through the incorporation of servomechanisms. A servomechanism is a device that accepts information from the system itself and then uses that information to adjust the system to maintain specific operating conditions. A servomechanism that controls the opening and closing of a valve in a process stream, for example, may use the pressure of the process stream to regulate the degree to which the valve is opened.

The stepper motor is another example of a servomechanism. Given a specific voltage input, the stepper motor turns to an angular position that exactly corresponds to that voltage. Stepper motors are essential components of disk drives in computers, moving the read and write heads to precise data locations on the disk surface.

Another essential component in the functioning of automated processes and servomechanisms is the feedback control systems that provide self-regulation and auto-adjustment of the overall system. Feedback control systems may be pneumatic, hydraulic, mechanical, or electrical in nature. Electrical feedback may be analogue in form, although digital electronic feedback methods provide the most versatile method of output sensing for input feedback to digital electronic control systems.

Background and History

Automation begins with the first artificial construct made to carry out a repetitive task in the place of a person. Early clock mechanisms, such as the water clock, used the automatic and repetitive dropping of a specific amount of water to accurately measure the passage of time. Water-, animal- or, wind-driven mills and threshing floors automated the repetitive action of processes that had been accomplished by humans. In many underdeveloped areas of the world, this repetitive human work is still a common practice.

With the mechanization that accompanied the Industrial Revolution, other means of automatically controlling machinery were developed, including self-regulating pressure valves on steam engines. Modern automation processes began in North America with the establishment of the assembly line as a standard industrial method by Henry Ford. In this method, each worker in his or her position along the assembly line performs a limited set of functions, using only the parts and tools appropriate to that task.

Servomechanism theory was further developed during World War II. The development of the transistor in 1951, and hence, digital electronics, enabled the development of electronic control and feedback devices. The field grew rapidly, especially following the development of the microcomputer in 1969. Digital logic and machine control can now be interfaced in an effective manner, such that today's automated systems function with an unprecedented degree of precision and dependability.

How It Works

An automated process is a series of repeated, identical operations under the control of a master operation or program. While simple in concept, it is complex in practice and difficult in implementation and execution. The process control operation must be designed in a logical, step-by-step manner that will provide the desired outcome each time the process is cycled. The sequential order of operations must be set so that the outcome of any one step does not prevent or interfere with the successful outcome of any other step in the process. In addition, the physical parameters of the desired outcome must be established and made subject to a monitoring protocol that can then act to correct any variation in the outcome of the process.

A plain analogy is found in the writing and structuring of a simple computer programming function. The definition of the steps involved in the function must be exact and logical, because the computer, like any other machine, can do only exactly what it is instructed to do. Once the order of instructions and the statement of variables and parameters have been finalized, they will be carried out in exactly the same manner each time the function is called in a program. The function is thus an automated process.

The same holds true for any physical process that has been automated. In a typical weld cell, for example, a set of individual parts are placed in a fixture that holds them in their proper relative orientations. Robotic welding machines may then act upon the setup to carry out a series of programmed welds to join the individual pieces into a single assembly. The series of welds is carried out in exactly the same manner each time the weld cell cycles. The robots that carry out the welds are guided under the control of a master program that defines the position of the welding tips, the motion that it must follow, and the duration of current flow in the welding process for each movement, along with many other variables that describe the overall action that will be followed. Any variation from this programmed pattern of movements and functions will result in an incorrect output.

The control of automated processes is carried out through various intermediate servomechanisms. A servomechanism uses input information from both the controlling program and the output of the process to carry out its function. Direct instruction from the controller defines the basic operation of the servomechanism. The output of the process generally includes monitoring functions that are compared to the desired output. They then provide an input signal to the servomechanism that informs how the operation must be adjusted to maintain the desired output. In the example of a robotic welder, the movement of the welding tip is performed through the action of an angular positioning device. The device may turn through a specific angle according to the voltage that is supplied to the mechanism. An input signal may be provided from a proximity sensor such that when the necessary part is not detected, the welding operation is interrupted and the movement of the mechanism ceases.

The variety of processes that may be automated is practically limitless given the interface of digital electronic control units. Similarly, servomechanisms may be designed to fit any needed parameter or to carry out any desired function.

Applications and Products

The applications of process automation and servomechanisms are as varied as modern industry and its products. It is perhaps more productive to think of process automation as a method that can be applied to the performance of repetitive tasks than to dwell on specific applications and products. The commonality of the automation process can be illustrated by examining a number of individual applications, and the products that support them.

"Repetitive tasks" are those tasks that are to be carried out in the same way, in the same circumstances, and for the same purpose a great number of times. The ideal goal of automating such a process is to ensure that the results are consistent each time the process cycle is carried out. In the case of the robotic weld cell described above, the central tasks to be repeated are the formation of welded joints of specified dimensions at the same specific locations over many hundreds or thousands of times. This is a typical operation in the manufacturing of subassemblies in the automobile industry and in other industries in which large numbers of identical fabricated units are produced.

Automation of the process, as described above, requires the identification of a set series of actions to be carried out by industrial robots. In turn, this requires the appropriate industrial robots be designed and constructed in such a way that the actual physical movements necessary for the task can be carried out. Each robot will incorporate a number of servomechanisms that drive the specific movements of parts of the robot according to the control instruction set. They will also incorporate any number of sensors and transducers that will provide input signal information for the self-regulation of the automated process. This input data may be delivered to the control program and compared to specified standards before it is fed back into the process, or it may be delivered directly into the process for immediate use.

Programmable logic controllers (PLCs), first specified by the General Motors Corporation in 1968, have become the standard devices for controlling automated machinery. The PLC is essentially a dedicated computer system that employs a limited-instruction-set programming language. The program of instructions for the automated process is stored in the PLC memory. Execution of the program sends the specified operating parameters to the corresponding machine in such a way that it carries out a set of operations that must otherwise be carried out under the control of a human operator.

A typical use of such methodology is in the various forms of CNC machining. CNC (computer numeric control) refers to the use of reduced-instruction-set computers to control the mechanical operation of machines. CNC lathes and mills are two common applications of the technology. In the traditional use of a lathe, a human operator adjusts all of the working parameters such as spindle rotation speed, feed rate, and depth of cut, through an order of operations that is designed to produce a finished piece to blueprint dimensions. The consistency of pieces produced over time in this manner tends to vary as operator fatigue and distractions affect human performance. In a CNC lathe, however, the order of operations and all of the operating parameters are specified in the control program, and are thus carried out in exactly the same manner for each piece that is produced. Operator error and fatigue do not affect production, and the machinery produces the desired pieces at the same rate throughout the entire working period. Human intervention is required only to maintain the machinery and is not involved in the actual machining process.

Servomechanisms used in automated systems check and monitor system parameters and adjust operating conditions to maintain the desired system output. The principles upon which they operate can range from crude mechanical levers to sophisticated and highly accurate digital electronic-measurement devices. All employ the principle of feedback to control or regulate the corresponding process that is in operation.

In a simple example of a rudimentary application, units of a specific component moving along a production line may in turn move a lever as they pass by. The movement of the lever activates a switch that prevents a warning light from turning on. If the switch is not triggered, the warning light tells an operator that the component has been missed. The lever, switch, and warning light system constitute a crude servomechanism that carries out a specific function in maintaining the proper operation of the system.

In more advanced applications, the dimensions of the product from a machining operation may be tested by accurately calibrated measuring devices before releasing the object from the lathe, mill, or other device. The measurements taken are then compared to the desired measurements, as stored in the PLC memory. Oversize measurements may trigger an action of the machinery to refine the dimensions of the piece to bring it into specified tolerances, while undersize measurements may trigger the rejection of the piece and a warning to maintenance personnel to adjust the working parameters of the device before continued production.

Two of the most important applications of servomechanisms in industrial operations are control of position and control of rotational speed. Both commonly employ digital measurement. Positional control is generally achieved through the use of servomotors, also known as stepper motors. In these devices, the rotor turns to a specific angular position according to the voltage that is supplied to the motor. Modern electronics, using digital devices constructed with integrated circuits, allows extremely fine and precise control of electrical and electronic factors, such as voltage, amperage, and resistance. This, in turn, facilitates extremely precise positional control. Sequential positional control of different servomotors in a machine, such as an industrial robot, permits precise positioning of operating features. In other robotic applications, the same operating principle allows for extremely delicate microsurgery that would not be possible otherwise.

The control of rotational speed is achieved through the same basic principle as the stroboscope. A strobe light flashing on and off at a fixed rate can be used to measure the rate of rotation of an object. When the strobe rate and the rate of rotation are equal, a specific point on the rotating object will always appear at the same location. If the speeds are not matched, that point will appear to move in one direction or the other according to which rate is the faster rate. By attaching a rotating component to a representation of a digital scale, such as the Gray code, sensors can detect both the rate of rotation of the component and its position when it is functioning as part of a servomechanism. Comparison with a digital statement of the desired parameter can then be used by the controlling device to adjust the speed or position, or both, of the component accordingly.

IMPACT ON INDUSTRY

Automated processes, and the servomechanisms that apply to them, have had an immeasurable impact on industry. While the term "mass production" does not necessarily imply automation, the presence of automated systems in an operation does indicate a significant enhancement of both production and precision. Mass production revolutionized the intrinsic nature of industry in North America, beginning with the assembly-line methods made standard by Ford. This innovation enabled Ford's industry to manufacture automobiles at a rate measured in units per day rather than days per unit.

While Ford's system represents a great improvement in productions efficiency, that system can only be described as an automatic system rather than an automated process. The automation of automobile production began when machines began to carry out some of the assembly functions in the place of humans. Today, it is entirely possible for the complete assembly process of automobiles, and of other goods, to be fully automated. This is not the most effective means of production, however, as even the most sophisticated of computers pales in comparison to the human brain in regard to intuition and intelligence. While the computer excels at controlling the mechanical function of processes, the human values of aesthetics and quality control and management are still far beyond computer calculation. Thus, production facilities today utilize the cooperative efforts of both machines and humans.

As technology changes and newer methods and materials are developed, and as machines wear out or become obsolete, constant upgrading of production facilities and methods is necessary. Process automation is essential for minimizing costs and for maximizing performance in the delivery of goods and services, with the goal of continuous improvement. It is in this area that process automation has had another profound effect on modern industry, identified by the various total quality management (TQM) programs that have been developed and adopted in many different fields throughout the world.

TQM can be traced to the Toyota method, developed by the Toyota Motor Corporation in Japan following World War II. To raise Japan's economy from the devastating effects of its defeat in that conflict, Toyota executives closely examined and analyzed the assembly-line methods of manufacturers in the United States, then enhanced those methods by stressing the importance of increasing efficiency and minimizing waste. The result was the method that not only turned Toyota into the world's largest and most profitable manufacturer of automobiles and other consumer goods, but revolutionized the manner in which manufacturers around the world conducted their businesses. Today, TQM programs such as Lean, 6Sigma, and other International Organization for Standardization (ISO) designations use process automation as a central feature of their operations; indeed, TQM represents an entirely new field of study and practice.

CAREERS AND COURSEWORK

Students looking to pursue a career that involves automated processes and servomechanisms can

Fascinating Facts About Automated Processes

- The ancient Egyptians used a remote hydraulic system to monitor the level of the water in the Nile River to determine the beginning and end of annual religious festivals.
- The water clock designed to measure the passage of a specific amount of time was perhaps the first practical, artificial automated process.
- Grain threshing mills powered by animals, wind, or water to automate the separation of grain kernels from harvested plants are still in use today.
- In relatively recent times, the turning of roasting spits in roadhouse kitchens was automated through the use of dogs called turnspits, which ran inside a treadmill wheel. A morsel of food suspended just beyond the reach of the dog was the "servomechanism" that kept the dog running.
- Control mechanisms of automated processes can use either analogue signal processing or digital signal processing.
- Computer control of physical operations is much more effective than human control, but human intervention is far superior for the subjective functions of management and quality control.
- Servomechanisms that perform some kind of automated control function can be found in unexpected places, including the spring-loaded switch that turns on the light in a refrigerator when its door is opened.
- Modern digital electronics permits extremely fine control of many physical actions and movements through servomechanisms.
- Process automation is as applicable to the fields of business management and accounting as it is to mass production in factories.

expect to take foundational courses in applied mathematics, physics, mechanics, electronics, and engineering. Specialization in feedback and control systems technology will be an option for those in a community college or pursuing an associate's, degree. Industrial electronics, digital technology, and machining and millwrighting also are optional routes at this level. More advanced levels of studies can be pursued through college or university programs in computer programming, mechanical engineering, electrical engineering, and some applied sciences in the biomedical field.

As may be imagined, robotics represents a significant aspect of work in this field, and students should expect that a considerable amount of their coursework will be related to the theory and practice of robotics.

As discussed, process automation is applicable in a variety of fields in which tasks of any kind must be repeated any number of times. The repetition of tasks is particularly appropriate in fields for which computer programming and control have become integral. It is therefore appropriate to speak of automated accounting practices, automated blood-sample testing, and other biomedical testing both in research and in treatment contexts, automated traffic control, and so on. One may note that even the procedure of parallel parking has become an automated process in some modern automobiles. The variety of careers that will accept specialized training geared to automated processes is therefore much broader than might at first be expected through a discussion that focuses on robotics alone.

Social Context and Future Prospects

While the vision of a utopian society in which all menial labor is automated, leaving humans free to create new ideas in relative leisure, is still far from reality, the vision does become more real each time another process is automated. Paradoxically, since the mid-twentieth century, knowledge and technology have changed so rapidly that what is new becomes obsolete almost as quickly as it is developed, seeming to increase rather than decrease the need for human labor.

New products and methods are continually being developed because of automated control. Similarly, existing automated processes can be re-automated using newer technology, newer materials, and modernized capabilities.

Particular areas of growth in automated processes and servomechanisms are found in the biomedical fields. Automated processes greatly increase the number of tests and analyses that can be performed for genetic research and new drug development. Robotic devices become more essential to the success of delicate surgical procedures each day, partly because of the ability of integrated circuits to amplify or reduce electrical signals by factors of hundreds of thousands. Someday, surgeons will be able to perform the most delicate of operations remotely, as normal actions by the surgeon are translated into the miniscule movements of microscopic surgical equipment manipulated through robotics.

Concerns that automated processes will eliminate the role of human workers are unfounded. The nature of work has repeatedly changed to reflect the capabilities of the technology of the time. The introduction of electric street lights, for example, did eliminate the job of lighting gas-fueled streetlamps, but it also created the need for workers to produce the electric lights and to ensure that they were functioning properly. The same sort of reasoning applies to the automation of processes today. Some traditional jobs will disappear, but new types of jobs will be created in their place through automation.

Richard M. Renneboog, M.Sc.

Further Reading

Bryan, L. A., and E. A. Bryan. *Programmable Controllers: Theory and Implementation.* Atlanta: Industrial Text, 1988. This textbook provides a sound discussion of digital logic principles and the digital electronic devices of PLCs. Proceeds through a detailed discussion of the operation of PLCs, data measurement, and the incorporation of the systems into more complex and centralized computer networks.

James, Hubert M. *Theory of Servomechanisms.* New York: McGraw-Hill, 1947. Discusses the theory and application of the principles of servomechanisms before the development of transistors and digital logic.

Kirchmer, Mathias. *High Performance Through Process Excellence.* Berlin: Springer, 2009. Examines the processes of business management as related to automation and continuous improvement in all areas of business operations.

Seal, A. M. *Practical Process Control.* Oxford, England: Butterworth-Heinemann/Elsevier, 1998. An advanced technical book that includes a good description of analogue and digital process control mechanisms and a discussion of several specific industrial applications.

Seames, Warren S. *Computer Numerical Control Concepts and Programming.* 4th ed. Albany, N.Y.: Delmar, Thomson Learning, 2002. Provides an overview of numerical control systems and servomechanisms, with an extended discussion of the specific functions of the technology.

Smith, Carlos A. *Automated Continuous Process Control.* New York: John Wiley & Sons, 2002. A comprehensive discussion of process control systems in chemical engineering applications.

Web Sites

Control-Systems-Principles
http://www.control-systems-principles.co.uk

See also: Computer-Aided Design and Manufacturing; Human-Computer Interaction.

B

BIOCHEMICAL ENGINEERING

FIELDS OF STUDY

Biochemistry; microbiology; biotechnology; cell biology; biology; chemical engineering; chemistry; genetics; molecular biology; pharmacology; medicine; agriculture; food science; environmental science; petroleum refinement; physiology; waste management.

SUMMARY

Biochemical engineers are responsible for designing and constructing those manufacturing processes that involve biological organisms or products made by them. Biochemical engineers take commercially valuable biological or biochemical commodities and design the means to produce those commodities effectively, cheaply, safely, and in mass quantities. They do this by optimizing the growth of organisms that produce valuable molecules or perform useful biochemical processes, establishing the most effective way to purify the desired molecules, and designing the operation systems that execute these processes, while adhering to a high standard of quality, purity, worker safety, and environmental cleanliness.

KEY TERMS AND CONCEPTS

- **Biofuels:** Solid, liquid, or gaseous fuels derived from biomass.
- **Biomass:** Plant materials and animal waste used especially as a source for fuel.
- **Bioreactor:** Device in which industrial biochemical reactions occur with the help of either enzymes or living cells.
- **Enzymes:** Particular proteins or ribonucleic acids (RNAs) that accelerate the rate of chemical reactions without being consumed or changed in the process.
- **Genetically Engineered Organisms:** Biological organisms that have had their endogenous deoxyribonucleic acid (DNA) altered, usually by the introduction of foreign DNA.
- **Hollow-Fiber Membrane Bioreactor (HFMB):** Cylindrical bioreactor that has a series of thin, porous, narrow, hollow tubes inside a plastic cylinder. Cultured cells grow in the spaces between the hollow tubes or fibers (extra-capillary spaces), while oxygen and nutrients continuously flow through the hollow fibers to the cells.
- **Hybridoma:** Cultured cell line that results from the fusion of an antibody-making B lymphocyte and a myeloma (B lymphocyte tumor cell) that secretes a monoclonal antibody.
- **Monoclonal Antibodies:** Proteins secreted by specific cells of the immune system that precisely bind to specific sites on the surface of foreign invaders and facilitate the destruction or neutralization of the foreign invaders.
- **Phage Display:** Test-tube selection technique that genetically fuses a protein to the outer-coat protein of a virus that infects bacteria, resulting in display of the fused protein on the outside of the virus. This allows screening of vast numbers of variants of the protein, each encoded by its corresponding DNA sequence.
- **Photobioreactor:** Translucent container that incorporates a light source and is used to grow small photosynthetic organisms for controlled biomass production.
- **Wave Bioreactor:** Disposable, sterile, plastic bag bioreactor that is mounted on a rocking platform, which creates wave action inside the bag to mix the culture that grows inside it.

DEFINITION AND BASIC PRINCIPLES

Biochemical engineering involves designing and building those industrial processes that use catalysts, feedstocks, or absorbents of biological origin. Industrial processes used in food, waste-management, pharmaceutical, and agricultural plants are often called unit operations. Those unit operations used in combination with biological organisms or molecules include heat and mass transfer, bioreactor design and operation, filtration, cell isolation, and sterilization.

One of the main tasks of bioengineers is to optimize the production of commercially valuable molecules by genetically engineered microorganisms. Biochemical engineers design culture containers known as bioreactors that accommodate growing cultures and maintain an environment that keeps growth at optimal levels. They also create the protocols that separate the cultured cells and their growth medium from the molecule of interest and purify this molecule from all contaminating components. Biochemical engineers do not make the genetically engineered organisms that produce or do valuable things, but instead they maximize the capacities of such organisms in the safest and most cost-effective ways.

Biochemical engineers also design systems that degrade organic or industrial waste. In these cases, bioreactors house biological organisms that receive and decompose waste. They select the right organism or mix of organisms for the job at hand, establish environments that allow these organisms to thrive, and design systems that feed waste to the organisms and remove the degradation products.

A branch of biochemical engineering called tissue engineering combines cultured cells with synthetic materials and external forces to mold those cells into organs that can serve as a replacement for diseased or damaged organs. Biochemical engineers determine the forces, materials, or biochemical cues that drive cells to form fully functional organs and then design the bioreactor and associated instrumentation to provide the proper environment and cues.

Background and History

Biochemical engineering is a subspecialty of chemical engineering. Chemical engineering began in 1901 when George E. Davis, its British pioneer, mathematically described all the physical operations commonly used in chemical plants (distillation, evaporation, filtration, gas absorption, and heat transfer) in his landmark book, *A Handbook of Chemical Engineering*.

Biochemical engineering emerged in the 1940's as advancements in biochemistry, the genetics of microorganisms, and engineering shepherded in the era of antibiotics. World War II created shortages in commonly used industrial agents; therefore, manufacturers turned to microorganisms or enzymes to synthesize many of the chemicals needed for the war effort. Growing large batches of microorganisms presented scaling, mixing, and oxygenation problems that had never been encountered before, and biochemical engineers solved these problems.

During the 1960's, advances in biochemistry, genetics, and engineering drove the creation of biomedical engineering, which is the application of all engineering disciplines to medicine, and separated it from biochemical engineering. During this decade, biochemical engineers developed new types of bioreactors and new instrumentation and control circuits for them. They also made breakthroughs in kinetics (the science that mathematically describes the rates of reactions) within bioreactors and whole-cell biotransformations.

The 1970's saw the development of enzyme technologies, biomass engineering, single-cell protein production, and advances in bioreactor design and operation. From 1980 to 2000 there was a virtual explosion in biochemical-engineering advances that had never been seen before. The advent of recombinant DNA and hybridoma technologies, cell culture, molecular models, large-scale protein chromatography, protein and DNA sequencing, metabolic engineering, and bioremediation technologies changed biochemical engineering in a drastic and profound way. These technologies also presented new challenges and problems, many of which are still the subject of intense research and development.

How It Works

Bioreactors. Bioreactors that utilize living cells are typically called fermenters. There are several different types of bioreactors: mechanically stirred or agitated tanks; bubble columns (cylindrical tanks that are not stirred but through which gas is bubbled); loop reactors, which have forced circulation; packed-bed reactors; membrane reactors; microreactors; and a variety of different types of reactors that are not easily classified (such as gas-liquid reactors and rotating-disk reactors). Biochemical engineers must choose the best bioreactor type for the desired purpose and outfit it with the right instrumentation and other features.

Bioreactor operation is either batch-wise or continuous. Batch-wise operation or batch cultures include all the nutrients required for the growth of cells prior to cultivation of the organisms. After inoculation, cell growth commences and ceases once the organisms have exhausted all the available nutrients in the culture medium. A modification of this type of operation is a fed-batch or semi-batch operation in which the reactants are continuously fed into the bioreactor, and the reaction is allowed to go to completion, after which the products are recovered. Continuously operated bioreactors, use "continuous culture systems" that continuously feed culture medium into the bioreactor and simultaneously remove excess medium at the same rate. Batch-culture bioreactors work best for fast-growing biological organisms. Slow-growing organisms usually require continuous-culture bioreactors.

Several factors influence the success of bioreactor-based operations. First, choosing the right strain to make the desired product is essential. Second, the culture medium and growth conditions must optimize the growth of the chosen organism. Third, supplying the culture with adequate oxygen requires the use of agitators or stirring equipment that must operate at high enough levels to aerate the culture without severely damaging the growing cells. Fourth, the bioreactor must have sensors to measure accurately the physical properties of the culture system, such as temperature, acidity (pH), and ionic strength. Fifth, the bioreactor should also be equipped with the means to adjust these physical properties as needed. Finally, the bioreactor must be integrated into a network of peripheral equipment that allows automated monitoring and adjustment of the culture's physical factors.

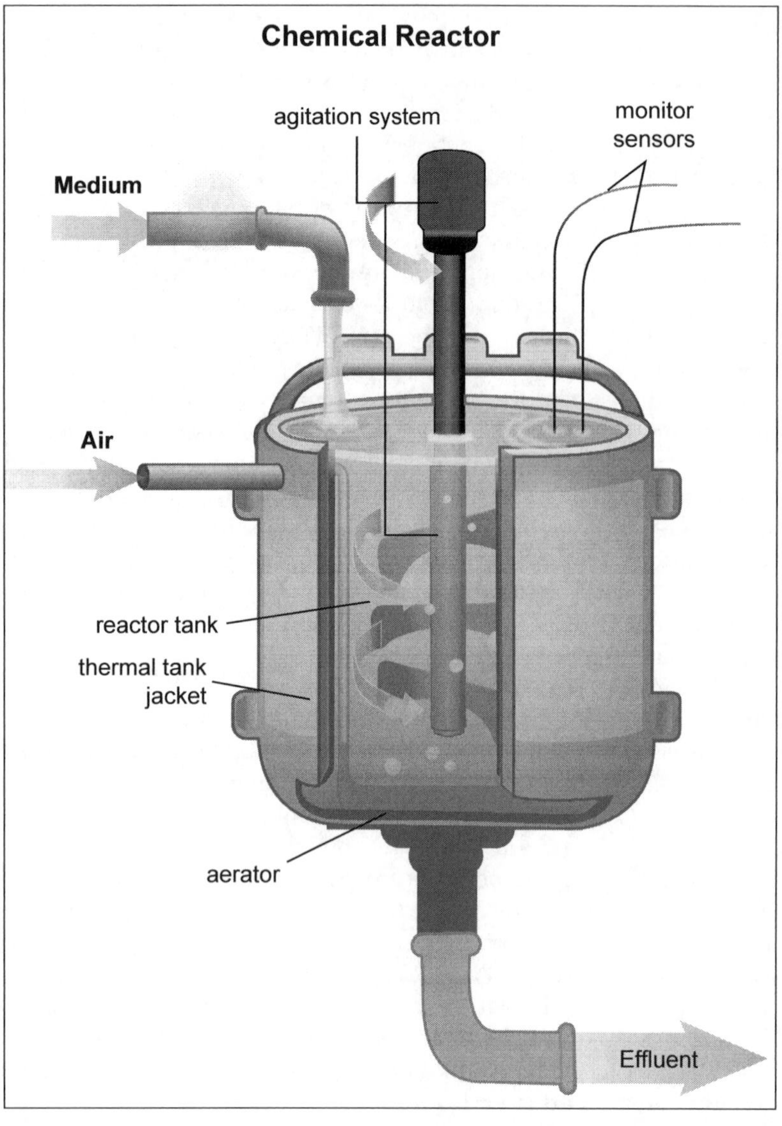

Chemical reactors are designed to contain chemical reactions such as emulsifying, solid suspension, and gas dispersion.

Separation. Once a bioreactor makes a product, separating this molecule or group of molecules from the remaining contaminants, byproducts, and other components is an integral part of preparing that molecule for market.

There are several different separation techniques. Filtration separates undissolved solids from liquids by passing the solid-liquid mixture through solids perforated by pores of a particular size (like a membrane). If the liquid is viscous or the particle size of the solid is too small for filtration, centrifugation can separate such solids from liquids. The liquid samples are loaded into centrifuges, which spin rotors at very high speeds. This process creates pellets from the solids and separates them from liquids. Neither filtration nor centrifugation can separate dissolved components from liquids.

Adsorption and chromatography can effectively separate dissolved molecules. Adsorption involves the accumulation of dissolved molecules on the surface of a solid in contact with the liquid. The solid in most cases consists of a resin made of porous charcoal, silica, polysaccharides (complex chains of sugars), or other molecules. Chromatography runs the liquid through a stationary medium packed into a cylindrical column that has particular chemical properties. The interaction between the desired molecules and the stationary medium facilitates their isolation. Other types of separation techniques include crystallization, in which the molecule of interest is driven to form crystals. This effectively removes it from solution and facilitates "salting out," in which gradually increased salt concentrations precipitate the molecules of interest, or contaminating molecules, from a liquid solution.

Sterilization. If a culture of genetically engineered organisms is used to produce a commercially useful product, contamination of that culture can decrease the amount of product or cause the production of harmful byproducts. Therefore, all tubes, valves, the bioreactor container, and the air supplied to it during operation must be effectively sterilized before the start of any production run.

Heat, radiation, chemicals, or filtration can sterilize equipment and liquids. One of the most economical means of sterilization is moist steam. Calculating the time it takes to sterilize something depends on the initial number of organisms present, the resilience of those organisms to killing with the chosen agent, the ability of the air or liquid to conduct the sterilizing agent, and length of time the organisms are exposed to the sterilizing agent.

Applications and Products

Pharmaceuticals. Hundreds of pharmaceuticals are proteins made by genetically engineered organisms. Because these reagents are intended for clinical use, they must be produced under completely sterile conditions and are usually grown in disposable (plastic), prepackaged, sterile bioreactor systems. A variety of wave bioreactors, hollow-fiber membrane bioreactors, and variations on these devices help grow the cells that make these products.

Some of the proteins made by genetically engineered cells are enzymes. Genentech, for example, makes dornase alfa, an enzyme that degrades DNA. This enzyme is made by genetically engineered Chinese hamster ovary (CHO) cells and is purified by filtration and column chromatography. Dornase alfa is administered as an inhalable aerosol to allay the symptoms of cystic fibrosis. Other therapeutic enzymes include clotting factors such as Helixate FS (native clotting factor VIII made by CSL Behring), NovoSeven (clotting factor VII made by Novo Nordisk) to treat hemophilia, and Fabrazyme or Replagal (agalsidase alfa) to treat Anderson-Fabry disease.

Other pharmaceuticals are peptide hormones. Serostim and Saizen are commercially available versions of recombinant human growth hormone. Both products are made with cultured mouse C127 cells in bioreactors. Human growth hormone is used to treat children with hypopituitary dwarfism or those who experience the chronic wasting associated with AIDS.

Therapeutic proteins are normally made in the human body under certain conditions, and synthetic versions of these proteins that are made in labs can be used as medicine. For example, human cells make a protein called interferon in response to viral infections, but synthetic interferon can also be used to treat multiple sclerosis. Two synthetic forms of interferon-1β, Rebif, which is made in CHO cells by EMD Serono, and Avonex, also made in CHO cells by Biogen Idec, serve as treatments for multiple sclerosis. Alefacept (brand name Amevive), which is made by Astellas Pharma, is a fusion protein that blocks the growth of specific T cells (immune cells). No such protein exists in the human body, but alefacept is used to treat psoriasis and various cancers.

These are only a few examples of the hundreds of pharmaceutical compounds made by genetically engineered organisms in bioreactors designed by biochemical engineers.

Monoclonal Antibodies. Monoclonal antibodies are Y-shaped proteins secreted by specific cells of the immune system that precisely bind to specific sites (epitopes) on the surface of foreign invaders, and act as guided missiles that facilitate the destruction or neutralization of the foreign invaders.

Immune cells called B lymphocytes secrete antibodies, and the fusion of these antibody-producing cells with myelomas (B-cell tumor cells) produces a hybridoma, an immortal cell that grows indefinitely in culture and secretes large quantities of a particular antibody. Antibodies made by hybridoma cells can bind to one and only one site on a specific target and are known as monoclonal antibodies.

Monoclonal antibodies are powerful clinical and industrial tools, and by growing hybridoma cell lines in bioreactors, biotechnology companies can produce large quantities of them for a variety of applications.

Mouse monoclonal antibodies end with the suffix "-omab." Tositumomab (brand name Bexxar) was approved by the Food and Drug Administration (FDA) for treatment of non-Hodgkin's lymphoma in 2003.

Chimeric or humanized monoclonal antibodies, and have the suffixes "-ximab" (chimeric antibodies that are about 65 percent human) or "-zumab" (humanized antibodies that are about 95 percent human). Cetuximab (Erbitux) is a chimeric antibody that was approved by the FDA in 2004 for the treatment of colorectal, head, and neck cancers. Bevacizumab (Avastin) is a humanized antibody approved by the FDA in 2004 that shrinks tumors by preventing the growth of new blood vessels into them.

Human monoclonal antibodies are made either by hybridomas from transgenic mice that have had their mouse antibody genes replaced with human antibody genes, or by a process called phage display. Human monoclonal antibodies end with the suffix "-mumab." The first human monoclonal antibody developed through phage display technologies was adalimumab (Humira), which was approved by the FDA to treat several immune system diseases.

Tissue Engineering. Making artificial organs for transplantation represents a unique challenge. Bioreactors tend to grow cells in two-dimensional cultures, but organs are three-dimensional structures. Thus, biochemical engineers have designed synthetic scaffolds that support the growth of cultured cells and mold them into structures that bear the shape and properties of organs. They have also designed special bioreactors that subject cells to the physical conditions that induce the cells to form the tissues that compose particular organs.

People often need cartilage repair or replacement, but bone and cartilage form only when their progenitor cells are subjected to mechanical stresses and shear forces. Biochemical engineers have grown bone by seeding bone marrow stem cells on a ceramic disc imbued with zirconium oxide and loading these discs into bioreactors with a rotating bed. Cartilage biopsies are taken from the nose or knee and grown in a bioreactor in which the cells are perfused into a complex sugar called glycosaminoglycan (GAG). This engineered cartilage is then used for transplantations. Such experiments have established that nasal cartilage responds to physical forces similarly to knee cartilage and might substitute for knee cartilage.

Heart muscle is grown in bioreactors that pulse the liquid growth medium through the chamber under high-oxygen tension. Blood vessels are grown in two-chambered bioreactors and contain a reservoir of smooth muscle cells and a chamber through which culture medium is repeatedly pulsed.

Food Engineering. Companies making foods that require fermentation by microorganisms or digestion of complex molecules by enzymes use bioreactors to optimize the conditions under which these reactions occur. Biochemical engineers design the industrial processes that manufacture, package, and sterilize foods in the most cost-effective manner.

Starch is a polymer of sugar made by plants and is a very cheap source of sugar. To convert starch into glucose, enzymes called amylases are employed. These enzymes are often isolated from bacteria or fungi, and some are even stable at high temperatures. Degrading starch at high temperatures often clarifies it and rids it of contaminating proteins.

Lactic acid fermentation metabolizes simple sugars to lactic acid and is commonly used in the production of yogurt, cheeses, breads, and some soy products. Cheese production begins with curdling milk by adding acids such as vinegar that separate solid curds from liquid whey and an enzyme mixture called rennet that comes from mammalian stomachs and coagulates the milk. Starter bacterial cultures then ferment the milk sugars into lactic acid. Yogurt is made from heat-treated milk to which starter cultures are added. The acidity of the culture is monitored, and when it reaches a particular point, the yogurt is heated to sterilize the culture for packaging.

Ethanol fermentation converts simple sugars to ethyl alcohol and is used in the production of alcoholic beverages. The most common organism utilized for ethanol fermentation is the baker's yeast, *Saccharomyces cerevisiae.* Malted barley is the sugar source in beer production, and grapes are used to make wine. Beer production involves the extraction of wort, a sugar-rich liquid from barley, which is treated with hops to add aroma and flavor and is then fermented by yeast to form beer. For wine production, the juice from crushed grapes is fermented by yeast for from five to twelve days to generate ethanol. For most red wines and some white wines, the mixture is fermented a second time by malolactic bacteria that degrade the malic acid in the wine, which has a rather harsh, bitter taste, to lactic acid. This lowers the acidity of the wine.

Biofuel Production. Burning of fossils fuels as an energy source is not sustainable, since the supply of these fuels is finite and their combustion generates greenhouse gases such as carbon dioxide (CO_2), sulfur dioxide (SO_2), and nitrogen oxides. First-generation biofuels (biodiesel and bioethanol) utilize biomass from cultivated crops such as corn, sugar beets, and sugar cane. This results in the unfortunate consequences of tying up large swathes of farmland for fuel production and raising food prices. Second-generation biofuels come from grasses, rice straw, and bio-ethers, which are economically superior to first-generation biofuels. Third-generation biofuels show the most ecological and economic promise and come from microalgae. The oil content of some microalgae can exceed 80 percent of their dry weight, and since they use sunlight as their energy source and atmospheric CO_2 as their carbon source, microalgae can produce substantial amounts of oil with little material investment.

Microalgae can be grown in open ponds, which ties up land, or special bioreactors called photobioreactors. The fast-growing microalgae are harvested and then liquefied by microwave high-pressure reactors. Oils extracted from the algal species *Dunaliella tertiolecta* at 340 degrees Celsius for 60 minutes had physical properties comparable to fossil fuel oil.

Waste Management. The removal of pollutants from air and water provides a large global challenge to environmental engineers. While there are nonbiological ways to degrade pollution, biological strategies represent some of the most innovative and potentially effective ways to remediate pollution.

To treat polluted air, it is piped through a biofilter, which consists of an inert substance called a carrier. Nutrients are trickled over the carrier, and consequently the carrier is colonized by biological organisms that can degrade the pollutant. Devices called bioscrubbers eliminate pollutants such as hydrogen sulfide (H_2S), which smells like rotten eggs, or SO_2, by dissolving the air pollutants in water and running the water into a bioreactor where the pollutants are degraded. For air pollutants that are poorly soluble in water, such as methane (CH_4) or nitric oxide (NO), hollow-fiber membrane bioreactor (HFMB) systems that house a robust population of biological organisms that can degrade gas-phase pollutants effectively treat air polluted with such molecules. Many of these same strategies can also treat polluted water.

Bioreactor landfills were designed to accelerate the degradation of municipal solid waste (MSW) in landfills. Bioreactor landfills use microorganisms to degrade solid wastes, but they also drain the water (leachate) that moves through the landfill, clean it, and recycle it back through the landfill in a process called leachate recirculation. The design of a bioreactor landfill requires extensive knowledge of the surroundings, the nature of the MSWs to be treated, and the quality of the water that becomes the leachate.

Impact on Industry

Government Agencies. Several government agencies regulate the work of biochemical engineers. The Environmental Protection Agency (EPA) enforces several regulations enacted by the U.S. Congress, which include the Clean Air Act, the Clean Water Act, and the Resource Conservation and Recovery Act (which amended the Solid Waste Disposal Act). In response to this legislation, the EPA established National Ambient Air Quality Standards, which set allowable ceilings for specific pollutants. The Clean Air Act listed almost two hundred chemicals the emissions of which had to be reduced or completely phased out. The EPA also established National Emissions Standards for Hazardous Air Pollutants, which specifies emission standards for pollutants such as asbestos, organic compounds, and heavy metals.

The National Institute for Occupational Safety and Health (NIOSH), part of the Centers for Disease Control, is responsible for conducting research and helping companies prevent work-related illness and injury. The Occupational Safety and Health Administration (OSHA) enforces the Hazardous Waste Operations and Emergency Response Standard, which specifies safe work conditions, safety training, and effective emergency-response plans.

All companies that produce medicines or food products for human consumption are subject to oversight by the FDA.

University Research. Several universities have biochemical engineering departments that engage in high-level research, much of which is supported by government funding bodies such as the National Science Foundation and the National Institutes of Health. Most of the research in biochemical engineering investigates bioreactor design, fermentation application, biotechnology/biomolecular engineering, nanopharmaceutics, pharmaceutical engineering, energy, and process systems. Some of the leading biochemical engineering programs in the United States are found at the following universities and colleges: University of California, Irvine; California Polytechnic State University, San Luis Obispo; Northwestern University; University of Maryland; Dartmouth College; Rutgers University; and Drexel University.

Industry and Business Sectors. Several industries make extensive use of biochemical engineers to design and maintain their production plants. Chemical industries use biological organisms to make useful chemicals. Since the vast majority of organic compounds are made from petroleum, the price of which keeps increasing, using microorganisms to make organic molecules is economically wise and environmentally sound. Energy industries, particularly those involved in cultivating alternative energy sources, use biochemical engineers to develop first-, second- and third-generation biofuels. Environmental firms also use bioreactors to detoxify and degrade pollutants. Food producers use a variety of enzymes and fermentations to produce their products. Pharmaceuticalindustries employ about 20 percent of all biochemical engineers. This industry utilizes genetically engineered organisms to make a variety of medicinal compounds ranging from antibiotics to over-the-counter drugs. Finally, the pulp and paper industry employs biochemical engineers to fashion cellulose, a biopolymer, into various types of products, such as stationery, cardboard, and recycled paper.

Major Corporations. In the pharmaceutical industry, some of the leading companies include Merck, Pfizer, Novartis, GlaxoSmithKline, Eli Lilly, Roche, AstraZeneca, Bristol-Myers-Squibb, Genentech, and Sanofi-Aventis. In the food industry, major companies include Unilever, Nestlé, PepsiCo, Diageo, and Mars. Some of the companies that use biochemical engineering to mitigate environmental pollution include Environmental Resources Management; Malcolm Pirnie; Versar; Ecology and Environment; EnSafe; First Environment; and Heritage-Crystal Clean. Biofuel and energy companies include Schlumberger, Dominion, E.ON, CenterPointEnergy, GE Power Systems, Sempra Energy, and Edison International. Chemical industries include Dow Chemical and DuPont. Finally, the paper and pulp industries that use biochemical engineering include Iogen Corporation, Eka Chemicals, Kalamazoo Paper Chemicals, and Plasmine Technology.

Fascinating Facts About Biochemical Engineering

- Capromab pendetide (ProstaScint), made by Cytogen Corporation, an engineered monoclonal antibody that specifically labels prostate cancers, was the first recombinant protein produced in hollow-fiber membrane bioreactors to be approved by the Food and Drug Administration in 1996.
- Heart muscle tissue grows in bioreactors only in the presence of high oxygen concentrations, but oxygen is poorly soluble in aqueous environments. To solve this problem, tissue engineers use an artificial oxygen carrier, such as perfluorocarbon, to act as a kind of artificial hemoglobin.
- A common component of household mildew, the fungus *Aspergillus niger*, converts sugars in beets or cane molasses to citric acid, which is used to acidulate foods, beverages, and candies.
- By injecting a gel made from seaweed (calcium alginate) into rabbit leg bones, tissue engineers were able to grow engineered bone that provided viable material for bone-transplant surgeries without affecting normal bone growth.
- Tempeh, a popular fermented food eaten as a meat substitute, is made in a solid-state fermentation bioreactor by cultivating the fungus *Rhizopus oligosporus* on cooked soybeans. The fungus binds the soybeans into compact cakes that are fried and packaged to sell to the public.
- Bioartificial hollow-fiber liver devices have cultured liver cells (hepatocytes) growing in between hollow fibers. The patient's blood is passed through the fibers, where it diffuses into the extra-fiber spaces, interacts with the hepatocytes, and returns to the patient. In laboratory tests, such devices maintained liver function for a few months.

Careers and Course Work

Foundational course work for biochemical engineering includes classes in biology, chemistry, physics, advanced mathematics, and computer programming. Additional course work in basic engineering concepts, like statics, dynamics, electronics, and thermodynamics are also necessary to train as a biochemical engineer. These would be followed by advanced courses in chemistry and chemical engineering. To work as a biochemical engineer, a bachelor's degree in biological, biochemical, or chemical engineering is required, as is becoming registered as professional engineer (PEng). Advanced degrees (M.S. or Ph.D.) are required for those who wish to work in academics or professional engineering research. In industry, advanced degrees are typically not required for entry-level jobs but are helpful for promotions to managerial or supervisory positions.

Biochemical engineering is a highly collaborative field, and an engineer must be able to work well with other professionals. Therefore, good communication skills are essential. Biochemical engineering is a highly analytical field that requires skill and an affinity for mathematics and chemistry as well as and problem solving. Since much of the problem solving occurs on a computer, the ability to visualize and design processes on computers is becoming an integral part of the field.

If a company employs biological organisms or enzymes for biological or chemical conversions, that company requires the expertise of a biochemical engineer. The majority of biochemical engineers work in chemical process industries (CPI), which include the chemical, gas and oil, food and beverage, textile, and agricultural sectors. Other enterprises that employ biochemical engineers include biotechnology and pharmaceutical companies, petroleum refining and oil sands extraction, waste management, paper and pulp manufacturing, farm machinery, construction and engineering design companies, and environmental companies and agencies. Government-employed biochemical engineers may work for the Departments of Energy or Agriculture, the EPA, or other such agencies.

Social Context and Future Prospects

Two aspects of biochemical engineering can be cause for concern to the general public. First, biochemical engineers work with genetically engineered organisms. Many people have never completely made peace with the use of such organisms, despite the fact that many of the items people consume on a regular basis, from seasonal flu vaccines and other medicines to the foods they eat, are made by genetically engineered organisms. Nevertheless, fear of genetically engineered organisms remains. For example, despite repeated tests establishing that genetically engineered foods are as safe as food from nongenetically modified crops, some people still feel the need to label genetically engineered food as Frankenfood. As long as this fear persists, the work of biochemical engineers will make some people uncomfortable. Second, biochemical engineers tend to work for large industries that are sometimes painted as inveterate polluters by environmental groups or as greedy, unconcerned capitalists by consumer-advocate groups. Since many companies abide by strict environmental standards and engage in humanitarian work, these accusations are somewhat unfair.

The development of new technologies in fields like genetic engineering, biomedicine, bioinstrumentation, biomechanics, waste management, and alternative energy development are driving new employment opportunities for biochemical engineers. According to the U.S. Department of Labor, Bureau of Labor Statistics, biochemical engineers are expected to have 21 percent employment growth during the period of 2006 through 2016, which is faster than the average for all occupations. Greater demands for more sophisticated medical equipment, procedures, and medicines will increase the need for greater cost-effectiveness. The growing interest in biochemical engineering is reflected in the tripling of the annual number of bachelor's degrees earned in biochemical engineering since 1991 and in the doubling of the annual number of graduate degrees earned in this field per year since 1991.

According to the U.S. Department of Labor, Bureau of Labor Statistics, the median annual salary for a biochemical engineer in 2009 was $73,930, with the lowest 10 percent earning $44,930 and highest 10 percent earning $116,330.

Michael A. Buratovich, B.S., M.A., Ph.D.

Further Reading

Katoh, Shigeo, and Fumitake Yoshida. *Biochemical Engineering: A Textbook for Engineers, Chemists, and Biologists.* Weinheim, Germany: Wiley-VCH Verlag,

2009. A basic, though rather technical, textbook of biochemical engineering by two prominent Japanese biochemical engineers that contains many tables of mathematical symbols and conversions, graphs that illustrate the application of the equations presented, and problems for interested students to solve.

McNamee, Gregory. *Careers in Renewable Energy: Get a Green Energy Job.* Masonville, Colo.: PixyJack Press, 2008. This highly readable summary of the renewable-energy job market includes more than just engineering jobs and discusses potential future employment opportunities in alternative energy sources.

Mosier, Nathan S., and Michael R. Ladisch. *Modern Biotechnology: Connecting Innovations in Microbiology and Biochemistry to Engineering Fundamentals.* Hoboken, N.J.: Wiley-AIChE, 2009. A very practical and richly illustrated and referenced guide to the advances in molecular biology for aspiring biochemical engineers.

Murphy, Kenneth M., Paul Travers, and Mark Walport. *Janeway's Immunobiology.* 7th ed. Oxford, England: Taylor & Francis, 2007. A standard immunology textbook that has an excellent section on the therapeutic use of antibodies.

Pahl, Greg. *Biodiesel: Growing a New Energy Economy.* 2d ed. White River Junction, Vt.: Chelsea Green, 2008. A popular guide to the advances in biodiesel technology and biofuel industries that also examines the food-for-fuel controversy and the issues surrounding genetically modified crops.

Vasic-Racki, Durda. "History of Biotransformations: Dreams and Realities." *Industrial Biotransformations,* edited by Andreas Liese, Karsten Seelbach, and Christian Wandrey. Weinheim, Germany: Wiley-VCH Verlag, 2000. This essay chronicles the history of using microorganisms and enzymes to synthesize commercially valuable products and the rise of biochemical engineering as an inevitable consequence of these developments.

Walker, Sharon. *Biotechnology Demystified.* New York: McGraw-Hill Professional, 2006. This is an introduction to the basics of and latest advances in molecular biology and the latest applications of these concepts to a range of discoveries, including new drugs and gene therapies.

Web Sites

Biohealthmatics
http://www.biohealthmatics.com/careers/PID00269.aspx

National Society for Professional Engineers
http://www.nspe.org

Sloan Career Cornerstone Center
http://www.careercornerstone.org/pdf/bioeng/bioeng.pdf

See also: Bioenergy Technologies; Biofuels and Synthetic Fuels; Food Preservation; Food Science; Genetically Modified Food Production; Industrial Fermentation.

BIOENERGY TECHNOLOGIES

FIELDS OF STUDY

Biology; chemistry; physics; chemical engineering; engineering technology; fuel resources and logistics; combustion technology; mathematical statistics and modeling; agricultural science; genetic engineering.

SUMMARY

The introduction of green bioenergy technologies may reduce pollution and dependence on finite fossil fuels. Bioenergy derived from biomass has the potential to provide renewable and sustained energy on both a local and a global scale. According to the International Energy Agency Bioenergy, the fundamental objective of bioenergy technology is to increase the use and implementation of ecologically sound, economically viable, and sustainable bioenergy that will help meet the world's increasing energy demands.

KEY TERMS AND CONCEPTS

- **Bioenergy:** Energy derived from biomass (living or recently living biological organisms); can be either a liquid (ethanol), gas (methane), or solid (biochar).
- **Biofuel:** Liquid fuels that are most commonly derived from plant material, such as sugar and starch crops (ethanol), or vegetable oils and animal fat (biodiesel) and used as a replacement for or additive to traditional fossil fuels (petrol).
- **Biomass:** Any renewable organic material derived from plants (such as crops or agricultural plant residue) or animals (such as human sewage or animal manure) that is used to produce bioenergy.
- **Feedstock:** Primary raw material used in the manufacture of a product or in an industrial process.
- **Fossil Fuels:** Deposits within the Earth's crust of either solid (coal), liquid (oil), or gaseous (natural gas) hydrocarbons produced through the natural decomposition of organic material (plants and animals) over many millions of years. These deposits contain high amounts of carbon, which can be burned with oxygen to provide heat and energy.
- **Peak Oil:** Point at which global oil production reaches its maximum point; it occurs when about half of the oil deposits in the Earth's crust have been extracted. After this, oil extraction falls into terminal decline. Production is thought to follow a bell curve, starting slowly, peaking, and falling at a relatively steady rate.
- **Pyrolysis:** Thermochemical process in which organic biomass is combusted at high temperatures in the absence of oxygen reagents to achieve decomposition and produce energy.
- **Renewable Energy:** Energy obtained from (usually) natural and sustainable sources, including the Sun (solar), ocean (marine), water (hydro), wind, geothermal sources, and organic matter (biomass); also known as alternative energy.

Definition and Basic Principles

Although they are closely related terms, bioenergy should not be confused with biomass. Fundamentally, bioenergy is energy derived from biomass, that is, energy derived from living (or recently living) biological organisms. Although fossil fuels are naturally occurring substances formed through the decomposition of biological organisms, the creation of these types of fuels takes millions of years, which means they are, on a human history scale, nonrenewable and unsustainable. The global demand for fossil fuels such as oil continues to increase. Oil has allowed human society to thrive because it has supplied seemingly endless cheap energy, creating diverse industries and employment. Scientific evidence continues to mount, however, in regard to its environmental impact. Oil is also a finite resource, and research has indicated that peak oil has either already occurred or will occur by the mid-2000's. A growing number of environmental and scientific organizations state that once the point of peak oil is reached, demand will far outstrip supply, and therefore, it is imperative that an alternative and sustainable source of fuel is found and implemented.

Background and History

The history of bioenergy is as long as the history of human civilization itself. The most basic form of bioenergy from biomass—the burning of wood for heat and light—has been used for thousands of years. Human society might rely heavily on fossil fuels for its energy needs, but bioenergy has been the world's primary energy source for most of human history and is still in use.

Wood and other combustibles such as corn husks are not the only sources of bioenergy with a long history of human use. One of the most popular biofuels is ethanol. Ethanol was first developed in the early to mid-1800's, before Edwin Drake's 1859 discovery of petroleum. The push for alternative fuels during that time was driven by the need to replace the whale oil used in lamps, as supplies dwindled and prices increased. By the late 1830's, ethanol mixed with naturally derived turpentine was the preferred and cheaper alternative.

In 1826, Samuel Morey invented the internal combustion engine. His engine was powered by a composite fuel of ethanol and turpentine, and although he was unable to find a suitable investor, his engine is considered to be his greatest and most progressive invention. It was not until 1860, however, that the German inventor Nicholas Otto independently invented the internal combustion engine (again fueled by ethanol) and achieved financial backing to develop the engine.

The next considerable leap in interest in bioenergy and in technology development occurred with the invention of the automobile in the early 1900's and Henry Ford's vision of an ethanol-fueled vehicle. As early as 1917, scientists knew that ethanol and other alcohols could be used as fuels. Because ethanol and other alcohols could be derived easily from any vegetable matter that undergoes fermentation, they could be cheaply and easily produced.

Despite this, however, ethanol fuels were not widely embraced as the fuel of choice for automobiles, particularly in counties such as the United States, where an ethanol tax made the alternative fuel more expensive than petrol. By 1906 when the tax was removed, gasoline fuel had developed an extensive infrastructure, and ethanol could not compete.

Because of rationing and shortages of petrol during World War II, ethanol and other vegetable-based fuels were used extensively. In particular, the use of vegetable-based fuels such as palm oil became common in European colonies in Africa so as to increase fuel self-sufficiency. During the oil crisis of the 1970's, oil prices skyrocketed and oil-dependent countries, such as the United States, became desperate to find a replacement for fossil fuels and to reduce their dependence on oil-producing countries. The 1974 oil embargo was influential in renewing interest in alternative fuels, such as ethanol, particularly in the United States.

Despite this, however, fossil fuels have continued to be used to a much greater degree than ethanol fuels. Although many automotive fuels are mixed with ethanol, there is still significant room for growth and technological and infrastructure development of bioenergy and biofuels. The potential of biofuels as a sustainable alternative to fossil fuels has once again increased interest in bioenergy. However, there are arguments against the use of bioenergy, particularly the fuel-versus-food debate. Some people believe that as long as large numbers of people worldwide do not get enough to eat, using crops for fuel instead of food is at best misguided and at worst highly unethical, and it contributes to the rates of starvation and malnutrition seen in many developing countries.

How It Works

Although scientific debate still surrounds human-influenced climate change and the timing of peak oil, many experts believe that developing sustainable fuels from renewable sources and implementing technology that helps reduce pollution is important and necessary. Bioenergy technologies play an important role in accelerating the adoption of environmentally sound bioenergy at a reasonable cost and in a sustainable manner, thereby helping meet future energy demands.

Bioenergy can be produced from many different biological materials, including wood and various crops, as well as human and animal waste. All these materials can be used to produce electricity and heat, and after coal, oil, and natural gas, biomass is the fourth largest energy resource in the world. About 14 percent of the world's electricity is produced from biomass sources. This percentage differs from country to country and is greater in developing countries, accounting for between 70 and 90 percent of the energy produced in some countries, while in developed countries such as the United States, bioenergy accounts for only between 3 and 4 percent. The United States is, however, the world's largest biopower generator and, according to the U.S. Energy Information Administration, it possesses more than half of the world's installed bioenergy capacity.

Since the 1990's, bioenergy technologies have experienced continuous development. Generally, these technologies can be divided into two main groups in relation to bioenergy production: energy crops and waste energy. Energy crops, which include trees, sugarcane, and rapeseed, are either combusted or fermented to produce high-energy alcohols such as ethanol and biodiesel that can be used as a replacement for petrol and liquid fuels. Certain waste, including organic waste from agriculture, human and animal effluent, food and plant waste, and industrial residue, can be used to produce methane gas. This methane gas can be combusted to produce steam, which can then be used to turn turbine generators and produce heat and electricity.

Most forms of bioenergy require combustion and thus the release of carbon dioxide into the atmosphere at some stage during their production. This release, however, is offset by the initial absorption of carbon dioxide by the fuel crops during the growing process. According to research, even accounting for all carbon dioxide released as a result of the planting, harvesting, producing, and transporting of bioenergy, net carbon emissions are significantly reduced (up to 90 percent).

Applications and Products

A number of different types of domestic biomass resources (also referred to as feedstocks) are used to produce bioenergy. These include biomass processing residues such as paper and pulp, agricultural and forestry wastes, urban landfill waste and gas, animal (including human) sewage and manure waste, and land and aquatic crops. There are basically two very important and useful applications of bioenergy: the production of electricity and the replacement of liquid fuels such as petrol.

Electricity Production. Biomass is capable of producing electricity in many different ways. The most commonly used methods include pyrolysis, cofiring, direct-fired/conventional stream method, biomass gasification, anaerobic digestion, and landfill gas collection.

The term "pyrolysis" is derived from the Greek words *pyro* (meaning fire) and *lysys* (meaning decomposition). Pyrolysis is a thermochemical conversion technology that involves the combustion of biomass at very high temperatures and its decomposition without oxygen. Although this process is energy consumptive and expensive, it can be used to produce electricity through the creation of pyrolysis oil, biochar, and syngas (oil, coke, and gas). These three products can be used for electricity production, as soil fertilizer, and for carbon storage. There are two types of pyrolysis—fast and slow. Fast, or flash, pyrolysis, which uses any organic material as a biomass source, takes place within seconds at temperatures of 300 to 550 degrees Celsius with rapid accumulation of biochar. In slow, or vacuum, pyrolysis, which uses any organic material as a biomass source, the combustion of the biomass occurs within a vacuum to reduce the boiling point and adverse chemical reactions.

Cofiring basically involves the combustion of solid biomass such as wood or agricultural waste with a traditional fossil fuel such as coal to produce energy. Many consider this form of bioenergy to be the most efficient in terms of the existing fossil fuel infrastructure, with its high dependence on coal. In addition, because the combustion of biomass is carbon neutral (the carbon absorbed during growth is equal to the amount released during combustion), mixing biomass with coal can assist in reducing net carbon emissions and other pollutants such as sulfur. The production of electricity by this method is considered advantageous because it is inexpensive and makes use of already existing power plants.

The direct-fired/conventional stream method is the most commonly used method of producing bioenergy. This process involves the direct combustion of biomass to produce steam, which then turns turbines that drive generators to produce electricity.

Biomass gasification is another type of thermochemical conversion technology, in which any organic biomass is converted into its gaseous form (known as syngas) and used to produce energy. The process relies on biomass gasifiers that heat solid biomass until it forms a combustible gas, which is then used in power production systems that merge gas and steam turbines to generate electricity. Although this technology is still in its infancy, it is hoped that gasification of biomass may lead to more efficient bioenergy production. Biomass gasification technologies basically can be categorized as fixed bed gasification, fluidized bed gasification, and novel design gasification.

Anaerobic digestion, a natural biological process, has a long history and involves the decomposition of organic biomass material such as manure and urban solid wastes in an air-deficient environment. Fundamentally, anaerobic digestion involves the production of methane gas through bacteria and archaea activity. This methane gas, also known as a type of biogas, is captured and then used to power turbines and produce electric and heat energy, as well as a soil-enhancement material called digestate. The advantage of this method is that it uses waste, such as wastewater sludge, to produce renewable energy and reduces the amount of greenhouse gas being released into the atmosphere.

The process of landfill gas collection is closely related to anaerobic digestion and involves the capture of gas from the decomposition of landfill urban wastes. This gas, which is about 50 percent methane, 45 percent carbon dioxide, 4 percent nitrogen, and 1 percent other gases, is then used to produce energy.

Fuel Production. Liquid biofuels are a significant alternative to petroleum-based vehicle and transportation fuels. Biofuels are attractive because they can be used in already existing vehicles with little modification required and also in the production of electricity. These fuels are estimated to account for almost 2 percent of the transportation fuels used in the world's vehicles.

Bioenergy production is generally divided into first-generation and second-generation fuels. The main distinction between these two types of fuels relates to the feedstock used. First-generation fuels are already in commercial production in many countries and are primarily made from edible grains, sugars, or seeds. The two most common biofuels are ethanol and biodiesel, both of which are already used in large quantities in many countries. Second-generation fuels, although not yet in commercial production, are considered superior to first-generation fuels because they are primarily made from nonedible whole plants or waste from food crops such as husks and stalks.

Ethanol, also known as ethyl alcohol, is an alcohol that can be used as a vehicle or transportation fuel. It is a renewable energy produced from sustainable agricultural feedstocks, particularly sugar and starch crops such as sugarcane, potatoes, and maize. Although it can be used as a direct replacement fuel, it is more commonly used as a fuel additive. Many vehicles use ethanol-petrol blends of 10 percent ethanol, which improve octane and decrease emissions. Ethanol is particularly popular in Brazil and the United States, which together produce almost 90 percent of the world's ethanol. Brazil has been at the forefront of ethanol use and as early as 1976 mandated the use of ethanol with fuels, eventually requiring a blend containing 25 percent ethanol (the most of any country). Although using arable land to grow crops to produce ethanol is somewhat controversial, the development of ethanol from cellulosic biomass (obtained from the cellulose of trees and grasses) is considered promising and may play a large role in the future of ethanol as a biofuel.

Biodiesel is a renewable energy produced from sustainable animal fat and vegetable oil feedstocks, such as soy, rapeseed, sunflowers, palm oil, hemp, and algae, and can be used as a vehicle or transportation fuel. As with ethanol, however, biodiesel is more often used as a diesel additive to reduce the levels of pollution emitted by traditional diesel engines. It is primarily produced through a process known as transesterification, which is the exchange or conversion of an organic acid ester into another ester.

Biobutanol can be used as fuel in internal combustion engines. It is usually produced from the fermentation of biomass, and because of its chemical properties, it is actually more similar to petrol than ethanol is. It can be produced from the same feedstocks used for ethanol production, such as corn, sugarcane, potatoes, and wheat. Despite its possible applications, however, this type of biofuel has not been produced commercially.

Impact on Industry

Many governments, universities, and organizations have been investigating bioenergy because of its potential for becoming one of the most important fuels of the future. Its ability to replace fossil fuels and its sustainability have pushed bioenergy into the forefront of fuel research.

Major Organizations. Many agencies and nongovernmental organizations (NGOs) are moving forward with bioenergy technology research and applications. A number of agencies and organizations are committed to increasing research and information exchange in an attempt to promote the use of bioenergy.

One of the earliest organizations established was the International Energy Agency (IEA). This autonomous body of the Organisation for Economic Cooperation and Development was founded in 1974 in response to the oil crisis. In 1978, the IEA established IEA Bioenergy to promote the development of new

Fascinating Facts About Bioenergy Technologies

- Organic material, such as plant and animal waste, can be used to produce both liquid fuel and electricity.
- Bioenergy can be produced from biomass, which is sustainable and renewable and offers a viable alternative to fossil fuels.
- The burning of biomass for fuel is carbon neutral–that is, the carbon absorbed during growth is equal to the amount released during combustion.
- Some experts are concerned that the use of biomass for fuel will create both social and environmental problems because it uses food for fuel and reduces biodiversity.
- The political climate in the twenty-first century has created renewed interest in bioenergy for the world's growing human population. The percentage of bioenergy used is much higher in less-developed countries than in developed nations.

green energy and improve international bioenergy research and information exchange, with a particular focus on economic and social development and environmental protection.

Given the global impact of fossil fuel consumption in terms of particulates and atmospheric pollution, many international organizations are taking an active interest in bioenergy technology and have released a number of statements regarding the adoption of bioenergy as an alternative fuel, including the Food and Agriculture Organization of the United Nations, which has stated that bioenergy may play a role in eradicating hunger and poverty and in ensuring environmental sustainability.

A number of organizations have been working toward the adoption of specific bioenergies. The International Biochar Initiative, formed in 2006 at the World Soil Science Congress, aims to encourage the development and use of biochar for carbon capture, soil enhancement, and energy production. The goal of the volunteer-led Biofuelwatch is to ensure that biofuels used within the European Union are obtained only from sustainable sources, thereby lessening the possibility of loss of biodiversity. Global support for organizations involved in bioenergy technology research and applications is provided by organizations such as the World Bioenergy Association, formed in 2008.

Government and University Research. Many countries have been expressing greater interest in bioenergy technology and conducting research. Some of the most significant research that has been undertaken on bioenergy has taken place in countries such as the United States, particularly within the Midwest. The oil embargo of 1973-1974 was influential in renewing interest in alternative fuels, particularly in countries such as the United States, which began intensive research and development of the dual fuel project at Ohio State University. Iowa State University has led the way in terms of academic research and education, with its Biorenewable Resources and Technology graduate program, the first of its kind in the United States. It not only offers advanced study in the use of plant and crop biomass for the creation of fuels and energy but also is home to the Bioeconomy Institute, one of the largest bioenergy institutes in the United States investigating the use of biorenewable resources as sustainable feedstocks for bioenergy.

The U.S. Department of Energy's Biomass Program, run by the office of Energy Efficiency and Renewable Energy, also aims to expand and enhance biomass power technology and increase production and use of biofuels such as ethanol and biodiesel. Specifically, the program seeks to produce diverse bioenergy products including electricity, liquid/solid/gaseous fuels, heat, and other bio-based materials. Bioenergy has become the second largest source of renewable energy used in the United States, and the president's Biofuels Initiative aims to make cellulosic ethanol economically competitive.

Many other countries, including those in the European Union, Australia, and the United Kingdom, have been increasing research into bioenergy technology and have contributed significantly to international knowledge. Australia, for example, established Bioenergy Australia in 1997 as part of its IEA Bioenergy collaborative agreement and to provide a forum to encourage and facilitate the expansion of domestic biomass energy development and use. The European Union has set goals for meeting 5.75 percent of transportation fuel requirements with biofuels. The United Kingdom Biomass Strategy, initially published in 2007, provides guidelines for the sustainable development of biomass for heat, fuel, and electricity applications. The Biomass Strategy, the Renewable

Energy Strategy, and the Bio-energy Capital Grants Scheme are among the United Kingdom's initiatives to achieve its 15 percent share of the 2020 renewable energy target established by the European Union.

Of most significance, perhaps, are the bioenergy initiatives of China and India. Although both of these rapidly developing countries use less energy per capita than most developed countries, such as the United States, Australia, and Canada (three of the top energy-consuming countries per capita), the sheer size of their populations means that they are among the world's highest energy consumers on a per country basis. Both China and India have been pursuing and researching bioenergy technology alternatives. China, the third-largest ethanol producer in the world, set a ten-year bioenergy target of generating 1 percent of its alternative energy from biomass sources. India has announced the development of bioenergy power plants to help meet growing energy demands and supply adequate energy to rural areas.

Careers and Course Work

Undergraduate and graduate courses in bioenergy technology are offered at many universities. Most students who follow this path have a strong background in science, agriculture, and fuel production technology. Following graduation, students studying bioenergy technology will understand methods for improving the efficiency of new energy technologies and have a solid understanding of environmentally friendly bioenergy concepts, theories, processes, and practices. These include biology and plant production, bioenergy electricity and fuel production from crops and waste, combustion science, sustainability in energy production, agricultural engineering, power-engine design, and emission-reduction techniques. They will also know how to integrate environmental issues and global economics into decision making. The primary purpose in bioenergy course work and research is to provide students with an understanding of energy as it relates to both economics and the environment. In addition, as with many modern sciences, students of bioenergy technology will require an understanding of computer modeling.

Bioenergy technology responds to the ever-changing needs of human society in relation to renewable and sustainable energy requirements, while focusing on environmental engineering in an internationally, socially, and ecologically responsible way. Students involved in bioenergy technology research and application can pursue various careers in environmental auditing and consulting, bioenergy education, agricultural and combustion engineering, sustainable agriculture, farm energy specializations, power plant and steam boiler engineering, pollution prevention and emissions reduction, waste treatment, renewable energy program management, and project and resource management. These careers span a wide range of industries and sectors, including, most prominently, the energy industry, private sector, nongovernmental organizations, specialized government organizations and agencies, and universities and institutions undertaking teaching and research.

Social Context and Future Prospects

The concept of bioenergy relies on the fact that such energy is sustainable. Although many believe bioenergy is synonymous with green energy, this is not always the case. Individual types of bioenergy produced in different ways and from various biomasses can have very diverse environmental impacts. Although the goal of bioenergy is to reduce the world's dependence on nonrenewable (to all intents and purposes) fossil fuels and thereby reduce greenhouse gas emissions and pollution, some forms of bioenergy can be equally harmful in terms of pollution or the energy expended to produce the bioenergy. As such, there have been significant movements since the 1980's to develop cleaner, greener, and move advanced biofuels and technologies. Many countries are investigating the potential of bioenergy, and many researchers believe that bioenergy will become a key contributor to sustainable global energy use.

Biofuels are, however, controversial, and some are calling for a moratorium on their use and advancement because of environmental and social concerns, particularly the loss of biodiversity and habitat destruction and the use of crops for fuel rather than food. First-generation fuels, already in commercial production in many countries, have been criticized because they are obtained from edible seeds and plants. However, because of the increasing problems of fossil fuel dependence, many of the world's governments and international organizations are stepping up research into bioenergy as a viable and sustainable alternative fuel. Many researchers believe that investigation and implementation of second-generation fuels, which are made from nonedible whole plants or waste from food crops, is the way of the future in terms of both efficiency and social responsibility. Biomass does offer many countries the opportunity to use fuels that are both sustainable and domestically sourced.

Christine Watts, Ph.D., B.App.Sc., B.Sc.

Further Reading

Geller, Howard. *Energy Revolution: Policies for a Sustainable Future.* Washington, D.C.: Island Press, 2003. Examines renewable energy, concentrating on energy patterns, trends and consequences, the barriers to sustainable energy use, and case studies on effective energy from different parts of the world.

Rosillo-Calle, Frank, et al., eds. *The Biomass Assessment Handbook: Bioenergy for a Sustainable Environment.* Sterling, Va.: Earthscan, 2008. Provides information on the supply and consumption of biomass and the skills and tools needed to understand biomass resource assessment and identify the effects and benefits of exploitation.

Scragg, Alan. *Biofuels: Production, Application, and Development.* Cambridge, Mass.: CAB International, 2009. Looks at biofuel production, concentrating on technological issues, benefits and problems, and existing and future forms of biofuels.

Silveira, Semida, ed. *Bioenergy: Realizing the Potential.* San Diego, Calif.: Elsevier, 2005. Investigates and integrates the fundamental technical, policy, and economic issues as they relate to bioenergy projects in both developed and developing countries. Focuses on biomass availability and potential and covers market development and technical and economic improvements.

Sims, Ralph. *The Brilliance of Bioenergy: In Business and Practice.* London: James and James, 2002. Examines the main biomass resources, technologies, processes, and principles of bioenergy production, with a focus on social, economic, and environmental issues in the form of small- and large-scale case studies in developed and developing countries.

Singh, Om V., and Steven P. Harvey, eds. *Sustainable Biotechnology: Sources of Renewable Energy.* London: Springer, 2009. An extensive collection of research reports and reviews, focuing on the progress and challenges involved in the use of sustainable resources for the production of renewable biofuels.

Web Sites

International Energy Agency Bioenergy
http://www.ieabioenergy.com

U.S. Department of Agriculture, Economic Research Service
Featuring Bioenergy
http://www.ers.usda.gov/features/bioenergy

U.S. Department of Energy
Bioenergy
http://www.energy.gov/energysources/bioenergy.htm

U.S. Energy Information Administration
Renewable and Alternative Fuels
http://www.eia.doe.gov/fuelrenewable.html

See also: Biofuels and Synthetic Fuels; Fuel Cell Technologies; Gasoline Processing and Production; Solar Energy; Wind Power Technologies.

BIOFUELS AND SYNTHETIC FUELS

FIELDS OF STUDY

Biology; microbiology; plant biology; chemistry; organic chemistry; biochemistry; agriculture; biotechnology; bioprocess engineering; chemical engineering.

SUMMARY

The study of biofuels and synthetic fuels is an interdisciplinary science that focuses on development of clean, renewable fuels that can be used as alternatives to fossil fuels. Biofuels include ethanol, biodiesel, methane, biogas, and hydrogen; synthetic fuels include syngas and synfuel. These fuels can be used as gasoline and diesel substitutes for transportation, as fuels for electric generators to produce electricity, and as fuels to heat houses (their traditional use). Both governmental agencies and private companies have invested heavily in research in this area of applied science.

KEY TERMS AND CONCEPTS

- **Biodiesel:** Biofuel with the chemical structure of fatty acid alkyl esters.
- **Biogas:** Biofuel that contains a mixture of methane (50-75 percent), carbon dioxide, hydrogen, and carbon monoxide.
- **Biomass:** Mass of organisms that can be used as an energy source; plants and algae convert the energy of the sun and carbon dioxide into energy that is stored in their biomass.
- **Ethanol:** Colorless liquid with the chemical formula C_2H_5OH that is used as a biofuel; also known as ethyl alcohol, grain alcohol, or just alcohol.
- **Fischer-Tropsch Process:** Process that indirectly converts coal, natural gas, or biomass through syngas into synthetic oil or synfuel (liquid hydrocarbons).
- **Fuel:** Any substance that is burned to provide heat or energy.
- **Gasification:** Conversion of coal, petroleum, or biomass into syngas.
- **Methane:** Colorless, odorless, nontoxic gas, with the molecular formula CH_4, that is the main chemical component of natural gas (70-90 percent) and is used as a biofuel.
- **Molecular Hydrogen:** Also known by its chemical symbol H_2, a flammable, colorless, odorless gas; hydrogen produced by microorganisms is used as a biofuel and is called a biohydrogen.
- **Synfuel:** A synthetic liquid fuel (synthetic oil) obtained via the Fischer-Tropsch process or methanol-to-gasoline conversion.
- **Synthesis Gas:** Synthetic fuel that is a mixture of carbon monoxide and H_2; also known as syngas.

Definition and Basic Principles

The science of biofuels and synthetic fuels deals with the development of renewable energy sources, alternatives to nonrenewable fossil fuels such as petroleum. Biofuels are fuels generated from organisms or by organisms. Living organisms can be used to generate a number of biofuels, including ethanol (bioethanol), biodiesel, biomass, butanol, biohydrogen, methane, and biogas. Synthetic fuels (synfuel and syngas) are a class of fuels derived from coal or biomass. Synthetic fuels are produced by a combination of chemical and physical means that convert carbon from coal or biomass into liquid or gaseous fuels.

Around the world, concerns about climate change and possible global warming due to the emission of greenhouse gases from human use of fossil fuels, as well as concerns over energy security, have ignited interest in biofuels and synthetic fuels. A large-scale biofuel and synthetic fuel industry has developed in many countries, including the United States. A number of companies in the United States have conducted research and development projects on synthetic fuels with the intent to begin commercial production of synthetic fuels. Although biofuels and synthetic fuels still require long-term scientific, economic, and political investments, investment in these alternatives to fossil fuels is expected to mitigate global warming, to help protect the global climate, and to reduce U.S. reliance on foreign oil.

Background and History

People have been using biofuels such as wood or dried manure to heat their houses for thousands of years. The use of biogas was mentioned in Chinese literature more than 2,000 years ago. The first biogas plant was built in a leper colony in Bombay, India, in the middle of the nineteenth century. In Europe, the first apparatus for biogas production was built in Exeter, England, in 1895. Biogas from this digester was used to fuel street lamps. Rudolf Diesel, the inventor of the diesel engine, used biofuel (peanut oil) for his engine during the World Exhibition in Paris in 1900. The version of the Model T Ford built by Henry Ford in 1908 ran on pure ethanol. In the 1920's, 25 percent of the fuels used for automobiles in the United States were biofuels rather than petroleum-based fuels. In the 1940's, biofuels were replaced by inexpensive petroleum-based fuels.

Gasification of wood and coal for production of syngas has been done since the nineteenth century. Syngas was used mainly for lighting purposes. During World War II, because of shortages of petroleum, internal combustion engines were modified to run on syngas and automobiles in the United States and the United Kingdom were powered by syngas. The United Kingdom continued to use syngas until the discovery in the 1960's of oil and natural gas in the North Sea.

The process of converting coal into synthetic liquid fuel, known as the Fischer-Tropsch process, was developed in Germany at the Kaiser Wilhelm Institute by Franz Fischer and Hans Tropsch in 1923. This process was used by Nazi Germany during World War II to produce synthetic fuels for aviation.

During the 1970's oil embargo, research on biofuels and synthetic fuels resumed in the United States and Europe. However, as petroleum prices fell in the 1980's, interest in alternative fuels diminished. In the twenty-first century, concerns about global warming and increasing oil prices reignited interest in biofuels and synthetic fuels.

How It Works

Biofuels and synthetic fuels are energy sources. People have been using firewood to heat houses since prehistorical time. During the Industrial Revolution, firewood was used in steam engines. In a steam engine, heat from burning wood is used to boil water; the steam produced pushes pistons, which turn the wheels of the machinery.

Biofuels and synthetic fuels such as ethanol, biodiesel, butanol, biohydrogen, and synthetic oil can be used in internal combustion engines, in which the combustion of fuel expands gases that move pistons or turbine blades. Other biofuels such as methane, biogas, or syngas are used in electric generators. Burning of these fuels in electric generators rotates a coil of wire in a magnetic field, which induces electric current (electricity) in the wire.

Hydrogen is used in fuel cells. Fuel cells generate electricity through a chemical reaction between molecular hydrogen (H_2) and oxygen (O_2). Ethanol, the most common biofuel, is produced by yeast fermentation of sugars derived from sugarcane, corn starch, or grain. Ethanol is separated from its fermentation broth by distillation. In the United States, most ethanol is produced from corn starch. Biodiesel, another commonly used biofuel, is made mainly by transesterification of plant vegetative oils such as soybean, canola, or rapeseed oil. Biodiesel may also be produced from waste cooking oils, restaurant grease, soap stocks, animal fats, and even from algae. Methane and biogas are produced by metabolism of microorganisms. Methane is produced by microorganisms called *Archaea* and is an integral part of their metabolism. Biogas produces by a mixture of bacteria and archaea.

Industrial production of biofuels is achieved mainly in bioreactors or fermenters of some hundreds gallons in volume. Bioreactors or fermenters are closed systems that are made of an array of tanks or tubes in which biofuel-producing microorganisms are cultivated and monitored under controlled conditions.

Syngas is produced by the process of gasification in gasifiers, which burn wood, coal, or charcoal. Syngas can be used in modified internal combustion engines. Synfuel can be generated from syngas through Fischer-Tropsch conversion or through methanol to gasoline conversion process.

Applications and Products

Transportation. Biofuels are mainly used in transportation as gasoline and diesel substitutes. As of the early twenty-first century, two biofuels—ethanol and biodiesel—were being used in vehicles. In 2005, the U.S. Congress passed an energy bill that required that ethanol sold in the United States for transportation be mixed with gasoline. By 2010, almost every fuel station in the United States was selling gasoline with a 10 percent ethanol content. The U.S. ethanol industry has lobbied the federal government to raise the ethanol content in gasoline from 10 to 15 percent. Most cars in Brazil can use an 85 percent/15 percent ethanol-gasoline mix (E85 blend). These cars must have a modified engine known as a flex engine. In the United States, only a small fraction of all cars have a flex engine.

Biodiesel performs similarly to diesel and is used in unmodified diesel engines of trucks, tractors, and other vehicles and is better for the environment. Biodiesel is often blended with petroleum diesel in ratios of 2, 5, or 20 percent. The most common blend is B20, or 20 percent biodiesel to 80 percent diesel fuel. Biodiesel can be used as a pure fuel (100 percent or B100), but pure fuel is a solvent that degrades the rubber hoses and gaskets of engines and cannot be used in winter because it thickens in cold temperatures. The energy content of biodiesel is less than that of diesel. In general, biodiesel is not used as widely as ethanol, and its users are mainly governmental and state bodies such as the U.S. Postal Service; the U.S. Departments of Defense, Energy, and Agriculture; national parks; school districts; transit authorities; public utilities; and waste-management facilities. Several companies across the United States (such as recycling companies) use biodiesel because of tax incentives.

Hydrogen power ran the rockets of the National Aeronautics and Space Administration for many years. A growing number of automobile manufactures around the world are making prototype hydrogen-powered vehicles. These vehicles emit only water, no greenhouse gases, from their tailpipes. These automobiles are powered by electricity generated in the fuel cell through a chemical reaction between H_2 and O_2. Hydrogen vehicles offer quiet operation, rapid acceleration, and low maintenance costs because of fewer moving parts. During peak time, when electricity is expensive, fuel-cell hydrogen automobiles could provide power for homes and offices. Hydrogen for these applications is obtained mainly from natural gas (methane and propane), through steam reforming, or by water electrolysis. As of 2010, hydrogen was used only in experimental applications. Many problems need to be overcome before hydrogen becomes widely used and readily available. The slow acceptance of biohydrogen is partly caused by the difficulty in producing it on a cost-effective basis. For hydrogen power to become a reality, a great deal of research and investment must take place.

Methane was used as a fuel for vehicles for a number of years. Several Volvo automobile models with Bi-Fuel engines were made to run on compressed methane with gasoline as a backup. Biogas can also be used, like methane, to power motor vehicles.

Electricity Generation. Biogas and methane are mainly used to generate electricity in electric generators. In the 1985 film *Mad Max Beyond Thunderdome*, starring Mel Gibson, a futuristic city ran on methane generated by pig manure. While the use of methane has not reached this stage, methane is a very good alternative fuel that has a number of advantages over biofuels produced by microorganisms. First, it is easy to make and can be generated locally, eliminating the need for an extensive distribution channel. Second, the use of methane as a fuel is a very attractive way to reduce wastes such as manure, wastewater, or municipal and industrial wastes. In farms, manure is fed into digesters (bioreactors), where microorganisms metabolize it into methane. There are several landfill gas facilities in the United States that generate electricity using methane. San Francisco has extended its recycling program to include conversion of dog waste into methane to produce electricity and to heat homes. With a dog population of 120,000, this initiative promises to generate a significant amount of fuel and reduce waste at the same time.

Heat Generation. Some examples of biomass being used as an alternative energy source include the burning of wood or agricultural residues to heat homes. This is a very inefficient use of energy, because typically only 5 to 15 percent of the biomass energy is actually used. Burning biomass also produces harmful indoor air pollutants such as carbon monoxide. On the positive side, biomass is an inexpensive resource whose costs are only the labor to collect it. Biomass supplies more than 15 percent of the energy consumed worldwide. Biomass is the number-one source of energy in developing countries; in some countries, it provides more than 90 percent of the energy used.

In many countries, millions of small farmers maintain a simple digester for biogas production to generate heat energy. More than 5 million household digesters are being used in China, mainly for cooking and lighting, and India has more than 1 million biogas plants of various capacities.

Impact on Industry

In 2009, the annual revenue of the global biofuels industry was $46.5 billion, with revenue in the United States alone reaching $20 billion. The United

States is leading the world in research on biofuels and synthetic fuels. Significant biofuels and synthetic fuels research has also been taking place in many European countries, Russia, Japan, Israel, Canada, Australia, and China.

Government and University Research. Many governmental agencies such as the U.S. Department of Energy (DOE), the National Science Foundation (NSF), and the U.S. Department of Agriculture provide funding for research in biofuels and synthetic fuels. The DOE has several national laboratories (such as the National Renewable Energy Laboratory in Golden, Colorado) where cutting-edge research on biofuels and synthetic fuels is performed. In addition, three DOE research centers are concentrated entirely on biofuels. These centers are the BioEnergy Science Center, led by Oak Ridge National Laboratory; the Great Lakes Bioenergy Research Center, led by the University of Wisconsin, Madison; and the Joint BioEnergy Institute, led by Lawrence Berkeley National Laboratory.

In 2007, DOE established the Advanced Research Projects Agency-Energy (ARPA-E) to fund the development and deployment of transformational energy technologies in the United States. Several projects funded by this agency are related to biofuels, such as the development of advanced or second-generation biofuels. Traditional biofuels such as biomass (wood material) and ethanol and biodiesel from crops are sometimes called first-generation biofuels. Second-generation biofuels such as cellulosic ethanol are produced from agricultural and forestry residues and do not take away from food production. Another second-generation biofuel is biohydrogen.

Biofuels such as butanol are referred to as third-generation biofuels. Butanol (C_4H_9OH) is an alcohol fuel, but compared with ethanol, it has a higher energy content (roughly 80 percent of gasoline energy content). It does not absorb water as ethanol does, is not as corrosive as ethanol is, and is more suitable for distribution through existing gasoline pipelines.

Scientists are trying to create "super-bugs" for superior biofuel yields and studying chemical processes and enzymes to improve existing bioprocesses for biofuels production. They also are working to improve the efficiency of the existing production process and to make it more environmentally friendly. Engineers and scientists are designing and developing new apparatuses (bioreactors or fermenters) for fuel generation and new applications for by-products of fuel production.

Fascinating Facts About Biofuels and Synthetic Fuels

- Scientists have discovered that *Gliocladium roseum*, a tree fungus, is able to convert cellulose directly into biodiesel, thus making transesterification unnecessary. The fungus eats the tree and, interestingly, keeps other fungi away from the tree by producing an antibiotic. Scientists are studying enzymes that will help this fungus eat cellulose.
- Termites can produce two liters of biofuel, molecular hydrogen, or H_2, by fermenting just one sheet of paper with microbes that live in their guts. Study of the biochemical pathways involved in H_2 production in termite guts may lead to application of this process industrially.
- In 2009, Continental Airlines successfully powered a Boeing 737-800 using a biodiesel fuel mixture partly produced from algae.
- Firewood can power an automobile, but its engine must be modified slightly and a trailer with a syngas generator (gasifier) must be attached to the automobile.
- An automobile that runs on diesel fuel can easily be modified to run on used cooking oil, which can be obtained for little or no charge from local restaurants.
- By attaching a water electrolyzer, an automobile can be modified to run partly on water. Electricity from the car battery splits water in the electrolyzer into hydrogen and oxygen. The hydrogen can be burned in the internal combustion engine and power the automobile.
- Modifying an automobile to run partly on methane is simple and definitely saves gasoline.

Industry and Business Sectors. The major products of the biofuel industry are ethanol, biodiesel, and biogas; therefore, research in industry has concentrated mainly on these biofuels. Some small businesses in the biofuel industry include startup research and development companies that study feedstocks (such as cellulose) and approaches for production of biofuels at competitive prices. These companies, many of which are funded by investment firms or government agencies, analyze biofuel feedstocks, looking for new feedstocks or modifying existing ones (corn, sugarcane, or rapeseed).

Big corporations such as Poet Energy, ExxonMobil, and BP are spending a significant part of their revenues on biofuel or synthetic fuels research. One area of biofuel research examines using algae to generate biofuels, especially biodiesel. More than fifty research companies worldwide, including GreenFuel Technologies, Solazyme, and Solix Biofuels, are conducting research in this area. Research conducted by the U.S. Department of Energy Aquatic Species Program from the 1970's to the 1990's demonstrated that many species of algae produce sufficient quantities of oil to become economical feedstock for biodiesel production. The oil productivity of many algae greatly exceeds the productivity of the best-producing oil crops. Algal oil content can exceed 80 percent per cell dry weight, with oil levels commonly at about 20 to 50 percent. In addition, crop land and potable water are not required to cultivate algae, because algae can grow in wastewater. Although development of biodiesel from algae is a very promising approach, the technology needs further research before it can be implemented commercially.

biofuel research examines using algae to generate biofuels, especially biodiesel. More than fifty research companies worldwide, including GreenFuel Technologies, Solazyme, and Solix Biofuels, are conducting research in this area. Research conducted by the U.S. Department of Energy Aquatic Species Program from the 1970's to the 1990's demonstrated that many species of algae produce sufficient quantities of oil to become economical feedstock for biodiesel production. The oil productivity of many algae greatly exceeds the productivity of the best-producing oil crops. Algal oil content can exceed 80 percent per cell dry weight, with oil levels commonly at about 20 to 50 percent. In addition, crop land and potable water are not required to cultivate algae, because algae can grow in wastewater. Although development of biodiesel from algae is a very promising approach, the technology needs further research before it can be implemented commercially.

Careers and Course Work

The alternative fuels industry is growing, and research in the area of biofuels and synthetic fuels is increasing. Growth in these areas is likely to produce many jobs. The basic courses for students interested in a career in biofuels and synthetic fuels are microbiology, plant biology, organic chemistry, biochemistry, agriculture, bioprocess engineering, and chemical engineering. Many educational institutions are offering courses in biofuels and synthetic fuels, although actual degrees or concentrations in these disciplines are still rare. Several community colleges offer associate degrees and certificate programs that prepare students to work in the biofuel and synthetic fuel industry. Some universities offer undergraduate courses in biofuels and synthetic fuels or concentrations in these areas. Almost all these programs are interdisciplinary. Graduates of these programs will have the knowledge and internship experience to enter directly into the biofuel and synthetic fuel workforce. Advanced degrees such as a master's degree or doctorate are necessary to obtain top positions in academia and industry related to biofuels and synthetic fuels. Some universities such as Colorado State University offer graduate programs in biofuels.

Careers in the fields of biofuels and synthetic fuels can take different paths. Ethanol, biodiesel, or biogas industries are the biggest employers. The available jobs are in sales, consulting, research, engineering, and installation and maintenance. People who are interested in research in biofuels and synthetic fuels can find jobs in governmental laboratories and in universities. In academic settings, fuel professionals may share their time between research and teaching.

Social Context and Future Prospects

The field of biofuels and synthetic fuels is undergoing expansion. Demands for biofuels and synthetic fuels are driven by environmental, social, and economic factors and governmental support for alternative fuels.

The use of biofuels and synthetic fuels reduces the U.S. dependence on foreign oil and helps mitigate the devastating impact of increases in the price of oil, which reached a record $140 per barrel in 2008. The production and use of biofuels and synthetic fuels reduces the need for oil and has helped hold world oil prices 15 percent lower than they would have been otherwise. Many experts believe that biofuels and synthetic fuels will replace oil in the future.

Pollution from oil use affects public health and causes global climate change because of the release of carbon dioxide. Using biofuels and synthetic fuels as an energy source generates fewer pollutants and little or no carbon dioxide.

The biofuel and synthetic fuel industry in the United States was affected by the economic crisis in 2008 and 2009. Several ethanol plants were closed, some plants were forced to work below capacity, and other companies filed for Chapter 11 bankruptcy protection. Such events led to layoffs and hiring freezes. Nevertheless, overall, the industry was growing and saw a return to profitability in the second half of 2009. Worldwide production of ethanol and biodiesel is expected to grow to $113 billion by 2019; this is more than 60 percent growth of annual earnings. One segment of the biofuel and synthetic fuel industry, the biogas industry, was not affected by recession at all. More than 8,900 new biogas plants were built worldwide in 2009. According to market analysts, the biogas industry has reached a turning point and may grow at a rate of 24 percent from 2010 to 2016. Research and development efforts in biofuels and synthetic fuels actually increased during the economic crisis. In general, the future of biofuels and synthetic fuels is bright and optimistic.

Sergei A. Markov, Ph.D.

Further Reading

Bart, Jan C. J., and Natale Palmeri. *Biodiesel Science and Technology: From Soil to Oil.* Cambridge, England: Woodhead, 2010. A comprehensive book on biodiesel fuels.

Bourne, Joel K. "Green Dreams." *National Geographic* 212, no. 4 (October, 2007): 38-59. An interesting discussion about ethanol and biodiesel fuels and their future.

Glazer, Alexander N., and Hiroshi Nikaido. *Microbial Biotechnology: Fundamentals of Applied Microbiology.* New York: Cambridge University Press, 2007. Provides an in-depth analysis of ethanol and biomass for fuel applications.

Mikityuk, Andrey. "Mr. Ethanol Fights Back." *Forbes,* November 24, 2008, 52-57. Excellent discussion about the problems and the hopes of the ethanol industry. Examines how the ethanol industry fought the economic crisis in 2008 and returned to profitability in 2009.

Probstein, Ronald F., and Edwin R. Hicks. *Synthetic Fuels.* Mineola, N.Y.: Dover Publications, 2006. A comprehensive work on synthetic fuels. Contains references and sources for further information.

Service, Robert F. "The Hydrogen Backlash." *Science* 305, no. 5686 (August 13, 2004): 958-961. Discusses the future of hydrogen power, including its maturity. Written in an easy-to-understand manner.

Wall, Judy, ed. *Bioenergy.* Washington, D.C.: ASM Press, 2008. Provides the information on generation of biofuels by microorganisms and points out future areas for research. Ten chapters focus on ethanol production from cellulosic material.

Web Sites

Advanced BioFuels USA
http://advancedbiofuelsusa.info

International Energy Agency Bioenergy
http://www.ieabioenergy.com

U.S. Department of Agriculture, Economic Research Service
Featuring Bioenergy
http://www.ers.usda.gov/features/bioenergy

U.S. Department of Energy
Energy Efficiency and Renewable Energy
http://www.energy.gov/energysources/index.htm

U.S. Energy Information Administration
Renewable and Alternative Fuels
http://www.eia.doe.gov/fuelrenewable.html

See also: Bioenergy Technologies; Hybrid Vehicle Technologies.

C

CAMERA TECHNOLOGIES

FIELDS OF STUDY

Cinematography; photography; mechanical and fluid engineering; electromechanical engineering; electrical engineering; computer programming; design and media management; electronics; cartography; computer science; mathematics; material science; digital photography; video production; film and digital production; design and media management; astrophotography.

SUMMARY

Camera technologies are concerned with the design, development, operation, and assessment of film and digital cameras. The field includes traditional photographic imaging as well as measuring and recording instruments in the fields of science, medicine, engineering, surveillance, and cartography. Camera technologies have widespread applications in almost every area of modern life, from digital video cameras used at banks to record information about customers to long-range space voyagers transmitting images back to scientists on Earth for study. Medical applications also have been developed to help surgeons "see" inside the human body and perform minimally invasive laparoscopic surgeries. Camera technologies continue to evolve and change rapidly as they integrate smaller digital technology for use in medicine, entertainment, industry, and science.

KEY TERMS AND CONCEPTS

- **Aperture:** Opening in the lens through which light passes.
- **Camera Obscura:** Darkened enclosure, such as a room, with a small aperture in one of the walls that allows light in and casts an inverted image on the opposite wall.
- **Exposure:** Total amount of light allowed to fall on the photographic medium (photographic film or image sensor) during the process of taking a photograph.
- **Focus:** Process of adjusting the camera lens in order to capture an image clearly.
- **Image Capture:** Process of taking a picture and storing it on film or an image sensor.
- **Lens:** Part of the camera that captures the light from the subject and brings it to photographic memory.
- **Negative Image:** Camera image with all the tones of the original scene reversed.
- **Photographic Film:** Sheet of plastic coated with light-sensitive pigments or chemicals used to capture a visible image.
- **Photographic Medium:** Photographic film or image sensor used to capture an image.
- **Pixel Count:** Resolution of an image captured by a digital camera.
- **Positive Images:** Standard image capture by a camera.
- **Processing:** Chemical means by which photographic film and paper are treated after photographic exposure to produce a negative or positive image.
- **Shutter:** Device that allows light to pass for a determined period of time for the purpose of exposing light to photographic film or a light-sensitive electronic sensor to capture a permanent image of a scene.
- **Stereoscopic:** Technique for enhancing a three-dimensional image that creates the illusion of depth by presenting two offset images separately to the left and right eye of the viewer.

Definition and Basic Principles

Camera technologies revolve around the science of capturing images. The field includes the design, development, operation, and assessment of both film and digital camera technologies. Film cameras use traditional photosensitive films to capture images by exposing the films to the right amount of light for the right length of time. Digital and video cameras use an array of photosensitive electronic devices or image sensors to capture images electronically and store them as a series of digital data.

Whether the camera is film, digital, or video, it typically consists of five basic parts: the body, the shutter and shutter-release button, the viewfinder, the camera lens complex, and the film or device that captures the image. The camera's body or casing is made of metal or high-grade plastic that holds the camera's other parts together and provides protection for the camera parts. The shutter captures the image, and the shutter-release button tells the camera to take the picture. The viewfinder is the window that shows the object or objects that compose the image that will be captured by the camera. The camera lens complex is made of several smaller pieces. Camera lens complexes control and create different image effects, such as the ability to zoom, focus, and distortion correction, which combine to control the appearance of the image being captured. The optical lens is the curved piece of glass located on the outside of the lenses that focuses light into the camera and onto the film or digital sensor. The aperture is controlled by an aperture ring, which controls the size of the opening and determines how much light goes into the lens. The aperture size is measured by an f-stop number; the smaller the f-stop, the more light is let in. The final component of the camera is the photographic medium used to capture and store the image, film or digital.

The series of scientific discoveries that bring us to the current array of camera technologies began with primitive temporary imaging devices, such as the camera obscura, and evolved to the various cameras using photographic film and digital image capturing. More advanced camera technologies, such as the Hubble Space Telescope and the laparoscope, have greatly furthered medical treatment and scientific knowledge. In both cases, the application of camera technology on a large and small scale allows human operators to direct actions and observe phenomena from a vantage point they cannot achieve with their eyes. The camera-technologies field is integrated in several other fields of science, medicine, engineering, surveillance, and cartography.

Background and History

The term camera comes from *camera obscura* (Latin for "dark chamber"), an ancient mechanism for projecting images. Camera obscura was a pinhole device used to produce temporary images on flat surfaces in dark rooms. Camera obscura was improved during the sixteenth and seventeenth centuries, but it was not until 1727, when German physician Johann Heinrich Schulze accidentally created the first photosensitive compound, that the next leap forward in camera technologies occurred. The early nineteenth century saw the application of Schulze's discovery to create development processes that translated temporarily viewed images to permanent images. By the mid-nineteenth century, direct positive images could be permanently affixed to glass (ambrotypes) or metal (tintypes or ferrotypes). In 1861, another development in camera technologies occurred when Scottish physicist James Clerk Maxwell demonstrated the first color photographs using a system of filters and slide projection called the color separation method. During this same era, the application of photos to record current events and new areas of the country occurred as photographers captured images taken during the American Civil War and of the West. The next innovation, by English physician Richard Leach Maddox in 1871, paved the way for a revolution in camera technologies with the development of the dry plate process—using an emulsion of gelatin and silver bromide on a glass plate. In the late nineteenth century, companies such as Kodak created commercially available cameras that used film to capture pictures. In 1907, the first commercial color film became available, and by 1914, the standard modern 24-by-36-millimeter (mm) frame and sprocketed 35 mm movie film were developed. In 1932, the idea of Technicolor for movies arose. In the Technicolor process, three black-and-white negatives were made in the same camera under different filters to create a colorful finished product. The rest of the twentieth century saw continued technical improvements and optimization of the quality and abilities of cameras and films, including development of multilayer color film, "instant" Polaroid film, underwater cameras, auto-focus cameras, the automatic diaphragm as well as introduction of the single-lens reflex (SLR) camera. The SLR camera changed photography by using a semiautomatic moving mirror system that permitted the photographer to see the exact image that would be captured by the film or digital imaging system. In 1972, Texas Instruments was granted the patent for an all-electric camera. In 1982, camera technologies moved forward again with the introduction of Sony's still video camera.

The ability to manipulate and transform images was then revolutionized with the 1990 release of Adobe Photoshop, which allowed nonprofessionals to transform photo images from their home computers. Also in the early 1990's, Kodak brought digital cameras to the general population by developing the photo CD system. The camera phone was introduced in 2000 in Japan. By 2003, more affordable digital cameras were made available and quickly became the dominant image-capturing format, so much so that Kodak ceased production of film cameras in 2004, with Nikon following suit with many of its film cameras as well, in 2006.

How It Works

The core element in camera technology is the capture of a desired image. The mechanism of capture varies depending on the type of camera used: film or digital. Film cameras capture images on a sheet of plastic coated with light-sensitive pigments (silver halide salts bonded by gelatin). The pigments contain variable crystal sizes that determine the sensitivity, contrast, and resolution of the film. When the film is sufficiently exposed to light or other electromagnetic radiation, it forms an image. The film is then developed using specialized chemical processes to create a visible image. Digital cameras use a special sensor or pixilated metal oxide semiconductors (photodiodes) made from silicon to convert the light that falls onto them into electrons in order to capture a desired image and store it in a memory device. The two most common types of sensors used in digital cameras are the charge-coupled device (CCD) and the complementary metal-oxide semiconductor (CMOS). A film recorded with a digital camcorder is captured in a similar way—saved as a series of frames rather than a single snapshot. In both traditional film and digital cameras, the image itself can be modified using different filters and lenses.

Applications and Products

Astronomy and Physics. Camera technologies have expanded the ability of astronomers to explore the universe far beyond manned space exploration or early telescopes. High-powered telescopes and cameras placed in space have provided photographic evidence of rare astronomical events and features that have resulted in revisions of scientific theory. For example, photographs from the Hubble Space Telescope led to the discovery of dark energy, the hypothetical and unexplained force that seems to be drawing galaxies away from each other. Other unique types of cameras, such as the near-infrared camera Lucifer 1, are powerful tools used to gain spectacular insights into universe phenomena such as star formation.

Medicine. Use of camera technologies has revolutionized medicine. Physicians can use cameras and robots to evaluate and treat distant patients as far away as the astronauts on the space station. Closer to home, medical professionals can use tiny cameras and image-capturing devices in pill form to evaluate the inner workings of their patients' intestines. Many surgeries can be performed laparoscopically, where surgeons use tiny cameras on medical instruments that project images of their patients' bodies. The surgeons then use images and controls attached to consoles to guide their surgical implements within the patients' bodies precisely. A major benefit of surgeries using camera technologies is that they are less invasive—smaller incisions need to be made because manipulators using cameras can be extremely narrow. Surgeries can also be performed remotely using imaging-incorporated technology such as the da Vinci Surgical System to conduct operations such as prostatectomy, cardiac surgery, bariatric surgery, and various forms of neurosurgery.

Journalism. The adage that "a picture is worth a thousand words" summarizes the incredible impact of images in conveying current events, news, and history. Photojournalists using camera technologies add depth to news stories that cannot always be conveyed through words alone. Examples of the impact of photography in journalism include photos of the famine in Ethiopia and the devastation wrought by Hurricane Katrina and other natural disasters.

Cartography. The field of cartography has been revolutionized through the use of camera technologies. The ability to take images of the geologic features of the Earth from the ground or from space satellites has increased the accuracy and speed of mapmaking. In fact, companies such as Google have traveled the United States to take street-level images of addresses that are then incorporated into maps and directions.

Engineering. Modern camera development has expanded the capabilities of engineering. Using computer technology and images, engineers can create models to predict a variety of outcomes from the impact of a head-on car collision to the minute changes in an electrical circuit. Modern images and specialized image-analysis systems can be used to enhance an observer's ability to make measurements from a large or complex set of images by improving accuracy, objectivity, or processing speed.

Entertainment. The development of camera technologies over time has provided entertainment options from the early moving pictures to specialized gaming systems. The use of advances in cameras can provide a more realistic or personalized experience. For example, a particular Wii fitness game comes with a motion-tracking camera that ensures the user is exercising in the optimal manner.

Surveillance and Security. The ability to use camera technologies to track and identify the movements of individuals or intruders often relies heavily on images. Photographic evidence can be used in court as proof of wrongdoing.

Military Operations. Satellite imaging using high-powered cameras has transformed war and peacetime military operations, as the images can provide real-time information on the movement of individuals, troops, and resources. Unmanned drone airplanes with cameras allow the military to explore locations that may be dangerous for humans to enter and provide needed information for military operations. The development of image-based technologies such as thermal and infrared cameras to detect changes in geologic features and topography are vital for planning military missions.

Exploration and Rescue. Sending robots and other unmanned machines with camera technologies to places too remote or dangerous for human beings to work in—outer space, great ocean depths, and disaster zones—is an important application. The space exploration robots Voyager 1 and 2 have been traveling through the solar system for nearly thirty-three years sending back images of distant planets, their moons, and the Earth itself. Images and cameras were also invaluable during the 2010 Deepwater Horizon oil spill, as underwater explorer robots equipped with cameras were able to send images of the damaged oil well as well as reports on how repairs were proceeding. Thermal-imaging cameras can also identify regions on fire even as they allow rescue robots to seek out injured humans trapped in fires or under rubble.

Impact on Industry

The total value of the camera-technologies industry is difficult to estimate as camera technologies are presently a startling cross-section of industries from photofinishing and retail camera industry to video surveillance to the motion picture and video industries. The camera technologies industry is, however, an expanding market. The video-surveillance market is expected to grow from $11.5 billion in 2008 to $37.7 billion in 2015 in the United States. According to market research company First Research, the U.S. photofinishing and retail camera industry is expanding and currently includes about 4,000 locations with combined annual revenue of about $6 billion. In another sector of the camera technologies, wage and salary employment in the motion picture and video industries is projected to grow 14 percent between 2008 and 2018, compared with 11 percent growth projected for wage and salary employment in all industries combined.

Medicine. The impact of cameras and camera technologies continues to revolutionize the medical field. From improvements in laparoscopic surgery and noninvasive imaging using tiny cameras, the use of camera technologies in medicine is expanding and improving the safety of medical procedures through knowledge of an individual's specific anatomy and medical issues. In addition, tiny, ingestible cameras can provide real-time imaging of body structure and function.

United States Military. The United States government, particularly the military, relies heavily on camera technologies to gather intelligence, perform safety inspections, and conduct military operations. Accordingly, they are one of the biggest sources of funding for camera research and development. The government has funded grants for projects such the improvement of optical communications and imaging systems and the use of magneto-optic imaging technology to detect hard-to-access corrosion on Air Force fighter planes. The purpose of these grants is to improve information gathering, military operations, security, and troop safety.

Industry and Business. The manufacturing industry is a large user of camera technologies, which help it to monitor industrial areas, cut costs, and increase efficiency. For example, many manufacturers use video-surveillance cameras to monitor areas that are intolerable to humans , as the cameras can be

Fascinating Facts About Camera Technologies

- One of the main missions of the Hubble Space Telescope was to look at the images and discover when the universe was born.
- The first photographic film was highly flammable
- The earliest cameras produced in significant numbers used glass plates to capture images.
- Barack Obama's official portrait was taken with a digital camera, a presidential first.
- The word photography is derived from the Greek words *photo* ("light") and *graphein* ("to write").
- American electrical engineer Steven Sasson, the inventor of the digital camera, was awarded the National Medal of Technology and Innovation in 2010. This is the highest honor bestowed by the U.S. government on scientists and inventors.
- As of 2011, about 80 percent of households in the United States had a digital camera.

designed to have a high resistance to weather and corrosion. Video cameras can also be used near heavy-duty machinery in an industrial environment to check the efficiency of the automated processes and safeguard the employees working there.

CAREERS AND COURSE WORK

There are many careers in a variety of industries that directly correlate to camera technologies, and entry-level requirements vary significantly by position. Given the wide spectrum of difference between the careers, a sampling of careers and course work follows.

The motion picture and video industries provide career options such as cinematographers, camera operators, and gaffers who work together to capture the scripted scenes on film and perform the actual shooting. Formal training can be an asset to workers in film and television production, but experience, talent, creativity, and professionalism are usually the most important factors in getting a job. In addition to colleges and technical schools, many independent centers offer training programs on various aspects of filmmaking, such as screen writing, editing, directing, and acting. For example, the American Film Institute offers training in directing, production, cinematography, screen writing, editing, and production design.

Another camera-technologies career option is photographer. Photographers produce and preserve images that paint a picture, tell a story, or record an event. They often specialize in areas such as portrait, commercial and industrial, scientific, news, or fine arts photography. Employers usually seek applicants with a "good eye," imagination, and creativity, as well as a good technical understanding of photography. Photojournalists or industrial or scientific photographers generally need a college degree. Freelance and portrait photographers need technical proficiency, gained through a degree, training program, or experience. Photography courses are offered by many universities, community colleges, and vocational and technical institutes. Basic courses in photography cover equipment, processes, and techniques. Learning good business and marketing skills is important and some bachelor's degree programs offer courses that focus on those. Art schools offer useful training in photographic design and composition. A good way for a photographer to start out is to work as an assistant to an experienced photographer.

Still another option is a mapmaker or geographer who works with cameras and images to create maps. Photogrammetrists interpret the more detailed data from aircraft to produce maps. Most mapmakers work for architectural and engineering services companies, governments, and consulting firms. Remote-sensing specialists and photogrammetrists often have at least a bachelor's degree in geography or a related subject, such as surveying or civil engineering. Classes in statistics, geometry, and matrix algebra also are useful. Many remote-sensing specialists have degrees in the natural sciences, including forestry, biology, and geology. They often take courses in remote sensing or mapping while earning these degrees. Not everyone working in this field has a bachelor's degree, however. People who have an associate's degree or a certificate in remote sensing or photogrammetry usually begin as assistants and gain additional skills on the job. Taking high school or college-level classes in mapping, drafting, and science can also lead to assistant jobs. Some employers hire entry-level workers who do not have college training but do have an aptitude for math and visualizing in three dimensions.

Social Context and Future Prospects

The ability to record and view the world will continue to change as new camera technologies allow individuals and companies to explore further. Future technological innovations in image capturing, size of cameras, storage of images, and viewing of images will change as the industry progresses.

Since the early days of camera technology, cameras have been used to bring the reality of different lives and social issues to the general population through books, exhibits, and newspapers. Early examples of this include Jacob Riis's 1890 publication *How the Other Half Lives*, which presented images of tenement life in New York City, and Lewis Hine's commission by the United States National Child Labor Committee to photograph children working in mills in 1909. A more recent example is the interest in the plight of Afghan women, which was stimulated in 1985 by photojournalist Steve McCurry's haunting photo of a young Afghan refugee named Sharbat Gula that appeared on the cover of *National Geographic*.

Employment in camera technologies is increasing for many careers. As an example, employment of photographers is expected to grow 12 percent over the period from 2008 to 2018, about as fast as the average for all occupations.

Dawn A. Laney, B.A., M.S., C.G.C., C.C.R.C.

Further Reading

Allan, Roger. "Robotics Give Doctors a Helping Hand." *Electronic Design*, June 19, 2008. Discusses the use of cameras and robots in medicine.

Ang, Tom. *Photography*. New York: Dorling Kindersley, 2005. Covers a number of aspects of photography, both historical and technical.

Chamberlain, P. "From Screen to Monitor." *Engineering & Technology* 3, issue 15 (2008): 18-21. A discussion of movie technology, images, and changes in viewing options.

Freeman, Michael. *The Photographer's Eye: Composition and Design for Better Digital Photos*. Burlington, Mass.: Focal Press, 2007. Discusses how to develop one's photographic eye and compose striking, effective photographs; with illustrations.

Gustavson, Todd. *Camera: A History of Photography from Daguerreotype to Digital*. New York: Sterling Innovation, 2009. An excellent history of the camera and photography.

Heron, Michal. *Creative Careers in Photography: Making a Living With or Without a Camera*. New York: Allworth Press, 2007. Provides information on numerous camera- and photography-related careers and includes a self-assessment tool.

Web Sites

American Photographic Artists
http://www.apanational.com

National Press Photographers Association
http://www.nppa.org

Professional Photographers of America
http://www.ppa.com

Women in Photography International
http://www.womeninphotography.org

See also: Cinematography; Photography.

CINEMATOGRAPHY

FIELDS OF STUDY

Physics of light; acoustics; holography; photography; computer graphics; computer animation; mathematics; chemistry; sensitometry; mechanical engineering; electronic engineering; perception.

SUMMARY

Cinematography is the science and practice of making motion pictures. It includes the technical processes behind all imaging formats, including film and analogue and digital video. Cinematography is concerned with the careful manipulation of technical and mechanical tools, such as lighting, exposure, lenses, filters, and special effects, to create a coherent visual expression of information, emotion or narrative. Besides the filmmaking and commercial industries, in which cinematography plays a central role, virtually any area where the production of a moving image is useful relies on the principles of this field of applied science. For example, cinematography has important applications in disciplines as diverse as military, education, scientific research, marketing, and medicine.

KEY TERMS AND CONCEPTS

- **Cut:** Transition between one camera shot and another without any effect in between, such as a dissolve or a fade; also called "edit."
- **Film Stock:** Traditional photographic film on which a moving image is captured, consisting of a base painted with a silver-halide emulsion that reacts to light. Common film stock gauges, or widths, are 16 and 35 millimeters.
- **High-Definition Technology:** Method of capturing and displaying images at a much higher resolution than standard-definition footage.
- **Interlaced Scanning:** Technology in which each frame of a motion picture is composed by scanning two separate fields and displaying them simultaneously.
- **Progressive Scanning:** Technology in which each frame of a motion picture is processed as a complete, separate image.
- **Sampling:** Process of repeatedly measuring an analogue signal at regular intervals and using this data to produce a digital signal; sampling allows motion pictures shot on film to be converted to a digital format.
- **Sensitometry:** Measurement of the sensitivity of various materials, especially photographic film, to light.
- **Widescreen:** Imaging format in which the ratio of the width of the image frame to its height (aspect ratio) is greater than 4:3. Widescreen formats appear much wider than they are tall.

Definition and Basic Principles

Cinematography is the science behind the techniques involved in creating a motion picture, as well as the practical application of those techniques. Basic cinematographic tools include film and video cameras; different lenses, filters, and film stock; the equipment involved with artificially lighting a set; the machinery used to mount and transport cameras and control their angles and movements; and a wide variety of computerized special effects that can be created or integrated into the motion picture in the postproduction stage. Unlike still photography, in which a complete product is composed of a single image, cinematography makes use of the relationships between quickly moving images to produce a narrative arc.

At heart, cinematography is based on an illusion caused by the interplay between technology and human perception. The typical motion-picture projection, whether the format is film, video, or digital, consists of twenty-four separate still frames per minute. These still frames are transformed into what appears to be a moving image through two related features of the human visual system known as persistence of vision and the phi phenomenon. Persistence of vision refers to the fact that when light hits the retina of the eye, the images it creates persist in the brain for a tiny fraction of a second longer than the physical stimulus itself. The phi phenomenon refers to the fact that when the eye is shown two separate images in rapid succession, the brain creates the appearance of seamless movement between the two frames. The combination of these two psychophysical characteristics makes it possible for people to ignore the tiny fractions of darkness that appear for a moment in between each still frame.

Background and History

Motion-picture technology was born in the late nineteenth century, when a research team led by American inventor Thomas Alva Edison and engineers working in England and France all independently developed a means of photographing still images at a rate fast enough to capture movement and of projecting successive images at a rate fast enough to create the illusion of seamless motion. Among the first motion-picture projection technologies were Edison's Kinetograph and Kinetoscope.

Early motion pictures were recorded in black and white (though they were often hand colored) and had no sound. Early in the twentieth century, various experiments with sound recording and amplification equipment were carried out, including a technique for using wax phonograph discs to record sound and an electronic loudspeaker for amplifying it in theaters. The Vitaphone system, created by the Warner Bros. studio, used these developments to produce a short musical known as *The Jazz Singer* (1927). By the late 1920's, studios had moved to recording sound optically on a separate reel of film stock. For the sound track of a film to be properly synchronized with the visual image, the speed of motion picture projection was standardized at twenty-four frames per second. Color cinematography was first achieved through a system of filters that recorded information from the red, green, and blue spectrums of light separately onto a strip of black-and-white film—later, color film stock was developed in which three layers of emulsion on the film served the same purpose.

In the 1990's, digital cinematography, which captures moving images in a digital, rather than analogue, format, began to take off. The first major cinematic release shot on digital video was George Lucas's *Star Wars Episode II: The Attack of the Clones* (2002).

How It Works

Cameras. The body of a traditional motion-picture camera consists of a sturdy, lightproof housing, a system of motors that control the movement of film and mechanical components such as the shutter, and a viewing system through which the camera operator can monitor the footage being shot. Onto the body of the camera is affixed a lens, which uses one or more convex glass elements to gather and focus rays of light onto the film. Depending on the type of framing the cinematographer wishes to achieve, standard, wide-angle, telephoto, or zoom lenses may be used. To produce a smooth, not jerky, moving image, cameras are usually mounted on sturdy supports during filming. A basic dolly mount consists of a large, heavy tripod on a sturdy base with casters. A mount has various adjustment levers that allow the camera operator to raise, lower, and tilt the camera. To raise the camera to an elevated height, a crane is added to the dolly; for long panning shots, the entire contraption is often moved along specially built rails that are laid on the floor. Steadicam systems, in which a camera is harnessed to the operator's body, reduce the jerkiness caused by his or her movements during shooting and allow for smoother handheld operation of cameras.

Film Cinematography. To capture a rapid succession of images, film is inserted behind the camera lens and automatically unrolled by an electric motor from a supply reel, through a gate, and onto a take-up reel. As the film travels through the gate, a motorized shutter lifts up and down at regular intervals, usually twenty-four times per second. Every time the shutter lifts, the film is exposed to light. Film stock is composed of a base, traditionally made of celluloid but later usually made of polyester or Mylar, painted with a layer or layers of silver-halide emulsion. This emulsion reacts when exposed to light. (Color films have three layers of emulsion, each of which reacts with only one of the three primary colors of light: blue, green, or red.) Once the film has been shot, a chemical known as a developer is used to process the exposed areas of the film and produce a negative image. With black-and-white film, a negative shows dark areas of the image as light areas and vice versa. With color film, a negative shows complementary colors—for example, red areas appear green. Further chemical processes make a positive print out of these negatives. When a positive print is projected onto a screen, the image appears in its correct form. A type of film known as reversal film is also available. It has the advantage of producing a positive rather than a negative image at the end of the developing process.

Video Cinematography. Although film remains an important tool for many cinematographers and motion-picture producers, other imaging formats have become increasingly relevant. With analogue video imaging, light gathered by the camera lens is not captured by the emulsions on film stock but instead is converted into electric signals by tiny photosensitive diodes on a component known as a charge-coupled device (CCD). In some cameras, three separate CCDs are used, with each capturing information about one of the three primary colors. This electric signal is then recorded as a pattern on a strip of magnetic tape. Analogue video imaging is used to produce news broadcasts and other television broadcasts. Because it is very inexpensive, it is also a popular tool for at-home or amateur motion-picture production. Digital video imaging also makes use of CCDs. With digital video, however, the electric signal from a CCD is not converted into a magnetic pattern on tape. Instead, it is translated into a digital signal or binary code consisting solely of 1's and 0's. Digital cameras can record onto hard disks, digital video discs (DVDs), flash drives, or any other digital format.

Applications and Products

Filmmaking. With the advent of motion-picture imaging came an entirely new form of visual communication, one that—unlike photographs—was able to depict a series of events, and thereby express a narrative rather than a single moment in time. In addition, the illusion of movement and the fidelity with which the camera records images lends viewers of motion pictures an irresistible sense that the events they are watching not only are somehow real but also are occurring in the present rather than the past. Arguably the most significant application of cinematography, then, at least in terms of its cultural impact, is the use of motion-picture imaging to create narrative, documentary, and newsreel films. Practicing cinematographers are able to apply the full range of cinematographic tools and techniques in different ways to achieve sophisticated effects. For instance, the use of a dolly mounted on tracks allows a camera operator to pan (turn) the camera in a horizontal plane. This is often used to produce the illusion that the camera's perspective is that of a character in the film. Although this application has largely been taken over by computer-generated imaging, cinematographic techniques also enable the creation of animated films. Animation is created by using a motion-picture camera to film a series of carefully constructed illustrations (or objects, such as clay puppets). When replayed, the still images appear to move.

Fascinating Facts About Cinematography

- The word cinematography is derived from two ancient Greek words that can be literally translated as "writing with motion"—the term distinguishes cinematography from photography, which comes from roots that mean "writing with light."
- An aerial shot is usually captured by a camera mounted on a helicopter. Before this became possible, filmmakers who wanted a shot from the air teamed camera men with stunt pilots to get the desired effect.
- In the 1939 film *The Wizard of Oz*, a scene in which the witch writes a message in the sky with her broom was filmed by a camera mounted beneath a glass tank. The "sky" was composed of a cloudy mixture of water and oil.
- The first motion-picture recordings were not shown on a theater screen but in a large machine the size of a cabinet. People would step into the machine and deposit a coin to watch the film.
- In 1903, projectionist Edwin S. Porter created the first motion picture that presented a narrative. Known as *The Great Train Robbery*, the 14-minute film contained the world's first motion-picture chase scene.
- When astronomers on the space shuttle *Atlantis* went up into space to repair the Hubble telescope in 2009, they carried a massive IMAX high-definition camera on board with which to shoot a film about the project. The camera required a mile of film to record 8 minutes of action.
- Before the development of computer-generated special effects, filmmakers used inventive techniques to create illusions. In the 1895 film *The Execution of Mary Queen of Scots*, the camera was stopped, the actress playing Mary ran off-screen and was replaced with a dummy, and the other actors froze in position. Then, the camera was restarted and the guillotine fell to "behead" Mary.

Education. Cinematography has made it possible for students of all subjects to engage in learning through direct observation without ever leaving their classrooms. Techniques such as time-lapse photography, in which still frames that were captured at a very slow rate are replayed at a much faster speed, enable teachers to demonstrate incredibly drawn-out and subtle processes, such as the development of vegetation from tiny seed to fruiting tree, or the movement of the stars across the sky. In the same way, high-speed photography, in which still frames that were captured at a very fast rate are replayed at a slower speed, enable students to see clearly the steps involved in processes that happen in the blink of an eye, such as the formation of droplets as a bead of water hits the surface of a pond. Time-lapse photography is created with a standard camera attached to a device set to trigger its shutter at extended, regular intervals—such as an hour apart or more. The technology behind high-speed photography is more complex and requires careful control over the precise timing of the camera's shutter and the amount of light the film or CCD is exposed to during the time the shutter is open. A shutter, for instance, may be triggered by a sound associated with the event the cinematographer is trying to capture, such as the firing of a gun.

Another cinematographic technology that has transformed both education and entertainment is the use of three-dimensional (3D) imaging and projection systems such as IMAX 3D. With three-dimensional cinematography, two camera lenses are used to capture two separate streams of images, one corresponding to the view a human observer would receive through his or her left eye, and one corresponding to the view through his or her right eye. In the theater, these two images are them projected onto the screen simultaneously. With the use of a special pair of three-dimensional glasses, viewers are able to perceive the dual cinematic streams as a single three-dimensional image. Three-dimensional imaging and projection is widely used in science museums as a means of giving museum-goers a more vivid entryway into presentations about topics such as space exploration and underwater habitats.

Medical Video Imaging. A host of specialized cinematographic techniques have found useful homes in the realm of clinical diagnosis and surgery. For example, stereo endoscopes make it far easier for physicians to perform minimally invasive surgeries. A stereo endoscope is an instrument that can be inserted into a patient's body through a small incision or down a natural orifice such as the throat. It is used to transmit a three-dimensional video image of a patient's internal parts to the surgeon as he or she works. A stereo endoscope consists of a tube, often flexible, containing a dual lens system and a bundle of optical fibers. These fibers bring light from an external light source into the patient's body, then transmit the video image from the lens back out to a large screen in the operating room. When cinematographic techniques are combined with tools from communications technology, another application of real-time video imaging in medicine emerges—Web-based conferencing, which enables collaborations between clinicians who are physically distant from each other.

Impact on Industry

Apart from the multibillion-dollar filmmaking industry, which is the biggest global user of cinematographic technology, many other industry sectors rely on this field of applied science to carry out a significant element of their business operations. The advertising industry, for example, uses the tools and visual conventions of motion-picture imaging to generate what are essentially short narrative films—commercials. Commercials promote virtually every consumer product on the market, from baby food to prescription drugs. The music industry has been revolutionized by the advent of cinematography in the form of music videos. These brief, lavishly produced motion pictures help recording studios showcase their performers as flesh-and-blood personalities rather than faceless voices, helping drive additional sales.

Cinematography also has important applications in the security industry. Both private security firms and public agencies make use of closed-circuit television (CCTV) technology—a system in which the video imaging signals from one or more cameras are transmitted to a restricted set of monitors—to keep a continuous record of what takes place in businesses such as banks, convenience markets, and hospitals, and public areas with a high record of criminal activity.

Major Corporations. Much of the research into new cinematographic techniques takes place within the context of the motion-picture industry rather than in academic or government research laboratories. In the early twenty-first century, for example, the Eastman Kodak company introduced a new generation of color motion-picture films that—because of a change in the chemistry of the silver-halide emulsion used to produce these films—was twice as sensitive to light as previous films, enabling camera operators to shoot in conditions with lower light without compromising the quality of the footage produced. Besides

Kodak, other large players in the field of cinematographic technology include Technicolor, Deluxe, and Fujifilm. The Arri Group and Panavision are among the largest suppliers of traditional film cameras to the motion-picture industry, and both groups have also entered the digital cinema market alongside camera-producing corporations such as Silicon Imaging and RED. The six most important corporations involved in filming and producing motion pictures in the United States are Paramount Pictures, Walt Disney Pictures/Touchstone Pictures, Warner Bros. Pictures, Twentieth Century Fox Film Corporation, Universal Pictures, and Columbia Pictures.

CAREERS AND COURSE WORK

Within the motion-picture industry, the role of the cinematographer is more creative than technical. Cinematographers work with the director of a film to decide on how each shot will be composed. Students interested in more hands-on career options may consider becoming camera operators, lighting technicians, sound engineers, set electricians, or postproduction editors.

Anyone interested in a career as a cinematographer or camera operator or who plans on pursuing some other kind of technical occupation related to cinematography such as lighting, sound recording, and postproduction and editing, should acquire a strong body of knowledge about the physics of light. Besides courses in optics, it is important to take specialized classes in the mechanical and electrical engineering of film, video, and digital cameras, and to understand how each type of cinematographic equipment, including lenses, interacts with light to produce an image. Other important educational requirements include an understanding of the chemistry of film emulsions and development and the mathematics of exposure.

Finally, it would be appropriate for any student of cinematography to take a course in human visual perception through a psychology or physiology department. Understanding how the eye and brain work together to process light and visual information allows those who work in motion-picture imaging to carefully manipulate those processes to achieve the desired emotional or visual effect. Many practicing cinematographers have graduated from a technical college rather than a traditional liberal arts university, or have pursued specialized training in motion-picture technology from a formal film school program. A degree in computer science or computer graphics would also be an appropriate starting point for someone wishing to work in special effects, animation, or video editing.

SOCIAL CONTEXT AND FUTURE PROSPECTS

It is hard to overestimate the social impact that cinematography and its associated technologies have had. One important cultural role for motion-picture imaging has been expanding the horizons of viewers beyond the concerns of their own lives, their own communities, and even their own countries—enabling them to see directly into entirely different worlds. In this way, many films have served as powerful mechanisms for social and humanitarian change. In the 1980's, for example, when documentary cameras captured dramatic footage of starving Ethiopian men, women, and children and delivered it to the eyes of the Western world, aid money poured into the impoverished nation. The 1993 film *Philadelphia*, which told the story of a lawyer infected with the human immunodeficiency virus and his partner, highlighted the growing devastation wrought by the acquired immunodeficiency syndrome (AIDS) epidemic in the United States and put a human face on what was still considered by many Americans to be a shameful disease.

The use of motion-picture technology by nonprofessionals has also had a profound cultural impact. In the second half of the twentieth century, the development of less expensive cameras for making home movies (particularly home video cameras) enabled individuals across the world to create a tangible archive of their personal experiences and domestic milestones. Unlike a photograph, which is often posed, the home movie preserves not just formal, fleeting glimpses of the past but rather ceremonies, events, and candid interactions. In addition, the home movie transports the viewer straight back to the moment when the footage was recorded.

One of the most promising emerging technologies in cinematography is a technique known as the digital intermediate (DI) process. This is a means of scanning motion pictures recorded on film—which many advocates argue is of a far superior visual quality than digital video—at an extremely high resolution, thus converting it into a digital format. The DI process may enable the motion-picture industry to take advantage of the best of both film and digital technologies to produce imaging that is stunningly crisp and clear as well as far easier and cheaper to distribute.

M. Lee, B.A., M.A.

Further Reading

Barclay, Steven. *The Motion Picture Image: From Film to Digital.* Boston: Focal Press, 2000. A narrative history of the development of cinematographic technology, as well as a practical overview of related technical issues. Contains a comprehensive index.

Brown, Blaine. *Cinematography: Theory and Practice.* Boston: Focal Press, 2002. A hands-on guide for the practicing cinematographer, heavily illustrated with photographs and diagrams. Includes a bibliography.

Meza, Philip E. *Coming Attractions? Hollywood, High Tech, and the Future of Entertainment.* Stanford, Calif.: Stanford Business Books, 2007. An examination of how technological developments in cinematography are acting to shape the U.S. entertainment market. Organized into a series of real-life business case studies.

Sawiki, Mark. *Filming the Fantastic: A Guide to Visual Effect Cinematography.* Boston: Focal Press, 2007. Includes hundreds of full-color photographs illustrating special-effects techniques.

Saxby, Graham. *The Science of Imaging: An Introduction.* Philadelphia: Institute of Physics Publishing, 2002. An overview of the fundamentals of physics and mathematics involved in creating television, holography, and other forms of imaging. Includes diagrams, glossaries, and numerous explanatory notes in the sidebars.

Web Sites

Eastman Kodak Company
The Essential Reference Guide for Filmmakers
http://motion.kodak.com/US/en/motion/Education/Publications/Essential_reference_guide/index.htm

U.S. Department of Labor, Bureau of Labor Statistics
Motion Picture and Video Industries
http://www.bls.gov/OCO/CG/CGS038.HTM

See also: Photography; Television Technologies.

COASTAL ENGINEERING

FIELDS OF STUDY

Marine/coastal science; marine biology; civil engineering; oceanography; meteorology; hydrodynamics; geomorphology and soil mechanics; numerical and statistical analysis; structural mechanics; fluid mechanics; electronics; geology; chemistry.

SUMMARY

Coastal engineering is a branch of civil engineering that aims to solve coastal zone problems such as shoreline erosion and the destruction caused by tsunamis through the development, construction, and preservation of mechanisms and structures. Coastal engineering also entails understanding the theories and processes of wave actions and forces, ocean currents, and wave-structure interactions. Given that coastal areas are often highly populated and especially vulnerable to human impact, coastal engineering attempts to blend conservation and management of the world's coastal zones with the development requirements of humans.

KEY TERMS AND CONCEPTS

- **Accretion:** Buildup of sediment (usually sand or shingle) through natural fluid flow processes on a beach or coastal area.
- **Estuary:** Semi-enclosed body of water with a mouth to the ocean; contains a mixture of fresh water from inland and salt water from the sea.
- **Hard Coastal Engineering:** Use of conventional tools and hard structures, including groins, seawalls, and breakwaters, to mitigate coastal erosion and flooding.
- **Land Reclamation:** Process of creating dry land by taking an area of land below sea level, pushing back the water, and preventing the water from returning with structures such as dykes, or introducing sediment and material into the sea, raising the level of the seabed.
- **Littoral (or Longshore) Drift/Transportation:** Movement of material or sediment (usually sand or shingle) along the coastline, caused by waves and currents.
- **Longshore Current:** Current formed or driven by waves breaking obliquely (diagonally) to the shore; flows in the same direction as the wave.
- **Shingle:** Coarse gravel, such as beach pebbles.
- **Soft Coastal Engineering:** Use of long-term, naturalistic tools, including beach nourishment and artificial dunes, to manage coastal areas.
- **Tsunami:** Very large and potentially damaging oceanic wave caused by underwater seismic activity, such as an earthquake or volcanic eruption.

Definition and Basic Principles

Coastal engineering is a branch of civil engineering involved in the development, design, construction, and preservation of structures in coastal zone areas. Its primary function is to monitor and control shoreline erosion, design and develop harbors and transport channels, and protect low-lying areas from tidal flooding and tsunamis. To achieve such objectives, coastal engineers must have a strong understanding of the sciences of engineering, oceanography, meteorology, hydrodynamics, geomorphology, and geology. They must also have a strong understanding of the theories and processes of wave motion and action, wave-structure interaction, wave-force forecasting, and ocean current prediction.

Coastal areas are socially, economically, and ecologically crucial. They are home to a significant proportion of the world's human population, serve as important breeding grounds for many animal species, and are important for tourism, aquaculture, fishing, transport, and trade. These multiple uses of a relatively small area have led to rapid environmental degradation and have created conflict and the need to achieve a sustainable balance.

Background and History

Although the specialty of coastal engineering emerged only in the latter half of the twentieth century, coastal engineering and management have been practiced for many centuries. Human beings have long lived in coastal zones and, for just as long, have attempted to control and manage the effects of tidal flooding and wave action and to engage in land reclamation. People have developed ports to make transportation and trading easier for thousands of years. The first river port—port of A-ur on the Nile—was built before 3000 B.C.E. and the first coastal port—the port on Pharos near Alexandria—about 2000 B.C.E. For centuries, many countries wanting to protect strategic coastal areas used structures such as embankments as sea defenses.

Early engineering projects mainly involved hard coastal engineering, which is the design and building of docks, ports, defenses, and walkways. Although coastal engineering has long been practiced, the actual coastal processes and their driving forces were little understood. Therefore, formal engineering development required much more comprehensive knowledge of the ocean and the processes and principles involved.

Traditionally, engineering projects in coastal zones were the province of civil and military engineers. However, because of the growing need for specific research in coastal engineering, the first International Conference on Coastal Engineering was held in 1950 in California, establishing the branch of applied science and introducing the terminology. The conference proceedings stated that coastal engineering was not a novel science or unconnected to civil engineering but rather was a branch of civil engineering that was strongly reliant on and influenced by sciences such as oceanography, meteorology, electronics, and fluid and structural mechanics.

By the 1960's, significant inroads had been made into understanding coastal processes and this, coupled with advances in modeling, highlighted the importance of incorporating such processes into design and construction. By the 1970's, the increase in understanding coastal processes led to the introduction of soft coastal engineering, which included beach nourishment and artificial dunes. In the following decades, significant progress was made in the development of coastal engineering techniques, such as modeling techniques for deep-water wave prediction, and the understanding of processes, including wave-structure interaction and coastal sediment erosion.

Essentially, coastal engineering has transformed from a science that simply responds to anthropocentric demands (docks, ports, and harbors) to one that aims to balance human requirements and environmental protection, with a particular focus on mitigating both natural and human-induced coastal erosion.

How It Works

It is the job of a coastal engineer to recognize the characteristics and natural processes of the shoreline environment and to apply fundamental engineering theories and philosophies to solving ecological problems. To be successful, a coastal engineer must have a strong understanding of other sciences such as oceanography, meteorology, geomorphology, structural and fluid mechanics, and hydrodynamics. In particular, this means clarifying and using theories of wave action, wave-structure interaction, wave-force forecasting, ocean current prediction, and beach profile modification.

Coastlines across the world are the natural division between the terrestrial landscape and the seas and oceans. The geological composition of these areas is distinctive, as are the natural and anthropomorphic processes that affect them. Coastal engineering techniques must take into account the morphological development of coasts, such as erosion and accretion, and relate this to shore protection through engineering approaches and structures.

The action of waves on the shoreline forms one of the fundamental areas of coastal engineering research. Waves contain a very large amount of energy formed by the action of wind over vast tracts of open water, the gravitational attraction of the Sun and Moon, and occasionally seismic activity (as with tsunamis). Although this energy is collected over very large areas, it is released along relatively small areas of coastline. This release of energy, in the form of breaking waves, has a strong influence on the currents, sediment, and geological structures of coastal areas. The shoreline or beaches, which are frequently composed of sediment such as sand, shingle, or gravel, are constantly battered and reshaped by the actions of wind and water.

Erosion is a natural process, and the movement of sediment is achieved by wave uprush, which transports sand onshore, and backwash, which transports sand offshore. The natural processes of erosion in coastal areas are very complex and involve flow fields created through not only the action of breaking waves but also erratic turbulent sediment transport in the water column and a shifting shoreline. Worldwide research aims to develop predictive models of this erosion process. Over time and under normal conditions, erosion is also a relatively cyclical process. Sediment can be carried offshore during one season and dumped back onshore during the next season, as well as moved obliquely along the shore over time. Erosion processes can, however, be significantly disrupted by human activity and structures or can be considered undesirable because of human requirements and desired aesthetics. Coastal engineering structures can be, in fact, both detrimental and beneficial in regard to coast erosion, and most coastal engineers believe that only an integrated and holistic approach to planning and design can generate long-term sustainability.

Within this energetic and process-driven natural boundary between land and sea, people have constructed coastal engineering structures such as ports and harbors, which have had both positive and negative effects on the environment. The design of such structures must predict wave dynamics and their impact on individual coastal zones and beach environments. Since the emergence of coastal engineering science, much greater emphasis has been placed on understanding coastal processes, such as wave action, and designing and developing policies and structures to protect coastal areas from erosion.

Although the study of coastal erosion is a fundamental concept for coastal engineering, coastal engineers must also recognize and study coastal protection measures and applications such as hard and soft shoreline protection structures, understand the effects of these structures on the morphology of the coastal areas, and develop effective coastal zone management plans and policies.

Applications and Products

Coastal engineering structures and activities vary significantly from country to country. They depend on social aspects, such as the country's history and development and its people's relationship to the ocean, and environmental and structural aspects, such as the nature of the ocean along the country's coast, geomorphological and geological conditions, the ecosystem, climate, extent of tectonic activity, currents, and wave action.

Generally speaking, however, there are two types of coastal engineering applications: hard and soft coastal engineering. Although hard stabilization techniques have been used for many years and are considered to be appropriate under certain circumstances, they can also be expensive and, in trying to rectify human impact, can actually disrupt and destroy natural shoreline processes and habitats, intertidal areas, and wetlands. In some cases, they can even increase erosion. As coastal engineering has evolved and become more sophisticated, engineers have moved away from constructing hard structures with little understanding of their impact on the environment and toward incorporating soft engineering and minimizing the ecological impact.

Hard Structures and Applications. Groins are solid structures, usually constructed from wood, concrete, or rocks, running perpendicular from the foreshore into the water (under normal wave levels). They aim to disrupt water flow and reduce the transportation of sediment from longshore drift. They trap sediment to extend or preserve the beach area on the up-drift side and reduce erosion on the down-drift side. The size and length of these structures is usually determined based on specific local conditions, including wave dynamics, beach slope, and environmental factors. Seawalls are very large, rigid, and usually vertical structures constructed from concrete. They are found at the transition between the low-lying beach and the higher mainland and run parallel to the shoreline. The main function of a seawall is to preserve the shoreline by preventing additional shoreline erosion and recession during direct wave-energy impact and flooding. Seawalls are, however, not effective in preventing longshore erosion.

Revetments are a concrete, rock, or stone "veneer," or facing, constructed on a beach slope with the aim of preventing erosion caused by wave action and storm surges. Unlike seawalls, which can assist in flood prevention, revetments are not usually constructed for this function. Revetments are always built as sloping structures, and although they may sometimes be completely solid and rigid, they are often built with interlocking slabs or stone and designed to be permeable. This permeability tends to enhance the strength of the revetment and the absorption of wave energy while reducing erosion and wave run-up. Revetments can also be made from sand-filled bags, interlocking tires, concrete-filled bags, and wire-mesh stone-filled gabions (sunken cylinders filled with earth or stones). As revetments tend to be passive structures, their application is often limited to areas that are already protected in some other way or by some other engineering structure.

Breakwaters are fixed or floating structures built parallel or at an angle to the shoreline. The main function of a breakwater is to protect the shore and activities along the coastline by reducing the impact of wave energy. The ability and impact of a breakwater depend on whether it is submerged or floating, its distance from the shoreline, its length and orientation, and if it is solid or segmented. The four main types of floating breakwaters are box, pontoon, mat, and tethered float; the three main types of detached breakwaters are offshore, coastal, and beach.

Soft Structures and Applications. Shore nourishment has become one of the most common soft coastal engineering applications. The three main types are backshore, beach, and shoreface nourishment. As the name suggests, nourishment is the action of artificially adding sand to the backshore (upper part of the beach), beach, or shoreface (usually the seaside of the bar) in an attempt to modify the effects of erosion. Although nourishment replaces sand in an eroded area and is considered to be a rather natural form of coastal engineering, it does not address the causes or processes of erosion nor does it reduce the impact of wave energy.

Sand dune stabilization is a relatively basic and common form of soft coastal engineering. It involves the planting of vegetation to stabilize and protect sand dunes along the coastline. The creation of artificial dunes is also considered to be an effective form of soft coast engineering protection, particularly in conjunction with shore nourishment.

Beach drainage, or beach dewatering, is a system of shore protection based on the concept of physically draining water from a beach. Drainpipes are installed and buried below the beach and parallel to the shoreline. These pipes collect seawater and transport it to a pumping station, where it is collected and either returned to the ocean or used. Beach drainage helps reduce erosion by lowering the watertable in the uprush zone and reducing the force of the backwash by increasing the volume of water that seeps into the beach.

Impact on Industry

The field of coastal engineering is being developed and researched by many countries and organizations all over the world. A definite shift from hard to soft engineering has occurred, and in some areas, hard engineering has been opposed and even actively prohibited. Many organizations and government agencies have become actively involved in coastal engineering research, particularly in holistic and soft approaches to coastal zone management.

Major Organizations. Because of the global interest in coastal management and the rather ubiquitous nature of coastal areas, many international organizations are interested in and are actively conducting research and contributing to the field of coastal engineering. Many countries have specific organizations that represent the specialist interests of coastal engineers, such as the National Committee on Coastal and Ocean Engineering (NCCOE), found within Engineers Australia. The NCCOE states that its priorities are to improve management of the coastal zone through understanding of the coastal environment, to develop strategies for hazards and risk in the coastal zone, and to establish a national coastal and near-ocean data program integrating advanced technologies such as satellite, airborne, and shore-based remote sensing.

Government and University Research. Many worldwide governments have a stake in the management of coastal areas. Numerous governments have increased their interest in monitoring and managing the land-sea interface through such methods as engineering projects, involving themselves not only in the economics of a project but also in its environmental aspects. Coastal engineering is important to the continued survival and sustainability of the world's shorelines. Although governments are increasingly aware of the importance of coastal engineering, many universities and organizations state that the funding of academic research in the field is still inadequate and affects the ability of researchers to conduct effective and extensive studies. Programs in coastal engineering are offered at many universities, including the University of California, Berkeley, the University of Florida, Oregon State University, the University of Queensland, Tokyo University, the University of Nottingham, and the Technical University of Denmark.

In many cases, no single federal department or agency is solely responsible for funding and managing coastal engineering research and education. In the United States, academic funding can come from many sources, including the National Science Foundation, the Sea Grant program of the National Oceanic and Atmospheric Administration (NOAA), the Office of Naval Research, and the U.S. Army Corps of Engineers. The NOAA is one of the world's largest governmental agencies involved in ocean management. The NOAA's Office of Ocean and Coastal Resource Management (OCRM) provides management and guidance for national, state, and territory coastal programs and research. The six divisions of the OCRM implement research and engineering techniques for shoreline protection. The Coastal Engineering Research Center, which was established in 1963 and is part of the U.S. Army Corps of Engineers' Coastal and Hydraulics Laboratory, is also at the forefront of coastal engineering research both nationally and internationally.

CAREERS AND COURSE WORK

Many universities offer undergraduate and graduate degrees in coastal engineering. Most commonly, students who wish to pursue a career in coastal engineering will study engineering or marine science as an undergraduate. By graduation, students should have a solid understanding of coastal engineering concepts, theories, processes, and practices, such as wave action, wave transformation, statistical and numerical analysis, tides and currents, beach dynamics and coastal structures, and the impact of natural and anthropogenic factors on the coastal environment. In addition, as coastal engineers need to provide the scientific base data used for coastal zone management, they are likely to require knowledge and experience in the use of modeling.

Students who study coastal engineering can enter such careers as consulting engineers, project managers, environmental consultants, and construction contractors and such fields as construction management, coastal and oceanographic engineering, water resource management, and hydrologic engineering. They may find employment in the private sector, with nongovernmental organizations, with specialized government organizations and agencies, or in universities undertaking teaching and research.

SOCIAL CONTEXT AND FUTURE PROSPECTS

The social, economic, and ecological consequences of coastal erosion are significant. Almost 60 percent of the world's population lives within 100 kilometers of a coast. Coastal areas not only provide significant economic benefits—fish and other maritime products, a means of transportation, and easier access to trading partners—but also offer numerous recreational opportunities. However, the economic and aesthetic appeal of coastal areas has created population pressure that has greatly increased the need for functional coastal structures, such as ports, wharfs, and jetties, and infrastructure, such as roads and sewerage facilities, which increase erosion and pollution. The erosion caused by human activities and structures can change beach dynamics and profiles, which in turn can have ecological effects, such as the loss of animal breeding grounds, and social effects, such as making the area unsafe for swimming and other recreational activities.

Research and climate change modeling has indicated that sea levels may rise significantly in the twenty-first century. This, coupled with an associated increase in coastal storm frequency and strength, will produce more floods and greater erosion, creating serious ecological repercussions for the coastal zones of the world, which are often densely populated and popular areas for tourism. The construction of such engineering structures as ports, harbors, recreational facilities, and resorts are considered necessary for

Fascinating Facts About Coastal Engineering

- Almost 650 million people living in low-lying coastal areas are at risk from the rising sea levels and associated erosion that are likely to be caused by climate change.
- Neskowin, a coastal town in northern Oregon that sits at 20 feet above sea level, began experiencing severe erosion of its beaches in the 1990's because of the rising sea level and high waves. Some relief has been provided by riprap (a stack of boulders) on the beach, but the future is uncertain.
- In 2008, a 45-foot-tall sheet pile ring and a rock apron were placed around the base of Morris Island Lighthouse to prevent further erosion. The lighthouse, built in 1875, was originally on high ground one-half mile from the Atlantic Ocean but has come to be located 2,000 feet offshore in 10 feet of water.
- The barrier island of Grand Isle, Louisiana, has experienced severe erosion problems since the 1980's. Numerous attempts to stop the erosion have had mixed results. In 2010, an artificial oyster reef was installed on the island's backside in an effort to end erosion.

continued economic and social development. Such engineering has, however, a very significant ecological impact on the coastal zone areas, particularly an increase in erosion. One of the primary goals in coastal engineering is to rectify erosion issues.

The difficulty for the future of coastal engineering is that some of the solutions to such environmental or aesthetic issues can be ineffective or ecologically unsound, thereby exacerbating the problem. As such, the future of coastal engineering requires integrated coastal zone management (ICZM), a holistic and integrated strategy and approach incorporating hard and soft engineering for the management of all aspects of coastal areas. It incorporates advances in modeling that are helping develop a framework for predicting coastal erosion hazards and processes. Many coastal engineers are advocating an integrated approach, stating that the future ecological and economic sustainability of the world's fragile and vulnerable coastal areas depends on its adoption. Such a strategy is optimal, as it considers such issues as climate change, the rise in sea level, navigational needs, the impact on plants and animals, and aesthetic considerations.

Christine Watts, Ph.D., B.App.Sc., B.Sc.

Further Reading

Dean, Robert G., and Robert A. Dalrymple. *Coastal Processes with Engineering Applications.* 2002. Reprint. New York: Cambridge University Press, 2004. Provides comprehensive information on coastal processes and management of shoreline erosion as well as an overview of the topic, the hydrodynamics of the coastal zone, and coastal responses to erosion and engineering applications.

Kamphuis, J. William. *Introduction to Coastal Engineering and Management.* 2d ed. London: World Scientific Publishing, 2009. Aimed toward undergraduate and graduate students, it discusses traditional and contemporary issues, methods, and practices in coastal engineering design, as well as environmental issues and climate change.

Kraus, Nicholas, ed. *History and Heritage of Coastal Engineering: A Collection of Papers on the History of Coastal Engineering in Countries Hosting the International Coastal Engineering Conference 1950-1996.* New York: American Society of Civil Engineers, 1996. Documents the history of coastal engineering in fifteen countries, including the United States, Canada, Australia, France, Denmark, Germany, South Africa, and Great Britain.

Reeve, Dominic, Andrew Chadwick, and Christopher Fleming. *Coastal Engineering: Processes, Theory and Design Practice.* New York: Spon Press, 2004. A comprehensive examination of coastal processes, morphology, design, and the effect of engineering structures for coastal defense, including modeling techniques and case studies.

Shibayama, Tomoya. *Coastal Processes: Concepts in Coastal Engineering and Their Application to Multifarious Environment.* Hackensack, N.J.: World Scientific Publishing, 2009. Designed for coastal engineering graduate students. Describes wave-induced problems, coastal processes, sediment transport, and coastal disasters.

Sorensen, Robert. *Basic Coastal Engineering.* 3d ed. New York: Springer, 2006. Provides students of coastal engineering with the basics on coastal processes and wave mechanics, theoretical and applied hydromechanics, and engineering design and examination.

Young, Kim, ed. *Handbook of Coastal and Ocean Engineering.* Hackensack, N.J.: World Scientific Publishing, 2010. Presents a thorough compilation of topics, such as wave and water fluctuations, coastal and offshore structures, sediment and erosion processes, modeling and environmental issues from more than seventy international authorities and experts on coastal engineering.

Web Sites

National Oceanic and Atmospheric Administration
Ocean and Coastal Resource Management
http://coastalmanagement.noaa.gov

U.S. Army Corps of Engineers
Coastal and Hydraulics Laboratory
http://chl.erdc.usace.army.mil

See also: Erosion Control; Flood-Control Technology; Ocean and Tidal Energy Technologies; Water-Pollution Control.

COATING TECHNOLOGY

FIELDS OF STUDY

Materials engineering; chemical engineering; chemistry; physics; high-efficiency devices; corrosion protection; thermal protection; resource management; optics; biotechnology; nanotechnology.

SUMMARY

A coating is a thin layer or film of a substance spread over a surface for protection or decoration. Coatings can significantly improve the performance of technology. Applications of coatings are far ranging, from corrosion protection of metals in vehicles to thermal protection in jet engine turbine blades. Functional coatings are also used to generate electricity in fuel cells and photovoltaic cells and can reduce thermal emissions from buildings through windows. There are numerous methods of applying these coatings, and each satisfies a different demand.

KEY TERMS AND CONCEPTS

- **Capacitor:** Storage device for electric charge.
- **Cataphoretic Dip Coating:** Electrophoretic process that coats substrates such as cars in a complete, adhesive, protective film.
- **Cathodic Protection:** Method by which an electrochemically inferior material is applied to a metal substrate, preventing the corrosion of the substrate.
- **Chemical Vapor Deposition:** Variety of deposition methods to create thin films by chemical reactions on substrates.
- **Colloid:** Small particles dispersed within a continuous medium in a manner that prevents them from agglomerating, settling, or being filtered.
- **Corrosion:** Chemical or electrochemical reaction of a material with its surroundings, leading to the disintegration of the material and the formation of new, often undesired materials, such as rust.
- **Creep:** Form of permanent plastic deformation in metals at elevated temperatures.
- **Dip Coating:** Method by which a substrate is fully immersed in a liquid either to chemically alter the substrate surface or to deposit a coating.
- **Doping:** Method by which tiny trace amounts of one element are mixed with another element, thereby significantly changing the properties of the resulting compound.
- **Electrophoresis:** Migration of charged colloidal particles or molecules through a conductive solution under the influence of an applied electric field.
- **Galvanization:** Method of applying a sacrificial anode such as zinc on the surface of a different metal to protect the latter from corrosion.
- **High Velocity Oxy-Fuel Process (HVOF):** Mixture of a combustible gas (such as propane) and oxygen ignited inside a combustion chamber to melt a coating material before it is accelerated via an inert carrier gas at high speeds toward a surface, where the hot material forms a coating.
- **Hydroxyapatite:** Naturally occurring mineral form of calcium apatite with the chemical formula $Ca_{10}(PO_4)_6(OH)_2$.
- **Line-Of-Sight Process:** Coating that is applied only on the surfaces that face the emission source of the coating material.
- **Non-Line-Of-Sight Process:** Coating that is applied equally to all surfaces of a material, even inside convoluted shapes.
- **Physical Vapor Deposition:** Methods of using low pressure to deposit thin films by the condensation or deposition of volatilized materials.
- **Proton Exchange Membrane Fuel Cell:** Polymer-based electrochemical device that converts various fuels and air into water, heat, and electric energy at low operating temperatures (70 to 140 degrees Celsius).
- **Refraction:** Turning or bending of any wave, such as light or sound, when it passes from one material into a distinctly different material.
- **Slurry:** Thick suspension of solids in a liquid that may separate and can be filtered.
- **Sodium Hypophosphite:** Catalytic agent in nickel plating that reduces dissolved metal ions and deposits them as metal atoms on a substrate surface; chemical composition is $NaPO_2H_2$-H_2O.
- **Solid Oxide Fuel Cell:** Ceramic electrochemical device that converts various fuels and air into water, heat, and electric energy at high operating temperatures (600 to 1,000 degrees Celsius).

- **Spallation:** Delamination or de-bonding of one section of a material from another.
- **Stent:** Metal cylinder that can be moved inside a clogged blood vessel, where it is permanently inflated and remains in place, allowing blood to circulate once more.
- **Thermal Barrier Coating:** Porous coating applied to turbine blades, reducing the temperature between the combustion chamber and the metal of the blade.
- **Thermal Spraying:** A variety of methods involving elevated temperatures to deposit materials on a surface by first melting or vaporizing them and then accelerating them toward the substrate.

Definition and Basic Principles

Coating technologies have been applied to the surface of materials for centuries. Corrosion of steels can cause structural degradation in buildings and vehicles. When exposed to the atmosphere, metals can corrode, forming ceramic materials such as oxides, which are much more brittle, and can lead to the mechanical failure of infrastructure. Coatings can significantly alter the performance of the substrates to which they are applied. Coating the surfaces of metals can prevent or at least delay the onset of detrimental corrosion, but modern coating technologies can do much more than that. Some examples of coatings on polymers and glasses include antireflective and ultraviolet protection on sunglasses as well as thermochromic windows, which reduce heat influx into buildings in summer and prevent heat emissions in winter. Electrochemically active thin coatings can act as photovoltaic cells, super capacitors, fuel cells, and as electrodes of batteries. Thin films of the right chemical composition and microstructure typically increase the efficiency of the device in which they are used. Patterned corrosion-protection coatings applied to semiconductors in photolithography prevent chemical etching and determine the size, shape, and functionality of the semiconductor device. Magnetic coatings are used as data-storage devices. Ceramic-glaze coatings create colors on pottery pieces. on polymers and glasses include antireflective and ultraviolet protection on sunglasses as well as thermochromic windows, which reduce heat influx into buildings in summer and prevent heat emissions in winter. Electrochemically active thin coatings can act as photovoltaic cells, super capacitors, fuel cells, and as electrodes of batteries. Thin films of the right chemical composition and microstructure typically increase the efficiency of the device in which they are used. Patterned corrosion-protection coatings applied to semiconductors in photolithography prevent chemical etching and determine the size, shape, and functionality of the semiconductor device. Magnetic coatings are used as data-storage devices. Ceramic-glaze coatings create colors on pottery pieces. Biomimetic coatings can prevent surgical implants from being attacked by the human immune system. Coating technologies can be physical, involving some type of melting or volatilization of a source material followed by deposition and solidification on a surface, or the solution or suspension of a coating material in a solvent, followed by the evaporation of the solvent after the deposition. Alternatively, the coating method can involve a chemical reaction that binds the coating to the substrate or chemically alters the substrate surface.

Background and History

Early humans developed pottery methods that still exists. By heating specific ceramic particles deposited from slurry on the surface of prefired clay, the pottery develops a dense glaze. This technology is one of the earliest surface-coating technologies. Paint, which was initially developed for artwork, has also been used during the last centuries as a protective coating.

Infrastructural corrosion of metallic structures in the United States alone amounts to an estimated $276 billion, or 4 percent of the United States gross domestic product, in damages each year. Coatings have been applied to infrastructure and machinery for hundreds of years in order to prolong the useful lifetime of the equipment. Early coatings included the application of paint, which often contained metal-oxide materials. Paints usually involve particles suspended in a colloidal solution. Once the paint is applied to a surface, the solvent evaporates over time, leaving behind a coating. These coatings often did not stick well to their substrates and were often permeable to the environment, leading to their flaking off. In the early 1820's, the English chemist Sir Humphry Davy discovered that combining steels with easily corroding materials such as zinc results in significantly reduced corrosion rates—at the cost of corroding the sacrificial zinc.

Modern coating technologies have been developed as a result of the need to have devices with higher efficiencies and longer lifetimes. Although vehicles produced until the 1990's still exhibit an affinity to form rust, modern vehicles are produced with improved corrosion-protection coatings and rarely rust.

How It Works

Coatings add an additional thin film between the surface of a material and the environment with which the material comes in contact. They can be applied as a design feature or with specific material properties in mind.

To apply a coating, the correct method of deposition has to be selected. Vehicle corrosion is addressed by applying several functional coatings, usually in a rapid, high-quality process called dip coating, in which the entire car components are immersed in an cataphoretic bath, which ensures complete, dense, and continuous coatings of all surfaces inside the structure. Further coatings may include color pigmentation and dyes, as well as a clear surface finish. Coatings can also be applied to much smaller substrates than vehicle parts. Small pigments such as mica and silica, with a size of a few micrometers, can change color if various oxide layers are applied by slurry casting to their surfaces, since the refraction of light is changed because of these surface oxide layers. For example, the chemical company Merck manufactures such coatings under the name Ronasphere for cosmetics and vehicles.

Whereas corrosion-protection coatings are supposed to be dense and not swell from water intake, other coatings must intentionally be porous. For example, the gas turbine engines, such as those in airplanes, can be made more efficient by increasing the temperature inside the combustion chamber. However, the turbine blades are rotating at a high velocity and high temperature, and consequently the metal surface of the blades may oxidize, and some deformation may occur because of creep. To prevent these detrimental effects, a 0.4-millimeter thin, porous layer of yttrium-stabilized zirconia (a type of beach sand) is applied to the surface of the turbine blades. This is usually performed in an open environment using a hot deposition gun that is capable of melting even ceramic zirconia, which has a melting point above 2,750 degrees Celsius. Heat is delivered by plasma spraying and high velocity oxy fuel, in which the ceramic material passes at high velocities through a flame to impact in liquid form and solidify on the surface of the turbine blades. The resultant coating is very porous, which is quite advantageous. Different materials expand at varying rates when heated, and dense ceramic coatings may develop such high stresses on the surface of metals, that they spall when exposed to varying temperatures. In porous coatings, on the other hand, these stresses are dispersed due to the porosity, and the coatings remain on the surface of the turbine blades. All these thermal-spraying methods are line-of-sight methods and cannot consequently be used if intricate three-dimensional structures that cannot be evenly positioned into the incoming material stream are to be coated evenly.

Similar hot spraying methods can also be used to create multiple graded functional layers in electrochemical devices such as batteries, fuel cells, solar cells, or capacitors. For example, the National Research Council Canada has developed a reactive spray deposition method, in which metals are dissolved in an organic solvent, vaporized at elevated temperatures, and sprayed through a short flame on various substrate surfaces as a deposition straight from the gas phase. This method allows very good control of the microstructure of the functional layers.

For high-quality thin film coatings, physical vapor deposition methods have been developed that volatilize materials under extreme high vacuum conditions from high-purity emission sources referred to as targets. The growth rates of coating layers resulting from such a deposition are slow, and it may take several hours to form a layer with a thickness of even one micrometer. However, the layers are very clean, pure, dense, and thin. Even something as simple as a potato chip bag can have a thin metallic coating on the inside to improve the quality of the sealed products.

Chemical vapor deposition is another method to deposit well-adhered thin layers. Here, a chemical reaction occurs on the usually heated surfaces of the substrates, and the desired coating forms. Some of these deposition methods involving chemical reaction may include harsh chemicals such as hydrofluoric acid.

Biomedical coatings can have several functions. First and foremost they must prevent the human body from rejecting an implant. Furthermore, implants have coatings that either reduce the growth of scar tissue in soft tissues or aid in tissue growth inside bones, facilitating a solid attachment between bones and the implant and consequently a longer lifetime of the implant. In heart surgery, for example, very thin (less than 0.01 millimeter), drug-laced hydroxyapatite coatings on stents can be used to create devices that will not become covered in human tissue, thus reducing the risk of difficult follow-up operations and replacements.

Applications and Products

Corrosion Protection. One of the largest application areas of coatings is corrosion protection. As a result of corrosion, fatal accidents can occur. For example, vessels that are under pressure or carry corrosive and abrasive liquids may rupture and corroding bridges may collapse. In 2010, a small oil pipeline in Wyoming ruptured because of corrosion, resultingin a spill of 85,000 gallons of crude oil. To prevent the corrosion of infrastructure, protective coatings are applied to the metal surfaces. In the production of steel, alloying elements such as nickel or chromium can be added to the liquid metals. Once exposed to oxygen, they form stable, protective oxide-layer coatings, which are very thin and invisible to the naked eye. This kind of metal is consequently called stainless steel. Although stainless steel is common, the alloying elements are more expensive than the iron-containing raw materials, and they are less often used in structural applications or in rebars within reinforced concrete. The most commonly produced stainless steel is called type 304; it contains 18 percent chromium and 8 percent nickel.

Once a type of steel has been produced, the metal surface can be protected by coatings that are applied externally. A method that was developed by French engineer Stanislas Sorel in 1837 is called galvanization. It involves dipping a metal part into a hot bath of zinc solution, leading to the formation of a visible zinc coating on the surface. Although steels are protected by this method, they degrade fast in corrosive environments—salted roads in winter, the ocean. An improved method to protect steels from corrosion results by cataphoretically coating the substrates. The entire substrate is hereby immersed in a conductive aqueous solution, and a dense, protective coating precipitates on the substrate surface due to an applied electric field. The German engineering firm Dürr works with many different industries worldwide using this coating method.

Fascinating Facts About Coating Technology

- Thermal barrier coatings on gas turbine blades are only about 100 micrometers thick, but they can reduce the temperature from the combustion chamber by 200 degrees Celsius so that the metal underneath does not degrade.
- Copper and zinc have been shown to be toxic to some aquatic animals and plants that would typically befoul parts of vessels in contact with the water. However, when in contact with steel, zinc may act as a sacrificial anode and protect the steel from corroding, thereby corroding much faster itself. With less zinc present, the vessels become more sustainable to befouling.
- Metallic coatings inside some chips bags aid in keeping the inert atmosphere from the packaging inside the bag and prevent the chips from touching the polymer of the bag directly.
- Some deposition methods involve high-temperature plasmas that can exceed 16,000 degrees Celsius. These methods can melt any material, even ceramics, very rapidly, to form a coating.
- A coating of hydroxyapatite fewer than 10 micrometers on the surface of heart stents can significantly reduce inflammation and reduce the risk of biofilm formation inside the blood vessel.
- Thinner electrolytes in electrochemical devices mean shorter pathways that the charge carriers have to travel. Consequently, significantly improved batteries, fuel cells, solar cells and super capacitors result from the use of thin coatings.
- Most clothing has been coated multiple times–in color dyes, dye leach protection, insect repellants, ultraviolet radiation protection for the fibers and the wearer, stain repellant nonstick coatings, and antistatic coatings. It is, therefore, always advantageous to wash clothes before wearing them for the first time.

Surfaces can also be coated without involving electricity. Electroless nickel plating, for example, involves pretreating the surface of any material, including nonconductive materials, with a catalyst such as sodium hypophosphite. This treated surface is then immersed in a heated nickel-phosphorous or nickel-boron solution. The metal ions from the solution are reduced to metal in contact with the catalyst and form a dense alloy layer on the treated surface.

Biomedical Coatings. Blood vessel stems are often-used implants. Early implants were plagued by inflammation around the implant and a resulting growth of tissue around the implant as a defense mechanism of the body to isolate the implant from the body. Companies such as MIV Therapeutics use the application of thin ceramic hydroxyapatite coatings to facilitate a better uptake of the implant in the body. These coatings are porous and can also be used to release anti-inflammatory drugs into the wound, locally and long term, without the need to medicate the entire patient with high doses of potentially dangerous drugs. In places where cellular growth is desired, such as in bone scaffolding and artificial joints, the outsides of the material in contact with the bone are coated with porous materials, such as stainless steel or titanium foams or beads that match the three-dimensional structure of the bone. It has been shown that these surfaces are overgrown and integrated into the bone much more easily than smooth metallic surfaces and constitute a significant improvement in implant lifetime. Because this is a large global market, there are a significant amount of international companies involved. Johnson & Johnson operates a subsidiary in North America for these implants called DePuy Orthopaedics. Other North American manufacturers include Stryker Orthopedics, Wright Medical Technologies, and Biomet.

Nonstick Coatings. Other coatings that are applied in artificial blood vessels and also in large engineering pipes carrying liquids are nonstick surfaces such as polytetrafluoroethylene, which is marketed under the trademark Teflon. Blood clotting inside the artificial blood vessel is prevented by the use of such a functional polymer coating. In engineering pipelines, liquid flow is slowed because of the friction of the liquid on the walls of the vessel. Nonstick coatings can significantly cut down the friction in the pipelines, reducing the power required to transport liquids, while simultaneously providing a chemical barrier coating between the liquid and the inside of the pipe. In addition to polymer nonstick coatings such as Teflon, the insides of pipes can be coated with thin layers of ceramic glass. Manufacturers such as the Swedish company Trelleborg provide coating and sheathing solutions for transoceanic pipelines that can carry a large variety of liquids, including oil products. Of course, one of the more well-known applications for nonstick coatings is cookware. Nonstick coatings reduce the likelihood of heated materials sticking to the inside of a metallic pan, since it does not chemically react with other materials. For this type of coating to stick to the inside of the pan, the metal is prepared with groves or porosity generated by sandblasting and coated with a porous primer.

Optical Coatings. Most ceramic glasses permit infrared radiation but will block ultraviolet radiation from the sun. As a consequence, in summer, the insides of buildings are heated, and energy-intensive air-conditioning is required to cool the building. Some ceramics such as vanadium oxide that has been mixed with small amounts of tungsten oxide can block infrared radiation, effectively preventing solar heat radiation from reaching the insides of buildings. Applied to the surface of glass, they automatically insulate the building from heat, silently and efficiently. However, this particular material is even smarter. Infrared radiation is blocked only above 29 degrees Celsius. In winter, the infrared radiation can pass into the building and help in heating. Polymers, on the other hand, seldom block ultraviolet radiation. Most eyeglasses and sunglasses are manufactured using polymers, which are coated with materials that block ultraviolet radiation to protect the eyes. Similarly, polymers and polymer fabrics can be protected from degradation due to ultraviolet radiation by applying thin coatings of zinc and titanium oxides. Other transparent coating materials, such as indium tin oxide, can be applied to glass surfaces to make them conductive, as is done in solar cells, where the transparent layer facing the sun acts as one of the electrodes. Schott and Carl Zeiss are Europe-based companies that manufacture high-quality specialty glasses with any number of functional coatings for the global market.

Magnetic Coatings. Videotapes, cassettes, floppy discs, and zip drives are all now technologically obsolete, but they have one thing in common: They used magnetic materials to store data. In the case of magnetic tapes, a polymer film was coated with magnetic material, for example, iron oxide. The chemical company BASF used to be one of the largest providers of magnetic tapes. Modern computer hard disks, which are still produced in large numbers, typically use cobalt-based alloys to store the data, as they allow for a faster and safer magnetization.

Impact on Industry

Government Research. Because coatings are involved in most technology applications, major research is produced in academia, industry, and government facilities. In Canada, for example, the National Research Council includes many major research facilities involved with coating technologies. The Institute for Fuel Cell Innovation in Vancouver has developed reactive spray-gas phase sublimation methods to deposit functional layers of fuel cells. In the United States, research facilities such as the Pacific Northwest Laboratories in Richland, Washington, are heavily involved in coatings research. Scientists there developed protective aluminide coatings for steel. Instead of forming a separate layer on the surface of the metal, the atoms of the coating diffuse into the underlying steel, thereby making it stronger and creating a protective coating that cannot chip or scratch. In Europe, one example of coatings research is the development of dye-sensitized solar cells at the Max Planck Institute for Physics in Munich, where researchers apply several thin layers onto a glass or polymer substrate to form a solar cell. Dense fluorine-doped tin oxide and titanium oxide layers are coated with titanium dioxide spheres that have, in turn, been coated with organic dyes to increase the efficiency of the solar cell. These are just a few examples of the broad research in coatings worldwide.

University Research. The Shanghai Museum contains a collection of very rare ancient Chinese pottery from the Neolithic times to the end of the Qing Dynasty. It shows that advanced ceramic coatings mimicking gold, wood, or gemstones in color and refraction have been known to humanity for several millennia.

Also situated in Shanghai are the Shanghai University School of Materials Science and Engineering and the Shanghai Institute of Ceramics at the Chinese Academy of Sciences. Both of these academic research facilities develop methods for ceramic-coating technologies. For example, researchers have developed nanometer-size tin additives for gas products that can significantly reduce the friction inside motors and pistons, thereby improving the efficiency and longevity of the movable parts of a motor. Other research groups in Shanghai have discovered novel applications for simple rust. They have created nanometer-size hollow spheres of iron oxide that can assist in wastewater treatment. Another example for the global scope of coating applications are silica nanosprings developed at the Washington State University. Because of their unique small size, these springs can be used to store hydrogen atoms for later use by clean-energy devices such as fuel cells.

Industry and Business. The largest global revenues in coatings come from industries involved in paints, often including corrosion protection. Although only about 2 percent ($2 billion) of the total financial turnover of the French petrochemical giant Total Fina Elf is produced by its subsidiary paint manufacturer SigmaKalon, the figures still give an idea of the large scale of the international paint coating market. Another large global company in the coating business is the Netherlands-based AkzoNobel, with coatings-related sales volumes of almost $7 billion. Large paint manufacturers in the United States include Pittsburgh-based PPG Industries, Cleveland-based Sherwin-Williams Corporation, Minneapolis-based Valspar Corporation, and the Wilmington, Delaware-based DuPont Coatings and Color Technologies Group. The annual sales volume is estimated at between $4 billion and $5.3 billion. The BASF Coatings and Glasurit research facilities in Münster, Germany; ICI Paints in Berkshire, England; and the Japanese Nippon Paint and Kansai Paint in Osaka generate gross sales of about $9 billion annually to the global paint market.

Careers and Course Work

Although it may seem trivial to pick up a paint roller to add color to a wall, the technological developments necessary to produce even a simple, stable paint are massive and have been ongoing for a long time. Detailed knowledge of the material properties, the chemistry of the solvents, and the fundamental physics of the optical, mechanical, thermal, and magnetic properties of coatings are essential to develop novel, innovative products for the global market. Career advancement in the global coatings technology typically requires an undergraduate degree, ideally with many industrial internships for practical experience.

SOCIAL CONTEXT AND FUTURE PROSPECTS

Vehicles manufactured as late as the 1980's were prone to corrosion, reducing not only the optical appearance of a vehicle, but also impairing the mechanical stability and safety. Since then, coating technologies have significantly advanced. All surfaces of a car body can now be coated, no matter what shape, and they are less permeable to air and absorb significantly less water. Coatings adhere better to surfaces and are dense and have higher impact and scratch resistance, and as a result vehicles manufactured in the early twenty-first century corrode far less frequently. This development also applies to aeronautic and marine applications. The coated devices, such as ships, have to be re-coated less frequently, reducing downtime and preventing fatal material failures while the device is in service. The same applies to biomedical devices. If, for example, a hip implant with a modern coating has to be renewed inside a patient every fifteen years instead of every five years, the result is a significant improvement in life quality for the patient.

Coating technologies are far from perfect, which is evidenced by the huge global research effort in the development of all types of coatings. Modern automotive coatings can typically involve more than 200 chemicals, and the exact physical and chemical interaction between all these chemicals can be evaluated only in small steps. Likewise, coating technology methods are constantly evolving. Consequently, there is a huge potential in coatings research and development with applications in almost every piece of technology in use as of 2011.

Lars Rose, M.Sc., Ph.D.

FURTHER READING

Birkmire, Robert. "Thin-Film Solar Cells and Modules." In *Solar Cells and Their Applications*, by Larry Partain and Lewis Fraas. 2d ed. Hoboken, N.J.: John Wiley & Sons, 2010. Birkmire, of the University of Delaware, has provided an excellent, detailed description of thin film solar cells.

Kumar, Challa, ed. *Nanostructured Thin Films and Surfaces.* Weinheim, Germany: Wiley-VCH, 2010. This assembly of research topics provides an overview of thin film coatings and nanoscale materials with a focus on their uses in the life sciences.

Lakhtakia, Akhlesh, and Russell Messier. *Sculptured Thin Films: Nanoengineered Morphology and Optics.* Bellingham, Wash.: SPIE, 2005. This very scientific book contains some good examples of how the morphology (the shape, size, and orientation) of thin surface films can significantly alter the properties of the material.

Lüth, Hans. *Solid Surfaces, Interfaces, and Thin Films.* 4th ed. Berlin: Springer-Verlag, 2001. Provides very detailed drawings and examples of different coating methods and the devices used to produce them.

Smeets, Stefan, Egbert Boerrigter, and Stephan Peeters. "UV Coatings for Plastics." *European Coatings Journal* 6 (2004): 42-48. Cytec (formerly Surface Specialties UCB) researchers' account of how to coat plastic surfaces with the help of ultraviolet radiation.

Weldon, Dwight G. *Failure Analysis of Paints and Coatings.* Rev. ed. John Wiley & Sons, 2009. There is a plethora of literature available on paints and paint technologies, but this book takes a unique look at how and why these coatings fail in service and how research can address these issues.

WEB SITES

American Coatings Association
http://www.paint.org

Chemical Coaters Association International
http://www.ccaiweb.com

Society for Protective Coatings
http://www.sspc.org

See also: Nanotechnology

COMPUTER-AIDED DESIGN AND MANUFACTURING

FIELDS OF STUDY

Three-dimensional modeling; aerospace engineering; architecture; civil engineering; computer animation; computer engineering; electrical engineering; graphic design; industrial design; landscaping; mechanical engineering; plant design; product design; ray tracing; textiles; texture mapping; virtual engineering.

SUMMARY

Computer-aided design and manufacturing is a method by which graphic artists, architects, and engineers design models for new products or structures within a computer's "virtual space." It is alternatively referred to by a host of other terms related to specific fields: computer-aided drafting (CAD), computer-aided manufacturing (CAM), computer-aided design and drafting (CADD), computer-aided drafting/computer-aided manufacturing (CAD/CAM), and computer-aided drafting/numerical control (CAD/NC). The generic term computer-aided technologies is, perhaps, the most useful in describing the computer technology as a whole. Used by professionals in many different disciplines, computer-aided technology allows designers in traditional fields, such as architecture, to render and edit two-dimensional (2-D) and three-dimensional (3-D) structures. Computer-aided technology has also revolutionized once highly specialized fields, such as video game design and digital animation, into broad-spectrum fields that have applications ranging from the film industry to ergonomics.

KEY TERMS AND CONCEPTS

- **Assembly Modeling:** Design that describes how parts and subassemblies can be put together into a single unit.
- **Critical Path Method:** Spreadsheet or activity diagram that schedules the time frame for each stage of production.
- **Gannt Chart:** Graphical representation (perhaps in bar chart or line format) that expresses a project schedule.
- **Kinematics:** Study of the movement of objects.
- **Numeric Control (NC):** Control of a machine's activity by the use of commands recorded in alphanumeric format.
- **Parametric Design:** Ability of a computer system to translate graphics (computer-drawn images) into mathematical calculations and calculations back into graphics.
- **Part Libraries:** Catalog or database of standard drawings of small parts or symbols used frequently in CAD designs.
- **Product Lifecycle Management (PLM):** Process describing the entire life cycle of a product, from concept to disposal.
- **Rapid Prototyping:** Function of computer-aided manufacturing that constructs physical objects by layer fabrication (for example, fabricating parts by melting layers of metal powder).
- **Vector Graphics:** Use of mathematically derived points, lines, and curves (or polygons based on the combination of lines and curves) to create computer graphic images.

Definition and Basic Principles

Computer-aided design and manufacturing is the combination of two different computer-aided technologies: computer-aided design (CAD) and computer-aided manufacturing (CAM). Although the two are related by their mutual reliance on task-specific computer software and hardware, they differ in how involved an engineer or architect must be in the creative process. A CAD program is a three-dimensional modeling software package that enables a user to create and modify architectural or engineering diagrams on a computer. CAD software allows a designer to edit a project proposal or produce a model of the product in the virtual "space" of the computer screen. CAD simply adds a technological layer to what is still, essentially, a user-directed activity. CAM, on the other hand, is a related computer-aided technology that connects a CAD system to laboratory or machine shop tools. There are many industrial applications for computer-directed manufacturing. One example of CAD/CAM might be when an automotive designer creates a three-dimensional model of a proposed design using a CAD program, verifies the physical details of the design in a computer-aided engineering (CAE) program, and then constructs a physical model via CAD or CAE interface with CAM hardware (using industrial lasers to burn a particular design to exact specifications out of a block of paper). In this case, the automotive designer is highly involved with the CAD program but allows the CAD or CAE software to direct the activity of the CAM hardware.

BACKGROUND AND HISTORY

CAD/CAM is, essentially, a modernization of the age-old process of developing an idea for a structure or a product by drawing sketches and sculpting models. Intrigued by the idea that machines could assist designers in the production of mathematically exact diagrams, Ivan Sutherland, a doctoral student at the Massachusetts Institute of Technology, proposed the development of a graphical user interface (GUI) in his 1963 dissertation "Sketchpad: A Man-Machine Graphical Communication System." GUIs allow users to direct computer actions via manipulation of images, rather than by text-based commands and sometimes requires the use of an input device (like a mouse or a light pen) to interact with the GUI images, also known as icons. Although it was never commercially developed, Sketchpad was innovative in that a user could create digital lines of a screen and determine the constraints and parameters of the lines via buttons located along the side of the screen. In succeeding decades, the CAD software package employed the use of a mouse, light pen, or touch screen to input line and texture data through a GUI, which was then manipulated by computer commands into a virtual, or "soft," model.

In the twenty-first century, CAM is more or less the mechanical output aspect of CAD work, but in the early days, CAM was developed independently to speed up industrial production on assembly lines. Patrick J. Hanratty, regarded as the father of CAD/CAM, developed the first industrial CAM using numerical control (NC) in 1957. Numerical control—the automation of machining equipment by the use of numerical data stored on tape or punched on cards—had been in existence since the 1940's, as it was cheaper to program controls into an assembly machine than to employ an operator. Hanratty's PRONTO, a computer language intended to replace the punched-tape system (similar to punch cards in that the automaton motors reacted to the position of holes punched into a strip of tape) with analogue and digital commands, was the first commercial attempt to modernize numerical control and improve tolerances (limiting the range of variation between the specified dimensions of multiple machined objects).

HOW IT WORKS

Traditionally, product and structural designers created their designs on a drawing board with pencils, compasses, and T squares. The entire design process was a labor-intensive art that demanded not just the ability to visualize a proposed product or structure in exact detail, but also demanded designers to possess extensive artistic abilities to render what, up until that point, had been only imagined. Changes to a design demanded either by a client's specific needs or by the limitations of the materials needed for construction frequently required the designer to "go back to the drawing board" and create an entirely new series of sketches. CAD/CAM was intended to speed up the design and model-making processes dramatically.

Computer-Aided Design (CAD). CAD programs function in the same way that many other types of computer software function: One loads the software into a computer containing an appropriate amount of memory, appropriate video card, and appropriate operating system to handle the mathematical calculations required by the three-dimensional modeling process. There are quite literally hundreds of different CAD programs for different applications available for trial use or purchase. Some programs are primarily two-dimensional drawing applications, while others allow three-dimensional drawing, shading, and rendering. Some are intended for use in construction management, incorporating project schedules, or for engineering, focused on structural design. There are a range of CAD programs of varying complexity for computer systems of different capabilities or differing operating systems.

Computer-Aided Manufacturing (CAM). CAM, on the other hand, tends to be a series of automated mechanical devices that interface with a CAD program installed in a computer system—typically, but not exclusively, a mainframe computer. The typical progress of a CAD/CAM system from design to soft model might involve an automotive designer using a light pen on a CAD system to sketch out the basic outlines of suggested improvements for an existing vehicle. Once the car's structure is defined, the designer can

Fascinating Facts About Computer-Aided Design and Manufacturing

- The earliest commercial uses of CAD, in the late 1960's, were in the automobile and aerospace industries. Ford, McDonnell-Douglas and Lockheed Martin all created proprietary CAD software programs.
- Will Wright, who designed *The Sims,* initially intended his engrossing people simulator to be a home creation and design tool.
- *Second Life,* another virtual life simulator, created by Linden Lab, had an active online player base of around 500,000 in 2007.
- *World of Warcraft,* developed by Blizzard Entertainment, had an active online player base of 10 million in 2008.
- Computer-aided tissue engineering is an example of the growing field of biotechnology. This specialty uses computers to assist in the development of artificial scaffolds for the growth of human tissues, such as ear and nose cartilage replacement, and the development of tissue models for artificial replacement of lost tissue.
- Fashion design is another industry where CAD is employed. The patternmakers who bring to fruition the designers' creations use CAD to determine the most efficient cuts of fabrics and to alter the pattern sizes easily.

then command the CAD program to overlay a surface, or "skin," on top of the skeletal frame. Surface rendering is a handy tool within many CAD programs to play with the use of color or texture in a design. When the designer's conceptual drawings are complete, they are typically loaded into a computer-aided engineering program to ascertain the structural integrity and adherence to the requirements of the materials from which the product will ultimately be constructed. If the product design is successful, it is then transferred to a computer-aided manufacturing program, where a product model, machined out of soft model board, can be constructed to demonstrate the product's features to the client more explicitly.

Advertisements for CAD programs tend to emphasize how the software will enhance a designer's imagination and creativity. They stress the many different artistic tools that a designer can easily manipulate to create a virtual product that is as close as possible to the idealized conceptual design that previously existed only in the designer's imagination. On the other hand, CAM advertisements stress a connection of the CAM hardware to real-world applications and a variety of alterable structural materials ranging from blocks of paper to machinable steel. The two processes have fused over the succeeding decades partly because of the complementary nature of their intended applications and partly because of clients' need to have a model that engages their own imaginations, but certain aspects of CAD (those technological elements most appropriate to the creation of virtual, rather than actual, worlds) may eventually allow the software package to reduce the need for soft models in terms of product development.

CAD in Video Games and Film Animation. CAD work intended for video games and film animation studios, unlike many other types of designs, may intentionally and necessarily violate one basic rule of engineering—that a design's form should never take precedence over a design's function. Industrial engineers also stress the consideration of ergonomics (the requirements of the human body are considered when a design for a product is developed) in the design process. In the case of the virtual engineer, however, the visual appeal of a video game or animation element may have the greater say in just how a design is packaged for a client. *Snow White and the Seven Dwarfs* was an animated film originally drawn between 1934 and 1937. Since animation was drawn by hand in those days—literally on drawing boards—with live models for the animated characters, the studio's founder, Walt Disney, looked for ways to increase the animated character's visual appeal. One technique he developed was to draw human characters (in this case, Snow White herself) with a larger-than-standard size head, ostensibly to make her appear to have the proportions of a child and, thus, to be more appealing to film viewers. In order that the artists might draw their distorted heroine more easily, the actress playing Snow White, Marjorie Bell, was asked to wear a football helmet while acting out sections of the story's plot. Unfortunately for Disney, Bell complained so much about the helmet that it had to be removed after only a few minutes. Ergonomics took precedence over design. Later animation styles, influenced by designers' use of CAD animation software, were able to be even more exaggerated without the need for a live model in an uncomfortable outfit. For example, the character of Jessica Rabbit in Amblin Entertainment's *Who Framed Roger Rabbit?* (1988) has a face and body that were designed exclusively for visual appeal, rather than to conform to external world reality. Form dominated function.

Applications and Products

CAD and CAD/CAM have an extensive range and variety of applications.

Medicine. In the medical field, for example, Abaqus 6.10-EF finite element analysis (FEA) allows designers to envision new developments in established medical devices (improving artificial heart valves, for example) or the creation of new medical devices. CAD software, ideally, allows a user to simulate real-world performance.

Aerospace. Femap and NEi Nastran are software applications used for aerospace engineering. It can simulate the tensile strength and load capacity of a variety of structural materials (aluminum, etc.) as well as the effect of weather on performance and durability.

Mechanical Engineering. Mechanical engineering firms sometimes use Siemens PLM Software NX suite for CAM or CAE functions because they require their software to create and revise structural models quickly.

Some forms of CAD have engineering applications that can anticipate a design's structural flaws to an even greater extent than an actual model might. One example of this application is SIMULIA, one of the simulation divisions of Dassault Systèmes that creates virtual versions of the traditional automobile safety tester—the crash-test dummy. The company's program of choice is Abaqus FEA, but it has altered the base software to use BioRID (Biofidelic Rear Impact Dummy) and WorldSID (World Side Impact Dummy) as well as other altered models for a wider variety of weight and height combinations. In the real world, crash-test dummies are rather complex mechanical devices, but their virtual simulacra have been declared as effective in anticipating structural failure as the original models. Because the nature of CAD is to make duplicating a successful element of a design quickly and easily, SIMULIA has been able to reduce the time spent on validating its simulations from four weeks to four days.

Animation. CAD programs have caused the rapid expansion and development of the animation industry—both in computer games and in film animation. Models of living creatures, both animal and human, are generated through the use of three-dimensional CAD software such as Massive (which was used by the graphic artists creating the battle sequences in Peter Jackson's *The Lord of the Rings* trilogy) and Autodesk's Maya (considered an industry standard for animated character designs). Autodesk's 3ds Max is one of the forerunner programs in the video game industry, since it allows a great deal of flexibility in rendering surfaces. Certain video games, such as Will Wright's *Sims* series, as well as any number of massively multiplayer online role-playing games (MMORPGs) such as *World of Warcraft*, *Second Life*, and *EverQuest*, not only have CAD functions buried in the design of the game itself but accept custom content designed by players as a fundamental part of individualization of a character or role. A player must design a living character to represent him or her in the video game including clothing, objects carried, and weapons used, and the process of creation is easily as enjoyable as the actual game play.

Avatar, released by Twentieth Century Fox in 2009, was envisioned by director James Cameron as being a sweeping epic in very much the same vein as George Lucas's *Star Wars* trilogy (1977, 1980, and 1983). *Avatar* was originally scheduled to start filming in 1997, after Cameron completed *Titanic*, but Cameron felt the computer-aided technology was not ready to create the detailed feature film he envisioned. Given that the film was also supposed to take advantage of newly developed three-dimensional stereoscopic film techniques, Cameron started work on the computer-aided technology aspects of his film only when he was sure the technology was up to the task—2006.

Even with an extensive three-dimensional software animation package at his command, the task of creating and populating an entire alien world was a daunting prospect. Most of the action of *Avatar* takes place on a distant planet, Pandora, which is intrinsically toxic to human beings. To get around this difficulty and successfully mine Pandora for a much-needed mineral (called, amusingly, Unobtainium), humans must use genetically created compatible bodies and interact with the native population, the Na'vi. Both the native creatures, who are ten feet tall with blue skin, and the hybrid human-Na'vi "avatar" bodies were a challenge for animators to render realistically even using computer-aided technology. The full IMAX edition was released on December 18, 2009, having ultimately cost its creators somewhere between $280 million and $310 million. Fortunately for Cameron, the film ended up rewarding its creators with gross revenues of more than $2 billion. Having created the highest-grossing film of all time, which also won Academy Awards for art direction, cinematography, and visual effects, Cameron was able to negotiate successfully with Fox for two planned sequels (echoing the original *Star Wars* trilogy).

Impact on Industry

There have been two schools of thought regarding the impact of CAD/CAM on industry. On one hand, the cost of producing new ideas for products or structures has dropped dramatically because of the quick processing facility of the computer processor and video card. Designers are no longer required to anticipate every angle of a new building or aircraft—it is possible to input just enough data for the computer to interpolate the views from other perspectives. CAD/CAM also saves money in the fabrication of models because it is able to control the cutting of angles and lines to .0001 inch—something that could not reliably be produced by a human hand. The ease of correcting flaws in a CAD-generated design allows for multiple drafts and more intensive collaboration between designer and client. As CAD programs dropped in price in the early twenty-first century, even small design companies could afford to upgrade to CAD, increasing the competition (and lowering the cost of a service) in a given market.

The other school of thought, mostly espoused by designers in the traditional fields of architecture and engineering, is the difficulty of combining the sometimes foreign concepts of three-dimensional modeling with traditional, human methods of expressing traditional design concepts. For example, some architects dislike CAD because the computer system tends to be based on three-dimensional solids modeling rather than traditional architectural terminology. Architecture evolved to use special terminology that reflects a particular understanding of buildings. When an architect designs a house using hand-drawn techniques and conceives of the "eaves" and "hip" of the roof, he or she must, because of the terminology used, see the roof as a series of peaks and valleys that visually describes spaces as well as surfaces. On the other hand, when a designer uses an older-generation CAD program for architectural purposes, he or she may be completely free from the strictures of the construction materials and the weight of what has already been built. As a consequence, the designer may conceive of the house as a cube and a pyramid that intersect at a particular height—visualizations of solid models that, while they may function adequately in the virtual world of the software, alienate the real-world sensibilities of the traditional drawing-board architect. The complaints occasionally raised by engineers and architects have caused some to question CAD's feasibility, at least at the initial creation stage, for the development of designs that rely on as deep an understanding of engineering as art.

To a greater or lesser degree, depending on the application, computer-aided technology has lowered the costs and increased the efficiency of designers at the drawing board—corrections can be made easily and quickly to product designs, and this can save an engineering or product-design firm time and money in the design and production stages. For some fields, however, the changes caused by computer-aided technology are immense and groundbreaking—as in the field of film visual effects. *Star Wars*, writer-director George Lucas's groundbreaking six-film science fiction classic, is a rare example of how computer-aided technology operates as a medium for the benefits of uninhibited imagination. The original film, released in 1977, was intentionally left incomplete because Lucas believed that computer-aided technology, then in its infancy, would eventually catch up so he could effectively realize the vision he had for his magnum opus. Several scenes were intentionally filmed without any attempt at visual effects, including one of actor Harrison Ford, in the role of Han Solo, trying to win over the confidence of the slug-like crime boss Jabba the Hutt (in the original shot, just a heavyset man in a tunic). Lucas retained these scenes, kept out of the 1977 version, and waited for the visual effects division of his Lucasfilm, Industrial Light & Magic, to integrate the new computer-animation technology he felt would change the film industry.

By 1997, Lucas felt that computer-aided technology had advanced enough to make the designed alterations to his original work. The unaltered scenes were extensively altered using CAD. The scene of Han

Solo and Jabba the Hutt was changed so dramatically that the original actor standing in the place of the future Jabba the Hutt cannot be seen under the CAD rendering. Apparently, Lucas has not yet declared his film epic to be complete. In 2010, he announced that all six *Star Wars* films would be remade with the computer-animation software used for *Avatar*.

Careers and Course Work

Careers in computer-aided design and manufacturing tend to be affected most obviously by the intended purpose of the designer's work. There could be significant differences in the course work required of a medical-device designer and that required of a film-animation specialist. Aside from course work in rapid visualization, CAD operation, graphic illustration, digital type and image manipulation, digital photography, new media, and three-dimensional modeling, those who wish to work with computer-aided technology should consider supplemental course work in the field most relevant to their area of design.

Social Context and Future Prospects

Computer-aided technologies, in general, are expected to keep developing and evolving. The possibilities of virtual reality are continuing to evolve as the need for structural and materials modeling increases. One common description of the future of computer-aided design and manufacturing is the term "exponential productivity." In other words, so much productive time in the fields of design has been wasted by the constant need to create by hand what can be better (and more quickly) produced by a computer. An architect or engineer might spend hours carefully working out the exact proportions of a given structure, more hours using a straight edge and compass to determine the exact measurements needed by the design, and even more time to clean up mistakes. On the other hand, computers can help that same architect draw straight lines, evaluate numerical formulas for proportions, and edit mistakes in a matter of minutes. When one considers the added benefits of a compatible computer-aided manufacturing program, one can further see the time saved in the quick and efficient production of a soft model.

Julia M. Meyers, M.A., Ph.D.

Further Reading

Duggal, Vijay, Sella Rush, and Al Zoli, eds. *CADD Primer: A General Guide to Computer-Aided Design and Drafting.* Elmhurst, N.Y.: Mailmax Publishing, 2000. An excellent introduction to CADD that is applicable to most CAD software programs.

Kerlow, Issac. *The Art of 3D Computer Animation and Effects.* 4th ed. Hoboken, N.J.: John Wiley & Sons, 2009. Written by a former Walt Disney production executive, this brings the traditional principles of animation into the computer age. The step-by-step guidelines to creating 3-D animation are written to work with any computer platform.

Lee, Kunwoo. *Principles of CAD/CAM/CAE Systems.* Reading, Mass.: Addison-Wesley, 1999. A good theoretical introduction to all computer-aided technologies for students of any level; also included is a case study that illustrates the product-development process.

Park, John Edgar. *Understanding 3D Animation Using Maya.* New York: Springer, 2005. An overview of computer animation using Autodesk's Maya, with hands-on tutorials and practical projects at the end of each chapter.

Schell, Jesse. *The Art of Game Design: A Book of Lenses.* Burlington, Mass.: Morgan Kaufmann, 2008. The author is the former chair of the International Game Developers Association, and his book is meant to inspire would-be designers and challenge them to look at their work through various "lenses": mathematics, music, design, film, among others.

Schodeck, Daniel, et al. *Digital Design and Manufacturing: CAD/CAM Applications in Architecture and Design.* Hoboken, N.J.: John Wiley & Sons, 2005. Comprehensive view of industrial design with case studies illustrating the work of architects Frank Gehry, Bernhard Franken, and Rafael Viñoly, among others.

Web Sites

Association for Computer Aided Design in Architecture (ACADIA)
http://www.acadia.org

Association for Computing Machinery's Special Interest Group on Computer Graphics and Interactive Techniques (ACM SIGGRAPH)
http://www.siggraph.org

Association of Professional Model Makers
http://www.modelmakers.org

International Game Developer's Association
http://www.igda.org

See also: Human-Computer Interaction

CRACKING

FIELDS OF STUDY

Chemistry; engineering; chemical engineering; chemical process modeling; fluid dynamics; heat transfer; distillation design; mechanical engineering; environmental engineering; control engineering; process engineering; industrial engineering; electrical engineering; safety engineering; plastics engineering; physics; thermodynamics; mathematics; materials science; metallurgy; business administration.

SUMMARY

In the petroleum industry, cracking refers to the chemical conversion process following the distillation of crude oil, by which fractions and residue with long-chain hydrocarbon molecules are broken down into short-chain hydrocarbons. Cracking is accomplished under pressure by thermal or catalytic means and by injecting extra hydrogen. Cracking is done because short-chain hydrocarbons, such as gasoline, diesel, and jet fuel, are more commercially valuable than long-chain hydrocarbons, such as fuel and bunker oil. Steam cracking of light gases or light naphtha is used in the petrochemical industry to obtain lighter alkenes, which are important petrochemical raw products.

KEY TERMS AND CONCEPTS

- **Catalytic Cracking:** Use of a catalyst to enhance cracking.
- **Coking:** Most severe form of thermal cracking.
- **Crude Oil:** Liquid part of petroleum; has a wide mix of different hydrocarbons.
- **Distillation:** Process of physically separating mixed components with different volatilities by heating them.
- **Fluid Catalytic Cracker (FCC):** Cracking equipment that uses fluid catalysts.
- **Fractionator:** Distillation unit in which product streams are separated and taken away; also known as fractionating tower, fractionating column, and bubble tower.
- **Fractions:** Product streams obtained after each distillation; also known as cuts.
- **Hydrocracking:** Special form of catalytic cracking that uses extra hydrogen to obtain the end products with the highest values.
- **Regenerator:** Catalytic cracker in which accumulated carbon is burned off the catalyst.
- **Residue:** Accumulated elements of crude oil that remain solid after distillation.
- **Steam Cracking:** Petrochemical process used to obtain lighter alkenes.
- **Thermal Cracking:** Oldest form of cracking; uses heat and pressure.

Definition and Basic Principles

Cracking is a key chemical conversion process in the petroleum and petrochemical industries. The process breaks down long-chain hydrocarbon molecules with high molecular weights and recombines them to form short-chain hydrocarbon molecules with lower molecular weights. This breaking apart, or cracking, is done by the application of heat and pressure and can be enhanced by catalysts and the addition of hydrogen. In general, products with short-chain hydrocarbons are more valuable. Cracking is a key process in obtaining the most valuable products from crude oil.

At a refinery, cracking follows the distillation of crude oil into fractions of hydrocarbons with different molecule chain lengths and the collection of the heavy residue. The heaviest fractions with the longest molecule chains and the residue are submitted to cracking. For petrochemical processes, steam cracking is used to convert light naphtha, light gases, or gas oil into short-chain hydrocarbons such as ethylene and propylene, crucial raw materials in the petrochemical industry.

Cracking may be done by various technological means, and the hydrocarbons cracked at a particular plant may differ. In general, the more sophisticated the cracking plant, the more valuable its end products will be but the more costly it will be to build. Being able to control and change a cracker's output to conform to changes in market demand has substantial economic benefits.

Background and History

By the end of the nineteenth century, demand rose for petroleum products with shorter hydrogen molecule chains, in particular, diesel and gasoline to fuel the new internal combustion engines. In Russia, engineer Vladimir Shukhov invented the thermal cracking process for hydrocarbon molecules and patented it on November 27, 1891. In the United States, the thermal cracking process was further developed and patented by William Merriam Burton on January 7, 1913. This doubled gasoline production at American refineries.

To enhance thermal cracking, engineers experimented with catalysts. American Almer McAfee was the first to demonstrate catalytic cracking in 1915. However, the catalyst he used was too expensive to justify industrial use. French mechanical engineer Eugene Jules Houdry is generally credited with inventing economically viable catalytic cracking in a process that started in a Paris laboratory in 1922 and ended in the Sun Oil refinery in Pennsylvania in 1937. Visbreaking, a noncatalytic thermal process that reduces fuel oil viscosity, was invented in 1939. On May 25, 1942, the first industrial-sized fluid catalytic cracker started operation at Standard Oil's Baton Rouge, Louisiana, refinery.

Research into hydrocracking began in the 1920's, and the process became commercially viable in the early 1960's because of cheaper catalysts such as zeolite and increased demand for the high-octane gasoline that hydrocracking could yield. By 2010, engineers and scientists worldwide sought to improve cracking processes by optimizing catalysts, using less energy and feedstock, and reducing pollution.

How It Works

Thermal Cracking. If hydrocarbon molecules are heated above 360 degrees Celsius, they begin to swing so vigorously that the bonds between the carbon atoms of the hydrocarbon molecule start to break apart. The higher the temperature and pressure, the more severe this breaking is. Breaking, or cracking, the molecules, creates short-chain hydrocarbon molecules as well as some new molecule combinations. Thermal cracking—cracking by heat and pressure only—is the oldest form of cracking at a refinery. In modern refineries, it is usually used on the heaviest residues from distillation and to obtain petrochemical raw materials.

Modern thermal crackers can operate at temperatures between 440 and 870 degrees Celsius. Pressure can be set from 10 to about 750 pound-force per square inch (psi). Heat and pressure inside different thermal crackers and the exact design of the crackers vary considerably.

Typically, thermal crackers are fed with residues from the two distillation processes of the crude oil. Steam crackers are primarily fed with light naphtha and other light hydrocarbons. After their preheating, often feedstocks are sent through a rising tube into the cracker's furnace area. Furnace temperature and feedstock retention time is carefully set, depending on the desired outcome of the cracking process. Retention time varies from fractions of a second to some minutes.

After hydrocarbon molecules are cracked, the resulting vapor is either first cooled in a soaker or sent directly into the fractionator. There, the different fractions, or different products, are separated by distillation and extracted.

In cokers, severe thermal cracking creates an unwanted by-product. It is a solid mass of pure carbon, called coke. It is collected in coke drums, one of which supports the cracking process while the other is emptied of coke.

Catalytic Cracking. Because it yields more of the desired short-chain hydrocarbon products with less use of energy than thermal cracking, catalytic cracking has become the most common method of cracking. Feedstocks are relatively heavy vacuum distillate hydrocarbon fractions from crude oil. They are preheated before being injected into the catalytic cracking reactor, usually at the bottom of a riser pipe. There they react with the hot catalyst, commonly synthetic aluminum silicates called zeolites, which are typically kept in fluid form. In the reactor, temperatures are about 480 to 566 degrees Celsius and pressure is between 10 and 30 psi.

As feedstock vaporizes and cracks during its seconds-long journey through the riser pipe, feedstock vapors are separated from the catalyst at cyclones at the top of the reactor and fed into a fractionator. There, different fractions condense and are extracted as separate products. The catalyst becomes inactive as coke builds on its surface. Spent catalyst is collected and fed into the regenerator, where coke deposits are burned off. The regenerated catalyst is recycled into the reactor at temperatures of 650 to 815 degrees Celsius.

Hydrocracking. The most sophisticated and flexible cracking process, hydrocracking delivers the

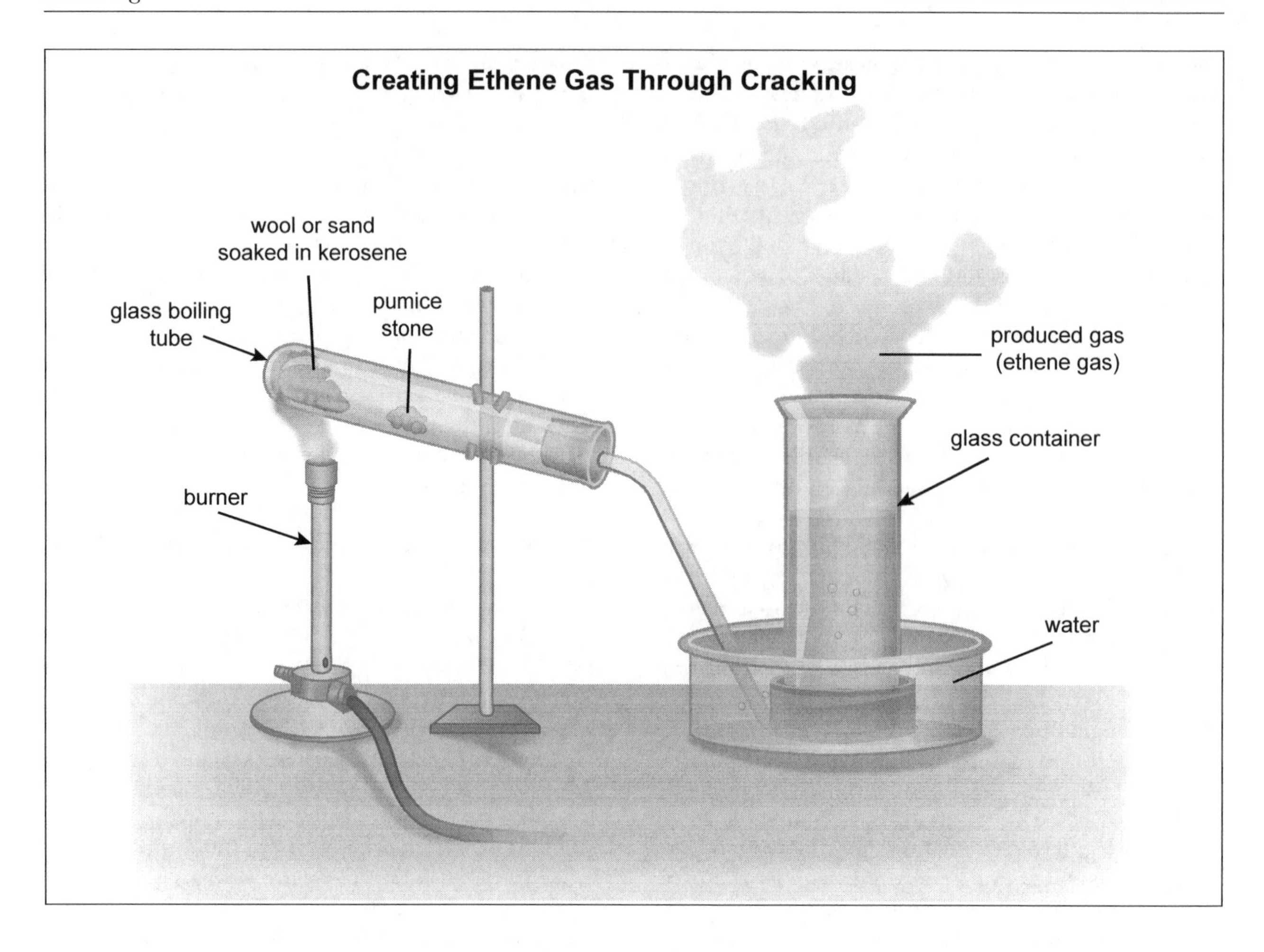

highest value products. It combines catalytic cracking with the insertion of hydrogen. Extra hydrogen is needed to form valuable hydrocarbon molecules with a low boiling point that have more hydrogen atoms per carbon atom than the less valuable, higher boiling point hydrocarbon molecules that are cracked. All hydrocracking involves high temperatures from 400 to 815 degrees Celsius and extremely high pressure from 1,000 to 2,000 psi, but each hydrocracker is basically custom designed.

In general, hydrocracking feedstock consists of middle distillates (gas oils), light and heavy cycle oils from the catalytic cracker, and coker distillates. Feedstock may also be contaminated by sulfur and nitrogen. Feedstocks are preheated and mixed with hydrogen in the first stage reactor. There, excess hydrogen and catalysts convert the contaminants sulfur and nitrogen into hydrogen sulfide and ammonia. Some initial hydrogenating cracking occurs before the hydrocarbon vapors leave the reactor. Vapors are cooled, and liquefied products are separated from gaseous ones and hydrogen in the hydrocarbon separator. Liquefied products are sent to the fractionator, where desired products can be extracted.

The remaining feedstock at the bottom of the fractionator is sent into a second-stage hydrocracking reactor with even higher temperatures and pressures. Desired hydrocracked products are extracted through repetition of the hydrocracking process. Unwanted residues can go through the second stage again.

Applications and Products

There are three modern applications of thermal cracking: visbreaking, steam cracking, and coking.

Visbreaking. Visbreaking is the mildest form of thermal cracking. It is applied to lower the viscosity (increasing the fluidity) of the heavy residue, usually from the first distillation of crude oil. In a visbreaker, the residue is heated no higher than 430 degrees Celsius. Visbreaking yields about 2 percent light gases such as butane, 5 percent gasoline products, 15 percent gas oil that can be catalytically cracked, and 78 percent tar.

Steam Cracking. Steam cracking is used to turn light naphtha and other light gases into valuable petrochemical raw materials such as ethene, propene, or butane. These are raw materials for solvents, detergents, plastics, and synthetic fibers. Because light feed gases are cracked at very high temperatures between 730 and 900 degrees Celsius, steam is added before they enter the furnace to prevent their coking. The mix remains in the furnace for only 0.2 to 0.4 second before being cooled and fractionated to extract the cracked products.

Coking. Coking is the hottest form of cracking distillation residue. Because it leaves no residue, it has all but replaced conventional thermal cracking. Cracking at about 500 degrees Celsius also forms coke. Delayed coking moves completion of the cracking process, during which coke is created, out of the furnace area. To start cracking, feedstock stays in the furnace for only a few seconds before flowing into a coking drum, where cracking can take as long as one day.

In the coke drum, about 30 percent of feedstock is turned into coke deposits, while the valuable rest is sent to the fractionator. In addition to coke, delayed coking yields about 7 percent light gas, 20 percent light and heavy coker naphtha from which gasoline and gasoline products can be created, and 50 percent light and heavy gas oils. Gas oils are further processed through hydrocracking, hydrotreating, or subsequent fluid catalytic cracking, or used as heavy fuel oil. The coke drum has to be emptied of coke every half or full day. To ensure uninterrupted cracking, at least two are used. The coke comes in three kinds, in descending order of value: needle coke, used in electrodes; sponge coke, for part of anodes; and shot coke, primarily used as fuel in power plants.

Flexicoking. Flexicoking, continuous coking, and fluid coking are technological attempts to recycle coke as fuel in the cracking process. Although these cokers are more efficient, they are more expensive to build.

Fluid Catalytic Cracking. Because of its high conversion rate of vacuum wax distillate into far more valuable gasoline and lighter, olefinic gases, fluid catalytic crackers (FCCs) are the most important crackers at a refinery. By 2007, there were about four hundred FCCs worldwide working continuously at refineries. Together, they cracked about 10.6 million barrels of feedstock each day, about half of which was cracked in the United States. FCCs are essential to meet the global demand for gasoline.

Design of individual FCCs, while following the basic principle of fluid catalytic cracking, varies considerably, as engineers and scientists continuously seek to improve efficiency and lessen the environmental impact. By 2010, there were five major patents for FCCs that arranged the reactor, where cracking occurs, and the regenerator for the spent catalyst side by side. One major patent places the reactor atop the generator.

The typical products derived from vacuum distillate conversion in a FCC are about 21 percent olefinic gases, often called liquefied petroleum gas; 47 percent gasoline of high quality; 20 percent light cycle (or gas) oil, often called middle distillate; 7 percent heavy cycle (gas) oil, often called heavy fuel oil; and 5 percent coke. Gasoline is generally the most valuable. Light cycle oil is blended into heating oil and diesel, with highest demand for these blends in winter. The more a FCC can change the percentages of its outcome, the higher its economic advantage.

Hydrocracking. Hydrocrackers are the most flexible and efficient cracker, but they have high building and operation costs. High temperatures and pressure require significant energy, and the steel wall of a hydrocracker reactor can be as thick as 15 centimeters. Its use of hydrogen in the conversion process often requires a separate hydrogen generation plant. Hydrocrackers can accept a wide variety of feedstock ranging from crude oil distillates (gas oils or middle distillate) to light and heavy cycle oils from the FCC to coker distillates.

Hydrocracker output typically falls into flexible ranges for each product. Liquefied petroleum gas and other gases can make up 7 to 18 percent. Gasoline products, particularly jet fuel, one of the prime products of the hydrocracker, can be from 28 to 55 percent. Middle distillates, especially diesel and kerosene, can make up from 15 to 56 percent. Heavy distillates and residuum can range from 11 to 12 percent.

Hydrocracking produces no coke but does have high carbon dioxide emissions. Its products have very low nitrogen and sulfur content.

Impact on Industry

Cracking is a core activity at any modern refinery. Without modern cracking proesses, the world's demand for gasoline, diesel, kerosene, and jet fuel, as well as for basic petrochemicals, could not easily be met from processing crude oil. About 45 percent of the world's gasoline, for example, comes from FCCs or related units in 2006. Significant public and private research is focusing on improving cracking efficiency.

Government Agencies. Throughout the industrialized and industrializing world, national governments and their agencies, as well as international entities, have promoted more environmentally friendly patterns of hydrocarbon production, processing, and consumption. This has increased the importance of and demand for cracking, as most of the heavy oil distillates and residues once used as heavy fuel oil in power plants are increasingly being replaced by cleaner burning natural gas. National legislation such as the United States Clean Air Act of 1963 with its crucial 1990 amendment imposes strict limits on emissions including those generated during the cracking process, as well as quality specifications for refined products such as cleaner burning gasoline and diesel. Crackers have come under governmental pressure to reduce their often high emissions of carbon dioxide and other gases through innovations or process changes, which means more costs.

Universities and Research Institutes. Particularly in industrialized societies, there is considerable public research into improving existing and developing new cracking processes. Of special interest are catalysts used by FCCs and hydrocrackers. Catalyst research looks into properties of and new materials for catalysts and experiments with optimizing their placement within the reactor, enhancing their recovery, and lengthening their life cycle.

However, a critical gap exists between what works in the laboratory or in small pilot plants and what can be implemented in an industrial setting. Occasionally, promising new catalysts have not risen to their perceived potential, and new reactor designs have proved commercially impractical. University and institute sciences tend to have a certain bias toward energy-saving and environmentally friendly discoveries, while industrial research tends to be more concerned with raising feedstock economy.

Industry and Business Sectors. Cracking provides a key process for the oil and gas and petrochemical industries. For example, in January, 2010, in the first step of atmospheric distillation, U.S. refineries processed a total of 17.8 million barrels of crude oil per stream day (almost identical to a calendar day). Of this, 8.5 million barrels were subjected to vacuum distillation. Among the distillates and residue from both steps, 2.5 million barrels (15 percent of the original crude oil entering the refinery) were processed in the thermal cracking process of delayed coking (fluid coking and visbreaking were negligible at U.S. refineries). Another 6.2 million barrels (35 percent of all products from the original crude oil) were submitted to an FCC for catalytic cracking, and 1.8 million barrels (10 percent of the base) were processed in a hydrocracker. Although the balance between FCC and hydrocracker processing is somewhat different in Europe and Asia, where demand for diesel and kerosene is higher than in the United States, the example is representative for the petroleum industry. Overall, about 60 percent of the crude oil that enters a refinery undergoes some form of cracking.

Major Corporations. Because crackers are integral parts of a refinery, require a very large initial investment, and have relatively high operating costs, they are typically owned by the large international or national oil and gas companies that own the refinery. Their construction, however, may be executed by specialized engineering companies, and their design can involve independent consultants. In addition, many cracking facilities are built under license from the company that holds the patent to their design.

Innovative or improved crackers can be added to existing refineries or built together with green field refineries that are increasingly erected in the Middle East, the source of much crude oil, or Asia, the source of much demand. Crackers follow economies of scale. In 2005, to build an FCC with a capacity of 10 million barrels of feedstock per stream day in a Gulf Coast refinery cost from $80 million to $120 million. An FCC with ten times more capacity, however, cost only between $230 million and $280 million. The economy-of-scale factor favors world-scale crackers at equally large refineries.

Fascinating Facts About Cracking

- If engineers had not invented cracking hydrocarbons, the world would need to produce almost double the amount of crude oil to meet the global demand of gasoline, diesel, and jet fuel.
- Crackers at refineries and petrochemical plants operate continuously for twenty-four hours and are typically staffed in three shifts.
- During World War II, newly invented catalytic cracking provided plenty of powerful high-octane gasoline for aircraft engines of the United States and Great Britain, giving their air forces an edge over German and Japanese air forces.
- Crackers are expensive chemical processing plants. The ethane steam cracker and high olefin fluid catalytic cracker constructed in 2009 at Rabigh Refining and Petrochemical Complex in Saudi Arabia cost $850 million to build.
- Reactors of catalytic and steam crackers are incredibly hot, reaching up to 900 degrees Celsius. At these temperatures, oil vapors are blown through the reactor in fractions of a second.
- New pharmaceuticals can contain some ingredients made from raw materials that came out of a steam cracker.
- Catalysts fundamentally support cracking processes. If operating staff mistreats or abuses catalysts, the whole cracker may have to be shut down for catalyst replacement.

Careers and Course Work

Cracking is a key refining process, so good job opportunities should continue in this field. Students interested in designing and constructing or operating and optimizing a cracker at a refinery or petrochemical plant should take science courses, particularly chemistry and physics, in high school. The same is true for those who want to pursue research in the field. There also are many opportunities for technicians in building, operationing, and maintaining a cracker.

A bachelor of science degree in an engineering discipline, particularly chemical, electrical, computer, or mechanical engineering, is excellent job preparation in this field. A bachelor of science or arts degree in a major such as chemistry, physics, computer science, environmental science, or mathematics is also a good foundation. An additional science minor is useful.

For an advanced career, any master of engineering degree is a suitable preparation. A doctorate in chemistry or chemical or mechanical engineering is needed if the student wants a top research position, either with a corporation or at a research facility. Postdoctoral work in materials science (engaged in activities such as searching for new catalysts) is also advantageous.

Because cracking is closely related to selecting crude oil for purchase by the refinery, there are also positions for graduates in business or business administration. The same is true for members of the medical profession with an emphasis in occupational health and safety. Technical writers with an undergraduate degree in English or communication may also find employment at oil and engineering companies. As cracking is a global business, career advancement in the industry often requires willingness to work abroad.

Social Context and Future Prospects

As fossilized hydrocarbons are a finite resource, they must be used as efficiently as possible. To this end, cracking at a refinery seeks to create the most valuable products out of its feedstocks derived from crude oil. This is not limited to fuels such as gasoline. The steam crackers of the petrochemical industry create raw materials for many extremely valuable and useful products such as pharmaceuticals, plastics, solvents, detergents, and adhesives, stretching the use of hydrocarbons for consumers.

At the same time, international concern with the negative environmental side effects of some hydrocarbon processes is increasing. Traditionally, crackers such as cokers or even hydrocrackers released a large amount of carbon dioxide or, in the case of cokers, other airborne pollutants as well. To make cracking more environmentally friendly, to save energy, and to convert feedstock efficiently are concerns shared by the public and the petroleum and petrochemical industries. This is especially so for companies when there are commercial rewards for efficient, clean operations.

The possible rise of alternative fuels, replacing some hydrocarbon-based fuels such as gasoline and diesel, would challenge refineries to adjust the output of their crackers. The more flexible hydrocrackers are best suited to meet this challenge.

R. C. Lutz, B.A., M.A., Ph.D.

Further Reading

Burdick, Donald. *Petrochemicals in Nontechnical Language.* 3d ed. Tulsa, Okla.: Penn Well, 2010. Accessible introduction. Covers use of petrochemical materials gained from steam cracking. Figures and tables.

Conaway, Charles. *The Petroleum Industry: A Nontechnical Guide.* Tulsa, Okla.: Penn Well, 1999. Chapter 13 provides a short summary of cracking processes.

Gary, James, et al. *Petroleum Refining: Technology and Economics.* 5th ed. New York: CRC Press, 2007. Well-written textbook. Good presentation of all types of cracking in the petrochemical industry. Includes economic aspects of cracking. Five appendixes, index, and photographs.

Leffler, William. *Petroleum Refining in Nontechnical Language.* Tulsa, Okla.: Penn Well, 2008. Covers all major aspects of cracking at a refinery, from chemical foundations to catalytic cracking and hydrocracking.

Meyers, Robert, ed. *Handbook of Petroleum Refining Processes.* 3d ed. New York: McGraw-Hill, 2004. Advanced-level technical compendium covers various industry methods for catalytic cracking, hydrocracking, visbreaking, and coking. In-depth information for those considering a career in this field.

Sadeghbeigi, Reza. *Fluid Catalytic Cracking Handbook.* 2d ed. Houston, Tex.: Gulf Publishing, 2000. Detailed, technical look at this key cracking process. Figures, conversion table, glossary.

Web Sites

American Association of Petroleum Geologists
http://www.aapg.org

American Petroleum Institute
http://www.api.org

National Petrochemical and Refiners Association
http://www.npra.org

National Petroleum Council
http://www.npc.org

See also: Detergents; Distillation; Gasoline Processing and Production.

CRYOGENICS

FIELDS OF STUDY

Astrophysics; cryogenic engineering; cryogenic electronics; nuclear physics; cryosurgery; cryobiology; high-energy physics; mechanical engineering; chemical engineering; electrical engineering; cryotronics; materials science; biotechnology; medical engineering; astronomy.

SUMMARY

Cryogenics is the branch of physics concerned with creation of extremely low temperatures and involves the observation and interpretation of natural phenomena resulting from subjecting various substances to those temperatures. At temperatures near absolute zero, the electric, magnetic, and thermal properties of most substances are greatly altered, allowing useful industrial, automotive, engineering, and medical applications.

KEY TERMS AND CONCEPTS

- **Absolute Zero:** Temperature measured 0 Kelvin (−273 degrees Celsius), where molecules and atoms oscillate at the lowest rate possible.
- **Cryocooler:** Device that uses cycling gases to produce temperatures necessary for cryogenic work.
- **Cryogenic Processing:** Deep cooling of matter using cryogenic temperatures so the molecules and atoms of the matter slow or almost stop movement.
- **Cryogenic Tempering:** Onetime process using sensitive computerization to cool metal to cryogenic temperatures then tempering the metal to enhance performance, strength, and durability.
- **Cryopreservation:** Cooling cells or tissues to subzero temperatures to preserve for future use.
- **Evaporative Cooling:** Process that allows heat in a liquid to change surface particles from a liquid to a gas.
- **Heat Conduction:** Technique where a substance is cooled by passing heat from matter of higher temperature to matter of lower temperature.
- **Joule-Thomson Effect:** Technique where a substance is cooled by rapid expansion, which drops the temperature. So named for its discoverers, British physicists James Prescott Joule and William Thomson (Lord Kelvin).
- **Kelvin Temperature Scale (K):** Used to study extremely cold temperatures. On the Kelvin scale, water freezes at 273 K and boils at 373 K.
- **Superconducting Device:** Device known for its electrical properties and magnetic fields, such as magnetic resonance imaging (MRI) in medicine.
- **Superconducting Magnet:** Electromagnet with a superconducting coil where a magnetic field is maintained without any continuing power source.
- **Superconductivity:** Absence of electrical resistance in metals, ceramics, and compounds when cooled to extremely low temperatures.
- **Superfluidity:** Phase of matter, such as liquid helium, that is absent of viscosity and flows freely without friction at very low temperatures.

Definition and Basic Principles

Cryogenics comes from two Greek words: *kryo*, meaning "frost," and *genic*, "to produce." This science studies the implications of producing extremely cold temperatures and how these temperatures affect substances such as gases and metal. Cryogenic temperature levels are not found naturally on Earth. The usefulness of cryogenics is based on scientific principles. The three basic states of matter are gas, liquid, and solid. Matter moves from one state to another by the addition or subtraction of heat (energy). The molecules or atoms in matter move or vibrate at different rates depending on the level of heat. Extremely low temperatures, as achieved through cryogenics, slow the vibration of atoms and can change the state of matter. For example, cryogenic temperatures are used in the liquefaction of atmospheric gases such as oxygen, nitrogen, hydrogen, and methane for diverse industrial, engineering, automotive, and medical applications.

Sometimes cryogenics and cryonics are mistakenly linked, but use of subzero temperatures is the only thing these practices share. Cryonics is the practice of freezing a body right after death to preserve it for a future time when a cure for fatal illness or remedy for fatal injury may be available. The practice of cryonics is based on the belief that technology from cryobiology can be applied to cryonics.

If cells, tissues, and organs can be preserved by cryogenic temperatures, then perhaps whole body can be preserved for future thawing and life restoration. Facilities exist for interested persons or families, although the cryonic process is not considered reversible as of this writing.

Background and History

The history of cryogenics follows the evolution of low-temperature techniques and technology. Principles of cryogenics can be traced to 2500 B.C.E., when Egyptians and Indians evaporated water through porous earthen containers to produce cooling. The Chinese, Romans, and Greeks collected ice and snow from the mountains and stored it in cellars to preserve food. In the early 1800's, American inventor Jacob Perkins created a sulfuric-ether ice machine, a precursor to the refrigerator. By the mid-1800's, William Thomson, a British physicist known as Lord Kelvin, theorized that extremely cold temperatures could stop the motion of atoms and molecules. This became known as absolute zero, and the Kelvin scale of temperature measurement emerged.

Scientists of the time focused on liquefaction of permanent gases. By 1845, the work of British physicist Michael Faraday accomplished liquefaction of permanent gases by cooling immersion baths of ether and dry ice followed by pressurization. Six permanent gases—oxygen, hydrogen, nitrogen, methane, nitric oxide, and carbon monoxide—still resisted liquefaction. In 1877, French physicist Louis-Paul Cailletet and Swiss physicist Raoul Pictet produced drops of liquid oxygen, working separately and using completely different methods. In 1883, S. F. von Wroblewski at the University of Krakow in Poland, discovered oxygen would liquefy at 90 Kelvin (K) and nitrogen at 77 K. In 1898, Scottish chemist James Dewar discovered the boiling point of hydrogen at 20 K and its freezing point at 14 K.

Helium, with the lowest boiling point of all known substances, was liquefied in 1908 by Dutch physicist Heike Kamerlingh Onnes at the University of Leiden. Onnes was the first person to use the word "cryogenics." In 1892, Scottish physicist James Dewar invented the Dewar flask, a vacuum flask designed to maintain temperatures necessary for liquefying gases, which was the precursor to the Thermos. The liquefaction of gases had many important commercial applications, and many industries use Dewar's concept in applying cryogenics to their processes and products.

The usefulness of cryogenics continued to evolve, and by 1934 the concept was well established. During World War II, scientists discovered that metals became resistant to wear when frozen. In the 1950's, the Dewar flask was improved with the multilayer insulation (MLI) technique for insulating cryogenic propellants used in rockets. Over the next thirty years, Dewar's concept led to the development of small cryocoolers, useful to the military in national defense. The National Aeronautics and Space Administration (NASA) space program applies cryogenics to its programs. Cryogenics can be used to preserve food for long periods—this is especially helpful during natural disasters. Cryogenics continues to grow globally and serve a wide variety of industries.

How It Works

Cryogenics is an ever-expanding science. The basic principle of cryogenics that the creation of extremely low temperatures will affect the properties of matter so the changed matter can be used for a number of applications. Four techniques can create the conditions necessary for cryogenics: heat conduction, evaporative cooling, rapid-expansion cooling (Joule-Thomson effect), and adiabatic demagnetization.

Creating Low Temperatures. With heat conduction, heat flows from matter of higher temperature to matter of lower temperature in what amounts, basically, to a transfer of thermal energy. As the process is repeated, the matter cools. This principle is used in cryogenics by allowing substances to be immersed in liquids with cryogenic temperatures or in an environment such as a cryogenic refrigerator for cooling.

Evaporative cooling is another technique employed in cryogenics. Evaporative cooling is demonstrated in the human body when heat is lost through liquid (perspiration) to cool the body via the skin. Perspiration absorbs heat from the body, which evaporates after it is expelled. In the early 1920's in Arizona during the summers, people hung wet sheets inside screened sleeping porches. Electric fans pulled air through the sheets to cool the sleeping space. In the same way, a container of liquid can evaporate, so the heat is removed as gas; the repetitive process drops the temperature of the liquid. An example is reducing the temperature of liquid nitrogen to its freezing point.

The Joule-Thomson effect occurs without the transfer of heat. Temperature is affected by the relationship between volume, mass, pressure, and temperature. Rapid expansion of a gas from high to low pressure results in a temperature drop. This principle was employed by Dutch physicist Heike Kamerlingh Onnes to liquefy helium in 1908 and is useful in home refrigerators and air conditioners.

Adiabatic demagnetization uses paramagnetic salts to absorb energy from liquid, resulting in a temperature drop. The principle in adiabatic demagnetization is the removal of the isothermal magnetized field from matter to lower the temperature. This principle is useful in application to refrigeration systems, which may include a superconducting magnet.

Cryogenic Refrigeration. Cryogenic refrigeration, used by the military, laboratories, and commercial businesses, employs gases such as helium (valued for its low boiling point), nitrogen, and hydrogen to cool equipment and related components at temperatures lower than 150 K. The selected gas is cooled through pressurization to liquid or solid forms (dry ice used in the food industry is solidified carbon dioxide). The cold liquid may be stored in insulated containers until used in a cold station to cool equipment in an immersion bath or with sprayer.

Cryogenic Processing and Tempering. Cryogenic processing or treatment increases the length of wear of many metals and some plastics using a deep-freezing process. Metal objects are introduced to cooled liquid gases such as liquid nitrogen. The computer-controlled process takes about seventy-two hours to affect the molecular structure of the metal. The next step is cryogenic or heat tempering to improve the strength and durability of the metal object. There are about forty companies in the United States that provide cryogenic processing.

Applications and Products

Early applications of cryogenics targeted the need to liquefy gases. The success of this process in the late 1800's paved the way for more study and research to apply to cryogenics to developing life needs and products. Examples include applications in the auto and health care industries and in development of rocket fuels and methods of food preservation. Cryogenic engineering has applications related to commercial, industrial, aerospace, medical, domestic, and defense ventures.

Superconductivity Applications. One property of cryogenics is superconductivity. This occurs when the temperature is dropped so low that the electrical current experiences no resistance. An example is electrical appliances, such as toasters, televisions, radios, or ovens, where energy is wasted trying to overcome electrical resistance. Another is with magnetic resonance imaging (MRI), which uses a powerful magnetic field generated by electromagnets to diagnosis

Fascinating Facts About Cryogenics

- American businessman Clarence Birdseye revolutionized the food industry when he discovered that deep-frozen food tasted better than regular frozen food. In 1923, he developed the flash-freezing method of preserving food at below-freezing temperatures under pressure. The "Father of Frozen Food" first sold small-packaged foods to the public in 1930 under the name Birds Eye Frosted Foods.
- In cryosurgery, super-freezing temperatures as low as −200 Celsius are introduced through a probe of circulating liquid nitrogen to treat malignant tumors, destroy damaged brain tissue in Parkinson's patients, control pain, halt bleeding, and repair detached retinas.
- Cryogenics can be used to save endangered species from extinction. Smithsonian researcher Mary Hagedorn is using cryogenics to establish the first coral seed banks: She's collecting thousands of sample species and freezing them for the future. Hagedorn refers to this as an insurance policy for natural resources.
- The Joule-Thomson effect, discovered in 1852 by James Prescott Joule and William Thomson (Lord Kelvin), is responsible for the cooling used in home refrigerators and air conditioners.
- Helium's boiling point, 173 Kelvin, is the lowest of all known substances.
- Surgical tools and implants used by surgeons and dentists have increased strength and resistance to wear because of cryogenic processing.
- Cryogenic processing is 100 percent environmentally friendly with no use of harmful chemicals and no waste products.
- In 1988, microbiologist Curt Jones, who studied freezing techniques to preserve bacteria and enzymes for commercial use, created Dippin' Dots, a popular ice cream treat, using a quick-freeze process with liquid nitrogen.

certain medical conditions. High magnetic field strength occurs with superconducting magnets. Liquid helium, which becomes a free-flowing superfluid, cools the superconducting coils; liquid nitrogen cools the superconducting compounds, making cryogenics an integral part of this process. Another application is the use of liquefied gases that are sprayed on buried electrical cables to minimize wasted power and energy and to maintain cool cables with decreased electrical resistance.

Health Care Applications. The health care industry recognizes the value of cryogenics. Medical applications using cryogenics include preservation of cells or tissues, blood products, semen, corneas, embryos, vaccines, and skin for grafting. Cryotubes with liquid nitrogen are useful in storing strains of bacteria at low temperatures. Chemical reactions needed to release active ingredients in statin drugs, used for cholesterol control, must be completed at very low temperatures (−100 degrees Celsius). High-resolution imaging, like MRI, depends on cryogenic principles for the diagnosis of disease and medical conditions. Dermatologists uses cryotherapy to treat warts or skin lesions.

Food and Beverage Applications. The food industry uses cryogenic gases to preserve and transport mass amounts of food without spoilage. This is also useful in supplying food to war zones or natural-disaster areas. Deep-frozen food retains color, taste, and nutrient content while increasing shelf life. Certain fruits and vegetables can be deep frozen for consumption out of season. Freeze-dried foods and beverages, such as coffee, soups, and military rations, can be safely stored for long periods without spoilage. Restaurants and bars use liquid gases to store beverages while maintaining the taste and look of the drink.

Automotive Applications. The automotive industry employs cryogenics in diverse ways. One is through the use of thermal contraction. Because materials will contract when cooled, the valve seals of automobiles are treated with liquid nitrogen, which shrinks to allow insertion and then expands as it warms up, resulting in a tight fit. The automotive industry also uses cryogenics to increase strength and minimize wear of metal engine parts, pistons, cranks, rods, spark plugs, gears, axles, brake rotors and pads, valves, rings, rockers, and clutches. Cryogenic-treated spark plugs can increase an automobile's horsepower as well as its gasoline mileage. The use of cryogenics allows a race car to race as many as thirty times without a major rebuild on the motor compared with racing twice on an untreated car.

Aerospace Industry Applications. NASA's space program utilizes cryogenic liquids to propel rockets. Rockets carry liquid hydrogen for fuel and liquid oxygen for combustion. Cryogenic hydrogen fuel is what enables NASA's workhorse space shuttle to get into orbit. Another application is using liquid helium to cool the infrared telescopes on rockets.

Tools, Equipment, and Instrument Applications. Metal tools can be treated with cryogenic applications that provide wear resistance. In surgery or dentistry, tools can be expensive, and cryogenic treatment can prolong usage. Sports equipment, such as golf clubs, benefits from cryogenics as it provides increased wear resistance and better performance. Another is the ability of a scuba diver to stay submerged for hours with an insulated Dewar flask of cryogenically cooled nitrogen and oxygen. Some claim musical instruments receive benefits from cryogenic treatment; in brass instruments, a crisper and cleaner sound is allegedly produced with cryogenic enhancement.

Other Applications. Other applications are evolving as industries recognize the benefits of cryogenics to their products and programs. The military have used cryogenics in various ways, including infrared tracking systems, unmanned vehicles, and missile warning receivers. Companies can immerse discarded recyclables in liquid nitrogen to make them brittle, then these recyclables can be pulverized or grinded down to a more eco-friendly form. No doubt with continued research, many more applications will emerge.

Impact on Industry

The science of cryogenics continues to impact the quality and effectiveness of products of many industries. Various groups have initiated research studies to support the expanded use of cryogenics.

Government and University Research. Government agencies and universities have conducted research on various aspects of cryogenics. The military have investigated the applications of cryogenics to national defense. The Air Force Research Laboratory (AFRL) Space Vehicles Directorate addressed applications of cryogenic refrigeration in the area of ballistic missile defense. The study looked at ground-based radars and space-based infrared sensors requiring cryogenic refrigeration. Future research targets the availability of flexible technology such as field cryocoolers. Such studies can be significant to a cost-effective national defense for the United States.

The health care industry has many possible applications for cryogenics. In 2005, Texas A&M University Health Science Center-Baylor College of Dentistry investigated the effect of cryogenic treatment of nickel-titanium on instruments used by dentists. Past research had been conducted on stainless-steel endodontic instruments with no significant increase in wear resistance. This research demonstrated an increase in microhardness but not in cutting efficiency.

Industry and Business. In 2006, the Cryogenic Institute of New England (CINE) recognized that although cryogenic processing was useful in many industries, research validating its technologic advantages and business potential was scant. CINE located forty commercial companies that provided cryogenic processing services and conducted telephone surveys with thirty of them. The survey found that some $8 million were generated by these deep cryogenic services in the United States, with 75 percent coming from the services and 25 percent from the sales of equipment.

The survey asked participants to identify the list of top metals they worked with in cryogenic processing. These were given as cast irons, various steels (carbon, stainless, tool, alloy, mold), aluminum, copper, and others. The revenue was documented by market application. Some 42 percent of the cryogenic-processing-plant market was in the motor sports and automotive industry, where the goal was treatment of engine components to improve performance, extend life wear, or treat brake rotors. Thirty percent of the application market fell into the category of heavy metals, tooling, and cutting; examples of these include manufacturing machine tools, dies, piping, grinders, knives, food processing, paper and pulp processing, and printing. Ten percent were listed in heavy components such as construction, in-ground drilling, and mining, while 18 percent of the market was in areas such as recreational, firearms, electronics, gears, copper electrodes, and grinding wheels.

The National Institute of Standards and Technology (NIST) initiated the Cryogenic Technologies Project, a collaborative research effort between industry and government agencies to improve cryogenic processes and products. One goal is the investigation of cryogenic refrigeration.

The nonprofit Cryogenic Society of America (CSA) offers conferences on related work areas such as superconductivity, space cryogenics, cryocoolers, refrigeration, and magnet technology. It also has continuing-education courses and lists job postings on its Web site.

Careers and Course Work

Careers in cryogenics are as diverse as the applications of cryogenics. Interested persons can enter the profession in various ways, depending on their field of interest. Some secure jobs through additional education, while others learn on the job. In general, the jobs include engineers, technologists or technicians, and researchers.

A primary career track for those interested in working in cryogenics is cryogenic engineering. To become a cryogenic engineer requires a bachelor's or master's degree in engineering. Course work may include thermodynamics, production of low temperatures, refrigeration, liquefaction, solid and fluid properties, and cryogenic systems and safety.

In the United States, some four hundred academic institutions offer graduate programs in engineering with about forty committed to research and academic opportunities in cryogenics. Schools with graduate programs in cryogenics include the University of California, Los Angeles; University of Colorado; Cornell University; Georgia Institute of Technology; Illinois Institute of Technology; Florida International University; Iowa State University; Massachusetts Institute of Technology; Ohio State University; University of Wisconsin-Madison; Florida State University; Northwestern University; and University of Southampton in the United Kingdom.

Social Context and Future Prospects

The economic and ecological impact of cryogenic research and applications holds global promise for the future. In 2009, Netherlands firm Stirling Cryogenics built a cooling system with liquid argon for the ICARUS project, which is being carried out by Italy's National Institute of Nuclear Physics. In China, the Cryogenic and Refrigeration Engineering Research Centre (CRERC) focuses on new innovations and technology in cryogenic engineering. Both private industry and government agencies in the United States are pursuing innovative ways to utilize existing applications and define future implications of cryogenics. Although cryogenics has proved useful to many industries, its full potential as a science has not yet been realized.

Marylane Wade Koch, M.S.N., R.N.

FURTHER READING

Hayes, Allyson E., ed. *Cryogenics: Theory, Processes and Applications.* Hauppauge, N.Y.: Nova Science Publishers, 2010. Details global research on cryogenics and applications such as genetic engineering and cryopreservation.

Jha, A. R. *Cryogenic Technology and Applications.* Burlington, Mass.: Elsevier, 2006. Deals with most aspects of cryogenics and cryogenic engineering, including historical development and various laws, such as heat transfer, that make cryogenics possible.

Schwadron, Terry. "Hot Sounds From a Cold Trumpet? Cryogenic Theory Falls Flat." *New York Times,* November 18, 2003. Explains how two Tufts University researchers studied cryogenic freezing of trumpets and determined the cold did not improve the sound.

Ventura, Gugliemo, and Lara Risegari. *The Art of Cryogenics: Low- Temperature Experimental Techniques.* Burlington, Mass.: Elsevier, 2008. Comprehensive discussion of various aspects of cryogenics from heat transfer and thermal isolation to cryoliquids and instrumentation for cryogenics, such as the use of magnets.

WEB SITES

Cryogenic Society of America
http://www.cryogenicsociety.org

Help Mary Save Coral
http://www.helpmarysavecoral.org/obe

National Aeronautics and Space Administration
Cryogenic Fluid Management
http://www.nasa.gov/centers/ames/research/technology-onepagers/cryogenic-fluid-management.html

National Institute of Standards and Technology
Cryogenic Technologies Project
http://www.nist.gov/mml/properties/cryogenics/index.cfm

See also: Electromagnet Technologies

CRYONICS

FIELDS OF STUDY

Cardiopulmonary resuscitation; cardiothoracic surgery; cryopreservation; nanotechnology; surgery.

SUMMARY

Cryonics is a theoretical life support technology, which involves stabilizing the condition of a terminally ill patient via freezing until a future date when technology will be able to revive that person and hopefully return him or her to a normal life. Storing a person at the temperature of liquid nitrogen (−196 degrees Celsius) can prevent further tissue damage indefinitely; however, the freezing process inflicts a degree of tissue damage that is not reversible by existing technology. Modern technology is successful in freezing sperm, ova (eggs), and embryos, which can later be thawed and restored to life. Human embryos have been thawed, implanted into a uterus, and have ultimately developed into a healthy newborn.

KEY TERMS AND CONCEPTS

- **Cryopreservation:** Process whereby cells and tissues are preserved by cooling to subzero temperatures.
- **Cryoprotectant:** Substance that protects biological tissue from freezing damage.
- **Ischemia:** Inadequate supply of oxygenated blood to a bodily organ or tissue.
- **Liquid Nitrogen:** Nitrogen, which is a gas at room temperature, can be maintained in a liquid state at a very low temperature (−196 degrees Celsius). It is used for the cryopreservation of sperm, ova (eggs), embryos, and adult human bodies.
- **Nanotechnology:** Technology that entails the manipulation of individual atoms or molecules, eventually to build or repair any physical object, including human cells and biological tissue.
- **Necrosis:** Cell death, which can be caused by factors such as ischemia or lack of oxygen.
- **Vitrification:** Process of converting something into a glass-like solid that is free of any crystal formation, which can cause tissue damage; vitrification lowers the freezing point of a solution.

Definition and Basic Principles

Cryonics (from the Greek *kryos*, which means "icy cold") involves the freezing, or cryopreservation, of an entire body or the brain until a future date when technology will be able to thaw the tissue and restore life. If only the brain is frozen, the ultimate goal is to grow a new body around the head. The brain is customarily frozen with the head to provide protection to the structure. In the United States, the process is begun when a patient's heart stops beating. In addition to humans, cryonics facilities preserve pets. Once frozen in liquid nitrogen, and with proper storage, a frozen body will not deteriorate further over time. As of 2011, technology is not available to thaw a cryopreserved individual and restore life. Therefore, any future thawing methods are purely hypothetical. As the technology improves, so will the chances of life restoration. A number of biological specimens have been cryopreserved, stored in liquid nitrogen, and revived. They include insects, eels, and organs. For example, in 2005, a cryopreserved rabbit kidney was thawed and transplanted to a rabbit; the organ functioned normally. The best examples of restoration of viability after cryopreservation of human tissue are ova (eggs), sperm, embryos, and ovarian tissue. A frozen ovum can be thawed and fertilized with previously frozen sperm, and a normal infant can develop. On a larger scale, human embryos can be cryopreserved, later thawed, and implanted into the uterus for growth. Infants derived from this process are no different in growth or intelligence than those who came into being under natural conditions.

Background and History

In 1866, Italian physician Paolo Mantegazza suggested that before soldiers departed for the battlefield, they should leave behind frozen sperm. He also suggested collection of sperm for freezing just before a mortally wounded soldier would die. The first successful human pregnancy using frozen sperm occurred in 1953; thirty years later, cryopreservation of oocytes was accomplished. In 1986, Australian physician Christopher Chen reported the world's first pregnancy using previously frozen oocytes. The rate of success of pregnancies using frozen oocytes, embryos, and fresh embryos are comparable.

The first cryonics organization, the Life Extension Society, was founded by Evan Cooper in 1963. Growth was slow over the next two years and most of the people interested in the procedure were wealthy celebrities. The first person to be cryogenically frozen was James Bedford, a seventy-three-year-old psychologist. He was cryopreserved in 1967, and his body is reportedly still in good condition at Alcor Life Extension Foundation. By the late 1970's, there were six cryonics companies operating in the United States. However, the preservation and subsequent maintenance of bodies indefinitely proved to be too expensive, and many of the companies ceased operations in the 1980's. Currently, seven companies offer cryonics services: Alcor Life Extension Foundation (Scottsdale, Arizona), American Cryonics Society (Cupertino, California), Cryonics Institute (Clinton Township, Michigan), Eucrio (Braga, Portugal), KrioRus (Alabychevo, Russia), Suspended Animation (Boynton Beach, Florida), and Trans Time (San Leandro, California). A facility in Australia is in the planning stages.

How It Works

Embryo Cryopreservation. Embryo cryopreservation has evolved to a technology with a high success rate. Embryos can be frozen at the pronuclear stage (one cell) up to the blastocyst stage (five to seven days after fertilization; 75 to 100 cells). Freezing and thawing of embryos is overseen by embryologists who are assisted by laboratory technicians. The embryos are mixed with a cryoprotectant and placed in straws before freezing. This allows for vitrification during the freezing process. The scientific definition of vitrification is the conversion of a liquid to a glass-like solid, which is free of any ice-crystal formation. The straw is placed in a cooling chamber for freezing. After freezing, the straw is placed into a carefully labeled metal cane, which is lowered into a liquid nitrogen tank with other frozen embryos. The thawing process involves warming the embryos to room temperature in 35 seconds. Over the next half hour, the embryos are incubated in decreasing concentrations of the cryoprotectant and increasing concentrations of water. These embryos are then transferred to a woman's uterus for growth. In the past, several embryos were transferred in the hope that at least one would survive. Existing technology has improved to the point that a single embryo has a high survival rate. The trend is to transfer a single embryo to prevent multiple births. Multiple births—even twins—have a much higher incidence of complications. Embryos stored for ten to twelve years have been thawed with subsequent development of a live infant. The cost for freezing an embryo is around $700. Storage costs vary depending on the length of storage.

Sperm and Ovary Cryopreservation. For men planning to have a vasectomy or men with testicular cancer, sperm can be cryopreserved, which will allow the future fathering of children. In addition to cryopreservation of embryos, women can have a portion of an ovary cryopreserved. This ovarian tissue can be thawed and transplanted at a later date. For women who must undergo chemotherapy for a malignancy, which can compromise their fertility, this is an extremely attractive option. The ovarian tissue contains thousands of eggs; furthermore, this tissue will produce female hormones.

Comparison of Embryo Cryopreservation to Cryonics. In contrast to the high success rate of human embryo transfer, cryopreservation of a fully formed individual is imperfect and theoretical. Embryos consist of a small number of cells, each of which has the potential to develop into any type of tissue or organ. Embryos are preselected and, if available, only high-quality embryos are used. The embryos are in good health and well oxygenated when frozen. The whole process from freezing to storage and ultimate thawing is done in an orderly process. An adult who wishes to be cryopreserved is usually of advanced age with multiple health problems, such as an advanced malignancy, severe heart disease, or Alzheimer's disease; the individual also is near death when the process is begun. Often, as death approaches, multiple organ failure occurs. The patient may have previously suffered a stroke or suffer one as death approaches. Unless the patient dies at a location and can be promptly instituted, a significant delay in the process can occur. Cryoprotectants must be infused into the body to circulate throughout it. With current technology, despite the use of cryoprotectants, the blood vessels are damaged. Even in cases where only the brain is frozen, blood-vessel damage occurs. Once frozen, no further damage will occur; however, all would agree, a cryopreserved human is significantly damaged. When thawed, the individual must be restored to life as well as good physical and mental health.

Cryonics. According to Alcor, when a patient who has enrolled in a cryonics program becomes critically ill, cryonics personnel will be placed on standby. When the heart stops beating, the patient is placed in an ice-water bath, and cardiopulmonary resuscitation is begun. Intravenous lines are established to infuse cryoprotectants, medications, and anesthetics. The medications maintain blood pressure. The cooling and the anesthetic lower the oxygen consumption to protect the brain and vital organs. If the patient is in a hospital that does not allow cryonics procedures, the patient is transferred to another facility; resuscitation and cooling are maintained during this process. The subsequent process includes surgically assessing the femoral (upper leg) arteries and veins and the patient is placed on cardiopulmonary bypass with a portable heart-lung machine. External resuscitation is discontinued. In the heart-lung machine, a heat exchanger works to lower the body temperature to a few degrees above the freezing point of water (0 degrees Celsius), and then blood is replaced with a cryoprotectant. The patient is then transferred to the cryonics facility (if not already there). Major blood vessels are accessed by thoracic surgery and attached to the perfusion circuit. Cryoprotectant is infused at nearly the freezing point of water for several minutes, which removes any residual blood. Then the cryoprotectant concentration is increased over a two-hour period to half the final target concentration. A rapid increase to the final concentration is then made. After cryoprotective perfusion, the patient is rapidly cooled under computer control by fans circulating nitrogen gas at a temperature of −125 degrees Celsius. The patient is then further cooled to −196 degrees Celsius over the next two weeks. The cost of the procedure and storage can exceed $160,000 for whole-body cryopreservation and $80,000 for the head-only option. The fee includes a basic cost at enrollment and usually requires an annual membership fee until death. Often, arrangements can be made to have these costs covered by life insurance.

Applications and Products

Cryoprotectants. A variety of cryoprotectants are available for cryopreservation of embryos and adults. Research is ongoing to improve these products. Conventional cryoprotectants in use include dimethyl sulfoxide (DMSO), ethylene glycol, glycerol, propylene glycol, propanediol (PROH), and sucrose. DMSO and glycerol have been used for decades by embryologists to cold-preserve embryos and sperm. Usually, a combination of ingredients is used because they have less toxicity and increased efficacy compared to a single substance. Cryoprotectants lower the freezing point of water; in addition, many cryoprotectants displace water molecules with hydrogen bonds in biological materials. This hydrogen bonding maintains proper protein and DNA function.

Cryonics Procedures. Cryonics companies employ specially designed resuscitation equipment to perform immediate cardiopulmonary resuscitation (CPR) immediately after death. Two such devices are the LUCAS Chest Compression System and Michigan Instruments' Thumper. These devices are powered by pressurized oxygen and restore blood flow much more efficiently than manual CPR. Later, a portable heart-lung machine is used to continue perfusion. After instillation of cryoprotectants, transport, ranging from a few miles to thousands, is needed. Cryonics facilities contain an operating room with a heart-lung machine as well as a variety of specialized equipment for instilling cryoprotectants and controlled cooling. The use of the operating room is sporadic at best. Months and sometimes years transpire between patients. Nevertheless, it must be maintained in a standby status that can be made fully operational in a matter of hours. Technicians and medical personnel, including doctors and nurses, are obviously not part of the regular staff. They are alerted and assembled when the death of a cryonics applicant is imminent. In some cases, sudden death occurs, and the facility must have a system in place to obtain the necessary personnel on short notice.

Embryo Culture. Specially designed media and incubators are used for the culture of embryos. Research is ongoing to improve both the media and incubators. Around day five of development, the embryo will be in the blastocyst stage, and it must hatch out from its "shell," a glycoprotein membrane called the zona pellucida, in order to implant in the uterus. If the embryo does not hatch, it cannot implant. Since the mid-1990's, assisted hatching has been performed to encourage embryo to come out. This is done by an embryologist, who makes a small hole in the zona of each embryo. The zona is not a living part of the embryo and making the hole does not harm the embryo. A laser beam is often used to make the hole.

Nanotechnology. Nanotechnology is focused on the manipulation of individual atoms or molecules with the ultimate goal of building or repairing any physical object, including human cells and biological tissue. Proponents of cryonics feel that, in the future, damaged tissue can be restored to a healthy state. This technology, if perfected, could repair damage from cryoprotectant toxicity, lack of oxygen, thermal stress (fracturing), ice-crystal formation in tissues that were not successfully vitrified, and reversal of the effects that caused the patient's death in the first place (heart attack, kidney failure, etc.). Advocates of freezing only the head feel that nanotechnology can regrow an entire body from the head. Theoretical revival scenarios describe repairs being accomplished via large numbers of devices or microscopic organisms. Repairs would be done at the molecular level before thawing.

Preservation of Mental Function and Memory. A successful cryonics outcome entails not only restoration of a healthy body but also a healthy mind, which is fully functional with an intact memory. It is well known that if the brain is deprived of oxygen for more than four to six minutes, ischemic changes occur, which result in brain damage and brain death. Proponents of cryonics claim that these ischemic changes may be reversible with future technology. They also claim that personality, identity, and long-term memory persist for some time after death because they are stored in resilient cell structures and patterns within the brain; thus, these features do not require continuous brain activity to survive. Another more radical cryonics concept is mind transfer. This entails a future technology that could scan the memory contents of a preserved brain.

Storage Facilities. Storage facilities for embryos are rooms containing tanks of liquid nitrogen, each containing many embryos. The tanks are, in essence, large thermoses. Cryonic storage tanks are obviously much larger; some are designed to contain a single individual while others can hold more than a dozen. Cryopreservation requires a constant source of liquid nitrogen because it evaporates, even in specially designed storage tanks. The tanks contain monitoring devices with alarms to indicate a drop in the nitrogen level. Some tanks have automated refilling devices attached. The nitrogen level in the air surrounding the tanks is also monitored. An increase in the nitrogen level would indicate leakage. The tanks are also inspected on a daily basis for any sign of leakage.

Fascinating Facts About Cryonics

- The most well-known cryopreserved patient is the baseball player Ted Williams.
- Science-fiction author Robert Heinlein wrote enthusiastically about cryonics; however, he was cremated after his death.
- The psychologist Timothy Leary, well known for his advocacy of psychedelic drugs such as LSD, was a cryonics advocate. However, shortly before his death, he changed his mind and was not cryopreserved.
- In a 1773 letter, Benjamin Franklin expressed regret that he lived "in a century too little advanced, and too near the infancy of science" that he could not be preserved and revived to fulfill his "very ardent desire to see and observe the state of America a hundred years hence."
- Arctic and Antarctic insects, fish, amphibians, and reptiles naturally produce cryoprotectants to minimize freezing damage during cold spells. Insects commonly use sugars or polyols (alcohols) as cryoprotectants. Arctic frogs use glucose, and Arctic salamanders produce glycerol in their livers for use as cryoprotectant.
- Woolly bear caterpillars can survive for ten months of the year frozen solid at temperatures as low as −50 degrees Celsius.
- Several novels based on contemporary cryonics have been published, including Bill Clem's *Immortal* (2008), which presents "a chilling look at the science behind cryonics," and James Halperin's *The First Immortal: A Novel of the Future* (1998), in which the protagonist suffers a massive coronary, is cryopreserved, and revived in 2072 by his great-grandson.

Impact on Industry

Cryonics has minimal impact on industry because of the small number of patients involved. As of 2010, about 200 individuals in the United States have undergone the procedure since it was first offered in 1962. Cryogenics research is also limited and is ongoing on a small scale at cryonics facilities such as the Cryonics Institute and Alcor. In 2006, the Cryonics Society, a nonprofit corporation, was formed. It is focused solely on the promotion and public education of the concept of cryonics. In the much broader field of cryogenics, considerable research is ongoing at facilities such as the National Institutes of Health (NIH) and universities. Furthermore, these facilities conduct research on nanotechnology and aging.

In contrast to cryonics, embryo cryopreservation and the associated areas of assisted reproduction technology (ART) have a significant impact on industry. ART facilities can be found in many metropolitan areas throughout the United States. These facilities are major consumers of medical equipment and supplies such as culture media, ultrasound equipment, and surgical instruments.

Careers and Course Work

Career opportunities are extremely limited in cryonics. The handful of cryonics facilities currently in existence employs a small staff of technicians to maintain the cryopreservation tanks and sales personnel. During a cryopreservation procedure, the staff expands tremendously for a brief period and includes physicians, nurses, and technicians. All temporary employees have full-time positions elsewhere in their areas of expertise.

Social Context and Future Prospects

The field of cryonics is speculative and controversial on scientific, sociological, and religious grounds. Proponents claim that it provides a possibility of a future existence to the finality of death. They suggest the possibility of immortality if revival is successful as well as the associated benefits that postponing or avoiding dying would bring. They tout that the cost of the procedure pales in comparison to the possible benefits. One of the arguments against the procedure is that it would change the concept of death. If life is restored at some time in the future, no friends or family would be left alive. Also, from a religious perspective, the soul would leave the body at the time of cryopreservation. Furthermore, religious beliefs often include a spiritual afterlife. Opponents state that the funds spent for cryopreservation could better be spent on worthwhile causes such as charities or providing funds for education of family members. They also claim that cryonics could lead to premature euthanasia to maximize the chances for revival. In fact, Alcor became involved in a lawsuit involving that topic. In 1987, an eighty-three-year-old woman who had opted for cryonics developed pneumonia after several years of poor health. Alcor personnel deemed that death was imminent and transferred the woman to their facility. She subsequently underwent a "neuro," a head preservation. After the local coroner received a headless body, he demanded the head, which Alcor refused. The coroner launched an investigation and accused Alcor of murder. According to Alcor, the case was settled out of court in the company's favor in 1991.

In its short history, several cryonics facilities have ceased operation. An economic downturn or a large lawsuit could bankrupt a company. If a suitable buyer could not be located to accept the ongoing maintenance costs, cryopreservation would not be maintained for the patients. Continuing the operations of a floundering company would occur only if the projected revenues from living individuals enrolled in the program would justify the costs. If, at some future date, technology exists to resuscitate cryopreserved patients and incorporate nanotechnology, the cost might be prohibitive.

Wide scientific acceptance of cryonics requires both cryopreservation and successful revival. This can be accomplished with animal experimentation. For example, dogs that are placed in a pound and scheduled for euthanasia after expiration of the adoption period could become subjects for cryonics experimentation.

Robin L. Wulffson, M.D., F.A.C.O.G.

Further Reading

Foster, Lynn E. *Nanotechnology: Science, Innovation, and Opportunity.* Upper Saddle River, N.J.: Prentice Hall, 2006. Experts in the field present information on where the industry currently stands, how it will evolve, and how it will impact the individual.

Immortality Institute. *The Scientific Conquest of Death: Essays on Infinite Lifespans.* Buenos Aires: Libros en Red, 2004. A collection of essays by nineteen doctors, scientists, and philosophers describing a positive perspective on cryonics.

Johnson, Larry, and Scott Baldyga. *Frozen: My Journey into the World of Cryonics, Deception, and Death.* New York: Vanguard Press, 2009. After a period of employment at Alcor, veteran paramedic Johnson became a whistleblower and described "horrific discoveries" that took place at the facility.

Perry, R. Michael. *Forever for All: Moral Philosophy, Cryonics, and the Scientific Prospects for Immortality.* Boca Raton, Fla.: Universal Publishers, 2000. The book follows recent immortalist thinking that places hope in future advances in humankind's understanding and technology.

Pommer, R. W., III. "Donaldson v. Van de Kamp: Cryonics, Assisted Suicide, and the Challenges of Medical Science." *Journal of Contemporary Health Law and Policy* Spring (1993): 589-603.Very interesting paper chronicling California mathematician Thomas Donaldson's constitutional right to premortem cryogenic suspension. Donaldson was diagnosed with an inoperable brain tumor.

Romain, T. "Extreme Life Extension: Investing in Cryonics for the Long, Long Term." *Medical Anthropology* 29 no. 2 (April, 2010): 194-215. Explores the possibility of biotechnology through cryonics; social aspect of anxiety about aging is also covered.

Shaw D. "Cryoethics: Seeking Life After Death." *Bioethics* 23 (November, 2009): 515-521. This article examines the ethical considerations of cryonic preservation.

Web Sites

Alcor Life Extension Foundation
http://www.alcor.org

Cryonics Institute
http://www.cryonics.org

The Cryonics Society
http://www.cryonicssociety.org

Sperm Bank, Inc.
http://www.spermbankcalifornia.com/embryo-egg-

D

DETERGENTS

FIELDS OF STUDY

Biochemistry; biology; biotechnology; engineering; chemistry; environmental science; chemistry; materials science; pharmacy; physics and kinematics; polymer chemistry; process design; systems engineering; textile design.

SUMMARY

Synthetic detergents enable otherwise immiscible materials (water and oil) to form homogeneous dispersions. In addition to facilitating the breakdown and removal of stains or soil from textiles, hard surfaces, and human skin, detergents have found widespread application in food technology, oil-spill cleanup, and other industrial processes, such as the separation of minerals from their ores, the recovery of oil from natural deposits, and the fabrication of ceramic materials from powders. Detergents are used in the manufacture of thousands of products and have applications in the household, personal care, pharmaceutical, agrochemical, oil and mining, and automotive industries, as well as in the processing of paints, paper coatings, inks, and ceramics.

KEY TERMS AND CONCEPTS

- **Critical Micelle Concentration (CMC):** Surfactant concentration at which appreciable micelle formation occurs, enabling the removal of soils.
- **Hydrophilic-Lipophilic Balance (HLB):** Value expressing hydrophilic (polar/water-loving) and lipophilic (nonpolar/oil-loving) character of surfactants.
- **Interface:** Surface that forms the boundary between two bodies of matter (liquid, solid, gas) or the area where two immiscible phases come in contact.
- **Micelle:** Aggregate or cluster of individual surfactant molecules formed in solution, whereby polar ends are on the outside (toward the solvent) and nonpolar ends are in the middle.
- **Surface Tension:** Property of liquids to preferentially contract or expand at the surface, depending on the strength of molecular association and attraction forces.
- **Surfactant:** Surface active agent containing both hydrophilic and hydrophobic groups, enabling it to lower surface tension and solubilize or disperse immiscible substances in water.

DEFINITION AND BASIC PRINCIPLES

Detergents comprise a group of synthetic water-soluble or liquid organic preparations that contain a mix of surfactants, builders, boosters, fillers, and other auxiliary constituents, the formulation of which is specially designed to promote cleansing action or detergency. Liquid laundry detergents are by far the largest single product category: Sales in the United States were reported at $3.6 billion for 2010. As of 2011, products are largely mixtures of surfactants, water softeners, optical brighteners and bleach substitutes, stain removers, and enzymes. They must be formulated with ingredients in the right proportion to provide optimum detergency without damaging the fabrics being washed.

Before the advent of synthetic detergents, soaps were used. Soaps are salts of fatty acids and made by alkaline hydrolysis of fats and oils. They consist of a long hydrocarbon chain with the carboxylic acid end group bonded to a metal ion. The hydrocarbon end is soluble in fats and oils, and the ionic end (carboxylic acid salt) is soluble in water. This structure gives soap surface activity, allowing it to emulsify or disperse oils and other water-insoluble substances. Because soaps are alkaline, they react with metal ions in hard water and form insoluble precipitates, decreasing their cleaning effectiveness. These precipitates became known as soap scum—the "gunk" that builds up and surrounds the bathtub and causes graying or yellowing in fabrics.

Once synthetic detergents were developed, this problem could be avoided. Detergents are structurally similar to soap and work much the same to emulsify oils and hold dirt in suspension, but they differ in their water-soluble portion in that their calcium, magnesium, and iron forms are more water soluble and do not precipitate out. This allowed detergents to work well in hard water, and thus, reduce the discoloration of clothes.

Background and History

Soap is the oldest cleaning agent and has been in use for 4,500 years. Ancient Egyptians were known to bathe regularly and Israelites had laws governing personal cleanliness. Records document that soap-like material was being manufactured as far back as 1500 B.C.E. Before soaps and detergents, clothes were cleansed by beating them on rocks by a stream. Plants such as soapwort and soapbark that contained saponins were known to produce soapy lather and probably served as the first crude detergent. By the 1800's, soap making was widespread throughout Europe and North America, and by the 1900's, its manufacture had grown into an industry. The chemistry of soap manufacturing remained primarily the same until 1907, when the first synthetic detergent was developed by Henkel. By the end of World War I, detergents had grown in popularity as people learned that they did not leave soap scum like their earlier counterparts.

The earliest synthetic detergents were short-chain alkyl naphthalene sulfonates. By the 1930's sulfonated straight-chain alcohols and long-chain alkyl and aryl sulfonates with benzene were being made. By the end of World War II, alkyl aryl sulfonates dominated the detergent market. In 1946, the use of phosphate compounds in combination with surfactants was a breakthrough in product development and spawned the first of the "built" detergents that would prove to be much better-performing products. Sodium tripolyphosphate (STPP) was the main cleaning agent in many detergents and household cleaners for decades. By 1953, U.S. sales of detergents had surpassed those of soap. In the mid-1960's, it was discovered that lakes and streams were being polluted, and the blame was laid on phosphate compounds; however, the actual cause was found to be branching in their molecular structure, which prevented them from being degraded by bacteria. Detergent manufacturers then switched from commonly used compounds such as propylene tetramer benzene sulphonate to a linear alkyl version. Detergent manufacturers are still grappling with sustainability issues and are focusing much attention on developing products that are safe for the environment as well as consumers.

How It Works

Just as forces exist between an ordered and disordered universe, so too do they between soil and cleanliness. People have been conditioned to believe that soil on the surface of an object is unwanted matter, but in reality, soil is being deposited continuously on all surfaces around us, and cleanliness itself is an unnatural, albeit desirable, state. In order to rid any surface of soil, one must work against nature and have an understanding of the concept of detergency—the act of cleaning soil from a surface (substrate).

The Function of Detergency. The cleaning action of detergents is based on their ability to emulsify or disperse different types of soil and hold it in suspension in water. The workhorse involved in this job is the surfactant, a compound used in all soaps and detergents. This ability comes from the surfactant's molecular structure and surface activity. When a soap or detergent product is added to water that contains insoluble materials like dirt, oil, or grease, surfactant molecules adsorb onto the substrate (clothes) and form clusters called micelles, which surround the immiscible droplets. The micelle itself is water soluble and allows the trapped oil droplets to be dispersed throughout the water and rinsed away. While this is a simplified explanation, detergency is a complex set of interrelated functions that relies on the diverse properties of surfactants, their interactions in solution, and their unique ability to disrupt the surface tension of water.

Surface Tension. The internal attraction or association of molecules in a liquid is called surface tension. However simple this may seem, it is a complex phenomenon, and for many students can be hard to grasp. Examining the properties of water and the action of surfactants may dispel any confusion.

Water is polar in nature and very strongly associated, such that the surface tension is high. This is because of its nonsymmetrical structure, in which the double-atom oxygen end is more negative than the single-atom hydrogen end is positive. As a result, water molecules associate so strongly that they are relatively stable, with only a slight tendency to ionize or split into oppositely charged particles. This is why their boiling point and heat of vaporization are very high in comparison to their low molecular weight.

The surface tension of water can be explained by how the molecules associate. Water molecules in the liquid state, such as those in the center of a beaker full of water, are very strongly attracted to their neighboring molecules, and the pull is equal in all directions. The molecules on the surface, however, have no neighboring molecules in the air above; hence, they are directed inward and pulled into the bulk of the liquid where the attraction is greater. The result is a force applied across the surface, which contracts as the water seeks the minimum surface area per unit of volume. An illustration of this is the fact that one can spin a pail of water around without spilling the contents. Surfactant compounds are amphiphilic, meaning their backbone contains at least one hydrophilic group attached to a hydrophobic group (called the hydrophobe), usually consisting of an 8 to 18 carbon-hydrocarbon chain. All surfactants possess the common property of lowering surface tension when added to water in small amounts, at which point the surfactant molecules are loosely integrated into the water structure. As they disperse, the hydrophilic portion of the surfactant causes an increased attraction of water molecules at the surface, leaving fewer sides of the molecule oriented toward the bulk of the liquid and lessening the forces of attraction that would otherwise pull them into solution.

Micelle Formation and Critical Micelle Concentration (CMC). As surface active agents, surfactants not only have the ability to lower surface tension but also to form micelles in solution, unique behavior that is at the core of detergent action. Micelles are aggregate or droplet-like clusters of individual surfactant molecules, whose polar ends are on the outside, oriented toward the solvent (usually water), with the nonpolar ends in the middle. The driving force for micelle formation is the reduction of contact between the hydrocarbon chain and water, thereby reducing the free energy of the system. The micelles are in dynamic equilibrium, but the rate of exchange between surfactant molecule and micelle increase exponentially, depending on the structure of each individual surfactant. As surfactant concentration increases, surface tension decreases rapidly, and micelles proliferate and form larger units.

The concentration at which this phenomenon occurs is known as the critical micelle concentration (CMC). The most common technique for measuring this is to plot surface tension against surfactant concentration and determine the break point, after which surface tension remains virtually constant with additional increases in concentration. The corresponding surfactant concentration at this discontinuity point corresponds to the CMC. Every surfactant has its own characteristic CMC at a given temperature and electrolyte concentration.

Fascinating Facts About Detergents

- According to biblical accounts, the ancient Israelites created a hair gel by mixing ashes and oil.
- Persil was the name of the first detergent made in 1907 by Germany's Henkel.
- Soap scum became immortalized with the ad campaign that coined the terms "ring around the tub" and "ring around the collar."
- Automatic dishwasher powders and liquid fabric softeners were invented in the 1950's, laundry powders with enzymes in the 1960's, and fabric softener sheets in the 1970's.
- Dawn dishwashing liquid was used to clean wildlife during the Exxon Valdez and BP oil spills in 1989 and 2010, respectively.
- The human lung contains a surface-active material called pulmonary surfactant that helps prevent the lung from collapsing after expiration.
- Enzymes used in detergents work much like they do in the body, since each has a personalized target soil it breaks down.

Applications and Products

The workhorse of detergents is the surfactant or more commonly, a mix of surfactants. The most important categories are the carboxylates (fatty acid soaps), the sulfates (alkyl sulfates, alkyl ether sulfates, and alkyl aryl sulfonates), and the phosphates.

Laundry Products. The primary purpose of laundry products is the removal of soil from fabrics. As the cleaning agent, the detergent must fulfill three functions: wet the substrate, remove the soil, and prevent it from redepositing. This usually requires a mix of surfactants. For example, a good wetting agent is not necessarily a good detergent. For best wetting, the surface tension need only be lowered a little, but it must be done quickly. That requires surfactants with short alkyl chain lengths of 8 carbons and surfactants with an HLB of 7 to 9. For best detergency, the surface tension needs to be substantially lowered and that requires surfactants with higher chain lengths of 12 to 14 carbons and an HLB of 13 to 15. To prevent particles from redepositing, the particles must be stabilized in a solution, and that is done best by nonionic surfactants of the polyethylene oxide type. In general, nonionics are not as effective in removing dirt as anionic surfactants, which is one reason why a mixture of anionic and nonionic surfactants is used. However, nonionics are more effective in liquid dirt removal because they lower the oil-water interfacial tension without reducing the oil-substrate tension.

Skin-Cleansing Bars. As of 2011, the bar soap being manufactured is called superfatted soap and is made by incomplete saponification, an improved process over the traditional method in which superfatting agents are added during saponification, which prevents all of the oil or fat from being processed. Superfatting increases the moisturizing properties and makes the product less irritating. Transparent soaps are like traditional soap bars but have had glycerin added. Glycerin is a humectant (similar to a moisturizer) and makes the bar transparent and much milder.

Syndet bars are made using synthetic surfactants. Since they are not made by saponification, they are actually not soap. Syndet bars are very mild on the skin and provide moisturizing and other benefits. Dove was the first syndet bar on the market.

Mining and Mineral Processing. Because minerals are rarely found pure in nature, the desired material, called values, needs to be separated from the rocky, unwanted material, called gangue. Detergents are used to extract metals from their ores by a process called froth flotation. The ore is first crushed and then treated with a fatty material (usually an oil) which binds to the particles of the metal (the values), but not to the unwanted gangue. The treated ore is submerged in a water bath containing a detergent and then air is pumped over the sides. The detergent's foaming action produces bubbles, which pick up the water-repellant coated particles or values, letting them float to the top, flow over the sides, and be recovered. The gangue stays in the water bath.

Enhanced Oil Recovery (EOR). This process refers to the recovery of oil that is left behind after primary and secondary recovery methods have either been exhausted or have become uneconomical. Enhanced oil recovery is the tertiary recovery phase in which surfactant-polymer (SP) flooding is used. SP flooding is similar to waterflooding, but the water is mixed with a surfactant-polymer compound. The surfactant literally cleans the oil off the rock and the polymer spreads the flow through more of the rock. An additional 15 to 25 percent of original oil in place (OOIP) can be recovered. Before this method is used, there is a great deal of evaluation and laboratory testing involved, but it has become a reliable and cost-effective method of oil recovery.

Ceramic Dispersions. Ceramic is a nonmetallic inorganic material. Ceramic dispersions are the starting material for many applications. The use of detergents or surfactants enhances the wetting ability of the binder onto the ceramic particles and aids in the dispersion of ceramic powders in liquids. As dispersants, they reduce bulk viscosity of high-solid slurries and maintain stability in finely divided particle dispersions. Bi-block surfactants help agglomeration of the ceramic particles. In wastewater treatment, detergents are used in ceramic dispersions to reduce the amount of flocculents.

Impact on Industry

Detergents are big business, not only in the United States, but around the world, and the industry and business sectors catering to it are expansive. These include the laundry and household products market; the industrial and institutional (I&I) market, which makes heavy-duty disinfectants, sanitizers, and cleaners; pharmaceuticals; the oil industry; food service and other service industries, such as hospitals and hospitality (hotels), as well as the personal-care industry.

The Global Picture. Worldwide, the detergent industry is estimated to be around $65 billion, $52 billion of which is the laundry-care category. Despite global cleaning products being a multibillion-dollar business, the global household products industry has seen slow growth over the last few years. The global I&I market is reported to be $30 billion.

Worldwide, the most growth for the detergent, household, and I&I sectors is expected to be in emerging markets, with China, India, and Brazil leading the pack. The three countries have been in the midst of a construction boom. China, for instance, has a new alkyl polyglucoside plant that is in full production. In contrast to the domestic market's slow-growth mode, experts say companies in these countries are seeing double-digit sales figures, while sales in other countries outside U.S. borders are rising by 5 to 10 percent. The consensus is that growth in the United States, Canada, Japan, and Western Europe will be slow compared with world averages. Some analysts predict that the growth in emerging markets will be explosive.

The Local Picture. The domestic market for detergents is not so rosy, say analysts. Market data for 2010 was mixed, as many companies are still feeling the pinch from the 2007-2009 recession. U.S. sales of liquid laundry detergents in food, drug, and mass merchant (FDMx) outlets (excluding Walmart) fell to $3.6 billion, a 3.14 percent drop from the previous year. However, growth in I&I as a whole rose 1.7 percent, amounting to about $11 billion. Projection data indicate that demand for disinfectants and sanitizers through 2014 will increase by more than 6 percent, outpacing the industry's overall growth for the same period. In the same vein, the personal-care sector saw FDMx sales of personal cleansers rise 22.22 percent, an impressive jump to nearly $143 million. A 2010 observational study sponsored by the American Society for Microbiology and the American Cleaning Institute (formerly the Soap and Detergent Association), indicates that 85 percent of adults now wash their hands in public restrooms—the highest percentage observed since studies began in 1996. FDMx soap sales reached $2.08 billion in 2010, which represents a 5 percent increase over the previous year.

The Supplier Side. Surfactants are the primary ingredient in detergents and are made from two different types of raw materials. Those derived from agricultural feedstocks are called oleochemicals, and those derived from petroleum (crude oil), synthetic surfactants. The global market for surfactants by volume size is more than 18 million tons per year, with 80 percent of the demand represented by just ten different types of compounds. The global share of surfactant raw materials represents only about 0.1 percent of crude-oil consumption.

Major Manufacturers. The top laundry detergent manufacturers in the United States are Procter & Gamble (P&G), Henkel, and Unilever, until 2008, when it sold its North American detergent business to Vestar Capital Partners, saying it could not compete with P&G. The former All, Wisk, Sunlight, Surf, and Snuggle brands are now part of a new company, the Sun Products Corporation. Vestar said Sun Products will have annual sales of more than $2 billion. Unilever NV of the Netherlands is still the largest detergent manufacturer in the United Kingdom.

P&G's detergent sales totaled $79 billion in 2010 and represent 70 percent of the company's annual sales. Tide continues to dominate the laundry detergent market, commanding more than 40 percent of sales and is the standout product of P&G's $1 Billion Club. Purex Complete was a big growth product for Henkel in the United States, but not in Europe. Besides the Purex brands, Henkel also markets Dial soaps, a brand it acquired in 2004. Henkel's worldwide sales grew 11 percent to $20.6 billion in 2010, but had stagnant sales in its U.S. division.

In the liquid laundry detergent category, the top five products by sales volume for the year ending October 31, 2010, were Tide (P&G), All (Sun Products), Arm & Hammer (Church & Dwight), Gain (P&G), and Purex (Henkel). Although ranked number eight, Cheer Brightclean (P&G) had the biggest jump in sales, reported at 41 percent. Reacting to the recession, P&G reduced the price of its line of laundry detergents by about 4.5 percent during 2010.

In the United States, Ecolab dominates the I&I sector. With a market share of more than 30 percent, it easily outdoes the number-two player, Diversey. Spartan Chemical is another big contender. Dial (Henkel), Purell (Gojo), and Gold Bond are top-selling brands in the personal cleanser and sanitizer market. Overall, in the personal-care market, private-label brands are by far the leader at mass merchandisers, with sales of $73.9 million in 2009.

Careers and Course Work

Science courses in the organic, inorganic, and physical chemistries, biology, and biochemistry, plus courses in calculus, physics, materials science, polymer technology, and analytical chemistry are typical requirements for students interested in pursuing careers in detergents. Other pertinent courses may include differential equations, instrumental analysis, statistics, thermodynamics, fluid mechanics, process design, quantitative analysis, and instrumental methods. Earning a bachelor of science degree in chemistry or chemical engineering is usually sufficient for entering the field or doing graduate work in a related area.

There are few degree programs in formulation chemistry, but students need to understand the chemistry involved, such as thermodynamics of mixing, phase equilibria, solutions, surface chemistry, colloids, emulsions, and suspensions. Even more important is how their dynamics affect such properties as adhesion, weather resistance, texture, shelf life, biodegradability, and allergenic response.

Students in undergraduate chemistry programs are encouraged to select a specialized degree track but are also being advised to take substantial course work in more than one of the primary fields of study related to detergent chemistry because product development requires skills drawn from multiple disciplines. Often a master's degree or doctorate is necessary for research and development.

There are a number of career paths for students interested in the detergent industry. Manufacturers of laundry detergents, household cleaning products, industrial and institutional (I&I) products, and personal-care products, as well as the ingredients suppliers of all these products, are the biggest employers of formulation chemists, technicians, and chemical operators. There are also career opportunities in research and development, marketing, or sales. Other areas where detergent chemists and technicians are needed include food technology, pharmaceuticals, oil drilling and recovery, mining and mineral processing, and ceramic powder production.

Social Context and Future Prospects

Around the world, sustainability and environmental protection remain the buzzwords for detergents in the twenty-first century. The industry is very dependent on the price and availability of fats and oils, since these materials are needed to make the fatty acids and alcohols used in the manufacture of surfactants. The turmoil in the oil industry is not only impacting how much is paid at the gas pump but is also directly related to the price paid to keep the environment clean.

While the detergents industry, on the whole, has traditionally been relatively recession-proof, the economic slowdown and conflicts in the Middle East have taken a toll on this huge market. Robert McDonald, chairman of P&G, alluded to changes in consumer habits, volatility in commodity costs, and increasing complexity of the regulatory environment as rationales for the slowdown. A spokesperson for Kline and Company, a market research firm reporting on the industry, stated: "The industry's mantra is greener, cleaner, safer," but goes on to say that the challenge is a double-edge sword, as the industry is battling to hold down costs while trying to produce environmentally sustainable products.

Sustainability and environmental protection are global issues, but insights are becoming more astute and action is being taken. Henkel published its twentieth sustainability report in conjunction with its annual report for 2010, saying the company met its targets for 2012 early, namely to commit to principles and objectives relating to occupational health and safety, resource conservation, and emissions reduction. Important emerging markets such as China are no exception in this battle. The Chinese government puts great emphasis on environmental regulations. Analysts say this is a clear message that to expand their sales to other parts of the world, China needs to hone in on the green demand and manufacture products that offer innovative and sustainable solutions without compromising on performance.

Barbara Woldin, B.S.

Further Reading

Carter, C. Barry, and M. Grant Norton. *Ceramic Materials: Science and Engineering*. New York: Springer, 2007. Covers ceramic science, defects, and the mechanical properties of ceramic materials and how these materials are processed. Provides many examples and illustrations relating theory to practical applications; suitable for advanced undergraduate and graduate study.

Myers, Drew. *Surfactant Science and Technology*. 3d ed. Hoboken, N.J.: John Wiley & Sons, 2006. Written with the beginner in mind, this text clearly illustrates the basic concepts of surfactant action and application.

Rosen, Milton J. *Surfactants and Interfacial Phenomena*. 3d ed. Hoboken, N.J.: Wiley-Interscience, 2004. Easy-to-understand text on properties and applications of surfactants; covers many topics, including dynamic surface tension and other interfacial processes.

Tadros, Tharwat F. *Applied Surfactants: Principles and Applications*. Weinheim, Germany: Wiley-VCH Verlag,

2005. Author covers a wide range of topics on the preparation and stabilization of emulsion systems and highlights the importance of emulsion science in many modern-day industrial applications; discusses physical chemistry of emulsion systems, adsorption of surfactants at liquid/liquid interfaces, emulsifier selection, polymeric surfactants, and more.

Zoller, Uri, and Paul Sosis, eds. *Handbook of Detergents, Part F: Production.* Boca Raton, Fla.: CRC Press, 2009. One of seven in the Surfactant Science Series, the book discusses state of the art in the industrial production of the main players in detergent formulation—surfactants, builders, auxiliaries, bleaching ingredients, chelating agents, and enzymes.

Web Sites

American Chemical Society
http://portal.acs.org/portal/acs/corg/content

American Chemistry Council
http://www.americanchemistry.com

American Oil Chemists' Society
http://www.aocs.org

Household and Personal Products Industry
http://www.happi.com

DIODE TECHNOLOGY

FIELDS OF STUDY

Electrical engineering; materials science; semiconductor technology; semiconductor manufacturing; electronics; physics; chemistry; nanotechnology; mathematics.

SUMMARY

Diodes act as one-way valves in electrical circuits, permitting electrical current to flow in only one direction and blocking current flow in the opposite direction. The original diodes used in circuits were constructed using vacuum tubes, but these diodes have been almost completely replaced by semiconductor-based diodes. Solid-state diodes, the most commonly used, are perhaps the simplest and most fundamental solid-state semiconductor devices, formed by joining two different types of semiconductors. Diodes have many applications, such as safety circuits to prevent damage by inadvertently putting batteries backward into devices and in rectifier circuits to produce direct current (DC) voltage output from an alternating current (AC) input.

KEY TERMS AND CONCEPTS

- **Anode:** More positive side of the diode, through which current can flow into a diode while forward biased.
- **Cathode:** More negative side of the diode, from which current flows out of the diode while forward biased.
- **Forward Bias:** Orientation of the diode in which current most easily flows through the diode.
- **Knee Voltage:** Minimum forward bias voltage required for current flow, also sometimes called the threshold voltage, cut-in voltage, or forward voltage drop.
- **Light-Emitting Diode (LED):** Diode that emits light when current passes through it in forward bias configuration.
- **P-N Junction:** Junction between positive type (p-type) and negative type (n-type) doped semiconductors.
- **Power Dissipation:** Amount of energy per unit time dissipated in the diode.
- **Rectifiers:** Diode used to make alternating current, or current flowing in two directions, into direct current, or current flowing in only one direction.
- **Reverse Bias:** Orientation of the diode in which current does not easily flow through the diode.
- **Reverse Breakdown Voltage:** Maximum reverse biased voltage that can be applied to the diode without it conducting electrical current, sometimes just called the breakdown voltage.
- **Thermionic Diode:** Vacuum tube that permits current to flow in only one direction, also called a vacuum tube diode.
- **Zener Diode:** Diode designed to operate in reverse bias mode, conducting current at a controlled breakdown voltage called the Zener voltage.

Definition and Basic Principles

A diode is perhaps the first semiconductor circuit element that a student learns about in electronics courses, though most early diodes were constructed using vacuum tubes. It is very simplistic in structure, and basic diodes are very simple to connect in circuits. They have only two terminals, a cathode and an anode. The very name diode was created by British physicist William Henry Eccles in 1919 to describe the circuit element as having only the two terminals, one in and one out.

Classic diode behavior, that for which most diodes are used, is to permit electric current to flow in only one direction. If voltage is applied in one direction across the diode, then current flows. This is called forward bias. The terminal on the diode into which the current flows is called the anode, and the terminal out of which current flows is called the cathode. However, if voltage is applied in the opposite direction, called reversed bias, then the diode prevents current flow. A theoretical ideal diode permits current to flow without loss in forward bias orientation for any voltage and prohibits current flow in reverse bias orientation for any voltage. Real diodes require a very small forward bias voltage in order for current to flow, called the knee voltage. The terms threshold voltage or cut-in voltage are also sometimes used in place of the term knee voltage. The electronic symbol for the diode signifies the classic diode behavior, with an arrow pointing in the direction of permitted current flow, and a bar on the other side of the diode signifying a block to current flow from the other direction.

Though most diodes are used to control the direction of current flow, there are many subtypes of diodes that have been developed with other useful properties, such as light-emitting diodes and even diodes designed to operate in reverse bias mode to provide a regulated voltage.

Background and History

Diode-like behavior was first observed in the nineteenth century. Working independently of each other in the 1870's, American inventors Thomas Alva Edison and Frederick Guthrie discovered that heating a negatively charged electrode in a vacuum permits current to flow through the vacuum but that heating a positively charged electrode does not produce the same behavior. Such behavior was only a scientific curiosity at the time, since there was no practical use for such a device.

At about the same time, German physicist Karl Ferdinand Braun discovered that certain naturally occurring electrically conducting crystals would conduct electricity in only one direction if they were connected to an electrical circuit by a tiny electrode connected to the crystal in just the right spot. By 1903, American electrical engineer Greenleaf Whittier Pickard had developed a method of detecting radio signals using the one-way crystals. By the middle of the twentieth century, homemade radio receivers using galena crystals had become quite popular among hobbyists.

As the electronics and the radio communication industries developed, it became apparent that there would be a need for human-made diodes to replace the natural crystals that were used in a trial-and-error manner. Two development paths were followed: solid-state diodes and vacuum tube diodes. By the middle of the twentieth century, inexpensive germanium-based diodes had been developed as solid-state devices. The problem with solid-state diodes was that they lacked the ability to handle large currents, so for high-current applications, vacuum tube diodes, or thermionic diodes, were developed. In the twenty-first century, most diodes are semiconductor devices, with thermionic diodes existing only for the rare very high-power applications.

How It Works

Thermionic Diodes. Though not used as frequently as they once were, thermionic diodes are the simplest type of diode to understand. Two electrodes are enclosed in an evacuated glass tube. Because the thermionic diode is a type of vacuum tube, it is often called a vacuum tube diode. The geometry of the electrodes in the tube depends on the manufacturer and the intended use of the tube. Heating one of the electrodes in some fashion permits electrons on that electrode to be thermally excited. If the electrode is heated past the work function of the material of which the electrode is fabricated, the electrons can come free of the electrode. If the heated electrode has a more negative voltage than the other electrode, then the electrons cross the space between the electrodes. More electrons flow into the negative electrode to replace the missing ones, and the electrons flow out of the positive electrode. Current flow is defined opposite to electron flow, so current would be defined as flowing into the positive electrode (labeled as the anode) and out of the negative electrode (labeled as the cathode). However, if the voltage is reversed, and the heated electrode is more positive than the other electrode, electrons liberated from the anode do not flow to the cathode, so no current flows, making the diode a one-way device for current flow.

Solid-State Diodes. Thermionic diodes, or vacuum tube diodes, tend to be large and consume a lot of electricity. However, paralleling the development of vacuum tube diodes was the development of diodes based on the crystal structure of solids. The most important type of solid-state diodes are based on semiconductor technology.

Semiconductors are neither good conductors nor good insulators. The purity of the semiconductor determines, in part, its electrical properties. Extremely pure semiconductors tend to be poor conductors. However, all semiconductors have some impurities in them, and some of those impurities tend to improve conductivity of the semiconductor. Purposely adding impurities of the proper type and concentration into the semiconductor during the manufacturing process is called doping the semiconductor. If the impurity has one more outer shell electron than the number of electrons in atoms of the semiconductor, then extra electrons are available to move and conduct electricity. This is called a negative doped or n-type semiconductor. If the impurity has one fewer electrons than the atoms of the semiconductor, then electrons can move from one atom to another in the semiconductor. This acts as a positive charge moving in the semiconductor, though it is really a missing electron moving from atom to atom. Electrical engineers refer to this as a hole moving in the semiconductor. Semiconductors with this type of impurity are called positive doped or p-type semiconductor.

What makes a semiconductor diode is fabricating a device in which a p-type semiconductor is in contact with an n-type semiconductor. This is called a p-n junction. At the junction, the electrons from the n-type region combine with the holes of the p-type region, resulting in a depletion of charge carriers in the vicinity of the p-n junction. However, if a small positive voltage is applied across the junction, with the p-type region having the higher voltage, then additional electrons are pulled from the n-type region and additional holes are pulled from the p-type region into the depletion region, with electrons flowing into the n-type region from outside the device to make up the difference and out of the device from the p-type region to produce more holes. As with the thermionic device, current flows through the device, with the p-type side of the device being the anode and the n-type side of the device being the cathode. This is the forward bias orientation. When the voltage is reversed on the device, the depletion region simply grows larger and no current flows, so the device acts as a one-way valve for the flow of electricity. This is the reverse bias orientation. Though reverse bias diodes do not normally conduct electricity, a sufficiently high reverse voltage can create electric fields within the diode capable of moving charges through the depletion region and creating a large current through the diode. Because diodes act much like resistors in reverse bias mode, such a large current through the diode can damage or destroy the diode. However, two types of diodes, avalanche diodes and Zener diodes, are designed to be safely operated in reverse bias mode.

Fascinating Facts About Diode Technology

- The term diode comes from the Greek *di* (two) and *ode* (paths) signifying the two possible ways of connecting diodes in circuits.
- Naturally occurring crystals with diode-like properties were used in amateur crystal radio sets purchased by millions of hobbyists in the middle of the twentieth century.
- Despite their widely known property of allowing electricity to pass in only one direction, some diodes, such as Zener diodes, are made to be operated in reverse bias mode.
- Light-emitting diodes (LEDs) emit light only when current passes through them in the right direction.
- The laser light produced in laser pointers is made using laser diodes.
- Diodes are often used in battery-operated devices to protect the electronics in the device from users accidentally inserting batteries backward.
- Diodes are used with automobile alternators to produce the DC voltage required for automobile electrical systems.

APPLICATIONS AND PRODUCTS

P-n junction devices, such as diodes, have a plethora of uses in modern technology.

Rectifiers. The classic application for a diode was to act as a one-way valve for electric current. Such a property makes diodes ideal for use in converting alternating current into direct-current circuits or circuits in which the current flows in only one direction. In fact, the devices were originally called rectifiers before the term diode was created to describe the function of these one-way current devices. Modern rectifier circuits consist of more than just a single diode, but they still rely heavily on diode properties.

Solid-state diodes, like most electronic components, are not 100 percent efficient, and so some energy is lost in their operation. This energy is typically dissipated in the diode as heat. However, semiconductor devices are designed to operate at only certain temperatures, and increasing the temperature beyond a specified range changes the electrical properties of the device. The more current that passed through the device, the hotter it gets. Thus, there is a limiting current that a solid-state diode can handle before it is damaged. Though solid-state diodes have been developed to handle higher currents, for the highest current and power situations, thermionic, or vacuum tube diodes, are still sometimes used, particularly in radio and television broadcasting.

Shottky Diodes. All diodes require at least a small forward bias voltage in order to work. Shottky diodes are fabricated by using a metal-to-semiconductor junction rather than the traditional dual semiconductor p-n junction used with other diodes. Such a construction allows Shottky diodes to operate with extremely low forward bias.

Zener Diodes. Though most diodes are designed to operate only in the forward bias orientation, Zener diodes are designed to operate in reverse bias mode. In such an orientation, they undergo a breakdown and conduct electric current in the reverse direction with a well-defined reverse voltage. Zener diodes are used to provide a stable and well-defined reference voltage.

Photodiodes. Operated in reverse bias mode, some p-n junctions conduct electricity when light shines on them. Such diodes can be used to detect and measure light intensity, since the more light that strikes the diodes, the more they conduct electricity.

Circuit Protection. In most applications of diodes, they are used to take advantage of the properties of the p-n junction on a regular basis in circuits. For some applications, though, diodes are included in circuits in the hope that they will never be needed. One such application is for DC circuits, which are typically designed for current to flow in only one direction. This is automatically accomplished through a power supply with a particular voltage orientation such as a DC source, power converter, or battery. However, if the power supply were connected in reverse or if the batteries were inserted backward, then damage to the circuit could result. Diodes are often used to prevent current flow in such situations where voltage is applied in reverse, acting as a simple but effective reverse voltage protection system.

Light-Emitting Diodes (LEDs). For diodes with just the right kind of semiconductor and doping, the combination of holes and electrons at the p-n junction releases energy equal to that carried by photons of light. Thus, when current flows through these diodes in forward bias mode, the diodes emit light. Unlike most lighting sources, which produce a great deal of waste heat in addition to light (with incandescent lights often using energy to produce more heat than visible light), most of the energy dissipated in LEDs goes into light, making them far more energy-efficient light sources than most other forms of artificial lighting. Unfortunately, large high-power applications of light-emitting diodes are somewhat expensive, limiting them to uses where their small size and long life characteristics offset the cost associated with other forms of lighting.

Laser Diodes. Very similar to light-emitting diodes are laser diodes, where the recombination of holes and electrons also produces light. However, with the laser diode, the p-n junction is placed inside a resonant cavity and the light produced stimulates more light, producing coherent laser light. Laser diodes typically have much shorter operational lifetimes than other diodes, including LEDs, and are generally much more expensive. However, laser diodes cost much less than other methods of producing laser light, so they have become more common. Most lasers not requiring high-power application are based on laser diodes.

Impact on Industry

The control of electric current is fundamental to electronics. Diodes, acting like one-way valves for electric current are, therefore, very important circuit elements in circuits. They can act as rectifiers, converting alternating current into direct current, but they have many other uses.

Government and University Research. Though the basic concept of the p-n junction is understood, research continues into new applications for the junction. This research is conducted in both public and private laboratories. Much research goes into an effort to make diodes smaller, cheaper, and capable of higher power applications.

Industry and Business. Most diodes are used in fabrication of devices. They are used in circuits and are part of almost every electronic device and will likely continue to play an important role in electronics. Specialized diodes, such as LEDs or laser diodes, have come to be far more important for their specialized properties than for the their current directionality. Laser diodes produce laser light inexpensively enough to permit the widespread use of lasers in many devices that would otherwise not be practical, such as DVD players. Laser diodes are easily modulated, so a great deal of fiber-optic communications use laser diodes. LEDs produce light much more efficiently than most other means of producing light. Many companies are working on developing LED lighting systems to replace more conventional lights.

Many uses of diodes exist outside of the traditional electronics industry. Diodes are, of course, used in electronic devices, but diodes are also used in automobiles, washing machines, and clocks. Almost all electronic devices use diodes, from televisions and radios to microwave ovens and DVD players.

Thermionic diodes are used in only a few industries, such as the power industry and radio and television broadcasting, where the extremely high-power applications would burn out semiconductor diodes. Semiconductor diodes are used in all other applications, since they are more efficient to operate and less expensive to produce.

CAREERS AND COURSE WORK

The electronics field is vast and encompasses a wide variety of careers. Diodes exist in some form in most electronic devices. Thus, a wide range of careers come into contact with diodes, and therefore a wide range of background knowledge and preparation exists for the different careers.

Development of new types of diodes requires considerable knowledge of solid-state physics, materials science, and semiconductor manufacturing. Often advanced degrees in these fields would be required for research, necessitating students studying physics, electrical engineering, mathematics, and chemistry. However, diode technology is quite well evolved, so there are limited job prospects for developing new diodes or diode-like devices other than academic curiosity. Most of this area of study is simply determining how to manufacture or include smaller diodes in integrated circuits.

Electronics technicians repair electrical circuits containing diodes. So, knowledge of diodes and diode behavior is important in diagnosing failures in electronic circuits and circuit boards. Sufficient knowledge can be gained in basic electronics courses. A two-year degree in electronic technology is sufficient for many such jobs, though some jobs may require a bachelor's degree. Likewise, technicians designing and building circuits often do not need to know much about the physics of diodes—just the nature of diode behavior in circuits. Such knowledge can be gained through basic electronics courses or an associate's or bachelor's degree in electronics.

Manufacturing diodes does not actually require much knowledge about diodes for technicians who are actually making semiconductor devices. Such technicians need course work and training in operating the equipment used to manufacture semiconductors and semiconductor devices, and they must be able to follow directions meticulously in operating the machines. An associate's degree in semiconductor manufacturing is often sufficient for many such jobs. Manufacturing circuit boards with diodes, or any other circuit element, does not really require much knowledge of the circuit elements themselves, save for the ability to identify them by sight, though it would be helpful to understand basic diode behavior. Basic course work in circuits would be needed for such jobs.

SOCIAL CONTEXT AND FUTURE PROSPECTS

Diodes exist in almost every electronic device, though most people do not realize that they are using diodes. Because electronics have been increasing in use in everyday life, diodes and diode technology will continue to play an important role in everyday devices. Diodes are very simple devices, however, and it is unlikely that the field will advance further in the development of basic diodes. Specialized devices using the properties of p-n junctions, such as laser diodes, continue to be important. It can be anticipated that additional uses of p-n junctions may be discovered and new types of diodes developed accordingly. Because the p-n junction is the basis of diode behavior and is the basis of semiconductor technology, diodes will continue to play an important role in electronics for the foreseeable future. LEDs produce light very efficiently, and work is proceeding to investigate the possibility of such devices replacing many other forms of lighting.

Raymond D. Benge, Jr., B.S., M.S.

FURTHER READING

Gibilisco, Stan. *Teach Yourself Electricity and Electronics.* 5th ed. New York: McGraw-Hill, 2011. Comprehensive introduction to electronics, with diagrams. A chapter on semiconductors includes a good description of the physics and use of diodes.

Held, Gilbert. *Introduction to Light Emitting Diode Technology and Applications.* Boca Raton, Fla.: Auerbach, 2009. A thorough overview of light-emitting diodes and their uses. The book also includes a good description of how diodes in general work.

Paynter, Robert T. *Introductory Electronic Devices and Circuits.* 7th ed. Upper Saddle River, N.J.: Prentice Hall, 2006. An excellent and frequently used introductory electronics textbook, with an excellent description of diodes, different diode types, and their use in circuits.

Razeghi, Manijeh. *Fundamentals of Solid State Engineering*. 3d ed. New York: Springer, 2009. An advanced undergraduate textbook on the physics of semiconductors, with a very detailed explanation of the physics of the p-n junction.

Schubert, E. Fred. *Light Emitting Diodes*. 2d ed. New York: Cambridge University Press, 2006. A very good and thorough overview of light-emitting diodes and their uses.

Turley, Jim. *The Essential Guide to Semiconductors*. Upper Saddle River, N.J.: Prentice Hall, 2003. A brief overview of the semiconductor industry and semiconductor manufacturing for the beginner.

Web Sites

The Photonics Society
http://photonicssociety.org

Schottkey Diode Flash Tutorial
http://cleanroom.byu.edu/schottky_animation.phtml

Semiconductor Industry Association
http://www.sia-online.org

University of Cambridge
Interactive Explanation of Semiconductor Diode
http://www-g.eng.cam.ac.uk/mmg/teaching/linearcircuits/diode.html

See also: Light-Emitting Diodes

DISTILLATION

FIELDS OF STUDY

Chemistry; chemical engineering; industrial studies; chemical hygiene; environmental chemistry.

SUMMARY

Distillation is a process for purifying liquid mixtures by collecting vapors from the boiling substance and condensing them back into the original liquid. Various forms of this technique, practiced since antiquity, continue to be used extensively in the petroleum, petrochemical, coal tar, chemical, and pharmaceutical industries to separate mixtures of mostly organic compounds as well as to isolate individual components in chemically pure form. Distillation has also been employed to acquire chemically pure water, including potable water through the desalination of seawater.

KEY TERMS AND CONCEPTS

- **Condensation:** Phase transition in which gas is converted into liquid.
- **Distilland:** Liquid mixture being distilled.
- **Distillate:** Product collected during distillation.
- **Forerun:** Small amount of low-boiling material discarded at the beginning of a distillation.
- **Fraction:** Portion of distillate with a particular boiling range.
- **Miscible:** Able to be mixed in any proportion without separation of phases.
- **Petrochemical:** Chemical product derived from petroleum.
- **Pot Residue:** Oily material remaining in the boiling flask after distillation.
- **Reflux:** To return condensed vapors (partially or totally) back to the original boiling flask.
- **Reflux Ratio:** Ratio of descending liquid to rising vapor during fractional distillation.
- **Theoretical Plate:** Efficiency of a fractionating column, being equal to the number of vapor-liquid equilibrium stages encountered by the distillate on passing through the column; often expressed as the height equivalent to a theoretical plate (HETP).
- **Vaporization:** Phase transition in which liquid is converted into gas.
- **Vapor Pressure:** Pressure exerted by gas in equilibrium with its liquid phase.

Definition and Basic Principles

Matter commonly exists in one of three physical states: solid, liquid, or gas. Any phase of matter can be changed reversibly into another at a temperature and pressure characteristic of that particular sample. When a liquid is heated to a temperature called the boiling point, it begins to boil and is transformed into a gas. Unlike the melting point of a solid, the boiling point of a liquid is proportional to the applied pressure, increasing at high pressures and decreasing at low pressures.

When a mixture of several miscible liquids is heated, the component with the lowest boiling point is converted to the gaseous phase preferentially over those with higher boiling points, which enriches the vapor with the more volatile component. The distillation operation removes this vapor and condenses it back to the liquid phase in a different receiving flask. Thus, liquids with unequal boiling points can be separated by collecting the condensed vapors sequentially as fractions. Distillation also removes nonvolatile components, which remain behind as a residue.

Background and History

Applications of fundamental concepts such as evaporation, sublimation, and condensation were mentioned by Aristotle and others in antiquity; however, many historians consider distillation to be a discovery of Alexandrian alchemists (300 B.C.E. to 200 C.E.), who added a lid (called the head) to the still and prepared oil of turpentine by distilling pine resin. The Arabians improved the apparatus by cooling the head (now called the alembic) with water, which allowed the isolation of a number of essential oils by distilling plant material and, by 800 C.E., permitted the Islamic scholar Jbir ibn Hayyn to obtain acetic acid from vinegar. Alembic stills and retorts were widely employed by alchemists of medieval Europe. The first fractional distillation was developed by Taddeo Alderotti in the thirteenth century. The first comprehensive manual of distillation techniques was *Liber de arte distillandi, de simplicibus*, by Hieronymus Brunschwig, published in 1500 in Strasbourg, France.

The first account of the destructive distillation of coal was published in 1726. Large-scale continuous stills with fractionating towers similar to modern industrial stills were devised for the distillation of alcoholic beverages in the first half of the nineteenth century and later adapted to coal and oil refining. Laboratory distillation similarly advanced with the introduction of the Liebig condenser around 1850. The modern theory of distillation was developed by Ernest Sorel and reduced to engineering terms in his *Distillation et rectification industrielles* (1899).

How It Works

Simple Distillation. A difference in boiling point of at least 25 degrees Celsius is generally required for successful separations with simple distillation. The glass apparatus for laboratory-scale distillations consists of a round-bottomed boiling flask, a condenser, and a receiving flask. Vapors from the boiling liquid are returned to the liquid state by the cooling action of the condenser and are collected as distillate in the receiving flask. For high-boiling liquids, an air condenser may be sufficient, but often a jacketed condenser—in which a cooling liquid such as cold water is circulated—is required. The design of many styles of condensers (such as Liebig and Wes) enhances the cooling effect of the circulating liquid. An adapter called a still head connects the condenser to the boiling flask at a 45-degree angle and is topped with a fitting in which a thermometer is inserted to measure the temperature of the vapor (the boiling point). A second take-off adapter is often used to attach the receiving flask to the condenser at a 45-degree angle so that it is vertical and parallel to the boiling flask. One should never heat a closed system, so the take-off adapter contains a side-arm for connection to either a drying tube or a source of vacuum for distillations under reduced pressure. The apparatus was formerly assembled by connecting individual pieces with cork or rubber stoppers, but the ground-glass joints of modern glassware make these stoppers unnecessary.

Fractional Distillation. When the boiling points of miscible liquids are within about 25 degrees Celsius, simple distillation does not yield separate fractions. Instead, the process produces a distillate whose composition contains varying amounts of the components, being initially enriched in the lower-boiling and more volatile one. The still assembly is modified to improve efficiency by placing a distilling column between the still head and boiling flask. This promotes multiple cycles of condensation and revaporization. Each of these steps is an equilibration of the liquid and gaseous phases and is, therefore, equivalent to a simple distillation. Thus, the distillate from a single fractional distillation has the composition of one obtained from numerous successive simple distillations. Still heads allowing higher reflux ratios and distilling columns having greater surface area permit more contact between vapor and liquid, which increases the number of equilibrations. Thus, a Vigreux column having a series of protruding fingers is more efficient than a smooth column. Even more efficient are columns packed with glass beads, single- or multiple-turn glass or wire helices, ceramic pieces, copper mesh, or stainless-steel wool. The limit of efficiency is approached by a spinning-band column that contains a very rapidly rotating spiral of metal or Teflon over its entire length.

Applications and Products

Batch and Continuous Distillation. Distilling very large quantities of liquids as a single batch is impractical, so industrial-scale distillations are often conducted by continuously introducing the material to be distilled. Continuous distillation is practiced in petrochemical and coal tar processing and can also be used for the low-temperature separation and purification of liquefied gases such as hydrogen, oxygen, nitrogen, and helium.

Vacuum Distillation. Heating liquids to temperatures above about 150 degrees Celsius is generally avoided to conserve energy, to minimize difficulties of insulating the still head and distilling column, and to prevent the thermal decomposition of heat-sensitive organic compounds. A vacuum distillation takes advantage of the fact that a liquid boils when its vapor pressure equals the external pressure, which causes the boiling point to be lowered when the pressure decreases. For example, the boiling point of water is 100 degrees Celsius at a pressure of 760 millimeters of mercury (mmHg), but this drops to 79 degrees Celsius at 341 mmHg and rises to 120 degrees Celsius at 1,489 mmHg. Vacuum distillation can be applied to solids that melt when heated in the boiling flask; however, higher temperatures may be required in the condenser to prevent the distillate from crystallizing. The term "vacuum distillation" is actually a misnomer, for these distillations are conducted at a reduced pressure rather than under an absolute vacuum. A pressure of about 20 mmHg is obtainable with ordinary water aspirators and down to about 1 mmHg with a laboratory vacuum pump.

Molecular Distillation. When the pressure of residual air in the still is lower than about 0.01 mmHg, the vapor can easily travel from the boiling liquid to the condenser, and distillate is collected at the lowest possible temperature. Distillation under high-vacuum conditions permits the purification of thermally unstable compounds of high molecular weight (such as glyceride fats and natural oils and waxes) that would otherwise decompose at temperatures encountered in an ordinary vacuum distillation. Molecular stills often have a simple design that minimizes refluxing and accelerates condensation. For example, the high-vacuum short-path still consists of two plates, one heated and one cooled, that are separated by a very short distance. Industrially, the distillate can be condensed on a rapidly rotating cone and removed quickly by centrifugal force.

Steam Distillation. Another method to lower boiling temperature is steam distillation. When a homogeneous mixture of two miscible liquids is distilled, the vapor pressure of each liquid is lowered according to Raoult's and Dalton's laws; however, when a heterogeneous mixture of two immiscible liquids is distilled, the boiling point of the mixture is lower than that of its most volatile component because the vapor pressure of each liquid is now independent of the other liquid. A steam distillation occurs when one of these components is water and the other an immiscible organic compound. The steam may be introduced into the boiling flask from an external source or may be generated internally by mixing water with the material to be distilled. Steam distillation is especially useful in isolating the volatile oils of plants.

Azeotropic Extractive Distillation. Certain nonideal solutions of two or more liquids form an azeotrope, which is a constant-boiling mixture whose composition does not change during distillation. Water (boiling point of 100.0 degrees Celsius) and ethanol (boiling point of 78.3 degrees Celsius) form a binary azeotrope (boiling point of 78.2 degrees Celsius) consisting of 4 percent water and 96 percent ethanol. No amount of further distillation will remove the remaining water; however, addition of benzene (boiling point 80.2 degrees Celsius) to this distillate forms a tertiary benzene-water-ethanol azeotrope (boiling point of 64.9 degrees Celsius) that leaves pure ethanol behind when the water is removed. This is an example of azeotropic drying, which is a special case of azeotropic extractive distillation.

Microscale Distillation. Microscale organic chemistry, with a history that spans more than a century, is not a new concept to research scientists; however, the traditional 5- to 100-gram macroscale of student laboratories was reduced one hundred to one thousand times by the introduction of microscale glassware in the 1980's to reduce the risk of fire and explosion, limit exposure to toxic substances, and minimize hazardous waste. Microscale glassware comes in a variety of configurations, such as Mayo-Pike or Williamson styles. Distillation procedures are especially troublesome in microscale because the ratio of wetted-glass surface area to the volume of distillate increases as the sample size is reduced, thereby causing significant loss of product. Specialized microscale glassware such as the Hickman still head has been designed to overcome this difficulty.

Analytical Distillation. The composition of liquid mixtures can be quantitatively determined by weighing the individual fractions collected during a carefully conducted fractional distillation; however, this technique has been largely replaced by instrumental methods such as gas and liquid chromatography.

Impact on Industry

Chemical and Petrochemical Industries. Distillation is one of the fundamental unit operations of chemical engineering and is an integral part of many chemical manufacturing processes. Modern industrial chemistry in the twentieth century was based on the numerous products obtainable from petrochemicals, especially when thermal and catalytic cracking is applied. Industrial distillations are performed in large, vertical distillation towers that are a common sight at chemical and petrochemical plants and petroleum refineries. These range from about 2 to 36 feet in diameter and 20 to 200 feet or more in height. Chemical reaction and separation can be combined in a process called reactive distillation, where the removal of a volatile product is used to shift the equilibrium toward completion.

Petroleum Industry. Distillation is extensively used in the petroleum industry to separate the hydrocarbon components of petroleum. Crude oil is a complex mixture of a great many compounds, so initial refining yields groups of substances within a range of

Fascinating Facts About Distillation

- The word "distill" in the late fourteenth century was originally applied to the separation of alcoholic liquors from fermented materials and comes from the Middle English "distillen," which comes from the Old French *distiller*, which comes from the Late Latin *distillare*, which is an alteration of *destillare* (*de* and *stillare*) meaning "to drip."
- In *Aristotelous peri geneses kai phthoras* (335-323 B.C.E.; *Meteoroligica*, 1812), Aristotle described how potable water was obtained as condensation in a sponge suspended in the neck of a bronze vessel containing boiling seawater.
- Distilled alcoholic beverages first appeared in Europe in the late twelfth century, developed by alchemists.
- The Celsius scale of temperature was originally defined in the eighteenth century by the freezing and boiling points of water at 0 and 100 degrees Celsius, respectively, at a pressure of one standard atmosphere (760 millimeters of mercury).
- The first vertical continuous distillation column was patented in France by Jean-Baptiste Cellier Blumenthal in 1813 for use in alcohol distilleries.
- The first modern book on the fundamentals of distillation, *La Rectification de l'alcool* (*The Rectification of Alcohol*), was published by Ernest Sorel in 1894.
- The gas lights of Sherlock Holmes's London burned coal gas from the destructive distillation of coal.
- In 1931, M. R. Fenske separated two isomeric hydrocarbons that boiled at only 3 degrees Celsius apart using a 52-foot fractionating column mounted in an airshaft at Pond Laboratory of Pennsylvania State College.

boiling points: natural gas below 20 degrees Celsius (C_1 to C_4 hydrocarbons), petroleum ether from 20 to 60 degrees Celsius (C_5 to C_6 hydrocarbons), naphtha (or ligroin) from 60 to 100 degrees Celsius (C_5 to C_9 hydrocarbons), gasoline from 40 to 205 degrees Celsius (C_5 to C_{12} hydrocarbons and cycloalkanes), kerosene from 175 to 325 degrees Celsius (C_{10} to C_{18} hydrocarbons and aromatics), gas oil (or diesel) from 250 to 400 degrees Celsius (C_{12} and higher hydrocarbons). Lubricating oil is distilled at reduced pressure, leaving asphalt behind as a residue

Destructive Distillation. When bituminous or soft coal is heated to high temperatures in an oven that excludes air, destructive distillation converts the coal into coke as gases and liquids form and are distilled over. The coke is chiefly employed in the iron and steel industry, and coal gas is used for heating. The liquid fraction, called coal tar, contains some compounds originally in the coal and others produced by chemical reactions during heating. Coal tar yields numerous organic compounds on repeated distilling and redistilling of the crude fractions. It provided the raw material for the early synthetic organic chemical industry in the nineteenth century. In the early twentieth century, coal-derived chemicals provided gas to light street lamps and were the main source of phenol, toluene, ammonia, and naphthalene during World War I. When wood is similarly heated in a closed vessel, destructive distillation converts the wood into charcoal and yields methyl alcohol (called wood alcohol), together with acetic acid and acetone as liquid distillates.

Essential Oils. Essential oils are generally obtained by steam distillation of plant materials (flowers, leaves, wood, bark, roots, seeds, and peel) and are used in perfumes, flavorings, cosmetics, soap, cleaning products, pharmaceuticals, and solvents. Examples include turpentine and oils of cloves, eucalyptus, lavender, and wintergreen.

Distilled Water and Desalination. Tap water commonly contains dissolved salts that contribute to its overall hardness. These are removed through distillation to provide distilled water for use in automobile batteries and radiators, steam irons, and other applications where pure water is beneficial. Specialized stills fed with continuously flowing tap water are used to prepare distilled water in chemistry laboratories; however, deionized water is becoming an increasingly popular and more convenient alternative. Seawater can be similarly distilled to provide potable drinking water. Desalination of seawater is especially important in arid and semi-arid regions of the Middle East, where the abundant sunshine is used in solar distillation facilities.

CAREERS AND COURSE WORK

The art of distillation is most commonly practiced by chemists whose college majors were chemistry or chemical engineering, the former distilling samples on a laboratory scale and the latter conducting distillations on larger pilot plant and considerably larger industrial scales. The difference between chemistry and chemical engineering majors occurs in advanced and elective course work. Chemistry majors concentrate on molecular structure to better understand the chemical and physical properties of matter, whereas chemical engineering students focus more on the properties of bulk matter involved in the large-scale, economical, and safe manufacture of useful products. Advanced graduate study can be pursued in both disciplines at the master's, and doctoral levels, the latter degree being very common among professors of chemistry and chemical engineering at colleges and universities.

Both chemistry and chemical engineering majors must have strong backgrounds in physics and mathematics and take a year of general chemistry followed by a year of organic chemistry and another year of physical chemistry. The ability to infer the relative boiling points of chemical substances by predicting the strength of intermolecular forces from molecular structure begins in general chemistry and is pursued in detail in organic chemistry. Student laboratories employ distillation to isolate and purify products of synthesis, often with very small amounts using microscale glassware. The theoretical and quantitative aspects of phase transitions are studied in depth in the physical chemistry lecture and laboratory. Participation in research by working with an established research group is an important way for students to gain practical experience in laboratory procedures, methodologies, and protocols. Knowledge of correct procedures for handling toxic materials and the disposal of hazardous wastes are vital for all practicing chemists.

Social Context and Future Prospects

The process and apparatus of distillation, more than any other technology, gave birth to modern chemical industry because of the numerous chemical products derived first from coal tar and later from petroleum. The continued role played by distillation in modern technology will depend on several factors including the sustainability of raw materials and energy conservation. Increased demand and diminishing supplies of raw materials, together with the accumulation of increasing amounts of hazardous waste, have made recycling economically feasible on an industrial scale, and distillation has a role to play in many of these processes. Likewise, the increasing cost of crude oil because of diminishing supplies of this finite resource encourages the use of alternate sources of oil such as coal (nearly 75 percent of total fossil fuel reserves), and distillation would be expected to play the same central role as it does in refining petroleum. However, distillation is also an energy-intensive technology in the requirements of both heating liquids to boiling and cooling the resulting vapors so that they condense back to liquid products. Thus, one would expect that the chemical industry of the future would seek alternate energy sources such as solar power as well as ways to conserve energy through improvements in distillation efficiency.

Martin V. Stewart, B.S., Ph.D.

Further Reading

Donahue, Craig J. "Fractional Distillation and GC Analysis of Hydrocarbon Mixtures." *Journal of Chemical Education* 79, no. 6 (June, 2002): 721-723. Demonstrates analytical and physical methods employed in petroleum refining.

El-Nashar, Ali M. *Multiple Effect Distillation of Seawater Using Solar Energy.* New York: Nova Science, 2008. Tells the story of the Abu Dhabi, United Arab Emirates, desalinization plant that uses solar energy to distill seawater.

Forbes, R. J. *Short History of the Art of Distillation from the Beginnings Up to the Death of Cellier Blumenthal.* 1948. Reprint. Leiden, Netherlands: E. J. Brill, 1970. Discusses the history of distillation, focusing on earlier periods.

Kister, Henry Z. *Distillation Troubleshooting.* Hoboken, N.J.: Wiley-Interscience, 2006. Examines distillation operations and how they are maintained and repaired.

Owens, Bill, and Alan Dikty, eds. *The Art of Distilling Whiskey and Other Spirits: An Enthusiast's Guide to the Artisan Distilling of Potent Potables.* Beverly, Mass.: Quarry Books, 2009. Photographs from two cross-country road trips illustrate this paperback guide to the small-scale distillation of whiskey, vodka, gin, rum, brandy, tequila, and liquors.

Stichlmair, Johann G., and James R. Fair. *Distillation: Principles and Practices.* New York: Wiley-VCH, 1998. This 544-page comprehensive treatise for distillation technicians contains chapters on modern industrial distillation processes and energy savings during distillation.

Towler, Gavin P., and R. K. Sinnott. *Chemical Engineering Design: Principles, Practice, and Economics of Plant and Process Design.* Boston: Elsevier/Butterworth-Heinemann, 2008. Chapter 11 examines continuous and multicomponent distillation, looking at the principles involved and design variations. Also discusses other distillation systems and components of a system.

WEB SITES

American Institute of Chemical Engineers
http://www.aiche.org

American Petroleum Institute
http://www.api.org

Fractionation Research
http://www.fri.org

Institution of Chemical Engineers
http://unified.icheme.org

DRUG TESTING

FIELDS OF STUDY

Biochemistry; biology; molecular biology; microbiology; chemistry; medical technology; pharmaceutical technology; pharmacology.

SUMMARY

Drug testing is done to ensure the safety of the general public, to maintain standards at schools and places of employment, and to make sure that athletes do not gain unfair advantage through the use of performance-enhancing drugs. The goal of these tests is detect whether a person has used drugs such as alcohol, marijuana, cocaine, amphetamines, barbiturates, benzodiazepines, lysergic acid diethylamide (LSD), opiates, phencyclidine (PCP), synthetic hormones, and steroids. Commonly used drug tests analyze a person's breath, urine, saliva, sweat, blood, or hair.

KEY TERMS AND CONCEPTS

- **Antibody:** Glycoprotein that binds to and immobilizes a substance that the cell recognizes as foreign.
- **Antigen:** Substance that triggers an immune response.
- **Barbiturate:** Class of drugs that depress the central nervous system, thereby inducing sleep and sedation.
- **Benzodiazepine:** Class of drugs that are structurally characterized by a seven-member ring containing two nitrogen atoms and that act as tranquilizers.
- **Beta Blocker:** Class of drugs that compete with beta-adrenergic receptor-stimulating agents.
- **Binding Assay:** Experimental method for selecting one molecule out of a number of possibilities by specific binding.
- **Creatine:** Substance that is a natural product of the kidney and is sometimes used by body builders as a performance-enhancing drug to quickly increase muscle mass.
- **Diuretic:** Any substance that increases the formation of urine.
- **Hormone:** Substance produced by endocrine glands and delivered by the bloodstream to target cells, producing a desired effect.
- **Masking Agent:** Any substance that can prevent the detection of a drug in a person's system.
- **Monoclonal Antibody:** Antibody produced from the progeny of a single cell and specific for a single antigen.
- **Tetrahydrocannabinol (THC):** The active ingredient in marijuana.

Definition and Basic Principles

Drug testing in the workplace and schools has become commonplace. A variety of tests are used to detect elevated levels of the most common drugs that can impair job performance or are illegal to use. The importance of drug testing has continued to increase since the Controlled Substances Act of 1970 placed all regulated drugs into five classifications based on their medicinal value, their potential to harm people, and their likelihood of being abused or causing addiction. Schedule I drugs have no known medical value and are most likely to be abused, while Schedule V drugs have little potential for abuse. These scheduled drugs are called controlled substances because their use, manufacture, sale, and distribution are subject to control by the federal government.

There are two general types of drug testing. Federally regulated drug testing, according to the National Institute on Drug Abuse (NIDA), requires testing for cannabinoids (THC, marijuana, hashish), cocaine, amphetamines, opiates (morphine, heroin, and codeine), and PCP. Nonfederally regulated drug testing is often used to test athletes in various sports for the use of creatine, hormones, steroids, and other performance-enhancing drugs. Additional tests are used to detect barbiturates and alcohol.

Urinalysis is typically used as a preliminary test because it is less expensive and more convenient than the other tests. Saliva tests and breathalyzers are commonly used. Blood tests, although less frequently employed because they are generally more expensive and invasive, are more dependable, as are hair strand tests. A preliminary positive test using urinalysis must be confirmed by diagnostic tests completed in an analytical laboratory setting, which can take several days to complete. These diagnostic tests include the analytical instruments of gas chromatography (GC), mass spectrometry (MS), ion scanning, high-pressure liquid chromatography (HPLC), immunoassay (IA), and inductively coupled plasma spectrometry (ICP-MS).

Background and History

The detection of ingested drugs in various body fluids first sparked the interest of the ancient alchemists. In 1936, Rolla N. Harger of Indiana University patented the Drunkometer, a breath test to measure a person's level of alcohol intoxication. In 1954, Robert F. Borkenstein of Indiana University invented the breathalyzer, which had the benefit of greater portability, to measure blood-alcohol content. However, it was not until the widespread use of recreational drugs in the 1960's that the National Institute of Drug Abuse was established to monitor drug use. With its creation, federal funding became available to researchers to develop drug testing methods, which led to rapid advances. In 1973, physician Robert L. DuPont was appointed director of the National Institute of Drug Abuse. As director, DuPont implemented the use of the urine test and further developed immunoassays to test for several controlled substances.

In 1981, an airplane crashed on the USS *Nimitz*, and the investigation into the incident revealed drug use to be a contributing factor. As a result, the United States Navy began random drug testing of all active-duty personnel in 1982. In the 1980's, the U.S. Department of Transportation began to test all of its employees. In September, 1986, President Ronald Reagan signed Executive Order 12564, making drug testing mandatory for federal employees and all employees in safety-sensitive positions, such as employees in the nuclear power industry. The National Institute of Drug Abuse extended this mandatory testing to include truck drivers working in the petroleum industry. This testing has come to be regulated by the Substance Abuse and Mental Health Services Administration (SAMHSA), which is part of the U.S. Department of Health and Human Services.

How It Works

Because of the commercial availability of so many masking agents, the most effective drug testing occurs when the subject has had no previous notification. Thus, random drug testing is very effective and has become common in the workplace, schools, and for athletes. The National Collegiate Athletic Association and the National Football League provide only one to two days notice before drug testing, and the United States Olympic Committee has a no-notice policy for drug testing.

The first commonly available drug testing method was the breathalyzer, followed by urinalysis. The usage of saliva tests continues to increase, while sweat tests remain the least-used testing method. Blood tests require additional medical staff and are also the most invasive testing method; therefore, although they are very accurate, they are not as commonly used as urinalysis. Hair tests are very accurate but do not detect drug use in the last four to five days. In terms of validity for legal purposes, any of these preliminary, or screening, tests must be confirmed by an analytical technique, most often gas chromatography/mass spectrometry, performed by trained personnel within a diagnostic laboratory.

Breath. Borkenstein received a patent in 1958 for the breathalyzer, which determines an individual's blood-alcohol content (BAC) from a breath sample. The ethanol in the breath of an individual reacts with the dichromate ion, which has a yellow-orange color, in the presence of acid to form the green chromate ion. This color change from pale orange to green can easily be observed. All fifty states and the District of Columbia have laws that forbid a person to drive with a BAC of 0.08 percent or greater, a level at which the individual is judged to be legally impaired.

Urine. Urinalysis became a common method of detecting drugs in the 1980's and has continued to be widely used. The urine sample is collected and sealed to ensure that it remains tamper-free. It is generally subjected to an immunoassay test first because this test is very fast.

Saliva. Testing of oral fluids is becoming increasingly common because of its convenience for random testing, and it is more resistant to adulteration than urine samples. Saliva testing can detect cocaine, amphetamine, methamphetamine, marijuana, bezodiazepines, PCP, opiates, and alcohol if the substance was ingested between six hours and three days before the test was administered.

Sweat. Although traditionally not considered to be as useful as the other methods because of the dilute sample obtained, patches that can be worn on the skin and collect samples over several hours are increasingly popular. This method of drug testing is preferred by government agencies such as parole departments and child protective services in which urine testing is not the method of choice.

Blood. Because blood tests are the most invasive and expensive method, requiring additional medical personnel, they are not as widely used as the other tests as a screening method. However, blood testing is very accurate and reliable, so it is often used to confirm a positive result from another type of drug test.

Hair. Hair samples from any part of the body can be used and are extremely resistant to any type of tampering or adulteration. Special fatty esters are permanently formed in the hair as a result of alcohol and drug metabolism, and therefore, this method is very reliable.

Gas Chromatography/Mass Spectrometry (GC/MS). This tandem analytical instrumentation must be done by trained personnel within a laboratory setting and therefore is not as convenient as the other testing methods. However, it is much more accurate and is used to confirm more rapid, preliminary tests. The gas chromatograph is able to separate molecules based on their attractive interactions with the material that packs a column. Molecules take varying amounts of time to travel, or elute, from the column, resulting in different retention times, or amounts of time retained on the packing material of the column. These separated molecules are then ionized in the mass spectrometer, which is able to produce molecular weight information.

APPLICATIONS AND PRODUCTS

Drug Test Dips. Test strips known as drug test dips, dip strips, drug test cards, or drug panels use a single immunoassay panel to test for several common drugs at once. Specific reactions between antibodies and antigens allow marijuana, cocaine, amphetamine, opiates, and methamphetamine to be detected in urine samples. These assay strips are so easy to use that the staff of many schools, sports clubs, and offices can use them. However, they must be used only as a preliminary test. Positive results should be confirmed using gas chromatography/mass spectrometry conducted by an independent diagnostic laboratory.

The test strip is removed from its protective pouch and allowed to equilibrate to room temperature. Meanwhile, a urine sample is obtained in a small cup and also allowed to equilibrate to room temperature. The test strip is dipped vertically, with the arrow on the test strip pointing down into the sample, and remains immersed in the urine sample for ten to fifteen seconds. Then the test strip is removed from the urine sample and placed on a flat, nonabsorbent surface. After five minutes, the test strip is checked for the appearance of any horizontal lines. The appearance of a colored line in the control region of the dip strip and a faded color line in the test region of the dip strip indicates a negative test and that the concentration of a drug is too low to be detected. If only one line appears in the control region, with no line visible in the test region, then the test is considered to be positive for the presence of drugs. The test is considered to be invalid if only one line appears in the test region or no line appears at all. An invalid test is usually the result of either not following the procedure correctly or not using a large enough urine sample.

The test strips for marijuana use a monoclonal antibody to detect levels of THC, the active ingredient in marijuana, in excess of 50 nanograms per milliliter (ng/ml), the level recommended by SAMHSA. Methamphetamine can be detected in urine samples for three to five days after usage by using a test strip equipped with a monoclonal antibody. A positive test indicates a level in excess of 1,000 ng/ml. A test strip detects the major metabolite of cocaine, benzoylecgonine, for up to twenty-four to forty-eight hours after use. Morphine in excess of 2,000 ng/ml can also be detected by using a test strip containing a specific antibody. Morphine is the primary metabolite product of heroin and codeine.

Kits. Drug tests used for fast, preliminary screening include easy-to-use kits that can test samples of urine, saliva, breath, hair, or sweat. Of these tests, the hair test for drugs is considered to be the most accurate but is still considered to be a preliminary test. To confirm preliminary results or to obtain results for legal purposes, a sample of saliva or urine must be sent to a laboratory when a more reliable test such as GC/MS must be performed. It can take three to seven days to obtain the results. Urine drug test kits are less expensive than other tests, provide instantaneous results, and are easy to store. However, because of variations in metabolism rate, there can be a three-day to one-month detection window, making these tests easier to adulterate than the saliva, hair, sweat, or blood tests.

Adulteration of Specimens. Adulteration generally refers to intentional tampering with a urine sample, and certain substances can be added to urine to create a false-negative test result. Adulteration of

Fascinating Facts About Drug Testing

- Each year in the United States, drug and alcohol use costs more than $100 billion in lost productivity.
- A single urine testing kit can detect up to twelve drugs.
- The drug detection window for testing urine samples is wider than for other methods for drug testing because it depends on metabolic rate, age, amount and frequency of use, urine pH, drug tolerance, body mass, and overall health. Therefore, urine testing is considered less accurate than other testing methods.
- Sweat patches for drug testing are about the size of a playing card and have a tamper-proof feature. They can be worn for up to seven hours to collect samples.
- The annual cost of conducting more than 100,000 drug tests to detect performance-enhancing drugs in athletes in the United States is about $30 million.
- There are more than four hundred commercial products readily available to mask urine samples.
- An aerosol detecting agent called DrugAlert can be sprayed on a paper towel that has been used to wipe any household surface to detect a variety of drugs on that surface.

urine samples is a common problem because four types of masking products—dilution substances, synthetic urine, cleaning substances, and adulterants—are readily available. More than four hundred readily available commercial products can mask urine samples. Dilution substances, including diuretics, lower the concentration of drug in a sample. An individual can either ingest one of these substances before submitting a urine sample or add the substance directly to the urine sample. Synthetic and dehydrated human urine can be bought and submitted for testing. An individual can also purchase a cleaning substance such as an herbal supplement for $30 to $70 and ingest the substance before submitting a sample. The herbal supplement reacts chemically with the drug to essentially nullify its active ingredient. Adulterants are chemicals that can actually react with the drugs, but these are actually added to the sample rather than ingested by the individual.

Methods to Detect Adulteration. Several methods can be used to detect the use of some type of adulterant. If the specific gravity of urine is outside the normal range of 1.003 and 1.030, then the sample may have been diluted. Another indication of adulteration via dilution is to test the level of creatine. If the level is too low (less than 5 milligrams, or mg, per deciliter) then dilution took place. Oxidants, such as pyridinium chlorochromate (PCC), bleach, or hydrogen peroxide, can react chemically with the drug to essentially nullify it. One commonly sold PCC adulterant is called Urine Luck. Tests can also detect the presence of any type of additional oxidant. If the pH (acidity-alkalinity) value of the urine is outside the normal range of 4.0 to 9.0, then an adulterant was added. Two common adulterants are sold under the names of Whizzies or Klear, and these react chemically with drugs in the urine by oxidizing the active ingredient in marijuana. Another chemical reaction occurs when an adulterant called Clear Choice or Urine Aid prevents enzyme activity in the test, which results in the presence of glutaraldehyde, which causes a false negative. Among the states that have passed laws to prevent the sale of masking agents are Florida, South Carolina, North Carolina, New Jersey, Maryland, Virginia, Kentucky, Oklahoma, Nebraska, Illinois, Pennsylvania, and Arkansas.

Impact on Industry

Drug testing was initially developed for military use and then mandated for federal employees. However, the widespread use of recreational drugs beginning in the 1960's caused private sector employers and then academic institutions to routinely administer drug tests. As a result, a huge market has developed not only for rapid, on-site testing methods but also for at-home testing methods, which many parents use to monitor their children. This demand produced explosive growth by companies that manufacture on-site or at-home drug testing kits, and the kits are sold by mass merchandisers and in drugstores such as Target, Walmart, Walgreens, and CVS. In addition, many of these inexpensive kits can be purchased on the Internet.

Careers and Course Work

An interest and aptitude for biology, chemistry, and quantitative classes are important prerequisites to pursuing a career in drug testing. Additional required characteristics include the ability to solve problems, pay close attention to detail, work under pressure, have good manual dexterity, and have normal color vision. Depending on the requirements of the state of residence, a person needs a certificate or license or an associate's degree in biology, chemistry, or medical technology to be employed as a clinical laboratory technician, medical technician, or clinical laboratory technician. Specific information regarding certification can be obtained from the board of registry of the American Association of Bioanalysts and the National Accrediting Agency for Clinical Laboratory Sciences. Typical job duties involve drug sample collection and storage and operation of automated analyzers, often wearing protective gloves and safety glasses. In May, 2008, the median annual salary for a technician was about $35,000.

A bachelor's degree is required to work as a technologist or scientist and earn a higher salary of about $50,000. In addition to a higher salary, a bachelor's degree allows a person to take on more responsibility and possibly advance into supervisory and managerial positions. The ideal bachelor's degree is medical technology, which requires courses in chemistry, biology, microbiology, statistics, computers, and mathematics. To pursue research or become a laboratory director, a master's or doctoral degree is necessary. Employment in the drug testing field is projected to grow at the rate of 14 percent through 2016, which is faster than the average, according to the U.S. Department of Labor. Typical employers include forensic science laboratories, research and development laboratories, and quality assurance laboratories in industry, government, schools, or hospitals.

Social Context and Future Prospects

Mandatory testing is regulated by SAMHSA, part of the U.S. Department of Health and Human Services. This mandatory testing does not yet test for semisynthetic opioids, such as oxycodone, oxymorphone, and hydrocodone, which are often used to relieve pain but have the potential to be abused. However, many employers, athletic organizations, and schools test for these drugs, and ongoing research is directed toward increasing the convenience and reliability of methods of detecting these drugs. Because so many masking agents are readily available, random testing without prior notification is the most effective method, although it is not without controversy. Schools are increasingly performing random drug testing, often leading to protests that the tests are an invasion of privacy and a violation of Fourth Amendment rights.

Organizations such as the International Olympic Committee, National Collegiate Athletic Association, National Basketball Association, and National Football League monitor athletes for the use of more than one hundred anabolic-androgenic steroids. Efforts are being made to eliminate the use of performance-enhancing drugs in all sports. The International Olympic Committee led a collective initiative in creating the World Anti-Doping Agency (WADA) in Switzerland in 1999. WADA created a code in an attempt to standardize regulation and procedures in all sporting countries and keeps a list of prohibited substances. Banned substances include anabolic steroids, hormones, masking agents, stimulants, narcotics, cannabinoids, glucocorticosteroids, and for some sports, alcohol and beta-blockers during competition. Also forbidden are methods of enhancing oxygen transfer (such as blood doping) and gene doping. The UNESCO International Convention Against Doping in Sport, which came into force in 2007, is a global treaty designed to help governments align their policies with the WADA code.

Jeanne L. Kuhler, B.S., M.S., Ph.D.

Further Reading

Jenkins, Amanda J., and Bruce A. Goldberger, eds. *On-Site Drug Testing*. Totowa, N.J.: Humana Press, 2002. Discusses on-site methods of testing for drugs in hospital, criminal, workplace, and school settings. Looks at many specific tests, discussing their efficacy and their underlying principles.

Karch, Stephen B., ed. *Workplace Drug Testing*. Boca Raton, Fla.: CRC Press, 2008. Examines regulations and mandatory guidelines for federal workplace drug testing and describes techniques. Provides sample protocols from the nuclear power and transportation industries.

Liska, Ken. *Drugs and the Human Body with Implications for Society*. Upper Saddle River, N.J.: Pearson/Prentice Hall, 2004. Simply describes the various classes of drugs and drug testing methods.

Mur, Cindy, ed. *Drug Testing*. Farmington Hills, Mich.:Greenhaven Press/Thomson Gale, 2006. A collection of essays on drug testing in schools and the workplace, discussing efficacy and ethical issues such as privacy.

Pascal, Kintz. *Analytical and Practical Aspects of Drug Testing in Hair.* Boca Raton, Fla.: CRC Press, 2006. Looks at advances in the use of strands of hair for drug testing in the workplace and in forensic crime laboratories and techniques for detecting specific drugs.

Thieme, Detlef, and Peter Hemmersbach. *Doping in Sports.* Berlin: Springer, 2010. Examines sports doping from its beginning, covering the use of anabolic steroids, erthyropoietin, human growth hormone, and gene doping in humans and the doping of race horses. Effects of the drugs, detection methods, and regulations are also discussed.

Web Sites

Drug and Alcohol Testing Industry Association
http://www.datia.org

Substance Abuse and Mental Health Services Administration
http://www.samhsa.gov

Substance Abuse Program Administrators Association
http://www.sapaa.com

U.S. Department of Labor
Drug-Free Workplace Adviser
http://www.dol.gov/elaws/drugfree.htm

World Anti-Doping Agency
http://www.wada-ama.org

See also: Forensic Science

EGG PRODUCTION

FIELDS OF STUDY

Poultry/animal science; reproduction; food technology; biology; physiology; business management.

SUMMARY

The egg production field includes farm production of shell eggs for direct consumption and further processing of eggs for use in products of the food industry. Egg production includes the development of highly productive strains of laying hens, advances in technology in the production and processing of eggs, and business models that permit the efficient production and marketing of eggs.

KEY TERMS AND CONCEPTS

- **Candling:** Inspecting the internal quality and embryonic development of eggs by shining a bright light through them.
- **Chalaza:** Stringlike attachment that anchors the yolk to the center of an egg.
- **In-Line Production:** Using a single location for production and packaging of eggs.
- **Line:** Group of related chickens that have similar production characteristics.
- **Off-Line Production:** Using different locations for the production and processing of eggs.
- **Pullet:** Immature female chicken destined for egg production.
- **Salmonella:** Genus of bacteria that can contaminate eggs, causing serious illness to humans who consume the eggs.
- **Vertical Integration:** Ownership by a single firm of multiple companies in order to cover all stages of egg production, from the raw materials through distribution, including feed mills, hens, buildings, egg-processing facilities, and transportation vehicles.

DEFINITION AND BASIC PRINCIPLES

Egg production in the United States has undergone a remarkable transformation. Before the twentieth century, hens ran loose around the farmyard, largely fending for themselves. Around the late 1800's, farm flocks came into being, and egg production became a serious part of the farm enterprise. Hens were given their own housing and provided with feeders, waterers, roosts, and nests, as well as a fenced-in yard. The farm flock system allowed for applying important management principles, such as proper feeding, breeding, and egg collection. The next advance took place around the 1960's with the emergence of farms that specialized in egg production. The farmer-manager could then focus entirely on egg production and use the latest in management and feeding techniques and production stock. Later in the twentieth century, egg producers became vertically integrated, with all aspects of production and marketing under the control of the same firm. The farmer-producer became just one part of the entire system.

Egg production involves genetic research to develop strains of highly productive hens; proper management of growing pullets to maximize their potential as laying hens; the use of advanced technology in buildings, equipment, feeding, and lighting for maximal egg production at minimal cost; and the development of new egg products for the consumer. It can also involve support services such as feed mills and transportation. Modern intensive production practices involving millions of birds have come under criticism as factory farming and have raised questions of animal welfare that must be addressed by the producer.

BACKGROUND AND HISTORY

Chickens were probably domesticated from red junglefowl in Southeast Asia. Genetic studies suggest multiple sites of domestication, including China and India. Archaeological studies indicate that chickens were present in the Americas before the time of the Spanish conquistadores.

The modern egg industry is a result of a series of technological advances. In the 1870's, incubators began to be used commercially to hatch chickens, rapidly increasing the number of commercial hatcheries. Poultry breeders applied scientific principles to develop improved breeds and strains of chickens for egg production. Land-grant colleges engaged in research in poultry nutrition and feeding. This led to improved management practices and more efficient production of eggs. Better understanding and treatment of diseases, together with improved sanitation and ventilation, allowed for the creation of confinement systems.

Improved distribution systems and the development of new egg products led to greatly increased consumption of eggs, reaching a maximum of 402 eggs per capita in 1945. Health concerns about the cholesterol content of eggs and changes in lifestyle led to declines in consumption to a low of 230 eggs per capita in 1991. However, after the publication of scientific studies that stated that consuming eggs does not raise blood cholesterol, consumption of eggs began to increase, reaching 248 eggs per capita in 2008.

How It Works

Egg production begins with the selection and development of breeding stock. Many breeds of chickens have developed over time, but for commercial purposes, the laying hen (layer) must be highly productive and efficient in converting feed into eggs. These criteria are met by the white leghorn breed, which is light in body weight, is highly active, and produces a white egg. A very few breeding companies dominate the supply of egg production chicks, and they have their own specialized lines or strains of breeders. The white leghorn has been overwhelmingly adopted by the egg industry, but other breeds are used in markets that prefer a brown egg. Traditionally, this has involved using heavy breeds, such as the New Hampshire or Rhode Island red. The development of specialized lines and crossbreeds has resulted in brown-egg layers that are almost as efficient in feed conversion as the white leghorn. In many countries, including European nations, brown eggs are preferred over white eggs. The breeders must be kept in floor management systems to facilitate the breeders' mating.

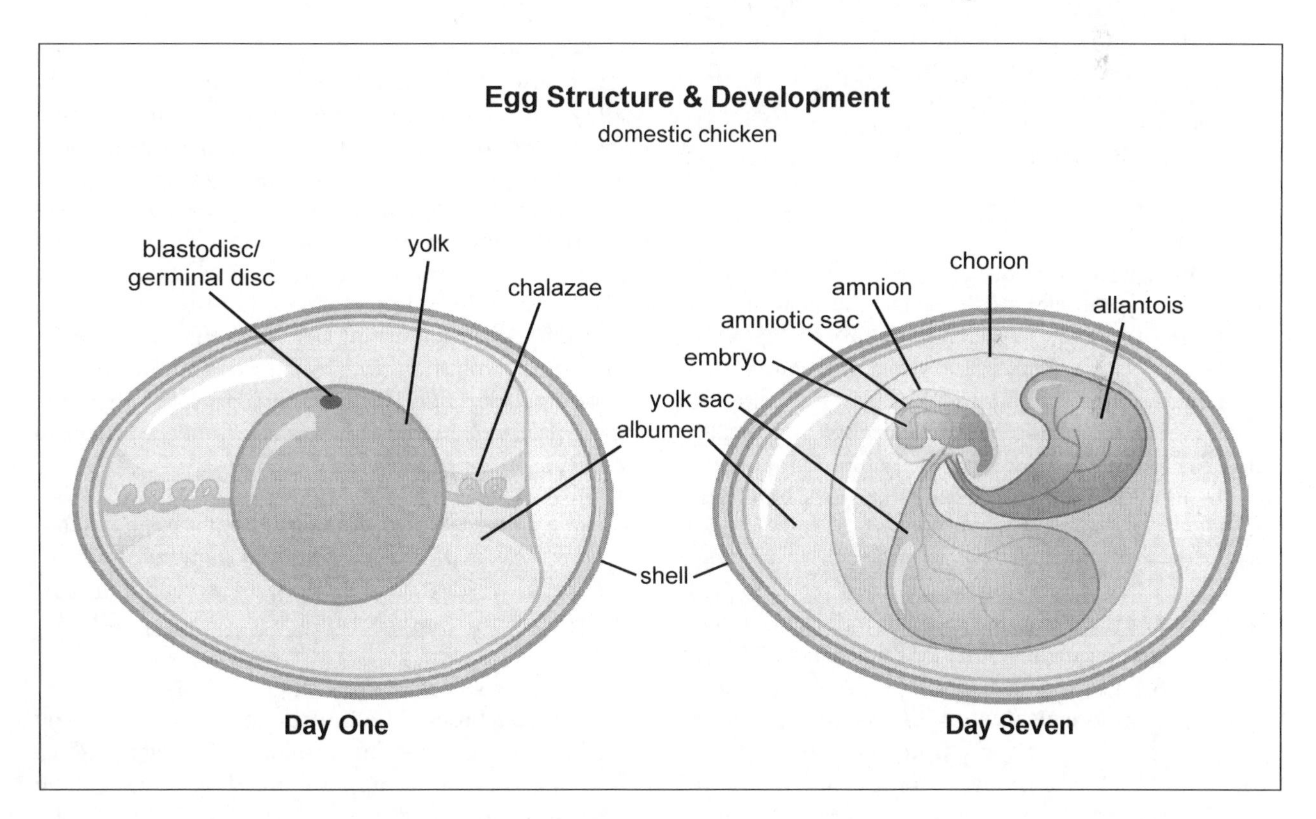

Fascinating Facts About Egg Production

- Eggs provide a unique source of balanced nutrients, including protein, essential fatty acids, vitamins, and minerals. The protein is of such high value that it is used as a standard to measure the quality of other food proteins.
- World consumption of eggs is increasing at about 8 percent per year because of higher living standards and the introduction of efficient production methods.
- Consolidation of egg farms has resulted in around three hundred producers supplying most of the nation's eggs. These producers are primarily located in the five top egg production states: Iowa, Ohio, Indiana, Pennsylvania, and California.
- Blood spots in egg yolk do not mean that the egg is fertilized. They are caused by a broken blood vessel on the surface of the yolk as the egg is forming.
- If a carton of eggs bears a U.S. Department of Agriculture grade, it must also have a Julian date, which is the date of packing. A sell-by date, if it appears, can be no more than thirty days after the date of packing, and a use-by date can be no more than forty-five days after packing.
- Pasteurized eggs have been exposed to heat to destroy bacteria. These are the best choice for recipes that call for partially cooked or raw eggs.

From Egg to Layer. Fertilized eggs are transported to commercial incubators for incubation and hatching. After a few days of incubation, the eggs are candled to test for fertility and for viable embryo development. An infertile egg is clear, and a developing embryo shows blood-vessel development. Typically, the eggs are moved to a separate hatching incubator for the final three days of incubation. After hatching, the chicks are vaccinated and sexed, as only the female chicks are useful for egg production. Debeaking (removal of part of the beak) is performed at this time, or after the chicks are seven to ten days old.

The pullets are raised in confinement either on the floor or in cages; outside range rearing is seldom used by commercial breeders. A lighting program is essential for proper development of the pullets. One-day-old chicks receive twenty-three hours of light per day, and for the rest of the growing period, they receive a minimum of ten hours of light per day. They are transferred to laying houses at around sixteen weeks of age. Hens usually begin to lay eggs when they are five months old and continue to lay for about twelve more months.

Egg Production. Several types of management systems are commonly used by egg producers: cages, floor systems, or free-range systems. Cages are used for more than 98 percent of production operations for a variety of reasons. They allow increased population density in the poultry houses, and they are more labor efficient, as feeding, watering, egg collection, and manure removal can all be mechanized. Floor or noncage systems keep hens on litter floors inside buildings that hold feeders, waterers, roosts, and nests. This was the most common management system before the adoption of cage systems. Free-range systems allow hens access to an outdoor yard when weather permits.

The term "organic eggs" refers not so much to a management system but to the feed the hens receive. The feed must be totally vegetarian, the grains used must be pesticide-free, and the hens must not receive hormones or antibiotics.

Because most laying hens produce eggs in windowless houses, artificial lighting is provided. In fact, in all systems, lighting is essential to stimulate the pituitary gland to secrete hormones that help initiate and sustain egg production. Various lighting programs have been developed, but a typical program increases lighting from ten hours at twenty-four weeks of age to seventeen hours at thirty-two weeks and maintains this lighting period until the end of the laying cycle. The length of the lighting period should never be decreased during the laying cycle. The number of eggs produced per hen during a laying cycle can range from 180 to 200 eggs in tropical climates to 250 to 300 eggs in more temperate climates.

In cage systems, after the eggs are laid, they are transported via a conveyor belt to an egg-processing facility, where they are washed, graded for size, and either packed in flats to be shipped to a retail store or broken for further processing.

APPLICATIONS AND PRODUCTS

Breeding Stock. The Institut de Sélection Animale (ISA) holds a dominant position in the egg production industry as it supplies breeding stock for 50 percent of the world's egg production industry. The company began as Hendrix in the Netherlands, where it still has its headquarters. ISA expanded by purchasing many well-known and respected laying-hen breeding companies, including Babcock, J. J. Warren, Kimber, Shaver, Dekalb, Hisex, and Bovans. Many of these companies began as family-owned businesses in the early part of the twentieth century. Many strains of white and brown egg layers under the names of the original companies are sold as day-old chicks. The chicks destined as breeders must have a good egg-production capability, but good fertility is essential.

Laying Stock. Laying stock is also sold by ISA and other breeders as day-old chicks. ISA has strains of white and brown egg layers that are companions to its breeding stock. High egg production and excellent feed efficiency are essential characteristics for these strains.

Ducks for Egg Production. Ducks have never been popular for egg production in the United States and, like quail eggs, are only a niche market. However, ducks are commonly used in Asia for egg production. The Khaki Campbell breed is best known for egg production, and Metzer Farms sells a hybrid duck that produces eggs at a rate similar to the best chicken egg strains. Duck eggs are larger, have a more deeply pigmented yolk, and have firmer albumen than chicken eggs. Compared with chicken eggs, duck eggs have a higher cholesterol content, tend to pick up off-flavors more readily, and are more susceptible to contamination

Shell Eggs. Eggs are most commonly marketed in the form in which they are laid, still in their shell. There is no difference in nutritional value between white and brown eggs, and although white eggs have a slightly thicker shell than brown eggs, brown eggshells have a stronger structure, so there is no difference in tendency to break. As the laying cycle nears its end, eggs tend to get bigger with thinner shells, leading to a greater tendency for breakage. When eggs are laid, they are coated with a protective layer called a cuticle. This cuticle is often removed during washing. The shell contains many pores, which nature intended for gaseous exchange for the developing embryo, but which also provide an entry point for bacteria.

The yolk consists of 32 to 36 percent lipids and around 16 percent protein. The lipids include triglycerides (fats), phospholipids, and cholesterol. Triglycerides contain various types of fatty acids. The fatty acid content of yolk can vary according to the diet fed to the hens. A popular modern egg product contains a high content of omega-3 fatty acids, typically 350 milligrams compared with a normal content of 60 milligrams. The eggs also have a lower content of saturated fat, as well as a somewhat lower content of cholesterol. The hens are fed flaxseed to produce these eggs. These eggs have purported health benefits and command a higher price.

Eggs are graded by weight and quality. Egg-processing machinery separates eggs by weight, which can range from jumbo to peewee. Eggs can be grade AA, A, or B in quality. Quality in eggs is determined by candling or breaking them out and measuring albumin height. Grade AA eggs are freshly laid, have a thick, cloudy albumin, and a small air cell. Most eggs in supermarkets are grade A because some time has passed since their laying. Grade A eggs have a larger air cell, and the albumin is clear but thinner. The yolk is more defined in candling but free of defects. Both AA and A eggs can be sold as shell eggs, while grade B eggs are used for further processing. Grade B eggs have poorer quality albumin and minor discoloration or minor blood or meat spots.

Liquid Egg Products. Grade B eggs or other eggs not needed for the shell egg market go to an egg-breaking plant. After breaking, the liquid products obtained include whole egg, egg white, and egg yolk. These products are destined for the food industry and are unlikely to be found in retail stores.

Dried Egg Products. The incentive for developing the technology for drying eggs in the United States began in the 1930's with the availability of large quantities of eggs from China at a very low cost. The industry got a boost during World War II when the military needed dried eggs. Dried eggs have several advantages over shell eggs or liquid eggs: They can be stored at low cost, take less space to store, are not susceptible to spoilage caused by bacteria, are easier to handle in a sanitary manner, and have lower transportation costs.

Dried eggs are used extensively in many products, including bakery foods and mixes, mayonnaise and salad dressings, ice cream, pastas, and convenience foods. Most dried egg products are obtained by spray drying, but before drying, the sugars are removed from the eggs by fermentation or enzymatic treatments. These processes are necessary to avoid reactions of glucose with proteins or phospholipids in the eggs that can result in poor baking qualities or off-flavors. The dried egg products are derived from egg white, egg yolk, whole egg, or blends of whole egg or yolk with carbohydrates such as sucrose or syrups.

Specialty dried egg products include a scrambled-egg mix that has good storage capability and low-cholesterol egg products. Most low-cholesterol egg products contain egg white, with nonfat milk, vegetable oil, and pigments substituting for yolk. The final composition is similar to that of a whole egg.

Impact on Industry

Worldwide egg production has increased considerably because of the establishment of intensive production systems in developing countries. A study in Southeast Asia has shown an annual growth rate in egg consumption of around 8 percent, with the greatest increase seen in the lower middle class. There is a direct association between increased per capita income and increased egg consumption. However, traditional extensive systems still account for 50 percent of production output and can provide a significant contribution to family nutrition.

Government Research. The U.S. Department of Agriculture (USDA) conducts poultry research at the Beltsville Agricultural Research Center in Maryland. Additional research is conducted at the Southeast Poultry Research Laboratory in Athens, Georgia. The USDA is responsible for determining the quality and grading standards for eggs, provides weekly summaries on national and international markets for eggs, and supplies educational and informational materials.

The most common research topics are feed ingredients, diseases, and management practices such as force molting, which reflect necessities in the egg production industry.

Although the Agricultural Research Service (ARS) does fewer poultry breeding studies than it once did, it developed a gentler line of laying hens that do not need debeaking to control cannibalism. Another interesting study related to breeding selection programs found that when breeders select for productivity traits, they should consider how these traits interact with behavioral traits. Application of such a breeding selection program resulted in dramatic improvements in productivity, livability, and welfare.

University Research and Extension. University research on egg production and egg products largely takes place in land-grant colleges and their associated research stations. Although some of the research is in basic science, much of it is applied research that can be used in a production setting. The results of this research are disseminated to producers or consumers through university extension programs, articles in the scientific and popular press, and electronic media.

Ongoing studies of cage versus alternative management systems indicate no inherent differences in mortality rates. Osteoporosis and other bone disorders may be greater in cage systems, although bone fractures may be greater in floor systems. Researchers have not arrived at definitive conclusions about animal welfare.

Industry Research. The development of specialized breeds and strains of laying hens has become the role of large poultry-breeding companies. The USDA and land-grant universities are less active in this field than previously and are inclined to focus on specialized studies. A few large feed companies still conduct nutrition research on laying hens, but quite often, they enter into contract with universities to conduct research. Poultry equipment manufacturers have played an important role in developing the modern, highly efficient egg industry. In some large facilities, all egg collection, grading, and packaging can be automated so that humans do not handle the eggs.

Careers and Course Work

Course work in poultry science is basic for students interested in pursuing a career in egg production. Because most poultry science departments have been absorbed into animal science departments at land-grant universities, the student should carefully search the curriculum, staff, and programs in these departments to determine if they can provide an adequate focus on egg production and technology.

Suitable undergraduate course work in the field of concentration could include poultry or laying hen production, poultry nutrition, poultry diseases, poultry anatomy and physiology, and reproduction (including breeding, embryonic development, and hatching). Students interested in the egg products industry should consider course work in food science. Supporting course work could include farm management, biology, and business courses such as economics and accounting.

Career opportunities for students with a bachelor's degree include poultry farm manager (as owner or employee), feed mill or hatchery manager, and salesperson for feeds, breeding or production stock, or equipment.

Students interested in a research career in poultry breeding or nutrition will need to complete graduate work leading to a doctorate degree. Advanced courses in animal breeding, statistics, endocrinology, genome analysis, genetics, animal breeding strategies, statistical methods, and biochemistry could be taken. Professional poultry scientists can obtain employment in academia or industry.

Social Context and Future Prospects

The modern cage system of egg production is a marvel of efficiency and low cost. However, the nature of the system has been brought to the attention of animal welfare activists. The hens are kept in very crowded conditions (typically 67 square inches per hen) and are not able to perform their natural or instinctive behaviors, such as sleeping on roosts, laying eggs in nests, and taking a dust bath. Animal activists say that this is not humane. However, egg producers reply that hens kept in cage systems are healthier than those raised in other systems, noting that their productivity is higher. Animal science departments have been aware of these criticisms and have developed a new field of farm animal welfare. Animal welfare can be studied scientifically in a manner that is objective, reliable, and reproducible. However, the demand for answers to animal welfare issues may be outpacing the results of scientific studies. This has resulted in legislation banning the use of cages for egg production in Europe and the passing of Proposition 2 in California. The California legislation will probably phase out cage use in the state, which producers say will increase production costs 40 to 70 percent and drive egg producers out of the state because they will no longer be competitive.

Egg consumption fell because eggs have a high level of cholesterol, but consistent research has shown that egg consumption will not increase blood cholesterol in healthy people. Persons with heart disease may want to consult their physician as their bodies may handle cholesterol differently. The image of eggs suffered, and egg producers must convince the public of the egg's nutritive value if egg consumption is to reach or approach its 1945 peak.

A problem with eggs is possible salmonella contamination. If the shells are contaminated with salmonella, proper washing can eliminate this hazard, but if hens become infected with salmonella during the growing period, the eggs are internally contaminated. In August, 2010, more than 500 million eggs produced by Wright County Egg and Hillandale Farms of Iowa were recalled because of possible salmonella contamination. Programs are being developed to certify hens in large flocks as being salmonella-free.

David Olle, B.S., M.S.

Further Reading

Bell, Donald D., William Daniel Weaver, and Mack O. North. *Commercial Chicken Meat and Egg Production.* 5th ed. Norwell, Mass.: Kluwer Academic, 2002. An essential guide for those interested in the poultry industry. This edition emphasizes managerial aspects.

Clancy, Kate. *Greener Eggs and Ham: The Benefits of Pasture-Raised Swine, Poultry, and Egg Production.* Cambridge, Mass.: Union of Concerned Scientists, 2006. The Union of Concerned Scientists looks at egg production, poultry, and pigs and presents an alternative to the intensive production methods in predominant use.

National Agricultural Statistics Service. *U.S. Broiler and Egg Production Cycles.* Washington, D.C.: USDA National Agricultural Statistics Service, 2005. A governmental document providing information on egg production cycles and chickens for those in the poultry industry.

Stedelman, William, and Owen Cotterill. *Egg Science and Technology.* 4th ed. New York: Haworth Press, 1995. Long recognized as the most comprehensive handbook on the egg-processing industry.

Web Sites

American Egg Board
http://www.aeb.org

American Poultry Association
http://www.amerpoultryassn.com

Institut de Sélection Animale
http://www.isapoultry.com

United Egg Producers
http://www.unitedegg.org

United Egg Producers Certified
http://www.uepcertified.com

U.S. Poultry and Egg Association

http://www.poultryegg.org

See also: Food Science

ELECTROMAGNET TECHNOLOGIES

FIELDS OF STUDY

Mathematics; physics; electronics; materials science.

SUMMARY

Electromagnetic technology is fundamental to the maintenance and progress of modern society. Electromagnetism is one of the essential characteristics of the physical nature of matter, and it is fair to say that everything, including life itself, is dependent upon it. The ability to harness electromagnetism has led to the production of most modern technology.

KEY TERMS AND CONCEPTS

- **Curie Temperature:** The temperature at which a ferromagnetic material is made to lose its magnetic properties.
- **Induction:** The generation of an electric current in a conductor by the action of a moving magnetic field.
- **LC Oscillation:** Alternation between stored electric and magnetic field energies in circuits containing both an inductance (L) and a capacitance (C).
- **Magnetic Damping:** The use of opposing magnetic fields to damp the motion of a magnetic oscillator.
- **Sinusoidal:** Varying continuously between maximum and minimum values of equal magnitude at a specific frequency, in the manner of a sine wave.
- **Toroid:** A doughnut-shaped circular core about which is wrapped a continuous wire coil carrying an electrical current.

Definition and Basic Principles

Magnetism is a fundamental field effect produced by the movement of an electrical charge, whether that charge is within individual atoms such as iron or is the movement of large quantities of electrons through an electrical conductor. The electrical field effect is intimately related to that of magnetism. The two can exist independently of each other, but when the electrical field is generated by the movement of a charge, the electrical field is always accompanied by a magnetic field. Together they are referred to as an electromagnetic field. The precise mathematical relationship between electric and magnetic fields allows electricity to be used to generate magnetic fields of specific strengths, commonly through the use of conductor coils, in which the flow of electricity follows a circular path. The method is well understood and is the basic operating principle of both electric motors and electric generators.

When used with magnetically susceptible materials, this method transmits magnetic effects that work for a variety of purposes. Bulk material handling can be carried out in this way. On a much smaller scale, the same method permits the manipulation of data bits on magnetic recording tape and hard-disk drives. This fine degree of control is made possible through the combination of digital technology and the relationship between electric and magnetic fields.

Background and History

The relationship between electric current and magnetic fields was first observed by Danish physicist and chemist Hans Christian Oersted (1777-1851), who noted, in 1820, how a compass placed near an electrified coil of wires responded to those wires. When electricity was made to flow through the coil, these wires changed the direction in which the compass pointed. From this Oersted reasoned that a magnetic field must exist around an electrified coil.

In 1821, English chemist and physicist Michael Faraday (1791-1867) found that he could make an electromagnetic field interact with a permanent magnetic field, inducing motion in one or the other of the magnetized objects. By controlling the electrical current in the electromagnet, the permanent magnet can be made to spin about. This became the operating principle of the electric motor. In 1831, Faraday found that moving a permanent magnetic field through a coil of wire caused an electrical current to flow in the wire, the principle by which electric generators function. In 1824, English physicist and inventor William Sturgeon (1783-1850) discovered that an electromagnet constructed around a core of solid magnetic material produced a much stronger magnetic field than either one alone could produce.

Since these initial discoveries, the study of electromagnetism has refined the details of the mathematical relationship between electricity and magnetism as it is known today. Research continues to refine understandings of the phenomena, enabling its use in new and valuable ways.

How It Works

The basic principles of electromagnetism are the same today as they have always been, because electromagnetism is a fundamental property of matter. The movement of an electrical charge through a conducting medium induces a magnetic field around the conductor. On an atomic scale, this occurs as a function of the electronic and nuclear structure of the atoms, in which the movement of an electron in a specific atomic orbital is similar in principle to the movement of an electron through a conducting material. On larger scales, magnetism is induced by the movement of electrons through the material as an electrical current.

Although much is known about the relationship of electricity and magnetism, a definitive understanding has so far escaped rigorous analysis by physicists and mathematicians. The relationship is apparently related to the wave-particle duality described by quantum mechanics, in which electrons are deemed to have the properties of both electromagnetic waves and physical particles. The allowed energy levels of electrons within any quantum shell are determined by two quantum values, one of which is designated the magnetic quantum number. This fundamental relationship is also reflected in the electromagnetic spectrum, a continuum of electromagnetic wave phenomena that includes all forms of light, radio waves, microwaves, X rays, and so on. Whereas the electromagnetic spectrum is well described by the mathematics of wave mechanics, there remains no clear comprehension of what it actually is, and the best theoretical analysis of it is that it is a field effect consisting of both an electric component and a magnetic component. This, however, is more than sufficient to facilitate the physical manipulation and use of electromagnetism in many forms.

Ampere's circuit law states that the magnetic field intensity around a closed circuit is determined by the sum of the currents at each point around that circuit. This defines a strict relationship between electrical current and the magnetic field that is produced around the conductor by that current. Because electrical current is a physical quantity that can be precisely controlled on scales that currently range from single electrons to large currents, the corresponding magnetic fields can be equally precisely controlled.

Electrical current exists in two forms: direct current and alternating current. In direct current, the movement of electrons in a conductor occurs in one direction only at a constant rate that is determined by the potential difference across the circuit and the resistance of the components that make up that circuit. The flow of direct current through a linear conductor generates a similarly constant magnetic field around that conductor.

In alternating current, the movement of electrons alternates direction at a set sinusoidal wave frequency. In North America this frequency is 60 hertz, which means that electrons reverse the direction of their movement in the conductor 120 times each second, effectively oscillating back and forth through the wires. Because of this, the vector direction of the magnetic field around those conductors also reverses at the same rate. This oscillation requires that the phase of the cycle be factored into design and operating principles of electric motors and other machines that use alternating current.

Both forms of electrical current can be used to induce a strong magnetic field in a magnetic material that has been surrounded by electrical conductors. The classic example of this effect is to wrap a large steel nail with insulated copper wire connected to the terminals of a battery so that as electrical current flows through the coil, the nail becomes magnetic. The same principle applies on all scales in which electromagnets are utilized to perform a function.

Applications and Products

The applications of electromagnetic technology are as varied and widespread as the nature of electromagnetic phenomena. Every possible variation of electromagnetism provides an opportunity for the development of some useful application.

Electrical Power Generation. The movement of a magnetic field across a conductor induces an electrical current flow in that conductor. This is the operating principle of every electrical generator. A variety of methods are used to convert the mechanical energy of motion into electrical energy, typically in generating stations.

Hydroelectric power uses the force provided by falling water to drive the magnetic rotors of generators. Other plants use the combustion of fuels to generate steam that is then used to drive electrical generators. Nuclear power plants use nuclear fission for the same purpose. Still other power generation projects use renewable resources such as wind and ocean tides to operate electrical generators.

Solar energy can be used in two ways to generate electricity. In regions with a high amount of sunlight, reflectors can be used to focus that energy on a point to produce steam or some other gaseous material under pressure; this material can then drive an electrical generator. Alternatively, and more commonly, semiconductor solar panels are used to capture the electromagnetic property of sunlight and drive electrons through the system to generate electrical current.

Material Handling. Electromagnetism has long been used on a fairly crude scale for the handling and manipulation of materials. A common site in metal recycling yards is a crane using a large electromagnet to pick up and move quantities of magnetically susceptible materials, such as scrap iron and steel, from one location to another. Such electromagnets are powerful enough to lift a ton or more of material at a time. In more refined applications, smaller electromagnets operating on exactly the same principle are often incorporated into automated processes under robotic control to manipulate and move individual metal parts within a production process.

Machine Operational Control. Electromagnetic technology is often used for the control of operating machinery. Operation is achieved normally through the use of solenoids and solenoid-operated relays. A solenoid is an electromagnet coil that acts on a movable core, such that when current flows in the coil, the core responds to the magnetic field by shifting its position as though to leave the coil. This motion can be used to close a switch or to apply pressure according to the magnetic field strength in the coil.

When the solenoid and switch are enclosed together in a discrete unit, the structure is known as a relay and is used to control the function of electrical circuitry and to operate valves in hydraulic or pneumatic systems. In these applications also, the extremely fine control of electrical current facilitates the design and application of a large variety of solenoids. These solenoids range in size from the very small (used in micromachinery) and those used in typical video equipment and CD players, to very large ones (used to stabilize and operate components of large, heavy machinery).

A second use of electromagnetic technology in machine control utilizes the opposition of magnetic fields to effect braking force for such precision machines as high-speed trains. Normal frictional braking in such situations would have serious negative effects on the interacting components, resulting in warping, scarring, material transfer welding, and other damage. Electromagnetic braking forces avoid any actual contact between components and can be adjusted continuously by control of the electrical current being applied.

Magnetic Media. The single most important aspect of electromagnetic technology is also the smallest and most rigidly controlled application of all. Magnetic media have been known for many years, beginning with the magnetic wire and progressing to magnetic recording tape. In these applications, a substrate material in tape form, typically a ribbon of an unreactive plastic film, is given a coating that contains fine granules of a magnetically susceptible material such as iron oxide or chromium dioxide. Exposure of the material to a magnetic field imparts the corresponding directionality and magnitude of that magnetic field to the individual granules in the coating. As the medium moves past an electromagnetic "read-write head" at a constant speed, the variation of the magnetic properties over time is embedded into the medium. When played back through the system, the read-write head senses the variation in magnetic patterns in the medium and translates that into an electronic signal that is converted into the corresponding audio and video information. This is the analogue methodology used in the operation of audio and videotape recording.

With the development of digital technology, electromagnetic control of read-write heads has come to mean the ability to record and retrieve data as single bits. This was realized with the development of hard drive and floppy disk-drive technologies. In these applications, an extremely fine read-write head records a magnetic signal into the magnetic medium of a spinning disk at a strictly defined location. The signal consists of a series of minute magnetic fields whose vector orientations correspond to the 1s and 0s of binary code; one orientation for 1, the opposite orientation for 0. Also recorded are identifying codes that allow the read-write head to locate the position of specific data on the disk when requested. The electrical control of the read-write head is such that it can record and relocate a single digit on the disk. Given that hard disks typically spin at 3,000 rpm (revolutions per minute) and that the recovery of specific data is usually achieved in a matter of milliseconds, one can readily grasp the finesse with which the system is constructed.

Floppy disk technology has never been the equal of hard disk technology, instead providing only a means by which to make data readily portable. Other technologies, particularly USB (universal serial bus) flash drives, have long since replaced the floppy drive as the portable data medium, but the hard disk drive remains the staple data storage device of modern computers. New technology in magnetic media and electromagnetic methods continues to increase the amount of data that can be stored using hard disk technology. In the space of twenty years this has progressed from merely a few hundred megabytes (10^6 bytes) of storage to systems that easily store 1 terabyte (10^{12} bytes) or more of data on a single floppy disk. Progress continues to be made in this field, with current research suggesting that a new read-write head design employing an electromagnetic "spin-valve" methodology will allow hard disk systems to store as much as 10 terabytes of data per square inch of disk space.

Impact on Industry

It is extremely difficult to imagine modern industry without electromagnetic technologies in place. Electric power generation and electric power utilization, both of which depend upon electromagnetism, are the two primary paths by which electromagnetism affects not only industry but also modern society as a whole. The generation of alternating current electricity, whether by hydroelectric, combustion, nuclear, or other means, energizes an electrical grid that spans much of North America. This, in turn, provides homes, businesses, and industry with power to carry out their daily activities, to the extent that the system is essentially taken for granted and grinds to an immediate halt when some part of the system fails. There is a grave danger to society in this combination of complacency and dependency, and efforts are expended both in recovering from any failure and in preventing failures from occurring.

That modern society is entirely dependent upon a continuous supply of electrical energy is an unavoidable conclusion of even the most cursory of examinations. The ability to use electricity at will is a fundamental requirement of many of the mainstays of modern industry. Aluminum, for example, is the most abundant metal known, and its use is essentially ubiquitous today, yet its ready availability depends on electricity. This metal, as well as magnesium, cannot be refined by heat methods used to refine other metals because of their propensity to undergo oxidation in catastrophic ways. Without electricity, these materials would become unavailable.

Paradoxically, the role of electromagnetism in modern electronic technology, although it has been very large, has not been as essential to the development of that technology as might be expected. It is a historical fact that the development of portable storage and hard disk drives greatly facilitated development and spawned whole new industries in the process. Nevertheless, it is also true that portable storage, however inefficient in comparison, was already available in the form of tape drives and nonmagnetic media, such as punch cards. It is possible that other methods, such as optical media (CDs and DVDs) or electronic media (USB flash drives), could have developed without magnetic media.

One area of electromagnetic technology that would be easy to overlook in the context of industry is that of analytical devices used for testing and research applications. Because light is an electromagnetic phenomenon, any device that uses light in specific ways for measurement represents a branch of electromagnetic technology. Foremost of these are the spectrophotometers that are routinely used in medical and technical analysis. These play an important role in a broad variety of fields. Forensic analysis utilizes spectrophotometry to identify and compare materials. Medical analytics uses this technology in much the same way to quantify components in body fluids and other substances. Industrial processes often use some form of spectrophotometric process to monitor system functions and obtain feedback data to be used in automated process control.

In basic research environments, electromagnetic technologies are essential components, providing instrumentation such as the mass spectrometer and the photoelectron spectrometer, both of which function strictly on the principle of electric and magnetic fiels. Nuclear magnetic resonance and electron

Fascinating Facts About Electromagnet Technologies

- It may soon be possible for hard disk drives to store as much as 10 terabytes of data per square inch of disk space.
- The giant electromagnet at work in a junkyard works on the same principle as the tiny read-write heads in computer disk drives.
- What came first: electricity or magnetism?
- An electron moving in an atomic orbital has the same magnetic effect as electric current moving through a wire.
- Before generators were invented to make electricity readily available for the refining of aluminum metal from bauxite ore, the metal was far more valuable than gold, even though it is the most common known metal.
- Electromagnetic "rail gun" tests have accelerated a projectile from 0 to 13,000 miles per hour in 0.2 seconds.
- Non-ionizing electromagnetic fields are being investigated as treatments for malaria and cancer.
- Coupling an electric field and a magnet produces a magnetic field that is stronger than either one alone could produce.
- In 1820, Oerstad first observed the existence of a magnetic field around an electrified coil.
- Because of the electrical activity occurring inside the human body, people generate their own magnetic fields.
- Mass spectrometers use the interaction of an electric charge with a magnetic field to analyze the mass of fragments of molecules.

spin resonance also are governed by those same principles. In the former case, these principles have been developed into the diagnostic procedure known as magnetic resonance imaging (MRI). Similar methods have application in materials testing, quality control, and nondestructive structural testing.

Careers and Coursework

Because electromagnetic technology is so pervasive in modern society and industry, students are faced with an extremely broad set of career options. It is not an overstatement to say that almost all possible careers rely on or are affected in some way by electromagnetic technology. Thus, those persons who have even a basic awareness of these principles will be somewhat better prepared than those who do not.

Students should expect to take courses in mathematics, physics, and electronics as a foundation for understanding electromagnetic technology. Basic programs will focus on the practical applications of electromagnetism, and some students will find a rewarding career simply working on electric motors and generators. Others may focus on the applications of electromagnetic technology as they apply to basic computer technology. More advanced careers, however, will require students to undertake more advanced studies that will include materials science, digital electronics, controls and feedback theory, servomechanism controls, and hydraulics and pneumatics, as well as advanced levels of physics and applied mathematics. Those who seek to understand the electromagnetic nature of matter will specialize in the study of quantum mechanics and quantum theory and of high-energy physics.

Social Context and Future Prospects

Electromagnetic technologies and modern society are inextricably linked, particularly in the area of communications. Despite modern telecommunications swiftly becoming entirely digital in nature, the transmission of digital telecommunications signals is still carried out through the use of electromagnetic carriers. These can be essentially any frequency, although regulations control what range of the electromagnetic spectrum can be used for what purpose. Cellular telephones, for example, use the microwave region of the spectrum only, while other devices are restricted to operate in only the infrared region or in the visible light region of the electromagnetic spectrum.

One cannot overlook the development of electromagnetic technologies that will come about through the development of new materials and methods. These will apply to many sectors of society, including transportation and the military. In development are high-speed trains that use the repulsion between electrically generated magnetic fields to levitate the vehicle so that there is no physical contact between the machine and the track. Electromagnetic technologies will work to control the acceleration, speed, deceleration, and other motions of the machinery. High-speed transit by such machines has the potential to drastically change the transportation industry.

In future military applications, electromagnetic technologies will play a role that is as important as its present role. Communications and intelligence, as well as analytical reconnaissance, will benefit from the development of new and existing electromagnetic technologies. Weaponry, also, may become an important field of military electromagnetic technology, as experimentation continues toward the development of such things as cloaking or invisibility devices and electromagnetically powered rail guns.

Richard M. Renneboog, M.Sc.

Further Reading

Askeland, Donald R. "Magnetic Behaviour in Materials." In *The Science and Engineering of Materials*, edited by Donald Askeland. New York: Chapman & Hall, 1998. Provides an overview of the behavior of magnetic materials when acted upon by electromagnetic fields.

Dugdale, David. *Essentials of Electromagnetism.* New York: American Institute of Physics, 1993. A thorough discussion of electromagnetism through electric and magnetic properties, from first principles to practical applications and materials in electrical circuits and magnetic circuits.

Funaro, Daniele. *Electromagnetism and the Structure of Matter.* Hackensack, N.J.: World Scientific, 2008. An advanced treatise discussing deficiencies of Maxwell's theories of electromagnetism and proposing amended versions.

Han, G. C., et al. "A Differential Dual Spin Valve with High Pinning Stability." *Applied Physics Letters* 96 (2010). Discusses how hard disk drives could include ten terabytes of data per square inch using a new read-write head design.

Sen, P. C. *Principles of Electric Machines and Power Electronics.* 2d ed. New York: John Wiley & Sons, 1997. An advanced textbook providing a thorough treatment of electromagnetic machine principles, including magnetic circuits, transformers, DC machines, and synchronous and asynchronous motors.

See also: Forensic Science; Magnetic Storage; Music Technology; Wind Power Technologies.

ELECTRONIC COMMERCE

FIELDS OF STUDY

Business information systems; computer science; computer programming; database administration; retailing; marketing; Web development; software engineering; Web design.

SUMMARY

Electronic commerce (e-commerce) refers to the buying, selling, and transfer of products and services using the Internet. It offers enormous advantages for supply-chain management and the coordination of distribution channels, in addition to being convenient for consumers. Convenience is a big selling point: In 2007, online retail generated $175 billion and is expected to climb to $335 billion by 2012.

KEY TERMS AND CONCEPTS

- **Bandwidth:** Difference between highest and lowest number of data that a medium can transmit, expressed in cycles per second or bits per second. The amount of data that can be transmitted increases as the bandwidth increases.
- **Browser:** Software program that is used to view Web pages on the Internet.
- **Download:** Copy digital data and transmit it electronically.
- **E-Book:** Electronic book that can exist with or without a printed form of the same content. E-books are downloaded from a Web site to an electronic reader (e-reader).
- **Encryption:** Any procedure used in cryptography to convert plaintext into ciphertext to prevent anyone but the intended recipient from reading the data.
- **Enterprise Resource Planning (ERP):** Software integration of projects, distribution, manufacturing, and employees.
- **Firewall:** Hardware device placed between the private network and Internet connection that prevents unauthorized users from gaining access to data on the network.
- **Hacking:** Process of penetrating the security of other computers by using programming skills.
- **Lead Time:** Time required for a product to be received from a supplier after an order has been placed.
- **Operating System:** Computer program designed to manage the resources (including input and output devices) used by the central processing unit and that functions as an interface between a computer user and the hardware that runs the computer.
- **Overhead Costs:** Daily operating expenses.
- **Server:** Computer that is dedicated to managing resources shared by users (clients).

Definition and Basic Principles

E-commerce refers to the communications between customers, vendors, and business partners over the Internet. The term e-business incorporates the additional activities carried out within a business using intranets, such as communications related to production management and product development. Many also view e-business as referring to collaborations with partners and e-learning organizations.

In addition to online purchases of goods and services, e-commerce involves bill payments, online banking, e-wallets, smart cards, and digital cash. E-commerce depends on secure connections to the Internet. Many precautions to ensure security are necessary to maintain successful e-commerce, including public key cryptography, the Secure Sockets Layer (SSL), digital signatures, digital certificates, firewalls, and antivirus programs. New technology is constantly being developed to protect against worms, viruses, and other cyber attacks.

Background and History

In 1969, the Advanced Research Projects Agency (now the Defense Advanced Research Projects Agency) of the U.S. Department of Defense proposed a method to link together the computers at several universities to share computational data via networks. This network became known as ARPANET, which was the precursor of the Internet. As a result, electronic mail (e-mail) was developed, along with protocols for sending information over phone lines in packets. The protocols for the transmission of these packets of data came to be known as Transmission Control

Protocol (TCP) and Internet Protocol (IP). Together these two protocols, known as TCP/IP, are still in use and are responsible for the efficient communication conducted through the network of networks referred to as the Internet.

Oxford University graduate Tim Berners-Lee initially created the World Wide Web in 1989 while working at CERN, the European Council for Nuclear Research, and made it available in 1991. During this time, Cisco Systems was growing to become the first company to produce the broad range of hardware products that allowed ordinary individuals to access the Internet. In 1993, Marc Andreessen and Eric Bina, employees at National Center for Supercomputing Applications (NCSA), created Mosaic, the first Web browser that supported clickable buttons and links and allowed users to view text and images on the same page. New software and programming-language developments rapidly followed, allowing ordinary consumers easy access to the Internet. As a result, companies saw the opportunity to gain customers, resulting in the creation of online businesses, including Amazon in 1994, eBay in 1995, and PayPal and Priceline in 1998.

How It Works

The engineers who developed ARPANET created the use of digital packets for transmitting data via packet switching. The general idea was that it would be faster and cheaper to transmit digital data using small packets that could be sent, or routed, to their destination in the most efficient way possible, even if the original message had to be split up into smaller packets that were then joined back together at their destination. In order to accomplish the packaging of data and transmission via the best routes, the engineers who developed ARPANET also developed Transmission Control Protocol (TCP). The first and most common access method to the Internet was through the wiring that has transmitted telephone calls. However, wireless Internet connections can now be made much faster from many locations—even from cell phones. The economy has become dependent on digital communication, and the companies that sell the most goods and services have a strong Web presence. A great deal of planning goes into the maintenance of an effective Web site, and some of the most important steps in establishing an e-commerce company are discussed next.

E-Commerce Business Establishment. After first developing a practical business plan, the process for establishing an online business that will be able to compete for sales successfully could be overwhelming. One way to start an e-business is by using a turnkey solution, which is essentially a prepackaged type of software specifically for a new business. An alternative is to use the services of an Internet incubator, which is a company that specializes in e-business development. Both eToys and NetZero used Internet incubators to help them get started. An Internet incubator typically obtains ownership of at least 50 percent of the business and may also enlist funding help from venture capitalists to get started. Web-hosting companies sell space on a Web server to customers and maintain enough storage space for the Web site and to provide support services as well. A domain name for the Web site must be chosen and registered. This domain name is to be used in the URL (uniform resource locator) for the Web site. The URL, or Web site's Internet address, consists of three parts: the host name, which is shown by the www for World Wide Web; the domain name, which is usually the name of the company; and lastly, the top-level domain (TLD), which describes the type of organization that owns the domain name, such as .com for a commercial organization, or .gov for a government organization. An initial public offering (IPO) of stock to assist with funding usually follows for enterprises that achieve a certain level of success.

Design of Markets and Mechanisms of Transactions. Initially the business-to-business (B2B) types of transactions were the primary e-commerce activities. These activities quickly expanded to include sales to consumers via electronic retailing (e-tailing), often called business-to-consumer (B2C). Since the late 1990's e-commerce has expanded to include consumer-to-consumer (C2C) Web sites, including eBay, and consumer-to-business (C2B), such as Priceline, where several airlines or hotels will compete for the purchase dollars of consumers. Each of these types of transactions can be completed within the general structure of one of the many different types of e-commerce models to generate revenue.

Automated Negotiation and Peer-to-Peer Distribution Systems. Auction models allow an Internet user to assume the role of a buyer using either the reverse-auction model (where the buyer sets a price and sellers have to compete to beat that price) or the reverse-price model (where the seller sets the minimum price that will be accepted). Auction sites, such as eBay, update listings, feature items, and earn submission and commission fees, but they leave the processes of payment and delivery up to the actual buyers and sellers.

Dynamic-pricing models include the name-your-price companies, such as Priceline.com, that use a shopping bot to collect bids from customers and deliver these bids to the providers of services to see if they are accepted. A shopping bot is a computer program that searches through vast amounts of information, then collects, summarizes, and reports the information. This is one example of intelligent agents, software programs that have been designed to gather information, used by e-businesses. Priceline's immense success is due in part to its use of this technology.

Network Resource Allocation: Electronic Data Interchange (EDI). Portal models present a whole variety of news, weather, sports, and shopping all on one Web page that allows a visitor to see an overview and then choose to obtain more in-depth information. Vertical portals are specific for a single item, while horizontal portals function as search engines with access to a large range of items. Storefront models require a product line to be accessible online via the merchant server so that customers can select items from the database of products and collect them in the order-processing technology called a shopping cart. Businesses use EDI as a standardized protocol for communication to monitor daily inventory, shipments, and payments. Standardized forms for invoices and purchase orders are routinely accessible via the use of extensible markup language (XML). Companies such as Commerce One and TIBCO Software were created with the sole purpose of helping companies to move their businesses to the Web via B2B techniques. The transition of traditional brick-and-mortar stores to click-and-mortar stores has helped to decrease lead time and has caused an increase of just-in-time (JIT) inventory management. JIT inventory management allows e-businesses to save money because the companies do not overbuy goods and create an inventory surplus that they then have to worry about storing and selling. JIT in turn decreases overhead costs.

Fascinating Facts About Electronic Commerce

- In 2010, e-commerce sales increased by 15.4 percent over 2009 e-commerce sales.
- E-book purchases in 2010 increased 150 percent over e-book purchases in 2009.
- By the year 2015 it is estimated that there will be more than 29.4 million e-books. Some believe that traditional book stores will eventually cease to exist. Borders bookstore is the largest national bookstore chain, and its financial difficulties (closing stores) could be an indication of this trend.
- The Apple iPhone is essentially a miniature, handheld computer, because it can download application programs ("apps") directly from the Internet, in addition to functioning as a phone. There are apps designed to perform just about any task, from accessing files on an office server to making the iPhone function as a flashlight. As of 2011 there were more than 100,000 apps available on the Apple App Store Web site.
- Almost 8 million Kindles were sold in 2010.
- Online sales for the 2010 holiday shopping season were estimated to total more than $36 billion.

Applications and Products

The development of computer technologies and the Internet has given rise to e-commerce, which, in turn, has spawned a variety of applications and products that facilitate e-business as well as enhance people's lives.

Consumer Products. Several tablets were among the most popular consumer digital purchases in 2010. Apple's iPad, Barnes & Noble's NookColor, Samsung's Galaxy Tab, Sony's Reader, and Amazon's Kindle are all capable of accessing the Internet and downloading magazines and books. The iPad is the most expensive of these tablets, but it is also capable of downloading video and audio files, while the Kindle is the least expensive and functions exclusively as an e-reader.

Both contact and contactless "smart cards," which resemble credit cards, have been developed to store much more information (banking, retail, identification, health care) because of a microprocessor embedded in the card. The E-ZPass, which is used by

New York and New Jersey commuters to pay tolls, is an example of a contactless smart card. Smart cards are more secure than credit cards because they are encrypted and password protected.

Security Applications and Products. Companies have been created to help merchants accept credit card payments online, which are called card-not-present (CNP) transactions. These companies, such as CyberCash and PAYware, offer services to facilitate the authentication and authorization processes through the Secure Socket Layer (SSL) using Secure Electronic Transaction (SET) technology to minimize fraud. Additional security features include firewalls, encryption, and antivirus software. Visa and other major credit card companies have introduced e-wallets that allow customers to save their shipment address and payment information securely in an online database so that purchases can be made with one click of the mouse, instead of having to reenter the same information each time. In 1999, the Electronic Commerce Modeling Language (ECML) emerged as the protocol for e-wallet usage by merchants. PayPal can be used to transfer payments between consumers securely by simply creating an account using an e-mail address and a credit card, which is used to pay for goods and services. PayPal is ideally suited for use on an auction site, such as eBay. PayPal is especially secure because credit card information is checked before the transaction actually begins. This allows for payment to take place in real time, minimizing the opportunity for fraud.

Applications Using Wireless Transactions. Mobile business (m-business) made possible by wireless technology will continue to grow in importance. The third generation, called 3G technology, is allowing wireless devices to transmit data more than seven times faster than the 56K modem, and 4G technology began to replace it in 2011. Sprint PCS provides access to the Internet using the Code Division Multiple Access (CDMA) technology. CDMA technology assigns a unique code to each transmission on a specific channel, which allows each transmission to use the entire bandwidth available for that channel, greatly decreasing the time it takes to complete an e-commerce transaction.

Impact on Industry

The developments in technology that have made the Internet more easily accessible to the average consumer have been made primarily by American companies. These companies have continued to grow and expand to reach consumers all around the world. Other nations have grown economically because of the explosion in e-commerce, and much of the e-commerce growth is expected to involve the BRIC nations (Brazil, Russia, India, and China).

Amazon.com was founded in Jeff Bezos's Seattle garage in 1994. Bezos was named *Time* magazine's Person of the Year in 1999, and the company that he started in his garage continues to grow at an amazing rate. In addition to books, products available on Amazon have come to include jewelry, sporting goods, shoes, digital downloads of music, videos, games, software, health and beauty aids, and just about everything else possible, including groceries. The company reported net sales of $7.56 billion for the third quarter of 2010, which was a 39 percent increase over the same period in 2009.

eBay.com has become the largest online auction site in the world. Visitors to the site can buy or sell just about anything, including iPods, laptops, digital cameras, tickets to concerts and sporting events, jewelry, books, antiques, crafts, sporting goods, pet supplies, and clothes. eBay is headquartered in San Jose, California, and revenue for the third quarter of 2010 was $2.2 billion, an increase of 1 percent over the same period of 2009. Its PayPal business, acquired in 2002, has grown by at least an additional 1 million accounts every month, and it can accept twenty-four different types of currencies worldwide.

PayPal was initially founded in 1998 in Palo Alto, California, and like many other original U.S.-based tech companies, it has long since expanded its operations worldwide. PayPal has locations in Berlin, Tel Aviv, Dublin, Luxembourg, and China. Its purpose is to facilitate the processing of online payments for various e-commerce businesses.

Priceline.com, located in Norwalk, Connecticut, developed its name-your-own-price system for its online auction type of business and has grown since it emerged in 1998. Visitors to the Web site can list the price they want to pay for travel-related services and items, including hotels, vacation packages, cruises, airplane tickets, and car rentals. Although it briefly tried to expand on its e-commerce activities to include home loans, long-distance telephone service, and cars, it discontinued these ventures in 2002 and has continued to excel in travel-related services. Its chief e-commerce competitors are Expedia.com, Travelocity.com, Orbitz.com, and Hotwire.com. As of 2011, Priceline.com was the leader. The company had more than three hundred employees, and its third-quarter 2010 revenue was reported to be $1 billion, a 37.1 percent increase over 2009.

FedEx is the leader in air and ground transportation and won recognition in *Fortune* magazine's "World's Most Admired Companies" in 2006, 2007, 2008, 2009, and 2010 and in *Business Week*'s "50 Best Performers" in 2006. In 2010, its second-quarter revenue was $9.63 billion, an increase of 12 percent from the same period in 2009. FedEx continues its commitment to innovation with its FedEx Institute of Technology at the University of Memphis, in the company's hometown. The institute focuses on nanotechnology, artificial intelligence, biotechnology, and multimedia arts.

While an undergraduate at Yale University in 1965, Fred W. Smith wrote a term paper describing the implementation of an airfreight system that could transport computer parts and medications in a timely fashion. Although at the time his idea was not viewed as feasible by Yale faculty, Smith later bought an interest in Arkansas Aviation Sales, which eventually grew to become FedEx. E-commerce has continued to grow because of the rapid and efficient transportation of goods made possible by FedEx and the United Parcel Service (UPS).

Careers and Course Work

The job titles, career paths, and salaries vary a great deal within the field of e-commerce. Some of the typical job titles include Web site developer, Web designer, database administrator, Web master, and Web site manager. Jobs related to Web content require skills in Web development tools and software languages, including hypertext markup language (HTML), extensible markup language (XML), Java, JavaScript, Visual Basic, Visual Basic Script, and Active Server Pages (ASP). Database administrators focus less on these Web tools and languages and more on database-related tools, such as structured query language (SQL), Microsoft Access, and Oracle. Knowledge of computer networks and operating systems is also very helpful.

Social Context and Future Prospects

Due in part to the easy accessibility of goods and services via the Internet, online businesses have become increasingly competitive, with more made-to-order goods being produced by companies such as Dell and corresponding decreases in the costs associated with the maintenance of a large inventory. Since so much more data are exchanged digitally via stock trades, mortgages, purchases of consumer goods, payment of bills, and banking transactions, it is conceivable that eventually digital cash and smart cards could replace traditional cash. Because consumers enjoy the comparison shopping among goods offered by companies all over the world, as well as the twenty-four-hour-a-day, seven-days-a-week convenience of online shopping, sales at the traditional brick-and-mortar stores will no doubt continue to decline, or at least migrate toward those goods that consumers prefer not to purchase online (such as those they wish to consider physically and those they wish to obtain immediately).

Hacking, identity theft, and other types of cyber theft have become problems, which have only increased the need for better security tools, such as digital certificates and digital signatures. Increased security needs will continue to fuel the ever-expanding Internet security industry.

Jeanne L. Kuhler, B.S., M.S., Ph.D.

Further Reading

Byrne, Joseph. *I-Net+ Certification Study System.* Foster City, Calif.: IDG Books Worldwide, 2000. This review guide provides technical information regarding hardware for networks, as well as software, important for e-commerce.

Castro, Elizabeth. *HTML, XHTML, and CSS, Sixth Edition: Visual Quick Start Guide.* Berkeley, Calif.: Peachpit Press, 2007. An introductory text describing how a novice can set up an individual Web site and includes plenty of helpful screen shots.

Deitel, Harvey M., Paul J. Deitel, and Kate Steinbuhler. *E-Business and E-Commerce for Managers.* Upper Saddle River, N.J.: Prentice Hall, 2001. This introductory textbook describes e-commerce in fairly nontechnical terms from the business perspective.

Longino, Carlo. "Your Wireless Future." *Business 2.0.* May 22, 2006. http://money.cnn.com/2006/05/18/technology/business2_wirelessfuture_intro/. Discusses the future of wireless technology in all sectors—business, entertainment, and communications.

Turban, Efraim, and Linda Volonino. *Information Technology for Management: Improving Performance in the Digital Economy.* 7th ed. Hoboken, N.J.: John Wiley & Sons, 2010. This introductory textbook provides both technical and nontechnical information related to e-commerce.

Umar, Amjad. "IT Infrastructure to Enable Next Generation Enterprises." *Information Systems Frontiers* 7, no. 3 (July, 2005): 217-256. Describes advances in network protocols and design.

Web Sites

E-Commerce Times
http://www.ecommercetimes.com

National Retail Foundation's Digital Division
http://www.shop.org

See also: Wireless Technologies and Communication

EPOXIES AND RESIN TECHNOLOGIES

FIELDS OF STUDY

Organic chemistry; polymer chemistry; industrial chemistry; chemical engineering; civil engineering; automotive engineering; watercraft design; adhesives; advanced composite technology; molding; mold making; extrusion manufacturing; prosthetics; packaging; environmental chemistry; materials science; materials recycling.

SUMMARY

Epoxies and resins are chemical systems, as opposed to single compounds, that are used in a variety of applications. Their value derives from their polymerization into three-dimensional or cross-linked polymeric materials when the components are combined and allowed to react. Epoxies are so-named because the principal component is a reactive epoxide compound. The combination of the epoxy compound with a second material that promotes the polymerization reaction is called a resin. The term also applies generally to any polymerizing combination of materials that is not epoxy based. Epoxies and resins are used primarily in structural composite applications, in which the combination of a reinforcement material (usually a specialized fiber) bound within a solid matrix of polymerized resin provides the advantages of high strength, low weight, and unique design capabilities. Resins are also used in injection molding and other molding operations, extrusion and pultrusion, prototype modeling, and as high-strength adhesives.

KEY TERMS AND CONCEPTS

- **Amine:** Organic compound of nitrogen derived from ammonia (NH_3) by replacement of one or more hydrogen molecules by a hydrocarbon radical. Amines can be primary (RNH_2), secondary (R_2NH), or tertiary (R_3N).
- **Cross-Linking:** Bonding of two molecules to each other from separate polymerization chain reactions; also the extent to which intermolecular bonds have formed between the otherwise separate molecular chains of a polymer.
- **Epoxide:** Compound whose molecules contain a three-membered ring structure of two carbon atoms and one oxygen atom; normally formed through the incomplete oxidation of a carbon-carbon double bond.
- **Epoxy:** Resinous material made up, in part, of epoxide compounds.
- **Fiber-Reinforced Plastic (FRP):** Structural composite consisting of a fibrous support material encased in a solid matrix of polymerized resin.
- **Functional Group:** Group of atoms that exhibits specific chemical behavior in a molecule.
- **Glass Transition Temperature (Tg):** Temperature range at which a thermoplastic material begins to lose the characteristic behavior of fracturing conchoidally (like glass) and becomes plastic; also the temperature range at which a thermosetting material begins to decompose.
- **Polymer:** Compound formed by sequential repeating reactions between simpler molecules; from "poly," meaning many, and "mer," meaning form.
- **Polymerization:** Process of forming a polymer; a type of reaction that produces very large molecules by the intermolecular bonding of smaller molecules to each other in a repeating manner.
- **Thermoplastic:** Describing a material, usually a polymer, that softens or becomes plastic with heating.
- **Thermosetting:** Describing a material, usually a resin, that polymerizes and hardens with heating.

Definition and Basic Principles

In the field of polymers, resin refers to the material or blend of materials that is specifically prepared to undergo a polymerization reaction. In this type of reaction, molecules add together sequentially to form much longer and larger molecules. A polymerization reaction can proceed in a linear manner to form long-chain single molecules whose bulk strength derives from physical entanglement of the molecules. It also can proceed with branching to form large, multiple-branch molecules that derive their bulk strength from their sheer size and complex three-dimensional interlinking bonds between the molecules.

The particular combination of materials used to prepare the resin for polymerization is chosen according to the extent and type of polymerization desired. Monomers containing only one reactive site or two functional groups can form only linear polymers. Three-dimensional polymers require the presence of three or more functional groups or reactive sites in at least one of the resin components. The polymerization reaction can proceed as a simple addition reaction, in which the single monomer molecules simply add together by forming chemical bonds between the reactive sites or functional groups on different molecules.

Polymerization reactions are generally driven to completion with heating, although the heat produced by exothermic reactions must be controlled to prevent overheating, decomposition, and dangerous runaway reactions from occurring. The polymeric product of the reaction may be thermoplastic, becoming soft, or plastic, with heating. Thermoplastics are characterized by this change of behavior at the glass transition temperature, *Tg*. Below this temperature, the material is solid and fractures in the characteristic conchoidal manner of glass rather than along any regular planes that would denote a regular crystal structure. At the *Tg*, the material begins to deform rather than to fracture. The *Tg* is always stated as a fairly broad temperature range, and at its higher value, the material has no resistance to deformation that would result in fracture, although it may not yet be entirely liquefied.

A thermosetting resin produces a polymer that does not soften with further heating and exhibits conchoidal fracture behavior at all temperatures at which it is stable. Such polymers will undergo thermal decomposition (also called thermolysis) when heated, as their *Tg* is at a higher temperature than the temperature at which they break down.

Epoxies are a specific type of resin in which one of the components is an epoxide compound. A second component, typically an amine, reacts irreversibly with the epoxide functional group, causing its three-membered ring structure to open up. The intermediate form produced reacts in a chainlike manner with other epoxide molecules to form complex, three-dimensional polymeric molecules.

Various technologies, methods, and applications are encompassed by the field of epoxies and other resins. These range from molecular design and testing in the chemical and material sciences laboratory, to injection molding and hand layup of fiber-reinforced plastics, and the repair of structures made from resin-based materials. The production of specific resin formulations on an industrial scale is particularly exacting because of the regulations governing certification of the materials for specific critical uses and concomitant purity requirements throughout the handling of the product. Specialized training and equipment is required for the safe production and transportation of the materials.

Background and History

Resins and their property of solidifying have been known and used since ancient times. During explorations of the New World and Asia, European explorers such as Christopher Columbus and Hernán Cortés found the indigenous peoples playing sports with balls that bounced and wearing clothing and footwear that had been waterproofed. The indigenous peoples used natural latex materials derived from plants in the making of these and other objects.

In the mid-nineteenth century, as the industrial sciences, especially chemistry, blossomed in Europe, these marvelous natural latex materials, known as caoutchouc and gutta percha, were imported and put to a variety of uses that capitalized on their unique properties. Gutta percha, for example, was used to make the corrosion-resistant coating and insulation for the first underseas telegraph cables laid across the English Channel between England and France. Other resins produced at this time were semisynthetic, chemical modifications of vegetable oils and latexes. The development of synthetic polymers such as Bakelite, especially after World War II, opened the way for untold applications. The unique and customizable properties of plastics and polymer resins served as the foundation of a very large and growing industry that has constantly sought new materials, new innovations, and new applications.

How It Works

Resins and Chain Reactions. The term "resin" was originally used to refer to secretions of natural origin that could be used in waterproofing. It has since come to mean any organic polymer that does not have a distinct molecular weight. Typically, organic polymers form through sequential addition reactions between small molecules that then form much larger molecules through a chain reaction mechanism. Once initiated, the progress of such a reaction chain becomes entirely random. Any particular reaction chain will proceed and continue to add monomer molecules to the growing polymer molecule as long as it encounters them in an orientation that permits the additional step to occur. Typically, this happens several thousand times before a condition, such as an errant impurity, is encountered that terminates the series of reactions. The exact molecular and chemical identity of any individual polymer molecule is determined precisely by the number of monomer molecules that have been combined to produce that particular polymer molecule. However, within a bulk polymerization process, billions of individual reaction chains progress at the same time, in competition for the available monomers, and there is no way to directly control any of the individual reactions. As a result, any polymerized resin contains a variety of homologous molecules whose molecular weights follow a standard distribution pattern. In thermoplastic resins, this composition, consisting of what is technically a large number of different chemical compounds, is the main reason that the *Tg* is characterized by softening and gradual melting behavior over a range of temperatures rather than as the distinct melting point typical of a pure compound.

PolymerizationReactionProcesses. Polymerization reactions occur in one of two modes. In one, monomer molecules add together in a linear head-to-tail manner in each single chain reaction. This occurs when only two atoms in the molecular structure function as reactive sites. In the other mode, there is more than one reactive site or functional group in each molecule. Polymerization reactions occur between reactive sites rather than between molecules. The presence of more than one reactive site in a molecule means that the molecule can take part in as many chain reaction sequences, with the resulting polymer molecules being cross-linked perhaps thousands of times and to as many different polymer chains. The result can, in theory, be a massive block of solid polymeric material composed of a single, large molecule.

Epoxy Resins and Cross-Linking. Polymerization and cross-linking bonds arise as the reactive site or sites of the molecules become connected by the formation of chemical bonds between them. As a bond forms from the atom at one end of a reactive site, the atom at the other end becomes able to form a bond to the reactive site of another molecule. When the reactive site is an epoxide ring structure, the resulting resin is called an epoxy. Epoxy resins are two-part reaction systems, requiring the mixture and thorough blending of the epoxide compound and the catalyst, a second compound that initiates the ring opening of the epoxide. This is typically an amine, and the relative amount of amine to epoxide controls the rate at which the polymerization occurs. This represents essentially all the control that can be exercised over the progress of a polymerization reaction. It is therefore critical to control the relative amounts of epoxide and catalyst in an epoxy resin blend.

Applications and Products

The value of epoxy and other resins is in their versatility; they are used in a wide variety of products and applications that have become central to modern society. Without epoxy resins and technology, many products would not exist, and modern society would be very different. Epoxy resins cure to a tough, resilient, and very durable solid that has high resistance to impact breakage, fracturing, erosion, and oxidation. They are also reasonably good thermal conductors that tolerate rapid temperature changes very well.

Aircraft. An excellent example of resin application is in aircraft technology, particularly in modern fighter jets. The fuselage and wing structures of many aircraft are constructed of fiber-reinforced plastics. The materials used in aircraft production must be able to tolerate drastic changes in temperature and pressure. For example, an aircraft may be stationed on the ground in a desert with surface temperatures in excess of 60 degrees Celsius, and less than one minute later, the aircraft may be in the air at altitudes where the air temperature is −35 degrees C or colder. That the structural materials of such an aircraft can repeatedly withstand abrupt changes of temperature and physical stresses says a great deal about the strength, toughness, and thermal properties of the epoxy resins used in its construction.

Electronic Devices. The thermal stability of epoxy resins is also evident in their use in the packaging material of modern integrated circuits, transistors, computer chips, and other electronic devices. The operation of these devices produces a great deal of heat because of the friction of electrons moving in the semiconductor material of the actual chip. Pushed to extremes, the devices can fail and burn out, but it is far more usual for the packaging material to adequately conduct and safely dissipate heat, allowing whatever process is running to continue uninterrupted. That may be something as trivial as some spare-time gaming, or as crucial as an emergency response call, the flight control program of an aircraft in the air, or an advanced medical procedure.

Structural Composite Applications. The applications of and the products produced from resins are numerous. The combinations of materials for the production of resins are essentially limitless, and each combination has specific qualities that make it suitable to particular applications. Thus, the varieties and possibilities in the field of epoxies and resin technologies are virtually limitless. A very significant area of application for epoxy resins and other types of resins is in the field of structural composites, particularly in fiber-reinforced plastics and as insulating or barrier foams. The particular application of a resin is determined as much by the desired properties of the product as by the properties of the polymerized resin. Resins that produce a hard, durable polymer such as those produced by epoxy resins are used in products of a corresponding nature. Resins that have good shape-retaining properties coupled with high compressibility, such as those used to produce urethane foams, are used in products such as furniture cushions, pillows, mattresses, floor mats, shoe insoles, and other applications in which the material provides protection from impact forces. Resins that exhibit high levels of expansion while forming a fairly rigid polymer with good thermal resistance are used in sealing and insulating applications, such as those for which urea-formaldehyde resin combinations are so useful.

Resin Production and Supply. A completely different set of technologies and applications is related to the supply and material processing of the resins themselves. Chemists and chemical engineers expend a great deal of effort and time in the development and testing of resins in order to identify new commercially valuable materials or to customize the properties of existing materials. When the new product leaves the laboratory for commercial applications, the system must then be established for the production and safe transport of the material from the supplier to the user. Systems and methods must also be established for the end user to prepare the intended products from the material. Resins for low-volume use can be packaged in cans and other small containers, while those for high-volume use may be transported by rail or in other types of large

Fascinating Facts About Epoxies and Resin Technologies

- The quest for a moldable material of chemical origin began with the need to find a replacement for ivory in the production of billiard balls. Ivory was considered to be the best material for billiard balls but was becoming scarce and expensive. In 1865, the company Phelan and Collender offered a reward of $10,000 in gold for a suitable substitute material.
- Nitrocellulose, or nitrated cellulose, was discovered accidentally by Christian Friedrich Schönbein in 1845 while he was carrying out chemical experiments with nitric acid and sulfuric acid in his kitchen. When he spilled a mixture of the two acids onto the linoleum floor, he used a clean cotton apron to mop up the spill. As it hung drying next to the hot stove, the apron burst into flame and vanished in a light puff of smoke. This was the origin of smokeless gun powder, called gun cotton.
- The first true synthetic polymer was a thermosetting resin compounded of phenol and formaldehyde called Bakelite. It was invented by Leo Hendrik Baekeland in 1907 in an effort to find a substitute for imported natural shellac.
- In theory, a single polymerization chain reaction initiated in a railway tank car full of an epoxy resin could turn the entire mass into a single, very large molecule.
- Variations in the composition of nitrocellulose led to its successful use as the basis of celluloid, the basic material of photographic film developed by George Eastman. The Kodak Brownie, a compact, portable camera (the body of which was made from Bakelite) was also introduced by Eastman, along with custom film-developing service.
- Epoxy resins represent some of the most advanced materials known, especially when used as the matrix material of a structural composite. In some modern aircraft, epoxy-based fiber-reinforced plastics have almost completely supplanted metals in structural components.

containers. Production methods must produce the resin material in a sufficiently pure state so that it will not polymerize en route. Methods of transport must also be such that the resin is protected against any contamination that could result in initiation of polymerization. This requires specialized applications in transportation technology. The end user of the resins will require the means to manipulate the resin, typically in spray-on applications or molding operations. The equipment used in the various molding operations also requires the creation of molds and forms appropriate to the product design. There is accordingly a very large sector of skilled support workers in industries and applications for resin usage.

Impact on Industry

The twentieth century has been referred to as the "synthetic century," an indication of the impact that plastics have had on both industry and society. Up until the mid-nineteenth century, goods produced by industry were restricted to traditional materials of natural origin, such as woods, metals, minerals, and animal by-products, and to rubber compounds from plant sources. Gasoline and diesel internal combustion engines had not yet been developed, and the "science" of chemistry lacked any theory of atomic structure that would permit its practitioners to predict atomic behaviors or to understand molecular structures and their interactions. As growing populations demanded more and more commercial goods and advances were made in the modes of transportation (such as railroads), industrial manufacturing capability increased, creating a need for cheap, accessible materials to replace those from natural sources that were diminishing in supply, such as ivory and exotic hardwoods. New materials were also wanted for use in rapid production methods, such as hot pressing. These production methods would cut costs by eliminating the large numbers of skilled artisans and craftspeople performing piecework in production settings. The first truly successful artificial material that fit these needs was the nitrocellulose known as collodion, and the first truly successful commercial application of that material was in the production of celluloid. As the base material of convenient, flexible photographic film for small personal cameras, celluloid revolutionized the practice of photography and ushered in the age of industrial plastics in the last years of the nineteenth century.

Atomic Theory and Explosive Industrial Growth. Around the turn of the twentieth century, modern atomic theory was developed, and chemistry became a mainstream science through which new materials could be produced. Each new material engendered new applications, and each new application played to a demand for still newer materials, mostly derived from coal tar, of which a ready supply existed. The final key requirement was the discovery and development of polymerization. The first completely synthetic polymer, compounded from phenol and formaldehyde, was developed in 1907 by Belgian chemist Leo Hendrik Baekeland. It proved to be the elusive material needed to expedite the mass production of consumer goods. Soon, many other new materials were created from polymerization, which led to the development of the modern plastics industry. These versatile resin materials were used in a variety of applications, from the synthetic fibers used to make cloth to essential structural components of modern space and aircraft.

Careers and Course Work

The demand for products of resin technologies is likely to increase in accord with the expanding human population. The facility with which large quantities of objects can be produced by epoxy and resin technologies ensures the continued growth of the field as it keeps pace with the needs of the population. The need for new or improved qualities in the materials being used in resin products means a need for materials scientists with advanced training and knowledge in organic and physical chemistry.

An individual who chooses to make a career of resin technologies must learn the chemical principles of polymerization by taking courses in advanced organic chemistry, physical chemistry, analytical chemistry, reaction kinetics, and specialty polymer chemistry. He or she will also need courses in mathematics, statistics, and physics. Specialized fields of engineering related to epoxy and resin technologies include chemical engineering, mechanical engineering, and civil engineering (in regard to certain special infrastructure applications). Aircraft maintenance engineers and technicians must go through specific training programs in the use of resins and epoxies as they apply to aircraft structural maintenance standards. These are hands-on training programs focused on the physical use and applications of the materials rather than courses of instruction in chemical theory. No special training is required for the layperson to make use of the materials, which are sold in the automotive and marine supply sections of many retail outlets and by certain hobby and craft suppliers.

Social Context and Future Prospects

A vast quantity of plastics is produced from resins. The strongest of these are the epoxy resins. In concert with the commercial and social benefits of epoxy and resin technologies are the logistical problems inherent in the materials themselves. The use of resin-based technologies carries with it the responsibility for the proper disposal of the used products. Thermoplastics are relatively easily managed because of their built-in ability to be reused. Because they can be rendered into a mobile fluid form simply by heating, used objects made from thermoplastic resins can be melted down and formed into new products. Thermosetting resins, however, cannot be reformed and must be processed for disposal in other ways. Thermoset plastics, such as the epoxy resins, are resistant to facile reprocessing as they are generally also impervious to solvents and all but the strongest oxidizing agents. Historically, and unfortunately, this has meant that the vast majority of goods made from thermosetting resins have been relegated to landfills, off-shore dumps, or just left as litter and refuse. Beginning in the late twentieth century, efforts began to be made to put such materials to other uses, the most common being simply to grind them up for use as bulk filling materials.

Epoxy and resin technologies and the plastics industry in general have had a huge impact on modern life since their inception, becoming essential to the infrastructure of modern society. Essentially every government and university research program, every industry and business sector, and every corporation that deals in material goods of any kind deals with resins and plastics in some way, and new ventures are established almost daily for the production of material goods designed specifically to be produced by epoxy or resin technologies.

Richard M. J. Renneboog, M.Sc.

Further Reading

Elias, Hans-Georg. *An Introduction to Plastics.* 2d ed. Weinheim, Germany: Wiley-VCH, 2003. Presents clear and usable descriptions of many aspects of the chemistry behind and the applications of resins and plastics, particularly in their manner of polymerization.

Fenichell, Stephen. *Plastic: The Making of a Synthetic Century.* New York: HarperBusiness, 1997. An extremely readable and entertaining account of the evolution of the plastics industry. Pays considerable attention to the social context and relevance of various materials.

Goodship, Vanessa. *Practical Guide to Injection Moulding.* Shrewsbury, England: Rapra Technology, 2004. Presents comprehensive information on the techniques and principles of injection molding, including troubleshooting of production machinery.

Green, Mark M., and Harold A. Wittcoff. *Organic Chemistry Principles and Industrial Practice.* 2003. Reprint. Weinheim, Germany: Wiley-VCH, 2006. Explains the principles of organic chemistry as related to epoxy resins and polymerization. Although the book is nicely presented in an understandable format, the level of chemistry is for the advanced student.

Lewis, Richard J., Sr., and Gessner G. Hawley. *Hawley's Condensed Chemical Dictionary.* 15th ed. Chichester, England: John Wiley & Sons, 2007. Entries include subjects such as trade names, chemical identities, and applications of diverse materials and methods.

Web Sites

European Resin Manufacturers Associaton
http://www.erma.org.uk

Plastics Institute of America
http://www.plasticsinstitute.org/index.php

EROSION CONTROL

FIELDS OF STUDY

Soil science; soil conservation; hydrology; hydrogeology; geology; agronomy; watershed management; water quality; air quality control.

SUMMARY

Preventing or eliminating the erosion of soil and rock protects water and air quality and the integrity and usability of public and private lands. As a result, erosion control requires both engineering and land-use management techniques. Erosion control is often seen as primarily an agricultural problem because the land area affected by farming practices is large. However, erosion control is also necessary in coastal regions and areas used for forestry, transportation, development, and recreational purposes because natural or human actions can change landscapes or soil covers and result in increased erosion.

KEY TERMS AND CONCEPTS

- **Conservation Tillage:** Sequence of farming practices that reduces loss of soil or water.
- **Desertification:** Gradual conversion of productive land into desert through loss of topsoil from abusive land-management practices.
- **Nonpoint Source Pollution:** Pollution from dispersed natural and human sources accumulated and carried by rainfall or snowmelt moving over and through the ground.
- **Overgrazing:** Pressure from travel and vegetation removal by grazing animals that results in exposure of bare soil surfaces.
- **Slope Stabilization:** Techniques or structures used to prevent movement of areas adjacent to excavations or in natural slopes forming cliffs, valley sides, and reservoirs that can have significant impacts.

Definition and Basic Principles

Erosion control is the practice of preventing the movement of soil or rock by the action of wind or water. Uncontrolled erosion by either natural or human actions can cause water or air pollution or damage to property. Erosion is a natural result of the action of water or wind; however, human activity can accelerate this process by removing protective vegetation or creating instabilities in existing soil and rock structures.

Wind erosion acts by selectively transporting soil particles; in other words, the higher the wind velocity, the larger the soil particle that can be transported. Water erosion works both as precipitation hits the ground surface and as water flows over the land. Erosion-control techniques and structures work to reduce the potential for soil or rock transport either by reducing the exposure of land surfaces to the effects of wind and water or by modifying the landscape or installing structures that increase the stability of the landform. For example, mulching, revegetation, and the application of geotextiles are all approaches that work to reduce the vulnerability of soil surfaces to erosion. Gabions, retaining walls, and terracing are examples of structures or modifications to landforms.

Lack of erosion control can result in declines in soil productivity for agricultural purposes and loss of land stability for land development and uses. Examples of the large-scale effects of erosion include landslides and desertification.

Background and History

Before 1920, erosion was noted during farming practices but not necessarily considered reparable, and eroded lands were abandoned in favor of new fields. This mind-set had its roots in the colonial days of the United States, where potential new farmland stretched to the horizon and was freely available to those willing to clear the land. Decades of intensive farming practice led to the loss of soil productivity and then the soil itself to water and wind erosion, with dramatic examples of wind erosion occurring during the Dust Bowl era of the 1930's.

In 1935, the United States Congress established the Soil Conservation Service (now the Natural Resources Conservation Service) to provide for ongoing work to conserve the nation's soils. This act established soil conservation as a national priority independent of agricultural programs, which had far-reaching advantages borne out by more recent concerns about erosion related to storm-water runoff, coastal wave and storm action, and river movement.

Erosion-control practices are required in many different circumstances. Agricultural practices for animal husbandry or to produce food crops can have significant adverse impacts on soil health if soil conservation techniques are not applied. Land development for residential housing or commercial and industrial facilities can have short-term erosion impacts during construction or long-term effects from rainwater or snowmelt on the sites. Recreational uses near lakes, streams, or oceans can affect the stability of the shoreline. Transportation systems may create unstable slopes with increased potential for landslides from cuts through hills for roads and rail lines.

HOW IT WORKS

Agriculture. Erosion-control practices in agriculture focus primarily on nonstructural techniques to retain and improve soil productivity. The intent of erosion-control measures is not only to protect the soil from raindrop impacts or wind transport but also to increase the infiltration capacity of the soil to lower the amount of water that runs off over land. One way that infiltration capacity is increased is to reduce the length of the slope that is available for water to travel by using terraces or contour plowing. Another way to increase infiltration is to slow the movement of water and physically protect the soil surface with mulch or vegetation, practices that also increase soil fertility by increasing organic content and biological activity.

Coastal Zone Management. Wave action and storm surges in coastal environments cause beach and headland erosion that is typically controlled through revegetation of coastal dunes and barrier islands or hardened structures like riprap, gabions, and retaining walls. An understanding of beach dynamics is needed, particularly of the seasonal transport of sand on- and offshore or the transverse migration of sediments along shorelines. Hard structural erosion-control features must be evaluated and designed to ensure that erosion is not increased elsewhere as an unintended consequence.

Transportation. The development and maintenance of transportation systems can increase erosion potential in landscape by altering the stability of landforms through either physically changing the landscape or by increasing water runoff. For example, road cuts can increase the potential for erosion and mass wasting by creating new steep slopes in previously stable hillsides. Unless control measures are used, for example, gabions, geotextiles, or terraces, erosional forces will work to reduce the steep slope to a more stable form. This erosive action can take the form of catastrophic landslides. Increased water runoff is also generated by road systems because of the impermeable nature of most road surfaces. The runoff, if not addressed, can increase erosion of road beds or nearby stream drainages, which can increase road-maintenance requirements or property damage.

Storm-Water Management. Residential, commercial, industrial, and public-facility uses generate erosion-control needs during both the development phase and long-term occupation of the site. Site preparation include activities such as tree and vegetation clearing and grading that expose soils and subsoils to erosion and alter the topography of the site. Erosion-control features during construction include phasing the work to minimize the amount of land exposed at any particular time, rapid revegetation of the site, and temporary approaches such as mulching.

Long-term land uses can also increase erosion potential if rainfall or snowmelt is not adequately accommodated on the site. In a similar manner to road systems, the impermeable surfaces created by roofs, driveways, parking areas, and walkways means that overland flow is increased and the infiltrative capacity of the site is reduced. Increased off-site erosion can result if excess overland flow is discharged to adjacent properties or the road system.

Recreation Management. The effects of recreational activities on landscapes may be dealt with through other disciplines and programs mentioned above. Effects of recreation often include soil compaction and vegetation removal through foot or vehicle travel. When the effects are limited in area, the overall effect on erosion potential can be slight; however, when the effects are widespread or occur on steep slopes or near bodies of water, the erosion potential can be high. Mitigating approaches may be to limit the area of travel, for example, designated trails or sites for activities, which will allow revegetation to occur naturally or through replanting. In some instances, the volume of traffic may be so great that structural approaches are needed. For example, a shoreline hardened with a retaining wall can allow

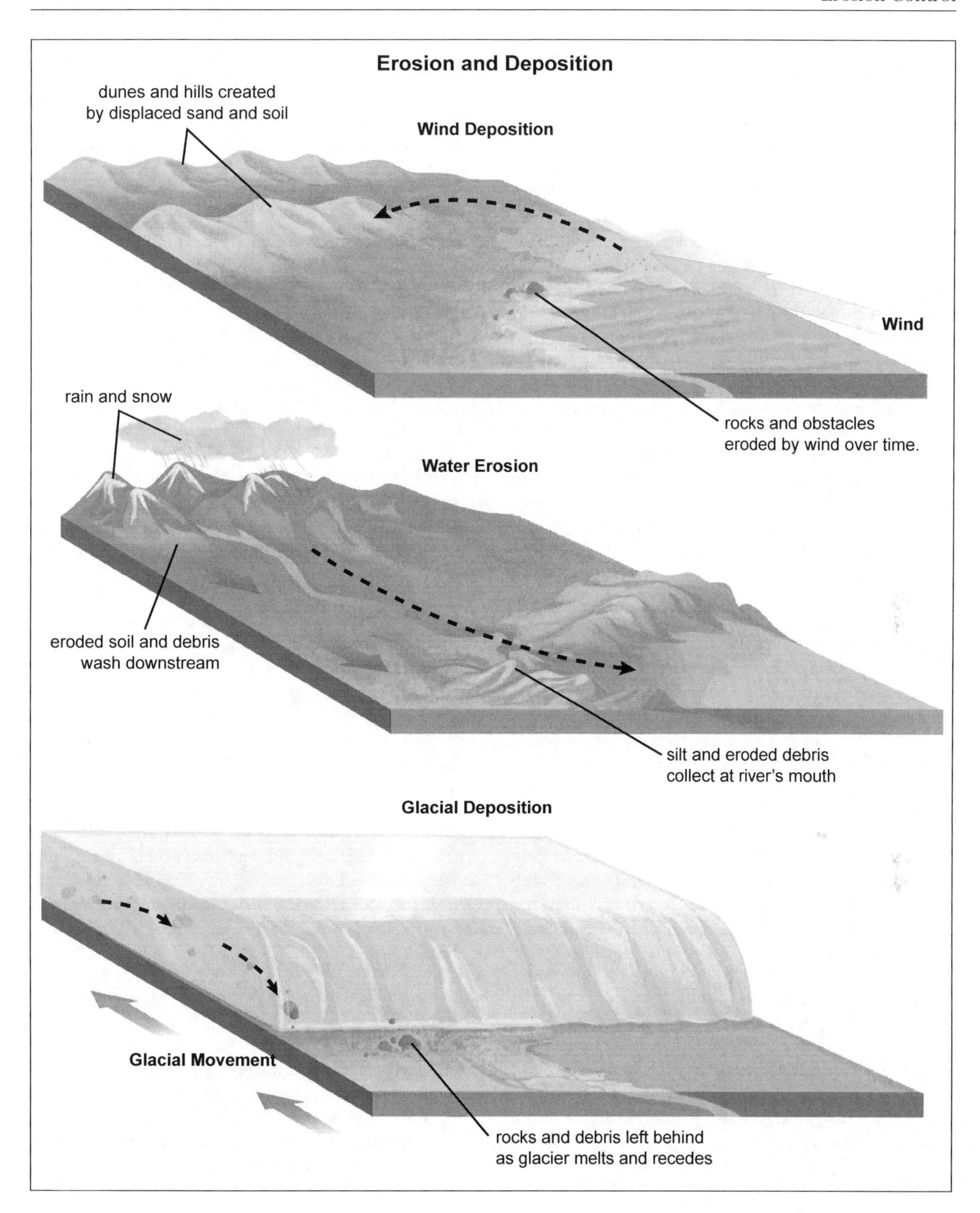
Erosion and Deposition
dunes and hills created by displaced sand and soil
Wind Deposition
Wind
rocks and obstacles eroded by wind over time.
rain and snow
Water Erosion
eroded soil and debris wash downstream
silt and eroded debris collect at river's mouth
Glacial Deposition
Glacial Movement
rocks and debris left behind as glacier melts and recedes

increased access for fishing when foot traffic would heavily impact a naturally vegetated shoreline.

APPLICATIONS AND PRODUCTS

Agriculture. Many agricultural erosion-control measures emphasize changing farming practices rather than necessarily installing a structure or feature on the site. For example, conservation tillage works to minimize soil tilling to leave as much crop residue in place as possible on the surface of the soil. This approach requires that the crop residue is not plowed into the soil at the end of the growing season and the new crop is planted in rows plowed through the residue at the beginning of the next season. The crop residue acts as a mulch to help stabilize the soil, retain soil moisture, and increase the soil's organic content. In this approach, chemical weed control is often used to eliminate undesirable plants or the crop from the previous year. Alternative methods of controlling weeds are available if a reduction in use of herbicides is needed.

Other nonstructural approaches to erosion control on agricultural land include contour plowing and crop or pasture rotation. Crop rotation is a practice where a sequence of crops are cultivated on the land over a series of seasons or years. Rotating crops improves soil structure and makes it more resistant to erosion, particularly when paired with conservation-tillage techniques. In addition, varying the crops grown can reduce reliance on herbicides and pesticides by, for example, changing the crop to one that creates field conditions that are less favorable to the weeds or pests of concern.

Structural erosion-control measures for agriculture include terraces and large windbreaks. Although windbreaks are not constructed, they usually are composed of trees planted in rows perpendicular to the prevailing wind direction. They do require planning for location and long-term maintenance to ensure continued effectiveness. Terraces are constructed to created areas of flat land in an otherwise sloped terrain. Terraces retain water behind a dyke or berm at the edge of the terrace and can help manage water on a site after heavy rains. Terraces cost more than other erosion-control measures in terms of labor and equipment to construct but can be effectively farmed with large equipment when properly designed, located, and constructed.

Coastal Zone Management. Erosion control in coastal zones can be necessary because the action of wind and waves erodes coastal lands so that it threatens structures or features such as roads, buildings, or navigation channels. Erosion control can also be needed when structures encroach into areas that are vulnerable to erosion or are otherwise unstable landforms. Examples include the construction of buildings on barrier islands or on exposed headlands.

Nonstructural erosion-control methods in coastal environments can be modeled after natural features such as coastal-dune environments or saltwater marshes to create buffers that absorb the energy of waves or storm surges. These approaches include limiting access for foot and vehicle traffic in areas vulnerable to erosion and revegetating with native plants. The plants trap wind-borne sediment and sand and anchor coastal soils with their root systems. Revegetation measures may include the use of geotextiles or mulches to protect the soil surface while the vegetation becomes established.

Structural erosion-control measures in coastal environments include riprap, gabions, and retaining walls. Riprap is rock placed to protect the toe of a slope or the base of a vertical wall. The rock absorbs the energy of wave action and prevents the undermining of existing structures. Gabions are large cages filled with rock, often the same rock that would be used for riprap, that armor the shoreline. Gabions can be stacked to create a retaining wall and can be used in place of riprap where space is limited. Gabions have an advantage over other armoring approaches in that the rock-filled cages drain water freely and are flexible enough to accommodate ground movement. In general, retaining walls need to be properly designed and installed to ensure that the pressure of the ground and subsurface water behind the wall can be withstood for the long term by creating weep holes for moisture to escape and using devices to prevent the soil load from transferring to the wall. Retaining walls can be constructed using either non-proprietary methods or proprietary products such as interlocking concrete blocks and grid materials.

Transportation. Transportation facilities, in particular roads and railways, are often constructed through hilly terrain using cuts into steep slopes. The resulting wall often creates an instability in the landscape that needs to be addressed to avoid damage to facilities and delays in shipping and travel.

Slope-stabilization techniques are commonly used in road cuts to prevent rock fall and landslides. These techniques include structural solutions (such as gabions, retaining walls, and spray-on concrete) and less structural approaches incorporating geotextiles and revegetation.

In addition, transportation facilities need to be designed and constructed to ensure that unintended erosion impacts are not created by runoff of storm water from impermeable surfaces. In these circumstances, facility designs need to include consideration of how storm water can be infiltrated.

Storm-Water Management. Construction and other site-development activities can significantly increase erosion potential on sites by removing vegetation and changing the topography by grading. Erosion control on construction sites can be achieved either temporarily during construction or for the long-term operation of the site after construction is complete. Temporary measures include phasing construction activities to minimize how much of the soil on the site is exposed at any one time and use of stabilizing materials such as mulches or geotextiles. Products such as sedimentation fences or storm-drain filters are not erosion-control devices, instead they control the movement of sediments that have already eroded to ensure they do not create off-site impacts to roads or adjacent lands.

Long-term storm-water management on developed sites includes the use of structural, such as retaining walls, or nonstructural approaches, such as revegetation, to stabilize sites and prevent erosion as the site is used.

Recreation Management. Recreation activities can be either dispersed or focused in terms of impacts to a site's erosion potential. Activities such as hiking or canoeing can have small impacts on sites if traffic is light. However, heavily used trails or boat launches can become significantly eroded. Other activities, for example off-road recreational vehicles, can create significantly de-vegetated areas and bare compacted soils vulnerable to wind and water erosion even with light usage. Erosion-control measures related to recreation areas will vary by need and facility. For example, boat launches for light craft in a rural area may consist of a graveled surface along the bank, whereas heavily used launches for heavy craft may consist of a concrete ramp extending into the water.

Fascinating Facts About Erosion Control

- Dust storms occurring during the 1930's in the Dust Bowl blackened skies from the Great Plains to as far east as Washington, D.C., and left more than a half million people homeless.
- Hurricane Hugo hit the east coast of the United States north of Charleston, South Carolina, in September, 1989. The nearly 20-foot-high storm surge inundated much of the barrier island system, which was less than 10 feet above sea level, and caused almost $6 billion in damages.
- The Grand Canyon is a dramatic example of naturally occurring erosion. It is more than a mile deep and fifteen miles wide.
- Development in hilly terrain and on steep hillsides increases the potential for landslides through a combination of factors, including the increased load on the slope from the weight of structures while increasing erosion potential by directing flow from roofs and pavement to surface drainages, and the loss of vegetation, the root systems of which anchor soils.
- Areas with high annual precipitation rates can be more vulnerable to landslides. For example, the state of Oregon averages more than $10 million in damages from landslides each year.
- The term "gabion" is derived from the Italian word for big cage, *gabbione*. Gabions have historically been used to create protective barriers for troops and equipment attacking a fortress or walled city.

Erosion control along river channels historically emphasized engineering solutions such as riprap, retaining walls, and channelization where the stream channel was lined with concrete. Because of the unintended erosion and flooding effects experienced as a result of these approaches, more recent stream-erosion projects have mimicked natural systems by reestablishing riparian vegetation and features such as floodplains and meanders. These approaches include removing barriers to river movement and using products such as coconut fiber bales and geotextiles to help revegetate areas and improve infiltration.

Impact on Industry

Erosion-control requirements affect a wide variety of public and private programs. Public agencies include environmental protection agencies concerned with protecting and enhancing water and air quality, public works or road departments responsible for constructing and maintaining roads and other transportation facilities, and agencies working on preparation for and the aftermath of natural hazards. Organizations in the private sector include technical consultants, land developers, construction contractors, and manufacturers of products used in erosion-control measures. The following provides examples of organizations working with erosion-control measures or regulations.

Government. The United States Environmental Protection Agency (EPA) administers the federal Clean Water Act, which requires protection of water quality. This act establishes water-quality standards, the foundation for storm-water management requirements and, through section 319, funds implementation projects to demonstrate the effectiveness of measures to control pollution from nonpoint sources.

The Natural Resources Conservation Service (NRCS), which is part of the United States Department of Agriculture, conducts assessments and research into the effectiveness of various soil-conservation approaches. The NRCS works with universities in each state to provide technical assistance on erosion control to land owners. The United States Geological Survey researches and compiles information on geologic hazards in each state, including landslide hazards. State agencies administer the Coastal Zone Management Program to respond to erosion-control needs in coastal environments. State and local agencies administer programs for public works and land-use management.

Industry and Business. Geotechnical and engineering firms provide professional services related to working in areas of geologic hazards such as areas prone to landslides or coastal environments or services related to design and construction of erosion-control structures or nonstructural installations. Land developers and construction contractors are responsible for ensuring that appropriate erosion-control measures are implemented during projects. Manufacturers and suppliers provide the products and materials required to construct or install erosion-control measures. Products include retaining-wall components, rock, geotextiles, or plants for revegetation.

CAREERS AND COURSE WORK

Course work that supports careers in erosion control or soil conservation are based in the physical and applied sciences. Degree programs include soil science, agronomy, hydrology, hydrogeology, and geology. Civil engineering is another alternative degree program, although course work from the other disciplines is required in order to create a focus on erosion control. A solid foundation of course work in soil science, geology, and hydrogeology is required to understand the basic physical processes involved in this subject in addition to course work in mathematics, physics, and engineering.

Students may focus on the engineering, land-use management, or regulatory approach to erosion control. A career oriented toward engineering or structural solutions to erosion-control requires an engineering degree, including course work in physics, mathematics, and engineering dynamics. This career may also require professional certifications, such as being certified as a Professional Engineer, which potentially requires additional study for the examination that must be passed to achieve the certification. A career focusing on land-use management or a regulatory approach to erosion control may be founded on any of the degree programs listed above and includes additional course work in land use and public-policy analysis.

A bachelor or master of science degree is typically required for career opportunities designing and implementing erosion-control practices. A degree is typically not required for a career as a construction contractor installing or building erosion-control structures or features, such as retaining walls or regrading and revegetation projects. However, certain professional certifications or licenses may be required for construction contractors depending on the jurisdiction.

SOCIAL CONTEXT AND FUTURE PROSPECTS

Erosion-control requirements affect many sectors of society and have become integrated into a wide variety of occupational disciplines. Given the potential for significant damage if erosion-control measures fail—for example, in areas with steep slopes or coastal zone environments—the emphasis on and expectations for high levels of success increase with increasing development pressures. Increasing development and market pressures can also affect food production in that farmers seek to maximize the productivity of their land while reducing costs. Soil-conservation measures can increase productivity while reducing the need for some herbicides and pesticides. Soil-conservation measures that increase the moisture-holding capacity of sites and make those sites less dependent on irrigation are also important for agriculture and urban or suburban areas where water supplies may be stressed by demands for domestic use.

Barbara J. Rich, B.S., M.A.

Further Reading

Baxter, Roberta. "Maintaining Vertical: Techniques for Slope Stabilization." *Erosion Control* 11, no. 2 (March/April, 2004). Provides an overview of slope stabilization.

Bell, F. G. *Basic Environmental and Engineering Geology.* Caithness, Scotland: Whittles Publishing, 2007. This book discusses geologic hazards, including soil erosion, in the context of the built environment and its effects on natural resources.

Coetzee, Ken. *Caring for Natural Rangelands.* Scottsville, South Africa: University of KwaZulu-Natal Press, 2005. Includes practical approaches for dealing with soil erosion on rangelands.

England, Gordon. "Implementing LID for New Development." *Stormwater* 11, no. 5 (July/August 2010). Covers the issues involved in low-impact development related to storm-water management.

Leposky, Rosalie E. "Retaining Walls: What You See and What You Don't." *Erosion Control* 11, no. 1 (January/February, 2004). Discusses manufactured materials and their use in stabilizing soil and managing groundwater to prevent mass wasting.

Lundgren, Lawrence W. *Environmental Geology.* 2d ed. Upper Saddle River, N.J.: Prentice Hall, 1999. Comprehensive overview of natural and anthropogenic causes of soil erosion, including discussion of land use, risk assessment, and surface water management.

Thompson, J. William, and Kim Sorvig. *Sustainable Landscape Construction: A Guide to Green Building Outdoors.* Washington, D.C.: Island Press, 2008. Provides information on land-development practices to preserve and restore vegetation and soil during and after construction.

Web Sites

Erosion Control Network
http://www.erosioncontrolnetwork.com

International Erosion Control Society
http://www.ieca.org

Natural Resources Conservation Service
http://www.nrcs.usda.gov

EXPLOSIVES TECHNOLOGIES

FIELDS OF STUDY

Chemistry; structural engineering; metallurgy; hazardous materials handling and transportation.

SUMMARY

Explosives, or the more generalized term "high-energy materials," are products that convert chemical energy into physical force through heat, blast, and compression. Explosives are useful in both creation and destruction. They are a key tool used by modern military forces and provide both destructive and propellant forces. A number of industries use explosives for both the separation of materials and the compression and shaping of processed products. There are also commercial applications for explosives, most notably in pyrotechnics and fireworks for both entertainment and commercial displays.

KEY TERMS AND CONCEPTS

- **Brisance:** Measurement of the shock output of an explosive product based on its ability to shatter a standardized material; from the French word for "break."
- **Deflagration:** Generating an explosive reaction by use of heat or fire; characteristic of low explosives.
- **Detonation:** Generating an explosive reaction by shock; characteristic of high explosives.
- **High Explosive:** Explosive substance that has a high rate of combustion, creating a large amount of blast.
- **Low Explosive:** Explosive substance with a low rate of combustion, creating a relatively small blast.
- **Nitroglycerin:** First advanced chemical explosive and the basis for all modern explosives developed in the last 150 years.
- **Oxidizer:** Product that promotes the propagation of oxygen consumption in an explosion, allowing the explosive product to achieve its most efficient consumption.

Definition and Basic Principles

Explosives are chemical compounds that contain high levels of stored chemical energy that release when the stored form is transformed by a catalyst into heat (often accompanied by visible flame) and blast (the sudden overpressurization of the atmosphere around the explosion). While some early explosives were very unstable in their resting form and prone to conversion by the smallest catalyst, modern explosives are relatively safe to handle and will explode only when initiated by specific means of detonation. The relative safety of explosives makes them ideal for uses ranging from constructive to lethal.

With the exception of nuclear reactions, explosives are chemical compounds that are not individually prone to detonation. When combined with other elements, however, the reaction between the chemicals, initiated by an outside promotion, causes the materials to break down their molecular bonds and release their stored potential energy in the form of blast and heat. Nuclear explosions are again an exception in that, besides heat and blast, they also release a large amount of radioactive particles not present in conventional explosions.

Unlike many products that burn or explode when exposed to heat or pressure, explosives are examples of the static potential of chemical energy in that they release their energy only when determined by the user. Instead of uncontrollable reactions, explosives can be manufactured, stored, and used when needed, a safe and predictable application for a potentially dangerous product. Because of their potential to convert mass into heat and energy, explosives are by definition a very hazardous product. The usefulness of their properties, however, means that explosives are constantly needed—and improved. Consequently, more powerful explosives constantly appear, and the methods for using them safely also keep pace.

Background and History

The earliest explosive, gunpowder, first appeared in ninth-century China. It was a combination of sulfur, saltpeter, and charcoal. The Chinese used gunpowder for limited military purposes, but its primary purpose was to provide pyrotechnics and fireworks for ceremonial and religious rites. By the thirteenth century, gunpowder was being widely used in European firearms and cannons. Because it was relatively weak, gunpowder had few industrial or commercial uses. In the 1840's, more potent and complex explosives, such as guncotton (cotton fibers seeped with nitric acid) and nitroglycerin, appeared and were used industrially, such as for mining. Both guncotton and nitroglycerin proved susceptible to spontaneous detonation, however, until mixed with a stabilizing ingredient. Swedish scientist and inventor Alfred Nobel was the first to create a safe, stable, and storable explosive in 1867 when he introduced dynamite: nitroglycerin stabilized by absorption of sawdust and containment in a waterproof wrapper.

These early products soon expanded into a wide range of acid-based explosives in the late nineteenth century. Cordite, an explosive commonly used in naval warfare, was a prime example. Unlike the relative instability of guncotton and nitroglycerin, cordite was inflammable, vibration resistant, and waterproof. Trinitrotoluene, better known as TNT, was another benchmark explosive. Very stable and of a consistency that allows it to be formed into virtually any shape, TNT remains in common use more than a century after its invention.

How It Works

Types of Explosives. Because of the myriad uses for explosive material, the term "explosives" is a general term and not always specific enough. The terminology for high-energy materials relates to the material's use and application. Explosive, as a general term, applies to any product that explodes, but it really refers to a substance intended to alter the form or physical properties of whatever is in its immediate vicinity when it explodes. A specifically placed explosive charge or an explosive used as a warhead in a shell or missile fits this definition. Explosive reactions can also be harnessed as a propellant to move projectiles toward a target. Simple explosives, like gunpowder used in a firearm, can be used as propellants, but more complex explosives are also used as fuel in missiles and rockets, where the propagation of energy must be sustained over an extended period. Propellants are also used to lift fireworks into the air and as pyrotechnics; this is the colorful burst of light that appears when the firework detonates. In addition to lifting the firework, the formulation of the bursting charge determines the size, pattern, and even the color of the resulting explosion.

Categories of Explosives. Low explosives have a relatively slow burning time and tend to generate more heat than blast. Low explosives, such as gunpowder, are commonly used in firearms because their chemical release is contained and will not burst the gun in which they are used. High explosives, which burn quickly and typically release more blast than heat, are used for open-air purposes because the blast needs to dissipate soon after explosion to avoid unnecessary or undesirable consequences. Instead of burning the application on which the high explosives are used, the blast is intended to shift, move, or remove something. The use of explosives to create mines and tunnels is a good example, as is the use of explosives to level buildings where conventional demolition is too dangerous or time consuming.

Fascinating Facts About Explosives Technologies

- In Greek, dynamite means "connected with power."
- The Chinese, the first innovators of explosives, used fireworks during religious ceremonies, believing the noise and flame produced by gunpowder drove away evil spirits.
- In the 1950's, the nuclear scientists at the Los Alamos Laboratory devised Project Orion, a proposed interplanetary vehicle propelled through space by riding the shockwave of a nuclear device exploded a distance behind the vehicle. Project Orion never got off the drawing board.
- Alfred Nobel, the inventor of dynamite, donated his vast fortune to the creation of the Nobel Prize, five international prizes, one of which was the peace prize, which would go to someone who "shall have done the most or the best work for fraternity between nations, for the abolition or reduction of standing armies and for the holding of peace congresses."
- The 21,000-pound GBU-43/B Massive Ordnance Air Blast Bomb (MOAB, also known as the Mother of all Bombs) is the largest and most powerful conventional explosive bomb in the world.
- In December, 1917, an ammunition ship, the *Mont Blanc*, exploded in the harbor at Halifax, Nova Scotia, with a force of roughly 3 kilotons, the largest conventional explosion in history. The damage from the explosion devastated most of the city and caused more than 2,000 deaths.

Characteristics of Explosives. How explosives transform into matter and energy is also an element of their type. Low explosives consume their mass through the process of deflagration. The initial ignition of the low explosive creates a concentrated charge of flame that breaks down the molecular bonds of the explosive matter, which in turn generates more flame and breaks down the next sequence of molecular bonds. Because the process must move from molecule to molecule, the resulting explosion takes more time and the low explosives do not dispel all of their stored energy at once, producing less blast. The generation of flame over mass is useful in applications that rely on heat, such as using explosives to burn through or sever metal parts. Because flame consumes materials in an irregular fashion, however, low explosives, especially gunpowder, often leave unburned residue as an unwanted byproduct. High explosives, on the other hand, consume their mass through the process of detonation. The molecular bonds of the explosive material are rapidly broken down by a shock wave passing through it. Because the detonation happens almost simultaneously, the resulting explosion largely occurs before flame can develop, and most of the explosive potential is released in the form of blast. This is particularly useful in applications where force transference is the goal, but damage by fire is not desirable. The variations in explosive characteristics permit a wider range of options in selecting the best explosive for a particular purpose. For many military applications, for instance, fire is a useful byproduct of an explosion, so low explosives are best. For purposes that prefer shock over flame, an explosive with a high brisance, or shattering, effect is the ideal option. Brisance is measured and standardized by the speed in which the explosion reaches its maximum force; the faster the propagation of force, the higher the brisance number. Maximum brisance is usually achieved with the aid of an oxidizer, an ingredient that burns at the start of the detonation to accelerate the charge to its maximum heat level at a rate faster than what the explosive would achieve on its own.

Initiation of Explosives. One of the early problems with explosives like nitroglycerin was their instability, which led to accidental explosions, property damage, and deaths. The development of later explosives removed this problem by creating stable explosives that could be safely stored, transported, and handled until detonated. Instead of casual handling, newer explosives required detonation by methods determined by the specialized purpose of the explosives themselves. Many military explosives are detonated by impact with a target that causes compression and initiation of the explosion. Friction is also a detonation method, whereby the heat generated by the movement of the explosion causes the explosion. For precisely timed explosions, electrical charges can also detonate high explosives. The electrical charge method allows the use of a timer mechanism or the initiation of a precisely timed sequence of explosions.

Applications and Products

Military Use of Explosives. The military use of explosives has been the primary driving factor in explosives research, as rival militaries attempt to create bigger and better explosions to gain military advantage of a potential enemy. Starting with crude firearms and cannons in the thirteenth century, gunpowder transformed the battlefield by transferring the emphasis of military power from man and animal muscle power to chemical power. Firearms permitted soldiers to attack and defend an area much larger than a hand-operated weapon, such as a spear or bow. The military applications of gunpowder were limited, however, by the relative weakness of gunpowder itself. Only a small percentage of the black powder actually propelled the bullet, while the rest remained in the gun as residue or went out the barrel in a large cloud of dense smoke. To improve their firearms, scientists in the late nineteenth century devised improved propellants, termed "smokeless" powders, that generated virtually no smoke, no residue, and used much more of the powder to propel the bullet downrange.

The invention of more complex and potent explosives in the late nineteenth century transformed war forever. The more lethal explosives, coupled with the new internal combustion engine, led to the tools of all modern armies: machine guns, grenades and rockets, heavy artillery, and tanks. Individual soldiers carried more personal firepower with each succeeding generation, and the accompanying heavy weapons also possessed greater ability to distribute explosives around the battlefield. Explosives also caused a decline in defensive warfare. Before gunpowder, a defender inside a stout fortification or castle could withstand the assault of a force several times the size of the defenders. Gunpowder reversed the advantage, as explosives, in sufficient amounts, could defeat any fixed defensive position, causing armies to pursue more mobile strategies. Explosives also went to sea, resulting in modern naval weapons such as the large-caliber guns on battleships, torpedoes, and mines. Aerial warfare also developed a wide range of explosive weapons, primarily warheads for bombs and missiles aimed at targets on the ground. Small hand grenades were hurled by pilots during World War I, and by World War II a single B-17 Flying Fortress bomber could carry up to four tons of bombs, while a Cold War-era B-52 Stratofortress could carry up to thirty-five tons. Modern cruise missiles carry a range of potential warheads, including low explosive fragmentation, high explosive blast, and nuclear warheads.

Civilian Use of Explosives. Explosives have found their way into a wide range of civilian applications. The earliest and perhaps best known is in the pyrotechnics industry. Fireworks displays are a common entertainment feature of many holidays and special events, and the evolution of explosives parallels the growing complexity and scale of modern fireworks companies. The railroad industry embraced the use of explosives: Confronted by mountains that defied construction, explosives were used to tunnel through them instead. The appearance of dynamite, the first safe and stable commercial explosive, appeared just as the expanding railroad system began moving into the mountainous western regions of North America, and its ability to create tunnels made explosives an important tool. The mining industry, also interested in boring holes through mountains, began to use explosives on a large scale. The blast component of explosives proved very valuable in creating tunnels, but also in crushing rock formations to ease the removal of the trace ores within.

The most visible industrial use of explosives is by the construction industry for the implosion of large structures. In circumstances where the traditional demolition of a building is too dangerous, time-consuming, or expensive, carefully placed explosives can topple a building so that it falls within a very precise footprint, preventing damage to nearby structures. The ability of explosives to topple buildings has, unfortunately, also attracted its use by criminal and terrorist organizations. Bombings, ranging from complex remote detonated bombs to crude "truck bombs," have become a common feature of life in unstable and contested regions of the world.

On a more benign level, explosives also play an unseen role in very common items. Small explosives charges, for instance, cause the rapid inflation of an automobile's air bag in case of a collision. On the scientific frontier, explosives in the form of propellants are the only means of lifting large cargoes into space, and heavy rockets consume tons of explosive material to maintain the International Space Station and the array of satellites in orbit.

Impact on Industry

Corporate Control. Because of its importance to national security, explosives production usually resided within only a limited number of easily monitored producers during most of the industrial age. Governments ran explosive factories themselves or, more usually, entrusted the production of explosives to large corporations that enjoyed virtual monopolistic control of the market and government contracts. In the United States, the DuPont company was the leading producer of explosives for both the U.S. military and industrial purposes. Founded in Delaware in 1802 by Eleuthère Irénée du Pont, who fled France to avoid the French Revolution, the company was a major producer of explosives and munitions throughout the nineteenth and early twentieth centuries. Although still a major component of the chemical industry, DuPont lost its control of the explosives industry to government production during World War I and II—and to producers who introduced new and more potent explosives into the market.

Research. As control of the explosives industry passed into the hands of a wider variety of manufacturers, the impetus to capture a larger portion of the market led to increased research into improving the characteristics, safety, and usefulness of high-energy materials. Instead of a one-size-fits-all approach, research has led to explosives for very specific purposes in very specific places. A good example is the mining industry. In open-air pit mining, traditional explosives are suitable because the blast can vent into open air. In shaft mining, however, the heat, blast, and gas are contained within the mine. Consequently, specific explosives that produce shorter blast areas and less heat (which can ignite other naturally occurring gases in the mine) have been developed. Explosives that produce less toxic gases and byproducts have appeared, permitting their use in traditional roles while creating less negative impact on the environment.

Government Regulation. Because of the expansive nature of the explosives industry and the potential damage from the irresponsible manufacture and distribution of explosives, explosives are heavily regulated throughout the industrialized world. In the United States, explosives are regulated by the Department of Justice through its Bureau of Alcohol, Tobacco, Firearms, and Explosives, known as the ATF. Congress granted the ATF the power to regulate explosives under the terms of the Organized Crime Control Act of 1970. Through its Explosives Industries Programs Branch, the ATF coordinates and oversees the production of explosive materials at more than 11,000 licensed manufacturers in the United States, ranging from small pyrotechnics companies to large defense contractors. The production of explosives also requires close cooperation between the ATF and the Environmental Protection Agency (EPA) to ensure the safe disposal of hazardous byproducts and degraded explosive material.

Careers and Course Work

Because of their complex chemical structures and potentially devastating potential, explosives technology requires very advanced and specific training and course work. Chemistry is the primary field of study leading to careers in the explosives industry, with jobs tending toward the development of new products that can be applied to new purposes. When using explosives in an occupational setting, knowing how explosives will react and change the environment in which they are used is essential. Structural engineering, for instance, provides knowledge on how explosives can affect or even destroy a structure or building. Conversely, structural engineering can also provide training on how to build structures resistant to explosive contact. If one is interested in propellants, a background in physics would prove useful, as would aerodynamics in the pyrotechnics industry. The potential hazards associated with explosives also require specialized training. The transport, storage, and disposal of explosives demand the adherence to strict safety protocols and the knowledge of how to use specialized equipment.

Social Context and Future Prospects

Explosives have a long past and a broad future. As a key element of industrialization over the past few centuries, explosives will continue to play the same role as other world economies continue to develop. As people's understanding of the physical world continues to change, new developments in explosives technology will lead to more potent and potentially lethal products for both civilian and military applications. Just as explosives joined with the internal combustion engine to revolutionize warfare in the past, explosives have begun joining with electronics to become smaller, smarter, and more lethal. Coupled with electronic aiming mechanisms, common explosive bombs became "smart" bombs capable of steering themselves to a target. As targeting got even more precise, military explosives reversed the trend of ever-larger explosive charges to become smaller and more discriminate, destroying the targeted adversary but leaving nearby civilian life and property unharmed, avoiding unwanted collateral damage. Instead of dropping tons of bombs by a B-52, the U.S. Air Force, for instance, has developed the diminutive 250-pound GBU-39 Small Diameter Bomb big enough to destroy a target yet small enough to avoid excessive blast or fire damage around the target.

Steven J. Ramold, Ph.D.

Further Reading

Agrawal, Jai Prakash. *High Energy Materials: Propellants, Explosives and Pyrotechnics.* Weinheim, Germany: Wiley-VCH, 2010. A comprehensive overview of explosives, both past and present, with an emphasis on their chemical construction and use.

Akhavan, Jacqueline. *The Chemistry of Explosives.* 2d ed. London: Royal Society of Chemistry, 2004. Similar to the Agrawal book, but with an emphasis on explosives developed since World War II.

Fant, Kenne. *Alfred Nobel: A Biography.* New York: Arcade, 1993. The best biography of the controversial inventor who created a product that changed the world, but who was haunted by its consequences.

Griffin, Roger D. *Principles of Hazardous Materials Management.* 2d ed. Boca Raton, Fla.: CRC Press, 2009. The book contains a lengthy discussion of the handling and transport of explosives, detailing how to move and employ safely even the most dangerous materials.

Persson, Per-Anders, Roger Holmberg, and Jaimin Lee. *Rock Blasting and Explosives Engineering.* Boca Raton, Fla.: CRC Press, 1994. A good example of how the results of explosives are as important as the explosives themselves, this study explains the use of explosives in the mining industry.

Web Sites

Bureau of Alcohol, Tobacco, Firearms, and Explosives
http://www.atf.gov

Institute of Makers of Explosives
http://www.ime.org

International Association of Bomb Technicians and Investigators
https://iabti.org

International Pyrotechnics Society
http://www.intpyro.org/Society.aspx

See also: Hazardous-Waste Disposal

FIBER TECHNOLOGIES

FIELDS OF STUDY

Chemistry; physics; mathematics; chemical engineering; polymer chemistry; agriculture; agronomy; mechanical engineering; industrial management; waste management; business management

SUMMARY

Fibers have been used for thousands of years, but not until the nineteenth and twentieth centuries did chemically modified natural fibers (cellulose) and synthetic plastic or polymer fibers become extremely important, opening new fields of application. Advanced composite materials rely exclusively on synthetic fibers. Research has also produced new applications of natural materials such as glass and basalt in the form of fibers. The current "king" among fibers is carbon, and new forms of carbon, such as carbon nanotubes, promise to advance fiber technology even further.

KEY TERMS AND CONCEPTS

- **Denier:** A unit indicating the fineness of a filament; a filament 9,000 meters in length weighing 1 gram has a fineness of 1 denier.
- **Roving:** Nonwoven fiber fabric whose strands all have the same absolute orientation.
- **Warp Clock:** A visual guide to the orientation of the warp of woven fiber fabrics, used in the laying-up of composite materials to provide a quasi-isotropic character to the final product.

Definition and Basic Principles

A fiber is a long, thin filament of a material. Fiber technologies are used to produce fibers from different materials that are either obtained from natural sources or produced synthetically. Natural fibers are either cellulose-based or protein-based, depending on their source. All cellulosic fibers come from plant sources, while protein-based fibers such as silk and wool are exclusively from animal sources; both fiber types are referred to as biopolymers. Synthetic fibers are manufactured from synthetic polymers, such as nylon, rayon, polyaramides, and polyesters. An infinite variety of synthetic materials can be used for the production of synthetic fibers.

Production typically consists of drawing a melted material through an orifice in such a way that it solidifies as it leaves the orifice, producing a single long strand or fiber. Any material that can be made to melt can be used in this way to produce fibers. There are also other ways in which specialty fibers also can be produced through chemical vapor deposition. Fibers are subsequently used in different ways, according to the characteristics of the material.

Background and History

Some of the earliest known applications of fibers date back to the ancient Egyptian and Babylonian civilizations. Papyrus was formed from the fibers of the papyrus reed. Linen fabrics were woven from flax fibers. Cotton fibers were used to make sail fabric. Ancient China produced the first paper from cellulose fiber and perfected the use of silk fiber.

Until the nineteenth century, all fibers came from natural sources. In the late nineteenth century, nitrocellulose was first used to develop smokeless gunpowder; it also became the first commercially successful plastic: celluloid.

As polymer science developed in the twentieth century, new and entirely synthetic materials were discovered that could be formed into fine fibers. Nylon-66 was invented in 1935 and Teflon in 1938. Following World War II, the plastics industry grew rapidly as new materials and uses were invented. The immense variety of polymer formulations provides an almost limitless array of materials, each with its own unique characteristics. The principal fibers used today are varieties of nylons, polyesters, polyamides, and epoxies that are capable of being produced in fiber form. In addition, large quantities of carbon and glass fibers are used in an ever-growing variety of functions.

HOW IT WORKS

The formation of fibers from natural or synthetic materials depends on some specific factors. A material must have the correct plastic characteristics that allow it to be formed into fibers. Without exception, all natural plant fibers are cellulose-based, and all fibers from animal sources are protein-based. In some cases, the fibers can be used just as they are taken from their source, but the vast majority of natural fibers must be subjected to chemical and physical treatment processes to improve their properties.

Cellulose Fibers. Cellulose fibers provide the greatest natural variety of fiber forms and types. Cellulose is a biopolymer; its individual molecules are constructed of thousands of molecules of glucose chemically bonded in a head-to-tail manner. Polymers in general are mixtures of many similar compounds that differ only in the number of monomer units from which they are constructed. The processes used to make natural and synthetic polymers produce similar molecules having a range of molecular weights. Physical and chemical manipulation of the bulk cellulose material, as in the production of rayon, is designed to provide a consistent form of the material that can then be formed into long filaments, or fibers.

Synthetic Polymers. Synthetic polymers have greatly expanded the range of fiber materials that are available, and the range of uses to which they can be applied. Synthetic polymers come in two varieties: thermoplastic and thermosetting. Thermoplastic polymers are those whose material becomes softer and eventually melts when heated. Thermosetting polymers are those whose the material sets and becomes hard or brittle through heating. It is possible to use both types of polymers to produce fibers, although thermoplastics are most commonly used for fiber production.

The process for both synthetic fibers is essentially the same, but with reversed logic. Fibers from thermoplastic polymers are produced by drawing the liquefied material through dies with orifices of the desired size. The material enters the die as a viscous liquid that is cooled and solidifies as it exits the die. The now-solid filament is then pulled from the die, drawing more molten material along as a continuous fiber. This is a simpler and more easily controlled method than forcing the liquid material through the die using pressure, and it produces highly consistent fibers with predictable properties.

Fibers from thermosetting polymers are formed in a similar manner, as the unpolymerized material is forced through the die. Rather than cooling, however, the material is heated as it exits the die to drive the polymerization to completion and to set the polymer.

Other materials are used to produce fibers in the manner used to produce fibers from thermoplastic polymers. Metal fibers were the first of these materials. The processes used for their production provided the basic technology for the production of fibers from polymers and other nonmetals. The best-known of these fibers is glass fiber, which is used with polymer resins to form composite materials. A somewhat more high-tech variety of glass fiber is used in fiber optics for high-speed communications networks. Basalt fiber has also been developed for use in composite materials. Both are available commercially in a variety of dimensions and forms.

Production of carbon fiber begins with fibers already formed from a carbon-based material, referred to as either pitch or PAN. Pitch is a blend of polymeric substances from tars, while PAN indicates that the carbon-based starting material is polyacrylonitrile. These starting fibers are then heat-treated in such a way that essentially all other atoms in the material are driven off, leaving the carbon skeletons of the original polymeric material as the end-product fiber.

Boron fiber is produced by passing a very thin filament of tungsten through a sealed chamber, during which the element boron is deposited onto the tungsten fiber by the process of chemical vapor deposition.

APPLICATIONS AND PRODUCTS

All fiber applications derive from the intrinsic nature of the material from which the fibers are formed. Each material, and each molecular variation of a material, produces fibers with unique characteristics and properties, even though the basic molecular formulas of different materials are very similar. As well, the physical structure of the fibers and the manner in which they were processed work to determine the properties of those fibers. The diameter of the fibers is a very important consideration. Other considerations are the temperature of the melt from which fibers of a material were drawn; whether the fibers were stretched or not, and the degree by which they were stretched; whether the fibers are hollow, filled, or solid; and the resistance of the fiber material to such environmental influences as exposure to light and other materials.

Structural Fibers. Loosely defined, all fibers are structural fibers in that they are used to form various structures, from plain weave cloth for clothing to advanced composite materials for high-tech applications. That they must resist physical loading is the common feature identifying them as structural fibers. In a stricter sense, structural fibers are fibers (materials such as glass, carbon, aramid, basalt, and boron) that are ordinarily used for construction purposes. They are used in normal and advanced composite materials to provide the fundamental load-bearing strength of the structure.

A typical application involves "laying-up" a structure of several layers of the fiber material, each with its own orientation, and encasing it within a rigid matrix of polymeric resin or other solidifying material. The solid matrix maintains the proper orientation of the encased fibers to maintain the intrinsic strength of the structure.

Materials so formed have many structural applications. Glass fiber, for example, is commonly used to construct different fiberglass shapes, from flower pots to boat hulls, and is the most familiar of composite fiber materials. Glass fiber is also used in the construction of modern aircraft, such as the Airbus A-380, whose fuselage panels are composite structures of glass fibers embedded in a matrix of aluminum metal.

Carbon and aramid fibers such as Kevlar are used for high-strength structures. Their strength is such that the application of a layer of carbon fiber composite is frequently used to prolong the usable lifetime of weakened concrete structures, such as bridge pillars and structural joists, by several years. While very light, Kevlar is so strong that high-performance automotive drive trains can be constructed from it. It is the material of choice for the construction of modern high-performance military and civilian aircraft, and for the remote manipulators that were used aboard the space shuttles of the National Aeronautics and Space Administration Kevlar is recognizable as the high stretch-resistance cord used to reinforce vehicle tires of all kinds and as the material that provides the impact-resistance of bulletproof vests.

In fiber structural applications, as with all material applications, it is important to understand the manner in which one material can interact with another. Allowing carbon fiber to form a galvanic connection to another structural component such as aluminum, for example, can result in damage to the overall structure caused by the electrical current that naturally results.

Fabrics and Textiles. The single most recognized application of fiber technologies is in the manufacture of textiles and fabrics. Textiles and fabrics are produced by interweaving strands of fibers consisting of single long fibers or of a number of fibers that have been spun together to form a single strand. There is no limit to the number of types of fibers that can be combined to form strands, or on the number of types

Fascinating Facts About Fiber Technologies

- In 1845, German-Swiss chemist C. F. Schonbein accidentally discovered the explosive properties of nitrocellulose fibers when he used his wife's cotton apron to mop up some nitric acid, then hung the apron by a stove to dry.
- Nitrocellulose became the first commercially successful plastic as celluloid, the basis of the photographic film industry.
- French chemist and microbiologist Louis Pasteur and French engineer Count Hilaire de Chardonnet tried to develop a synthetic alternative for silk. The cellulose fiber rayon was the result, patented by Chardonnet in 1885.
- Any material that can be melted, even the volcanic lava known as basalt, can be formed into a usable fiber.
- Many materials, such as PET bottles, which have long been resigned to landfills after a single use, are now being recycled as feedstock for synthetic fiber production.
- A fiber with a textile weight of 1 denier is 9,000 meters long, but weighs only 1 gram.
- The synthetic fiber industry has an economic value of about US$100 billion annually, worldwide.

of strands that can be combined in a weave.

The fiber manufacturing processes used with any individual material can be adjusted or altered to produce a range of fiber textures, including those that are soft and spongy or hard and resilient. The range of chemical compositions for any individual polymeric material, natural or synthetic, and the range of available processing options, provides a variety of properties that affect the application of fabrics and textiles produced.

Clothing and clothing design consume great quantities of fabrics and textiles. Also, clothing designers seek to find and utilize basic differences in fabric and textile properties that derive from variations in chemical composition and fiber processing methods.

Fibers for fabrics and textiles are quantified in units of deniers. Because the diameter of the fiber can be produced on a continuous diameter scale, it is therefore possible to have an essentially infinite range of denier weights. The effective weight of a fiber may also be adjusted by the use of sizing materials added to fibers during processing to augment or improve their stiffness, strength, smoothness, or weight. The gradual loss of sizing from the fibers accounts for cotton denim jeans and other clothing items becoming suppler, less weighty, and more comfortable over time.

The high resistance of woven fabrics and textiles to physical loading makes them extremely valuable in many applications that do not relate to clothing. Sailcloth, whether from heavy cotton canvas or light nylon fabric, is more than sufficiently strong to move the entire mass of a large ship through water by resisting the force of wind pressing against the sails. Utility covers made from woven polypropylene strands are also a common consumer item, though used more for their water-repellent properties than for their strength. Sacks made from woven materials are used worldwide to carry quantities of goods ranging from coffee beans to gold coins and bullion. One reason for this latter use is that the fiber fabric can at some point be completely burned away to permit recovery of miniscule flakes of gold that chip off during handling.

Cordage. Ropes, cords, and strings in many weights and winds traditionally have been made from natural fibers such as cotton, hemp, sisal, and manila. These require little processing for rough cordage, but the suppleness of the cordage product increases with additional processing. Typically, many small fibers are combined to produce strands of the desired size, and these larger strands can then be entwined or plaited to produce cordage of larger sizes. The accumulated strength of the small fibers produces cordage that is stronger than cordage of the same size consisting of a single strand. The same concept is applied to cordage made from synthetic fibers.

Ropes and cords made from polypropylene can be produced as a single strand. However, the properties of such cordage would reflect the properties of the bulk material rather than the properties of combined small fibers. It would become brittle when cold, overly stretchy when warm, and subject to failure by impact shock. Combined fibers, although still subject to the effects of heat, cold, and impact shock, overcome many of these properties as the individual fibers act to support each other and provide superior resistance.

Impact on Industry

Industries based on fiber technologies can be divided into two sectors: those that produce fibers and those that consume fibers; both are multibillion dollar sectors.

Fiber production industries are agricultural and technical. Forestry and other agricultural industries produce large amounts of cellulosic fiber each year. Production of cellulosic fiber peaked at some 3 million metric tons in 1982, representing 21 percent of the fiber market share, but this heavy production has steadily declined since 1982. By 2002, cellulosic fiber production had decreased to 6 percent of the world fiber-market share. Production in Eastern Europe dropped from 1.1 million metric tons to just 92,000 metric tons, and production in Asia increased by 660,000 metric tons over the same period, accounting for fully 69 percent of global production in 2002. Most of the cellulosic fiber that is produced by the forestry industry is used in the manufacture of paper and paper products, while the other cellulosic fiber types—cotton, hemp, sisal, and manila—are used primarily for fabrics, textiles, and cordage.

Synthetic fiber production is the principal driving force behind the decline in cellulosic fiber production. Industrial polymerization processes provide a much greater variety of fiber-forming materials with a lower requirement for process control. In the period from 1982 to 2002, manufactured fiber production increased by 155 percent, with the rise attributed to synthetic fiber as the production of cellulosic fiber decreased. Fibers from synthetic materials account for 94 percent of total global fiber production.

A great deal of research has been expended in the field of polymerization chemistry to identify effective catalysts and processes for the reactions involved, so it has become ever easier to control the nature of the materials being produced. Polymerization reactions have become controllable such that their products now span a much narrower range of monomer weights. This specificity of control has enabled the production of more specific fiber materials, which has in turn driven research and development of more specific uses of those fibers.

The greatest increase in synthetic fiber materials has been an increase in the polyesters, which now account for almost two-thirds of total synthetic fiber production. It must be remembered that terms such as "polyester" and "polyamide" refer to broad classes of compounds and not to specific materials. Each class contains untold thousands of possible variations in molecular structure, both from the chemical identity of the monomers used and from the order in which they react during polymerization.

Technical industries that produce fibers from both natural and synthetic materials make use of several classes of machinery. One class of machinery (for example, pelletizers and masticators) manipulates raw materials into a form that is readily transported and usable in the fiber-forming process. Another class of machinery (for example, injection molders and spinnerets) carries out the fiber-forming process, while still another class of machinery (for example, sizers and mercerizers) uses the raw fiber to provide a finished fiber product ready for consumers.

Industries that consume finished fibers include weaving mills that produce fabrics and textiles from all manner of fibers, including glass, basalt, and carbon. These fibers are then used accordingly, for clothing, composites, or other uses. Cordage manufacturers use finished fibers as an input feedstock to produce every type of cordage, from fine thread to coarse rope.

A more recent development in the fiber industry is the utilization of recycling wastes as feedstock for fiber-forming processes. Synthetic materials such as nylon, polyethylene, and polyethylene terephalate (PET), which have historically been condemned to landfills, are now accepted for reprocessing through recycling programs and formed into useful fibers for many different purposes. The versatility of the materials themselves lends to the development of new industries based on their use.

Careers and Coursework

Careers in textile production depend on a sound basic education in chemistry, physics, mathematics, and materials science. Students anticipating such a career should be prepared to take advanced courses in these areas at the college and university level, with specialization in organic chemistry, polymer chemistry, and industrial chemistry. Postsecondary programs specializing in textiles and the textile industry also are available, and provide the specialist training necessary for a career in the manufacture and use of fibers and textiles. The chemistry of color and dyes is another important aspect of the field, representing a distinct area of specialization.

Composite-materials training and advanced composite-materials specialist training can be obtained through only a limited number of public and private training facilities, although considerable research in this field is carried out in a number of universities and colleges. Private industries that require this kind of specialization, particularly aircraft manufacturers, often have their own patented fabrication processes and so prefer to train their personnel on-site rather than through outside agencies.

Social Context and Future Prospects

One could argue that the fiber industry is the principal industry of modern society, solely on the basis that everyone wears clothes of some kind that have been made from natural or synthetic fibers. As this is unlikely ever to change, given the climatic conditions that prevail on this planet and given the need for protective outerwear in any environment, there is every likelihood that there will always be a need for specialists who are proficient in both fiber manufacturing and fiber utilization.

Richard M. Renneboog, M.Sc.

Further Reading

Fenichell, Stephen. *Plastic: The Making of a Synthetic Century.* New York: HarperCollins, 1996. A well-researched account of the plastics industry, focusing on the social and historical contexts of plastics, their technical development, and the many uses for synthetic fibers.

Morrison, Robert Thornton, and Robert Nielson Boyd. *Organic Chemistry..* 5th ed. Newton, Mass.: Allyn & Bacon, 1987. Provides one of the best and most readable introductions to organic chemistry and polymerization.

Selinger, Ben. *Chemistry in the Marketplace.* 5th ed. Sydney: Allen & Unwin, 2002. The seventh chapter of this book provides a concise overview of many fiber materials and their common uses and properties.

Weinberger, Charles B. "Instructional Module on Synthetic Fiber Manufacturing." Gateway Engineering Education Coalition: 30 Aug. 1996. This article presents an introduction to the chemical engineering of synthetic fiber production, giving an idea of the sort of training and specialization required for careers in this field.

Web Sites

University of Tennessee Space Institute
http://www.utsi.edu/research/carbonfiber

U.S. Environmental Protection Agency
http://www.epa.gov

See also: Forestry

FISHERIES SCIENCE

FIELDS OF STUDY

Ecology; biology; chemistry; marine biology; limnology; ichthyology; oceanography; ethology; hydrology; anthropology; law; economics; political science; genetic engineering; mechanical engineering; electrical engineering; systems engineering; veterinary medicine; ethology.

SUMMARY

Fisheries science is an interdisciplinary study concerned with the hunting and farming of aquatic organisms in oceans and bodies of fresh- or saltwater as part of the ongoing effort to feed the world's population. Until recently, most aquatic organisms from fisheries were wild creatures. Farming and the use of genetic-engineering techniques to produce desirable domesticated aquatic food organisms are widespread, but the majority of fisheries science still involves the management of fish stocks for sustainability of the resource.

KEY TERMS AND CONCEPTS

- **Allowable Biological Catch (ABC):** Maximum amount of fish stock that can be harvested without adversely affecting recruitment or other biological components of the stock.
- **Benthos:** All those animals and plants living on or in sediments at the bottom of the sea; benthic animals are usually described by their position in the sediment relative to the surface, and their size, either living within (burrowing in) the sediments, or living at, near, or on the sediment surface.
- **Biomass:** Amount or mass of some organism, such as fish.
- **Bycatch:** Fish, other than the primary target species, that are caught incidental to the harvest of the primary target species.
- **Derby:** Fishery in which the total allowable catch is fixed and participants do not have individual quotas; participants attempt to maximize their harvest as quickly as possible, before the fishery is closed for the season.
- **Epipelagic:** Pertaining to the community of suspended organisms inhabiting an aquatic environment between the surface and a depth of 200 meters.
- **Individual Fishing Quota (IFQ):** Fishery management tool that allocates a certain proportion of the total allowable catch to individual vessels, fishermen, or other eligible recipients based on initial qualifying criteria.
- **Input Controls:** Fishery management measures that seek to limit the amount or effectiveness of effort in a fishery, including licenses limiting numbers of fishermen, gear restrictions, and time limits on fishing activities.
- **Output Controls:** Fishery management measures (including total allowable catch and individual fishing quotas) intended to limit the amount of catch or harvest in a fishery.
- **Pelagic:** Of or pertaining to the open waters of the sea.
- **Recruitment:** Number or percentage of fish that survive from birth to a specific age or size.
- **Total Allowable Catch (TAC):** Total catch permitted to be caught from a stock in a given period, typically a year; usually this amount is less than the allowable biological catch.

Definition and Basic Principles

Fisheries science is concerned with the continued extraction of aquatic organisms from marine, brackish, and freshwater environments for subsistence, commercial, or recreational purposes. As such, fisheries science necessarily involves issues of yield (the numbers of fish harvested from a given stock) and sustainability (the numbers of fish that must not be harvested if the particular fish stock is to demonstrate continued productivity). This seemingly simple biological situation, however, is made much more complex by ecological, economic, and social considerations.

No aquatic species—not even farmed fish—exists in isolation, unconnected to the food web of its surrounding ecosystem. To the extent that fishing or aquaculture alters the population of one species, it also inevitably alters the populations of other species in the same environment. The health of a fishery is dependent as much on the health of the entire ecosystem as it is on any particular species. Sustainability, then, must ultimately take into account not just numbers of a particular fish stock harvested or left behind but also the health of that fish stock's entire ecosystem.

Fisheries science is further complicated by the fact

that fish, particularly in the sea, have long been assumed to be a limitless resource, available to all for the taking. Depleted stocks and increasing competition for declining numbers of fish have led to an understanding that even fish in the sea are also a limited resource. This has led to the search for new ways to manage marine fisheries and to produce more fish aquaculturally through fish farming of various types. Basic, applied, and developmental research in fisheries science is deeply involved with aquaculture and fisheries management. These are fields in which biological sustainability and productivity must be balanced with the economic productivity of those employed in the fishing and fish-farming industries and with the social productivity of communities who have long been dependent on the harvest of aquatic organisms.

Background and History

Although evidence of aquaculture (particularly the raising of fish in ponds or estuaries) goes back at least 5,000 years, and evidence of the hunting and gathering of aquatic organisms goes back significantly more than 100,000 years. It was only in the twentieth century that fisheries science developed out of the biological study of aquatic organisms. The development of the discipline can be traced to the fact that, although there had been occasional spot depletions of fish stocks (particularly in freshwater environments or among large sea mammals such as whales), widespread depletions of fish stocks did not become a concern until the twentieth century.

These depletions and eventual collapses of fisheries resulted from technological advances that increased both the efficiency in how a given fish stock was harvested and the rate at which that stock was exploited. Through the first half of the twentieth century these technological advances were primarily mechanical: More powerful steam and then diesel engines allowed larger and sturdier fishing vessels, hauling larger and stronger fishing gear, to travel farther and, along with better refrigeration, stay at sea longer. During the second half of the twentieth century, the technological advances were primarily electronic: sonar and the Global Positioning System (GPS), along with many other informational technologies, made it possible to find the fish more accurately and extract them.

Over the course of the twentieth century, each new technological advance allowed fishers to eat deeper into the natural capital of a given fish stock. Short-term-oriented market forces, driving an extractive technological arms race within the context of open-access fisheries, often reached the point that the fish stocks in question collapsed entirely. The need to manage fish stocks more rationally became increasingly obvious, and fisheries science has begun to serve this need.

How It Works

The core of fisheries science is biological. It is concerned with understanding the life cycles of individual aquatic organisms, including growth rates, ages of sexual maturity, longevity, predation, and mortality. How these factors affect estimates of fish-stock population sizes and long-term population management of fish stocks are also all of key importance in fisheries science. Fisheries science also requires practical understanding of fish tagging and fish marking, the particular fishing gear used in specific fisheries, habitat improvement and bioremediation, fishways, screens, and guiding devices, the role of hatchery-raised fish and stocking, and small-pond, floating-enclosure, or net-pen management (these last being particularly important in aquacultural contexts).

Whether in aquacultural or traditional fishing approaches, however, fisheries scientists find they must go beyond a simply biological understanding of fish stocks. Fisheries science is profoundly influenced by the context of the cultural and legal framework within which fisheries management takes place. An important part of this framework is the common-law doctrine of the "public trust," particularly the idea that the resources of the rivers and seas within a nation's jurisdiction belong to the people, and the government holds them in trust for the public.

Fisheries as a Public Trust. The idea of the public trust has its roots in the principle of ancient Roman law that things such as the air, running water, the sea, and the shores of the sea are incapable of private ownership. In English law, and later in U.S. law, the fish and wild beasts were added to the list of "common property" that the government holds in trust for the public. Resources relating to the public good cannot be given away to private interests because public trust resources are inalienable—that is, they cannot be given over or transferred by the government to other entities or individuals. The government also must, by law, exercise continuing stewardship responsibility and authority over public trust resources, including fisheries.

In conjunction with improved extractive technologies and largely unregulated market forces, the open-access nature of fisheries in many cases resulted in the degradation of those fisheries. It thus became increasingly clear that government, as trustee of the fisheries, ought to exercise its stewardship authority over these public trust resources. In the United States, this stewardship responsibility is seen most clearly in the Magnuson-Stevens Fishery Conservation and Management Act of 1976 (which extended to 200 miles offshore the nation's jurisdiction over the sea as an "exclusive economic zone" for fishing) and the Sustainable Fisheries Act of 1996 (which called for further analysis of sustainability and closed to fishing many depleted fisheries, in hopes of their recovery).

The places where aquatic organisms (whether wild or farmed) live and from which they are to be harvested or protected from harvest will likely remain waters held primarily in public trust. As a result, the fisheries scientists will continue to serve most often in the role of knowledge expert, assessing fish stocks and advising government, industry, and the public how best to use the fish stocks that are the common property of a nation's citizens.

Assessing Estimated Population Size and Ecological Risk: Traditional Fisheries. To estimate yield, to understand basic changes in population number and composition, and as a basis for sound management, the fisheries scientist must be able to estimate fish population reliably. Like much of the rest of fisheries science, population assessment is based in mathematical and systems-analysis approaches. Although opportunities do exist for direct counting of fish, most statistics on fish populations are estimates. Some examples of methods for estimating population include area density, mark-recapture or "single census," and catch-effort.

The area-density approach to estimating population involves counting the number of animals in a series of sample strips or plots distributed randomly or systematically throughout the total environmental area in which the fish stock population is to be determined. The sample count, once taken, is expanded to an estimate of the entire population by multiplying the aggregate sample count by a particular fraction (total area divided by the sum of sample areas). Because a subarea can also be sampled for time instead of space, the same equations applied to area-density can also be applied to partial-time coverage.

In the mark-recapture method of estimating fish population, a sample of fish is collected, marked, and released, and then at a later time a second or recapture sample is taken, which includes both marked and unmarked fish. This approach is based on the assumption that the proportion of marked fish recovered is to the total catch in the second sample as the total number of marked fish released is to the total fish population.

The catch-effort method of estimating fish population depends on the premise that all individual fish in a sample have the same chance of being caught and that the effort made by fishermen to catch the fish is constant. This approach works best when the population is closed, when chance of capture and constancy of effort remains unchanged from sample to sample, and when there are enough fish removed so that the population shows a decline.

Assessing Estimated Ecological Risk: Aquaculture. In aquaculture, population is much more controlled, so assessment emphasizes questions of ecological risk rather than determination of population. Because of the intensive nature of fish-farming practices, these questions of ecological risk include near-field and far-field effects of increased organic loading (mainly from uneaten fish feed, fish fecal material, and decomposing dead farm fish), increased inorganic loading (mainly nitrogen, phosphorus, trace elements, and vitamins in fish excretory products and uneaten feed), residual heavy metals (mainly zinc compounds in uneaten feed and fish feces), the transmission of disease organisms (often increased because of the high density of fish per volume of water in net-pen enclosures), and residual therapeutants (from biomedical treatments of the fish performed in response to the increased transmission of disease organisms resulting from crowded net-pen conditions).

Other ecological risks to be assessed for aquaculture include biological interaction of farmed fish that have escaped into wild populations and vice versa (with the potential not only for breeding but also for cross-infection with parasites and pathogens), the impact on marine habitat of fish-farm enclosures (including entanglements involving nets, anchors,

Fascinating Facts About Fisheries Science

- Wild fisheries generated more than $63 billion in household incomes worldwide in 2009.
- Of the world's total fish stocks, 80 percent are either fully exploited or have collapsed.
- One billion residents of Asia depend on seafood for the majority of their animal protein.
- One in five Africans depends on seafood for the majority of his or her animal protein.
- The United States, Japan, and the countries of the European Union are net importers of seafood.
- Between 1986 and 2009, the world population of hammerhead sharks declined by 89 percent—largely from "finning" to make soup.
- Between 1965 and 2009, the Russian sturgeon population, exploited for caviar, declined by 90 percent.
- Overfishing by bottom trawlers along the New Zealand coast has resulted in an 80 percent decline in orange roughy numbers.
- Ghost fishing—nets, lines, and traps that continue to catch fish despite having been lost or abandoned—is a particular problem of derby fisheries.

and moorings), control of natural predators in the fish-farm environment, and increasing pressure on shoaling small pelagic fish populations to be fed to the farmed fish.

Applications and Products

The primary role of most fisheries scientists continues to be consultative. That consultative role usually falls into three categories: assessor (the person responsible for making reliable estimates of fish populations, predicting what harvest levels those populations can support, and the environmental impacts of both fishing and fish-farming approaches), adviser (the person responsible for communicating to government, industry, and public all findings relevant to the health of a fishery and its environment), and educator (the person responsible for increasing governmental, industrial, and popular awareness of both the economic and ecological importance of aquatic organisms).

Some fisheries scientists, however, already find themselves involved in efforts with direct applications and products. These include new ways of growing or harvesting aquatic organisms, some as revolutionary as genetic modification of farmed fish, some as evolutionary as the development of more efficient fishing gear.

Aquatic and Marine Food. Fisheries scientists, as experts in the health of individual fish species and overall fishery populations, help industry provide high-quality seafood products for the consumer—an important activity, considering that seafoods are one of the world's primary sources of high-quality protein. Fisheries scientists are involved in product development, physicochemical principles, and process technology for aquatic food and marine bioproduct utilization, as well as in examining and improving aquatic and marine products and manufacturing processes.

In the aquacultural context, fisheries science helps to increase the aquacultural contribution to the food supply while also developing new methods of production and improved cultural practices for selective species. These production and cultural concerns include environmental, ecological, and disease considerations, selective breeding, feeding, processing, and marketing.

Aquatic and Marine Nonfood Products. In addition to foodstuffs, however, many other products are derived from aquatic and marine organisms. Fish oil, which contains omega-3 fatty acids and anti-inflammatory eicosanoids, are important for a healthy diet. Fish meal, a high-protein food supplement used in aquaculture, is a by-product of rendering and processing fish for fish oil. Fish emulsion, a fertilizer, is produced from the fluid remainder of fish already processed for fish oil and fish meal. Fish skins, swim bladders, and bones are boiled to produce fish glue for specialized uses, while isinglass, a form of collagen obtained from dried swim bladders, is used for clarifying and refining wine and beer, for preserving eggs, and for conserving parchment. Kelp is steadily growing in popularity as a fertilizer and has long been a major source of iodine. Traditional royal or Tyrian purple is derived from sea snails of the *Murex* genus. Pearls and mother-of-pearl are key components of lustrous jewelry.

Fisheries scientists continue to explore aquatic and marine bioresources for pharmaceuticals, nutraceuticals, and novel biomaterials as well as investigate distribution and biodiversity of marine organisms important to industrial utilization. In finding innovative uses for aquatic and marine products and thereby increasing the value of specific fish stocks, fisheries scientists also make more likely the prospect that a given stock will be more sustainably harvested over the long term.

Aquaculture. In the aquaculture context, fisheries scientists not only examine the impacts of aquaculture practices on the environment (including habitat alteration, release of drugs and chemicals, interaction of cultured and wild organisms, and related environmental and regulatory issues) but also propose and evaluate methods for reducing or eliminating those impacts, including modeling, siting, and monitoring of aquaculture facilities and the use of polyculture and water-reuse systems. Fisheries scientists propose the design criteria, provide operational analysis, and develop management strategies for selected species in water-reuse systems.

Fisheries scientists provide in-depth understanding of the natural and social ecology of aquaculture ecosystems, applying principles of systems ecology to the management of the world's aquaculture ecosystems. They also evaluate the nature, causes, and spread of diseases limiting the success of freshwater and marine aquaculture projects. They provide diagnoses of diseases affecting hatchery management and other aquacultural contexts, as well as define appropriate prevention, control, and treatment strategies for those diseases.

Traditional Fisheries. For traditional or wild fisheries, much fisheries science involves advising and educating government, industry, and the public on the biology of aquatic-resource animals, as well as assessing fisheries' populations, stock abundance, and overall ecological health. Fisheries scientists also help frame the debate concerning aquatic resource management, conservation legislation, rehabilitation of depleted fisheries, and socioeconomic considerations involved in national and international fishery issues, practices, patterns, and public policy.

Fisheries scientists also are involved in the development of new fish-catching methods and technologies, including the development and assessment of electronic enhancements to fishing and fishing-vessel operation that have increased fishing power. Fisheries science contributes to the application of these new methods to scientific sampling, to commercial harvesting, to recreational and subsistence fishing. Fisheries scientists not only contribute to advancements and innovations in fishing-gear construction, maintenance, and operation, but also to the evaluation—through empirical, theoretical, model scaling, and statistical-analysis techniques—of the behavior and performance of fish-capture systems.

Impact on Industry

The best current estimates conclude that global fisheries contribute between $225 billion and $240 billion per year to the worldwide economy. Marine fisheries alone provide fifteen percent of the animal protein consumed by humans annually. Despite the importance of fisheries, however, management of fisheries worldwide lags far behind international guidelines recommended to minimize the effects of exploitation. Many researchers agree that conversion of scientific advice into policy, through a participatory and transparent process, is at the core of achieving fisheries' sustainability built on science-based management.

Achieving such sustainability is no easy task, however. Conservation and economic growth often tend to work against each other, at least in the short term. According to a recent National Oceanic and Atmospheric Administration (NOAA) study, in 2009 heavy and effective controls on American fishermen's efforts (focused on limiting fishermen's days at sea and barring access to certain fishing grounds at certain times of the year) drove down landings of fish by 6 percent in the United States—but also moved the nation's fisheries closer to sustainability for the tenth straight year.

A particular challenge that faces fisheries scientists in their role as advisers to government, industry, and the public is that prediction of fish stocks in fisheries and of risk assessment in aquaculture tend to be complicated by the fact that most of the factors involved are interactive. It is therefore important that the fisheries scientist, in communicating findings and making recommendations, ought to qualify and quantify the uncertainty associated with both the findings and the recommendations.

Despite the inherent uncertainties, important actions do occur as a result of the advice of fisheries scientists. For example, more and more of the world's major fisheries are no longer open access. Derby or race-for-fish situations—in which there is a total allowable catch in a fishery but no limitation on fishing by any individual fisherman—are being phased out in most major fisheries. Input controls limiting fisher effort and output controls limiting catch or harvest have become increasingly important tools for managing fisheries as a public trust resource.

Also on the advice of fisheries scientists, the duty to prevent overfishing, attention to marine resources used for noncommercial purposes, and the broader ecological context of fisheries have all increasingly become legally enforceable obligations. This more responsible approach has in many cases resulted in the reduction of bycatch and fish discarding as well as reduced the adverse impacts of fishing activity on critical fish habitat.

Now widely implemented in U.S. fisheries, the catch-share system is a transferable version of individual fishing quota, which grants fishermen in sector cooperatives an assigned share of a total allowable catch for each species and encourages fishermen to buy, sell, or trade shares with colleagues or outside investors. The catch-share approach, in removing the race for fish, has reduced the incentive to fish during unsafe conditions and to buy ever-larger vessels and more equipment (overcapitalization). Under this approach, fresh fish are available to consumers over longer periods during the year, and fishermen have also been able to maintain higher-quality product while reducing bycatch and gear conflicts.

Economic and socioeconomic downsides to the approach do exist, however. Fish imported into the United States from unsustainable Asian and European fisheries are cheaper than sustainably caught U.S. fish. There are also important concerns about the equity of gifting a public trust resource, the elimination of vessels and reductions in crews, and the consolidation of shares (and therefore economic power) in fewer hands and into larger businesses. The effects of this last not only on jobs but also on the fishing culture the industry has long made possible in many locales.

Government and University Research. As part of the public trust nature of the fisheries involved, many U.S. governmental entities make use of the advice provided by fisheries scientists—departments and agencies as diverse as NOAA, the United States Geological Survey (USGS), the Food and Drug Administration (FDA), the Department of Commerce and the National Marine Fisheries Service (NMFS), the National Academy of Science and National Research Council, the eight regional fisheries management councils, and the scientific and statistical committees that provide scientific advice to them. Individual state and even county departments of fish and game make use of such expertise, as do supranational entities like the United Nations Food and Agriculture Organization (FAO), the International Council for the Exploration of the Sea (ICES), the North Atlantic Fisheries Organization (NAFO), and the International Pacific Halibut Commission (IPHC).

Universities and related research institutions employ nearly as many fisheries scientists as do governments. Massachusetts' Woods Hole Oceanographic Institution and its marine biological laboratory, several campuses in the University of California system, the University of Maryland system, the University of Florida system, and the University of Alaska system all have substantial research programs in fisheries science.

Industry, Business, and Nongovernmental (NGO) Sectors. Although most participants in the fishing industry continue to rely on statistics produced by governmental agencies and university researchers, fisheries scientists are also being employed in increasing numbers by industry groups ranging from aquaculture firms such as AquaBounty (creators of the AquAdvantage line of genetically modified salmon), to fishermen's associations, to large fish-processing concerns such as the At Sea Processor's Association.

In response to the changing landscape in the fishing and aquaculture industries, nongovernmental organizations such as the Center for Science in the Public Interest, the Alliance for Natural Health, the World Wildlife Fund, People for the Ethical Treatment of Animals (PETA), the Marine Stewardship Council, Earth Island Institute, and many others are also increasingly finding a need for fisheries scientists and their advice.

Careers and Course Work

Career titles in fisheries science involve both traditional wild fisheries and aquaculture. They include aquarium director, biologist, conservation ecologist, conservation officer, environmental consultant, fish culturist, fish-processing manager, fisheries manager, fisheries technician, fisheries biologist, fisherman, hatchery manager, lab technician, museum or aquarium curator, marine biologist, natural-resource specialist, nature interpreter, lake or pond manager, researcher, vessel captain, and wildlife manager.

Courses in ecology, biology, chemistry, and mathematics are foundational for students wishing to pursue careers in fisheries science. A fisheries technician, fish-hatchery manager, fish-processing manager, lab technician, or nature interpreter may need little more than this background, together with some basic skills in mechanical, electrical, or systems engineering.

The pursuit of the master's and doctoral degrees—which are often the necessary minimum qualification for more advanced academic, governmental, or industrial career in fisheries science—generally requires more specialized course work beginning at the upper division of undergraduate studies or the beginning of graduate studies. More specialized courses may include genetics, marine biology, limnology, ichthyology, oceanography, and veterinary medicine.

Although fisheries science is primarily biological at its root, it is also strongly interdisciplinary and advisory, so background in fields such as hydrology, anthropology, law, economics, and political science can also prove very helpful.

Social Context and Future Prospects

As more and more of the world's fisheries become overfished or fished out, and wild fish stocks are threatened globally, there is increased pressure to make the changeover from the mechanized hunting-gathering approach of commercial fishing to aquaculture. This must be done in decades, rather than the millennia involved in the analogous changeover to agriculture. Concerns about potential ecological risks, ranging from aquaculture's inherently dense and intense practices to genetic modification of fish species, have also increased as the rapidity of this changeover has increased.

One of the most pivotal roles of fisheries science in the future will be the assessment of how much of the near-shore oceans should remain wild and how much should be intensively farmed. This will involve not only the assessment of ecological risk associated with the growing aquaculture industry but also the determination of how fully and how rapidly depleted or damaged fisheries can recover. Marine reserves, aquatic refuges, and bioremediation all have roles to play here, as do concerns associated with reef die-offs, pollution and climate change, and economic and political pressures on the remaining 20 percent of fisheries worldwide that are not yet fully exploited.

During the twentieth century, the survival or extinction of many populations of aquatic organisms will depend on how the growing population of a terrestrial organism with a fondness for seafood—the human species—interacts with its global environment. Fisheries scientists, in their roles as assessors and advisers, are key to helping shape that interaction.

Howard V. Hendrix, B.A., M.A., Ph.D.

Further Reading

Clover, Charles. *The End of the Line: How Overfishing Is Changing the World and What We Eat.* Berkeley: University of California Press, 2008. An investigative journalist passionately (and sometimes polemically) writes about his personal research into the crisis in global fisheries.

Committee to Review Individual Fishing Quotas, et al. *Sharing the Fish: Toward a National Policy on Individual Fishing Quotas.* Washington, D.C.: National Academy Press, 1999. Report of committee empaneled by Congress to report on individual fishing quotas.

Everhart, W. Harry, and William D. Youngs. *Principles of Fisheries Science.* 2d ed. Ithaca, N.Y.: Cornell University Press, 1981. Classic introductory textbook on the subject, providing a broad overview of the field.

Iudicello, Suzanne, Michael Weber, and Robert Wieland. *Fish, Markets, and Fishermen: The Economics of Overfishing.* Washington, D.C.: Island Press, 1999. Emphasizes the importance of human economic behavior in the exploitation of fisheries and their management.

McGoodwin, James R. *Crisis in the World's Fisheries: People, Problems, and Policies.* Stanford, Calif.: Stanford University Press, 1990. An anthropological perspective on fisheries, emphasizing indigenous and small-scale fishers.

Molyneaux, Paul. *Swimming in Circles: Aquaculture and the End of Wild Oceans.* New York: Thunder's Mouth Press, 2007. A thoroughly researched, critical journalistic account of salmon aquaculture in Maine and shrimp aquaculture in Mexico.

Walters, Carl J., and Steven J. D. Martell. *Fisheries*

Ecology and Management. Princeton, N.J.: Princeton University Press, 2004. Upper division and graduate textbook for classes in fisheries assessment and management, with an emphasis on quantitative modeling methods.

Web Sites

American Fisheries Society
http://www.fisheries.org

Food and Agriculture Organization of the United Nations
Fisheries and Aquaculture Department
http://www.fao.org/fishery/en

National Oceanic and Atmospheric Administration
National Marine Fisheries Service
http://www.nmfs.noaa.gov

National Oceanic and Atmospheric Administration
Office of Sustainable Fisheries
http://www.nmfs.noaa.gov/sfa/statusoffisheries/SOSmain.htm

See also: Anthropology

FLOOD-CONTROL TECHNOLOGY

FIELDS OF STUDY

Hydrology; civil engineering; fluvial geomorphology; watersheds; climatology; meteorology; coastal erosion; stream hydraulics; flood forecasting; dam building; physical geography.

SUMMARY

Flood-control technology deals with the myriad techniques that can be employed to deal with water that overflows stream banks, thereby leading to deaths and injuries, property and crop damage, and severe erosion. The magnitude of the flood damage varies considerably, as it depends on the duration of the storm and the amount of precipitation. Each watershed has different physical features, such as size, slope, basin relief, impoundments, flood history, soil types, and drainage characteristics.

KEY TERMS AND CONCEPTS

- **Flash Flood:** Occurs when storms cause a river to rise rapidly.
- **Flood Control:** Refers to various measures that could reduce flood damage.
- **Flood Frequency:** Average amount of time between floods that are the same or greater than a selected magnitude.
- **Floodplain:** Low-lying area of varying width that could be on one or both sides of a river.
- **Flood Probability:** Statistical determination that flood events of a particular size would be less or the same during a particular period of time.
- **Flood Stage:** Occurs when the river overflows its banks and moves onto the floodplain.
- **Flood Wall:** Structure designed to reduce flooding.
- **Floodway:** Large channel built to divert floods away from populated areas.
- **Hydrograph:** Graphic representation of the amount of water passing a given point.

Definition and Basic Principles

Flood-control technology includes various structural and nonstructural measures that play a substantial role in mitigating flood damages. The methods vary from place to place, given the enormous heterogeneity of stream discharge, precipitation frequency and amount, and watershed factors that include slope, soil permeability, infiltration characteristics, degree of urbanization, varying land-management practices, governmental interest, and land-ownership practices.

The structural or physical measures that are employed include levees or flood walls made of earth or concrete, channel modifications such as deepening or widening the stream itself, building dams or reservoirs to hold additional water, watershed-improvement techniques that include reforestation and planting of vegetative cover in bare areas, flood proofing in lower-risk floodplain areas that include dikes, elevating the buildings, and waterproofing.

The nonstructural or preventative measures that could be adopted include floodplain regulations such as zoning laws that stipulate where development may or may not occur, building codes that put restrictions on basements and the permissible amount of impervious cover allowed on the site, open-space requirements for large-scale developments, and designing flood warning systems to provide advance notice about impending problems.

Background and History

The beginnings of speculation about water movement (the hydrologic cycle) began in ancient times by such renowned philosophers as Homer, Thales, Plato, and Aristotle in Greece, and later on by Lucretius, Seneca, and Pliny in ancient Rome. The field of hydrometry that pertains to stream-flow measurement began in the seventeenth century and improved techniques increased with each century. For example, flow formulas, measuring instruments, and stream-gauging procedures became better established in the eighteenth and later centuries. Some of the many advances include civil engineer Theodore G. Ellis's and surveyor William Gunn Price's stream current meters in 1870 and 1885, respectively, Irish engineer Robert Manning's flow formula in 1891, civil engineer Allen Hazen's comments on increasing the use of statistics in flood studies in 1930, and German statistician E.J. Gumbel's suggestion for using frequency analysis for floods in 1941.

Major changes at the governmental level in the United States assumed increasing importance as several hydrologic agencies were created in the nineteenth century: the Army Corps of Engineers in 1802, the Weather Bureau in 1870 (now called the National Weather Service), the U.S. Geological Survey, and the Mississippi River Commission in 1879.

How It Works

Floods. In the hydrologic cycle, atmospheric precipitation falls to Earth and is either evapotranspired back to the atmosphere, absorbed by vegetation, or moved downslope as overland flow that eventually becomes large enough to form stream channels. Water generally stays within the channel for most of the year. However, if the precipitation is heavy enough, the channel cannot transport all of this water and the stream overflows its banks, thereby creating floods.

Floodplains are low areas that can be found on one or both sides of a stream and are common in humid regions. Small watersheds with steep slopes often have flash floods that move very quickly with enough turbulence to damage buildings and vehicles easily. Since the flash floods occur so rapidly, people cannot be readily warned and drownings can occur.

Stream Discharge. In the United States, stream flow is measured in cubic feet per second at most gauging stations operated by the U.S. Geological Survey (USGS). Most of the larger streams have higher discharges as they flow downstream. There are some exceptions, such as the Colorado River in the arid southwestern United States, where the demand for irrigation water and public supply literally dries up the stream at its mouth.

Flood Frequency. It would be wonderful if flood prediction were reasonably exact. However, the probability of knowing when floods would occur turns out to be a markedly difficult task as it involves frequency analysis, probability theory, data that need to be homogeneous and independent, and selected assumptions that existing stream-flow data would be similar to future flows. This fallacious assumption implies that there will be no changes in future land use and climate.

The longer the period of record, the better the estimation of future flood flows. Accordingly, a "100-year flood" does not mean that the next one of that magnitude will occur 100 years in the future. It does mean that there is a 1 percent chance that a 100-year flood could occur in any year. Accordingly, a 100-year flood could occur this year and again the following year, although the odds are against it. In hindsight, it would have been better if the term "100-year flood" would have been expressed as a "1-in-100 chance flood" in order to reduce public confusion.

Floodplain Development. The stream channel itself occupies a relatively narrow area within a floodplain. The channel is bordered by low-lying land that is called a floodway. As the distance away from the river increases, the slope of the land gradually increases and a slightly higher landscape called a floodway fringe is created. A cross-sectional profile of this floodplain would then show a channel with a floodway on one or both sides of the channel and a floodway fringe farther away from the stream.

In a nonregulated development scenario that was common in earlier years and still exists in some areas, residential and commercial buildings were constructed in the floodway zone, which lead to higher flood levels, structural damage, and potential loss of life. Low bridges that cross the stream in this zone could partially block the flow and thereby increase flood levels.

The situation would be quite different in a regulated floodplain. Ideally, the floodway would be zoned for as much open space as possible, such as parks, picnic areas, and golf courses. This would facilitate easier passage of flood waters and restrict buildings to higher ground wherever possible.

Floodplain Extent. It has been estimated that 100-year floodplains make up from 7 to 10 percent of the total land area of the United States. The floodplains with the largest areas are located in the southern portions of the country; those with large populations are located along the north Atlantic coast, the Great Lakes, and California.

Types of Floods. Damaging floods occur in varying locations. Flash floods are associated with quickly developing thunderstorms in mountainous areas. The rapid downslope movement of water, even if relatively shallow, can quickly move vehicles and their occupants to the extent that 50 percent of the flood deaths in the United States are caused by flash floods. If floodwater is flowing at a depth of only 2 feet, a trapped vehicle experiences a lateral force of 1,000 pounds and a buoyant force of 1,500 pounds, more than enough to flip the vehicle over and drown the passengers.

Regional floods are associated with level land that experiences long periods of heavy rain. The resulting damage to homes and properties in the floodplain can be enormous, as evidenced in the August 1993 flood in the lower Missouri and upper Mississippi rivers. About 69 percent of the levees built along the upper Mississippi River to protect the floodplain occupants were overtopped, drowning forty-eight persons, submerging seventy-five towns, destroying 50,000 homes, and even excavating more than 700 coffins from a cemetery in Missouri.

Storm surges powered by tropical storms and hurricanes can wreak havoc along low-lying coasts. The settlers who founded New Orleans in 1718 along the lower Mississippi River experienced their first major flood the same year. Over time, many other floods have occurred in New Orleans and the lower reaches of the Mississippi, aided by the slow sinking of the floodplain as river sediments compact over time.

Applications and Products

Structural Flood-Control Objectives. Most structural flood-control measures are physical and expensive. They include construction of large reservoirs, diversion structures, levees and flood walls, channel alterations, modifications to bridges and culverts, and tidal barriers.

Flood-control reservoirs are built so that excess water from storms can be stored and released at a later time, thereby reducing peak discharge. The economic value of protecting property that may be damaged in floods can justify the construction costs of a reservoir. In addition, the stored water can also be used for hydropower, water-supply purposes, and recreation. Diversion structures are designed to reduce peak flows by forcing floods to go to another location via channels or tunnels. Levees and flood walls can physically keep water away from floodplains, thereby providing good potential for damage reduction. Levees are usually made of earth and flood walls of concrete. Both structures are usually set parallel to the stream.

Channel alterations to the stream can lower the height of water so that peak discharges are reduced at one point, but they can be increased downstream. The cost of these measures is substantial, but they can be justified in certain situations by their potential protection for valuable property. Possible long-term negative effects of these structural measures include aggradation or degradation of the downstream channel and sediment deposition in downstream bypass channels or tunnels. Bridge and culvert modifications are useful when the structures have inadequate capacity to handle flood flows. Repairing or raising the structures over the channel can lower damages by allowing more water to flow downstream. Tidal barriers along the coast can prevent high tides from moving upstream and damage developed areas. They are expensive to build and are generally used if large urbanized areas need protection. Note that the potential for tidal flooding is expected to increase as sea level rises because of climate change.

Nonstructural Flood-Control Objectives. These measures are employed in order to lower flooding susceptibility, thereby decreasing possible damage. They include flood warning, flood proofing, and a variety of land-use control procedures. Flood warning can reduce possible loss of life, or even better, eliminate that possibility. Flood-proofing procedures include rearranging the working space in buildings, waterproofing outside walls, and raising the height of buildings that occupy particularly susceptible locations. Land-use controls include many kinds of action within the floodplain that can reduce flood-hazard potential. These controls include proper building codes, purchasing flood insurance, zoning restrictions, or purchase of land and property in the less vulnerable portions of the floodplain.

Climate Change. There is growing concern that global climate change could substantially impact flooding in coastal locations. For example, since 1870 sea level has increased about 5.9 inches. The overall sea-ice cover in the Arctic has been decreasing about 3 percent per decade and by 7 percent per decade in the summer. Estimates indicate that average global surface temperatures for the period 1990-2100 will increase between 3.1 and 10.5 degrees Fahrenheit. As a result, sea level could increase between 8.3-18.5 inches by the end of the twenty-first century. In addition, future peak precipitation events are expected to increase in intensity. To make matters worse, some of the climatic models employed suggest that rates of sea-level rise will double along the shorelines of certain portions of the eastern United States and the western North American and Arctic coasts.

Flood Mapping. It would be very useful to be able to generate flood maps in advance of approaching storms. Experimental work has been developed by

Fascinating Facts About Flood-Control Technology

- The Black Sea was a freshwater lake during the last glacial advance. As the glaciers receded, global sea level rose and Mediterranean seawater flowed through the Bosporus Strait in Turkey into the Black Sea. More than 40,000 square miles of farmland was flooded, displacing an untold number of people.
- Glacial recession at the end of the last Ice Age allowed enormous volumes of meltwater to scour out land in North America to depths as great as 1,600 feet, thereby forming the Great Lakes, which hold 18 percent of all of the freshwater on the Earth's surface.
- Peak stream flow on the Mississippi is generally highest from February to June. However, the greatest flood that was ever recorded on the upper Mississippi River at St. Louis occurred during the summer of 1993, due mostly to a wet winter that was followed by an even wetter summer (double the normal precipitation). About two-thirds of the 1,576 levees were damaged enough to allow water into the floodplain.
- Hurricane Katrina in August 2005 holds the record as the most damaging storm to strike the United States The high winds (category 3) and heavy precipitation of eight to ten inches, in conjunction with a storm surge of twenty-four to twenty-eight feet, resulted in a death toll of 1,836 and property losses of more than $100 billion. Floodwaters with depths approaching twenty feet covered 80 percent of New Orleans.
- Paleoflood hydrology is the study of ancient floods. It is generally used for events within the past 10,000 years, as earlier times experienced very different global climatic changes, namely, the Ice Age.
- Major floods that occur in historical times can be exceeded centuries later: Florence, Italy, was devastated on November 4, 1333, when the Arno River flooded the city to depths of 14 feet. Almost 633 years to the day, on November 3, 1966, another flood occurred, covering the city with depths of 20 feet.

the USGS and the National Weather Service (NWS) to make forecasting and mapping of potential floods readily available to local officials. The goal is to have NWS provide storm forecasts for a particular area, thereby enabling the USGS to generate maps showing potential flooded areas on the floodplain, arrival times, and the depth of the flood itself. The combination of new methodologies and techniques, such as light detection and ranging (LIDAR), advanced computer programs, and geographic information systems (GIS) have led to the strong possibility of creating useful flood-forecast maps.

Flood Warning Systems. A useful addition to the regulation of floodplain development occurs when flood-warning systems are employed by local governments. These systems use radar, rainfall, and stream-flow gauging that are connected by satellite transmitters to relay real-time data to computers at some central site, which in turn make the data available to interested parties via the Internet. Automatic warnings are then dispatched to emergency-management officials who can then institute procedures, based on the predicted flood levels, that range from selective road closures to complete evacuation.

Impact on Industry

U.S. Government Research. Several agencies of the federal government are involved with flooding issues. The U.S. Army Corps of Engineers (Corps) is the oldest agency that was assigned to deal with increasing flooding problems, particularly on the Mississippi. Flood Control Acts of 1928, 1936, and 1944 gave additional authority to the Corps to reduce flooding.

The major source of stream-flow data for the United States has been primarily gathered by the USGS. Its first gauging station was built on the Rio Grande in New Mexico in 1889. The early gauges were mechanical-current devices that measured the velocity of flowing water in conjunction with measurement of the stream's cross-sectional area. Improvements since 1995 have included hydroacoustic meters that have come to constitute about 30 percent of the discharge gauges that now number more than 7,400. The new techniques have lowered the average discharge measurement time from 96 to only 18 minutes. The National Water Information System (NWIS), part of the USGS, provides real-time data for selected surface water, groundwater, and water-quality sites around the country with maps and links to the sites.

Other federal and regional agencies that record

specialized stream-flow data include the Tennessee Valley Authority (TVA), Bureau of Reclamation, Natural Resources Conservation Service (NRCS), U.S. Forest Service, and the Agricultural Research Service (ARS).

Government Regulations. The National Flood Insurance Program (NFIP) was created by the United States Congress in 1968, and is now under the control of the Federal Emergency Management Agency (FEMA). FEMA became part of the U.S. Department of Homeland Security (DHS) in March 1, 2003, and employs more than 3,700 workers. The purpose of this agency is to identify and map flood-hazard areas, disseminate this information to those living in floodplains, and provide information about flood insurance. Local governments are required by the NFIP to adopt regulations that would deter proposed housing developments that would not be in accordance with national standards. USGS maps are used by the NFIP to delineate 100- and 500-year floodplains. Banks that lend money for homes and commercial property in the 100-year floodplain require the prospective owners to purchase flood insurance. Developments in the 500-year floodplain do not necessarily have to have flood insurance. The probability of a 100-year and 500-year flood occurring in any 1-year period is 1 percent and 0.2 percent, respectively.

Association of State Floodplain Managers (ASFPM). This organization of 13,000 professionals in the United States was formed in 1976. The major concerns that interest ASFPM include flood-hazard mitigation, floodplain management, the NFIP, and just about anything else that involves flooding issues.

Careers and Course Work

There are a surprising number of varied courses and future jobs that are included in the field of flood-control technology. The most obvious ones include civil engineering for construction of dams and levees, hydraulic engineering to handle channel deepening where appropriate, and surveying. However, there are many other skills that are needed, and it would benefit those interested in this field to study geology, physical geography, hydrology, meteorology (storms), and water-resources management. Other useful courses of study include environmental planning, computer science, economics, and environmental law.

Many governmental organizations, including the Army Corps of Engineers, American Water Resources Association, and FEMA, have job postings on their Web sites that encompass a range of employment opportunities. These agencies employ civil engineers, scientists, natural-resource specialists, administrators, hydrologic technicians, and ecologists, among others.

Social Context and Future Prospects

Floods have occurred on numerous occasions over many millennia. Societies have designed ways to store some of this water in special collecting basins or small reservoirs, and in selected cases by creating diversion canals. In other cases, the floods have been very damaging to settlements and farms. What makes the situation even worse is the population growth on floodplains, the inhabitants of which have the mistaken notion that floods can be contained by structural techniques. Too many damaging floods have occurred in too many countries that renders naive the presumption that one is safe behind the levee.

One step in the right direction has been the growing recognition that floodplain development carries a substantial risk: predicting flood heights is fraught with uncertainty because of a complicated mix of storm tracks, soil-moisture conditions, available upstream reservoir storage, and the elevations of commercial facilities and homes. One can now recognize the benefits of nonstructural measures, such as zoning lands that are in the 100-year floodplain as not suitable for development. To make matters more interesting, and more complicated, how does one factor in shifts that could occur over many decades, such as climate change, river behavior, and sea-level rise that would impact coastal areas as well as the existing homeowners in the floodplain who simply enjoy living close to rivers?

Two things will certainly help out in this situation. One would simply be better explanation of the flooding danger as given by the probability of certain events occurring in any year. These odds could, of course, change as floodplain development continues or decreases. The second element is simply better discussion of future weather patterns, as unpredictable as they are, that could occur on a global scale.

Robert M. Hordon, B.A., M.A., Ph.D.

Further Reading

Bedient, Philip B., Wayne C. Huber, and Baxter E. Vieux. *Hydrology and Floodplain Analysis.* 4th ed. Upper Saddle River, N.J.: Prentice Hall, 2008. In addition to chapters on hydrology, floodplain hydraulics, and flood management, the book contains seven pages of hydrology-related Internet links.

Cech, Thomas V. *Principles of Water Resources: History, Development, Management, and Policy.* 3d ed. Hoboken, N.J.: John Wiley & Sons, 2010. A highly readable and very informative book on hydrology, flood events, and federal water agencies with useful diagrams, maps, and photos.

Dunne, Thomas, and Luna B. Leopold. *Water in Environmental Planning.* New York: W. H. Freeman, 1978. A classic text in the water field; contains useful information regarding runoff processes, flood hazards, and human occupancy of flood-prone areas.

Dzurik, Andrew A. *Water Resources Planning.* 3d ed. Lanham, Md.: Rowman and Littlefield, 2003. A readable book on hydrology, federal water agencies and related legislation, and floodplain management.

Mays, Larry W. *Water Resources Engineering.* 2d ed. Hoboken, N.J.: John Wiley & Sons, 2010. A detailed textbook that contains numerous examples and diagrams involving hydrology, surface runoff, and flood control.

Strahler, Alan. *Introducing Physical Geography.* 5th ed. Hoboken, N.J.: John Wiley & Sons, 2010. An excellent textbook with very useful diagrams, maps, and color photos that cover many aspects of hydrology and stream flow.

Viessman, Warren, Jr., and Gary L. Lewis. *Introduction to Hydrology.* 5th ed. Upper Saddle River, N.J.: Prentice Hall, 2003. Includes many chapters on floods and floodplains with good diagrams.

Web Sites

American Water Resources Association
http://careers.awra.org/jobs

Federal Emergency Management Agency
http://www.fema.gov

U.S. Army Corps of Engineers
http://www.usace.army.mil

U.S. Geological Survey
Water Resources
http://www.usgs.gov/water

FLUIDIZED BED PROCESSING

FIELDS OF STUDY

Chemistry; engineering; chemical engineering; chemical process modeling; corrosion engineering; control engineering; process engineering; industrial engineering; mechanical engineering; electrical engineering; safety engineering; physics; thermodynamics; mathematics; materials science; food processing; cryogenics.

SUMMARY

Fluidized bed processing is a technique that mixes solid particles with either a liquid or a gaseous stream to form a mix that behaves like a fluid. This mix can be made to flow, keep a uniform temperature, and facilitate interaction among solid and fluid particles, supporting heat and mass transfers. The early 1940's saw its first major application in fluid catalytic cracking units that increase a refinery's yield of high-quality gasoline and aviation fuel. The technique has become widely used in the process industry, including synthesis of polyethylene and polypropylene, key feedstocks of the chemical industry. It is also used in the food and pharmaceutical industries, at power plants, and for waste incineration.

KEY TERMS AND CONCEPTS

- **Aggregative Fluidization:** As a gaseous liquid fluidizes the bed of solids, bubbles form, making the process look like boiling water.
- **Bed:** Area of a processing unit where solid particles are fluidized.
- **Distributor:** Grid, or a plate with holes, through which the fluid is injected into the bed.
- **Entrainment:** Transport of solid particles up out of the bed in the fluid mix.
- **Fluid:** In physics, either a liquid or a gas.
- **Fluidization:** Physical process that converts the state of solid particles to a fluidlike state as a fluid is sent through the solid material.
- **Minimum Fluidization:** Point at which the solid particles of the bed begin to behave like a fluid.
- **Minimum Fluidizing Velocity:** Minimum superficial velocity with which the fluid has to enter the bed to fluidize the solid particles.
- **Particulate Fluidization:** Bed of solids expands continuously and uniformly, typically caused by a liquid fluidization medium; the opposite of aggregative fluidization.
- **Superficial Velocity:** Roughly, the mean velocity with which the fluid hits the solids.

Definition and Basic Principles

Fluidized bed processing is a technique developed from the fact that if small solid particles encounter a quickly moving fluid of liquid or gaseous nature, the resulting mix including the solids behaves like a fluid itself. This provides numerous advantages in process engineering.

For fluidized bed processing to work properly, the solids have to exist in the form of tiny particles, often called granules, ranging in size from 1 or 2 millimeters to as little as 0.02 millimeter. The smaller the particles are, the easier it is to get them into a fluidized state. The only exceptions are food particles, which can be much larger because of their low density.

To achieve fluidization and to make use of this physical conversion process, a fluid medium is injected into a bed of solid particles. As the fluid rises at its own superficial velocity—roughly meaning its own speed, which is measured by dividing the volumetric flow rate of the fluid by the cross-sectional area of the bed of solids—it encounters the solid particles. Once the rising fluid exerts a drag force greater than the net weight of the solid particle, which is called minimum fluidization, the solid particles begin to behave like the rest of the fluid. Typically, they become entrained in the fluid and rise with it to the top of the unit. Because of the laws of thermodynamics, the solids quickly share the same temperature as the fluid medium. Because of their large shared surface area, solids and fluid medium particles can interact very efficiently.

Background and History

On December 16, 1921, German chemist Fritz Winkler was the first to use the principle of fluidized bed processing in a coal-gasification experiment based on small-grained lignite. By 1926, the German company BASF that employed Winkler began commercial coal gasification based on Winkler's new technique. Fluidized bed processing has remained an important part of coal gasification and has been applied to gasification of biomass as well.

The second most important discovery was the use of fluidized bed processing for the catalytic cracking of long hydrocarbons at petroleum refineries. This was done to obtain more valuable gasoline and aviation fuel from crude oil distillate. Four American scientists, Donald Campbell, Homer Martin, Eger Murphree, and Charles Tyson filed their patent for the process on December 27, 1940. It was granted on October 19, 1948. Based on their research, a pilot plant started in Baton Rouge in May, 1940, and the commercial plant began operations on May 25, 1942. Since then, a fluid catalytic cracker has been at the heart of every modern refinery.

In each subsequent decade, fluidized bed processing has found new applications in the chemical, petrochemical, pharmaceutical, and food-processing industries. The technique is important in the quest for cleaner power generation—and even the production of carbon nanotubes.

How It Works

Bed Preparation. To begin the process, solid particles are placed or injected onto a bed in a holding unit. This is called a packed or a fixed bed, until fluid is applied. The bed is typically in a reactor, boiler furnace, or another part of a processing unit, such as the catalytic riser of a refinery's fluid catalytic cracker. There are different kinds of beds, the most common being stationary or bubbling beds or more complex circulating beds.

Typically, the solid particles of the bed are distinguished by their size and density. They are commonly classified into four categories, proposed first by English chemical engineer and professor Derek Geldart in 1973, and known as Geldart groups. The larger and denser the particles are, the more energy is needed to fluidize them.

In the case of fluidized bed combustion, there is a difference between solid bed materials (which will not be burned), solid sorbents, and the solid particles that serve as fuel. Solid particles can be inserted into the bed in a continuous process, even if there is already a fluid mix. This is very commonly done, rather than individual batch processing, to save time and energy.

Fluidization. Once the solids are in the bed, fluid is injected into the bed at a high velocity through a distributor, which is commonly either a grid or a porous plate at the bottom of the bed. The fluid can be both in liquid or gas form, as defined in physics. Very often, superheated steam is used or, in the case of fluid catalytic cracking, evaporated hydrocarbon molecule chains.

Because of the velocity with which the fluid flows through the bed to the top of the containing unit, it will exert a drag force on the solids in the bed. Minimum fluidization begins once the solids start to rise with the fluid medium and become entrained in it as they leave the bed for the top. If a gaseous liquid is used as fluid medium, bubbles will form in the bed, giving it the characteristic look of boiling water in a process called aggregative fluidization. If a liquid fluid is used, the bed of solids will expand continuously and uniformly, as solids rise within the liquid fluid to the top. This is called particulate fluidization.

To measure and control fluidization, process engineers measure the so-called superficial velocity with which the fluid charges through the bed. It is obtained by dividing the volumetric flow rate of the fluid by the cross-sectional area of the bed and corresponds roughly to the speed with which the fluid medium is pressed through the bed. It is typically given in meters per second. Each type of solid has its own minimal fluidization velocity that must be reached for the solids to begin to fluidize.

Processing. In the fluid state, processing of solids and fluid occurs as designed for each application. This is aided by the uniform heat of the fluid mix, its transportability through pipes, and the high rate of surface interaction between solid and fluid medium particles. At the end of desired processing, remaining solids and fluid medium are typically separated.

Applications and Products

Fluid Catalytic Cracker (FCC). Traditionally, this is the key industrial application of fluidized bed processing. Both atmospheric and vacuum distillation of crude oil at a refinery leave behind a large percentage of low-value residue instead of desired gasoline, diesel, or aviation fuel. This distillation residue leaves behind long hydrocarbon molecule chains.

Cracking creates much more of the shorter hydrocarbon chains, which yield desirable fuel such as gasoline. The American invention of the FCC in 1942 quickly spread to refineries all over the world after

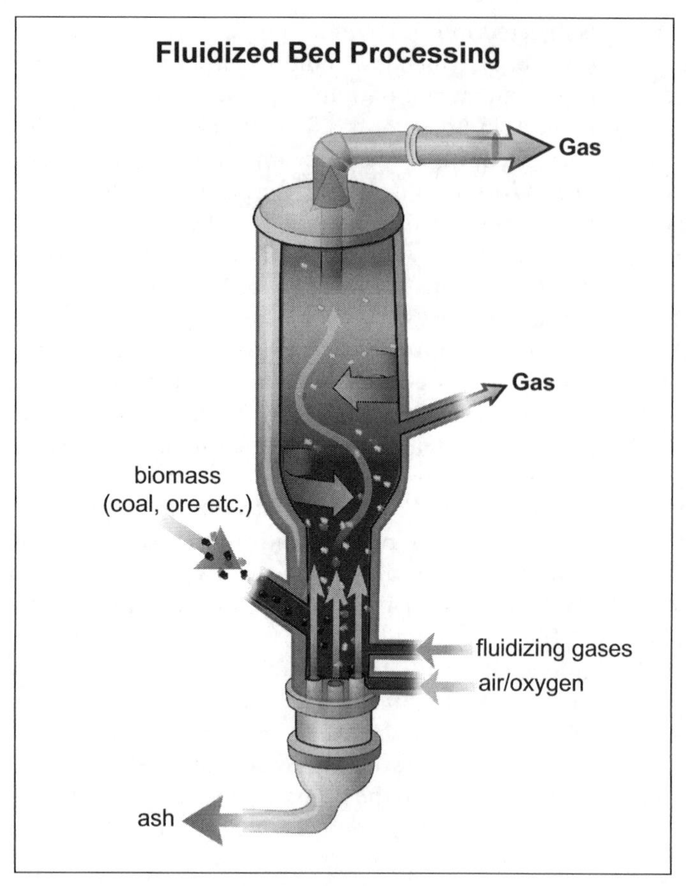

Fluidized bed processing is a technique that mixes solid particles with either a liquid or a gaseous stream to form a mix that behaves like a fluid.

World War II ended in 1945

The great technical advantage of the fluid catalytic cracker is that the long hydrocarbon chains, typically light fuel (or gas) oil, come into contact with the solid catalyst particles in a fluidized, rather than a stationary, bed environment. That means that the total surface area between catalyst and oil particles is much larger than possible if the catalyst were just fixed to the floor or sides of the reactor. In the cracker's catalytic riser, tiny catalyst particles are completely surrounded by the oil particles and swim in a fluid stream together for a few crucial fractions of a second.

In this case, it is the fluid medium, the fuel oil preheated to its evaporation point, that undergoes the value-adding conversion. Its long hydrocarbon chains are quickly cracked when encountering the catalyst particles in the fluid mix. The fluid is sent to a reactor. Cyclones at the top of the reactor separate the gaseous hydrocarbon fluid from the spent catalyst and transport it into a distillation unit. The solid catalyst particles are collected at the bottom of the reactor and sent to a regenerator, from where they are fed back into the catalytic riser in a continuous process.

Fluidized Bed Combustion (FBC). This is the oldest application that has gained new attention in the quest for cleaner power generation and waste incineration. Its basic goal is to convert the energy of solid fuels, such as coal, biomass, or waste materials, through gasification, steam generation, or incineration using fluidized bed combustion.

An advanced power plant can use FBC to increase its fuel efficiency and lower harmful emissions. Inside the combustor, there is a bed of inert material that will not be burned itself, typically a form of sand. To the solid fuel particles that are added to the bed is mixed a solid sorbent, which absorbs potentially harmful substances released during combustion. Typically, when coal is used as fuel, limestone or dolomite is used as sorbent to absorb sulfur that would otherwise react and be emitted as sulfur oxide, a strong atmospheric pollutant. The solid particles are fluidized by the injection of a fluid, typically heated air. There are two kinds of arrangements for the bed of solids that is hit by the fluid, either a bubbling fluid bed or a circulating fluid bed.

The two great advantages of the FBC are that it can be fired by a wide variety and mix of solid fuel particles, including coal, solid wastes, biomass, or natural gas; and it generates much less pollution because of the use of sorbent in the fluid to catch sulfur. It is able to burn fuel at lower temperatures where oxygen and nitrogen of the air fluid do not yet react to form the pollutant nitrogen oxide.

A simple FBC operates at atmospheric pressure as the solid fuel particles are burned in the fluidized state and the sorbent particles bind potentially harmful by-products. In a more advanced design, the fluid of generally hot air is pressurized. Now, combustion creates a hot pressurized gas flow that fires a gas turbine, while steam generated during combustion fires a steam turbine for maximum energy yield

Fascinating Facts About Fluidized Bed Processing

- The invention and industrial application of fluid catalytic cracking is credited by some historians for contributing significantly to the Allied victory in World War II because of the wide availability of quality automotive and aviation fuel.
- In food processing, solid particles of the fluidized bed can be as large as a potato chip, while the smallest particles used in the chemical industry are as tiny as 20 micrometers.
- It takes less than one second for a catalyst to crack a long hydrocarbon molecule chain in the catalytic riser of a fluid catalyst cracker at the heart of a modern refinery.
- The four American scientists who invented fluid catalytic cracking, Donald Campbell, Homer Martin, Eger Murphree, and Charles Tyson, had to wait almost eight years, from December, 1940, to October, 1948, for their U.S. Patent No. 2,451,804 to be granted after filing it.
- For many years, the chemical engineering department at the University College London used to show its students how fluidization of a bed of sand made a buried light plastic duck rise from the bottom to the top and exchange places with the brass duck that was sinking down.
- Fluidized bed combustion holds some promise for use within zero-emission power generation.

in what is called a combined-cycle power plant. This accounts for one of the most efficient and least polluting modes of power generation.

Chemical Applications. Fluidized bed processing has become widely used in the chemical industry. It is important in particular for the synthesis of polyethylene and polypropylene, key basic plastics used for packaging, textiles, and plastic components. Fluidized bed reactors are used also for the industrial production of monomers such as vinyl chloride or acrylonitrile, which are both used to make plastics. These reactors are also employed to produce polymers such as synthetic rubber and polystyrene. The advantages of uniform heat transfer, great surface interaction, and transportation as fluid, whether in liquid or gaseous form, have made fluidized bed processing very valuable for contemporary chemical industry processes.

Pharmaceutical and Food Processing. Because of its excellent properties facilitating material and heat transfer, fluidized bed processing has become an important application for the coating or drying of pharmaceuticals. It is also used for batch granulation of pharmaceuticals. In the food-processing industry, the technique is used especially for creating individually quick-frozen (IQF) products. Because of their low density, individual food particles as large as bite-size diced vegetables can be fluidized and frozen quickly to maintain their taste. Complete food packages, as well as individual food particles, are also frozen, blanched, cooked, roasted, or heat sterilized in fluidized bed processing.

Mineral Processing. This older application dating back to the 1950's is used to decompose or purify ores through calcination or the roasting and pre-reduction of ores. Fluidized bed processing is also used in cement manufacture.

Impact on Industry

The vital global importance of fluidized bed processing for petroleum refining, cleaner and more efficient power generation, as well as often innovative applications in the chemical, food, and pharmaceutical industries have made this a field of considerable research and industry interest. There is both academic and industry inquiry to learn and fully understand all theoretical scientific underpinnings of the technique as well as practical developments to optimize processes and come up with new applications or design solutions.

Government Research. As of 2011, national governments have taken a lead in promoting both basic and applied research into fluidized bed processing in the quest to develop more environmentally friendly and energy-efficient modes of power generation. For example, China promotes research into coal-fired cogeneration plants using fluidized bed processing in its coal-particle combustion units. A particular concern is discovering economically viable alternatives (other than coal washing or flue-gas scrubbing) to reduce the massive sulfur oxide-emissions of older-style Chinese power plants, especially since coal washing also causes water pollution.

The United States Department of Energy sponsors many research projects, particularly into next-generation pressurized fluid bed combustion combined-cycle power plants. The goal is to design plants with a net system efficiency of more than 50 percent, extremely low sulfur and nitrogen oxide emissions well below 2010 emission limits, and at a power-generation cost of three-quarters by a conventional coal-fired power plant. The European Union similarly sponsors research in this area, as does Japan and other developed or developing countries.

Universities and Research Institutes. Universities and research institutes are interested in promoting both primary research and innovative process designs for fluidized bed processes for the same reason governments and industry are interested. As the quest to limit human-made carbon and other emissions intensifies, the promise of efficiency and low-emission output gained from fluidized bed processing motivates much research for which government grants tend to be available. In addition, catalyst research affecting FCC performance is a vibrant area, and often finds partners in industrial users. There has been consistent growth internationally in patents awarded for fluidized bed process optimizations resulting from the work of university and other institutional research teams.

Industry and Business. Fluidized bed processing is vital in the petroleum, chemical, petrochemical, pharmaceutical, and other process industries. Its impact is considerable, particularly in the petroleum and power industry.

In refining alone, all over the world in 2007, about four hundred fluid catalytic crackers cracked about 10.6 million barrels of feedstock each day, with half of it occurring in the United States. In 2010, FCCs produced from 35 to 45 percent of the gasoline gained at different U.S. refineries. The exact designs of an FCC, and the catalysts used therein, have been closely held corporate proprietary information.

Major Corporations. Because of the capital-intensive nature of units operating fluidized bed processing in most cases, there is a concentration of major corporations active in the field. All the international oil and gas companies employ engineers and scientists seeking to improve operations at their fluid catalytic crackers. Many FCCs are built under license by internationally active companies such as Kellogg Brown and Root (KBR) and Universal Oil Products (UOP). These companies hold patents for the unique design of their specialized fluidized bed reactors.

All major engineering companies who provide utility companies with fluidized bed combustion units, such as Germany's Lurgi GmbH or France's Technip, for example, need specialists to design, build, and improve these units. A pharmaceutical or food-processing corporation can gain a significant competitive edge if it develops a new or improved fluidized bed process for its manufacturing process.

Careers and Course Work

Students interested in a career in designing, developing, or supervising the operation of fluidized bed processes should take high school courses in the natural and computer sciences and mathematics. Refineries, power plants, and chemical and pharmaceutical companies employ skilled technicians for their operations of fluidized bed processes, and an associate's degree in science or engineering is a good foundation. Universities and research institutes also hire technicians to assist their researchers in the field.

A bachelor of science or arts degree in physics or chemistry, and perhaps a double major in computer science, is very useful for a career in the field. Students interested in food processing could also major in biology.

An engineering degree is an excellent foundation for an advanced career in the field, especially in chemical, electrical, mechanical, or computer engineering. Students should make sure to take classes in process engineering, chemical process modeling, control and safety engineering, and have a good understanding of thermodynamics.

Any master of engineering or doctoral degree in chemistry, physics, or chemical engineering is good preparation for a high-level career. A Ph.D. in one of the sciences can lead to an advanced research position either with a university, company, or government agency.

For a career in private industry, some sense of economics is also helpful. For those who are interested in applications in the refining or the power-generation industry, a willingness to work outside the United States is a plus as many contemporary Greenfield plants using the latest technology are built abroad. For this reason, a semester or two of study abroad is advised.

SOCIAL CONTEXT AND FUTURE PROSPECTS

As global demand for high-end petroleum products such as gasoline, diesel, and jet fuel continues, optimization of fluidized bed processes in FCCs will lose none of their importance. The drive to develop more efficient catalysts and processes, as well as to increase desirable product yield, lower energy consumption, and emissions, will provide for an exciting field of research in this technique.

The quest for a zero-emission power-generation plant would bring enormous prestige as well as business benefits for its developer. As of 2011, engineers and scientists in the power industry seek to develop a commercially viable chemical-looping combustion process that would use two fluidized beds. In the first, oxygen for combustion would be gained from a metal oxide bed. Then the metal would be reoxidized in a second bed, ready for return into the first one. This process could greatly revolutionize the power industry.

As the chemical and petrochemical industry has been growing, so have applications for well-designed, specialized fluidized bed processes in their production processes. The same holds true for the pharmaceutical and food-processing industries, where qualitative process advantages have been sought as well as new modes for innovative processes.

R. C. Lutz, B.A., M.A., Ph.D.

FURTHER READING

Basu, Prabir. *Combustion and Gasification in Fluidized Beds.* Boca Raton, Fla.: CRC Press, 2006. Comprehensive account of all aspects of this important application of fluidized bed processing.

Froment, Gilbert, Juray DeWilde, and Kenneth Bischoff. *Chemical Reactor Analysis and Design.* 3d ed. Hoboken, N.J.: John Wiley & Sons, 2011. Includes description of chemical reactors using fluidized bed processing; for undergraduate and graduate students, comprehensive scope.

Occelli, Mario L., ed. *Advances in Fluid Catalytic Cracking: Testing, Characterization, and Environmental Regulations.* Boca Raton, Fla.: CRC Press, 2010. Series of essays concerning technological advances in this key application of fluidized bed processing.

_______. *Fluid Catalytic Cracking VII: Materials, Methods and Process Innovation.* Oxford, England: Elsevier, 2007. Seventeen individual chapters cover key aspects of fluidized bed processing in fluid catalytic cracking at refineries, focus on catalysts used, process optimization, cleaner processes, and future developments.

Smith, P. G. *Application of Fluidization to Food Processing.* Oxford, England: Blackwell Science, 2007. Excellent survey of basic principles of fluid bed processing and the technique's growing application in the food processing industry. Somewhat technical but accessible; glossary.

Tsotsas, Evangelos, and Arun S. Mujumdar, eds. *Modern Drying Technology: Experimental Techniques, Volume 2.* Weinheim, Germany: Wiley-VCH, 2009. Chapter 5.4 focuses on the application of fluidized bed processing in granulation of detergents and batch granulation of pharmaceuticals and methods to measure attrition dust and over-spray in continuous spray processes.

WEB SITES

American Petroleum Institute
http://www.api.org

International Energy Agency
http://www.iea.org

National Petrochemical and Refiners Association
http://www.npra.org.

FOOD PRESERVATION

FIELDS OF STUDY

Chemistry; food science; food technology; nutrition; agriculture; animal and dairy science; agricultural engineering; veterinary science; microbiology; botany; biology; horticulture; industrial engineering; mechanical engineering; electrical engineering; computer science; robotics.

SUMMARY

Food preservation involves the application of scientific methods to prevent agricultural resources, whether eaten raw or further processed, from becoming contaminated or spoiling. Preservation ensures that people have sufficient supplies of food to survive. The availability of preserved foods makes it easier for people to migrate and settle in new areas. Explorers and military forces rely on preserved foods for nourishment. Preservation methods also enable perishable foods to be traded worldwide.

KEY TERMS AND CONCEPTS

- **Aseptic Packaging:** Preservation method in which food is sealed in pathogen-free containers in a sterile environment.
- **Canning:** Preservation method in which food is sealed in airtight containers by using pressure or other scientific techniques.
- **Freeze-Drying:** Preservation method in which food is dried by being quickly frozen and then vacuum processed.
- **Irradiation:** Preservation method in which food is exposed to ionizing radiation to slow decay.
- **Modified Atmosphere Packaging:** Preservation method in which the gas in food containers is altered to improve shelf life; typically, the amount of oxygen is reduced to slow oxidation and the growth of aerobic organisms.
- **Pasteurization:** Preservation methods that use heat, chemicals, or radiation to kill microorganisms and inactivate some enzymes.
- **Shelf Life:** Duration that a food can be stored and still be suitable for consumption.
- **Thermal Processing:** Preservation methods using different types and degrees of heat.
- **Vacuum Packaging:** Preservation method in which air is extracted from sealed containers holding food.

DEFINITION AND BASIC PRINCIPLES

Food preservation is the science of destroying or impeding the growth of harmful microorganisms, slowing oxidation, and controlling chemical reactions that cause foods to decay and become inedible. Food technologists and scientists realize that food will inevitably decay but that preservation strategies can extend its freshness. Food preservation professionals develop procedures and technologies to decrease spoilage, which ruins between one-tenth and one-third of food produced globally. They improve packaging, which keeps out hazardous contaminants, and devise innovative storage and transportation modes to maintain food quality, facilitating the manufacture and distribution of more food varieties worldwide.

Preservation is essential to ensure a consistent and sufficient supply of safe food to nourish global populations. Manufacturing processes to protect foods from contaminants typically involve exposing them to temperature extremes, using chemicals, applying high pressures, placing foods in a vacuum, or passing them through radiation or diverse energy sources. Removing moisture and drying foods is basic to many food preservation methods. Food and beverage packaging makes use of protective films, including edible coatings; of gases injected into or removed from food containers; and of aseptic environments in which to package foods.

Critics of food preservation allege that many of the preservatives added to foods are dangerous and suggest that nonchemical alternatives be developed and employed. They also protest the use of irradiation and other controversial technologies that they believe may be carcinogenic or pose other health risks to consumers. The U.S. Food and Drug Administration (FDA) responds to these criticisms by investigating potential threats involving food preservation and setting standards for those processes, such as establishing permitted irradiation dosages for various foods.

Background and History

Since ancient times, people have preserved surplus food to eat when fresh foods are not readily available. Early methods included drying or cooling foods and using easily obtainable preservatives such as honey, vinegar, alcohol, fats, sugar, and oils. Hanging foods to dry reduced their moisture content and minimized decay. Foods could be cooled in caves, snow, and streams. Standard preservation methods included salting, pickling, and fermenting. Many of these basic preservation strategies are incorporated into modern methods.

In the late eighteenth century, scientists became aware that microorganisms existed but did not know that microbes could cause food to spoil. By the early nineteenth century, confectioner Nicolas François Appert had devised a technique to preserve food in glass bottles, which served as the foundation for modern industrialized canning. Food preservation techniques improved as people learned how microorganisms made food spoil. In the mid-nineteenth century, French bacteriologist Louis Pasteur developed the pasteurization process, which uses heat to destroy microbes. This process, named after its inventor, became intrinsic to food preservation. In the 1850's, inventor Gail Borden, Jr., used vacuum technology to remove air from containers to create condensed milk, which remained fresh for extended periods while shipping or in storage. In the 1880's, Gustavus F. Swift, founder of a meatpacking company, created refrigerated railroad cars to preserve meat being transported over long distances. In the 1920's, businessman and inventor Clarence Birdseye patented a process to quick freeze foods.

As food preservation methods (some harmful) and patent medicines proliferated with little or no regulation, the U.S. government enacted legislation to ensure the quality of foods and drugs. The Pure Food and Drug Act of 1906 allowed the federal government to inspect meat products and prohibited the "manufacture, sale, or transportation of adulterated or misbranded or poisonous or deleterious foods, drugs, medicines, and liquors." The law's focus was on the accurate labeling of products. Although the FDA can trace its history to 1848, it began acting as a consumer protection agency after the passage of the 1906 act. Many additional laws regarding food and drug safety followed.

How It Works

Maintaining the freshness of food as it goes from market to the table is accomplished through chemicals and other substances that act as preservatives, processes that prevent or delay food spoilage, and packaging that guards against the chemical reactions that initiate decay.

Preservatives. Food technologists use many preservatives, both chemical and natural. Natural food preservatives include antimicrobials that attack microorganisms. Enzymes that are effective antimicrobials include glucose oxidase, lactoperoxidase, and lysozymes. Antioxidants, such as vitamins C and E, butylated hydropxytoluene (BHT), and butylated hydroxyyanisole (BHA), block spoilage caused by oxidation. Other food preservation strategies involve adding acids to lower pH levels, which disrupts microbe activity and slows food deterioration.

Thermal Processes. Heat damages or destroys most microbes and harmful enzymes in foods. During pasteurization, heat penetrates food and eliminates most microorganisms. Food technologists adjust the duration and intensity of heat applied to foods depending on the process and the food and microbes or enzymes involved. Many manufacturers use high-temperature short-time (HTST) or ultra-high-temperature (UHT) methods. Foods are transported on conveyor belts into contained areas, where they are heated to the proper temperature and for the appropriate length of time. Automated canning and bottling systems produce thousands of cans or bottles per hour. The bottles or cans are filled with liquid or food and sealed with lids as they move along conveyor belts into heating chambers.

Aseptic food preservation processes differ from canning in that foods and containers are both sterilized separately before packaging occurs. The foods are flash pasteurized then cooled before being packaged and sealed in sterile containers in a sterile environment. This type of packaging, often used for soy milk and drink boxes, allows foods—with no added preservatives—to be kept for long periods without refrigeration. Other thermal food preservation methods include using energy from radio frequencies and microwaves to heat foods during manufacturing.

Nonthermal Processes. Because exposure to heat alters some foods' chemical properties, food technologists have devised alternative methods. Nonthermal techniques control microorganisms with the application of electricity, radiation, pressure, or optics, individually or in combination. Some food manufacturers employ high-pressure processing (HPP), placing foods in chambers that can be pressurized at varying intensities. The ultra-high-pressure processing (UHP) technique pasteurizes foods, typically vegetables and fruits, by spraying them with powerful cold-water jets that remove most microbes.

Irradiation exposes foods to electron beams emitted by gamma rays, X rays, or other radioactive sources producing ions for times ranging from several seconds to an hour. The FDA and U.S. Department of Agriculture regulate irradiation dosages to disinfect specific foods. Other nonthermal food preservation methods include electronic pasteurization or the use of a pulsed electric field (PEF), pulsed magnetic field (PMF), or high intensities of pulsed light (PL) to inactivate microbes. Ultrasound preservation processes inactivate hazardous spores, enzymes, and microorganisms in foods.

Cooling and Freezing Processes. Reducing food temperatures is an effective way to slow biodeterioration because chilling foods alters the kinetic energy of microorganisms. In these processes, food is placed on conveyor belts and moved past chilled metal plates, coils holding refrigerants, compressors, and evaporators. Temperatures of 0 degrees Celsius (freezing point of water) or less kill most microbes or impede their movement. The type of freezing machinery and chemicals used determine how long the freezing process takes; some methods instantly freeze foods and others slowly transform them. Some food manufacturers use cool water to chill vegetables and fruits to preserve their freshness.

Vacuum and Dehydration Processes. Removing air from environments surrounding foods by creating a vacuum in a contained space helps minimize oxidation and other chemical reactions that cause spoilage. Engineers design industrial machinery to remove water efficiently from large quantities of food. Food manufacturers often use mechanical dryers to blow streams of warm, dry air over food. Fans circulate this air, which carries water molecules evaporated from food. Osmotic drying uses chemical solutions containing sodium chloride, sucrose, or other agents. Freeze-drying involves ice sublimating from frozen food in a vacuum chamber.

Packaging. Food preservation relies on protective containers or wrappings to block contaminants. Films impede oxygen and moisture from contacting foods. Packaging technology includes films containing antimicrobial preservatives and coatings that are safe to eat. The modified atmosphere packaging process involves machinery that places foods in containers that are injected with gases, including nitrogen, oxygen, carbon dioxide, or noble gases such as argon. Variations of modified atmosphere packing include controlled-atmosphere packaging, which uses scrubber technology to attain more precise gas levels and vacuum packaging at higher pressures.

Applications and Products

Nutrition. People require consistent access to sufficient nutritious foods to survive. Preservation strategies help provide adequate nourishment to populations worldwide. Aseptic processing and packaging contributes significantly to saving agricultural surpluses, which otherwise would spoil or be discarded. Preservation techniques enable people to have a safe source of food and water when emergency situations interrupt the normal flow of goods. Relief organizations distribute bulk shipments of preserved foods that do not require refrigeration. French manufacturer Nutriset and U.S. producer Edesia Global Nutrition Solutions produce Plumpy'nut, a high-calorie, vitamin-rich paste preserved in foil pouches, which alleviates malnourishment in areas affected by famine or disasters.

Travel, Transportation, and Trade. Food preservation has aided the movement of people for centuries. Travelers, including nomads and migrant workers, often rely on preserved foods to sustain them on journeys. Transportation of preserved foods takes place over the land and seas and through the air. Trucks move preserved foods and beverages from manufacturers to wholesalers and retailers. Engineers have designed special shipping cartons and vehicles to protect foods. Boxes of food are often wrapped with plastic. Technicians remove air between the plastic and the food, inserting gases to create a modified atmosphere to stabilize the food until delivery. Films placed around egg packages absorb shock. Refrigerated trucks keep foods frozen or chilled, and fans circulate air within cargo areas to keep perishables fresh. Many of those trucks contain microprocessors or are regulated by satellites to assure refrigeration is consistent while traveling.

Many railroad companies use refrigerated cars to move foods. These cars are designed to transport specific preserved foods and can be removed from trains and hauled by trucks to their final destinations. Rail yards near ports often have stacks of thousands of containers waiting to be transported inland or to sea. Cargo ships transport large tanks filled with several million gallons of preserved foods. Aseptic preservation processes enable worldwide shipments of vegetables and fruit products. Many preserved foods are shipped by airplane, with Federal Express carrying much of that cargo worldwide. Innovations in transporting preserved foods have enhanced international trade. Unique foods associated with specific geographical areas can be shipped and sold in distant markets within days of harvests, expanding sales of native produce.

Exploration. The availability of preserved foods often determines whether an expedition to remote areas is successful. Historically, explorers traveling for extended periods across land masses and bodies of water relied on salted, cured, canned, and other preserved food to provide them energy and nutrients. Modern explorers also carry preserved foods. The National Aeronautics and Space Administration (NASA) asked Pillsbury Company food scientists to help them develop preserved foods that would not produce crumbs. Mercury, Gemini, and early Apollo program crews consumed foods contained in tubes. NASA arranged for Oregon Freeze Dry to use its freeze-drying technology to preserve more elaborate meals for astronauts to eat during later Apollo and space shuttle missions.

NASA encourages commercial businesses to adapt its food preservation and packaging technology, granting permission to those businesses that want to use these innovations for preserving and packaging their food products. In the 1980's, Sky-Lab Foods used NASA technology to manufacture freeze-dried meals in bags. Oregon Freeze Dry produces freeze-dried military rations, fruits, vegetables, meats, and pet treats as well as a line of products for backpackers under the brand name Mountain House. Reflective metals used in space films were adapted for use in packages to protect and insulate foods.

In 2010, NASA sponsored a project examining the feasibility of growing vegetables in space to feed astronauts on possible missions to Mars or conolists of a Moon base. Orbitec Technologies assisted NASA scientists in determining ways to create gardens on spacecraft. This program supplemented ongoing plant experiments on the International Space Station and attempts by Martek Biosciences in the 1990's to cultivate algae to create oil with polyunsaturated fatty acids for lengthy missions.

Military. Since ancient times, armies have preserved foods to serve as military rations. These foods boost morale by providing soldiers familiar meals similar to those they consumed at home. Quartermasters distribute preserved food provisions to wounded soldiers in military hospitals and to prisoners of war. During the late twentieth century, military forces began supplying troops with Meals, Ready-to-Eat (MRE). These preserved foods and drinks, resembling those eaten in space, are packaged in pouches. MRE manufacturers process foods, such as

Fascinating Facts About Food Preservation

- Philip E. Nelson received the 2007 World Food Prize in recognition of his technological contributions to preserving food. He has received numerous patents for aseptic process and packaging inventions.
- In 1938, British biochemist Jack Drummond unsealed canned mutton that Captain William Parry and his crew had taken on their 1824 search for the Northwest Passage. The canned mutton had been left behind when the crew abandoned their icebound ship. Drummond found that the salvaged food was still safe to eat.
- In 1955, France began circulating a twelve-franc postal stamp featuring Nicolas François Appert and his food-preserving bottling process.
- Aseptic packaging ranges in size from small pudding cups and juice boxes commonly found in grocery stores to square tanks, standing as high as a six-story building and having an equivalent width, which can hold more than 1 million gallons.
- In August, 2010, Al and Susie Howell opened a can of Cougar Gold cheese that the Washington State University Creamery had produced in 1987 using technology that the school had created for packaging foods in tin containers. Although the can was twenty-three years old, the cheese retained its flavor.

beef stew, barbecued chicken, and cultural favorites such as curry and vegemite, with special techniques to preserve them for consumption under the adverse conditions associated with military deployments. Soldiers often combine MREs to concoct mixtures that have become part of military tradition.

The U.S. Department of Defense secures MREs from several manufacturers, including International Meals Supply, Wornick, Sopakco, and Ameriqual. International Meals Supply makes MREs for military units worldwide and produces MREStars for civilians. The other MRE manufacturers also make versions for nonmilitary customers. Applications of MRE manufacturing technology include producing energy foods for athletes and survival foods for use in emergencies. Governments, hospitals, and safety and relief agencies acquire MRE for use in crises.

Impact on Industry

In 2008, the global food preservation industry's value exceeded $500 billion. The food preservation market consists of manufacturers that produce technology to perform food preservation processes in factories, businesses, and homes and that develop and distribute synthetic and natural preservatives. Researchers throughout North America, Europe, Asia, Africa, Australia, and New Zealand contribute food preservation advances in academic, government, and industrial settings.

Government and University Research. The USDA provides the majority of government funding for food preservation work in the United States. Researchers at academic laboratories and agricultural experiment stations conduct studies relevant to developing new preservation procedures and identifying natural antimicrobials to counter microorganisms in food. Many researchers participate in interdisciplinary projects, using knowledge and insights from colleagues with expertise in related fields to envision, refine, and design technology that can maintain the quality of food for longer periods and to provide effective packaging to protect foods. Modeling techniques project how microbes will react to preservation methods and how various processes and intensities might affect spoilage rates. Researchers address problems associated with microorganisms becoming resistant to food preservation technologies. They assess if new food preservation techniques are compatible with food safety requirements and how they affect the environment.

American universities with food science departments making significant contributions to food preservation research include Purdue University and Washington State University. In 2000, the USDA's Cooperative State Research, Education, and Extension Service provided funds to establish the National Center for Home Food Preservation based at the University of Georgia. The center's primary mission is to educate consumers and teachers regarding scientific aspects of food preservation and new techniques introduced by researchers. Notable European food preservation work occurs at Wageningen University's Agrotechnological Research Institute in the Netherlands. In 2010, the journal *Applied and Environmental Microbiology* reported on research conducted by German scientists at the University of Applied Sciences into the application of acids to kill the norovirus.

Industry and Business Sectors. According to the Business Communications Company, U.S. industries focusing on food preservation generated more than $250 billion in 2008. Worldwide, food preservation contributed to almost three-fourths of the food manufacturing industry's profits. A 2003 report issued by the company attributed growth in the food preservation industry to successful incorporation of research to minimize spoilage and increase the shelf life of fresh foods.

Manufacturers produce equipment that incorporates food preservation in the process of converting agricultural materials into consumer products. Examples of the successful industrial use of food preservation technology include Fresherized Foods, which outfitted its plants with high-pressure processing machines in 1997 to manufacture guacamole for a variety of commercial foods. Japanese food manufacturers had begun implementing high-pressure processing technology in the 1980's. By the early twenty-first century, about fifty-five businesses worldwide were equipped with this technology.

Businesses manufacture and sell home food preservation devices, such as FoodSaver products, which use a vacuum process to seal foods in bags. Other basic food preservation equipment used in homes and retail businesses includes pressure cookers, canning jars, and seals.

The food preservation industry's packaging sector thrives because 98 percent of food is sold in some form of plastic, polymer, or metal wrapping or container. The U.S. food packaging industry generates more than $110 billion yearly. A 2009 report by the research firm Frost & Sullivan suggested that more food manufacturers are likely to incorporate nanotechnology approaches to food preservation, producing packaging films containing natural antimicrobial agents such as enzymes and cultures with antibiotic properties. A 2009 report released by Food Technology Intelligence noted that methods involving fungal amylases, low pH levels, and zero-oxygen packaging extend shelf live, increasing profits for manufacturers that implement these strategies.

Careers and Course Work

People seeking careers associated with food preservation can pursue several educational options depending on their professional interests and goals. Most industrial and government food preservation career paths require employees to have a bachelor of science degree. Food preservation professionals can find positions with both industrial and academic employers wanting to improve techniques to protect foods. Undergraduate classes focusing on food science, nutrition, agricultural engineering, or animal and dairy science provide students with a fundamental comprehension of how sciences are applied to cultivating, harvesting, and preserving agricultural resources.

An interdisciplinary approach expands students' employability. Study of related scientific and technological fields, especially microbiology, chemistry, and engineering, can qualify students for entry-level positions at industries using food preservation techniques or prepare them for further study. Food manufacturing industries recruit people with engineering degrees and work experience who can design machinery to preserve food or can incorporate robotics and automation into packaging processes. Computer expertise is needed to write programs monitoring preservation processes.

Graduate degrees, usually a master's degree and sometimes a doctorate, in specialized fields are usually needed for research positions involving food preservation work at industrial, government, or academic laboratories. Veterinary science degrees prepare professionals focusing on food preservation issues associated with livestock products. Faculty positions enable qualified personnel to teach students or perform research at university experiment stations. Government agencies, especially the USDA and FDA, hire employees with food preservation knowledge to work in diverse research, education, and administrative roles.

Social Context and Future Prospects

International humanitarian and agricultural groups emphasize the need for continued food preservation research to mitigate hunger and malnutrition. In 2010, the United Nations stated that 925 million people worldwide suffer from chronic hunger. Aid workers teach impoverished people basic food preservation techniques so that they can stockpile foods. The recession in the early twenty-first century made bulk buying of foods attractive to many people, who then became interested in preserving these foods to store them for longer periods. The number of people growing gardens or purchasing produce from farmers markets increased, making home food preservation more popular. These people were motivated not only by economic considerations but also by a desire to control the quality of their food.

Manufacturers expect food technologists to improve preservation techniques so that their foods remain fresher longer than those of their competitors. Packaging research and creative technological advances, such as aseptic methods and antimicrobial films, represent a growing component of the food preservation industry. Industries also have asked scientists to use more natural preservatives and avoid synthetic chemicals because many consumers refuse to purchase foods containing any additives that might be detrimental to their health. Food preservation researchers strive to advance and patent unique methods incorporating bacteriophages (bacteria-eating viruses), enzymes, and other innovative scientific concepts to combat microorganisms while producing affordable, nutritious foods with long shelf lives.

Elizabeth D. Schafer, Ph.D.

Further Reading

Belasco, Warren, and Roger Horowitz, eds. *Food Chains: From Farmyard to Shopping Cart.* Philadelphia: University of Pennsylvania Press, 2009. A collection of case studies describing how various agricultural products—including hogs, poultry, and seafood—are turned into food products.

Branen, A. Larry, et al., eds. *Food Additives.* 2d ed. New

York: Marcel Dekker, 2002. Contains chapters on regulations governing additives, hypersensitivity, and children.

Kurlansky, Mark. *Salt: A World History.* New York: Walker, 2002. Explores salt's role as a food preservative and its place in culture, economics, and the military.

Lelieveld, Huub L. M., Servé Notermans, and Sjoerd W. H. De Haan, eds. *Food Preservation by Pulsed Electric Fields: From Research to Application.* Cambridge, England: Woodhead, 2007. Explains how pulsed electric fields are used to preserve foods.

Nelson, Philip E., ed. *Principles of Aseptic Processing and Packaging.* 3d ed. Lafayette, Ind.: Purdue University Press, 2010. A technical discussion of aseptic manufacturing methods and applications and government regulations establishing guidelines.

Rahman, Mohammad Shafiur, ed. *Handbook of Food Preservation.* 2d ed. Boca Raton, Fla.: CRC Press/Taylor & Francis Group, 2007. Comprehensive guide examines traditional techniques and modern developments in technology and processes to preserve foods, suggesting possibilities for future methods.

Shephard, Sue. *Pickled, Potted, and Canned: How the Art and Science of Food Preserving Changed the World.* New York: Simon & Schuster, 2000. Illustrated history discusses methods used by people at various places and times to preserve foods, providing contemporary quotations and food preservatives references.

Web Sites

American Council for Food Safety and Quality
http://agfoodsafety.org

American Frozen Foods Institute
http://www.affi.com

International Association for Food Protection
http://www.foodprotection.org

National Aeronautics and Space Administration
Space Food
http://spaceflight.nasa.gov/living/spacefood/index.html

National Center for Home Food Preservation
http://www.uga.edu/nchfp

See also: Food Science; Pasteurization and Irradiation.

FOOD SCIENCE

FIELDS OF STUDY

Organic chemistry; inorganic chemistry; biochemistry; cell biology; molecular biology; genetics; tissue engineering; microbiology; nutrition; bacteriology; agricultural science; genetics; physiology; medicine; pharmaceutics; horticulture; phytochemistry; engineering; mathematics; calculus; statistics; physics.

SUMMARY

Food science is a field concerned with studying the biological, chemical, and physical properties of food. Food scientists make use of the tools of science, technology, and engineering to develop effective ways of producing and preserving a safe, healthy food supply for communities and nations. Issues addressed by food science include the safe cultivation and harvest of food plants; the healthy and humane breeding and slaughter of livestock; the optimal preservation of food as it is processed, stored, packaged, and distributed; and the manufacture of new food products with maximal nutritional value. Food science is also the discipline responsible for identifying the inherent nutritional properties of various foods and the ways in which foods interact with human biological systems once they have been consumed.

KEY TERMS AND CONCEPTS

- **Aeration:** Process of introducing air into a mixture by chemical, biological, or mechanical means.
- **Commercial Sterility:** Level of sterility in which only one out of every 10,000 products of a certain kind can be expected to contain bacteria or other microorganisms.
- **Cross-Contamination:** Transfer of a harmful microorganism to a food item through contact with food or another object.
- **Denaturization:** Change in molecular structure that occurs when proteins experience an external stress such as heat or contact with an acid, salt, or organic solvent.
- **Emulsifier:** Substance that enables the smooth combining of two liquids, such as milk and oil, which otherwise could not form a homogenous mixture.
- **Pasteurization:** Process of heating a food to a particular temperature and keeping it there for a specified time before cooling it, so as to destroy pathogenic microorganisms.
- **Phytochemical:** Chemical substance derived from a plant.
- **Refrigeration:** To preserve a food product at a temperature lower than 40 degrees Fahrenheit.
- **Shelf Life:** Time it takes for a given food to become chemically degraded.
- **Standard Plate Count (SPC):** Test for detecting the level of microorganisms in food.
- **Sterilization:** Process of heating a food or other material to the point where most or all microorganisms have been killed.
- **Triglyceride:** Fat molecule (or lipid) composed of three fatty acids linked chemically to one molecule of glycerol.
- **Xantham Gum:** Microorganism produced by fermenting corn sugar; used as a thickening and emulsification agent.
- **Zymurgy:** Science and practice of fermentation, such as in beer or wine making.

DEFINITION AND BASIC PRINCIPLES

Food science is a multidisciplinary field in which principles from biology, chemistry, and engineering are applied to the study of the chemical, physical, and microbiological properties of food. Food scientists investigate the elements contained in foods, factors involved in the physical and chemical deterioration of food, and how different foods interact with human physiology. The specific application of knowledge from food science to practical issues such as preserving, processing, manufacturing, packaging, and distributing food, as well as questions of food safety and quality, is sometimes called food technology.

The fact that foods are chemical systems is an important fundamental principle of food science. The same basic elements that form the cells and tissues of the human body are the ones that make up the majority of the foods people eat. These include carbon, oxygen, hydrogen, nitrogen, sulfur, and calcium. The chemical composition of foods and the ways in which the atoms and molecules within a food are arranged and bonded to each other determine properties such as a food's flavor, texture, color, and nutritional value. Food scientists are also concerned with the processes of digestion—how foods are broken down and absorbed by the body. Digestion occurs through the action of enzymes (biological substances that facilitate chemical reactions) found in saliva, pancreatic juices, and the lining of the small intestine. Nutrition—the study of how the chemical composition of different foods contributes to human health, growth, and disease—is another important subfield of food science. The six basic nutritional components of the human diet are carbohydrates, proteins, fats or lipids, vitamins, minerals, and water. Each of these components serves different vital functions within the human body.

BACKGROUND AND HISTORY

Human beings have been working to identify the properties of different foods and experimenting with ways of processing and preserving foods since time immemorial. In ancient China, Rome, Greece, Egypt, and India, fermentation was commonly used to produce alcoholic beverages, and archaeological finds suggest that in the Middle East, as long ago as 12,000 B.C.E., fruits, vegetables, and meats were deliberately laid out in the hot sun in order to preserve them. Many ancient civilizations also had a rudimentary understanding of the nutritional and medicinal properties of specific foods. The Roman writer Pliny the Elder, for instance, asserted that consuming cabbage was good for vision, could relieve headaches and stomach problems, and could even prevent hangovers.

Food science as a formal scientific discipline has a much more recent history. Beginning in about the nineteenth century, researchers working in fields such as biology and chemistry often took an interest in questions of food science. For example, the French bacteriologist Louis Pasteur introduced the technique of pasteurization in 1864, and in 1847, the German chemist Justus von Liebig published his seminal work, *Researches on the Chemistry of Food.* Liebig was among the first scientists to clearly outline the principle that foods are metabolized, or broken down by the body to produce energy. He also understood the nutritional importance of chemicals such as nitrogen and sulfur.

It was not until the twentieth century, however, that food science and the application of technology to the processing and packaging of food became a recognized field of study in its own right. One of the major achievements of twentieth-century food scientists was the invention of vacuum packaging, which helps preserve perishable foods longer by removing air from the containers in which they are kept, thus preventing the action of bacteria. Other breakthroughs were the development of quick-freezing (pioneered by the frozen vegetable manufacturer Clarence Birdseye), the use of thermal processing to improve the safety of canned foods (which had been around since the late eighteenth century), and the use of radiation to kill microorganisms in foods. Dedicated food science departments were created at universities around the United States and the world, and professional organizations such as the Institute of Food Technologists and the European Federation of Food Science and Technology gathered practicing food scientists together to share their knowledge.

By the twenty-first century, food science had progressed to a point where it was enabling researchers to tinker with the production of various foods in ways that had never before been thought possible. For example, in the making of cheese, rennet (curdled milk) has largely been replaced by a synthetically engineered enzyme called chymosin, which is more chemically consistent and pure. Also, advances in food science allow growers to use genetic engineering rather than breeding to create plants with more desirable attributes—such as tomatoes whose DNA is altered so that they take longer to ripen and reach supermarket shelves just at the point when they are ready to eat.

HOW IT WORKS

There are three chemical compounds that form the building blocks of food: carbohydrates, fats (or lipids), and proteins. Carbohydrate molecules, which are found in fruits, vegetables, starches, and dairy products, consist of atoms of carbon, hydrogen, and oxygen, chemically bonded in a ratio of 1:2:1. Monosaccharides and disaccharides such as glucose, fructose, and sucrose have just one or two molecules of this kind and are known as simple sugars. Polysaccharides like starch, glycogen, and cellulose (an important component of dietary fiber) have several carbohydrate molecules and are known as complex carbohydrates. Food science is able to reveal the chemistry behind the behaviors of carbohydrates in different foods. For example, because sugar molecules tend to crystallize (form solid geometric structures) at low temperatures, ice cream with too much milk in it can be gritty because of the crystallization of lactose, a sugar found in milk whey.

Protein molecules, which are found in meats, nuts, eggs, beans, and dairy, are composed of long chains of amino acids. An amino acid is an organic compound that contains at least one carboxyl group (-COOH) and one amino group (-NH_2). The amino acids in a protein are linked to each other by peptide bonds and folded into particular shapes. By identifying how a protein denatures (how the shape of its folded amino acids changes) when heat or chemicals are applied to it, food science can demonstrate how different methods of cooking affect foods and why. For example, when heat is applied to meat, the tidy folds created by the chains of amino acids it contains collapse, causing the protein molecules to shrink and release water. This is why a steak seared on a pan is smaller after cooking.

Lipids, such as animal fats and vegetable oils, are large molecules that are not soluble in water but are soluble in organic solvents such as acid. There are many different kinds of naturally occurring lipids, but most are largely composed of various fatty acids and glycerol. Glycerol is a type of alcohol, and a fatty acid is a compound that contains carbon and hydrogen atoms linked together in a long line and ending in a carboxyl group. Again, food science is concerned with how the chemical structure of lipids contributes to their physical characteristics. For example, the way in which the fatty acids in a lipid molecule are arranged determines whether a given fat will be solid at room temperature (such as lard) or liquid (such as vegetable oil).

Contemporary food scientists use a variety of sophisticated instruments to analyze the chemical composition and physical properties of foods. For example, a tool known as a spectrophotometer is used to detect how much light is absorbed by the atoms and molecules in a given sample of food and how much passes through it. Another food analysis technique is chromatography, which passes a sample of a food substance—either in liquid or gaseous form—through a medium that allows different components of the sample to travel at various rates. Spectrophotometry, liquid chromatography, and gas chromatography are all methods of analysis that enable food scientists to determine exactly what percentage of a specific food is made up of components such as fatty acids, amino acids, cholesterol, and carbohydrates. They also allow researchers to test for the presence of particular vitamins and minerals.

Digestion and Nutrition. The chemical composition of foods affects how they are broken down, or digested, by the body. The primary chemical reaction involved in digestion is called aerobic respiration, which involves tearing apart the bonds between carbon, oxygen, and hydrogen, the elements found in all three major types of food compounds. When aerobic respiration occurs, the energy contained within these bonds is released, and the individual atoms can be rearranged into different forms.

By analyzing the chemical characteristics of foods and how these properties relate to their digestion, food science can reveal why certain foods affect people the way they do. For example, because aerobic respiration is able to act directly on molecules of simple sugars—commonly found in soft drinks or colas—they are very quickly converted into energy. This results in an intense burst of energy that rapidly fades away—the so-called sugar high that a sweet drink can provide. In contrast, complex carbohydrates, commonly found in whole grains such as oats and brown rice, must first be broken down into simpler form by enzymes before aerobic respiration can take place. These foods take longer to digest and provide longer lasting energy, which is why a bowl of oatmeal in the morning can make a person feel full for hours. The structure of proteins is still more complex than that of whole grains, so they provide an even longer lasting source of energy. Lipid molecules, the hardest of all to digest, provide the body with its most long-lasting source of energy.

Although lipids take the longest time to digest, once they have been broken down, they provide the most energy per weight of food. For this reason, excess energy is stored within the body as fat deposits under the skin and in the abdomen. These enable the vital activities of cells to continue even when there is a temporary shortage of food. However, excess fat deposits are also stored inside organs and blood vessels. Here, they can block the normal flow of blood and lead to serious health problems such as heart failure. The question of how people can maximize the positive physiological effects of foods and avoid the harmful effects of an unhealthy diet is one of the primary concerns of food scientists. Among other pieces of knowledge, research has revealed that fiber, a type of carbohydrate found in fruits, vegetables, grains, and legumes, is essential for the proper functioning of the bowels and adheres to fat molecules traveling through the digestive system so they can be more easily disposed of as waste. Food science has also shown that certain kinds of unsaturated fatty acids, including omega-3 fatty acids, are extremely helpful in promoting heart health and preventing the buildup of cholesterol in the blood vessels.

Although it may not satiate a person's hunger, water is one of the most important nutrients required by the body. Water is the major component of every cell in the body and the environment within which every chemical reaction in the body takes place. It serves as a medium of transport for nutrients and waste, and it helps maintain a steady body temperature. Other chemical components found in food and important for human nutrition include vitamins and minerals. These nutrients serve a variety of essential functions. For example, vitamin E is an antioxidant, a substance that inhibits oxidizing reactions that can damage cells, and vitamin C helps the body process amino acids and fats. Calcium and magnesium are both minerals that are important in the formation of strong, healthy bones.

APPLICATIONS AND PRODUCTS

Preservation Techniques. One of the most important applications of food science is the development of preservation techniques that lengthen the time that a food can remain safe to consume and palatable. Among the many factors that cause foods to deteriorate are being exposed to microorganisms such as bacteria, molds, and yeast; experiencing changes in moisture content; being exposed to oxygen or light; undergoing the action of natural enzymes over time; being contaminated by industrial chemicals; and being attacked by insects or animals such as ants or rodents. Any of these factors may cause physical or biochemical reactions in foods that result in changes in texture, color, and taste, or that make them unsafe to eat.

Heating food is an effective method of preservation because high temperatures destroy both microorganisms and enzymes. Too much heat, however, can cause detrimental changes in the flavor, texture, and nutritional content of foods. Two commonly used methods of mild heat treatment are pasteurization and blanching. Pasteurization, which is most often applied to milk but also used to preserve fruit juices, beer, and eggs, involves heating the food to a temperature of 161 degrees Fahrenheit for just a few seconds. Blanching, most often used for vegetables destined to be frozen, dried, or canned, involves briefly dipping the food in water of about 212 degrees Fahrenheit. The most common severe heat treatment is canning. First, the food is placed inside a cylindrical steel or aluminum container and the air drawn out of it using a vacuum. Then, the lid is sealed in place and heat of about 240 to 250 degrees Fahrenheit is applied to the can. The process of canning ensures that the food in question reaches the point of commercial sterility and that it does not contain any live bacterium of the species *Clostridium botulinum.* This potentially deadly pathogen, if ingested, causes a kind of poisoning known as botulism.

Refrigeration and freezing are the two major types of cold preservation. Refrigeration, which takes place at temperatures ranging from 40 to 45 degrees Fahrenheit, does not destroy microorganisms or enzymes but somewhat inhibits the reactions they cause that result in spoiling. In and of itself, refrigeration is not a long-term method of preservation for most foods. Freezing, which takes place at temperatures ranging from 0 to 32 degrees Fahrenheit, is a more effective inhibitor of biochemical reactions than refrigeration and can preserve foods for longer periods of time. However, extremely cold temperatures can cause undesirable chemical changes, such as crystallization, in foods. If the water molecules in a food crystallize, they will rupture cell walls and cause the food to be softer and more liquid when it is eventually thawed.

Besides thermal processing and the use of cold temperatures, food scientists have developed a host of other techniques to combat the deterioration of food, such as dehydration, radiation, fermentation, and the use of natural and artificial preservative agents such as sugars, salts, acids, and inorganic chemicals like sodium benzoate and sulfur dioxide. Preservation is also aided by careful control over the characteristics of the atmosphere in which a food is stored.

Genetically Modified Foods. Genetically modified (GM) foods, also known as genetically engineered foods, are those whose existing genetic structure has been changed by the introduction of a new gene from a different organism. The technology that enables scientists to do this is called gene splicing, and the new genetic information is known as recombinant DNA. The purpose of genetic modification is usually to achieve some specific trait that will increase the food's usefulness for either producers or consumers. For example, DNA from bacteria has been incorporated into many strains of plants to enable them to resist attacks from insects and other pests. Some plants, such as soybeans, have been genetically modified so that they no longer produce particular substances, such as certain proteins, that can cause allergic reactions in people. According to some estimates, more than half of the foods for sale on supermarket shelves in the United States contain some ingredient from a genetically modified, or transgenic, plant.

Food products from genetically modified animals have not been available in the United States; however, in 2009, the Food and Drug Administration determined its requirements and regulations for genetically engineered animals, opening the door for food products from such animals. Food science research in this area has begun but largely remains in the experimental stages. Potential applications of this technology to animals raised for food include pigs with a genetic modification that causes them to produce omega-3 fatty acids, or salmon or chickens that grow much faster than usual.

Manufactured Foods. In vitro meat, also known as cultured meat, is animal muscle tissue that is grown outside of a living organism. It represents a relatively new and very specialized application of food science that makes use of tissue engineering techniques borrowed from cell biology and biotechnology. The production of in vitro meat involves harvesting either muscle cells or stem cells (cells that are pluripotent, or able to give rise to any number of different cell types) from a live animal, such as a chicken, cow, or pig. Alternatively, cells from a slaughtered animal can be used. Then, the cells are cultured within a medium that provides them with a large quantity of the nutrients required for growth. Typically, this includes amino acids, vitamins, minerals, and glucose. In this environment, the cells multiply rapidly. To encourage the cultured cells to fuse and form the three-dimensional structures that make up muscle fibers and tissues, they are placed on a scaffold, usually made of collagen. They may also be stretched or electronically stimulated to help them form the correct structures. At the end of this process, a substance similar to ground meat is produced.

Beverages and Snacks. Food science principles lie behind almost every beverage and snack food found on supermarket shelves. Carbonation, for example, enables the production of nonalcoholic soft drinks such as colas and sparkling water. Carbonation is simply the introduction of carbon dioxide gas into a liquid. It takes place at high pressures and low temperatures, because both of these factors increase the solubility of carbon dioxide. Alcoholic beverages such as beer and wine are produced through a process called fermentation. A substance—barley in the case of most beers, grapes in the case of wines—is chemically broken down by the action of microorganisms such as bacteria or yeast. It takes place under anaerobic conditions (in the absence of oxygen), because this is what causes the microorganisms to react with the carbohydrates in the substance to be fermented. In the process, sugars present in the original substance are converted into ethyl alcohol, the flavor and texture of the beverage is markedly changed, and its shelf life is prolonged.

Candy and other sweet confections are also the products of food science principles. To control the taste and texture of their products, candy manufacturers use a number of clever techniques designed to manipulate the behavior of sugar. For instance, many different candies begin with the same two ingredients: sugar and water. By changing the ratio of sugar to water, changing the boiling temperature of the sugar-water solution, controlling the time it takes for the boiling mixture to cool down, and adding various interfering agents such as butter, gelatin, cocoa, or pectin to the mixture, candy manufacturers can create a host of different textures from the same basic foundation: creamy and smooth, hard and brittle, moist and chewy, or smooth and transparent.

Impact on Industry

The application of food science technologies can have a tremendous impact on the amount of money a country spends on the production and distribution of food. For example, although it does not tell the whole story, one of the reasons consumers in the United States spend the lowest percentage of their per capita income on food when compared with consumers in other countries is that farms in the United States are able to take advantage of state-of-the-art technologies for growing, harvesting, distributing, and processing foods, such as advanced chemical fertilizers and preservatives that keep fruits and vegetables fresh over long journeys. These techniques help lower the costs of food production.

Government and University Research. In the United States, several large government agencies are involved in funding or directly conducting food science and technology research. Among them are the Food Safety and Inspection Service, a subsidiary of the U.S. Department of Agriculture, and the Food and Drug Administration. Government research scientists are largely concerned with conducting scientific studies related to specific issues of food safety and quality, such as how to prevent the spread of bacteria, such as salmonella or *Escherichia coli*, that can cause serious disease outbreaks within the population. Academic researchers may tackle a slightly different set of questions, some of which may be more basic and less problem-driven, such as investigations into the chemistry of different species of fruits or nuts.

Some academic research is funded by industry partners within agriculture. For example, the grain-farming sector may provide support for university laboratories conducting research into certain crop species that possess a natural genetic resistance to soil-borne viruses, or the dairy industry for research into the health benefits of consuming milk and yogurt. Strict regulations over conflict of interest issues ensure that the integrity of such scientific research is maintained. The outcomes of studies such as these can have broad economic impacts on the agriculture industry.

Industry and Business. Food science treats foods as complex chemical systems whose components have specific properties and give rise to specific nutritional effects. The packaged foods industry has capitalized on the discoveries of food science to create products that are marketed as having particular health benefits. For example, several butter substitutes on the market are made with plant-derived sterols, which are substances that inhibit the body's ability to absorb cholesterol and thus improve heart health, reduce the risk of stroke, and combat inflammation in the body. Sports drinks containing stimulants, amino acids, vitamins, and other active ingredients are marketed as providing athletes and other consumers with a quick, safe, and effective burst of energy. Thousands of such functional food and beverage items fill grocery store shelves, allowing food retailers to capitalize on the unique properties of these products. Customers may not be able to effectively differentiate amongGranny Smith apples from three separate orchards, but they are likely to have an opinion as to whether they want to buy regular flour or flour that has been fortified with additional beneficial nutrients. The restaurant industry is another major business sector that makes use of food science principles. For example, many restaurants around the world are using a type of cooking known as molecular gastronomy. Restaurants use dozens of high-tech gadgets to transform the chemical and physical structure of foods, creating dishes such as chocolate in the form of a jelly and tomato and cheese in the form of foam.

Careers and Course Work

Food science research and related industries such as agriculture, food processing and manufacturing, and food safety and inspection offer a wide variety of career opportunities. For example, food chemists analyze, create, and modify the biochemical compounds and chemical processes involved in the synthesis or development of food products. They may work for food manufacturing companies, biotechnology corporations, government institutions, environmental consulting companies, and agricultural universities. Food bacteriologists, who often work in similar settings, carry out research into the microbiological and molecular basis of food-borne pathogens. Their work supports the safe production, packaging, and distribution of food plants and animals. Food safety inspectors are employed in settings such as meat-processing plants, fisheries, large farms, restaurants, and food production facilities. They may also work for federal agencies. Food safety inspectors make use of their knowledge of biology and chemistry to ensure that food products such as dairy, grains, fishes, fruits, vegetables, meats, and poultry properly conform to national or industry-based standards of sanitation.

Taking high school courses in biology, chemistry, physics, and mathematics provides a good early foundation for an eventual career in food science.

At the undergraduate level, courses in organic and inorganic chemistry, nutrition, cell and molecular

> **Fascinating Facts About Food Science**
>
> - Food scientists have gone far beyond freeze-dried ice cream when it comes to creating high-tech food for astronauts. Salt and pepper are turned into liquids so that they do not float around in zero gravity, and perishable fruits and meat can be thermostabilized (processed with heat) to allow them to stay fresh for long space journeys.
> - Nutrition can come not just from the food that people eat but also from the vessels in which they cook it. Cooking foods in a cast-iron pot rather than a nonstick or glass pan, for instance, increases the amount of iron—an essential mineral found in hemoglobin, or blood molecules—in the finished dish.
> - Food science reveals that plants have an incredibly complex chemical composition. Raw spinach, for instance, contains calcium, iron, magnesium, phosphorus, potassium, sodium, zinc, copper, manganese, and selenium—and that is just the minerals. It also contains at least eighteen vitamins and nearly twenty amino acids.
> - Freezer burn is caused by the escape of water molecules from food stored at very cold temperatures. When moisture is lost from a food in this way, it becomes dried out and covered in frost.
> - Pickled vegetables can be more nutritious than fresh ones. The bacteria that are involved in fermentation break down molecules in the vegetables and, as a by-product, create vitamins.
> - Food and beverage manufacturers in the United States are constantly working to combine existing ingredients into new formulations. In total, they come up with nearly 19,000 new products for supermarket shoppers every year.

biology, biochemistry, genetics, bacteriology, microbiology, and agricultural science are especially relevant. For the student who is especially certain of his or her ambitions, a number of colleges and universities offer specialized undergraduate degrees in food science or food technology. To acquire a position as a technician, an associate's degree or certification in an area such as chemical or biochemical technology may be all that is required. Positions conducting independent research at an academic or government institution or food science research in private laboratories require the completion of a bachelor's degree in science and, usually, a master's or doctoral degree in biochemistry, bacteriology, chemical technology, or a related field.

Social Context and Future Prospects

The potential impact of the field of food science is both far reaching and profoundly positive. For example, the advent of in vitro meat has both environmental and ethical implications. If it became widely used, cultured meat could result in huge energy and water savings over traditional methods of bringing meat to market, reduce the amount of pollution produced by farming, and end the raising of animals in factory farms. Similarly, the gene splicing technology used to create genetically modified foods has already been applied to the pressing problem of global poverty. To increase the nutritional content of rice for consumption in developing countries, where rice often makes up a huge portion of the diet of the poor, researchers have created strains of genetically modified rice. The new rice contains large amounts of beta-carotene, a precursor for the synthesis of vitamin A, which contributes to the iron content of blood and helps maintain the structure and functioning of eyes. The product, called Golden Rice, has proven to be beneficial in places where malnutrition often leads to problems such as anemia and blindness.

Emerging food technologies such as these, however, are still surrounded by a cloud of skepticism and debate. Many consumers and advocacy groups argue that the health safety risks and environmental impact of products such as genetically modified foods have not been properly assessed and that these products could have devastating unintended consequences. For example, some people worry that plants containing genes from foreign species may turn out to be allergenic or even toxic to humans. Others are concerned that genetically modified organisms may interact with the environment in unpredictable ways. One study of monarch butterfly larvae found that feeding them on leaves dusted with bioengineered corn pollen caused their growth to be stunted and, in some cases, resulted in their deaths. Nevertheless, many food scientists point out that attempts to modify the genetic information in both plants and animals is hardly new; the process of selective breeding, they say, pursues essentially the same goal and has been going on for centuries.

M. Lee, B.A., M.A.

Further Reading

Brown, Amy. *Understanding Food: Principles and Preparation.* 4th ed. Pacific Grove, Calif.: Brooks/Cole, 2010. An introductory textbook largely organized around specific food types, such as meat, cereals, and fats. Also discusses career options within the field.

Heldman, Dennis R., and Daryl B. Lund, eds. *Handbook of Food Engineering.* 2d ed. Boca Raton, Fla.: CRC Press/Taylor & Francis, 2007. Includes chapters on topics such as cleaning and sanitation, heating and cooling processes, food dehydration, and thermal processing of canned foods. Contains figures, relevant equations, and a comprehensive index.

Hui, Yiu H., et al., eds. *Food Biochemistry and Food Processing.* Ames, Iowa: Blackwell Publishing, 2006. Covers the biochemistry of food and the biotechnology involved in food processing. Includes numerous figures, diagrams, and tables illustrating essential concepts.

Wansink, Brian. *Marketing Nutrition: Soy, Functional Foods, Biotechnology, and Obesity.* 2006. Reprint. Urbana: University of Illinois Press, 2007. An analysis of food science, nutrition, and labeling from a business and marketing perspective. Contains a list of suggested further readings.

Web Sites

European Federation of Food Science and Technology
http://www.effost.com

Institute of Food Technologists
http://www.ift.org

U.S. Department of Agriculture
Food Safety and Inspection Service
http://www.fsis.usda.gov

U.S. Food and Drug Administration
Food
http://www.fda.gov/Food/default.htm

See also: Egg Production; Food Preservation; Genetically Modified Food Production; Pasteurization and Irradiation; Plant Breeding and Propagation.

FORENSIC SCIENCE

FIELDS OF STUDY

Chemistry; biology; biochemistry; mathematics; microbiology; physics.

SUMMARY

Forensic science is commonly defined as the application of science to legal matters. Although forensic science incorporates numerous disciplines, ranging from accounting to psychology, in the traditional sense, forensic science refers to the scientific analysis of evidence collected at crime scenes, which is also known as "criminalistics." Pattern evidence, such as fingerprints, bullets, and tool marks, is often compared visually, and chemical evidence (such as illicit drugs) and biological evidence (such as DNA, blood, and bodily fluids) are analyzed and compared using scientific instruments.

KEY TERMS AND CONCEPTS

- **Class Evidence:** Evidence that can be identified as belonging to a group containing many members, all with similar characteristics or features.
- **Criminalistics:** Refers to the analysis of pattern, chemical, and biological evidence; often used interchangeably with forensic science.
- **DNA (Deoxyribonucleic Acid):** Nucleic acid that contains the genetic code and is present in nearly every cell in the body.
- **Illicit Drug:** Substance or drug that is prohibited by federal or state laws because of its undesirable effects or high risk of abuse.
- **Impression Evidence:** Evidence, such as fingerprints, tire tracks, and footprints, formed when an object leaves behind a characteristic marking on a surface.
- **Individualizing Evidence:** Evidence that can be identified as belonging to a group containing only itself as a member.
- **Latent Print:** Fingerprint residue that is not easily visible to the naked eye and must be treated, either physically or chemically, to be observed.
- **Toxicology:** Study of drugs and poisons and their effect on the body.
- **Trace Evidence:** General term for microscopic pieces of evidence that are transferred by contact between people and objects. Examples include hairs and fibers, as well as fragments of paint and glass.

Definition and Basic Principles

Forensic science is the application of scientific principles to the analysis of numerous types of evidence, most commonly evidence collected at a crime scene. Crime scene investigators, usually police officers, collect evidence at the crime scene and submit it to a crime laboratory for analysis by forensic scientists.

Crime laboratories contain different sections, each of which specializes in a particular type of analysis, such as controlled substances, DNA, firearms and tool marks, latent prints, questioned documents, toxicology, and trace evidence. The type of analysis conducted depends on the type of evidence as well as the circumstances of the crime. A single piece of evidence may be analyzed in more than one section. For example, a firearm may be analyzed in the latent prints and DNA sections, as well as the firearms and tool marks section.

Following analysis, forensic scientists may be summoned to present their findings in a court of law. Forensic scientists present their analysis and interpretation of the evidence before a judge and jury, who are charged with determining the guilt or innocence of the defendant. The unbiased, accurate analysis presented by the forensic scientist is an integral part of the criminal proceedings.

Background and History

Forensic science aims to determine identifying or individualizing characteristics to link people, places, and objects. In the late 1880's, French criminologist Alphonse Bertillon developed a method of identifying humans based on eleven physical measurements including height, head width, and foot length. However, limitations in this method soon became apparent. In 1880, Scottish scientist Henry Faulds published an article in *Nature* that discussed the use of fingerprints as a means of identification. In 1892, Sir Francis Galton published *Fingerprints*, proposing a system of classifying fingerprint patterns. That same year, an Argentine police officer, Juan Vucetich, used fingerprint evidence that resulted in the arrest and conviction of a murder suspect. From 1896 to 1925, Sir Edward Henry, a police official in British India, developed the Henry Classification System for fingerprints, which was based on the pattern on each finger and the two thumbs.

In the late nineteenth and early twentieth centuries, advances were being made in other areas that have become integral to forensic science. Spanish-born French scientist Mathieu Joseph Bonaventure Orfila, often considered as the pioneer of forensic toxicology, is credited with developing and improving methods for the detection of arsenic (a common poison in the nineteenth century) in the body. French scientist Edmond Locard developed the hypothesis that "every contact leaves a trace," which implies that whenever two objects make contact, there is an exchange between them. This hypothesis became known as Locard's exchange principle and is the foundation of modern trace evidence analysis. In the 1920's, the comparison microscope, which analyzes side-by-side specimens, was developed by American chemist Philip Gravelle and popularized by forensic scientist Calvin Goddard. This microscope enabled significant advances in many areas of forensic science, particularly firearms, tool marks, and trace evidence.

A major scientific breakthrough in the 1980's revolutionized the field of forensic DNA analysis. British geneticist Sir Alec Jeffreys developed DNA profiling, enabling individuals to be identified from samples of blood and other body fluids left at a crime scene. The development of the polymerase chain reaction by American biochemist Kary Mullis in 1983 allowed DNA profiling to be conducted on degraded and very small samples of DNA, making it possible for forensic scientists to test a wider range of evidence.

As the field of forensic science evolves, newly developed technologies and instrumentation allow evidence to be analyzed and compared in an increasingly rapid, objective, and reliable manner. Forensic science is a truly dynamic field, constantly seeking further improvements and advancements in its analytical methodologies.

How It Works

Forensic science incorporates numerous subdisciplines, but the most common types of analysis conducted by crime laboratories are the analysis of illicit drugs, biological evidence, latent prints, firearms, footprints, tire marks, tool marks, and trace evidence. Latent prints, footprints, tire marks, and tool marks are considered pattern evidence. The patterns of an unknown sample (usually from the crime scene) and a known sample are visually compared to find similarities between the two. Samples can also be analyzed, either chemically or biologically, with scientific instruments. Some of the more common methods of testing are infrared spectroscopy, ultraviolet/visible microspectrophotometry, gas chromatography-mass spectrometry, and electrophoresis.

Infrared Spectroscopy. In infrared spectroscopy, the chemical structure of a sample is determined based on how the sample interacts with infrared radiation. Chemical bonds can absorb infrared radiation of a specific energy, which causes the bond to vibrate. Additionally, each bond can vibrate in different ways. Therefore, when infrared radiation is introduced, chemical bonds within the sample absorb different energies, and the results are shown in the form of an infrared spectrum. The spectrum is essentially a graph of radiation transmitted versus wave number, which is related to the energy of the radiation. Additionally, transmission can be mathematically converted to absorbance such that the spectrum can be displayed as absorbance versus wave number. The infrared spectrum of a sample displays numerous absorptions, each corresponding to a particular type of chemical bond and a particular type of vibration. The infrared spectrum of a sample is unique to that sample, and therefore, this technique can be used to definitively identify compounds.

Ultraviolet/Visible Microspectrophotometry. Infrared spectroscopy and ultraviolet/visible microspectrophotometry are both based on the principle of the interaction of radiation with a sample. However, ultraviolet/visible microspectrophotometry is typically used to compare the dye or pigment composition of samples. The technique is used to determine the color of a sample and identify subtle differences in color that cannot be seen with the naked eye.

A microspectrophotometer consists of a microscope with a spectrometer attached, which allows the analysis of microscopic pieces of evidence. The sample is viewed under the microscope, and ultraviolet and/or visible radiation is introduced.

Depending on the chemical structure of the sample, wavelengths of light will be absorbed, reflected, or transmitted. The transmitted light is collected in the spectrophotometer, and the intensity of each wavelength is measured. Results are displayed in the form of a spectrum that is a graph of transmittance (or absorbance) versus wavelength. Subtle differences in color between two samples are observed as differences in wavelengths of light transmitted or absorbed in the corresponding spectra. Such differences are caused by differences in chemical composition between the two samples, and therefore, comparison of the resulting spectra can be used to determine if the two samples are similar in color.

Gas Chromatography-Mass Spectrometry. In any chromatography technique, sample mixtures are separated based on differences in interaction between a mobile phase and a stationary phase. In gas chromatography (GC), the mobile phase is a gas, and the stationary phase is a liquid coated on the inner walls of a very thin column. Liquid samples are typically introduced into the system and carried, in the mobile phase, through the stationary phase. Sample components that have a stronger attraction for the stationary phase spend longer in that phase, and components with less attraction spend less time in that phase and move more quickly through the system. The time it takes for sample components to travel through the system and reach the detector is known as the retention time.

In gas chromatography-mass spectrometry (GC-MS), the detector is the mass spectrometer, which contains three major components: the ion source, the mass analyzer, and the detector. On emerging from the GC column, sample components enter the ion source, where each component is first ionized. The resulting ion is known as the molecular ion. This ion is unstable because of its high energy, so it breaks down, or fragments, into smaller ions. Molecular ions and fragment ions then enter the mass analyzer, where the ions are separated according to their mass-to-charge ratio. The separated ions enter the detector, where the number of ions of each mass-to-charge ratio is counted. Results are displayed in the form of a mass spectrum, which is a graph of intensity versus the mass-to-charge ratio. Because molecules break down, or fragment, in a predictable manner, the mass spectrum can be used to determine the structure of the original sample component. Furthermore, because the fragmentation pattern is unique to a molecule, the mass spectrum can be used to definitively identify the component.

On analyzing a sample by GC-MS, two pieces of information are obtained. First, from gas chromatography, a chromatogram is obtained, which is a graph of detector response versus retention time. Each separated component in the sample mixture is shown as a peak on the chromatogram. Components that take longer to reach the detector have greater attraction for the stationary phase and have longer retention times. Additionally, for each separated component, the mass spectrum is also obtained, which can be used to definitively identify the component.

Electrophoresis. Although electrophoresis is also used to separate sample mixtures, the technique is not considered a chromatographic technique because no mobile phase is involved. Instead, sample mixtures are separated based on differences in migration under the influence of an applied electrical potential. Therefore, electrophoresis is used for the analysis of samples that have an electric charge.

Although there are different types of electrophoresis, capillary electrophoresis is most commonly used for DNA profiling purposes. In this technique, a capillary column is filled with a polymer, and the ends of the column are immersed in reservoirs containing a buffer solution. The reservoirs also contain electrodes to allow the application of the electric potential. The sample is introduced to one end of the column, and the sample components move through the column under the influence of the applied potential. Separation occurs based on differences in the migration rate of the components through the column, which depends on size and charge. Separated components pass through a detector at the other end of the column, producing an electropherogram. The electropherogram shows the migration time of the separated components. Smaller components move more quickly, reaching the detector before larger components and have shorter migration times.

Applications and Products

The major role of the forensic scientist is to analyze submitted evidence for the purposes of characterization and identification. For example, a blue fiber collected from the scene may be submitted to the trace evidence section, where forensic scientists characterize the fiber (for example, by its dimensions, color, cross-sectional shape) and then identify the type of fiber (for example, nylon, polyester, acrylic). Furthermore, when a known sample is available (such as fibers from the suspect's clothing), forensic scientists compare it with the unknown sample (collected from the crime scene) to determine if the two most likely originated from a common source. This process of characterization, identification, and comparison requires multiple stages of analysis, ranging from visual examination to instrumental analysis.

Infrared Spectroscopy. The technique of infrared spectroscopy is commonly used in the controlled substance and the trace evidence sections of the crime laboratory. This technique can identify illicit drugs present in unknown samples, the type of fiber found at a crime scene or on a person, the polymer present in a paint chip, or the organic compounds present in explosive residues. The evidence is prepared for analysis in several ways, depending on the type of sample.

Solid samples of illicit drugs can be mixed with potassium bromide and pressed into a pellet, which is then placed in the spectrometer. Infrared radiation is passed through the sample, which will absorb at characteristic energies depending on its chemical structure. The transmitted radiation is collected and the infrared spectrum is generated. Because potassium bromide does not absorb infrared radiation, the subsequent infrared spectrum shows only contributions from any drug present in the sample.

For opaque samples, such as fibers or paint chips, attenuated total reflectance-infrared (ATR-IR) spectroscopy is more commonly used. The sample is positioned over a crystal, and pressure is applied to ensure good contact between the sample and crystal. Infrared radiation is passed through the crystal, and because of the close contact, the radiation penetrates a small depth into the sample. Certain energies are absorbed depending on the chemical bonds within the sample, resulting in the characteristic spectrum of the sample.

The infrared spectrum of the questioned sample can be compared to a database containing infrared spectra for known standards (drugs, fibers, paints, and so on) to identify the unknown sample. However, care must be taken when comparing a spectrum to spectra in a database. Although the spectrum of a given compound is unique, it can vary slightly depending on the instrument used to analyze the sample and standard. Rather than relying on a database search, it is often preferable to analyze the unknown sample and known standards on the same day, using the same instrument, to allow for a direct comparison of spectra.

Although samples can be rapidly analyzed using infrared spectroscopy, the technique works best for relatively pure samples. If impurities are present in the sample and they also absorb infrared radiation,

Fascinating Facts About Forensic Science

- Forensic entomology involves the study of insects that invade a body after death to determine the time that has elapsed since the person's death.
- The saliva on a discarded cigarette contains enough DNA to identify the person who smoked it.
- The first use of fingerprint evidence to solve a criminal case was recorded in Argentina in 1892. Police official Juan Vucetich used a bloody fingerprint found at the crime scene to prove that two boys were murdered by their own mother.
- "Forensic" comes from the Latin *forensic*, which means "of the forum." The "forum" relates to the law courts in ancient Rome. In modern times, forensic science is defined as science relating to the law.
- In 1981, a German publishing company purchased what were thought to be Adolf Hitler's diaries. However, forensic document examiners proved that the diaries were fake, based on the presence of a paper-whitening agent that was not used in paper manufacturing until at least 1954.
- Marie Lafarge was the first person to be found guilty of murder based on toxicology evidence. Although she poisoned her husband with arsenic, initial testing did not find any arsenic in his body. However, when French scientist Mathieu Joseph Bonaventure Orfila repeated the tests, he found arsenic in the man's body and proved that the initial testing was inaccurate.
- In 1995, O. J. Simpson was cleared of murdering his wife, Nicole Brown, and her friend Ronald Goldman, despite DNA evidence identifying blood at the crime scene as belonging to the former football player. Furthermore, DNA from Simpson, Brown, and Goldman was found in a leather glove found at the scene.

the resulting spectrum contains contributions from both the sample and the impurities. This can complicate interpretation of the spectrum and subsequent identification of the sample.

Microspectrophotometry. The comparison and analysis of colored samples is often undertaken using microspectrophotometry. This technique is used in the trace evidence and questioned documents sections to compare the dye or pigment composition of fibers, paints, and inks.

Methods for sample preparation vary depending on the type of sample to be analyzed. Fibers are flattened and mounted on a microscope slide with a drop of immersion oil. Paint samples require more involved preparation, particularly for transmission spectra. The paint chip must be cut into a section so thin that light can be transmitted through it. Spectra of inks can be obtained directly if the paper is sufficiently thin to allow transmission. Otherwise, the ink must be removed from the document. This can be done by removing a small sample of the paper containing the ink and immersing the paper in a solvent to extract the ink. The resulting ink solution is placed on a microscope slide, and the solvent is allowed to evaporate, leaving a residue of ink for analysis. However, this is a destructive procedure because the document is damaged in removing the sample. Alternatively, a piece of clear tape can be placed on an area of the document that contains the ink. When the tape is lifted off, particles of ink adhere to the tape. These particles can be removed from the tape and transferred to a microscope slide for analysis. The document is minimally damaged using this procedure.

Although microspectrophotometry offers a rapid means to investigate the dye or pigment composition of certain samples, no extensive spectral databases are readily available. Therefore, the technique is more useful when known samples are available and the color of the unknown and known samples can be compared directly, based on spectral interpretation.

Gas Chromatography-Mass Spectrometry. As with infrared spectroscopy, gas chromatography-mass spectrometry is commonly used in the controlled substances and trace evidence sections, as well as in the toxicology section, for the determination of drugs and poisons in body fluids.

GC-MS is advantageous over infrared spectroscopyin that samples containing impurities can still be identified because of the separation abilities of gas chromatography. For example, gas chromatography analysis of a drug mixture containing methamphetamine and caffeine separates the two components. In the resulting chromatogram, two peaks are observed: one for methamphetamine and one for caffeine. The mass spectrum of each peak is also obtained, which can be used to definitively identify each component.

In most cases, samples must be in liquid form for GC-MS analysis. This is achieved by adding a suitable solvent to the sample and analyzing the resulting solution. For body fluid or tissue samples, a solid phase extraction or liquid-liquid extraction is necessary to isolate any drugs and poisons from additional components present in the fluids or tissues.

Solid samples can be analyzed using pyrolysis GC-MS. In this case, a pyrolysis unit is attached to the gas chromatography inlet. Solid samples (for example, paint chips or fiber fragments) are placed in a small quartz tube and introduced into the pyrolysis unit, which rapidly heats the sample to a very high temperature. The sample is broken down and vaporized in the pyrolysis unit, and then carried in the flow of carrier gas onto the gas chromatography column, where the sample components are separated.

Before analyzing the sample, it is important to demonstrate that the GC-MS system is free from contamination. This is usually done by injecting a volume of the solvent used to prepare the sample. If the solvent and instrument are not contaminated, the resulting chromatogram should show no peaks. For pyrolysis GC, the empty quartz tube is analyzed to demonstrate that there is no contamination in the tube or instrument.

Because the mass spectrum rather than the retention time is unique to a sample component, the spectrum of an unknown sample is compared to a suitable database of spectra. However, there may be slight differences between the database spectrum and the spectrum obtained for the unknown sample, depending on the instrument used to collect the spectra. It is often preferable to prepare and analyze a known standard in the same way as the unknown sample and then compare the corresponding mass spectra.

Electrophoresis. DNA profiling makes the most use of electrophoresis. Typically, blood, semen, saliva, or another body fluid from the crime scene is used to generate a DNA profile, which is compared with profiles generated from known samples. If known samples are not available, the generated DNA profile can be compared to a database of profiles. The Federal Bureau of Investigation (FBI) maintains a database of DNA profiles submitted by crime laboratories across the United States. This database, the combined DNA index system (CODIS), contains profiles from crime scenes, convicted criminals, and missing persons.

Modern DNA profiling is based on the characterization of short tandem repeats (STRs) that are regions (loci) on the chromosome that repeat at least twice within the DNA. For profiling, the number of repeats at each location on the chromosome is determined. To do this, the DNA is first amplified via the polymerase chain reaction (PCR), in which the double-stranded DNA is split into two single strands and a mixture of enzymes and primers are used to replicate specific STR regions of the DNA. In the United States, STRs at thirteen loci are typically considered. The reaction is repeated many times, generating exact copies of the STRs. Because of this amplification procedure, profiles can be obtained from very small samples of DNA.

The STRs are analyzed using electrophoresis, most commonly capillary electrophoresis, which allows rapid and automated analysis. The STR mixture is separated based on differences in migration rate through the capillary column, which is related to the size of the STR. The resulting electropherogram displays a series of peaks that correspond to the STRs at each loci. Additionally, for each STR, there are two variants, one inherited from the mother and one from the father; therefore, the electropherogram actually shows a pair of peaks at each loci. A match in the number of STRs for both variants at all loci is considered strong evidence that the unknown and known samples originate from the same person. Because DNA is unique to an individual, this is one type of evidence that is considered individualizing rather than class evidence.

Impact on Industry

Local, State, and Federal Laboratories. The majority of forensic science laboratories in the United States are funded by the local government, the state, or the federal government. These laboratories offer a variety of services to their customers, who are typically police departments and other law enforcement agencies. The actual services offered vary depending on the size of the laboratory and the geographical area that it covers. Typical services include analysis of illicit drugs, body fluids (often for DNA), fingerprints, firearms, tool marks, and trace evidence. Within any state, there may be only one laboratory that offers all services, with a number of smaller laboratories throughout the state offering two or three services.

The federal government operates numerous specialized forensic science laboratories. The U.S. Department of Justice oversees the FBI, the Drug Enforcement Administration (DEA), and the Bureau of Alcohol, Tobacco, Firearms, and Explosives (ATF), all of which have forensic science laboratories. These laboratories offer analytical services to local and state law enforcement agencies and conduct research to develop new analytical tools and technologies to advance forensic science.

Within the U.S. Department of the Treasury, the Treasury Inspector General for Tax Administration operates a forensic science laboratory that principally analyzes suspected counterfeit documents, using fingerprint and handwriting analyses, along with digital image enhancement procedures. The U.S. Postal Inspection Service operates a forensic science laboratory with the role of analyzing evidence from postal-related crimes. This laboratory mainly conducts fingerprint, document, and chemical analyses, along with digital image enhancement.

The U.S. Fish and Wildlife Forensics Laboratory is operated through the Department of the Interior and is the only laboratory worldwide that focuses solely on crimes against wildlife. Laboratory expertise in genetic and chemical analysis, as well as firearms, trace evidence, latent prints, and pathology is used to analyze wildlife evidence to identify species and determine cause of death.

Private Laboratories. A number of private forensic science laboratories operate throughout the United States. The majority of these laboratories offer expertise in one or two areas (for example, analysis of illicit drugs and trace evidence) rather than a full range of services. The vast majority of these laboratories focus on DNA analysis, particularly paternity testing. Because these laboratories are privately funded, their services are offered to the general public and are not limited to law enforcement agencies.

Federal Funding. The National Institute of Justice

(NIJ) is the largest funding agency for forensic science within the United States. The agency funds research that improves methods for the collection, analysis, and interpretation of forensic evidence. Funds for research, development, and evaluation go to projects that improve the analytical tools and technologies available to forensic scientists, from developing new methods for the comparison of evidence to developing new analytical instrumentation for forensic analyses. Forensic laboratory enhancement funds are for projects that improve sample throughput in laboratories, enabling evidence to be analyzed in a timely manner. The agency also awards research fellowships to individuals to conduct specific research that will improve or enhance existing practices in forensic science. In 2009, the institute awarded a total of $284 million.

Careers and Course Work

At a minimum, a forensic scientist must have a bachelor's degree in a natural science, such as biology or chemistry, or in a related discipline, such as microbiology or biochemistry. It is also possible to obtain a bachelor's degree in forensic science; however, care should be taken to ensure that the degree meets minimum credit hour requirements in either biology or chemistry. Successful completion of the bachelor's degree ensures that students have a strong background in the appropriate science for their future field of work.

The popularity of forensic science has made the field highly competitive, and many forensic scientists have a master's degree in forensic science. Although as of 2010, there were more than forty such degree programs in the United States, only thirteen were accredited by the American Academy of Forensic Sciences. Accreditation ensures that rigorous standards have been met in terms of the content and quality of the degree program. In most cases, the degree obtained is a master of science in forensic science with a concentration in a specific discipline, for example, forensic biology, forensic chemistry, or forensic toxicology.

Forensic scientists are not limited to careers in local and state crime laboratories. Many federal agencies, such as the FBI, the DEA, and the ATF, employ forensic scientists. In addition, forensic scientists can be employed by private forensic laboratories (such as paternity testing or sport testing laboratories) or as independent consultants.

Social Context and Future Prospects

Although great advances have been made in forensic science, many more have yet to be achieved. In 2009, the National Research Council published *Strengthening Forensic Science in the United States: A Path Forward*, a report on forensic science in the United States. The report highlighted several deficiencies in the field and recommended improving education, training, and certification for forensic scientists as well as developing standardized procedures and protocols for evidence analysis and reporting. Additionally, the report recommended research into the reliability and validity of many of the procedures used for evidence analysis. The report concluded that more research is necessary, not only to improve existing practices but also to develop new technologies that can be implemented in forensic science laboratories. It called for the development of a national institute of forensic science that would have many objectives, including the development of standards for certification for forensic scientists and accreditation of forensic laboratories, along with improving education and research in the field.

Ruth Waddell Smith, Ph.D.

Further Reading

Bertino, Anthony J., and Patricia N. Bertino. *Forensic Science: Fundamentals and Investigations.* Mason, Ohio: South-Western Cengage Learning, 2009. Examines the tests and techniques used for the scientific analysis of various evidence types, including hairs and fibers, DNA, handwriting, and soil.

Brettell, Thomas A., John M. Butler, and José R. Almirall. "Forensic Science." *Analytical Chemistry* 79, no. 12 (2007): 4365-4384. A review of forensic science applications used in common disciplines.

Embar-Seddon, Ayn, and Allan D. Pass, eds. *Forensic Science.* 3 vols. Pasadena, Calif.: Salem Press, 2008. Extensive coverage of forensics, including historical events, famous cases, and types of investigations, evidence, and equipment.

Houck, Max M., and Jay A. Siegel. *Fundamentals of Forensic Science.* 2d ed. Burlington, Mass.: Academic Press, 2010. An introduction to forensic science and common techniques used for the analysis of physical, biological, and chemical evidence.

James, Stuart H., and Jon J. Nordby, eds. *Forensic Science: An Introduction to the Scientific and Investigative Techniques.* 3d ed. Boca Raton, Fla.: CRC Press, 2009.

Discusses mass spectrometry techniques in relation to forensic applications, including forensic toxicology, controlled substance identification, and DNA analysis.

Kobilinsky, Lawrence, Thomas F. Liotti, and Jamel Oeser-Sweat. *DNA: Forensic and Legal Applications.* Hoboken, N.J.: Wiley-Interscience, 2005. Presents an overview of DNA analysis, including the historical perspective, scientific principles, and laboratory procedures.

Rudin, Norah, and Keith Inman. *An Introduction to Forensic DNA Analysis.* 2d ed. Boca Raton, Fla.: CRC Press, 2002. Contains an overview of DNA analysis, beginning with its history and examining the principles on which it is based.

Saferstein, Richard. *Criminalistics: An Introduction to Forensic Science.* 10th ed. Upper Saddle River, N.J.: Prentice Hall, 2011. Provides an introduction to forensic science, detailing the techniques to analyze physical, biological, and chemical evidence.

Web Sites

American Academy of Forensic Sciences
http://www.aafs.org

American Forensic Association
http://www.americanforensics.org

Association of Forensic DNA Analysts and Administrators
http://www.afdaa.org

National Institute of Justice
Forensic Sciences
http://www.ojp.usdoj.gov/nij/topics/forensics/welcome.htm

See also: Vehicular Accident Reconstruction

FORESTRY

FIELDS OF STUDY

Botany; ecology; accounting; geographical information systems; civil engineering; statistics; agriculture; policy; applied mathematics; law.

SUMMARY

Forestry is the management of forests for human resource development, extraction, utilization, regeneration, and conservation. It relies heavily on a range of sciences, mathematics, engineering, and administrative procedures. Typical forest products are paper, joinery timbers, composite boards, cardboard, firewood, drinking water, and carbon sequestration. The guiding principle is that both the trees and the supporting resources (such as soil nutrients and hydrology) should not be depleted but be used in such a way that they can persist in perpetuity.

KEY TERMS AND CONCEPTS

- **Afforestation:** Establishment of forest on land previously vacant for a considerable period.
- **Allometric:** Pertaining to the relationship, usually mathematical, between the size of one component of a tree to the size of another of its components.
- **Clearfell:** Felling prescription, or state, of a forest, in which all vegetation types are either felled or pushed over; also known as clearcut.
- **Coppice:** Practice of cutting a trunk near ground level to allow the tree to regenerate with multiple stems.
- **Diameter At Breast Height (DBH):** Diameter of a tree measured at a person's breast height, a universal measurement used in allometrics.
- **Mensuration:** Science of measuring tree and forest stand attributes, especially wood volume, through other parameters, such as diameter at breast height, height, and stand density.
- **Selective Logging:** Felling prescription in which either individual trees or groups of trees are felled, and the demographics of the remainder form a viable population.
- **Senescent Tree:** Living tree of advanced years, beyond its prime mass, with significant crown decay or hollowing of the trunk; senescent trees remain valuable for biodiversity, catchment water, and carbon storage.
- **Silviculture:** Planning and execution of forest yield maintenance through such practices as regeneration, cultivation, thinning, and various harvesting methods.

Definition and Basic Principles

Forestry is the professional management of forests. It consists of several interlinked processes relying heavily on a range of basic and interdisciplinary sciences (such as soil science, forest ecology, entomology, remote sensing, geographic information systems, and statistics), on engineering (road design and harvesting machinery), on labor (stand inventory, tree felling, haulage, log grading, firefighting, and prescribed burning), on budgeting and marketing, and on product technologies (mill technology and wood processing chemistry). Forestry professionals are guided by government policies, laws, cultural mores, and advances in science.

The purpose of the forestry profession is to provide the long-term sustainability of forest-product yields and financial returns, in contrast to one-time plundering and gradual forest degradation. Sustainability of a variety of forest attributes and products constitutes good planning in forest management. Timber, in some form, is the most common and marketable of forestry outputs. Timber yield and quality is therefore of primary concern to most foresters, with a consequent drive for maximization of timber yields (especially of the more valuable products) and for forecasting those yields.

Background and History

Forests have been used unsustainably for millennia. Forests have been converted to pasture for livestock, used to smelt ores, or to manufacture timber, tar, and other products. Trees have been turned into charcoal and removed from mountain tops to mine the ores below. Forestry originated in attempts to prevent the overuse of forests. For example, in the eleventh century, William the Conqueror established forest laws governing the usage of vegetation and wild game.

Until the eighteenth century, forest regeneration was accomplished through regrowth, replanting, or coppicing. In the early 1700's, several French scientists and naturalists, Jacques Roger, Henri-Louis Duhamel du Monceau, and René Antoine Ferchault de Réaumur, as well as British naturalist John Turberville Needham, made scientific discoveries in the areas of tree breeding and cultivation, wood strength, and yields. Later that century, the German agriculturalist Georg Ludwig Hartig focused his research on sustained yield and founded the first German school of forestry.

During the nineteenth century, the first forest preservation programs were established in the United States. In 1887, the American Forestry Association created a movement to preserve the forests of the United States. In 1891, the Yellowstone Park Timberland Reserve was created by President William Henry Harrison. Seven years later, Cornell University became the first American college to offer college-level education in forestry. The Organic Administration Act of 1897 established the first national forests; however, they were to be working forests, designed to improve water flows and to provide timber.

The U.S. Forest Service, part of the U.S. Department of Agriculture, was established in 1905 to make sure that the forests continued to provide timber and water. Several national forests, most in the eastern United States, were created by the Weeks Law of 1911, which was designed to reforest areas that had been logged or cleared for farming. In the 1970's, after the start of the environmental movement, forestry began to expand its horizons and encompass environmental concerns. Ecological forestry, although still lacking a standard definition, is generally used to refer to forestry that seeks to make use of forests in a sustainable manner, considering the whole ecosystem and the social and economic environments, as well as incorporating scientific research.

How It Works

Stock and Yield. Forecasting timber yields requires the measurement of existing stock (making an inventory of the forest) and estimating its growth. Growth is estimated either by repeated measurements over many years or by sampling different sites of similar productivity and different times since harvest. A common measure of productivity is the site index, or the mean stand height at the age of fifty years. Yield tables are created, which cross-reference site index and time since harvest, to give harvestable timber volume.

Estimating the standing timber volume requires on-ground measurement of properties such as height and diameter at breast height. Such basic measurements are combined in allometric formulas to provide trunk taper, tree volume, or merchantable wood volume. A technology used at some locations is ground-based lidar (light distance and ranging), a scanning laser providing a map of the trunk surface and, therefore, the volume of several trees per scan.

Statistics are taken and estimates made at several stages.

Stratification. For efficiency, rather than measuring the whole forest estate, inventory is taken of only portions, or stratums, often selected by their site index. This method relies on matching various environmental attributes such as slope and aspect, precipitation, fire history, soil nutrients, and soil depth. Good stratification provides more reliable interpolation between measured strata and thereby improves overall stock and yield estimates.

Stratification is facilitated by on-ground sampling, remote sensing, and geographic information systems (GIS). Relevant remote-sensing platforms include aerial photography, airborne lidar, and satellites. After stratification, the annual yields of the entire forest can be tallied from the yield tables.

Sustainability. Two main time frames are considered in sustainability: harvesting cycles and the long term.

The primary considerations for each harvesting cycle include maintenance or regeneration of the ecosystem type (for native forests). For fauna, this most often refers to the landscape level, as local demographics often depend on time since harvest. Another consideration at this level is the maintenance of near-optimum catchment-level water flows and water-table levels.

Long-term considerations include measures to ensure that there is no downward trend in yields of primary forest products (notably timber), the financial returns from the forest estate, and the levels of vital nutrients such as calcium and phosphorus. Such depletions can occur if successive harvests are too frequent—that is, if they exceed the replacement time needed by nutrient- and site-dependent ecological

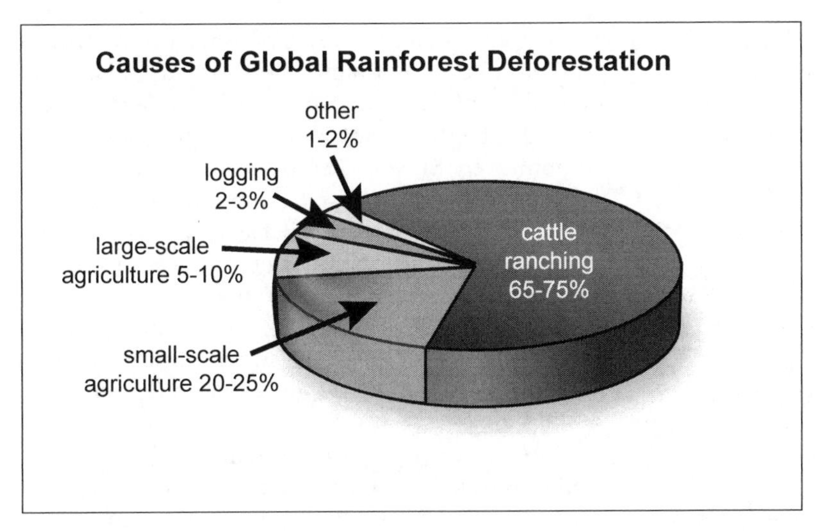

Sustainable logging and ranching are two approaches to reducing rainforest deforestation worldwide.

processes. Another long-term consideration is the total carbon stock of the wider forest estate; the forest management activity must not constitute a net carbon emission. This consideration has come to the forefront with the rise of concerns over global climate change; this requirement is readily met for plantations on old-field farmlands but is more difficult for native forests newly opened to logging and harvest cycles.

Harvesting Plans. Detailed harvest planning is part of legitimate forest management, entailing, for example, a local inventory, the silvicultural prescription, mapping drag lines and buffer zones, yield expectations, and plan approval from government departments. It ensures that on-ground operations meet all prerequisites, such as application of appropriate engineering expertise in forest areas newly opened to industry and local environmental regulations. The plan also forms part of the site-history record for future forestry operations.

Production Mechanisms. Silviculturalists grow forests by various techniques, such as collecting seed from local stock before harvest, coppicing, genetically engineering saplings, and prescribed burns sometimes followed by aerial seeding. Follow-up treatment to promote growth may be unnecessary or include herbivore and weed control and stand thinning. Growth models, combined with mapped environmental variability, produce catchment-level forecasts of wood volume and carbon sequestration. Models help in forecasting responses to management efforts, such as spacing of plantings, pruning, commercial thinning, fertilizer addition, provenance selection, genetic improvement, clearfelling, prescribed burns, as well of less controllable affects such as drought and wildfire. Linear programming is one method of optimizing finances, timber yields, sustainability, and carbon sequestration over multiple harvesting areas (coupes) within the forest estate. This allows development of harvesting plans and their approval several years before implementation, which in turn allows employment stability through prior assignment of logging teams and mill deliveries. Forest managers generally prefer error margins in yields to be less than 10 percent. Uncertainties are higher for previously unlogged native forests and mature stands because of buttressing and senescence but lower for more intensively managed forests and plantations.

Product Sourcing. Exotic species form much of the global plantation stock. For example natural rubber comes from the Amazonian tree *Hevea brasiliensis,* which is cultivated extensively in Southeast Asian countries. *Eucalyptus globulus* from Australia is cultivated widely in southern Europe and South America. *Pinus radiata* from California is grown widely in Australia. This translocation reduces browsing by herbivores that evolved alongside the tree species, and some exotic plantation species can out-compete indigenous species, producing higher growth rates than they do in their place of origin. Such plantations form a major source of products consumed at a high rate, such as paper. In some regions, native forests are used unsustainably for such products. The more obvious environmental effects can be observed and unsustainable use can be proven through the same fundamentals of forest sciences used in sustainable yield calculations.

Applications and Products

Internal Products. Some of forestry's outputs are consumed within the industry. These include

Fascinating Facts About Forestry

- Although the tallest recorded individual trees are no longer standing, scientists determined from taper formulas calibrated on shorter, living specimens that the tallest flowering plant, *Eucalyptus regnans*, a hardwood tree in southeast Australia, can reach 120 meters.
- Since medieval times, coppicing of broadleaf species was practiced in England, often with stands of varying ages, producing timber for infrastructure, weaving, charcoal production, smelting, tools, and tanneries. It is continued in some places for conservation of wildlife that had adapted to the coppices and for craftspeople.
- Nearly every year, cyclists climb Mount Ventoux, a 6,273-foot bald mountain in the Provence region, during the Tour de France. The mountain was once covered by a forest, but it was stripped of trees beginning in the twelfth century to provide wood for shipbuilders in Toulon.
- Cork is produced by peeling the bark from the branches of cork oak trees in the Mediterranean basin at about ten-year intervals, when the tree is between thirty and two hundred years old. It is traditionally used for flooring, bottle stoppers, and buoyancy devices. As the trees are not felled, cork oak forests are an unusual wildlife habitat.
- A variety of remote-sensing platforms are used to study forests. Hyperspectral and lidar combined in a light aircraft provide the most detailed information, such as forest health, species distribution, canopy structure, and individual tree height.
- Urban forests in the northern United States may serve as seed sources for southern species during their northern migration, necessary for their conservation under climate change, but the urban trees may also enable invasion by exotic species.

computer software developed by foresters to aid in forestry operations and budgeting, harvesting machinery such as felling mechanisms, virtual reality harvesting simulators, stock valuation and carbon assessments, mill residues to fuel kiln drying of milled timber, and felled trees for construction of bridges on haulage roads.

External Sales. Volume quantities of wood products delivered from the forest can be converted directly to monetary values in terms of dollars per metric ton or cubic meter, depending on the type of product and its destination. Other outputs are not so easily converted to monetary equivalents. These include remediation of land through afforestation, unmonitored firewood collection, maintenance of biodiversity, water quality, salinity remediation, and reduced air pollution. Such calculations are, however, becoming more frequent in the scientific literature and even in government reports.

Mill Products. Two main mill products from logged timber are pulpwood and sawlogs, sometimes from the same forest tree or stand (integrated harvesting) or from entirely separate forests. The proportion of the original tree volume that becomes cut lumber is usually no more than 55 percent and often lower. Consequently, optimization of sawmilling is financially crucial. Taper formulas, which model tree shape, indicate not only total wood volumes but also volumes of different mill products. Computer-automated milling to optimize the cutting pattern within individual logs can reduce wastage. Final product yields are fed back to aid future revenue forecasting. Woodchip and pulp mills are coarser processors, with throughput up to 400 metric tons of woodchips per hour, equivalent to one fully laden log truck every two and one-half minutes. These rates are driven by the global market developed for paper products.

Product Variety. In terms of volume, the material products of forestry are dominated by lumber and paper, but other products such as firewood, charcoal, cork, sandalwood oil, cinnamon, and palm oil can be regionally significant. Some outputs such as firewood, although measurable, can be from forestry harvests or simply collected from forests without forest management, but both come under the heading of forest products in national accounts. Broader-scale outputs of nontraditional forestry include salinity remediation, regeneration of semiarid landscape functionality, amelioration of urban air pollution, and attempts to address climate change. Reducing the emissions from deforestation and forest degradation is an area of increasing investment and requires adaptation of traditional forest sciences to different forest components (such as nonmerchantable tree components, woody debris, and soil) and to a greater range of time scales.

Impact on Industry

Application of Forestry Principles. Sustainability regulations in forestry vary between private and public land. The distinction between government and industry may be unclear, with forest management on public land operated by corporation-like government agencies. Sustainability principles are applied differently for various forest attributes, such as wood yield and wildlife habitat, and the legality of operations also varies. Therefore, the proportion of annual yields retrieved through application of sustainability principles is difficult to ascertain. Nevertheless, some experts have tried to estimate the amount of illegally sourced lumber. Policies are being implemented to curtail this illegal harvesting, with a view to attenuating environmental damage and establishing civil law; however, one side effect will be a rise in global timber prices.

Global Trade. Trade in forestry is generally perceived in terms of forest products. Annual global trade in forest products in 2009 was about 140 million cubic meters of sawn timber, about 130 million cubic meters of roundwood (premilled timber), 130 million metric tons of paper, 80 million cubic meters of wood panels, and 45 million metric tons of woodpulp. The country with the largest output of roundwood is Russia, and the major importers are Finland, China, Japan, and Korea. Other major exporters of roundwood are the United States, Germany, New Zealand, and Canada. Canada is the lead exporter of both sawn timber and pulpwood, although the United States produces more pulpwood. Africa is the largest producer of firewood. Japan is the leading importer of woodchips, importing hardwood chips principally from Australia, South Africa, and Chile. Within Australia, Tasmania is the leading state-exporter of hardwood chips, most of which are sourced from native forests, although increasingly they are being sourced from plantations.

In 2006, the companies with the largest net sales of fully processed paper products were U.S.-based International Paper ($21.2 billion), Finland's Stora Enso ($16.2 billion), and U.S.-based Procter & Gamble ($12.0 billion). In early 2010, the Nippon Paper Group reported net sales of $11.2 billion and a net profit of $334 million. Paper consumption per capita is highest in the United States, although China and India have the highest growth rates of consumption.

Forestry Research. Scientific research related to forestry is undertaken in colleges and universities, in dedicated research institutes, private companies, and consortiums of research bodies. Funding sources are primarily government and forest agencies, large corporations, educational institutions, and industries. Research areas include primary wood production, manufacturing wood products, and the environment. Most funds from the private sector are devoted to research on manufacturing. Most scientific research data and advances in forestry reach the public arena, although some data, conducted by the forest agencies, forest product companies, or even government departments remain confidential. Demand for comprehensive, high-quality data on forests is intense, as it can advance new research and investment frontiers, especially climate change mitigation.

Careers and Course Work

Trained foresters are in demand for forestry management of both native and plantation forests. The two main drivers are production of wood products and conservation of the environment. The traditional route for those interested in pursuing careers as foresters has been to gain a bachelor's degree at a special forestry school or university department, often with a focus on wood production. Often these studies are combined with environmental sciences or geography, as part of a more general environmental science degree. Nevertheless, dedicated forestry departments are still part of many universities, especially those in major capital cities or those where the forest products industry has regional presence and a substantial stake in the workforce. Some universities link with others to provide a more comprehensive training.

Students who wish to pursue forestry in college should take science classes and at least one mathematics class while in high school. A forestry degree can be combined with biology, botany, accounting, economics, geography, or resource management. A forestry degree will often include at least one class in forest ecology. For more advanced studies in forestry, statistics is necessary at the undergraduate level.

Social Context and Future Prospects

Sustainability of key environmental attributes remains problematic, even with rigorous protocols and planning because advances in forest science must be comprehensively proven and written into policy before integration with demands for productivity. For example, evidence and modeling suggest that in many forests, heat stress from higher summer temperatures and droughts will outweigh the carbon dioxide fertilization and longer growing season accompanying climate change, requiring recalculation of harvesting quotas.

A question remains as to where and how forestry should be implemented. Many plantations, such as rubber and palm oil plantations in Southeast Asia, required clearing of the original forest and have since been reforested. Afforestation can have environmental benefits, such as providing a reprieve for native forests, land rehabilitation, and climate change mitigation. In 2010, although progress had been made toward recycling paper and curtailing illegal logging, the paper industry and some forestry professionals participated in an initiative designed to get people to print more e-mails.

Application of forestry know-how is increasingly sought in the newer, nontraditional industries such as mine-site and rangeland rehabilitation and in urban forestry. Forestry principles are also used in climate change mitigation efforts such as carbon budgeting of practices and calculating emission offsets.

Christopher Dean, B.Sc., Ph.D.

Further Reading

Bekessy, Sarah A., et al. "Modelling Human Impacts on the Tasmanian Wedge-Tailed Eagle (*Aquila audax fleayi*)." *Biological Conservation* 142 (2009): 2438-2448. Discusses a case in which forestry and other societal influences affected an endangered species. Notes the steps for conservation and graphs the decline of the population.

Cox, Thomas R. *The Lumberman's Frontier: Three Centuries of Land Use, Society, and Change in America's Forests.* Corvallis: Oregon State University Press, 2010. Examines the history of the lumber industry in the United States and the changing attitudes over time.

Hays, Samuel P. *Wars in the Woods: The Rise of Ecological Forestry in America.* Pittsburgh: University of Pittsburgh Press, 2007. Examines forestry practices and the conflict between ecological forestry and the more traditional approach, which he calls commodity forestry.

Hicke, Jeffrey A., et al. "Spatial Patterns of Forest Characteristics in the Western United States Derived from Inventories." *Ecological Applications* 17, no. 8 (2007): 2387-2402. Shows how continental-scale forest inventory can be combined with remote sensing and modeling to provide a history of management effects on carbon stocks.

Maser, Chris, Andrew W. Claridge, and James M. Trappe. *Trees, Truffles, and Beasts: How Forests Function.* New Brunswick, N.J.: Rutgers University Press, 2008. Describes examples of ecosystem processes and players in forest productivity in the disparate forests of the northern United States and southern Australia.

Pretzsch, Hans. *Forest Dynamics, Growth, and Yield: From Measurement to Model.* New York: Springer, 2009. Presents a modern and detailed review of methods for forest growth and yield calculations, including comparisons with ecological modeling. Provides for carbon accounting. Builds from basics to the professional level.

Web Sites

International Union of Forest Research Organizations
http://www.iufro.org

National Association of State Foresters
http://www.stateforesters.org

Oregon Forest Resources Institute
http://www.oregonforests.org

Society of American Foresters
http://www.safnet.org

U.S. Forest Service
http://www.fs.fed.us

See also: Silviculture

FUEL CELL TECHNOLOGIES

FIELDS OF STUDY

Physics; chemistry; electrochemistry; thermodynamics; heat and mass transfer; fluid mechanics; combustion; materials science; chemical engineering; mechanical engineering; electrical engineering; systems engineering; advanced energy conversion.

SUMMARY

The devices known as fuel cells convert the chemical energy stored in fuel materials directly into electrical energy, bypassing the thermal-energy stage. Among the many technologies used to convert chemical energy to electrical energy, fuel cells are favored for their high efficiency and low emissions. Because of their high efficiency, fuel cells have found applications in spacecraft and show great potential as sources of energy in generating stations.

KEY TERMS AND CONCEPTS

- **Anode:** Electrode through which electric current flows into a polarized electrical device.
- **Carnot Efficiency:** Highest efficiency at which a heat engine can operate between two temperatures: that at which energy enters the cycle and that at which energy exits the cycle.
- **Cathode:** Electrode through which electric current flows out of a polarized electrical device.
- **Cogeneration:** Using a heat engine to generate both electricity and useful heat simultaneously.
- **Electrocatalysis:** Using a material to enhance electrode kinetics and minimize overpotential.
- **Electrode:** Electrical conductor used to make contact with a nonmetallic part of a circuit.
- **Electrolyte:** Substance containing free ions that make the substance electrically conductive.
- **Electron:** Subatomic particle carrying a negative electric charge.
- **In Situ:** Latin for "in position"; here, it refers to being in the reaction mixture.
- **Proton:** Subatomic particle carrying a positive electric charge.

Definition and Basic Principles

Fuel cells provide a clean and versatile means to convert chemical energy to electricity. The reaction between a fuel and an oxidizer is what generates electricity. The reactants flow into the cell, and the products of that reaction flow out of it, leaving the electrolyte behind. As long as the necessary reactant and oxidant flows are maintained, they can operate continuously. Fuel cells differ from electrochemical cell batteries in that they use reactant from an external source that must be replenished. This is known as a thermodynamically open system. Batteries store electrical energy chemically and are considered a thermodynamically closed system. In general, fuel cells consist of three components: the anode, where oxidation of the fuel occurs; the electrolyte, which allows ions but not electrons to pass through; and the cathode, which consumes electrons from the anode.

A fuel cell does not produce heat as a primary energy conversion mode and is not considered a heat engine. Consequently, fuel cell efficiencies are not limited by the Carnot efficiency. They convert chemical energy to electrical energy essentially in an isothermal manner.

Fuel cells can be distinguished by: reactant type (hydrogen, methane, carbon monoxide, methanol for a fuel and oxygen, air, or chlorine for an oxidizer); electrolyte type (liquid or solid); and working temperature (low temperature, below 120 degrees Celsius, intermediate temperature, 120 degrees to 300 degrees Celsius, or high temperature, more than 600 degrees Celsius).

Background and History

The first fuel cell was developed by the Welsh physicist and judge Sir William Robert Grove in 1839, but fuel cells did not receive serious attention until the early 1960's, when they were used to produce water and electricity for the Gemini and Apollo space programs. These were the first practical fuel cell applications developed by Pratt & Whitney. In 1989, Canadian geophysicist Geoffrey Ballard's Ballard Power Systems and Perry Oceanographics developed a submarine powered by a polymer electrolyte membrane or proton exchange membrane fuel cell (PEMFC). In 1993, Ballard developed a fuel-cell-powered bus and later a PEMFC-powered passenger car. Also in the late twentieth century, United Technologies (UTC) manufactured a large stationary fuel cell system for the cogeneration power plant, while continuously developing the fuel cells for the U.S. space program. UTC is also developing fuel cells for automobiles. Siemens Westinghouse has successfully operated a 100-kilowatt (kW) cogeneration solid oxide fuel cell (SOFC) system, and 1-megawatt (MW) systems are being developed.

How It Works

Polymer Electrolyte Membranes or Proton Exchange Membrane Fuel Cells (PEMFCs). PEMFCs use a proton conductive polymer membrane as an electrolyte. At the anode, the hydrogen separates into protons and electrons, and only the protons pass through the proton exchange membrane. The excess of electrons on the anode creates a voltage difference that can work across an exterior load. At the cathode, electrons and protons are consumed and water is formed.

For PEMFC, the water management is critical to the fuel cell performance: Excess water at the positive electrode leads to flooding of the membrane; dehydration of the membrane leads to the increase of ohmic resistance. In addition, the catalyst of the membrane is sensitive to carbon monoxide poisoning. In practice, pure hydrogen gas is not economical to mass produce. Thus, hydrogen gas is typically produced by steam reforming of hydrocarbons, which contains carbon monoxide.

Direct Methanol Fuel Cells (DMFCs). Like PEMFCs, DMFCs also use a proton exchange membrane. The main advantage of DMFCs is the use of liquid methanol, which is more convenient and less dangerous than gaseous hydrogen. As of 2011, the efficiency is low for DMFCs, so they are used where the energy and power density are more important than efficiency, such as in portable electronic devices.

At the anode, methanol oxidation on a catalyst layer forms carbon dioxide. Protons pass through the proton exchange membrane to the cathode. Water is produced by the reaction between protons and oxygen at the cathode and is consumed at the anode. Electrons are transported through an external circuit from anode to cathode, providing power to connected devices.

Solid Oxide Fuel Cells (SOFCs). Unlike PEMFCs, SOFCs can use hydrocarbon fuels directly and do not require fuel preprocessing to generate hydrogen prior to utilization. Rather, hydrogen and carbon monoxide are generated in situ, either by partial oxidation or, more typically, by steam reforming of the hydrocarbon fuel in the anode chamber of the fuel cell. SOFCs are all-solid electrochemical devices. There is no liquid electrolyte with its attendant material corrosion and electrolyte management problems. The high operating temperature (typically 500-1,000 degrees Celsius) allows internal reforming, promotes rapid kinetics with nonprecious materials, and yields high-quality byproduct heat for cogeneration. The total efficiency of a cogeneration system can be 80 percent—far beyond the conventional power-production system.

The function of the fuel cell with oxides is based on the activity of oxide ions passing from the cathode region to the anode region, where they combine with hydrogen or hydrocarbons; the freed electrons flow through the external circuit. The ideal performance of an SOFC depends on the electrochemical reaction that occurs with different fuels and oxygen.

Molten Carbonate Fuel Cells (MCFCs). MCFCs use an electrolyte composed of a molten carbonate salt mixture suspended in a porous, chemically inert ceramic matrix. Like SOFCs, MCFCs do not require an external reformer to convert fuels to hydrogen. Because of the high operating temperatures, these fuels are converted to hydrogen within the fuel cell itself by an internal re-forming process.

MCFCs are also able to use carbon oxides as fuel. They are not poisoned by carbon monoxide or carbon dioxide, thus MCFCs are advanced to use gases from coal so that they can be integrated with coal gasification.

Applications and Products

Hydrogen Fuel Cell Vehicles. In recent years, both the automobile and energy industries have had great interest in the fuel cell powered vehicle as an alternative to internal combustion engine vehicles, which are driven by petroleum-based liquid fuels. Many automobile manufacturers, such as General Motors, Renault, Hyundai, Toyota, and Honda, have been developing prototype hydrogen fuel cell vehicles. Energy industries have also been installing prototype hydrogen filling stations in large cities, including Los Angeles; Washington, D.C.; and Tokyo.

The first hydrogen fuel cell passenger vehicle for a private individual was leased by Honda in 2005. However, public buses provide better demonstrations of hydrogen fuel cell vehicles compared with passenger vehicles, since public buses are operated and maintained by professionals and they have more volume for the hydrogen fuel storage than passenger vehicles. A number of bus manufacturers such as Toyota, Man, and Daimler have developed hydrogen fuel cell buses and they have been in service in Palm Springs, California; Nagoya, Japan; Vancouver; and Stockholm.

Despite many advantages of hydrogen fuel cell vehicles, this technology still faces substantial challenges such as high costs of novel metal catalyst, safety of hydrogen fuel, effective storage of hydrogen onboard, and infrastructure needed for public refueling stations.

Stationary Power Plants and Hybrid Power Systems. Siemens Westinghouse and UTC have produced a number of power plant units in the range of about 100 kW by using SOFCs, MCFCs, and phosphoric acid fuel cells (PAFCs). Approximately half of the power plants were MCFC-based plants. They showed that these fuel cell systems have exceeded the research-and-discovery level and already produced an economic benefit. These systems generate power with less fossil fuel and lower emissions of greenhouse gases and other harmful products. Just a small number PEMFC-based power plants were built as the cost of fuel cell materials was prohibitive. In many cases, the fuel-cell-based stationary power plants are used for heat supply in addition to power production, enabling so-called combined heat and power systems. Such systems increase the total efficiency of the power plants and offer an economic benefit.

More recently, many efforts to develop hybrid power plants combining fuel cells and gas turbines were made. While the high-temperature fuel cells, such as SOFCs and MCFCs, produce electrical power, the gas turbines produce additional electrical power from the heat produced by the fuel cells' operation. At the same time, the gas turbines compress the air fed into the fuel cells. The expected overall efficiency for the direct conversion of chemical energy to electrical energy is up to 80 percent.

Fascinating Facts About Fuel Cell Technologies

- In 2003, U.S. president George Bush launched the Hydrogen Fuel Initiative (HFI), which was later implemented by legislation through the 2005 Energy Policy Act and the 2006 Advanced Energy Initiative. President Bush stated that "the first car driven by a child born today could be powered by hydrogen and pollution free."
- The Department of Energy is the largest funder of fuel cell science and technology in the United States.
- As of 2011, 191 states have signed Kyoto Protocol, which is a legally binding international agreement to reduce greenhouse-gas emissions by 5.2 percent of 1990 levels by the year 2012.
- In 2008, Boeing announced that it has, for the first time in aviation history, flown a manned airplane powered by hydrogen fuel cells. The Fuel Cell Demonstrator Airplane used a proton exchange membrane fuel cell and lithium-ion battery hybrid system to power an electric motor, which was coupled to a conventional propeller.
- In 2002, typical fuel cell systems cost $1,000 per kilowatt. But, by 2009, the fuel cell system costs had been reduced with volume production (estimated at 500,000 units per year) to $61 per kilowatt.
- Top international universities built their own hydrogen fuel cell racing vehicle to compete against one another on a mobile track in a race called Formula Zero Championship. More advanced races are planned for 2011 Street Edition: The race class will scale up to hydrogen racers, which will compete globally on street circuits in city centers. In the 2015 Circuit Edition, full-size hydrogen fuel cell racing cars built by car manufacturers will compete on racing circuits around the world.

Small Power Generation for the Portable Electronic Devices. At the end of the twentieth century, the demand for electricity continued to increase in many applications, including portable electronics. Batteries have seen significant advances, but their power density is still far inferior to combustion devices. Typically, hydrocarbon fuels have 50 to 100 times more energy storage density than commercially available batteries. Even with low conversion efficiencies, fuel-driven generators will still have superior energy density. There is considerable interest in miniaturizing thermochemical systems for electrical power generation for remote sensors, micro-robots, unmanned vehicles (UMVs), unmanned aerial vehicles (UAVs), even portable electronic devices such as laptop computers and cell phones.

Much work on such systems has been developed by the military. The Defense Advanced Research Projects Agency (DARPA) has initiated and developed many types of portable power concepts using the fuel cells. Industries such as Samsung, Sony, NEC, Toshiba, and Fujitsu have developed fuel cells based portable power generation. Most (about 90 percent) devices were based on PEMFCs or DMFCs, which require lower operating temperatures than SOFC. However, development of SOFC-based portable power generation under the DARPA Microsystems Technology Office showed the feasibility of employing high-temperature fuel cells with appropriate thermal management.

The Military. In addition to the portable power generation for the foot soldiers, the military market has been interested in developing medium-size power plants (a few hundred watts) for recharging various types of storage batteries and high stationary power plants (more than a few kW) for the auxiliary power units.

Military programs in particular have been interested in the direct use of logistic fuel (for example, Jet Propellant 8) for the fuel cells, because of the complexities and difficulties of the re-forming processes. While the new and improved re-forming processes of logistic fuel were being developed to feed hydrogen into the fuel cells, direct jet-fuel SOFCs were also demonstrated by developing new anode materials that had a high resistance to coking and sulfur poisoning.

IMPACT ON INDUSTRY

Government and University Research. One of the biggest sources of funding for fuel cell research in the United States is the Department of Energy (DOE). DOE has developed many programs for the fuel cells and hydrogen. For example, DOE formed Solid State Energy Conversion Alliance (SECA) in 1999 and formulated a program with funding of $1 billion for 10 years. Other government agencies, such as the Department of Defense (DOD), DARPA, Air Force Office of Scientific Research (AFOSR), Office of Naval Research (ONR), and Army Research Laboratory (ARL) have also funded a number of the fuel cell projects taking place in academic and corporate settings to bring about the transfer of the energy technologies to those fighting wars.

Professional societies have also noticed the importance of the energy security and advanced energy technologies. In 2003, the American Institute of Aeronautics and Astronautics (AIAA) and American Society of Mechanical Engineers (ASME) brought in new international conferences: AIAA International Energy Conversion Engineering Conference (IECEC) and ASME International Fuel Cell Science, Engineering, and Technology Conference. The conferences' goals are to expand international cooperation, understanding, promotion of efforts, and disciplines in the area of energy conversion technology, advanced energy and power systems and devices, and the policies, programs, and environmental impacts associated with the development and utilization of energy technologies.

As of 2011, the Korean fuel cell market is in a nascent stage and is expected to witness rapid growth as a result of government-supported policies. Korea has nine fuel cell units installed in various regions. The major driver behind the future development of the hydrogen and fuel cell industry in Korea is the country's need to achieve energy security.

Industry and Business. Almost every car manufacturer has developed a fuel cell vehicle powered by PEMFCs or SOFCs. They hope that the fuel cell vehicles double the efficiency of internal combustion engine vehicles. Some (Honda, General Motors, Toyota) are using their own developed fuel cells, but most companies buy the fuel cell systems from the fuel cell manufacturers such as UTC and De Nora. Many electronic companies such as Motorola, NEC, Toshiba, Samsung, and Matsushita are rushing to develop their own small fuel cells that will provide power up to ten times longer on a single charge than conventional batteries for small portable electronic devices.

The U.S. fuel cell market is growing rapidly. By the end of 2009, 620 fuel cell power units were installed. Government-supported promotion of clean energy is responsible for this, and the tax credits permitted under Energy Policy Act of 2005 continue to drive the U.S. fuel cell market. Fuel cell manufacturers are also working on development of small fuel cell power systems intended to be used in homes and office buildings. For example, a 200 kW PAFC system was installed to power a remote police station in New York City's Central Park.

CAREERS AND COURSE WORK

Courses in chemistry, physics, electrochemistry, materials science, chemical engineering, and mechanical engineering make up the foundational requirements for students interested in pursuing careers in fuel cell research. Earning a bachelor of science degree in any of these fields would be appropriate preparation for graduate work in a similar area. In most circumstances, either a master's or doctorate degree is necessary for the most advanced career opportunities in both academia and industry.

Careers in the fuel cells field can take several different shapes. Fuel cell industries are the biggest employers of fuel cell engineers, who focus on developing and manufacturing new fuel cell units as well as maintaining or repairing fuel cell units. Other industries in which fuel cell engineers often find work include aviation, automotive, electronics, telecommunications, and education.

Many fuel cell engineers prefer employment within the national laboratories and government agencies such as the Pacific Northwest National Laboratory, the National Renewable Energy Laboratory, the Argonne National Laboratory, the National Aeronautics and Space Administration (NASA), DOE, and DARPA. Others find work in academia. Such professionals divide their time between teaching university classes on fuel cells and conducting their own research.

SOCIAL CONTEXT AND FUTURE PROSPECTS

In the future, it is not likely that sustainable transportation will involve use of conventional petroleum. Transportation energy technologies should be developed with both the goal of providing an alternative to the petroleum-based internal combustion engine vehicles. People evaluate vehicles not only on the basis of fuel economy but also performance. Vehicles using an alternative energy source should be designed with these parameters.

One of the most promising energy sources for the future will be hydrogen. The hydrogen fuel cell vehicles face cost and technical challenges, especially the fuel cell stack and onboard hydrogen storage.

For the fuel cell power plants, the economic and lifetime related issues hinder the acceptance of fuel cell technologies. Such problems were not associated with fuel cells but with auxiliary fuel cell units such as thermal management, reactant storage, and water management. Therefore, the auxiliary units of fuel cell systems should be further developed to address these issues.

The fundamental problems of fuel cells related to electrocatalysis also need to be addressed for improvement in performance, as highly selective catalysts will provide better electrochemical reactions.

Lastly, once the new fuel cell technologies are successfully developed and meet the safety requirements, the infrastructure to distribute and to recycle fuel cells will also be necessary.

Jeongmin Ahn, B.S., M.S., Ph.D.

FURTHER READING

Bagotsky, Vladimir S. *Fuel Cells: Problems and Solutions.* Hoboken, N.J.: John Wiley & Sons, 2009. Provides extensive explanations of the various types of fuel cells operation.

Hoogers, Gregor, ed. *Fuel Cell Technology Handbook.* Boca Raton, Fla.: CRC Press, 2003. Recognizes the part played by the change in Gibb's potential.

Kotas, T.J. *The Exergy Method of Thermal Plant Analysis.* Malabar, Fla.: Krieger Publications, 1995. Proves that the fuel chemical exergy and the lower calorific value of the fuel, with different units, are numerically equal.

Larminie, James, and Andrew Dicks. *Fuel Cell Systems Explained.* 2d ed. Hoboken, N.J.: John Wiley & Sons, 2000. This text provides construction details of the various types of fuel cells.

O'Hayre, Ryan P., et al. *Fuel Cell Fundamentals.* 2d ed. Hoboken, N.J.: John Wiley & Sons, 2008. Includes extensive discussions on thermodynamics, transport science, and chemical kinetics in the early chapters with a supporting appendix on quantum-mechanical issues. Also addresses modeling and characterization of fuel cells and fuel cell systems and their environmental impact.

Reddy, Thomas B., ed. *Linden's Handbook of Batteries.* 4th ed. New York: McGraw-Hill, 2011. Includes detailed technical descriptions of chemistry, electrical characteristics, construction details, applications, and pros and cons charts.

WEB SITES

Battery Council International
http://www.batterycouncil.org

Fuel Cell and Hydrogen Energy Association
http://www.fchea.org

Fuel Cell Europe
http://www.fuelcelleurope.org

See also: Hybrid Vehicle Technologies

G

GASOLINE PROCESSING AND PRODUCTION

FIELDS OF STUDY

Chemistry; engineering; chemical engineering; chemical process modeling; fluid dynamics; heat transfer; distillation design; distillation processes; unit operation; corrosion engineering; environmental engineering; control engineering; process engineering; industrial engineering; mechanical engineering; electrical engineering; safety engineering; physics; thermodynamics; mathematics; materials science; computer science; business administration.

SUMMARY

About one-quarter of the crude oil extracted globally is processed to produce gasoline, the fuel for the internal combustion engine that powers most cars, light trucks, motorcycles, and piston engine aircraft. To meet the world's demand for quantity and quality of gasoline, refineries perform several processing steps, beginning with desalting and distillation of crude oil into separate fractions. Selected fractions undergo desulfurization, cracking, reforming, and other processes. The different components gained are blended with further additives to produce various grades of gasoline. Gasoline can also be produced as synthetic fuel from coal, oil sands, natural gas, and biomass.

KEY TERMS AND CONCEPTS

- **Additive:** Special chemical component added to improve the quality of gasoline.
- **Alkylation:** Catalytic combination process to gain high-octane isoparaffins to add to the gasoline pool.
- **Aviation Gasoline:** Gasoline for use by only piston engine aircraft, not by jet engines.
- **Blending:** Physical process of creating various grades of gasoline for consumption.
- **Catalytic Reforming:** Chemical process used on gasoline components to increase their octane ratings.
- **Cracking:** Breaking of long-chain hydrocarbon molecules into short-chain hydrocarbon molecules, some of which are suitable as gasoline components.
- **Crude Oil:** Liquid part of petroleum, the raw material from which gasoline, among many other products, is processed.
- **Desulfurization:** Process of cleansing gasoline components of unwanted sulfur.
- **Distillation:** Process in which a mix of components with different volatilities is heated to create physical separation of the components.
- **Isomerization:** Chemical process to rearrange the atoms in a molecule to form another molecule, done to create a higher octane rating for a gasoline blend.
- **Octane Rating:** Unit to measure quality of gasoline; the higher the octane rating, the less likely the gasoline will knock, or prematurely ignite in the engine.
- **Polymerization:** Chemical process used to increase the octane rating by forming longer molecules from propene and butane as gasoline stocks.
- **Refining:** Processing crude oil by various means to form different end products such as gasoline.

Definition and Basic Principles

Gasoline is the most effective fuel for the internal combustion engine that powers most automobiles and light trucks as well as piston engine aircraft. It is generally gained from processing crude oil at a refinery. The gasoline sold to consumers consists of a complex blend of hydrocarbons that have boiling ranges from about 38 to 205 degrees Celsius (100 to 400 degrees Fahrenheit).

The first step in processing at a refinery is generally the desalting of crude oil. Then the oil is heated, partially vaporized, and sent to a distillation tower operating at atmospheric pressure. There it condenses into separate fractions that are extracted. The yield of light straight-run (LSR) gasoline from crude oil after atmospheric distillation generally consists of only up to 10 percent. This is much too low to satisfy the global demand for gasoline. Therefore, additional, different fractions from distillation are processed into suitable components of gasoline.

Heavy straight-run (HSR) naphtha gained from atmospheric distillation and from vacuum distillation, followed by thermal, catalytic, or hydrocracking, is fed into a catalytic reformer to become reformate, a gasoline component. To boost the octane rating of the final gasoline blend, which is a key indicator of its quality as engine fuel, gasoline components also undergo such chemical processes as isomerization, polymerization, and alkylation. Special additives gained in other refining processes are mixed in to form high-quality gasoline.

Gasoline can also be produced by processing feedstock such as coal, oil sands and oil shale, natural gas, or even biofuels. These processes are far more expensive than producing gasoline from crude oil. They are usually undertaken only in special circumstances, such as periods of abundance of alternative feedstock and lack of crude oil, during wars, or as the result of a political commitment to alternative fuels.

BACKGROUND AND HISTORY

In 1856, in Poland and Romania, the first refineries were built to process crude oil into more useful products through distillation into different fractions. The first refinery opened in the United States in 1861. What would later be known as gasoline in the United States, a naphtha-based hydrogen compound, called petrol in Great Britain, was one of the different refinery products. The rise of the internal combustion engine in the late nineteenth century, particularly as the motor of the newly invented automobile, led to a tremendous increase in the demand for gasoline as its fuel.

To satisfy this demand, methods of increasing the yield of gasoline from crude oil were developed. Thermal cracking of other distillation fractions was invented separately by Vladimir Shukhov in Russia in 1891 and by William Merriam Burton in the United States in 1913. Thermal cracking doubled the yield of gasoline in the United States.

In 1930, the invention of thermal reforming boosted the octane rating of gasoline and lessened the stress on engines burning gasoline. In 1932, hydrogenation came into use to lower the undesirable sulfur content of gasoline, and coking created additional base stocks for gasoline out of heavier distillation fractions. In 1935, catalytic polymerization further boosted octane ratings.

The big breakthrough in producing more gasoline with higher octane ratings came with the invention of catalytic cracking in 1937 and fluid catalytic cracking in 1942. French engineer Eugene Jules Houdry is generally given credit for the first and a consortium of American university scientists and oil industry researchers for the second. Other important steps were introduction of visbreaking in 1939, alkylation and isomerization in 1940, catalytic reforming in 1952, and hydrotreating in 1954. Research to improve refinery processes to optimize gasoline yield and quality and to make processes and products more environmentally friendly continues into the twenty-first century.

Producing gasoline synthetically from coal was invented in 1913 by German chemist Friedrich Bergius. By 2010, SASOL, a South African company, had become the leading producer of gasoline from coal. In 1978, commercial production of synthetic crude oil from oil sands began in Canada; in 2009, this accounted for 13 percent of Canada's petroleum product consumption including gasoline.

HOW IT WORKS

Gasoline is processed and produced from crude oil at a refinery. As refineries are very complex installations, processing a wide variety of crude oils, each refinery follows its own, customized process. However, the following are the most common processes.

Desalting. As crude oil arrives at a refinery, it is generally desalted to remove suspended salt and solid contaminants. Crude oil is heated and mixed with hot water. Salt and contaminants are washed out either by adding chemicals or by application of an electric field. Another method is to filter heated crude oil through diatomaceous earth.

Distillation. Desalted crude oil is heated for fractional distillation between 343 and 399 degrees Celsius (650 to 750 degrees Fahrenheit). The resulting vapor and liquid mix is sent to the first distillation tower that operates at atmospheric pressure. Because of the different boiling points of the different hydrocarbon molecules of crude oil, the hydrocarbons can be separated into different fractions (also called cuts). This occurs at the distillation tower, which can be as tall as 50 meters (164 feet). The heaviest hydrocarbons remain at the bottom, while the middle and lighter ones are extracted.

From the lightest hydrocarbons collected from distillation, light straight-run (LSR) gasoline (sometimes called light naphtha) is extracted at a gas separation plant that commonly contains a hydrodesulfurization unit. Among the lighter gases, butane is used for blending into the final gasoline products or as feedstock for the alkylation unit.

Atmospheric distillation also yields lower-boiling heavy straight-run (HSR) gasoline (or HSR naphtha). Often after hydrotreating, this fraction is processed further at the catalytic reformer.

Cracking. Distillation also yields residue. Among its fractions is what is commonly called gas oils, middle distillate, or wax distillate. After hydroprocessing, these gas oils are sent to a cracker at the refinery to convert them into more valuable products including gasoline components. If a hydrocracker is used, no prior hydroprocessing occurs.

Cracking is the chemical conversion process used to break down longer-chain hydrocarbon molecules into shorter-chain ones that are more valuable. The naphtha gained from various cracking processes is used for further processing into gasoline components with high octane ratings.

The mildest form of thermal cracking is visbreaking. It breaks the longest hydrocarbon molecules to eventually yield also more gasoline. Delayed coking is severe thermal cracking, heating the gas oil feedstock to 500 degrees Celsius (930 degrees Fahrenheit). The product gained for gasoline processing is called coker naphtha.

Fluid catalytic cracking (FCC) is a key process to gain components for gasoline blending from the heavier feedstocks processed at a FCC unit. Fluid catalytic cracking converts nearly half of the heavy feedstocks into naphtha for gasoline production and accounts for 35 to 45 percent of the volume of the gasoline produced at U.S. refineries as of 2010. The naphtha gained from cracking is typically divided. The light fraction is used directly for gasoline blending. The heavy fraction is sent to the catalytic reformer and functions as octane booster.

Hydrocracking is the most sophisticated, flexible, and expensive form of cracking. Hydrocrackers are less common at American refineries than at European and Asian ones. This is because gasoline is the most in-demand product in the United States, and a hydrocracker's maximum yield of gasoline components is only 8 percent more than that of a much cheaper fluid catalytic cracker. Hydrocracking combines catalytic cracking with the insertion of hydrogen. Light and heavy naphtha is gained for gasoline blending and processing.

Catalytic Reforming. To improve the octane rating of the blended gasoline, heavy straight-run naphtha and naphtha gained from cracking is sent to a catalytic reformer. During catalytic reforming, the feedstock naphtha has its hydrocarbon molecules restructured by light cracking in the presence of a platinum-based catalyst. Catalytic reforming creates a desirable high octane gasoline stock. Its aromatics that are responsible for its high octane rating have come under environmental scrutiny.

Isomerization, Polymerization, and Alkylation. There are some further processes to increase the octane rating of gasoline. Light straight-run gasoline often has at least some of its components isomerized, so that it consists of some molecules with a different structure but with the same number of atoms. Polymerization is a cost-effective way to boost octane ratings by combining the very light gases propene and butane into longer-chain olefin molecules. Alkylation is an effective but expensive process to increase the octane rating of light components of the gasoline blending pool. It refers to the addition of an alkyl group to isobutene at low temperatures of 5 to 38 degrees Celsius (41 to 100 degrees Fahrenheit) with a catalyst of sulfuric or hydrofluoric acid.

Blending and Additives. In the end, all gasoline components are blended together. At a modern refinery, this is done in a computer-controlled process. Additives are used primarily to boost octane rating. Lead was widely used once but was banned in 1996 in the United States, as well as in many other nations. With the increasing prohibition of methyl-tert-butyl-ether (MTBE) in many U.S. states, use of alternatives such as tertiary amyl methyl ether (TAME) or ethyl tert-butyl ether (ETBE) increased. There are many governments that mandate the blending of ethanol, which is done after gasoline leaves the refinery.

Gasoline from Synthetic Fuels. Gasoline can be produced from oil shale, coal, biomass, or natural gas. These processes are invariably more expensive than processing and producing gasoline from crude oil. They are employed in specific locations such as in South Africa from coal and in Canada and Venezuela from huge oil sand deposits.

Applications and Products

Fuel Gasoline. By 2009, worldwide, 676 refineries produced 7.7 billion barrels of gasoline, or 325 billion gallons. This meant that globally, about one-quarter of the 30.8 billion barrels of crude oil refined was converted to gasoline. In the United States, because of a higher demand for gasoline, 141 operating refineries produced 3.2 billion barrels (135 billion gallons) of gasoline, amounting to 46 percent of all refinery output. This U.S. percentage for gasoline production has been basically stable since the 1990's, with a slight dip to 44 percent in 2008 because of the recession.

Over 90 percent of gasoline produced in the United States is used to fuel automobiles and light trucks. All over the world, gasoline as fuel is essential for transportation in any industrial and industrializing society. Its desired qualities are strong resistance to premature ignition in the engine, facilitating easy start, warm-up, and acceleration of the engine. Further desirable qualities include high mileage for the fuel consumed, prevention of vapor lock and deposit build-up in the engine, and as few polluting emissions as possible.

Gasoline Components. To optimize its qualities as fuel for the internal combustion engine, gasoline is blended from different components. Out of concern for the environment and for protection from cancer-causing components, national governments often regulate the composition of gasoline. An example is the prohibition of lead as an additive for fuel gasoline in the United States and the European Union and increasingly in many other nations.

In the United States, gasoline sold to consumers consists primarily of naphtha gained from catalytic cracking at 38 percent of the total, reformate from catalytic reforming at 27 percent, alkylate at 12 percent, and light straight-run gasoline and its isomeric form at 7 percent. Smaller contributors are the light component normal butane at 3 percent, light naphtha from hydrocracking at 2.4 percent, light coker naphtha at 0.7 percent, and polymers at 0.4 percent. Other additives, especially ethanol, account for the remainder.

Out of environmental concerns in the United States, the concept of reformulated gasoline (RFG) has been developed. This refers to a blend of gasoline that burns at least as cleanly as high methanol-content alternative fuels. As a result of federal and state regulations, in 2009, about one-third, or 1.1 billion gallons, of gasoline produced by U.S. refineries was blended with ethanol to become reformulated gasoline.

Measuring Gasoline Quality. For the consumer, the most important indicator of gasoline quality is its octane rating. The higher the octane rating, the less likely is the gasoline blend to ignite prematurely in the engine, damaging it in an event commonly called knocking. There are many different ways to calculate the octane rating. The research octane number (RON) is derived from testing the gasoline blend in a laboratory engine. The gasoline's ability to burn by controlled ignition and not ignite prematurely is related to the respective quality of a mix of iso-octane and heptanes. A RON rating of 95, for example, indicates that this gasoline burns as well as a mix of 95 percent iso-octane and 5 percent heptanes would. The motor octane number (MON) uses the same comparison but places the test engine under more stress to simulate actual driving situations. For this reason, the MON octane rating is between 8 to 10 points lower than the RON. Different nations use different octane ratings. In the United States, an average of RON and MON is used and posted at gas station pumps. It can be called PON (posted octane number), or (R+M)/2, and is also called the anti-knocking index (AKI).

In the United States, gasoline is typically available as regular unleaded gasoline, with a PON of generally 85, and premium gasoline ranges from a PON of 89 to 93. California typically offers the three grades of 87, 89, and 91 PON. Premiums are branded under different names. Because lower atmospheric pressure at high altitudes reduces pressure in the engine's combustion chamber and lessens the danger of premature ignition, gasoline sold at high elevations in the Rocky Mountains states typically have a PON of 85 up to 91, signifying that gasoline has to be blended to account for the environment where it is burned as fuel. Special gasoline for automobile racing in the United States can have a PON of 100 or higher.

Two other indicators of gasoline quality are as important but less visible to consumers than the octane rating. They are Reid vapor pressure (RVP) and boiling range of the blended gasoline. Low RVP, or low volatility, and higher boiling ranges mean the gasoline is less likely to evaporate too quickly, causing vapor lock in the engine. They also prevent higher evaporation losses that lower mileage gained from gasoline. A high RVP, or high volatility, and lower boiling ranges make the engine start easier and warm up more quickly. Outside temperature influences gasoline behavior. As a result, gasoline is blended differently for use during the summertime, with a RVP of about 7.2 pounds per square inch (psi), or 49.6 kilo Pascal (kPA), and during the wintertime with a RVP of about 13.5 psi (93.1 kPa).

E 85. Ethanol, or grain alcohol, is blended into gasoline before sale to consumers because it is a very clean-burning fuel from renewable resources. However, it gets only 70 percent of the mileage of undiluted gasoline. Several states have mandated a blend of at least 5.9 percent of ethanol into gasoline, and many gas pumps state that the gasoline sold may contain up to 10 percent ethanol. An alternate fuel, marketed as E 85, contains 85 percent ethanol to only 15 percent gasoline. Its lack of fuel efficiency is made up by a federal tax subsidy trying to encourage its consumption.

Avgas. For use as fuel for aircraft with piston engines, refineries produce relatively small quantities of aviation gasoline, called avgas. This is very different from jet fuel. In 2009, U.S. refineries produced about 211 million gallons of avgas, about 0.1 percent of their products, and down from 261 million gallons in 2004. This decline occurred because avgas, most commonly of the 10011 variety, contains a small amount of lead in the form of tetraethyl lead (TEL). As public pressure increases to completely phase out leaded gasoline, scientists have looked for an alternative to lead in aviation gasoline. Avgas must have a high octane rating and cannot lead to engine vapor lock at the low pressure encountered during flight. By 2010, there had been some promising experiments with a nonleaded aviation gasoline called G100UL.

Impact on Industry

Because gasoline is the key fuel for the internal combustion engine that powers most of the world's automobiles and a significant number of small buses and light trucks, its impact on global industry is extremely significant. This begins with the quest for the raw materials from which the majority of gasoline is produced, in particular crude oil from petroleum. Exploration, extraction, and transport to processing of crude oil is a major industrial enterprise. However, the oil industry is significant not only in itself but also as a customer of highly specialized operating equipment and customized plants. Transporting gasoline from the refinery to the customer requires a large logistics effort involving pipelines, tanker trucks, and a network of gas stations. The huge costs of producing gasoline are justified only by its high value.

Government Agencies. Governments of nations with exploitable stock of petroleum benefit hugely, primarily from royalties for extraction concessions on national land or territorial waters. Gasoline taxes also make up a major revenue stream for many industrialized countries. Government agencies such as the U.S. Environmental Protection Agency (EPA) or those linked to ministries of transportation or the environment all over the industrial and industrializing world also affect processing and production of gasoline, particularly in regard to gasoline standards. A key example was the long, drawn-out fight to end the addition of lead to gasoline in the United States, in which EPA scientists became involved. This was mirrored in other nations, where governments set similarly strict standards for gasoline components.

Government agencies also set standards for transportation, storage, and distribution of gasoline, as well as for specifications of the internal combustion engines and their exhaust mechanisms, all to lessen the negative environmental impact of gasoline combustion. The requirement for catalytic converters to treat gasoline engine exhaust has been set by the United States, the European Union, Japan, and many other nations.

National governments have been concerned with safeguarding the supply of gasoline for their industries and citizens. This has led to conflicts triggered in part by disputes and fears over petroleum extraction, such as arguably the United Nation's successful termination of Iraq's occupation of Kuwait (1990-1991).

Universities and Research Institutes. Universities educate the chemists, engineers, and other scientists and professionals involved in all steps of gasoline processing and production or in petroleum extraction and production as gasoline's prime feedstock. Universities also conduct research into fuel alternatives to gasoline that are more environmentally friendly. Research institutes employ scientists and researchers to engage in the continuing project to improve the quality, cost efficiency, and the environmental impact of gasoline at all steps of production and consumption. Trade associations such as the American Petroleum Institute also conduct public relations and see themselves as an interface between the oil-gasoline industry and the general public.

Industry and Business Sectors. As part of the chemical industry, gasoline production belongs to the oil and gas industry sector. There, it is part of the midstream and downstream sector that engages in the processing, blending, transport, and marketing of gasoline to the consumer and oversees its final distribution at a gas station.

Gasoline production also affects the financial industry as its key raw material, crude oil, is one of the world's most widely traded commodities. Shock waves from swings in the oil price can affect the global economy. Manufacturers, particularly automakers and their suppliers, are also affected by gasoline production. The petrochemical industry is affected as its feedstocks such as ethane compete with the production of gasoline at refineries.

The transportation industry is involved in bringing gasoline to its individual points of sale at a network of gas stations. Gas stations, like most of the 162,000 operating in the United States in 2010, often also serve as small retail outlets. The advertising industry earns revenue from creating major gasoline brands and public relation campaigns for oil companies.

Major Corporations. Overall, gasoline processing and production tends to be highly integrated into the business of large private or national oil and gas companies. Because of significant economies of scale, the number of refineries has declined. Smaller refineries independent of big oil companies have become the exception. Although huge new world-scale refineries have been built, particularly in the Middle East, the source of much crude oil, or Asia, source of increasing demand for gasoline, the number of operable U.S. refineries has shrunk by half. By 2010, the number had dropped to 148 from 301 in 1982. However, U.S. refinery capacity was virtually the same in 2010 as in 1982, at about 17.5 million barrels per day, which indicates consolidation of refining operations

In 2007, the world's largest oil and gas company was Saudi Aramco in Saudi Arabia. Its four sole-owned refineries in Saudi Arabia processed almost 1 billion barrels of crude oil, and a new refinery opened in 2008.

In 2010, ExxonMobil was the world's largest private oil company and operated the two largest refineries in the United States. Its Baytown, Texas, refinery had a capacity of 205 million barrels per year, and its Baton Rouge, Louisiana, refinery was a close second at 184 million barrels.

Royal Dutch Shell and BP were other leading international private oil companies heavily engaged in the refining business as part of their integrated oil and gas operations. In 2009, Shell held interests in forty refineries around the world processing

Fascinating Facts About Gasoline Processing and Production

- Americans are the world's top consumers of gasoline. In 2009, the American population of about 305 million used 377 million gallons of gasoline per day, more than one gallon for every person.
- Japan is the only industrialized nation with virtually no domestic raw materials to produce gasoline and must import all of its crude oil for processing at its thirty-one refineries.
- Gasoline production is the largest of all U.S. basic industries, exceeding the output of steel or lumber.
- South Africa is the world's leading producer of gasoline from coal, an abundant natural resource in that nation, following a process invented in Germany in 1913.
- Antarctica is the world's only continent without a refinery.
- Because of heavy taxation, gasoline is more expensive in the European Union than in the United States. On June 30, 2010, a gallon of unleaded gasoline cost on average $2.78 in the United States but cost $7.19 in the Netherlands.
- Both Norway and Venezuela produce more gasoline than they consume. However, while Norwegian taxes raise the price of a gallon of unleaded gasoline to $7.65, Venezuelan subsidies lower it to the world's cheapest price of $0.09 (as of June, 2010).
- During World War II, Japanese occupying forces requisitioned much-needed rice from Vietnamese farmers to convert into synthetic gasoline for army trucks and aircraft, exacerbating Vietnam's severe famine of 1944 to 1945.
- Aviation gasoline is artificially colored so that it can be quickly distinguished from automobile gasoline by sight. The most common 10011 variety shines a bright blue.

about 1.5 billion barrels of crude. Similarly, worldwide, in 2009, BP held an interest in sixteen refineries that processed 1.35 billion barrels of crude oil, with BP's share being 973 million barrels.

Careers and Course Work

Students interested in a career in the gasoline processing and production industry should take courses in science, mathematics, and economics in high school. Refineries employ many skilled technicians for their operations and need laboratory analysts, occupations for which a two-year associate degree is helpful. A refinery also has positions for firefighters and employs members of the medical profession, from paramedics to physicians.

A bachelor's degree in chemistry, physics, computer science, mathematics, or environmental science is very useful for a career in the actual processing and production field. The same is true for an engineering degree, whether at the undergraduate or postgraduate level, especially in chemical engineering, or electrical, mechanical or computer engineering. Such degrees create good employment prospects. For a career in the purchasing department of a refinery, selecting the different crude oils for processing, for instance, a business major or master of business administration is helpful and can also lead to a higher management position.

Any master of engineering or doctoral degree in chemistry or chemical engineering serves as good preparation for a top-level career. A doctorate in a science field can lead to an advanced research position either with a company or a government agency. A research career linked to the subject of gasoline production, whether at a university, institute, or corporation, also benefits from postdoctoral work in chemistry or materials science.

Particularly for a career in private industry, the cyclical nature of the oil industry does not guarantee employment during industry downturns. As many new refineries are built outside the United States and many operate in different places of the world, global mobility is of significant advantage when pursuing an advanced career.

Social Context and Future Prospects

As the most efficient fuel for the internal combustion engine, gasoline has vastly increased private and public mobility of the world's people, especially in industrialized societies. The nineteenth century increased humanity's mobility through the steam engines of railroad trains, and the twentieth century saw the rise of gasoline to enable private automobile transportation. As such, gasoline significantly affects the lives of almost every person in a developed country.

However, the emissions caused by both the production of gasoline and its use as fuel have had a negative impact on the environment that has been fiercely debated and publicly discussed. Of particular concern are carbon dioxide emissions and the side effects of many gasoline additives. The wide availability of gasoline also promotes the manufacture of automobiles and other vehicles, which generates more greenhouse gases and sometimes hazardous wastes, especially during the vehicle disposal process. Oil, gasoline, and automobiles have been attacked by some activists as the main causes of human-made environmental degradation, although they have granted freedom of mobility to a vast number of people on a scale unthinkable in the nineteenth century.

Because of environmental concerns, research seeks to minimize the aromatics content of gasoline, as additives such as benzene are particularly harmful if handled carelessly. Research is ongoing to make gasoline burn cleaner and increase the cost efficiency of its production. To improve the economics of gasoline production, very large world-class capacity refineries have been built to use economics of scale and take advantage of proximity to either raw materials or markets.

Research is under way into alternative fuels to gasoline that would offer similar fuel efficiency at a lesser environmental cost. The rise of the petrochemical industry as a competitor for gasoline's raw material, crude oil, led to more refineries increasing their output of ethane, a very light gaseous hydrocarbon, at the expense of gasoline production, as ethane is a prime and valuable petrochemical feedstock.

R. C. Lutz, B.A., M.A., Ph.D.

Further Reading

Duffield, John. *Over a Barrel: The Costs of U.S. Foreign Oil Dependence.* Stanford, Calif.: Stanford University Press, 2008. Notes that as demand for gasoline far outstrips domestic resources, U.S. economic and foreign policy is forced to address this problem in specific ways, including military options. Concludes with a proposal to lower the cost of dependence. Tables, figures, maps.

Gary, James, et al. *Petroleum Refining: Technology and Economics.* 5th ed. New York: CRC Press, 2007. A well-written textbook that details all the refining steps of gasoline processing and production. Five appendixes, index, photographs.

Leffler, William. *Petroleum Refining in Nontechnical Language.* Tulsa, Okla.: Penn Well, 2008. Covers all major aspects of gasoline processing and production at a refinery, from the chemical foundations to the processes themselves.

Meyers, Robert, ed. *Handbook of Petroleum Refining Processes.* 3d ed. New York: McGraw-Hill, 2004. Advanced-level technical compendium covers various industry methods for gasoline production. In-depth information for those considering a career in this field.

Raymond, Martin, and William Leffler. *Oil and Gas Production in Nontechnical Language.* Tulsa, Okla.: Penn Well, 2005. Accessible introduction to the process of producing gasoline from crude oil. Covers all major aspects. Light-hearted but very informative style.

Yeomans, Matthew. *Oil: Anatomy of an Industry.* New York: The New Press, 2004. Critical look at the oil industry, particularly in the United States. Good historical overview; addresses issues of America's dependency on oil caused by its gasoline demand, world conflicts caused by oil, and the question of alternatives like hydrogen fuel.

WEB SITES

Global Petroleum Research Institute
http://www.pe.tamu.edu/gpri-new/home

Independent Petroleum Association of America
http://www.ipaa.org

National Petrochemical and Refiners Association
http://www.npra.org

National Petroleum Council
http://www.npc.org

Society of Petroleum Engineers
http://www.spe.org/index.php

U.S. Department of Energy
Energy Information Administration, Petroleum
http://www.eia.doe.gov/oil_gas/petroleum/info_glance/petroleum.html

U.S. Environmental Protection Agency
Gasoline Fuels
http://www.epa.gov/otaq/gasoline.htm

See also: Biofuels and Synthetic Fuels

GAS TURBINE TECHNOLOGY

FIELDS OF STUDY

Mechanical engineering; thermodynamics; aerodynamics; fluid dynamics; heat transfer; environmental engineering; control engineering; industrial engineering; electrical engineering; safety engineering; physics; mathematics; materials science; metallurgy; ceramics; process engineering; engineering; electronics; aeronautics; business administration.

SUMMARY

Gas turbine technology covers design, manufacture, operation, and maintenance of rotary conversion engines that generate power from the energy of the hot, pressurized gas they create. Key components are air and fuel intake systems, compressor, combustor, the gas turbine itself, and an output shaft in those gas turbines that are not designed to provide thrust alone as jet engines. Primary applications of gas turbines are jet engines and generators for industrial power and electric utilities. Sea and land vehicles can also be propelled by gas turbines.

KEY TERMS AND CONCEPTS

- **Airfoil:** Wing-shaped blades used in compressors and turbines to interact with gas stream.
- **Brayton Cycle:** Term for the scientific principle behind the operation of the gas turbine.
- **Cogeneration:** Purposeful use of the remaining heat from a gas turbine exhaust stream.
- **Combined Cycle:** Using a gas turbine in combination with a steam turbine to increase its fuel efficiency.
- **Combustor:** Unit where fuel mixes with pressurized air and is ignited and burned.
- **Compressor:** Unit that increases the pressure of intake air by compressing its volume.
- **Free Power:** Portion of power generated by the turbine not needed to drive the compressor.
- **Jet Engine:** Gas turbine designed to release its free power as thrust to lift, propel, and stop an aircraft.
- **Nozzle:** Apparatus through which the gas stream is injected into the turbine; also used to inject fuel into combustor.
- **Spool:** Industry name for the output shaft of the turbine conveying its power.
- **Turbine:** Rotary engine that can be fueled by a gas stream.
- **Turbofan:** Jet engine designed to provide stronger thrust by driving a fan instead of relying on exhaust.
- **Turboprop:** Jet engine designed to provide stronger thrust by driving a propeller instead of relying on exhaust.

Definition and Basic Principles

Gas turbine technology concerns itself with all aspects of designing, building, running, and servicing gas turbines, which provide jet engines for airplanes and represent the heart of many contemporary power plants, among other uses. Strictly speaking, the gas turbine itself is only one part of a complex engine assembly commonly given this name.

A gas turbine employs the physical fact that thermal energy can be converted into mechanical energy. Its basic principle is often called the Brayton cycle, named after George Brayton, the American engineer who developed it in 1872. The Brayton cycle involves compression of air, its heating by fuel combustion, and the release of the hot gas stream to expand and drive a turbine. The turbine creates both power for air compression and free power. The free power gained can be used either as thrust in traditional jet engines or as mechanical power driving another unit such as an electric generator, pump, or compressor.

The gas turbine is closely related to the steam turbine. The gas turbine got its name from the fact that it operates with air in gaseous form. A wide variety of fuel can be used to heat the gaseous air.

Background and History

The oldest known reference to an apparatus utilizing the physical principles of a gas turbine is a design by Leonardo da Vinci from 1550 for a hot-air-powered roasting spit. In 1791, British inventor John Barber obtained a patent for the design of a combined gas and steam turbine. However, the lack of suitable materials to withstand the heat and pressure needed for a working gas turbine impeded any practical applications for a long time.

In 1872, German engineer Franz Stolze designed a gas turbine, but driving its compressor used more energy than the turbine generated. In 1903, the first gas turbine with a surplus of power was built by Norwegian inventor Aegidius Elling. The idea to use gas turbines to build jet engines was pioneered by British aviation engineer and pilot Sir Frank Whittle by 1930. At the same time, German physicist Hans von Ohain developed a jet engine on his own. On August 27, 1939, the German Heinkel He 178 became the first flying jet airplane. That year, the first power plant using a gas turbine became operational in Switzerland. Since that time, the twin use of gas turbine technology either to propel jet airplanes or serve as a source for generating power on the ground, particularly electricity, has been subject to many technological advances.

How It Works

Brayton Cycle. A gas turbine engine is a device to convert fuel energy via the compression and subsequent expansion of air. Its working process is commonly called a Brayton cycle. The Brayton cycle begins with the compression of air, in gaseous form, for which energy is expended. For this reason, a gas turbine assembly needs an air intake and a compressor where air is pressurized. Next, more energy is added to the compressed air through the heat from the combustion of fuel. Fuel is burned, most commonly, internally in the engine's combustor. However, in a variation called the Ericsson cycle, the fuel is burned externally and the generated heat is relayed to the compressed air. The third step of the cycle comes with the release of the heated gas stream through nozzles into the gas turbine proper. The gas expands as it loses some of its pressure and cools off. The energy released by gas expansion is captured by the gas turbine, which is driven in a rotary fashion as its blades are turned by the exiting hot gas stream. In the last step, the expanding air releases its leftover heat into the atmosphere.

Power Generation. As the hot gas stream flows along stator vanes to hit the airfoil-shaped rotor blades arranged on a disk inside the gas turbine, it drives the blades in rotary fashion, creating mechanical power. This power is captured by one or more output shafts, called spools. There are two uses for this power. The first is to drive the compressor of the gas turbine assembly, feeding power back to the first step of operations. Any free power remaining can be used to drive external loads or to provide thrust for jet engines, either directly through exhaust or by driving a fan or propeller.

The engineering challenge in gas turbine technology is to gain as much free power as possible. Attention has been focused both on materials used inside the gas turbine, looking for those that can withstand the most heat and pressure, and in arranging the individual components of the gas turbine to optimize its output. The metal of single-crystal cast alloy turbine blades can withstand temperatures of up to 1,940 degrees Fahrenheit. Ingenious air-cooling systems enable these blades to deal with gas as hot as 2,912 degrees Fahrenheit. Top compressors can achieve a 40:1 ratio. At the same time, designing gas turbines with multiple shafts to drive low- and high-pressure compressors or adding a power turbine behind the first turbine used only to gather sufficient power for the compressor has also improved efficiency. While early gas turbines used between 66 and 75 percent of the power they generated from fuel to drive their own compressor, leaving only 25 to 34 percent of free power, contemporary gas turbines for industrial power can achieve up to 65 percent of free power.

Fuel. Fuel for gas turbines is variable and ranges from the hydrocarbon product kerosene for jet engines to coal or natural gas for industrial gas turbines. The engineering challenges have been to optimize fuel efficiency and to lower emissions, particularly of nitrogen oxides.

Operation and Maintenance. Gas turbines require careful operation as they are very responsive, and malfunctions in either the compressor or turbine can happen in fractions of a second. The primary control systems are handled by computer and are hydromechanical and electrical. From start and stop and loading and unloading the gas turbine, operating controls cover speed, temperature, load, surges, and output. The control regimen ranges from sequencing to routine operation and protection control.

Gas turbines have some accessories to facilitate their operation. These include starting and ignition systems, lubrication and bearings, air-inlet cooling and injection systems for water, and steam or technical gases such as ammonia to control nitrogen oxide emissions.

To facilitate maintenance, gas turbines used for

aircraft and those modeled after these have a modular design so that the individual components such as compressor, combustor, and turbine can be taken out of the assembly individually. Those gas turbines allow also for a borescopic inspection of their insides by an optical tube with lens and eyepiece. Heavy industrial gas turbines are not designed for borescopic inspection and must be dismantled for inspection and maintenance. Preventive maintenance is very important for gas turbines.

Applications and Products

Jet Engines. The first jet engines were designed to use the free power of gas turbines exclusively for thrust. This gave them a speed advantage over piston- engine aircraft but at the price of very low fuel economy. As a result, gas turbine technology developed more economical alternatives. For helicopters and smaller commercial airplanes, the turboprop system was developed. Here, the gas turbine uses its free power to drive the propeller of the aircraft.

The turbofan jet engine is used in 90 percent of contemporary medium to large commercial aircraft. Efficiency is increased by the addition of a fan that acts like a ducted propeller and that is driven by the free power of the gas turbine. In contemporary high-bypass turbofan jet engines in use since the 1970's and continuously improved since, much of the air taken in bypasses the compressor and is directly propelled into the engine by the turbine-driven fan, up to a bypass ratio of 5:1. Only a smaller part of air is taken into a low- and then a high-pressure compressor, joined with fuel burned in the combustor and driving both a high- and low-pressure turbine. The net resulting thrust, primarily from the fan, is achieved with high fuel economy and relatively little noise.

Military aircraft, the only ones flying at supersonic speed, also use afterburners with their gas turbines. This is done by adding another combustor behind the turbine blades and before the exhaust nozzle, creating extra thrust at the expense of much fuel.

Aero-Derivative Turbines. Because of their relatively low weight, gas turbines based on jet engine design have been used on the ground for power generation and propulsion. Especially with the contemporary trend toward turbofan jet engines, an aero-derivative turbine that uses one or more spools, or output shafts, to provide a mechanical drive needs very little adaptation from air to ground use. There are also hybrid gas turbines that use an aero-derivative design but replace the jet engine's lighter roller and antifriction ball bearings with hydrodynamic bearings typical of the heavy industrial gas turbine.

With higher shaft speed but lower airflow through the turbine, aero-derivative turbines require less complex and shorter maintenance than other ground gas turbines. They are often used in remote areas, where they are employed to drive pumps and compressors for pipelines, for example. Because of their quicker start, stop, and loading times, aero-derivative gas turbines are also used for flexible peak load power generation and for ground propulsion.

Heavy Industrial Gas Turbines. Gas turbines for use on the ground can be built more sturdily and larger than jet engines. These heavy units have been generally used for power generation or to drive heavy industrial pumps or compressors, with power generation increasingly important. Gas turbine technology has experimented with a variety of designs for these turbines. One decision is whether to place the output shaft (spool) at the "hot" end where the gas stream exits the turbine, or at the "cold" end in front of the air intake. "Cold" end drives are easier to access for service and do not have to withstand the hot environment at the turbine end, but their position has to be carefully designed so as not to disturb the air intake. If the output shaft would cause a turbulence or vortex in the air flowing into the compressor this could lead to a surge potentially destroying the whole engine.

There are also design differences regarding the numbers of shafts (spools) and turbines within a contemporary heavy industrial gas turbine. The basic form has one output shaft rotating at the speed of the compressor and turbine. At the output, this speed can be geared up or down depending on the speed desired for the application the gas turbine is driving. This design is almost exclusively used for power generation. An alternative to minimize gear losses is to put a second, free-power turbine behind the first gas turbine driving the compressor. This means that the speed of the free-power turbine can be regulated independently of the turbine speed needed to drive the compressor, which makes it an attractive design when pumps or compressors are driven by the gas turbine. This design is only possible with a "hot" end configuration. Finally, there are gas turbine designs that use more than one shaft (spool). A dual spool split output shaft gas turbine, for example, employs three output shafts to operate independently with a high-pressure and low-pressure turbine-compressor assembly as well as a free- power turbine.

Because a single-cycle, stand-alone gas turbine has a fuel efficiency of as little as 17 percent, meaning 83 percent of the energy created is used for the compressor, engineers have combined gas turbines for power generation in cogeneration or combined-cycle power plants. In a cogeneration plant, the remaining heat that exits the gas turbine is used for industrial purposes, such as heating steam for a refinery. In a combined-cycle power plant, the heat from typically two gas turbines fuels a steam turbine. This can create fuel efficiencies ranging from 55 to 65 percent or more. These gas turbines typically create about 250 to 350 megawatts of electrical power each.

Marine and Tank Propulsion. Aero-derivative gas turbines are also used to propel ships, particularly military vessels. Military requirements of high speed outweigh the fuel and construction cost disadvantages that make gas turbines too expensive for commercial ships. Gas turbines can also be used as tank engines, for example in the American M1A1 Abrams tank or the Russian T-80. However, their high fuel use provides an engineering challenge, particularly at idle speed. The M1A1 tanks have been retrofitted with batteries to use for idling, and the Russian T-80 was replaced by the diesel-engine-powered T-90.

Turbochargers. Their low fuel efficiency makes gas turbines unsuitable for car propulsion. However, small gas turbines working as turbochargers are commonly added to increase the power of diesel car engines. The power from the turbine is used to compress the air taken in by the diesel engine, increasing its performance.

Fascinating Facts About Gas Turbine Technology

- Gas turbines are small mechanical wonders, with up to 4,000 individual parts often handled and fitted by hand. Maintaining them requires the same great technical skills as building them.
- With the Messerschmitt Me 262, Germany developed the world's first jet fighter, but by the end of World War II, the British had their own with the Gloster Meteor. The two jets never fought each other, though.
- Industry analysts expect that by 2018, about 7,000 state-of-the-art jet engines will be built and commissioned every year, the vast majority of the turbofan design.
- The AGT1500 gas turbine power plant can propel the American M1A1 Abrams battle tank up to 41.5 miles per hour across level terrain. However, even the U.S. Army balked at the fuel costs and retrofitted the engine with a battery pack to power the tank while idling.
- Since the late eighteenth century folk craftsmen in Germany's Erz Mountains have used the principle of the gas turbine to move small wooden Christmas ornaments: They revolve on disks powered by rotor blades on a shaft driven by hot air rising from candles.
- Once engineers designed ways to utilize the remaining heat from the exhaust of a gas turbine employed for power generation, overall system efficiency almost doubled to as much as 65 percent.
- A contemporary gas turbine airfoil blade is grown as a single crystal from a molten super alloy to achieve great heat resistance in the turbine.

Impact on Industry

Gas turbine technology has a vast global impact in the fields of electricity, industrial power generation, and civil and military aviation. Because of the capital and research-intensive nature of the field, gas turbines for most applications are designed and built by a few large global companies with headquarters in North America, Western Europe, Russia, Japan, India, and China. However, gas turbines are used worldwide, both on the ground and powering most of the world's commercial airliners and military aircraft. The key engineering challenges that drive research across public and private institutions are to make gas turbines more efficient and operate with fewer emissions and noise.

Government Agencies. Traditionally, governments of industrialized countries at war or in prolonged political antagonism toward another have been prime sponsors of gas turbine technology, especially in the field of the jet engine. World War II and the Cold War saw key government support for jet engine and gas turbine design to gain a military advantage in the air and on water. In the West, gas turbine technology was one of the key products of the military-industrial complex, as Western governments sponsored research by private industrial corporations, and these military products were later adapted for civilian use. Governmental focus in industrial nations has generally shifted away from military applications and toward sponsoring research in cleaner, "greener," and more efficient gas turbines for electric utilities and power generation. These governments funnel considerable sums into public and commercial research and set policy guidelines for cleaner power generation that drives respective innovations.

Universities and Research Institutes. Public research in almost all industrially advanced nations is focused on improving efficiency, emissions, and innovation in gas turbine technology. One new research avenue concerned micro gas turbines which, if they could be constructed at affordable cost and with reasonable efficiency, may become a source of decentralized and customized power generation. There was primary research to discover a technologically feasible way for gas turbines to run on hydrogen, which would constitute a technological quantum leap. There was also great research in material sciences, particularly ceramics, to create new components for gas turbines that could cope with even higher temperatures and pressure. Research also centered on integrating gas turbines in ever-more efficient power plant assemblies. Gas turbine technology was also pursued in aeronautics for improved jet engine designs.

Industry and Business. The power industry has represented a huge market for gas turbines. From 2000 to 2009, the total global electric output of gas turbines rose from 2.5 million megawatts to 3.7 million megawatts per year and is expected to rise to 5 million megawatts by 2020. This created a huge demand for the design and manufacture of new gas turbines as well as their operation and maintenance around the world. Even though the annual market value for new gas turbines was reduced by almost half during the 2007-2009 recession (from about $20 billion to just above $10 billion), economic recovery increased this value again.

In the aviation industry, the market for jet engines has been very robust. For the period from 2007 to 2016, market analysts have predicted an annual global value of new engine sales ranging from a conservative $18 billion to an enthusiastic $30 billion. Of this, 85 percent are for civilian and 15 percent for military aircraft. As all operating jet engines have to be maintained regularly, there was an annual global market value of about $25 billion for engine maintenance, repair, and operations (MRO) in 2005, which analysts expected to grow as the world added more operating aircraft.

Major Corporations. America's General Electric has been the market leader both in manufacturing of gas turbines for power generation (at more than 40 percent market share throughout the 2000's) and in jet engine manufacturing (holding about 30 percent market share during the same time). In the power industry, Germany's Siemens company, Japan's Mitsubishi Heavy Industries, France's Alstom, and India's state-owned Bharat Heavy Electricals have been major players.

England's Rolls-Royce Group has been a close contender for market-share leader in jet engines, achieving this feat with a 32 percent share in 2001. Another key jet engine manufacturing company is America's Pratt & Whitney. Other niche players include Germany's Daimler, Russia's NK Engines Company, Japan's Kawasaki Heavy Industries, or mainland China's Shenyang Aeroengine Research Institute. Cooperation and partnership among companies in the gas turbine manufacturing industry have been very common.

Careers and Course Work

Gas turbine technology has been a key to two of the world's leading industries, power and aviation, so job demand in the field should remain very strong. Students interested in the field should take science courses in high school, particularly physics, as well as mathematics and computer science. An associate's degree in an engineering or science field (engineering or industrial technology) will provide a good entry.

A bachelor of science in an engineering discipline, particularly mechanical, electrical, or computer engineering, is excellent preparation for an advanced job. A B. S. in physics or mathematics would point to a more theoretically informed career, perhaps in design. A bachelor of arts in environmental studies is also useful, as emission control is becoming a major part of gas turbine technology. A minor in any science is always beneficial.

If one's career focus is on advanced work, a master of science in mechanical, electrical, and computer engineering, or in environmental science and management, could be chosen. For top scientific positions, a Ph.D. in these disciplines, together with some postdoctoral work, is advisable.

Because of the global nature of the field, students should maintain a general openness to work abroad or in somewhat remote locations, with the exception of those purely interested in design. The field can also be attractive to students with expertise in support functions, including those who have earned a B.A. in English, communications, economics, or biology. A master of business administration would serve as preparation for the business end of the field.

Social Context and Future Prospects

As more nations industrialize and global development continues, the demand for power and mobility, including air travel, is expected to increase. Especially with the key applications of jet engines and gas turbines for power generation, the field of gas turbine technology is likely to keep its great relevancy. The quest for more efficient gas turbines that combust their fuel with as little emissions as possible will continue to motivate major developments in the field. If gas turbines can become an ever-more efficient and low-emission source of generating power they have the potential, like fuel cells, to become part of the next generation of power sources. There is also much promise linked to micro gas turbines as a source of efficient, affordable, clean, and decentralized power.

For jet engines, the design challenge is to reduce fuel consumption and noise and increase power. There is ongoing research to employ gas turbines in combination with electric hybrid car engines to lower overall carbon dioxide emissions from personal transport. This will grow in importance as more people in developing nations acquire cars.

R. C. Lutz, B.A., M.A., Ph.D.

Further Reading

Boyce, Meherwan P. *Gas Turbine Engineering Handbook.* 3d ed. Burlington, Mass.: Gulf Professional Publishing, 2006. Focuses on design, components, operation, and maintenance issues with emphasis on gas turbines used in combined cycle power plants. Each chapter has own bibliography.

Giampaolo, Tony. *Gas Turbine Handbook: Principles and Practices.* 4th ed. Boca Raton, Fla.: CRC Press, 2009. Comprehensive coverage of technology and all applications. Clear organization, figures, bibliography.

Kehlhofer, Rolf, et al. *Combined-Cycle Gas and Steam Turbine Power Plants.* 3d ed. Tulsa, Okla.: PennWell, 2009. Application of gas turbine as part of cutting-edge power-generation technology. Presents concepts, components, applications, operations of these plants. Each chapter has illustrative figures.

Peng, William W. *Fundamentals of Turbomachinery.* Hoboken, N.J.: John Wiley & Sons, 2008. Chapter 8 discusses gas turbines and covers thermodynamics, design, efficiency, and performance, as well as applications; illustrated.

Rangwala, A. S. *Turbo-Machinery Dynamics: Design and Operation.* New York: McGraw-Hill, 2005. Comprehensive textbook, very useful presentation of the field. Covers all applications, component design, materials, and manufacture for gas turbines. Excellent illustrations.

Soares, Claire. *Gas Turbines: A Handbook of Air, Land and Sea Applications.* Burlington, Mass.: Butterworth-Heinemann, 2008. Good, comprehensive overview of all aspects of the field from a practical viewpoint.

Web Sites

Gas Turbine Association
http://www.gasturbine.org

Gas Turbine Builders Association
http://www.gtba.co.uk

International Gas Turbine Institute
http://igti.asme.org

GENETICALLY MODIFIED FOOD PRODUCTION

FIELDS OF STUDY

Genetic engineering; genetic manipulation; gene technology; recombinant DNA technology; biotechnology; food production; agriculture.

SUMMARY

Genetically modified food production is a subset of biotechnology and genetic engineering. This developing field offers both hope and concern for global food production. Genetically modified food production is the direct result of the development of genetically modified organisms. Conventional plant breeding is a slow process, and it take several years to develop plants with desirable traits. Advances in genetic engineering have allowed scientists to speed up the process of developing plants with the most desirable traits and with greater predictability. These plants help farmers increase production and obtain higher yields. However, the use of genetically modified organisms in food production is controversial in some countries because the effect of these organisms on humans has not been assessed.

KEY TERMS AND CONCEPTS

- **Biotechnology:** Technology that uses biological systems, living organisms, or derivatives thereof to create or modify organisms, existing products, or processes.
- **DNA (Deoxyribonucleic Acid):** Nucleic acid found in all cells that contains the genetic material for the growth and functioning of an organism.
- **Gene:** Piece of the DNA code that regulates biological processes in living organisms and contains the information needed to make a specific protein in that organism; it is the functional and physical unit of heredity passed from a parent to its offspring.
- **Genetically Modified Organism (GMO):** Organism that has been modified or altered using genetic engineering techniques for a specific purpose; also known as a genetically engineered organism.
- **Genetic Engineering:** Technique used to remove, modify, or add genes to a DNA molecule of a living organism to change its genetic contents and enable it to make different proteins or perform different functions.
- **Protein:** Large molecule containing one or more chains of amino acids arranged in a specific order; formed according to information coded in a gene.
- **Recombinant DNA:** Form of DNA that has been created by combining two or more DNA sequences that do not usually occur together.
- **Recombinant Protein:** Protein derived from recombinant DNA.
- **Transgenic Plant Or Animal:** Plant or animal that has been altered by the introduction of a gene from another species.

Definition and Basic Principles

Genetically modified food production is the creation of food products using genetically modified organisms. Some of the food production problems that can be addressed using genetically modified organisms are limiting or eliminating the damage caused by pests, weeds, and diseases, and providing tolerance to specific herbicides, extreme temperatures, and drought, as well as other production-related issues.

One of the initial goals of developing genetically modified plants was to achieve higher crop yields by creating versions of plants such as corn and soybeans that offered greater resistance to diseases caused by pests and viruses. Genetically modified foods, also known as genetically engineered food, are developed, produced, and marketed mainly because they present an advantage over traditionally produced food to either the farmer/producer or the consumer. The benefit to the producers comes from increased productivity and the reduction of lost crops. The consumer benefits from genetically engineered food by having access to food with better nutritional value and, in some cases, lower prices.

Background and History

People began harvesting plants and domesticating animals around 10,000 B.C.E., and they soon were selecting the best seed and breeding the best animals through a process of trial and error. In the latter part of the nineteenth century, the monk Gregor Mendel used peas to demonstrate heredity in plants, thereby laying the foundation for modern plant breeding and genetics. The early work of Mendel and others led to the development of commercial crops by the 1930's. The field of molecular biology, which emerged in the twentieth century, has led to a better understanding of the cells and molecular processes of living organisms. This understanding has allowed researchers to develop genetically engineered plants and animals to address and solve many of the problems faced by farmers in crop and livestock production.

The first field trials of crops developed using genetically modified organisms took place in 1990, and by 1992, the first genetically modified corn was approved for use by the U.S. Food and Drug Administration (FDA). In the 1990's, additional advances led to genetically engineered vaccines and hormones as well as to the cloning of animals. Since their introduction in the United States, genetically modified crop plants have been widely used by American farmers, and the percentage of acreage planted with genetically modified crops has been steadily increasing, reaching more than 90 percent for some crops. Soybeans and cotton genetically modified to tolerate herbicides are the two most widely adopted genetically modified crops. Cotton and corn with insect resistance are the next most common crops grown by American farmers.

How It Works

Genetic engineering allows scientists to insert a gene from one organism into another, resulting in a genetically modified organism. Therefore, genetic engineering begins with the identification and isolation of a gene that expresses a desirable trait. This gene can be found in a relative of the target species or in a completely unrelated species. A recipient plant or animal is selected, and the gene is inserted and incorporated into the genome of the recipient. The desired gene is inserted by various techniques such as using *Agrobacterium* as a vector or using a gene gun. The newly inserted gene becomes part of the genome of the recipient and is regulated in the same way as its other genes. Genetic engineering confers a new ability on the organism that has received the new gene. One advantage of genetic engineering is that genes can be introduced in a plant even if they do not occur in the genome of the target plant.

The use of genetically modified food production, a new and valuable tool in agriculture, fisheries, and forestry production, has allowed significant improvements in food production to meet the needs of the ever-expanding world population. With conventional plant breeding techniques, researchers crossbred plants, taking five to seven years to generate a plant with the desired traits. In conventional breeding, half of an individual's genes come from each parent, but in genetically modified organisms, one or more genetically desirable traits has been added to the genetic material of the desired plant. One of the main differences between conventional breeding and genetic engineering is that in the conventional process, crosses are possible only between close relatives, whereas with genetic engineering, scientists can transfer genes between plants that are not related and might not be able to crossbreed in nature.

For example, in the case of Bt-corn, a gene from a naturally occurring soil bacterium, *Bacillus thuringiensis,* was inserted into corn to provide resistance to the corn borer. The gene from the bacterium produces a protein, Bt delta endotoxin, which kills the European and southwestern corn borer larvae. Bt-corn eliminates the need to spray insecticides to control corn borers. Although planting these crops reduces the amount of pesticides released into the environment, the long-term effects of Bt-corn on human health and the environment are not known.

Companies involved in the research and development of plant and animals derived from genetic engineering patent these new products and processes. The patent allows them to protect their investment; however, it costs farmers who use the seed. A contract between the farmers and these companies prohibits the farmers from saving seed for use the following year, reselling seed to a third party, or exchanging seed with other farmers.

Scientists envisage numerous future applications of genetically modified food. For example, food could be used to produce drugs to address human health problems including infectious diseases. One of the crops that is being considered for such an application is bananas, which could be used to produce a human vaccine.

Applications and Products

Over the years, many biotechnology firms and well-established companies have become involved in

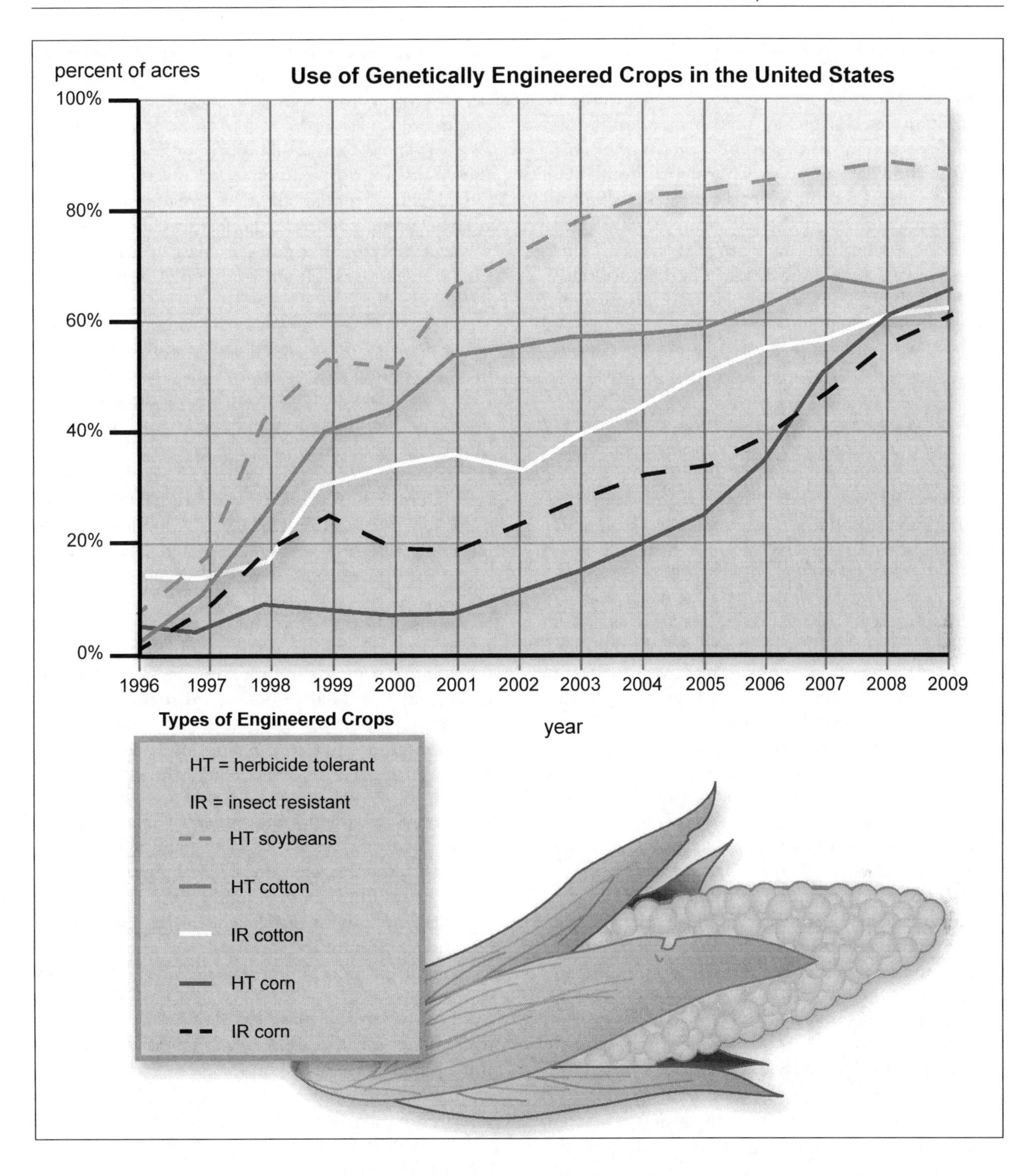

research and development using genetically modified organisms. The genetic engineering technique and biotechnology are likely to have many potential commercial uses. Genetically modified organisms have many applications in food production and livestock. Many products, including plants and animals, have been developed using genetic engineering techniques.

Insect Resistance. The use of biotechnology to achieve insect resistance in food plants destined for human or animal consumption is accomplished by incorporating a gene into a particular food plant such as corn. In the case of Bt-corn, the gene for toxin production from the bacterium *B. thuringiensis* was incorporated into the corn plant. This toxin is an insecticide commonly used in agriculture and is safe for human consumption. Genetically modified plants that permanently produce this toxin have been shown to require lower quantities of insecticides in specific situations, for example, where pest pressure is high.

Virus Resistance. Virus resistance is achieved through the introduction of a gene from certain viruses that cause disease in plants. Virus resistance makes plants less susceptible to diseases caused by such viruses, resulting in higher crop yields.

Herbicide Tolerance. Herbicide tolerance is achieved through the introduction of a gene from a bacterium conveying resistance to some herbicides. In situations where weed pressure is high, the use of such crops has resulted in a reduction in the quantity of herbicides used. For example, corn resistant to the popular Monsanto weedkiller Roundup (glyphosate) has been developed so that farmers can treat their crops with Roundup without damaging their corn crop. This means that farmers can reduce the number and amount of herbicides used in any given year and situation.

Genetically Modified Animals. Genetic engineering of animals has taken place for different purposes and using a number of different species such as cattle, sheep, goats, rabbits, pigs, chickens, and fish. For example, well-performing bulls have been cloned to create better breeding stock, and animals have been used to produce useful human proteins. Research using genetic engineering in cattle production is trying to produce cows that are resistant to mad cow disease or that have the capacity to produce milk with higher levels of protein.

Scientists have been able to genetically engineer a variety of salmon that grows at twice the rate of Atlantic salmon. The fish is no bigger than the Atlantic salmon, but it reaches that size in half the time. Scientists inserted a Chinook salmon growth hormone gene into a fertilized egg of an Atlantic salmon. To ensure that the gene remains active all year round, scientists added a "switch" from the ocean pout to the Atlantic salmon. This genetically engineered fish reduces the production cost of salmon. AquaBounty Technologies, which developed the AquAdvantage salmon applied to the FDA for permission to market the fish. In September, 2010, the FDA concluded that the fish was safe to eat but felt that more scientific research was needed, particularly on the possible environmental impact of the modified salmon.

Possible Risks. Most scientists and regulators agree that the use of genetically modified organisms in food production may pose some risks. These potential risks usually fall into two basic categories: the effects on human and animal health and the impact on the environment where the genetically modified organisms are grown. Scientists and regulators advise that care must be exercised to reduce these risks, especially the possibility of transferring toxins or allergenic compounds used in genetically modified organisms to ordinary plants or animals. This cross-contamination could result in unexpected allergic reactions in humans and animals. One of the major risks to natural resources and the environment is the possibility of outcrossing, the transfer of genes from genetically modified organisms to regular crops or related wild species. For example, the use of herbicide-resistant corn and soybeans raises the possibility of outcrossing, which could lead to the development of more aggressive weeds or herbicide-resistant wild relatives of these cultivated plants. This outcrossing could upset the balance of the natural ecosystem. The introduction of genetically modified plants could also lead to a loss of biodiversity as traditional varieties of plants are displaced by a smaller number of genetically modified varieties.

Genetically modified organisms are not always advantageous: The cost of research and development can be prohibitively high, much cheaper ways to control the undesired pests or diseases may exist, and the still unknown effects on humans and the environment may potentially result in lawsuits against the developers of the plant or animal.

IMPACT ON INDUSTRY

The private sector is responsible for the major investments in research and development in the field of genetically modified food production. These investments have targeted mostly agriculture plants and livestock to address the quality of these plants and animals and their vulnerability to disease.

Government and University Research. The U.S. government was quick to create a research and regulatory framework under which modified organisms could be used in food production. Under this framework, corporations have funded research at American universities that has played a major role in the development of genetically modified food. However, some academics are concerned about the power of these corporations. The U.S. Department of Agriculture (USDA), the agency trusted with regulation of genetically modified food production, does not conduct any tests of its own to assess the risks posed by these crops but instead relies on data provided by the industry. Some scientists have pointed out that as long as the USDA relies on these companies to provide data on the risks and benefits of genetically modified crops, the agency cannot be sure that the reports it is receiving are not biased. In addition, university researchers have complained of company intervention in their research. In at least one case, several companies did not grant a researcher permission to conduct research using their seeds. In another case, a researcher was asked to sign an agreement that he would not test the seed if he wanted access to it. Companies have control over the seed of crops derived from genetic engineering because of the patent rights they hold.

Industry and Business Sector. The industry and business sector has been working very hard to develop new products and techniques, as well as solutions to some unsolved problems. For example, the seed industry is developing new techniques and products to prevent disease from destroying crops, to make crops tolerant to specific herbicides, and to improve the nutritional quality of certain foods using genetic engineering. These new crops are helping solve hunger problems in various countries around the world.

Major Corporations. U.S. and European corporations have played a major role in the research and development of genetically modified crops. These corporations are primarily seed companies and producers of agricultural chemicals.

Syngenta, a global seed and agricultural chemicals company based in Switzerland, has created corn resistant to the European corn borer by inserting the *cry1Ab* gene from *B. thuringiensis*, corn resistant to rootworm by transforming corn with a modified *cry3A* gene, and insect-resistant and herbicide-tolerant maize by conventionally crossbreeding genetically modified parents. Aventis CropScience, a France-based crop production and technologies company, developed maize that is both resistant to insects and tolerant of glufosinate ammonium herbicide by inserting genes encoding Cry9C protein from *B. thuringiensis* subspecies tolworthi and phosphinothricin acetyltransferase (PAT) from *Streptomyces hygroscopicus*. It also created an insect-resistant and herbicide-tolerant maize by conventional crossbreeding of genetically modified parental lines.

Pioneer Hi-Bred International, an Iowa-based developer and supplier of plant genetics, created maize

Fascinating Facts About Genetically Modified Food Production

- As of 2010, pest-resistant Bt-cotton was produced on about 93 percent of the acreage devoted to cotton production in the United States.
- Genetically engineered soybeans resistant to the herbicide glyphosate (Roundup) occupied about 93 percent of the soybean acreage in the United States in 2010.
- Vitamin-enriched corn has been derived from a South African white corn hybrid (M37W). The new corn has bright orange kernels that have 169 times more beta carotene, 6 times more vitamin C, and twice the folate content of the regular corn.
- Rice, a major staple in Asian countries, has been modified by the introduction of three new genes, two from daffodils and one from a bacterium, to create Golden Rice, which is high in beta carotene, a precursor to vitamin A.
- It is estimated that genetic engineering has increased U.S. farmers' income by $1.5 billion a year.
- The Grocery Manufacturers Association estimates that about 80 percent of processed foods in the United States contain genetically modified organisms.
- In New Zealand, cows were genetically modified to produce milk with a higher protein content, which allows for the production of more cheese from the same amount of milk.

that is male-sterile and tolerant of glufosinate ammonium herbicide by inserting genes encoding DNA adenine methylase and PAT from *Escherichia coli* and *Streptomyces viridochromogenes*, respectively. In collaboration with Pioneer, Dow Agrosciences, an Indiana-based subsidiary of Dow Chemical, produced a rootworm-resistant maize by inserting the *cry34Ab1* and *cry35Ab1* genes from *B. thuringiensis*.

Monsanto, a U.S.-based agricultural biotechnology company, produced maize resistant to the European corn borer by inserting the *cry1Ab* gene from *B. thuringiensis* subspecies kurstaki and a rootworm-resistant maize by inserting the *cry3Bb1* gene from *B. thuringiensis* subspecies kumamotoensis. It used conventional crossbreeding of genetically modified parent plants to create an insect-resistant and enhanced lysine content maize and an insect-resistant and glyphosate-tolerant maize. In 1998, Monsanto bought Dekalb Genetics and acquired rights to an herbicide-tolerant maize and a insect-resistant and herbicide-tolerant maize that Dekalb Genetics had developed.

International Organizations. Several international organizations address issues related to genetically modified food production. Among them are the World Trade Organization, the World Health Organization, the Food and Agriculture Organization of the United Nations, the Codex Alimentarius Commission, the World Organisation for Animal Health, and the International Plant Protection Convention. These organizations address genetically modified food production from different angles: its environmental impact, the effect on human health, and the regulatory aspects.

Over the years and with the involvement of the World Trade Organization, the opportunities for global agricultural trade have increased dramatically. The organization has mainly concerned itself with reducing tariffs and subsidies, but in 1994, it adopted the Agreement on the Application of Sanitary and Phytosanitary Measures, which establishes that countries retain their right to ensure that the food, animal, and plant products that they import are safe. The agreement also states that countries should use internationally agreed standards developed by the Codex Alimentarius Commission for food safety, the World Organisation for Animal Health for animal health, and the International Plant Protection Convention for plant health.

Careers and Course Work

Career pathways in agricultural production, biotechnology, food production, gene technology, genetic engineering, genetic manipulation, and recombinant DNA technology offer great opportunities for students interested in issues dealing with environmental challenges, human health, and reducing world hunger. These fields are all rapidly changing with new applications, techniques, and innovations being developed all the time. These fields of study can be loosely termed "biotechnology." Students can prepare themselves for a career in biotechnology in many ways. The opportunities in the field include positions as government research scientists, corporate scientists, laboratory technicians, engineers, process technicians, and maintenance and instrumental technicians. Students can chose among a great many private and public sector employers and a wide range of work environments and types of jobs.

A biotechnology degree can lead to employment in the private sector, with drug and chemical companies, seed and agricultural companies, or environmental remediation companies; in government agencies such as the USDA's Agricultural Research Services or the Food and Drug Administration; or in educational institutions, teaching and doing research.

Social Context and Future Prospects

The social debate about the use of genetically modified food centers on the level of risk to health and the environment that they present and whether these types of food are necessary. The possible risks to human health presented by genetically modified food fall into two categories, direct and indirect. Direct effects include toxicity, an allergic reaction (also called allergenicity), and negative nutritional effects, and indirect effects include the stability of the inserted gene, outcrossing, and unintended effects as a result of the gene insertion. Therefore, genetically modified foods must be tested very carefully and thoroughly to ensure that the benefits of such plants outweigh their risks and the hidden costs of developing them.

The debate about whether genetically modified food needs to be created to deal with hunger among people in the developing world is taking place in many venues, both political and scientific. Some scientists opposed to genetically engineered food argue that there is more than enough food in the world and that the hunger crisis in some countries is the result of problems with food distribution and the politics in those countries rather than production levels and systems. They also argue that offering food with unknown levels of risk to those in need is unethical. This argument assumes that genetically modified foods have risks not present in traditional foods; however, the proponents of genetically modified food argue that traditional foods are not devoid of risk. However, as production of genetically modified foods has increased and no major adverse effects have emerged, some of the earlier criticism has disappeared. However, a consensus does not exist among U.S. scientists on the risks and the safety of genetically modified plants and animals.

Lakhdar Boukerrou, B.Sc., M.S., Ph.D.

Further Reading

Fedoroff, Nina, and Nancy Marie Brown. *Mendel in the Kitchen: A Scientist's View of Genetically Modified Food.* 2004. Reprint. Washington, D.C.: Joseph Henry Press, 2006. Fedoroff, an expert in plant molecular biology and genetics, teams with a science writer to dispel many misconceptions about genetically modified foods and describes them as an "environmentally conservative" way to increase the food supply.

Henningfeld, Diane Andrews, ed. *Genetically Modified Food.* Detroit: Greenhaven Press, 2009. Contains a collection of essays that looks at the issues surrounding genetically modified food from both sides, pro and con.

King, Robert C., William D. Stansfield, and Pamela Khipple Mulligan. *A Dictionary of Genetics.* 7th ed. New York: Oxford University Press, 2006. Provides information on all aspects of genetics in alphabetical order by topic.

Lurquin, Paul F. *High Tech Harvest: Understanding Genetically Modified Food Plants.* Boulder, Colo.: Westview Press, 2002. A research biologist examines genetically modified plants, explaining the science behind them. He is generally supportive of biotechnology.

Weasel, Lisa. *Food Fray: Inside the Controversy over Genetically Modified Food.* New York: American Management Association, 2009. Provides a historical perspective on the debate over genetically modified food. Has some very useful information on the use of genetically modified food in places such as India and Africa as well as on the future of genetically modified food.

Web Sites

Food and Agriculture Organization of the United Nations
Biotechnology
http://www.fao.org/biotech/stat.asp

GMO Compass
http://www.gmo-compass.org/eng/home

U.S. Department of Agriculture, Economic Research Service
Adoption of Genetically Engineered Crops in the United States
http://www.ers.usda.gov/Data/BiotechCrops

U.S. Food and Drug Administration
Genetically Engineered Foods
http://www.fda.gov/NewsEvents/Testimony/ucm115032.htm

World Health Organization
Twenty Questions on Genetically Modified (GM) Foods
http://www.who.int/foodsafety/publications/biotech/en/20questions_en.pdf

See also: Food Preservation; Food Science; Plant Breeding and Propagation.

GLASS AND GLASSMAKING

FIELDS OF STUDY

Optics; physics; mechanical engineering; electrical engineering; materials science; ceramics; chemistry; architecture; mathematics; thermodynamics; microscopy; spectroscopy.

SUMMARY

Glassmaking is a diverse field with applications ranging from optics to art. Most commercially produced glass is used to make windows, lenses, and food containers such as bottles and jars. Glass is also a key component in products such as fiber-optic cables and medical devices. Because many types of glass are recyclable, the demand for glass is expected to increase. Most careers in glassmaking require significant hands-on experience as well as formal education.

KEY TERMS AND CONCEPTS

- **Ceramic:** Material made from nonorganic, nonmetallic compounds subjected to high levels of heat.
- **Crystal:** Solid formed from molecules in a repeating, three-dimensional pattern; also refers to a high-quality grade of lead glass.
- **Glazier:** Professional who makes, cuts, installs, and replaces glass products.
- **Pellucidity:** Rate at which light can pass through a substance; also known as transparency.
- **Silica:** Silicon dioxide, the primary chemical ingredient in common glass.
- **Tempered Glass:** Glass strengthened by intensive heat treatment during manufacturing.
- **Viscosity:** Rate at which a fluid resists flow; sometimes described as thickness.
- **Vitreous:** Being glasslike in texture or containing glass.

Definition and Basic Principles

Glass is one of the most widely used materials in residential and commercial buildings, vehicles, and many different devices. Glass windows provide natural light while protecting indoor environments from changes in the outside temperature and humidity. The transparency of glass also makes it a good choice for lightbulbs, light fixtures, and lenses for items such as eyeglasses. A nonporous and nonreactive substance, glass is an ideal material for food packaging and preparation. Most types of glass can be recycled with no loss in purity or quality, a factor that has increased its appeal.

Although glass resembles crystalline substances found in nature, it does not have the chemical properties of a crystal. Glass is formed through a fusion process involving inorganic chemical compounds such as silica, sodium carbonate, and calcium oxide (also known as lime). The compounds are heated to form a liquid. When the liquid is cooled rapidly, it becomes a solid but retains certain physical characteristics of a liquid, a process known as glass transformation.

Background and History

The earliest known glass artifacts come from Egypt and eastern Mesopotamia, where craftspeople began making objects such as beads more than five thousand years ago. Around 1500 B.C.E., people began dipping metal rods into molten sand to create bottles. In the third century B.C.E., craftspeople in Babylon discovered that blowing air into molten glass was a rapid, inexpensive way to make hollow shapes.

Glassmaking techniques quickly spread throughout Europe with the expansion of the Roman Empire in the first through fourth centuries. The region that would later be known as Italy, led by the city of Venice, dominated the glass trade in Europe and the Americas for several hundred years.

With the development of the split mold in 1821, individual glass objects no longer needed to be blown and shaped by hand. The automation of glass manufacturing allowed American tinsmith John L. Mason to introduce the Mason jar in 1858. Over the next few decades, further innovations in glassmaking led to a wide range of glass applications at increasingly lower costs.

One of the most noteworthy advancements in glass use came in the 1960's and 1970's with the design and rollout of fiber-optic cable for long-range communications.

How It Works

Most types of glass have translucent properties, which means that certain frequencies, or colors, of light can pass through them. Transparent glass can transmit all frequencies of visible light. Other types of glass act as filters for certain colors of light so that when objects are viewed through them, the objects appear to be tinted a specific shade. Some types of glass transmit light but scatter its rays so that objects on the other side are not visible to the human eye. Glass is often smooth to the touch because surface tension, a feature similar to that of water, binds its molecules together during the cooling process. Unless combined with certain other compounds, glass is brittle in texture.

There is a widespread but incorrect belief that glass is a liquid. Many types of glass are made by heating a mixture to a liquid state, then allowing it to reach a supercooled state in which it cannot flow. The process, known in glassmaking as the glass transition, causes the molecules to organize themselves into a form that does not follow an extended pattern. In this state, known as the vitreous or amorphous state, glass behaves like a solid because of its hardness and its tendency to break under force.

Chemical Characteristics. The most common type of glass is known as soda-lime glass and is made primarily of silica (60 to 75 percent). Sodium carbonate, or soda ash, is added to the silica to lower its melting point. Because the presence of soda ash makes it possible for water to dissolve glass, a third compound such as calcium carbonate, or limestone, must be added to increase insolubility and hardness. For some types of glass, compounds such as lead oxide or boric oxide are added to enhance properties such as brilliance and resistance to heat.

Manufacturing Processes. The melting and cooling of a mixture to a liquid, then supercooled state is the oldest form of glassmaking. Most glass products, including soda-lime, are made using this type of process. As the liquid mixture reaches the supercooled state, it is often treated to remove stresses that could weaken the glass item in its final form. This process is known as annealing. The item's surface is smoothed and polished through a stream of pressurized nitrogen. Once the glass transformation is complete, the item can be coated or laminated to increase traits such as strength, electrical conductivity, or chemical durability.

Specialized glass can be made by processes such as vapor deposition or sol-gel (solution-gel). Under vapor deposition, chemicals are combined without being melted. This approach allows for the creation of thin films that can be used in industrial settings. Glass created through a sol-gel process is made by combining a chemical solution, often a metal oxide, with a compound that causes the oxide to convert to a gel. High-precision lenses and mirrors are examples of sol-gel glass.

Applications and Products

Glass is one of the most widely used materials in the world. Some of the most common uses for glass include its incorporation into buildings and other structures, vehicles such as automobiles, fiberglass packaging materials, consumer and industrial optics, and fiber optics.

Construction. The homebuilding industry relies on glass for the design and installation of windows, external doors, skylights, sunrooms, porches, mirrors, bathroom and shower doors, shelving, and display cases. Many office buildings have floor-to-ceiling windows or are paneled with glass. Glass windows allow internal temperatures and humidity levels to be controlled while still allowing natural light to enter a room. Light fixtures frequently include glass components, particularly bulbs and shades, because of the translucent quality of glass and its low manufacturing costs.

Automotive Glass. Windshields and windows in automobiles, trucks, and other vehicles are nearly always made of glass. Safety glass is used for windshields. Most safety glass consists of transparent glass layered with thin sheets of a nonglass substance such as polyvinyl butyral. The nonglass layer, or laminate, keeps the windshield from shattering if something strikes it and protects drivers and passengers from injury in an accident. A car's side and back windows generally are not made of safety glass but rather of tempered glass treated to be heat resistant and to block frequencies of light such as ultraviolet rays.

Fiberglass. Russell Games Slayter, an employee of Owens-Corning, invented a glass-fiber product in 1938 that the company named Fiberglas. The product was originally sold as thermal insulation for buildings and helped phase out the use of asbestos, a carcinogen. The tiny pockets of air created by the material's fabriclike glass filaments are the source of its insulating properties. Fiberglass is also used in the manufacturing of vehicle bodies, boat hulls, and sporting goods such as fishing poles and archery bows because of its light weight and durability. Because recycled glass makes up a significant portion of many fiberglass products, the material is considered environmentally friendly.

Packaging. Processed foods and beverages are often sold in glass bottles and jars. The clarity of glass allows consumers to see the product and judge its quality. Glass is a nonporous, nonreactive material that does not affect the flavor, aroma, or consistency of the product it contains. Consumers also prefer glass because it can be recycled. A disadvantage of glass packaging is the ease with which containers can be shattered. Glass also is heavier than competing packaging materials such as plastic, aluminum, and paper.

Consumer Optics. Glass lenses can be found in a wide range of consumer-oriented optical products ranging from eyeglasses to telescopes. The use of glass in eyeglasses is diminishing as more lenses are made from specialized plastics. Cameras, however, rarely use plastic lenses, as they tend to create lower quality images and are at greater risk of being scratched. Binoculars, microscopes, and telescopes designed for consumer use are likely to contain a series of glass lenses. These lenses are often treated with coatings that minimize glare and improve the quality of the image being viewed.

Industrial Optics. For the same reasons that glass is preferred in consumer optics, it is the material of choice for industrial applications. Many specialized lenses and mirrors used for telescopes, microscopes, and lasers are asymmetrical or aspherical in shape. The manufacture of aspherical lenses for high-precision equipment was difficult and expensive until the 1990's, when glass engineers developed new techniques based on the processing of preformed glass shapes at relatively low temperatures. These techniques left the surface of the glass smooth and made it more cost-effective to manufacture highly customized lenses.

Fiber Optics. The optical fibers in fiber-optic communication infrastructures are made from glass. The fibers carry data in the form of light pulses from one end of a glass strand to the other. The light pulses then jump through a spliced connection to the next glass fiber. Fiber-optic cable can transmit information more quickly and over a longer distance than electrical wire or cable. Glass fibers can carry more than one channel of data when multiple frequencies of light are used. They also are not subject to electromagnetic interference such as lightning strikes, an advantage when fiber-optic cable is used to wire offices in skyscrapers.

Fascinating Facts About Glass and Glassmaking

- The oldest known glassmaking handbook was etched onto tablets and kept in the library of Ashurbanipal, king of Assyria in the seventh century B.C.E.
- In 1608, settlers in Jamestown, Virginia, opened the first glassmaking workshop in the American colonies. It produced bottles used by pharmacists for medicines and tonics.
- Nearly 36 percent of glass soft drink and beer bottles and 28 percent of all glass containers were recycled in 2008. About 80 percent of recovered glass is used for new bottles.
- Toledo, Ohio, is nicknamed the Glass City because of its industrial heritage of glassmaking. In the early twentieth century, Toledo began providing glass to automobile factories in nearby Detroit.
- China manufactures about 45 percent of all glass in the world. Most Chinese glass is installed in automobiles made locally.
- The first sunglasses were built from lenses made out of quartz stone and tinted by smoke. The glasses were worn by judges in fourteenth-century China to hide their facial expressions during court proceedings.
- The National Aeronautics and Space Administration has developed NuStar (Nuclear Spectroscopic Telescope Array), a set of two space telescopes carrying a new type of glass lens made from 133 glass layers. NuStar is about one thousand times as sensitive as space telescopes such as Hubble.

IMPACT ON INDUSTRY

The worldwide market for glass is divided by regions and product types. By its nature, glass is heavy and fragile. It is subject to breaking when shipped over long distances. For this reason, most glass is manufactured in plants located as close to customers as possible.

Flat glass is one of the largest glass markets

globally. An estimated 53 million tons of flat glass, valued at $66 billion, were made in 2008. The majority of flat glass, about 70 percent of the total market, was used for windows in homes and industrial buildings. Furniture, light fixtures, and other interior products made up 20 percent of the market, while the rest (10 percent) was used primarily in the automotive industry. The manufacturing of flat glass requires significant capital investment in large plants, so the market is dominated by a few large companies. In 2009, the world's leading makers of flat glass were the NSG Group (based in Japan), Saint-Gobain (France), Asahi Glass (Japan), and Guardian Industries (United States). The market is growing at about 4 to 5 percent each year. Much of this growth is fueled by demand from China, where the construction and automotive industries are expanding at a faster rate than in North America and Europe. New flat glass products incorporate nonglass materials such as polymers, which enhance safety and the insulating properties of windows.

The glass packaging industry worldwide is led by bottles used for beverage products. In 2009, glass container manufacturers produced more than 150 billion beer bottles and more than 60 billion bottles for spirits, wines, and carbonated drinks. Research firm Euromonitor International predicts that demand for glass packaging will grow, but growth rates across package types will vary significantly. Glass bottles face competition from polyethylene terephthalate (PET) bottles, which are much lighter and more durable. A rise in median consumer income across the Asia-Pacific region in the 2000's has increased demand for processed food products sold in jars and bottles.

Niches such as glass labware are also facing pressure from items made with nonglass materials. The global market for glass labware could be as high as $3.87 billion by 2015, according to Global Industry Analysts. New plastics developed for specialized applications such as the containment of hazardous chemicals may slow down the rate at which laboratories purchase glass containers. However, glass will continue to be a laboratory staple as long as it remains inexpensive and widely available.

Fiber-optic cable continues to fuel growth for glass manufacturing. In late 2010, many telecommunications firms were preparing to invest in expanding their fiber-optic networks. Telecom research firm CRU estimated that companies worldwide ordered more than 70 million miles of new fiber-optic cable between July, 2009, and June, 2010. The growth in demand in Europe was nearly twice that of the United States in 2009 and 2010.

Careers and Course Work

The skills required to work in the glassmaking industry depend on the nature of the glass products being made. Manufacturing and installing glass windows in buildings is a career within the construction industry, while the design of glass homewares involves the consumer products industry. The making of lenses for eyeglasses is a career track that differs significantly from the production of specialty lenses for precision equipment such as telescopes.

Glaziers in construction gain much of their knowledge from on-the-job training. Many glaziers enter apprenticeships of about three years with a contractor. The apprentice glazier learns skills such as glass cutting by first practicing on discarded glass. Tasks increase in difficulty and responsibility as the apprentice's skills grow. This work is supplemented by classroom instruction on topics such as mathematics and the design of construction blueprints. Similarly, makers of stemware and high-end glass products used in homes learn many of their skills through hands-on work with experienced designers.

Careers in optical glass and lens design, on the whole, require more formal education than other areas of glass production do. Several universities offer course concentrations in glass science, optics, or ceramics within undergraduate engineering programs. Graduate programs can focus on fields as narrow as electro-optics (fewer than ten American schools offer this option).

Students planning careers in optical glass take general classes such as calculus, physics, and chemistry. Course work on topics such as thermodynamics, microscopy, spectroscopy, and computer-aided design are also standard parts of an optical engineer's education.

Social Context and Future Prospects

Demand for glaziers in the construction industry rises and falls with the rate of new buildings being built. The early 2000's saw a sharp increase in demand for glaziers, particularly those with experience in installing windows in new houses. This trend reversed with the end of the construction boom in the late 2000's. At the same time, the development of window glass with energy-efficient features led to the upgrading of windows in office buildings and homes, which has increased the need for skilled glaziers. The U.S. Bureau of Labor Statistics predicts that jobs in this field will grow at about 8 percent from 2008 to 2018, an average rate compared to all other occupations.

Within the field of optical glass, manufacturers have developed a wide range of glass types for specialized applications such as lasers. The need for increasingly precise lenses has grown steadily with developments in fields ranging from microsurgery to astronomy.

Julia A. Rosenthal, B.A., M.S.

Further Reading

Beretta, Marco. *The Alchemy of Glass: Counterfeit, Imitation, and Transmutation in Ancient Glassmaking.* Sagamore Beach, Mass.: Science History Publications/USA, 2009. Examines the history of glassmaking, from Egypt and Babylonia to modern times, looking at the effects of technology.

Hartmann, Peter, et al. "Optical Glass and Glass Ceramic Historical Aspects and Recent Developments: A Schott View." *Applied Optics* (June 1, 2010): D157-D176. A brief but informative overview of optical glass from the Italian Renaissance to the modern times.

Le Bourhis, Eric. *Glass: Mechanics and Technology.* Weinheim, Germany: Wiley-VCH, 2008. Explores the properties of glass and their physical and chemical sources.

Opie, Jennifer Hawkins. *Contemporary International Glass.* London: Victoria and Albert Museum, 2004. Photographs and artistic statements from sixty international glass artists, based on pieces in the collection of the Victoria and Albert Museum in London.

Shelby, J. E. *Introduction to Glass Science and Technology.* Cambridge, England: Royal Society of Chemistry, 2005. A college-level textbook with detailed information on the physical and chemical properties of glass.

Skrabec, Quentin, Jr. *Michael Owens and the Glass Industry.* Gretna, La.: Pelican, 2007. A biography of American industrialist Michael Owens, whose mass production techniques for glass bottles and cans led to major changes in the lives of consumers.

Web Sites

American Scientific Glassblowers Society
http://www.asgs-glass.org

Association for the History of Glass
http://www.historyofglass.org.uk

Corning Museum of Glass
http://www.cmog.org

Society of Glass Technology
http://www.societyofglasstechnology.org.uk

GRAPHICS TECHNOLOGIES

FIELDS OF STUDY

Mathematics; physics; computer programming; graphic arts and animation; anatomy and physiology; optics.

SUMMARY

Graphics technology, which includes computer-generated imagery, has become an essential technology of the motion picture and video-gaming industries, of television, and of virtual reality. The production of such images, and especially of animated images, is a complex process that demands a sound understanding of not only physics and mathematics but also anatomy and physiology.

KEY TERMS AND CONCEPTS:

- **Anti-Aliasing:** A programming technique for producing smooth curve lines in an image by designating the color of an edge pixel according to how much of the pixel extends beyond the defined curve.
- **Morphing:** The process of converting one image object into another through a series of tweened images.
- **Resolution:** The fineness of detail that can be displayed in an image, determined by the number of pixels contained within a specified area.
- **Sprite:** A small static image, like a desktop icon, that is used to animate subjects that have a constant appearance.
- **Tweening:** The process of creating images to fill in the space between key images of a sequence, used in animations.

Definition and Basic Principles

While graphics technologies include all of the theoretical principles and physical methods used to produce images, it more specifically refers to the principles and methods associated with digital or computer-generated images. Digital graphics are displayed as a limited array of colored picture elements (pixels). The greater the number of pixels that are used for an image, the greater the resolution of the image and the finer the detail that can be portrayed. The data that specifies the attributes of each individual pixel are stored in an electronic file using one of several specific formats. Each file format has its own characteristics with regard to how the image data can be manipulated and utilized.

Because the content of images is intended to portray real-world objects, the data for each image must be mathematically manipulated to reflect real-world structures and physics. The rendering of images, especially for photo-realistic animation, is thus a calculation-intensive process. For images that are not produced photographically, special techniques and applications are continually being developed to produce image content that looks and moves as though it were real.

Background and History

Imaging is as old as the human race. Static graphics have historically been the norm up to the invention of the devices that could make a series of still pictures appear to move. The invention of celluloid in the late nineteenth century provided the material for photographic film, with the invention of motion picture cameras and projectors to follow. Animated films, commonly known as cartoons, have been produced since the early twentieth century by repeatedly photographing a series of hand-drawn cels. With the development of the digital computer and color displays in the last half of the century, it became possible to generate images without the need for hand-drawn intermediaries.

Computer graphics in the twenty-first century can produce images that are indistinguishable from traditional photographs of real objects. The methodology continues to develop in step with the development of new computer technology and new programming methods that make use of the computing abilities of the technology.

How It Works

Images are produced initially as still or static images. Human perception requires about one-thirtieth of a second to process the visual information obtained through the seeing of a still image. If a sequential series of static images is displayed at a rate that exceeds the frequency of thirty images per second, the images are perceived as continuous motion. This is the basic principle of motion pictures, which are nothing more than displays of a sequential series of still pictures. Computer-generated still images (now indistinguishable from still photographs since the advent of digital cameras) have the same relationship to computer animation.

Images are presented on a computer screen as an array of colored dots called pixels (an abbreviation of "picture elements"). The clarity, or resolution, of the image depends on the number of pixels that it contains within a defined area. The more pixels within a defined area, the smaller each pixel must be and the finer the detail that can be displayed. Modern digital cameras typically capture image data in an array of between 5 and 20 megapixels. The electronic data file of the image contains the specific color, hue, saturation, and brightness designations for each pixel in the associated image, as well as other information about the image itself.

To obtain photorealistic representation in computer-generated images, effects must be applied that correspond to the mathematical laws of physics. In still images, computational techniques such as ray tracing and reflection must be used to imitate the effect of light sources and reflective surfaces. For the virtual reality of the image to be effective, all of the actual physical characteristics that the subject would have if it were real must be clearly defined as well so that when the particular graphics application being used renders the image to the screen, all of the various parts of the image are displayed in their proper positions.

To achieve photorealistic effects in animation, the corresponding motion of each pixel must be coordinated with the defined surfaces of the virtual object, and their positions must be calculated for each frame of the animation. Because the motions of the objects would be strictly governed by the mathematics of physics in the real world, so must the motions of the virtual objects. For example, an animated image of a round object bouncing down a street must appear to obey the laws of gravity and Newtonian mechanics. Thus, the same mathematical equations that apply to the motion and properties of the real object must also apply to the virtual object.

Other essential techniques are required to produce realistic animated images. When two virtual objects are designed to interact as though they are real, solid objects, clipping instructions identify where the virtual solid surfaces of the objects are located; the instructions then mandate the clipping of any corresponding portions of an image to prevent the objects from seeming to pass through each other. Surface textures are mapped and associated with underlying data in such a way that movement corresponds to real body movements and surface responses. Image animation to produce realistic skin and hair effects is based on a sound understanding of anatomy and physiology and represents a specialized field of graphics technology.

Applications and Products

Software. The vast majority of products and applications related to graphics technology are software applications created specifically to manipulate electronic data so that it produces realistic images and animations. The software ranges from basic paint programs installed on most personal computers (PCs) to full-featured programs that produce wireframe structures, map surface textures, coordinate behaviors, movements of surfaces to underlying structures, and 360-degree, three-dimensional animated views of the resulting images.

Other types of software applications are used to design objects and processes that are to be produced as real objects. Computer-assisted design is commonly used to generate construction-specification drawings and to design printed-circuit boards, electronic circuits and integrated circuits, complex machines, and many other real-world constructs. The features and capabilities of individual applications vary.

The simplest applications produce only a static image of a schematic layout, while the most advanced are capable of modeling the behavior of the system being designed in real time. The latter are increasingly useful in designing and virtual-testing such dynamic systems as advanced jet engines and industrial processes. One significant benefit that has accrued from the use of such applications has been the ability to refine the efficiency of systems such as production lines in manufacturing facilities.

Hardware. The computational requirements of graphics can quickly exceed the capabilities of any particular computer system. This is especially true of PCs. Modern graphics technology in this area makes use of separate graphics processing units (GPUs) to

Fascinating Facts About Graphic Technologies

- Graphic technologies began with simple cave drawings and has progressed to full-motion, photorealistic, CGI motion pictures.
- The brain is tricked into seeing a rapid sequential series of still images as motion by the physiological process called persistence of vision.
- Jaggies are steplike edges that result when rectangular pixels are used to portray curves in images.
- Ray-tracing applies the mathematics of optical physics to a text file to produce realistic lighting effects in a corresponding image.
- A fractal is an infinitely detailed, self-similar shape that results when an image operation is repeated many times.
- Graphics technology is rapidly progressing to the point in which printed books will be able to incorporate video images displayed as part of the printed page.

handle the computational load of graphics display. This allows the PC's central processing unit (CPU) to carry out the other computational requirements of the application without having to switch back and forth between graphic and nongraphic tasks.

Many graphics boards also include dedicated memory for exclusive use in graphics processing. This eliminates the need for large sectors of a PC's random access memory (RAM) to be used for storing graphics data, a requirement that can render a computer practically unusable.

Another requirement of hardware is an instruction system to operate the various components so that they function together. For graphics applications, with the long periods of time they require to carry out the calculations needed to render a detailed image, it is essential that the computer's operating system be functionally stable. The main operating systems of PCs are Microsoft Windows, its open-source competitor Linux, and the Apple Mac OS. Some versions of other operating systems, such as Sun Microsystems' Solaris, have been made available but do not account for a significant share of the PC market.

The huge amounts of graphics and rendering required for large-scale projects such as motion pictures demand the services of mainframe computers. The operating systems for these units have a longer history than do PC operating systems. Mainframe computers function primarily with the UNIX operating system, although many now run under some variant of the Linux operating system. UNIX and Linux are similar operating systems, the main difference being that UNIX is a proprietary system whereas Linux is open-source.

Motion Pictures and Television. Graphics technology is a hardware- and software-intensive field. The modern motion picture industry would not be possible without the digital technology that has been developed since 1980. While live-action images are still recorded on standard photographic film in the traditional way, motion picture special effects and animation have become the exclusive realm of digital graphics technologies. The use of computer generated imagery (CGI) in motion pictures has driven the development of new technology and continually raised the standards of image quality. Amalgamating live action with CGI through digital processing and manipulation enables film-makers to produce motion pictures in which live characters interact seamlessly with virtual characters, sometimes in entirely fantastic environments. Examples of such motion pictures are numerous in the science-fiction and fantasy film genre, but the technique is finding application in all areas, especially in educational programming.

Video Gaming and Virtual Training. The most graphics-intensive application is video gaming. All video games, in all genres, exist only as the graphic representation of complex program code. The variety of video game types ranges from straightforward computer versions of simple card games to complex three-dimensional virtual worlds.

Many graphics software applications are developed for the use of game designers, but they have also made their way into many other imaging uses. The same software that is used to create a fictional virtual world can also be used to create virtual copies of the real world. This technology has been adapted for use in pilot- and driver-training programs in all aspects of transportation. Military, police, and security personnel are given extensive practical and scenario training through the use of virtual simulators. A simulator uses video and graphic displays of actual terrain to give the person being trained hands-on experience without endangering either personnel or actual machinery.

Impact on Industry

Graphics technology has revolutionized the way in which industry functions by streamlining processes for speed and efficiency. Barcoding is one very simple graphics technology by which this has been achieved. By scanning a barcode with a laser barcode-reader, the pertinent information about that object is available and automatically updated in just a few seconds. Barcode technology is universally applicable and has been used for tasks ranging from pricing goods in stores and cataloging library books, to animal identification tattooing and monitoring goods in transport.

Graphics technology in product design has greatly facilitated the development of new products and made existing processes both more efficient and cost effective. Printed circuit-board layout is one area in which graphics technology is important. Using a software application designed for the purpose, an electronics engineer or technologist can lay out the circuitry and component placement of a new circuit board in a short time and then use the same software to run virtual tests of the performance of the new design. The preparation and assembly of an actual circuit board, even a simple one, is a material- and time-consuming process when carried out for the production of a small number of test cases. Virtual testing of the graphic design bypasses that aspect of development and enables the rapid prototyping of the design in development.

Process modeling is closely related to virtual testing. With the use of process modeling, graphic software applications, engineers, and designers can create complex systems in the virtual world and then animate them to see how they work. The complex mathematics of the physics involved in the operation of the system is built into the model to make it as accurate as possible. This allows designers to make changes to the system to optimize its function before it is built rather than having to disassemble and reconstruct various components afterward.

The economic value of graphics technology is demonstrated by the motion picture industry and the video-gaming industry. The top-grossing films have historically been science-fiction and fantasy titles that use CGI graphics. The cost of production of any one of these films is measured in tens of millions of dollars, and in some cases, in hundreds of millions of dollars. The gross proceeds of these same films are often ten times as much, and in a few cases have moved into the billion-dollar range. The special effects field is a multibillion dollar industry in its own right, intimately linked to the computer hardware and software industries.

The video-gaming industry is the second area of note. A survey conducted by the industry trade magazine *Game Developer* and published in April, 2011, reported that more than 48,000 game developers were active throughout North America alone. This is a low number in actuality because it does not include developers in other nations, many of whom produce more video game titles than are produced in North America. The survey presented data for the associated professions of programming, art and animation, and design within the video game industry. Of those, programmers garnered an average annual salary of $85,733; artists and animators earned $71,354 annually, on average; and game designers earned an annual average of $70,223. If one carries out the simple calculation using the overall average of these groups ($75,770), then game developers represent a combined annual economic value of at least $3.6 billion.

Video game development and graphics have also impacted the course of postsecondary education. *Game Developer*, for example, routinely carries advertisements from several colleges and universities for dedicated programs of study in video game design at both undergraduate and graduate levels. The magazine also carries recruitment ads for several video-game production companies.

Careers and Coursework

Many colleges and universities now offer undergraduate and graduate degree programs in applied graphics technology, and in many other traditional programs, graphics technology holds a central position as a tool of the trade. Students interested in pursuing studies aimed at the development of graphics technology must have an advanced understanding of real-world physics and mathematics as a minimum requirement. While it is possible to acquire computer programming skills and expertise with graphics software programs without formal training, it is unlikely that anything less than formal training in those areas will be acceptable to established game-development organizations and other graphics-technology users. Students should therefore expect to take courses in advanced programming to become proficient in assembly language and the Java and C/C++ programming languages. All of these areas are utilized in the production of graphics programming.

Depending on the desired area of specialization, students should also expect to study human and animal anatomy, physiology, kinesiology (the study of movement), and general graphic arts.

Students interested in the technical side of graphics technology should take courses of study in electronics engineering and technology and in computer science. These areas also require a basic knowledge of physics and mathematics and will include a significant component of design principles.

Social Context and Future Prospects

Graphics technology is inextricably linked to the computer and digital electronics industries. Accordingly, graphics technology changes at a rate that at minimum equals the rate of change in those industries. Since 1980, graphics technology using computers has developed from the display of just sixteen colors on color television screens, yielding blocky image components and very slow animation effects, to photorealistic full-motion video, with the capacity to display more colors than the human eye can perceive and to display real-time animation in intricate detail. The rate of change in graphics technology exceeds that of computer technology because it also depends on the development of newer algorithms and coding strategies. Each of these changes produces a corresponding new set of applications and upgrades to graphic technology systems, in addition to the changes introduced to the technology itself.

Each successive generation of computer processors has introduced new architectures and capabilities that exceed those of the preceding generation, requiring that applications update both their capabilities and the manner in which those capabilities are performed. At the same time, advances in the technology of display devices require that graphics applications keep pace to display the best renderings possible. All of these factors combine to produce the unparalleled value of graphics technologies in modern society and into the future.

Richard M. Renneboog, M.Sc.

Further Reading

Abrash, Michael. *Michael Abrash's Graphics Programming Black Book.* Albany, N.Y.: Coriolis Group, 1997. Consisting of seventy short chapters, the book covers graphics programming up to the time of the Pentium processor.

Brown, Eric. "True Physics." *Game Developer* 17, no. 5 (May, 2010): 13-18. This article provides a clear explanation of the relationship between real-world physics and the motion of objects in animation.

Jimenez, Jorge, et al. "Destroy All Jaggies." *Game Developer* 18, no. 6 (June/July, 2011): 13-20. This article describes the method of anti-aliasing in animation graphics.

Oliver, Dick, et al. *Tricks of the Graphics Gurus.* Carmel, Ind.: Sams, 1993. Provides detailed explanations of the mathematics of several computer graphics processes.

Ryan, Dan. *History of Computer Graphics: DLR Associates Series.* Bloomington, Ind.: AuthorHouse, 2011. A comprehensive history of computer graphics and graphics technologies.

See Also: Computer-Aided Design and Manufacturing; Lithography; Photography; Television Technologies.

HAZARDOUS-WASTE DISPOSAL

FIELDS OF STUDY

Biology; chemistry; sanitary and sewage engineering; emergency response; environmental science; geology; hydrology; law; toxicology.

SUMMARY

Hazardous-waste disposal involves transporting hazardous materials to appropriate facilities for treatment, recycling, and possible storage in a manner that will protect the environment and public health. Materials that are radioactive, toxic, corrosive, ignitable, irritating, poisonous, and infectious are examples of hazardous wastes that threaten human health. Moreover, some hazardous wastes, such as mercury, present a danger because they accumulate in the environment when they are improperly managed. Although manufacturing companies are most often thought to be the entities that generate hazard wastes, there is a diversity of generators from farms to nuclear power plants to households.

KEY TERMS AND CONCEPTS

- **Corrosive Wastes:** Acidic wastes that are able to corrode metal storage containers.
- **Generators:** Entities that produce hazardous materials from physical processes or disposal of nonrenewable resources.
- **Hazardous Wastes:** Wastes that exhibit one of four characteristics: ignitability, corrosivity, reactivity, or toxicity.
- **Household Hazardous Wastes:** Consumer products that have the characteristics of hazardous wastes.
- **Ignitable Wastes:** Wastes that can spontaneously combust and have low flash points.
- **Reactive Wastes:** Unstable wastes that can explode or release toxic gases when heated, compressed, or mixed with water.
- **Toxic Materials:** Materials harmful to humans through ingestion or absorption and to the environment through groundwater pollution.
- **Transporter:** Company that removes and transports hazardous wastes for generators.

Definition and Basic Principles

Hazardous-waste disposal is the cradle-to-grave management of materials that threaten human health and the environment. Hazardous wastes are ubiquitous and their proper disposal is a global issue. In fact, industrialized nations are still involved in cleanup of hazardous wastes that contaminated land and waters before disposal laws were adopted and enforced. One method of recycling hazardous waste sites is to clean up the wastes and designate the land as a brownfield site that can be reused by commercial and industrial enterprises.

Arriving at effective methods for disposal of hazardous wastes is difficult, because these wastes come in many forms, including solid, liquid, sludge, and gas. Governments regulate hazardous wastes and require their disposal in designated facilities. Hazardous wastes can be stored in sealed containers in the ground or recycled into new products such as fertilizers. The methods used and locations for disposing of hazardous wastes are not the same as those employed in solid waste disposal. Nonhazardous solid wastes or trash can be disposed of in nonsecure landfills, while hazardous wastes should be deposited in secure landfills so they do not contaminate the groundwater.

Background and History

Disposal of wastes has existed since ancient cultures burned, recycled, and buried their trash. After World War II, toxic materials were manufactured in greater quantities, and industrialized nations allowed uncontrolled dumping, burning, and disposal of hazardous wastes until the public demanded regulation because of human health and environmental threats. In the United States, Congress charged the Environmental Protection Agency (EPA) with regulating hazardous-waste disposal through the Resource Conservation and Recovery Act (RCRA) of 1976, which amended the Solid Waste Disposal Act of 1965. The EPA is authorized to control hazardous wastes from cradle to grave under RCRA, which means regulating generators, transporters, and the facilities involved in treatment, recycling, storage, and disposal.

From the 1940's through the 1960's research showed a relationship between hazardous-waste disposal and contamination of groundwater. By employing hydrology studies of underground water movement, scientists were able to pinpoint the sources of hazardous waste contamination. In 1980, the EPA established the Superfund program for the clean up of sites contaminated with hazardous wastes.

How It Works

Manifest Disposal System. In the United States, hazardous-waste disposal is governed by the RCRA manifest system. Generators must prepare a form that discloses the type and quality of wastes to be transported to an off-site facility for treatment, storage, recycling, or disposal subject to the Hazardous Materials Transportation Act (HMTA). Because the generator remains liable for proper disposal, a copy of the manifest form, signed by all handlers of the hazardous wastes, is returned to the generator to verify its delivery. However, some wastes are not transported but are treated and disposed of on-site.

Disposal methods for hazardous wastes vary depending on the dangers posed by the wastes. The goal is to remove the characteristics of the wastes that are harmful to human health and the environment. Some methods include treatment, disposal in secure landfills, and incineration.

Treatment. Treatment involves neutralization of chemical, biological, and physical characteristics that make a waste hazardous to the environment. For example, acids that cause corrosion can be neutralized with basic substances, reagents, or through pH adjustment. Cement has also been used to decrease toxicity of some hazardous wastes through stabilization, often those found in sludge. Industrial hazardous wastes may receive either chemical treatments, such as chlorination, oxidation, and chemical bonding, or the application of physical techniques not limited to distillation and filtration. Medical wastes are considered biohazards that can cause the spread of disease and present different treatment issues. Treatment techniques for these biological wastes include steam sterilization and chemical decontamination before disposal.

Landfill Disposal. Hazardous materials that are not stored on the surface can be disposed of in hazardous-waste landfills. The materials must first be sequestered from nonhazardous wastes and be treated. Moreover, because hazardous wastes can interact with each other they must be segregated and stored by type before they are disposed of in a landfill. The hazardous wastes are then placed in secure storage containers that are buried in landfills with plastic and clay liners that are thicker than the ones used in solid-waste landfills to prevent leaching into the groundwater.

Incineration. Incinerating some hazardous wastes, such as oils and solvents, at high temperatures will reduce the amount of hazardous material through destruction. Although gases released through incineration can generate energy, incinerating facilities must comply with the Clean Air Act. Starved air incineration, which controls the combustion rate of hazardous wastes, is one of the newer technologies that aids in reducing the amount of air pollution caused by incineration. Another treatment method known as pyrolysis may be used to destroy concentrated organic wastes such as polychlorinated biphenyls (PCBs) and pesticides. Pyrolysis is different from high-temperature incineration in that it employs ultrahigh temperatures in the absence of oxygen, often under pressure.

Applications and Products

Energy. Hazardous wastes such as solvents or used oil are often burned directly to produce heat or electricity. This disposal application is regulated throughout the world, and in the United States, the EPA requires the combustion units employed in the burning of hazardous materials to comply with specific standards that prevent harmful air pollution. Some companies, however, have developed proprietary processes that treat the hazardous materials using plasma-enhanced waste-recovery systems that turn hazardous wastes into nontoxic, synthetic, gas-alternative energy sources. Unlike direct burning, these new applications claim to destroy all hazardous components that might pollute the air.

Storage. Some hazardous wastes are disposed of through storage. Storage facilities must be able to withstand natural disasters such as seismic events. Storage drums and containers must be secure from fire and water intrusion, and an entire industry has developed to provide appropriate containers that will not leak into the groundwater or explode on impact. Hazardous wastes must first be characterized before they are disposed of in storage containers. Containers used to store hazardous wastes must be designed to accommodate diverse characteristics such as toxicity, corrosivity, and ignitability, and the stored materials should not react with the container or be mixed with other incompatible materials. Storage containers must be sealed and appropriately labeled with the contents and characteristics of the stored hazardous wastes, and a permit is required in the United States to allow for permanent storage. Liquid hazardous wastes, however, are often disposed of in underground injection wells. The EPA regulates the construction, operations, and closure of such wells, which are also subject to regulation under the Safe Drinking Water Act.

Remediation. Many companies specialize in hazardous-waste remediation by providing a variety of disposal services and equipment, not limited to catalytic oxidizers and carbon adsorption systems. Remediation includes reduction and cleanup of hazardous wastewater by using oil-water separators to separate oil and solids from wastewater effluents of petroleum-based industries. Other hazardous-waste remediation services include demolition and removal of hazardous materials such as asbestos, removal of leaky fuel tanks, cleanup of contaminated soils, and the construction of slurry walls to aid in the remediation of groundwater that is polluted with hazardous waste. Bioremediation is also a growing field that involves disposal of medical and laboratory wastes often by using microorganisms to break down the hazardous materials.

Recycling and Reclamation. Recycling and reclamation are important disposal applications because they help reduce hazardous wastes and the amount of raw materials that are consumed. Some materials that may be classified as hazardous wastes can be recycled and used again, such as rechargeable batteries and heating and air-conditioning thermostats. Spent solvents are an example of a product that can be made pure and reused for the same purpose, as can used oil, which can also be processed to create new oil-based products. In addition, the EPA allows some hazardous materials to be treated, recycled, and disposed of on the land in fertilizers or as an ingredient in asphalt. Components of hazardous wastes can also be reclaimed for use in new products. Examples include mercury reclaimed from thermometers, silver from photographic fixers, and scrap metal from auto bodies or manufacturing processes.

Telephones, televisions, audiovisual equipment, computers, computer components, circuit boards, and handheld devices that store music and books are all examples of products that are disposed of by businesses and individual consumers as electronic

Fascinating Facts About Hazardous-Waste Disposal

- Annual hazardous-waste generation in the United States could fill the Louisiana Superdome more than 1,500 times.
- Estimates of annual e-waste disposal include more than 100 million phones in European nations and 30 million computers in the United States. These wastes contain hazardous materials such as lead, cadmium, mercury, and chromium. Computer monitors contain about four to five pounds of lead.
- The Environmental Protection Agency (EPA) has estimated that about 1.6 million tons of household hazardous wastes are generated in America annually, and include paints, pesticides, cleaning products, and used motor oil.
- Billions of batteries are thrown away each year, resulting in a majority of the mercury and cadmium in public landfills.
- When household hazardous-waste products are poured down sinks into storm sewers or onto the ground, it takes very little of the wastes to contaminate groundwater and soil.
- A single gallon of motor oil can contaminate 250 gallons of drinking water.
- About 5 million tons of oil are dumped or spilled into the world's oceans each year.
- Golf courses generate many hazardous wastes including pesticides, herbicides, and gasoline and engine oil.

wastes (e-wastes) when they become obsolete. E-wastes, however, contain hazardous materials such as lead, cadmium, and some precious metals that can be reclaimed and reused. Recycling of e-wastes is an important hazardous-waste disposal application throughout the developed world to mitigate the growth of this type of waste.

Hazardous Material (Hazmat) Products. Hazardous-waste disposal is not always carried out effectively, so an entire industry has grown out of spills that need to be cleaned up. Hazmat team members will wear differing types of hazmat suits, often with ventilators, depending on the characteristics of the hazardous material to be cleaned up. Specially manufactured boots, gloves, and socks are also part of the necessary clothing. Those involved in hazmat-cleanup applications will use spill kits that enable first responders to determine whether the hazardous material presents a chemical, biological, or radioactive hazard. These products are employed in many situations that include not only hazardous-material spills, but also disposal of illegal drugs and abandoned storage drums and cleanup of any dump site that contains suspicious materials. After the hazardous wastes have been cleaned up and properly disposed of, decontamination is necessary. Decontamination products include temporary shelters, showers, and cleaning solutions.

Impact on Industry

Although there are global initiatives to ensure adequate hazardous-waste disposal, the United States has some of the strictest laws dealing with the subject, while developing nations such as China, are struggling with environmental damage due to lack of environmental laws and improper disposal of hazardous materials.

Government Agencies. In the United States, both the EPA and the Nuclear Regulatory Commission (NRC) play significant roles in enforcing laws pertaining to hazardous-waste disposal. The NRC is instrumental in regulating radioactive-waste disposal. Each state has its own agency to regulate hazardous-waste disposal, and municipalities throughout the country are involved not only with regulation, but also with many hazardous-waste disposal activities including collection and recycling. The Occupational Safety and Health Administration (OSHA) is involved in emergency preparedness and response to hazardous-waste accidents. In addition, the Departments of Defense and Energy play a role in ensuring legal disposal of hazardous materials.

University Research. During the 1990's, when the EPA was involved in cleanup of contaminated sites throughout the United States, universities were actively involved in developing new techniques for effective hazardous-waste disposal and aiding in remediation of polluted sites. As of 2011, universities have established hazardous-waste research facilities, but much of the research is concerned with RCRA's mission of mitigating hazardous wastes and better recycling of wastes and reusable components to reduce hazardous-waste disposal in landfills. Moreover, educational institutions are working with local governments to educate the public in the proper recycling of household hazardous wastes.

International Organizations and Treaties. The United Nations is involved in treaties concerning the international disposal of hazardous materials and certifying containers that may be used in the transportation of hazardous wastes by airplanes, ground vehicles, and ships. The United States is one of several nations that disposes of some of its hazardous waste in other countries. These countries usually welcome the additional revenue that compensates for such disposal, but they have little expertise in proper disposal techniques. Some nations have entered into treaties concerning hazardous-waste imports and exports to prevent developed nations from taking advantage of some of the underdeveloped countries. One of these agreements is known as the Bamako Convention, which bans the import of hazardous wastes into Africa and regulates trans-boundary movement of hazardous wastes within Africa. The Basel Convention, which preceded the Bamako Convention, also regulates hazardous-waste imports and exports but not radioactive wastes, which are often exported to the less developed countries.

Generators. There are many types of businesses and industries that generate hazardous wastes. These include manufacturers, oil refineries, professional offices, commercial facilities such as dry cleaners, service industries including beauty salons, automobile repair shops, and exterminators and medical facilities, hospitals, and laboratories. Based on the RCRA manifest system, these generators have cradle-to-grave responsibility for the proper disposal of the hazardous wastes in the United States.

Transporters. Hazardous-waste transportation is a growing business. In the United States, the Hazardous Materials Transportation Act (HMTA) requires strict compliance with federal laws and applies not only to transporters of hazardous materials but also to generators who engage the services of such transportation companies. Transporters must be issued an identification number and are required to use the RCRA manifest system. The Department of Transportation is involved in ensuring compliance with the HMTA.

Careers and Course Work

Hazardous-waste disposal involves interdisciplinary course work such as engineering courses concerning the characteristics and sources of hazardous wastes and their treatment, destruction, and recycling; biology and chemistry classes, especially related to human-health issues and the environment; and law and economics studies that consider disposal regulations and their costs. Certifications that may be necessary for a career in this field include environmental hazardous materials technology and emergency response.

Career opportunities are growing in this field including positions with agencies at all levels of government and with waste-management firms and engineering companies involved in transporting, treating, recycling, and disposing of hazardous wastes and remediating and cleaning up sites where there has been contamination. Hazardous-waste generators, such as utility companies, hospitals, and manufacturers, often hire in-house hazardous-waste experts. Educational institutions and laboratories also employ educators and researchers with doctoral degrees in engineering and hazardous-waste management.

Those interested in careers as chemical, environmental, sanitary, and sewage engineers will need at least a bachelor's degree to obtain a job as a hazardous-waste management specialist. Careers that require a minimum of an associate's degree and some additional training and certifications include workers involved in collecting, transporting, treating, and destroying hazardous wastes, in addition to emergency response and cleanup. Common job titles for hazardous-waste workers are hazardous-waste technician, field technician, and environmental technician.

Social Context and Future Prospects

Nations must continue to strive for hazardous-waste disposal that proactively reduces the amount of hazardous wastes and, where possible, recycles these wastes to conserve the consumption of raw materials. Although globally some progress has been made in ensuring proper hazardous-waste disposal, the funds needed for education of the public and cleanup of improperly managed hazardous-waste sites are not always available. International cooperation is also necessary to resolve several hazardous-waste disposal issues. Among these matters are the disposal of radioactive and nuclear wastes and the exporting of hazardous materials by developed nations to underdeveloped countries that are incapable of proper disposal. As of 2011, it is estimated that less than one-quarter of global e-wastes are recycled. Appropriate laws, disposal systems, and international treaties are needed to manage these rapidly growing e-wastes, which are mostly disposed of in landfills.

Disposal of household hazardous wastes (HHWs), such as used batteries, cleaning products, and medications, continues to be problematic because they contaminate groundwater when disposed of in landfills. Although the United States federal government exempts HHWs from regulation, some municipalities require the separation of household hazardous wastes from other solid wastes for collection or disposal at a municipal facility. Better public education and enhanced laws will help mitigate the damage caused by unregulated HHW disposal.

Carol A. Rolf, B.S.L.A., M.B.A., M.Ed., J.D.

Further Reading

Blackman, William C., Jr. *Basic Hazardous Waste Management.* 3d ed. Boca Raton, Fla.: CRC Press, 2001. Overview of hazardous-waste management technologies with discussion concerning disposal of radioactive and biomedical wastes.

Cabaniss, Amy D., ed. *Handbook on Household Hazardous Waste.* Lanham, Md.: Government Institutes, 2008. Comprehensive discussion concerning household hazardous-waste disposal that encourages responsible disposal.

LaGrega, Michael D., Philip L. Buckingham, and Jeffrey C. Evans. 2d ed. *Hazardous Waste Management.* Long Grove, Ill.: Waveland Press, 2010. Text written for students interested in resolving hazardous-waste problems based on case studies and treatment solutions.

Miller, Jeffrey G., and Craig N. Johnston. *Law of Hazardous Waste Disposal and Remediation: Cases, Legislation, Regulations, Policies.* 2d ed. St. Paul, Minn.: Thomson/West, 2005. Introductory text discussing federal and state laws pertaining to hazardous-waste disposal and cleanup.

Spellman, Frank. *Transportation of Hazardous Materials Post-9/11.* Lanham, Md.: Rowman & Littlefield, 2007. Useful information concerning current hazardous materials, transportation regulations, and management of security threats.

Voyles, James K. *Managing Your Hazardous Wastes: A Step-by-Step RCRA Compliance Guide.* 2d ed. Rockville, Md.: Government Institutes, 2002. Guide for developing management plans that comply with laws regulating the generation and processing of hazardous wastes.

Web Sites

Environmental Protection Agency
Wastes–Hazardous Wastes
http://www.epa.gov/epawaste/hazard/index.htm

Nuclear Regulatory Commission
Radioactive Waste
http://www.nrc.gov/waste.html

Occupational Safety and Health Administration
Safety and Health Topics: Hazardous Waste
http://www.osha.gov/SLTC/hazardouswaste/index.html

Status Clean
Hazardous Wastes
http://www.statusclean.com/waste-management/waste-disposal/hazardous-waste.aspx

HEAT-EXCHANGER TECHNOLOGIES

FIELDS OF STUDY

Thermodynamics; fluid mechanics; heat transfer; mechanics of materials; calculus; differential equations.

SUMMARY

A heat exchanger transfers thermal energy from one flowing fluid to another. A car radiator transfers thermal energy from the engine-cooling water to the atmosphere. A nuclear power plant contains very large heat exchangers called steam generators, which transfer thermal energy out of the water that circulates through the reactor core and makes steam that drives the turbines. In some heat exchangers, the two fluids mix together, while in others, the fluids are separated by a solid surface such as a tube wall.

KEY TERMS AND CONCEPTS

- **Closed Heat Exchanger:** Exchanger in which the two fluids are separated by a solid surface.
- **Counterflow:** Fluids flowing in opposite directions.
- **Cross-Flow:** Fluids flowing perpendicular to each other.
- **Deaerator (Direct-Contact Heater):** Open heat exchanger in which steam is mixed with water to bring the water to its boiling point.
- **Open Heat Exchanger:** Exchanger in which the two fluids mix together and exit as a single stream.
- **Parallel Flow (Co-Current):** Both fluids flowing in the same direction.
- **Plate Heat Exchanger:** Exchanger in which corrugated flat metal sheets separate the two flowing fluids.
- **Shell:** Relatively large, usually cylindrical, enclosure with many small tubes running through it.
- **Shell And Tube Heat Exchanger:** Exchanger composed of many relatively small tubes running through a much larger enclosure called a shell.
- **Tube Sheet:** Flat plate with many holes drilled in it. Tubes are inserted through the holes.

Definition and Basic Principles

There are three modes of heat transfer. Conduction is the method of heat transfer within a solid. In closed heat exchangers, conduction is how thermal energy moves through the solid boundary that separates the two fluids. Convection is the method for transferring heat between a fluid and a solid surface. In a heat exchanger, heat moves out of the hotter fluid into the solid boundary by convection. That is also the way it moves from the solid boundary into the cooler fluid. The final mode of heat transfer is radiation. This is how the energy from the Sun is transmitted through space to Earth.

The simplest type of closed heat exchanger is composed of a small tube running inside a larger one. One fluid flows through the inner tube, while the other fluid flows in the annular space between the inner and outer tubes. In most applications, a double-pipe heat exchanger would be very long and narrow. It is usually more appropriate to make the outer tube large in diameter and have many small tubes inside it. Such a device is called a shell and tube heat exchanger. When one of the fluids is a gas, fins are often added to heat-exchanger tubes on the gas side, which is usually on the outside.

Plate heat exchangers consist of many thin sheets of metal. One fluid flows across one side of each sheet, and the other fluid flows across the other side.

Background and History

Boilers were probably the first important heat exchangers. One of the first documented boilers was invented by Hero of Alexandria in the first century. This device included a crude steam turbine, but Hero's engine was little more than a toy. The first truly useful boiler may have been invented by the Marquess of Worcester in about 1663. His boiler provided steam to drive a water pump. Further developments were made by British engineer Thomas Savery and British blacksmith Thomas Newcomen, though many people mistakenly believe that James Watt invented the steam engine. Watt invented the condenser, another kind of heat exchanger. Combining the condenser with existing engines made them much more efficient. Until the late nineteenth century boilers and condensers dominated the heat-exchanger scene.

The invention of the diesel engine in 1897 by Bavarian engineer Rudolf Diesel gave rise to the need for other heat exchangers: lubricating oil coolers, radiators, and fuel oil heaters. During the twentieth century, heat exchangers grew rapidly in number, size, and variety. Plate heat exchangers were invented. The huge steam generators used in nuclear plants were produced. Highly specialized heat exchangers were developed for use in spacecraft.

How It Works

Thermal Calculations. There are two kinds of thermal calculations–design calculations and rating calculations. In design calculations, engineers know what rate of heat transfer is needed in a particular application. The dimensions of the heat exchanger that will satisfy the need must be determined. In rating calculations, an existing heat exchanger is to be used in a new situation. The rate of heat transfer that it will provide in this situation must be determined.

In both design and rating analyses, engineers must deal with the following resistances to heat transfer: the convection resistance between the hot fluid and the solid boundary, the conduction resistance of the solid boundary itself, and the convection resistance between the solid boundary and the cold fluid. The conduction resistance is easy to calculate. It depends only on the thickness of the boundary and the thermal conductivity of the boundary material. Calculation of the convection resistances is much more complicated. They depend on the velocities of the fluid flows and on the properties of the fluids such as viscosity, thermal conductivity, heat capacity, and density. The geometries of the flow passages are also a factor.

Because the convection resistances are so complicated, they are usually determined by empirical methods. This means that research engineers have conducted many experiments, graphed the results, and found equations that represent the lines on their graphs. Other engineers use these graphs or equations to predict the convection resistances in their heat exchangers.

In liquids, the convection resistance is usually low, but in gases, it is usually high. In order to compensate for high convection resistance, fins are often installed on the gas side of heat-exchanger tubes. This can increase the amount of heat transfer area on the gas side by a factor of ten without significantly increasing the overall size of the heat exchanger.

Hydraulic Calculations. Fluid friction and turbulence within a heat exchanger cause the exit pressure of each fluid to be lower than its entrance pressure. It is desirable to minimize this pressure drop. If a fluid is made to flow by a pump, increased pressure drop will require more pumping power. As with convection resistance, pressure drop depends on many factors and is difficult to predict accurately. Empirical methods are again used. Generally, design changes that reduce the convection resistance will increase the pressure drop, so engineers must reach a compromise between these issues.

Strength Calculations. The pressure of the fluid flowing inside the tubes is often significantly different from the pressure of the fluid flowing around the outside. Engineers must ensure that the tubes are strong enough to withstand this pressure difference so the tubes do not burst. Similarly, the pressure of the fluid in the shell (outside the tubes) is often significantly different from the atmospheric pressure outside the shell. The shell must be strong enough to withstand this.

Fouling. In many applications, one or both fluids may cause corrosion of heat-exchanger tubes, and they may deposit unwanted material on the tube surfaces. River water may deposit mud. Seawater may deposit barnacles and other biological contamination. The general term for all these things is fouling. The tubes may have a layer of fouling on the inside surface and another one on the outside. In addition to the two convection resistances and the conduction resistance, there may be two fouling resistances. When heat exchangers are designed, a reasonable allowance must be made for these fouling resistances.

Applications and Products

Heat exchangers come in an amazing variety of shapes and sizes. They are used with fluids ranging from liquid metals to water and air. A home with hot-water heat has a heat exchanger in each room. They are called radiators, but they rely on convection, not radiation. A room air conditioner has two heat exchangers in it. One transfers heat from room air to the refrigerant, and the other transfers heat from the refrigerant to outside air. Cars with water-cooled engines have a heat exchanger to get rid of engine heat. It is called a radiator, but again it relies on convection.

Boilers. Boilers come in two basic types: fire tube and water tube. In both cases, the heat transfer is between the hot gases produced by combustion of fuel and water that is turning to steam. As the name suggests, a fire-tube boiler has very hot gas, not actually fire, inside the tubes. These tubes are submerged in water, which absorbs heat and turns into steam. In a water-tube boiler, water goes inside the tubes and hot gases pass around them. Water-tube boilers often include superheaters. These heat exchangers allow the steam to flow through tubes that are exposed to the hot combustion gases. As a result the final temperature of the steam may reach 1,000 degrees Fahrenheit or higher. An important and dangerous kind of fouling in boilers is called scale. Scale forms when minerals in the water come out of solution and form a layer of fouling on the hot tube surface. Scale is dangerous because it causes the tube metal behind it to get hotter. In high-performance boilers, this can cause a tube to overheat and burst.

Condensers. Many electric plants have generators driven by steam turbines. As steam leaves a turbine it is transformed back into liquid water in a condenser. This increases the efficiency of the system, and it recovers the mineral-free water for reuse. A typical condenser has thousands of tubes with cooling water flowing inside them. Steam flows around the outsides, transfers heat, and turns back into liquid water. The cooling water may come from a river or ocean. When a source of a large amount of water is not available, cooling water may be recirculated through a cooling tower. Hot cooling water leaving the condenser is sprayed into a stream of atmospheric air. Some of it evaporates, which lowers the temperature of the remaining water. This remaining water can be reused as cooling water in the condenser. The water that evaporates must be replaced from some source, but the quantity of new water needed is much less than when cooling water is used only once. When river water or seawater is used for cooling, there may be significant fouling on the insides of the tubes. Because the steam leaving a turbine contains very small droplets of water moving at high speed, erosion on the outsides of the tubes is a problem. Eventually a hole may develop, and cooling water, which contains dissolved minerals, can leak into the condensing steam.

Steam Generators. In a pressurized-water nuclear power plant, there is a huge heat exchanger called a steam generator. The primary loop contains water under high pressure that circulates through the reactor core. This water, which does not boil, then moves to the steam generator, where it flows inside a large number of tubes. Secondary water at lower pressure surrounds these tubes. As the secondary water absorbs heat it turns into steam. This steam is used to drive the turbines. Steam generators are among the largest heat exchangers in existence.

Deaerators. Because the condensers in steam systems operate with internal pressures below one atmosphere, air may leak in. Some of this air dissolves in the water that forms as steam condenses. If this air remained in the water as it reached the boiler, rapid rusting of boiler surfaces would result. To prevent this, an open heat exchanger, called a deaerator, is installed. Water is sprayed into the deaerator as a fine mist, and steam is also admitted. As the steam and water mix, the water droplets are heated to their boiling point but not actually boiled. The solubility of air in water goes to zero as the water temperature approaches the boiling point, so nearly all air is forced out of solution in the deaerator. Once the air is in gaseous form, it is removed from the system.

Feedwater Heaters. Leaving the deaerator, the water in a steam plant is on its way to the boiler. The system can be made more efficient by preheating the water along the way. This is done in a feedwater heater. Steam is extracted from the steam turbines to serve as the heat source in feedwater heaters. Feedwater flows inside the tubes and steam flows around the tubes. Feedwater heaters are often multipass heat exchangers. This means that the feedwater passes back and forth through the heat exchanger several times. This makes the heat exchanger shorter and fatter, which is a more convenient shape.

Intercoolers. Many diesel engines have turbochargers that pressurize the air being fed to the cylinders. As air is compressed, its temperature rises. It is desirable to lower the temperature of the air before it enters the cylinders, because that means a greater mass of air can occupy the same space. More air in the cylinder means more fuel can be burned, and more power can be produced. An intercooler is a closed heat exchanger between the turbocharger and the engine cylinders. In this device, air passes around the tubes, and cooling water passes inside them. There are usually fins on the outsides of the tubes to provide increased heat transfer area.

Industrial air compressors also have intercoolers. These compressors are two-stage machines. That means air is compressed part way in one cylinder and the rest of the way in another. As with turbochargers, the first compression raises the air temperature. An intercooler is often installed between the cylinders.

Compressed air flows through the intercooler tubes. Either atmospheric air or cooling water flows around the outside. Cooling the air before the second compression reduces the power required there.

Impact on Industry

By 1880 the use of boilers was widespread, but there was little or no regulation of their design and construction. As a result, boiler explosions were commonplace, and many lives were lost in that way every year. The American Society of Mechanical Engineers (ASME) was founded that year to develop safety standards for boilers.

Early in the twenty-first century, an industry group called Tubular Exchanger Manufacturers Association (TEMA) published standards for shell-and-tube heat exchangers. This organization certifies heat exchangers that are built in accordance with its standards. Purchasers of heat exchangers with TEMA certification are assured that these devices are properly designed and safe to operate. More than twenty manufacturers are members of TEMA. Standards for shell-and-tube heat exchangers are also published by ASME, Heat Transfer Research, and the American Petroleum Institute.

Manufacturers of small heat exchangers include Exergy LLC and Springboard Manufacturing. These heat exchangers are small enough to hold in one's hand. Basco/Whitlock manufactures medium-size shell-and-tube heat exchangers from two inches to eight inches in diameter and eight inches to ninety-six inches in length. The shells are made of brass or stainless steel, and the tubes are made of copper, Admiralty metal (an alloy of copper, zinc, and tin), or stainless steel.

Westinghouse, Combustion Engineering, and Babcock & Wilcox have built many of the huge steam generators used in U.S. nuclear power plants. Heat Transfer Equipment Company manufactures large steel heat exchangers. It specializes in heat exchangers for the oil and gas industry. In Japan, Mitsubishi Heavy Industries has manufactured about one hundred steam generators for use in nuclear power plants there.

The Electric Power Research Institute, an organization composed of companies that generate electricity, funds and conducts research on all aspects of electric power generation, including the heat exchangers that are involved. Much of the basic research on convection resistance in heat exchangers was conducted in university laboratories during the middle of the twentieth century. Although this remains an active research area, the level of activity is lower in the early twenty-first century.

Fascinating Facts About Heat-Exchanger Technologies

- Heat exchangers that have seawater flowing through them usually have small zinc blocks in the flow. These blocks help protect the exchanger surfaces from corrosion.
- Steam condensers operate below atmospheric pressure. Any leakage that occurs is air going in rather than steam coming out. These heat exchangers have air-removal devices called air ejectors connected to them.
- Heat exchanger tubes are usually attached to tube sheets by "rolling," a process that expands the tube to make it fit tightly in its hole in the tube sheet.
- Shell and tube heat exchangers have baffle plates that direct the fluid flowing over the outsides of the tubes. The baffle plates also help support the tubes.
- Electronic sensing devices called eddy-current sensors are passed through the tubes of a shell heat exchanger to detect potential problems before tubes fail.
- Steam generators in nuclear power plants can be as tall as 70 feet and weigh as much as 800 tons when empty. These are among the largest heat exchangers in existence.
- The slang "crud" is said to be an acronym of Chalk River unidentified deposits, which refers to radioactive fouling found in the steam generators of the Chalk River Laboratories in Canada.
- Radiators in cars and home-heating systems make use of convection heat transfer, which refers to the transfer of heat into a fluid. In both cases, heat is transferred to air. They do not make significant use of radiation heat transfer.
- Heat exchangers in spacecraft make use of radiation heat transfer, because there is no air or water to absorb the heat.

Careers and Course Work

Heat exchangers are usually designed by mechanical engineers who hold bachelor or master of science degrees in this field. Students of mechanical engineering take courses in advanced mathematics, mechanics of materials, thermodynamics, fluid mechanics, and heat transfer. An M.S. degree provides advanced understanding of the physical phenomena involved in heat exchangers. Research into the theory of heat transfer is normally carried out by mechanical engineers with doctoral degrees. They conduct research in laboratories, at universities, private research companies, or large corporations that build heat exchangers. As mentioned earlier, convection heat transfer calculations rely on equations that are derived from extensive experiments. Much research work continues to be devoted to improving the accuracy of these equations.

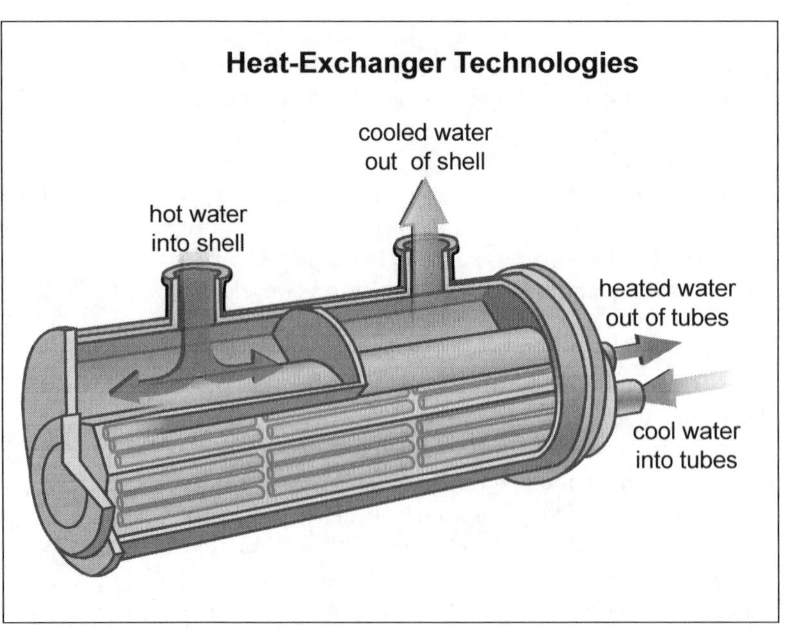

Construction of heat exchangers is executed by companies large and small. The work is carried out by skilled craftsmen using precise machine tools and other equipment. Machinists, welders, sheet-metal, and other highly trained workers are involved. Students who pursue such careers may begin with vocational-technical training at the high school level. They become apprentices to one of these trades. During apprenticeship, the workers receive formal training in classrooms and on-the-job training. As their skills develop they become journeymen and then master mechanics.

Workers who operate, maintain, and repair heat exchangers have a variety of backgrounds. Some have engineering or engineering technology degrees. Others have vocational-technical and on-the-job training. At nuclear power plants, the Nuclear Regulatory Commission requires a program of extensive testing of the vital heat exchangers. This is carried out by engineers with B.S. or M.S. degrees, assisted by skilled craftsmen.

SOCIAL CONTEXT AND FUTURE PROSPECTS

Although heat exchangers are not glamorous, they are an essential part of people's lives. Every home has several, as does every car and truck. Without heat exchangers, people would still be heating their homes with fireplaces, and engines of all types and sizes would not be possible. Heat exchangers are essential in all manner of industries. In particular, they play a key role in the generation of electricity.

The design of heat exchangers is based on empirical methods rather than basic principles. While empirical methods are reasonably effective, design from basic principles would be preferred. In the early twenty-first century, extensive research projects are under way with the goal of solving the very complicated equations that represent the basic principles of heat transfer. These projects make use of very powerful computers. As the cost of computers continues to drop and their power continues to increase, heat exchangers may come to be designed from basic principles.

Edwin G. Wiggins, B.S., M.S., Ph.D.

FURTHER READING

Babcock and Wilcox Company. *Steam: Its Generation and Use.* Reprint. Whitefish, Mont.: Kessinger Publishing, 2010. This easy-to-read book provides extensive information about all manner of fossil-fueled boilers as well as nuclear steam generators.

Blank, David A., Arthur E. Bock, and David J. Richardson. *Introduction to Naval Engineering.* 2d ed. Annapolis, Md.: Naval Institute Press, 2005. Intended for use by freshmen at the U.S. Naval Academy, this book provides a simple, nonmathematical discussion of heat exchangers used on naval ships.

McGeorge, H. D. *Marine Auxiliary Machinery.* 7th ed. Burlington, Mass.: Butterworth-Heinemann, 1995. Excellent descriptions of shell-and-tube heat exchangers, plate heat exchangers, and deaerators are provided in very readable form, including a simple presentation of heat exchanger theory.

Thurston, Robert Henry. *A History of the Growth of the Steam-Engine.* Ithaca, N.Y.: Cornell University Press, 1939. Although very old, this book is useful because it provides a comprehensive history of the development of steam engines, boilers, and condensers.

Web Sites

American Petroleum Institute
http://www.api.org

American Society of Mechanical Engineers
http://www.asme.org

Electric Power Research Institute
http://my.epri.com/portal/server.pt?

Heat Transfer Research, Inc.
http://www.htri.net

Tubular Exchanger Manufacturers Association
http://tema.org.

HEAT PUMPS

FIELDS OF STUDY

Civil engineering; mechanical engineering; renewable-energy engineering; physics; geology; environmental science; heating, ventilation, and air-conditioning (HVAC); refrigeration; thermodynamics.

SUMMARY

A heat pump is a thermal energy exchanger that transfers energy from one medium to another in order to provide heating or cooling. This technology is rapidly developing as a renewable-energy technology to provide moderate space heating and cooling or water heating in residential, commercial, and institutional settings. The benefit of the heat-pump design over conventional energy systems is that it condenses and uses existing thermal energy, rather than fossil fuels, to generate new thermal energy. As a renewable-energy technology, heat pumps rely on the relatively constant temperature just below the Earth's surface to act as a source or sink for thermal energy.

KEY TERMS AND CONCEPTS

- **Closed-Loop Subsystem:** Earth connection subsystem in which a fluid, typically water or a refrigerant, is circulated continuously through a system and not released.
- **Coefficient Of Performance (COP):** Measure of energy performance applied to electric, vapor-compression heat pumps that describes the number of kilowatts of thermal energy output per kilowatt electricity input.
- **Earth Connection Subsystem:** Component of a ground source heat-pump system that collects thermal energy from or distributes thermal energy to the Earth below its surface; also known as a "loop."
- **Geoexchange:** Refers to ground-source heat-pump systems.
- **Heat Distribution Subsystem:** Component of a ground-source heat-pump system that collects thermal energy from or distributes thermal energy to the inside of a building.
- **Open-Loop Subsystem:** Earth-connection subsystem in which a fluid, typically water, is drawn in, circulated, and released.
- **Temperature Lift:** Increase in temperature achieved by a heat pump; measures the difference between the temperatures inside the evaporator and condenser.

DEFINITION AND BASIC PRINCIPLES

In the simplest sense, a heat pump takes in thermal energy at a low temperature (the source) and releases it at a higher temperature (the sink), working against the temperature gradient using a small amount of electricity or other energy. This thermal energy is carried between two spaces by a liquid refrigerant. The temperature at which a liquid will change into a gaseous state is related to the atmospheric pressure it is under. For example, water will normally change state from liquid to gas at 212 degrees Fahrenheit (F) at sea level, or under 1 atmosphere of pressure. However, under one-tenth of normal atmospheric pressure water will boil at 98 degrees F.

Thermal energy naturally moves down a temperature gradient, from a warmer medium to a cooler medium, in accordance with the second law of thermodynamics. This law characterizes the trend of the universe toward disorganization, or entropy, and explains how heat is transferred to or from the heat-pump system. While the refrigerant in a heat pump is in the sink, or evaporation stage, it absorbs ambient heat, which turns the low-pressure liquid into a higher-energy gas. This gas is then compressed to a higher pressure, and it releases that energy into the sink and condenses into a high-pressure liquid. These are the basic principles under which heat pumps, refrigerators, and air-conditioning units operate.

BACKGROUND AND HISTORY

The vapor-compression cycle was first used by French engineer Nicolas Léonard Sadi Carnot in 1824. Then in 1832, American inventor Jacob Perkins was the first to demonstrate a compression cooling technology that used ether as a refrigerant. But it was in 1852 that Scottish engineer William Thomson, also known as Lord Kelvin, conceptualized the first heat pump system, dubbed the "heat multiplier." He recognized that this early predecessor of the air-source heat pump could provide both heating and cooling. Thomson estimated that his machine could generate an equal amount of heat as that produced through direct heating but using only 3 percent of the energy required.

In the early nineteenth century, skeptics maintained that electricity would never be a practical energy source for generating heat in large amounts. Just a handful of scientists looked at the possibility of converting electrical energy to mechanical energy as a means of generating heat, and then pumping that heat from a lower temperature to a higher temperature. But in 1927, British engineer T. G. N. Haldane demonstrated that the heat pump could operate both as a heating and a cooling apparatus using a vapor-compression cycle refrigerator. When it functioned normally, the refrigerator provided cooling, but when operated in reverse, it could provide heat.

How It Works

There are two main types of heat pumps: vapor-compression pumps and absorption pumps. Vapor-compression heat pumps are the more common of the two.

Vapor-Compression Pumps. The vapor-compression refrigeration cycle removes heat from the air or water inside a building and enables this energy to be transferred to another medium. Vapor-compression pumps are filled with a heat-conducting refrigerant, such as Freon, and include four main components: a motor, a compressor, an evaporator, and a condenser. The motor uses electricity to draw the refrigerant in its vaporous form from the evaporator through the compressor, where it is compressed to increase pressure and temperature. The vapor then passes on to the condenser for cooling and condensing. At this point, the vapor surrenders much of the thermal energy it contains to the air or liquid it is designed to heat and returns to a liquid phase. The liquid refrigerant then continues through the expansion valve to the evaporator. The expansion valve allows only as much refrigerant to pass through as will vaporize. As this occurs, the pressure of the refrigerant falls rapidly and lowers its boiling point. The lower boiling point allows it to vaporize and accept thermal energy from the air or liquid. The refrigerant then moves on to the compressor to repeat the process.

Absorption Heat Pumps. Absorption heat pumps operate using thermal energy, rather than electrical energy, and function like an air-source heat pump. Just a small amount of electricity may be required by the solution pump. Fossil fuels, solar energy, or another energy source may be used to power the pump. Absorption heat-pump components consist of an absorber, a solution pump, a generator, an evaporator, a condenser, and an expansion valve. While vapor-compression heat pumps use only one fluid, the refrigerant, absorption heat pumps require a working fluid and an absorbent. The working fluid must have the ability to vaporize under operating conditions, and the absorbent must be able to absorb this vapor. Most systems employ either water as the working fluid and lithium bromide as the absorbent or ammonia as the working fluid and water as the absorbent. Inside the evaporator, low-pressure vapor from the working fluid is absorbed by the absorbent, resulting in the production of heat. The solution pump then increases the pressure of the resulting solution, and the heat added to the system by the external heat source causes the working fluid to boil out of the solution. The vaporous working fluid is cooled and condensed back into liquid form in the condenser, and the absorbent is diverted back to the absorber by the expansion valve. As in the vapor-compression heat pump, heat is accepted by the working fluid inside the evaporator and released inside the condenser; however, in an absorption heat pump, heat is also released inside the absorber.

Applications and Products

Many people unknowingly use heat pumps every day. The refrigerator is probably the most common example of a heat pump. The refrigerant inside this type of heat pump is selected to have a vaporizing temperature that is below the desired temperature inside the refrigerator, so that heat from inside the refrigerator is captured by the refrigerant following the temperature gradient. The refrigerant flows to the evaporator, where is turns into a gas. But the refrigerant must also have a condensing temperature that is higher than room temperature, so that the refrigerant draws thermal energy away from the condenser and changes to a liquid, releasing the heat into the air surrounding the refrigerator. Water chillers, air conditioners, and dehumidifiers are other common forms of heat pumps that operate similarly. All of these types of heat pumps are designed to allow heat to move in only one direction.

Three kinds of heat-pump systems can be used for indoor space heating and cooling: air source, water source, and ground source. Since they must perform both heating and cooling, these systems are designed both to produce and receive thermal energy.

Air-Source Heat Pumps. Air-source systems may be used in place of central air-conditioning systems, as a means of cooling indoor air. These are typically installed on rooftops or other building areas where access to circulating air is available. In addition to the four main components of a heat pump, air-source heat pumps also have a four-way reversing valve that is capable of reversing the flow of refrigerant through the heat pump while maintaining the same direction of flow through the compressor. The expansion valve in an air-source heat pump is also modified to control bidirectional flow of refrigerant. When it is performing space cooling, an air-source heat pump operates exactly like an air conditioner. Refrigerant under pressure leaves the compressor at a higher temperature than the outside air, causing heat to follow the temperature gradient from the coil to the outside air as the refrigerant condenses. When the refrigerant passes out of the condenser, it is in a pressurized liquid form, and the expansion valve releases only as much refrigerant as can be vaporized when it receives thermal energy from the indoor air. The pressure of the refrigerant falls as it passes through the expansion valve and its temperature decreases, causing it to receive more heat from the indoor air and vaporize the remaining refrigerant. Once the pressure of the refrigerant has completely fallen, the refrigerant passes through the reversing valve and is directed back to the compressor to repeat the process. If the direction of the reversing valve is switched to allow for heating, the process moves in the opposite direction. Air-source heat pump systems may be connected to water-heating systems to provide hot water during the cooling cycle.

One limitation of air-source heat pump systems is the need for relatively mild winter temperatures, between 25 and 35 degrees F. Otherwise, supplemental heating may be required. One solution may be to include an electric heating device on the air outflow inside the building.

Water-Source Heat Pumps. Water-source systems transfer thermal energy to and from water, rather than air. During the winter, a fossil-fuel-powered water heater produces heat that is condensed by the heat pump and distributed throughout the building by water-filled pipes. In warmer temperatures, the indoor air heats the water in the pipes and the warm water is pumped up to an evaporative cooling tower outside the building, typically located on the roof. The water inside the pipes remains at a steady temperature of between 60 and 90 degrees F.

Since water is more heat conductive than air because of its higher specific heat, these distribution systems require only 25 percent of the energy that would be required for an air-filled duct distribution system, and because water is denser than air, water-filled pipes take up less space inside a building than ducts. For these reasons, water-source systems are typically preferred in large, multistory buildings. Because many such buildings also have sprinkler systems, the distribution pipes may be integrated into the sprinkler system to reduce overall costs.

Fascinating Facts About Heat Pumps

- When the temperature outside is 50 degrees Fahrenheit (F), the coefficient of performance (COP) of an air-source heat pump is approximately 3.3. When the temperature falls to 17 degrees F, the COP decreases to approximately 2.3, because the heat pump must perform more work to extract thermal energy from colder air.
- Because the outside air temperature in countries such as the United States, Canada, Japan, and Germany is more variable than the temperature underground or in lakes or ponds, ground-source heat pumps can operate in a temperature range closer to their optimal COP and yield the greatest energy savings.
- Electric vapor-compression heat pumps typically have a COP of 2.5 to 5.0, indicating that they produce up to five times as much energy as they require to operate.
- Integrating space heating and cooling and water-heating systems can reduce water-heating costs by up to 50 percent.
- The potential energy cost savings achieved by switching to a heat-pump system depends both on the energy performance of the new system and the fuel source used by the system currently in place.

Ground-Source Heat Pumps. Ground-source heat pumps take advantage of the relatively consistent temperature underground or beneath the surface of bodies of water such as lakes and ponds as a source or sink for thermal energy. Ground-source systems include three major components: the Earth connection subsystem, the heat-distribution system, and the heat pump. The Earth connection subsystem, or "loops," consists of pipes buried approximately six to ten feet below the Earth's surface. Loops may be buried underground in the soil, placed in an underground aquifer, or located in a lake or pond. Loops may be oriented vertically or horizontally, based on the available space and the difficulty associated with vertical drilling or excavation. Earth connection subsystems may be open or closed to the surrounding soil, groundwater, or surface water. The heat-distribution system consists of air-filled ducts or fluid-filled pipes that transport thermal energy through a building, either to or from the heat pump at any given time.

Unlike air-source heat pumps, ground-source heat pumps are not exposed to fluctuating outdoor temperatures because the temperature below the Earth's surface remains relatively consistent year-round. Consequently, ground-source systems may be more appropriate where both heating and cooling is needed and seasonal temperatures fall below 25 degrees F to 35 degrees F. Supplemental heating systems may still be required to maintain desired indoor temperatures during extreme temperatures. Where loops are submerged in ponds or lakes, it is important to ensure that freezing will not occur.

Air cooling and water heating may be achieved simultaneously by connecting the heat-distribution system and heat pump to the hot-water system inside a building. Instead of transferring thermal energy from the heat-distribution system into the ground, this energy can be put to use for water heating. In this way, energy use is being reduced from both space cooling and water heating.

Rising energy prices and awareness about the environmental impacts of energy use have contributed to the popularity of heat-pump systems. Heat-pump systems offer a potentially lower-emission option for space heating and cooling and water heating because of their relative efficiency under suitable conditions and their reliance on electricity as an energy source. Where prevailing climatic conditions are moderate and electricity is a lower-emission energy source than natural gas, heat-pump systems reduce greenhouse gas emissions from residential, commercial, or industrial buildings. Since the heat pump is using existing thermal energy rather than attempting to generate it from another source, such as fossil fuels, heat-pump systems require energy inputs only to operate the motor and compressor. For this reason, heat pumps require only 20 kilowatts (kW) to 40 kW of electricity to generate 100 kW of thermal energy; some highly efficient heat pumps use even less electricity for a similar heat output.

While it is often easier and more cost effective to incorporate a heat-pump system into a new building design, retrofits can also be tailored to existing site and building characteristics. Electric heat-pump systems tend to be most appropriate for single-family dwellings, while engine-driven systems can be developed for larger condominium-style, commercial, or institutional buildings.

Industrial Applications. Industrial applications of heat pumps are currently limited, but heat pumps may be used for space heating, process heating and cooling, water heating, steam generation, drying, evaporation, distillation, and concentration. Waste heat is typically captured for space heating, and recapture is minimal; however, heat reuse is common in drying, evaporation, and distillation.

Industrial heat pumps may use conventional heating from fossil fuels or waste-heat capture to provide space heating inside greenhouses or other facilities. These systems typically employ vapor-compression heat pumps.

Operations that require hot water for cleaning typically require water temperatures of between 100 degrees F and 190 degrees F. Heat-pump systems for these applications may simultaneously provide space cooling and water heating. Vapor-compression heat pumps are typically employed, but absorption heat pumps may be adapted to these uses. Vapor-compression heat-pump systems can also provide steam at various pressures for process water heating at temperatures between 210 degrees F and 390 degrees F.

Pulp and paper, lumber, and food processing industries may use heat pumps to provide drying and dehumidification at temperatures up to 210 degrees F. In these applications, heat pumps have demonstrated superior performance and product quality compared with conventional drying systems. Drying systems are typically closed and produce fewer odor emissions.

There are several types of heat pumps that are currently feasible for industrial applications, including mechanical vapor-compression systems, closed-cycle mechanical heat pumps, absorption heat pumps, heat transformers, and reverse Brayton-cycle heat pumps.

Heat pumps can provide valuable energy savings and lower costs associated with industrial energy use by capturing waste steam that is at a temperature too low to be useful for heating and increasing its temperature, allowing for reuse. The exact dollar value of energy savings can be calculated based on the temperature lift, the increase in temperature achieved by the heat pump. Achieving a higher lift requires a more powerful pump, which in turn requires a larger amount of energy.

Impact on Industry

With the wide variety of heat-pump technology applications available around the world, there are countless businesses involved in engineering, manufacturing, and selling heat pumps; and an even greater number that design and install air-source, water-source, and ground-source heat-pump systems. Annual growth of the global market for ground-source heat pumps has grown by at least 10 percent since 2000. The majority of market expansion has occurred in North America and Europe. The United States boasts most of the world's installed ground-source heat-pump capacity, followed by Sweden, Germany, Switzerland, and Canada.

Government and University Research. The International Energy Association (IEA) has several annexes under its heat pump program dedicated specifically to studying and promoting applications of heat-pump systems. Annexes are country-specific activities that occur under the IEA banner. IEA activities focus primarily on ground-source heat pumps and include examination of barriers to heat-pump implementation.

The International Ground-Source Heat Pump Association (IGSHPA) is a nonprofit organization housed within Oklahoma State University in Stillwater. Through its affiliation with the IGSHPA, Oklahoma State University's Building and Environmental Thermal Systems Research Group is involved in considerable research focused on the development, improvement, and application of heat-pump technology.

A number of countries, including Austria, Canada, China, Denmark, Japan, the Netherlands, Norway, Sweden, Switzerland, and the United States, have also established national heat-pump research institutes or government programs supporting heat-pump research.

Natural Resources Canada provides online access to RETScreen International, a standardized renewable energy project analysis software program that features case studies and free ground-source heat-pump project modeling. The Canadian Office of Energy Efficiency provides information about heat-pump applications.

The Heat Pump and Thermal Storage Technology Center of Japan is a national research organization that is linked with the IEA through its research into heat-pump cooling. Its researchers explore thermal storage systems and technologies and actively promote adoption of heat-pump technologies. Japan's New Energy and Industrial Technology Development Organization has also been developed as a collaborative institution involving government, academia, and industry, with a mandate to advance environmental and renewable-energy issues. Offices are located in Japan as well as China, Europe, Thailand, India, and the United States.

In the United States, the Geo-Heat Center is operated out of the Oregon Institute of Technology and provides access to research, case studies, periodic newsletters, and a directory of heat-pump consultants and manufacturers. The institute also owns a database of known groundwater wells and springs with high potential for ground-source heat-pump applications, fluid chemistry characteristics of these sites, and known sites currently using ground-source heat pumps, which includes information for sixteen western states. The U.S. Department of Energy (DOE) has also conducted a significant amount of research focused on heat pumps and promotes residential, commercial, and institutional applications through its Geothermal Technologies Program.

Industry and Business. The Institute of Electrical and Electronics Engineers (IEEE) is an international, nonprofit, member-driven organization that works to advance electricity-related technology. The organization promotes innovation, develops standards, and provides members with newsletters and networking opportunities. Its membership currently includes more than 400,000 engineers, among whom are experts working in the field of heat pump technologies.

Industry associations for heat pump-related businesses have been established in Austria, Canada, China, Finland, France, Germany, Italy, Japan, the Netherlands, Romania, Sweden, Switzerland, the United Kingdom, and the United States

The Earth Energy Society of Canada is an organization that was established by a small group of industry members to promote ground-source heat-pump systems. The organization provides information about the technology, a list of members, and a directory of International Ground-Source Heat Pump Association-accredited system contractors located across the country.

The Canadian Geoexchange Coalition has both a national presence and provincial branches. The organization offers training and accreditation for system designers and installers, information about installation in new and existing buildings, links to government resources and grant programs, case studies, upcoming events, and a list of members.

The United Kingdom's Heat Pump Association is dedicated to providing information about domestic, commercial, and industrial heat-pump applications to the public, as well as case studies, and a list of members. The Ground-Source Heat Pump Association is a more specialized industry group also operating out of the United Kingdom that focuses on providing members with services such as news, events, and grant opportunities.

In the United States, the Geothermal Heat Pump Consortium encompasses all aspects of the ground-source heat-pump industry, including contractors, manufacturers, drillers, installers, engineers, and system designers. It maintains a list of professionals in Canada and the United States and provides information about case studies, current projects, incentive programs, and offers training and events.

Careers and Course Work

The design, manufacturing, and installation of heat-pump components and technologies often requires an academic background in mechanical engineering. Specific courses required may include thermodynamics, heat transfer, fluid dynamics, mechanical design, thermal systems design, control systems, and manufacturing processes. Professional licensure, achieved through a recognized engineering association, such as the National Society of Professional Engineers, may be required for some positions. Regional, state, or federal requirements may dictate that an accredited professional engineer must approve the installation of ground-source heat-pump systems.

Many technical and trade schools offer programs geared toward installing renewable energy technology systems. Engineering systems and energy technology programs teach applied skills for ground-source heat-pump installation. Careers in heating, refrigeration, and air conditioning may require either a bachelor's degree in mechanical engineering or a technical diploma in a related field. Diploma programs in heating, refrigeration, and air conditioning enable individuals to install, service, maintain, and upgrade heat pumps and systems for these applications. Additional certificates may be required. A diploma in plumbing may be required for installing, servicing, or upgrading heat-distribution systems involving water-source or ground-source heat pumps. Where drilling is required, as in the case of an Earth connection subsystem, a diploma in drilling may be required for installing loops.

While many university or technical programs provide the foundations for understanding heat pumps and associated systems, industry associations offer additional, specialized training and accreditation. It is important to check regional, state, or federal requirements for professionals in the specific field of interest.

Social Context and Future Prospects

Tighter environmental regulations for building construction and industrial processes are forecast to increase the use of heat-pump technologies to reduce energy use and greenhouse-gas emissions. However, it is important that the source of electricity used to power the heat pump does not generate more greenhouse-gas emissions on a life-cycle basis than the fossil fuels required to generate the same amount of heat in a conventional energy system. One advantage of heat-pump technologies is that they permit switching to electricity produced from renewable-energy sources, such as hydro or wind, as these energy sources become available, while a heating system that relies solely on fossil fuels, such as natural gas, does not. Therefore, as more renewable energy is supplied to the grid, greenhouse-gas emissions associated with heat-pump systems will continue to fall.

Residential, Commercial, and Institutional Buildings. Heat pumps can decrease building energy consumption by up to 70 percent in cold climates, such as those of Nordic countries. Leadership in Energy and Environmental Design (LEED) and net zero emissions facility design have led to the incorporation of heat pumps into green-building design. Retrofits to existing buildings and advanced heat-pump systems will present new challenges to heat-pump system designers and installers. For example, in member-countries of the international Organisation for Economic Co-operation and Development (OECD), new housing is not projected to grow rapidly, and the majority of opportunities for heat-pump systems are expected to be retrofits to existing buildings.

Historically, heat-pump applications have been limited by high initial cost, difficulty optimizing retrofitted heat-pump systems in existing buildings, and inadequate performance of air-source heat pumps in seasonally variable climates. Technological advancements in heat-pump engineering have made significant progress toward eliminating these issues. While the growing popularity of ground-source heat-pump systems has led to an increase in the number of installations and decreasing system and installation costs, research as of 2011 is focusing on further cost reduction and system optimization.

Ultrahigh efficiency heat pumps are being developed to achieve space cooling using a fraction of the energy of the heat pumps available in the second decade of the twenty-first century. As the standard of living continues to rise in the developing world, demand for space cooling is forecasted to increase; thus the market for ultrahigh efficiency heat pumps is projected to grow considerably as the technology matures.

Industry. The applicability of heat pumps in the industrial sector depends largely on the temperatures of the heat source and sink and the temperature lift required. Temperature lift is influenced by the design of the heat pump and the properties of the refrigerant. While previously developed refrigerants, such as Freon, performed well in technological applications, the longer-term environmental consequences of their use were unacceptable. Research has turned to developing improved refrigerants that are capable of producing greater temperature lift for industrial applications. Research from the International Energy Agency is focusing on developing refrigerants capable of producing temperature lift in the range between 170 degrees and 300 degrees F.

Ngaio Hotte, B.Sc. M.F.R.E.

Further Reading

Egg, Jay, and Brian Howard. *Geothermal HVAC: Green Heating and Cooling.* New York: McGraw-Hill, 2011. Provides a guide for contractors and consumers about the benefits of commercial and residential heating, ventilation, and cooling using heat pumps. Includes practical information about the types of systems available, how they work, costs and efficiencies, and troubleshooting.

Herold, Keith E., Reinhard Radermacher, and Sanford A. Klein. *Absorption Chillers and Heat Pumps.* Boca Raton, Fla.: CRC Press, 1996. This text provides a detailed investigation of absorption chillers and heat pumps, with particular attention to thermodynamic properties and movement of working fluids.

Kavanaugh, Stephen P., and Kevin Rafferty. *Ground-Source Heat Pumps: Design of Geothermal Systems for Commercial and Institutional Buildings.* Atlanta: American Society of Heating, Refrigerating, and Air-Conditioning Engineers, 1997. Written and published by a respected professional organization, this guide is a resource for engineers working with ground-source and water-source heat pumps.

Langley, Billy C. *Heat Pump Technology.* 3d ed. Upper Saddle River, N.J.: Prentice Hall, 2002. This book serves as both a text and reference manual. Fundamental concepts of heat pump operation and service are written for a student audience.

Radermacher, Reinhard, and Yunho Hwang. *Vapor Compression Heat Pumps with Refrigerant Mixtures.* Boca Raton, Fla.: CRC Press, 2005. The author provides a more advanced examination of refrigerant mixtures and working fluids, including thermodynamic aspects, refrigerant cycles, and heat transfer.

Silberstein, Eugene. *Heat Pumps.* Clifton, N.Y.: Thomson/Delmar Learning, 2003. Designed as a reference text for technicians working with heat pumps, this text provides an introduction to vapor-compression heat-pump technology and applications. Includes several diagrams and uses straightforward explanations of components and processes.

U.S. Department of Energy, Energy Efficiency and Renewable Energy, Industrial Technologies Program. "Industrial Heat Pumps for Steam and Fuel

Savings." June 2003. http://www1.eere.energy.gov/industry/bestpractices/pdfs/heatpump.pdf . Explains heat-pump applications for industrial processes and includes diagrams, types of heat pumps, benefits, and typical applications by type.

WEB SITES

American Society of Heating, Refrigerating and Air-Conditioning Engineers
http://www.ashrae.org

Geothermal Technologies Program
http://www1.eere.energy.gov/geothermal/heatpumps.html

Heat Pump Centre
http://www.heatpumpcentre.org/en

Institute of Electrical and Electronics Engineers (IEEE)
http://www.ieee.org

International Energy Agency
http://www.iea.org

National Society for Professional Engineers
http://www.nspe.org

U.S. Department of Energy
http://www.energy.gov

HERBOLOGY

FIELDS OF STUDY

Medicine; nutrition; naturopathy; biochemistry; botany; pharmacy; physiology.

SUMMARY

Herbology is the study of plants for use in the prevention and treatment of health conditions and disease. Herbs have been used throughout history for medicinal purposes. Herbology is an essential part of traditional Chinese medicine and has been incorporated into Western naturopathy. Growing interest in natural remedies in North America has increased the popularity of training programs in herbology. Although some cultures have used herbal treatments for thousands of years, their efficacy and quality have been questioned, and some people have called for stricter regulation of herbal products.

KEY TERMS AND CONCEPTS

- **Adaptogen:** Herb that treats physical and emotional stress.
- **Alterative:** Herb that restores normal bodily function.
- **Astringent:** Tannin containing herb used for contraction of tissues.
- **Bitter:** Digestive herb.
- **Carminative:** Herb for relief of gas.
- **Cathartic:** Laxative herb.
- **Cholagogue/Choleretic:** Herb to increase bile production.
- **Decoction:** Tea made from the fibrous part of plants.
- **Demulcent:** Mucilaginous herb that treats mucous membranes.
- **Diaphoretci:** Herb to promote perspiration.
- **Emmenagogue:** Herb to stimulate menstruation.
- **Extract:** Concentrated liquid made from herbs.
- **Galactagogue:** Herb with sugar and nonsugar composition.
- **Hepatic:** Quality of supporting liver function.
- **Herbalist:** Practitioner trained in the medicinal use of plants.
- **Infusion:** Herbal tea made from the soft parts of plants such as flowers and leaves.
- **Tincture:** Concentrated liquid made with water, alcohol, and herbs.
- **Vunerary:** Herb to promote wound healing.

Definition and Basic Principles

Herbology, also known as phytotherapy, is the study of plants and plant extracts for medicinal use. Herbology is an integral part of Chinese traditional medicine, folk medicine, and naturopathic medicine. Herbology is often used by practitioners who also favor alternative medical treatments such as acupuncture. An herbalist is a person who specializes in the use of plants and plant extracts for the prevention and treatment of diseases and conditions. Some herbalists also use naturally occurring nonplant products such as minerals and animal products.

Herbology is based on the knowledge that phytochemicals (chemical substances within plants) have pharmaceutical properties. Many herbal remedies are based on traditional practices, and many scientific studies have been done to verify the properties of herbal products. In 1978, the German Commission E, a governmental regulatory agency, was formed to evaluate the safety and efficacy of herbs available for general use. The commission produced 380 monographs. Although the commission is no longer in existence, the monographs are available in *The Complete German Commission E Monographs* published by the American Botanical Council in 1998. These monographs are one of the most thorough collections of scientific data on the subject of herbology. Since the commission was disbanded, numerous scientific studies have been published by other groups regarding the effectiveness of various herbs in the treatment and prevention of disease.

As of 2010, herbalists were not required to be licensed in the United States and Canada. Training programs in herbology are available through a number of organizations. The American Herbalists Guild was founded in 1989 to serve as an educational organization for herbalists in the United States. The American Botanical Council is a nonprofit organization that was founded in 1988 to provide information to herbalists and consumers.

Herbal remedies are not regulated by the U.S. Food and Drug Administration (FDA) and are not investigated by the FDA unless there are reports of adverse reactions. A similar situation exists in Canada. As more and more herbal products have become available, reports of adverse reactions due to product misuse or product contaminants have been increasing. The National Center for Complementary and Alternative Medicine (NCCAM), a branch of the National Institutes of Health (NIH), takes a scientific approach. It provides research-based information on the safety and efficacy of herbs and offers training in herbology.

A wide variety of medical practitioners uses herbology. In addition to traditional herbalists, medical herbalists, natural healers, holistic medical doctors, naturopaths, and practitioners of alternative medicine (such as traditional Chinese medicine) use herbs. In a practice called complementary medicine, some conventional doctors and other medical professionals treat patients using herbology and other alternative therapies in addition to conventional Western medicine. According to the National Institutes of Health, in 2007, more than 38 percent of adults in the United States reported using complementary or alternative medicine, including herbs.

Background and History

Plants and plant extracts have been used in the prevention and treatment of medical conditions for many centuries. Ancient Egyptians used opium, garlic, and other plants, and the Old Testament includes references to the use of herbs. The use of herbology has been an integral part of traditional Chinese medicine for more than a thousand years.

Many pharmaceutical products are derived from plants. A well-known example is digitalis, which is derived from the foxglove plant and is used in the treatment of heart conditions. Another example is morphine, which was originally extracted from poppies. Aspirin is derived from salicin, which is present in the bark and leaves of the willow tree. As herbal remedies have become more popular, the manufacture and sale of herbal products including combinations of herbs in tablet or extract forms has increased. Many products are readily available at health food stores and pharmacies.

Although the practice of herbology relies on traditional knowledge, a number of efforts have been made to verify the efficacy and safety of herbal remedies. The most well-known source for scientific information about herbs is the monographs produced by the German Commission E. The National Center for Complementary and Alternative Medicine also provides information on and training in herbology.

How It Works

Herbalists use herbal remedies made from plants. These remedies are based on the knowledge that phytochemicals have pharmacologic properties. These phytochemicals can be brought out by preparing the plant material in a variety of ways, depending on the remedy.

Herbal medicines are often prepared as a tea, which is properly called an infusion or decoction. A decoction is typically made by placing the stems, roots, or bark of a specific plant in water, and boiling the water until the volume is reduced. Infusions are made by steeping plant parts in hot or cold water. Infusions are usually made from the softer parts of plants such as the flowers or leaves. Decoctions and infusions extract water-soluble phytochemicals.

Herbal remedies may also be prepared as a tincture or extract. Tinctures are made by macerating the plant material in water and alcohol. The ratios of plant material, water, and alcohol will vary depending on the herbal remedy and dosages. Extracts will usually be more concentrated. For some extracts, glycerin is used as a solvent.

Tablets, lozenges, and dried herbs are also available. Some herbal products can also be found in creams or ointments.

Commercially available products are not regulated by the FDA, and the contents of some products have differed from the product label. Others have been contaminated with other herbs, metals, or pesticides. To ensure safety, NCCAM suggests that patients become informed about the herbal product, discuss it with their primary care physician, read the label carefully, and remember that natural does not always mean safe.

Consumers and practitioners should be aware that herbal products can cause adverse reactions or result in negative outcomes for a number of reasons. First, consumers or herbalists may not recognize the seriousness of a medical condition, which might require more aggressive medical treatment than herbs can provide. Second, problems can arise from allergic reactions, misidentification of herbal products, intentional or unintentional contamination of herbal products with heavy metals or other harmful substances, or mistakes in the dosage. Third, herbs and conventional medications may interact and exacerbate an existing condition. Finally, unsubstantiated herbal remedies may be totally ineffective and allow a condition to remain basically untreated.

Herbology is practiced in different ways according to the philosophy of treatment. Some dosage recommendations are derived from herbal traditions such as traditional Ayurvedic or Chinese medicine. Many books are available on the subject of herbal remedies, and some commercially available products will have recommended dosages on the label.

Although there are no standards for training in herbology, training programs are available through NCCAM and other organizations.

APPLICATIONS AND PRODUCTS

Traditional Chinese Medicine. Herbology involves the use of plants and plant extracts in the treatment and prevention of disease. The application of herbal remedies may depend on the medical tradition followed by the herbalist. For example, a herbalist following traditional Chinese medicine may recommend herbs to restore a person's body to a balanced state, based on an examination of aspects such as yin and yang, as well as qi (chi) and blood. Yin and yang are opposite aspects, defined as light and dark, hot and cold, and moist and dry, respectively. Qi is often described as energy, or a dynamic essence. Blood, transformed from the essence of food, is important as well. Chinese medicine also determines treatment according to the five elements–wood, fire, earth, metal, and water–which correspond to specific organs, senses, and bodily functions. Chinese herbal products range from specific foods such as wheat or grapes to combinations of herbs in a tablet or a tea.

Western Herbology. Some Western herbalists subdivide herbal remedies according to the organs or systems affected or the action of the herb. Herbal actions fall into several categories. Adaptogens are herbs used to treat emotional and physical stress. Tonics provide energy. Vulnerary herbs promote healing, and nervines work on the nervous system. Astringents contract tissues. Diaphoretics promote perspiration, and febrifuge is used to treat a fever. Alterative herbs restore normal function and may be further divided into hepatic, digestive, or antimicrobial. The gallbladder can be treated with a cholagogue or choleritic to promote secretion of bile. Bitters and carminative herbs are used to support digestion. Purgative and carthartic herbs are used for a laxative effect, and stomachic is an agent to promote appetite. Demulcents are mucilaginous agents that sooth mucous membranes. Menstrual flow is enhanced by emmenagogues. Phytoestrogens enhance the estrogen system. Galactagogues promote lactation.

Applications. Herbology is used in a number of settings. Herbalists may be solo practitioners or may work in a health food store. Nutritionists or conventional physicians may incorporate herbology in their practices. Herbal remedies are readily available in pharmacies, and many pharmacists will have some knowledge about the benefits and dosing of herbal remedies. Herbal remedies are often used along with other forms of alternative medicine such as yoga, acupuncture, and meditation.

IMPACT ON INDUSTRY

Herbal products, used by more than 38 percent of Americans and millions of people worldwide, have become big business. Billions of dollars of herbal products are sold each year. According to some estimates, Americans spent more than $20 billion on vitamins, herbs, and other supplements in 2007. Health Canada reports that more than 71 percent of Canadians regularly take natural health products. Suppliers of herbal products range from small companies that specialize in a few products to major companies, including some major pharmaceutical companies. The growth in sales of herbs and supplements slowed after deaths were linked to the supplement ephedra, which was banned in 2004. However, herbal products continue to generate large revenues.

Organizations. Herbology organizations range from governmental agencies designed to educate consumers to associations made up of people in the industry. The National Center for Complementary and Alternative Medicine, part of the NIH, provides research-based information to consumers as well as training in herbology. The American Botanical Council provides information about herbal medicine to consumers and herbalists. The American Herbal Products Association represents growers, importers, manufacturers, marketers, and corporations in the herbal industry. The American Herbalists Guild represents herbalists specializing in the medicinal use of plants.

Governmental Regulation. According to the Dietary Supplement Health and Education Act of

1994, companies marketing supplements in the United States cannot make specific therapeutic claims about their products. Companies can make nutritional claims if they include a disclaimer that the claims have not been evaluated by the FDA. Reports of adverse reactions from herbal products are investigated by the FDA. In Canada, supplements can be treated as either foods or drugs. Supplements regulated as prescription drugs have a drug identification number (DIN) that appears on the product label. The National Health Products Regulations, which came into effect in January 1, 2004, defined natural health products as over-the-counter drugs. Those evaluated by Health Canada are given a natural product number (NPN).

Government Research. Research on herbal products has been conducted by governmental agencies such as the German Commission E and Canada's Natural Health Products Research Program (2003-2008). National Center for Complementary and Alternative Medicine funds a number of research programs and, in cooperation with the NIH's Office of Dietary Supplements, runs the Botanical Research Centers Program. Thousands of scientific articles have been published on herbal products and peer-reviewed articles are available through NIH's National Library of Medicine.

It can be challenging to study herbal remedies because of large variations in the practice of herbology, including variable dosing and the practice of using a mixture of herbs. Another problem is the placebo effect. In blind studies, up to 30 percent of those taking placebos (a substance that resembles the product but does not have the active ingredient) report benefits. In addition, some researchers feel that those who take herbs are more likely to have a healthier lifestyle than those who do not take herbal products. Researchers have produced conflicting reports about a number of products, including St. John's wort, which has been used to treat depression. Later data seem to indicate that St. John's wort may not be effective and can have serious drug interactions. The efficacy of ginkgo biloba, a common herb taken to combat memory loss, has also been called into question.

Industry and Business. Herbology involves a variety of sectors, both small and large. Herbology practices range from individuals working solo to those who are part of a larger alternative group practice. Herbal products are comonly sold in health food stores that specialize in the sales of herbs and supplements. These stores may be mom-and-pop establishments or branches of large corporate chains. Herbal products are also sold by several direct-sales companies.

The industry also contains companies that grow, import, manufacture, and distribute herbal products. This segment of the industry includes large and small companies dedicated to herbal and other supplements as well as major pharmaceutical companies with an herbal product line. The distribution and sales of these herbal products in small and large retail chain stores generate billions of dollars in revenue every year.

Fascinating Facts About Herbology

- Herbs have been used as medicine for thousands of years. Perhaps the most expensive herb is saffron, which costs between $500 and $1,000 per pound. Some herbalists recommend saffron for memory enhancement and antiaging.
- Garlic is both a common remedy and a common ingredient in foods. The ancient Greeks and Romans used garlic to ward off disease. Garlic, which may reduce cholesterol, has been used to prevent heart disease, cancer, colds, and the flu. It is considered both a vegetable and a herb and is a member of the onion family. Garlic has been used as a natural mosquito repellent and was once believed to ward off vampires.
- Red clover has been reported to have magical powers, including the ability to protect against witchcraft and evil spirits. Four-leaf variants have been used as lucky charms.
- A flavenoid called quercetin present in apples may protect the brain, lending support for the adage "An apple a day keeps the doctor away."

Careers and Course Work

For those wishing to become a practicing herbalist, programs are readily available in the United States and Canada. Herbology training usually involves studying botany, herbal medicine, and the history and philosophy of herbal medicine, and receiving specialized training in the use of herbs in pregnancy and childhood.

The Association of Accredited Naturopathic Medical Colleges consists of seven North American schools that provide naturopathic training. These schools offer a variety of programs leading to a degree or certificate, including a bachelor of science in herbal sciences, a certificate in Chinese herbal medicine, and a doctor of naturopathic medicine degree. The requirements for entry and completion of these programs vary depending on the chosen degree. These naturopathic programs are built around the six fundamental principles of healing: the body has an inherent ability to create, maintain, and restore health (the healing power of nature), the physician should not interfere with this healing power (first do no harm), the cause of an illness must be identified for its treatment, the physician should heal the whole person, the physician should educate and encourage the patient, and the physician should help the patient achieve health through preventative medicine. There are also a variety of online programs available for training in herbology.

For careers in the business side of herbology, business training or industry experience may be helpful. Researchers in herbology often have advanced degrees in related fields such as biochemistry, botany, chemistry, or physiology. Grants are available through the National Center for Complementary and Alternative Medicine for additional training in herbology.

Social Context and Future Prospects

The use of plants and plant extracts for therapeutic treatment and prevention of disease will continue to be practiced worldwide. With the development of government organizations, degree programs, and medical school courses in alternative medicine, it is likely that someday mainstream medicine will incorporate the study of herbology.

Consumer interest in herbal products continues to increase and to spur growth in the herbal industry. As the number of herbal products expands and more and more people use them, governments are likely to increase oversight of these products to ensure consumer safety and prevent fraudulent practices. The FDA may introduce licensing or other measures to control the production and use of herbal products.

Continued research into herbal remedies is likely to lead to improved understanding of the benefits of herbs as well as possible side effects and interactions. The phytochemicals responsible for the herbal effects will probably lead to the development of conventional pharmaceuticals as well.

Ellen E. Anderson Penno, M.D., M.S., F.R.C.S.C., A.B.O.

Further Reading

Blumenthol, Mark, et al., eds. *The Complete German Commission E Monographs: Therapeutic Guide to Herbal Medicines.* Austin, Tex.: American Botanical Council, 1998. Still considered a valuable resource for science-based information about herbs for therapeutic use.

Boon, Heather, and Michael Smith. *Fifty-five Most Common Medicinal Herbs.* 2d ed. Toronto: Robert Rose, 2009. A detailed description of the fifty most commonly used herbs. Includes an introduction with information about herbal preparations and legislation of herbal products.

Hurley, Dan. *Natural Causes: Death, Lies, and Politics in America's Vitamin and Herbal Supplement Industry.* New York: Broadway Books, 2006. Hurley criticizes the supplements industry for its excesses and discusses the politics behind regulatory decisions.

PDR for Nonprescription Drugs, Dietary Supplements, and Herbs. 30th ed. Montvale, N.J.: Thomson Healthcare, 2008. The physician's desk reference provides information on dietary supplements and herbs.

Tierra, Michael. *The Way of Chinese Herbs.* New York: Pocket Books, 1998. A very thorough compendium of Chinese herbal remedies with descriptions, including the philosophy of treatment underlying the Chinese herbal tradition.

Web Sites

American Botanical Council
http://abc.herbalgram.org

American Herbalists Guild
http://www.americanherbalistsguild.com

American Herbal Products Association
http://www.ahpa.org

Health Canada
Natural Health Products
http://www.hc-sc.gc.ca

Herb Research Foundation
http://www.herbs.org

National Center for Complementary and Alternative Medicine, NIH
http://www.nccam.nih.gov

Office of Dietary Supplements, NIH
http://ods.od.nih.gov

See also: Horticulture

HOLOGRAPHIC TECHNOLOGY

FIELDS OF STUDY

Optics; photography; computer science; laser technology; chemistry; electrical engineering; electronics; mathematics; and physics.

SUMMARY

Holographic technology employs beams of light to record information and then rebuilds that information so the reconstruction appears three-dimensional. Unlike photography, which traditionally produces fixed two-dimensional images, holography re-creates the lighting from the original scene and results in a hologram that can be viewed from different angles and perspectives as if the observer were seeing the original scene. The technology, which was greatly improved with the invention of the laser, is used in various fields such as product packaging, consumer electronics, medical imaging, security, architecture, geology, and cosmology.

KEY TERMS AND CONCEPTS

- **Diffracted Light:** Light that is modified by bending when it passes through narrow openings or around objects.
- **Emulsion:** Light-sensitive mixture that forms a coating on photographic film, plates, or paper.
- **Incandescent Light:** Light that is white, glowing, or luminous as a result of being heated.
- **Monochromatic Light:** 1. Having or appearing to have one color or hue. 2. Consisting of or composed of radiation of a single wavelength.
- **Object Beam:** Beam of light that first reaches the object and is then reflected off the object.
- **Reference Beam:** Beam of light that does not reflect off an object but interferes with the object beam to produce an interference pattern on the emulsion.

Definition and Basic Principles

Holography is a technique that uses interference and diffraction of light to record a likeness and then rebuild and illuminate the likeness. Holograms use coherent light, which contains waves that are aligned with one another. Beams of coherent light interfere with one another as the image is recorded and stored, thus producing interference patterns. When the image is re-illuminated, diffracted light allows the resulting hologram to appear three dimensional. Unlike photography, which produces a fixed image, holography re-creates the light of the original scene and yields a hologram, which can be viewed from different angles and different perspectives just as if the original subject were still present.

Several basic types of holograms can be produced. A transmission hologram requires an observer to see the image through light as it passes through the hologram. A rainbow hologram is a special kind of transmission hologram: Colors change as the observer moves his or her head. This type of transmission hologram can also be viewed in white light, such as that produced by an incandescent lightbulb. A reflection hologram can also be viewed in white light. This type allows the observer to see the image with light reflected off the surface of the hologram. The holographic stereogram uses attributes of both holography and photography. Industry and art utilize the basic types of holograms as well as create new and advanced technologies and applications.

Background and History

Around 1947, Hungarian-born physicist Dennis Gabor, while attempting to improve the electron microscope, discovered the basics of holography. Early efforts by scientists to develop the technique were restricted by the use of the mercury arc lamp as a monochromatic light source. This inferior light source contributed to the poor quality of holograms, and the field advanced little throughout the next decade. Laser light was introduced in the 1960's and was considered stable and coherent. Coherent light contains waves that are aligned with one another and is well suited for high-quality holograms. Subsequently, discoveries and innovations in the field began to increase and accelerate.

In 1960, American physicist Theodore Maiman of Hughes Research Laboratories developed the pulsed ruby laser. This laser used rubies to operate and generated powerful bursts of light lasting only nanoseconds. This laser, which acted much like a camera's flashbulb, became ideal for capturing images of moving objects or people.

In 1962, deciding to improve upon Gabor's technique, scientists Emmett Leith and Juris Upatnieks, working at the University of Michigan, produced images of three-dimensional (3-D) objects–a toy bird and train. These were the first transmission holograms and required an observer to see the image through light as it passed through the holograms.

Additionally in 1962, Russian scientist Yuri N. Denisyuk combined his own work with the color photography work of French physicist Gabriel Lippmann. This resulted in a reflection hologram that could be viewed with white light reflecting off the surface of a hologram. Reflection holograms did not need laser light to be viewed.

In 1968 electrical engineer Stephen Benton developed the rainbow hologram–as the observer moves his head, he sees the spectrum of color as in a rainbow. This type of hologram can also be viewed in white light.

Holographic art appeared in exhibits beginning in the late 1960's and early 1970's, and holographic portraits, made with the pulsed ruby laser, found some favor beginning in the 1980's. Advances in the field have continued, and many varied types of holograms are utilized in many different areas of science and technology, while artistic applications have lagged in comparison.

How It Works

A 3-D subject captured by conventional photography becomes stored on a medium, such as photographic film, as a two-dimensional (2-D) scene. Information about the intensity of the light from a static scene is acquired, but information about the path of the light is not recorded. Holographic creation captures information about the light, including the path, and the whole field of light is recorded.

A beam of light first reaches the object from a light source. Wavelengths of coherent light, such as laser light, leave the light source "in phase" (in sync) and are known collectively as an object beam. These waves reach the object, are scattered, and then are interfered with when a reference beam from the same light source is introduced. A pattern occurs from the reference beam interfering with the object waves. This interference pattern is recorded on the emulsion. Re-illumination of the hologram with the reference beam results in the reconstruction of the object light wave, and a 3-D image appears.

Fascinating Facts About Holographic Technology

- Larry Siebert of the Conductron Corporation produced the first pulsed hologram of a person when he created a self-portrait on October 31, 1967.
- The Cranbrook Academy of Art in Michigan held the first holographic art exhibition in 1968.
- Dennis Gabor received the 1971 Nobel Prize in Physics for his discovery of and work in holography.
- Kiss II, a well-known multiplex hologram, was produced in 1974 by Lloyd Cross. A 3-D image of Pam Brazier floats in the air–she blows a kiss and winks to passersby.
- *National Geographic* crafted an entire magazine cover made of an embossed hologram in 1988.
- The first holoprinter was created by Dutch Holographic Laboratory in 1992.

Light Sources. An incandescent bulb generates light in a host of different wavelengths, whereas a laser produces monochromatic wavelengths of the same frequency. Laser light, also referred to as coherent light, is used most often to create holograms. The helium-neon laser is the most commonly recognized type.

Types of lasers include all gas-phase iodine, argon, carbon dioxide, carbon monoxide, chemical oxygen-iodine, helium-neon, and many others.

To produce wavelengths in color, the most frequently used lasers are the helium-neon (for red) and the argon-ion (for blue and green). Lasers at one time were expensive and sometimes difficult to obtain, but modern-day lasers can be relatively inexpensive and easier to use for recording holograms.

Recording Materials. Light-sensitive materials such as photographic films and plates, the first resources used for recording holograms, still prove useful. Since the color of light is determined by its wavelength, varying emulsions on the film that are sensitive to different wavelengths can be utilized to record information about scene colors. However, many different types of materials have proven valuable in various applications.

Other recording materials include dichromated gelatin, elastomers, photoreactive polymers, photochromics, photorefractive crystals, photoresists, photothermoplastics, and silver-halide sensitized gelatin.

Applications and Products

Art. Holographic art, prevalent in the 1960's through the 1980's, still exists, although fewer artists practice holography solely. Many artistic creations contain holographic components. A modicum of schools and universities teach holographic art.

Digital Holography. Digital holography is one of the fastest-growing realms and has applications in the artistic, scientific, and technological communities. Computer processing of digital holograms lends an advantage, as a separate light source is not needed for re-illumination.

Digital holography first began to appear in the late 1970's. The process initially involved two steps: the writing of a string of digital images onto film and then converting the images into a hologram. Around 1988, holographer Ken Haines invented a process of creating digital holograms in one step.

Digital holographic microscopy (DHM) can be utilized noninvasively to study changes in the cells of living tissue subjected to simulated microgravity. Information is captured by a digital camera and processed by software.

Display. Different types of holograms can be displayed in store windows, as visual aids to accompany lectures or presentations, in museums, at art, science, or technology exhibits, in schools, libraries, or at home as simple decorations hung on a wall and lit by spotlights.

Embossed Holograms. Embossed holograms, which are special kinds of rainbow holograms, can be duplicated and mass produced. These holograms can be used as means of authentication on credit cards and driver's licenses as well as for decorative use on wrapping paper, book covers, magazine covers, bumper stickers, greeting cards, stickers, and product packaging.

Holograms in Medicine and Biology. The field of dentistry provided a setting for an early application in medical holography. Creating holograms of dental casts markedly reduced the space needed to store dental records for Britain's National Health Service. Holograms have also proved useful in regular dental practice and dentistry training.

The use of various types of holograms prove beneficial for viewing sections of living and nonliving tissue, preparing joint replacement devices, noninvasive scrutiny of tumors or suspected tumors, and viewing the human eye. A volume-multiplexed hologram can be used in medical-scanning applications.

Moving Holograms. Holographic movies created for entertaining audiences in a cinema are in development. While moving holograms can be made, limitations exist for the production of motion pictures. Somewhat more promising is the field of holographic video and possibly television.

Security. A recurring issue in world trade is that of counterfeit goods. Vendors increasingly rely on special holograms embedded in product packaging to combat the problem. The creation of complex brand images using holographic technology can offer a degree of brand protection for almost any product, including pharmaceuticals.

Security holography garners a large segment of the market. However, makers of security holograms, whose designs are utilized for authentication of bank notes, credit cards, and driver's licenses, face the perplexing challenge of counterfeit security images. As time progresses, these images become increasingly easier to fake; therefore, this area of industry must continually create newer and more complex holographic techniques to stay ahead of deceptive practices.

Stereograms. Holographic stereograms, unique and divergent, use attributes of both holography and photography. Makers of stereograms have the potential of creating both very large and moving images. Stereograms can be produced in color and also processed by a computer.

Non-Optical Holography. Types of holography exist that use waves other than light. Some examples include acoustical holography, which operates with sound waves; atomic holography, which is used in applications with atomic beams; and electron holography, which utilizes electron waves.

Impact on Industry

As time progresses, holography increasingly appears in more and varied industries, even branching into non-optical fields. In addition to those mentioned above, applications can also be found in industries such as aeronautics, architecture, automotive, the environment and weather, cosmology, geology, law enforcement, solar energy, and video-game design and programming. As of 2011, commercial, technological, and scientific applications outnumber applications in fine art.

Government Research. United States Army and Air Force research offices have funded holographic work leading to the development of technologies including 3-D mapping, risk reduction for helicopter flight and landing, and improving visualization and interface technology during military operations.

Acoustic holography and sonar technology have been used in research by the United States Navy, which has also worked with holographic cameras capable of operating underwater.

Schools and Universities. A few schools are breaking ground with new developments in optics, holographic television, holographic sensors, and recording materials. Some companies have collaborated with a few progressive schools and universities, providing art students with innovative holographic systems for the purpose of experimental and creative work. Such work evolves the field of holography and opens doors to future applications, which can ultimately influence industry. Among the schools that have made valuable contributions to the holographic frontier as well as to the many related disciplines are Adelphi University; Cambridge University and De Montfort University in the United Kingdom; Korean National University of Art and Holocenter Korea; Academy of Media Arts (KHM) Cologne in Germany; Kun Shan University in Taiwan; University of Arizona; and University of New South Wales College of Art in Australia.

Major Corporations. Holographic applications in industry have become, over time, increasingly diverse and valuable. Among the global corporations that have made significant contributions to the industry are Colour Holographic and De La Rue in the United Kingdom, Dutch Holographic Laboratory in the Netherlands, Geola in Italy, Holographics North in Vermont, Illumina in California, Rabbitholes Media in Canada, SAIC in Virginia, and Zebra Imaging in Texas.

Careers and Course Work

Outside of fine arts, careers that include holographic applications are varied. While studies should be concentrated in a particular industry or area of interest, students should have a basic foundation in chemistry, electronics, mathematics, and physics. Advanced courses in these subjects should be taken according to the requirements of the overall industry of interest. Course work in computer science, software engineering, electrical engineering, laser technology, optics, and photography should also be given consideration. Students should expect to earn at least a bachelor of science in their chosen field. For advanced career positions, generally a master's or doctorate degree will be required.

Social Context and Future Prospects

Holography in one form or another is prevalent in modern society, whether as a security feature on a credit card, a component of a medical technique, or a colorful wrapping paper. Holograms have been interwoven into daily life and will likely continue to increase their impact in the future.

Next-generation holographic storage devices have been developed, setting the stage for companies to compete for future markets. Data is stored on the surface of DVDs; however, devices have been invented to store holographic data within a disk. The significantly enlarged storage capacity is appealing for customers with large storage needs who can afford the expensive disks and drives, but some companies are also interested in targeting an even larger market by revising existing technology. Possible modification of current technology, such as DVD players, could potentially result in less expensive methods of playing 3-D data.

Upcoming technology for recording and displaying 3-D images in another format is that of 3-D television. This idea has been around for awhile; however, advances have been slow but potentially promising.

Glenda Griffin, B.A., M.L.S.

Further Reading

Ackermann, Gerhard K., and Jürgen Eichler. *Holography: A Practical Approach.* Weinheim, Germany: Wiley-VCH, 2007. Based on university laboratory courses, and contains more than 100 problems with solutions. Also discusses new developments in holography.

Hariharan, P. *Basics of Holography.* Cambridge, England: Cambridge University Press, 2002. This resource introduces the basics of holography for students of science and engineering as well as for people with an interest in holography and a basic understanding of physics.

Harper, Gavin D. J. *Holography Projects for the Evil Genius.* New York: McGraw Hill, 2010. Explains the basics of holography and provides do-it-yourself projects with easily accessible materials. Glossary, index, and supplier's index included.

Johnston, Sean F. "Absorbing New Subjects: Holography as an Analog of Photography." *Physics in Perspective* vol. 8, issue 2 (2006): 164-188. Provides the history and background in the development of holography and examines the cultural influences on the field.

Saxby, Graham. *Practical Holography.* 3d ed. Bristol, England: Institute of Physics Publishing, 2004. Explains the basics of holography and provides easy steps to making holograms. Offers ideas on creating a holographic studio.

Yaroslavsky, Leonid. *Digital Holography and Digital Image Processing: Principles, Methods, Algorithms.* Norwell, Mass.: Kluwer Academic, 2010. A basic introduction to digital holography including valuable information on signal processing.

Web Sites

American Institute of Physics
http://www.aip.org

International Hologram Manufacturer's Association
http://www.ihma.org

National Science Foundation
http://www.nsf.gov

See also: Laser Technologies; Photography.

HORTICULTURE

FIELDS OF STUDY

Plant science; crop science; botany; agronomy; agriculture; biochemistry; ecology; entomology; genetics; landscape architecture; soils engineering.

SUMMARY

Horticulture is both a science and an art and includes the propagation and cultivation of plants often associated with gardening. Trees, bushes, grasses, and fungi are typical naturally occurring plant materials. Cultivated plants include not only trees, shrubs, grasses, and fungi but also vegetables, fruits, nuts, herbs, spices, ornamentals, and flowers. Although horticulture has existed since ancient times, modern scientific advances in the field have resulted in many practical uses of plant materials, including the development of pharmaceuticals. Genetics plays a significant role in modern-day horticultural practices, especially in areas such as food production.

KEY TERMS AND CONCEPTS

- **Botany:** Branch of biology that involves the scientific study of plants.
- **Genetic Engineering:** Technique used to modify the genes of a plant so that it has a desired trait.
- **Hydroponics:** Method of growing plants in water, without soil, by adding nutrients to the water.
- **Micropropagation:** Cloning of multiple plants from the tissues of another plant.
- **Plant Biotechnology:** Using biological systems to genetically engineer hardier plants and enhance their productivity.
- **Plant Structures:** Parts of plants used most often in propagation, including runners, tubers, corms, bulbs, and rhizomes.
- **Propagation:** Sexual production of plants from seeds or asexual production through cuttings, grafts, and tissue cultures.
- **Xeriscaping:** Method of growing gardens in arid conditions by using drought-resistant plants and water conservation techniques.

Definition and Basic Principles

Horticulture is the art of aesthetically arranging plant materials and the science of breeding, propagating, and growing plants. Horticulture requires a knowledge of landscape design, plant science, and individual plants. The ability to identify plants and their growing needs is necessary for determining how to use plant materials in landscape designs and also in plant breeding and propagation.

Plants are classified as herbaceous or woody. Woody plants include deciduous and evergreen trees, shrubs, and smaller plant materials. Deciduous plants become dormant at the end of the growing season, and plants classified as evergreens will retain their foliage–needles and broad leaves–throughout the year.

Horticulture is not the same as agriculture, which involves growing crops on a much larger scale than the typical horticultural venture and may also include animal production. Horticulture can be pursued either as a hobby or as a commercial pursuit. Those involved in horticulture need to be versed in many disciplines. Design, planning, graphics, and construction engineering are important for the art of horticulture, and a background in math and biochemistry are needed for horticultural science.

Background and History

Horticulture is associated with ancient botany, which began as early as the fourth century B.C.E. Plants were selected and domesticated, and early propagation practices such as grafting can be seen in artwork found in ancient Egyptian tombs. Classical philosophers, such as Aristotle, defined plant types and developed some of their early uses for food, drugs, and fuel. Although an interest in botanical sciences waned during the Middle Ages, monasteries were known for their gardening prowess. During the eighteenth and nineteenth centuries, gardening became important as an art form, especially in Europe and Asia, and horticulture as a science began to expand.

One of the first modern horticulturalists in the United States was Liberty Hyde Bailey, a graduate of Michigan Agricultural College. As is true of many horticulturalists, Bailey wrote many texts on botany and horticulture. Bailey also worked with President

Theodore Roosevelt's administration to improve rural life through horticulture. Early landscape architects, such as Frederick Law Olmstead, were also instrumental in advancing the field of horticulture.

How It Works

A horticulturalist must be cognizant of soil, light, temperature, and water conditions to propagate and cultivate plants. Horticulture begins with the preparation of soil through tilling and fertilization, unless the plant is grown without soil as in hydroponics. Light and temperature can be controlled by beginning plant cultivation inside greenhouses and transplanting plants to the outdoors when light and temperature are optimal for growth. Installation of water collectors and irrigation systems, which have been part of horticulture since ancient times, is also necessary. Xeriscaping with modified drought-resistant plant materials, however, has reduced the reliance on water in some locations. Control of pests and diseases is also important, although many pest- and disease-resistant plant materials exist because of modern technology. Some horticulturalists no longer rely on chemical fertilizers and pesticides, and organic gardening is becoming profitable because of the public demand for naturally grown foods. Horticulture may also involve modern harvesting and handling practices that use more mechanical methods as opposed to manual labor.

Propagation. Propagation techniques vary from microscopic techniques such as gene splicing and producing tissue cultures to plant breeding and employing natural and cultivated plant structures and cuttings to generate new plants. The most common form of plant propagation is with seeds. Vegetables and flowers are typically propagated in this manner.

Clonal propagation can be done naturally or achieved through scientific methods involving plant structures rather than seeds. Some examples of natural clonal propagation through plant structures include strawberry runners, potato tubers, crocus corms, daffodils bulbs, and Calla lily rhizomes. Scientific propagation can be as simple as placing cuttings of shoots, roots, and leaves into water or moist soil. Shoots, such as willows, will regenerate roots. Roots, such as sweet potatoes, will grow new shoots, and leaves, such as African violets, will grow both new shoots and roots.

More complex clonal propagation methods include grafting techniques that physically join the parts from two plants together (often used in the propagation of fruit trees) and micropropagation, which requires the horticulturalist to grow plant cells and tissues artificially. Micropropagation is used in the production of modified plants that may have desirable qualities such as pest and disease resistance. It enables the production of a greater number of hardier plants in a shorter amount of time than could be produced from seeds or other propagation methods in the same amount of time.

Cultivation. Cultivation involves the direct planting of seeds and plant structures or the transplanting of plants that were started elsewhere. Cultivation of plant materials requires pruning and weeding to enhance future growth and improve on a plant's appearance. Ornamental cultivation techniques might include training plants so that they are reoriented to better conditions, such as light. Dwarfing is an ornamental cultivation practice that is often used in Asian cultures. Ornamental horticulture is important for those involved in floriculture and landscape architecture.

Applications and Products

Plants are everywhere and are necessary to sustain life. The practice of horticulture, therefore, has multiple applications that are important to society. In addition, modern technology has resulted in a greater diversity in plant-based products.

Food Production. Plant biotechnology includes genetic engineering, which allows the exchange of genes between plant species, usually ones that are closely related. More than 70 percent of processed foods contain some ingredients that have been genetically engineered, and biotech crops, such as corn and soybeans, are on the rise. New horticultural technologies and research are especially important in global regions, where horticultural production is restricted because of marginal climate conditions, poor soils, lack of water for irrigation, and reduced land resources for cultivation. DNA plays a major role in plant breeding of both quantitative and qualitative traits. Developing nations must especially depend on horticultural research to breed plants with pest- and drought-resistant characteristics to expand productivity on less acreage.

The global food market is changing. The ability

to transport food products longer distances and the improved shelf life of such food are the result of horticultural advances that have opened up new markets for foods that may be exported by small producers in developing nations. Moreover, the methods of producing food are changing, and horticulturalists are finding ways to meet society's desire for more natural and organic food products while sustaining and protecting the environment.

Pharmaceuticals. Herbal remedies and medicinal plants have been used by healing practitioners since ancient times. Although scientists have questioned the curative powers of some herbs, horticulture continues to play an important role in the breeding of plants used as alternative medicines and in the development of plant-based mainstream drugs. Horticulturalists are also involved in synthesizing plant compounds into new medicines. Rainforest plants have become especially important in this research because of the many naturally occurring chemical defenses against pests and diseases that these plants exhibit. Horticulturalists are on the forefront in conducting research concerning the curative properties of naturally occurring chemical compounds found in plants. These chemicals, known as pytochemicals, may provide new drugs for fighting cancer.

Flavors and Fragrances. The cultivation of plants such as flowers and fruits is necessary to produce many diverse products. Among these are cosmetics, perfumes, and food flavorings. Plant breeding and genetic engineering have enhanced the quality of plant-based flavorings, thus making some foods more palatable.

Farms and Nurseries. Horticultural practices play an important role in any enterprise that involves the growing of plants for commercial, public, or private uses. Tree farms range from farms specializing in the production of Christmas trees to those providing trees for forest conservation, which protects the environment by providing flood control and carbon sequestration. Other farms may produce vegetables, flowers, or turf, and golf course, park, and garden designers depend on farms and nurseries to produce plant materials to implement their plans.

Energy Production. Horticulture plays a role in the production and breeding of plants that can be converted into biofuels. Although the biofuel industry is in its infancy, substantial global research is taking place. For example, Clemson University is involved in research to turn discarded peaches into hydrogen gas, and nations such as India are producing biodiesel from nonedible plant oil seeds. Sugarcane and maize have also been used to produce ethanol. Because diverse plants can be used to produce biofuels, developing nations may be able to create biofuels cheaply by using naturally growing plant materials and thus create new industries to sustain population growth. However, horticulturalists must ensure that sufficient plants are available for both food production and these emerging biofuel industries.

Education. Development of healthy eating habits has become an important goal in many developed nations, such as the United States, where obesity is on the increase. Educators are employing gardening and horticulture in their lesson plans to teach children about nutrition. In less-developed countries, where poor nutrition is caused by a lack of horticultural crops, international organizations such as the United Nations provide education not only on advanced production and cultivation practices including irrigation and pest control but also on food safety, conservation, and marketing.

Therapy. Horticulture and gardening are therapeutic for those with learning, mental health, and physical disabilities. The American Horticultural Therapy Association is one of several organizations involved in this growing practice. Horticultural therapists incorporate gardens into wellness programs, and this type of therapy has become popular in programs to treat alcohol and substance abusers.

Impact on Industry

The Global Horticulture Initiative, known as GlobalHort, is an international consortium of institutions, governmental agencies, and nongovernmental organizations involved in horticultural research, training, and information sharing. Located in both Belgium and Tanzania, GlobalHort focuses on improving international health and nutrition through the production and promotion of various horticultural crops. Partners include the Food and Agriculture Organization of the United Nations, the World Health Organization, and the World Bank.

Government Agencies. In the United States, one of the primary agencies involved in horticultural research and education is the U.S. Department of Agriculture (USDA). State and local governments also have active horticultural research and public education programs, including those associated with county and university agricultural extension services. The USDA is in partnership with many researchers at land-grant universities and sponsors the National Agricultural Library, which collaborates with public, private, and nonprofit agencies to provide education, often concerning horticultural aspects of food production.

University Research. Much of horticultural research is conducted at universities. In the United States, many of these institutions are state universities that started as land-grant colleges. University research is concentrated on new plant science, especially related to propagation and cultivation. Many land-grant universities have botanical gardens that provide ample opportunities for field research with new plant materials and research trials to enhance plant characteristics. Often university research is conducted at experimental stations; research ranges from agroforestry experimentation at the University of Missouri to urban horticulture research at the University of California Cooperative Extension.

Institutions and Organizations. Great Britain's Royal Horticultural Society, which has focused on gardening for about 150 years, continues to provide horticultural advice and education, especially through its published journal. Horticultural research and education has been provided internationally since 1864 by the International Society for Horticultural Science, made up of scientists representing about 150 nations. The American Society for Horticultural Science provides a similar service in the United States. Within the Society of Chemical Industry, an international private organization with offices in the United States, United Kingdom, and Australia, is a horticultural group concerned with transferring horticultural science into business practices, especially those associated with agriculture and food. In the United States, the American Horticultural Society is an important organization for providing private gardeners with horticultural information. Online networking is possible through the nonprofit organization, Society for Advancement of Horticulture, which connects researchers, growers, and commercial enterprises associated with horticulture.

Growers and Producers. Commercial and private growers and producers associate with one another to learn the latest technologies in horticulture and to share information. Growers and producers cultivate a variety of plant materials not limited to commercial produce, such as vegetables, fruits, nuts, and decorative trees, shrubs, grasses, and flowers. Some horticultural specialties include growing grapes and berries for wine, aquatic gardening, and producing unusual species such as carnivorous plants, cacti and succulents, and bonsai and dwarf varieties. Flower growers often specialize in one species such as roses, orchids, or lilies.

Related Industry and Business Sectors. Growers and producers must depend on industry to supply them with new horticultural products and the latest scientific advances, including seeds and plant structures, machinery, greenhouse and irrigation systems, and fertilizers and pesticides. Many of the newer companies are associated with the growth in organic and natural plant production. Once commercial products–everything from vegetables to flowers–are produced, the growers and producers must depend

Fascinating Facts About Horticulture

- Approximately 25 percent of Western medicines and drugs are derived from plants. The U.S. National Cancer Institute believes that more than 70 percent of tropical plants have medicinal properties that could help fight cancers.
- The oldest known living tree is a Norway spruce found in Sweden. Although the exposed part of the tree is only 13 feet tall, scientists believe its root system is about 9,550 years old.
- Most plants contain both male and female reproductive parts, thus simplifying pollination. Some plants, however, such as hollies, have separate male and female plants, and the female plant cannot produce red berries unless a male plant, which produces the pollen, is nearby so that pollination can take place.
- Plants such as ferns do not flower and must reproduce through spores, which are located on the underside of the plant's fronds.
- Bamboo not only is the world's tallest grass but also grows quickly. The plant can reach heights of 130 feet and grow 4 feet in a day.
- There are more than one hundred species of carnivorous plants that capture insects and digest them for their nutrients.

on the business sector to process plant products safely and to market them.

Careers and Course Work

Horticultural careers include gardening and landscaping, which may require little education, and scientific careers, such as plant breeding and genetic engineering, for which advanced degrees are required. A bachelor of science degree in horticulture or a related field prepares a student for careers including plant and crop cultivator and plant disease or pesticide specialist. Governmental careers such as extension work are also popular. A dual bachelor's degree in business administration and horticulture would support work in a greenhouse, nursery, garden center, or golf course. Those interested in garden and landscape design might seek a bachelor's degree in landscape architecture.

Students interested in plant biotechnology, one of the fastest growing fields, can work in research laboratories as technicians or research scientists. Many who have gained a doctorate degree in biotechnology work in educational settings and continue to conduct horticultural research. Biotechnologists are sought by governmental agencies and private industry, especially companies associated with food production and pharmaceuticals.

Basic courses for students seeking a career in horticulture include plant biology, botany, and plant materials. More advanced course work includes entomology, plant physiology, breeding and propagation, plant nutrition, and plant pathology. Specialized educational courses might include floriculture for those who want to work with flowers or arboriculture for those interested in caring for trees. Students interested in an advanced degree, such as biotechnology, must expand their course work to include biochemistry, genetics, and molecular science.

Social Context and Future Prospects

As the worldwide population continues to increase, horticultural advances through biotechnology can benefit society by ensuring an adequate food supply. Genetically engineered plants are more pest and disease resistant, and modified plants that withstand cold temperatures and are more tolerant of drought and salt allow for longer growing seasons and more numerous habitats. Plants are being engineered to have superior nutrient properties that could reduce the quantity of food necessary to nourish the world's growing population. Furthermore, many of the plants created through biotechnology and genetic engineering are playing a greater role in the production of beneficial medicines and drugs; some experts predict that society may one day have access to edible vaccines.

However, these scientific advances are raising concerns. Scientists are debating whether genetically engineered foods could produce gene mutations in organisms that consume them, create new food allergens, or cause environmental damage because of the antibiotic resistance of genetically engineered foods. Cross-pollination has also become an issue, as it cannot be prevented or stopped. Some fear the creation of super weeds that cannot be controlled after naturally occurring weeds cross-pollinate with genetically modified pesticide-resistant plants.

Carol A. Rolf, B.S.L.A., M.Ed., M.B.A., J.D.

Further Reading

Adams, Charles R., Katherine M. Bamford, and Michael P. Early. *Principles of Horticulture.* 5th ed. Burlington, Mass.: Butterworth-Heinemann, 2008. Introductory text dealing with all aspects of horticulture care and propagation, pest control, and soils, water, and nutrition.

DiSabato-Aust, Tracy. *The Well-Tended Perennial Garden: Planting and Pruning Techniques.* 2d ed. Portland, Oreg.: Timber Press, 2006. Useful guide for planning a perennial garden with a section for notes and personal documentation.

Garner, Jerry. *Careers in Horticulture and Botany.* 2d ed. New York: McGraw-Hill, 2006. Lists many job opportunities in the field of horticulture and describes the necessary education and training.

Pauly, Philip J. *Fruits and Plains: The Horticultural Transformation of America.* Cambridge, Mass.: Harvard University Press, 2008. Discusses how horticultural practice in the United States helped shape the nation.

Reiley, Edward H., and Carroll L. Shry, Jr. *Introductory Horticulture.* 7th ed. Florence, Ky.: Delmar Cengage Learning, 2006. Easy-to-read textbook that includes information on the latest technologies and on finding a job in horticulture.

Toogood, Alan. *American Horticultural Society Plant Propagation: The Fully Illustrated Plant-by-Plant Manual of Practical Techniques.* New York: DK Adult, 1999. Good reference for propagating more than 1,500 plants from existing specimens with illustrations and explanations of plant biology.

Web Sites

American Horticultural Society
http://www.ahs.org

American Horticultural Therapy Association
http://www.ahta.org

American Society for Horticultural Science
http://www.ashs.org

Global Horticulture Initiative
http://www.globalhort.org

International Society for Horticultural Science
http://www.ishs.org

U.S. Department of Agriculture
Horticultural Crops Research Unit
http://www.ars.usda.gov/main/site_main.htm?modecode=53-58-10-00

See also: Herbology; Hydroponics; Silviculture.

HUMAN-COMPUTER INTERACTION

FIELDS OF STUDY

Computer science; computer graphics; software engineering; systems analysis; graphic design; industrial design; ergonomics; mechanical engineering; information science; information architecture; robotics; artificial intelligence; cognitive science; psychology; social psychology; linguistics; neurobiology; psychophysics; social neuroscience; anthropology; scientific computing; data visualization; typography; anthropometrics.

SUMMARY

Human-computer interaction (HCI) is a field concerned with the study, design, implementation, evaluation, and improvement of the ways in which human beings use or interact with computer systems. The importance of human-computer interaction within the field of computer science has grown in tandem with technology's potential to help people accomplish an increasing number and variety of personal, professional, and social goals. For example, the development of user-friendly interactive computer interfaces, Web sites, games, home appliances, office equipment, art installations, and information distribution systems such as advertising and public awareness campaigns are all applications that fall within the realm of HCI.

KEY TERMS AND CONCEPTS

- **Accessibility:** Extent to which an interface can be used by people with visual, auditory, cognitive, or physical impairments.
- **Direct Manipulation:** Interacting with a graphic representation of an object to accomplish a task.
- **Ethnography:** Process of observing, interviewing, and analyzing the activities of people in their everyday environments in order to gain the perspective of a user.
- **Graphical User Interface (GUI):** Means of interacting with an electronic device based on images rather than text.
- **Heuristic:** Guideline or rule of thumb used to quickly evaluate an interface or product.
- **Information Architecture:** Study and practice of organizing data so that they can be found and used efficiently.
- **Likert Scale:** Method of ascribing quantitative value to qualitative data, used in questionnaires; typically asks respondents to rate how much they agree or disagree with a statement.
- **Ubiquitous Computing:** Model of computing that sees technology as being fully integrated into every aspect of daily life rather than limited to the functionality of a machine on a desktop.
- **Usability:** Extent to which a product can easily, efficiently, and effectively be used to achieve a certain goal.
- **Wayfinding:** Ways in which users orient themselves and navigate from place to place within an interface.
- **Widget:** Interactive component on a Web site, such as one that allows the user to click through a list of options and choose one.
- **Wire Frame:** Skeleton version of a Web site that leaves out visual design elements and focuses on how pages will be linked.

DEFINITION AND BASIC PRINCIPLES

Human-computer interaction is an interdisciplinary science with the primary goal of harnessing the full potential of computer and communication systems for the benefit of individuals and groups. HCI researchers design and implement innovative interactive technologies that are not only useful but also easy and pleasurable to use and anticipate and satisfy the specific needs of the user. The study of HCI has applications throughout every realm of modern life, including work, education, communications, health care, and recreation.

The fundamental philosophy that guides HCI is the principle of user-centered design. This philosophy proposes that the development of any product or interface should be driven by the needs of the person or people who will ultimately use it, rather than by any design considerations that center around the object itself. A key element of usability is affordance, the notion that the appearance of any interactive element should suggest the ways in which it can be manipulated. For example, the use of shadowing around a button on a Web site might help make it look three-dimensional, thus suggesting that it can be pushed or clicked. Visibility is closely related to affordance; it is the notion that the function of all the controls with which a user interacts should be clearly mapped to their effects. For example, a label such as "Volume Up" beneath a button might indicate exactly what it does. Various protocols facilitate the creation of highly usable applications. A cornerstone of HCI is iterative design, a method of development that uses repeated cycles of feedback and analysis to improve each prototype version of a product, instead of simply creating a single design and launching it immediately. To learn more about the people who will eventually use a product and how they will use it, designers also make use of ethnographic field studies and usability tests.

Background and History

Before the advent of the personal computer, those who interacted with computers were largely technology specialists. In the 1980's, however, more and more individual users began making use of software such as word-processing programs, computer games, and spreadsheets. HCI as a field emerged from the growing need to redesign such tools to make them practical and useful to ordinary people with no technical training. The first HCI researchers came from a variety of related fields: cognitive science, psychology, computer graphics, human factors (the study of how human capabilities affect the design of mechanical systems), and technology. Among the thinkers and researchers whose ideas have shaped the formation of HCI as a science are John M. Carroll, best known for his theory of minimalism (an approach to instruction that emphasizes real-life applications and the chunking of new material into logical parts), and Adele Goldberg, whose work on early software interfaces at the Palo Alto Research Center (PARC) was instrumental in the development of the modern graphical user interface.

In the early days of HCI, the notion of usability was simply defined as the degree to which a computer system was easy and effective to use. However, usability has come to encompass a number of other qualities, including whether an interface is enjoyable, encourages creativity, relieves tension, anticipates points of confusion, and facilitates the combined efforts of multiple users. In addition, there has been a shift in HCI away from a reliance on theoretical findings from cognitive science and toward a more hands-on approach that prioritizes field studies and usability testing by real participants.

How It Works

Input and Output Devices. The essential goal of HCI is to improve the ways in which information is transferred between a user and the machine he or she is using. Input and output devices are the basic tools HCI researchers and professionals use for this purpose. The more sophisticated the interaction between input and output devices–the more complex the feedback loop between the two directions of information flow–the more the human user will be able to accomplish with the machine.

An input device is any tool that delivers data of some kind from a human to a machine. The most familiar input devices are the ones associated with personal computers: keyboards and mice. Other commonly used devices include joysticks, trackballs, pen styluses, and tablets. Still more unconventional or elaborate input devices might take the shape of head gear designed to track the movements of a user's head and neck, video cameras that track the movements of a user's eyes, skin sensors that detect changes in body temperature or heart rate, wearable gloves that precisely track hand gestures, or automatic speech recognition devices that translate spoken commands into instructions that a machine can understand. Some input devices, such as the sensors that open automatic doors at the fronts of banks or supermarkets, are designed to record information passively, without the user having to take any action.

An output device is any tool that delivers information from a machine to a human. Again, the most familiar output devices are those associated with personal computers: monitors, flat-panel displays, and audio speakers. Other output devices include wearable head-mounted displays or goggles that provide visual feedback directly in front of the user's field of vision and full-body suits that provide tactile feedback to the user in the form of pressure.

Perceptual-Motor Interaction. When HCI theorists speak about perceptual-motor interaction, what they are referring to is the notion that users' perceptions—the information they gather from the machine—are inextricably linked to their physical actions, or how they relate to the machine. Computer systems can take advantage of this by using both input and output devices to provide feedback about the user's actions that will help him or her make the next move. For example, a word on a Web site may change in color when a user hovers the mouse over it, indicating that it is a functional link. A joystick being used in a racing game may exert what feels like muscular tension or pressure against the user's hand in response to the device being steered to the left or right. Ideally, any feedback a system gives a user should be aligned to the physical direction in which he or she is moving an input device. For example, the direction in which a cursor moves on screen should be the same as the direction in which the user is moving the mouse. This is known as kinesthetic correspondence.

Another technique HCI researchers have devised to facilitate the feedback loop between a user's perceptions and actions is known as augmented reality. With this approach, rather than providing the user with data from a single source, the output device projects digital information, such as labels, descriptions, charts, and outlines, on the physical world. When an engineer is looking at a complex mechanical system, for example, the display might show what each part in the system is called and enable him or her to call up additional troubleshooting or repair information.

Applications and Products

Computers. At one time, interacting with a personal computer required knowing how to use a command-line interface in which the user typed in instructions–often worded in abstract technical language–for a computer to execute. A graphical user interface, based on HCI principles, supplements or replaces text-based commands with visual elements such as icons, labels, windows, widgets, menus, and control buttons. These elements are controlled using a physical pointing device such as a mouse. For instance, a user may use a mouse to open, close, or resize a window or to pull down a list of options in a menu in order to select one. The major advantage graphical user interfaces have over text-based interfaces is that they make completing tasks far simpler and more intuitive. Using graphic images rather than text reduces the amount of time it takes to interpret and use a control, even for a novice user. This enables users to focus on the task at hand rather than to spend time figuring out how to manipulate the technology itself. For instance, rather than having to recall and then correctly type in a complicated command, a user can print a particular file by selecting its name in a window, opening it, and clicking on an icon designed to look like a printer. Similarly, rather than choosing options from a menu in order to open a certain file within an application, a user might drag and drop the icon for the file onto the icon for the application.

Besides helping individuals navigate through and execute commands in operating systems, software engineers also use HCI principles to increase the usability of specific computer programs. One example is the way pop-up windows appear in the word-processing program Microsoft Word when a user types in the salutation in a letter or the beginning item in a list. The program is designed to recognize the user's task, anticipate the needs of that task, and offer assistance with formatting customized to that particular kind of writing.

Consumer Appliances. Besides computers, a host of consumer appliances use aspects of HCI design to improve usability. Graphic icons are ubiquitous parts of the interfaces commonly found on cameras, stereos, microwave ovens, refrigerators, and televisions. Smartphones such as Apple's iPhone rely on the same graphic displays and direct manipulation techniques as used in full-sized computers. Many also add extra tactile, or haptic, dimensions of usability such as touchscreen keyboards and the ability to rotate windows on the device by physically rotating the device itself in space. Entertainment products such as video game consoles have moved away from keyboard and joystick interfaces, which may not have kinesthetic correspondence, toward far more sophisticated controls. The hand-held device that accompanies the Nintendo Wii, for instance, allows players to control the motions of avatars within a game through the natural movements of their own bodies. Finally, HCI research influences the physical design of many household devices. For example, a plug for an appliance designed with the user in mind might be deliberately shaped so that it can be inserted into an outlet in any orientation, based on the understanding that a user may have to fit several plugs into a limited amount of space, and many appliances have bulky plugs that take up a lot of room.

Increasingly, HCI research is helping appliance designers move toward multimodal user interfaces. These are systems that engage the whole array of human senses and physical capabilities, match particular tasks to the modalities that are the easiest and most effective for people to use, and respond in tangible ways to the actions and behaviors of users. Multimodal interfaces combine input devices for collecting data from the human user (such as video cameras, sound recording devices, and pressure sensors) with software tools that use statistical analysis or artificial intelligence to interpret these data (such as natural language processing programs and computer vision applications). For example, a multimodal interface for a GPS system installed in an automobile might allow the user to simply speak the name of a destination aloud rather than having to type it in while driving. The system might use auditory processing of the user's voice as well as visual processing of his or her lip movements to more accurately interpret speech. It might also use a camera to closely follow the movements of the user's eyes, tracking his or her gaze from one part of the screen to another and using this information to helpfully zoom in on particular parts of the map or automatically select a particular item in a menu.

Workplace Information Systems. HCI research plays an important role in many products that enable people to perform workplace tasks more effectively. For example, experimental computer systems are being designed for air traffic control that will increase safety and efficiency. Such systems work by collecting data about the operator's pupil size, facial expression, heart rate, and the forward momentum and intensity of his or her mouse movements and clicks. This information helps the computer interpret the operator's behavior and state of mind and respond accordingly. When an airplane drifts slightly off its course, the system analyzes the operator's physical modalities. If his or her gaze travels quickly over the relevant area of the screen, with no change in pupil size or mouse click intensity, the computer might conclude that the operator has missed the anomaly and attempt to draw attention to it by using a flashing light or an alarm.

Other common workplace applications of HCI include products that are designed to facilitate communication and collaboration between team members, such as instant messaging programs, wikis (collaboratively edited Web sites), and videoconferencing tools. In addition, HCI principles have contributed to many project management tools that enable groups to schedule and track the progress they are making on a shared task or to make changes to common documents without overriding someone else's work.

Fascinating Facts About Human-Computer Interaction

- Children working with scientists at a University of Maryland HCI laboratory created their own toy: a set of animal blocks that each plays a recorded factoid about an animal and reveal its name. When the blocks are jumbled up, they create a nonsensical animal using syllables from each animal's name.
- Roomba, a robotic vacuum cleaner, appeals so well to users' emotions (through features such as "cute" chirping alert noises when the robot finds itself stuck in a corner) that many people find themselves naming and talking to their robots.
- To print a file using a command-line interface rather than a graphical user interface, a user would have to type in the instruction "print [/d:Printer] [Drive:][Path] FileName [. . .]."
- HCI researchers have coined the term critical incident to describe times when a piece of technology makes a person feel frustrated, angry, or confused. Some 80 percent of computer users say frustrating experiences with their machines have led them to shout expletives out loud.
- One common application of HCI research in Japan is smart toilets with automatic seat warmers and speakers that emit flushing sounds to mask bathroom noises.
- The mouse was invented in 1964 and began its life as a simple wooden box with rolling wheels.

Education and Training. Schools, museums, and businesses all make use of HCI principles when designing educational and training curricula for students, visitors, and staff. For example, many school districts are moving away from printed textbooks and toward interactive electronic programs that target a variety of information-processing modalities through multimedia. Unlike paper and pencil worksheets, such programs also provide instant feedback, making it easier for students to learn and understand new concepts. Businesses use similar programs to train employees in such areas as the use of new software and the company's policies on issues of workplace ethics. Many art and science museums have installed electronic kiosks with touchscreens that visitors can use to learn more about a particular exhibit. HCI principles underlie the design of such kiosks. For example, rather than using a text-heavy interface, the screen on an interactive kiosk at a science museum might display video of a museum staff member talking to the visitor about each available option.

Impact on Industry

Cutting-edge HCI research is taking place at technology centers all over the globe, but Japan is perhaps the country in which usability has become most established as a cornerstone of product design. With robotic receptionists so realistically human and efficient that they are almost capable of fooling visitors, Japan is a world leader in the goal of making technology ever more able to anticipate the needs of the user.

Government and University Research. Although many HCI technologies have emerged from privately funded corporations, both government institutes and university research laboratories are equally important contributors to the field. For example, the United States' Defense Advanced Research Projects Agency (DARPA) invests heavily in military-related HCI applications such as simulation equipment used to train soldiers for the battlefield and airplane pilots for flying. The number of universities that have set up academic research institutes focusing on HCI has been growing rapidly. Among them are Carnegie Mellon University, Northeastern University, Stanford University, and the University of Maryland.

Industry and Business. The major industry sectors related to HCI are those that directly involve the development of technological devices for personal and business use, including computer and software manufacturers and electronic appliance manufacturers. Companies such as Apple, Microsoft, IBM, Hewlett-Packard, and Xerox are among the major corporations whose research and development departments have been at the forefront of new HCI applications for many years. However, because the use of technology of all forms is so deeply embedded in contemporary life, HCI design principles have come to influence virtually every other commercial sector.

Marketers of all stripes, for instance, use HCI theories to create Web sites that make browsing for, choosing, and purchasing products as easy as possible. Amazon.com's one-click ordering system is a good example of an HCI-influenced application. It shortens the distance between the decision to buy something and the point of purchase to a single click, making it less likely that a customer will be turned off by a lengthy ordering process or change his or her mind halfway through the process. Similarly, several travel Web sites have begun using a search interface that allows users to type in their desired parameters (departure and destination airports, dates of travel) just once, then simply click on a button to use the same search terms with any of a dozen different airline Web sites–a far less laborious task than visiting each of those sites individually and using their internal search functionalities. Increasingly, it has become common for businesses to create site elements that are not immediately related to the purchase of a product but that add value to customers' experiences on the Web site and develop brand loyalty. For instance, a company that sells cooking and baking equipment for home cooks might add a forum to their site in which users can share recipes and discuss the foods they love. By including on their site a piece of technology that allows potential customers to build a virtual community around a shared interest, the company can encourage repeat visits and future sales.

Careers and Course Work

The paths toward becoming a HCI professional are extraordinarily varied. Bachelor's degrees in cognitive science, neuroscience, computer science, graphic design, psychology, engineering, art, and many other fields could serve as appropriate preparation for a career as someone who uses HCI principles. No matter which concentration an aspiring HCI researcher or student chooses, it is important to acquire basic programming skills, a broad understanding of human psychophysiology, and some practical experience or training with either graphic design or product design. Common areas of work include developing Web sites; computer operating systems; interfaces for consumer appliances such as cell phones, printers, or cameras; and educational materials such as interactive employee training courses, advertising campaigns, or any other applications that demand accessible, learnable, and usable computer systems. Although a graduate degree is not required for entering the field, many universities offer specialized master's programs in HCI.

SOCIAL CONTEXT AND FUTURE PROSPECTS

As HCI moves forward with research into multimodal interfaces and ubiquitous computing, notion of the computer as an object separate from the user may eventually be relegated to the archives of technological history, to be replaced by wearable machine interfaces that can be worn like clothing on the user's head, arm, or torso. Virtual reality interfaces have been developed that are capable of immersing the user in a 360-degree space that looks, sounds, feels, and perhaps even smells like a real environment–and with which they can interact naturally and intuitively, using their whole bodies. As the capacity to measure the physical properties of human beings becomes ever more sophisticated, input devices may grow more and more sensitive; it is possible to envision a future, for instance, in which a machine might "listen in" to the synaptic firings of the neurons in a user's brain and respond accordingly. Indeed, it is not beyond the realm of possibility that a means could be found of stimulating a user's neurons to produce direct visual or auditory sensations. The future of HCI research may be wide open, but its essential place in the workplace, home, recreational spaces, and the broader human culture is assured.

M. Lee, B.A., M.A.

FURTHER READING

Bainbridge, William Sims, ed. *Berkshire Encyclopedia of Human-Computer Interaction: When Science Fiction Becomes Fact.* 2 vols. Great Barrington, Mass.: Berkshire, 2004. Contains more than a hundred HCI topics, including gesture recognition, natural-language processing, and education. Each article includes further reading listings and may contain sidebars, figures, tables, or photographs.

Helander, Martin. *A Guide to Human Factors and Ergonomics.* 2d ed. Boca Raton, Fla.: CRC Press/ Taylor & Francis, 2006. Discusses the cognitive and physical aspects of human capabilities that inform product development, and includes an appendix examining the use of checklists to improve safety and effectiveness in human factors design.

Sears, Andrew, and Julie A. Jacko, eds. *The Human-Computer Interaction Handbook: Fundamentals, Evolving Technologies, and Emerging Applications.* 2d ed. New York: Lawrence Erlbaum Associates, 2008. Prominent researchers in the field address issues in design, development, testing, and evaluation. Contains hundreds of explanatory figures and tables.

Sharp, Heken, Yvonne Rogers, and Jenny Preece. *Interaction Design: Beyond Human-Computer Interaction.* 2d ed. Hoboken, N.J.: John Wiley & Sons, 2007. Each chapter on interaction design contains an outline, summary, subsections, and text boxes highlighting case studies and other points of focus. Includes a comprehensive bibliography.

Thatcher, Jim, et al. *Web Accessibility: Web Standards and Regulatory Compliance.* New York: Springer, 2006. Presents an overview of laws, policies, and technical standards for creating accessible Web sites.

Tufte, Edward R. *The Visual Display of Quantitative Information.* 2d ed. Cheshire, Conn.: Graphics Press, 2007. A seminal work on ways in which complex information can be clearly and simply displayed in visual forms such as graphics, maps, charts, and tables.

WEB SITES

Human Factors and Ergonomics Society
Internet Technical Group
http://www.internettg.org

Special Interest Group on Human Computer Interaction
http://www.sigchi.org

See also: Graphics Technologies; Wireless Technologies and Communication.

HYBRID VEHICLE TECHNOLOGIES

FIELDS OF STUDY

Electrical engineering; mechanical engineering; chemistry; ecology.

SUMMARY

Hybrid vehicle technologies use shared systems of electrical and gas power to create ecologically sustainable industrial and passenger vehicles. With both types of vehicles, the main goals are to reduce hazardous emissions and conserve fuel consumption.

KEY TERMS AND CONCEPTS

- **Driving Range:** Distances, both in city driving and open-road driving, that can be covered by HEVs before refueling or recharging is necessary.
- **Fuel Cells:** Very advanced form of electrochemical cell (instead of a conventional battery) that produces electricity following a reaction between an externally supplied fuel, frequently hydrogen, which becomes positively charged ions following contact with an oxidant, frequently oxygen or chlorine, which "strips" its electrons.
- **Internal Combustion Engine (ICE):** Produces power by the pressure of expanding fossil fuel gases after they are ignited in cylinder chambers.
- **Lithium-Ion Battery:** Lightweight rechargeable battery with energy-efficient qualities tied to the fact that lithium ions travel from the negative to positive electrode when the battery discharges, returning to the negative during charging.
- **Parallel System Hybrid:** Main power comes from the gas engine; electric motor steps in only for power to accelerate.
- **Series System Hybrid:** Power is delivered to the wheels from the electric motor; the gas engine's role is to generate electricity.
- **Third-Generation Hybrid:** Improved vehicle models using newer technology after the previous (first and second) generations.

Definition and Basic Principles

As the word "hybrid" suggests, hybrid vehicle technology seeks to develop an automobile (or, more broadly defined, any power-driven mechanical system) using power from at least two different sources. Before and during the first decade of the twenty-first century, hybrid technology emphasized the combination of an internal combustion engine working with an electric motor component.

Background and History

Before technological development of what is now called a hybrid vehicle, the automobile industry, by necessity, had to have two existing forms of motor energy to hybridize–namely internal combustion in combination with some form of electric power. Early versions of cars driven with electric motors emerged in the 1890's and seemed destined to compete very seriously with both gasoline (internal combustion engines) and steam engines at the turn of the twentieth century.

Although development of commercially attractive hybrid vehicles would not occur until the middle of the twentieth century, the Austrian engineer Ferdinand Porsche made a first-series hybrid automobile in 1900. Within a short time, however, the commercial attractiveness of mass-produced internal combustion engines became the force that dominated the automobile industry for more than a half century. Experimentation with hybrid technology as it could be applied to other forms of transport, especially motorcycles, however, continued throughout this early period.

By the 1970's the main emerging goal of hybrid car engineering was to reduce exhaust emissions; conservation of fuel was a secondary consideration. This situation changed when, in the wake of the 1973 Arab-Israeli War, many petroleum-producing countries supporting the Arab cause cut exports drastically, causing a nationwide gas shortage and worldwide fears that oil would be used as a political weapon.

Until 1975 government support for research and development of hybrid cars was tied to the Environmental Protection Agency (EPA). In that year (and after at least two unsatisfactory results of EPA-supported hybrid car projects), this role was shifted to the Energy Research and Development Administration, which later became the Department of Energy (DOE).

During the decade that followed the introduction of Honda's Insight hybrid car in 1999, the most widely recognized commercially marketed hybrid automobile was Toyota's Prius. Despite some setbacks in sales in 2010 following major recalls connected with (among other less dangerous problems) the malfunctioning anti-lock braking system and accelerator devices, the third-generation Prius still held a strong position in total hybrid car sales globally going into 2011.

How It Works

"Integrated motor assist," a common layperson's engineering phrase borrowed from Honda's late-1990's technology, suggests a simple explanation of how a hybrid vehicle works. The well-known relationship between the electrical starter motor and the gas-driven engine in an internal combustion engine (ICE) car provides a (technically incomplete) analogy: The electric starter motor takes the load needed to turn the crankcase (and the wheels if gears are engaged) until the ICE itself kicks in. This overly general analogy could be carried further by including the alternator in the system, since it relieves the battery of the job of supplying constant electricity to the running engine (recharging the battery at the same time).

In a hybrid system, however, power from the electric motor (or the gas engine) enters and leaves the drivetrain as the demand for power to move the vehicle increases or decreases. To obtain optimum results in terms of carbon dioxide emissions and overall fuel efficiency, the power train of most hybrid vehicles is designed to depend on a relatively small internal combustion engine with various forms of rechargeable electrical energy. Although petroleum-driven ICE's are commonly used, hybrid car engineering is not limited to petroleum. Ethanol, biodiesel, and natural gas have also been used.

In a parallel hybrid, the electric motor and ICE are installed so that they can power the vehicle either individually or together. These power sources are integrated by automatically controlled clutches. For electric driving, the clutch between the ICE and the gearbox is disengaged, while the clutch connecting the electric motor to the gearbox is engaged. A typical situation requiring simultaneous operation of the ICE and the electric motor would be for rapid acceleration (as in passing) or in climbing hills. Reliance on the electric motor would happen only when the car is braking, coasting, or advancing on level surfaces.

It is extremely important to note that one of the most vital challenges for researchers involved in hybrid-vehicle technology has to do with variable options for supplying electricity to the system. It is far too simple to say that the electrical motor is run by a rechargeable battery, since a wide range of batteries (and alternatives to batteries) exists. A primary and obvious concern, of course, will always be reducing battery weight. To this aim, several carmakers, including Ford, have developed highly effective first-, second-, and even third-generation lithium-ion batteries. Many engineers predict that, in the future, hydrogen-driven fuel cells will play a bigger role in the electrical components of hybrids.

Selection of the basic source of electrical power ties in with corollary issues such as calculation of the driving range (time elapsed and distances covered before the electrical system must be recharged) and optimal technologies for recharging. The simplest scenario for recharging, which is an early direct borrowing from pure-electric car technology, involves plugging into a household outlet (either 110 volt or 220 volt) overnight. But, hybrid-car engineers have developed several more sophisticated methods. One is a "sub-hybrid" procedure, which uses very lightweight fuel cells, mentioned above, in combination with conventional batteries (the latter being recharged by the fuel cells while the vehicle is underway). Research engineers continue to look at any number of ways to tweak energy and power sources from different phases of hybrid vehicle operation. One example, which has been used in Honda's Insight, is a process that temporarily converts the motor into a generator when the car does not require application of the accelerator. Other channels are being investigated for tapping kinetic-energy recovery during different phases of simple mechanical operation of hybrid vehicles.

Applications and Products

Some countries, especially Japan, have begun to use the principle of the hybrid engine for heavy-duty transport or construction-equipment needs, as well as hybrid systems for diesel road graders and new forms of diesel-powered industrial cranes. Hybrid medium-power commercial vehicles, especially urban trolleys and buses, have been manufactured, mainly in Europe and Japan.

Important for broad ecological planning, several countries, including China and Japan, have incorporated hybrid (diesel combined with electric) technology into their programs for rail transport. The biggest potential consumer market for hybrid technology, however, is probably in the private automobile sector.

By the second decade of the twenty-first century, a wide variety of commercially produced hybrid automobiles were on European, Asian, and American markets. Among U.S. manufacturers, Ford has developed the increasingly popular Escape, and General Motors produces about five models ranging from Chevrolet's economical Volt to Cadillac's more expensive Escalade. Japanese manufacturers Nissan, Honda, and Toyota have introduced at least one, three, and two standard hybrid models, respectively, to which one should add Lexus's RX semi-luxury and technologically more advanced series of cars. Korea's Hyundai Elantra and Germany's Volkswagen Golf also competed for some share of the market.

One of the chief attractions to Toyota's hybrid technology has usually been its primary goal of using electric motors in as many operational phases as possible. The closest sales competitor in the United States to the Prius (mainly for mileage efficiency) was Chevrolet's Volt.

At the outset of 2011, Lexus launched an ambitious campaign to attract attention to what it called its full hybrid technology(as compared with mild hybrid) in its high-end RX models. A main feature of the full hybrid system, according to Lexus, is a combination of both series and parallel hybrid power in one vehicle. Such a combination aims at transferring a variable but continuously optimum ratio of gas-engine and electric-motor power to the car. Another advance claimed by Lexus's full hybrid over parallel hybrids is its reliance on the electric motor only at lower speeds.

Early in 2011, Mercedes-Benz also announced its intention to capture more sales of high-end hybrids by dedicating, over a three-year period, more research to improve the technology used in its S400 model. Audi, a somewhat latecomer, unveiled plans for its first hybrid, the Q5, to appear in European markets late in 2011.

As fuel alternatives continue to be added to the ICE components of HEVs, advanced fuel-cell technology could transform the technological field that supplies electrical energy to the combined system.

Impact on Industry

Given the factors of added cost associated with designing and producing hybrid vehicles, private companies, both manufacturers and research institutions, are more likely to enter the field if they can receive some form of governmental financial assistance. This can be in the form of direct subsidies, tax reductions, or grants. Similarly, institutions of higher learning, both public and private, frequently seek outside sources of funding for specially targeted research activities. Major national laboratories that are not tied to academic institutions, such as the Argonne National Laboratory near Chicago, also submit research proposals for grants. Private foundations that favor ecological research, particularly reduction of carbon dioxide emissions and alternative methods for producing electrical energy, can also be approached for seed money.

In the United States in 2010, hybrid vehicles made up only 3 percent of the total car sales. If this is a valid indicator, only a much bigger sales potential is likely to induce manufacturers to fund research that could bring to the market more fuel-efficient hybrid-technology cars at increasingly attractive prices. Major vested interests on the potentially negative side are: the gigantic ICE automobile industry itself, which is resistant to changes that require major new investment costs, and the fossil-fuel (petroleum) industry, which holds a near monopoly on fuels supplying power to automobiles, including diesel, and, in some cases, ethanol, around the world.

Future expansion in the number of hybrid cars might also cause important changes in the nature of equipment needed for various aspects of hybrid refueling and recharging–equipment that will eventually have to be made available for commercial distribution. At an even more local level, hiring specialized and, perhaps, higher-paid mechanics capable of dealing with the more advanced technical components of hybrid vehicles may also become part of automotive shops' planning for as-yet unpredictable new directions in their business.

Careers and Course Work

Academic preparation for careers tied to HEV technology is, of course, closely tied to the fields of electrical and mechanical engineering and, perhaps to a lesser degree, chemistry. All of these fields demand course work at the undergraduate level to develop familiarity not only with engineering principles but with basic sciences and mathematics used, especially those used by physicists. Beyond a bachelor's degree, graduate-level preparation would include continuation of all of the above subjects at more advanced levels, plus an eventual choice for specialization, based on the assumption that some subfields of engineering are more relevant to HEV technology than others.

The most obvious employment possibilities for engineers interested in HEV technology is with actual manufacturers of automobiles or heavy equipment. Depending on the applicant's academic background, employment with manufacturing firms can range from hands-on engineering applications to more conceptually based research and design functions.

Employment openings in research may be found with a wide variety of private- sector firms, some involving studies of environmental impact, others embedded in actual hybrid-engineering technology. These are too numerous to list here, but one outstanding example of a major private firm that is engaged on an international level in environmentally sustainable technology linked to hybrid vehicle research is ABB. ABB grew from late-nineteenth-century origins in electrical lighting and generator manufacturing in Sweden (ASEA), merging in 1987 with the Swiss firm Brown Boveri. ABB carries on operations in many locations throughout the world.

Internationally known U.S. firm Argonne National Laboratory not only produces research data but also serves as a training ground for engineers who either move on to work with smaller ecology-sensitive engineering enterprises or enter government agencies and university research programs.

Finally, employment with government agencies, especially the EPA, the DOE, and Department of Transportation, represents a viable alternative for applicants with requisite advanced engineering and managerial training.

Fascinating Facts About Hybrid Vehicle Technologies

- The basic technology that is being used, albeit in perfected form, in post-2000 hybrid vehicles was first used to manufacture a working hybrid car more than one hundred years ago.
- Consideration of total weight of a hybrid vehicle is so important that engineers devote major attention to possible innovations for any and all hybrid electric vehicle (HEV) components. The most obvious component that undergoes changes from one generation of HEV to the next involves ever-more-efficient modes of supplying electrical energy.
- Technological use of hybrid power systems need not, probably will not, be limited to land transportation. It is possible that aviation–a transport sector that went from conventional to jet engines in the middle of the 20th century–could become considerably more economical and ecological by a combination of power sources.
- Many major cities, especially in Europe (most notably Paris) have fleets of municipal bicycles that can be checked in and out for inner-city use by individuals physically capable and desirous of peddling. It is to be hoped that a next stage–hourly rentals of small HEVs, especially those with major electrical power sources–will follow when production of "basic" (markedly less expensive) hybrid vehicles becomes feasible.
- As more and more sophisticated hybrid-power procedures are developed, the possibility of using different forms of fuel, and eventually bypassing dependence on petroleum, and even biofuels, is a long-range goal of hybrid technology.
- Using regenerative braking, engineers are able to recover electric energy from the magnetic field created when braking results and store it in the HEV's battery for future use.

SOCIAL CONTEXT AND FUTURE PROSPECTS

Although obvious ecological advantages can result as more and more buyers of new vehicles opt for hybrid cars, a variety of potentially negative socioeconomic factors could come into play, certainly over the short to medium term. The higher sales price of hybrids that were available toward the end of 2010 already raised the question of consumer ability (or willingness) to pay more at the outset for fuel-economy savings that would have to be spread out over a fairly long time frame–possibly even longer than the owner kept the vehicle. It is nearly impossible to predict the number of potential buyers whose statistically lower purchasing ability prevents them from paying higher prices for hybrids. Continued unwillingness or inability to purchase hybrids would mean that a proportionally large number of used older-model ICE's (or brand-new models of older-technology vehicles) would remain on the roads. This socioeconomic potentiality remains linked, of course, to any investment strategies under consideration by industrial producers of cars.

How is one to know which companies worldwide are developing new, economically attractive applications for forthcoming hybrid cars?

The European digital news service EIN News, established in the mid-1990's, provides (among dozens of other categories of information) a specific subsection on hybrid vehicle technology and marketing events, including exhibitions, to its subscribers.

Subscribers from all over the world can obtain up-to-date information on hybrid technology from the Detroit publication *Automotive News* and *Automotive News Europe* and *Automotive News China*. There are, of course, many different marketing congresses (popularly labeled automotive shows) all over the globe, where the latest hybrid technology is introduced and different manufacturers' models can be compared.

In the United States, the Society of Automotive Engineers (SAE) is an important source of up-to-date information for ongoing hybrid vehicle research for both engineering specialists and well-informed general readers.

Byron D. Cannon, M.A., Ph.D.

Further Reading

Bethscheider-Kieser, Ulrich. *Green Designed: Future Cars.* Ludwigsburg, Germany: Avedition, 2008. Presents European estimates of technologies that should be compared with the hybrid gas-electric approach to fuel economy.

Clemens, Kevin. *The Crooked Mile: Through Peak Oil, Hybrid Cars and Global Climate Change to Reach a Brighter Future.* Lake Elmo, Minn.: Demontreville Press, 2009. As the title suggests, issues of hybrid car technology need to be placed in a very broad ecological context, where even bigger issues (downward decline in world oil reserves, climate change) may necessitate emphasis on new possible technological solutions.

Lim, Kieran. *Hybrid Cars, Fuel-cell Buses and Sustainable Energy Use.* North Melbourne, Australia: Royal Australian Chemical Institute, 2004. Provides an idea of technologies and programs imagined in other countries.

Society of Automotive Engineers. *1994 Hybrid Electric Vehicle Challenge.* Warrendale, Penn.: Society of Automotive Engineers, 1995. Reports published by thirty American and Canadian college and university engineering laboratories on their respective HEV research programs.

Web Sites

Electric Auto Association
http://www.electricauto.org

Electric Drive Transportation Association
http://www.electricdrive.org

Society of Automotive Engineers
http://www.sae.org

U.S. Department of Energy
Clean Cities
http://www1.eere.energy.gov/cleancities

See also: Biofuels and Synthetic Fuels

HYDROPONICS

FIELDS OF STUDY

Horticulture; agriculture; agricultural engineering; agronomy; botany; biology; biochemistry; biological engineering; chemistry; plant physiology; microbiology; fisheries and allied aquacultures; food technology; food science; nutrition; floriculture; hydrology; genetics; agricultural economics; rural sociology; industrial engineering; electrical engineering; mechanical engineering; computer science; physics; robotics.

SUMMARY

Hydroponics uses varied scientific and technological processes to cultivate plants without soils usually associated with agriculture. Throughout history, humans have delivered nutrients directly to plant roots with water. Based on that principle, modern hydroponics has diverse applications for commercial and utilitarian agriculture. Hydroponics supplies food to military personnel in places where agricultural resources are limited because of climate and terrain. Astronauts eat fresh vegetables grown with hydroponics in space. Food security is bolstered by the availability of substantial yields year-round assured by hydroponics, providing people access to nutrients and relief from hunger. Agribusinesses sell hydroponic crops and equipment to consumers. Many scientific educational curricula incorporate hydroponic lessons.

KEY TERMS AND CONCEPTS

- **Aeroponics:** Nutritional material sprayed in fine droplets onto roots.
- **Aquaponics:** Using hydroponic techniques for plant cultivation simultaneously with fisheries production.
- **Controlled Environment Agriculture:** Cultivation providing plants optimum growing temperatures, moisture, and aeration.
- **Macronutrients:** Elements that plants require significant quantities of to thrive, including nitrogen, phosphorus, sulfur, potassium, calcium, and magnesium.
- **Micronutrients:** Elements that plants need in lesser amounts such as iron, chlorine, manganese, boron, zinc, copper, and molybdenum.
- **Nutrient Solution:** Combination of water, minerals, and elements mixed in specific proportions for each plant type.
- **Rock Wool:** Inorganic fibers created by melting limestone, coke, and volcanic rock at extremely high temperatures to enhance water retention.
- **Substrate:** Material used to support roots.

Definition and Basic Principles

Hydroponics is the scientific use of chemicals, organic and inorganic materials, and technology to grow plants independently of soil. Solutions composed of water and dissolved minerals and elements provide essential macronutrients and micronutrients and supplement oxygen and light necessary for plant growth. Plant roots absorb nutrients that are supplied through various methods. Some hydroponic systems involve suspending roots into liquid solutions. Other hydroponic techniques periodically wash or spray roots with solutions. Methods also utilize containers filled with substrates, such as gravel, where roots are flooded with nutrient solutions. Hydroponics is practiced both in greenhouses, where temperatures and lighting can be regulated, and outdoors, where milder climates pose few natural detriments to plants.

Hydroponics enables agriculturists to grow crops continually without relying on weather, precipitation, and other factors associated with natural growing seasons. These systems permit agricultural production in otherwise unsuitable settings for crop cultivation such as congested cities, deserts, and mountains. Hydroponics is convenient, producing foods in all seasons. Plants can be grown closely together because root growth does not spread like soil-based plant roots extending to seek nutrients and water. Agriculturists can grow crops that are not indigenous to areas, such as tropical fruits. Growth typically occurs more quickly with hydroponics than in soil, because plants invest energy in maturing rather than competing for resources, resulting in large yields. Many hydroponic systems recycle water not absorbed by roots to use for other purposes. Crops cultivated with hydroponic systems are usually safer for consumers than field-grown crops because their exposure to soil-transmitted diseases has been minimized.

Negative aspects of hydroponics include costs associated with acquiring equipment and supplies. Automation and computerized systems require substantial investments in machinery, software, and training personnel to operate them.

BACKGROUND AND HISTORY

Records indicate people cultivated plants using water instead of soil in ancient Mesopotamia, Egypt, and Rome. Aztecs in Mexico innovated floating barges, *chinampas*, for growing food because they lacked suitable agricultural land. In the mid-nineteenth century, German botanist Julius von Sachs and German agrochemist Wilhelm Knop experimented with combining minerals with water to nourish plants.

During the 1920's, University of California, Berkeley, plant nutrition professor Dennis R. Hoagland studied how roots absorb nutrients. William F. Gericke, also a professor at the University of California, Berkeley, cultivated tomatoes in tanks of mineral-rich solution and discussed his research in a February, 1937, *Science* article, noting colleague William A. Setchell referred to that process as hydroponics, representing Greek vocabulary: *hydro* (water) and *ponos* (work).

In the 1940's, the United States military utilized hydroponics to provide sustenance to World War II soldiers in the Pacific. Oil companies built hydroponic gardens on Caribbean islands to feed employees extracting natural resources in that region. The United States Army established a hydroponics branch to supply troops serving in the Korean War in the early 1950's, growing eight million pounds of food.

By mid-century, researchers began incorporating plastic equipment in hydroponic systems. Engineers innovated better pumps and devices, automating some hydroponic processes with computers. By the 1970's new methods included drip-irrigation systems and the nutrient film technique (NFT) created by English scientist Allen Cooper, which helped commercial hydroponics expand globally. The United Nations' Food and Agriculture Organization (FAO) funded hydroponic programs in areas experiencing food crises. Scientists continued devising new techniques, such as aerohydroponics, in the late twentieth century.

Fascinating Facts About Hydroponics

- Diners eating at restaurants in Disney World can purchase vegetables, including squash and tomatoes, resembling Mickey Mouse, which are specially cultivated with hydroponics using plastic molds in Epcot Center's Land Pavilion.
- Mukesh Ambani, the world's fifth wealthiest person in 2010 according to *Forbes*, cultivates food with hydroponic systems placed throughout his twenty-seven-story tall, $2 billion Mumbai, India, residence. He values hydroponic plants' role in conserving energy by cooling that structure through adsorption of solar heat.
- Japanese agriculturists grow Koshihikari rice using hydroponic techniques in a renovated, subterranean bank vault equipped with fans to ventilate the underground paddy.
- OrganiTech in Israel utilizes automation technology to cultivate vegetables unable to thrive naturally in the country's arid conditions. The company's Grow-tech 2000, a hydroponic machine, uses sensors that alert robots when to plant and harvest crops.
- Hydroponic technology at McMurdo Station, located on Ross Island in Antarctica, produces about 250 pounds of vegetables and herbs monthly.
- In 1992, teacher Vonneke Miller and her students at Peterson Middle School in Sunnyvale, California, built a hydroponics laboratory they called ASTRO 1 because it resembled a spacecraft capsule. The vegetables they cultivated, including tomatoes and lettuce, were incorporated into meals prepared in the cafeteria.

HOW IT WORKS

Hydroponics. Hydroponic processes represent examples of controlled environment agriculture in which plant cultivation involves technology such as greenhouses enabling growers to stabilize conditions. Most hydroponic systems function with basic components that supply oxygen and nutrients necessary to sustain plants until they have matured for harvest. Electrical or solar-powered lights, fans, heaters, and pumps regulate temperatures, ventilation for plant respiration, water flow, and photosynthesis impacting plant growth. Each hydroponic system incorporates variations of equipment and methods according to growers' resources and goals. Styrofoam, wood, glass, and plastic are materials used to construct hydroponic systems.

Basic hydroponic procedures involve placing seeds in substrates that consist of organic materials such as coconut fibers, rice hulls, sawdust, peat moss or inorganic mediums including gravel, pumice, perlite, rock wool, or vermiculite. After roots emerge during germination, growers keep seedlings in substrates or remove them depending on which hydroponic method is selected for cultivation. Roots undergo varying durations of exposure to nutrient solutions to absorb macronutrients and micronutrients. Most hydroponic processes utilize either an open, or nonrecirculating, system, or a closed system, referred to as recirculating, depending on whether nutrient solution contacts roots once and is discarded or is kept for consistent or repeated use.

Water-Culture Techniques. These hydroponic methods, which are frequently used to cultivate plants that quickly attain maturity, involve roots constantly being suspended in a nutrient solution. Water-culture hydroponic techniques are often utilized to grow lettuce crops. For the raft culture technique, growers place plants on platforms drilled with holes to pull roots through so roots can be submerged in pools of nutrient solution on which the platforms float. In the dynamic root floating technique, roots closest to the plant are kept dry so they can supply oxygen to the plant. The lower roots are constantly exposed to nutrient solutions and absorb those minerals and elements to nourish the plant.

Pumps and air stones oxygenate nutrient solutions so that roots are aerated. Lighting is essential for plants to undergo photosynthesis above the solution surface. Growers monitor nutrient solutions' pH levels and the presence of any algae, which might harm roots, interfere with their adsorption of nutrients, and impede plant growth. Growers also replenish fluids lost to evaporation.

Nutrient Solution Culture (NSC) Methods. Several forms of NSC are utilized to feed plants. Continuous-flow NSC involves nutrient solutions being poured into a trough and constantly moving through roots. Nutrient solutions contact roots less frequently in intermittent-flow NSC. The drip NSC technique delivers nutrient solutions through tubing and emitters that dispense water on the substrate near roots. Some drip systems recycle nutrient solution. The wick system utilizes strings that extend from substrates to a reservoir filled with nutrient solution.

In the ebb-and-flow method, nutrient solution contacts roots in cycles after flooding trays containing roots and substrates then draining and returning to a tank to store for additional delivery. Timers control pump mechanisms, which move nutrient solutions. Aquaponics systems transport water from ponds or greenhouses where fish tanks are kept to greenhouses where plants are grown so that wastes from the fish can provide nutrients for plants.

Nutrient Film Technique (NFT). This closed system continually pumps nutrient solutions into a channel placed at an angle in which roots hang under plants supported from above by platforms or other equipment. No substrates are used. The solution contacting roots is delivered as a watery film to assure roots will receive sufficient oxygen. The hydroponic trough system uses a reservoir, which includes a filtering device to strain contaminants from nutrient solutions. Resembling NFT methods, aeroponics does not rely on substrates. Sprayers attached to timers continually dispense nutrient solutions on roots suspended in air below plants.

Applications and Products

Nutrition. Hydroponics enables ample production of food supplies that meet nutritional needs for vitamins, antioxidants, and amino acids crucial for maintaining people's bodies. These techniques aid hunger relief in arid regions where climate change is associated with expanding desertification and loss of arable land, threatening food security. Hydroponic cultivation provides both rural and urban populations access to affordable, fresh, healthy food despite loss of access to traditional agricultural supplies due to political, economic, or military crises; natural disasters; or famines. Various hydroponic techniques can be applied to produce crops with increased levels of calcium, potassium, and other elements essential to sustain health. Hydroponic processes can be designed to grow food with appealing tastes, textures, and appearances.

Agribusiness. Hydroponics generates profits for commercial sellers of crops, manufacturers of hydroponic equipment, nutrient solutions, and supplies and wholesalers and retailers that distribute hydroponic merchandise to consumers. Agribusinesses create and market hydroponic greenhouses of varying sizes, including small growing containers such as AeroGarden for use inside homes, to consumers. Many florists grow stock cultivated with hydroponics at their stores. Internationally, the number of hydroponic businesses has expanded on all continents except Antarctica, contributing to countries' economies. By 2008, Advanced Nutrients, one of the most successful hydroponic businesses internationally, sold its merchandise to customers from forty-one countries. Some hydroponic companies develop and sell smart phone applications to perform hydroponic functions, such as General Hydroponics' calculator for preparing nutrient solutions.

Education. Students at various levels, from elementary through graduate school, often study hydroponics in science classes. Some courses may discuss hydroponics to explain basic scientific principles such as how roots absorb nutrients, while others may focus on special topics such as genetics. Students frequently investigate aspects of hydroponics for science-fair competitions or projects for the Future Farmers of America. Teachers instructing Advanced Placement biology courses often encourage students to develop hydroponic systems to comprehend concepts associated with plant growth and nutrition. Some school cafeterias use foods grown on their campuses or students sell products cultivated with hydroponic techniques for fund-raisers. Universities sometimes award funds to students' innovative hydroponic projects, especially those with humanitarian applications. The Denver Botanic Gardens and other botanical centers offer hydroponic classes.

Military and Exploration. Military troops benefit from the establishment of hydroponic systems near bases and battlefields to produce fresh vegetables for rations regardless of soil and climate conditions in those areas. Hydroponic applications for military usage enable crews on vessels undergoing lengthy sea voyages to grow foods when they are between ports. Veterans with hydroponic experience or who complete Veterans Sustainable Agriculture Training or similar programs are often sought out for employment in that field. The ability to grow foods without soil nourishes people traveling by submarine, whether for military or scientific reasons. Workers in remote locations, such as off-shore oil and natural-gas rigs, eat meals incorporating hydroponic produce grown at those sites.

Scientists conducting research at Antarctic stations rely on hydroponics for sustenance and as a method to recycle, purify, and store water. The South Pole Food Growth Chamber, designed by the Controlled Environment Agriculture Center at the University of Arizona, uses NFT methods and is automated with an Argus climate-control system. The National Aeronautics and Space Administration (NASA) funds projects such as Controlled Ecological Life Support Systems (CELSS), in which hydroponic plants remove carbon dioxide and pollutants while producing food on spacecraft. Researchers are investigating using hydroponics for future missions of long duration.

Urban Planning. Some twenty-first-century architecture incorporates hydroponics as a strategy to feed increasing populations, particularly in urban areas. Rooftops are popular sites for hydroponic systems in places where land is unavailable for gardens. These urban farms grow large yields of basic vegetable crops and supply fresh produce to residents who might otherwise not have access to those foods. Vertical farming techniques inspired proposals applying hydroponics. In New York City, Dickson Despommier, a Columbia University microbiologist, introduced his idea to renovate almost two thousand empty structures with hydroponic equipment. The Seoul Commune 2026 in South Korea presented another vertical farming proposal. This project will involve covering skyscrapers, some fifty floors high, with supports for plants. Nutrients delivered by fog machines and irrigation technology to roots will nourish plants growing on those garden buildings.

Tourism. Some hotels, especially in exotic locales, apply hydroponics for agricultural and aesthetic uses. On Anguilla in the Caribbean, the CuisinArt Resort and Spa grows hydroponic herbs, vegetables, and flowers. Guests can tour areas with hydroponic equipment to see where food served in the hotel's restaurant is grown. These businesses sometimes sell hydroponic products, often identified by resorts' brands, to cruise ships docking nearby or to markets. Visitors can tour hydroponic displays at the Hampshire Hydroponicum in England and Epcot's Land Pavilion in Orlando, Florida. Zoos occasionally utilize hydroponic processes to cultivate grain and grass to feed animals.

Impact on Industry

Organic Monitor declared the organic agriculture industry, including hydroponics, generated $50.9 billion globally in 2008. Hydroponic systems used for production of legal crops represented part of the $26.6 billion of organic products sold during 2009 in North America. In 2006, hydroponic retail sales totaled $55 million, a $40-million increase since 2002. Internationally, scientists and engineers, working for universities, industries, and governments, conduct research to improve the hydroponic industry for humanitarian, utilitarian, and commercial applications by manufacturers, businesses, and agriculturists.

Government and University Research. Most hydroponic research is supported with funds from academic institutions, governmental agencies, or professional organizations. Hydroponic projects conducted at land-grant universities typically occur at experiment stations where scientists and engineers contact experts in other fields so they can incorporate interdisciplinary approaches to resolving problems and innovating new technologies. Academic and research laboratories host studies investigating procedures to advance hydroponics by designing plants and equipment compatible with hydroponic processes. Some hydroponics researchers utilize computer models to test hypotheses. Scientists can observe distant hydroponic facilities such as those on Antarctica with Internet and satellite technology to gather data.

The Ohio Agricultural and Research Development Center initiated notable research in the early twenty-first century regarding hydroponics used to grow lettuce. Scientists perform experiments relevant to hydroponic systems designs, substrates, and rate of nutrient flow to determine how those factors impact yields. Horticulturists study which plant species are best suited for cultivation using hydroponic methods. Some genetics researchers at the University of Wisconsin focused on developing smaller versions of plants, such as dwarf wheat, to grow in limited spaces. Scientists also investigate increasing maturation of plants so they can be harvested more quickly. Starting in 2008, Pakistan's government funded BioBlitz, a hydroponic program that grows 50,000 kilograms of tomatoes weekly on five acres. Officials projected hydroponics could enable Pakistan to earn $500 million annually by cultivating 1,000 acres.

Environmental impact of hydroponics is a universal concern among researchers. The United States Environmental Protection Agency (EPA) promoted hydroponic research to increase use of inorganic substrates to minimize agriculturists' reliance on chemicals in fields. New York's State Energy Office encouraged hydroponic cultivation to decrease the amount of energy needed to grow and transport food.

Industry and Business. Agribusinesses invest in hydroponic research to enhance the profitability of their companies by offering improved technology and products that appeal to customers, whether individuals, governments, or corporations.

Industry experts project that the hydroponic industry will continue growing in the twenty-first century. This is concurrent with the market for organic foods, which increased eleven percent from 2008 to 2009. The hydroponic industry will expand to produce equipment for new applications as they are envisioned and introduced. Businesses will recognize investment opportunities hydroponics systems offer because that technology can expand food production beyond the limited quantities grown from natural resources.

Partnerships link hydroponic agriculture with existing food manufacturing methods. For example, in India, Himalya International, which manufactures frozen foods, contracts with the British company New World Paradigm to produce crops using hydroponic systems for Himalya International to process for distribution to markets. New World Paradigm also supplies hydroponic produce to food manufacturers in Europe and North America. Alliances between companies, such as American Hydroponics and Rimol Greenhouse Systems, are financially benefited by centralizing hydroponics resources for consumers to purchase.

Successful hydroponic technology manufacturers include Orbitec, which created the Biomass Production System that can be used to grow dwarf plants developed by researchers. The Orbitec hydroponics technology regulates nutrients, moisture, lighting, and temperatures. DuPont researchers innovated materials that are biodegradable for growing plants hydroponically. Companies design and produce greenhouses of various sizes and shapes, including geodesic domes. Manufacturers build machinery incorporating computerized functions to perform tasks. OrganiTech developed the automated hydroponic Grow-tech 2000 equipment, securing revenues of $3.5 million in 2005, including a $2.73

million contract to construct a 100,000-square-foot greenhouse in Russia.

Careers and Course Work

Students interested in professions associated with hydroponics can complete diverse educational programs to pursue their career goals. Many entry-level hydroponic positions are available to people with high school educations or associate's degrees. Some employers seeking qualified workers to build and maintain hydroponic systems expect candidates to have completed basic horticultural courses at technical schools, community colleges, or universities, preparing them to cultivate plants and assemble equipment. Experience working for landscaping businesses, farms, or other positions that involve tending plants enhances one's employability. One can sometimes find available positions at gardening businesses that use hydroponics to grow crops and ornamental plants to sell to consumers and markets. Resorts, botanical gardens, and theme parks hire people with educational and work experience to establish and maintain hydroponic gardens.

Government, academic, and industrial employers that staff scientific and technological positions focusing on hydroponics usually require the minimum of a bachelor of science degree in a related field. Candidates can acquire basic knowledge for plant cultivation by studying horticulture, botany, agriculture, or subjects applicable to hydroponics. Those seeking research positions typically need to earn advanced degrees–a master's or doctorate–in relevant subjects, acquiring expertise that will benefit the quality of their employers' services and hydroponic products. Agricultural engineering, computer science, or robotics courses prepare employees for positions designing hydroponic structures, machinery, and automation software. Candidates with advanced education or hydroponic experience have credentials for many positions as administrators or educators in schools, experiment stations, extension services, or government agencies.

Social Context and Future Prospects

Throughout the twenty-first century, hydroponics will continue to provide humanitarian and commercial benefits. The Hydroponic Merchants Association stated in 2004 that hydroponic greenhouses grew 55,000 acres of vegetables internationally, of which 5,800 acres in North America produced tomatoes, peppers, and cucumbers valued at $2.4 billion. That organization estimated the hydroponic industry will continue growing ten percent yearly because of increasing demand and advances in hydroponic technology. Industry experts suggest that hydroponics, universal to diverse cultures, will continue to expand for several reasons, including depletion of arable lands caused by natural disasters and global warming, expenses associated with machinery and operation of conventional agriculture; and public disapproval of bioengineering associated with field crops.

The early twenty-first-century economic recession motivated consumers to use hydroponic equipment because many cannot afford produce sold in stores. Some domestic hydroponic growers sell their products to earn money while they are unemployed or to supplement incomes. Many people practice hydroponics when grocery stores in their communities close because of financial problems, resulting in those consumers lacking access to fresh food. Hydroponics presents food-security solutions to the increasing population, which is estimated to reach nine billion people by 2050. Some experts speculate hydroponics will eventually surpass mainstream agriculture to produce the most food worldwide.

Elizabeth D. Schafer, Ph.D.

Further Reading

Despommier, Dickson. *The Vertical Farm: Feeding the World in the Twenty-First Century.* New York: St. Martin's Press, 2010. Emphasizing humanitarian and environmental benefits, narrative promotes development of urban hydroponic farms to increase food supplies.

Giacomelli, Gene A., et al. "CEA in Antarctica: Growing Vegetables on 'the Ice.'" *Resource: Engineering & Technology for a Sustainable World* 13, no. 1 (January/February, 2006): 3-5. Discusses how hydroponic systems serve both scientists in polar regions and researchers monitoring those processes remotely with communications technology.

Hansen, Robert, Jeff Balduff, and Harold Keener. "Development and Operation of a Hydroponic Lettuce Research Laboratory." *Resource: Engineering & Technology for a Sustainable World* 17, no. 4 (July/August, 2010): 4-7. Text and photographs document Ohio Agricultural and Research Development Center cultivation experiments' growth sequence, results, and impact on growers.

Jones, J. Benton, Jr. *Hydroponics: A Practical Guide for the Soilless Grower.* 2d ed. Boca Raton, Fla.: CRC Press, 2005. Agronomist addresses diverse aspects of cultivating crops with nutrient solutions. Chapter concentrating on hydroponics in education outlines ideas and resources for projects; includes CD-ROM.

Lloyd, Marion. "Gardens of Hope on the Rooftops of Rio." *The Chronicle of Higher Education* 52, no. 8 (October 14, 2005): A56. Describes hydroponic work conducted by students in a Brazilian shantytown to teach residents how to grow fresh, nutritious foods.

WEB SITES

Food and Agriculture Organization of the United Nations
http://www.fao.org

National Aeronautics and Space Administration
Farming for the Future
http://www.nasa.gov/missions/science/biofarming.html

National Plant Data Center
http://npdc.usda.gov

See also: Food Science; Horticulture.

I

INDUSTRIAL FERMENTATION

FIELDS OF STUDY

Biology; microbiology; biochemistry; organic chemistry; biotechnology; bioprocess engineering; chemical engineering.

SUMMARY

Industrial fermentation is an interdisciplinary science that applies principles associated with biology and engineering. The biological aspect focuses on microbiology and biochemistry. The engineering aspect applies fluid dynamics and materials engineering. Industrial fermentation is associated primarily with the commercial exploitation of microorganisms on a large scale. The microbes used may be natural species, mutants, or microorganisms that have been genetically engineered. Many products of considerable economic value are derived from industrial fermentation processes. Common products such as antibiotics, cheese, pickles, wine, beer, biofuels, vitamins, amino acids, solvents, and biological insecticides and pesticides are produced via industrial fermentation.

KEY TERMS AND CONCEPTS

- **Antioxidant:** Chemical that prevents the oxidation of other chemicals.
- **Biomass:** Mass of organisms; traditionally, this term refers to the biomass of plants and microorganisms.
- **Bioreactor:** Apparatus for cell growth with practical purposes under controlled conditions. Bioreactors are closed systems and vary in size from the small laboratory scale (five to ten milliliters) to the large industrial scale (more than 500,000 liters).
- **Enzymes:** Biological catalysts made of proteins.
- **Fermentation:** In biology, the metabolic reactions necessary to generate energy in living (mainly microbial) cells; in industry, any large industrial process based on living things is called fermentation.
- **Fermenter:** Type of traditional bioreactor (stirred or nonstirred tanks) where cell fermentation takes place. Fermenters can be operated as continuous or batch-culture systems. In a continuous culture, nutrients are continuously fed into the fermentation vessel. This allows the cells to ferment indefinitely.
- **Probiotics:** Microorganisms and substances that promote the development of healthy intestinal microbial communities.
- **Substrate:** Molecule that is broken down by fermentation.

Definition and Basic Principles

Industrial fermentation is the use of living organisms (mainly microorganisms), typically on a large scale, to produce commercial products or to carry out important chemical transformations. The goal of industrial fermentation is to improve biochemical or physiological processes that microbes are capable of performing while yielding the highest quality and quantity of a particular product. The development of fermentation processes requires knowledge from disciplines such as microbiology, biochemistry, genetics, chemistry, chemical and bioprocess engineering, mathematics, and computer science. The major microorganisms used in industrial fermentation are fungi (such as yeast) and bacteria. Fermentation is performed in large fermenters or other bioreactors often of several thousand liters in volume. Industrial fermentation is a part of many industries, including microbiology, food, pharmaceutical, biotechnology, and chemical.

Background and History

Traditional fermentations such as those for making bread, cheese, yogurt, vinegar, beer, and wine had been used by people for thousands of years before its microbial nature was understood. Brewing beer was one of the first applications of fermentation in ancient Egypt as long as 10,000 years ago. The exact origins of dairy products are unknown—it may have been as early as 8000 B.C.E. It was probably nomadic Turkish tribes in Central Asia who invented cheese and yogurt making. Traditionally, dairy fermentation was a means of milk preservation. The scientific understanding of fermentation began only in the nineteenth century after French scientist Louis Pasteur published the results of his studies on the microbial nature of wine making.

The first industrial fermentation bioprocesses based on knowledge of microbes appeared in the early twentieth century. Russian biochemist Chaim Weizmann is considered to be the father of industrial fermentation. Weizmann used the bacterium *Clostridium acetobutylicum* for the production of acetone from starch in 1916. Acetone was used to make explosives during World War I.

Significant growth of this field began in the middle of the twentieth century, when the fermentation process for the large-scale production of antibiotic penicillin was developed. The goal of industrial-scale production of penicillin during World War II led to development of fermenters by engineers that were working together with biologists from the pharmaceutical company Pfizer. The fungus *Penicillium* grows and produces an antibiotic much more effectively under controlled conditions inside the fermenter. Continuous progress in industrial fermentation technology in the twentieth century has followed the development of genetic engineering. Genetic engineering allows gene transfer between species and creates possibilities to generate new products from genetically modified microorganisms that are grown in fermenters.

The twenty-first century has been characterized by the introduction of biofuels, which are made by industrial fermentation processes. Once again, past and future developments in fermentation technology require contributions from a wide range of disciplines, including microbiology, genetics, biochemistry, chemistry, engineering, mathematics, and computer science.

How It Works

Industrial fermentation is based on microbial metabolism. Microbes produce different kinds of substances that they used for growth and maintenance of their cells. These substances can be useful for humans. The goal of industrial fermentation technology is to enhance the microbial production of useful substances.

Process of Fermentation. In biology, fermentation is a process of harvesting energy of organic molecules in oxygen-free conditions. Sugars are a prime example of what can be fermented, although, there are many other organic molecules that can be used. Different fermentations are known and are categorized by the substrate metabolized or the type of the product.

In industry, any large microbiological process is called fermentation. Thus, the term fermentation has a different meaning than in biology. Most industrial fermentations require oxygen.

Industrial Fermentation Organisms. Different organisms, such as bacteria, fungi, and plant and animal cells, are used in industrial fermentation processes. An industrial fermentation organism must produce the product of interest in high yield, grow rapidly on inexpensive culture media available in bulk quantities, be open to genetic manipulation, and be nonpathogenic (does not cause any diseases).

Fermentation Media. To make a desired product by fermentation, microorganisms need nutrients (substrates). Nutrients for microbial growth are known as media. Most fermentation requires liquid media or broth. General media components include carbon, nitrogen, oxygen, and hydrogen in the form of organic or inorganic compounds. Other minor or trace elements must also be supplied, for example, iron, phosphorus, or sulfur.

Fermentation Systems. Industrial fermentation takes place in fermenters, which are also called bioreactors. Fermenters are closed vessels (to avoid microbial contamination) that reach vast volumes, as many as several hundred thousand liters. Designed by engineers, the main purpose of a fermenter is to provide controllable conditions for growth of microbial cells or other cells. Parameters such as pH, temperature, nutrients, fluid flow, and other variables are controlled. There are two kinds of fermenters, those for anaerobic processes (oxygen-free) and aerobic processes. Aerobic fermentation is the most common in industry. Anaerobic fermenters can be as simple as stainless-steel tanks or barrels. Aerobic fermenters are more complicated. The most critical part in these systems is aeration. In a large-scale fermenter transfer of oxygen is very important. Oxygen transfer and dispersion are provided by stirring with impellers or oxygen (air) sparging.

Fermentation Control and Monitoring. Industrial fermentation control is very important to ensure that organisms behave properly. In most cases computers are used for controlling and monitoring the fermentation process. Computers control temperature, pH, cell density, oxygen concentration, level of nutrients, and product concentration.

Applications and Products

There are a wide range of industrial fermentation products and applications.

Food, Beverages, Food Additives, and Supplements. Industrial fermentation plays a major role in the production of food. Food products traditionally made by fermentation include dairy products (cheeses, sour cream, yogurt, and kefir); food additives and supplements (flavors, proteins, vitamins, and carotenoids); alcoholic beverages (beer, wines, and distilled spirits); plant products (bread, coffee, soy sauce, tofu, sauerkraut); and fermented meat and fish (pepperoni and salami).

Industrial Fermentation. The primary and largest industry revolves around food products. Milk from cows, sheep, goats, and horses have traditionally been used for the production of fermented dairy products. These products include cheese, sour cream, kefir, and yogurt. More recently so-called probiotics appeared and have been marketed as health-food drinks. Dairy products are produced via fermentation using lactic bacteria such as *Lactobacillus acidophilus* and *Bifidobacterium*. Fungi are also involved in making some cheeses. Fermentation produces lactic acid and other flavors and aroma compounds that make dairy products taste good.

Many products of industrial fermentation are added into food as flavors, vitamins, colors, preservatives, and antioxidants. These products are more desirable than food additives produced chemically. Many of the vitamins are made by microbial fermentations including thiamine (vitamin B_1), riboflavin (vitamin B_2), cobalamin (vitamin B_{12}), and vitamin C (ascorbic acid). Vitamin C is not only a vitamin but is also an important antioxidant that helps to prevent heart diseases. Carotenoids are another effective antioxidant. They are also used as a natural food color for butter and ice cream. Carotenoids are red, orange, and yellow pigments produced by bacteria, algae, and plants.

Food preservatives are yet another product of industrial fermentation. Organic acids, particularly lactic and citric acids, are extensively used as food preservatives. Some of these preservatives (such as citric acid) are used as flavoring agents. A mixture of two bacterial species (*Lactobacillus* and *Streptococcus*) is usually used for industrial production of lactic acid. The mold *Aspergillus niger* is used for citric acid manufacturing. Another common preservative is the protein nisin. Nisin is produced via fermentation by the bacterium *Lactococcus lactis*. It is employed in the dairy industry especially for production of processed cheese.

Antibiotics and Other Health Care Products. Antibiotics are chemicals that are produced by fungi and bacteria that kill or inhibit the growth of other microbes. They are the second most significant product of industrial fermentation. Most antibiotics are generated by molds or bacteria called actinomycetes. More than 4,000 antibiotics have been isolated from microorganisms, but only about 50 are produced regularly. Among them, beta-lactams, such as penicillins and tetracyclines, are most common. Penicillin is produced by the mold *Penicillium chrysogenum* via corn fermentation in bioreactors of up to 200,000 liters.

The other major health care products produced with the help of industrial fermentation are bacterial vaccines, therapeutic proteins, steroids, and gene therapy vectors. There are two categories of bacterial vaccines: living and inactivated vaccines. Living vaccines consist of weakened, also known as attenuated, bacteria. Examples of living vaccines include those for diseases such as anthrax, which is caused by *Bacillus anthracis*, and typhoid fever, which is caused by *Salmonella typhi*. Inactivated vaccines are composed of bacterial cells or their parts that have been inactivated by heat or formaldehyde. Examples of these vaccines are those for meningitis, whooping cough, and cholera. Vaccine production takes place in fermenters no bigger than 1,000 liters in volume. It requires highly controlled operations to avoid the release of bacteria into the environment. All exhaust gases pass through sterilization processes.

Therapeutic proteins include growth hormone, insulin, wound-healing factors, and interferon. Previously, such compounds were made from animal tissues and were very expensive to manufacture. Genetic engineering now allows their production by fermentation from bacteria. Human growth hormone is synthesized in the human brain and controls growth. Too little growth hormone can cause some cases of dwarfism. The American company Genentech started production of human growth hormone from genetically modified *Escherichia coli* by fermentation in

1985. Insulin is an animal and human hormone that is involved in the regulation of blood sugar. The body's inability to make sufficient insulin causes diabetes. Insulin extracted from pigs had been used to treat diabetes, but it has been replaced by insulin produced by industrial fermentation from genetically modified bacteria.

Chemicals. Numerous chemicals, such as amino acids, polymers, organic acids (citric, acetic, and lactic), and bioinsecticides are produced by industrial fermentation. Amino acids are used as a food and animal feed, as well as in the pharmaceutical, cosmetic, and chemical industries. Bacteria such as *Micrococcus luteus* and *Corynebacterium glutamicum* are used for industrial fermentation to produce chemicals. Bacterial toxins are effective against different insects. Since the 1960's, preparations of the bacteria *Bacillus thuringiensis* have been produced by fermentation as a biological insecticide.

Enzymes. Enzymes are used in many industries as catalysts. Microorganisms are the favored source for industrial enzymes. Seventy percent of these enzymes are made from *Bacillus* bacteria via fermentation. Most commercial microbial enzymes are hydrolases, which break down different organic molecules such as proteins and lipids. The enzyme glucose isomerase is important in the production of fructose syrups from corn and is widely used in the food industry.

Biomass Production. During biomass production by fermentation, the cells produced are the products. Biomass is used for four purposes: as a source of protein for human food or animal feed, in industry as fermentation starter cultures, in agriculture as a pesticide or fertilizer, and as a fuel source.

One major product of this application of industrial fermentation is baker's yeast biomass. Baker's yeast is required for making bread, bakery products, beer, wine, ethanol, microbial media, vitamins, animal feed, and biochemicals for research. Yeast is produced in large aerated fermenters of up to 200,000 liters. Molasses is used as a nutrient source for the cells. Yeast is recovered from fermentation liquid by centrifugation and then dried. It can then be sold as compressed yeast cakes or dry yeast.

Many bacteria have been considered as potential sources of protein to fulfill the food needs in some countries of the world. As of 2011, only a few species are cultivated around the world as a source of food and feed. Among them, cyanobacteria are the most popular. The protein level of *Cyanobacterium Spirulina* can be as high as those found in meat, nuts, and soybeans, from 50 to 70 percent. This cyanobacterium has been used as a human food for millennia in Asia, parts of Africa, and in Mexico.

Apart from yeast and bacteria, people are also using the biomass of algae. Algae are a source of animal feed, plant fertilizer, chemicals, and biodiesel. Because light is necessary to grow algae, the biomass is produced in open ponds or in transparent tubular glass or plastic bioreactors, called photobioreactors.

Fascinating Facts About Industrial Fermentation

- Yeast (*Saccharomyces cerevisiae*) was one of the earliest domesticated organisms. It was used by ancient Egyptians for making bread and beer. People used yeast for thousands of years without knowing about it until the nineteenth century.
- The holes in some cheeses, such as Swiss cheese, are a result of gas production during bacterial fermentation. Bacteria *Propionibacterium* are used during ripening (aging) of cheese to add flavor.
- The most common mushroom in grocery stores is *Agaricus bisporus.* Since 1810, it has been cultivated using wheat straw and horse manure as food sources. *Agaricus* cultivation represents an example of the most efficient conversion of plant and animal wastes into edible food.
- Vinegar is produced via industrial fermentation of alcohol (ethanol) by acetic acid bacteria. These bacteria are not welcomed by winemakers since bacteria tend to spoil wine by converting it into vinegar. The word vinegar is derived from the French words *vin* (wine) and *egre* (sour).
- San Francisco has extended its recycling program to include fermentation of dog waste into methane. Methane is used to fuel electrical generators to produce electricity and to heat homes. San Francisco's dog population is 120,000, and this initiative promises to generate a significant amount of fuel while reducing a huge amount of waste.
- Chaim Weizmann, who is considered to be the father of industrial fermentation, became the first president of Israel. Weizmann's life path from the industrial fermentation to becoming a head of state illustrates the importance of industrial fermentation.

Biofuels. Industrial fermentation is used in the production of biofuels, mainly ethanol and biogas. These two biofuels are produced by the action of microorganisms in bioreactors. Fermentation can also be used for generation of biodiesel, butanol, and biohydrogen. Biofuels are considered, by many, as a future substitute for fossil fuels. Pollution from fossil fuels affects public health and causes global climate change due to the release of carbon dioxide (CO_2). Using biofuels as an energy source generates less pollutants and little or no CO_2.

Production of ethanol is a process based on fungal or bacterial fermentation of a variety of materials. In the United States, most of the ethanol is produced by yeast (fungal) fermentation of sugar from cornstarch. Sugar is extracted using enzymes, and then yeast cells convert the sugar into ethanol and CO_2. Ethanol is separated from the fermentation broth by distillation. Brazil, the second largest ethanol producer after the United States, uses sugarcane fermentation to generate ethanol. The Brazilian production of ethanol from sugarcane is more efficient than the American corn-based ethanol.

Biogas is produced during the anaerobic (non-oxygen) fermentation of organic matter by communities of microorganisms (bacteria and *Archaea*). There are different types of biogas. One type contains a mixture of methane (50 to 75 percent) and CO_2. Another type is composed primarily of nitrogen, hydrogen, and carbon monoxide (CO) with trace amounts of methane. Methane is generated by microorganisms called *Archaea* and is an integral part of their metabolism. For practical use, biogas is generated from wastewater, animal waste, and "gas wells" in landfills.

Impact on Industry

Industrial fermentation plays a major role in a number of multibillion-dollar industries, including food, pharmaceutical, microbiological, biotechnological, chemical, and biofuel. The United States maintains a dominant position in the world in industrial fermentation. In fact, the first really large-scale industrial fermentation process to produce the antibiotic penicillin was developed in the United States. Many other developed and developing countries use industrial fermentation to produce varieties of products. Some countries have made industrial fermentation a prime national interest. Brazil, for example, is using the industrial fermentation for ethanol production. This country has the largest and most successful ethanol for fuel program in the world. As the result of this successful program, Brazil reached complete self-sufficiency in energy supply in 2006.

Government and University Research. Governmental agencies such as the National Science Foundation (NSF), the United States Department of Energy (DOE), and the United States Department of Agriculture provide funding for research in the industrial fermentation area. Currently, a vast majority of research is concentrated on biofuel generation by microorganisms and environmental applications.

Industry and Business. Scientists in the industry traditionally carry out a significant load of research in fermentation area. Companies such as Pfizer and Merck were pioneers in industrial production of the first antibiotic penicillin produced by fungal fermentation. A significant proportion of industrial fermentation research in industry has been directed to health care products (such as antibiotics).

Major Corporations. Examples of major corporations in food and beverage industries are Kraft Foods, Dannon, Coors, Guinness, and Anheuser-Busch. In the biofuel industry, major companies are Archer Daniels Midland, Poet Energy, Abengoa Bioenergy, and VeraSun Energy.

Careers and Course Work

There are several career options for people who are interested in being trained in industrial fermentation. Food, biotechnology, microbiology, pharmaceutical, chemical, and biofuel companies are the biggest employers in the area. Students who are interested in conducting research in industrial fermentation can find jobs in university, government, and industry laboratories.

When choosing a career in industrial fermentation, one should be prepared for an interdisciplinary science. Students should obtain skills in microbiology, molecular biology, bioengineering, plant biology, organic chemistry, biochemistry, agriculture, bioprocess engineering, and chemical engineering.

Most professionals in industrial fermentation have a bachelor's degree in biology, microbiology, or biotechnology. Individuals who have managerial responsibilities often have a master's or doctorate in biology, microbiology, fermentation, molecular biology, biochemistry, biotechnology, bioprocess or chemical engineering, or genetics. Some universities, including the University of California, Davis, and Oregon State University, offer degrees in fermentation.

A career in industrial fermentation presents a variety of work options such as research, process development, production, technical services, or quality control. Some industrial fermentation specialists may be considered as genetic engineers (using DNA techniques to modify living organisms), while others are classified as bioprocess or chemical engineers (optimizing bioreactors and biochemical pathways for a desired product).

Social Context and Future Prospects

Industrial fermentation plays a major role in providing food, chemicals, and fuels. End users are consumers, farmers, medical doctors, and industrialists. Industrial fermentation is changing the course of history. People have made food by fermentation for centuries. In the twentieth century, the development of antibiotics and their production by industrial fermentation had the most significant impact on the practice of medicine than any other development. The growth of the industrial fermentation field is continuing rapidly. Since the beginning of the twenty-first century, industrial fermentation underwent an unprecedented growth and expansion due to biofuel introduction. This record growth is particularly visible in the U.S. ethanol industry. In 1980, the U.S. ethanol industry produced 175 million gallons of ethanol by fermentation, and in 2009, 10.6 billion gallons.

The role of industrial fermentation in human society is likely to expand in the future because of increasing requirements for resources.

Sergei A. Markov, Ph.D.

Further Reading

Bailey, James E., and David F. Ollis. *Biochemical Engineering Fundamentals.* 2d ed. New York: McGraw-Hill, 1986. Classic textbook on biochemical engineering.

Bourgaize, David, Thomas R. Jewell, and Rodolfo G. Buiser. *Biotechnology: Demystifying the Concepts.* San Francisco: Benjamin/Cummings, 2000. Classic text on biotechnology.

Doran, Pauline M. *Bioprocess Engineering Principles.* San Diego: Academic Press, 1995. Describes various bioreactors and fermenters used for industrial fermentation.

Glazer, Alexander N., and Hiroshi Nikaido. *Microbial Biotechnology: Fundamentals of Applied Microbiology.* 2d ed. New York: Cambridge University Press, 2007. In-depth analysis of application of microorganisms in industrial fermentation.

Lydersen, Bjorn K., Nancy A. D'Elia, and Kim L. Nelson, eds. *Bioprocess Engineering: Systems, Equipment and Facilities.* New York: John Wiley & Sons, 1994. Describes equipment and facilities for industrial fermentation.

Wright, Richard T., and Dorothy F. Boorse. *Environmental Science: Toward a Sustainable Future.* 11th ed. Boston: Benjamin/Cummings, 2011. This textbook describes several bioprocesses used in waste treatment and pollution control.

Web Sites

Biotechnology Industry Organization
http://www.bio.org

Society of Bioscience and Technology
http://www.socbioscience.org

United States Department of Agriculture
Biotechnology
http://www.usda.gov/wps/portal/usda/usdahome?navid=BIOTECH

See also: Food Preservation; Food Science.

ION PROPULSION

FIELDS OF STUDY

Inorganic chemistry; general physics; plasma physics; astrophysics; mechanical engineering; electronics; electrical engineering; systems engineering; mathematics; statistics; material science.

SUMMARY

Ion propulsion is the ejection of charged particles from a space ship, causing the ship to move in the opposite direction from the ejected particles. Chemical rocket engines move ships in much the same way, but the speed of their exhaust is limited by the energy released in the chemical reaction. Ions, however, can be accelerated by electric and magnetic fields to far higher speeds, so ion engines require a far smaller mass of fuel than chemical engines.

KEY TERMS AND CONCEPTS

- **Burn Time:** Time it takes to burn up an amount of rocket fuel, and by extension, the time it takes an ion engine to eject a given amount of mass.
- **Geostationary Earth Orbit (GEO):** Geosynchronous orbit, 35,785 kilometers above the equator, so that the object in orbit appears to stay at a fixed point relative to the Earth's surface at all times.
- **Hall Effect:** Generation of a difference in potential across a conductor carrying an electric current that is exposed to an external magnetic field applied at a right angle.
- **Hall-Effect Ion Engine:** Ion engine that uses magnetic fields to accelerate ions.
- **Ion Engine:** Device that ionizes reaction mass, accelerates it with either electric or magnetic fields, and expels it from the rear of the engine, thereby driving the engine forward; generally refers to an engine that uses electric fields to accelerate ions.
- **Low Earth Orbit (LEO):** Orbit 160 to 2,000 kilometers above the Earth's surface.
- **Reaction Mass:** Mass ejected from the rear of a rocket that drives the rocket forward.
- **Space Charge:** Concentration of charge in a region that is large enough to significantly affect other charges moving toward that region.
- **Thrust:** Force exerted on a rocket by its motor; equals the mass ejected from the motor per unit time multiplied by the exhaust velocity.

Definition and Basic Principles

The fundamental principle of rocket propulsion is Sir Isaac Newton's third law of motion: For every action, there is an equal and opposite reaction. The mass of fuel ejected rearward from a rocket results in the rocket being propelled forward. More precisely, the mass of fuel multiplied by the exhaust velocity equals the force pushing the rocket forward multiplied by the burn time. It follows then that the greatest amount of mass possible must be ejected each second at the highest possible speed to provide maximum acceleration to a rocket. The maximum exhaust speed for a chemical rocket is about 3 to 5 kilometers per second, but the exhaust speed for an ion engine can be 30 to 50 kilometers per second or greater.

The best thrust of an ion engine is only a small fraction of its weight, so ion engines cannot be used to lift a space ship into orbit. Chemical engines must perform that task. However, once the space ship is in orbit, ion engines can be very useful. Short bursts from an ion engine are ideal for station keeping (maintaining position and orientation with respect to another craft or object). Because ion engines use so little mass, they may be fired for days or months instead of the minutes that a chemical engine can burn. The low thrust of an ion engine fired for a long time can make major changes in the orbit of a space ship. For example, an ion engine took the National Aeronautics and Space Administration's Deep Space 1 from low Earth orbit to fly by asteroid 9969 Braille and Comet Borrelly.

Background and History

Anyone working with electric fields and charged particles soon discovers that it is easy to accelerate charged particles to high speeds. Rocket pioneers Robert Goddard (in 1906) and Konstantin Tsiolkovsky (in 1911) discussed the idea of an ion engine. In 1916, Goddard built an ion engine and demonstrated that it produced thrust. In 1964, National Aeronautics and Space Administration

(NASA) scientist Harold R. Kaufman built and successfully tested an ion engine that used mercury as reaction mass in the suborbital flight of the Space Electric Rocket Test 1 (SERT 1).

During the 1950's and 1960's, both the United States and the Soviet Union (now Russia) worked on a Hall-effect ion engine that used magnetic fields to accelerate ions. The United States continued to work on the electric field ion engine but dropped out of the competition for the Hall-effect engine. Russia eventually developed Hall-effect engines and began using them in space in 1972.

In 1992, the West adopted some of the Russian technology and started using Hall-effect engines as well as the electric field ion engine. In 1998, NASA launched Deep Space 1, which used an ion engine and flew by asteroid 9969 Braille and Comet Borrelly. The Deep Space 1 used an ion engine as its main source of propulsion, and its success paved the way for future missions powered by ion engines. In the 2000's, NASA, the European Space Agency (ESA), and the Japan Aerospace Exploration Agency launched spacecraft that used versions of ion engines.

How It Works

Heavy atoms are usually selected as the reaction mass for an ion engine because heavy ions increase the mass ejected per second. It is also desirable to use a substance that requires relatively little energy to ionize it. Mercury, cesium, xenon, bismuth, and argon have all been used for reaction mass. However, mercury and cesium are poisonous and require special handling, so the substance most commonly used for reaction mass is xenon. Xenon can be stored under pressure as a liquid for long periods of time. As a gas, it is easy to transfer from storage to the rocket motor, the molecules are relatively heavy and easily ionized, and it is chemically inert, so it will not corrode the engine. Ion engine bodies are often made of boron nitride, a good insulator that can withstand the operating conditions of the engine.

A small amount of the reaction mass is leaked at a controlled rate into a plasma chamber, where it is bombarded with electrons to ionize it. Fields from permanent magnets are used to confine and control the resulting plasma.

The walls of the plasma chamber are maintained at a high positive voltage (such as +1,100 volts) so that free electrons migrate to the walls and are absorbed. Two grids or wire screens are at the rear of the plasma chamber. The first grid is at a lower voltage, such as +1,065 volts, so that positive ions from the chamber are attracted to it. A second grid is placed 2 or 3 millimeters beyond the first grid and electrified at −180 volts. Positive ions will be accelerated by the electric field in the gap between the two grids and will gain 1,245 electron volts of energy. This means that xenon ions will exit the engine at an impressive 43 kilometers per second.

There are three areas that have to be addressed for ion propulsion to work. First, because the craft is ejecting positive charge into space, it is becoming more negative. If nothing were done about this, the craft would eventually become so negative that the positive ion exhaust would be attracted back to the ship, and there would be no propulsion. To deal with this, a special electrode sprays negative charge onto the ion plume behind the craft to keep the craft's charge neutral. Second, a positive charge being accelerated between the grids constitutes a space charge density. This space charge repels approaching positive charge, and this limits the ion current to fairly small values, which in turn limits the thrust of an ion engine to small values. These limits on thrust must be carefully calculated and considered. Third, high-energy ions strike electrodes and erode them. Careful attention to the electrode design and the materials used can extend electrode lives to 10,000 hours.

Hall-Effect Thrusters. The Hall effect (named for Edwin Hall, who discovered it in 1879) states that if a conductor carrying a current is placed in a magnetic field perpendicular to the current, the magnetic field will cause a new current to flow perpendicular to both the magnetic field and the original current.

As an example, consider a horizontal tube lying along the bottom of a page. Let the tube contain a vertical electric field pointing from the bottom to the top of the page. Suppose further that there is a horizontal magnetic field pointing out of the paper. If electrons and ionized xenon atoms are introduced at the left end of the tube, electrons will be propelled by the electric field to sink to the bottom side of the tube along the lower edge of the page. Because of the greater mass of the xenon ions, their progress toward the upper side of the tube will be far slower. A charged particle moving in a magnetic field experiences a force perpendicular both to its velocity and to the magnetic field. This causes the electrons that were

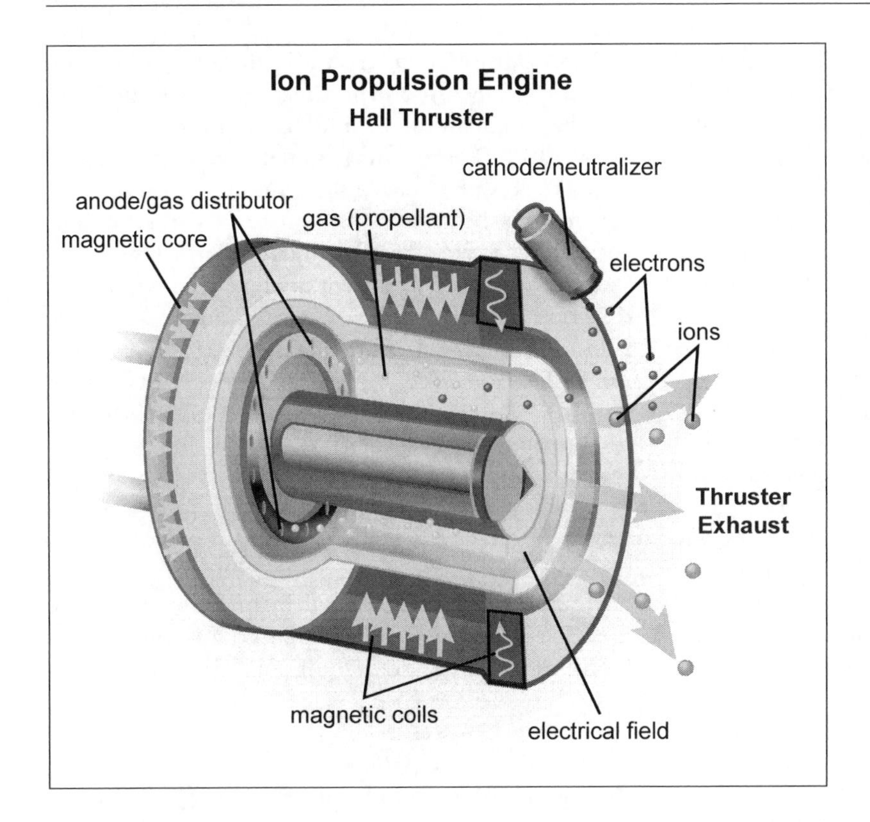

sinking to the bottom of the tube to move from left to right down the axis of the tube. These electrons will strike other electrons and ions, forcing them to move down the tube as well. This plasma now moving along the tube's axis will become the exhaust jet that propels the spacecraft. The mass exhausted is not limited by space charge and can be as large as the device can handle.

VASIMR. The VASIMR (Variable Specific Impulse Magnetoplasma Rocket) project of the Ad Astra Rocket Company in Webster, Texas, seems to hold the key to more powerful ion engines. The VASIMR engine uses microwaves to ionize and heat the propulsive gas. The temperature of the gas can be controlled by increasing or decreasing microwave intensity. The absence of electrodes means they cannot be eroded by hot plasma and suffer degraded performance as happens in a normal ion engine. Magnetic fields confine, direct, and accelerate the plasma to exhaust speeds up to 50 kilometers per second. Magnetic fields form a rocket nozzle inside the metal nozzle. Gas introduced between the metal nozzle and the plasma is heated by the plasma and adds to the thrust. This thrust can be modified by changing the microwaves heating the plasma, by changing the rate at which reaction mass is delivered to the engine, or by changing the size of the magnetic field nozzle. Engines using 100 and 200 kilowatts have been built and ground-tested. One day it may be possible to build a 200-megawatt engine powered by a small nuclear reactor. A craft powered by such an engine could make the trip from Earth to Mars in only 39 days, a considerable improvement over the 255 days required by conventional rockets.

Applications and Products

Numerous ion engines have been used for station keeping, and an increasing number of missions are using them as main engines.

SERT 1. Launched by NASA in 1964, SERT 1 showed that an ion engine actually functioned as expected in space. In 1970, SERT 2 was sent into space to test two ion engines using mercury as reaction mass. The engines operated 2,011 hours and 3,781 hours respectively and were restarted about three hundred times.

Deep Space 1. Launched by NASA in 1998, Deep Space 1 tested twelve high-risk technologies, including the NSTAR ion engine. The engine drew up to 2.5 kilowatts from solar cells and produced 0.092 newtons of thrust. It carried 81.5 kilograms of xenon reaction mass, enough to last 678 days. Each day, the engine added 24 to 32 kilometers per hour to the craft's speed. It flew by asteroid 9969 Braille and Comet Borrelly. Because of a software crash, the pictures of Braille were not as good as had been hoped, but the pictures of Comet Borrelly were outstanding. A camera had to be reprogramed to take over tracking duties when the star tracker failed, but most of the other instruments worked well. The mission ended with engine shutdown on December 18, 2001.

Artemis. The European Space Agency launched the Artemis in 2001, but the telecommunications satellite failed to reach its intended orbit. It used its remaining chemical fuel to transfer to a higher orbit, then used its xenon ion engine for eighteen months to raise it to the intended geostationary Earth orbit.

Hayabusa. The Japan Aerospace Exploration Agency launched Hayabusa in 2003. It reached the asteroid Itokawa in 2005, but the lander module MINERVA flew by the asteroid instead of landing as planned. Astronomers have long suspected that some asteroids are rubble piles held together by only self-gravitation. Images sent from Hayabusa seem to show exactly that. Hayabusa successfully landed and activated a collection capsule. The craft was propelled by four xenon ion engines that amassed a combined total of 31,400 operating hours. These were the first engines in space to use microwaves to form and heat the xenon plasma. Unfortunately, the engines failed one by one. None was operational to bring Hayabusa home, but operators were able to use the ion generator from one engine with the electron gun neutralizer from another engine to make one working engine. The sample capsule returned to Earth on June 13, 2010. It had several grains of dust inside, and if they prove to be from the asteroid, it will be the first such sample ever obtained.

SMART-1. The European Space Agency launched SMART-1 (Small Missions for Advanced Research in Technology 1) in 2003. A Hall-effect ion engine, using xenon as the reaction mass, allowed SMART-1 to travel from low Earth orbit to lunar orbit. SMART-1 surveyed the chemical elements on the lunar surface, and then was driven into the Moon in a controlled crash in 2006. Although the Russians had used Hall-effect engines for station keeping for many years, this was the first use of a Hall-effect engine as the main engine.

Dawn and GOCE. In 2007, NASA launched Dawn to explore the asteroid Vesta in 2011 and the dwarf planet Ceres in 2015. The spacecraft is propelled by three xenon ion engines but typically uses only one at a time. The European Space Agency launched GOCE (Gravity Field and Steady-State Ocean Circulation Explorer) in 2009. GOCE orbits only 260 kilometers above the Earth's surface so that it can map small changes in the Earth's field. It uses a xenon ion engine to make up for loses caused by air drag.

GSAT-4. The Indian Space Research Organization launched the GSAT-4, a geostationary satellite for navigation and communications, in April, 2010. It had an ion engine for station keeping, but the third stage failed to ignite, and the satellite was lost. The ion engine would have extended GSAT-4's normal ten-year life to fifteen years.

LISA. Although gravity waves have never been detected directly, theory predicts their existence, and they best explain the orbital decay of binary neutron stars. LISA (Laser Interferometer Space Antenna) is a very sensitive device that can detect the waves if they exist. The device will consist of three satellites containing special masses, mirrors, and lasers. The satellites will be 5×10^6 kilometers apart, each at a different corner of an equilateral triangle. The lasers should be able to measure small changes in the positions of the satellites caused by a gravity wave. LISA and LISA Pathfinder are joint projects of NASA and the European Space Agency. LISA Pathfinder will test several techniques to be used on LISA, including colloid thrusters. These thrusters are ion engines that use charged liquid drops for reaction mass. Their thrust is very small, perhaps only 20×10^{-6} newtons. Such small forces are useful in making tiny adjustments to the speed or position of a satellite. LISA Pathfinder is set for launch in 2011.

Impact on Industry

Ever since the Chinese developed gunpowder rockets in the eleventh century, the word "rocket" has usually referred to a device powered by explosive chemical reactions. It soon became apparent to those trying to improve rockets that increasing the exhaust speed of the reaction mass was the key to faster rockets. Therefore, in 1906, Robert Goddard conceived of an ion engine as a solution. One of the best chemical rocket fuel combinations is liquid oxygen and liquid hydrogen. When burned in an efficient engine, the resultant steam has an exhaust velocity of 4.5 kilometers per second, but singly ionized xenon atoms accelerated through 1,000 volts in an ion engine have an exhaust velocity of 38 kilometers per second. Modern rocket designers have increasingly turned toward using the ion engine.

Government Development. Because ion engines are roughly ten times as efficient as chemical engines, all the major players in space exploration have adopted ion engines wherever possible. The Soviet Union pioneered Hall-effect engines in the 1960's, and they were gradually adopted for station keeping and later for powering spacecraft. Missions using ion engines have been launched by the Russian Federal Space Agency (Roscosmos), NASA, the Indian Space Research Organization, the Japan Aerospace Exploration Agency, and the European Space Agency (its member nations are Austria, Belgium, the Czech Republic, Denmark, Finland, France, Germany, Greece, Ireland, Italy, Luxembourg, the Netherlands, Norway, Portugal, Spain, Sweden, Switzerland and the United Kingdom).

Industrial Development. Private companies are increasingly involved in the creation of engines to power spacecraft and satellites as well as the satellites themselves. The Ad Astra Rocket Company in Webster, Texas, is developing the VASIMR engine, and the British company QinetiQ is preparing engines to power the European Space Agency's BepiColombo to the planet Mercury. The Satellite Business Development Office of NEC built the four ion engines for Japan's Hayabusa. The Boeing Satellite Development Center bills itself as the world's leading manufacturer of commercial communications satellites, a major supplier of spacecraft and equipment, and as the supplier for weather satellites for the United States and Japan.

Careers and Course Work

Any job in design and development is likely to be exciting and interesting to those who like to solve problems and see ideas made into reality. A strong background in the physical sciences—at least a bachelor's degree in physics, engineering, electrical engineering, mechanical engineering, materials science, or chemistry—is required to design and develop ion engines or other space hardware. A high school student should take all available courses in physics, chemistry, and mathematics. The same is true for college students, who will need at least one year of basic chemistry, at least one year of calculus-based physics with laboratory practice, and mathematics through differential equations and matrices. Classes in statistics and computer programing may be helpful.

Other useful classes include writing and speech (for reports and presentations) and a simple business course. Those wishing to work in research and development need a feel for how things work and some level of creativity. Those working in quality control may need to design tests that demonstrate that a device works and that it will continue to work reliably.

Social Context and Future Prospects

As Hall-effect ion engines and VASIMR engines become more powerful, rocket fuel will account for less of the spacecraft's total weight. This will allow for larger payloads and enable robot missions to conduct more scientific experiments. If high-powered engines can be developed, the only source capable of providing enough power is a small nuclear reactor. Fission reactors have been used in space by Russia, which has used more than thirty to power satellites and by the United States, which flew a test reactor, the SNAP-10A, in 1965. With a nuclear reactor for power and with a large ion engine, a spacecraft could travel from Earth to Mars in 39 days instead of the 255 days that it would take a conventionally powered vessel. This may make a manned mission possible

Fascinating Facts About Ion Propulsion

- If space travel is ever to become much more common, it seems likely that it will be powered by ion engines.
- The bluish color of the exhaust from an ion engine comes from the recombination of electrons with xenon ions.
- NASA's Evolutionary Xenon Thruster (NEXT) generates 2.5 times as much thrust as the NSTAR engine used on the Dawn mission. It provides a record-breaking 10 million Newton-seconds of total impulse (the overall acceleration available to a spacecraft).
- Microwaves will heat the plasma in the VASIMR engine to millions of Kelvins. Magnetic fields hold the plasma away from the rocket's walls so they do not melt.
- The VASIMR VX-200 ion engine produces a thrust of 5 newtons at 5,000 seconds specific impulse. In comparison, the NEXT engine produces 0.327 newtons at 4,300 seconds specific impulse and the NSTAR engine 0.092 newtons at 3,300 seconds specific impulse.
- The Australian National University designed and built the dual-stage four-grid (DS4G) ion thruster for the European Space Agency. An exhaust speed of 210 kilometers per second was recorded during testing in January, 2006. This engine, however, remains in the prototype stage.
- To reduce the risk of radioactive contamination, nuclear reactors used to power ion engines will not be turned on until they are in space.

because of the reduction in days of exposure to cosmic rays. Such an engine might make it possible to fly to an asteroid and exploit its minerals. (A single metallic asteroid would most likely contain more nickel than has ever been mined on Earth.) An additional possible benefit is that people's apprehensions about using nuclear reactors for power on Earth might lessen if nuclear reactors in space built up a good safety record.

Charles W. Rogers, B.A., M.S., Ph.D.

FURTHER READING

Bruno, Claudio, ed. *Nuclear Space Power and Propulsion Systems.* Reston, Va.: American Institute of Aeronautics and Astronautics, 2008. Essays examine the use of nuclear power on spacecraft. One chapter specifically examines ion thrusters and nuclear power.

Doody, Dave. *Deep Space Craft: An Overview of Interplanetary Flight.* New York: Springer, 2009. Contains a chapter on propulsion and discusses ion propulsion in the greater context of spacecraft.

Gilster, Paul. *Centauri Dreams: Imagining and Planning Interstellar Exploration.* New York: Copernicus Books, 2004. Includes a good discussion of various means of propulsion, including ion engines.

Goebel, Dan M., and Ira Katz. *Fundamentals of Electric Propulsion: Ion and Hall Thrusters.* Hoboken, N.J.: John Wiley & Sons, 2008. Provides a guide to the technology and physics of Hall-effect and ion thrusters. Contains photographs of successful engines and figures and tables that show thruster and cathode schematics.

Turner, Martin J. L. *Rocket and Spacecraft Propulsion: Principles, Practice and New Developments.* 2d ed. Chichester, England: Praxis, 2006. Contains general discussions of propulsion methods, including ion engines. Excellent diagrams and pictures.

WEB SITES

Ad Astra Rocket Company
http://www.adastrarocket.com

European Space Agency
Propulsion, Advanced Concepts Team
http://www.esa.int/gsp/ACT/pro/pp/DS4G/background.htm

National Aeronautics and Space Administration
Glenn Research Center
http://www.nasa.gov/centers/glenn/home/index.html

NEC
Hayabusa's Seven-Year Journey
http://www.nec.com/global/ad/hayabusa

Propulsion Engineering Research Center
http://www.psu.edu/dept/PERC

Propulsion Research Center
http://prc.uah.edu/Propulsion_Research_Center/Welcome.html

See also: Jet Engine Technology

J

JET ENGINE TECHNOLOGY

FIELDS OF STUDY

Physics; chemistry; thermodynamics; gas dynamics; aerodynamics; heat transfer; materials science; metallurgy.

SUMMARY

Jet engines are machines that add energy to a fluid stream and generate thrust from the increase of momentum and pressure of the fluid. Jet engines, which usually include turbomachines to raise the pressure, vary greatly in size. Applications include microelectromechanical gas turbines for insect-sized devices; mid-sized engines for helicopters, ships, and cruise missiles; giant turbofans for airliners; and scramjets for hypersonic vehicles. Jet engine development pushes technology frontiers in materials, chemical kinetics, measurement techniques, and control systems. Born in the desperation of World War II, jet engines have come to power most modern aircraft.

KEY TERMS AND CONCEPTS

- **Cycle Efficiency:** Ratio of the useful work delivered by a thermodynamic system to the heat put into the system.
- **Equivalent Exhaust Velocity:** Thrust divided by the mass flow rate of propellant through the exhaust.
- **Fuel-to-Air Ratio:** Ratio of the mass of fuel burned per unit of time to the mass of air passing through the burner per unit of time.
- **Overall Pressure Ratio:** Ratio between the highest pressure in the propulsion system and the lowest pressure, typically the atmospheric pressure at the altitude at which the vehicle is flying.
- **Propulsive Efficiency:** Ratio of work done by the propulsion system on the vehicle to the kinetic energy imparted by the system to the working fluid.
- **Specific Impulse:** Equivalent exhaust velocity divided by the standard value of acceleration due to gravity; expressed in seconds.
- **Stage Pressure Ratio:** Ratio of the stagnation pressure at the end of a single stage of a turbomachine to the stagnation pressure ahead of the stage.
- **Thermal Efficiency:** Ratio of work done by the propulsion system to the heat put into the system.
- **Thrust:** Force exerted by the propulsion system on the vehicle.
- **Thrust-Specific Fuel Consumption:** Weight of fuel consumed per unit thrust force, per hour.
- **Turbine Inlet Temperature:** Highest temperature of the working fluid, typically reached at the end of heat addition and before work is taken out in the turbine.

Definition and Basic Principles

The term "jet engine" is typically used to denote an engine in which the working fluid is mostly atmospheric air, so that the only propellant carried on the vehicle is the fuel used to release heat. Typically, the mass of fuel used is only about 2 to 4 percent of the mass of air that is accelerated by the vehicle.

Background and History

British engineer Frank Whittle and German engineer Hans Pabst von Ohain independently invented the jet engine, earning patents in Britain in 1932 and in Germany in 1936, respectively. Whittle's engine used a centrifugal compressor and turbine. The gas came in near the axis, was flung out to the periphery, and returned to the axis through ducts. Ohain's engine used a combination of centrifugal and axial-flow turbomachines. The gas direction inside the engine was mostly aligned with the axis of the engine, but it underwent small changes as it passed through the compressor and turbine. The axial-flow engine had higher thrust per unit weight of the engine and smaller frontal area than the centrifugal machine. Initially, the axial-flow machine, which had a large number of stages and blades, was much more prone to failure than the centrifugal machine, which had sturdier blades and fewer moving parts. However, these problems were resolved, and most modern jet engines use purely axial flow.

Several early experiments with jet engines ended in explosions. Ohain's engine was successfully used to power a Heinkel He 178 aircraft on August 27, 1939. Ohain's engine led to the Junkers Jumo 004 axial turbojet engine, which was mass-produced to power the Messerschmitt Me 262 fighter aircraft in 1944. The British Gloster E.28/39, which contained one of Whittle's engines, flew in 1941. In September, 1942, a General Electric engine powered the Bell XP-59 Airacomet, the first American jet fighter aircraft.

The de Havilland Comet jet airliner service began in 1952, powered by the Rolls-Royce Avon. In 1958, a de Havilland Comet jet operated by British Overseas Airways Corporation flew from London to New York, initiating transatlantic passenger jet service. In the mid-1950's, the Rolls-Royce Conway became the first turbofan in airliner service when it was used for the Boeing 707. The Rolls-Royce Olympus afterburning turbofan powered the supersonic Concorde in 1969. The majority of aircraft have come to be powered by jet engines, mostly turbofans, the largest, as of 2010, being the General Electric GE90-115, which produces nearly 128,000 pounds (rated at 115,000 pounds, or 570,000 Newtons) of thrust.

From 1903 to the 1940's, the power per unit weight of engines increased from 0.05 to about 0.8 horsepower per pound (hp/lb), with the largest engines producing 4,000 hp. With jet engines, the power per unit weight has risen to more than 20 hp/lb, with the largest engines producing more than 100,000 hp. The compressor pressure ratio has risen from 3:1 for initial engines to more than 40:1 for modern engines.

How It Works

Jet engines operate by creating thrust through the Brayton thermodynamic cycle. In a process called isentropic compression, a mixture of working gases is compressed, with no losses in stagnation pressure. Heat is chemically released or externally added to the fluid, ideally at constant pressure. A turbine extracts work from the expanding gases. This work runs the compressor and other components such as a fan or propeller, depending on the engine type and application. The gas leaving the turbine expands further through a nozzle, exiting at a high speed.

Jet engine developers try to maximize the pressure and temperature that the engine can tolerate. The thermal efficiency of the Brayton cycle increases with the overall pressure ratio; therefore, designers try to get the highest possible pressure at the end of the compression. However, the temperature rise accompanying the pressure rise limits the amount of heat that can be added before the temperature limit of the engine is reached. Thrust is highest when the greatest net momentum increase is added to the flow. The propulsive efficiency of the engine is highest when the speed of the jet exhaust is close to the flight speed of the vehicle. These considerations drive jet engine design in different directions depending on the application. For very-high-speed applications (typically military engines), engine mass and frontal area must be kept low, so a smaller amount of air is ingested and accelerated through a large speed difference. For engines such as those used on airliners, a large amount of air is ingested using a large diameter intake and accelerated through a small speed difference for best propulsive efficiency. The major components of a jet engine are the inlet, diffuser, compressor, fan, propeller, combustor, turbine, afterburner, nozzles, and gearbox.

Inlet and Diffuser. The engine inlet is designed to capture the required airflow without causing flow separation. An aircraft flying at supersonic speeds has a supersonic inlet in which a series of shocks slows down the flow to subsonic speeds with minimal losses in pressure. Once the flow is subsonic, it is slowed further to the Mach number of about 0.4 needed at the face of the compressor or fan. A supersonic inlet and diffuser may lose 5 to 10 percent of the stagnation pressure of the incoming flow when used with aircraft flying in excess of Mach 2.

Compressor, Fan, and Propeller. In 1908, René Lorin patented a jet engine in which a reciprocating piston engine compressed the fluid. In 1913, he patented a supersonic ramjet engine in which enough compression would occur simply by slowing the flow down to subsonic speeds. In 1921, Maxime Guillaume patented a jet engine with a rotating axial-flow compressor and turbine. Most modern engines use the turbomachine in some form. The compressor is built in several stages, with the pressure ratio of each stage limited to prevent flow separation and stall. A centrifugal compressor stage consists of a rotor that imparts a strong radial velocity to the flow, and the flow

Fascinating Facts About Jet Engine Technology

- Hans Pabst von Ohain's first demonstrator jet engine used hydrogen as fuel, and the first Heinkel jet engine patent application was for a hydrogen combustion system.
- The Pratt & Whittney J-58 engines of the Lockheed SR-71 Blackbird aircraft started as afterburning turbojets but morphed into ramjets to allow the aircraft to reach more than three times the speed of sound.
- A turbojet engine built with microelectromechanical systems (MEMS) technology at the Lincoln Laboratories in Boston has a 4 millimeter-diameter turbine that spins at more than 1 million revolutions per minute.
- The General Electric GE90-115 turbofan engine for the Boeing 777, which produces 128,000 pounds of thrust, has a diameter of 3.43 meters, nearly as large as the cabin of the Boeing 737 aircraft.
- The blue light seen shooting out of jet engine nozzles indicates incomplete combustion, with opportunities for further gains in efficiency.
- In 1985, twin-engine airliners were allowed to fly transatlantic routes because jet engine reliability was determined to be so high that failures of other critical aircraft components were far more likely.
- Ramjet engines proposed for vehicles designed to explore the atmosphere of the planet Jupiter would use atmospheric hydrogen as the working fluid and add oxygen as fuel for combustion.
- The Bussard intersteller ramjet engine concept, first proposed in the 1960's, involves capturing hydrogen ions from interstellar gas using an electromagnetic field, compressing the flow in a diffuser, and heating it using an nuclear thermal reactor.
- The Pratt & Whitney F135 and the competing General Electric F136 engines for the F-35 joint strike fighter use fans and compressors that are attached to turbine stages that spin in opposite directions, saving the weight of the cooled turbine stator disks that would otherwise be needed.
- Modern jet engine turbine blades are grown as single crystals, or the entire blade-disk assembly of a stage is cast as a single piece, hollow inside.
- Supersonic combustion ramjet engines for hypersonic vehicles must provide enough distance for fuel to ignite. An ignition delay of 1 millisecond, for a flow moving at 2,000 meters per second, means an ignition distance of 2 meters.

is flung out at the periphery of the blades with added kinetic energy. The flow is then turned and brought back near the axis by a diffuser stage in which the static pressure rises and the flow speed decreases. In an axial compressor, work is added to the flow in several stages, as many as fifteen in some engines. In each stage, a spinning rotor wheel with many blades that act as lifting wings, imparts a swirl velocity to the flow. This added energy is then converted to a pressure rise in stator blade passages, bringing the flow back to being axial, but with increased pressure. Some newer compressors have counter-rotating wheels in each stage instead of a rotor and a stator. Shock-in-rotor stages use blades moving at supersonic speed relative to the flow to create large pressure rise because of shocks. Supersonic through-flow compressor designs are being developed for future high-speed engines.

The fans of most engines are extensions of the first, or low-pressure, compressor stages. Fans have only one or two stages, and fewer, larger blades than the compressor stages.

Propellers are used in turboprop engines to produce a portion of the engine's thrust.

Combustor. The combustor is designed to mix and react the fuel with the air rapidly and to contain the reaction zone within an envelope of cooling air. At the exit of the combustor, these flows are mixed to ensure the most uniform temperature distribution across the gases entering the turbine. Older combustors were either several cans connected by tubes, arranged around the turbomachine shaft, or an annular passage. Some modern combustors are arranged in a reverse-flow geometry to enable better mixing and reaction. Ideally, a jet engine combustor must add all the heat that can be released from the fuel, at a constant pressure, with minimal pressure losses because of flow separation and turbulence. Heat addition must also be done at the lowest flow Mach number possible.

Turbine. The critical limiting temperature in a jet engine is the temperature at the inlet to the first turbine stage. This is usually tied to the strength of the blade material at high temperatures. A turbine has only a few stages, typically four or fewer. The highest mass flow rate through the engine is limited to the flow rate at which the passages at the final turbine stage are choked, in other words, when the Mach number reaches 1 at these passages. Turbine stage disks and blades are integrated into blisks and can be made as a single piece, with cooling passages inside the blades, using powder metallurgy.

The turbine is directly attached to the compressor through a shaft. To enable starting the compressor and better matching the requirements of the different stages, a twin-spool or three-spool design is used, where the outer (low-pressure, lower rotation speed) stages of the compressor, fan, and turbine are connected through an inner shaft, and the inner (high-pressure) stages are connected through a concentric, outer shaft.

Afterburner. Older military engines use an afterburner (also known as reheat) duct attached downstream of the turbine, where more fuel is added and burned, with temperatures possibly exceeding the turbine inlet temperature. Afterburners are highly inefficient but produce a large increment of thrust for short durations. Therefore, afterburners are used at takeoff, in supersonic dashes, or in combat situations.

Nozzles. For jet engines without afterburners, the exit velocity is at most sonic and the nozzle is just a converging duct. If the exhaust is expanded to supersonic speeds, the nozzle has a convergent-divergent contour. Thrust-vectoring nozzles either rotate the whole nozzle, as in the case of the Harrier or the F-22, or use paddles in the exhaust, as in the case of the Sukhoi-30.

Gearbox. When the engine must drive a rotor, propeller, or counter-rotating fan, a gearbox is used to reduce the speed or change the direction of rotation. This usually adds considerable weight to the engine.

Applications and Products

Ramjet and Scramjet. Ramjets are used to power vehicles at speeds from about Mach 0.8 to 4. The diffuser slows the flow down to subsonic speeds, increasing the pressure so much that thrust can be generated without a mechanical compressor or turbine. Beyond Mach 4, the pressure loss in slowing down the flow to below Mach 1 is greater than the loss due to adding heat to a supersonic flow. In addition, if such a flow were decelerated to subsonic conditions, the pressure and temperature rise would be too high, either exceeding engine strength or leaving too little room for heat addition. In this regime, the supersonic combustion ramjet, or scramjet, becomes a better solution.

Turbojet. The turbojet is the purest jet engine, with a compressor and turbine added to the components of the ramjet. The turbojet can start from rest, which the pure ramjet cannot. However, since the turbojet converts all its net work into the kinetic energy of the jet exhaust, the exhaust speed is high. High propulsive efficiency requires a high flight speed, making the turbojet most suitable near Mach 2 to 3. Because jet noise scales as the fifth or sixth power of jet speed, the turbojet engine was unable to meet the noise regulations near airports in the 1970's and was rapidly superseded by the turbofan for airliner applications.

Turbofan. The turbine of the turbofan engine extracts more work than that required to run the compressor. The remaining work is used to drive a fan, which accelerates a large volume of air, albeit through a small pressure ratio. The air that goes through the fan may exit the engine through a separate fan nozzle or mix with the core exhaust that goes through the turbine before exiting. Because the overall exhaust speed of the turbofan is much lower than that of the turbojet, the propulsive efficiency is high in the transonic speed range where airliner flight is most efficient, yet airport noise levels are far lower than with a turbojet. Turbofan engines are used for most civilian airliner applications and even for fighter and business jet engines.

Turboprop. In the turboprop engine, a separate power turbine extracts work to run a propeller instead of a fan. The propeller typically has a larger diameter than a fan for an engine of comparable thrust. However, the rotating speed of a propeller, constrained by the Mach number at the tip, is only on the order of 3,000 to 5,000 revolutions per minute, as opposed to turbomachine speeds, which may be three to ten times higher. Therefore, a gearbox is required.

Turboshaft. Instead of a propeller, a helicopter rotor or other device may be driven by the power turbine. Automobile turbochargers, turbopumps for rocket propellants, and gas turbine electric power generators are all turboshaft engines.

Propfan. Propfans are turbofans in which the fan has no cowling, so that it resembles a propeller and has a larger capture area, but the blades are highly swept and wider than propeller blades.

Air Liquefaction. The high pressures encountered in high-speed flight make it possible to liquefy some of the captured and compressed air at lower altitudes, using heat transfer to cryogenic fuels such as hydrogen. The oxygen from this liquid can be separated out and stored for use as the vehicle reaches the edge of the atmosphere and beyond. Turboramjet engines using this technology could enable routine travel to and from space, with fully reusable, single-stage vehicles.

Impact on Industry

The jet engine allowed airplanes to fly efficiently, well above the weather and the troposphere in the smooth air of the stratosphere. The smooth stratosphere allowed airplanes to travel at high speeds without exceeding structural strength limits posed by air turbulence. This in turn has enabled intercontinental flights in jet-smooth comfort and the construction of very large airliners. An immense expansion of commerce has resulted from these developments. Engine size and power have increased far beyond what was possible with the internal combustion engine. Outside the aerospace industry, gas turbines are used to power ships and in most thermal power plants.

Careers and Course Work

Jet engine development and manufacturing are parts of a highly specialized industry. Those interested in jet engine technology must have a very good basic understanding of thermodynamics and dynamics and can specialize in combustion, turbomachine aerodynamics, gas dynamics, or materials engineering. Airlines operate engine test cells and employ many technical workers to diagnose and repair problems and ensure proper maintenance and operating procedures. The National Aeronautics and Space Administration (NASA) and the U.S. Department of Defense offer many opportunities in all aspects of jet engine technology. Engine developers include the very large corporations that supply airliner engines and the smaller companies that develop engines for business jets, cruise missiles, and other applications. Companies doing research, development, and manufacture of jet engines are in the United States, Europe, Russia, Japan, and China. Many other nations also produce jet engines under license.

Social Context and Future Prospects

Jet engines have developed rapidly since the 1940's and have come to dominate the propulsion market for atmospheric flight vehicles. Because the air that makes up more than 95 percent of the working fluid is free and does not have to be carried up from the ground, air-breathing propulsion offers a huge increase in specific impulse over rocket engines for flight in the atmosphere. As technology advances to enable rotating machinery to tolerate higher temperatures, pressures, and stresses, jet engines can become substantially lighter and more efficient per unit of thrust. Hydrogen-fueled engines can operate much more efficiently than hydrocarbon-fueled engines. Turbo-ramjet engines may one day enable swift and inexpensive access to space. Helicopter engines and the lift fans developed for vertical-landing fighter planes bring personal air vehicles closer to reality. Supersonic airline travel using hydrogen fuel is much closer to becoming routine. At the other extreme, micro jet engines are finding use in devices to power actuators for many applications, including surgical tools and devices to control stall on wings and larger engines. These advances create the prospect for revolutionary advances in human capabilities. Jet engine development promises to remain a leading-edge technological field for many years to come.

Narayanan M. Komerath, Ph.D.

Further Reading

Constant, Edward W. *The Origin of the Turbojet Revolution.* Baltimore: The Johns Hopkins University Press, 1980. Traces the history of the turbojet.

Conway, Erik M. *High-Speed Dreams: NASA and the Technopolitics of Supersonic Transportation, 1945-1999.* 2005. Reprint. Baltimore: The Johns Hopkins University Press, 2008. Written by a NASA historian, this book traces the early development of engines for military supersonic flight and then analyzes the economic and political factors that led to the failure of three attempts to develop an American supersonic airliner.

Golley, John, Frank Whittle, and Bill Gunston. *Whittle: The True Story.* Washington, D.C.: Smithsonian Institution Press, 1987. Biography of Whittle, giving the British perspective on the invention and development of early jet engines.

Hill, Philip Graham, and Carl R. Peterson. *Mechanics*

and Thermodynamics of Propulsion. 2d ed. Reading, Mass.: Addison-Wesley, 2010. This popular textbook provides succinct presentations of the theory and applications of propulsion.

Hünecke, Klaus. *Jet Engines: Fundamentals of Theory, Design and Operation.* 1998. Reprint. London: Zenith Press, 2005. Begins with the history of jet engines and continues with the basics of how they operate and their structure.

Mattingly, Jack. *Elements of Gas Turbine Propulsion.* 1996. Reprint. Reston, Va.: American Institute of Aeronautics and Astronautics, 2005. Contains large sections on turbomachines and fighter engines. The authoritative foreword by Hans Pabst von Ohain provides historical insight and perspective.

Web Sites

American Institute of Aeronautics and Astronautics
http://www.aiaa.org

General Electric Aviation
http://www.geae.com/engines/index.html

National Aeronautic Association
http://www.naa.aero

National Aeronautics and Space Administration
Gas Turbine Propulsion
http://www.grc.nasa.gov/WWW/K-12/airplane/turbine.html

See also: Vertical Takeoff and Landing Aircraft

L

LASER INTERFEROMETRY

FIELDS OF STUDY

Control engineering; electrical engineering; engineering metrology; interferometry; laser science; manufacturing engineering; materials science; mechanical engineering; nanometrology; optical engineering; optics; physics.

SUMMARY

Laser interferometry includes many different measurement methods that are all based on the unique interference properties of laser lights. These techniques are used to measure distance, velocity, vibration, and surface roughness in industry, military, and scientific research.

KEY TERMS AND CONCEPTS

- **Beam Splitter:** Partially reflecting and partially transmitting mirror.
- **Constructive Interference:** Addition of two or more waves that are in phase, leading to a larger overall wave.
- **Destructive Interference:** Addition of two or more waves that are out of phase, leading to a smaller overall wave.
- **Heterodyne Detection:** Mixing of two different frequencies of light to create a detectable difference in their interference pattern.
- **Homodyne Detection:** Mixing of two beams of light at the same frequency, but different relative phase, to create a detectable difference in their interference pattern.
- **Monochromatic:** Containing a single wavelength.
- **Spatial Coherence:** Measure of the phase of light over a defined space.
- **Temporal Coherence:** Measure of the phase of light as a function of time.

Definition and Basic Principles

Laser interferometry is a technique that is used to make extremely precise difference measurements between two beams of light by measuring their interference pattern. One beam is reflected off a reference surface and the other either reflects from or passes through a surface to be measured. When the beams are recombined, they either add (constructive interference) or subtract (destructive interference) from each other to yield dark and light patterns that can be read by a photosensitive detector. This interference pattern changes as the relative path length changes or if the relative wavelength or frequency of the two beams changes. For instance, the path lengths might vary because one object is moving, yielding a measurement of vibration or velocity. If the path lengths vary because of the roughness of one surface, a "map" of surface smoothness can be recorded. If the two beams travel through different media, then the resulting phase shift of the beams can be used to characterize the media.

Lasers are not required for interferometric measurements, but they are often used because laser light is monochromatic and coherent. It is principally these characteristics that make lasers ideal for interferometric measurements. The resulting interference pattern is stable over time and can be easily measured, and the precision is on the order of the wavelength of the laser light.

Background and History

The interference of light was first demonstrated in the early 1800's by English physicist Thomas Young in his double-slit experiment, in which he showed that two beams of light can interact like waves to produce alternating dark and light bands. Many scientists believed that if light were composed of waves, it would require a medium to travel through, and this medium (termed "ether") had never been detected. In the late 1800's, American physicist Albert Michelson designed an interferometer to measure the effect of the ether on the speed of light. His experiment was considered a failure in that he was not able to provide proof of the existence of the ether. However, the utility of the interferometer for measuring a precise distance was soon exploited. One of Michelson's first uses of his interferometer was to measure the international unit of a meter using a platinum-iridium metal bar, paving the way for modern interferometric methods of measurement. Up until the mid-twentieth century, atomic sources were used in interferometers, but their use for measurement was limited to their coherence length, which was less than a meter. When lasers were first developed in the 1960's, they quickly replaced the spectral line sources used for interferometric measurements because of their long coherence length, and the modern field of laser interferometry was born.

How It Works

The most common interferometer is the Michelson interferometer, in which a laser beam is divided in two by use of a beam splitter. The split beams travel at right angles from each other to different surfaces, where they are reflected back to the beam splitter and redirected into a common path. The interference between the recombined beams is recorded on a photosensitive detector and is directly correlated with the differences in the two paths that the light has traveled.

In the visible region, one of the most commonly available lasers is the helium-neon laser, which produces interference patterns that can be visually observed, but it is also possible to use invisible light lasers, such as those in the X-ray, infrared, or ultraviolet regions. Digital cameras and photodiodes are routinely used to capture interference patterns, and these can be recorded as a function of time to create a movie of an interference pattern that changes with time. Mathematical methods, such as Fourier analysis, are often used to help resolve the wavelength composition of the interference patterns. In heterodyne detection, one of the beams is purposefully phase shifted a small amount relative to the other, and this gives rise to a beat frequency, which can be measured to even higher precision than in standard homodyne detection. Fiber optics can be used to direct the light beams, and these are especially useful to control the environment through which the light travels. In this case, the reflection from the ends of the fiber optics have to be taken into account or used in place of reflecting mirrors. Polarizers and wave-retarding lenses can be inserted in the beam path to control the polarization or the phase of one beam relative to the other.

While Michelson interferometers are typically used to measure distance differences between the two reflecting surfaces, there are many other configurations. Some examples are the Mach-Zehnder and Jamin interferometers, in which two beams are reflected off of identical mirrors but travel through different media. For instance, if one beam travels through a gas, and the other beam travels through vacuum, the beams will be phase shifted relative to the other, causing an interference pattern that can be interpreted to give the index of refraction of the gas. In a Fabry-Perot interferometer, light is directed into a cavity consisting of two highly reflecting surfaces. The light bounces between the surfaces multiple times before exiting to a detector, creating an interference pattern that is much more highly resolved than in a standard Michelson interferometer. Several other types of interferometers are described below in relation to specific applications.

Applications and Products

Measures of Standards and Calibration. Because of the accuracy possible with laser interferometry, it is widely used for calibration of length measurements. The National Institute of Standards and Technology (NIST), for example, offers measurements of gauge blocks and line scales for a fee using a modified Michelson-type interferometer. Many commercial companies also offer measurement services based on laser interferometer technology. Typical services are for precise measurement of mechanical devices, such as bearings, as well as for linear, angular, and flatness calibration of other tools such as calipers, micrometers, and machine tools. Interferometers are also used to measure wavelength, coherence, and spectral purity of other laser systems.

Dimensional Measurements. Many commercial laser interferometers are available for purchase and can be used for measurements of length, distance, and angle. Industries that require noncontact measurements of complex parts use laser interferometers to test whether a part is good or to maintain precise positioning of parts during fabrication. Laser interferometers are widely used for these purposes in the automotive, semiconductor, machine tool, and medical- and scientific-parts industries.

Vibrational Measurements. Laser vibrometers make use of the Doppler shift, which occurs when one laser beam experiences a frequency shift relative to the other because of the motion of the sample. These interferometers are used in many industries to measure vibration of moving parts, such as in airline or automotive parts, or parts under stress, such as those in bridges.

Optical Metrology. Mirrors and lenses used in astronomy require high-quality surfaces. The Twyman-Green and Fizeau interferometers are variations on the Michelson interferometer, in which an optical lens or mirror to be tested is inserted into the path of one of beams, and the measured interference pattern is a result of the optical deviations between the two surfaces. Other industrial optical testing applications include the quality control of lenses in glasses or microscopes, the testing of DVD reader optical components, and the testing of masks used in lithography in the semiconductor industry.

Ring Lasers and Gyroscopes. In the last few decades, laser interferometers have started to replace mechanical gyroscopes in many aircraft navigation systems. In these interferometers, the laser light is reflected off of mirrors such that the two beams travel in opposite directions to each other in a ring, recombining to produce an interference pattern at the starting point. If the entire interferometer is rotated, the path that the light travels in one direction is longer than the path length in the other direction, and this results in the Sagnac effect: an interference pattern that changes with the angular velocity of the apparatus. These ring interferometers are widely available from both civilian and military suppliers such as Honeywell or Northrop Grumman.

Ophthalmology. A laser interferometry technique using infrared light for measuring intraocular

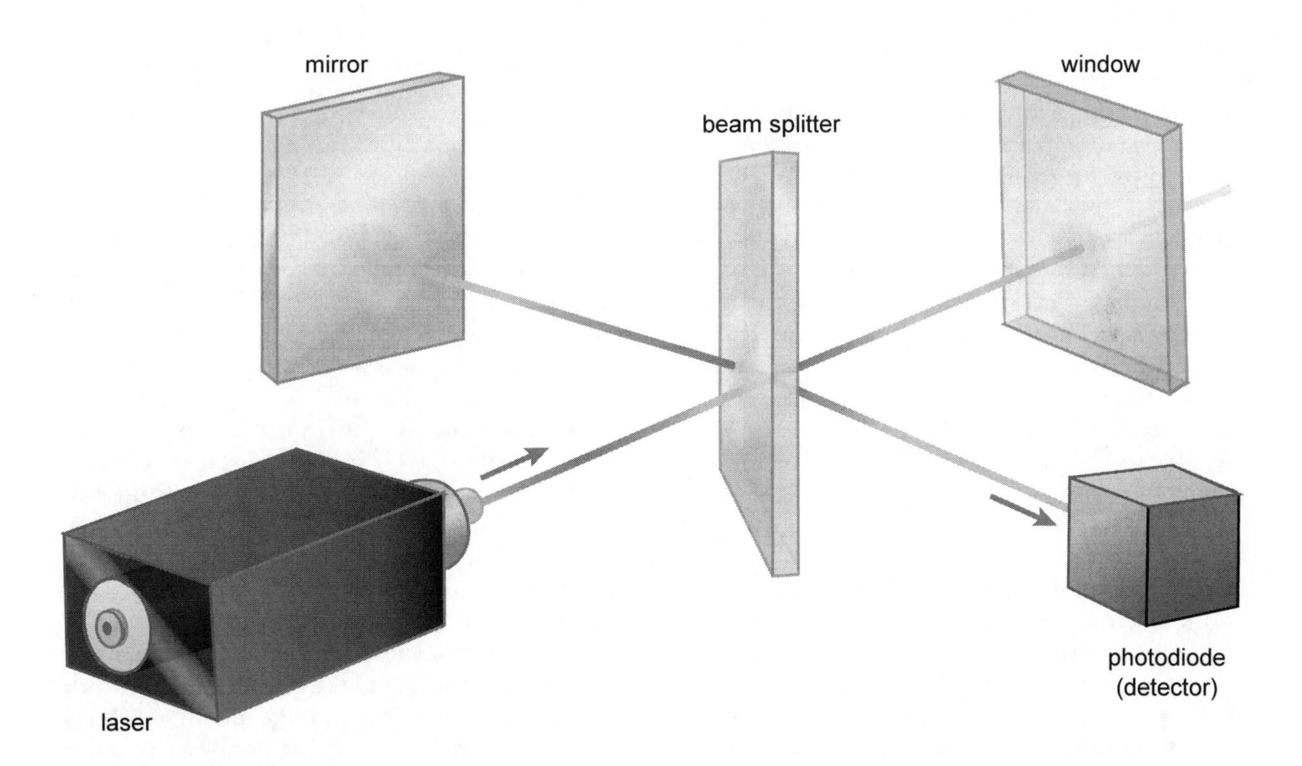

distances (eye length) is widely used in ophthalmology. This technique has been developed and marketed primarily by Zeiss, which sells an instrument called the IOL Master. The technique is also referred to as partial coherence interferometry (PCI) and laser Doppler interferometry (LDI) and is an area of active research for other biological applications.

Sensors. Technology based on fiber-optic acoustic sensors to detect sound waves in water have been developed by the Navy and are commercially available from manufacturers such as Northrop Grumman.

Gravitational Wave Detection. General relativity predicts that large astronomical events, such as black hole formation or supernova, will cause "ripples" of gravitational waves that spread out from their source. Several interferometers have been built to try to measure these tiny disturbances in the local gravitational fields around the earth. These interferometers typically have arm lengths on the order of miles and require a huge engineering effort to achieve the necessary mechanical and vibrational stability of the lasers and the mirrors. There are currently efforts to build space-based gravity-wave-detecting interferometers, which would not be subject to the same seismic instability as Earth-based systems.

Research Applications. Laser interferometers are used in diverse form in many scientific experiments. In many optical physics applications, laser interferometers are used to align mirrors and other experimental components precisely. Interferometers are also used for materials characterization in many basic research applications, while ultrasonic laser interferometers are used to characterize velocity distributions and structures in solids and liquids. A more recent technology development is interferometric sensors, which are used to monitor chemical reactions in real time by comparing laser light directed through waveguides. The interference pattern from the two beams changes as the chemical reaction progresses. This technique is often referred to as dual-polarization interferometry.

Impact on Industry

The invention of the laser in 1960 opened up the field of interferometry to a huge range of applications, allowing measurements both at long distances and with extremely high precision. The following decades saw the initial application of laser interferometry to vibration and dimensional measurements in industrial processes. The invention of cheap semiconductor diode lasers in the last decade has further decreased the price of laser interferometers and has widened the range of accessible wavelengths. Many laser interferometer systems are now available commercially, and they are used in a large segment of the semiconductor, automotive, and measurement industries. Systems now sold by many companies give compact, reliable ways to measure surface roughness, calibrate mechanical components, align or position parts during manufacturing, or for precision machining. Lasers are a multibillion-dollar industry, and laser interferometers are a constantly growing segment of this market.

Laser interferometers have hugely affected the quality control of lenses in the optical-manufacturing industries. Interferometers are the method of choice to measure curvature and smoothness of lenses

Fascinating Facts About Laser Interferometry

- When the first laser was demonstrated in 1960, it was called a "solution looking for a problem."
- The Laser Interferometer Gravitational-Wave Observatory (LIGO) is sensitive to disturbances in the local gravitational fields that are caused by astronomical events as far back in time as 70 million years.
- Sensors based on laser interferometers are being developed to detect acoustic signals for surveillance. A buried sensor can detect the sound of footsteps from as far away as 30 feet.
- The Laser Interferometer Space Antenna (LISA) will consist of three spacecraft orbiting at 5 million kilometers apart and will provide information on the growth and formation of black holes and other events never before seen.
- Laser interferometers will be used to measure the optical quality of the mirror used in the James Webb Space Telescope, scheduled to launch in 2014. The surface smoothness must be unprecedented: If the mirror were scaled up to a size of 3,000 miles across, the surface height would be allowed to vary only by a foot at most.
- Quantum interferometers have been built to recreate the classic double-slit interference experiment, using single atoms instead of a light beam.

used in microscopes, telescopes, and eyeglasses. Companies specializing in optical measurements are available for on- or off-site testing of optical components. Astronomy, in particular, has benefited from the precision testing of polished surfaces made possible by laser interferometers; mirrors such as the ones used in the Hubble Space Telescope would not be possible without a laser interferometry testing system.

The National Science Foundation (NSF) has invested substantially in ground-based interferometers for the measurement of gravity waves to test the predictions of general relativity. In the 1990's, the NSF funded the building of two very large-scale Michelson interferometers in the United States–one in Hanford, Washington, and the other in Livingston, Louisiana. Together they make up the Laser Interferometer Gravitational-Wave Observatory (LIGO) and consist of interferometer arms that are miles long. Though gravity waves have not yet been detected, government-funded efforts around the world continue with large-scale interferometers currently operating in a half-dozen countries.

Sensor technology is a growing field in laser interferometry applications. Sensors based on interference of laser light using fiber optics have been developed to detect environmental changes, such as temperature, moisture, pressure, strain on components, or chemical composition of an environment. These types of sensors are not yet widely commercially available but are in development by industry and by the military for applications such as chemical-agent detection in the field or for use in harsh or normally inaccessible environments. Combined with wireless technology, they could be used, for instance, to monitor environmental conditions far underground or within the walls of buildings.

Careers and Course Work

Basic research on laser interferometry and its applications is conducted in academia, in many metrology industries, and in government laboratories and agencies. For research careers in new and emerging interferometry methods in academia or as a primary investigator in industry, a doctorate degree is generally required. Graduate work should be in the area of physics or engineering. The undergraduate program leading into the graduate program shouldinclude classes in mathematics, engineering, computer, and materials science.

For careers in industries that provide or use commercial laser interferometers but do not conduct basic research, a master's or bachelor's degree would be sufficient, depending on the career path. Senior careers in these industries involve leading a team of engineers in new designs and applications or guiding a new application into the manufacturing field. In this case, the focus of course work should be in engineering. Mechanical, electrical, optical, or laser engineering will provide a solid background and an understanding of the basic theory of interferometer science. Additional courses should include physics, mathematics, and materials science. A bachelor's degree would also be required for a marketing position in laser interferometry industries. In this case, focus should be on business, but a strong background in engineering or physics will make a candidate much more competitive. Technical jobs that do not require a bachelor's degree could involve maintenance, servicing, or calibration of laser interferometers for measurements in industry. They could involve assembly of precision optomechanical systems or machining of precision parts.

Social Context and Future Prospects

The development of increasingly precise interferometers in the last few decades, such as for gravitational-wave measurement, has spurred corresponding leaps in mechanical and materials engineering, since these systems require unprecedented mechanical and vibrational stability. Laser interferometers are beginning to be used in characterization of nanomaterials, and this will push the limits of resolution of laser interferometers even further. As the cost of lasers and optical components continues to decrease, the use of laser interferometers in many industrial manufacturing applications will likely increase. They are an ideal measurement system in that they do not contain moving parts, so there is no wear on parts, and they do not mechanically contact the sample being measured.

Active research is conducted in the field of laser interferometric sensors, with potential applications in military and manufacturing industries. Oil and gas companies may also drive development of sensors for leak and gas detection during drilling. In addition, commercial applications for laser sensors will open up in areas of surveillance as acoustic laser interferometry technology is developed.

Research continues in academic labs in areas that may someday become real-world applications, such as using laser interferometers for the detection of seismic waves, or in the area of quantum interferometry in which single photons are manipulated to interfere with each other in a highly controlled manner.

Corie Ralston, B.S., Ph.D.

Further Reading

Beers, John S., and William B. Penzes. "The NIST Length Scale Interferometer." *Journal of Research of the National Institute of Standards and Technology* 104, no. 3 (May/June, 1999): 225-252. Detailed description of the original NIST Michelson interferometer, with historical context.

Halsey, David, and William Raynor, eds. *Handbook of Interferometers: Research, Technology and Applications.* New York: Nova Science Publishers, 2009. A comprehensive text on current interferometry technologies. Designed for readers with a strong physics background.

Hariharan, P. *Basics of Interferometry.* 2d ed. Burlington, Mass.: Academic Press, 2007. Covers most of the current interferometer configurations and includes sections on applications.

_______. *Optical Interferometry.* 2d ed. San Diego: Academic Press, 2003. Overview of the basic theory of interferometry methods, including sections on laser interferometry.

Malacara, Daniel, ed. *Optical Shop Testing.* 3d ed. Hoboken, N.J.: John Wiley & Sons, 2007. An introduction to interferometers used in testing optical components.

Sirohi, Rajpal S. *Optical Methods of Measurement: Wholefield Techniques.* 2d ed. Boca Raton, Fla.: CRC Press, 2009. Covers the basics of wave equations and interference phenomena. Includes multiwavelength techniques of interferometry.

Tolansky, Samuel. *An Introduction to Interferometry.* 2d ed. London: Longman, 1973. Includes many interferometry methods that do not use lasers.

Web Sites

Laser Interferometer Gravitational-Wave Observatory
http://www.ligo.caltech.edu

National Aeronautics and Space Administration
James Webb Space Telescope
http://www.jwst.nasa.gov

National Institute of Standards and Technology
http://www.nist.gov

See also: Laser Technologies

LASER TECHNOLOGIES

FIELDS OF STUDY

Cosmetic surgery; dermatology; fiber optics; laparoscopic surgery; medicine; ophthalmology; optical disk manufacturing; physics; printer manufacturing; spectroscopy; surgery.

SUMMARY

The term "laser" is an acronym for light amplification by stimulated emission of radiation. A laser device emits light via a process of optical amplification; the process is based on the stimulated emission of photons (particles of electromagnetic energy). Laser beams exhibit a high degree of spatial and temporal coherence, which means that the beam does not widen or deteriorate over a distance. Laser applications are numerous and include surgery, skin treatment, eye treatment, kidney-stone treatment, light displays, optical disks, bar-code scanners, welding, and printers.

KEY TERMS AND CONCEPTS

- **Fiber Optics:** Transmission of light through extremely fine, flexible glass or plastic fibers.
- **Laparoscopic Surgery:** Minimally invasive surgery that is accomplished with a laparoscope and the use of specialized instruments, including laser devices, through small incisions.
- **Laser Guidance:** Guidance via a laser beam, which continuously illuminates a target; missiles, bombs, or projectiles (for example, bullets) home on the beam to strike the target.
- **Optical Disk:** Plastic-coated disk that stores digital data, such as music images, or text; tiny pits in the disk are read by a laser beam.
- **Photon:** Electromagnetic energy, which is regarded to be a discrete particle that has zero mass, no electrical charge, and an indefinitely long lifetime.
- **Spectroscopy:** Analysis of matter via a light beam.

Definition and Basic Principles

A laser consists of a highly reflective optical cavity, which contains mirrors and a gain medium. The gain medium is a substance with light-amplification properties. Energy is applied to the gain medium via an electrical discharge or an optical source such as another laser or a flash lamp. The process of applying energy to the gain medium is known as pumping. Light of a specific wavelength is amplified in the optical cavity. The mirrors ensure that the light bounces repeatedly back and forth in the chamber. With each bounce, the light is further amplified. One mirror in the chamber is partially transparent; amplified light escapes through this mirror as a light beam.

Many types of gain media are employed in lasers, they include: gases (carbon dioxide, carbon monoxide, nitrogen, argon, and helium/neon mixtures); silicate or phosphate glasses; certain crystals (yttrium aluminum garnet); and semiconductors (gallium arsenide and indium gallium arsenide). A basic concept of laser technology is "population inversion." Normally, most of the particles comprising a gain medium lack energy and are in the ground state. Pumping the medium places most or all the particles in an excited state. This results in a powerful, focused laser beam. Lasers are classified as operating in either continuous or pulsed mode. In continuous mode, the power output is continuous and constant; in pulse mode, the laser output takes on the form of intermittent pulses of light.

Background and History

In 1917, Albert Einstein theorized about the process of stimulated emission, which makes lasers possible. The precursor of the laser was the maser (microwave amplification by stimulated emission of radiation). A patent for the maser was granted to American physicists Charles Hard Townes and Arthur L. Schawlow on March 24, 1959. The maser did not emit light; rather it amplified radio signals for space research. In 1958, Townes and Schawlow published scientific papers, which theorized about visible lasers. Also in 1958, Gordon Gould, a doctoral student at Columbia University under Townes, began building an optical laser. He did not obtain a patent until 1977. In 1960, the first gas laser (helium-neon) was invented by Iranian-American physicist Ali Javan. The device converted electrical energy to light, and this type of laser has many practical applications such as laser surgery. In 1962, American engineer Robert Hall invented the semiconductor laser, which has many applications for communications systems and electronic appliances. In 1969, Gary Starkweather, a Xerox researcher, demonstrated the use of a laser beam to print. Laser printers were a marked improvement over the dot-matrix printer. The print quality was much better and it could print on a single sheet of paper rather than a continuous pile of fan-folded paper. In September, 1976, Sony first demonstrated an optical audio disk. This was the precursor to compact discs (CDs) and digital versatile discs (DVDs). In 1977, telephone companies began trials using fiber-optic cables to carry telephone traffic. In 1987, New York City ophthalmologist Stephen Trokel performed the first laser surgery on a patient's eyes.

How It Works

Laser Ablation. Laser ablation involves removing material from a surface (usually a solid but occasionally a liquid). Pulsed laser is most commonly used; however, at a high intensity, continuous laser can ablate material. At lower levels of laser energy, the material is heated and evaporates. At higher levels, the material is converted to plasma. Plasma is similar to gas, but it differs from gas in that some of the particles are ionized (a loss of electrons). Because of this ionization, plasma is electrically conductive.

Laser Cutting. Laser cutting uses laser energy to cut materials either by melting, burning, or vaporizing. Laser cutting is extremely focusable to about 25 microns (one-quarter the width of a human hair); thus, a minimal amount of material is removed. The three common types of lasers used for cutting are: carbon dioxide (CO_2), neodymium (Nd), and neodymium-yttrium aluminum garnet (Nd-YAG).

Laser Guidance. Laser guidance involves the use of a laser beam to guide a projectile (a bomb or a bullet) to a target. In its simplest form, such as a beam emitted from a rifle, the shooter points the laser beam so that the bullet will hit the target. A much more complex process is the guidance of a missile or a bomb. In some cases, the missile contains a laser homing device in which the projectile "rides" the laser beam to the target. More commonly, a technique referred to as semi-active laser homing (SALH) is employed. With SALH, the laser beam is kept pointed at the target after the projectile is launched. Laser energy is scattered from the target, and as the missile approaches the target, heat sensors home in on this energy. If a target does not reflect laser energy well, the beam is aimed at a reflective source near the target.

Laser Lighting Displays. The focused beam emitted by a laser makes it useful for light shows. The bright, narrow beam is highly visible in the night sky. The beam can also be used to draw images on a variety of surfaces, such as walls or ceilings, or even theatrical smoke. The image can be reflected from mirrors to produce laser sculptures. The beam can be moved at any speed in different directions by the use of a galvanometer, which deflects the beam via an electrical current. The variety of vivid colors available with lasers enhances the visual effects.

Laser Printing. Laser printing involves projecting an image onto an electrically charged, rotating drum that is coated with a photoconductor. When exposed to light, the electrical conductivity of the photoconductor increases. The drum then picks up particles of dry ink (toner). These particles are picked up by varying degrees depending on the amount of charge. The toner is then applied to a sheet of paper. The process involves rapidly "painting" the image line by line. Fast (and more expensive) laser printers can print up to 200 pages per minute. Printers are both monochrome (one color–usually black) or color. Color printers contain toners in four colors: cyan, magenta, yellow, and black. A separate drum is used for each toner. The mixture of the colors on the toners produces a crisp, multicolored image. Duplex laser printers are available, which print on both sides of a sheet of paper. Some duplex printers are manual devices, which require the operator to flip one or more pages manually when indicated. Automatic duplexers mechanically turn each sheet of paper and feed it past the drum twice.

Optical Disks. Optical disks are flat, circular disks that contain binary data in the form of microscopic pits, which are non-reflective and form a binary value of 0. Smooth areas are reflective and form a binary value of 1. Optical disks are both created and read with a laser beam. The disks are encoded in a continuous spiral running from the center of the disk to the perimeter. Some disks are dual layer; with these disks, after reaching the perimeter, a second spiral track is etched back to the center. The amount of data storage is dependent on the wavelength of the laser beam. The shorter the wavelength, the greater the storage capacity (shorter-wavelength lasers can read a smaller pit on the disk surface). For example, the high-capacity Blu-ray Disc uses short-wavelength blue light. Laser can be used to create a master disk from which duplicates can be made by a stamping process.

Laser 3-D Scanners. Laser 3-D scanners analyze an object via a laser beam. The collected data is used to construct a digital, three-dimensional object of the model. In addition to shape, some scanners can replicate color.

Applications and Products

Laser technology includes a vast number of applications for business, entertainment, industrial, medical, and military use. The reflective ability of a laser beam has one major drawback: It can inadvertently strike an unintended target. For example, a reflected laser beam could damage an eye, so protective goggles are worn by laser operators. Products range from inexpensive laser pointers and CDs to surgical, industrial, and military devices costing hundreds of thousands of dollars. The following applications and products are a representative sample and are by no means comprehensive.

Optical Disks. Most optical disks are read-only; however, some are rewritable. They are used for the storage of data, computer programs, music, graphic images, and video games. Since the first CD was introduced in 1982, this technology has evolved markedly. Optical data storage has in a large part supplanted storage on magnetic tape. Although optical storage media can degrade over time from environmental factors, they are much more durable than magnetic tape, which loses its magnetic charge over time. Magnetic tape is also subject to wear as it passes through the rollers and recording head. This is not the case for optical media in which the only contact with the recording surface is the laser beam. CDs are primarily used for the storage of music: A five-inch CD can hold an entire recorded album, replacing the vinyl record, which was subject to wear and degradation. A limitation of the CD is its storage capacity: 700 megabytes of data (eighty minutes of music). Actually, three years before the introduction of the CD, the larger LaserDisc appeared for home video use. However, it never attained much popularity in the United States. The DVD, which appeared in 1996, rapidly gained popularity and soon outpaced VHS tape for the storage of feature-length movies. The DVD can store 4.7 gigabytes of data in single-layer format and 8.5 gigabytes in dual-layer format. The development of high-definition (HD) television fueled the development of higher-capacity storage media. After a format war of several year's duration, the Blu-ray DVD won out over the HD disc. The Blu-ray Disc can store about six times the amount of data as a standard DVD: 25 gigabytes of data in single-layer format and 50 gigabytes of data in dual-layer format.

Medical Uses. Medical applications of laser

Fascinating Facts About Laser Technologies

- In early 1942, long before laser guidance systems appeared, psychologist B. F. Skinner at the University of Minnesota investigated weapons guidance by trained pigeons. The birds were trained with slides of aerial photographs of the target. They were placed in a harness inside the guidance system and kept the target in the crosshairs by pecking. Skinner never overcame official skepticism, and the project was abandoned.
- The radial keratotomy procedure originated in Soviet Russia with physician Svyatoslav Fyodorov in the 1970's. Fyodorov treated a boy whose glasses had been broken in a fight and cut his cornea. After the boy recovered, his eyesight had improved—his myopia was significantly lessened.
- Physician Stephen Trokel, who patented the excimer laser for vision correction, adapted the device from its original use: etching silicone computer chips.
- In 1960, American inventor Hildreth Walker, Jr., developed a ruby laser, which precisely measured the distance from the Earth to the Moon during the Apollo 11 mission.
- The original Hewlett Packard (HP) LaserJet printer retailed for more than $3,000 and could print eight pages per minute at a resolution of 300 dots per inch. The only font available was Courier, and it could print only a quarter-page of graphics at that resolution. As of 2011, much smaller laser printers with a multitude of features, including higher speed and print quality, retail for around $100.
- Laboratory lasers can measure distances accurate to less than a thousandth of the wavelength of light, which is less than the size of an atom.

technology exist for abdominal surgery, ophthalmic surgery, vascular surgery, and dermatology. The narrow laser beam can cut through tissue and cauterize small blood vessels at the same time. Laser can be used with an open surgical incision as well as laparoscopic procedures in which the surgery is performed through small incisions for passage of the laparoscope and surgical instruments. The surgeon and his or her assistants can view images on a video monitor. Mirrors can be used to deflect the laser beam in the desired direction. However, the surgeon must be extremely careful when directing a laser beam at an internal structure–inadvertent reflection of the beam can damage tissue (for example, puncture the bowel).

A common ophthalmic procedure is the radial keratotomy. With this procedure, fine laser cuts are made in a radial fashion around the cornea. These precision cuts can correct myopia (nearsightedness) and hyperopia (farsightedness). Laser is also used to treat a number of eye problems such as a detached retina (separation of the imaging surface of the retina from the back of the eye), glaucoma (increased pressure within the eyeball), and cataracts (clouding of the lens).

Atherosclerosis (a narrowing of the blood vessels due to plaque formation) is a common disease. Laser can be used to vaporize the plaque. It also can be used to bore small holes within the heart to improve circulation within the heart muscle.

Dermatology procedures can be performed with the laser–many of which are cosmetic procedures done to improve appearance. Some laser dermatologic applications are: acne treatment and acne scar removal, removal of age spots, skin resurfacing, scar removal, tattoo removal, spider vein removal, and hair removal.

Innovative uses of laser for medical purposes are being reported frequently. For example, in December, 2010, French researchers reported a noninvasive technique to diagnose cystic fibrosis prenatally. Cystic fibrosis is a genetic disease in which thick mucus secretions form in the lungs and glands such as the pancreas, which result in progressive deterioration of the affected organs. The technology involves the identification of affected fetuses by laser microdissection of a fetal cell, which was circulating in the mother's bloodstream. The procedure avoids the risk of a miscarriage when either chorionic villus sampling or amniocentesis is used. Another innovative use of laser was also reported in December, 2010. Medical devices, which are made of plastic or metal, are surgically placed within the body. Examples of medical devices are stents, which are inserted to improve blood flow to the heart, and hip-replacement prosthetics. These devices can become coated with biofilm, which is of bacterial origin. This biofilm is resistant to antibiotics; however, a laser technique was reported to be successful in removing biofilm.

Industrial Uses. Lasers can make precision cuts on metal or other materials. A minimal amount of material is removed, leaving a smooth, polished surface on both sides of the cut. It also can weld and heat-treat materials. Other industrial uses include marking and measuring parts.

Business Uses. The supermarket bar-code scanner was one of the earliest laser applications–it appeared in 1974. Laser printers are ubiquitous in all but the smallest business offices. They range in price from less than $100 to more than $10,000. The higher-priced models have features such as duplexing capability, color, high-speed printing, and collating.

Military Uses. In addition to marking targets and weapons guidance, laser is used by the military for defensive countermeasures. It also can produce temporary blindness, which can temporarily impair an enemy's ability to fire a weapon or engage in another harmful activity. The military also uses laser guidance systems. Defensive countermeasure applications include small infrared lasers, which confuse heat-seeking missiles to intercept lasers, and power boost-phase intercept laser systems, which contain a complex system of lasers that can locate, track, and destroy intercontinental ballistic missiles (ICBMs). Intercept systems are powered by chemical lasers. When deployed, a chemical reaction results in the quick release of large amounts of energy.

Laser Lighting Displays. Laser light displays are popular worldwide. The displays range from simple to complex and are often choreographed to music. A popular laser multimedia display is Hong Kong's Symphony of Lights, which is presented nightly. The display is accentuated during Tet, the Vietnamese New Year. The exteriors of 44 buildings on either side of Victoria Harbour are illuminated with a synchronized, multicolored laser display, which is accompanied by music. More than four million visitors and locals have viewed the display.

Impact on Industry

Laser technology is a major component of many industries, including electronics, medical devices, data storage, data transmission, research, and entertainment. Laser products enjoy significant repeat business. Over time the devices fail and need to be repaired or replaced. Often, before device failure occurs, the consumer will be attracted to an improved rendition of the product. For example the original LaserDisc, which never achieved popularity in the United States, has evolved to the Blu-ray Disc, which boasts vastly improved storage capacity as well as image and sound quality. In addition, many laser devices come with a hefty price tag–ranging from several thousand dollars to well in excess of 1 million dollars.

Industry and Business. The average home contains laser devices such as CDs (discs and players) and DVDs (discs and players); and home computers, televisions, and telephones that are often connected to a fiber-optic cable network. Many homes and businesses use one or more laser printers. Virtually every retail establishment has a laser bar-code scanner. The military uses laser devices ranging from inexpensive pointers to guidance systems, which can cost many thousands of dollars. These expensive devices offer a continuous revenue stream for a vendor as they are only used once. The United States automobile industry has been using laser to cut the steel for its vehicles since the 1980's. As of 2011, many manufacturing industries use laser for the cutting and welding of metal parts. The medical-device industry is responsible for many high-end products. Laparoscopic surgery is currently popular, and laser instruments are well-adapted to operating by remote control through small incisions. Ophthalmologists are heavy users of laser instruments. Dermatologic procedures, particularly in the realm of cosmetic surgery, often employ expensive laser devices. Scientific procedures such as spectroscopy are commonplace in all industrialized nations.

Government and University Research. One of the biggest sources of funding for laser research in the United States is the Defense Advanced Research Projects Agency (DARPA). It is also the biggest client for certain kinds of laser applications. The agency is primarily concerned with lasers with a military focus, such as guidance systems, missile countermeasures, and laser-based weapons. Another source of funding for laser research is the United States Department of Energy (DOE).The DOE is currently focused on energy efficiency and renewable energy.

DARPA and the DOE supply funds to many universities in the United States for laser research and development. For example, the two major universities in the Los Angeles area, University of California, Los Angeles (UCLA), and the University of Southern California (USC), are actively engaged in highly technical laser research. UCLA is the home to the Plasma-Accelerator Group, which publishes papers such as "Highest Power CO_2 Laser in the World" (2010). The USC Center for Laser Studies contains a number of research groups: Optical Devices Research Group, Research at High Speed Technology, Semiconductor Device Technology, Solid State Lasers, and Theoretical Research.

Careers and Course Work

The laser technology industry offers careers ranging from entry-level positions, such as lighting display operators, to high-level and highly technical positions, which require a scientific or engineering degree. The high-level positions require at least a bachelor's degree with course work in several fields related to laser technology: engineering, physics, computer science, mathematics, and robotics. Many of the positions require a master's or doctorate degree. Positions are available for individuals with a degree in laser engineering in both the government and private sectors. The ability to be a team player is often of value for these positions because ongoing research is often a collaborative effort. University research positions are also available. In this arena, the employee is expected to divide his or her time between research and teaching.

Laser technicians are needed in a variety of fields including medical, business, and entertainment. Many of these positions require some training beyond high school at a community college or trade school. Dermatologists and other medical specialists often employ technicians to perform laser procedures. If a company employs a number of laser technicians, supervisory positions may be available.

Social Context and Future Prospects

As of 2011, laser applications have become innumerable and ubiquitous. Many aspects of daily living that are taken for granted, such as ringing up groceries, making a phone call, or playing a video, are dependent on this technology. Virtually every branch of laser technology is continually evolving. Furthermore, existing technologies are being implemented in increasing numbers. For example, use of the CD for data storage is being replaced with the much higher capacity DVD. Research in the realm of military applications is particularly vigorous. The science-fiction weapon, the laser blaster, will soon become a reality. A primary obstacle to developing lasers as weapons has been the generation of enough power to produce a laser blast with a sufficient level of destruction. That power level is rapidly approaching. In March, 2009, Northrop Grumman first fired up their 100 kilowatt (100,000 watts) weapons-grade laser. (One hundred kilowatts is enough energy to power about six U.S. homes for a month). Tests are ongoing for the device. Grumman announced in late 2010 that the device would soon be transferred to the White Sands Missile Range in New Mexico for further testing. Medical applications for laser are expanding rapidly. The consumer, however, must be aware that adding the "laser" adjective to a procedure does not necessarily make it superior to previous techniques. The laser scalpel, for example, is basically just another means of cutting tissue.

Robin L. Wulffson, M.D., F.A.C.O.G.

FURTHER READING

Bone, Jan. *Opportunities in Laser Technology Careers.* New York: McGraw-Hill, 2008. Offers a comprehensive overview of career opportunities in laser technology; includes salary figures as well as experience required to enter the field.

Hecht, Jeff. *Understanding Lasers: An Entry-Level Guide.* 3d ed. Piscataway, N.J.: IEEE Press, 2008. This introductory text is suitable for students at the advanced high school level.

Lele, Ajey. *Strategic Technologies for the Military: Breaking New Frontiers.* Thousand Oaks, Calif.: Sage, 2009. Describes the nuances of technological development in a purely scientific manner and provides a social perspective to their relevance for future warfare and for issues such as disarmament and arms control, as well as their impact on the environment.

Sarnoff, Deborah S., and Joan Swirsky. *Beauty and the Beam: Your Complete Guide to Cosmetic Laser Surgery.* New York: St. Martin's, 1998. Provides information for the consumer on laser dermatology procedures and aids the reader in selecting the safest and most experienced practitioner. It also describes what each surgical procedure entails and details costs in different regions.

Silfvast, William Thomas. *Laser Fundamentals.* 2d ed. New York: Cambridge University Press, 2004. Covers topics from laser basics to advanced laser physics and engineering.

Townes, Charles H. *How the Laser Happened: Adventures of a Scientist.* New York: Oxford University Press, 1999. A personal account by a Nobel laureate that describes some of the leading events in twentieth-century physics.

WEB SITES

American Society for Laser Medicine and Surgery
http://www.aslms.org

Defense Advanced Research Projects Agency
http://www.darpa.mil

USC Center for Laser Studies
http://www.usc.edu/dept/CLS/page.html

See also: Printing

LIGHT-EMITTING DIODES

FIELDS OF STUDY

Electrical engineering; materials science; semiconductor technology; semiconductor manufacturing; electronics; physics; chemistry; mathematics; optics; lighting; environmental studies; physics.

SUMMARY

Light-emitting diodes (LEDs) are diodes, semiconductor devices that pass current easily in only one direction, that emit light when current is passing through them in the proper direction. LEDs are small and are easier to install in limited spaces or where small light sources are preferred, such as indicator lights in devices. LEDs are also generally much more efficient at producing visible light than other light sources. As solid-state devices, when used properly, LEDs also have very few failure modes and have longer operational lives than many other light sources. For these reasons, LEDs are gaining popularity as light sources in many applications, despite their higher cost compared with other more traditional light sources.

KEY TERMS AND CONCEPTS

- **Anode:** More positive side of the diode, through which current can easily flow into the device when forward biased.
- **Cathode:** More negative side of the diode, through which current can flow out of the device when forward biased.
- **Color Temperature:** Temperature of a blackbody radiating thermal energy having the same color as the light emitted by the LED.
- **Forward Bias:** Orientation of the diode in which current most easily flows through the device.
- **Photon:** Quantum mechanical particle of light.
- **P-N Junction:** Junction between positive type (p-type) and negative type (n-type) doped semiconductors on which all diodes are based.
- **Radiant Efficiency:** Ratio of optical power output to the electrical power input of the device.
- **Reverse Bias:** Orientation of the diode in which current does not easily flow through the device.
- **Reverse Breakdown Voltage:** Maximum reverse biased voltage that can be applied to the device before it begins to conduct electricity, often in an uncontrolled manner; sometimes simply called the breakdown voltage.
- **Thermal Power Dissipation:** Rate of energy per unit time dissipated in the device in the form of heat.

Definition and Basic Principles

Diodes act as one-way valves for electrical current. Current flows through a diode easily in one direction, and the ideal diode blocks current flow in the other direction. The very name diode comes from the Greek meaning two pathways. The diode-like behavior comes from joining two types of semiconductors, one that conducts electricity using electrons (n-type semiconductor) and one that conducts electrons using holes, or the lack of electrons (p-type semiconductor). The electrons will try to fill the holes, but applying voltage in the proper direction ensures a constant supply of holes and electrons to conduct electricity through the diode. The electrons and holes have different energies, so when the electrons combine with holes, they release energy. For most diodes this energy heats the diode. However, by adjusting the types and properties of the semiconductors making up the diodes, the energy difference between holes and electrons can be made larger or smaller. If the energy difference corresponds to the energy of a photon of light, then the energy is given off in the form of light. This is the basis of how LEDs work.

LEDs are not 100 percent efficient, and some energy is lost in current passing through the device, but the majority of energy consumed by LEDs goes into the production of light. The color of light is determined by the semiconductors making up the device, so LEDs can be fabricated to make light only in the range of wavelengths desired. This makes LEDs among the most energy-efficient sources of light.

Background and History

In 1907, H. J. Round reported that light could be emitted by passing current through a crystal rectifier junction under the right circumstances. This was the ancestor of the modern LED, though the term diode had not yet been invented. Though research continued on these crystal lamps, as they were called, they were seen as impractical alternatives to incandescent and other far less expensive means of producing light. By 1955, Rubin Braunstein, working at RCA, had shown that certain semiconductor junctions produced infrared light when current passed through them. Scientists Robert Biard and Gary Pittman, however, managed to produce a usable infrared LED, receiving a patent for their device. Nick Holonyak, Jr., a scientist at General Electric, then created a red LED–the viable and useful visual spectrum LED–in 1961. Though these early LEDs were usable, they were far too expensive for widespread adoption. By the 1970's, Fairchild Semiconductor had developed inexpensive red LEDs. These LEDs were soon incorporated into seven segment numeric indicators for calculators produced by Hewlett Packard and Texas Instruments. Red LEDs were also used in digital watch displays and as red indicator lights on various pieces of equipment.

Early LEDs were limited in brightness, and only the red ones could be fabricated inexpensively. Eventually, other color LEDs and LEDs capable of higher light output were developed. As the capabilities of LEDs expanded, they began to see more uses. By the early twenty-first century, LEDs began to compete with other forms of artificial lighting based on their energy efficiency.

How It Works

An LED is a specific type of solid-state diode, but it still retains the other properties typical of diodes. Solid-state diodes are formed at the junction of two semiconductors of different properties. Semiconductors are materials that are inherently neither good conductors nor good insulators. The electrical properties of the materials making up semiconductors can be altered by the addition of impurities into the crystal structure of the material as it is fabricated. Adding impurities to semiconductors to achieve the proper electrical nature is called doping. If the added impurity has one more electron in its outermost electron shell compared with the semiconductor material, then extra electrons are available to conduct electricity. This is a negative doped, or n-type, semiconductor. However, if the impurity has one fewer electron in its outermost electron shell compared with the semiconductor material, then there are too few electrons in the crystal structure, and an electron can move from atom to atom to fill the void. This results in a missing electron moving from place to place and acts like a positive charge moving through the semiconductor. Engineers call this missing electron a hole, and semiconductors in which holes dominate are called positive doped, or p-type, semiconductors.

The P-N Junction. To make a diode, a device is fabricated in which a p-type semiconductor is placed in contact with a n-type semiconductor. The shared boundary between the two types of semiconductors is called a p-n junction. In the vicinity of the junction, the extra electrons in the n-type region combine with the holes of the p-type region. The results in the removal of charge carriers in the vicinity of the p-n junction and the area of few charge carriers is called the depletion region.

When a voltage is applied across the p-n junction, with the p-type region having the higher voltage, then electrons are pulled from the n-type region and holes are pulled from the p-type region into the depletion region. Additionally, electrons are pulled into cathode (the exterior terminal connecting to the n-type region) replenishing the supply of electrons in the n-type region, and electrons are pulled from the anode (the exterior terminal connecting to the p-type region) replenishing the holes in the p-type region. This is the forward-bias orientation of the diode, and current flows through the diode when voltage is applied in this manner. However, when voltage is applied in the reverse direction, electrons are pulled from the n-type region and into the p-type region, resulting in a larger depletion region and fewer available charge carriers. Electric current does not flow through the diode in this reverse-bias orientation.

Electroluminescence. Electrons and holes have different energy levels. When the electrons and holes combine in the depletion region, therefore, they release energy. For most diodes, the energy difference between the p-type holes and the n-type electrons is fairly small, so the energy released is correspondingly small. However, if the energy difference is sufficiently large, then when an electron and hole combine, the amount of energy released is the same as that of a photon of light, and the energy is released in the form of light. This is called electroluminescence. Different wavelengths or colors of light have different energies, with infrared light having less energy than visual light, and red light having less energy than other forms of visual light. Blue light has more energy than other forms of visual light. The color of light emitted by the recombination of electrons and holes is determined by the energy difference between the electrons and holes. Different semiconductors have different energies of electrons and holes, so p-n junctions made of different kinds of semiconductors with different doping result in different colors of light emitted by the LED.

Efficiency. Most light sources emit light over a wide range of wavelengths, often including both visual and nonvisual light as well as heat. Therefore, only a portion of the energy used goes into the form of light desired. For an idealized LED, all of the light goes into one color of light, and that color is determined by the composition of the semiconductors making the p-n junction. For real LEDs, not all of the light makes it out of the material. Some of it is internally reflected and absorbed. Furthermore, there is some electrical resistance to the device, so there is some energy lost in heat in the LED–but nowhere near as much as with many other light sources. This makes LEDs very efficient as light sources. However, LED efficiency is temperature dependent, and they are most efficient at lower temperatures. High temperatures tend to reduce LED efficiency and shorten the lifetime of the devices.

Applications and Products

LEDs produce light, like any other light source, and they can be used in applications where other light sources would have been used. LEDs have certain properties, however, that sometimes make their use preferable to other artificial light sources.

Indicator Lights. Among the first widespread commercial use of LEDs for public consumption was as indicator lights. The early red LEDs were used as small lights on instruments in place of small incandescent lights. The LEDs were smaller and less likely to burn out. LEDs are still used in a similar way, though not with only the red LEDs. They are used as the indicator lights in automobile dashboards and in aircraft instrument panels. They are also used in many other applications where a light is needed and there is little room for an incandescent bulb.

Another early widespread commercial use of LEDs was the seven-segment numeric displays used to show digits in calculators and timepieces. However, LEDs require electrical current to operate, and calculators and watches would rather quickly discharge the batteries of these devices. Often the display on the watches was visible only when a button was pressed to light up the display. However, the advent of liquid crystal displays (LCDs) has rendered these uses mostly obsolete since they require far less energy to operate, and LEDs are needed to light the display at night only.

Replacements for Colored Incandescent Lights. Red LEDs have become bright enough to be used as brake lights in automobiles. Red, green, and yellow LEDs are sometimes used for traffic lights and for runway lights at airports. LEDs are even used in Christmas-decoration lighting. They are also used in message boards and signs. LEDs are sometimes used for backlighting LCD screens on televisions

Fascinating Facts About Light-Emitting Diodes

- Most of the device often called a light-emitting diode is really just the packaging for the LED. The actual diode is typically very tiny and embedded deep inside the packaging.
- The fist inexpensive digital watches marketed to the public used red LEDs for displays.
- Most remote controls for televisions, DVD players, and similar devices use an infrared LED to communicate between the remote control and the devices.
- LEDs operate more efficiently when they are cold than when they are hot.
- Many automobile manufacturers use LEDs in brake lights, particularly the center third brake light.
- An early term for the LED was "crystal lamp." This term eventually gave way to the term "light-emitting diode," later commonly abbreviated LED.
- LEDs, if properly cared for and operated under specified conditions, can last for upward of 40,000 hours of operation.
- The first LEDs produced were infrared-emitting diodes. The first mass-produced visible-light LEDs emitted red light.
- LEDs switch on and off far more quickly than most other light sources, often being able to go from off to fully on in a fraction of a microsecond.

and laptops. Colored LEDs are also frequently used in decorative or accent lighting, such as lighting in aquariums to accentuate the colors of coral or fish. Some aircraft use LED lighting in their cabins because of energy efficiency. Red LEDs are also used in pulse oximeters used in a medical setting to measure the oxygen saturation in a patient's blood.

The biggest obstacle to replacing incandescent lights with LEDs for room lighting or building lighting is that they produce light of only one color. Several strategies are in development for producing white light using LEDs. One strategy is to use multiple-colored LEDs to simulate the broad spectrum of light produced by incandescent lights or fluorescent lights. However, arrays of LEDs produce a set of discrete colors of light rather than all colors of the rainbow, thus distorting colors of objects illuminated by the LED arrays. This is aesthetically unpleasing to most people. Another strategy for producing white light from LEDs is to include a phosphorescent coating in the casing around the LED. This coating would provide the different colors of light that would mimic the light of fluorescent bulbs; however, such a strategy removes much of the efficiency of LEDs. Research continues to produce a pleasing white light from LEDs.

Despite the color problems and the high initial cost of LEDs, LEDs have many properties that make them attractive replacements for incandescent or fluorescent lights. LEDs typically have no breakable parts and being solid-state devices are very durable and have low susceptibility to vibrational damage. LEDs are very energy efficient, but they tend to be less efficient at high power and high light output. LEDs are slightly more efficient, and far more expensive, than high-efficiency fluorescent lights, but research continues.

Nonvisual Uses for LEDs. Infrared LEDs are often used as door sensors or for communication by remote controls for electronic devices. They can also be used in fiber optics. The rapid switching capabilities of LEDs makes them well suited for high-speed communication purposes. Ultraviolet LEDs are being investigated as replacements for black lights for purposes of sterilization, since many bacteria are killed by ultraviolet light.

Impact on Industry

Government and University Research. Research continues to produce less expensive and more capable LEDs. The nature of LEDs makes them potentially among the most energy-efficient light sources. The United States Department of Energy is a significant driving force behind development of energy-efficient solid-state lighting systems, specifically the development of LED technology for widespread use. University researchers, particularly faculty in engineering and materials science, receive government grants to study improving LED performance.

Industry and Business. There are two different ways that the private sector is impacted by advances in LEDs and LED uses. Major corporations, particularly those in semiconductor manufacturing, are doing research alongside government and university researchers to develop more efficient LEDs with greater capabilities. The increased demand for LED lighting from energy-conscious consumers spurs companies to compete to develop new LEDs for consumer use.

However, no matter what types of new LEDs are developed, they must be used in products that consumers want to purchase in order to be commercially viable. In the early twenty-first century, environmentally conscious consumers began looking for more energy-efficient lights. A great many companies began to make products that used LED lights in place of incandescent lights for these energy-conscious consumers who were willing to pay more for a product that was perceived as being more environmentally friendly and energy efficient. A prime example of this sort of development was the advent of LED Christmas lights; however, many other applications have been developed. The long operational life of LEDs is also attractive to consumers, since they are less likely to need a replacement compared with other light sources. The continued development of increasingly inexpensive LEDs with more different colors of light possible is spurring more companies to develop more products using LEDs.

Careers and Course Work

LEDs are used in many industries, not just in electronics, which means that there are many different degree and course-work pathways to working with LEDs.

The development of new types of LEDs requires

detailed understanding of semiconductor physics, chemistry, and materials science. Typically, such research requires an advanced degree in physics, materials science, or electrical engineering. Such degrees require courses in physics, mathematics, chemistry, and electronics. The different degrees will have different proportions of those courses.

Utilization of LEDs in circuits, however, requires a quite different background. Technicians and assembly workers need only basic electronics and circuits courses to incorporate the LEDs into circuits or devices.

Lighting technicians and lighting engineers also work with LEDs in new applications. Such careers could require bachelor's degrees in their field. New LED lamps are being developed and LEDs are seen as a possible energy-efficient alternative to other types of lighting. They also have long operational lives, so there is continual development to include LEDs in any type of application where light sources of any sort are used.

Social Context and Future Prospects

At first, LEDs were a niche field, with limited uses. However, as LEDs with greater capabilities and different colors of emitted light were produced, uses began to grow. LEDs have evolved past the point of simply being indicator lights or alphanumeric displays. Developments in semiconductor manufacturing have driven down the cost of many semiconductor devices, including LEDs. The reducing cost combined with the energy efficiency of LEDs has led these devices to become more prominent, particularly where colored lights are desired. Research continues to produce newer LEDs with different colors, different power requirements, and different intensities. Newer techniques are being developed to produce white light using LEDs. These technological developments will make LEDs even more practical replacements for current light sources, despite their higher initial up-front costs.

Research continues on LEDs to make them more commercially and aesthetically viable as alternatives to more traditional light sources. However, research is also continuing on other alternative light sources. The highest-efficiency fluorescent lights have similar efficiencies to standard LEDs, but they cost less and are able to produce pleasing white light that LEDs do not yet produce. LEDs will continue to play an increasing role in their current uses, but it is unclear if they will eventually become wide-scale replacements for incandescent or fluorescent lights.

Raymond D. Benge, Jr., B.S., M.S.

Further Reading

Held, Gilbert. *Introduction to Light Emitting Diode Technology and Applications.* Boca Raton, Fla.: Auerbach, 2009. A comprehensive overview of light-emitting diode technology and applications of LEDs.

Mottier, Patrick, ed. *LEDs for Lighting Applications.* Hoboken, N.J.: John Wiley & Sons, 2009. A detailed book about LEDs, the manufacture of LEDs, and use of the devices in artificial lighting.

Paynter, Robert T. *Introductory Electronic Devices and Circuits: Conventional Flow Version.* 7th ed. Upper Saddle River, N.J.: Prentice Hall, 2006. An excellent and frequently used introductory electronics textbook, containing several sections on diodes, with a very good description of light-emitting diodes.

Razeghi, Manijeh. *Fundamentals of Solid State Engineering.* 3d ed. New York: Springer, 2009. An advanced undergraduate textbook on the physics of semiconductors, with a very detailed explanation of the physics of the p-n junction, which is the heart of diode technology.

Schubert, E. Fred. *Light-Emitting Diodes.* 2d ed. New York: Cambridge University Press, 2006. A very good and thorough overview of light-emitting diodes and their uses.

Žukauskas, Artūras, Michael S. Shur, and Remis Gaska. *Introduction to Solid-State Lighting.* New York: John Wiley & Sons, 2002. A fairly advanced and very thorough treatise on artificial lighting technologies, particularly solid-state lighting, such as LEDs.

Web Sites

LED Magazine
http://www.ledsmagazine.com

The Photonics Society
http://photonicssociety.org

Schottkey Diode Flash Tutorial
http://cleanroom.byu.edu/schottky_animation.phtml

University of Cambridge
Interactive Explanation of Semiconductor Diode
http://www-G.eng.cam.ac.uk/mmg/teaching/linearcircuits/diode.html

U.S. Department of Energy
Solid-State Lighting
http://www1.eere.energy.gov/buildings/ssl/about.html

See also: Diode Technology

LIQUID CRYSTAL TECHNOLOGY

FIELDS OF STUDY

Electronics; chemistry; organic chemistry; physics; optics; materials science; mathematics

SUMMARY

Liquid crystal devices are the energy efficient, low-cost displays used in a variety of applications in which information or images are presented. The operation of the devices is based on the unique electrical and optical properties of liquid crystal materials.

KEY TERMS AND CONCEPTS

- **Active Matrix Liquid Crystal Display:** A display with transistors to control voltage at each pixel, which allows for a sharper picture.
- **Anisotropic:** Molecules that are not similar in all directions but have a longer axis in certain planes, so certain measurements may depend on direction.
- **Calamitic Liquid Crystal:** An anisotropic liquid crystal with a longer axis in one direction, giving it a rodlike shape (also referred to as prolate).
- **Discotic Liquid Crystal:** An anisotropic liquid crystal with one axis shorter than the other two, giving it a disclike shape (also referred to as oblate).
- **Lyotropic Liquid Crystal:** Liquid crystals achieved when the molecule is combined to a high enough concentration with a solvent (rather than through temperature change). Soap is one example.
- **Nematic State:** Liquid crystals that are not distinctly layered but can be oriented and aligned by a charge.
- **Smectic State:** Liquid crystals that are somewhat disordered but maintain distinct layers.
- **Thermotropic Liquid Crystals:** Liquid crystals in which the specific properties are maintained within a specific temperature range.
- **Thin Film Transistor (TFT):** Transistors in a liquid crystal display (LCD) that help control the voltage, and therefore, the coloring, at individual pixels.
- **Twisted-Nematic Effect (TNE):** The effect in which, in a turned-on LCD, light travels through the display changing from being polarized in one direction to being twisted 90 degrees and polarized in that direction.
- **Twisted-Nematic Liquid Crystal Display (TN LCD):** The dominant type of LCD display, which utilizes the TNE.

Definition and Basic Principles

Liquid crystal technology is the use of a unique property of matter to create visual displays that have become the standard for modern technology.

Originally discovered as a state existing between a solid and a liquid, liquid crystals were later found to have applications for visual display. While liquid crystals are less rigid than something in a solid state of matter, they also are ordered in a manner not found in liquids. As anisotropic molecules, liquid crystals can be polarized to a specific orientation to achieve the desired lighting effects in display technologies.

Liquid crystals themselves can exist in several states. These range from a well-ordered crystal state to a disordered liquid state. In-between states are known as the smectic phase, which have layering, and the nematic phase, in which the separate layers no longer exist but the molecules can still be ordered.

Liquid crystal displays (LCDs) at this point typically use crystals in the nematic state. They also use calamitic liquid crystals, whose rodlike shape and orientation along one axis allow the display to lighten and darken.

Over time, researchers have made advances in the materials used for LCDs and the route of power for manipulating the crystals, allowing for the low-cost, high-resolution, energy-efficient displays that have become the dominant technology for displays such as computers and television sets. Future research should allow for improvements in the response time of the displays and for better viewing from different angles.

Background and History

Liquid crystals were discovered by Austrian botanist and chemist Friedrich Reinitzer in 1888. While working with cholesterol, he discovered what appeared to be a phase of matter between the solid (crystal) state and the liquid state. While attempting to find the melting point, Reinitzer observed that within a certain temperature range he had a cloudy mixture, and only at a higher temperature did that mixture become a liquid. Reinitzer wrote of his discovery to his friend, German physicist Otto Lehmann, who not only confirmed Reinitzer's discovery—that the liquid crystal state was unique and not simply a mixture of solid and liquid states—but also noted some distinct visual properties, namely that light can travel in one of two different ways through the crystals, a property known as birefringence.

After the discovery of liquid crystals, the field saw a lengthy period of dormancy. Modern display applications have their roots in the early 1960s, in part because of the work of French physicist and Nobel laureate Pierre-Gilles de Gennes, who connected research in liquid crystals with that in superconductors. He found that applying voltages to liquid crystals allowed for control of their orientation, thus allowing for control of the passage of light through them.

In the early 1970s, researchers, including Swiss physicist and inventor Martin Schadt at the Swiss company Hoffman-LaRoche, discovered the twisted-nematic effect—a central idea in LCD technology. (The year of invention is typically said to be 1971, although patents were awarded later.) The idea was patented in the United States at the same time by the International Liquid Xtal Company (now LXD), which was founded by American physicist and inventor James Fergason in Kent, Ohio. (Fergason was part of the Liquid Crystal Institute at Kent State University. The institute was founded by American chemist Glenn H. Brown in 1965.) Licensing the patents to outside manufacturers allowed for the production of simple LCDs in products such as calculators and wristwatches.

In the 1980s, LCD technology expanded into computers. LCDs became critical components of laptop computers and smaller television sets. With research continuing on liquid crystals into the twenty-first century, LCD televisions overtook cathode ray tubes (CRTs) as the dominant technology for television sets.

How It Works

LCDs have a similar structure, whether in a digital watch or in a 40-inch television. The liquid crystals are held between two layers of glass. A layer of transparent conductors on the liquid crystal side of the glass allows the liquid crystal layer to be manipulated. Polarized film layers are placed on the outside ends of the glass, one of which will face the viewer and the other will remain at the back of the display.

Polarizers alter the course of light. Typically, light travels outward in random directions. Polarizers present a barrier, blocking light from traveling in certain directions and preventing glare. The polarizers in an LCD are oriented at 90-degree angles from each other. With the polarizers in place alone, all light would be blocked from traveling through an LCD, but the workings of liquid crystals allow that light to come through.

The electrical current running through the liquid crystals controls their orientation. The rodlike crystals, without voltage, are oriented perpendicular to the glass of the screens. In this state, the crystals do not alter the direction of the light passing through. As voltage is applied, the crystals turn parallel to the direction of the screen. Like the polarized films, the conductors are oriented at 90-degree angles to each other, as are the crystals next to each screen (that is, crystals on one end are at a 90-degree angle from those on the other end). Between, however, the crystals orient in a twisting pattern, so light polarized in one direction will be redirected and turned 90 degrees when it emerges at the other end of the display. This is known as the twisted-nematic effect.

Thus, the voltage applied to the crystals controls the light coming through the LCD screen. At lower voltage levels, some, but not all, light is allowed through the display. By manipulating the intensity of the incoming light, the LCD can display in a gray scale.

Because liquid crystals are a state of matter, they exist only at a certain temperature: between the melting and freezing points of the material. Thus, LCD displays may have trouble working in extreme heat or extreme cold. One of the primary challenges of LCD display is finding materials that remain in liquid crystal forms at the temperatures in which the devices are likely to be used. Some of the challenge also lies in finding materials that may display better color or allow for lower energy consumption. However, materials that do one of these things better may make other features of a display worse. In some cases, a mixture of compounds for the liquid crystals in a device may be used.

Simple LCDs. The simplest LCD displays, such as calculators and watches, typically do not have their own light sources. Instead, they have what is known as passive display. In back of the LCD display is a reflective surface. Light enters the display and then bounces off the reflective surface to allow for the screen display. Simple LCDs are monochromatic and have specific areas (typically bars or dots) that become light or dark. While these devices are lower-powered, some do still use a light source of their own. Alarm clocks, for example, have light-emitting diodes (LEDs) as part of their display so that they can be seen in the dark.

Personal Computers and Televisions. For larger monitors that display complex images in color, the setup for an LCD becomes more complicated. Multicolor LCDs need a significant light source at the back of the display.

The glass used for more sophisticated LCD displays will have microscopic etchings on the glass plates at the front and back of the display. As with the polarizing filters and the conductors, the etchings are at 90-degree angles from each other, vertical on one plate and horizontal on the other. This alignment forms a matrix of points in each location where the horizontal and vertical etchings cross, resulting in what are known as pixels. Each pixel has a unique "address" for the electronic workings of the display. Many television sets are marketed as having 1080p, referring to 1,080 horizontal lines of pixels.

An active matrix (AM) display will have individual thin film transistors (TFTs) added at each pixel to allow for control of those sites. Three transistors are actually present at each pixel, each accompanied by an additional filter of red, green, or blue. Each of those transistors has 256 power levels. The blending of the different levels of those three colors (256^3, or 16,777,216 possible combinations) and the number of pixels allows for the full-color LCD displays.

While light is displayed as a combination of red, green, and blue on screens, printing is typically done on a scale that uses cyan, magenta, yellow, and black as base colors. This accounts for some discrepancy between colors that appear on screen and those that show up on paper.

Applications and Products

Liquid crystals are used in displays for a number of products. Early uses included digital thermometers, digital wristwatches, electronic games, and calculators. As the power needed for an LCD display and resolution improved, LCDs came to be used in computer monitors, television sets, car dashboards, and cellphones.

Calculators and Digital Watches. Watches and calculators use what is known as a seven-segment display, wherein each of the seven segments that make up a number are "lit" or "unlit" to represent the ten digits. Looking closely at an LCD will reveal that most numbers come from seven segments, which can be lit to display the ten different digits. Without the polarizing layer, the display would not work. Placing the polarized layers in parallel on the surface would, for example, cause the outlines of all the numbers and other areas on the display to illuminate (appearing as 8s) and leave as blanks the rest of the display.

Early electronic games also used a segment display. Fixed places on the display would be either lit or unlit, allowing game characters to appear to move across the screen.

Temperature Monitors. Because of their sensitivity to heat, liquid crystals have been studied for their use as temperature monitors. Molecules in the smectic liquid crystal state rotate around their axes, and the angle at which they rotate (the pitch) can be temperature sensitive. At different temperatures, the wavelength of light given off will change. Some liquid crystal mixtures are fairly temperature sensitive, and so the mixture of the colors will change with relatively small changes in the temperature. Because of this, they can be used for displays such as infrared or surface temperatures.

Computer Monitors. LCDs have been, and will likely remain, the standard for laptop computers. They have been used for monitors since the notebook computer was introduced. Because of the low power consumption and thinness of the monitor, their use is likely to continue.

Television Sets. The workings of LCDs in televisions have been outlined in the foregoing section. As will be discussed further in the next section, LCDs have had a marked impact on television displays, both in the quality of displays in the home and in industry overall.

There are several developments that could affect LCD technology in the near future. One example is the development of LEDs for use as backlighting for LCDs. By using LEDs rather than a fluorescent bulb, as LCD technology now uses, LCDs can manifest greater contrast in different areas of the screen. Other areas of LCD development include photoalignment and supertwisted nematic (STN) LCDs.

Grooves are made in glass used for LCDs, but this has raised some concern about possible electric charges, reducing the picture quality. Additionally, photoalignment—a focus on the materials used to align the liquid crystals in the display—should ultimately allow for liquid display screens that are flexible or curved, rather than rigid (as are glass panels).

STN LCDs are modified versions of TN LCDs. Rather than twisting the crystals between the layers a total of 90 degrees, STN LCDs rotate the crystals by 270 degrees within the display. This greater level of twisting allows for a much greater degree of change in the levels of brightness in a display. At the same time, it presents a challenge because the response time for the screen is significantly slower.

Fascinating Facts About Liquid Crystal Technology

- Although it was the dominant display technology in televisions for more than sixty years, cathode ray tubes were discovered close to twenty years after liquid crystals.
- Partly for his work with liquid crystals, Pierre-Gilles de Gennes was awarded the Nobel Prize in Physics in 1991.
- The world's largest LCDs are in Cowboys Stadium in Arlington, Texas. Each screen of the four-sided display is 72 feet high and 160 feet wide. The screens are backlit by a total of 10.5 million LEDs, and altogether, the display weighs 1.2 million pounds.
- The first LCD television, the black-and-white Casio TV-10, came to market in 1983.
- Introduced in 1982, the Epson HX-20, generally considered the world's first standard-sized laptop, utilized an LCD.
- The discovery of liquid crystals was made by Friedrich Reinitzer while working with the compound cholesteryl benzoate. In the time since its discovery, thousands of other organic compounds have been found to have a liquid crystal state.
- In addition to use in displays, liquid crystals have some application in polymers. Because they can be oriented, they can produce stronger fibers. As a result, they are used in the production of Kevlar.

IMPACT ON INDUSTRY

The introduction of LCDs allowed for the replacement of CRTs, which had been the standard in the industry. LCDs require less energy than CRTs, and because they do not need the space for the cathode ray, the physical sets are lighter and take up less space for a similar-sized screen.

In larger displays, LCDs compete in the consumer market with plasma televisions, which utilize small compartments of gas for the light on their screens. There are still a number of reasons a consumer might choose plasma over LCD, or vice versa.

Because of their polarized light, LCDs do not have the problems with glare that plasma displays may have. However, because LCDs have polarized films on the screen, LCD images vary greatly when they are not viewed straight on. While plasma televisions may have a slightly faster response rate than LCD televisions, LCD televisions consume less energy. While there is some competition between LCDs and plasma screens in the large television market, plasmas are made only in larger sizes, so smaller television sets and other applications such as computer monitors continue to use LCDs.

Political Change. With the proliferation of high-definition television sets, whether in plasma or LCD form, came a corresponding need for higher quality signals for broadcast. Cable and satellite television providers have offered packages of channels that are distributed at higher quality.

Additionally, the presence of these televisions in homes was partly responsible for the industry's decision to move to a different broadcast format for television and to free up parts of the broadcast spectrum. Starting in 2005 and 2006 (depending on the size of the set), television sets were required to have digital receivers. As of June 12, 2009, all television signals were broadcast at a high definition (HD) frequency.

CAREERS AND COURSEWORK

Much of liquid crystal technology is oriented toward the production of display screens, but the complexity of the subject leaves a number of career path options. Master's degrees and doctorates are available in the area of liquid crystal research. Programs typically involve interdisciplinary study in chemistry and physics, and in other potentially relevant areas. Liquid crystal technology builds off basic knowledge of physics, chemistry, and organic chemistry.

Some research in the area of liquid crystals focuses on the material of the crystals themselves and in the development of crystals that improve upon current LCD displays. A background in chemical analysis and optics is important. Design of the products themselves involves knowledge about the design of circuits and backlighting.

There are a number of areas for prospective research and product development. These include the design of LCDs themselves, design of the manufacturing process, and the process of creating the molecules used in the displays.

Social Context and Future Prospects

Some of the concerns and problems with LCDs are being confronted by society as a whole. One concern is the high energy consumption of fluorescent lamps used by LCDs. In contrast, LED lights, which use less energy, are being used more and more in LCDs. There is concern, however, about the environmental hazards LEDs may create when they are disposed of in landfills. Another possibility is the use of carbon nanotubes, which would provide LCD backlighting but would use even less energy than LEDs.

Durability concerns may also come to play a role. The grooves in the glass necessary for high definition LCDs also lead to physical wear and tear on the product. Refining the technology further may produce more durable sets while also alleviating some of the concerns about electronics disposal. Future work on LCDs also will involve altering components to overcome picture quality and durability concerns. Given the prominence of the products that utilize liquid crystals, the technology is likely to be important for development for the foreseeable future.

Joseph I. Brownstein, M.S.

Further Reading

Chandrasekar, Sivaramakrishna. *Liquid Crystals.* 2d ed. New York: Cambridge University Press, 1992. Originally written in 1977, this work is considered one of the classic textbooks in the field and provides early history and an overview of work in liquid crystals.

Chigrinov, Vladimir G., Vladimir M. Kozenkov, and Hoi-Sing Kwok. *Photoalignment of Liquid Crystalline Materials.* Chichester, England: John Wiley & Sons, 2008. This book covers some areas of development in improving screens for LCD devices.

Collings, Peter J., and Michael Hird. *Introduction to Liquid Crystals.* New York: Taylor & Francis, 1997. As an introduction to the field, this book goes through the basics of liquid crystals and then some of the applications, using less technical language than many other texts on the subject.

Delepierre, Gabriel, et al. "Green Backlighting for TV Liquid Crystal Display Using Carbon Nanotubes." *Journal of Applied Physics* 108, no. 4 (September 2010). This article examines the possibility of using carbon nanotubes to backlight LCDs, potentially reducing both production and energy costs.

Web Sites

Kent State University, Liquid Crystal Institute
http://www.lcinet.kent.edu

Nobel Prize Foundation
.http://www.nobelprize.org

See also: Diode Technology; Graphics Technologies; Light-Emitting Diodes; Television Technologies.

LITHOGRAPHY

FIELDS OF STUDY

Printing; photolithography; photography; physics; chemistry; mathematics; calculus; optics; mechanical engineering; material science; graphic design; electromagnetics; microfabrication; semiconductor manufacturing; laser imaging.

SUMMARY

Lithography is an ink-based printing process that was first used in Europe at the end of the eighteenth century. Unlike an older printing press, in which individual pieces of raised type were pressed down onto sheets of paper, lithography uses a flat plate to transfer an image to a sheet of paper. Nearly all books, newspapers, and magazines being published are printed using lithography, as are posters and packing materials. A specialized subfield of lithography known as photolithography is also used in the making of semiconductors for computers. Career opportunities in lithography are growing in specialized areas but overall are neither increasing or decreasing because of the rise of electronic publishing and marketing.

KEY TERMS AND CONCEPTS

- **Emulsion:** Mixture of two chemicals; in lithography, often used on plate surfaces.
- **Hydrophilic:** Chemical property on a plate's surface that attracts and holds a water-based ink; the opposite is hydrophobic.
- **Image:** Words, pictures, or both on a printing plate.
- **Imagesetter:** Device that transfers an image from a computer directly to a plate without the use of photographic negatives.
- **Offset:** Transfer of an image from a plate to a secondary surface, often a rubber mat, that reverses it before final printing.
- **Photolithography:** Process that uses high-precision equipment and light-sensitive chemicals to make products such as semiconductors.
- **Photomask:** Flat surface into which holes have been cut to allow light to pass through; used in photolithography.
- **Plate:** Printing surface, made of metal or stone, on which areas have been chemically treated to attract or repel ink.

DEFINITION AND BASIC PRINCIPLES

Lithography is the process of making an image on a flat stone or metal plate and using ink to print the image onto another surface. Areas of the plate are etched or treated chemically in order to attract or repel ink. The ink is then transferred, directly or indirectly, to the surface where the final image appears.

Unlike a process such as letterpress, where raised letters or blocks of type are coated with ink and pressed against a surface such as paper, lithographic printing yields a result that is smooth to the touch. Lithography differs from photocopying in that plates must be created and ink applied before prints can be made. Photocopying uses a process known as xerography, in which a tube-shaped drum charged with light-sensitive material picks up an image directly from a source. Laser printing is another application of xerography and is not the same as lithography.

Photolithography is a process that imitates traditional lithography in several ways but is not identical. Its high level of precision–a photolithographic image can be accurate down to the level of a micrometer or smaller–is useful in applications such as the manufacturing of computer components.

BACKGROUND AND HISTORY

Lithography was invented in 1798 by Alois Senefelder, a German playwright. Senefelder, who was looking for a way to publish his plays cheaply, discovered that printing plates could be made by writing on a flat stone block with grease pencil and etching away the stone surface around the writing. Eventually Senefelder developed a process by which ink adhered only to the parts of a flat surface not covered by grease. He later expanded the process to include multiple ink colors and predicted that lithography would one day be advanced enough to reproduce works of fine art.

German and French printers in the early 1800's made additional innovations. A patent was issued in 1837 to artist Godefroy Engelmann in France for a process he called chromalithography, in which colors were layered to create book illustrations. Interest in lithography and color printing also spread to North America, where printers in Boston invented new technologies that made the mass production of lithographic prints both high quality and economical. The process quickly spread from books to greeting cards, personal and business cards, posters, advertisements, and packaging labels. Lithography is still the leading process by which mass-produced reading material and packaging are printed.

How It Works

Lithography in the context of printing follows a different set of steps than photolithography as used to make microprocessors.

Offset Lithography. While there are many ways to print on paper or packaging using lithographic techniques, most items involve a process known as offset lithography. The term "offset" refers to the fact that the printing plate does not touch the paper or item itself.

In offset lithography, a plate is first created with the image to be printed. The plate may be made of metal, paper, or a composite such as polyester or Mylar. Lithographic printing plates were flat at one time, but modern printing presses use plates shaped like cylinders, with the image on the outside. To transfer the image to the plate, the surface of the plate is roughened slightly and covered with a light-sensitive chemical emulsion. A sheet of photo film with a reverse, or negative, of the image is laid over the emulsion. When an ultraviolet light is shone on the negative, the light filters through the image only in the areas where the negative is translucent. The result is a positive image–essentially, a negative of the negative–left on the printing plate.

The plate is treated again with a series of chemicals that make the darker areas of the image more likely to pick up ink, which is oil based. The lighter areas of the image are made to be hydrophilic, or water loving. Because oil and water do not mix, water blocks ink from being absorbed by these areas. A water-based mixture called fountain solution is applied to the surface of the plate and is picked up by the hydrophilic areas of the image. Rollers then coat the plate with ink, which adheres only to the hydrophobic (water-fearing) areas that will appear darker on the final image. Once the plate is inked, the press rolls it against a rubber-covered cylinder known as a blanket. The ink from the plate is transferred to the blanket in the form of a negative image. Excess water from the ink as well as fountain solution is removed in the process. The blanket is rolled against the sheet of paper or other item that will receive the final image. Finally, the paper carrying the newly inked image passes through an oven, followed by a set of water-chilled metal rollers, to set the image and prevent the ink from smudging.

Photolithography. Like lithography, the process of photolithography depends on the making of a plate coated with a light-sensitive substance. The plate is known as the substrate, while the light-sensitive chemical is known as the photoresist. Instead of a photo negative with the image to be printed, a photomask is used to shield the photoresist from light in some areas and expose it in others.

The similarities to traditional lithography end here, however. In photolithography, the substrate—rather than a sheet of paper or packaging material—is the final product. Once the image is transferred through the photo mask onto the photoresist, the substrate is treated with a series of chemicals that engraves the image into the surface. In lithographic printing, the image is never engraved directly onto the plate. Unlike printing plates and blankets, which are cylindrical, substrates are always flat. The result is a thin sheet of silicon, glass, quartz, or a composite etched precisely enough to be used as a microprocessing component.

Applications and Products

Lithography as a printing technology has developed in multiple, almost opposing, directions throughout its history. Because lithographic plates can be used to make large numbers of impressions, the development of lithography allowed for printing of images and type on a mass scale that was commercially viable, a major change from the letterpress and intaglio methods of printing that came earlier. Over time, lithography came to be associated with lower-cost editions of books and other printed matter intended to be short-lived, such as newspapers, magazines, and catalogs. Lithography has also evolved as a method of artistic printmaking that can produce works of great beauty and high value. On the photolithography side, the technology has kept pace with the needs of generations of computers.

Web-Fed Offset Printing. Large numbers of copies of a printed work–in the range of 50,000 copies and up–require printing processes that can run quickly and efficiently. Web-fed offset printing takes its name from the way in which paper is fed into the press. A web press uses a roll of paper, known as a web, which is printed and later cut into individual sheets. The largest web presses stand nearly three stories tall, print images on both sides of a sheet at once, and can print at a rate of 20,000 copies per hour. Major newspapers and magazines as well as best-selling books with high print runs are printed on web presses. One of the disadvantages of using a web press is that post-print options, such as folding and binding, are limited. Page sizes are highly standardized and cannot be changed easily to meet the needs of an individual print run. Image quality also is not as high as that offered by other types of lithographic presses.

Sheet-Fed Offset Printing. As its name suggests, sheet-fed offset printing uses a paper supply of individually cut sheets rather than a paper roll. Each press has a mechanism that feeds paper sheets into the machine, one at a time. This process is less efficient than web-fed printing and can lead to a higher rate of mechanical problems, such as damage to the rubber blanket when more than one sheet is fed into the press in error. However, sheet-fed printing allows for a greater degree of customization for each printing job. The size and type of the paper can be changed, as can the area of the page on which each image is to be printed. A paper of heavier, higher-quality grade may be used in a sheet-fed printer. A wider range of post-print options are also available. Sheet-fed print runs can be bound using a number of different methods, including lamination and glue. These features make it more suitable for products such as sales brochures, corporate annual reports, coffee-table books, and posters.

Lithography in Art. When it first appeared in the United States in the mid-1800's, lithography was associated with high-quality printing, particularly reproductions of works of art. The later introduction of technologies such as photogravure printing eventually made lithographic illustrations in mass-produced printed matter obsolete. At the same time, a number of artists on both sides of the Atlantic Ocean were making advances in lithographic printing as an art form. Henri Toulouse-Lautrec depended on lithography to achieve the bold lines and fields of color in his iconic posters for the Moulin Rouge and other French cabarets in the late 1800's. Another surge of interest in lithography came in the 1920's with works from painters Wassily Kandinsky, Georges Braque, and Pablo Picasso. In some cases, such as Toulouse-Lautrec's posters, these works were originally commercial in nature and intended to be reproduced in large print runs. Artists who experimented with lithography in the twentieth century were more likely to be drawn to the medium for its visual characteristics and possibilities for expression, not for its ability to generate copies. Paris was a major center of lithographic art until World War II, at which point many artists relocated to New York. A revival of the technique emerged in the 1950's with new prints from artists such as Sam Francis, Jasper Johns, and Robert Rauschenberg. Lithography is taught in many fine-arts schools. Some artists prefer to work directly with the stone or metal printing plates, while others draw or paint images and rely on third parties to transfer the work from the page to the plate.

Semiconductor Manufacturing. Photolithography has been used to manufacture semiconductors and microprocessing components for about fifty years. When it was first developed, photolithography depended on the use of photomasks that came into direct contact with the photoresist. This contact often damaged the photomasks and made the manufacturing process costly. Next, a system was developed in which photomasks were suspended a few microns above the photoresist without touching it. This strategy reduced damage, but also lowered the precision with which a photomask could project an image. Since the 1970's manufacturing plants have used a system known as projection printing, in which an image is reflected through an ultrahigh-precision lens onto a photoresist. This technology has allowed manufacturers to fit increasingly higher numbers of integrated circuits onto a single microchip. In 1965, Gordon Moore, a technology executive who would go on to cofound Intel, predicted that the number of transistors that could be placed on a microchip would double about every two years. The prediction has been so accurate that the principle is now known as Moore's law.

IMPACT ON INDUSTRY

Commercial Printing. The market for commercial lithographic printing is a mature one and is not expected to see significant growth in the next several years. Unlike many other areas of technology, lithography in printing does not receive government research funding or have programs at academic institutions devoted to its study.

The commercial printing industry in North America is divided into tiers by company size and niche. As a market, lithographic printing on a large scale is led by RR Donnelley followed by Quad/Graphics. These firms and their competitors dominate market segments such as books, magazines, directories, catalogs, and direct-mail marketing pieces. Beyond corporations such as these, however, the commercial printing industry is made up primarily of small businesses with local clientele. According to the U.S. Bureau of Labor Statistics, seven out of ten companies offering lithographic printing services have fewer than ten employees. Taken as a whole, commercial printers earned about $100 billion in revenue in 2009. The roughly 1,600 daily newspapers in the United States make up another major market segment in lithographic printing. Because most newspapers own and operate their own printing facilities, they are seen as belonging to a related but separate industry, and their revenues are not included in most printing-industry estimates.

Most of the innovations in large-scale lithographic printing that have occurred since the 1990's involve the use of digital technology. Rather than using film and photo negatives to create reverse images, computers allow typesetters and printers to transfer an image directly onto the surface of a printing plate. However, many developments in commercial printing methods largely involve technologies that do not rely on lithography. When this trend is taken into account, along with the migration of many books and news sources from print formats to electronic ones, the future of commercial lithographic printing seems very limited.

Photolithography. Prospects for photolithography and semiconductor manufacturing are much brighter. The semiconductor industry reported sales worldwide of $298.3 billion in 2010, reflecting a 32 percent increase over 2009. Much of the growth was due to a rise in microchip purchases by customers in the Asia Pacific region, which makes up slightly more than half of the global market by volume. Microchip buyers in this context are not consumers, but rather companies that manufacture computers, mobile phones, and other types of hardware. The leading semiconductor manufacturers also reflect the global nature of this industry. Intel tops the list by volume. Its competitors include Analog Devices, Texas Instruments, and Micron Technology in the United States; STMicroelectronics and NXP Semiconductors in Europe; and, Samsung, Toshiba, NEC Electronics, and Taiwan Semiconductor Manufacturing in Asia.

For many years, industry sources have predicted that photolithography would be replaced by other technologies because of an increasing need for precision in the making of microchips. Innovations such as excimer laser technology have allowed lithography to become so precise that features smaller than a single wavelength of light can be printed accurately. The vast expense of microtechnology on this scale prevents many nonprofit institutions such as universities from devoting significant resources to its study. Instead, advances are most likely to come directly from manufacturing companies themselves, which reinvest about 15 percent of sales into research and development each year. Research funding is also supplied by government sources such as the National Science Foundation, the National Institute of Standards and Technology, the U.S. Department of Energy, and the U.S. Department of Defense.

Careers and Course Work

The course work required for a career in lithography varies widely with the nature of the product and the stage in the printing process.

In traditional lithography, one major professional area is media printing. Books, magazines, and newspapers must be designed and laid out page by page before lithographic plates can be created. Many of the professionals who hold these jobs have earned bachelor's or master's degrees in academic areas such as art, graphic design, industrial and product design, and journalism. A background of this type could include course work in typography, color theory, digital imaging, or consumer marketing. Students seeking opportunities in media design also pursue internships with publishers and other companies in their fields of interest.

The mechanical process of lithography has become highly automated. Fewer employees are needed in printing plants than before. Most lithographic press operators receive their training on the job and through apprenticeships. Formal education is offered through postsecondary programs at community colleges, vocational and technical schools, and some universities. Students take courses in mechanical engineering and in the maintenance and repair of heavy equipment. Additional course work may include mathematics, chemistry, physics, and color theory.

Lithography as an artistic printing technique is taught in many college and university art departments. While it is considered too specialized by most institutions for a degree, artists may choose to use lithographic printing to create visual works on paper and other materials.

Photolithography is a highly specialized area of technological manufacturing. Its course work and career track are notably different from those in traditional lithography. Professionals working in photolithography have undergraduate or graduate degrees in fields such as engineering, physics, mathematics, and chemistry. An extensive knowledge of microtechnology and the properties of light-sensitive materials is needed. Because most of the world's semiconductor manufacturing takes place outside North America, careers in photolithography can involve frequent travel to areas such as Asia.

SOCIAL CONTEXT AND FUTURE PROSPECTS

As a broad category, lithographic printing offers very limited job growth. Consumers are increasingly concerned about the environmental impact of paper use in catalogs and other sources of bulk mail. In an effort to respond to these concerns, many companies have reduced their use of paper-based marketing campaigns. This change has lowered the demand for commercial offset lithography. A similar trend has affected the printing of checks and invoices, which are being replaced by electronic systems and online banking.

The growth of electronic media, from the Internet to handheld e-book readers, has also lowered the need for lithography in the publishing industry. Newspapers are reducing the circulation and length of their paper editions and shifting their publishing efforts to Web sites and news feeds. While the demand for paper-based books is not likely to disappear in the near future, sales of new books in electronic formats are growing at a more rapid pace than their print counterparts. In the print segment, new technology is boosting the use of print-on-demand systems for books, which use digital printing techniques rather than lithography.

The prospects for growth in photolithography are more optimistic. Photolithography continues to be one of the most effective and precise ways to make semiconductors. Until it is replaced by a new technology, the field is expected to keep growing with new demand for smaller, faster computers.

Julia A. Rosenthal, B.A., M.S.

Fascinating Facts About Lithography

- Commercial lithography changed the way in which companies marketed products to consumers, starting in the late 1800's. Inexpensive, mass-produced pictures launched the catalog industry. Product advertisements and labels became more vivid and colorful.
- Christmas cards printed by L. Prang and Company with images of Santa Claus and evergreen trees first appeared in England in 1873 and in the United States in 1874. Affordable and attractive, the cards were popular instantly and launched the tradition of sending holiday cards to friends and family.
- Alois Senefelder, the inventor of lithography, taught himself to write backwards quickly in order to make his first printing plates.
- The idea of using a mixture of wax and ink written on stone came to Senefelder by accident. He was interrupted in his work one day by his mother, who needed him to write a bill for a washerwoman waiting at the door. Senefelder scrawled the information onto a new stone printing plate. The result inspired him to develop the process that became lithography.
- Senefelder's first play was called *Die Maedchenkenner* (*The Connoisseur of Girls*). It is believed to have been his only commercial success as a playwright.
- In 1846 in Boston, inventor Richard M. Hoe redesigned a lithographic flatbed press by putting the plates onto a rotary drum. The new machine could print six times faster, earning it the nickname "the lightning press."

FURTHER READING

Devon, Marjorie. *Tamarind Techniques for Fine Art Lithography.* New York: Abrams, 2009. A hands-on manual of techniques for artists seeking to produce lithographic prints, written by the director of the Tamarind Institute of Lithography at the University of New Mexico in Albuquerque.

Landis, Stefan, ed. *Lithography.* Hoboken, N.J.: Wiley-I STE, 2010. A new detailed textbook on lithography as applied to the design and manufacturing of microtechnology.

Meggs, Philip B., and Alston W. Purvis. *Meggs' History of Graphic Design.* 4th ed. Hoboken, N.J.: John Wiley & Sons, 2006. A broad history of graphic design and printing processes throughout the world, including the role of lithography.

Senefelder, Alois. *Senefelder on Lithography: The Classic 1819 Treatise.* Mineola, N.Y.: Dover Publications, 2005. A reproduction of an essay published nearly two centuries ago by the founder of lithography.

Suzuki, Kazuaki, and Bruce W. Smith, eds. *Microlithography: Science and Technology.* 2d ed. Boca Raton, Fla.: CRC Press, 2007. A series of essays by several contributors on the processes behind microlithography, one of the manufacturing techniques used to make semiconductors.

Wilson, Daniel G. *Lithography Primer.* 3d ed. Pittsburgh: GATF Press, 2005. An illustrated overview of each step in the lithographic printing process. Includes chapters on topics such as plate imaging, inks, and papers.

Web Sites

American Institute of Graphic Arts (AIGA)
http://www.aiga.org

National Association of Litho Clubs
http://www.graphicarts.org

Tamarind Institute
http://tamarind.unm.edu/index.html

See also: Photography; Printing.

LONG-RANGE ARTILLERY

FIELDS OF STUDY

Physics; ballistics; chemistry; mathematics; geometry; trigonometry; military science; mechanical engineering; materials science; political science; military history.

SUMMARY

Artillery is any weapon that projects power from afar, striking an enemy over distance with decisive destruction at great rates of fire. Historically, artillery has been the technological advancement driving the evolution of warfare. Over the centuries, artillery in one form or another has been the weapons system either deciding a battle's outcome or offering the necessary support required to win a victory.

KEY TERMS AND CONCEPTS

- **Battery:** Three to six artillery pieces grouped according to caliber and type.
- **Breechloader:** Gun in which shells are loaded through the breech at the back of the weapon.
- **Corning:** Compressive process of formulating granular gunpowder.
- **Gun:** Long-range, long-rifled barrel artillery cannon.
- **Howitzer:** Cannon capable of both high-angle and low-angle firing.
- **Mortar:** Muzzle-loading, high-angle cannon capable of delivering ordnance at very short ranges.
- **Muzzle Loader:** Gun in which the propellant and round are loaded down the barrel muzzle.

Definition and Basic Principles

In war, the ability to strike an enemy from afar is paramount: If an army can project power from a distance, its soldiers can stay farther from harm's way. Artillery evolved as a means to this end, and until the beginning of the twentieth century, the role of explosive-force artillery–as field, siege, naval, or fortress guns–was largely unchanged. In World War I, additional specialized artillery was developed. Explosive propellant artillery pieces are classified according to their projectile trajectories: Mortars lob objects in high arc parabolas; guns tend to have straight, high-velocity trajectories; and howitzers are a compromise, propelling large shells at slower speeds, over moderate distances.

The history of artillery is the quest for precision and accuracy. Ballistic science is concerned with the properties of classical physical mechanics governing the motions of bodies under force. With artillery, these motions involve the mechanics of gun machinery, the dynamics of propellants, and the trajectory of discharged projectiles. The basic dynamics of artillery fire–whether bow and arrow, catapult, howitzer, or railroad gun–are based on Newton's second law of motion: Net force is the product of the mass times the acceleration. Traditionally, for artillery this has meant that the amount of destruction was equal to weight of the projectile times how fast it could be propelled. In modern warfare, this destructive force is multiplied by adding explosives and submunitions to the projectile.

Newton's third law of motion–for every action there is an equal and opposite reaction–also plays a role in the history and development of artillery. Explosions propelling projectiles out of barrels result in recoil, moving the entire gun, carriage and all, backward. Until barrel recoil could be countered, guns had to be aimed after each shot, disrupting their accuracy and consistency to target. Accuracy requires the ability to calculate the trajectory, path geometry, and position of an object over time. Knowledge of projectile weight, the force of acceleration, the forces of gravity, the Earth's curvature, atmospheric conditions, and the effects of friction are necessary to accurately predict the parabolic arc and subsequent impact point of a projectile. The result is the proper angle of elevation required to propel an object to a specified location at a certain distance.

Background and History

Before the invention of gunpowder and its practical application in launching projectiles, four types of mechanical advantage were used to increase the range of missiles. During the Middle Ages, volley-fired archery dominated battlefields and great siege engines were required to attack large fortified cities and castles. The first form of practical artillery used in battle was the bow and arrow. The bow is a tension device capable of launching an arrow farther than a person could throw a dart. Arrows became the fastest, farthest flying, and most accurate of historical artilleries, delivering great destructive power, but they were incapable of bringing down fortifications.

Large mechanical devices were used to overcome the limitations of archery. Although these machines could not match the distance and accuracy of the bow and arrow, they brought massive destructive power to the battlefield. Counterweight lever weapons had a short range and slow velocity, but could propel massive missiles. The best-known counterweight lever weapon is the trebuchet, the most powerful weapon of the Middle Ages. Spring-powered artillery used either an oversized double-stave bow or a series of laminated leaf springs to store energy that was released when the missile was shot. The spring's power was limited to the elastic strength of the materials used to build it. The most well-known spring artilleries are the giant crossbow, the spring engine catapult, and the spring engine strike-hurler. Torsion artillery used the elasticity of a twisted bundle of fiber to store energy until it was released to hurl an object. Torsion artillery–such as engine catapults, ballistas, and onagers–throw objects with great accuracy and high velocity.

Using the explosive force of gunpowder to launch projectiles was the greatest step forward in weapons technology. Although gunpowder was invented in China during the eleventh century, it was not exploited as a weapons propellant until the early thirteenth century in Europe. Early gunpowder mixtures burned slowly, resulting in less explosive force. The development of grained gunpowder by corning produced a fast-burning powder with more explosive force, which equated to greater velocity, more range, and more smashing power. Since the inception of propellant-based artillery, engineers have sought to improve it by maximizing the rate of fire, lengthening the flight distance, perfecting accuracy, and increasing the lethality of munitions. They have also striven to minimize discharge signatures, barrel fouling and overheating, recoil, and concussion.

The first explosive-force artillery were muzzle-loading guns. These required a charge of powder to be pushed to the bottom of the gun barrel, followed by a wad, then a projectile. An ignition source applied to a vent would explode the powder and propel the missile. Muzzle-loaded cannons had a relative short range and were aimed directly at their targets. The smooth bore of muzzle-loading guns required the shot to be round, causing the shot to be unstable and tumble in flight, resulting in relative inaccuracy. However, sustained bombardment of massed ranks of infantry and cavalry by multiple muzzle-loading guns produced devastating effects. Maintaining accuracy with muzzle-loading guns requires readjusting them after each firing to compensate for recoil. In the 1860's, the rifling of cannon barrels allowed muzzle-loading guns to put spin on projectiles, giving them gyroscopic stability. Point-first ammunition was designed to take advantage of this stability, providing for larger, heavier shells that could fly greater distances with high accuracy.

The next step forward in artillery technology was the breech-loading mechanism: a system of sealing blocks, screws, or interrupters designed to allow loading from the rear of a gun and prevent the escape of propellant gases on firing. When combined with self-contained ammunition (in which a metal casing containing explosive propellant, fuse, and projectile was loaded into the gun's breech), it allowed for rapid firing. As quickly as one round was fired and its casing ejected, another round could be placed into the gun and discharged. To exploit the advantages afforded by breech-loading guns, recoil had to be countered. The development of a reliable recoil-suppression mechanism in combination with breech-loading technology resulted in artillery capable of sustained, accurate firing beyond the line of sight. In the 1890's, French engineers designed a 75-millimeter field gun with a hydropneumatic recoil-suppression system: Rates of fire for artillery doubled from ten rounds per minute to twenty. Furthermore, recoil suppression meant the gun did not have to be readjusted after each round to stay on target.

During World War I, the application of indirect artillery fire on static enemy positions transformed war forever: Mass bombardments and indirect artillery barrages became the driving force in ground warfare until the end of the twentieth century. World War I also saw the development of three new types of artillery designed to counter advances in military technology and doctrine: antiaircraft guns, mechanized armor (tanks and tracked artillery), and antitank guns.

In World War II, artillery's role changed again. Fast-moving operations equipped with mobile infantry and armored targets became difficult to destroy with indirect bombardment. Artillery technology changed to reflect a need for direct-fire guns to take out designated targets. The United States began to establish centers to coordinate artillery operations, allowing multiple gun batteries to accurately bombard targets designated by forward observers either on the ground or in the air. Since World War II, automatic data processing, digital communications, lasers, radar, Global Positioning Systems (GPS), and satellite technology have made it possible for a targeting sensor to directly communicate with artillery units and pinpoint fire. These technological applications reduce the time needed for targeting and increase accuracy.

How It Works

Artillery pieces are large, unwieldy, and complex, with multiple components, making their operation a team effort. Specialized training, organization, and team cooperation are required to effectively deploy, target, and fire artillery. Modern artillery bombardments are mostly indirect fire support of ground operations. For gunnery teams to strike their target requires a series of coordinated actions organized and timed for maximum efficient use of valuable ammunition and manpower. Forward observers on the ground or in the air, usually senior artillery officers, determine targets and communicate them to a fire direction center. The center sets priorities, selects ordnance type, directs fire control, and selects the battery units to complete the mission against priority targets. The gun battery calculates firing data needed to hit the target with the proper gun or rocket launcher. The forward observer gives the order to fire to the battery and can communicate needed corrections to register the target. The battery continues to fire until a cease-fire order is given.

Applications and Products

Throughout history, artillery guns have been designed to accomplish specific tasks. Field guns, such as howitzers, are designed to accompany a military force on campaign. They were towed by horses and then by trucks and later became self-propelled gun platforms. In World War I, arms manufacturer Friedrich Krupp designed and built the Big Bertha (*Dicke Bertha*), a massive howitzer, so that the German army could destroy forts along the Belgian frontier. Krupp also built the massive traversing railroad cannon, the Paris Gun (*Paris-Geschütz*), which bombarded Paris from 75 miles away. The gun had poor accuracy but was never meant to destroy Paris but rather to terrorize the populace. After World War I, antiaircraft artillery became a highly specialized weapons system with the sole purpose of destroying aircraft, and this type of artillery has been largely replaced by guided missiles.

Traversing, stabilized, high-velocity, quick-firing rifled cannons were developed for armored fighting vehicles, specifically tanks, and specialized oversized high-velocity rifled guns were developed to counter tanks. Large-caliber naval rifles were designed to defeat the armor of battle fleets before airpower rendered their use obsolete. Multiple launch rocket systems were developed as a cost-saving means of providing indirect artillery fire. The ballistic missile is self-propelled artillery that can reach any place on the planet. As of the early twenty-first century, the most advanced artillery gun was the U.S. M777 howitzer, which uses a digital fire-control system to shoot a 155-millimeter GPS-guided M982 Excalibur fire-and-forget projectile. When fired from a distance of 24 miles, the round will land within 30 feet of its target.

Impact on Industry

Military technology reacts to existing needs, and military doctrine dictates what weapons systems are needed. Warfare became the driving force of the Industrial Revolution as the need for mass-produced weapons inspired manufacturers to discover ways to meet the demands of warring nations. The quicker the military could be armed with the most powerful and reliable weapons, the better the chance of victory and the greater the arms manufacturer's profits. The need for bigger, stronger cannons to hold and direct the explosive forces necessary to launch larger projectiles drove the creation of new metal casting techniques and new designs to perform precision machine turning of rifled cannon barrels. The need for advanced artillery design resulted in the development of high-tensile steel alloys to withstand high compression and heat from propellants used in larger rifle-barreled long-range field guns.

Through World War II, one of the most prolific

Fascinating Facts About Long-Range Artillery

- From the mid-eighteenth to mid-nineteenth centuries, half of all battlefield casualties during war resulted from artillery bombardment.
- Artillery accounted for 70 percent of all deaths in World War I.
- The British fired 1 million artillery shells in the first week of the Battle of the Somme, which began on July 1, 1916, and lasted until November 18.
- The German Big Bertha howitzer was named after Alfred Krupp's daughter, Bertha Krupp von Bohlen und Halbach, heiress to the Krupp steel and weapons manufacturing empire.
- German traversing railroad guns of both world wars required gun crews of eighty men to operate efficiently.
- In World War II, artillery was so valued as a means to achieving victory on the battlefield that between 30 and 40 percent of Allied armies were made up of artillery gunnery teams.
- In 1953, the United States built and test fired Atomic Annie, an artillery gun intended to shoot a tactical 280-millimeter shell with a fission warhead to yield a 15-kiloton explosion.
- Modern mortars are the last surviving muzzle-loading artillery, and along with direct-fire recoilless guns, are no longer classified as field artillery but as infantry weapons.
- Since the 1990's, technological advances, such as laser, radar, and Global Positioning Systems, have been used to guide smart bombs and fire-and-forget missile systems, which have lessened the importance of field artillery.
- Crews manning large-scale artillery guns usually fire only three rounds before being replaced by a new crew because extensive, prolonged, or continuous exposure to the gun's percussive forces can cause internal organ damage.

and successful artillery makers was Friedrich Krupp (which became part of Thyssen Krupp in 1999), a German company that produced advanced alloy steels, ammunition, and armaments, especially high-quality artillery pieces.

Careers and Course Work

Careers in artillery are limited to the military and weapons research, development, and testing. The U.S. Army Field Artillery School teaches members of the military to use cannons, rockets, and missiles in coordination with combined arms operations. Nonmilitary careers in artillery design and testing require degrees in engineering, physics, or materials sciences. Most jobs are with large defense contractors such as Lockheed-Martin, BAE Systems Bofors, or Raytheon, or governmental agencies such as the Defense Advanced Research Projects Agency (DARPA), part of the U.S. Department of Defense.

Social Context and Future Prospects

The modern design and application of artillery is a direct by-product of the interdependence of political and military doctrine and advances in technology. Substantial technological advances since the end of World War II have improved the accuracy and lethality of artillery fire and increased the mobility of both guns and support crews. Stronger and lighter alloys have been adopted in the manufacture of modern artillery, reducing weight and allowing for greater mobility. Advances in the chemical composition of propellant charges and the development of self-propelled rounds has increased missile velocities, equating to longer range. These chemical changes, as well as the introduction of caseless ammunition, result in less-corrosive barrel wear, increasing barrel life. Innovations in recoil suppression reduce fatigue to gun parts and allow for increased rates of fire with minimal aiming corrections. The use of laser systems to measure range has created nearly pinpoint artillery accuracy. Laser, electro-optical, infrared, GPS, radio-beam, and radar target acquisition and designation systems eliminate nearly all error in target allocation, increasing the likelihood of first-round accuracy. The use of computer fire-control systems allows modern artillery to be electronically aimed, with computers making any necessary corrections in range, elevation, azimuth, and depression. This increases accuracy and reduces the time between target acquisition and firing.

A large portion of modern artillery consists of mounted, self-propelled weapons, which reflects the modern military's preference for fast, mobile fighting forces. Modern field artillery can advance, stop, set up, target, fire, and move on in minutes, all before it can be located and counterattacked. The United States military is testing the non-line-of-sight (NLOS) cannon, a lightweight, self-loading, highly computerized, self-propelled gun that requires only two people to operate it. The gun can fire four shells in sequence at differing trajectories and land them all on target simultaneously. Future advances in artillery technology will most likely be linked to changes in mobile rocket launchers, electronic targeting systems, projectiles, propellants, self-propelled gun carriages, and tactical changes on the battlefield. Many long-range artillery missions are being replaced by precision-guided aerial munitions, or smart bombs. As the accuracy of aerial attack munitions grows, and mechanized armor and armored personal carriers increase their speeds over the battlefield, mobile artillery will need to keep pace technologically if it wishes to maintain a place in the modern arsenal.

Randall L. Milstein, B.S., M.S., Ph.D.

Further Reading

Bailey, B. A. *Field Artillery and Firepower.* Rev. ed. Annapolis, Md.: U.S. Naval Institute Press, 2003. Documents the changing technology, tactics, and strategy of artillery usage and then analyzes artillery effectiveness under multiple combat conditions.

Bidwell, Shelford, Brian Blunt, and Tolley Taylor, eds. *Brassey's Artillery of the World: Guns, Howitzers, Mortars, Guided Weapons, Rockets, and Ancillary Equipment in Service with the Regular and Reserve Forces of All Nations.* 2d ed. Elmsford, N.Y.: Pergamon Press, 1981. A classic work on the characteristics of modern field artillery pieces.

Foss, Christopher F., ed. *Jane's Armour and Artillery, 2010-2011.* Alexandria, Va.: Jane's Information Group, 2010. An authoritative, essential recognition guide to tanks and artillery for armies of all nations.

Grice, Michael D. *On Gunnery: The Art and Science of Field Artillery from the American Civil War to the Dawn of the Twenty-first Century.* North Charleston, S.C.: Booksurge, 2009. Documents American artillery doctrine from the Civil War through Operation Iraqi Freedom.

Grossman, Dave. *On Killing: The Psychological Cost of Learning to Kill in War and Society.* New York: Little, Brown, 2009. Although it focuses on the Vietnam War, the book reflects on the devastating effect prolonged exposure to intense artillery bombardment had on the psychological health of soldiers in World War I and World War II, as well as the artillery men whose guns killed from great distances.

Hogg, Ian. *Twentieth-Century Artillery: Three Hundred of the World's Greatest Artillery Pieces.* New York: Friedman-Fairfax, 2001. Hogg, one of the world's leading experts on guns of all types, documents some of the most famous and widely used artillery pieces.

Manchester, William. *The Arms of Krupp: The Rise and Fall of the Industrial Dynasty That Armed Germany at War.* 1968. Reprint. Boston: Little Brown, 2003. A biography of the Krupp family, innovators and manufacturers of weapons that shaped modern warfare.

Manson, M. P. *Guns, Mortars, and Rockets.* London: Brassey's United Kingdom, 1997. Documents the historical development of artillery, emphasizing three types of missile launch platforms–guns, mortars, and rockets.

Miller, Henry W. *The Paris Gun: The Bombardment of Paris by the German Long-Range Guns and the Great German Offensives of 1918.* 1930. Reprint. East Sussex, England: Naval and Military Press, 2003. A detailed account of the construction and use of the Paris gun; richly illustrated with photographs and maps.

Web Sites

American Artillery Association
http://www.americanartillery.org

U.S. Army Field Artillery School
http://sill-www.army.mil/USAFAS

U.S. Field Artillery Association
http://fieldartillery.org

M

MAGNETIC STORAGE

FIELDS OF STUDY

Computer engineering; computer networking systems; electrical engineering; information technology; materials engineering.

SUMMARY

Magnetic storage is a durable and non-volatile way of recording analogue, digital, and alphanumerical data. In most applications, an electrical current is used to generate a variable magnetic field over a specially prepared tape or disk that imprints the tape or disk with patterns that, when "read" by an electromagnetic drive "head," duplicates the wavelengths of the original signal. Magnetic storage has been a particularly enduring technology, as the original conceptual designs were published well over a century ago.

KEY TERMS AND CONCEPTS

- **Biasing:** Process that pre-magnetizes a magnetic medium to reproduce the magnetic flux from the recording head more exactly.
- **Bit:** Storage required of one binary digit.
- **Byte:** Basic unit, composed of eight bits, of measuring storage capacity and the basis of larger units.
- **Ferromagnetic Media:** Iron-based media sometimes used as a source of tape, hard, or flexible disks.
- **Magnetic Anisotrophy:** Tendency, in certain substances, to hold onto a magnetically induced pattern even without the presence of electrical current.
- **Magnetization:** Property of an object determining whether it can be affected by magnetism.
- **Random Access Memory (RAM):** Property of certain kinds of storage media where all elements of a sequence may be accessed in the same length of time; also called direct access.
- **Sequential Access Memory:** Property of certain kinds of storage media where remote elements require a longer access time than elements located in the immediate vicinity.

Definition and Basic Principles

Magnetic storage is a term describing one method in which recorded information is stored for later access. A magnetized medium can be one or a combination of several different substances: iron wire, steel bands, strips of paper or cotton string coated with powdered iron filings, cellulose or polyester tape coated with iron oxide or chromium oxide particles, or aluminum or ceramic disks coated with multiple layers of nonmetallic alloys overlaid with a thin layer of a magnetic (typically a ferrite) alloy. The varying magnetic structures are encoded with alphanumerical data and become a temporary or permanent nonvolatile repository of that data. Typical uses of magnetic storage media range from magnetic recording tape and hard and floppy computer disks to the striping material on the backs of credit, debit, and identification cards as well as certain kinds of bank checks.

Background and History

American engineer Oberlin Smith's 1878 trip to Thomas Edison's laboratory in Menlo Park, New Jersey, was the source for Smith's earliest prototypes of a form of magnetic storage. Disappointed by the poor recording quality of Edison's wax cylinder phonograph, Smith imagined a different method for recording and replaying sound. In the early 1820's, electrical pioneers such as Hans Ørsted had demonstrated basic electromagnetic principles: Electrical current, when run through a iron wire, could generate a magnetic field, and electrically charged wires affected each other magnetically. Smith toyed with the idea, but did not file a patent—possibly because he never found the time to construct a complete, working model. On September 8, 1888, he finally published a description of his conceptual design, involving a cotton cord woven with iron filings passing through a coil of electrically charged wire, in *Electrical World* magazine. The concept in the article, "Some Possible Forms of Phonograph," though theoretically possible, was never tested.

The first actual magnetic audio recording was Danish inventor Valdemar Poulsen's telegraphone, developed in 1896 and demonstrated at the Exposition Universelle in Paris in 1900. The telegraphone was composed of a cylinder, cut with grooves along its surface, wrapped in steel wire. The electromagnetic head, as it passed over the tightly wrapped iron wire, operated both in recording sound and in playing back the recorded audio. Poulsen, trying to reduce distortion in his recordings, had also made early attempts at biasing (increasing the fidelity of a recording by including a DC current in his phonograph model) but, like Oberlin Smith's earlier model, his recorders, based on wire, steel tape, and steel disks, could not easily be heard and lacked a method of amplification.

Austrian inventor Fritz Pfleumer was the originator of magnetic tape recording. Since Pfleumer was accustomed to working with paper (his business was cigarette-paper manufacturing), he created the original magnetic tape by gluing pulverized iron particles (ferrous oxide) onto strips of paper that could be wound into rolls. Pfleumer also constructed a tape recorder to use his tape. On January 31, 1928, Pfleumer received German patent DE 500900 for his sound record carrier (lautschriftträger), unaware that an American inventor, Joseph O'Neill, had filed a patent—the first—for a device that magnetically recorded sound in December, 1927.

How It Works

The theory underlying magnetic storage and magnetic recording is simple: An electrical or magnetic current imprints patterns on the magnetic-storage medium. Magnetic tape, magnetic hard and floppy disks, and other forms of magnetic media operate in a very similar way: An electric current is generated and applied to a demagnetized surface to vary the substratum and form a pattern based on variations in the electrical current. The biggest differences between the three dominant types of magnetic-storage media (tape, rigid or hard disks, and flexible or floppy disks) are the varying speeds at which stored data can be recovered.

Magnetic Tape. Magnetic tape used to be employed extensively for archival computer data storage as well as analogue sound or video recording. The ferrous- or chromium-impregnated plastic tape, initially demagnetized, passes at a constant rate over a recording head, which generates a weak magnetic field proportional to the audio or video impulses being recorded and selectively magnetizes the surface of the tape. Although fairly durable, given the correct storage conditions, magnetic tape has the significant disadvantage of being consecutively ordered—the recovery of stored information depends on how quickly the spooling mechanism within the recorder can operate. Sometimes the demand for high-density, cheap data storage outweighs the slower rate of data access. Large computer systems commonly archive information on magnetic tape cassettes or cartridges. Despite the archaic form, advances in tape density allow magnetic tape cassettes to store up to five terabytes (TB) of data in uncompressed formats.

For audio or video applications, sequential retrieval of information (watching a movie or listening to a piece of music) is the most common method, so a delay in locating a particular part is regarded with greater tolerance. Analogue tape was an industry standard for recording music, film, and television until the advent of optical storage, which uses a laser to encode data streams into a recordable media disk and is less affected by temperature and humidity.

Magnetic Disks. Two other types of recordable magnetic media are the hard and floppy diskettes—both of which involve the imprinting of data onto a circular disk or platter. The ease and speed of access to recorded information encouraged the development of a new magnetic storage media form for the computer industry. The initial push to develop a nonlinear system resulted, in 1956, with the unveiling of IBM's 350 Disk Storage Unit—an early example of what is currently known as a hard drive. Circular, ferrous-impregnated aluminum disks were designed to spin at a high rate of speed and were written upon or read by magnetic heads moving radially over the disk's surface. Hard and floppy disks differ only in the range of components available within a standard unit. Hard disks, composed of a spindle of disks and a magnetic read-write apparatus, are typically located inside a metal case. Floppy disks, on the other hand, were packaged as a single or dual-density magnetic disk (separate from the read-write apparatus that encodes them) inside a plastic cover. Floppy disks, because they do not contain recording hardware, were intended to be more portable (and less fragile) than hard disks—a trait that made them extremely popular for home-computer users.

Another variant of the disk-based magnetic storage technology is magneto-optical recording. Like an optical drive, magneto-optical recording operates by burning encoded information with a laser and accesses the stored information through optical means. Unlike an optical-storage medium, a magneto-optical drive directs its laser at the layer of magnetic material. In 1992, Sony released the MiniDisc, an unsuccessful magneto-optical storage medium.

Applications and Products

Applications for magnetic storage range from industrial or institutional uses to private-sector applications, but the technology underlying each of these formats is functionally the same. The technology that created so many different inventions based on electrical current and magnetic imprinting also had a big impact on children's toys. The Magna Doodle, a toy developed in 1974, demonstrates a simple application of the concept behind magnetic storage that can shed light on how the more complex applications of the technology also work. In this toy, a dense, opaque fluid encapsulates fine iron filings between two thin sheets of plastic. The upper layer of plastic is transparent and thin enough that the weak magnetic current generated by a small magnet encased in a cylinder of plastic (a magnetic pen) can make the iron filings float to the surface of the opaque fluid and form a visible dark line. Any images produced by the pen are, like analogue audio signals encoded into magnetic tape, nonvolatile and remain visible until manually erased by a strip of magnets passing over the plastic and drawing the filings back under the opaque fluid.

Magnetic Tape Drives. It is this basic principle of nonvolatile storage that underlies the usage of the three basic types of magnetic storage media—magnetic tape, hard disks, and floppy disks. All three have been used for a wide variety of purposes. Magnetic tape, whether in the form of steel bands, paper, or any one of a number of plastic formulations, was the original magnetic media and was extensively used by the technologically developed nations to capture audio and, eventually, video signals. It remains the medium of choice for archival mainframe data storage because large computer systems intended for mass data archival require a system of data storage that is both capable of recording vast amounts of information in a minimum of space (high-density) and is extremely inexpensive—two qualities inherent to magnetic tape.

Early versions of home computers also had magnetic tape drives as a secondary method of data storage. In the 1970's, IBM offered their own version of a magnetic cassette tape recorder (compatible with its desktop computer) that used the widely available cassette tape. By 1985, however, hard disks and floppy disks had dominated the market for computer systems designed to access smaller amounts of data frequently and quickly, and cassette tapes became obsolete for home-computer data storage.

Hard Disk Drives. In 1955, IBM's 350 Disk Storage Unit, one of the computer industry's earliest hard drives, had only a five megabyte (MB) capacity despite its massive size (it contained a spindle of fifty twenty-four-inch disks in a casing the size of a large refrigerator). However, the 350 was just the first of a long series of hard drives with ever-increasing storage capacity. Between the years of 1950 and 2010, the average area density of a hard drive has doubled every few years, starting from about three megabytes to the current high-end availability of three terabytes. Higher-capacity drives are in development, as computer companies such as Microsoft are redefining the basic unit of storage capacity on a hard drive from 512 bytes (IBM's standard unit, established during the 1980's) to a far larger 4 kilobytes (KB). The size of the typical hard drive made it difficult to transport and caused the development, in 1971, of another, similar form of magnetic media—the floppy disk.

Early hard drives such as IBM's 350 were huge (88 cubic feet) and prohibitively expensive (costing about $15,000 per megabyte of data capacity). Given that commercial sales of IBM computers to nongovernmental customers were increasing rapidly, IBM wanted some way to be able to deliver software updates to clients cheaply and efficiently. Consequently, engineers conceived of a way of separating a hard disk's components into two units—the recording mechanism (the drive) and the recording medium (the floppy disk).

Floppy Disk Drives. The floppy disk itself, even in its initial eight-inch diameter, was a fraction of the weight and size needed for a contemporary hard disk.

Because of the rapidly increasing storage needs of the most popular computer programs, smaller disk size and higher disk density became the end goal of the major producers of magnetic media—Memorex, Shugart, and Mitsumi, among others. As with the hard drive, floppy disk size and storage capacity respectively decreased and increased over time until the 3.5-inch floppy disk became the industry standard. Similarly, the physical dimensions of hard drives also shrunk (from 88 cubic feet to 2.5 cubic inches), allowing the introduction of portable external hard drives into the market. Floppy disks were made functionally obsolete when another small, cheap, portable recording device came on the market—the thumb or flash drive. Sony, the last manufacturer of floppy disks, announced that, as of March, 2011, it would stop producing floppy disks.

Magnetic Striping. Magnetic storage, apart from tape, hard and floppy disk, is also widely used for the frequent transmission of small amounts of exclusively personal data—namely, the strip of magnetic tape located on the back of credit, debit, and identification cards as well as the ferrous-impregnated inks that are used to print numbers along the bottom of paper checks. Since durability over time is a key factor, encasing the magnetic stripe into a durable, plastic card has become the industry standard for banks and other lending institutions.

Impact on Industry

At the base of any industry is the need to store information. Written languages, ideographic art, and even musical notation were developed for the purpose of passing knowledge on to others. At the most basic, any individual, company, or government division that uses a computer system relies on some form of magnetic storage to encode and store data.

Government Use. Few government entities in even moderately technologically developed countries can operate without using some form of computer system and the requisite magnetic storage of either magnetic tape, internal or external hard drive, or both.

In 1942, the Armour Institute of Technology sold literally thousands of wire sound recorders to the American military, since the Army intelligence service was experimenting with technology that intercepted foreign radio transmissions and wished to record those transmissions for the purpose of decoding them.

In June, 1949, IBM envisioned a new kind of magnetic storage device that could act as a data repository for another new invention—the computer. On May 21, 1952, the IBM 726 Tape Unit with the IBM 701 Defense Calculator was unveiled. On September 13, 1956, a IBM announced their creation of the 305 RAMAC (Random Access Memory Accounting System) computer. These devices (and other, similar inventions) allowed both the government and the private sector to start phasing out punch-card-based computer storage systems. In the following decades, governmental agencies were able to switch from the massive boxes of punch cards to the more efficient database system based on external magnetic tape cassettes and internal magnetic disks. In the United States, for example, the Internal Revenue Service, which records and processes financial data on every working individual in the country identified by a Social Security Number, is one of many governmental agencies that must store massive amounts of sensitive information. As a result, the federal government purchases roughly a million reels every year.

Business Use. Besides the magnetic media used in computers, magnetic storage media in the form of magnetic tape has had a dramatic impact on the industrial and business sectors of the developed countries of the world. Miniature cassette recorders, introduced in 1964 by the Philips Corporation, became a common sight in college classrooms, executive offices, and journalists' hands for recording lectures, memos, or interviews because of their small size and adjustable tape speed. These recorders suggested the importance of knowing exactly what another individual has said for reasons of accuracy and truth in reporting.

Entertainment Use. The invention of magnetic storage allowed for a more precise system of audio and video recording. Dramatic shifts in popular entertainment and the rise of the major music labels such as Victor or RCA occurred precisely because there were two alternative methods of audio recording that, ultimately, introduced music into vastly different venues and allowed individuals to take an interest in music that was as lucrative to the music company as it was pleasurable to the consumer. Phonograph recordings made on wax, lacquered wood, or vinyl brought music into a family's living room and millions of teenagers' bedrooms, while electromagnetic recordings on tape allowed music to

Fascinating Facts About Magnetic Storage

- One of the earliest magnetic recording devices, Valdemar Poulsen's telegraphone, was not only intended for the recording of telephone calls but was actually constructed out of telephone components.
- In 2011, the highest-capacity desktop hard drive was the three-terabyte Hitachi Deskstar 7K1000. It contains the highest number of hard disk platters—five—and rotates them at 7,200 rpm.
- In 2011, the highest-capacity floppy disk was 3M's 3.5-inch LS-240 SuperDisk, which could store up to 240 megabytes of data.
- The oldest form of magnetic storage media, magnetic tape, still has the highest capacity of all of the current forms of magnetic data storage, at 5 terabytes per cassette.
- Forrest Parry, under contract to the United States government, invented the first security card in 1960—a plastic card with a strip of magnetic tape affixed to the back. "Magstripes" (typically three tracks per stripe) are universally used to record identifying data for a variety of security and financial purposes: credit cards, banking identification cards, gift cards, and government benefit cards.
- Twenty-one states in the United States and six provinces in Canada use driver licenses that have magnetic striping to carry identifying data about the individual user depicted on the card. Some passports also have a magnetic stripe for easier handling by immigration officials.

be played as background music in vehicles and workplaces and to be carried from place to place in portable devices. Competition between the rival manufacturers of records (such as Columbia and EMI) and tapes (such as BASF and TDK) was fierce at times, even inspiring copyright-infringement campaigns by such organizations as the British Phonographic Industry (BPI) against the manufacturers and users of blank cassette tapes in the 1980's. Regardless, the portability and shock resistance of magnetic tape-driven music players ensured that magnetic storage was unlikely to be made obsolete by either phonograph technology or music-industry politics.

Consumer Use. Magnetic-storage media probably would not have the impact it did on Western society if it had not been so enthusiastically embraced by the willingness of the private individual to purchase and use the technology. For example, although magnetic media gained a reputation for illicit use because of music- and film-industry complaints about some consumers' misuse of blank magnetic media to record illegal (bootleg) copies of phonograph recordings, television programs, and movies broadcast over cable or satellite systems, magnetic-storage media has also given consumers the ability to use blank media as a vehicle for the recording of their own audio and video creations. The technology was robust enough to make a durable recording but flexible enough to allow repeated erasures and rerecordings on a single tape.

Careers and Course Work

Individuals interested in further developing the technology of magnetic media should study computer engineering, electrical engineering, or materials engineering depending on whether they wish to pursue increasing storage capacity of the existing electromagnetic technology or finding alternative methods of using electromagnetic theory on the problems of data storage. Some universities are particularly sensitive to the related nature of computer-component design and electrical-system design and allow for a dual degree in both disciplines. Supplemental course work in information technology would also be helpful, as an understanding of how networks transmit data will more precisely define where future storage needs may be anticipated and remedied.

Social Context and Future Prospects

The current social trend, both locally and globally, is to seek to collect and store vast amounts of data that, nevertheless, must be readily accessible. An example of this is in meteorology, which has been attempting over the past few decades to formulate complex computer models to predict weather trends based on previous trends observed and recorded globally during the past century. Weather and temperature records are increasingly detailed, so computer storage for the ideal weather-predicting computer will need to be able to keep up with the storage requirements of a weather-system archive and meet reasonable deadlines for the access and evaluation of a century's worth of atmospheric data.

Other storage developments will probably center

on making more biometric data immediately available in identification-card striping. With surveillance, rather than incarceration, of petty criminals being considered as a method by which state and federal government might cut expenses, an improvement in the design and implementation of identification cards might be one magnetic-storage trend for the future. Another archival-related goal might be the increasing need for quick comparisons of DNA evidence in various sectors of law enforcement.

Julia M. Meyers, M.A., Ph.D.

Further Reading

Bertram, H. Neal. *Theory of Magnetic Recording.* Cambridge, England: Cambridge University Press, 1994. A complete and thorough handbook on magnetic recording theory; goes well beyond the typical introductory-style discussion of magnetic and strives for a more advanced discussion of magnetic recording and playback theory.

Hadjipanayis, George C, ed. *Magnetic Storage Systems Beyond 2000.* Dordrecht, The Netherlands: Kluwer Academic, 2001. This book is the collected papers presented at the "Magnetic Storage Systems Beyond 2000" Conference held by the NATO Advanced Study Institute (ASI) in Rhodes, Greece. Although the papers are technical discussions of the limitations of magnetic technology (such as magnetic heads, particulate media, and systems still in development), the speculative emphasis often describes new areas of possible development. Each paper has a tutorial, which provides an introduction to the ideas discussed.

Mee, C. Denis, and Eric D. Daniel. *Magnetic Recording Technology.* 2d ed. New York: McGraw-Hill, 1996. An older handbook, but one with particularly clear description and analysis of many of the emergent trends in the last few decades in the field of data recording and storage.

National Research Council of the National Academies. *Innovation in Information Technology.* Washington, D.C.: National Academies Press, 2003. A good handbook for the study of storage media and other issues in the field of information technology.

Prince, Betty. *Emerging Memories: Technologies and Trends.* Norwell, Mass.: Kluwer Academic, 2002. A good primer for understanding the development of memory in computing systems. Has application both to data storage and volatile memory forms.

Wang, Shan X., and Alexander Markovich Taratorin. *Magnetic Information Storage Technology.* San Diego: Academic Press, 1999. Provides the basic principles of magnetic storage and digital information and describes the technological need for data recording and the resulting push for increased capacity, faster access rates, and greater durability of media.

Web Sites

IEEE Magnetics Society
http://www.ieeemagnetics.org

Information Technology Association of America
http://www.itaa.org

International Disk Drive Equipment and Materials Association
http://www.idema.org.

MAPS AND MAPPING

FIELDS OF STUDY

Cartography; geography; geology; graphic arts; mathematics; statistics; geographic information systems; remote sensing.

SUMMARY

Mapping encompasses methods for representing geospatial information in paper and digital forms as well as newer digital technologies such as remote sensing and geographic information systems. Maps are needed in areas ranging from civil engineering to regional planning. Computer technologies and the Internet have brought profound changes to the ways maps are created and used, and a growing number of mapping applications have been developed for portable devices such as car navigation systems.

KEY TERMS AND CONCEPTS

- **Cartography:** Art and science of making maps.
- **Digital Orthophoto Quadrangle:** Aerial photograph that has been rectified to remove camera distortion and tied to a system of coordinates such as latitude and longitude.
- **Digitizing:** Process of converting analogue information into a digital or machine-readable format.
- **Geographic Information System (GIS):** Computer-based system for storing, analyzing, and visualizing geographic information.
- **Global Positioning System (GPS):** Satellite-based positioning system developed and operated by the U.S. Department of Defense.
- **Map Projection:** Mathematical method for transferring information from the spherical Earth to a flat surface.
- **Scale:** Relationship in measurement units between a map and the real-world location it represents.
- **Thematic Map:** Map representing data corresponding to a single theme such as rainfall or wheat production.
- **Virtual Map:** Temporary map displayed on a computer screen.

Definition and Basic Principles

Maps assume many forms, including spherical globes, folded paper charts, wall-sized murals, and images displayed on tiny electronic screens. As scale models of reality, maps show a selection of information. In addition to describing features on the surface of the Earth, maps can represent underwater areas, the interiors of caves, and celestial bodies such as planets. The term "cartography" refers to the process of creating maps and related geographic products. According to the International Cartographic Association, cartography is the art, science, and technology of making maps.

Maps can be divided into two categories: general-purpose and thematic maps. General-purpose maps show a variety of information, including lakes, rivers, roads, cities, and administrative or political boundaries. Most general-purpose maps, such as road and topographic maps, are designed to serve as reference tools. In contrast, the purpose of thematic maps is to show patterns and distributions corresponding to a single topic or theme such as rainfall or corn production.

Background and History

For more than 4,000 years, maps have been used to record and communicate information about the Earth. The ancient Greeks created maps to characterize the Earth's spherical shape, while legions of Roman soldiers used them as tools in the conquest of new territories. Drawn by hand, early maps showed landscape features using pictorial symbols to represent mountains, rivers, and other physical features. Later technologies such as printing eliminated the need for maps to be painstakingly copied and increased the speed at which maps could be reproduced and disseminated. Exploration, especially during the fifteenth and sixteenth centuries, created a demand for maps to assist in ocean navigation and to document newly discovered locations.

Modern maps play many roles in society, from guiding aircraft to providing reference information within school textbooks. Under most circumstances, maps are superior to charts and graphs in their ability to represent distances, directions, and the relative sizes of objects over space. In addition to showing locations of features, maps are able to efficiently represent geographic patterns and the spatial extent of physical entities such as rivers and mountains. They are also useful for showing cultural features such as property boundaries and political or administrative areas. As archives of spatial information, maps are used extensively for recording the locations of historical events as well as spatially referenced scientific data. In addition to their use in storing information, maps serve as important research tools for visualizing geographic data needed for evaluating hypotheses about spatial distributions.

How It Works

Scale and Projection. All maps of the Earth's surface require some level of reduction. Map scale is an important concept to maps and mapping because it determines the level of map reduction necessary. Scale is the ratio of a distance measured on a map to the distance it represents in the real world, using the same units of measurement. For example, 1 inch measured on a map with a scale of 1:24,000 represents 24,000 inches (or 2,000 feet) in the real world. The scale used on any given map determines the size of the area that can be represented as well as the level of detail that can be depicted. In addition to the representative fraction (for example, 1:100,000), scale can be shown using a verbal statement such as "1 inch equals 1 mile" or a graphic or bar scale printed directly on the map.

Cartographers draw a distinction between small- and large-scale maps. Small-scale maps represent very large areas of the world and are useful for representing entire continents. In contrast, large-scale maps are capable of showing very small areas at higher levels of detail. City road maps and topographic maps are examples of large-scale maps. Given that mapmaking involves reductions, the cartographer must make choices about the features to be shown and the symbols used to represent those features. This process, called selection, must be closely tied to the purpose of the map. In addition, complex objects such as coastlines must be simplified to reduce unnecessary detail.

Another important consideration is the choice of projection for displaying mapped information. Projection refers to a mathematical transformation of geographic information from the spherical Earth to a flat or developable surface. Because spherical surfaces cannot be flattened without distortion, the cartographer must select the most important properties to maintain during this transformation. For example, on a world map, it may be desirable to maintain the relative sizes of countries for making comparisons. Although maintaining the property of equivalence, this type of projection distorts the shapes of countries near the poles. The well-known Mercator projection is useful for ocean navigation because it preserves shapes as well as lines of constant geographic direction as straight lines. However, a disadvantage of the Mercator projection is that it severely distorts the relative sizes of polar areas, making the island of Greenland in the North Atlantic look larger than the continent of South America, which is actually fifteen times larger than Greenland.

Geographic Grids. Cartographers use geographic grids, or coordinate systems, for the placement of points, lines, and areas to systematically organize geographic information. Most maps are created using either the latitude-longitude coordinate system or a rectangular coordinate system. Latitude-longitude and rectangular grids use Cartesian coordinates (x,y). The point where the prime meridian (0 degrees longitude) and the equator (0 degrees latitude) cross serves as an origin against which all other locations are referenced. Smaller geographic areas are often mapped with rectangular coordinates that use an origin located outside and to the southwest of the area of interest to make all coordinate values positive.

Map Design. Map design is a systematic process involving the selection and arrangement of map elements in a way that facilitates a user's correct interpretation of ideas and concepts. Cartographic design has benefited from developments in both graphic design and cognitive psychology. Because the principal objective of mapmaking is communication, it is important for cartographers to select visual elements and arrangements that unambiguously present concepts and spatial relationships. The design process can require several stages as alternatives are tested and modified. Basic graphic elements manipulated in the design process include points, lines, and area symbols. In addition, the selection and placement of text is an important design element. As suggested by French cartographer Jacques Bertin, graphic elements used in cartography can be made more or less prominent through changes in hue (color), value (darkness or lightness), texture, orientation, size, and shape. Design is affected not only by the choice and design of symbols but also by the arrangement of map elements. In the design process, cartographers must be careful to use space efficiently and to avoid symbols that create unwanted attention or noise. Other important concepts within map design include achieving balance in the arrangement of map elements and creating a visual hierarchy that draws the map reader's attention to features most important to the map's purpose.

Map Output. Developments in map reproduction have followed innovations in printing, beginning with Johann Gutenberg's invention of the printing press in 1440. Early printing technology employed a process called letterpress, in which ink was applied to raised portions of wooden blocks that were then pressed against paper. Introduced in the early 1700's, engraving involved the application of ink to depressions in a metal printing plate. Engraving improved map reproduction because it enabled the use of finer line widths and improved gray tones and patterns. Lithographic reproduction, a technique based on the incompatibility of grease and water, was introduced in the 1800's. Most modern maps are reproduced using lithographic plates, which are applied to a rotating drum to transfer an image to a moving sheet of paper. Four-color maps use lithographic plates corresponding to the subtractive primary colors of cyan, magenta, and yellow plus black. A wide variety of colors can be created by combining percentages of each primary color. Modern printing presses use computer files to create plates needed to produce color maps, with sheets of paper passing through the press once for each plate.

Maps can also be reproduced in smaller quantities using low-cost output devices such as laser printers and ink-jet plotters. Electrostatic (or xerographic) copying is a common method for reproduction. Laser and ink-jet printers have also become a popular method for reproducing small numbers of maps. The readability of map lettering and the sharpness of lines and grey tones is a function of the printer's resolution as represented in dots per inch (dpi). Low-end printers typically offer 300-dpi output, meaning that each square inch has 90,000 dots. Higher-end printers and plotters offer resolutions exceeding 2,000 dpi.

The introduction of personal communications devices with high-resolution screens has led to an

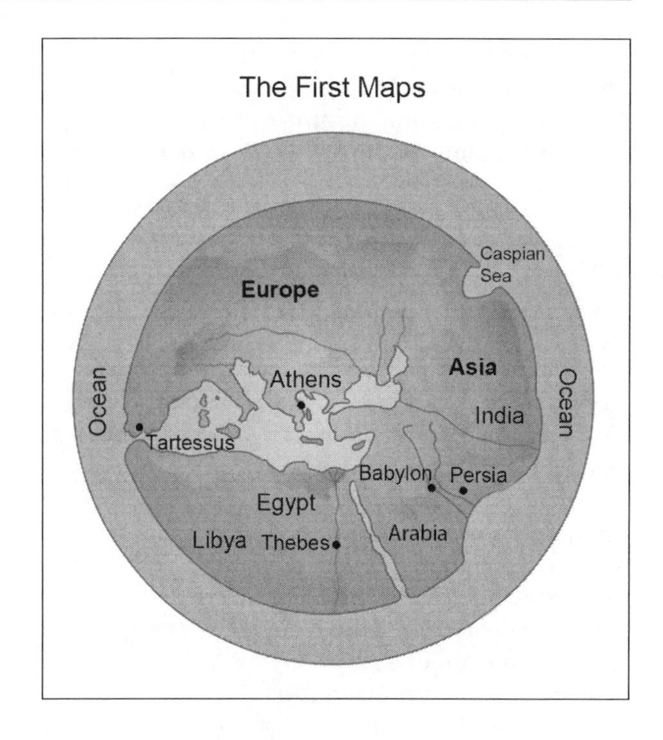

increasing number of maps that are never printed. These virtual maps may be used for checking weather conditions or finding a street address. Internet-based Google Maps has been an innovator in making virtual maps, aerial photographs, and satellite images accessible to the public. Google Maps and Google Earth enable the simultaneous viewing of aerial photographs and map information in the form of roads, contour lines, and political boundaries. Mapping tools enable the user to zoom in or out, pan in any direction, and select specific map features and points of interest to be displayed.

Computers and Mapmaking. The single most significant change to cartography during the twentieth century was the introduction of computers that enabled tasks previously requiring hours to be carried out in seconds. Computer cartography, or the use of computers in map production, has been particularly beneficial in enabling cartographers to experiment. Using mapping software, the cartographer can resize and reposition map elements such as text labels, scale bars, and legend boxes. Computers also facilitate mathematical and statistical transformations of tabular data needed for making thematic maps. Some programs enable map images to be combined with other spatial data such as aerial photographs or satellite images. Increasingly, digital databases are replacing paper maps as the principal method for storing and retrieving geographic information.

Before geographic information can be manipulated by a computer, it must be converted to a digital form. The process of capturing points, lines, and areas shown on a paper map in a digital format is called digitizing. Analogue information can be digitized by tracing features of interest individually using a tablet or table-sized digitizer. Digitizing can also be accomplished by scanning an entire map sheet. Before scanned data are useful, they must be transformed into a system of coordinates such as latitude and longitude. This process is called geocoding or georeferencing.

APPLICATIONS AND PRODUCTS

Since 1985, a significant amount of data used in computer mapping has been converted into digital format. For example, the U.S. GeoData digital line graphs are computer files containing transportation and other line information that has been captured from topographic maps maintained by the United States Geological Survey (USGS). The set of digital files called digital raster graphics includes information scanned from USGS topographic maps. USGS digital elevation model data files contain grids of elevation values captured every 30 meters and are useful for constructing contour maps, three-dimensional terrain diagrams, and computer simulations such as aerial fly-bys.

Computer technology also makes it possible to create new types of cartographic visualizations such as animated maps. Maps that use a sequence of digital images are called temporal animations. This type of map is useful for representing changes over time, such as the movement of military fronts during a battle.

Computer technologies have influenced maps and mapping through developments in closely related fields such as aerial photography, satellite remote sensing, and geographic information systems (GIS). Remote sensing involves the capture of information about the Earth using sensors and is commonly divided into photographic methods such as aerial photography and nonphotographic imaging that include sensors looking outside visible portions of the electromagnetic spectrum.

Digital orthophoto quadrangles (DOQs) have emerged as an important source of information for constructing maps. DOQs are aerial photographs in digital form that have been rectified (corrected) for camera distortion. In addition to serving as a source from which to extract the location of physical and cultural features, they can be used as background images for maps. Multispectral scanners used by Landsat and other satellites capture large amounts of data using scanners that are sensitive to narrow wavelengths of energy emitted from the Earth's surface. Such data can be useful for revealing information that cannot be seen in photographic images, including stress or disease in vegetation.

Radio detection and ranging (radar) is another nonphotographic imaging system that can be used for mapping surface features, underground areas, and atmospheric conditions. Radar uses transmitted energy that is reflected by objects and then captured and converted into an image. Imaging radar used

Fascinating Facts About Maps and Mapping

- Faculty in the Department of Geography at the University of California, Santa Barbara, designed a mapping system for blind persons that uses audio signals coupled with GPS guidance.
- Geographic information systems and remote sensing were used extensively after Hurricane Katrina to assist in directing emergency responders and organizations involved with recovery efforts. Geographic information systems were also important in the recovery effort following the September 11, 2001, attack on the World Trade Center in New York City.
- The U.S. military is using geographic information systems to provide battlefield commanders with tactical views of personnel and weapons systems to improve situational awareness.
- Assisted by French cartographer Louis-Alexander Berthier, George Washington used hinged overlay maps during the Battle of Yorktown to study troop movements.
- Most modern cell phones contain GPS receivers that can guide emergency responders to the telephone's location.
- The destruction wrought by the 1906 earthquake and resulting fire in San Francisco was first documented in aerial photographs taken by cameras carried by kites.

on space shuttle missions has become an important source of elevation data used in a variety of mapping applications. Light detection and ranging (lidar) is a innovation for mapping the Earth by measuring the properties of scattered light as it comes into contact with distant objects. Lidar images are created by measuring the time needed for laser pulses to reach the Earth's surface. Some applications for lidar include three-dimensional mapping of urban areas and assessments following natural disasters

Impact on Industry

Computer technology has not only affected the way maps are created but also introduced new ways of accessing, analyzing, visualizing, and sharing geographic information. Among the most important developments has been the growth of geographic information systems (GISs). GIS has been defined as a system for storing, analyzing, and visualizing spatial information. Although there is no well-defined boundary between computer cartography and GIS, the former is concerned with the production of maps, while the later focuses on the analysis and visualization of geographic data. Examples of GIS applications include surface-water modeling, tracking the spread of invasive plant and animal species, determining optimal locations for bus stops, and developing strategies to improve emergency response following natural and human-caused disasters. An important area of GIS development has been precision agriculture, in which GIS maps of soil properties and other site-specific factors are used to guide tractors moving through fields as they apply fertilizers, insecticides, or herbicides.

Another innovation that has greatly influenced mapping is the Global Positioning System (GPS). GPS uses Earth-orbiting satellites to provide a stream of data that can be used by portable receivers for pinpointing Earth locations within less than a meter. GPS receivers have introduced possibilities for the rapid collection of geographic coordinates corresponding to points, lines, and areas. GPS receivers determine latitude, longitude, and elevation above sea level by measuring the time needed for a radio signal to be received from four or more satellites. Data captured using GPS receivers can be used to update cartographic data sets or develop real-time mapping applications that track the location and speed of cars, trucks, boats, or aircraft.

Careers and Course Work

College and university courses in cartography are most often taught within departments of geography. Students seeking careers in the mapping sciences should complete college course work in geography, mathematics, statistics, and technical writing. Depending on the type of maps they will be involved in producing, they may also take course work in graphic arts, computer programming, physics, remote sensing, computer-aided drafting, database management, planning, or geology. A wide variety of agencies and organizations employ cartographers, GIS analysts, and remote-sensing specialists. The U.S. Geological Survey employs cartographers to maintain both paper and digital map products. Its National Imagery and Mapping Agency supports mapping and imagery needs for the military and policy making. The U.S. Department of Agriculture's Natural Resources Conservation Service maintains the Soil Survey Geographic Database, an important spatial data set defining soil boundaries within the United States. The Central Intelligence Agency provides mapping products used internally and by other intelligence agencies. State and local government agencies and private organizations also employ geospatial experts. State agencies produce road maps used for promoting tourism, and local government departments maintain property and planning maps. Maps and proprietary data used in mapping are also created by private companies.

Social Context and Future Prospects

Maps and other cartographic products are important within areas ranging from civil engineering and planning to the humanities. For example, maps are used in weekly magazines, in newspapers, and as part of Web sites. Journalistic maps typically show the countries, cities, or accident sites that are the focus of news events.

Transportation maps such as nautical and aviation charts are of critical importance to the safe movement of ships and aircraft. Within the United States, the National Ocean Service of the National Oceanic and Atmospheric Administration (NOAA) publishes four types of nautical charts: general, sailing, coast, and harbor charts. The Federal Aviation Administration publishes several aviation charts, including the world aeronautical charts at a scale of 1:1,000,000 and sectional charts that represent smaller areas in greater detail at a scale of 1:500,000. Nautical and aviation charts must be updated frequently to reflect changes that can affect safety. Other types of transportation maps include street and highway maps, road atlases, maps showing airline or train routes, and bicycle or walking trail maps.

Topographic maps show rivers, lakes, mountains, and other surface features, as well as cultural features such as roads, cities, levees, and towers. Elevations on topographic maps are shown using contours defined as lines that join points of equal elevation. Bathymetric or hydrographic charts are a related type of map that uses lines to represent underwater depths. The U.S. Geological Survey maintains topographic maps as well as maps showing geological cross sections, groundwater areas, earthquake faults, and seismically active areas. Geologic maps have a wide range of uses, from mineral exploration to civil engineering. Most geologic maps use colors, gray tones, patterns, and sometimes letter codes to represent different rock types. In addition to government agencies, energy and mining companies also maintain maps showing fossil fuels, mineral reserves, and areas with the potential to generate power using solar or wind energy.

Maps are important for recording the location of property boundaries and for planning applications. Cadastral maps are maintained by local government agencies to show ownership boundaries as a basis for property taxes. This type of map is also important to insurance companies. Many local government planning agencies maintain maps showing the city or county infrastructure, such as water resources, sewers, electric lines, and gas lines. City and regional planners use maps for zoning changes and to maintain 911 emergency response systems.

The NOAA's National Weather Service is the principal federal agency responsible for disseminating information about weather conditions. Weather maps are used by forecasters and others to monitor or predict atmospheric conditions including temperature, precipitation, wind direction and speed, humidity, cloud cover, and anomalies such as thunderstorms, hurricanes, and tornadoes. Weather maps may show a single parameter such as surface pressure or multiple types of information. Climate maps include some of the same information shown on weather maps but for longer timeframes. Examples include maps depicting monthly temperature highs and lows or average yearly rainfall.

Widespread availability of the Internet has created possibilities for the publication of new types of mapping products such as electronic atlases that allow users to generate customized maps by selecting the data to be shown. Electronic atlases enable users to search on place-names and easily locate cities and find other information. An added benefit of the Internet is its ability to incorporate updates as soon as new information becomes available.

Thomas A. Wikle, Ph.D.

Further Reading

Cartwright, William, Michael P. Peterson, and Georg F. Garner, eds. *Multimedia Cartography.* 2d ed. New York: Springer, 2007. Examines the growing field of interactive mapping that has gained interest as a result of the Internet's expansion.

Dodge, Martin, Rob Kitchin, and Chris Perkins, eds. *Rethinking Maps: New Frontiers in Cartographic Theory.* New York: Routledge, 2009. Discusses the changing meaning of maps and how their construction and use has evolved. Chapters are written by theorists who examine topics ranging from sustainable mapping to the role of maps in fields such as anthropology.

Kraak, Menno-Jan, and Ferjan Ormeling. *Cartography: Visualization of Spatial Data.* 3d ed. New York: Prentice Hall, 2010. Examines map design, statistical mapping, and data acquisition.

Okada, Alexandra, Simon J. Buckingham Shum, and Tony Sherborne, eds. *Knowledge Cartography: Software Tools and Mapping Techniques.* London: Springer-Verlag London, 2008. Knowledge cartography refers to the mapping of intellectual landscapes. Topics include visual languages, the role of hypertext maps, and methods for evaluating maps.

Robinson, Arthur H., et al. *Elements of Cartography.* 6th ed. New York: John Wiley & Sons, 1995. A comprehensive, well-illustrated text on cartography.

Wilford, John N. *The Mapmakers: The Story of the Great Pioneers in Cartography from Antiquity to the Space Age.* London: Pimlico, 2002. Looks at the origins of mapmaking, beginning with Ptolemy's crude maps and extending through technological achievements that enable scientists to map distant planets.

Web Sites

Federal Aviation Administration
World Aeronautical Charts
http://www.naco.faa.gov/index.asp?xml=naco/catalog/charts/vfr/world

Global Positioning System
http://gps.gov

Google Maps
http://maps.google.com

National Oceanic and Atmospheric Administration
National Ocean Service, Office of Coastal Survey
http://www.nauticalcharts.noaa.gov/staff/chartspubs.html

National Resources Conservation Service
Soil Survey Geographic Database
http://soils.usda.gov/survey/geography/ssurgo

U.S. Geological Survey
Maps, Imagery, and Publications
http://www.usgs.gov/pubprod

See also: Air Traffic Control; Navigation; Remote Sensing.

MARINE MINING

FIELDS OF STUDY

Mining engineering; petroleum engineering; geological engineering; ocean engineering; environmental engineering; geological oceanography; economic geology; applied science; environmental science; materials science; mineralogy; metallurgy, gemology.

SUMMARY

Marine mining is the mining of minerals and other natural resources from the ocean. Valuable resources in the near-shore zone include sand and gravel for construction, salt, phosphate deposits needed for fertilizers, coral for marine aquariums, carbonate sands, and diamonds, gold, tin, and other minerals in placers. Farther out are oil and natural gas deposits, sulfur for industry, and gas hydrates. On the seafloor are nodules and crusts containing strategically important metals such as manganese and cobalt and sulfide deposits around the black smokers in the rift zones, which contain gold, silver, and other useful metals.

KEY TERMS AND CONCEPTS

- **Black Smoker:** Seafloor opening that emits jets of superheated, mineral-rich water.
- **Dynamic Positioning:** Computer-controlled way to keep a ship directly over a drilled hole.
- **Gas Hydrate:** Gas-rich layer of ice found in ocean sediments.
- **Manganese Nodule:** Metal-rich lump found on the deep seafloor.
- **Placer:** Accumulation of mineral grains washed by rivers into the sea.
- **Salt Dome:** Salt mass that rises through sediments because of salt's lesser density.
- **Semisubmersible Rig:** Deep-water drilling rig supported by pontoons.
- **Sulfide Deposit:** Metal-rich crust found surrounding a back smoker.

DEFINITION AND BASIC PRINCIPLES

Marine mining deals with the extraction of economically valuable substances from the ocean. These may be gases, liquids, minerals, rocks, base and precious metals, or ornamental stones such as pearls, diamonds, and coral. Near-shore mining has been done for years, and drilling for oil and gas at sea began in the 1890's, but the mineral deposits on the seafloor have been largely ignored because of the technological difficulty in obtaining them. The substances obtained by marine mining fall into three categories: rocks and minerals found in shallow, near-shore waters; oil and gas deposits found in waters up to thousands of feet deep; and metal-rich deposits found on the seafloor, two miles or more down. Near-shore sand and gravel and the minerals in placers can be scooped or suctioned up by giant dredges, salt can be evaporated in shallow coastal ponds, and pearls and ornamental coral for aquariums can be obtained by divers. Oil wells have been drilled in the ocean for more than one hundred years, and during that time, marine oil and gas production has grown so rapidly that by the twenty-first century, thirty-four percent of the world's oil and twenty-eight percent of the world's natural gas come from below the seafloor. Someday the gas hydrates, manganese nodules, and sulfide deposits may be tapped as well.

BACKGROUND AND HISTORY

The ocean has provided salt for human dietary purposes since ancient times, so evaporation of salt from seawater probably represents one of the earliest examples of marine mining. Another early example was the collecting of minerals washed into the sea by rivers to form placers. Shoreline tin was mined in Indonesia in the early 1900's using dredges, and diamonds carried into the sea by the Orange River were mined along the coast of South Africa at about the same time. Oil has been produced on land using dug or drilled wells for hundreds, if not thousands, of years, but the first offshore oil wells were not drilled until the early 1890's, using piers or pilings driven into shallow water.

Not until the 1930's did oil drillers venture more than a mile from shore, and the immense floating platforms used in the twenty-first century were not introduced until the 1960's. The first manganese nodules were discovered on the seafloor by the HMS *Challenger* expedition in the 1870's, but the techniques for mining them and the related sulfide deposits are still being developed.

How It Works

Near-Shore Resources. Humans must have found early on that the salt crystals in dried-up lagoons near the ocean were useful as a seasoning and for the curing and preserving of foods. The construction of evaporating ponds was a logical next step, and these are common in countries that have access to the ocean and lack underground salt deposits. The mining of placers uses large, floating dredges that either suck the minerals up in long tubes or scoop them up in a series of circulating buckets. Other valuable substances found in shallow water include phosphate for fertilizer, pearls for jewelry, live coral for aquariums, and pure carbonate sand. Although phosphate deposits are widely present in the ocean, they are not mined because phosphate deposits on land can be tapped at a lower cost. The pearls are grown from oysters in pearl farms, which consist of floating wooden rafts or posts driven into the bottom of the sea. A few pearls are still brought up by divers. Divers also bring up the live coral so prized by marine aquarium hobbyists. In the Bahamas, shallow-water shoals of pure carbonate sand are mined as a source of lime for industry.

Oil and Natural Gas. To extract oil and natural gas, a fixed or floating platform is built so that a hole may be drilled into the seabed. The first wells used fixed platforms, either built on pilings at the ends of piers or on artificial islands. The first true offshore well in the Gulf of Mexico was spudded in 1937, one mile from shore in the then-intimidating water depth of 14 feet. The first well located out of sight of land in the Gulf of Mexico was drilled in 1947; it was 10 miles from shore and in 18 feet of water. In the 2010's, companies drill wells in the Gulf of Mexico that are 50 miles out and in nearly 10,000 feet of water. For wells in moderate water depths, the drilling platform rests on a concrete structure or a steel tower. Deeper waters necessitate the use of semisubmersible rigs in which the platform rests on underwater pontoons, moored to the bottom by steel cables or kept in place by dynamic positioning.

Related Products. Sulfur is commonly associated with oil deposits, especially when salt domes are present. As the salt moves upward because of its lesser density, it arches up the sediments, and chemical reactions between the salt and the sediments may form a cap rock that is rich in sulfur. The sulfur is mined by drilling into the cap rock, melting the sulfur with high-pressure steam, and bringing it to the surface as a liquid for processing. The methane gas hydrates in the ocean, which contains a large amount of natural gas, and has not been mined yet. In these hydrates, the gas is bonded with water in the form of ice instead of being associated with oil.

Deep-Sea Resources. Technology for mining the manganese nodules and sulfide deposits found on the seafloor is still in the developmental stages. The nodules are lumps about the size of hamburgers and are scattered about. Someday it may be possible to vacuum them up with suction hoses or scoop them up with a series of buckets. The sulfide deposits around the black smokers present a greater challenge. They are welded to the seafloor and would have to be broken loose.

Applications and Products

Sand and Gravel. Sand and gravel for construction purposes are in short supply in many parts of the world. When available deposits on land have been exhausted, the only other supplies are in the ocean. This situation is particularly true for some of the older countries in Europe, such as England, where construction has gone on for hundreds of years and real estate values are so high that the few remaining deposits on land are prohibitively expensive. Deposits of sand and gravel in the ocean have the advantage of being well sorted, with grains all of the same size, so they are suitable for building purposes once the salt has been washed out of them. Sand and gravel are mined along the coast of the United States to rebuild beaches that have lost their sand because of erosion. The sand is pumped directly onto the beach from the offshore zone in a process known as beach nourishment. The cost of nourishing a beach may exceed several million dollars for each mile of beach treated, and a significant problem is that if the sand is fine grained, it may wash away quickly in storms. In the Bahamas, shallow-water shoals of carbonate sand are mined as a source of lime for making cement and for separating iron from its ore in blast furnaces.

Placers. Gold, tin, and diamonds are the most sought-after substances found in placers, but platinum, iron minerals, and other gemstones may also be found in them. The gold, platinum, and gems are used in the manufacture of jewelry, with the gold also being needed for dentistry, coinage, and monetary purposes. Both gold and platinum have electrical properties that are useful for industry, and platinum is required in the manufacture of catalytic converters for automobiles. Tin was highly prized in ancient times because it was needed for the manufacture of bronze and pewter. In modern times, it is used in solder and for the tin plating of steel to prevent corrosion. The familiar so-called tin can used to package food is simply a steel container with a thin, protective coating of tin.

Salt. Because of its dietary importance and usefulness for food preservation, salt has been valued by humans throughout history. It is also widely used in the chemical industry as a water softener, for snow and ice removal, and by anyone who makes homemade ice cream. However, only 86 percent of the dissolved material in the ocean is table salt. The remainder includes bromine, which is used in medicines, chemistry, and antiknock gasoline; magnesium, an important light-weight metal used in fireworks and flares because it is flammable; potassium, which is used in fertilizers; and calcium sulfate (gypsum), an important component of wallboard.

Oil and Natural Gas. By far the most valuable of the resources obtained from the sea are oil and natural gas. In addition to being used for fuels, such as gasoline, diesel fuel, kerosene, fuel oil for heating homes, and jet fuel, oil is the raw material that is the basis of grease, motor oil and other lubricants, asphalt and tar, and a variety of chemical products, including nylon fibers for stockings. Even the paraffin wax used to seal jars of homemade jelly comes from oil. The natural gas found with oil is used domestically for cooking and heating, in the power industry to generate electricity, and in the production of ammonia for making fertilizer. The methane gas hydrates deposited on the seafloor, which have not yet been exploited commercially, may someday have similar uses.

Sulfur. Although sulfur is not derived from the seafloor in the twenty-first century, it has been in the past. Sulfuric acid, which is widely used in chemistry and industry, is its most important product. Because of its toxicity to bacteria, sulfur is an ingredient in medicines, insecticides, fungicides, and pesticides. Sulfur is also flammable and is used in the manufacture of matches and gunpowder.

Manganese Nodules. Although the manganese nodules and crusts found on the seafloor are not being mined commercially, the United States has an interest in them because of the metals they contain. Three of the metals—iron, needed for making steel; copper, used in wiring; and nickel, important in the manufacture of stainless steel—are already available on land. The other two metals, strategically important in wartime, are not available in the United States and must be imported. They are manganese, used in making armor plate, and cobalt, which is needed for the high-temperature steel used in tools and aircraft engines.

Sulfide Deposits. The sulfide deposits found on the seafloor have not been mined yet but are of interest because of their metal content. They contain not only the five metals found in the manganese nodules but also zinc, which is needed for galvanizing and for alloys such as brass; lead, which is used in batteries, scuba divers' weight belts, and as shielding for atomic reactors; silver, which is needed for coinage and in industry; and gold, with its monetary and industrial uses.

IMPACT ON INDUSTRY

Governmental agencies and numerous nongovernmental agencies play an active role in overseeing and regulating marine mining, partly because it is done in the sea.

Governmental Agencies. The International Seabed Authority (ISA), headquartered in Kingston, Jamaica, was established in 1982 under the United Nations Convention of the law of the sea. The 159 nations that ratified the law of the sea make up the membership of the ISA. The organization's mission is to supervise and control activities regarding the seabed and the subsoil below it whenever these activities lie beyond the limits of any single nation's jurisdiction. The law of the sea gives each nation full jurisdiction over its territorial waters, which extend twelve nautical miles out from the nation's shores. In addition, each nation has an exclusive economic zone (EZZ) that extends offshore for two hundred nautical miles, or even more if the edge of the continental shelf lies farther out than two hundred nautical miles. Within the boundaries of its economic zone, each nation is able to regulate fishing policies, the extraction of mineral resources, the control of pollution, and scientific research.

The ISA is developing regulations that will govern commercial prospecting for minerals in the part of the seafloor outside the various member nations' exclusive economic zones. The ISA has the power to make licensing decisions for this area, and part of the profits from mining operations in this area will be used for the benefit of the world's developing nations. Although the United States participated in the writing of the law of the sea and signed the treaty that implemented it, the U.S. Senate has never ratified the treaty. As a result, the United States is not a member nation of the ISA and therefore has only observer status at the organization's meetings.

Industry and Business. An industry-sponsored organization concerned with marine mining is the International Marine Minerals Society, headquartered in Honolulu. The society sponsors an international forum, the Underwater Mining Institute, where leaders from universities, industry, and government agencies can exchange ideas and recommend policies regarding marine mining. The society also publishes a scientific journal, *Marine Georesources and Geotechnology*, in which researchers can share findings related to seafloor resources.

Another example of an industry-sponsored organization in marine mining is the American Petroleum Institute, the leading trade association for the oil and gas industry, founded in 1919 and headquartered in Washington, D.C. The institute has nearly four hundred corporate members from all segments of the industry: producing companies, refiners, suppliers such as drillers and electric log companies, pipeline operators, and marine transporters. Its mission is to represent the oil and gas industry to the public, the U.S. Congress, the executive branch, state governments, and the media. It also negotiates with regulatory agencies regarding rules and fees and represents the industry in legal proceedings. Among the services it provides for members are sponsoring research projects, collecting and publishing statistics regarding oil and gas production, setting standards for the industry and recommending approved procedures, certifying equipment as safe and individuals as competent, and running educational programs.

Careers and Course Work

The only marine mining done in the United States as of 2010 is the dredging of sand and gravel for beach nourishment and oil and gas production in the Gulf of Mexico, along the Southern California coast, and off the north slope of Alaska. Dredging jobs are limited because the work is intermittent and seasonal and takes place largely in resort areas where beach erosion is a problem. Advanced training and professional degrees are not required. A captain is needed to run the dredge, a hydrographic surveyor has to chart the bottom and locate the deposits of sand and gravel, a mechanic has to keep the equipment running, and deckhands are required to do routine tasks. By contrast, a wide variety of jobs are available in the oil and gas industry, both on land and at sea.

A good place for the beginner to start is with an undergraduate, college-level course in geology. This could lead to graduate work in geology and a career in oil and gas exploration, or to a degree in

Fascinating Facts About Marine Mining

- Salt is used to make homemade ice cream because salt lowers the freezing point of water. This is also the reason that salt is put on roads in the northern part of the United States for snow removal and why northern harbors do not freeze when temperatures dip below 32 degrees Fahrenheit.
- Offshore sand is frequently used to replace the sand eroded from a beach in a process known as beach nourishment. The cost of dredging this sand can be high—as much as several million dollars a mile—and there is always a danger that the next storm will wash the sand away.
- De Beers Marine Namibia is mining diamonds off the shores of Africa with specialized ships that serve as mines.
- According to the U.S. Geological Survey, there may be more organic carbon in methane hydrate than in all the coal, oil, and nonhydrate natural gas in the entire world.
- Scientists believe that some large sulfide deposits on land may have originally been formed in the deep sea and thrust up by land movements. The ancient Romans got their copper from the thirty large sulfide deposits on the island of Crete.
- Nautilus Minerals, a Canadian company, is planning to mine sulfide deposits on the ocean floor near Papua New Guinea for copper, gold, silver, and zinc. Its actions have been the focus of some criticism by scientists.

petroleum engineering and a job designing, constructing, and maintaining offshore drilling platforms. The oil industry also offers jobs doing seismic surveying, drilling and logging wells, and producing, refining, and marketing the oil that is produced. Management positions also are available, so courses in business management are also recommended

Social Context and Future Prospects

As glaciers and ice sheets melt because of global warming, the water added to the sea by this melting will result in rising sea levels. This, in turn, will increase the amount of beach erosion around the world and consequently the need to nourish beaches in resort areas. Jobs in the dredging industry should continue to be available as a result. Jobs in the oil and gas industry will continue to be available as well. More than three-quarters of world oil production goes into powering cars, trucks, ships, and planes, so the demand for oil is likely to grow as more and more people in the developing nations are able to afford to buy cars.

One problem that may lie ahead in marine mining involves the manganese nodules and sulfide crusts found on the seafloor. The developed nations are already investigating ways to mine these deposits. They are the only nations with the technological expertise to bring up these nodules and crusts. However, the law of the sea states that the profits from mining minerals in the area in which the nodules and crusts are found must be shared with the developing nations, and the developed nations may find it hard to agree to this condition.

Donald W. Lovejoy, B.S., M.S., Ph.D.

Further Reading

Drew, Lisa W. "The Promise and Peril of Seafloor Mining: Can Minerals Be Extracted from the Seafloor Without Environmental Impact?" *Oceanus* 47, no. 3 (2009): 8-14. Summarizes the way in which seafloor mineral deposits form, the metals they contain, how they can be mined, and the environmental damage that might result.

______. "Who Regulates Mining on the Seafloor?" *Oceanus* 47, no. 3 (2009): 15. Describes the role played by the International Seabed Authority and the International Marine Minerals Society in overseeing seafloor mining and determining the rules that govern it.

Garrison, Tom. *Oceanography: An Invitation to Marine Science.* Belmont, Calif.: Thomson Brooks/Cole, 2007. An excellent introduction to the world's oceans; chapters cover the seafloor and its structures, the many marine resources, and the environmental concerns regarding the extraction of these resources.

McLachlan, Anton, and Alec Brown. *The Ecology of Sandy Shores.* Burlington, Mass.: Elsevier/Academic Press, 2006. Discusses beach erosion and the problems caused by mining the beach and the offshore zone; has an extensive treatment of beach nourishment and its environmental effects.

Rice, Tony. *Deep Ocean.* Washington, D.C.: Smithsonian Institution Press, 2009. An excellent introduction to the deep-ocean environment: profusely illustrated, and with sections on oil and gas as well as the metal-rich nodules and sulfide crusts.

Welland, Michael. *Sand: The Never-Ending Story.* Berkeley: University of California Press, 2009. A good summary of the ways sand is important; several examples of beach nourishment are given, as well as examples of the problems that beach nourishment may cause.

Web Sites

American Petroleum Institute
http://www.api.org

International Marine Minerals Society
http://www.immsoc.org

International Seabed Authority
http://www.isa.org.jm

Underwater Mining Institute
http://www.underwaterming.org

MICROLITHOGRAPHY AND NANOLITHOGRAPHY

FIELDS OF STUDY

Physics; chemistry; mathematics; calculus; optics; mechanical engineering; materials science; thermodynamics; electrical engineering; microtechnology; nanotechnology; electromagnetics; semiconductor manufacturing; radiology; laser imaging.

SUMMARY

Microlithography and nanolithography are two closely related fields in the area of electronics manufacturing. Microlithography is a process used in making integrated circuits in semiconductors. The process relies on the projection of an image onto a light-sensitive plate. Circuits made by microlithographic techniques can carry features as small as 100 nanometers in width. Nanolithography refers to the making of circuits and other features even smaller and finer than those possible through traditional microlithography. New developments in nanolithography are making it possible to apply the technology in areas beyond semiconductors, such as nanoelectromechanical systems (NEMS) in highly sensitive measuring devices.

KEY TERMS AND CONCEPTS

- **Exposure:** Measured amount of light shone onto a light-sensitive surface to create an image.
- **Integrated Circuit:** Circuit manufactured as a single, rigid object; also known as a microchip.
- **Lithography:** Printing process in which an image is transferred from a flat plate to another surface.
- **Photomask:** Template used to project an image onto a light-sensitive surface.
- **Resist:** Light-sensitive material found in a thin layer on the surface of a microchip.
- **Self-Assembly:** Manufacturing method used in nanolithography in which individual chemical molecules are used to create highly precise images.
- **Substrate:** Rigid surface that carries the resist layer on a microchip.
- **Wafer:** In electronics, a sheet of material, most commonly silicon and as thin as 200 microns, carrying up to several hundred microchips.

Definition and Basic Principles

Some sources say that the difference between microlithography and nanolithography has to do with the size and precision of the electronic components being made. Microlithography is associated with a level of precision in manufacturing that can be measured in microns, while nanolithography is associated with nanometers.

In reality, the differences are more complicated than feature size. Microlithography is a term associated with the making of integrated circuits and semiconductors. It refers to a process of projecting images from a master pattern onto a light-sensitive surface. The process shares some basic traits with traditional lithography, a form of ink-based printing in which an image is transferred from a flat plate to a surface such as paper or cardboard.

Nanolithography refers not just to microlithography on a smaller scale. It involves a wide range of technologies, including the uses of electron beams and scanning probes, to make components in products ranging from video screens to cultures for the growing of organic tissue used in transplant surgery. While the terms are not used interchangeably, the fields are often grouped together, such as in the *Journal of Micro/ Nanolithography, MEMS, and MOEMS.* (MEMS stands for microelectromechanical systems and MOEMS for micro-opto-electromechanical systems.)

Background and History

The roots of microlithography can be found in a program funded from 1957 to 1963 by the U.S. Army Signal Corps. The Micro-Module Program was launched shortly after the discovery of the transistor by scientists at Bell Telephone Laboratories in 1947. Before transistors, electronic components relied on vacuum tubes, which were large, broke easily, and generated heat. Transistors offered the possibility of making much smaller and more durable devices that needed less energy to run. However, transistors presented a number of manufacturing problems. They were made from numerous parts consisting of different materials that had to be assembled and soldered by hand. The Micro-Module Program encouraged the development of newer technologies in which components could be made with built-in wiring and joined together to form circuits. Texas Instruments engineer Jack Kilby discovered a way to make integrated circuits in a more cost-effective way in 1958. At nearly the same time, Robert Noyce at Fairchild Semiconductor made a related breakthrough using silicon transistors. Microlithography grew rapidly with the demand for microchips in electronics.

One of the earliest advances in the field of nanolithography was a 1983 NATO Advanced Research Workshop in Rome. Leading experts from a range of fields, including X-ray and electron-beam technology, met for a cross-disciplinary discussion of the ways in which nanolithography could be applied to technological problems. The field has since evolved beyond semiconductors to areas such as chemistry and medicine.

How It Works

Microlithography is a process used specifically for the manufacturing of microchips. It allows a series of circuits to be placed on a hard surface with a tremendous amount of precision. Improvements in technology have made it possible for microchips to carry increasing numbers of circuits in new combinations and patterns.

Microlithography begins with the creation of a template, or pattern, known as a photomask. The photomask plays a role similar to that of a plate on a printing press. It carries an image of the circuits to be built on the surface of a chip. The image on the photomask is projected onto the chip's surface with the help of a lithographic lens. This lens is one of the most accurate and advanced pieces of equipment needed to make a microchip. An individual lens can be made out of more than thirty components and weigh almost a half ton. A source of intense light, such as a high-pressure mercury arc lamp or a laser, is shone through a series of refraction elements and eventually through the photomask itself. The light projects the photomask's image onto the surface of a silicon wafer by shining through the lithographic lens, which condenses the image down to a suitably small size for placement on the surface of the wafer. The wafer's surface has been coated with a thin layer of a light-sensitive substance. When the image is shone onto this layer, known as a photoresist or simply a resist, the intensity of the light burns the image into the resist's surface within a fraction of a second. This procedure is known as an exposure.

A single microchip goes through many exposures in the manufacturing process. At each step, the accuracy of the procedure must be absolute or the microchip will be unusable. The tolerance for errors in focusing and placing the image from the photomask is below 1 percent, or a few nanometers. Wafers are held in place by a vacuum system, and their positions are verified by lasers before each exposure occurs. For a manufacturing process to be profitable, errors must be as low as possible while wafers pass through each stage quickly. A standard system might process fifty to one hundred or more wafers in an hour.

Unlike microlithography, there are a large number of manufacturing methods used in nanolithography, each with its own steps and specialized equipment. Some of these methods rely on radiation such as X rays or ultraviolet light rather than lasers. Electron projection lithography falls into a similar category. It relies on an electronic lens and electric and magnetic fields to guide charged particles into specific shapes and patterns with the help of a mask. Many of these particles move on wavelengths short enough to pass directly through a mask. In these cases, the masks deflect the particles away from the resist rather than stopping the energy entirely. Other methods do not use photomasks at all, which are expensive and difficult to produce for small manufacturing batches, but focus a beam of light directly onto the resist surface to make the needed patterns.

These methods belong to a category of nanolithography known as top-down. In contrast, bottom-up manufacturing methods use combinations of chemical solutions to assemble images directly on a surface. These images are so precise that the assembly process takes place on the scale of individual molecules.

Applications and Products

Semiconductors. Virtually all manufacturing that involves microlithography is used to make semiconductors. Like printing with lithographic plates, microlithography allows a highly complex pattern of electronic circuits to be transferred to many individual microchips at the same time. This process makes it possible to invest a great deal of time and financial resources into the creation of a sophisticated technological design, then duplicate the design on a broad scale precisely and quickly.

The process of microlithography is only one set of steps in a chain that leads to a finished microchip. Before the surface of a wafer is ready to receive a microlithographic image, the wafer itself must be formed from highly refined silicon. After refining, silicon is formed into single-crystal rods known as ingots and sliced into wafers. The refining and wafer-slicing process requires such specialized technology that this stage is often handled by companies devoted exclusively to it.

The silicon wafers are then purchased by semiconductor manufacturers, which prepare the wafers for microlithography by adding silicon dioxide (a form of sand) and elements with specific electrochemical properties such as boron, phosphorus, or arsenic. Silicon dioxide serves as an insulating base for the other additives, which are formed on the surface of the chip into transistors and circuits through repeated exposures from a photomask in the process of microlithography. The transistors are connected to each other through a metallization process that uses aluminum, tungsten, or a number of other metals and their compounds. The same metallization process is also used to make bonding pads on the surface of the chip. These pads are the points through which the chip communicates electronically with other chips and with the machine in which it will be installed.

Once the image on the chip's surface is complete, the surface itself is smoothed chemically and covered by a protective layer. The chip's transistors and circuits are tested before the chip is sold by the manufacturer. At this stage, the chip may be altered to a larger product's specifications, a process known as die cutting, before being installed inside another device.

Carbon Nanotubes. In contrast to microlithography, nanolithography as a manufacturing process has applications in a wide range of fields. Nanolithography has been found to be useful in the making of nanoelectromechanical systems (NEMS). One type of structure that shows the most promise is the carbon nanotube. Through nanolithography, carbon atoms are bonded together into molecules shaped like long, hollow tubes with closed ends. The result is no wider than a few nanometers but can be up to several inches long. Carbon nanotubes belong to a family of carbon molecules known as fullerenes, a group that also includes structures such as buckyballs. Because carbon molecules are strong conductors of electricity, nanotubes have many potential uses as components in transistors and circuits. They can also be used in electrical sensors that need a high level of precision and sensitivity, such as those used to track changes in gases.

The potential uses of carbon nanotubes are still being discovered. The presence of carbon atoms in compounds such as steel makes the resulting product lighter and stronger. A company in Finland has developed a way to add carbon nanotubes to the blades of windmills. The nanotubes make the windmill blades twice as light as those made from glass fibers and several times stronger, which allows for larger blades that move more efficiently. Other applications are still in the research stage. Scientists have noted, for example, that carbon nanotubes resemble tiny needles and may be used to carry antibodies and pharmaceuticals to highly targeted areas within the body, such as cancerous tumors.

Biosensors and Cell Biology. Dip-pen nanolithography (DPN) has shown great promise in the field of cell biology. DPN allows a chemical pattern of molecules, such as those found in the DNA of a cell, to be copied onto the surface of a microchip. DPN makes it possible to copy this pattern not once but thousands of times within a single manufacturing process. Unlike microlithography, which depends on light being shone through a photomask, DPN deposits a chemical agent directly onto the surface of a chip with the help of a "dip pen"—the highly precise tip of an atomic-force microscope. Direct contact between the pen and the surface means a higher possibility of problems, such as contaminating agents.

However, this challenge is outweighed by the possibilities presented by the technology in developing new kinds of cell-based therapies and other medical applications. Scientists at Northwestern University in Chicago have found that DPN may be used to replicate electrodes in DNA patterns. With the information from these patterns, customized biosensors could be developed that could, in theory, be reintroduced into living cells. These sensors could then transmit data that would be used to monitor a body's vital functions. The sensors could also track the progress of a disease or drug therapy.

Advancements beyond the technology of DPN are already being pursued. Polymer-pen lithography (PPL) uses many of the same processes as DPN,

Fascinating Facts About Microlithography and Nanolithography

- Microlithography makes up about one-third of the manufacturing cost of each microchip.
- Microchips become more complex as they go through each microlithographic procedure. The most advanced designs have fifty or more images layered on top of one other.
- Semiconductor expert Gordon Moore states in Moore's law that the number of devices that can be installed on a single microchip will double every year. Moore published his theory in an article in 1965. Though the cycle is now closer to a year and a half, the rule has held true for more than forty years.
- Silicon wafers are made into semiconductors in areas known as "clean rooms." Microlithograpic procedures are so precise that even dust and airborne contaminants can destroy the etching process on the surface of a wafer. Employees working in clean rooms wear garments over their clothes called "bunny suits."
- In 1999 an application of nanolithography known as dip-pen nanolithography was developed by Chad Mirkin at Northwestern University. The technology allows patterns at the nanoscale level, such as certain electronic circuits and biological cell material, to be copied.
- Texas Instruments engineer Jack Kilby, credited with developing one of the first designs for an integrated circuit, came up with his idea in July, 1958, while he was a new employee. He had not yet earned vacation time, so he was alone in the office while most of his colleagues were taking time off. Kilby received the Nobel Prize in Physics in 2000.

but it involves larger arrays of pens—up to 11 million, by an estimate from Northwestern—as well as the ability to push the pens with varying amounts of force against the writing surface. These features allow many details to be transferred to the writing surface during a single procedure, which makes the manufacturing process faster and more efficient. Beam-pen lithography (BPL) blends lithographic techniques with the technology behind scanning electron microscopes. These approaches are still in the early research stages, but they present possibilities for developing new treatments in fields ranging from genetics to heart disease.

Impact on Industry

Microlithography's role in semiconductor manufacturing will continue to be critical in the early twenty-first century. New research in the field is primarily focused on finding ways to make manufacturing processes more efficient. Because of this, there is not a significant amount of federal or academic research funding devoted to finding new applications of the technology. However, institutes at schools such as the Georgia Institute of Technology and Rice University have built facilities devoted to microlithography research.

Nanolithography is receiving more attention and support from governments, academia and the private sector. Major efforts include:

Government Agencies and Initiatives. One of the leading providers of U.S. government agency support for new work in nanolithography is the National Nanotechnology Initiative (NNI). Launched in 2000, the NNI brings together more than twenty federal agencies and helps to coordinate their work in the field. The NNI serves as a communications hub as well as a source of financial support for research. It also tracks educational programs, particularly for college and graduate students, and is a good resource for learning more about careers in the field.

The NNI has also been instrumental in supporting nanotechnology development in other countries, primarily through the backing of cross-border efforts. Its international equivalent is the Global Issues in Nanotechnology (GIN) working group. Other programs led or supported by NNI include the Organisation for Economic Co-operation and Development (OECD) Working Party on Manufactured Nanomaterials and the OECD Working Party on Nanotechnology. Countries such as Japan, Australia, and the United Kingdom have also established government agencies or specially funded efforts to boost the development of nanotechnology. Nanolithography is too specialized a field to receive significant attention at the level of national government, but it is most commonly included under the banner of nanotechnology.

Universities and Research Institutions. Nanolithography is being explored in a variety of forms at universities and research institutions around the world. Because nanolithography has applications in more than one field, work contributing to its advancement can be carried out in departments such as electrical engineering, mechanical engineering, physics, chemistry, and others. In the United States, Northwestern and Stanford universities are leading contributors to new research in nanolithography. Institutions making significant contributions outside the United States include the University of Toronto, the University of Twente in the Netherlands, and the University of Strathclyde in Scotland.

Private Sector. The applications of nanolithography within the private sector involve many industries spread across the globe. One of the best ways to understand current developments is to look at the activities of cross-sector forums and consortiums. The memberships of these groups are made up of organizations of all kinds, but they include manufacturers that make nanolithography equipment as well as those that use the technology itself. Some of the most visible and active international groups are the Global Nanotechnology Network (GNN), the International Council on Nanotechnology (ICON), the European Nano Forum, and the Asia Pacific Nanotechnology Forum (APNF). These groups host conferences and publications supporting the sharing of information about nanotechnology.

Careers and Course Work

Microlithography and nanolithography are highly specialized areas when it comes to developing a career. At the same time, they involve the intersection of many different academic disciplines, such as engineering, physics, and chemistry.

Students interested in working in microlithography and nanolithography are likely to pursue bachelor's degrees with majors in fields such as electronics engineering, electrical engineering, or materials science. A student with a background in a related field such as mechanical engineering, physics, or computer science would also be well-positioned for a job in microlithography or nanolithography. Relevant course work starts with a foundation in the physical sciences and mathematics. Depending on the institution, advanced course work can be highly specialized. More than ten schools in the United States alone offer bachelor's degrees with majors in nanotechnology.

Due to the complex and specialized knowledge required to work in microlithography or nanolithography, many job candidates complete master's degrees or doctorates. Advanced degrees improve the earning potential of graduates in the field as well as prepare one for higher-level positions such as research team leader. Among the degree programs tracked by the National Nanotechnology Initiative, none specializes exclusively in microlithography or nanolithography. However, some large research universities such as Georgia Institute of Technology host interdisciplinary centers that offer students and faculty members the opportunity to gain experience working with microlithography and nanolithography applications.

Outside of designing new systems and applying specialized knowledge, the job market for semiconductor manufacturing is very limited. The U.S. Bureau of Labor Statistics finds that the number of jobs in the field has fallen sharply in the first decade of the twenty-first century due to increased efficiencies in manufacturing processes. Most factory workers are required to wear full protective suits and to work in clean rooms to avoid introducing contaminants, which can affect the high precision of the components being made.

Social Context and Future Prospects

The demand for new applications of microlithography and nanolithography is expected to continue for the foreseeable future. Microlithography remains one of the most precise and cost-effective ways to manufacture integrated circuits. An increasing number of consumer devices rely on microchips to function, ensuring that the need for inexpensive components will stay in place for many years. However, some sources say that further innovations in the field of microlithography are not expected because of the limitations of the technology. Without major changes to manufacturing processes, microchip makers are likely to continue to move their plants to markets around the world where manual labor is least expensive. A significant share of semiconductor manufacturing already takes place in regions such as Asia for cost-related reasons.

The outlook for nanolithography is more optimistic. As new applications for the technology are discovered, there is an increasing need for specialists. At present, some of the most promising areas of opportunity are in biotechnology, chemistry, and electronics outside of traditional integrated circuits.

There is debate within the field about the level of precision that can be achieved in manufacturing through nanolithography. If the process is no longer cost-effective below a certain point, a next-generation technology will need to be developed.

Julia A. Rosenthal, B.A., M.S.

Further Reading

Cao, Guozhong, and Ying Wang. *Nanostructures and Nanomaterials: Synthesis, Properties, and Applications.* 2d ed. Singapore: World Scientific Publishing, 2011. Chapter 7 examines photolithography and nanolithography in detail as used in making semiconductors.

Guo, Zhen, and Li Tan. *Fundamentals and Applications of Nanomaterials.* Norwood, Mass.: Artech House, 2009. Chapter 6 places nanolithography into the context of nanotechnology as a field.

Levinson, Harry J. *Principles of Lithography.* 3d ed. Bellingham, Wash.: SPIE Press, 2011. Contains a detailed but readable explanation of the process of microlithography in making semiconductors.

Prasad, Paras N. *Nanophotonics.* Hoboken, N.J.: John Wiley & Sons, 2004. Excellent reference that covers photonics on a nano scale, written by a leader in the field.

Suzuki, Kazuaki, and Bruce W. Smith, eds. *Microlithography: Science and Technology.* 2d ed. Boca Raton, Fla.: CRC Press, 2007. Provides a thorough examination of microlithography and its applications, including technologies using electron projection and extreme ultraviolet light.

Web Sites

Global Nanotechnology Network (GNN)
http://www.globalnanotechnologynetwork.org

Journal of Micro/Nanolithography, MEMS, and MOEMS
http://spie.org/x865.xml

National Nanotechnology Initiative (NNI)
http://www.nano.gov

See also: Electromagnet Technologies; Nanotechnology; Transistor Technologies.

MICROWAVE TECHNOLOGY

FIELDS OF STUDY

Communications; electronics; radar technology; radio technology; radio astronomy; spectroscopy; telecommunications.

SUMMARY

A microwave is an electromagnetic wave the wavelength of which ranges from one meter to one millimeter. Microwave energy has a frequency ranging from 0.3 gigahertz (GHz) to 300 GHz. The high frequency of microwaves provides the microwave band with a very large information-carrying capacity; the band has a bandwidth thirty times that of the rest of the radio spectrum below it. Microwave signals propagate in straight lines; thus, they are limited to line-of-sight transmission. Unlike lower-frequency radio waves, they cannot pass around hills or mountains and are not refracted or reflected by atmospheric layers. In addition to the familiar microwave oven, applications include communications, radar, radio astronomy, navigation, and spectroscopy.

KEY TERMS AND CONCEPTS

- **Amplification:** Process of increasing the strength of an electronic transmission or a sound wave.
- **Amplitude:** Refers to the height of a radio wave.
- **Antenna:** Device that either converts an electric current into an electromagnetic radiation (transmitter) or converts electromagnetic radiation into an electric current (receiver).
- **Modulation:** Process of varying one or more properties of an electromagnetic wave. Three parameters can be altered via modulation: amplitude (height), its phase (timing), and its frequency (pitch). Two common forms of radio modulation are amplitude modulation (AM) and frequency modulation (FM).
- **Radio Frequency:** An oscillation of a radio wave in the range of 3 kilohertz (kHz) to 300 GHz.
- **Radio Wave:** Electromagnetic radiation that travels at the speed of light; radio waves are of a longer wavelength than infrared light.

Definition and Basic Principles

In contrast to sound waves, which require a medium such as air or water for propagation, microwaves can travel through a vacuum. In a vacuum such as outer space they travel at the speed of light (299,800 kilometers per second). Microwaves travel in a straight line, limiting them to line-of-sight applications. In space, microwaves conform to the inverse-square law: The power density of a microwave is proportional to the inverse of the square of the distance from a point source. All microwaves weaken as they travel a distance. At some point, depending on the strength of the signal, the microwave will no longer be discernible. Interference can weaken or destroy a radio signal. Other microwave transmitters in the same frequency range produce interference; however, their small wavelength allows small antennae to direct the electromagnetic energy in narrow beams, which can be pointed directly at the receiving antenna. This feature allows nearby microwave equipment to broadcast on the same frequencies without interfering with each other, as lower-frequency radio waves do.

Microwave ovens pass radiation, usually at a frequency of 2.45 GHz, through food. Energy from the microwaves is absorbed by fat, water, and other substances in the food through a process known as dielectric heating. Water and many other molecules have a partial positive charge at one end and a partial negative charge at the other. They rotate in an attempt to align themselves with the alternating electric field of the microwaves. This molecular movement produces heat.

Background and History

Electromagnetic waves were discovered in 1877 by the German physicist Heinrich Hertz, whose name is used to describe radio frequencies in cycles per second. Eight years later, American inventor Thomas Edison obtained a patent for wireless telegraphy by discontinuous (intermittent) wave. A far superior system was developed in 1894 by the Italian inventor Guglielmo Marconi. Marconi initially transmitted telegraph signals over a short distance on land. Subsequently, an improved system was capable of transmitting signals across the Atlantic Ocean.

Much of microwave technology was developed

during World War II for radar applications. The technology was developed secretly; it became available for public use only after the war. In 1951,AT&T's new microwave radio-relay skyway carried a telephone call via a series of 107 microwave towers that were spaced about 30 miles apart; this was the first microwave application that could carry telephone conversations across the United States via radio (as opposed to wire or cable). The system could also carry television signals; three weeks after the first telephone call, at least 30 million people saw and heard President Harry Truman open the Japanese Peace Treaty Conference in San Francisco. Then, in 1946, Percy Spencer, an engineer at the Raytheon Corporation, developed the now-ubiquitous microwave oven.

How It Works

Communication. Because of the short wavelength, microwave radio transmission employs small, highly directional antennae, which are smaller and therefore more practical than ones used for longer wavelengths (lower frequencies). Considerably more bandwidth is available in the microwave spectrum than in lower frequencies. This wider bandwidth is suitable for the transmission of video and audio. Since microwave transmission is line of sight, distance is limited by the curvature of the Earth. A higher antenna can transmit over a greater distance. Much greater transmission distances between the Earth's surface and an orbiting satellite are possible because the only limitation is the attenuation that occurs from the atmosphere; however, even in the vacuum of outer space, the signal degrades over a distance.

Energy Transmission. The concept of using microwaves for the transmission of power emerged following World War II; high-power microwave emitters, also known as cavity magnetrons, were developed. These emitters can transfer electrical energy from a power source to a target without interconnecting wires. Wireless transmission is reserved for cases in which interconnecting wires are inconvenient, dangerous, or impossible. To be effective as well as economical, a large part of the energy sent out by the generating plant must arrive at the receiver(s). The short wavelengths of microwave radiation can be made more directional than lower radio frequencies; thus, allowing power beaming over longer distances. A rectenna (rectifying antenna) at the target can convert the microwave energy back into electricity. Conversion efficiencies of more than 95 percent have been achieved with rectennae.

Microwave Ovens. A microwave oven passes non-ionizing microwave radiation, usually at a frequency of 2.45 GHz, through food. Water, fat, and other substances in the food absorb energy from the microwaves in a process called dielectric heating. Many molecules (such as those of water) are electric dipoles, which means they have a partial positive charge at one end and a partial negative charge at the other. They rotate in order to align themselves with the alternating electric field of the microwaves. This movement represents heat, which is then dispersed as the rotating molecules strike each other.

Navigation. Global Positioning Systems (GPSs) operate in microwave frequencies ranging from about 1.2 GHz to 1.6 GHz. GPS is a space-based global navigation satellite system, which provides accurate location and time information at any place on Earth where there is an unobstructed line of sight to four or more GPS satellites. GPS can function under any weather conditions and anywhere on the planet. Since it cannot function underwater, a submarine must surface to use a GPS system. The technology depends upon triangulation, just as a land-based systems do, to locate a discrete point. GPS is composed of three segments: the space segment; the control segment; and the user segment. The U.S. Air Force operates and maintains both the space and control segments. The space segment is made up of satellites, which are in medium-space orbit. The satellites broadcast signals from space, and a GPS receiver (user segment) uses these signals to calculate a three-dimensional location (latitude, longitude, and altitude). The signal transmits the current time, accurate within nanoseconds.

Radar. Radar is an acronym for radio detecting and ranging. The device consists of a transmitter and receiver. The transmitter emits radio waves, which are deflected from a fixed or moving object. The receiver, which can be a dish or an antenna, receives the wave. Radar circuitry then displays an image of the object in real time. The screen displays the distance of the object from the radar. If the object is moving, consecutive readings can calculate the speed and direction of the object.

If the object is airborne, and the device is so

equipped, the altitude is displayed. Radar is invaluable in foggy weather when visibility can be severely reduced.

Radio Astronomy. Radio astronomy is a subfield of astronomy that examines celestial objects, which emit radio frequencies. Much of radio astronomy is focused on the microwave band.

Spectroscopy. Spectroscopy involves the use of spectrometers and spectroscopes to analyze the distribution of atomic or subatomic particles in a system, such as a molecular beam. Microwave spectroscopy is a form of spectroscopy in which information is obtained on the structure and chemical bonding of molecules and crystals via measuring the wavelengths of microwaves emitted or absorbed by them. Microwave spectroscopy can be conducted only on gases.

Applications and Products

Communication. Microwaves are commonly used by communication systems on the Earth's surface, in satellite communications, and in deep-space radio communications. Microwaves are commonly used by television news media to transmit audio and video from a specially equipped van to a television station. Mobile telephone networks operate in the lower end of the microwave band, while others operate at frequencies just beneath the microwave band. Networks of microwave relay links have largely been replaced by fiber-optic networks. Wireless transmission employing local area network (LAN) protocols such as Bluetooth operate in the 2.4 GHz microwave band. Other LAN protocols operate at higher microwave frequencies. Wireless Internet services operate in the 3.5-4.0 GHz range.

Energy Transmission. Although wireless power transmission via microwaves is well proven, many applications are still in the experimental stage. In 1964, a miniature helicopter propelled by microwave power was demonstrated. In 2008, 20 watts of power were transmitted 92 miles from a mountain on Maui to the big island of Hawaii. The U.S. military is currently using microwave energy transmission as a form of sublethal weaponry. The application, known as an Active Denial System, uses microwaves to heat a thin layer of human skin to an unbearable temperature, which forces the recipient to move away from the energy source. The skin can be heated to a temperature of 54 degrees Celsius at a depth of 0.4 mm with a two-second burst of a 95-GHz-focused beam.

Microwave Ovens. A microwave oven is a common household appliance. It rapidly heats frozen foods, pops popcorn, bakes potatoes, and boils water. Its compact size makes it beneficial when space is at a premium. It is used on commercial airlines for meal preparation and is found in many other locations, including offices. In addition, microwave heating is employed in many industrial processes for drying and curing products.

Navigation. GPS applications, which operate in the microwave band, are used for navigation over land and water, and numerous military, civilian, and commercial applications exist. GPS devices are used for navigation on military and commercial vessels, on pleasure boats, and in automobiles. The device gives a visual display of the vehicle's position on a map overlay. Useful information such as distance to the next turn and the destination are given visually and aurally, if the GPS is equipped with an audio system. Some vehicle GPS devices also provide information such as the nearest gasoline stations, rest areas, restaurants, and hospitals. GPS units for off-road use are also available. The motorist on a budget can purchase an inexpensive handheld GPS; however, these devices are popular with campers and hikers. Many small aircraft and virtually all large aircraft contain a GPS. GPS is also used for space navigation. Another application of GPS is an ankle monitor, which tracks the location of individuals under house arrest.

Radar. Radar is an essential component of any vessel that operates offshore. Navy vessels and many pleasure boats have radar. Larger vessels often have several. Aircraft are guided by land-based radar installations, known as airport surveillance radar (ASR), located at civilian and military airfields. An ASR tracks airport positions and weather conditions in the vicinity of the airport.

Radio Astronomy. Most radio astronomy applications operate in the microwave spectrum. Usually, naturally occurring microwave radiation is observed; however, radio astronomy has been used to measure distances precisely within the solar system. Radio astronomy has also been employed to map the surface of Venus, which is not visible via optical telescopes because of its dense cloud cover. The technology has expanded astronomical knowledge and has led to the discovery of new objects, including radio galaxies, pulsars, and quasars. Radio astronomy allows objects that are not detectable with an optical telescope to be seen. These objects are some of the most extreme and energetic physical processes that exist in the universe.

Since microwaves penetrate dust, radio astronomy techniques can study the dust-shrouded environments where stars and planets are born. Radio astronomy is also used to trace the location, density, and motion of the hydrogen gas, which constitutes about 75 percent of the ordinary matter in the universe.

Spectroscopy. Microwave radiation is used for electron paramagnetic resonance, also known as electron spin resonance, analysis, which has been crucial in the development of the most fundamental theories in physics, including quantum mechanics, the special and general theories of relativity, and quantum electrodynamics. Microwave spectroscopy is an essential tool for the development of scientific understanding of electromagnetic and nuclear forces.

Impact on Industry

Microwave technology has had a significant impact on industry. As of 2011, cell phones are widely used; in fact, some individuals rely on a cell phone rather than a landline telephone for communication. Cell phone service providers and manufacturers are currently experiencing significant growth. Wireless microwave applications such as Bluetooth are widespread and increasing in popularity as are GPS devices for military, commercial, and personal use. Radar is an essential navigation tool. Microwave ovens represent a significant segment of the appliance industry. Microwave devices also support a repair industry, and the constantly evolving technology fuels replacement of these devices with an upgraded model often before the device fails.

Regulation. Radio transmission is regulated and closely monitored by government agencies. In the United States, it is regulated by the Federal Communications Commission (FCC). The FCC is responsible for regulating interstate and international communications by radio, television, wire, satellite, and cable in all fifty states, the District of Columbia, and U.S. possessions.

Industry and Business. Although the civilian population is a significant consumer of microwave equipment, commercial and military applications represent a large market share. In the commercial sector, the civilian aircraft and shipping industries are major purchasers of microwave equipment. The National Aeronautics and Space Administration (NASA) requires extremely precise, complex, and expensive navigational and communications equipment for its missions. All branches of the military—Army, Navy, Air Force, Marines, and Coast Guard—use microwave equipment extensively.

Government and University Research. One of the biggest sources of funding for microwave research in the United States is the Defense Advanced Research Projects Agency (DARPA). It is also the biggest client for certain kinds of microwave and navigation applications. The agency is primarily concerned with

Fascinating Facts About Microwave Technology

- Science fiction author Robert Heinlein foresaw the development of the microwave oven in two of his novels. His 1948 novel *Space Cadet* mentioned the uses of "high frequency heating" to prepare food. His 1950 *Farmer in the Sky* described the preparation of a twenty-first century meal: "I grabbed two Syntho-Steaks out of the freezer and slapped them in quickthaw, added a big Idaho baked potato for Dad . . . and stepped up the gain on the quickthaw so that the spuds would be ready when the steaks were."
- In the 1926, Japanese researcher Hidetsugu Yagi and his colleague Shintaro Uda designed a directional array antenna, which they named the Yagi antenna. Although the antenna did not prove to be particularly useful for power transmission, it is used widely throughout the broadcasting and wireless telecommunications industries because of its excellent performance characteristics.
- Radio gained acceptance more rapidly in the United States among amateur radio operators than the general public. In 1913, 322 amateurs were licensed, and by 1917 there were 13,581—primarily boys and young men. Many older individuals considered radio to be a fad. They reasoned that listening to dots and dashes or the occasional experimental broadcast of music or speech over earphones was a worthless endeavor.
- The microwave oven was a by-product of radar research. In 1945, Raytheon engineer Percy Spencer was testing a new vacuum tube called a magnetron and found that a candy bar in his pocket had melted. Deducing that electromagnetic energy was responsible, he placed some popcorn kernels near the tube; the kernels sputtered and popped. His next trial was with an egg, which exploded in short order.

radio systems with a military focus, such as guidance systems. It also conducts research on satellite communication. Another source of funding for radio research is the U.S. Department of Energy (DOE).The DOE is currently focused on energy efficiency and renewable energy. The DARPA and the DOE supply funds to many universities in the United States for microwave research and development. For example, as far back as 1959, DARPA began work with Johns Hopkins Applied Physics Laboratory to develop the first satellite positioning system.

Careers and Course Work

Many technical and nontechnical careers are available in microwave technology. The technical fields require a minimum of a bachelor's degree in engineering or other scientific field; however, many also require a master's or doctorate degree. Course work should include mathematics, engineering, computer science, and robotics. Positions are available in both the government and private sector. The ability to be a team player is often of value for these positions because ongoing research is often a collaborative effort.

Technicians are needed in a variety of fields for equipment repair. Many of these positions require some training beyond high school at a community college or trade school. If a company employs a number of technicians, supervisory positions may be available. Numerous opportunities are available for sales.

Social Context and Future Prospects

Microwave technology is an integral component of everyday life in all but the most primitive societies. In view of the continuous advances in radio technology, including microwave, further advances are extremely likely.

The microwave oven demonstrates the fact that electromagnetic waves are a form of energy, which is capable of heating, and thus, damaging tissue. Microwave ovens are shielded to prevent exposure; however, other microwave devices are not. For example, if one stands near or touches a microwave transmitter, severe burns can result. The heating effect of an electromagnetic wave varies depending on its power and the frequency. The Institute of Electrical and Electronics Engineers (IEEE) as well as many national governments have established safety limits for exposure to various frequencies of electromagnetic energy. Controversy still exists as to whether microwave energy can be harmful to humans and animals. Low levels of microwave radiation have no proven harmful effect. Examples of low-level microwave radiation are the small amount of leakage from a microwave oven and microwave transmission from a cell phone. Some experts express concern that prolonged low-level exposure (for example, holding a cell phone to one's ear for extended periods of time) might be harmful. Countering this concern is research that has involved exposing multiple generations of animals to microwave radiation at the cell phone intensity or higher, and no health issues have been found. High levels are definitely harmful. An example is U.S. military's Active Denial System, which can heat human skin to an intolerable level.

Robin L. Wulffson, M.D., F.A.C.O.G.

Further Reading

Balbi, Amedeo. *The Music of the Big Bang: Cosmic Microwave Background and the New Cosmology.* Berlin: Springer-Verlag, 2010. Focuses on how the exploration of the cosmic background radiation has shaped the picture of the universe, leading even the non-specialized readers toward the frontier of cosmological research, helping them to understand the mechanisms behind the universe.

Hallas, Joel R. *Basic Radio: Understanding the Key Building Blocks.* Newington, Conn.: American Radio Relay League, 2005. An introduction to radio that discusses its components: receivers, transmitters, antennae, propagation, and their applications to telecommunications, radio navigation, and radio location.

Rudel, Anthony. *Hello, Everybody! The Dawn of American Radio.* Orlando, Fla.: Houghton Mifflin Harcourt, 2008. Addresses early radio, covering the entrepreneurs, evangelists, hucksters, and opportunists who exploited the new technology.

Scott, Alan W. *Understanding Microwaves.* Hoboken, N.J.: Wiley-Interscience, 2005. Covers terminology, devices, and systems, and shows how the technology works to make communications, navigation, and radar equipment. Aimed at engineers, technicians, managers, and students.

Web Sites

Institute of Electrical and Electronics Engineers
http://www.ieee.org

National Association of Broadcasters
http://www.nab.org

Society for Applied Spectroscopy
http://www.s-a-s.org

Society of Amateur Radio Astronomers
http://www.radio-astronomy.org

See also: Telecommunications

MUSIC TECHNOLOGY

FIELDS OF STUDY

Acoustic science; computer programming; music theory; analogue signal processing; digital signal processing; MIDI recording; mixing and mastering, music synchronization; composition; music history.

SUMMARY

Music technology is the application of computer software, electronic instruments, and other sound-manipulation equipment in the composition, performance, and recording of music. The boundaries of personal expression through music are expanding with computer-instrument hybrids, such as those based on the piano and the guitar. Software facilitates the composition of scores with automatic notation and immediate playback features. Technology also allows the precise reproduction and efficient storage of music as well as advanced editing, mixing, and mastering capabilities. It even makes music more accessible, such as the creation of ring tones for cellular phones and electronic music effects for video games.

KEY TERMS AND CONCEPTS

- **Acoustic Sound:** Natural sound that is not produced or enhanced electronically.
- **Analogue Signal:** Information is transmitted as a continuous signal that is responsive to change.
- **Digital Signal:** Information is transmitted as a string of binary code without distortion.
- **Electronic Sound:** Sound that is produced or enhanced with technological manipulation, such as with computer software.
- **Equalization:** Boosting or weakening bass or treble frequencies to achieve a desired sound.
- **MIDI:** Musical instrument digital interface.
- **Modulation:** Altering the frequency or amplitude of a sound wave.
- **Oscillator:** Electronic circuit that generates a regularly repeating sound wave.
- **Synthesizer:** Electronic apparatus that combines signals of different frequencies to produce music.
- **Voice:** Musical sound that is produced from thecombined output of multiple oscillators.

DEFINITION AND BASIC PRINCIPLES

Music technology is the integration of personal expression through instrumental and vocal music and cutting-edge applied science that enhances rather than replaces traditional music theory, composition techniques, and philosophy. It also facilitates the invention of novel instrumental and choral sounds and styles of music. Music technology easily incorporates mathematic underpinnings to create synergy in composition and performance among electronic instruments, such as synthesizers, sequencers, and samplers, and classic instruments. Musicians now have access to technical equipment that enables them to make stronger harmonic decisions and more fully manifest their artistic visions. The social culture of music has changed with the advent of personal MP3 players and downloadable open-sharing files; more people have immediate access to new music from around the world than ever before.

Music technology is often confused with sound technology. Sound technology involves the reproduction of sounds with adjustments for volume and clarity, and is applied in the development of hearing aids and acoustic halls. It also differs from audio or recording engineering, which is concerned with the capturing, editing, and storing of digital music files. Music technology takes into account aesthetic composition and innovative performance, as well as reproduction with fidelity.

BACKGROUND AND HISTORY

In 1961, engineering physicist Robert Moog revived the theremin, the archetypal electronic instrument invented by Russian engineer Léon Thérémin in 1920 and patented in 1928. It consists of two loop antennae that control oscillators; one regulates amplitude (volume) and the other regulates frequency (pitch). The antennae sense movements of the performer's hands without actual physical contact to create changes in the electronic signal and the resulting eerie music is amplified and transmitted through speakers. Moog studied the theremin and applied some of its principles to the invention of the Moog synthesizer, introduced in 1964. His invention used electronic circuit boards to alter the sound of music produced by a keyboard. It quickly caught on with the advent of rock and funk music styles; it was used in concerts by the Beatles and the Rolling Stones in the late 1960's. Moog is acknowledged to be an American pioneer in electronic music.

The development of hyperinstruments began in 1986 in the Massachusetts Institute of Technology Media Lab under the direction of renowned composer and professor of media arts and sciences Tod Machover. The goal of the project was to find ways to use technology to increase the power and finesse of musical instruments for superior performers. Since 1992, the focus of this project has shifted toward creating innovative interactive musical instruments for use by nonprofessional musicians and students, thus making music technology more accessible to the general public.

How It Works

Composition. Aesthetic music compositions are typically layers of patterns that can be translated into mathematical expressions. Computer programs are able to perform this translation quickly, and the results can then be applied as objective parameters for developing new musical creations. Software can assist with rapid, clear musical notation and real-time playback during the composing period as well as with subsequent complex harmonious orchestration. Songwriting is similarly facilitated and appropriate vocal keys, harmonies, and arrangements can be easily derived.

Performance. Traditional instruments are enhanced with computer parts that create novel yet related voices. Electronic instruments are also being introduced into performances. A music synthesizer is an electronic keyboard instrument that produces electrical signals of various frequencies to generate new voices, some of which imitate other traditional instruments. Synthesizers operate on programmed algorithms and these synthesis techniques may be additive, subtractive, modulate frequency, or cause phase distortion. A music sequencer is a software program or an electronic device for recording, editing, and transmitting music in a musical instrument digital instrument (MIDI) format. A music sampler is similar to a synthesizer, except that instead of generating music, it plays back preloaded music samples, such as those from a sequencer, to compose as well as perform. Like synthesizers, samplers can modify the original sound signals to generate new voices and many can generate multiple notes simultaneously.

Recording. In the recording process, music may be modified in its analogue form or converted to digital data before alteration. Sound mixing is the process of blending sounds from multiple sources into a desired end product. It often starts with finding a balance between vocal and instrumental music or dissimilar instruments so that one does not overshadow the other. It involves the creation of stereo or surround sound from the placement of sound in the sound field to simulate directionality (left, center, or right). Equalizing adjusts the bass and treble frequency ranges. Effects such as reverberation may be added to create dimension. Auditory electrical signals may then be sent to speakers, where they are converted into acoustical signals to be heard by an audience. Digital signals may be broadcast in real time over the Internet as streaming audio. Otherwise, the processed signals may be stored for future reproduction and distribution. Analogue signals may be stored on magnetic tape. Digital signals may be stored on a compact disc or subjected to MP3 encoding for storage on a computer or personal music player.

Applications and Products

Hyperinstruments. Music technology products include hyperinstruments, a term coined in 1986 for electroacoustic instruments. They were originally intended for virtuosos; customized instruments have been fabricated for Yo-Yo Ma and Prince. However, their accessibility is being expanded for use by nonprofessional musicians and children. The theremin is an archetypal electronic musical instrument that is played without actual physical contact. The performer uses fine motor control to wave his hands in proximity to a pair of antennae to control the pitch and volume produced. The instrument responds to every movement, whether deliberate or unintentional. A chameleon guitar has interchangeable soundboards in the central cavity that allow the same body, neck, and frets to have a familiar feel while producing various voices that imitate other instruments. Hyperpianos yield MIDI data that are augmented as solo music or keyboard accompaniment carefully matched to other instruments and voices. With the hypercello, the bow pressure, string contact, wrist measurements, and neck fingering are measured, the data are processed mathematically, and an enhanced signal is generated. A hyperviolin makes no sound of its own but creates electronic output when it is played with a hyperbow. The speed, force, and position of the hyperbow are measured wirelessly, the resulting data are adjusted, and the sound is appropriately enhanced for consistently intentional expression. The hyperviolin and hyperbow may be seen in performances by virtuoso Joshua Bell.

Innovative Musical Productions. The technology that has emerged from the Massachusetts Institute of Technology Media Lab has been channeled by MIT professor Tod Machover into a new form of opera in the broadest sense, telling a story through music. The first project, *The Brain Opera*, debuted in 1996. Each night's event occurred in two parts. In the first half, the audience is invited to experiment with electronic instruments such as the Rhythm Tree (by punching a node on the tree, a thump comes from a nearby speaker), Gesture Walls (advanced theremins), and Harmonic Driving devices (video games that people drive to make musical choices). In the second half, the music created in the first half is incorporated into a multimedia presentation for a unique show every night. Its story is the psychological exploration of how we think about music. The project toured worldwide, finding a permanent home in Vienna in July 2000. The second and most recent project, *Death and the Powers*, debuted in September 2010. This one-act opera tells the story of an inventor, Simon Powers, who builds an electronic system into which he downloads his memories and humanity, which is then expressed by robots, animated bookcases, and a musical chandelier. Typically inanimate objects become personified with electronic music and voices in what Machover calls "disembodied performance." Music technology is responsible for connecting audiences with the emotion of the story.

Fascinating Facts About Music Technology

- The Institut de Recherche et Coordination Acoustique/Musique (IRCAM) in Paris was founded to study the intersection of music and technology. Its mission is to assist composers in manifesting their artistic vision through technology.
- Music technology has been used to create a virtual castrato, reproducing the distinctive high-pitch quality of castrated boy singers in eighteenth century Europe. After hearing a careful mix of a soprano and a countertenor, even the singers could not discern their own voices.
- Music technology can serve corporations in producing subtly effective advertising. The automobile manufacturer Renault once sought expertise from a music technology institute to fine-tune the musicality of car doors closing in a commercial to support its marketing image.
- Barcodas, an iPhone app created by Dutch media artist Leo van der Veen, allows the user to scan any standard Universal Product Code bar code and translate it into a musical phrase. While entertaining, it is primarily intended as an electronic muse for composing.
- The audience-interactive musical production *The Brain Opera*, created by MIT Media Lab professor Tod Machover, was the technological forerunner of the *Guitar Hero* and *Rock Band* video games.
- The first commercially available digital music was the ABBA CD *The Visitors* in 1982.
- Record albums, even in a digital format, may be becoming obsolete. In 2009, digital single downloads outsold physical and digital albums by 250 percent.

Impact on Industry

International Institutes. The International Federation of the Phonographic Industry (IFPI), with headquarters in London and regional offices in Brussels, Hong Kong, Moscow, Beijing, Zurich, and Miami, unites 1,400 members in sixty-six countries with affiliated industry associations in forty-five countries to uphold its mission to "promote the value of recorded music, safeguard the rights of record

Pro Tools digital audio workstation platform; Propellerhead, the Swedish developer and manufacturer of Record and Reason music software applications; and Apple, with products such as iPod and iTunes. Leading instrument manufacturers include Yamaha and Korg. The major music labels are Sony Music Entertainment, Warner Music Group, EMI Music, and Universal Music Group.

Careers and Course Work

In the United States, music technology degrees are offered at such prestigious schools as New York

University and the Juilliard School in New York City and Massachusetts Institute of Technology in Cambridge. These degrees are also offered at universities across the country, many located in metropolitan areas, including Wayne State University in Detroit and Northwestern University in Evanston, Illinois.

The Cork Institute of Technology, Cork School of Music, in Cork, Ireland, offers a master's degree in music and technology as of 2009.

Students of music technology may pursue various levels of education. Although some associate's and doctoral degree programs are available, most are bachelor's and master's degree programs. Admission into these programs often requires the ability to read music, the ability to play a traditional instrument (piano is preferred and voice may be considered), some experience with computer hardware and software, and some experience with music recording. Nearly all programs offer courses in music history, traditional music theory, contemporary music theory, music appreciation and critical evaluation, analogue and digital signal processing, recording, mixing, mastering, synchronization, songwriting, orchestration and ensemble performance, and music business. Many colleges and universities arrange internships with local recording studios and radio stations to provide practical application.

Graduates may pursue many career paths in music technology. One is studio production, engineering, and recording for film, television, radio, video, Internet, and record labels. Another is composing, arranging, and orchestration for film, television, theater, church, advertising, and video games. Some choose to perform in live music venues. Some work in education and research, studying music design based on how the mind interprets pitch, rhythm, melody, and harmony. Similarly, some become music critics who must stay abreast of the current trends in both popular music and advancing technology. Another option is multimedia collaboration, integrating audio and video. This collaboration extends to Web design, digital video editing, and interactive media. A similar option is equipment design, creating the next generations of synthesizers, sequencers, and samplers. Another popular career path is music business and administration involving sales, marketing, and management for music retailers, production companies, and record labels.

Social Context and Future Prospects

The Recording Industry Association of America estimates that recorded music is a $10.4 billion industry in the United States. In 2008, the United States Bureau of Labor Statistics reported that there were 53,600 music directors, arrangers, and composers (with a projected 10 percent increase in the subsequent decade), 114,600 broadcasting and sound engineering professionals (with a projected 8 percent increase in the subsequent decade), and 186,400 commercial musicians and singers (with a projected 8 percent increase in the subsequent decade).

Music technology is going in several exciting and novel directions beyond those of audio engineering. While audio engineering is developing music information retrieval systems, advanced digital signal processing, and mobile recording studios, music technology is pursuing more intimate avenues that will strengthen the bond between the artists and the audience.

One such area is music cognition. Technology is developing the capability to interpret and map expressive gestures made by professional musicians to personify electronic music. Similarly, researchers are mapping the brain's responses to different pieces of music to predict the effect of new compositions on an audience. Conversely, bioengineers are working to make brain waves audible so that happy, sad, fearful, or angry thoughts can be heard as musical patterns and subsequently evaluated by mental health professionals.

Another emerging area is interactive music systems, which take the audience from listening passively to thinking actively about how they react to music as they create it by making choices. This ranges from live musical performances, in which the audience is deliberately invited to participate, to blind research in which musical stairs are placed in proximity to escalators, and observers measure the rate at which pedestrians progressively choose to take the stairs, thus getting more exercise for the reward of music. Interactive music systems overall make music an inclusive experience.

Bethany Thivierge, B.S., M.P.H.

Further Reading

Ballora, Mark. *Essentials of Music Technology.* Upper Saddle River, N.J.: Prentice Hall, 2003. This book discusses five broad areas: the sound of music, computer software and hardware, MIDI, digital audio signals, and additional equipment.

Brown, Andrew R. *Computers in Music Education: Amplifying Musicality.* New York: Routledge, 2007. This book presents an easily understood overview of music technology to teachers and students, including sections on notation software, MIDI files, and downloading music.

Hosken, Dan. *An Introduction to Music Technology.* New York: Routledge, 2011. This covers the basics for music students to enhance their composition and performance with computer hardware and software.

Katz, Mark. *Capturing Sound: How Technology Has Changed Music.* Rev. ed. Berkeley: University of California Press, 2010. Describes how technology "has changed the way we listen to, perform, and compose music."

Middleton, Paul, and Steven Gurevitz. *Music Technology Workbook: Key Concepts and Practical Projects.* Burlington, Mass.: Focal Press, 2008. This workbook offers commonly encountered problems and their practical solutions, while discussing a variety of relevant software programs.

Roads, Curtis. *The Computer Music Tutorial.* Cambridge, Mass.: MIT Press, 1996. A comprehensive reference written by an acknowledged expert and respected professor of music technology.

Williams, David Brian, and Peter Richard Webster. *Experiencing Music Technology.* 3d ed. Belmont, Calif.: Cengage Learning, 2008. This paperback comes with a DVD and presents a complete overview of this rapidly changing field.

Web Sites

Institut de Recherche et Coordination Acoustique/Musique
http://www.ircam.fr/?L=1

International Federation of the Phonographic Industry (IFPI)
http://www.ifpi.org

MIDI Manufacturers Association
http://www.midi.org

Recording Industry Association of America
http://www.riaa.com

N

NANOTECHNOLOGY

FIELDS OF STUDY

Agriculture; artificial intelligence; bioinformatics; biomedical nanotechnology; business; chemical engineering; chemistry; computational nanotechnology; electronics; engineering; environmental studies; mathematics; mechanical engineering; medicine; molecular biology; microelectromechanical systems; molecular scale manufacturing; nanobiotechnology; nanoelectronics; nanofabrication; molecular nanoscience; molecular nanotechnology; nanomedicines; pharmacy; physics; toxicology.

SUMMARY

Nanotechnology is dedicated to the study and manipulation of structures at the extremely small nano level. The technology focuses on how particles of a substance at a nanoscale behave differently than particles at a larger scale. Nanotechnology explores how those differences can benefit applications in a variety of fields. In medicine, nanomaterials can be used to deliver drugs to targeted areas of the body needing treatment. Environmental scientists can use nanoparticles to target and eliminate pollutants in the water and air. Microprocessors and consumer products will also benefit from the use of nanotechnology, as components, and associated products, become exponentially smaller.

KEY TERMS AND CONCEPTS

- **Bottom-Up Nanofabrication:** The creation of nanoparticles that will combine to create nanostructures that meet the requirements of the application.
- **Buckyballs (Fullerenes):** Molecules made up of carbon atoms that make a soccer ball shape of hexagons and pentagons on the surface. Named for the creator of the geodesic dome, Buckminster Fuller.
- **Carbon Nanotubes:** Molecules made up of carbon atoms arranged in hexagonal patterns on the surface of cylinders.
- **Mechanosynthesis:** Placing atoms or molecules in specific locations to build structures with covalent bonds.
- **Molecular Electronics:** Electronics that depend upon or use the molecular organization of space.
- **Moore's Law:** Transistor density on integrated circuits doubles about every two years. American chemist Gordon Moore attributed this effect to shrinking chip size.
- **Nano:** Prefix in the metric system that means "10^{-9}" or "1 billionth." The word comes from the Greek *nanos*, meaning "dwarf." The symbol is *n*.
- **Nanoelectronics:** Electronic devices that include components that are less than 100 nanometers (nm) in size.
- **Nanoionics:** Materials and devices that depend upon ion transport and chemical changes at the nanoscale.
- **Nanolithography:** Printing nanoscale patterns on a surface.
- **Nanometer:** 10^{-9} or 1 billionth of a meter. The symbol is *nm*.
- **Nanorobotics:** Theoretical construction and use of nano-scaled robots.
- **Nanosensors:** Sensors that use nanoscale materials to detect biological or chemical molecules.
- **Scanning Probe Microscopy:** The use of a fine probe to scan a surface and create an image at the nanoscale.
- **Scanning Tunneling Microscopy (STM):** Device that creates images of molecules and atoms on conductive surfaces.
- **Self-Assembly:** Related to bottom-up nanofabrication. A technique that causes functionalized nanoparticles to chemically bond with or repel other atoms or molecules to assemble in specified patterns without external manipulation.
- **Top-Down Nanofabrication:** Manufacturing technique that removes portions of larger materials to create nano-sized materials.

Definition and Basic Principles

Nanotechnology is the science that deals with the study and manipulation of structures at the nano level. At the nano level, things are measured in nanometers, or 1 billionth of a meter (10^{-9}). The symbol for nanometers is *nm*. Nanoparticles can be produced using a process known as top-down nanofabrication, which starts with a larger quantity of material and removes portions to create the nano-sized material. Another method being developed is bottom-up nanofabrication, in which nanoparticles will create themselves when the necessary materials are placed in contact with one another.

Nanotechnology is based upon the discovery that materials behave differently at the nano scale, less than 100 nm in size, than they do at slightly larger scale. For instance, gold is classified as an inert material because it neither corrodes nor tarnishes. However, at the nano level, gold will oxidize in carbon monoxide. It will also appear as colors other than the yellow for which it is known.

Nanotechnology is not simply about working with materials like gold at the nano level. It is about taking advantage of these differences at the nanoscale to create markers and other new structures that are of use in a wide variety of medical and other applications.

Background and History

In 1931, German scientists Ernst Ruska and Max Knoll built the first transmission electron microscope (TEM). Capable of magnifying objects by a factor of up to one million, the TEM made it possible to see things at the molecular level. The TEM was used to study the proteins that make up the human body. It was also used to study metals. The TEM made it possible to view these particles smaller than 200 nm by focusing a beam of electrons to pass through an object, rather than focusing light on an object, as was the case with traditional microscopes.

In 1959 the noted American theoretical physicist Richard Feynman brought nanoscale possibilities to the forefront with his talk "There's Plenty of Room at the Bottom," presented at the California Institute of Technology in 1959. In this talk, he asked the audience to consider what would happen if they could arrange individual atoms, and he included a discussion of the scaling issues that would arise. It is generally considered that Feynman's reputation and influence brought increased attention to the possible uses of structures at the atomic level.

In the 1970s scientists worked with nanoscale materials to create technology for space colonies. In 1974 Tokyo Science University professor Norio Taniguchi coined the term "nano-technology." As he defined it, nanotechnology would be a manufacturing process whose materials were built by atoms or molecules.

In the 1980s the invention of the scanning tunneling microscope (STM) led to the discovery of diffullerenes in 1986. The carbon nanotube was discovered a few years later. In 1986, Eric Drexler's seminal work on nanotechnology, *Engines of Creation*, was published. In this work, Drexler used the term "nanotechnology" to describe a process that is now understood to be molecular nanotechnology. Drexler's book explores the positive and negative consequences of being able to manipulate the structure of matter. Included in his book are ruminations on a time when all the works in the Library of Congress can fit on a sugar cube and when nanoscale robots and scrubbers can clear capillaries or whisk pollutants from the air. Controversy continues as to whether the view of a world with nanotechnology envisioned by Drexler is even attainable.

In 2000 the U.S. National Nanotechnology Initiative was founded. Its mandate is to coordinate federal nanotech research and development. Great growth in the creation of improved products using nanoparticles has taken place since that time. The creation of smaller and smaller components—which reduces all aspects of manufacture, from the amount of materials needed to the cost of shipping the finished product—is driving the use of nanoscale materials in the manufacturing sector. Furthermore, the ability to target delivery of treatments to areas of the body needing those treatments is spurring research in the medical field.

The true promise of nanotechnology is not yet known, but this multidisciplinary science is widely viewed as one that will alter the landscape of fields from manufacturing to medicine.

How It Works

Basic Tools. Nanoscale materials can be created for specific purposes, but there exists also natural nanoscale material, like smoke from fire. To create nanoscale material and to be able to work with it requires specialized tools and technology. One essential piece of equipment is an electron microscope. Electron microscopy makes use of electrons, rather than light, to view objects. Because these microscopes have to get the electrons moving, and because they need several thousand volts of electricity, they are often quite large.

One type of electron microscope, the scanning electron microscope (SEM), requires a metallic sample. If the sample is not metallic, it is coated with gold. The SEM can give an accurate image with good resolution at sizes as small as a few nanometers.

For smaller objects or closer viewing, a transmission electron microscope (TEM) is more appropriate. With a TEM, the electrons pass through the object. To accomplish this, the sample has to be very thin, and preparing the sample is time-consuming. The TEM also has greater power needs than the SEM, so that in most cases the SEM is used, reserving the TEM for times when a resolution of a few tenths of a nanometer is absolutely necessary.

The atomic force microscope (AFM) is a third type of electron microscope. Designed to give a clear image of the surface of a sample, this microscope uses a laser to scan across the surface. The result is an image that shows the surface of the object, making visible the object's "peaks and valleys."

Moving the actual atoms around is an important part of creating nanoscale materials for specific purposes. Another type of electron microscope, the scanning tunneling microscope (STM), not only images the surface of a material in the same way as the AFM; the tip of the probe, which is typically made up of a single atom, also can be used to pass an electrical current to the sample. This charge lessens the space between the probe and the sample. As the probe moves across the sample, the atoms nearest the charged atom move with the charged atom. In this way, individual atoms can be moved to a desired location in a process known as quantum mechanical tunneling.

Molecular assemblers and nanorobots are two other potential tools. The assemblers would use specialized tips to form bonds with materials that would make specific types of materials easier to move. Nanorobots might someday move through a person's blood stream or through the atmosphere, equipped with nanoscale processors and other materials that enable them to perform specific functions.

Bottom-Up Nanofabrication. Bottom-up nanofabrication is one approach to nanomanufacturing. This process builds a specific nanostructure or material by combining components of atomic and molecular scale. Creating a structure this way is time-consuming, so scientists are working to create nanoscale materials that will spontaneously join to assemble a desired structure without physical manipulation.

Top-Down Nanofabrication. Top-down nanofabrication is a process in which a larger amount of material is used at the start. The desired nanomaterial is created by removing, or carving away, the material that is not needed. This is less time-consuming than bottom-up nanofabrication, but it produces considerable waste.

Specialized Processes. To facilitate the manufacture of nanoscale materials, a number of specialized processes are used. They include nanoimprint lithography, in which nanoscale features are stamped or printed onto a surface; atomic layer epitaxy, in which a layer that is only one atom thick is deposited on a surface; and dip pen lithography, in which the tip of an atomic force microscope writes on a surface after being dipped into a chemical.

Applications and Products

Smart Materials. Smart materials are materials that react in ways appropriate to the stimulus or situation they encounter. Combining smart materials with nanoscale materials would, for example, enable scientists to create drugs that would respond when encountering specific viruses or diseases. They could also be used to signal problems with other systems, like nuclear power generators or pollution levels.

Sensors. The difference between a smart material and a sensor is that the smart material will generate a response to the situation encountered and the sensor will generate an alarm or signal that there is something that requires attention. The capacity to incorporate sensors at a nanoscale will greatly enhance the ability of engineers and manufacturers to create structures and products with a feedback loop that is not cumbersome. Nanoscale materials can easily be incorporated into the product.

Medical Uses. The potential use for nanoscale materials in the field of medicine is one that is of particular interest to researchers. Theoretically, nanorobots could be programmed to perform functions that would eliminate the possibility of infection at a wound site. They could also speed healing. Smart materials could be designed to dispense medication

Fascinating Facts About Nanotechnology

- Approximately 80,000 nanos equal the width of one strand of human hair.
- Gold is considered an inert material because it does not corrode or tarnish. At the nano level, this is not the case, as gold will oxidize in carbon monoxide.
- The transmission electron microscope (TEM) allowed the first look at nanoparticles in 1931. The TEM was built by German scientists Ernst Ruska and Max Knoll.
- German physicist Gerd Binning and Swiss physicist Heinrich Rohrer, colleagues at IBM Zurich Research Laboratory, invented the scanning tunneling microscope in 1981. The two scientists were awarded the Nobel Prize in Physics in 1986 for their invention.
- Nanoparticles can be enclosed in an outer covering or coat.

through, but others, such as mold or moisture, cannot. With this level of control, living conditions can be designed to meet the specific needs of different categories of residents.

Extending the life of batteries and prolonging their charge has been the subject of decades of research. With nanoparticles, researchers at Rutgers University and Bell Labs have been able to better separate the chemical components of batteries, resulting in longer battery life. With further nanoscale research, it may be possible to alter the internal composition of batteries to achieve even greater performance.

Light-emitting diode (LED) technology uses 90 percent less energy than conventional, non-LED, lighting. It also generates less heat than traditional metal filament light bulbs. Nanomanufacture would make it possible to create a new generation of efficient LED lighting products.

Electronics. Moore's law states that transistor density on integrated circuits doubles about every two years. With the advent of nanotechnology, the rate of miniaturization has the potential to double at a much greater rate. This miniaturization will profoundly affect the computer industry. Computers will become lighter and smaller as nanoparticles are used to increase everything from screen resolution and battery life while reducing the size of essential internal components, such as capacitors.

Impact on Industry

Nanotechnology will have a profound impact on industry. Reducing the size of components to nanoscale will mean that the internal circuits of a computer will be etched on a surface too small to be seen by the naked eye. With circuit boards and other components reduced in size, finished products will be similarly reduced in scale.

Industries such as the automotive industry, which must have products of a certain size to function, also will take advantage of nanoparticles and materials. Their use will result in components that weigh less and are stronger because of the manufacturing use of structures such as the carbon nanotube.

The use of lighter coatings of materials on the components of vehicles destined for space will result in significantly lighter loads and reduced fuel requirements. Even space suits will be improved by the use of nanoparticles in their manufacture, as layers of bulky material are replaced with trimmer fabrics providing the same or better protection.

In the medical field, experimentation is already underway to deliver medication to tumors in places that are not readily accessible through conventional means. Researchers are also experimenting with coatings to protect tissue as medication moves through the human body and with markers to guide medication to the proper treatment site.

Governments and universities around the world have shown a strong interest in nanotechnology. Alliances are being formed to create solutions to problems such as air and water pollution. One such application under development is the use of nanostructures to create products that will clean oil spills before widespread damage is done to wildlife and their habitats. The National Nanotechnology Initiative is involved in education and research. To support its project, the organization's Web site clearly explains nanotechnology. The site also includes discussion of the expected benefits of nanotechnology. The initiative is also involved in setting standards for nanotechnology and is legislating for policies and regulations for this new branch of science.

The private sector is also involved in

nanotechnology-related research, as individuals and private companies work to develop ideas and products they can patent and produce. For example, major corporations are investigating the use of nanoparticles such as carbon nanotubes to strengthen the materials they use in the components that make up cars and airplanes.

Nanotechnology is already a technology with broad applications for the business community. Commercialization of research findings is a priority. The funding from the commercialization will pay for future research as consumers take advantage of the positive effects of nanotechnology. Because nanotechnology is rising to prominence during the information age, which includes the Internet and web, research findings and writings are readily available. Another factor advancing progress in this field is the global reach of research and information dissemination.

Careers and Coursework

Engineering and architecture, manufacturing, and the health care industry are just a few of the career fields that will be touched by advances in nanotechnology and its applications.

For those pursuing a career in engineering, nanoscale materials will result in stronger, lighter materials that can be created specifically for their function. This will result in lighter, more durable airplanes, ships, spacecraft, and other vehicles. Nanotechnology will also be a factor in the selection of materials used by civil engineers for roads, bridges, and dams. Architects will have a new range of materials available as they design homes and commercial buildings.

In manufacturing, advances in the use of nanotechnology will result in stronger materials of lighter weight that are better suited for their purpose. Products that now must be of a certain size to accommodate internal components will not have this constraint in the future. In a move analogous to the switch from televisions with cathode ray tubes to those using liquid crystal or plasma displays, advances in the production of nanocomponents will alter design constraints on a variety of products. Those pursuing manufacturing careers will need to stay abreast of those changes to produce products at the greatest level of efficiency and at the lowest cost.

Professionals in the health industry, from personal trainers to surgeons, will feel the effect of advances in nanotechnology. From the use of nanomaterials to speed patient recovery to the use of nanostructures in prosthetics and related aids, nanotechnology will change current practices at a rapid pace.

Because nanotechnology is multidisciplinary, coursework applying to nanotechnology can be found in a variety of disciplines. Because nanotechnology is such a new field, much research is reported in journals and on reputable Web sites dedicated to the field. Keeping informed of developments and advances will benefit those seeking careers that touch on the application of nanotechnology.

Social Context and Future Prospects

Whether nanotechnology will be good or bad for the human race remains to be seen. There is tremendous potential associated with the ability to manipulate individual atoms and molecules, to deliver medications to a disease site, and to build products such as cars that are lighter yet stronger than ever. There also exists the persistent worry that humans will lose control of this technology and face what one writer called "gray goo," a scenario in which self-replicating nanorobots take over and ultimately destroy the world.

Gina Hagler, M.B.A.

Further Reading

Drexler, K. E. *Engines of Creation.* London: Fourth Estate, 1990.

Ratner, D., and M. A. Ratner. *Nanotechnology and Homeland Security: New Weapons for New Wars.* Upper Saddle River, N.J.: Prentice Hall/PTR, 2004.

Ratner, M. A., and D. Ratner. *Nanotechnology: A Gentle Introduction to the Next Big Idea.* Upper Saddle River, N.J.: Prentice Hall, 2003.

Web Sites

Arizona State University, Arizona Initiative for Nano-Electronics
http://www.asu.edu/aine/app_nanoionics.htm

Carnegie Mellon NanoRobotics Lab
http://nanolab.me.cmu.edu/

NanoScience Instruments
http://www.nanoscience.com/index.html

Nanotechnology Research Foundation
http://www.nanotechnologyresearchfoundation.org/

National Nanotechnology Initiative
http://www.nano.gov/

University of Illinois at Urbana-Champaign Imaging Technology Group
http://virtual.itg.uiuc.edu/training/AFM_tutorial/

See also: Liquid Crystal Technology; Microlithography and Nanolithography; Transistor Technologies.

NAVIGATION

FIELDS OF STUDY

Aeronautics; aviation; cartography; celestial navigation; electronic navigation systems; geodesy; geoinformatics; space exploration; ship piloting.

SUMMARY

Navigation is the process of plotting an object from one point to another. This is accomplished by determining one's exact position on Earth by visual or electronic means and determining the course to the exact position of the intended destination. Specific points are often defined as measurements of longitude and latitude.

KEY TERMS AND CONCEPTS

- **Celestial Navigation:** Method of locating one's position by the angular measurement between celestial bodies (the Sun, Moon, stars, or planets).
- **Global Positioning System (GPS):** Space-based global navigation satellite system that provides accurate location and time information for any point on Earth where there is an unobstructed line of sight to four or more GPS satellites.
- **Gyrocompass:** Compass that points true north (rather than magnetic north) by using a fast-spinning gyroscope. Unlike a magnetic compass, a gyrocompass is unaffected by magnetic fields.
- **Latitude:** Angular measurement (degrees, minutes, and seconds) of a location on the Earth north or south of the equator. The North Pole has a latitude of 90 degree north, the South Pole has a latitude of 90 degrees south, and the equator has a latitude of 0 degree.
- **Longitude:** Angular measurement (degrees, minutes, and seconds) of a location relative to the prime meridian, which runs from the North to the South Pole through the original site of the Royal Observatory in Greenwich, England. Measurements range from 180 degrees east to 180 degrees west.
- **Magnetic Compass:** Device that contains a magnetized needle that points to magnetic north.
- **Radar:** Acronym for radio detecting and ranging; a system that emits a narrow beam of extremely high-frequency pulses that are reflected back from objects (buoys, ships, and landmasses). A radar screen updates in real time, yielding the speed and direction of moving objects.
- **Triangulation:** Determining the position of an object by calculation from two known quantities; the distance between two fixed points and the angles formed between the line described by those points and the line between a third point.

Definition and Basic Principles

Location of one's position is based on trigonometry and the process of triangulation. Triangulation involves determining the location of a point by measuring angles to it from two or more known points. The intersection of the lines from the known points represents one's position.

Prior to the development of electronic instruments such as radio, radar, and Global Positioning System (GPS), the determination of one's position was accomplished using a sextant. The sextant was able to pinpoint one's location accurately by measuring the angular distance between the horizon and a celestial object. Repeated sightings were plotted on a nautical chart. This method was suitable for ships and propeller aircraft.

After the development of the radio, a position could be determined by triangulating radio stations with known positions. With the development of radar, images of known landmasses or buoys could be used for triangulation.

GPS also involves triangulation. The device electronically determines one's position by triangulation of orbiting satellites and marks the position on a map overlay.

Navigation also involves correction for ocean currents and obstacles such as landmasses and other vessels. Experienced seamen past and present, aided by experience, can predict current with reasonable accuracy. Wind affects currents, which can be estimated by the sailor or measured with an anemometer (wind gauge). Airplanes are affected by currents known as jet streams. Spacecraft are affected by gravitational forces, which increase in power as the craft approaches. That force can be used to change the direction of the craft substantially.

Background and History

Human migration and discovering new lands by navigation of the oceans was accomplished by many ancient cultures, including the Phoenicians, ancient Greeks, the Norse, the Persians, and the Polynesians. Primitive navigation depended on knowledge of ocean currents and the position of celestial objects.

The magnetic compass was first used in China around 200 B.C.E. The ancient compasses used a naturally magnetic lodestone, which pointed to the magnetic North Pole. In eighth-century China, the lodestone was replaced by a magnetized needle.

Navigation charts first appeared in Italy at the close of the thirteenth century. Called portolan charts, they were rough maps based on accounts of sailors plying the coastlines of the Mediterranean and Black seas. The octant, precursor to the sextant and developed around 1730, made it possible to determine latitude accurately but not longitude. The sextant added the ability to determine longitude. At the close of the nineteenth century, radios that transmitted and received Morse code began to appear on oceangoing vessels.

The prototype of the modern radar was installed on the USS *Leary* in 1937. In 1942, the first long-range navigation system (loran) was installed at various points along the Atlantic coast of the United States. Loran is based on using the intersection of two radio waves to determine one's position. The subsequent proliferation of satellites, after the Soviet Union launched Sputnik, the first Earth-orbiting satellite, in 1957, led to the development of satellite technology and the highly accurate GPS.

How It Works

Compasses. The magnetic compass needle points to magnetic north and floats over a 360-degree compass face, which displays the cardinal points: north (0 degree), east (90 degrees), south (180 degrees), and west (270 degrees). The deviation (declination) of magnetic north is not a constant and is updated annually. Nautical charts display the amount of needed correction from the baseline, which is the declination value of the year the chart was published. The magnetic compass is a simple, reliable instrument; however, its readings can be affected by any magnetic field in the vicinity. Marine electronics located near the compass generate magnetic fields; ferrous (iron) material often has a magnetic field. To improve accuracy, the compass can be calibrated by small magnets or other devices to correct for extraneous magnetic fields. The compass must be recalibrated with the addition of any new electronics. A gyrocompass points to true north and is unaffected by stray magnetic fields. The gyrocompass employs a rapidly spinning gyroscope, which is motor driven. A disadvantage of the gyrocompass is that electrical or mechanical failure can render it useless.

Loran. Loran is an acronym for long-range navigation. The system relies on land-based low-frequency radio transmitters. The device calculates a ship's position by the time difference between the receipt of signals from two radio transmitters. The device can display a line of position, which can be plotted on a nautical chart. Most lorans convert the data into longitude and latitude. Since GPS became available, the use of loran has markedly declined.

Radar. Radar consists of a transmitter and receiver. The transmitter emits radio waves, which are deflected from a fixed or moving object. The receiver, which can be a dish or an antenna, receives the wave. Radar circuitry then displays an image of the object in real time. The screen displays the distance of the object from the radar. If the object is moving, consecutive readings can calculate the speed and direction of the object. If the object is airborne, and the radio is so equipped, the altitude is displayed. Radar is invaluable in foggy weather when visibility can be severely reduced.

Sonar. Sonar, which is an acronym for sound navigation ranging, transmits and receives sound waves for underwater navigation by submarines. Passive sonar is a related technology in which the equipment merely listens for underwater sound made by vessels. Active sonar emits a pulse of sound (a "ping") then listens for a reflection or echo of the pulse. The distance of the object is determined by the time difference from transmission to reception of the ping (the speed of sound in water is a constant). To measure the bearing, two or more separated transmitter/receivers are used for triangulation of the object. Sonar can also be used by both submarines and surface vessels to determine the depth and contour of the ocean (or other body of water). By consulting a nautical chart, which is marked with depth gradients, the distance from the shoreline can be calculated.

Global Positioning System (GPS). GPS is a space-based global navigation satellite system that provides accurate location and time information for any point on Earth where there is an unobstructed line of sight to four or more GPS satellites. GPS can function under any weather condition anywhere on the planet. It cannot function underwater—a submarine must surface to use a GPS system. The technology depends on triangulation, just as a land-based system such as loran employs. GPS is composed of three segments: the space segment, the control segment, and the user segment. The U.S. Air Force operates and maintains both the space and control segments. The space segment is made up of satellites, which are in medium-space orbit. The satellites broadcast signals from space, and a GPS receiver (user segment) uses these signals to calculate a three-dimensional location (latitude, longitude, and altitude). The signal transmits the time, accurate within nanoseconds.

Navigational Aids. According to the U.S. Coast Guard, a navigational aid is a device external to a vessel or aircraft specifically intended to assist navigators in determining their position or safe course, or to warn them of dangers or obstructions to navigation. Navigational aids include buoys, lighthouses, fog signals, and day beacons. Buoys are used worldwide to mark nautical channels. They are color coded: Red buoys mark the right (starboard) side of the channel for a vessel returning to a port; green buoys mark the port (left) side of the channel; and red-and-green-striped buoys mark the junction of two channels. Buoys often contain gongs, which are activated by wave motion and solar-powered lights. Some buoys do not mark safe channels and are referred to as "nonlateral markers." They contain shapes identifying their purpose. Squares depict information, such as food, fuel, and repairs. Diamonds warn of dangers such as rocks. Circles mark areas of reduced speed. Crossed diamonds indicate off-limits areas such as dams and places where people may be swimming.

Applications and Products

Navigational aids have many applications, which are tailored to underwater, water surface, terrestrial, atmospheric, and space use.

Water-Surface Navigation. Any vessel, which is capable of venturing offshore, must contain some basic navigation equipment. A magnetic compass is mandatory, as is a marine radio, which can be used to call for help. A knot meter, which consists of an underwater paddle wheel connected to a dial, can give the relative speed over water. The vessel should contain nautical charts of the area, which are prepared by the National Oceanic and Atmospheric Administration (NOAA). Another essential is an accurate timepiece (chronometer). By combining the readings of speed registered on the knot meter, the compass heading, and chronometer, one can plot the vessel's approximate position on the nautical chart. All inboard marine engines are equipped (or can be equipped) with a tachometer, which registers engine speed in revolutions per minute (rpm). A prudent skipper can run his boat over a known distance (between measured mile markers) and construct a chart of the vessel's speed at various rpm. If the vessel does not have a knot meter, or it fails, the approximate boat speed can be determined via the tachometer. Another relatively inexpensive piece of equipment is a simple depth sounder, which displays depth (more expensive models display a graphic representation of the seabed). The depth sounder can aid in determining the vessel's position, and more importantly, alert the skipper to shallow water, which could damage the hull. Binoculars, which can help identify shoreline features, are another inexpensive necessity. The sextant, which can pinpoint one's position, is another inexpensive navigation aid; however, most recreational boaters are unfamiliar with its use.

Many recreational vessels of modest size (thirty-two to forty-two feet long) have an automatic pilot, radar, and GPS in addition to the aforementioned items. Some utilize a loran receiver, which may be less expensive than a GPS, and it can accurately display the ship's position. An automatic pilot adjusts the rudder(s) to maintain a desired heading. Radar can be adjusted for the distance it displays; ranges of less than one mile and more than thirty miles are commonplace. In many situations, the shortest range is the most valuable because it can clearly depict an approaching vessel or a small landmass (an islet or an offshore oil rig). Even in clear weather, radar can provide invaluable navigation information. Ship-mounted GPS devices can be purchased and installed for as little as $1,000 and are becoming commonplace on recreational vessels.

Large commercial and naval vessels (and many luxury yachts) contain sophisticated navigation equipment. Large vessels also boast onboard computers, which interpret data from radar, GPS, and the tachometers to display detailed navigation information. These

Fascinating Facts About Navigation

- About 1,500 years ago, Tahitians began navigating back and forth between Tahiti and Hawaii in large, double-hulled canoes without the aid of a compass. Their techniques are not completely understood because they did not have a written language; however, their skills were based primarily on celestial navigation augmented by wave patterns, bird sightings, and even the light from active volcanoes on the Hawaiian Islands.
- At the time of his 1492 sailing, Christopher Columbus's maps showed that Japan was 2,700 miles from Europe on the other side of the Atlantic Ocean. Japan was actually 12,200 miles to the west, and North and South America are in the way.
- Native Americans were referred to as "Indians" because when Columbus first landed on the North American continent he thought that he had come ashore in India.
- In early 1942, long before laser guidance systems appeared, behavioral psychologist B. F. Skinner at the University of Minnesota investigated weapons guidance by trained pigeons. The birds were trained with slides of aerial photographs of the target, placed in a harness inside the guidance system, and kept the target in the crosshairs by pecking. Skinner never overcame official skepticism, and the project was eventually abandoned.
- The Mars Climate Orbiter, launched in December 1998, crashed on Mars nine months later because of a computational error. The orbiter's approach altitude to the planet was set at 160 kilometers, when, in fact, it was just under 60 kilometers. This is an example of human, not computer, error.

computers also replace the automatic pilot found on smaller vessels. The computer can adjust the rudders and throttles to maintain the ship's heading and time of arrival. Depth sounders with graphic displays and sonar are components of most large vessels as well as fishing boats, which use them for locating schools of fish.

Underwater Navigation. Unlike surface vessels, submarines must navigate in three dimensions. Active sonar is employed by distance to a target and depth. Transmission of sound from two separate sources can give the submarine's heading. Submarines are equipped with GPS, which can be used to determine position only when the vessel is on the surface. When below the surface, an inertial navigation system (INS) is employed. The INS is composed of precise accelerometers and gyroscopes to record every change in the submarine's speed and direction. The data from these instruments is fed to a computer, which determines the vessel's position. Over time, small errors accrue; these errors are corrected when the submarine surfaces and the GPS is activated.

Terrestrial Navigation. When traveling on a road, a motorist can navigate using street and highway signs as well as a map of the area. Many motor vehicles are equipped with a GPS. The GPS gives a visual display of the vehicle's position on a map overlay. Useful information such as distance to the next turn and the destination are given. The GPS in most automobiles has an audio system, which informs the motorist of upcoming turns. The GPS of some vehicles also gives additional information such as gasoline stations, rest areas, local restaurants, and hospitals. GPS units for off-road use are also available. The motorist on a budget can purchase an inexpensive handheld GPS, devices that are also popular with campers and hikers. Foot travelers and off-roaders also rely on compasses for navigation. Celestial navigation with a sextant can also be utilized. Being able to locate Polaris (the North Star), which is the brightest star in the constellation Ursa Minor (the Little Dipper or Little Bear), is a necessary skill for all off-road travelers. The rising and setting sun or other visual aids such as the glow from a city at night can also aid in navigation.

Atmospheric Navigation. Aircraft navigate in three-dimensional space, so latitude, longitude, and elevation must be known. All aircraft contain basic instrumentation, which includes: a magnetic compass; an altimeter, which displays the altitude; an airspeed indicator; a heading indicator, which is similar to a gyrocompass; a turn indicator, which displays direction and rate of turn; a vertical-speed indicator, which displays the rate of climb or descent; and an attitude indicator. The attitude indicator, which is also known as an artificial horizon, is an invaluable navigation instrument. It contains a gyroscope that reflects both the horizontal and vertical alignment of the aircraft. When visibility is reduced, the pilot is unable to visualize the horizon. Many small aircraft and virtually all large aircraft contain a GPS. Other navigation equipment commonly found on larger aircraft include a course deviation indicator (CDI) and a radio magnetic indicator (RMI). However, a GPS is much superior to either a CDI or RMI. The CDI displays an aircraft's lateral position in relation to a track, which is provided by a very high frequency (VHF) omnidirectional range (VOR) or an instrument landing system. A VOR is a ground-based station that transmits a magnetic bearing of the ship from the station. An instrument landing system is a ground-based system that can provide precision guidance of the aircraft to a runway. It includes radio signals, and often, high-intensity lighting arrays. An RMI is usually coupled to an automatic direction finder (ADF); the RMI displays a bearing to a nondirectional beacon (NDB), which is a ground-based radio transmitter.

Unmanned aircraft (drones) and guided missiles must also be navigated. In the case of an unmanned drone, a pilot sits at a console on the ground and directs the flight via radio. Guided missiles are either internally or externally guided, sometimes both. Some missiles contain inertial navigation systems similar to those found on submarines. The system is programmed to strike a specified target. External control can be via radio waves or laser. In some cases of laser guidance, the missile contains a laser homing device in which the projectile "rides" the laser beam to the target. More commonly, a technique referred to as semi-active laser homing (SALH) is employed. With SALH, the laser beam is kept pointed at the target after the projectile is launched. Laser energy is scattered from the target, and as the missile approaches the target, heat sensors home in on this energy.

Space Navigation. Space navigation is a complex and highly technical science. In addition to involving three dimensions it also requires plotting a course between two moving objects (the Earth and a space station or an orbiting space shuttle and the Earth). Space navigation also entails calculating the gravitational force of celestial objects such as the Moon and planets. Space navigation requires the collaboration between computers (both ground and ship based) and complex instrumentation. It also requires collaborative effort between highly skilled personnel and either astronauts or navigation equipment within unmanned spacecraft.

Impact on Industry

Navigation has a significant industrial impact. Navigation is required for mundane activities such as driving to the local market as well as more complex ones such as directing a space shuttle. It is a component of daily living. Although a market still exists for simple navigational aids such as magnetic compasses and nautical charts, the thrust of the market is focused on high-end, sophisticated electronics such as GPS. Since GPS was first developed for military use in 1978, it has spread into many civilian applications for navigation on the Earth's surface and the air. Electronics evolves rapidly, and new, improved products regularly reach the marketplace. Marine electronics is particularly vulnerable because it functions in an environment where corrosion from salt is an ongoing threat.

Industry and Business Sector. Although navigation equipment for owners of recreational boats and aircraft generates significant business, commercial and military applications represent the lion's share of the market. In the commercial sector, the civilian aircraft and shipping industries are major purchasers of navigation equipment. The National Aeronautics and Space Administration (NASA) requires extremely precise, complex, and expensive navigational equipment for its missions. All branches of the military—Army, Navy, Air Force, Marines, and Coast Guard—are major purchasers of navigation equipment. Many automobiles are equipped with a GPS or a compass. Although a compass is rudimentary when compared to a GPS, it can be helpful for finding one's way. Handheld GPS devices are popular and inexpensive. They are used not only by pedestrians but also can be taken along in an automobile or boat.

Government and University Research. One of the biggest sources of funding for navigation research in the United States is the Defense Advanced Research Projects Agency (DARPA). It is also the biggest client for certain kinds of navigation applications. The agency is primarily concerned with navigation systems with a military focus, such as inertial navigation systems. It also conducts research on satellite navigation. Another source of funding for navigation research is the U.S. Department of Energy (DOE).The DOE is focused on energy efficiency and renewable energy. DARPA and the DOE supply funds to many universities in the United States for navigation research and development. As far back as 1959, DARPA began working with Johns Hopkins Applied Physics Laboratory to develop the first satellite positioning system.

Careers and Course Work

Any navigation course requires knowledge of trigonometry. Whether navigation is based on visual or electronic information, triangulation to fix a position is required. The military offers many careers and course work involving navigation. Following military service, this training can be applied to many civilian job opportunities, up to and including piloting an aircraft or navigating a ship. The U.S. Coast Guard Auxiliary and the U.S. Power Squadrons offer free boating-safety courses, which include navigational skills. Courses are available throughout the United States as well as online. For individuals interested in more advanced courses, such as ones for preparation for the U.S. Coast Guard captain's license, reasonably priced courses are available throughout the U.S. and online. For those not interested in a military career, navigation training is available at civilian institutions. These include California Maritime Academy in Vallejo; Great Lakes Maritime Academy in Traverse City, Michigan; Maine Maritime Academy in Castine; Massachusetts Maritime Academy in Buzzards Bay; State University of New York Maritime College in the Bronx; Texas Maritime Academy, which is part of Texas A&M University at Galveston; and the United States Merchant Marine Academy in Kings Point, New York.

Although the basic principles of navigation can be learned and applied by any high school graduate, more advanced topics require a college degree and often a postgraduate degree such as a master of arts (M.A.) or doctorate (Ph.D.). An M.A. requires one year of study after four years of college; however, an academically aggressive student can earn an M.A. concurrently with a bachelor's degree. A Ph.D. requires two to three additional years of study followed by submission of a thesis or dissertation. Course work should include mathematics, engineering, computer science, and robotics. Positions are available for individuals with a degree in laser engineering in both the government and private sector. The ability to be a team player is often of value for these positions because ongoing research is often a collaborative effort.

Social Context and Future Prospects

Although the basic concepts of navigation have remained unchanged for centuries, modern navigation technology is highly advanced and continues to evolve. The frontier of navigation research lies in military and extraterrestrial applications. The military is focused on guidance systems for missiles and drones. Extraterrestrial applications range from the navigation of Earth-orbiting shuttles to interplanetary (and beyond) navigation. Navigation is an essential component of daily life—both civilian and military.

Robin L. Wulffson, M.D., F.A.C.O.G.

Further Reading

Burch, David. *Emergency Navigation: Find Your Position and Shape Your Course at Sea Even if Your Instruments Fail.* 2d ed. New York: McGraw-Hill, 2008. Burch, a veteran sailor and the founder of the Starpath School of Navigation, provides all manner of how to navigate without electronics: celestial, wind, swells, and even using airliners' contrails.

Burns, Bob, and Mike Burns. *Wilderness Navigation: Finding Your Way Using Map, Compass, Altimeter and GPS.* 2d ed. Seattle: Mountaineers Books, 2004. An invaluable aid for the backpacker, hiker, and camper.

Cutler, Thomas J. *Dutton's Nautical Navigation.* 15th ed. Annapolis, Md.: Naval Institute Press, 2004. An essential textbook and reference for anyone venturing offshore in any size vessel.

Launer, Donald. *Navigation Through the Ages.* Dobbs Ferry, N.Y.: Sheridan House, 2009. Covers navigation history from antiquity to the present.

Lele, Ajey. *Strategic Technologies for the Military: Breaking New Frontiers.* Thousand Oaks, Calif.: Sage Publications, 2009. Describes the nuances of technological development in a purely scientific manner and provides a social perspective to the relevance of future warfare and issues such as disarmament and arms control and their impact on the environment.

Web Sites

Boating Safety Resource Center
http://www.uscgboating.org/safety/boating_safety_courses_.aspx

National Aeronautics and Space Administration
http://www.nasa.gov

U.S. Department of Energy
http://www.energy.gov

U.S. Power Squadrons
http://www.usps.org

See also: Maps and Mapping; Sonar Technologies.

NIGHT VISION TECHNOLOGY

FIELDS OF STUDY

Vision science; optics; photonics; physics.

SUMMARY

Night vision technology is used to allow for better night vision than is possible with the human eye alone. Night vision technology uses light amplification and thermal-imaging components incorporated into goggles, cameras, binoculars, and other devices to improve vision under low-light conditions.

KEY TERMS AND CONCEPTS

- **Cones:** Photoreceptors in the human retina responsible for color vision and vision in bright light.
- **Infrared:** Electromagnetic energy with wavelengths of 750 nanometers to 1 millimeter, not visible by the human eye.
- **Mesopic Vision:** Vision under medium-light conditions, such as twilight.
- **Microchannel Plate:** Device made from coated glass that consists of an array of tiny glass tubes used to accelerate electrons while maintaining the entering pattern or image of the electrons.
- **Phosphor:** Fluorescent coating used in night vision technology that emits green light when electrons strike the coating.
- **Photocathode:** Coated metallic electrode that emits electrons in response to light.
- **Photoelectric:** Phenomenon of light in the form of photons striking a metallic surface, which causes electrons to be released.
- **Photonics:** Science of light and light particles.
- **Photopic Vision:** Vision under bright-light conditions.
- **Photoreceptor:** Cell in the human retina that responds to light.
- **Rods:** Photoreceptor cells in the human retina responsible for vision under dim light.
- **Scotopic Vision:** Vision under low light.

DEFINITION AND BASIC PRINCIPLES

Night vision technology is the use of light-amplifying and thermal-imaging devices to enhance human vision performance in low light. These devices can take the form of cameras, goggles, binoculars, and spotting scopes. This technology takes ambient light and amplifies it through photoelectric techniques or thermal imaging that takes advantage of the energy released in the infrared spectrum in the form of heat.

Night vision devices use a photocathode that collects photons, which are light particles present even in dim light. These photons strike a photocathode, which then emits electrons. Photocathodes can be made of a variety of coated metallic materials. These electrons are multiplied by a microchannel plate and then transformed back into green light using a phosphor screen. Green light works well because of the sensitivity of the human eye to these wavelengths. There are variations on this technology, including early night vision systems that project infrared light and then amplify the reflected light.

BACKGROUND AND HISTORY

The groundwork for the development of night vision technology was laid by early scientists such as Heinrich Hertz who described the photoelectric effect in 1887. The discovery that electrons are emitted when light strikes metal was further developed by German physicists Max Planck and Albert Einstein in the early twentieth century. Their work confirmed the particle nature of light and provided the foundation for future applications, which included night vision technology.

William E. Spicer was a cofounder of the Stanford Synchotron Radiation Lightsource and was instrumental in the development of light amplification. His work paved the way for the first generation of night vision goggles and had applications in medical-imaging technology. Spicer's work provided the basis by which light in the infrared spectrum, which is not visible to the human eye, can be detected, amplified, and transformed into visible green light. All of the night vision devices rely on this basic technology.

As a result of the research done by Spicer night vision goggles were developed for use by the military in World War II in the 1940's. England, Germany, and the United States all developed sniper scopes using infrared cathodes. These devices used an infrared beam to generate reflected light from the surroundings that were then amplified by the scope. These devices had the disadvantages of low range and the ability of the enemy to detect the infrared beam. Early devices using an infrared beam to create reflected light are called active night vision devices and are referred to as generation zero.

Militaries around the world continued to work on improved night vision technology. Generation one devices, the next iteration, improved on the light amplification so that ambient light could be used without the need to use an infrared beam. These systems did not work well on very dark or cloudy nights. Early night vision devices were large and created distortion of images. The Starlight scope used in Vietnam is an example of this generation of devices. Generation zero and generation one night vision devices are now available to the general public.

As technology advanced, the next generation of night vision devices became more sensitive by the addition of microchannel plates, which further amplified the signal. A microchannel plate is manufactured from lead oxide cladding glass. Generation two devices have less distortion and increased brightness. Generation three night vision technology incorporates gallium arsenide cathodes, which further increases sensitivity. Generation four devices, which are typically used for military applications, incorporated changes to the microchannel and added gating. Gating is a system that switches on and off to allow for rapid response to changes in light. For example, if night vision goggles are on and then a light is suddenly switched on, the user will be then able to see under the lighted conditions.

Thermal imaging has been made possible with improved sensitivity and light amplification and also creates images using infrared wavelengths that are emitted as heat. Not all night vision devices are able to detect thermal energy.

How It Works

To understand how night vision technology works, it is important to have a basic understanding of light and of how the human eye responds to light. Before the twentieth century, there was an ongoing debate as to whether light was a wave or a particle. Sir Isaac Newton favored a particle theory, which was later substantiated by Henrich Hertz, Max Planck, and Albert Einstein. However, modern understanding of light is that it behaves like both a wave and a particle. For the purpose of understanding night vision technology, it is the photoelectric effect that forms the basis for these devices. When particles of light called photons strike metal, electrons are emitted. Specialized photocathodes are coated with various materials to make them more sensitive. The technical specifications of the photocathodes have improved over the generations since the 1940's, in part because of the use of different materials and coatings. The function of the photocathode in a night vision device is to convert the light into electrons. In low-light conditions the night vision devices are able to detect infrared light that is not detectable by the human eye.

The electrons are then converted into visible light by a phosphor screen, which then converts the electrons back into green light visible to the human eye. Later devices added a microchannel plate, which serves to amplify the electron energy while preserving the pattern or image. The microchannel plate is an array of tiny glass tubes. The electrons enter and are confined in each tube as they travel through, which results in the preservation of their entering pattern. While traveling through the microchannel plate, the electrons are further amplified by the application of voltage across the microchannel plate. This allows for more energy entering the phosphor screen and a subsequently brighter image. Infrared light travels from the environment to the photocathode, where it is translated into electrons, which in turn enter the microchannel plate. The amplified signal then strikes the phosphor screen, which turns the energy into green light that the viewer can see.

The human eye is most sensitive to visible light with wavelengths of about 400 to 700 nanometers (nm). Infrared light is in the 700 nm to 1 millimeter (mm) range. Infrared is further divided into near-infrared IR-A with 750 to 1400 nm wavelength range, medium wavelength IR-B of 1,400 to 3,000 nm range, and long wavelength or far IR-C with wavelengths of 3,000 nm to 1 mm. The long wavelengths are used in thermal-imaging devices. Infrared light is not detected by the human eye, so night vision devices are used to transcribe this light into visible green light. The human eye is particularly sensitive to green light. For example, 0.001 watt of green light will appear bright, while 0.001 watt of blue light will appear dim.

Applications and Products

Military Applications. Military organizations used night vision technology in World War II and continue to be at the forefront of new developments. Military applications include night vision goggles for military personnel, sniper scopes, reconnaissance, and vehicle navigation. The advances that led to thermal imaging were on display in the media during the Gulf War in 1991. Those who may have watched the coverage of this war on television will remember the pictures with greenish images and periodic flashes of bright green corresponding to tracers and explosions.

Thermal forward-looking imaging (FLIR) devices are installed on vehicles and helicopters. Night vision devices are available to personnel for survival purposes even in a downed aircraft. This technology has continued to be employed in weapons-aiming devices. Data collection and communications technology have been added to some night vision devices in order to improve military communication and reconnaissance.

Law Enforcement. Law-enforcement applications are similar to military applications and include surveillance, weapons aiming, recording, and identification of suspects in situations of low light. Thermal imaging is used to identify illegal marijuana-growing operations, which are sometimes located in ordinary urban neighborhoods. The heat lamps used in growing the plants make it possible for law enforcement to identify these operations by air. A helicopter equipped with thermal-imaging equipment can detect an increased heat signature coming from the roof of the house that contains the growing operations. FLIR is also used on law-enforcement vehicles. Night vision technology is also used in search-and-rescue operations by law enforcement and other agencies.

Photography. Some photographers are using night vision cameras to create artistic images. To address the green images created by this technology, the photographers employ digital-editing techniques. The resulting images are unique works of art.

Recreational Use. Recreational use of night vision technology has expanded as the older generation of devices has become less expensive. Newer generations are still mostly used by the military and law enforcement because of the higher costs of these advanced devices. Spotting scopes, binoculars, and cameras are used by hunters, campers, hikers, and fisherman. Night vision devices are used for wildlife viewing and photography.

A unique activity that makes use of night vision goggles is dining in the dark. The servers use night vision goggles to provide a meal for diners who do not have the night vision goggles. The idea is to make the meal more of an adventure and to enhance the dining experience. Some companies use this as a team-building activity.

Scientific Research. Scientists use night vision devices to study nocturnal animals and other phenomena that might not otherwise be visible to the human eye. This has opened up a new area of study for wildlife biologists. In some parks, night vision technology is used to study wildlife and vehicle

Fascinating Facts About Night Vision Technology

- William E. Spicer, who is credited with the development of light amplification, suffered from speech difficulties and dyslexia as a child. He credits his dyslexia to his later success, since he believed it allowed him to think about problems in a different way.
- In some of the Spider-Man comics, this superhero has night vision capabilities. Night vision goggles have also been featured in movies such as *Iron Man* (2008) and *Watchmen* (2009).
- Thermal imaging can be used in accident reconstruction since thermal images can be detected even when there are no visible skid marks.
- Cosmic rays can be detected by astronauts as flashes of light. Cosmic rays are subatomic particles that originate in space and resemble particles produced by particle accelerators.
- The dark-adapted human eye undergoes physiological and biochemical changes to become 100,000 times more sensitive than when in bright light. Exposure to bright light reverses these changes and will reduce night vision until the eye is dark adapted again. The retinal cells responsible for night vision are rods, which become most sensitive after thirty or forty minutes.
- Infrared video cameras can detect the infrared signal from TV remote controls and may also reveal images behind tinted glass. Thermal cameras are also used for home-energy audits and for detection of mold spores.

collisions in order to determine ways to reduce these incidents, which are dangerous to both humans and animals.

Astronomical research has also benefited from the use of night vision technology. The National Aeronautics and Space Administration (NASA) has used night vision technology to acquire images with the Hubble Space Telescope and the Mars Rovers. This technology is also being offered to amateur astronomers to enhance the images that can be acquired.

IMPACT ON INDUSTRY

The invention of night vision devices has led to an industry of device manufacturers and researchers. Although a large part of the market continues to be in military applications and law enforcement, night vision technology is being marketed widely to the general public.

In addition to the creation of an industry based on the manufacture and sales of night vision devices, the technology of the photocathode and light amplification has impacted numerous fields, including television, astronomy, and medicine. NASA has used infrared imaging to explore the solar system. There are images available online from the Hubble Space Telescope and from the Mars Rovers that were generated using the same techniques that are used for night vision devices.

CAREERS AND COURSE WORK

Manufacturers and distributors of night vision technology utilize a variety of personnel. For technical jobs in this field, a solid mathematics background is necessary. Understanding of physics, electronics, optics, photonics, and software is important for some of the career paths.

Automatic data processing equipment technicians (ADPE) are used in the night vision technology industry. This type of technician will require two years of technical training. ADPE technicians work with engineers and other personnel in a range of activities including assembly, design, and troubleshooting.

Software engineers should have a bachelor of science degree in a field such as physics, mathematics, or engineering. Some jobs require a master's degree. Strong computer programming skills are also required.

Engineers in a variety of areas are employed in the night vision technology industry. Electrical engineers and manufacturing engineers are two examples. A bachelor's degree and a strong understanding of mathematics are requirements. Fields of study will vary between the various engineering programs.

Researchers in the night vision technology will often have a master's degree or doctorate in physics, engineering, mathematics, or other related field. Research in this field is done by universities, industry, and by the military.

SOCIAL CONTEXT AND FUTURE PROSPECTS

The development of night vision technology has changed the way wars are fought. Before this technology was available most militaries avoided night operations. Militaries have competed to stay on the forefront of night vision technology research in order to maintain a tactical advantage. Night vision technology has been credited with the success of Desert Storm in 1991, giving the U.S. military an advantage in the conflict.

As this technology advances into solid-state formats, additional communications and analysis features will be added to allow real-time communication between soldiers. Remote surveillance and reconnaissance using thermal imaging is becoming more widely used. Night vision technology is already being used in the acquisition of astronomic images. NASA is already using thermal and infrared imaging in their missions to Mars.

Thermal-imaging systems are now being marketed for night driving, heavy equipment operators, maritime applications, and pilots. As the costs of these systems decline they will be more widely available for the general public and possibly may eventually become a standard option in passenger vehicles.

Ellen E. Anderson Penno, M.D., M.S., F.R.C.S.C., Dip. A.B.O.

FURTHER READING

American Academy of Ophthalmology. *Clinical Optics.* San Francisco: American Academy of Ophthalmology, 2006. This volume covers the fundamental concepts of optics as it relates to lenses, refraction, and reflection. It also covers the basic optics of the human eye and the fundamental principles of lasers.

Hobson, Art. *Physics: Concepts and Connections.* 5th ed. Boston: Pearson Addison-Wesley, 2010. Includes chapters on light, geometric optics, wave nature of light, and a section on night vision imaging.

Kakalios, James. *The Physics of Superheroes.* 2d ed. New York: Gotham Books, 2009. Uses comic-book references to cover basic physics theory. Includes chapters on mechanics, energy (heat and light), and modern physics.

Newell, Frank W. *Ophthalmology:Principles and Concepts.* 5th ed. St. Louis: Mosby, 1982. Covers basic eye anatomy, optics, and retinal physiology and biochemistry.

Tipler, Paul A., and Gene Mosca. *Physics for Scientists and Engineers.* 6th ed. New York: W. H. Freeman, 2008. Paul Tipler's physics text has been a staple for introductory university physics courses for many years. Chapters cover basic physics concepts including the basic physics of optics and the dual wave and particle nature of light.

Web Sites

HubbleSite
http://hubblesite.org

Jet Propulsion Laboratory
Mars Exploration Rover Mission
http://marsrover.nasa.gov

Night Vision and Electronic Sensors Directorate
http://www.nvl.army.mil

NOISE CONTROL

FIELDS OF STUDY

Acoustics; architectural acoustics; audiology; electrical engineering; mechanical engineering; physics.

SUMMARY

Scientifically, noise is defined as an intermittent random oscillation of the air that may be perceived by humans. Because the effects on humans are usually unwelcome, often quite disturbing, or even dangerous, psychologically, noise may be defined as any unwanted sound. During the twentieth century, as humans became increasingly aware of the dangers of noise, many techniques for mitigating the problem evolved. These included sound absorbing materials for buildings, barriers to reduce traffic noise near residential areas, reducing noise at its source, preventing noise incursion into ears, and most recently, active noise control by sound cancellation.

KEY TERMS AND CONCEPTS

- **Decibel (dB):** Measure of the intensity of a sound source; as the dB level increases sounds are perceived as being louder. Humans can process sounds from 0 dB (threshold of hearing) to 120 dB (threshold of feeling).
- **Hertz (Hz):** Frequency of a sound wave, that is, the number of vibrations per sec.
- **Noise Reduction Coefficient (NRC):** Single number, obtained as an average over low and high frequencies, rating the sound absorption of a material.
- **Presbycusis:** Age-related hearing loss characterized by progressively decreasing sensitivity to high frequencies.
- **Reverberation:** Repeated reflection of sound waves in an enclosure that is perceived as a continuously decreasing prolongation of the original sound.
- **Sound:** Vibration of the air within the frequency range of 20 Hz to 20,000 Hz, the normal range of human hearing.

Definition and Basic Principles

A noise problem can arise from one of three sources: It may be airborne, structure-borne, or a combination of the two. Airborne noise control is achieved either by reducing noise in the environment of a listener, or reducing noise at its source, for example, by designing quieter machinery or enclosing loud machines. Enclosing a noise source to reduce propagated sound presents the somewhat daunting challenge of balancing contradictory factors; the enclosure must be sufficiently airtight to contain noise while allowing for ventilation, power cable access, and material flow. Structure-borne noise can be suppressed by shock absorbers to isolate the machinery from the enclosure. Ductwork that penetrates the isolation walls must be resiliently supported on flexible hangers. When extremely quiet environments are mandatory, such as for a recording studio, it is easier to isolate the room from its environment rather than attempt to suppress all possible sources of impinging noise. This is best achieved by forming an acoustically and vibrationally isolated room within the building. This requires not only soundproof walls and ceiling but also a resiliently supported floor with isolated walls and ceiling to attenuate structure-borne vibration. Protecting individual listeners from dangerous noise levels is best achieved by using earplugs or tight-fitting acoustical earmuffs.

Background and History

Attempts to control noise by legislation have occurred for centuries. Bern, Switzerland, for example, has over the centuries imposed dozens of noise ordinances, including regulations against singing and shouting in the streets (1628), noisy conduct at night (1763), industrial woodworking noise (1886), and loud automobile sounds (1913). With increasing technology has come increasing noise. By the 1960's, it was realized that exposure to loud industrial sounds will eventually cause employee hearing loss. The Walsh-Healey Act of 1960 attempted to control this noise, but standards were not specified until 1966 after the Bureau of Labor Standards accumulated data indicating that daily exposure to sounds above 85 dB would eventually cause permanent hearing impairment and that at higher sound levels the detrimental effects occurred sooner and more severely.

By the 1970's, a heightened environmental awareness recognized noise as an insidiously dangerous and pervasive epidemic pollutant, not confined to the workplace. Even the moderate levels experienced in the nonwork environment will have a cumulative effect, eventually causing hearing loss. Noise occurring during sleep, even though it may not awaken the sleeper, can result in a less soporific slumber by interfering with sleep stages, and adversely affecting physical and mental health. In the first concerted effort to control environmental noise, Congress enacted the Noise Control Act in 1972.

How It Works

Effects of Noise on People. Noise can impair performance and adversely affect human health by increasing stress, interfering with sleep, producing anxiety, and raising blood pressure. Chronic exposure may also cause increased hormone excretion and psychological depression. Continuous exposure to sounds 90 dB or greater will causes permanent hearing loss, particularly in the higher frequencies. Sick people are particularly susceptible to excessive environmental noise; it can lengthen convalescence and necessitate additional medical treatments. It is also known that long-term exposure to moderate sound levels (85 dB or less) is the leading cause of presbycusis in the elderly.

Legislation for Noise Control. Pressured by environmental organizations and concerned citizens, Congress enacted the Noise Control Act (1972), which set noise emission standards for industrial machinery, commercial products, major appliances, aircraft, and motor vehicles. The act also required the U.S. Environmental Protection Agency (EPA) to coordinate all existing federal programs relating to noise research, control, and regulation. For the first time, environmental noise-impact studies would be required for new industrial sites and interstate highways. The EPA was also mandated to provide technical assistance to state and local governments planning to address noise mitigation and hazards. Since 1983, the Occupational Safety and Health Administration (OSHA) has regulated industrial noise exposure for workers; this limits the permissible exposure to 90 dB over an eight-hour work day and mandates regular audiometric testing.

Subsequent to federal legislation, municipalities plan zoning to keep housing developments isolated from industrial regions. Noise ordinances, for day and night, have been enacted for most residential regions as well as industrial sites. Highway noise is abated in residential areas by concrete noise-reducing barriers installed along interstate highways.

Design for Quiet. It is considerably easier to design noise control into new buildings than to attempt a retrofit when unanticipated problems appear. Incorporated noise control does not substantially increase construction costs, while later corrections can be quite expensive.

In apartment buildings, noise transmitted through walls is always problematic; the porous insulation typically used to retard heat flow does not effectively absorb transmitted sounds. Effective absorption of transmitted noise is best achieved with massive materials (such as brick), but this is at variance with modern construction practice using lightweight materials. Living spaces located near busy freeways should be designed to reduce the transmitted noise to no more than 60 dB.

Concert halls designed for metropolitan areas present unique challenges. Automotive traffic noise is ubiquitous, and low-jet flyovers are not uncommon. The interior must be isolated from conversation in the foyer as well as from ventilation and plumbing sounds if one is to hear the nuances of a performance. Shielding from external noise is best achieved by constructing a box-within-a-box—an outer shell enclosing a spacious foyer with several well-isolated performance halls located in the interior area. Ventilation noise is controlled by careful design of ducts to minimize turbulence. This design has been successfully employed for the Kennedy Center in Washington, D.C., where the interior background noise never exceeds 30 dB despite the extremely noisy location.

Applications and Products

There are three places where noise can be reduced: at its source, during transmission, and at the receiving end.

Controlling Noise at Its Source. Traffic noise, the bane of modern civilization, is best reduced at its source by improved engine enclosures and better tire design. When interstate highways pass near residential areas, the next best means of controlling traffic noise is to block it by lining the highway with concrete barriers or sound-absorbing vegetation. Internal automotive noise is best decreased by reducing air turbulence with more aerodynamically efficient shapes, employing additional sound isolation, and suppressing vibration.

Fascinating Facts About Noise Control

- Active noise control earphones use a technology whereby external sounds are recorded, phase inverted, and reemitted with the original sound to effectively cancel it by destructive interference. This ensures a quiet background for hearing recorded music in noisy surroundings.
- Aircraft traveling faster than the speed of sound create a sonic boom, a large pressure wave experienced as an explosion. This startles people, breaks windows, and damages structures. Commercial aircraft are therefore prohibited from flying at supersonic speeds over land areas.
- A noise dosimeter is a small (cellphone-size) device to measure noise exposure. It uses a small lapel microphone that records the local sound field and sends it to a pocket-held processor to store intensity levels and their time duration.
- Research indicates that for most cases presbycusis is a result of one's lifetime exposure to noise. Members of the Mabaan tribe in Sudan, living in primitive but extremely quiet environments, have virtually no cases of presbycusis among the elderly.
- If a tree falls and there is no one there to hear it, does it produce a noise? This centuries-old conundrum is readily answered, depending on one's definition of noise. Defined physically, then yes, the air will vibrate. If noise is defined psychologically, then no: Without an ear to hear it, there can be no sound.
- A bionic ear is an electronic device surgically implanted in the inner ear to provide a sense of hearing for those with severely damaged cochleas.

In the vicinity of airports jet noise is increasingly problematic because ever-more powerful engines create more sound. To reduce the radiated noise, two methods are employed: surrounding the engine's moving parts by acoustic linings to attenuate high-frequency jet whine and reducing the number of rotor blades and stator vanes, which may diminish power.

Construction equipment produces notoriously high decibel levels, both in the operator's cab and outside it. If the cab is open, the operator has no recourse but to use earplugs. Equipment manufacturers, however, have begun producing machines with enclosed cabins carefully designed to attenuate the interior noise level, but reducing external noise has been considerably less successful. On occasion, rubber tractor treads have been employed instead of metal; they produce less noise, but they are considerably less durable.

If noise cannot be suppressed at its origin, it can often be contained in a sound-attenuating structure. Inert and relatively massive walls reduce sound transmission, the actual decibel loss being dependent on frequency, sound absorption, and the tightness of the enclosure. An effective enclosure must also minimize structure-borne noise. This is best achieved by mounting vibrating machinery on resilient supports and supporting penetrating ducts on spring hangers. If the duct connects to the machine, as in the case of ventilation fans, the duct must be vibrationally isolated from the fan housing by a flexible hose.

Controlling Transmitted Noise. Propagated noise can be either airborne or structure-borne. Airborne noise is generated by vibration, impact, or airflow turbulence. In closed rooms sound is both reflected from the surfaces and transmitted through the enclosures, the material determining the proportion reflected and transmitted. Absorbent walls reflect less energy, reducing the interior sound level, but transmitted sound may be increased. Hard surfaces reflect most of the sound, creating the phenomenon termed reverberation. A long reverberation time is detrimental in any room where speech intelligibility is important; reverberation is best controlled by applying acoustical absorbing materials to interior surfaces. Transmitted structure-borne noise is extremely difficult to control and is best managed at the source.

Airborne transmitted noise may arise from two different sources: external noise penetrating the structure or internal noise transmitted through walls, ducts, or openings. For external noise, barriers or massive walls must be employed to attenuate the sound. If the noise is produced in the same building and transmitted through walls or ducts, specific procedures must be implemented. Noise transmitted through apartment walls is best prevented by relatively massive walls; if this is not possible another method of control is to use staggered studs. The framing studs on one side of the wall are staggered with those constituting the other side. The two interior walls do not share the same supporting studs, and there is airspace between the walls, reducing transmitted sound. Noise transmitted through air ducts is a pervasive problem. The first line of defense is the source; ventilation equipment should be designed to be as quiet as possible, while still performing its desired function. Turbulence in the ducts, heard as a hissing sound, is best controlled by keeping the airflow slow and eliminating sharp turns in the ducts. When turns are necessary, turning vanes consisting of curved surfaces mounted inside the duct smoothly direct the airflow around corners. There is no excuse for sound-leak openings around conduit and pipe penetrations, and such are typical of shoddy workmanship. The optimum noise reduction acceptable for a room depends on the room's intended purpose. The requirements for music rehearsal space are much more stringent than the requirements for many other spaces.

Controlling Noise at the Receiver. When noise cannot be adequately controlled at its source or reduced during transmission, the last resort is to attenuate it at the ears of the receiver. Hearing protection was first instituted for outdoor airport workers exposed to the extremely high levels of noise generated by jet aircraft. Gradually, hearing protection became mandatory for noisy industrial environments, with the type of protection dependent on the noise spectrum. It is generally less expensive for industry to reduce noise at the source or to isolate workers from noisy environments then to provide adequate protective devices. The cost of these devices usually exceeds the original capital outlay. Employees must be trained in their correct usage, monitored to ensure they are being worn, and possibly compensated when hearing loss occurs because of noncompliance.

There are three primary types of ear-protection devices: earplugs, earmuffs, and ear caps. To be effective they must form an acoustic seal across a broad frequency range. Earplugs achieve this by being inserted into the ear canal. Typically, they are made from malleable dense foam that tightly fills the ear canal when inserted, attenuating noise by about 20 dB across the frequency spectrum, but they are unable to prevent sound from reaching the inner ear through bone conduction by the skull. Earmuffs cover the ear forming an acoustic seal around the ear. Being external to the ear they also reduce bone-conduction sound. The attenuation provided by the muff depends on its construction and how well it seals the ear. Ear caps fit over the entrance to the ear canal and protrude only slightly into the interior. They are held in position by a tension band going over the head or under the chin.

Electronic noise-reduction earmuffs use the principle of adding two out-of-phase signals to reduce the sound level. A small microphone on the earmuff surface records the surrounding noise, inverts the signal phase and adds it, through small speakers in the earphones, to the original signal, effectively canceling the sound. Although expensive, these active noise-suppression systems are quite effective.

Impact on Industry

Construction and Traffic Noise Control. Outdoor construction noise in cities is being contained by "sound curtains," flexible noise-reducing curtains connected with Velcro seals, which enclose the job site. These curtains block exterior noise and absorb noise created within the curtains. Typically the curtains reduce noise by 10 to 15 dB. In the European Union, a large consortium of leading industries, named Euro Noise Control, has formed to address noise issues and offer solutions by manufacturing means of reducing noise in urban areas.

Traffic noise, both inside and outside of vehicles, is a difficult problem to solve because of the sheer number of vehicles on the road. The first line of defense is to reduce noise at its source by creating better engine enclosures, mufflers, and tire design. The next method is to block the noise by the constructing concrete barriers or planting sound-absorbing vegetation on either side of the roadway. Internal automobile noise has been greatly abated by designing more aerodynamically efficient vehicles to reduce air turbulence, using better sound isolation materials, and improving vibration isolation.

Aircraft Noise Control. Aircraft noise, particularly in the vicinity of airports, is a serious global problem exacerbated by the fact that as modern airplanes have become more powerful, the noise generated has risen concomitantly. The European Union approaches aircraft noise control by four principal means: reduction of noise at the source, land-use management, noise-abatement procedures, and operating restrictions on airplanes. New methods are constantly being investigated to reduce jet-engine noise without sacrificing power. Unfortunately, it takes several decades for new technology to become standard in most of the operating fleets.

Residential annoyance by aircraft noise is best controlled by zoning restrictions prohibiting housing developments near airports. It may also be possible to relocate runways to reduce flights over nearby homes. Noise-reflecting screens and active noise-suppression procedures, although expensive, are under serious deliberation as possible means of airport noise abatement.

Special noise problems occur when aircraft travel faster than the speed of sound (supersonic), as this creates a sonic boom, which startles people, breaks windows, and damages structures. Because of this, commercial aircraft are prohibited from flying at supersonic speeds over land areas.

Reduction of aircraft noise at its source, the airplane or airport, has become increasingly problematic because the major potentials for improvement have already been implemented, leaving only minor possible adjustments. Manufacturers attempt to produce airplanes that are more economical to fly; although unplanned, this also involves an associated noise reduction. As new commercial aircraft become quieter, overall airport noise reduction is best realized by withdrawing the noisier older airplanes from service.

Careers and Course Work

Noise control, like other environmental problems, requires people with broad scientific, technical, and creative skills. Because the global soundscape is not becoming less noisy, despite heightened awareness, pubic protests, and new legislation, there is an ever-increasing need for qualified people in noise control engineering and audiology. Vibration and means of suppressing it are best learned by obtaining a bachelor's degree in mechanical engineering or physics. If one wishes to work in the field of noise reduction by means of signal modification, this is best achieved by a degree in electrical engineering. Audiologists not only assess and diagnose hearing problems but also design hearing-conservation programs for industry and initiate basic research on hearing loss and its prevention by noise control. Becoming a certified audiologist requires a bachelor's degree in communication sciences and a minimum of a master's degree in audiology. The U.S. Bureau of Labor Statistics has predicted that job opportunities for audiologists will increase by about 25 percent by 2018 due to hearing problems in an aging population as well as federal legislation mandating more audiologists be recruited for public schools.

Social Context and Future Prospects

Engineers working in industry can provide valuable assistance in developing quieter products and designing sound-insulating areas to keep industrial workers separated by from noisy machines. In the area of noise control engineering there is a need for innovative ideas and new technologies for noise and vibration measurement and control in industrial and commercial applications.

As noise-induced hearing loss becomes more problematic with an aging population subjected to a lifetime of excessively loud sounds, there is a concomitant need for audiologists to assist those with hearing loss by fitting hearing aids. Manufacturers of hearing aids are continuously researching how to produce smaller, more accurate, and higher-quality devices for the hearing impaired. Considerable research effort is also being extended on ear-covering devices to attenuate incoming sound before it reaches the eardrum.

George R. Plitnik, B.A., B.S., M.A., Ph.D.

Further Reading

Baron, Robert Alex. *The Tyranny of Noise.* New York: St. Martin's Press, 1970. Somewhat biased, but very readable, tirade against the costs and politics of noise pollution; methods of alleviation are also included.

Behar, Alberto, Marshall Chasin, and Margaret Cheesman. *Noise Control: A Primer.* San Diego: Singular, 2000. Relevant information for people working in fields concerned with noise exposure; a short but readable exposition.

Rettinger, Michael. *Acoustic Design and Noise Control: Noise Control.* Vol. 2. New York: Chemical Publishing, 1977. Multiple aspects of noise and noise reduction for designers and environmentalists, including a plethora of charts, graphs, and practical examples.

Rossing, Thomas, ed. *Environmental Noise Control.* College Park, Md.: American Association of Physics Teachers, 1979. Collection of essays, from elementary to advanced, on the many and diverse methods for controlling noise. Also included is a resource letter annotating hundreds of books and journal articles concerned with the theory and practice of noise control.

Strong, William J., and George R. Plitnik. *Music, Speech, Audio.* 3d ed. Provo, Utah: Brigham Young University Academic Publishing, 2007. A comprehensive text, written for the layperson, with noise covered in Chapter 17, "Hearing Impairments and Hazards" and Chapter 18, "Noise in the Environment."

Vér, István L., and Leo L. Beranek, eds. *Noise and Vibration Control Engineering: Principles and Applications.* 2d ed. Hoboken, N.J.: John Wiley & Sons, 2006. Details sound absorption, passive silencers, enclosures, vibration isolation and damping, machinery and heating, ventilation, and air-conditioning (HVAC) noise control, and active noise suppression.

Web Sites

Acoustical Society of America
http://asa.aip.org

American Physical Society
http://www.aps.org

American Society of Mechanical Engineers
http://www.asme.org

Institute of Noise Control Engineering
http://www.inceusa.org.

NUCLEAR TECHNOLOGY

FIELDS OF STUDY

Energy; chemistry; physics; medicine; environmental science; government; international politics.

SUMMARY

Nuclear technology focuses on the particles composing the atom to produce reactions and radioactive materials that have practical use in such areas as agriculture, industry, medicine, and consumer products, as well as in the generation of electrical power and the construction of nuclear weapons.

KEY TERMS AND CONCEPTS

- **Alpha Decay:** Decay of matter emitting an alpha particle composed of two protons and two neutrons, a property of unstable radioactive elements.
- **Atom:** Microscopic building block of ordinary matter, composed of negatively charged electrons, positively charged protons, and neutrons, which have no charge.
- **Beta Decay:** Radioactive decay of matter emitting a beta particle when a proton is converted into a neutron in the atom's nucleus and released as an electron.
- **Fission:** Energy-generating process by which the nucleus of a heavy atom is split into two or more lighter atoms by the absorption of a neutron, and in which a dense number of heavy atoms will generate an atom-splitting chain reaction.
- **Fusion:** Process by which two light atoms are combined at extraordinarily high temperatures into a single, heavy nucleus, resulting in the release of much greater amounts of energy than produced by nuclear fission.
- **Gamma Decay:** Release of gamma rays; electromagnetic radiation in the form of energy, not matter.
- **Laser:** Device producing light in the form of electromagnetic radiation.
- **Plutonium:** Heavy, manmade radioactive element useful in the production of both nuclear energy and nuclear weapons.
- **Radioactivity:** Spontaneous emission of radiation from an atom's nucleus.
- **Radium:** Highly radioactive element found in uranium ores.
- **Uranium:** Radioactive metallic substance, and one of the most abundant elements found in nature, used to produce the fissile isotope uranium-235 (U-235), the principal fuel of nuclear reactors.
- **X Ray:** High-frequency electromagnetic radiation emitted when atomic electrons drop to lower energy levels.

Definition and Basic Principles

One of most significant developments of the twentieth century, nuclear technology focuses on practical applications of the atomic nuclei reactions that result in the Earth from the decay of uranium and from the artificial stimulation of uranium particles, the atomic nuclei of which are normally separate from one another because they contain positive electrical charges that cause them to repel one another.

The scientific principles employed in nuclear technology grow out of the initial research on radium conducted during the early twentieth century by Henri Becquerel and Marie and Pierre Curie and their daughter, Irène Joliet-Curie. This research involved the alpha, beta, and gamma ray activity of radium. The subsequent research of their successors focused on manipulating the relationship between the proton, neutron, and electron properties of the atom of radioactive elements to produce chain reactions and radioactive isotopes of value in numerous fields. Most of the scientific principles involving the peaceful use of nuclear technology relate to the process of fission, an energy- and neutron-releasing process in which the nucleus of an atom is split into two relatively equal parts. The production of nuclear weapons also builds on these principles but in addition to fission, it involves the fusion of several small nuclei into a single one, the mass of which is less than the sum of the small nuclei used in its creation.

Since the use of atomic weapons in World War II, a second set of principles relating to the technology has also emerged—one designed to govern its application in a manner beneficial to humanity. The most important of these applied principles pertain to the beneficial and responsible use of nuclear technologies in a manner mindful of human and environmental safety, the technology's continued improvement in terms of efficiency and safety, and the securing of research information and nuclear material from acquisition by those who might use them for destructive purposes.

Background and History

Most accounts of the nuclear age's roots begin with the 1896 work by the French physicist Henri Becquerel, who is credited with discovering radioactivity while exploring uranium's phosphorescence. The research center in the field remained in France for decades thereafter, most notably in the groundbreaking work of Pierre Curie, and that of his wife, Marie, and their daughter Irène Joliot-Curie following Pierre's death in 1906. It is Pierre Curie who is credited with coining the term "radioactivity," and it was his and his wife's work on the properties of decaying uranium that provided the foundation for the research on nuclear fission and nuclear fusion, which eventually led to the creation of the atomic bomb, nuclear electricity-generating power plants, and the radioisotopes so abundantly useful in medicine, industry,, and daily life

By the 1940's, research conducted a decade earlier by British physicist James Chadwick, Italian physicist Enrico Fermi, German chemist Otto Hahn, and others on the unstable property of the atom had progressed sufficiently for scientists to envision the development of nuclear weapons that could, through an induced chain reaction, release enormous amounts of destructive energy. A wartime race to produce the atomic bomb ensued between Germany and the eventual winners, the German, British, and American scientists who collaborated in the United States' Manhattan Project under the direction of American physicist Robert Oppenheimer. Using reactors constructed in Hanford, Washington, to create weapons-grade uranium, U-238 and U-235, the project's scientists tested the first atomic bomb on July 16, 1945. Shortly thereafter, detonation of atomic bombs on the Japanese cities of Hiroshima (August 6) and Nagasaki (August 9) led to Japan surrendering unconditionally, ending World War II. The resultant

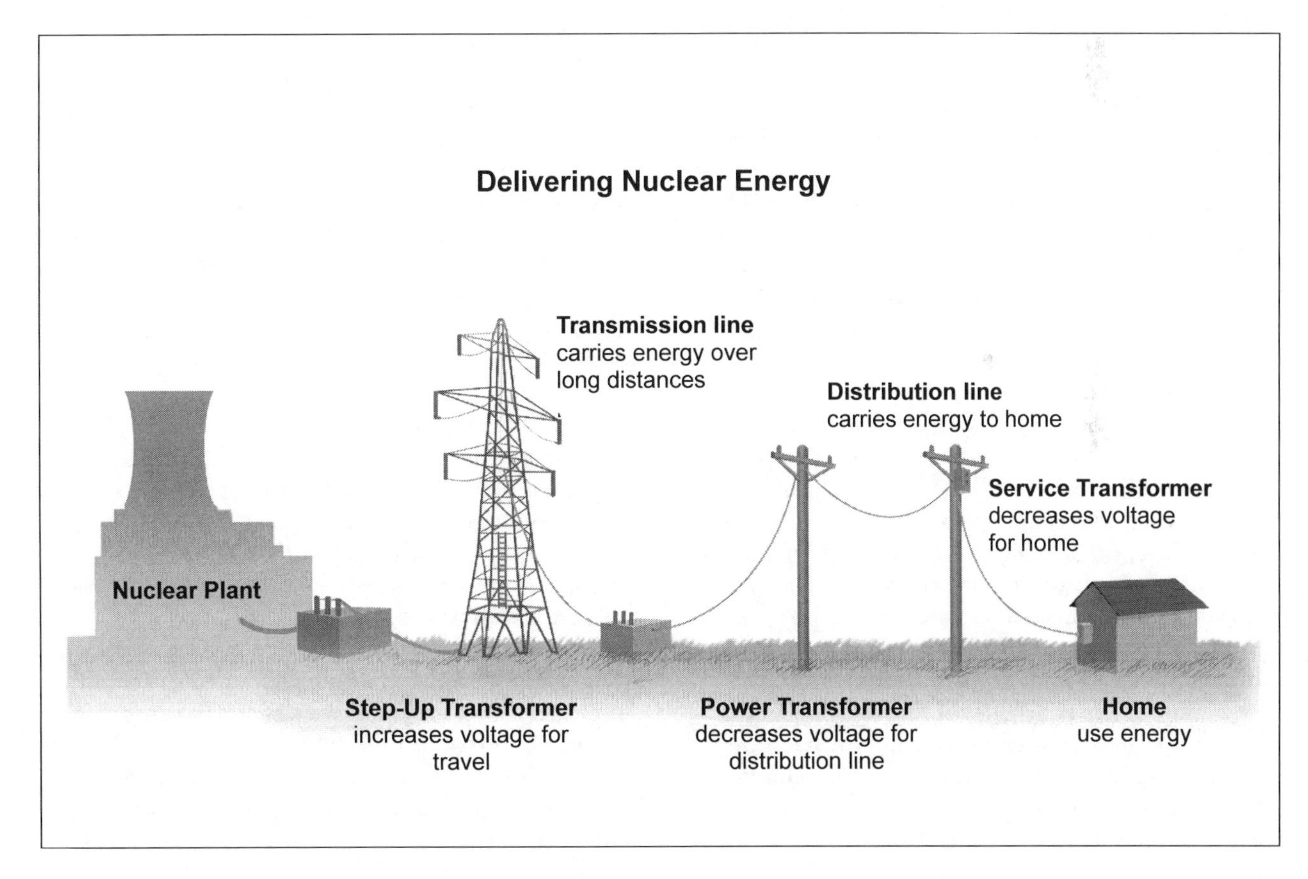

peace, however, was short lived. By 1947, the wartime alliance between the United States and the Soviet Union had dissolved into an intense competition for global influence. When the Soviet Union exploded its first atomic weapon in 1949, it began a nuclear arms race between the two countries that both threatened the safety of the world and kept it in check out of the mutual recognition by the two superpowers that given such weaponry, no nuclear war could be "won" in any meaningful sense.

Meanwhile, research in the field of nuclear technology began to focus on the use of nuclear energy to generate electricity. This captured the attention of world leaders concerned with their ability to meet the anticipated postwar demand for electricity in their growing cities. That feat was first accomplished on a test basis near Arco, Idaho, on December 20, 1951. The first nuclear power station went online in the Soviet Union on June 27, 1954, and two years later the first commercial nuclear power station opened in Sellafield, England. The United States joined the world, commercially producing nuclear power in December, 1957, with the opening of the Shippingport Atomic Power Station in Pennsylvania. By then, the United Nations (UN) had already convened in 1955 a conference to explore the peaceful use of nuclear technology, six Western European countries were banding together to form the European Atomic Energy Community (EURATOM), a supranational body committed to the cooperative development of nuclear power in Europe. The UN had created the International Atomic Energy Agency (IAEA) to encourage the peaceful use of nuclear technology on an even wider basis.

In retrospect, it appears that the early zeal in opening nuclear energy plants inadvertently paved the way for the technology's declining appeal by the end of the century. The early emphasis was on constructing and making operational a growing number of power plants, with a resultant neglect of safety issues in the choice of reactor design, the construction of the plants, and the training of the technicians charged with operating them. The United States Atomic Energy Commission (AEC), for example, was charged with both promoting and regulating the commercial use of nuclear energy. It was a dual mandate in which the first charge invariably got the better of the second, most notably demonstrated in 1979's Three Mile Island accident in Middletown, Pennsylvania, when the rush to get the plant online before all tests were performed resulted in a construction-flaw-induced accident compounded by human error that did much to dampen the appeal of nuclear power in the United States. Seven years later, human error merged with the flawed design of the Soviet nuclear power station in Chernobyl to undercut the appeal of nuclear power throughout most of Western Europe.

The military utility of the atom demonstrated during World War II also encouraged its continued pursuit in the military field in the United States, where nuclear reactors were harnessed to propel fleets of nuclear submarines and aircraft carriers that remain a mainstay of the U.S. Navy. Nonetheless, the real diffusion of nuclear technology has occurred in the civilian field, and as a result of scientific and political events separated by nearly a generation. In 1934, Irène Joliot-Curie and her husband, Frédéric Joliot-Curie, discovered that radium-like elements could be created by bombarding materials with neutrons ("induced radioactivity"), a discovery that eventually led to the inexpensive production of radioisotopes.

After World War II ended, though, there was a concerted postwar effort by governments to control tightly all research pertaining to nuclear technology.

It was not until 1953, when President Dwight Eisenhower proposed a broad sharing of information in his "Atoms for Peace" speech at the UN. That was a significant declassification of information after which the fruits of research in the field of nuclear technology became widely available and began to be utilized in myriad areas.

How It Works

The means by which nuclear technology is applied in the various arenas surveyed below varies from sector to sector. In general, however, nuclear technology produces its benefits either by altering the activity and/or weight of nuclear particles, or by exposing nonnuclear matter to radiation.

The nuclear power plants that generate electrical power, for example, like those that burn fossil fuel, function by heating water into steam in order to turn the turbines that produce electricity. The fuel consists of uranium oxide (commonly known as yellowcake) processed into solid ceramic pellets and packaged into long vertical tubes that are inserted into reactors

Fascinating Facts About Nuclear Technology

- J. Robert Oppenheimer, who headed the Manhattan Project that produced the atomic bomb, was later denied security clearance when he lobbied for international control of atomic energy and openly opposed the further development of nuclear weapons and the dangers he saw lining that path.
- It is widely believed that it was the Balance of Terror–the likelihood that a hot war between the United States and the Soviet Union would result in absolute mutually assured destruction (the MAD system)–that kept the peace between the two superpowers throughout the Cold War.
- Caught up in the optimism of the 1950's atomic age, in 1958 the Ford Motor Company unveiled the design of its future Nucleon line of atomic cars, which were to be powered by small reactors in their trunks.
- Outdoor malls across the United States often feature clocks with dials lit by radiation emanating from the low level of nuclear material that they contain, but not so low a level that Homeland Security does not fret every time one disappears, lest its radiation be used to create a dirty bomb.
- By the twenty-first century the products of nuclear technology had become so pervasive that most Western homes unknowingly contained them in such common, everyday devices as smoke detectors and DVD players.

to produce a controlled fissile chain reaction. Either pressure or cold water is utilized to control reactor heat and the intensity of the reaction.

By contrast, atomic weapons rely on generating a uranium chain reaction of an intentionally uncontrolled nature for maximum destructive effect, with the principle of fusion (the compressing of the atom into a smaller particle in order to produce energy) being exclusively utilized in the production of the more powerful thermonuclear bombs.

Elsewhere, the industrial use of radioisotopes rests on the fact that radiation loses energy as it passes through substances. Manufacturers have consequently been able to develop gauges to measure the thickness and density of products and, using radioisotopes as imaging devices, to check finished products for flaws and other sources of weakness. For their part, the fossil fuel industries involved both in mining and oil and gas exploration are using radioactive waves that measure density to search for resource deposits beneath the soil and sea. The medical community, the agriculture industry, and the producers of consumer goods that use nuclear technology largely rely on radioisotopes—more specifically, on exposing selected "targets" to radioisotope-containing chemical elements that can either be injected into a patient's body to "photograph" how an organ is functioning or employed to destroy undesirable or harmful elements.

Applications and Products

Although for some the mention of nuclear technology is most likely to conjure up threatening images of mushroom clouds or out-of-control nuclear power plants, nuclear technology has become a daily part of the lives of citizens in much of the developed world.

Nuclear Power Industry. The nuclear-based power industry that has emerged in the United States, United Kingdom, France, and more than twenty other countries around the globe encompasses more than 440 nuclear power plants and produces nearly one-fifth of the world's electrical output. Moreover, despite a slowdown in the construction of new plants, the slightly more than 100 U.S. nuclear power stations still generate more electricity than any fuel source and have assisted with the electrical needs of a steadily growing population.

Space Exploration. Space exploration has also substantially profited from nuclear technology—in particular, the development of the radioisotope thermoelectric generators (RTGs) that have used plutonium-generated heat to produce electrical power for unmanned spaced travel ever since the launch of Voyager I in 1977.

Medicine. Apart from the growing area of laser optical surgery, the existing applications of nuclear technology in medicine principally involve the use of positron emission tomography (PET) scans, other forms of imagining, and X rays in diagnostics, and radiation in the treatment of cancer.

Industry. The centerpiece of nuclear technology in the industrial field revolves around the diagnostic use of lasers and radioisotopes in order to improve the quality of goods, including the quality of the steel used in the automotive industry and the detection of flaws in jet engines.

Agriculture and the Pharmaceutical Industry. Nuclear technology is also used in these sectors to test the quality of products. The U.S. Food and Drug Administration (FDA), for example, requires testing of all new drugs, and 80 percent of that testing employs radioisotopes. But radiation is also widely used to treat products, especially in agriculture, where an irradiation process exposes food to gamma rays from a radioisotope of cobalt 60 to eliminate potentially harmful or disease-causing elements. Even livestock products are covered. Like its counterparts in at least ten other countries, the FDA approves the use of irradiation for pork, poultry, and red meat as well as for fruits, vegetables, and spices in order to kill bacteria, insects, and parasites that can lead to such diseases as salmonella and cholera.

Mining, and Oil and Gas Exploration. The process of searching for valuable natural resources has been radically altered in the last generation by the introduction of radiation wave-based exploratory techniques. Nuclear technology is also important to resource recovery and transportation in these industries. Lateral drilling, for example, relies on radiation wave directives to tap into small oil deposits, and construction and pipeline crews routinely use radiation waves to test the durability of welds and the density of road surfaces.

Consumer Products. Virtually every American home contains several consumer products using nuclear technology, from nonstick pans treated with radiation to prolong the life span of their surfaces, to photocopiers and computer disks that use small amounts of radiation to eliminate static, to cosmetics, bandages, contact-lens solutions and hygiene products sterilized with radiation to remove allergens.

Impact on Industry

In the case of nuclear power, the technology has created an industry that has found at least a small home on every continent and continues to be an attractive environmental alternative to coal-fired electrical plants. In other areas, the use of the technology has revolutionized the way of doing things—most notably in the field of oil and gas exploration, where the use of radium wave devices has essentially replaced the century-old exploration system that relied on geological formations and exploratory drilling to locate gas and oil deposits.

Elsewhere its impact has perhaps been less stark but often no less dramatic. Imagine devices that allow medical personnel to detect tumors, blocked arteries, and other life-impairing problems earlier, when they are more treatable. The use of radioactive materials to treat cancer has saved innumerable lives, especially when surgery has not been an option. The food that is available where irradiation is utilized is safer, and for others life is just easier or better, from frying bacon in nonstick pans to driving on roads with stronger foundations and fewer potholes. The nuclear age has become a lot more pervasive and diffused than anyone imagined three generations ago, and its imprint on industry and life is etched—albeit almost undetectably—virtually everywhere.

Careers and Course Work

Given the breadth of the applications of nuclear technology, career options lie in almost every field, but almost all require a college degree involving specific technical training, especially in fields such as nuclear chemistry, nuclear physics, nuclear engineering, and nuclear medicine. For the more specialized areas, a career in either nuclear-technology research or in development and application tends to require one or more advanced degrees. Given the complex, cutting-edge nature of such work, graduate training is often important for even nuclear-technician positions. Jobs nonetheless remain plentiful in most sectors for those who have that training, both in government (maintaining and operating the Navy's nuclear fleet) and in the private sector. The power industry routinely advertises its need for design engineers, process control engineers, technical consultants, civil, mechanical, and electrical nuclear engineers, and nuclear work planners; the medical community constantly seeks radiologists and other personnel trained in nuclear technology; and nuclear technicians remain in high demand in consumer product manufacturing, mining, and agriculture.

Public administration careers should not be ignored. National and local government entities such as the U.S. Department of Energy, the Nuclear Regulatory Commission, oversight agencies at the state level, and their counterparts in other countries are also career outlets for those combining business administration or public administration training with a knowledge of nuclear technologies.

Social Context and Future Prospects

The application of nuclear technology is basically progressing on three pathways, the first of which is of serious global concern.

Students of international affairs have long been concerned with the problem of "runaway" nuclear proliferation: the acquisition of nuclear weapons by so many states that others will also feel the need to acquire them, trebling the number of nuclear-armed states in a short time and making an accidental or intentional nuclear war more likely. The presence of stateless terrorist organizations who are willing to engage in extremist activity involving high kill numbers has significantly elevated this concern. Until the 2000's, the pace of proliferation was incremental and those who acquired the weaponry were sometimes loath to publicize the fact. The acquisition of atomic weapons by Pakistan and North Korea, and the apparent pursuit of nuclear arms by Iran, have been highly publicized and may have pushed the world to that "runaway" point in which countries with fast breeder reactors—whose recycled nuclear fuel can be brought to weapons-grade quality—may be encouraged to develop nuclear weapons themselves.

The second track holds considerably more potential for good: a renewed interest in nuclear power to meet the world's growing electrical needs. As a source of electrification, nuclear power fell largely out of fashion during the late twentieth century in much of the world as a result of: the cost of building and maintaining nuclear power plants compared with the cheap cost of imported energy between 1984 and 2003; the antinuclear movement and the public's concern about the construction of nuclear power plants in their backyards; and the appeal of environmentally friendly, renewable green energy sources during the era of rising oil prices that followed the U.S.-led invasion of Iraq in 2003.

That noted, the prospect for employment in the field of nuclear power remains good for three reasons. First, research and development activity has resulted in the application of techniques that have prolonged the life span of existing nuclear power plants well beyond their intended use cycle. Trained personnel are needed at all levels to continue that research and safely operate those nuclear power plants. Second, the green technologies being pursued are unlikely to be able to power the giant electrical grids that are increasingly being demanded by the megacities that are emerging, especially in third world countries. Such technologies will not be able to meet the existing global demand for ever more electrical power, which is increasing at about 1 percent per year in the United States, and at a far higher rate in developing areas. Large-scale nuclear power plants can meet those needs, while also adhering to the environmental standards to which northern and southern hemisphere states have committed themselves.

Finally, there is the broad, umbrella area of civilian societal applications, where a continuing, high demand for nuclear technology in the field of medicine, virtually every area of industry, agriculture, and consumer products can be predicted with a far greater assurance than the future demand for nuclear power as a source of electrification. In fact, so assured is the presumption of a steadily growing demand for nuclear-based products in medicine alone that it is driving much of the interest in constructing new reactor facilities just to produce the materials used in radiation-based therapies.

Joseph R. Rudolph, Jr., Ph.D.

Further Reading

Angelo, Joseph A., Jr. *Nuclear Technology.* Westport, Conn.: Greenwood Press, 2004. Basic follow-up reading on the history and impact of, and future for, nuclear technology.

Morris, Robert C. *The Environmental Case for Nuclear Power: Economic, Medical and Political Considerations.* St. Paul, Minn.: Paragon House, 2000. A concise (two-hundred-page) argument for continuing the construction of nuclear power plants based on the contributions that technology makes to multiple sectors.

Shackett, Peter. *Nuclear Medicine Technology: Procedures and Quick Reference.* 2d ed. Philadelphia: Lippincott, Williams and Wilkins, 2008. Although aimed at medical students, this text offers a useful, alphabetized catalog of applications of nuclear technology in disease diagnosis and treatment.

Stanculescu, A., ed. *The Role of Nuclear Power and Nuclear Propulsion in the Peaceful Exploration of Space.* Vienna: International Atomic Energy Agency, 2005. Slightly more than one hundred pages, the work contains an immensely interesting set of easy-to-understand essays on the topic.

United States Congress, House Committee on Foreign Affairs, Subcommittee on Terrorism,

Nonproliferation, and Trade. *Isolating Proliferators, and Sponsors of Terror: The Use of Sanctions and the International Financial System to Change Regime Behavior.* Washington, D.C.: Government Printing Office, 2007. An examination of the dangers posed by the international movement of weapons-grade materials and the personnel capable of constructing nuclear weapons out of them.

Yang, Chi-Jen. *Belief-Based Energy Technology Development in the United States: A Comparative Study of Nuclear Power and Synthetic Fuel Policies.* Amherst, N.Y.: Cambria Press, 2009. An excellent study of the importance of political clout as well as technological feasibility in explaining energy choices in the American political process.

Web Sites

American Nuclear Society
http://www.aboutnuclear.org

International Atomic Energy Agency
http://www.iaea.org

Nuclear Energy Institute
http://nei.org

U.S. Department of Energy
Nuclear Security
http://energy.gov/nationalsecurity/nuclearsecurity.htm

O

OCEAN AND TIDAL ENERGY TECHNOLOGIES

FIELDS OF STUDY

Engineering; oceanography; fluid mechanics; hydrodynamics; atmospheric physics; energy conversion; computer control systems.

SUMMARY

Every continent on the planet is surrounded by a cleaner, safer, more efficient energy resource. As conventional energy supplies are depleted, means are being developed, and in some cases are in actual operation, to convert the energy found in waves, tidal currents, ocean and river currents, ocean thermal gradients, and offshore wind into usable electric power for utility-scale grids, independent power producers, and the public sector.

KEY TERMS AND CONCEPTS

- **Barrage:** Artificial obstruction in a watercourse, such as a bridge or dam.
- **Biomass:** Mass of living organisms, such as kelp, from which energy can be obtained.
- **Greenhouse Gases:** Atmospheric gases thought to contribute to global warming.
- **Ocean Temperature Energy Conversion (OTEC):** Obtains energy from water-temperature differences.
- **Reversing Tidal Power Plant:** Power plant that generates power both on rising and falling tides.
- **Spring Tide:** High tide that is unusually high, occurring twice monthly because of the cycle of the Moon.
- **Tidal Range:** Vertical distance between the highest and lowest stands of the tide.
- **Tidal Stream:** Fast-moving ocean or estuary current generated by the tides.
- **Tide Mill:** Grinding mill for grain that uses tidal power to turn a water wheel.

DEFINITION AND BASIC PRINCIPLES

The tides were the earliest source of obtaining power from the ocean. The requirements were simple: a dam to contain a head of water that was brought in by high tide and a water wheel to turn as the water was let out, thus generating the power. The obtaining of power from crashing waves was another dream of oceanographers, and it has finally been realized in a variety of modest projects that generate power in a number of ingenious ways. Another dream of oceanographers has been deriving power from fast-moving currents in estuaries on in the ocean itself. Several so-called in-stream devices have now been developed, but the ultimate goal of obtaining power from Florida's famous Gulf Stream, the fastest-moving ocean current in the world, is still elusive although study is under way. Another interesting project was the famous ocean temperature energy conversion (OTEC) project to obtain power from the temperature difference between warm- and cold-ocean waters. This project had a brief success in Hawaii during the 1980's, was dropped, but is being looked at again by several countries. And a spectacular newcomer for obtaining power from the ocean is the development of offshore wind turbines. These giant windmills, whirling high on stilts, are now enjoying great success in the coastal waters of Europe, although development in the United States as of 2011 has been slow.

BACKGROUND AND HISTORY

Power has been generated from the tides in Europe since at least the Middle Ages, and tide mills were common in Great Britain, France, Ireland, and along the east coast of the United States until the middle of the nineteenth century. As of 2011, the world has four power stations generating electricity from the tides. The largest is located on the Rance River estuary in France and is rated at 240 megawatts. Others are on the Annapolis River in Nova Scotia, Canada (rated at up to twenty megawatts), the Kislaya Guba, Russia, plant (rated at one to two megawatts), and the Xiamen, China, plant (rated at up to three megawatts). The generation of power from the waves has been a more recent development. Small-scale installations are now generating power in Scotland and at other locations, and projects involving various new methods are under way.

How It Works

The model for a successful tidal barrage plant is on the Rance River estuary in Brittany, France. Here the tide range is about forty feet. As the rising tide passes through the circular openings in the barrage, which is located near the mouth of the estuary, rotors spaced at regular intervals in the circular openings generate power as the water level rises. After six hours, the tide will have turned and the flow of water is in the opposite direction. Then the rotors are turned 180 degrees so that the plant can generate power as the tide flows out to sea.

Wave Power. Many ingenious devices have been designed to obtain electric power from waves. One type is the point absorber, a bottom-mounted or floating structure that can absorb wave energy coming from all directions. A second type is the terminator, which reflects or absorbs all the wave energy coming at it. Another type is the linear absorber, which is oriented parallel to the direction of the oncoming waves. It is composed of interlocking sections, and the pitching and yawing of these sections, because of the waves, pressurizes a hydraulic fluid that turns a turbine. A fourth device is an oscillating water-column terminator, which is a partially submerged chamber with air trapped above the water's surface. As entering waves push the air column up, the compression of the air will act as a piston to drive a turbine. All of these devices are in the developmental stage, but no single design has been judged the best. Factors that must be considered are the corrosive and occasionally violent marine environment and biofouling, which begins the moment any device is placed in the ocean.

Power from Tidal and Ocean Currents. The dream of obtaining power from the ocean's fast-moving currents may finally be nearing realization. Tidal generators are being tested in several estuaries and ocean channels, using either bottom-mounted rotors or rotors suspended from floating barges. Power generation is nearly continuous, with the rotors turning for both flood and ebb currents. The greatest challenge will be harnessing power from the ocean's famous Gulf Stream. A group of researchers at Florida Atlantic University in Boca Raton, Florida, plans to mount one or more giant rotors in this current as it flows between Florida and the Bahamas. Water depths approach 2,500 feet there, so attaching the rotors to the seabed will be a challenge. In addition, there will be the usual problems of biofouling, corrosion, and getting the power to shore. The rotors will constitute a potential hazard for passing ships, submarines, and large marine mammals, such as whales, so these concerns will also have to be addressed as well. The amount of power generated might well approach 10,000 megawatts.

Ocean Thermal Energy Conversion (OTEC). An OTEC plant operated successful in Hawaii during the 1980's and is being considered by several European countries, despite its high cost. This plant derives power from the differential between 40-degree-Fahrenheit deep water and 80-degree-Fahrenheit surface water. The 40-degree-Fahrenheit water, drawn from thousands of feet down, is used to condense ammonia, which is then brought in contact with 80-degree-Fahrenheit surface water, causing it to vaporize explosively, driving a turbine. One problem is the disposal of the 40-degree-Fahrenheit water after it has been warmed in the vaporization process. It cannot be returned to the ocean, where it would kill reefs and tropical fish, so the solution in Hawaii was to pipe it through the soil where it fooled cool-weather crops such strawberries and asparagus into growing in a tropical climate.

Wind Power. The latest method for obtaining energy from the ocean is the giant turbines turning in the near-shore, shallow waters of Europe, where more than 1,500 megawatts of power is being generated. These huge turbines function much like windmills on land, except that they are firmly anchored to the seafloor, often in plain sight of coastal residents. Environmental and aesthetic concerns have so far held up the installation of such turbines along the east coast of the United States, but construction has finally been approved for a group of wind turbines off the coast of Massachusetts.

Applications and Products

The number of tidal barrage power plants in the world is extremely limited because of the large tidal range required—a minimum of a ten-foot rise between low tide and high tide. Few coasts in the world have that great a rise in a body of water narrow enough to be dammed, and even fewer have it in an area that is sufficiently populated to provide a market for the power generated. In addition to the four tidal power plants already mentioned, an additional plant has been proposed at the Severn Estuary in southwestern England (rated at 1,200 to 4,000 megawatts). One problem faced by all tidal barrage plants is that the power generated is intermittent. The time suitable for power generation shifts steadily as the Moon orbits around the Earth. This means that the supply of power and the demand for power will not always coincide. For several nights customers will have ample power to cook their supper and enjoy evening activities, but the next few nights they will have no power at all.

Power Generation from Waves. Installations generating a total of about 4 megawatts of power have now been created worldwide, mostly on an experimental or demonstration project basis, but one that has been generating power successfully since 2000 is the Limpet, the world's first commercial, wave power station, on the rockbound coast of the Island of Islay in Scotland. The plant is of the oscillating water column design with a fortress-like exterior fronting the waves just at sea level along the rocky shore. When a breaking wave enters the long, concrete tube, with its opening just below water level, it drives air in and out of a pressure chamber through a specially designed air turbine, generating electricity. The Limpet produces 0.5 megawatts of power, which is fed into the island's power grid. The design makes the Limpet easy to build and install, and its low visible profile does not intrude on the coastal landscape or the ocean views.

Power from Currents. Besides the proposed giant rotors in the Gulf Stream, which would be an open-ocean device, a number of projects have been designed for obtaining power from tidal currents in estuaries and other constricted passages using bottom- or surface-mounted turbines. These turbines do not require construction of an expensive barrage. The flow of water simply turns the turbine as the tide comes in, and, if the turbine is reversed 180 degrees, it can also generate power as the tide goes out. An experimental array of six turbines was installed in New York City's East River in 2007 and has thus far proved successful. Similar arrays are being installed in Ireland, the United Kingdom, Italy, Korea, and Canada. The turbines strongly resemble torpedoes, with the rotor at one end, and they stand on a stout pedestal firmly attached to a channel bed or they can be suspended from floating barges. Power is transmitted to shore by means of cables.

Wind Farms. The major components for a wind turbine are a tower, a rotor with hub and blades, a gear box, and a generator. Offshore systems are larger than those on land because of the greater cost to install and service them. Most are rated at three to five megawatts, compared with one and one-half to three megawatts for those on land. They are usually fixed in water depths of fifteen to seventy-five feet and require a foundation driven deep into the seabed to support the weight of the tower and the rotor. Frequently the turbines are arranged in arrays to reduce maintenance and cabling costs. The power generated is sent to shore through a high-voltage cable buried in the seabed. A service area is always required, with boats and a hoist crane for repairs and maintenance. By the end of 2008, 1,471 megawatts of turbines have been installed worldwide, primarily in the United Kingdom, Germany, Denmark, the Netherlands, and Sweden. The U.S. projects will be located along the North and Central Atlantic coasts, and several are in the permit process. Wind farms in the Great Lakes, the Gulf of Mexico, and eastern Canada will probably come next.

Impact on Industry

It is estimated that marine renewable energy projects have the potential to provide 15 percent of the United Kingdom's electrical needs and about 7 percent of the United States' electrical power needs. Research is most active in the countries with the most access to wave, tidal, and current power, namely the United Kingdom, Ireland, Portugal, and the United States. Jointly operated projects will be the rule because the marine environment is a harsh one and significant engineering and environmental challenges must be overcome.

Government Agencies. The Federal Energy Regulatory Commission (FERC) has broad authority to issue licenses for hydroelectric projects located in the navigable waters of the United States. These licenses can carry a term of up to fifty years. Projects cannot harm the waterway and must be in the public's interest. Environmental concerns must also be taken into consideration. FERC has 130 applications for permits under consideration. For proposals relating to the outer continental shelf, generally defined as beginning three nautical miles from shore, the Minerals

Management Service (MMS) has jurisdiction to authorize and manage projects. As of 2011, the MMS is presently considering proposals for six wind farms, two wave projects, and three ocean current projects. On October 6, 2010, Secretary Ken Salazar of the Department of the Interior and Cape Wind Associates signed the nation's first lease for a wind farm. It will be located on the outer continental shelf in Nantucket Sound, off the Massachusetts coast, and will generate 468 megawatts of power.

University Research. Coastal universities frequently partner with industry and government agencies to test, model, and develop marine-energy generators. Oregon State University and Ocean Power Technologies have developed kilowatt-scale wave converters off the West Coast of the United States. In New England, the Maine Maritime Academy has proposed a small-scale tidal energy conversion test facility, and the Northwest National Marine Renewable Energy Center, working with Oregon State University and the University of Washington, is developing a full-scale test of this facility. The University of Hawaii runs a test center for the U.S. Department of Energy, and in 2005, the Florida Atlantic University Center for Ocean Energy began investigating ways of generating electrical power from the Gulf Stream.

Industry. Companies are involved in renewable ocean energy projects as well. In the United States, Ocean Renewable Power Company is partnering with the Coast Guard station near Westport, Connecticut, to demonstrate a sixty-kilowatt tidal turbine, which will be the largest ocean tidal energy generator thus far constructed in the United States. Ocean Power Technologies, a Nasdaq listed company with the symbol OPTT, is conducting trials of a large floating-wave energy converter to be attached to the seabed. The United Kingdom's E.ON UK, which has stakes in twenty wind farms across the country, has just begun generating power at Scroby Sands, off the Norfolk coast, the United Kingdom's largest wind farm so far. Its thirty two-megawatt turbines were built by the Danish wind turbine company Vestas.

Fascinating Facts About Ocean and Tidal Energy Technologies

- In its World Energy Outlook for 2008, the International Energy Agency predicts that if present trends continue, the world's overall energy demands will almost double by 2050.
- As of 2011, only about 7 percent of the U.S. energy supply comes from renewable sources, and most of this is derived from biomass and hydropower. The rest of U.S. energy needs come from oil, coal, natural gas, and nuclear power. Marine renewable energy sources accounted for only about 0.5 percent of U.S. energy needs in 2008.
- Thirty states in the United States border either an ocean or one of the Great Lakes. These states generate and consume 75 percent of the nation's electricity, but the development of renewable marine power resources by these states so far is practically nil.
- The fastest growing renewable energy source in the world as of 2011 is wind energy, with a yearly growth rate of 20 to 30 percent. Germany and the United Kingdom are two of the largest wind-power generators, and Germany's Innogy Nordsee 1 project, being developed in the North Sea, is expected to provide power for nearly 1,000,000 homes.
- Limpet, the world's first commercial wave power station, generates 0.5 megawatt of power–enough to light about 500 homes.

CAREERS AND COURSE WORK

College-level students seeking careers in marine renewable energy are advised to take a general oceanography course in order to familiarize themselves with the characteristics of the environment in which they will be working. In addition, they should supplement this study with basic courses in math and physics so that they can understand the technology involved in the design of the various marine-energy projects. For those students planning to go on to graduate study, engineering courses are highly recommended, especially for those students interested in designing and building ocean energy generators. As of 2011, the jobs available in the United States for students seeking opportunities in the field of marine-renewable energy are still somewhat limited because most of the American marine energy projects operating are just getting under way or are still in developmental stages. The country's first wind farm project was approved for Massachusetts in late 2010, but construction has not yet begun. Additional wind farm projects are planned but the permit process for them is a lengthy one. Several coastal wave energy projects are now operating, as well as a few tidal stream projects in rivers and estuaries with strong tidal flows, but these projects are all in developmental stages. Once they mature into full-time operations, job opportunities in these areas should begin to appear.

Social Context and Future Prospects

For many years the United States has been totally dependent on traditional energy sources, such as oil, natural gas, coal, water power, and atomic energy. Attention is turning to the generation of electricity from marine renewable energy sources. Projects in this field are still in the developmental stages, with tidal stream and wind farm installations showing the most promise. The potential for them to make a significant contribution to U.S. energy needs is great because nearly half the states have access to the oceans along their borders. Although the United States has ample supplies of coal, the burning of coal harms the environment, and U.S. reserves of oil and natural gas are on a downward trend. Attention to marine-energy sources cannot help but increase in the coming years. The recent approval of a wind farm for Massachusetts is an encouraging sign. Wind farms are already making a significant contribution to energy needs in Europe, and coastal conditions along the Atlantic and Gulf coasts of the United States are equally favorable. The cost of these projects is large, but they will provide many jobs for the installation and maintenance of the equipment, as well as for the manufacture of the turbines.

Donald W. Lovejoy, B.S., M.S., Ph.D.

Further Reading

Bedard, Roger, et al. "An Overview of Ocean Renewable Energy Technologies." *Oceanography* 23, no. 2 (June, 2010): 22-31. Highlights the development status of the various marine-energy conversion technologies. Contains many helpful color photographs and explanatory diagrams.

Boon, John D. *Secrets of the Tide: Tide and Tidal Current Application and Prediction, Storm Surges and Sea Level Trends.* Chichester, England: Horwood, 2004. Describes the so-called tide mills used in many locations along the East Coast of the United States in the eighteenth century for grinding grain into flour or meal.

Carter, R. W. G. *Coastal Environments: An Introduction to the Physical, Ecological, and Cultural Systems of Coastlines.* San Diego: Academic Press, 1988. Includes an outstanding analysis of tidal power, with a description of the mechanics of power generation and a discussion of the environmental and ecological effects.

Segar, Douglas A. *Introduction to Ocean Sciences.* 2d ed. New York: W. W. Norton, 2007. A well-respected textbook with solid foundational information.

Sverdrup, Keith A., and E. Virginia Armbrust. *An Introduction to the World's Oceans.* 10th ed. New York: McGraw-Hill, 2009. Provides information on how energy is now being obtained, or may someday be obtained, from ocean waves, currents, and tides. Many useful diagrams.

Thresher, R., and W. Musial. "Ocean Renewable Energy's Potential Role in Supplying Future Electrical Needs." *Oceanography* 23, no. 2 (June, 2010): 16-21. A summary of the nation's and the world's energy needs and the sources from which these energy needs have been met and will be met in the future.

Ulanski, Stan. *Gulf Stream: Tiny Plankton, Giant Bluefin, and the Amazing Story of the Powerful River in the Atlantic.* Chapel Hill: University of North Carolina Press, 2010. Describes the world's fastest ocean current, summarizing the physical characteristics of location and speed that make this current a prime candidate for energy generation.

Web Sites

American Wind Energy Association
http://www.awea.org

National Hydropower Association
http://hydro.org

National Oceanographic and Atmospheric Administration
Earth System Research Laboratory
http://www.esrl.noaa.gov/research/renewable_energy

OPTICAL STORAGE

FIELDS OF STUDY

Optical disk manufacturing; optical disk reader and writer manufacturing.

SUMMARY

Optical storage refers to a variety of technologies that are used to read and write data. It employs special materials that are selected for the way they interact with light (an optical, or visible, medium). As of 2011, most optical storage devices being manufactured are digital; however, some are analogue. Both the computer and entertainment industries offer numerous practical applications of optical storage devices. Common optical storage applications are compact discs (CDs), digital versatile discs (DVDs), and Blu-ray discs (BDs). A variety of data can be stored optically, including audio, video, text, and computer programs. Data are stored in binary form.

KEY TERMS AND CONCEPTS

- **Blu-Ray Disc (BD):** Disc commonly used for storage of high-definition video and audio that can hold twenty-five to fifty gigabytes of information. BDs can also contain other forms of information, such as computer programs or data backup.
- **Compact Disc (CD):** Disc that can store 700 megabytes, or eighty minutes of music, in audio format.
- **Compact Disc Read-Only Memory (CD-ROM):** Compact disc that contains data that can be read by a computer. It can hold any form of data; however, it is commonly used for computer programs, games, and multimedia applications.
- **Digital Versatile Disc (DVD):** Optical storage disk that can hold from 4.7 to 8.7 gigabytes of data.
- **Optical Disk:** Plastic-coated disk that stores digital data, such as music, images, or text. Tiny pits in the disk are read by a laser beam.
- **Optical Disk Drive:** Device that can read one or more types of optical disks.

Definition and Basic Principles

Optical storage differs from other data storage technologies such as magnetic tape, which stores data as an electrical charge. Optical disks are flat and circular and contain binary data in the form of microscopic pits, which are non-reflective and have a binary value of 0. Smooth areas are reflective and have a binary value of 1. Optical disks are both created and read with a laser beam. The disks are encoded in a continuous spiral running from the center of the disk to the perimeter. Some disks are dual layer: With these disks, after reaching the perimeter, a second spiral track is etched back to the center. The amount of data storage is dependent upon the wavelength of the laser beam. The shorter the wavelength, the greater the storage capacity (shorter-wavelength lasers can read a smaller pit on the disk surface). For example, the high-capacity Blu-ray Disc uses short-wavelength blue light. Lasers can be used to create a master disk from which duplicates can be made by a stamping process.

Optical media is more durable than electromagnetic tape and is less vulnerable to environmental conditions. With the exception of Blu-ray Discs, the speed of data retrieval is considerably slower than that of a computer hard drive. The storage capacity of optical disks is significantly less than that of hard drives. Another, less common form of optical storage is optical tape, which consists of a long, narrow strip of plastic upon which patterns can be written and from which the patterns can be read back.

Background and History

Optical storage originated in the nineteenth century. In 1839, English inventor John Benjamin Dancer produced microphotographs with a reduction ratio of 160:1. Microphotography progressed slowly for almost a century until microfilm began to be used commercially in the 1920's. Between 1927 and 1935, more than three million pages of books and manuscripts in the British Library were microfilmed by the Library of Congress. Newspaper preservation on film had its onset in 1935 when Kodak's Recordak division filmed and published the *New York Times* on thirty-five-millimeter (mm) microfilm reels.

Analogue optical disks were developed in 1958 by American inventor David Paul Gregg, who patented the videodisc in 1961. In 1969, physicists at the Netherlands-based Royal Philips Electronics began experimenting with optical disks. Subsequently, Philips and the Music Corporation of America (MCA) joined forces to create the laser disc, which was first introduced in the United States in 1978. Although the laser disc achieved greater popularity in Asia and Europe than it did in the United States, it never successfully competed with VHS tape. In 1980, Philips partnered with Sony to develop the compact disc (CD) for the storage of music. A few years later, the CD had evolved into a compact disc read-only memory (CD-ROM) format, which in addition to audio, could store computer programs, text, and video. In November 1966, the digital versatile disc (DVD) format was first introduced by Toshiba in Japan; it first appeared in the United States in March, 1997; in Europe in October, 1998; and in Australia in February, 1999. A format war between two higher-capacity data storage technologies emerged in 2006 when Sony's Blu-ray and Toshiba's HD DVD players became commercially available for the recording and playback of high-definition video. Two years later, Toshiba conceded to Sony; Blu-ray was based on newer technology and had a greater storage capacity.

How It Works

CDs, DVDs, and BDs are produced with a diameter of 120 mm. The storage capacity is dependent upon the wavelength of the laser: the shorter the wavelength, the greater the storage capacity. CDs have a wavelength of 780 nanometers (nm), DVDs have a wavelength of 650 nm, and BDs have a wavelength of 405 nm. Some disk drives can read data from a disk while others can both read and write data.

Optical System. In a disk reader, the optical system consists of pickup head (which houses a laser), a lens for guiding the laser beam, and photodiodes that detect the light reflection from disk's surface. The photodiodes convert the light into an electrical signal. An optical disk drive contains two main servomechanisms: One maintains the correct distance between the lens and the disk and also ensures that the laser beam is focused on a small area of the disk; the other servomechanism moves the head along the disk's radius, keeping the beam on a continuous spiral data path. The same servomechanism can be used to position the head for both reading and writing. A disk writer employs a laser with a significantly higher power output. It burns data onto the disk by heating an organic dye layer, which changes the dye's reflectivity. Higher writing speeds require a more powerful laser because of the decreased time the laser is focused on a specific point. Some disks are rewritable—they contain a crystalline metal alloy in their recording layer. Depending on the amount of power applied, the substance may be melted into a crystalline form or left in an amorphous form. This enables the creation of marks of varying reflectivity. The number of times the recording layer of a disk can be reliably switched between its crystalline and amorphous states is limited. Estimates range from 1,000 to 100,000 times, depending on the type of media. Some formats may employ defect-management schemes to verify data as it is written and skip over or relocate problems to a spare area of the disk. A third laser function is available on Hewlett-Packard's LightScribe disks. The laser can burn a label onto the side opposite the recording surface on specially coated disks.

Double-Layer Media. Double-layer (DL) media has up to twice the storage capacity of single-layer media. DL media have a polycarbonate first layer with a shallow groove; a first data layer, a semi-reflective layer; a second polycarbonate spacer layer with a deep groove; and a second data layer. The first groove spiral begins on the inner diameter and extends outward; the second groove starts on the outer diameter and extends inward. If data exists in the transition zone, a momentary hiccup of sound and/or video will occur as the playback head changes direction.

Disk Replication. Most commercial optical drives are copy protected; in some cases, a limited number of copies can be made. Disks produced on home or business computers can be readily copied with inexpensive (or included) software. If two optical drives are available, a disk-to-disk copy can be made. If only one drive is available, the data is first stored on the computer's hard drive and then transferred to a blank disk placed in the same read-write drive. For copying larger numbers of disks, dedicated disk-duplication devices are available. The more expensive ones incorporate a robotic disk-handling system, which automates the process. Some products incorporate a label printer. Industrial processes are used for mass replication of more than 1,000 disks, such as DVDs, CDs, or computer programs. These disks are manufactured from a mold and are created via a series of industrial processes including pre-mastering, mastering, electroplating, injection molding, metallization, bonding, spin coating, printing, and advanced quality control.

Applications and Products

Numerous applications and products are focused on optical storage. Most applications are geared toward the computer and entertainment industries.

Optical Disks. Most optical disks are read-only; however, some are rewritable. They are used for the storage of data, computer programs, music, graphic images, and video games. Since the first CD was introduced in 1982, this technology has evolved markedly. Optical data storage has in large part supplanted storage on magnetic tape. Although optical storage media can degrade over time from environmental factors, they are much more durable than magnetic tape, which loses its magnetic charge over time. Magnetic tape is also subject to wear as it passes through the rollers and recording head. This is not the case for optical media, in which the only contact with the recording surface is the laser beam. CDs are commonly used for the storage of music: A CD can hold an entire recorded album and has replaced the vinyl record, which was subject to wear and degradation. A limitation of the CD is its storage capacity: 700 megabytes of data (eighty minutes of music). The DVD, which appeared in 1996, rapidly gained popularity and soon outpaced VHS tape for the storage of feature-length movies. The DVD can store 4.7 gigabytes of data in single-layer format and 8.5 gigabytes in dual-layer format. The development of high-definition television fueled the development of higher-capacity storage media. The Blu-ray (BD) Disc can store about six times the amount of data as a standard DVD: 25 gigabytes of data in single-layer format and 50 gigabytes of data in dual-layer format. The most recent evolution of the BD disc is the 3-D format; the increased storage capacity of this medium allows for the playback of video in three dimensions.

Computer Applications. Almost all computers, including basic laptops, contain one or more optical drives, most of which house a read-write drive. The drive is used to load computer programs onto a hard drive, data storage and retrieval, and entertainment. The inclusion of Blu-ray readers as well as writers (burners) is increasing; this is a result of significant drops in price of these devices as well as blank media since their introduction to the marketplace. Most internal drives for computers are designed to fit in a 5.25-inch drive bay and connect to their host via an interface. External drives can be added to a computer; they connect via a universal serial bus (USB) or FireWire interface.

Entertainment Applications. Optical disk players are a common form of home entertainment. Most play DVDs as well as CDs. As with computer applications, the presence of Blu-ray devices is rapidly increasing. Some load a single optical disk at a time; others load a magazine, which holds five or six optical disks; others have a capacity for several hundred disks. The higher-capacity players often have a TV interface in which the user can make a selection from a list of the disks contained within. Often, the list contains a thumbnail image of the CDs or DVDs. These devices contain audio and video outputs to interface with home entertainment systems. If attached to an audio-video receiver (AVR), surround-sound audio playback can be enjoyed. Some DVD players have an Ethernet interface for connection to the Internet. This allows for streaming video from companies such as Netflix and Web surfing. Many automobiles contain a CD or DVD drive. DVD video is displayed to backseat passengers. Vehicle navigation systems often contain an optical drive (CD or DVD) containing route information. Portable players are also available. These range from small devices that can be strapped on an arm to small battery-operated players similar in size to a laptop computer. Many laptop owners use their devices for viewing DVD or BD movies.

Games. Although a wealth of games can be played on a computer, a number of devices in the marketplace are designed strictly for game playing. Most of these devices attach to a television set and are interactive; thus, the player can immerse oneself in the action. Some accommodate more than one player. Although earlier devices had proprietary cartridges for data storage, the trend is toward DVDs and BDs. Three-dimension consoles, which have added realism, are also available.

Impact on Industry

Optical storage is a major component of the computer and entertainment industries. Technology is rapidly evolving in both segments, fueling the purchase of upgraded items. The Blu-ray Disc is increasing in popularity. Fueling this increase is a drop in price of players, recorders, and media. New

Fascinating Facts About Optical Storage

- The precursor of the DVD and BD, the laser disc was almost twelve inches in diameter and could store an hour of standard-definition analogue video on each side. The BD is 4.7 inches in diameter and can readily store more than two hours of high-definition digital video on a single side. The smaller track width of the BD is one factor; however, a larger factor is the compression of both video and audio. The video is in a compressed format and is uncompressed during playback.
- Microfilm was first used by the military during the Franco-Prussian War. Communication between Paris and the provincial government in Tours was by pigeon post. Since the pigeons were unable to carry paper dispatches, the Tours government photographed paper dispatches and compressed them to microfilm, which was carried to Paris by homing pigeons.
- On occasion, a traveler to a foreign country has purchased a DVD only to find upon returning home that the disc cannot be played. That is because two different formats are in use: National Television System Committee (NTSC) and Phase Alternating Line (PAL). The NTSC format runs thirty frames per second, and each frame includes 525 scan lines. The PAL format runs twenty-five frames per second, and each frame includes 625 scan lines. Because PAL has more scan lines, it is considered to have a higher resolution and therefore a higher-quality image. NTSC is used in the United States, while PAL is prevalent in Europe.
- In the 1990's, optical tape was forecast to be the standard for high-capacity, high-speed computer data storage format; however, this prediction did not come to pass.

features, such as 3-D video, spur new sales. When recorders or players fail, the device is either replaced or repaired. Backward compatibility improves the sales of new devices. For example, a 3-D BD player can also play DVDs and CDs. When a computer is replaced, a new one is purchased, usually housing one or more new optical drives. Gaming consoles are a rapidly evolving segment of this industry. A variety of consoles compete for market share and each manufacturer regularly releases upgrades of a current product line or introduces an all-new device.

Industry and Business. The average home contains optical devices and disks. Many will have one or more computers, one or more optical disk players, and gaming consoles, particularly if children live in the home. A wealth of educational material is available on optical media. Some are designed for an optical disk player while others are designed for use on a home computer. Those designed for computer use are often interactive, allowing feedback on the child's progress. Software is available for transferring video from a camcorder to a computer for editing and production of a DVD or Blu-ray video. Software ranges from basic, easy-to-use programs for home use to sophisticated programs for business and commercial applications. In some cases, an amateur videographer will purchase high-end software to develop high-quality home videos. Business uses of optical storage include the development of multimedia presentations with varying degrees of sophistication ranging from high-quality video and audio to simple slide shows. Playback of multimedia presentations requires either a computer with an optical drive or an optical player. It also requires a video monitor or projector, and often a sound system. The military uses optical storage in a manner similar to that of businesses.

Government and University Research. A source of funding for optical storage research in the United States is the Defense Advanced Research Projects Agency (DARPA). It is also a client for certain kinds of optical-storage applications. The agency is primarily interested in optical storage with a military focus. Another source of funding for optical storage research is the U.S. Department of Energy (DOE). The DOE is focused on energy efficiency and renewable energy. The DARPA and the DOE supply funds to many universities in the United States for optical-storage research and development.

CAREERS AND COURSE WORK

The optical-storage industry offers careers ranging from entry-level positions such as assemblers to high-level and highly technical positions, which require a scientific or engineering degree. Management positions are also available. The high-level positions require at least a bachelor's degree with course work in several fields related to laser technology:

engineering, physics, computer science, mathematics, and robotics. Many of positions require a master's or doctorate degree. Management positions are more accessible to individuals with a master of business administration (M.B.A.) degree. Positions are available for individuals with scientific degrees in both the government and private sector. The ability to be a team player is often of value for these positions because ongoing research is often a collaborative effort. University research positions are also available; in this arena, the employee is expected to divide his or her time between research and teaching.

Technicians are needed in a variety of fields for equipment repair. Many of these positions require some training beyond high school at a community college or trade school. If a company employs a number of technicians, supervisory positions may be available.

Social Context and Future Prospects

Merely three decades after optical-storage devices appeared on the market, laser applications have become innumerable and ubiquitous. Most households contain one or more computers with optical drives. Many have a DVD burner. DVD and CD players are common household items. Optical disks with games that can be played on a computer or specialized gaming consoles are popular not only with children but also with adults. Criticism has been directed at gaming because some are fearful that children and even adults will devote excessive amounts of time at the expense of other, more productive activities. Many computer games are violent. The goal of these games is to kill game inhabitants; often a "kill" is graphically displayed. Critics complain that this could lead to antisocial behavior, particularly in teens.

The trend for optical storage has progressed from CD to DVD to BD. High-capacity media are being used not only for entertainment, such as high-definition movies, but also for data storage. Many computer programs are marketed in DVD format. This is cost-effective because the production cost of three or more CDs is much higher than that of a single DVD, which can hold all of the data. Another trend is toward video streaming and hard-disk storage. Music can be downloaded from a Web site onto a hard drive or solid-state memory. Movies are being streamed directly to a computer, DVD player, or gaming console from companies such as Netflix. CD and DVD "jukeboxes," which can house up to several hundred disks are giving way to hard-drive devices (either computer or stand-alone). Some automobiles can be equipped with a hard-drive player of audio and sometimes video. These devices can be programmed with playlists to suit particular musical tastes. In contrast to a jukebox, the transition between selections is swift without any annoying "clanking." These devices are commonly loaded from CDs or DVDs, which can be stored for future use in the event of a hard-drive failure.

Robin L. Wulffson, M.D., F.A.C.O.G.

Further Reading

Bunzel, Tom. *Easy Creating CDs and DVDs.* 2d ed. Indianapolis: Que, 2005. A book and CD-ROM edition, which explains how to burn CDs and DVDs.

McDonald, Paul. *Video and DVD Industries.* London: British Film Institute, 2008. Examines the business of video entertainment and details divisions of the video business. It outlines industry battles over incompatible formats, from the Betamax-VHS war to competing laser disc systems, alternatives such as video compact disc or Digital Video Express, and the introduction of HD DVD and Blu-ray high-definition systems.

Taylor, Jim, Mark R. Johnson, and Charles G. Crawford. *DVD Demystified.* 3d ed. New York: McGraw-Hill, 2006. Covers basic technology as well as production and authoring processes. Describes the vast variety of competing video, audio, and data formats and explains how DVD standards and specs dovetail or clash with related digital-media standards.

Taylor, Jim, et al. *Blu-ray Disc Demystified.* New York: McGraw-Hill, 2009. Provides a detailed overview of Blu-ray technology, its uses, and its shortcomings.

Web Sites

Association for Computing Machinery
http://www.acm.org

Blu-ray Disc Association
http://www.blu-raydisc.com/en.html

Optical Storage Technology Association
http://www.osta.org.

P

PASTEURIZATION AND IRRADIATION

FIELDS OF STUDY

Food engineering; food science; microbiology; chemistry; medicine; nuclear science; mathematics.

SUMMARY

Pasteurization and irradiation are processes that partially sterilize food in order to make it safe to eat, without substantially altering its nutritional content, structure, and taste. Pasteurization uses mild heat treatment, whereas irradiation makes use of ionizing radiation. Both reduce the levels of pathogenic (disease-causing) and spoilage microorganisms to a level that renders the food safe to eat provided that it is stored appropriately for no longer than the prescribed time. Irradiation can also be used on fresh fruits and vegetables to kill insects and to retard biological processes, such as ripening. At high levels, irradiation will fully sterilize food, packing material, and disposable medical items.

KEY TERMS AND CONCEPTS

- **Gram-Negative Bacteria:** Bacteria that do not stain violet after a treatment developed by Danish bacteriologist Hans Gram; these include *Campylobacter, Escherichia coli, Salmonella,* and *Vibrio.*
- **Gram-Positive Bacteria:** Bacteria that do stain violet after a treatment developed by Hans Gram; these include *Clostridia* and *Listeria.*
- **Gray (Gy) Unit:** Unit of absorbed irradiation, where one Gy is the absorption of one joule of energy per kilogram of food.
- **Ionizing Radiation:** Subatomic particles or electromagnetic waves of sufficient energy to dislodge electrons from atoms or molecules.
- **Microorganism:** Any small plant or animal organism, especially one visible only with an optical or electron microscope, such as a bacterium, protozoan, or virus.
- **Pathogenic:** Causing disease.
- **Radioisotope:** Unstable isotope of an element that decays generating alpha particles, electrons, positrons, or gamma rays.
- **Radiolysis:** Breakdown of molecules struck by ionizing radiation.
- **Spore-Forming Bacteria:** Bacteria that form thick-walled seedlike structures that are highly resistant to destruction; these include *Clostridia.*
- **Sterilization:** Destruction of all living organisms present.
- **Vegetative Cell:** Cell that is in the resting phase of its life cycle, when it is not involved in reproduction.

Definition and Basic Principles

Pasteurization is the process of using mild heat to treat food, whereas irradiation, sometimes called radiation pasteurization or cold pasteurization, uses ionizing radiation. The primary purpose of each is to destroy microorganisms that would be pathogenic to human consumers, without significantly changing the food's attributes. In addition, these processes can be used to destroy microorganisms or enzymes that spoil food, leading to a longer shelf life and less waste. Irradiation can also be used on fresh fruits and vegetables to kill insects and to delay germination, ripening, or sprouting.

Pasteurization, used primarily with liquid foods, such as milk, fruit juices, and beer, refers to heat treatments that do not exceed 100 degrees Celsius (C), whereas heat sterilization (such as canning) uses temperatures of 100 degrees C or higher. In general, high-temperature short-time (HTST) pasteurization destroys undesirable organisms while minimizing deleterious effects on the food. In milk, comparable killing of microorganisms can be achieved by conventional pasteurization at 63 degrees C for thirty minutes or by HTST at 72 degrees C for fifteen seconds or 88 degrees C for one second. Sterilization of milk at ultrahigh temperature (UHT) typically requires 138 degrees C for two seconds.

For irradiation, the ionizing radiation used are

gamma rays, generated from the decay of radioisotopes cobalt 60 or cesium 137, X rays, and electrons, the latter two generated by machines for such purposes. The operators of equipment involving ionizing radiation need to be protected from its effects. The ability subsequently to pasteurize or irradiate food should not compensate for best practices to minimize contamination of food before treatment. Moreover, treated food also needs to be protected from subsequent contamination.

Background and History

Thermal and nonthermal processes have long been used to ensure the safety and storage of food. Cooking and smoking food were practiced in prehistoric times and likely permitted a survival advantage. Subsequently, drying, salting, and pickling were also used.

Pasteurization was developed in 1862 by French scientists Louis Pasteur and Claude Bernard, initially for the preservation of beer and wine. The pasteurization of milk became widespread in the early 1900's.

X rays were discovered by German physicist Wilhelm Röntgen in 1895, and radiation emitted from uranium and other radioactive elements was discovered by French physicists Henri Becquerel, Marie Curie, and Pierre Curie shortly thereafter. Patents were issued for food preservation using ionizing radiation in 1905 and for the use of X rays to destroy *Trichinella* in pork in 1921.

By 2005, food irradiation was widely used in Asia, less often used in America and Eastern Europe, and rarely used in Western Europe, where consumer resistance is high. The primary use of irradiation is for herbs, spices, and dry vegetables, followed by root crops (such as potatoes and garlic, to inhibit sprouting), grains and fruits, meat, and seafood. Food irradiation is strictly regulated within each country, and all irradiated foods must be labeled with words such as "Treated by irradiation" and the international Radura logo. Global trade would be enhanced by harmonization of national regulations.

How It Works

Pasteurization. Heat kills pathogenic and spoilage microorganisms by disrupting their cellular structure and metabolism. In pasteurization, sufficient heat is applied to the food being treated to kill undesirable organisms, but without damaging the food itself. In this balance, not all undesirable organisms are destroyed, but they are reduced to such a level that the product, if stored appropriately, will be safe for consumption until its use-by date.

Although some liquid foods, such as beer and fruit juices, may be pasteurized after filling containers (with warm water or steam applied to raise the temperature appropriately), most are pasteurized in a vat process or a continuous-flow process and then packaged. The vat (or batch) process involves heating in a well-agitated tank for the required time and temperature. The vat process is suitable for relatively small-scale operations.

The continuous flow process became possible when plate heat exchangers were developed in the late 1920's and has been enhanced by the development of concentric-tube heat exchangers. It is particularly well-suited to HTST and large-scale operations. In a typical system, the liquid to be pasteurized flows in a continuous tube from a cooled holding tank, through a preheater, through the heater (which heats the fluid to the required temperature), through a holding tube (whose size coupled with the flow rate determines the length of time that the liquid is held at the specified temperature), and then cooled down for storage. In practice, the preheater acts as a precooler, extracting heat from the heated liquid, before it is further cooled, permitting 85 to 90 percent of the heat to be reclaimed. The temperature in the holding tube must be monitored to ensure that the desired temperature has been maintained and, if not, the flow must be automatically diverted back to the starting tank.

In both vat and continuous-flow processes, the system must be thoroughly cleaned between uses, but a single use of the latter may last for many hours.

Irradiation. Radiation kills pathogenic and spoilage organisms, as well as insects. It also retards germination, ripening, and sprouting. It does so by disrupting cell structure, cell metabolism, and, most importantly, DNA molecules, preventing further growth and reproduction. Irradiation exerts its effects by direct action of the radiation or indirectly, principally via the radiolysis of water, which leads to the generation of highly reactive chemical species, such as hydroxyl radicals and hydrogen peroxide. Smaller organisms are more resistant than larger ones, for instance, viruses compared with bacteria. Spores of species such as *Clostridia* that cause botulism are more resistant than vegetative cells. Gram-negative bacteria, including primary food pathogens *Escherichia coli* and *Salmonella*, are more sensitive than gram-positive bacteria. Gamma rays are more effective than X rays, which in turn are more effective than electrons. These differences relate to the penetrating power of the radiation, with electron irradiation only suitable for treatment of surfaces or thin packages.

In the process of irradiation, the product is brought in line with the source of radiation for the requisite period of time. Electrons or X rays are generated by machines for these purposes and can be turned on and off as required. Because gamma rays result from radioactive decay, they cannot be turned on or off; when not needed, the sources of radioactivity are stored in a large water tank that absorbs the radiation. Irradiation with electrons and X rays is well suited to a conveyor-belt system that brings the product into the radiation beam. With gamma rays, the use of an overhead rail system is preferred. The packages of product to be irradiated are suspended from that system and moved so that the package can be bombarded from various sides and angles to ensure uniformity of treatment. In all cases, a dosimeter (or dose meter) must be periodically included to ensure that the material has received the required dosage.

In food irradiation, sufficient radiation is applied to destroy undesirable organisms (including insects) or to inhibit a biological process, without adversely affecting the nutritive value and sensory characteristics of the food. As with pasteurization, organisms may not be completely eliminated but are reduced to a safe level provided the food is stored appropriately for no longer than the permitted time. Radiation is able to penetrate packaging materials, which reduces the risk of contamination after treatment. On the other hand, packaging materials can be affected by the ionizing radiation generating radiolysis that may migrate to the food and affect its taste. Careful choice of packaging material, as well as adhesives and printed material, must be made to avoid such problems. The fats in foods are susceptible to breakdown, forming products with unacceptable taste, but this effect is minimized by irradiating foods high in fat while frozen. Irradiation in the absence of oxygen minimizes the generation of byproducts that can affect the color and taste of the food.

Applications and Products

Pasteurized Products. Pasteurization is typically applied to liquids—milk is the best-known example. Almost all milk consumed around the world is pasteurized or heat sterilized. Before the development of milk pasteurization, more than 25 percent of food-borne diseases were attributed to milk and milk products. Many microorganisms, including pathogenic ones, survive well in milk. Combined with aseptic packaging technology, pasteurization makes milk less prone to spreading disease and less perishable. Fruit juices and beers may be flash pasteurized (HTST) to minimize spoilage. A few wines are pasteurized; wines with less than 14 percent alcohol are sometimes pasteurized (or ultra-filtered) to stop any further fermentation. Flash pasteurization is also used to make wines acceptable to strict Orthodox Jews. Liquid eggs can be similarly pasteurized; those in shell can also be pasteurized in a series of warm-water baths.

Nonliquid pasteurized products include cheese, almonds, smokeless tobacco, crabmeat, bread, and ready-to-eat meals. Pasteurized cheese is made from pasteurized milk, so the liquid is treated in this process, although the cheese is subjected to heat treatments as well. Almonds can be pasteurized with a steam treatment, designed to kill any microorganisms on the outside of the nut; pasteurization of almonds can also refer to their treatment with propylene oxide, but that treatment is more appropriately called chemical fumigation. Smokeless (or chewing) tobacco is pasteurized by heating to 85 degrees C. Crabmeat labeled as pasteurized is heated to 113 degrees C for one minute in sealed cans or plastic containers. Because this temperature is higher than 100 degrees C, it is not properly termed pasteurization. Nevertheless, this treatment does not kill all pathogens present, it merely reduces them to a safe level, and the product must be stored at refrigerated temperatures for no longer than prescribed to ensure its safety. Bread and ready-to-eat meals are usually pasteurized by microwaves, which generate heat in the product being treated.

Irradiated Products. The extent of irradiation of a food will vary according to the desired end point and nature of the substance being irradiated. The absorbed doses are expressed in units of gray (Gy) and kilogray (kGy). Low doses (less than one kGy) will inhibit sprouting of potatoes, garlic, onions and other root foods, will disinfect insects on fruits, grains and dry foods, will delay ripening of fresh fruits and vegetables, and will inactivate parasites on pork and fresh fish. Medium doses (one to ten kGys) will extend shelf life of strawberries, mushrooms, fresh fish, and meat (if stored at between 0 and 4 degrees C), destroy parasites in meats, control molds on fresh fruit, and destroy pathogenic and spoilage microorganisms in spices, raw or frozen poultry, meat, and shrimp. High doses (greater than ten kGys) will sterilize herbs, spices, meat, poultry, seafood, food additives (such as enzymes and natural gums), packaging materials (such as wine corks), disposable medical items (such as syringes, tubing, and gloves), and hospital food (especially for immune-compromised patients). In the United States, the only exception to a maximum of thirty kGys is for sterilizing frozen packaged meats for National Aeronautics and Space Administration (NASA) space flights. When in space, American astronauts have been eating irradiated foods, such as beef, pork, smoked turkey, and corned beef, since the beginning of the space program. Interestingly, milk and milk products are not good prospects for irradiation because it generates undesirable flavors.

In irradiation, gamma rays are used more often than electrons or X rays, largely because of their greater penetrating power. Irradiation of fresh fruits and vegetables to destroy any mature or immature insects obviates any need to quarantine these products. In the past, such disinfection was done with methyl bromide, but its use is being phased out because it is an ozone-depleting chemical. The costs of food irradiation include high capital costs and modest operating costs. Irradiation facilities must protect workers from the radiation used; this involves thick walls and fail-safe design that prevents accidental radiation exposure to employees when in operation. Several methods are available to determine if a food has been irradiated, but biological methods that enumerate dead and live microorganisms of the species of concern are particularly useful because they not only show how many survived, but the total burden before irradiation, providing a check on good handling practices before treatment.

Impact on Industry

Pasteurization is practiced worldwide and is important in ensuring the safety of the food supply. Pasteurization is an established and well-accepted technology. Microwave technology is likely to become more important as the source of the heat for pasteurization in the future, particularly if uneven heating can be overcome. The dairy and beverage industries are main users of pasteurization. Major companies in the United States include Dean Foods for packaged fluid milk, Kraft Foods for cheese, Tropicana Products for fruit juice and Anheuser-Busch InBev for beer. National Pasteurized Eggs is the main producer of pasteurized shell eggs in the country.

Irradiation is underused commercially, despite extensive research, especially since the 1960's, and its advantage as a nonthermal process is that it is less likely to alter the nutritional content, structure, and taste of food. Worldwide, the International Atomic Energy Agency provides general guidance for the safety and use of irradiation technology. It supervises research and collates results from research laboratories around the world. It often does so in partnership with the United Nations' Food and Agriculture Organization. Irradiation is approved for use in more than fifty-five countries for more than 200 food items. Many countries, however, restrict it to herbs and spices. By contrast, more than 80 foods are approved for irradiation in South Africa. Worldwide, almost 400,000 metric tons of food was irradiated in 2005: 183,000 in Asia, 116,000 in the Americas, 70,000 in Eastern Europe and 15,000 in Western Europe. The three countries irradiating the most food were China, the United States, and Ukraine, amounting to three quarters of the world total. Herbs, spices, and dry vegetables accounted for 46 percent of the food irradiated; garlic, potatoes, and other root crops, 22 percent; grains and fruits, 20 percent; and meat and seafood, 8 percent.

In the United States, responsibility for food irradiation is shared by the Food and Drug Administration (FDA), the U.S. Department of Agriculture (USDA), and the Environmental Protection Agency (EPA). The rules are published in the Code of Federal Regulations (citation: 21CFR179.26). In response to new research and outbreaks of food-borne diseases, the rules are amended, as they were to permit irradiation to control *Salmonella* in fresh shell eggs in 2000 and *E. coli* in ground beef in 1999. In 2003, the USDA announced that it would allow school districts to purchase irradiated beef for school cafeterias, but in 2004 it was explicitly forbidden from requiring them to do so or to subsidize such meats. Meat producers have been reluctant to introduce irradiated beef because of increased cost, possibility of color and taste changes, and consumer resistance.

Fascinating Facts About Pasteurization and Irradiation

- Pasteurization was originally developed to preserve beer and wine.
- In the United States, millions become ill and as many as 5,000 die annually from food-borne infections, most of which could have been prevented by pasteurization or irradiation. The problem is more acute in the developing world.
- Pasteurization and irradiation require a balance between minimizing undesirable organisms and retaining desirable characteristics of the food.
- High-temperature short-time pasteurization of milk at 88 degrees Celsius for one second is as effective in killing pathogenic bacteria as conventional pasteurization at 63 degrees Celsius for thirty minutes and is less damaging to the value of the food.
- In the United States, about 80 percent of disposable medical items (such as syringes, tubing, and gloves) are sterilized by irradiation and yet little of the food supply is so treated.
- Irradiation of herbs and spices is widely practiced because many are grown and harvested by small landowners in tropical countries where insects and disease organisms are prevalent. Previously used chemical fumigants, such as ethylene oxide, are banned because they deplete atmospheric ozone.
- Prospects for irradiating seafood should be good because of its short shelf life.

Two companies, Sterigenics and Steris, operate most of the contract irradiation facilities in the United States. They have at least thirty sites, primarily for medical applications, clustered in areas with a substantial manufacturing base for disposable medical products, as around Chicago, New York City, and Los Angeles. Most are equipped for irradiation with gamma rays, two with electron beams, and one with X rays; the latter is under contract by the U.S. Postal Service to sanitize mail for several federal departments and agencies. Food Technology Service has a gamma-ray operation in Florida that specializes in food irradiation. However, in the food industry, irradiation is best incorporated into the food-processing system, rather than as a stand-alone facility. Accordingly, other companies specialize in the design and construction of customized irradiation facilities, Gray Star for gamma rays and SADEX for electron beams. Hawaii Pride has its own electron-beam facility to destroy fruit flies on papaya and other tropical products destined for the mainland.

Careers and Course Work

Pasteurization and irradiation are integral to the food industry, ensuring the safety of food from its source on a farm to its consumption by people. To design or oversee operations involving these processes, one should possess undergraduate or graduate degrees in food science or food engineering; knowledge of the structure and value of food must be coupled with knowledge of engineering processes. In addition, one should possess a strong knowledge of microbiology and chemistry. One should understand the medical consequences of ingesting pathogenic microorganisms and, for irradiation, the fundamentals of nuclear reactions and nuclear safety. Mathematics is an essential foundation for careers in these fields.

Bachelor of science majors appropriate for positions in these fields would be food science or food engineering, with course work in mathematics, chemistry, microbiology, nutrition, food chemistry, food and industrial microbiology, and food processing. Such degrees prepare one for oversight positions in the food industry and for subsequent advanced degrees. Appropriate master of science and doctoral degrees in food science or food engineering focus on research in the area of interest, with course work to supplement that research. Such advanced degrees prepare one for positions in applied or basic research into pasteurization or irradiation processes in the food industry, government laboratories, or universities.

Social Context and Future Prospects

Pasteurization is a well-accepted technology. Nevertheless, proponents of raw milk contend that pasteurization is unnecessary if milk is kept clean from the udder of the cow to the consumer and that it destroys some desirable components in milk. Replacing hand milking with milking machines and direct transfer of raw milk to a cooling tank have made contamination from microorganisms from the environment less likely but do not eliminate the risk. Recall that milk is a good medium for growing many microorganisms, including pathogens. With regard to components of milk that are destroyed in pasteurization, the only substantial loss is some enzymes, such as alkaline phosphatase. But, on eating, enzymes are inactivated and digested in the stomach and intestines of consumers and do not provide any demonstrable health benefits aside from being a source of amino acids. Public health and medical organizations promote the pasteurization of milk in the interest of food safety.

Irradiation is not as accepted, largely because of public concerns about nuclear radiation. Gamma irradiation does not increase radioactivity in food over what occurs naturally. Although it does produce radiolytic products in foods, animal testing indicates that irradiated food is safe and high doses of radiolytic products have no adverse effects. Public health and medical organizations, such as the American Medical Association, attest to the safety of irradiation in protecting the food supply. Public education at the point of sale has been demonstrated to be effective in overcoming resistance to irradiated foods.

James L. Robinson, Ph.D.

Further Reading

Berk, Zeki. *Food Process Engineering and Technology.* Burlington, Mass.: Academic Press, 2009. Contains chapters on thermal and nonthermal food preservation, includes periodic word problems with answers to illustrate concepts and practical applications.

Fellows, P. J. *Food Processing Technology: Principles and Practice.* 3d ed. Boca Raton, Fla.: CRC Press, 2009. A comprehensive treatment of the technology that includes separate chapters on pasteurization and irradiation.

Komolprasert, Vanee, and Kim M. Morehouse, eds. *Irradiation of Food and Packaging: Recent Developments.* Washington, D.C.: American Chemical Society, 2004. Excellent coverage of irradiated foods and foods contaminated with pathogens and the effects of irradiation on food packaging and additives.

Ortega-Rivas, Enrique, ed. *Processing Effects on Safety and Quality of Foods.* Boca Raton, Fla.: CRC Press, 2010. A readable text that includes sections on pasteurization, irradiation, and other food-processing technologies.

Stewart, Eileen. "Food Irradiation: More Pros than Cons?" *The Biologist* 51, no. 2 (June, 2004): 91-96. The first part of an impartial review of the advantages and disadvantages of food irradiation, illustrated with graphs, tables, and pictures.

_______. "Food Irradiation: More Pros than Cons? Part 2." *The Biologist* 51, no. 3 (September, 2004): 141-144. The second half of Stewart's review of the value of food irradiation. Many graphs, tables, and pictures are included.

Web Sites

Food and Agriculture Organization of the United Nations
http://www.fao.org

Food and Drug Administration
Food Facts
http://www.fda.gov/Food/ResourcesForYou/Consumers/ucm079516.htm

Idaho State University
Radiation Information Network
http://www.physics.isu.edu/radinf

U.S. Department of Agriculture
Food Safety and Inspection Service
http://www.fsis.usda.gov/Fact_Sheets/Irradiation_and_Food_Safety/index.asp

See also: Food Preservation; Food Science.

PHOTOGRAPHY

FIELDS OF STUDY

Chemistry; physics; optics; art and architecture; visual media; visual design; environmental science; history and aesthetics of images; journalism; computer science; computer graphics.

SUMMARY

Photography is the process of forming images on surfaces that are sensitive to electromagnetic radiation. These surfaces are usually silver halide-based film or an electronic photosensor, and the radiation recorded is usually visible light, infrared radiation, ultraviolet radiation, or X rays. Photography is an important technological tool for examining and documenting the natural and human-built world. For more than 150 years, photography has added vital knowledge to the physical sciences, environmental sciences, biological sciences, medical sciences, forensic sciences, materials sciences, and engineering sciences.

KEY TERMS AND CONCEPTS

- **Charge-Coupled Device (CCD):** Type of sensor that converts electromagnetic radiation into electric charges that are processed to form a digital image.
- **Color Temperature:** Color that appears when a hypothetical body, called a blackbody, is heated to a specific temperature.
- **Complementary Metal-Oxide Semiconductor (CMOS):** Type of sensor that converts electromagnetic radiation into electric charges that are processed to form a digital image.
- **Dynamic Range:** Range of light contrast in a scene.
- **Electromagnetic Radiation:** Wave energy emitted by any substance that possesses heat. The temperature of the substance determines the wavelength of emission; for visible light, the wavelengths range from about 0.4 to 0.7 micrometers (μm).
- **High Dynamic Range (HDR):** Combination of two or more images of the same scene but with different exposures to form a single image that more faithfully represents the scene's dynamic range.
- **Pixel:** Picture element; the smallest unit that contains light-sensitive information on a digital imaging sensor or display screen.
- **Silver Halide:** Light-sensitive compound used in photographic film and paper.
- **Single Lens Reflex (SLR):** Mechanical mirror system in a film or digital camera that reflects the light entering the lens so that the view through the camera's eyepiece matches the scene recorded on the film or digital sensor.

Definition and Basic Principles

Photography is the process of forming images on surfaces that are sensitive to electromagnetic radiation. Usually the surfaces are either a silver halide-based film or an electronic (digital) photosensor that has been designed to be sensitive to visible light. Depending on the application, the film or electronic sensor may be designed to form images in response to infrared radiation, ultraviolet radiation, or X rays.

Background and History

The word "photography" derives from the Greek words *photos* and *graphos*, which together mean "light drawing." Hercules Florence is believed to be the first person to introduce the word photography in 1833 while living in Brazil, where he experimented with light-sensitive silver salts to produce prints of drawings. Florence's pioneering work remained largely unknown until 1973. While Florence was conducting his experiments in Brazil, work on permanently fixing an image with light-sensitive materials was being conducted by Nicéphore Niépce and Jacques Daguerre in France and independently by William Fox Talbot in England. It was Talbot's work, however, that led preeminent scientist John Herschel to coin the word "photography" in 1839 to describe Talbot's process for fixing an image. Because of Herschel's broad connections to the scientific community, he is traditionally credited with introducing the words "photography" and "photograph."

The invention of photography was announced to the world on January 7, 1839, at a meeting of the French Academy of Sciences in Paris, France. François Arago, a physicist and secretary of the academy, announced that Daguerre had employed the action of light to permanently fix an image with the aid of a camera obscura, a device used by artists to assist in drawing. Daguerre's image was made on a highly polished, light-sensitized silver plate, which produced a one-of-a-kind image called a daguerreotype.

Several months after Arago's announcement of the daguerreotype, Talbot announced his process for fixing an image by the action of light. In contrast to Daguerre's process, which relied on a light-sensitized silver plate, Talbot's process relied on light-sensitized paper. Talbot's process, which he called photogenic drawing, produced a stable paper negative from which multiple prints could be made. Talbot went on to improve his process to produce images called calotypes, after the Greek word *kalos*, meaning beautiful. Although the daguerreotype was sharper and produced greater detail than the calotype, the paper-based calotype is more similar to the modern photograph.

How It Works

The camera is a device that basically consists of three components: a light-tight box containing a light-sensitive surface (such as film or a digital sensor), a lens for focusing the light onto the surface, and a means for controlling the exposure, such as a mechanical or electronic shutter. With a film-based camera, image processing takes place outside the camera, often in a darkroom or image-processing laboratory. With a digital camera, image capture, processing, and storage all take place within the camera.

Film-based cameras are designed in a variety of formats, although the three main formats are 35 millimeter (mm), medium format, and large format view. The type of camera format plays an important role in image quality; typically, larger format cameras produce higher quality images. Because each format differs sharply in portability, the choice of format depends largely on the application. In most film-based cameras, the film is coated with silver halide, a light-sensitive compound that is suspended in a gelatin. The film characteristics, such as grain, light sensitivity, and overall image quality, depend on the size, shape, and distribution of the silver-halide crystals. The sensitivity of the film to light is denoted by its International Organization for Standardization (ISO) number. Film characterized as slow has a low ISO number, such as ISO 50, whereas film characterized as fast has a high ISO number, such as ISO 400. The film grain becomesincreasingly apparent as the ISO number increases. Correspondingly, the image quality tends to decrease as the ISO number increases.

Digital cameras are designed with a light-sensitive electronic sensor that records image information. The sensor is made up of an array of photosensors that convert light to electric signals. The signals are proportional to the intensity of the light and are assigned numbers, which the image processor uses to create individual picture elements called pixels. The pixels contain information on brightness (luminance) and color (chrominance). The digital information is then processed by the camera's computer chips to produce the image that is stored in the camera's memory.

Digital cameras generally use two types of digital sensors: the charge-coupled device (CCD) and the complementary metal-oxide semiconductor (CMOS). The CCD and CMOS sensors differ primarily in how they convert charge to voltage. In the CMOS sensor, the conversion takes place at each photosensor, whereas in the CCD sensor, the conversion takes place at a common output amplifier. The advantages of the CCD sensor over the CMOS are that it has a higher dynamic range and very low noise levels. The advantages of the CMOS sensor over the CCD are that it is less costly to produce and requires less power.

Digital cameras are designed with image sensors of varying sizes. The size of the sensor and the number of pixels contained on the sensor play an important role in image quality. Rapid technological advances in sensor design have resulted in dramatic improvements in the dynamic range recorded by the sensors and the reduction of digital noise in long exposure, low-light conditions.

Digital photography has several advantages over film photography. For example, image composition and exposure can be reviewed immediately; if necessary, images can be deleted to free up file space on the camera's memory card. There is no need for film or chemical processing in a darkroom. Color temperature can be changed with different lighting situations to ensure color fidelity on the image. Images can be transferred almost instantly and wirelessly to remote locations. Digital images are easily edited using image processing software. Despite a number of advantages, there are some disadvantages to digital photography. For example, digital camera equipment, ranging from the camera to the computer for postprocessing, is costly. Imaging systems and software are rapidly changing, often requiring costly upgrades of equipment and additional training to operate the software. Overall, the advantages of digital photography generally outweigh the disadvantages, a fact that has led to the ever-increasing popularity of digital photography over film photography.

Applications and Products

Aerial Photography. First accomplished with balloons in the nineteenth century and with aircraft and satellites in the twentieth century, aerial photography has enabled scientists to acquire information from platforms above the Earth's surface. That information is used to characterize and track changes in land use, soil erosion, agricultural development, water resources, vegetation distribution, animal and human populations, and ecosystems. It is also used to detect water pollution, monitor oil spills, assess habitats, and provide the basis for geologic mapping. Because aerial photographs can record wavelengths of electromagnetic radiation that are invisible to the human eye, such as thermal infrared radiation, plant canopy temperatures can be measured and displayed on an aerial photograph that characterize the plant's stress due to environmental conditions.

By applying photogrammetric methods, whereby spatial relationships on an aerial photograph are related to spatial relationships on Earth's surface, analysts can relate distances on the photograph to distances on the ground. Object heights and terrain elevations can also be obtained by comparing photographs made from two different vantage points, each with a different line of sight.

Additional information can be gleaned from aerial photographs by examining tonal changes and shadow distributions within the photograph. Tonal changes are related to surface texture, which can be used to distinguish between vegetation type, soil type, and other surface features. Because the shapes of shadows change with time of day and are unique to particular objects, such as bridges, trees, and buildings, the shadows can be used to aid in the identification of the objects.

Environmental Photography. As the environment undergoes changes because of human activities and natural forces such as floods and earthquakes, documenting those changes has become increasingly important. Photography has played, and will continue to play, an important role in that documentation process. For example, scenes photographed in the nineteenth century by such preeminent photographers as William Henry Jackson, Timothy O'Sullivan, and Carleton Watkins as part of government-sponsored surveys of the American West were used to acquire scientific data about the geology and geomorphology of the land. In the twenty-first century, those same scenes are being photographed from the same vantage points and under similar lighting conditions to document the environmental changes that have occurred since the mid-nineteenth century. Environmental photographs made from land-based and elevated platforms, such as airplanes and satellites, provide valuable visual information for monitoring present-day and future environmental changes involving ecological systems, snowpacks, forests, deserts, soils, water sources, and the human-built landscape. Such information aids in habitat restoration, land-use planning, and environmental policy making.

Medical Photography. Photography has been an integral part of medical science since the mid-nineteenth century. In 1840, one year after the invention of photography was announced to the world, Alfred Donné, a Paris-based physician, used a microscope-daguerreotype to photograph bone and dental tissue. In 1865, in a presentation to the Royal Society, British physician Hugh Welch Diamond advocated the use of photography to document mental patients for later analysis. In the late nineteenth century, Frederick Glendening, a London-based pioneer in medical photography, used clinical photographs of the human body to assist in the diagnosis of disease. At around the same time, Thomas R. French and George Brainerd collaborated to produce the first photographs of the larynx, while W. T. Jackman and J. D. Webster are believed to have made the first photographs of the human retina. The first X-ray photographs were made by Wilhelm Conrad Röntgen in 1895. Since then, X-ray photographs have become a routine tool for medical diagnostics.

Medical photography, also known as biomedical photography, has become a highly specialized field that requires precision and accuracy to be an effective diagnostic tool. Because medical photography is deeply rooted in digital technology, involving, for example, X rays, magnetic resonance imaging (MRI), and positron emission tomography (PET), practitioners must be well versed in digital imaging techniques. Additional training in the biological sciences, medical sciences, photogrammetry, and photomicrography may be required depending on the area of specialization.

Scientific Photography. In addition to aerial photography, environmental photography, and medical photography, there are many other specialties that fall under scientific photography. Kirilian photography involves using electrophotography to form a contact print, for example, by applying high-voltage but low current to an object on a photosensitive surface. Some uses are primarily for record keeping or documentation, including archaeological photography (images of past human remains and artifacts taken on-site at excavations or in laboratories), forensic photography (documentary images that can be used in the legal process or admitted as evidence in court), and time-lapse photography (images taken at specified time intervals to show movement or change over time. Botanical photography involves taking images of plants, paying particular attention to their botanical and morphological adaptations, to aid in the identification and study of environmental stresses, diseases, and their interactions with other plants and animals. In biological photography, which sometimes uses optical and electron microscopy, images are taken of plant and animal specimens from micro to macro scales and in land and aquatic environments. In underwater photography, waterproof cameras or traditional cameras in unmanned or manned submersibles are used to photograph underwater scenes. Astrophotography uses both Earth-based and satellite platforms to reproduce images of celestial objects.

Fascinating Facts About Photography

- The mammoth camera built in 1900 for the Chicago and Alton Railway to photograph trains—along with the holder for the 8-by-4.5-foot photographic plates—weighed 1,400 pounds and is believed to be the largest camera ever built.
- In 2008, about 100 million digital cameras were sold, 100 billion photographic images were made, and more than 1 trillion digital images were posted on the World Wide Web.
- The first image taken by Wilhelm Conrad Röntgen, who discovered X rays in 1895, was of his wife's hand. It showed her opaque flesh, bones, and rings.
- In 1890, Austrian physiologist Sigmund Exner gazed through a microscope and photographed through the amputated eye of a firefly. The photomicrograph showed a dreamlike image of a barn framed by a window. Exner's groundbreaking photograph provided proof for a then-controversial theory that the multiple images produced by the firefly's compound eye are resolved into a single upright image.
- In 1872, Leland Stanford, former governor of California, president of the Central Pacific Railroad, and founder of Stanford University, commissioned English photographer Eadweard Muybridge to solve a debate regarding whether all four hooves of a galloping horse were ever in the air at the same time. The momentary airborne state of the horse, not visible to the human eye, was revealed by Muybridge's pioneering high-speed photography.
- Photographic manipulation was commonplace in the nineteenth century, when images were combined in the darkroom to overcome the inability of the photographic emulsions to resolve detail in both the sky and the landscape.

IMPACT ON INDUSTRY

The impact of photography on industry is pervasive and global in scope and spans research and development, manufacturing, and marketing. Digital camera sales have grown annually worldwide since 2000, with sales of film-based cameras occupying an ever-shrinking part of the market. During 2008, 106 million digital cameras were shipped to distributors, about four times the number of film-based cameras. In 2009, shipments decreased slightly because of the global economic downturn. Sales of digital cameras are expected to rebound as the global markets expand, particularly in Asia. The growth in digital camera sales will spur sales of related accessories and technologies, including camera lenses, image processing software, image storage media, and digital printing devices. Corporations such as Canon, Kodak, and Nikon are expected to continue as industry leaders in the development, manufacture, and distribution of digital camera technology and related products.

Government and University Research. Research and development in photography is mostly carried out by major corporations, government laboratories, and universities. Research involving photography generally follows one of two avenues: improving existing camera technologies and developing new ones or using existing camera technology to document and carry out research in science, engineering, and medicine.

Research on camera technology is competitive and ongoing, as consumers, industry, businesses, engineering firms, and scientists continue to demand new and improved innovations. Advances continue to be made in camera resolution, noise reduction, capacity of memory cards, dynamic range, battery life, auto focusing, wireless transmission of images, face recognition and detection, and the incorporation of still images captured with high-definition video in digital single-lens reflex (SLR) cameras. Lenses have evolved along with advances in digital camera technology. Autofocusing has become faster, and some lenses have motion sensors that provide image stabilization, a technology that adjusts for camera shake to provide sharper images. Dramatic advances in digital printing technology also have occurred. High-quality archival prints are easily made without resorting to a professional laboratory or working with chemicals in a darkroom.

Application of digital camera technology to research is far-reaching. Digital cameras have become an integral part of nearly every research laboratory. Digital cameras provide immediate sharing of visual information, documentation of experiments for later analysis, and the production of high-quality images for posting on the World Wide Web and for publication in scientific and industry journals. Because digital cameras can be very small and still produce high-quality images, they are easily transported to remote sites, where images can be recorded and quickly sent back to base operations through conventional land lines or wireless communications.

Corporations and Businesses. Corporations and businesses are commercial organizations that provide services and sell goods to consumers, other businesses, and government entities. They use photography for advertising and documenting the workplace, from operations to products and services. Larger corporations and businesses often use in-house photographers, as they have an intimate knowledge of the company's products, operations, and philosophy. Photography also is used for security monitoring and public relations. As corporations and businesses continue to expand to new global markets, the need to share visual information efficiently and quickly also expands, leading to an increasing demand for new camera technologies and the skills to operate and manage those technologies.

Careers and Course Work

Photography is a broad field with many specialty areas and applications. The best preparation for a career in photography is to earn a bachelor of fine arts degree or a bachelor of science degree with elective courses in fields related to specific photographic specializations. Beyond a core of liberal arts courses, general course work should include classes in the materials and processes of photography, general art classes, the history of art and architecture, studio photography, studio drawing, two-dimensional design, and computer science. Although a master's degree is not required for most careers in photography, the additional graduate-level training may be beneficial for some areas of specialization, such as fine arts photography, architectural photography, and biomedical photography.

For photographic specialties in the physical, medical, or engineering sciences, additional course work is recommended in chemistry, physics, mathematics, applied computer science, data analysis, imaging systems, and technical writing.

For a photography career specializing in the environmental sciences, additional course work in meteorology, climate science, soil sciences, ecology, and hydrology is recommended. For a photography career in the biological sciences, perhaps photographing wildlife, plants, or insects, additional courses in wildlife management, biology, entomology, and related fields would be beneficial. A career in forensic photography, which focuses on documentation of accident and crime scenes for law enforcement and disaster scenes for the insurance industry, may require additional training in underwater photography, the principles of photogrammetry, and criminal justice.

Because academia, industry, and government will continue to have an increasing demand for accurate visual information and visual communication skills, career opportunities in photography should continue to grow for those individuals who are best prepared academically.

Social Context and Future Prospects

Since Daguerre "imprisoned" light in the mid-nineteenth century, photography has had a profound influence on art and science. In the hands of the artist, the camera has heightened awareness of the aesthetic qualities of space and light while revealing hidden truths about culture and society. From centuries-old experiments in optics and chemistry to the digital revolution, the camera has relied on science for its development while also serving as an essential scientific tool for probing and documenting the natural and human-built world.

In its relatively short life, photography has evolved rapidly and profoundly. During the nineteenth century, the slow exposure times associated with the daguerreotype and calotype processes were greatly improved by the newer wet plate collodian process. Film was eventually introduced, shortening the exposure times even more. The shorter exposure times combined with improved lens optics enabled scientists of the nineteenth century to study phenomena that were too quick for the unaided eye to see. Photographs of galloping horses, humans in motion, birds in flight, lightning, and distant galaxies were among the phenomena studied.

At the beginning of the twenty-first century, digital photography largely supplanted film photography. Digital photography continues on a trajectory of rapid growth and development. It has enabled nearly immediate sharing of visual information, which has greatly aided in the monitoring of the environment, the diagnostics of diseases and health problems, the documentation of crime scenes by law enforcement, and the dissemination of images by the news media. In the coming years, digital cameras will continue to diminish in size and improve in resolution and noise reduction. The ease of digital image making, especially as digital cameras become smaller and less conspicuous, is likely to raise issues concerning personal privacy and image making in public venues.

Terrence R. Nathan, Ph.D.

Further Reading

Frankel, Felice. *Envisioning Science: The Design and Craft of the Science Image.* Cambridge, Mass.: The MIT Press, 2004. Examines scientific photography from the physical and biological sciences, nanotechnology, materials science, and engineering.

Hunter, Fil, Steven Biver, and Paul Fuqua. *Light, Science and Magic: An Introduction to Photographic Lighting.* 3d ed. Burlington, Mass.: Focal Press, 2007. Covers the basic principles of the physics of light with applications to available and artificial lighting in photography.

Keller, Corey, ed. *Brought to Light: Photography and the Invisible, 1840-1900.* New Haven, Conn.: San Francisco Museum of Modern Art in association with Yale University Press, 2008. Looks at scientific photography from 1840 to 1900, with many extraordinary scientific photographs.

Mante, Harald. *The Photograph: Composition and Color Design.* Santa Barbara, Calif.: Rocky Nook and Verlag Photographie, 2008. Covers the principles of visual design as applied to photography.

Peat, F. Davis. "Photography and Science: Conspirators." In *Photography's Multiple Roles: Art, Document, Market, Science,* edited by Terry A. Neff. Chicago: Museum of Contemporary Photography, 1998. Examines the relationship between photography and science, which is supported by analysis of photographs of magnetism, laboratory still lifes, and medical imaging technology.

Rosenblum, Naomi. *A World History of Photography.* New York: Abbeville Press, 2007. Presents a comprehensive review of the historical and technological developments in photography.

Thomas, Ann. *Beauty of Another Order: Photography in Science.* New Haven, Conn.: Yale University Press, 1997. Contains chapters on the search for patterns, medical photography before 1900, photography of movement, and photographing the universe.

Web Sites

International Environment Photographers Association
http://theiepa.org

North American Nature Photography Association
http://www.nanpa.org

Professional Aerial Photographers Association
http://www.papainternational.org

Professional Photographers of America
http://www.ppa.com

See also: Optical Storage

PLANT BREEDING AND PROPAGATION

FIELDS OF STUDY

Agronomy; horticulture; botany; genetics; molecular biology; statistics.

SUMMARY

The science of plant breeding and propagation is the controlled, systematic identification and multiplication of useful plant varieties. It is critical to human survival. Of all discoveries furthering the advance of civilization, improvements in food production have arguably been the most significant. Breeding food crops for higher productivity, improved nutritional content, and greater resistance to stress and disease; preserving rare species and reintroducing them into the natural environment; and creating plant varieties that act as factories for complex pharmaceuticals are among the most active fields for the plant breeder. Genetic engineering techniques are revolutionizing the industry.

KEY TERMS AND CONCEPTS

- **Biopharmaceutical:** Complex drug produced in managed living systems, with or without genetic modification.
- **Bioreactor:** Closed system for batch growing of microbial or plant populations.
- **Explant:** Small piece of plant vegetative tissue used to establish tissue cultures.
- **Genetically Modified Organism (GMO):** Organism whose genotype has been transposed through use of genetic engineering technology.
- **Germ Plasm Bank:** Facility for storing and perpetuating plant stocks, including rare species and heritage cultivars.
- **Molecular Farming:** Growing genetically modified crops for production of a chemical compound.
- **Somatic Embryogenesis:** Production of embryonic plants from vegetative tissues.
- **Terminator Technology:** Genetic modification of food crops to produce seed sterility.

Definition and Basic Principles

Plant breeding and propagation science encompasses any systematic attempt to create or identify and select useful varieties of plants and multiply them for commercial production. Historically, the raw materials came from naturally occurring variations. Using increased knowledge of genetics and the mechanisms of inheritance, modern plant breeders have been able to crossbreed strains to produce hybrids with specific suites of characters. Breeders may use radiation or chemical mutagens to increase variability. Beginning in the 1980's, genetic engineering has enabled plant scientists to insert specific genes into genomes of cultivated plants.

Having produced and selected a desirable strain, the breeder must then produce sufficient numbers of plants to test for trait stability and determine optimal conditions for commercial production. Unlike annuals that produce abundant seed, perennials with long generation times such as forest trees are a challenge to the plant breeder. For species such as conifers, advances in tissue-culturing techniques have transformed production processes in both the development and the marketing phases.

The advent of genetic engineering in agriculture has produced a number of hotly debated issues concerning the safety, long-term environmental impact, and wisdom of introducing nonplant genes into human dietary staples on a global scale. This debate has slowed the spread of this technology.

Background and History

Humans have been selecting and propagating useful varieties of plants for more than 10,000 years. The domestication of wheat, rice, and barley in the Old World and maize and potatoes in the New World was critical to the development of the earliest civilizations. When plant breeding became systematic enough to constitute science is difficult to pinpoint. Roman agricultural science, which drew heavily on Greek and Mesopotamian antecedents, included the deliberate introduction and propagation of novel plant varieties. Contemporary Chinese were at least as advanced as the Romans in agronomy and horticulture, and the sophisticated agricultural techniques and variety of cultivars in pre-Columbian Mexico and Peru argues for the existence of a scientific approach in those cultures.

Agricultural science came into renewed prominence in the eighteenth century as educated landowners brought the tools of the enlightenment to bear. Private botanical gardens became establishments for introducing, testing, and propagating exotic species and varieties of food and ornamental plants. By midcentury, breeders recognized that flowering plants did have a sexual cycle and that the principles of animal breeding applied to them.

Beginning with the concept of genes pioneered by Gregor Mendel in the nineteenth century, the science of genetics progressed steadily through the twentieth century, giving plant breeders increasing understanding of the processes underlying their work. Plant breeders have contributed a great deal to genetics in general.

How It Works

Conventional Plant Breeding. The plant breeder seeks to combine the desirable traits from different strains into a single variety. Sources for the genes producing these traits include existing commercial cultivars, field collections or germ plasm banks, closely related wild plants, and spontaneous or induced mutations. Under carefully controlled conditions, the researcher cross-pollinates two varieties, plants the resulting seed, and evaluates the progeny for the desired combination of characteristics. To produce uniform seed of the resulting hybrid for sale, a company will develop breeding stocks of the two parent strains, crossing them in production plots. A gene for male sterility is often incorporated into one of the parents, preventing self-pollination.

Tremendous improvements in methods of vegetative propagation have enabled it to be used for commercial production of perennials with a generation time of many years. There have also been significant advances in methods for evaluating desirable characteristics at an earlier stage in a plant's life cycle. If the biochemical basis for a mature trait is known, precursors can often be detected in very young seedlings. Alternatively, if a mature trait is closely linked genetically to a trait that is expressed earlier in the life cycle, the presence of that marker can be used for screening.

Many plants form viable interspecific crosses. Often these are sterile, which results in seedless fruits. Seedless varieties must either be hybrids of seed-producing parents or propagated vegetatively. Sometimes breeders use colchicine to induce polyploidy, resulting in a sexually reproducing interspecific hybrid with a double chromosome complement.

Genetic Engineering. Certain viruses and viruslike plasmids invade cells and attach their DNA to that of the host. Any DNA attached to the virus or plasmid will also be incorporated. Plant genetic engineering uses a plasmid from a phytopathogenic bacterium, *Agrobacterium tumefaciens*, to transform plant cells. Because transformation rates are very low, geneticists incorporate an antibiotic-resistance gene to facilitate screening. Plant cells in undifferentiated tissue culture are exposed to transformed *Agrobacterium* plasmids and transferred to an antibiotic-containing medium. The surviving cells are then grown out as plants.

In theory, the ability to synthesize any biologically produced compound can be transferred to any plant species by this method. In practice, the absence of activators or presence of inhibitory genes stymies many attempts.

Once engineered into a plant variety, a gene propagates normally from generation to generation and, under field conditions, into other populations of the same species. A gene for herbicide resistance, introduced into a crop to facilitate management by chemical means, can backfire if it spreads to closely related weeds. Of serious concern is so-called terminator technology, a sterility gene introduced into genetically engineered seed by seed companies to prevent patent infringement.

Modern Methods of Plant Propagation. A key feature of modern plant propagation is the use of explants and their proliferation under sterile conditions to produce large numbers of genetically identical, pathogen-free plant starts. Undifferentiated plant cells are grown in liquid or solid media promoting rapid growth. To produce a mature plant, subcultures are then subjected to growth regimes promoting differentiation and maturation. For biopharmaceutical production, plant cells may be grown indefinitely in liquid-culture bioreactors and harvested in the same manner as microbial populations, often heterotrophically.

Applications and Products

Seed. The primary use of plant breeding is to improve the productivity, cultivation characteristics, stress tolerance, and nutritional content of food crops, particularly cereal grains. The green revolution stressed productivity under modern agricultural methods. Later, more attention has been paid to tailoring crops to specific environmental and social conditions. An example is yellow rice, genetically engineered for high vitamin A content, to combat vitamin A deficiency in rice-dependent populations.

Preservation of Biodiversity. Tissue culture methods for propagation have been a great boon for the proliferation of medicinal plants, heritage varieties, and stock for habitat restoration. Orchids and cycads, which are difficult to propagate from seed, are under tremendous pressure from collectors who encourage poaching in nature preserves. Modern nursery propagation helps protect these vulnerable species.

Molecular Farms. The use of genetically engineered plants to produce exotic organic compounds is still in the development phase but shows promise. Researchers have successfully produced a strain of *Arabis* (a mustard) that synthesizes hirudin, an anticlotting agent, from leeches. Molecular-farmed pharmaceuticals of plant origin are safer than those extracted from animals or human plasma. Another promising line of development is edible vaccines—food plants synthesizing proteins that provoke an antigen response to a particular disease-causing bacterium.

Reforestation. With tissue culture and rapid mass multiplication of stocks of woody plants, reforestation following disturbance has become more targeted. The forester can readily access seedlings adapted to the site, with built-in insect and disease resistance. Species such as the American chestnut, eliminated from most of its original range by blight in the early 1900's, are being successfully reintroduced as highly selected resistant strains.

Virus-Indexed Plants. Production of disease-free stock of potatoes and bananas has always been a challenge. The ability to produce commercial quantities of plant starts from a small amount of sterilized tissue greatly reduces the spread of viruses and other pathogens in these vegetatively propagated crops.

Impact on Industry

The Seed Industry. One result of technological advances in plant breeding has been the increased centralization and globalization of the seed industry. Ten large corporations control most of the worldwide production of commercial seed. Their efforts, which are extremely competitive, are geared to intensive large-scale agriculture. Positive results include overall increased productivity and a decrease in prices to the consumer. However, traditional farming methods have been a casualty, even in the developed world.

Because of germ plasm banks, global diversity of food crops is no longer plummeting, but the diversity of field production continues to decline, with vast acreages of critical crops devoted to monocultures of very narrow genotypes emanating from one or two suppliers. The southern corn leaf blight epidemic of 1970 highlighted the vulnerability of such genetic monocultures. The maize crop in the United States succumbed to an outbreak of a fungus that attacked plants carrying the Texas male sterile gene, used by all seed producers to facilitate hybridization.

Although the proportion is declining, 75 percent of grain farmers worldwide still save seed for planting. They may be liable to lawsuit if the variety is patented and, in the case of genetically modified crops, barred from planting their own seed if their crop cross-pollinates with a neighbor's genetically modified crop. Farmers growing patented seed on contract for multinational corporations can earn healthy profits but are vulnerable to changing regulations and the vagaries of international markets.

Commercial production of fruit trees and ornamental woody plants traditionally involved rootstocks grown from seed and buds or scions (grafts) from stock plants maintained by individual nurseries. Now both rootstocks and grafts are typically produced via tissue culture in large centralized operations, and grafting is done in these plant factories at a very early stage in development. Retail nurseries seldom do their own grafting. The number of local varieties available has declined, and standardization for the needs of the largest markets has increased.

Government, University, and Industry Research. In the United States, federal and state agricultural experiment stations and state land-grant colleges sponsored the bulk of plant cultivar development for much of the twentieth century. With a decline in public funding, the emphasis has shifted to private industry and large corporations. These corporations conduct their own research and development but also fund university-based research projects, skewing efforts toward their bottom lines, which may not correspond with public interest.

Developing Countries. The challenges of tropical agriculture and the pressing needs of rapidly expanding populations have prompted India, China, Mexico, the Philippines, Brazil, and a number of other countries to mount aggressive state-supported plant breeding programs employing genetic engineering and sophisticated propagating techniques. In many ways, the developing world is ahead of Europe and the United States in embracing the challenges and potential of the new technology.

Careers and Course Work

In the United States, a person intending to go into the production end of the plant propagation industry as a contract seed grower or operator of a large retail nursery will probably need a four-year degree in agronomy or horticulture, with a strong dose of business management course work as well as preparation in plant science. For employment in corporate research and development or in a university or government laboratory, an advanced degree is usually necessary. Individuals intending to do genetic engineering will need extensive graduate level course work in genetics, molecular biology, and computerized data management. They also will probably need to complete a dissertation and obtain a doctoral degree specifically in genetic engineering. As with most scientific fields, obtaining the credentials is no guarantee of permanent employment at a living wage. Most university positions in the United States are filled on a temporary basis with graduate students or recent graduates classified as interns and research fellows, and the situation in government laboratories is not much better. Corporate employment is typically better paid, but the number of positions available is not encouraging.

Social Context and Future Prospects

Advances in plant breeding have provoked many controversies, the most pressing of which are the contribution of improved crop varieties to unsustainable population growth, ownership of rights to germ plasm and to the products of genetic engineering, the safety of transgenic plants as human food, the risks of the unplanned spread of modified genes, and the adverse effects on traditional agriculture in developing countries. Concern over safety has led to a patchwork of national laws that inhibit but fail to effectively regulate global commerce in the products of genetic engineering.

The relationship of increased crop productivity to exponential population growth, already noted by William Malthus in 1798, was a feature of the green revolution in the twentieth century. Gains from improved technology or new crops are quickly canceled out unless population growth slows. Rural populations may actually end up worse off because improved technology favors large-scale operations and

Fascinating Facts About Plant Breeding and Propagation

- Thomas Jefferson and King George III of England may have been at odds politically, but they shared a common passion for plant breeding and improvement.
- Barbara McClintock, whose discovery of transposons earned her the Nobel Prize in Physiology or Medicine, specialized in maize.
- Cereal crop seeds bearing genes for insect resistance derived from the bacterium *Bacillus thuringensis* or genes for resistance to a specific herbicide are the principal uses of gene-splicing technology.
- Scientists developed a variety of maize whose protein content was enhanced by splicing genes from Brazil nuts but halted commercial production of the strain (developed as stock feed) when they discovered that it triggered severe nut allergies.
- Nearly all of the bananas grown commercially worldwide belong to a single variety, Cavendish, and are effectively clones.
- Natural foods enthusiasts have seized on the term "Frankenstein plants," or "Frankenfoods," to characterize food crops produced by interspecific genetic engineering.
- Wheat is a hexaploid containing three sets of chromosomes from two genera of grasses. A wheat cell contains about twice as much DNA as a human cell. The crossings that produced the parent of all commercial wheat occurred in prehistoric times.
- Chinese herbal medicine has provided a powerful incentive for developing novel plant propagation techniques. Many key species have never been brought into cultivation and have become rare in the wild because of overharvesting and habitat destruction

displaces farmers. Third World-based institutes such as the International Maize and Wheat Improvement Center (Centro Internacional de Mejoramiento de Maíz y Trigo, or CIMMYT) are working to ensure that the new round of agricultural technology will have a more beneficial human impact.

New varieties of plants under patent can be propagated only under license from the originator, which prevents farmers from saving seed. This increases costs and forces small operations to use older, less productive strains. When genetically engineered strains cross with crops in adjacent fields, the farmer of the traditional crop may be unable to sell it for human consumption. There are conflicting claims over the ownership of rights to improved varieties whose base stock came from field collections in developing nations.

If these objections can be overcome, the potential for the new technology is tremendous. The combination of greater productivity and improved nutritional content in staple crops could be of immense benefit. Incorporating resistance to disease and stress should lower the need for pesticides and herbicides and make farming on marginal lands more environmentally friendly. Using genetically engineered green plants to produce pharmaceutical and other complex organic chemicals ought to lower their cost and increase availability. The ability to propagate a species from vegetative portions of a few individuals and reintroduce it into its native habitat will undoubtedly be a boon in preserving endangered species, restoring native vegetation, and promoting sustainable ecologically friendly landscaping. As long as science and the public do not lose sight of potential social costs and maintain adequate safeguards, plant breeding will continue to serve humankind well.

Martha A. Sherwood, Ph.D.

Further Reading

Hrazdina, Geza, ed. *The Use of Agriculturally Important Genes in Biotechnology.* Washington, D.C.: NATO Scientific Affairs Division, 2000. A collection of symposium papers on genetic engineering, genome analysis, plant transformation, breeding for disease and stress resistance, and legal aspects of biotechnology.

Kalloo, G., and J. B. Chowdhury, eds. *Distant Hybridization of Crop Plants.* Berlin: Springer-Verlag, 1992. Describes the techniques and reasons for creating hybrids between closely related species of plants, at a time when genetic engineering was still in its infancy.

Liang, Geoge H., and Daniel Z. Skinner. *Genetically Modified Crops: Their Development, Uses, and Risks.* New York: Food Products Press, 2004. Written for the nonspecialist, the book is favorable to the technology and adopts the point of view that benefits outweigh risks.

Parekh, Sarad R. *The GMO Handbook: Genetically Modified Animals, Microbes, and Plants in Biotechnology.* Totowa, N.J.: Humana Press, 2004. A collection of papers including an overview, a chapter on biosafety and ethical issues, and a chapter dealing specifically with food plants.

Slater, Adrian, Nigel W. Scott, and Mark Fowler. *Plant Biotechnology: The Genetic Manipulation of Plants.* New York: Oxford University Press, 2008. A thorough and comprehensive text with much detail on the mechanisms and techniques of genetic engineering; general information is highlighted and segregated so as to be more accessible to nonspecialists.

Web Sites

International Maize and Wheat Improvement Center
http://www.cimmyt.org

National Association of Plant Breeders
http://www.plantbreeding.org/napb/index.htm

National Council of Commercial Plant Breeders
http://www.nccpb.org

U.S. Food and Drug Administration
Genetically Engineered Foods
http://www.fda.gov/NewsEvents/Testimony/ucm115032.htm

World Health Organization
Twenty Questions on Genetically Modified (GM) Foodshttp://www.who.int/foodsafety/publications/biotech/en/20questions_en.pdf

See also: Erosion Control; Food Science; Forestry; Genetically Modified Food Production; Hydroponics; Viticulture and Enology.

PRINTING

FIELDS OF STUDY

Graphic arts; graphic communications; computer science; graphic design; electronics; chemistry; color theory and optics; physics; geometry; keyboard systems; algebra; technical writing; computer design; information systems; chromatics; metallurgy; journalism; calligraphy; database management; typography; accounting; art and drawing; marketing; English; mechanical engineering; software design.

SUMMARY

Printing includes a wide variety of technologies, increasingly computerized, engaged in the transference and manipulation of images—both words and illustrations—from one surface to another. Surfaces range from paper to glass to metal and even apparel. With the emerging demands of the information age, in which literally millions of agents are engaged in the process of moving information and maintaining accurate records, printing technology has evolved into a cornerstone industry: Commercial and business printing employ an estimated 750,000 people in the United States alone.

KEY TERMS AND CONCEPTS

- **Desktop Publishing:** Process of formatting and designing a page of text (images and words) using computer software.
- **Flexography:** Highly efficient rotary printing process in which the image carrier is a flexible plate, usually rubber.
- **Gravure:** Industrial printing process in which an image is cut into a plate and then transferred from the sunken surface.
- **Ink-Jet Printing:** Printing process in which the image is produced by spraying individual drops of ink through an aperture onto the intended surface.
- **Letterpress:** Process of printing from an inked elevated surface.
- **Lithography:** Process of transferring an image from any surface.
- **Offset Printing:** Process in which an image is transferred from a plate to a rubber blanket that, in turn, presses the image onto a sheet.
- **Output:** Second phase of the printing process, specifically when the image is transferred.
- **Post-Press:** Third stage of the printing process, specifically when printed material is prepared for distribution usually through folding, cutting, trimming, packaging, or binding.
- **Prepress:** First stage of the printing process, specifically the arrangement, organization, and selection of text for printing an acceptable image.
- **Reprography:** Transferring technical and industrial images via electronic means.
- **Screen Printing:** Transferring an image by letting ink seep through an opening, usually a stencil.
- **Typography:** Art and industry, both commercial and private, of manipulating movable type to produce images.

Definition and Basic Principles

Printing refers to the applied technologies of reproducing texts, specifically the storing and distribution of data. In many ways, printing is an invisible technology despite being surrounded by evidence of its application. It might be assumed that given the rapid rise in digital technology, specifically the unprecedented storage capability of digital technologies and the ubiquitous presence of personal computers, that printing has become obsolete. Although the computer has revolutionized printing, the demand for the organized and controlled transference of information still relies on printing. From publishing (books, magazines, newspapers) to commercial applications (labels, manuals, banking checks, forms, letterhead, as well as glassware and even clothing), printing is central to a culture that relies on maintaining records and disseminating information. Unlike more prominent applied technologies—such as steel manufacturing or auto making—printing maintains its low profile because, unlike those industries, printing concerns are more likely to be smaller, privately owned, independent industrial facilities that seldom employ more than fifty workers.

Nearly all of the commercial nondigital printing is involved in one of five basic processes: lithography, screen, flexography, letterpress, or gravure. Over the last generation, however, digital technologies have begun to reshape printing, and microtechnology has allowed printing to diversify into data-management and Web-based systems. In fact, with the advent of desktop publishing and laser printing, text manufacture and information display have become part of personal computing with texts that, unlike those produced a generation earlier by the typewriter, are professional looking.

Background and History

Given the impractical and unreliable methods of the ancient oral cultures (memory and verbal communication), the desire to print information is as old as humanity itself: Papyrus, the first practical printing system, dates to 6000 B.C.E. The earliest developments in printing, woodblocks used to transfer images and symbols to cloth, date to first century China. In addition, advances in paper production in the Far East were brought back to Europe in the twelfth century as were advances in movable type, efficient systems of reproducing text using revolving stone tablets and numbered boxes of handset type, most often made of wood. In the early fifteenth century, German metalworker Johannes Gutenberg perfected an efficient system of movable type technology using durable type pieces made of a lead and tin alloy.

It is impossible to overestimate the impact of movable type. It enabled information to be more rapidly shared, and science, philosophy, and literature enjoyed unprecedented prosperity. The printing press has been credited with making the Renaissance possible. Knowledge could accumulate across generations: Scientists could build on each others' work; books could circulate widely; governments could maintain public records; education was readily accessible; writers had authority over their work; reading itself became a private act; and languages became part of a national culture. Over three centuries, the basic technology of Gutenberg's printing press remained unchanged, save for being driven by progressively more energy-efficient means: first steam then electricity. In the closing decades of the twentieth century, however, printing began to redefine image reproduction and text manufacturing to accommodate the rapid strides in information gathering and display offered by digital technologies.

How It Works

Offset. By far the most common process for commercial printing is offset lithography, which includes the mass-quantity, high-speed production of mail-order catalogs, newspapers, directories, weekly magazines, and books. The process is designed for competent and speedy production. It involves spools of paper that weigh upwards of a ton and machines capable of printing more than 1,000 impressions every minute.

First, the text to be printed must be converted to plates, a process that involves photo emulsion. This is the same process as photography in that negatives of the text are exposed to controlled light that passes through the negative and in turn triggers a chemical reaction that transfers the transparent image to a thin flat plate, most often made of zinc or aluminum. The process moves to the actual press run. Because oil and water do not mix, the plate surface is broken down into two areas, the actual image area, which repels water (and thus remains dry and can absorb ink), and the nonimage area, which accepts water and is not part of the reproduced image.

Ink is distributed to the prepared plates through high-efficiency rollers. The plates are rolled first with water and then with ink. The ink adheres to the image, and the water rollers keep the ink off the nonimage area. The plate is then pressed to a rubber blanket, transferring the image quickly and cleanly. High-speed presses then roll the paper across the rubber blanket; the paper never touches the actual plate, thus the term offset. Because the paper, once through the press, is wet, the paper then passes through a gas oven at a temperature near 400 degrees Fahrenheit to dry it, and then it is quickly run through a refrigeration unit to cool the paper to avoid smudging.

Rotogravure Printing. The other most widely used industrial printing process (accounting for roughly 20 percent) is rotogravure, or intaglio, printing, in which the image or text is cut into the surface of a rounded (hence "roto") printing plate. The image is actually formed from the sunken surface on the plate. Each cylinder can hold the image of up to sixteen pages of text. As the cylinder is turned at high speeds, the roll of paper (or fabric or foil or plastic) is fed continuously across its surface, in which computerized stressors maintain the registration controls necessary to transfer the images. The paper is passed across the image cylinder. After that, the paper is passed through high-temperature driers to help set the ink quickly.

Although the process is more technologically sophisticated and more expensive than offset, gravure is actually more cost efficient (state-of-the-art presses can run as fast as 3,000 feet per second), and the reproduction is more consistent, better detailed, and of finer resolution. The tempered plates are more durable than the rubber blankets used in offset. The only drawback to gravure from an industry standpoint is the time it takes to prepare the cylinder plates, a process called diffusion etching, which involves a thick copper mask and an acid bath to achieve the appropriate cut in the cylinder.

Applications and Products

Printing applications figure in virtually every industry. These industries are routinely held together through the reproduction and dissemination of printed materials. In the 1950's, Marshall McLuhan, one of the pioneer theorists in communication and language, predicted the inevitable end of the age of print and the evolution of the paperless office, and despite the innovations of the last generation of printers driven by digital technology, reproducing and storing data through printing processes has only increased in application.

Lithography. Offset lithography, introduced more than a century ago, is the predominant technology for commercial printing. Sheet-fed offset printing is cost effective and produces reliably clean images and text because the rubber blanket that actually transfers the image adheres tightly and precisely to the texture of the printing surface. Color- control reproduction is entirely directed by computerized settings that blends the inks together at tremendous speed. Although the color reproduction is not fine point, the types of printing that use lithography offset—daily and weekly newspapers, most weekly magazines, advertising and informational brochures, and most paperback books—are not designed for long-term usage and do not require precise image reproduction.

Gravure Printing. Given its history (the process of cutting designs into soft metallic plates was first used by Renaissance artisans interested in fine and expensive engravings), it is not surprising that gravure printing is used for printing projects that, while requiring a significant number of rungs, have more of an aesthetic element than those created with offset printing applications. These include commercial bags and boxes that bear the trademark logo of a company or store; decorative wrapping paper; labels; mass-mailing glossy inserts; Sunday-newspaper inserts, high-end glossy commercial and department-store catalogs, and monthly magazines; specialty limited-run printing, such as ornamental wallpaper rolls, art prints, publicity posters, or commemorative postage stamps; and, finally, printing runs that involve delicate fabric. Because of the durability of the cut plates, gravure printing is also used in the production of long-term forms and applications.

Flexography. Developed in the early 1900's but not widely used until the invention of cellophane during the Depression created a new industry for food packaging, flexography is a variation of rotary printing using an image carrier made of flexible rubber or a soft plastic plate with raised image designs. It accounts for just about 15 percent of commercial printing applications. Flexography is particularly attractive for high-demand printing work that is less ornamental and more functional. It is based on a simple three-stage technology: an in-feeding unit feeds materials such as cardboard or film; a printing unit delivers high-speed drying ink, which is spread by passing the medium through a series of distribution rollers; the out-feed unit rewinds the printed material to be sent to the appropriate plants. Candy wrappers, for instance, would be rewound and sent to the factory. Flexographic printing is versatile in its applications. It is used widely in the production of cardboard products, product wrapping, corrugated board, medical, sanitary, and food packaging, and plastic film. Because printing on cellophane using the gravure process proved prohibitively expensive, flexography made cellophane printing cost effective.

Screen Printing. As with other printing technologies, screen printing is based on a simple principle that dates back to ancient China. The idea is to transfer an image by passing ink through openings in a precut stencil. A stencil is attached to a piece of fabric stretched on a metal or wooden frame. A squeegee-like device is used to force ink through the stencil opening, and then the material, most often paper but also metal, plastic, or wood, is fed and pressed under the stencil and then removed for drying. Although originally used for producing elaborate wallpaper designs and shop signs in the 1800's, screen printing is now used for short-run projects that require high-quality resolution: photograph reproduction, clothing lines and silk-screen work, cardboard cartons, and even printing microcircuits for computer systems.

Ink-Jet Printing. With the rise of personal computers and the emergence of digital printing in offices, ink-jet printing is becoming the most familiar and most versatile application of printing technology. Given the resources of computer programs and the speed at which such software is marketed (any description of existing ink-jet printing technology will quickly become obsolete), the applications of ink-jet printing are bound only by the resources of the technological imagination. The basic premise—the transfer of images and text using digital data memory and on-screen formatting to create individual and personalized printing—puts printing skills in the hands of anyone with a computer. The effects are clear and readable but relatively low resolution. High-quality resolution is seldom the aim.

In ink-jet printing, computers create the desired effect by controlling the spraying of individual drops of ink by using electronic impulses. Although used most frequently with paper for high-speed reproduction of information within a large or small office network, ink-jet printing can also be used with corrugated board, plastic, and fabric.

IMPACT ON INDUSTRY

Commercial printing is the third-largest employer in the world. In the information age, workers skilled in the printing process, particularly the burgeoning field of digital printing, have virtually reshaped every industry. Printing is still at the core of the twentieth-century corporate model. Significant development in the field of computerized printing processes, particularly developing secure identification cards, secure banking and medical data, more efficient and better resolution mass printings, and developing environmentally friendly, cost-effective food packaging, has been spearheaded in the United States, most notably by the international conglomerate Toppan America, headquartered in New Jersey. Important printing technology has also been pioneered in Japan and Great Britain.

Government and University Research. The largest producer of paperwork—forms, applications, manuals, brochures, contracts, informational publications, legislative records, as well as currency and bank notes—is the United States government. The United States Government Printing Office, which is only responsible for records of the legislative and executive branches, alone produces more than a half million separate printed documents weekly. At the state and local levels, government services routinely require significant accumulations of printed material, not only stored as records but disseminated as useful information about legislation and regulations. The medical industry is second in the production of printed material and third is

Fascinating Facts About Printing

- The United States alone prints 27.8 million tons of paper annually. This does not include 12 million tons for newsprint. To keep up with printing needs worldwide, 4 billion trees must be harvested each year. In the United States alone, each person uses an average of a little more than 750 pounds of paper per year. That is up 400 percent from just fifty years ago. The United States, which has only 5 percent of the world's population, produces the most printed material: 30 percent.
- The dominant producer of printed materials for the United States government is the Bureau of Engraving and Printing, a division of the Treasury Department, which each day produces 38 million notes with a face value of $750 million. This involves the world's most secure printing process.
- The rich tones of offset printing are achieved through the digital combination of four basic colors: cyan, magenta, yellow, and black.
- Centuries before Gutenberg, Bibles were transcribed by hand by generations of monastics who would labor over a single illustrated text. Gutenberg's first design was based on a winepress. Gutenberg's Bible, the first book produced by movable type, was massive—1,282 double-column pages—that took a little more than a year to set. Of the 180 Bibles that Gutenberg produced, only 48 are still in existence.
- Statistical projections based on readership declines and the upgrading of news services on the Internet and on television indicate that the last newspaper will most likely be printed in 2044.
- Project Gutenberg is a cooperative archival computer system the mission of which is to digitize all public-domain books, histories, and historic documents. It was begun in 1971, when computer students at the University of Illinois had the idea to digitize the Declaration of Independence. As of 2010, this digital library has archived 34,000 titles.

public education. From day-care facilities to graduate schools, educational systems routinely contract out or print on-site nearly 3 million separate documents annually, according to the commercial printing sources.

Research Direction. Two concerns—government and business—direct research in the printing field. Commercial printing is spearheading efforts to become more environmentally friendly. Many printing processes rely on petroleum-based inks, organic emissions, and lubricants that create disposal problems and long-term environmental hazards. Given that printing businesses are seldom major industrial plants but rather are small neighborhood ventures that average fewer than fifty employees, regulating that environmental hazard is a growing concern. Paper itself represents a special challenge to the printing industry. Worry about the availability of raw materials for paper production have encouraged recycling campaigns, but researchers in printing processes are investigating technologies that would free data—images and words—from paper altogether. That, however, creates a second area of concern within the printing field. Printing processes, relying increasingly on paperless technology, need to be brought into compliance with post-September 11 security realities. Given the volume of data recorded by computer imaging and transmitting through, among others, government offices, law firms, banks, and hospitals, information relayed through the printing medium needs to be secured in ways that would have seemed extreme before September 11, 2001.

Careers and Course Work

Courses in mechanical engineering, chemistry, physics, computer science, mathematics, and graphic communication account for the foundational requirements for students interested in careers in printing. Although there are an estimated 70,000 printing establishments in the United States alone, career opportunities and requirements for long-term success have dramatically changed in just the last twenty years. The United States Department of Labor cites increasing computerization and the declining volume of printed matter due to the availability of computerized data storage as reasons for the decline. The most promising career opportunities in larger commercial printing firms are geographically concentrated: California, Illinois, Massachusetts, Texas, Michigan, and the Washington, D.C., Chicago, and New York City metropolitan areas account for more than one-third of such facilities.

Training for most hourly wage printing jobs (technicians and operators) is on-site and job specific. Production workers, printing machine operators, and manual laborers who load and maintain the presses typically have associate's degrees, specific vocational training, or bachelor's degrees (most often in engineering). However, significant career advancement is best secured through knowledge of computer skills, desktop publishing, journalism, and digital graphic design. In fact, skilled press operators are expected to be the most securely employed in the printing industry through 2025. Upper-level-salary jobs require familiarity with business administration and marketing.

Social Context and Future Prospects

A casual glance at a contemporary department store or shopping mall, an office complex or a bank, a hospital or a local government building confirms the place of print technology. Although seldom acknowledged, the printing industry defines a twenty-first century consumer culture. The advent of computerized printing techniques has reshaped the employment profile away from manual work into prepress design and computerized formatting, and digitalized data banks have drastically cut into the volume of printed material produced. Printing and printed material are still central to virtually every service industry and business enterprise.

Newspapers have suffered dramatically because of competition from Internet news sources and television. Despite the advent of electronic books and electronic readers, the book industry is solid—more than 135,000 new titles are published each year—and magazine publishing maintains a healthy market share. Researchers have estimated that before the end of the twenty-first century, the enormous volume of data required by an information-hungry global community will be entirely digitalized, virtually eliminating the need for paper and printing for data retrieval. That same technology, however, will increase the efficiency, reliability, and safety of printing presses able to provide the other, wide-ranging printing applications.

Joseph Dewey, B.A., M.A., Ph.D.

Further Reading

Adams, J. Michael, and Penny Ann Dolin. *Printing Technology*. 5th ed. Albany, N.Y.: Delmar, 2002. This text includes step-by-step explanations of each of the basic printing technologies alongside helpful and detailed illustrations.

Howard, Nicole. *The Book: The Life Story of a Technology*. Westport, Conn.: Greenwood Press, 2005. Careful history of the evolution of book printing that includes a comprehensive look at printing technologies in the digital era. Particularly helpful in detailing early printing techniques and presses before 1900.

Johnson, Harald. *Mastering Digital Printing*. 2d ed. Boston: Thomson Course Technology, 2005. Comprehensive summary of printing technologies in the last twenty years, written in accessible terms with handsome illustrations.

Needham, Paul, and Michael Joseph. *Adventure and Art: The First One Hundred Years of Printing*. New Brunswick, N.J.: Rutgers University Press, 1999. Handy and concise overview of original printing techniques that are very much the basic principles of modern printing.

Young, Sherman. *The Book Is Dead. Long Live the Book*. New South Wales, Australia: University of New South Wales Press, 2007. Highly readable examination of the crossroads position of printing in the digital age with particular emphasis on the development of electronic books and the rise of the Internet.

Web Sites

Amalgamated Lithographers Union
http://www.litho.org

Association for Suppliers of Printing, Publishing and Converting Technologies (NPES)
http://www.npes.org

Book Manufacturers' Institute
http://www.bmibook.com

IDEAlliance
http://www.idealliance.org

National Association for Printing Leadership
http://www.napl.org

Printing Industries of America
http://www.printing.org

Specialty Graphic Imaging Association
http://www.sgia.org

See also: Graphics Technologies

R

RECYCLING TECHNOLOGY

FIELDS OF STUDY

Environmental studies; earth science; physics; mechanical engineering; electrical engineering; systems engineering; design; mathematics; human-computer interaction; field robotics; civil engineering; materials science; science and technology studies; sustainability studies.

SUMMARY

The environmental movement, which has been growing steadily since the 1940's, has brought increasing attention to the need to change the public and industrial practices that have enormous environmental costs. Managing the vast amounts of waste produced in a consumer society is a prodigious challenge that is exacerbated by the closing and non-replacement of landfill-waste dump sites. One important aspect of the green attitude is to reduce the level of consumption of virgin planetary resources by reusing and recycling products. While citizens are being called upon to reduce, reuse, and recycle, it is generally recognized that efficient recycling requires the application of new technologies. The major targets for residential recycling programs, which now exist in most cities of the developed world, have traditionally been glass and aluminum containers, paper products, certain types of plastics, and plant and yard waste. However, as technologies are further developed and industries find profitable ways of recycling and reusing waste, recycling programs are being extended to include rubber, a variety of metals, asphalt shingles, and electronics.

KEY TERMS AND CONCEPTS

- **Downcycling:** Converting valuable protects into low-value raw materials.
- **Ecological Footprint:** Measure of human demand on the Earth's ecosystems and of ecological capacity needed to regenerate the resources a human population consumes, as well as the capacity to absorb and render harmless the corresponding amount of waste generated by human practices.
- **E-Waste:** Electronic waste; discarded electronics, such as computers, monitors, televisions, traditionally thought nonrecyclable and disposed of in landfills.
- **Generate:** Capture useful waste material for use as energy; includes methane collection, gasification, and digestion.
- **Incinerate:** Destroy material by burning it at very high temperatures.
- **Landfill Site:** Site for the disposal of waste materials by burial; the oldest method of garbage disposal.
- **Printed Circuit Board (PCB):** Thin plate where chips and other electronic components reside.
- **Recycle:** Break down old materials and products into raw materials for remanufacture into new goods.
- **Reduce:** Buy less and use less; includes limiting new purchases, turning off lights, taking shorter showers.
- **Upcycling:** Converting low-value materials into high-value products.
- **Waste Hierarchy:** Refers to the three caveats of the green movement, reduce, reuse, recycle, which classify waste-management strategies according to their desirability.

Definition and Basic Principles

Recycling is the process of returning previously manufactured products to a raw material state to be utilized afresh for the manufacture of new products. The purpose of recycling technologies is to discover new ways to extend this practice into new material fields. In keeping with the mantra of the global environmental movement—reduce, reuse, recycle—recycling is a crucial aspect of the global environmental justice movement because recycling reduces pollution by reducing the amount of landfill waste and the amount of waste destroyed in incinerators, both of which add to greenhouse gases. Recycling conserves natural resources, which would otherwise be relied upon for the manufacture of new products. Recycling also reduces pollution by reducing manufacturing energy consumption, since less energy is required to recycle a product than to make a new one from virgin materials.

Background and History

The creation of industries focused on reusing previously manufactured products grew out of an increasing awareness of the effects of industrialization and urbanization on the planet's atmosphere, waters, and land. The launching of the global environmental justice movement is credited to Rachel Carson, an American marine biologist and conservationist whose book *Silent Spring* (1962) stimulated public awareness of the effects of unrestricted pesticide use in the United States.

People have always reused their possessions, finding new ways to put old things to fresh uses. However, reuse of products had been waning in affluent societies, since the notion of "planned obsolescence" rendered voracious consumption a virtue. The idea was first introduced by New York real estate broker Bernard London and promoted in his 1932 paper, "Ending the Depression Through Planned Obsolescence." London argued that science and business were successfully producing products in factories and fields; inadequate consumption practices were to blame for the economic woes of the country. Consumers were not buying enough, but London had the answer: Products must be designed to serve only a specific, limited life span. Increased consumption offered the preventative against future economic depressions. Against the traditional Protestant ethic of extreme frugality, London posited that an insatiable appetite for goods was the mark of the good citizen, acting in support of the nation's economic health.

During World War II, Americans had experimented with conservation and recycling as a function of national security. However, in the economic boom following the war, Western middle-class life unapologetically adopted an ethos of consumption, where product value was directly tied to newness and abundance. As environmental problems grew, in the United States and across the planet, the environmental justice movement gained increasing credence among middle-class Americans, and the wastefulness of modern consumer societies became increasingly obvious, intolerable, and shameful to more and more citizens. Across the industrialized West, people began to call upon their governments for regulation of polluting industries, and they began to integrate more sound practices into their private lives. The cultural and social impact of the new public awareness culminated in the establishment of Earth Day in 1970.

Recycling is an important aspect of the modern environmentalism movement, but, in the light of the vast environmental damage done by industry all over the planet, recycling can be undervalued by some environmental activists as having little actual environmental impact. However, one crucial caveat of the environmental justice movement has been to encourage consumers to do their part and institute small lifestyle changes as their personal contribution to the health of the planet. A minor shift in general human consumption behavior, in the context of a globalizing consumerism, suggests that significant benefits can be reaped by reducing, reusing, and recycling old products. The change in public attitudes has forced environmental concerns to the forefront of political debate and contributed to a new ethic that values restraint in consuming. Recycling is a crucial aspect of the new ecological awareness. Recycling has become the fashion, and industry has responded to the new attitudes, finding new applications for the new ethic, creating new business opportunities for profit and development.

How It Works

When people and communities recycle, used materials are set aside in recycling bins, collected, and converted into new products. Simple societies have always been avid recyclers, since their access to new products is limited compared with consumer societies. They tend to produce very little waste and place a high premium on the human creativity that reuses old products for new tasks and designs new products from old discards. Most cities in the developed world support the reduce, reuse, and recycle ethic by collecting used products as part of their curbside trash-collection services. The materials may be sorted as they are placed on the recycling trucks or once collected, they may be sorted in substations, where the used products and materials are then sold to dedicated recycling companies to be made into newly valuable commodities for sale in the global marketplace. Some recycling programs ask residents to separate their recyclables from their trash, which may be collected on different days. To encourage recycling, many Canadian communities have the policy of charging residents for the trash put out for collection but not for their recycling. Residents may be asked to sort and separately bind their recycling into various types of materials—paper products, plastics, glass, Styrofoam, corrugated cardboard, and boxes.

Applications and Products

Most recycling programs target for collection previously used household items that are made from paper, cardboard, metal, glass, plastic, and compostable yard wastes. Recycling recovered metals and glass involves a relatively simple procedure: The items are melted down and sold to processing plants that refill the newly formed containers or to manufacturing plants that create new products from the raw materials. Yard waste can be composted with very little investment in equipment, and many families have created their own composting systems. Paper is the most important recycled material, because recycling is simple and saves harvesting new trees. Paper is recycled by mixing with water, and if necessary, deinking the pulp before reusing it. Plastics recycling, on the other hand, entails an expensive process, whereby differing resins are separated. Since plastics are made from the Earth's dwindling stores of petroleum, they are a favorite target of recycling programs, despite the more complicated processes involved. Plastics can be recycled into a wide range of secondary products, from textiles, such as fiberfill and polyester-like fibers, to plastic toys, recycling bins, and plastic furniture. Recycling of e-waste began in the late twentieth century. Prior to that, television sets, computer monitors, computers and laptops, printers, and microwave ovens had to be disposed of in landfill sites, where they presented a number of health and safety risks, because of the dangerous materials they contain (barium, beryllium, flame retardants, cadmium, hexavalent chromium, lead, mercury, plastics). Moreover, the precious metals used in electronic boards are rare, and thus highly coveted, in their natural state. New industries have cropped up to salvage useful and valuable materials from these sources.

Impact on Industry

Recycling affects a broad spectrum of industries that used to rely solely on newly harvested raw materials. Since the green ethos has elevated recycling to the status of a virtue, recycling technologies have burgeoned as well. As university and corporate research churns out new ideas about how to recycle not only the products consumers discard, but metal from airplanes and ships, electronic products, asphalt from existing roadways and roofing materials, rubber from automobile tires, and methane gas from the decaying refuse in landfills. While recycling is a term generally reserved for metals, glass, papers, and like materials, water can be recycled as well. Water recycling reuses treated wastewater for agricultural and landscape irrigation, industrial processes, toilet flushing, and replenishing groundwater basins. Water is sometimes recycled and reused on-site in industrial facilities, for example, for cooling processes.

The business community works hand in hand with research teams to put the new ideas into practice and to develop new products from recycled materials. In turn, purchasing products stamped with the label, "made from recycled materials," allows consumers to attach themselves to the ethically significant global movement, which promises to save the planet. Recycling stimulates industry at every level, from the curbside-pickup industry to the dedicated recycling companies that break down the various products to their raw-material states to the manufacturing industries that implement the new technologies for fashioning new products from the recycled materials. The search for the most innovative new technology is always a constant challenge in the recycling industry. Companies, research scientists, and environmental groups are racing to come up with better recycling processes and machines, as well as to create new schemes for recycling previously nonrecyclable materials.

Government and University Research. Green technologies represent heavily funded arenas for research. Most research funding comes from two major sources: corporations, motivated mainly by profit, and government, seeking innovative answers to its waste, energy, and pollution problems. Grants for green technologies are primarily awarded to universities or specialized government agencies. A third, more limited, funding source is nonprofit organizations.

Industry and Business. Full-service recycling companies exist in most towns and cities across the

Fascinating Facts About Recycling Technology

- More than one-half million trees could be saved if every household in the United States replaced just one roll of paper towels with recycled ones.
- Twenty million tons of electronic waste are disposed of in landfills each year. One ton of scrap from discarded computers contains more gold than can be produced from seventeen tons of gold ore.
- Nine cubic yards of landfill can be saved by recycling one ton of cardboard.
- The global recycling industry is worth more than $160 billion and employs more than 1.5 million people.
- It takes five percent of the energy to recycle aluminum compared with mining and refining new aluminum.
- Each year, the United States alone generates more than 160 million metric tons of solid waste.
- In the United States alone, recycling is estimated to save more than 18 million tons of CO_2 each year, equivalent to reducing the number of cars on the roads by approximately 5 million.
- Environmental groups often head the search for new recycling methods and products. In 2006, Earth First engineered a new proprietary tire processing system, which salvages steel, carbon, high-energy gas and oil, as well as effectively recycles tires. The benefit of the new method is that the tires are burned at a fraction of the temperature needed for pyrolysis, which satisfies very strict emissions regulations while effectively preserving the tire components.
- New recycling technologies, developed by scientists at University College Dublin in 2006, use a bacteria that eats polystyrene foam, commonly known as Styrofoam, and turns it into a usable plastic. The bacteria offer an efficient scheme for recycling the more than 14 million metric tons of Styrofoam produced annually, which has traditionally ended up in landfills.
- In 2009, Motorola unveiled the world's first-carbon neutral cell phone, made from recycled water bottles. The company promised to offset the carbon emissions created during the phone's manufacture by investing in reforestation and renewable resources.
- New York-based Ecovative Design has developed a substitute for pink fiberglass insulation named Greensulate, which is made from rice hulls (agricultural garbage), mushroom fibers (very inexpensive), and recycled paper (readily available). Greensulate repels water, prevents fire, and is resistant to temperature changes. As of 2011, the product still needs to be tested to determine whether it is mold resistant.

developed world. Some of these remain municipally owned but private industries appear to be taking the lead in assuming the responsibility for the residential collection of recyclable materials. RecycleInAmerica is one of the leading recyclers of plastics in the United States, with outlets in all fifty states. Republic Services is another recycling industry leader, providing services for commercial, industrial, municipal, and residential customers through more than 375 collection companies in forty states and Puerto Rico. Republic also owns or operates more than 223 transfer stations and seventy-eight recycling facilities, serving millions of residential customers in more than 2,800 municipalities. Many industry leaders are joining forces to reach their green goals: The Consumer Electronics Association is teaming up with manufacturers and retailers, such as Panasonic, Best Buy, Sony, and Toshiba, for an industry-wide initiative in e-waste recycling, with a goal of recycling one billion pounds by 2016. This represents triple the e-waste recycled in 2010. The effort focuses on educating consumers about electronics recycling, increasing e-waste recycling collection centers, transparency in reporting progress, and supporting third-party recycler certification.

Major Corporations. Recyclingcompanies.com lists forty-seven global recycling companies and 173 new recycling organizations around the world. Busch Systems of Barrie, Ontario, Canada, is the world's largest supplier of recycling, waste, and compost containers, and has been the industry leader for more than twenty-five years. Based in Connecticut, the Canusa Hershman Recycling Company manages monthly totals of more than 100,000 tons of fiber and 3 million pounds of plastics and other materials. It has a dozen offices and processing facilities in North America and offers a wide range of services to generators and consumers of secondary recyclable materials. Greenstar Recycling is a American paper recycling company that serves a global market. Salvage America is an example of a multi-material recycling company that services industries and municipalities to recycle a great spectrum of materials, including cardboard and paper products, wood pallets, scrap lumber and tree brush, plastics, construction materials, such as concrete, brick, block, and dirt, as well as all metals, ferrous and nonferrous.

Careers and Course Work

On the theoretical side, undergraduate and graduate degrees are offered in the fields of sustainability and environmental studies, which grants students the opportunity to grasp the fundamental scientific concepts that underpin sustainability, understand the policy framework and the political environments in which environmental policy is enacted, create new economic motivations for developing sustainable practices, and analyze and evaluate the consequences of adopting the new technologies. Environmental ethics is a popular course offered in many philosophy departments, business schools, earth science programs, and science and technology studies departments. On the practical side, recycling technologies is an aspect of engineering studies, with varied applications in mechanical, electrical, civil, and systems engineering. Courses in design and materials science look to the creation of new recycling methods and new materials to be targeted for recycling, while civil engineering addresses problems of landfill elimination, road surface recycling, and asphalt shingle recycling.

Social Context and Future Prospects

Throughout the middle decades of the twentieth century, public attitudes in industrialized nations attached virtue to consumption, believing heavy consuming would stimulate the economy and guard societies against economic depressions. Western middle-class life began to locate its identity in products. Through the phenomenon of branding, companies capitalized on this identifying tendency, and advertising convinced consumers that newer, bigger, and more plentiful was better. However, as environmental problems grew more and more evident, in the United States and across the planet, the environmental justice movement gained increasing authority among middle-class Americans, and the wastefulness of modern consumer societies came to be seen as deplorable, irresponsible, and shameful. Across the industrialized West, people began to call upon their governments for regulation of polluting industries, and they began to integrate more sound practices into their private lives. The cultural and social impact of the new public awareness led to the creation of recycling programs in many communities.

Recycling is an important aspect of the modern environmental justice movement. The planet is one component of the triple bottom line (planet, people, and profit), which is a cornerstone of global ethics and the Corporate Social Responsibility movement. In the light of the vast environmental damage still being done by industry all over the planet, recycling is sometimes undervalued by environmental activists, who feel that the biggest offenders are actually governments that need to regulate polluting practices and industries that need to implement environmentally sound manufacturing methods. There is good reason to agree with this activist charge. However, since overconsumption and consumer wastefulness are enormous aspects of the environmental problem, the claim that recycling fails to target the biggest culprits is hardly helpful in changing the public's attitude about its extravagant lifestyle, which is fast being exported to the developing world. Recycling, as a lifestyle ethos, can and has begun to spill out into the business world, stimulating new recycling industries at every level of public life. It is precisely this change in public attitudes that has forced environmental concerns to the forefront of political debate and contributed to public policy changes in favor of industrial pollutant regulation and recycling in the public realm. Recycling programs in the eleven U.S. states in which deposit legislation forces bottle-return practices have been enormously successful. Recycling is a crucial aspect of the ecological justice movement. Scientists and other experts agree that the future of recycling technologies consists in placing more of various recycling technology sites around the country and throughout the world. As of 2011, it is generally agreed that the process of developing new recycling technologies will move very quickly because so many scientists and business leaders are on board.

However, the biggest boost to recycling technologies comes when governments get onboard, and politicians raise recycling as an issue of importance, which then draws to the subject both media coverage and research funding. The United States, traditionally a slow participant in global ecological concerns, is leading new efforts to explore the possible recycling of commercial nuclear waste, as a potential fuel product. Dangerous materials, such as plutonium, require enhanced security in transporting to, and managing in, recycling plants, but these matters are currently under investigation as recycling technologies spread into broader arenas of application.

Wendy C. Hamblet, Ph.D.

Further Reading

Crampton, Norman. *Green House: Eco-Friendly Disposal and Recycling at Home.* Lanham, Md.: M. Evans, 2008. Crampton, a career worker in the field of waste prevention, is the widely acclaimed expert on practical information for adopting green habits of disposal.

Friedman, Lauri S., ed. *Garbage and Recycling.* Farmington Hills, Mich.: Greenhaven Press, 2009. Offers comprehensive overview of recycling that is useful in teaching novices about the field and introduces debate and sorts out misinformation about the subject.

Kintisch, Eli. "Congress Tells DOE to Take Fresh Look at Recycling Spent Reactor Fuel." *Science* 310, no. 5753 (December 2, 2005): 1406. This article reports on renewed U.S. plans to recycle commercial nuclear waste into fuel, an increasingly attractive option as a petroleum alternative, but one fraught with many dangers.

Lund, Herbert F. *McGraw-Hill Recycling Handbook.* 2d ed. New York: McGraw-Hill, 2001. Lund offers answers to hundreds of questions about recycling, focusing on how to develop an effective recycling program, given current recycling technologies, strategic goals, public awareness levels, and legislation.

Mancini, Candice, ed. *Garbage and Recycling: Global Viewpoints.* Farmington Hills, Mich.: Greenhaven Press, 2010. Provides international perspective on recycling technologies through a broad collection of primary-source information, including government documents and essays from international magazines and journals. A good resource for comparative studies of recycling attitudes and programs; annotated table of contents, world map, and country and subject indices.

Porter, Richard C. *The Economics of Waste.* Washington, D.C.: RFP Press, 2002. Porter applies economic and ethical analysis to the dangerous aspects of waste disposal and recycling, especially landfills, incineration, illegal disposal, international trade in waste, and disposal and recycling of hazardous materials, revealing the true costs and risks of recycling policies and the complex problems associated with meeting strategic recycling goals, while still protecting human health and the environment.

Scott, Nicky. *Reduce, Reuse, Recycle: An Easy Household Guide.* White River Junction, Vt.: Chelsea Green Publishing, 2007. This book offers a helpful A-to-Z listing of household items, explanations on how to recycle discarded products, suggestions on how to make discards profitable, and information on how to draw upon local resources to get more involved in recycling projects and opportunities in local communities.

Web Sites

Charitable Recycling Program
http://www.charitablerecycling.com/CR/home.asp

Construction Materials Recycling Association
http://www.cdrecycling.org

Grass Roots Recycling Network
State Recycling Organizations
http://www.grrn.org/resources/sros.html

National Recycling Coalition
http://www.nrcrecycles.org

REMOTE SENSING

FIELDS OF STUDY

Mathematics; physics; chemistry; computer graphics; environmental science; electrical engineering; geographical information systems.

SUMMARY

Remote sensing is a technology-based, interdisciplinary field of science involving the measurement of some characteristic of a physical entity, using a device (sensor) not in contact with that entity. The sensor, or the entity, may be on Earth, below the Earth's surface, in the atmosphere, or in outer space. Data that characterize the entity are transmitted to the sensor, carried by electromagnetic radiation or acoustic waves. The data are usually stored in the sensor or downloaded to a second device for storage, then processed, analyzed, and interpreted. Typical applications are resource extraction, environmental assessment, biomedical imaging, weather monitoring and prediction, and military surveillance.

KEY TERMS AND CONCEPTS

- **Band:** Specific wavelength range, or bandwidth. Remotely sensed images often consist of several bands.
- **Band Indices:** Mathematical combinations of bands developed to detect pixels corresponding to particular features in raster imagery, such as water or living vegetation.
- **Digital Elevation Model (DEM):** Topographic map in the form of a raster image; the land surface is represented by coordinate triplets $[x, y, z]$.
- **Geographical Information System (GIS):** Software used in processing and analyzing remotely sensed data.
- **Global Positioning System (GPS):** System that uses orbiting satellites to provide terrestrial location coordinates. Used in ground truthing and calibrating remotely sensed imagery. Differential-GPS (D-GPS) provides more accurate coordinates.
- **Hyperspectral Imagery:** Imagery collected from dividing the electromagnetic spectrum into narrow bandwidths; can be used to form images of terrain.
- **Light Distance And Ranging (Lidar):** Type of scanning radar in which the timing of returns from the pulsed laser yields a target's shape. Full-waveform lidar allows detection of objects between the forest canopy and floor.
- **Multispectral Imagery:** Imagery collected from the electromagnetic spectrum from the ultraviolet through the infrared region.
- **Rectification:** Process of referring pixel row and column numbers of an image to real-world coordinates to remove sensor distortion. Orthorectification, use of three-dimensional coordinates, can correct for an uneven target.
- **Synthetic Aperture Radar (SAR):** Radar that uses movement of the sensor (for example, in an aircraft) to provide an artificially larger antenna and therefore higher spatial resolution.

Definition and Basic Principles

Remote sensing is the process of measuring some property of an entity, such as its size, constitution, or condition, through use of a sensor that is not in contact with the entity and that leaves it essentially unaltered after the measurement. Entities and the properties that might be measured include a flood (its extent and flow rate), the sea (distribution of a chemical contaminant), the landscape (change of vegetation cover), and the Sun (timing and size of solar flares). Some biomedical imaging, such as optical coherence tomography (high-resolution three-dimensional internal scans), qualifies as remote sensing, whereas interplanetary X-ray diffractometry does not, as a sample must be collected and prepared. Information about the entity is commonly carried to the sensor by either light or sound waves. When energy is applied to the entity to collect data, it is termed "active sensing," otherwise, it is called "passive sensing." Data received by the sensor are stored for processing; thus, personal observation alone is not remote sensing. The location of the data source is important and requires a spatial frame of reference to be recorded with the remotely sensed data.

Background and History

In the early seventeenth century, Galileo used a form of remote sensing to make groundbreaking observations in astronomy, including the geometry and movement of sunspots. During the same century, dowsing for minerals and water were commonplace, and this method is still used for finding water. Although dowsing meets the geometric requisite of remote sensing, it is not scientific in that it collects an interpretation rather than data that can be independently reexamined.

Aerial remote sensing was founded in the nineteenth century by Félix Nadar, who photographed the Earth's surface from a hot-air balloon, showing the value of remote sensing in surveying and mapmaking. Spectral remote sensing was reported in 1957 by E. S. Artsybashev and S. V. Belov, who were studying forests. Many technological advances in remote sensing came about as the result of military applications and were later developed for civilian use. In the nineteenth and early twentieth centuries, the military mounted cameras on carrier pigeons, kites, and rockets. World War II saw the development of radar, and satellites were developed during the Cold War. Conflicts in the Mideast produced miniaturized autonomous aircraft. Early satellites used photographic film, but modern satellites transmit digital imagery.

How It Works

Target Affects. Waves have three main components: amplitude (intensity), frequency, and wavelength. For example, light ranges from gamma rays (frequencies of 10^{20} to 10^{24} hertz and wavelengths of less than 10^{-11} meters) to radio waves (frequencies of 3×10^{11} hertz and wavelengths of greater than 1 millimeter). Wavelength is inversely related to frequency and directly related to the size of the object that can be detected. Generally at least one wave characteristic, when received at the sensor, has been influenced by the target, giving it a characteristic pattern or signature because of selective absorption, transmittance, reflection, or irradiation. Other influences from the target may include altered frequency by relative movement of the target, change in polarization of electromagnetic radiation (direction of oscillation), change in phase, and wavelength-angle-of-incidence affects (for example, bidirectional reflectance in forest aerial photography). Some of these additional effects may be helpful in characterization of the target or may simply complicate its measurement.

Spectral Data. Remotely sensed spectral data—color information stored as multiple measurements across the range of wavelengths of visible light—are often two-dimensional. Raster images, or bitmaps, are data structures representing a rectangular grid of pixels. Each point in the raster, or pixel, is characterized by its position and by intensities of different wavelengths associated with it. In a typical color photograph, each pixel refers to only three wavelengths (red, green, and blue). Such three-band triplets can also be selected from multispectral and hyperspectral images and rendered as false-color images for visual interpretation. Quantitative analysis, however, uses any number of bands or mathematical combinations of them.

Nonspectral Data. Remote sensing that relies only on reflectance of a single wavelength from an object produces an image of the object's outline. The image produced may be a surface relief (for example, with first-return lidar) or a cross section through the target (for example, with ground-penetrating radar, seismic imagery, or full-waveform lidar). Both two-dimensional and three-dimensional images can be created by repeated scans. Remote-sensing systems that typically provide such information are radar, lidar, sonar, and seismic systems. Radar systems usually use an antenna in place of a camera's aperture and a photosensitive detector. Shorter wavelengths and longer antennas produce data of greater spatial resolution. Dual antennas and examination of phase differences can produce high-resolution surface-relief maps. Lidar uses the inherently thin beam of a laser, along with short pulse durations and fast scans, to provide high-resolution data, down to plus-or-minus 10 centimeters for high-resolution systems.

Image Processing. Frequently the collected data require quantitative processing before interpretation. Two primary processing steps are removal of interference and distortions from objects between the target and sensor (such as atmospheric scattering and absorption) and artifacts introduced by the sensor and image rectification. The three-dimensional coordinates of recognizable data points (geocontrol points) are used in image rectification. These may be obtained through triangulation, GPS, or D-GPS. A third layer of processing is mathematical operations on the intensities of different bands assigned to each pixel. Different mathematical band combinations highlight various types of objects (for example, vegetation or water). A higher level of processing is feature detection, done through mathematical groupings of pixel neighborhoods based on properties such as spectral classification, texture, size, and shape.

Applications and Products

Numerous sensor types, platforms, and applications are being used in remote sensing. Some of the more prominent applications are bathymetry, emergency response, natural resources management, meterological or other mapping, military reconnaissance, and subterranean investigations. Lidar and X-ray remote sensing have multiple applications.

Lidar. Applications for lidar include surveying, forest inventorying, and mapping of power lines, structure and biomass, ice shelves, and open mine pits and stockpiles. Platforms can be helicopters, airplanes, semiautonomous aircraft, satellites, and terrestrial systems on tripods. Lidar can be combined with other data types, such as hyperspectral, to yield fused imagery for more comprehensive analyses of issues such as forest health.

X Rays. X-ray remote sensing has a wide variety of uses, particularly biomedical applications, security (where the target provides data through relative absorption of different materials), and interplanetary imaging from orbiting satellites. The remote sensing for interplanetary imaging is passive, as the X rays originate from the planet's surface because of fluorescence in response to solar X rays, bombarding charged particles, and magnetospheric solar wind induction. The sensor is an imaging spectrometer, producing hyperspectral, raster imagery in geochemistry and astronomy.

Bathymetry. Bathymetry involves the charting of the underwater terrain and measuring the water's depth. Coastline mapping (such as recording the sea-level rise resulting from climate change) requires high-resolution relief data, such as from lidar, for both onshore measurement and shallow bathymetry. Bathymetry in deeper water (such as in exploring for oil and gas) uses multibeam sonar systems with roughly 100-meter resolution, mounted on ships or functioning as autonomous submarines, to chart the seafloor. Deeper water is penetrable from the surface only by lower frequency sonar, which means longer wavelengths and hence lower resolution. Very coarse-resolution bathymetry (of about 1 kilometer) can be obtained from radar satellites (altimeters) that measure changes in the height of the sea surface caused by the gravitational effects of seafloor topography, and then sonar systems can be used to map specific areas of interest.

Emergency Response. In the twenty-first century, many satellite-mounted sensors from different nations and consortiums have been deployed in disaster monitoring and relief work after severe earthquakes, tidal waves, oil spills, and volcanic eruptions. These sensors allow detection of changes in natural systems and in infrastructure (vegetation and leveled buildings), landslides, altered water-surface properties, and the presence of aerosol clouds. In more stable metropolitan settings, multiangle, orthorectified high-resolution imagery is used by emergency services to locate individuals.

Natural Resource Management. The Landsat series of satellites has been operational since the 1970's, providing initially 80-meter and then 25-meter resolution multispectral imagery. The French SPOT (satellite probatoire de l'observation de la terre) series began a decade later, providing 20-meter then 10-meter resolution multispectral imagery. Applications for both have principally been in land-use change and agricultural and vegetation mapping, often through development of band indices and time-series imagery. Later satellites for these applications include the experimental ASTER (advanced spaceborne thermal emission and reflection radiometer) and MODIS (moderate resolution imaging spectroradiometer) satellites, the GMES Sentinels series, and the hyperspectral AVIRIS (airborne visible/infrared imaging spectrometer), EOS (earth-observing system), Hyperion, and HyMAP.

The German company RapidEye provides terrestrial multispectral imagery with 6.5-meter resolution from the world's first commercial Earth-observing satellite constellation. Specific applications include agricultural insurance, change detection, risk management, digital elevation modeling production, and disaster relief. The company is increasingly offering image processing for specific turnkey applications, such as a forest health and inventory service, using traditional ground-based inventory combined with remote-sensing image classification and GIS.

Meteorological. Meteorological remote sensing includes mapping phenomena such as cloud coverage (spectral), wind speed and direction (by sodar, or sonic detection and ranging), and rainfall (radar). Sensors are on both terrestrial and satellite platforms.

Since the 1990's, AVHRR (advanced very high resolution radiometer) satellites have provided low spatial resolution data on sea-surface temperature, volcanic cloud drift, and floods, among other meteorological and natural phenomena. The MODIS and the geostationary Meteosat satellites support similar applications.

Military. Military remote sensing was developed to obtain, with minimal detriment to the observer, information such as the enemy's location, troop size, and types of weaponry to determine its strengths, weaknesses, and intentions. The need for detailed and up-to-date information led the drive for higher spatial resolution and the ability to detect a range of frequencies, such as infrared for night vision. For the military to maintain a technological advantage over its enemies, public access to sensors and imagery developed for military purposes has often been limited, resulting in delayed usage for nonmilitary purposes. For example, from 1959 to the early 1970's, the Corona satellites provided 2- to 8-meter resolution stereophotography (for uses such as missile counting), but the images could not be used for environmental applications (such as mapping changes in biomass) until they were declassified in 1995. Observer safety has been heightened by the military's use of autonomous, miniaturized aircraft to obtain data. Development of the propulsion and navigation systems for these aircraft was conducted by universities, so some information on these systems, although not the imagery obtained, is in the public domain.

Subterranean. Numerous techniques are used in subterranean remote sensing. For example, the data from aerial hyperspectral sensors are compared against spectral signature libraries for detection of mineral ores. Electromagnetic induction coils are used to detect changes in conductivity caused by groundwater, buried objects, or chemical contaminants. Handheld, aerial, or spaceborne ground-penetrating radar is used for mapping tree roots, finding buried hazardous waste and unexploded ordinance, and assisting in archaeological digs. Marine sub-bottom seismic reflection is used to explore for oil and gas.

Impact on Industry

Sensor and Platform Development. Research and production of remote-sensing systems and image products is increasingly being performed by cross-sector consortiums, involving universities, government agencies, and private industry. For example, in 2010, the National Aeronautics and Space Administration (NASA) and Old Dominion University in Virginia jointly developed a commercial helium airship suited to remote sensing of Earth and carrying weaponry. Although the majority of research in remote sensing is still for military applications, the time before these discoveries can be used in civilian applications is decreasing, in part because of the privatization of research and higher commercial returns from civilian applications in mining, medicine, meteorology, agriculture, and other areas.

Numerous agencies in China, Europe, India, Israel, Japan, Russia, and the United States design and operate remote-sensing satellites but deploy them using international facilities with rocket-launching capability. Most of these agencies are governmental, although some are partially funded by private industry investment. The 2010 budget for the European Space Agency (ESA), an international organization with eighteen member states, was 3.7 billion euros, or about $5.2 billion. Earth observation, ESA's largest program, encompasses a variety of remote-sensing satellites and spacecraft such as the European Remote Sensing (ERS) satellites, the Envisat spacecraft, and the GMES Sentinels that monitor meteorological conditions, natural disasters, and changes in the land, atmosphere, oceans, and ice caps. Some ESA member countries run their own space agencies and support independent remote-sensing platforms (for example, France's long-running SPOT series and Germany's Boeing-747-based astronomical telescope). In 2009, total annual spending worldwide on space agencies was about $68 billion, with roughly equal portions allotted to military and civilian projects.

Image Processing. Image processing is an integral part of remote sensing. Innovation in platform-independent image processing occurs mostly in universities and private industries rather than in governmental agencies. These innovations are typically published in public-domain scientific journals and can then be used in applications. Military organizations often operate their own image-processing and interpretation services; however, the military often purchases end-product remote-sensing imagery from private industry for emergency response

Fascinating Facts About Remote Sensing

- By 2030, infrared space interferometry is expected to develop to the point that it can assess whether environmental conditions on other planets can support life.
- Identification of minerals on other planets by hyperspectral imaging relies on spectral libraries built using minerals known on Earth. Therefore, some minerals remain unidentified although some of their components, such as bound water, can be identified.
- Regular arrays of objects that are smaller than the wavelengths of light with which their constituents interact can appear invisible because of substantial absorption of that wavelength. These objects, therefore, can be used in military applications for concealment from radar, although they would still cast a shadow or form a silhouette.
- To prevent imprecision resulting from movement, precise measurements of a person's face (to better than 0.5 millimeters) for use in maxilla-facial surgery can be calculated from imagery taken within 25 nanoseconds, using a pulsed laser and holographic imaging.
- Satellite radar can penetrate clouds to allow measurement of deformations at the millimeter level from interferograms (records of optical interference); this allows remote detection of landslides and blocked roads after earthquakes.
- Telescopes based on neutrinos (a penetrating elementary particle) rather than light have been built deep underground and have yielded information on the interiors of astrophysical objects such as the Sun.
- Satellite imagery is used to map deforestation for grazing in Australia. Illegal clearing can result in prosecution; only a few court proceedings have resulted in convictions with fines issued.
- Quickbird imagery has facilitated identification of prehistoric roads used to transport the Easter Island statues. Soil compaction and depression, and therefore surface-water retention, allowed more vegetative growth on these roads, leading to their visibility.

and navigation. Private industry involved in remote sensing is partially dependent on the military, as it relies on the U.S. military's maintenance of GPS satellites.

Software that interprets or manages remotely sensed imagery so that it better fits users' needs has been developed by many private companies. For example, Trimble's eCognition (developed by Definiens) is a software suite that analyzes images for geospatial applications by extracting data from remote-sensing imagery obtained from various sources, including satellites, aerial photography, radar, and hyperspectral data. Its applications include urban planning, forestry, agriculture, energy, and security. Other companies offering similar geospatial software include ERDAS, which has the ER Mapper (used in oil and gas exploration) and the IMAGINE suite (used for remote-sensing analysis and spatial modeling), and ERSI, which offers ArcGIS (geographical information systems, used by insurance companies, financial institutions, law enforcement, the military, and emergency services). Open-source packages (not-for-profit software developed by users) is also available to process remotely sensed images.

Image processing and application solutions are increasingly being undertaken by consortiums of institutions, private companies, and governmental agencies, including the National Centre for Earth Observation (United Kingdom) and the Cooperative Research Centre for Spatial Information (Australia). These efforts are designed to increase collaboration between academic researchers and private industry.

Turnkey Systems. Leading biomedical imaging companies, such as General Electric, Hitachi, Philips, Siemens, and Toshiba, produce both sensors and turnkey image-processing systems for X-ray, MRI, and computed tomography (CT) scans. For example, one product from Siemens includes a CT scanner and processed high-resolution (to about 0.33 millimeters), color-coded, volume-filling images. Similarly, some earth-observing remote-sensing companies offer integrated systems. For example, Colorado-based DigitalGlobe offers a satellite constellation and online processed imagery for a range of applications.

CAREERS AND COURSE WORK

Prerequisites for professions in remote sensing depend on the area in which the individual is interested: the hardware technology, the image processing, or the solving of applied issues. All three

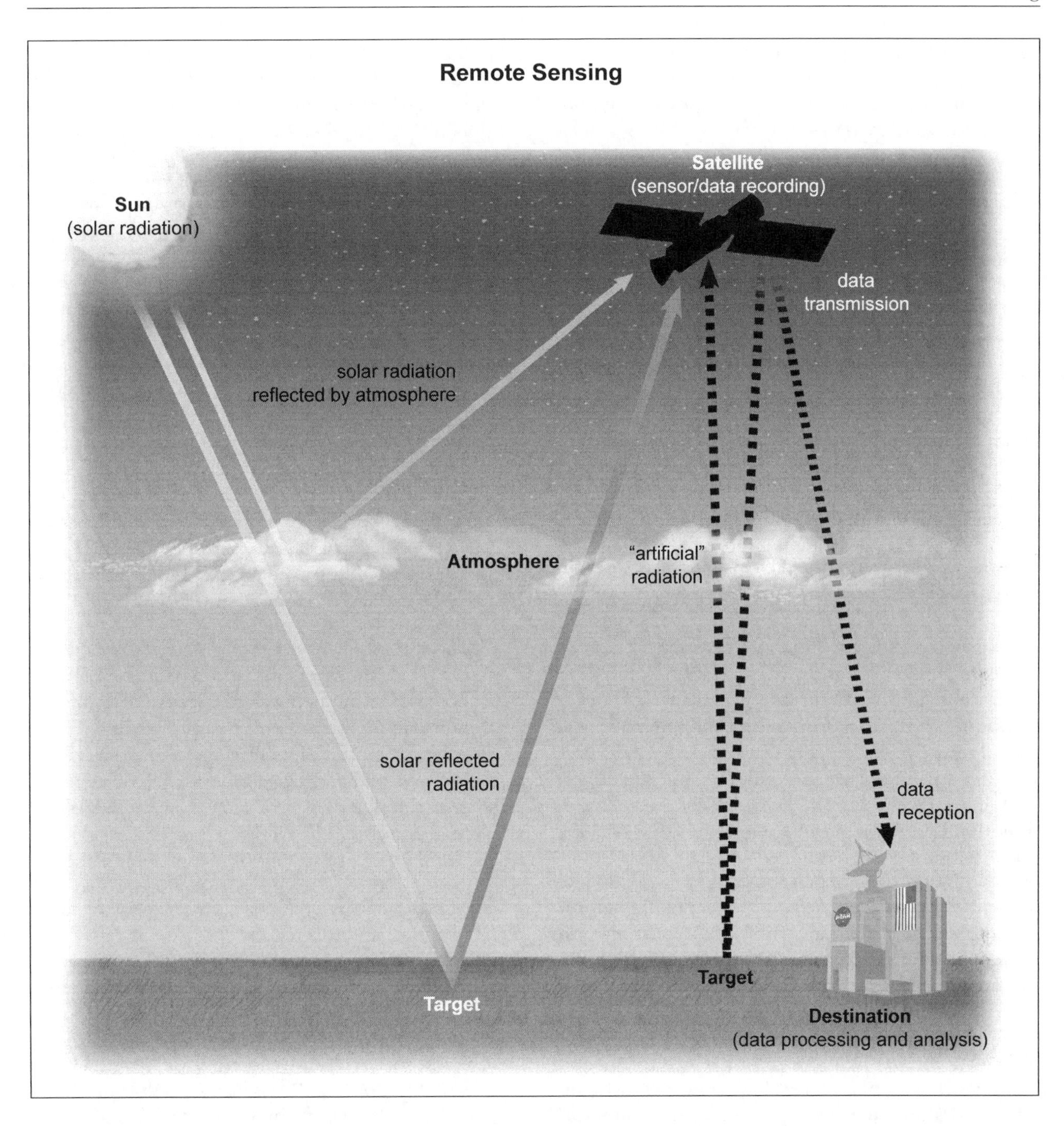

areas require mathematics and physics at the high school level.

Developers of remote-sensing technology usually require a degree in electrical engineering, materials science, computer science, computer programming, physics, or geophysics. Image processing, which includes photogrammetry, spectral manipulation, and feature extraction, requires a strong background in applied mathematics, computer programming, computer graphics, and possibly physics. Working with processed imagery is specific to the particular application. Therefore, training in the specific application is necessary, whether it is archaeology, astronomy, geography, geology, or medicine. Medical students would touch on the physics of imaging in biophysics courses. Many universities offer courses specifically in remote sensing as part of geography or environmental studies, areas in which remote sensing can be a core component. Advanced image processing requires an understanding of computer graphics and possibly pure mathematics, regardless of the application. Advanced studies in applied remote sensing is undertaken in relevant departments, such as geography, geology, and climatology.

Increasingly, remote sensing is being used by professionals in nontraditional fields. For example, field biologists may be equipped with computers whose background display is a processed or partially processed remotely sensed image; tracking of individual animals may be by office-based remote sensing. Fieldworkers may be asked to take measurements for calibration in image processing (ground truthing). Virtually all outdoor scientific or resource management work now requires use of a GPS, and this means understanding coordinate systems—the link to imagery.

Social Context and Future Prospects

Remote sensing has many applications, including monitoring Earth's resources and climate conditions, providing medical information, and ensuring national security. It is being used to assess the pressures of the world's increasing population and the effects of the industrialization of land and sea, such as pollution and global climate change. In 2010, specific satellites were being developed and deployed to measure atmospheric concentrations of carbon dioxide, carbon monoxide, ozone, and methane. Biomedical remote sensing provides unique information and functions as a surgical aid. Technological advances in remote sensing will increasingly aid global security surveillance and discoveries in astronomy.

Remote-sensing devices and usage have a sizable carbon footprint through resource extraction and energy usage in their manufacturing, deployment, and operation. For example, rare-earth minerals are mined for sensors. In addition, remote sensing is used for some activities that may not be environmentally sustainable, such as the extraction of fossil fuels and iron ores. However, it can also be applied to monitoring and attenuation of unsustainable practices such as fauna habitat depletion.

The demand continues for more real-time applications in remote sensing, especially in biomedical, military, and emergency service applications. These are likely to be met through further improvements in imaging and data retrieval by way of improved radiometric stability, precision manufacturing, preprocessing algorithms, higher performance computing, and creative innovation. Progress in virtual reality is likely to allow new methods of data interpretation.

Christopher Dean, B.Sc., Ph.D.

Further Reading

Cracknell, Arthur P., and Ladson Hayes. *Introduction to Remote Sensing*. 2d ed. Boca Raton, Fla.: CRC Press, 2007. A comprehensive, concise review of remote sensing that explains modern applications.

Delalieux, Stephanie, et al. "Hyperspectral Reflectance and Fluorescence Imaging to Detect Scab-Induced Stress in Apple Leaves." *Remote Sensing* 1 (2009): 858-874. Provides an example of hyperspectral remote sensing using band indices, applied to orchard health.

Dhome, Michel, ed. *Visual Perception Through Video Imagery*. Hoboken, N.J.: John Wiley & Sons, 2009. Shows how the three-dimensional environment, as perceived by humans, can be recreated as a mathematical model using computer processing of video imagery or multiple still cameras (a developing application of photogrammetry).

Kalacska, Margaret, and G. Arturo Sanchez-Azofeifa, eds. *Hyperspectral Remote Sensing of Tropical and Subtropical Forests*. Boca Raton, Fla.: CRC Press, 2008. Provides comprehensive examples of hyperspectral remote sensing used to characterize a variety of tropical and subtropical ecosystems.

Schowengerdt, Robert A. *Models and Methods for Image Processing*. 3d ed. San Diego: Academic Press, 2007. Shows how to get the most out of imagery through a range of corrections and enhancements.

Stein, Alfred, Ben Gorte, and Freek van der Meer, eds. *Spatial Statistics for Remote Sensing*. Boston: Kluwer Academic, 2002. A guidebook on reliability that shows where errors occur and how to experiment effectively, and explains interpolation, interpretation, and decision support.

Web Sites

American Society for Photogrammetry and Remote Sensing
http://www.asprs.org

European Space Agency
Earth Observing
http://www.esa.int/esaEO/index.html

Geoscience and Remote Sensing Society
http://www.grss-ieee.org

National Oceanic and Atmospheric Administration
National Geophysical Data Center, Earth Observation Group
http://www.ngdc.noaa.gov/dmsp/dmsp.html

U.S. Department of Defense
National Geospatial-Intelligence Agency
http://www.nga.mil

U.S. Geographical Survey
Aerial Photographs and Imagery
http://www.usgs.gov/pubprod/aerial.html

See also: Air-Quality Monitoring; Coastal Engineering; Erosion Control; Forestry; Maps and Mapping; Sonar Technologies.

S

SILVICULTURE

FIELDS OF STUDY

Forestry; agroforestry; horticulture; plant physiology; soils engineering; botany; ecology; biology; entomology; genetics; geology; microbiology.

SUMMARY

Silviculture is a branch of forestry. The term is based on *silva*, the Latin root for "woodland." Silviculturalists are concerned with the protection, restoration, and management of forest resources, often for timber production but also for wildlife habitats, recreation, and other uses. Silviculturalists are involved in harvesting the forest for its timber value or to reduce wildfire risk and in reestablishing the forest in ways that ensure that it is sustainable. Controlling tree growth, managing the composition of tree stands, protecting trees from pathogens, and ensuring the quality of trees are some of the activities that make up the field of silviculture.

KEY TERMS AND CONCEPTS

- **Clearcutting:** Cutting of all trees in an area at the same time.
- **Coppice:** Reforestation method that uses stump sprouts or root suckers of harvested trees.
- **Dominant Trees:** Trees with crowns that receive the most sunlight within a stand because of their height and size.
- **Even-Aged Stand:** Group of trees that are of the same age.
- **Overstory:** Tallest trees in the forest; also known as the canopy.
- **Prescribed Burning:** Forest habitat manipulation through systematic use of fire.
- **Silviculture Prescription:** Proposed plan for implementing silviculture activities prepared by a licensed silviculturalist.
- **Stand:** Grouping of trees made up of one or many species with uniformity in age, composition, and planting site qualities.
- **Understory:** Lower-level forest plants, including seedlings and saplings from the overstory trees.

Definition and Basic Principles

Silviculture is the art and science of establishing and caring for forests to meet the needs and management objectives of landowners and society. Although timber production sustainability is often the most important objective of silviculture, other benefits of a well-managed forest are preservation of wildlife habitats, water conservation, and aesthetics. Silviculture relies on ecology and natural processes, including fire, to control forest composition and growth. However, too much interference in the forests by silviculturalists can negatively affect the forest ecosystem.

Silviculture is practiced during all stages of the life of a tree stand to meet the dual goals of sustaining forest growth and satisfying market demands and landowner objectives. A silviculturalist's management decisions concern not only the present forest and timber crop but also future forest quality and growth. Silviculturalists use many methods, such as selectively cutting trees while retaining some of the high-grade tree growth for the future. Other practices include cutting of all trees except the ones with the best genetics for producing seeds to regenerate the forest and cutting out all undesirable trees in a stand to allow for natural regeneration of a desirable species whenever possible. In both cases, the harvesting methods allow for better growth because the new trees are in full sunlight rather than being blocked by the overstory.

Background and History

Society has been removing trees to satisfy its needs since ancient times. Early settlers thought little about the value of the forests and clearcut them to allow for agriculture, development, and wood production with little thought of the future, as the supply of trees appeared to be limitless. The practice of silviculture, however, became a necessity in Europe by the 1600's to restore the timber used in everyday life for fuel and building materials. Timber production also became important to the economy of many societies in the eighteenth to twentieth centuries, and silviculture concentrated on ensuring that there would be future growth available for harvesting.

Early silviculture involved practical experience, but during the 1700's, Henri-Louis Duhamel du Monceau, a French botanist, was instrumental in applying scientific methods to forest regeneration, including experimenting with plant physiology. In the mid-1800's, Martin Faustmann, a German forester, published the Faustmann formula for capital evaluation, which became the basis for modern forest valuation. By the end of the 1800's, silviculturalists had developed a theory of tolerance, and forest experimental stations had been established to gain a better understanding of the relationship between forests, organisms within the forest, and soil conditions. In the 2000's, agencies such as the U.S. Forest Service are concerned not only with timber production and harvesting but also ecology and sustainability.

How It Works

The practice of silviculture is systematic and involves many different operations including preparing the site, cultivating, regenerating, thinning, harvesting, clearcutting, pruning, fertilizing, and prescribed burning. Before any silviculture applications are implemented, however, silviculturalists must conduct surveys of the forest to develop silviculture plans that specify not only the goals and objectives of silviculture activities but also the means of achieving them. These plans are known as silvicultural prescriptions. Often Global Positioning Systems (GPS) and geographical information systems (GIS) mapping is used, and survey data are entered into databases for analysis of forest resources, such as wildlife habitats and watersheds. Two types of plans that a silviculturalist might develop include regeneration and thinning plans.

Regeneration Plans. Trees are lost through harvesting, wind, fire, insects, and disease. Silviculture activities include implementing methods for regenerating tree stands. Among the methods are tree seeding, coppice, clearcutting, shelterwood, selection cutting, and variable retention. When trees are grown from seed, the seeds are collected from parent trees in the forest that are located close to the planting site. Before planting takes place, data are analyzed to ensure that the biophysical environment of the planting site will allow the tree seedlings to survive. Coppice techniques, in which trees are cut and then regenerated from sprouts and suckers, can result in even-aged tree stands, although the initial cutting can reserve some of the original trees to develop a two-aged stand. Clearcutting is used to generate new, even-aged, or two-aged forest stands either through natural seeding or replanting methods. Shelterwood regeneration results in new, even-aged stands, as silviculturalists reserve mature trees and then manage new tree growth that develops in the undergrowth and is sheltered by the reserve trees. Selection cutting involves the removal of older trees, which enhances the growth and health of the remaining trees, while variable retention involves retaining of some trees in a harvested stand until the stand undergoes the next harvest in order to maintain specific tree species on a site.

Thinning Plans. Trees that are planted too close together are deprived of sun, water, and nutrients needed for growth. Therefore, a silviculturalist must engage in planning for tree thinning to allow for sustainable growth. Thinning provides for tree stands that are more resistant to fire, wind, pests, and disease. Thinning techniques may serve commercial as well as noncommercial purposes and include crown, low, free, and selective thinning. In crown thinning, dominant trees are removed from a tree stand to encourage growth of codominant trees that may be more desirable. When employing the low thinning method, the silviculturalist removes dominant trees from the understory to enhance growth in the overstory. Free and selective thinning techniques allow the silviculturalist to remove less desirable trees in the understory or overstory, often to establish an even-aged tree stand made up of one species. Once the thinning is completed, however, a regeneration plan must be implemented.

Other Silviculture Practices. Before regeneration of a harvested site can take place, the planting site must be prepared through activities such as prescribed burning and trenching. A planted forest that consists of genetically enhanced seedlings will regenerate faster than a naturally seeded forest, and a planted forest will also receive more care by the silviculturalist, including removal of naturally occurring competing plants, application of pesticides to combat insects and diseases, and fertilization to enhance growth.

Applications and Products

Silviculture involves maximizing multiple uses of the forest by private and public landowners in a manner that preserves and sustains the forest resources for the future. Silviculture is based on the natural science of silvics, which concerns the study of environmental factors, such as soil conditions, climate, competing plants, and other living organisms, and how they affect the vegetative biomass. From this study, the silviculturalist applies scientific methods to ensure adequate growth of the trees used in multiple applications and products.

Timber Forest Products. Forests provide many marketable wood products, from lumber used in building and construction industries to timber that is made into wooden panels, barrels, and gift items. Commercial harvesting also supplies the timber needed by pulp and paper companies. Often a silviculturalist is involved in growing a specific species of tree because of its marketability, such as for furniture production. Silviculturalists achieve this objective by using herbicides to control competing species and planting high-quality seedlings of the plant variety that is desired.

Nontimber Forest Products. Forests are managed through silviculture to produce more than timber. Examples include maple products, nutmeats, and shell products that have economic viability through the silviculture practice of single-species planting. Silviculturalists also manage undergrowth vegetation in forests to produce products such as edible ferns and mushrooms. Trees produce important oils, including the cedar oil used in many cedar-based products such as insecticides. Timber branches and leaves are used in wreaths and garlands, dried and sold for decorating products, or woven together into baskets and screens.

Wood Energy. Forests are also harvested as a biomass for energy, including the production of firewood and charcoal. Silviculture practices for this application usually involve management for hardwood species through thinning, quality stand rehabilitation, and effective harvesting to mitigate the environmental effects of harvesting on soils, organic matter, and wildlife habitats. The growing market in wood energy production has resulted in the growth of the silviculture contracting industry.

Christmas Trees. Christmas trees are an important seasonal crop for both retail and wholesale economies. Silviculture is practiced by cultivating one species of tree, such as balsam fir. Single species are maintained through weeding of competitive species and shearing, fertilizing, and treating the Christmas trees with insecticides until they are harvested.

Restoration Ecology. Silviculturalists accelerate recovery of ecosystems in areas that have been disturbed or destroyed by natural disasters or human beings. They ensure a sustainable and productive forest through reforestation methods and restore and expand wildlife habitats, even creating corridors that link existing and newly expanded habitats by planting the plant species that support specific animal habitats. Restoration ecology also involves the use of erosion control methods that are necessary after floods, fires, and human disturbances, such as mining, to ensure adequate soil for plant growth. These methods include planting fast-growing species that hold the soils in place.

Hydrology and Watershed Protection. Silvicultural activities, such as timber harvesting, can greatly affect watersheds, stream flow, and the underlying hydrology by causing pollution and disrupting existing hydrologic processes. The pollution is often caused by soil erosion created by logging and associated activities, such as road building, and is not easy to control, as it usually does not come from a single source. Because entire watersheds can be negatively affected by poorly managed timbering operations, state and federal laws have been adopted to ensure that silviculturalists engage in timber harvest planning before cutting down trees.

Climate Control. Deforestation can affect climate, as trees soak up fossil fuel emissions. Therefore, silviculturalists must understand how their activities affect climatic changes and ultimately the economies in many countries. Silviculture is being employed as a climate mitigation method by increasing the size of silviculture plantations, especially in areas where drought has become a serious problem. However, expanding the size of plantations to control climate is not always possible because of the need to develop land to meet growing global social needs.

Control of Invasive Plants. Throughout history, nonnative plants have been introduced intentionally or by accident for purposes such as food, medicine, and aesthetics, sometimes with negative results, such as lost revenues. Once introduced, about one out of every ten of these nonnative plants becomes difficult to control and can destroy ecosystems and habitats. Some of the ways that silviculture methods can be used to overcome these negative outcomes is to plant sterile hybrids of the invasive species and to reintroduce native species into an area.

Pest and Disease Control. Silviculturalists act as plant doctors throughout the globe when they apply methods to protect plants from insect and disease hazards. Because insecticides may have side effects and do not remove the cause of the insect invasion, silviculturalists use methods such as altering forest conditions, preventing injuries to trees, selective cutting to remove physiologically weakened trees, and accelerating regeneration in areas where there has been an outbreak of insect pests or diseases. When establishing a new tree stand, a silviculturalist will introduce tree species that have been able to grow in the area before without devastation from pests and diseases.

Urban Forestry. Silviculture activities are found around cities. Urban forestry is a type of silviculture that involves planning, cultivating, and maintaining trees and greenspaces in developed communities. Urban forestry can help meet many objectives, including improving public health and enhancing environmental education in urban settings.

IMPACT ON INDUSTRY

The United States leads the world in the production and consumption of wood products. Therefore, silviculture is important to the future of industries that rely on healthy forests. Silviculture is being practiced throughout the globe, from single-species plantation silviculture in countries such as New Zealand, Brazil, and South Africa and in the southern United States to silviculture education in the historic forests of Europe, where the field of silviculture was born. In addition, forested wetlands and coasts in locations such as Australia and the United Kingdom are maintained through silviculture practices. Silviculturalists are necessary participants in meeting industrial objectives to eradicate insect pests and plant diseases that affect commercial forest lands.

Government Agencies. The U.S. Forest Service, a division of the Department of Agriculture, is one of the primary agencies involved in silviculture and manages a program that sells timber to private companies. The National Park Service is also active in forest management and silviculture, as are state and local governments, usually to serve recreational and aesthetic purposes. Every state has an agency responsible for its forests, and silviculturalists are needed to carry out the forest management policies set by the government.

University Research. Many universities have active forestry programs involved in silviculture research. Much of the early research began in European universities, but in the United States, many of the land-grant institutions are involved in such work. The National Association of University Forest Resources Programs (NAUFRP) represents more than sixty universities, scientists, extension service specialists, and international programs involved in forest resource education and research. The association's goal is to sustain forests through research, education, and outreach. Many universities also conduct research through experimental stations, such as Yale University's Silas Little Experimental Station and the Bent Creek Experimental Forest south of Asheville, North Carolina, which is the oldest experimental station in the United States and is made up of partners from universities, governmental agencies, and nonprofit organizations.

Nonprofit Entities, Conservation Organizations, and Land Trusts. The Nature Conservancy International is concerned with protecting important biological features worldwide, including forests that help control climate and land where significant ecosystems are threatened. Other nonprofit and conservation organizations that are active in sustaining forests and preserving wildlife habitats include EarthTrust, EarthWatch, the Forests Forever Foundation, the National Audubon Society, the Sierra Club, and the Land Trust Alliance. Silviculture education to promote tree care, conservation, and the planting of trees by individuals and organizations is the goal of the National Arbor Day Foundation in the United States. Rainforest conservation groups include the Rainforest Alliance and the Rainforest Action Network. Silviculture practice is also represented by the nonprofit Society of American Foresters, which is the largest professional organization representing forestry and related professions in the world. One of its goals is to advance the applied science of silviculture as it relates to the practice of forestry.

Timber Companies and Private Landowners. Two-thirds of the forests in the United States are classified as timberland, which means their purpose is for the production of commercial wood products. A little less than three-fourths of all timberland is owned by private landowners, often private land trusts. Because of the conservation movement, many of the private land trusts have taken on the responsibility of being stewards of the forests by managing them for wildlife habitats and sanctuaries and for recreational purposes, rather than harvesting them for their commercial value. The largest private timber company in the United States with the most diverse holdings is the Plum Creek Timber Company. This company owns timberland in more than a dozen states throughout the country. Plum Creek is typical of other large timber companies that employ silviculture and sustainable practices to conserve natural resources and offer recreational opportunities, such as hunting and fishing, while selling timber to companies that produce lumber and manufacture paper products.

Fascinating Facts About Silviculture

- Trees, the longest-living organisms on Earth, do not die from old age but rather from disease or the influences of humans and insects.
- In 1872, J. Sterling Morton founded Arbor Day in Nebraska for the purpose of encouraging individuals and organizations to plant trees to replace what society had destroyed. About 1 million trees were planted on the first Arbor Day celebration.
- Since adoption of the National Forest Management Act of 1976, all federal forests harvested in the United States are to be replanted within five years.
- Some trees, such as willows, are able to fight disease without the assistance of a silviculturalist by emitting a chemical that communicates to other trees that an insect attack is underway, thus warning the surrounding trees to pump tannin into their leaves to prevent further invasion.
- Dutch elm disease, named after the Dutch scientist who identified the fungus that causes the disease, spread to the United States in crates made of diseased elm wood.
- Trees grow from the top rather than the bottom. Thus branches on tree trunks remain mostly in the same location for the life of a tree, even though the tree gets taller.
- It takes about three trees to generate the amount of oxygen needed by each person to survive on Earth.

Careers and Course Work

Basic course work for a career in silviculture includes botany, biology, ecology, and forestry. Silviculture, however, is a multidisciplinary field, and students must take courses in more than one field. Students can concentrate their studies in specific areas, including botany, genetics, and zoology for those interested in stand development, plant composition, habitat management, and ecology. Soils engineering, geology, and hydrology are important if site preparation, forest growth, and water quality protection are of interest. If a student wants to work with insect pests and diseases, then courses in plant physiology, microbiology, and entomology are a must. Finally, for students interested in the business of timber production, courses in economics, statistics, sociology, planning, and business administration will be helpful.

Entry-level positions for silviculture workers do not usually require a college degree. These positions include laborers involved in such jobs as tree planting, harvesting, stand thinning, and fire control. Those involved in more complex silviculture practices—such as silviculture technicians, regeneration planners and surveyors, producers of genetically enhanced plants, forest and ecosystem managers, and applicators of chemical herbicides and fertilizers—require training, certification, and usually college degrees. Master's and doctoral degrees are held by many silviculturalists, especially those involved in research, as well as certified and licensed silviculturalists, who are able to prepare silviculture prescriptions.

Jobs are available with private timber companies and at all levels of government. Some silviculturalists find work as consultants to private landowners. The future is bright for silviculture because of the many demands placed on the forests, their multiple uses, and the wood products they produce.

Social Context and Future Prospects

The growing demand for multiple-use forests and harvesting trees for wood products conflicts with the importance of preserving diverse forest ecosystems. Forests are complex, and silviculturalists recognize that by using scientific methods of tree removal and management, forests can adapt to natural changes and those caused by humans. Advancements in technology allow silviculturalists to use best management practices to harvest, regenerate, and tend forests to ensure their sustainability. These methods include advance regeneration before harvesting trees, maintaining soil quality and stability before and after tree harvest, genetically improving planting stock, improving the use of chemical fertilizers and herbicides, and protecting water quality by preventing the loss of nitrogen into streams and water bodies through the use of plant buffers.

Silviculturalists must educate the public about the many facets of silviculture to destroy the myth that it simply promotes timber production. In the future, silviculture will be useful to people for purposes such as reducing fire risks in areas where residential development coexists with forests and in abating specific types of insect pestilence. Silviculturalists can also manipulate the landscape to restore former plant communities or establish new plant materials that provide benefits such as habitats for endangered or threatened wildlife species. However, silviculturalists must work with members of other disciplines throughout the world, including economists and social scientists, to ensure that the future needs and objectives related to silviculture are met and that private landowners and the timber industries are constantly receiving education concerning the benefits of silviculture in managing forests.

Carol A. Rolf, B.S.L.A., M.Ed., M.B.A., J.D.

Further Reading

Andrews, Ralph. *This Was Logging.* Atglen, Pa.: Schiffer, 1997. A pictorial history of logging in the Pacific Northwest from 1890 to 1925 before society recognized the need to implement systematic silviculture methods.

Burton, L. Devere. *Introduction to Forestry Science.* 2d ed. Clifton Park, N.Y.: Delmar Cengage Learning, 2007. Comprehensive textbook that uses an applied science approach including a discussion of silviculture and a list of career opportunities.

Nyland, Ralph. *Silviculture: Concepts and Applications.* Long Grove, Ill.: Waveland Press, 2007. Provides a good discussion and actual examples concerning the conflicts between ecology and economy in managing forests, and explains how a silviculturalist can integrate these multiple needs and objectives.

Puettmann, Klaus, Christian Messier, and K. David Coates. *A Critique of Silviculture: Managing for Complexity.* Chicago: Island Press, 2008. Considers silviculture as it is practiced and its future based on the pressure to provide wood products while simultaneously preserving forest ecosystems.

Sinclair, Wayne, and Howard Lyon. *Disease of Trees and Shrubs.* 2d ed. Ithaca, N.Y.: Cornell University Press, 2005. Examines tree pathology and diseases at length.

Walsh, Ann, and Kathleen Cook Waldron. *Forestry A-Z.* Olympia, Wash.: Orca Book, 2008. An alphabetical listing of facts accompanied by pictures related to the forestry industry, from heavy logging equipment to artistic uses of trees.

Web Sites

British Columbia, Ministry of Forests, Mines, and Lands
http://www.for.gov.bc.ca

National Association of University Forest Resources Programs
http://www.naufrp.org

Society of American Foresters
http://www.safnet.org

U.S. Forest Service
http://www.fs.fed.us

Western Forestry and Conservation Association
http://www.westernforestry.org

See also: Forestry; Horticulture; Plant Breeding and Propagation; Wildlife Conservation.

SOLAR ENERGY

FIELDS OF STUDY

Energy; electrical engineering; mechanical engineering; physics.

SUMMARY

Solar energy is the energy from the Sun, captured and used to heat homes or provide electricity. The three main types of solar energy systems are passive, in which solar energy is stored without using any other energy source; active, in which electricity is used to capture the Sun's energy; and photovoltaic, which directly converts sunlight into electricity. Although solar energy is free in that costs are not involved in generating it, it is not constant and must be captured and stored. Also, the systems used to capture solar power remain expensive.

KEY TERMS AND CONCEPTS

- **Active Solar:** Type of solar energy system that uses energy to capture solar energy.
- **Alternate Energy:** Any sustainable nonfossil fuel source of energy, including solar, wind, and geothermal power.
- **Passive Solar:** Type of solar energy system that is powered only by the Sun.
- **Photoelectric Effect:** Electric current produced when two dissimilar materials pressed together are exposed to light.
- **Photovoltaic (PV) Cell:** Semiconducting material, such as crystalline silicon or cadmium telluride, that converts solar radiation into direct current electricity using the photoelectric effect.
- **Photovoltaic Module:** Configuration of photovoltaic cells laminated between a clear glaze and a solid substrate.
- **Photovoltaic System:** One or more photovoltaic modules arranged into an array, with associated control and storage devices.

Definition and Basic Principles

Solar energy is light from the Sun that has been converted into heat energy or electricity. The three most common conversion methods are passive systems, which collect and store solar energy without the use of any other source of energy; active systems, which collect and store energy by employing electric energy; and photovoltaic systems (PV), which convert sunlight into electricity. Both passive and active systems use glass to admit sunlight and prevent heat from escaping, and mass to store the heat collected. The four types of passive systems are direct gain, indirect gain, attached gain, and thermosyphon. Active systems either collect sunlight directly on flat surfaces or use parabolic reflectors to achieve high temperatures by focusing the light. Either air or water may be used to transfer the heat from the collector to storage.

PV systems use arrays of photocells to transform solar radiation directly into DC (direct current) electricity. The photocells, typically semiconducting silicon crystals, act as insulators until illuminated by radiant energy. The material then conducts electricity, effectively making each cell a small battery. By connecting photocells into large modular arrays, sufficient electric energy can be generated to power homes or, when sufficient modules are present, to produce electricity for a centralized power plant. Small PV systems may use battery storage or tie into the electric grid to allow energy to be withdrawn when needed or fed into the grid when not necessary.

Although solar energy is free energy in that there is no cost to generate it, it has several disadvantages. It is diffuse and intermittent, and it must be stored. Also, active collection devices are constructed of expensive nonrenewable resources such as aluminum and copper.

Background and History

Solar energy has always provided, directly or indirectly, virtually all of humanity's energy. Ancient Greek homes were oriented toward the south, and early Chinese architecture incorporated solar design to heat interior spaces. By the first century B.C.E., Rome had added clear glass windows to Greek solar designs to trap the heat, thus creating the first true passive solar design. The Sun was also used to heat water entering the huge public baths indigenous to Roman society.

After the fall of Rome, solar architecture was

forgotten until the sixteenth century, when greenhouses were used to grow exotic fruits and vegetables in northern Europe. By the eighteenth century, large glass windows enabled the construction of better greenhouses, which evolved in the nineteenth century into ostentatious conservatories for displaying exotic plants.

Active systems that focused sunlight to produce high temperatures were developed in the nineteenth century. Domestic hot water systems were first built and marketed in the early twentieth century. By midcentury, active systems using air to heat homes appeared, but their acceptance was limited because of their high costs.

Photovoltaic (PV) trace their origin to 1880 when Charles Fritts developed a solar electric cell using germanium crystals, but commercial development stagnated until the 1950's when Bell Laboratories produced a viable but costly silicon-based system to power remote communication devices. The National Aeronautics and Space Administration, needing lightweight reliable energy sources for its nascent space program, adapted these PV systems for its first satellites, launched in the late 1950's.

In the 1970's, because of the oil embargoes and rise in fuel prices, solar energy began to capture the interest of the public. However, the decline in oil prices in the 1980's produced a drop in the interest in solar power. Since the mid-1990's, solar water heating systems have grown at an average annual rate of 20 percent, rendering solar water heating the most widely deployed solar technology of the early twenty-first century.

How It Works

Although less than half of the solar radiation that reaches the Earth is available for human use (some is absorbed by the atmosphere, land, and oceans, and some is radiated back to space), this amount is prodigious enough to provide for all human energy needs if it could be efficiently captured. Because solar radiation is dilute and noncontinuous, large collector areas are necessary, and storage devices must be integrated. Photovoltaic systems convert radiation directly into electricity, and solar thermal units collect energy for interior spaces or water heating. Passive systems convert sunlight directly into interior space heating, and active heating systems require electricity to power pumps or fans. Active systems may be subdivided into those that use flat-panel stationary collectors and those that focus incoming solar rays to achieve temperatures high enough to create steam.

Photovoltaic. Photovoltaic cells transform solar radiation directly into electricity. The cells consist of two types of silicon crystals in which bound electrons are energized into a conducting state when irradiated by light. The freed electrons cross the junction between the two crystals more easily in one direction than the other, thus creating negative and positive surfaces, the basis of a battery. This photobattery provides direct current (DC) electricity. The brighter the irradiating light, the greater the current. By connecting large arrays of such cells, a solar module, which typically can provide 170 watts per square meter of surface area at 14 percent efficiency, is created.

Solar panels used to power homes and businesses are typically made from modules holding about forty cells. A typical house requires an array of ten to twenty solar panels to provide sufficient power. The panels are mounted at a fixed angle facing south, or they can be mounted on a tracking device that follows the sun, allowing them to maximize solar energy capture. For large electric facility or industrial applications, hundreds of solar arrays are interconnected to form a large utility-scale photovoltaic system.

Passive Solar Heating. Passive solar heating systems use south-facing glass windows to collect solar energy and a room's interior mass to store energy and regulate temperature swings. The three main types of passive systems are direct, indirect, and attached gain. Direct-gain systems incorporate ample interior mass for storage, and indirect-gain systems require a massive wall positioned directly behind the south-facing glass. Attached-gain systems consist of a greenhouse, accessible to the house, attached to an exterior southern wall. When the greenhouse is warm, the doors can be opened to heat the house. A fourth system, the thermo-siphon, uses flat-plate collectors to heat water and a storage tank located above the collector top. Heated water rises by natural convection into the storage tank, creating a siphon effect that keeps the fluid circulating. Because no electricity is used, this constitutes a passive system.

A well-designed passive system, in addition to double-paned south-facing glass and interior mass, includes movable insulation to cover the windows at night, overhangs above the windows to keep out the summer sun, and sufficient house insulation to minimize heat loss.

In direct-gain systems, the thermal mass, incorporated into a floor or wall, is typically brick, tile, or concrete. The mass must be sized to the total area of south-facing glass—the greater the area of glass, the greater the mass required to prevent overheating the room. Indirect gain systems use a massive wall of brick or barrels of water located in proximity to the south-facing glass. The outside-facing surface of the mass is painted black to effectuate solar gain stored in the mass. The heat is released through vents into the interior living space by natural convection; at night, the vents are closed, preventing convective heat loss, while the mass radiates heat into the interior space.

Attached gain, or greenhouse, systems are usually entirely glass with concrete or soil serving as the mass. When properly designed, an attached greenhouse can be used to provide food as well as heat during the winter. A different application of passive solar is the thermal chimney ventilation system, consisting of an interior vertical shaft vented to the exterior. When the chimney warms, the enclosed air is heated, causing an updraft that draws air through the building.

Active Solar Heating. Active solar heating systems transfer a sun-heated fluid (air or water) from an exterior south-facing collector to the point of use or to a storage facility. In air systems, the storage facility is a bed of rocks, and in water systems, tanks of water.

In addition to the collector and storage feature, active systems include a pump or fan to circulate the fluid and a differential thermostat that regulates fluid flow to those times when the collector is at least 10 degrees Fahrenheit hotter than the storage facility. Active systems may be used to heat interior space or to produce domestic hot water. The circulating fluid for domestic hot-water systems is always water, and the storage unit is a water tank. Space heating units may circulate either air or water, but air systems are more common.

Solar flat-plate collectors consist of a rectangular box containing a black metallic plate covered by nonreflecting tempered glass. In water systems, tubes to conduct water are soldered to the plate, and in air systems, small channels direct the air. When the sun strikes the black surface, light is changed into heat, which is transferred to the fluid moving across the heated surface. For maximum efficiency, collectors should be oriented directly south and set at an angle (from horizontal) equal to the local latitude (for domestic hot-water systems) or latitude plus 10 degrees (for space heating systems).

Both space heating and domestic hot-water systems use water mixed with propylene glycol antifreeze as the circulating fluid, propelled by a pump. Heat from the fluid is transferred to the hot-water storage tank through a heat exchanger, to be used for domestic hot water or as preheated fluid for hydronic baseboard heating systems.

When air is used as the working fluid, excess heat is stored either in smooth rocks or in a phase-change material. Rock storage consists of a 280-cubic-foot bin of 1-inch-diameter smooth rocks weighing 7 tons. When the collector is warmer than the house, a fan pumps the air directly from the collectors into the house. When heat is not required, air is directed through the rock bin, transferring the heat into the rocks. At night, air from the house can be circulated through the rocks to reclaim the stored heat.

Focusing Collectors. Concentrated sunlight is realized in one of two ways: troughs of parabolic mirrors that focus sunlight to an oil-filled tube positioned along the focal line, or huge assemblies of mirrors that reflect sunlight from a large area to a small central receiver. Either method may be used in a solar thermal power plant, where the concentrated radiation is used to produce high-temperature steam that drives a turbine to create electricity. A solar furnace is a type of focusing collector employing parabolic curved mirrors to concentrate sunlight to a focal point to generate extremely high temperatures.

Applications and Products

Photovoltaic Systems. Because PV cell arrays are expensive, the cost of the electricity produced in photovoltaic systems has traditionally exceeded the cost of fossil fuel electricity. Since the mid 1970's, however, the cost has consistently decreased so that by 2010, the cost per kilowatt of photovoltaic electricity was comparable to fossil fuel costs. Low market penetration and insufficient economies of scale have inhibited even lower costs for PV systems, but prices are projected to continue to drop as research raises the conversion efficiency of PV cells. It is projected that by the year 2020, the cost of installed PV units will be half the 2010 cost. As one of the most rapidly growing alternate energy sources with production doubling every two years since about the 1980's, it is projected that by 2020, there will be 500,000 additional installations and a cumulative world capacity of 1,500 gigawatts.

Although individual household PV modules are more expensive per kilowatt than large centralized PV power plants, individual units become cost competitive when distribution costs are eliminated. From 1995 to 2009, solar module costs per installed watt declined at 5 to 6 percent annually, a trend projected to continue, particularly in regions with ample sunlight, expensive fossil fuel electricity, and government incentives.

Traditionally solar cells have been made from pure crystalline silicon doped with boron or phosphorus. The manufacturing process is not inexpensive, and the conversion efficiency rarely exceeds 15 percent. Research on using amorphous silicon has led to less expensive PV cells but these cells have considerably lower efficiencies of less than 6 percent. Nevertheless, PV laminates composed of thin nonreflective layers of amorphous silicon photocells coated on flexible plastic have been made into roofing material. In 2010, Solarmer Energy, a Los Angles-based manufacturer of plastic solar panels, achieved a recordbreaking high efficiency of 7.6 percent for plastic PV panels. With such panels, a dual function is served: Roofs are weather-protected with material that generates electricity whenever the Sun shines. Because PV modules have no moving parts and low maintenance, these systems are projected to last at least thirty years, the typical lifetime of quality roofing shingles. If the excess energy is stored in rechargeable batteries, it would be possible for homeowners to eliminate their reliance on the grid. Alternately, if the PV system is integrated with the grid, the need for a large bank of storage batteries is eliminated. Excess electricity is sent back into the grid for credit, and the grid provides for nighttime or cloudy weather requirements.

Worldwide, the trend through the 2000's has been toward ever larger scale centralized PV power plants, as typified by a 14-megawatt plant in Nevada and a 20-megawatt facility in Beneixama, Spain, both completed in 2007. Since batteries are not practical as a backup supply in large-scale applications, storage is achieved using excess energy to pump water from a lower elevation reservoir to a higher one; the energy is reclaimed by releasing the water through a hydroelectric generator. By judiciously pairing PV systems with wind energy and biogas generators, a twenty-four-hour supply of renewable electricity can be virtually guaranteed. Such a system has been successfully pilot tested by the Institute for Solar Energy Supply Technology at the University of Kassel, Germany.

Passive Solar Heating. Daylighting systems collect and disperse sunlight into interior spaces using skylights, clerestory windows, and light tubes. Physiological and psychological benefits accrue when natural lighting replaces artificial, and the necessity of summer air-conditioning to eliminate waste heat from incandescent bulbs is reduced. Properly implemented systems can reduce lighting related energy consumption by 25 percent.

Integrated passive systems that combine solar heating, ventilation, and lighting, tailored to the local climate, create well-lit spaces maintaining a comfortable temperature with minimal use of fossil fuel energy. For agriculture, greenhouses have been superseded by less expensive tunnels of polyvinyl covering rows of crops to support winter growth.

Another application, still in the experimental stage, is the solar pond, a pool of saturated saltwater at least 6 feet deep that collects and stores solar energy. The concentration of salt increases with depth, preventing convection currents and allowing the temperature to increase with depth. An experimental pond near the Dead Sea was able to achieve temperatures approaching 200 degrees Fahrenheit at its bottom layer. When used to drive a heat engine to produce electricity, the overall efficiency was 2 percent.

Solar cookers use solar radiation for cooking, drying, and pasteurization. The simplest solar cooker consists of an insulated container with a transparent cover that can achieve temperatures as high as 300 degrees Fahrenheit. More elaborate cookers, using focusing mirrors, can achieve temperatures of 600 degrees Fahrenheit in direct sunlight.

Active Solar Heating. Solar thermal technologies can be used for water heating, space heating, air-conditioning, and process heating. The most common types of solar water heaters are glazed flat-plate collectors, evacuated tube collectors to achieve higher temperatures, and unglazed flat-plate collectors used to heat swimming pools. As of 2007, the capacity of these systems totaled 154 gigawatts, with China being the greatest consumer. Over 90 percent of homes in

Israel and Cyprus use solar domestic hot-water systems, while in the United States and Australia, the main application is as heaters for swimming pools.

Solar distillation, operating by passive, active, or hybrid modes, is used to make saltwater or brackish water potable. Water for household use or storage may be easily disinfected by exposing water-filled bottles to sunlight for several hours. More than 2 million people in developing countries disinfect their daily drinking water by this method. In small-scale sewage treatment plants, solar radiation is an effective means of treating wastewater in stabilization ponds without employing chemicals or using electricity.

Phase-change materials, such as Glauber's salt (sodium sulfate decahydrate), store energy by transforming from solid to liquid at a temperature of about 85 degrees Fahrenheit. Heat from the Sun is absorbed by melting the salt; when the temperature drops below the melting point, the salt resolidifies, releasing the stored heat.

Concentrating Collectors. Hybrid solar lighting systems use sun-tracking focusing mirrors and optical fiber transmission to provide interior lighting. Typically half of the incident sunlight can be transmitted to rooms, where it replaces or supplements conventional lighting.

The Solar Kitchen, located in Auroville, India, uses a stationary spherical reflector to focus light to a linear receiver, perpendicular to the sphere's interior surface, where steam, used for kitchen process heat, is produced.

A solar concentrating device developed by Wolfgang Scheffler in 1986 produces temperatures between 850 and 1,200 degrees Fahrenheit at a fixed focal point by means of flexible parabolic dishes that track the Sun's diurnal motion and adjust curvature seasonally. By 2008, more than 2,000 large Scheffler cookers had been built, most used for cooking meals. The world's largest system, in Rajasthan, India, can cook up to 35,000 meals daily.

Another application, developed by Sandia National Laboratories, combines high temperatures from focusing collectors with a catalyst to decompose carbon dioxide into oxygen and carbon monoxide. The carbon monoxide can then be reacted with hydrogen to produce hydrocarbon fuels.

In the United States, the first commercial concentrating system, Solar Total Energy Project, was developed in Shenandoah, Georgia. This system uses a field of 114 parabolic dishes to provide 50 percent of the process heating, air-conditioning, and electrical requirements for a clothing factory. In Southern California, a system of parabolic trough collectors heats oil in tubes along the focal line. The heated oil is used to produce steam to power a generator. This system supplies up to 350 megawatts of power at a 25 percent conversion efficiency at a price competitive with fossil fuel electricity.

Central receivers, or power towers, use an extended assembly of moveable Sun-tracking mirrors to reflect sunlight to a small region on top of a tower, where temperatures between 1,000 and 2,700 degrees Fahrenheit provide the motive power to produce electricity. The first large-scale demonstration facility, constructed in southern California in 1982, was a 10-megawatt plant, later increased to 200 megawatts, at a cost competitive with fossil fuel plants. Shortly after this plant proved its feasibility, additional commercial units in the 30- to 50-megawatt range were constructed in the southwestern United States, Spain, Italy, Egypt, and Morocco. By the end of 2006, fifteen large thermal solar generating stations were operational in the United States alone—ten in California and five in Arizona. Thermal storage is provided by molten nitrate salts pumped from a cold reservoir to a hot reservoir by excess solar energy. The stored energy is used to produce superheated steam for the electric generation system when the tanks are emptied. The high-temperature storage increases the efficiency of electrical conversion, making these systems competitive with coal-burning plants.

The world's largest solar furnace, constructed in 1970 in the French Pyrenees, where annual sunlight exceeds three hundred days, consists of an array of 63 flat moveable mirrors that reflect sunlight into a huge curved mirror. The mirror, covering one entire side of a multistoried building, consists of 9,600 curved glass reflectors totaling an area of 20,000 square feet. This mirror focuses the light onto an area of about 1 square foot, where the 1,000 kilowatts of power delivered creates a temperature in excess of 5,400 degrees Fahrenheit. Furnaces of this type are primarily used for research in the high-temperature properties of metal oxides or in exposing materials to intense thermal shock.

Impact on Industry

Electricity from fossil fuels, which accounts for

one-third of the world's primary energy, experienced escalating price increases during the first decade of the twenty-first century. During the same time period, the average growth of PV systems exceeded 30 percent, with even more rapid growth predicted for the future as modules become more efficient and costs decrease. As the world's most rapidly growing energy technology, the total global power production from PV had exceeded 15 gigawatts by the end of 2008. Because technological innovations and increased efficiencies are projected for the next two decades, PV solar energy is uniquely posed to become the dominant provider of electric power by mid-century.

United States. The premier U.S. laboratory for renewable energy research and development is the National Renewable Energy Laboratory based in Golden, Colorado. Its mission is to develop innovative alternate energy technologies and facilitate the production of cost-effective marketable products through public-private partnerships. Although almost all alternate energy systems are being investigated, a significant portion of research is dedicated to photovoltaics, particularly designing buildings with integrated PV panels.

To help maintain a global competitive advantage, the U.S. government created the National Center for Photovoltaics. The center uses the expertise of the National Renewable Energy Laboratory and Sandia National Laboratories to oversee and coordinate other PV research organizations, including Brookhaven National Laboratory, the Strategic Energy Initiative at Georgia Institue of Technology, and the University of Delaware Institute of Energy Conversion. The center's main goals are to enhance PV performance, boost reliability, simplify installation, and ignite technological innovation.

Sandia's PV division provides technical support and validation for federal agencies and institutions planning to install PV systems, focusing on evaluation of performance, reliability, and cost-effectiveness. In addition to basic PV research, Brookhaven National Laboratory offers health and safety information to homeowners interested in PV installations.

International Research and Development. Germany has become the world leader in renewable energy production; electricity from renewable sources increased from 6.3 percent in 2000 to about 16 percent in 2009, with a target goal of 27 percent by 2020. By the end of 2007, Germany had PV energy systems capable of producing 3,800 megawatts of PV energy and was aiming at 4,500 megawatts by 2011. In 2006, the world's largest PV system, at 12 megawatts, began producing electricity in Arnstein. It was soon superseded by a 40-megawatt thin-film system, the Waldpolenz Solar Park, which became fully operational in 2008.

The Fraunhofer Institute for Solar Energy Systems, located in Freiburg, Germany, conducts research on solar thermal and PV systems, as well as on chemical energy conversion and energy storage. To increase PV conversion efficiency, a solar module that concentrates incoming radiation by a factor of five hundred has been developed. This procedure, combined with newly invented tiny extremely efficient solar cells, reduces the area of the semiconductor material required to achieve a recordbreaking efficiency of 41 percent.

In 2009, Michael Grätzel, director of the Laboratory of Photonics and Interfaces at Ecole Polytechnique Fédérale de Lausanne in Switzerland, invented dye-sensitized solar cells. These cells have an excellent price-to-performance ratio, positioning them as a possible future replacement for silicon PV cells. Constructed from low-cost materials and easy to manufacture, Grätzel cells will be particularly important for low-cost large-scale systems.

The Solar Energy Research Institute of Singapore is a national institute for both theoretical and applied solar energy research. Its focus is on developing materials, components, processes, and systems for PV electricity generation and recommending energy-efficient building strategies for Southeast Asia.

Concentrated solar energy can separate water directly into oxygen and hydrogen at temperatures in excess of 4,000 degrees Fahrenheit, but an alternate approach uses concentrated solar energy to remove hydrogen from natural gas. This process, termed SOLZINC, developed at the Weizmann Institute of Science in Israel, employs a 1-megawatt solar furnace to create the 2,200 degrees Fahrenheit necessary to separate zinc oxide into its constituent elements. The pure zinc product can be reacted with water to separate hydrogen and oxygen.

India, with rising power demands and high levels of solar radiation, is committed to expanding its PV capacity from its 2010 level of 30 megawatts to 20 gigawatts by 2022. Non-Indian firms, particularly in the United States and Taiwan, will be the immediate beneficiaries of this program, as India is not yet manufacturing PV cells. India has, however, been developing concentrating solar thermal units to create electricity by steam turbines, the goal being a capacity of 1 gigawatt by 2013 rising to 20 gigawatts by 2022.

A passive method of preheating ventilation air uses unglazed transpired collectors and perforated sun-facing walls to raise incoming air temperatures by up to 70 degrees Fahrenheit, thus delivering temperatures between 110 and 140 degrees Fahrenheit. By 2003, more than eighty systems, including one in Costa Rica for drying coffee beans and one in India for drying marigolds, with a total area of 350,000 square feet, had been installed worldwide.

Cooperative Ventures. In 2010, GrafTech International entered into a partnership with SENER Engineering and Systems and the University of California, Berkeley, to develop a high-efficiency solar thermal plant storage system by raising the operating temperature of the high-temperature reservoir. This system will contain a new graphite material capable of tolerating higher temperatures than the traditional molten salts, thus raising the plant's electric energy conversion efficiency.

In January, 2010, the Association for Electrical, Electronic and Information Technologies (one of the largest scientific associations in Europe), the Fraunhofer Institute for Solar Energy Systems, and the Solar Energy Research Institute of Singapore entered into a joint venture. The consortium will provide a complete range of testing services, including safety testing, for any PV modules and module components. Solar companies in tropical regions will benefit from the consortium's expertise by realizing shorter certification times for new products.

Also in 2010, Natcore Technology (Red Bank, N.J.) entered into a joint venture with a Chinese consortium consisting of government-supported Zhuzhou Hi-Tech Industrial Development Zone in Hunan Province, and Chuangke Silicon, a polycrystalline silicon producer. The consortium will manufacture equipment that grows an antireflective silicon film on a substrate in a room-temperature chemical bath. Existing technology uses an energy-intensive high-temperature vacuum furnace to grow the coating, which requires considerably more silicon than necessary because of the high firing temperatures. The new method is simpler and less energy-intensive, thus having the potential to significantly lower manufacturing costs. This joint venture is an example of the opportunities for collaboration between private companies and government as a means of reducing PV module costs as well as attenuating environmental pollution.

Careers and Course Work

By the middle of the twenty-first century, solar power is likely to be the dominant global energy resource; consequently, numerous new career opportunities await those with technological interests and skills. Although most universities do not offer an undergraduate major in solar energy, those wanting to enter the field can major or minor in electrical engineering, mechanical engineering, or physics. One of the nation's best developed solar programs, leading to bachelor's or graduate degrees in solar engineering, can be found at the University of Massachusetts, Lowell. Washington State University, Olympia, offers an associate degree for a position as a solar energy technician.

The Research Laboratory of the University of Central Florida (Cocoa), in addition to researching PV materials, conducts solar thermal systems testing. Other U.S. research programs are found at Georgia Institute of Technology, North Carolina State University, and the Universities of Wisconsin, Texas, Delaware, Oregon, and Arizona.

Social Context and Future Prospects

Every day, the Earth receives 10,000 times more energy from the Sun than humans consume from fossil fuels. As fossil fuels are depleted and the pollution produced by them becomes increasingly problematic, sustainable alternate energies, with the Sun as a major provider, will become the only viable energy future. The increased use of solar energy will require two economic shifts, supply and demand. First, the pressure to shift to clean, renewable energy supplies is mandated by the increased costs to society of continued reliance on polluting nonrenewable fuels. Second, a move away from large centralized power plants to increased reliance on smaller locally generated energy providers is anticipated. Several indirect benefits of the world's transition to a solar economy include the creation of wealth in underdeveloped countries rich in solar resources, improved homeland security through reductions in energy imports, reduced pollution and lessened

Fascinating Facts About Solar Energy

- The JOOS Orange portable solar cell phone charger, first marketed in 2010, delivers 2.5 hours of talk time for every hour of charging, twenty times the capacity of existing devices. The 24-ounce device can charge on cloudy or rainy days and costs less then $100.
- Greendix (Taiwan) has produced a soccer ball in which the traditional black pentagon-shaped leather patches have been replaced with solar cells. The mini solar panels power built-in motion sensors and an audio device, which emits a tracking sound when the ball is kicked, thus enabling the visually impaired to play soccer.
- The World Solar Challenge is a biannual solar-powered car race of 1,877 miles across central Australia. When the race was founded in 1987, the winner's speed averaged 42 miles per hour, but in 2007, the winner's speed had increased to 56 miles per hour.
- Roof-mounted PV panels can effectively be employed on sunny days to air-condition a car's interior without using engine power.
- The use of PV panels for auxiliary power in passenger boats was first introduced in 1995 but has become routine. The first solar-powered crossing of the Pacific Ocean occurred in 1996; the first solar crossing of the Atlantic took place during the winter of 2006-2007.
- The Gossamer Penguin made the first piloted flight using only PV energy in 1980. In 1990, a PV-powered flight from California to North Carolina was completed in twenty-one segments. The Helios, in 2001, set an altitude record for a non-rocket propelled aircraft at 96,864 feet, and in 2007, the Zephyr achieved a fifty-four-hour flight.
- Solar sails, consisting of large membrane mirrors that would utilize solar radiation pressure, have been proposed for spacecraft propulsion. Unlike rockets, the thrust is small, but because solar sails require no fuel and operate continuously when exposed to sunlight, significant speeds can be achieved when they are deployed in space.

effects on the global climate, and the ready availability of potable water through desalination plants. The European Union, for example, is exploring the concept of collecting solar energy in the Sahara Desert, converting it to electricity, and importing it via undersea power lines to Europe. Transportation costs would be offset by the greater and more reliable energy available in the Sahara. This arrangement would benefit both parties: energy-hungry Europe receiving ecologically friendly power and the less-developed nations of North Africa acquiring investments, infrastructure, technological expertise, and employment opportunities.

During the first decade of the twenty-first century, it became apparent that several crises were converging. As world population increases, obtaining the basic necessities of life, such as food and water, becomes progressively more problematic for underdeveloped nations. At the same time, the global demand for energy is accelerating as fossil fuel resources are being depleted and global warming threatens ultimately to render Earth uninhabitable. Arguably, the best course for humanity is to convert to sustainable food production and energy use. Solar energy, in all its myriad forms, is uniquely positioned to accomplish this transition, if people have the fortitude to endure temporary deprivation so as to ultimately abide harmoniously with the natural environment.

George R. Plitnik, B.A., B.S., M.A., Ph.D.

Further Reading

Bradford, Travis. *Solar Revolution: The Economic Transformation of the Global Energy Industry.* 2006. Reprint. Cambridge, Mass.: MIT Press, 2008. Uses economic forecasting models to predict that solar energy will become the best and least expensive future source of energy.

Chiras, Daniel. *The Solar House: Passive Heating and Cooling.* White River Junction, Vt.: Chelsea Green, 2002. A practical guide to selecting, designing, and installing cost-effective alternate energy heating systems using environmentally friendly materials.

Fisk, Marion, and H. C. William Anderson. *Introduction to Solar Technology.* New York: Addison-Wesley, 1982. Explores both theory and applications of solar technology.

Hough, Tom P., ed. *Trends in Solar Energy Research.* New York: Nova Science, 2006. Examines research in solar energy, including building applications.

Kut, David, and Gerald Hare. *Applied Solar Energy.* 2d ed. London: Butterworth, 1983. Practical advice on understanding and designing solar heating systems, including domestic how-water systems and solar greenhouses.

Ramsey, Dan. *The Complete Idiot's Guide to Solar Power for Your Home.* New York: Penguin Alpha Books, 2003. A detailed guide to financing and building a new solar home or incorporating a solar addition into an existing home.

Scheer, Hermann. *The Solar Economy: Renewable Energy for a Sustainable Global Future.* Sterling, Va.: Earthscan, 2002. Arguments supporting the immediate replacement of fossil fuels by various solar technologies as the only means of avoiding a future economic collapse.

Solar Energy International. *Photovoltaics: Design and Installation Manual.* Gabriola Island, B.C.: New Society, 2004. A resource manual to assist the layperson in comprehending PV modules so as to successfully design, install, and maintain a home system.

WEB SITES

American Solar Energy Society
http://www.ases.org

National Renewable Energy Laboratory
http://www.nrel.gov

Solar Energy Industries Association
http://www.seia.org

See also: Fuel Cell Technologies; Wind Power Technologies.

SONAR TECHNOLOGIES

FIELDS OF STUDY

Acoustics; computer science; signal processing; marine engineering; electronics and electrical engineering; oceanography.

SUMMARY

Sonar technology embraces the use of sound waves to facilitate travel and use of waterways, primarily the ocean. Through the use and detection of sound waves, sonar enables ships to avoid underwater obstacles and discover the ocean's resources. Naval forces use sonar to locate, target, and destroy hostile forces. Sonar started out as simply listening to sound in the ocean, but as technology improved sonar became a complex system of emitters, receivers, and processors that enabled a much wider use in bodies of water of all sizes.

KEY TERMS AND CONCEPTS

- **ASDIC:** Acronym for the developers of the first military application of sonar technology, Anti-Submarine Detection and Investigation Committee, created during World War I.
- **Biosonar:** Naturally occurring sonar used by aquatic animals, such as whales and dolphins, and some mammals, such as bats, to communicate and navigate.
- **Piezoelectricity:** Electrical charge generated in solid materials by the introduction of mechanical strain.
- **Towed Array:** Cable with a number of underwater microphones used to detect underwater objects across a range of aquatic environments.
- **Transducer:** Device that converts one type of energy into another. In the case of active sonar, the transducer converts electrical power into sound waves generated into the water.
- **Ultrasonic:** General term for sound waves generated in air, instead of water, used for various medical and industrial purposes.
- **Variable Depth Sonar:** Similar to a towed array, but a variable depth sonar has a motorized "fish" at the end of the array to permit the array to move through a range of depths and water characteristics.

Definition and Basic Principles

The idea of using sound to locate submerged objects came from nature. Many marine animals, most notably whales and dolphins, generate sounds of various frequencies (the high-frequency dolphin "squeak" or the low-frequency whale "song") that they use to communicate, navigate, and locate food. Humans soon copied this natural phenomenon. In its most common form, sonar is a means of acquiring data from bodies of water. Sonar, an acronym for sound navigation ranging, takes advantage of the sound-amplifying qualities of water to distinguish objects hidden beneath the surface. Usually mounted on a ship or submarine, sonar systems convert electrical energy into sound energy, propagate the sound energy into the water in the form of a wave, measure the time interval between the transmission of the wave and its return to the ship and conversion back into electrical energy, and convert that measurement into processed data. The term sonar, therefore, applies not only to the generated sound wave, but also to the apparatus that generates it. Because water tends to amplify and transmit sound to a greater degree than air, sonar is the primary means of detecting objects in areas where light and clear visibility is often in short supply. Depending on circumstances, water will transmit sounds hundreds of times farther than open air and at a greater speed. Modern sonar is an especially useful tool for exploring the nautical realm because it can both search and record data on a wide area and can also be focused to provide detail on specific areas. Consequently, sonar has become the preferred tool in mapping the vast reaches of the ocean.

Background and History

In the early twentieth century, scientists developed the first acoustical systems to study the ocean floor. Using sound emitters, oceanographers could determine the depth of ocean waters and the contours of the bottom by measuring the time interval between the emitting of a sound wave and its return to the emitter. One of the first uses of acoustical devices was a means to detect icebergs, developed in the aftermath of the 1912 sinking of the RMS *Titanic*. Such systems proved their military value during World War I, when German submarines wrecked havoc on

British shipping. To counter the submarines, the British developed a device known as ASDIC that allowed them to locate submerged submarines before the German vessels could get within firing distance of their torpedoes. When the German submarine menace reappeared during World War II, sonar (the American term for ASDIC) again became a useful tool in stopping the destruction of merchant ships.

In the aftermath of World War II, the military use of sonar expanded. As nuclear power made submarines capable of faster speeds and diving to deeper depths, sonar became the only reliable means to target hostile threats. Advances in sonar also aided submarines in finding their enemy forces and directing weapons against them by providing electronic-targeting information instead of relying on visual inputs from binoculars or the periscope. Sonar also became effective at detecting the new generation of mines that rested on the bottom of the ocean as well as locating underwater obstacles and navigating choke points.

Postwar sonar advances also had civilian applications. Sonar proved a great asset to the fishing and oil industries, as it could efficiently detect resources, permitting more efficient use of those resources and avoiding the risks of sending deep-sea divers into dangerous circumstances.

How It Works

Types of Sonar. Sonar falls into the two categories of active and passive use. Active sonar uses a transducer to propagate a sound wave into the water, and a receiver, either on the emitting transducer or at another location, detects the echo of the sound wave bouncing off objects nearby and determines their depth and range by measuring the time interval between the initiation of the sound wave and its return. The data collected by the receiver must take various circumstances into context as water conditions affect the power of the sonar sound wave and the speed at which it moves through the water. Passive sonar relies on listening for sounds in the ocean rather than using a transducer-produced sound wave to generate data. Passive sonar is useful for determining conditions in the water without disturbing or altering the condition by introducing active sonar waves. Passive sonar also allows the user to remain silent and hidden, a useful element for the military, in which stealth is important.

Transmission and Reception. The method of acquiring information from a sonar system depends on the type of sonar employed. Passive sonar relies entirely on reception systems without an active transmission component. The first passive system was the use of a hydrophone, essentially a microphone placed into the water. First introduced during World War I, hydrophones were a cheap and basic means of detecting hostile submarines. The hydrophone operator simply listened to sounds emanating through the water, trying to sort artificial man-made noise from the ambient sea sounds. The main handicap was the motion of the ship, as water moving past the hydrophone and propeller noise muffled the sound of distant objects. Modern passive sonars work on the same principle of reception-based detection but use advanced receptor arrays positioned at various locations to counter the noise generated by the ship itself. The early hydrophones used carbonized paper as baffle to transform the sound wave energy into electrical signals heard in the operator's headset, but modern systems employ fiber-optic receptors that decode the incoming sound. Active systems have a standard receptor and an active emitter to generate sound in the water. The first systems used quartz crystal oscillators to generate sound, but later systems used more powerful and adaptive piezoelectric materials, such as lead magnesium niobate. The more powerful the electrical charge, the farther the detection range of the sonar system, but transmitting sound itself is not sufficient for object detection.

Environmental Challenges. Although modern power plants on ships and submarines, especially nuclear ones, are capable of generating vast amounts of power to their sonar systems, the environment in which they operate limits the capability of the sonar to transmit and receive sound. The temperature of the water affects sound propagation, as does the salinity level, with sonar capability seriously degrading in freshwater compared with saltwater. The amount of marine life also affects sonar effectiveness. Large amounts of algae or other microscopic life can, like a fog, cause different levels of water density. Marine animals that employ biosonar can also generate enough sound in a specific locality to drown out man-made sonar emissions. But the biggest obstacles to effective sonar use are geography and temperature. Sonar works best in deep water, where reflections off the shore or bottom do not cause multiple echoes or reflections. In shallow water, there are multiple reflective surfaces, and sonars have trouble discerning between the multiple inputs to determine whether a sound is artificial or natural. As water depths increase, the declining level of sunlight also causes various strata or layers of water with different temperatures, known as thermal layers. Sonar-emitted sound tends to bounce off these different layers and hide an object or become warped, which distorts the sonar beam and provides an inaccurate measurement of a target's location.

Data Processing. Acquiring data from sound propagation is one thing, but converting it into useful information is another. Unless the received data is converted into a useful form, the sonar is worthless. Early sonar sets used the human ear to discern useful information from the ambient ocean sounds, a process fraught with difficulties ranging from changing conditions to human error. The advancement of electronics and computers, however, provided a means to turn random noise into coherent sonar images. Advanced data processing permitted sonar operators to "scrub" radar returns of unwanted clutter as well as enhance weak or scattered returns to provide otherwise useless information. Computers also allowed sonar to take additional forms aimed at overcoming geographical and conditional limitations. Towed arrays, long cables with many attached hydrophones, provide multiple sonar sources and permit the operator to triangulate a sound source or concentrate on the particular hydrophones that are providing the clearest data. To overcome the limitations of thermal layers, a vessel might employ a variable depth sonar (VDS), a towed array with a small maneuvering craft, something like a small submarine, on the end of the array. By passing signals down the array, the operator can direct the maneuvering craft through the thermal layers, concealing nothing. The most modern sonar sets feature synthetic aperture sonar, which uses Doppler-shift differences in sonar return to convert the incoming data into a three-dimensional image.

Applications and Products

Military Use of Sonar. Although originally created for scientific purposes, sonar owes its development to the immediate military need to hunt German submarines during World War I. Starting with the basic sonar sets and hydrophones that equipped Allied sub hunters in World War I, sonar continued its development as an antisubmarine weapon through the years after the conflict. By World War II, sonar had become a key electronic aid to all major navies, as submarines continued to develop as a fighting force. By the 1950's, the arrival of faster and more capable nuclear-powered submarines made early detection even more vital. Sonar systems evolved to provide precise location of potential submerged threats and data for new weapons to counter them. In addition to improved shipboard systems with long-range and greater data-processing power, sonar took on new forms. The U.S. Navy developed the Sound Surveillance System (SOSUS), massive hydrophone arrays planted on the seafloor, to monitor geographical choke points for passing Soviet submarines during the Cold War. Complementing the shipboard sonars, airborne systems began to use sonar. Helicopters equipped with dunking sonar, a sonar transducer on the end of a cable that a hovering helicopter could drop into the water, greatly expanded a ship's sonar-coverage area. Helicopters and fixed-wing aircraft could also drop sonobuoys, a buoy that is able to detect underwater sounds and transmit them via radio, within a general area identified by another sonar system. As nuclear-powered submarines could remain submerged for months at a time, sonar became the primary means of navigating without external visual cues. Sonar also provided targeting information for submarine weapons, such as torpedoes and cruise missiles.

Civilian Use of Sonar. Sonar plays a significant role in a variety of industries. The fishing industry makes widespread use of sonar to locate schools of fish in the vast stretches of the ocean. Locating fish by sonar allows the fishing boat to sweep a specific area of water, as opposed to pre-sonar fishing, where operators had to drop their nets in spots likely to contain fish but which might not. Targeted fishing saves both time and fuel by making the process more efficient. The use of sonar in fishing has even filtered down to the recreational sportsman: Small, portable "fish finders" now abound. The other major industry to employ sonar is resource exploration. Sonar permits the mapping of the ocean floor and the exploration of likely locations of valuable materials. Oil exploration in particular employs sonar. Prior to exploratory drilling, sonar is used to determine the geological formations present on the ocean floor most likely to contain pockets of oil.

The nature of high-frequency sound waves has contributed to other industrial applications, even out of water. Unlike radar waves, sound waves are not radioactive and latently lethal. Although there is a risk of hearing loss due to lengthy exposure to high-frequency sound, sound waves generated into the air, known as ultrasonic waves, do not pose the health risks of radiation emitters such as radar. Consequently, sound waves are used in industrial processes such as materials inspection, where they are bounced off a product to determine its structural soundness and uniformity. The most recognizable use of commercial sound waves, however, is the prenatal ultrasound. As radiation might harm an unborn child, physicians use sound waves to generate, through a synthetic aperture system, a three-dimensional view of the child for diagnostic and preventive-care measures.

IMPACT ON INDUSTRY

In addition to the various industries and uses identified for sonar and sound waves, the application of sonar helps to support and encourage research in a significant number of other industries and scientific fields. Power for sonar systems is vital, but space on ships or submarines is limited, so sonar development drives the pursuit of compact power sources. The effectiveness of the sonar depends on the power and focus of the beam produced by the transducer, which in turn promotes the exploration of more efficient and effective transmitting materials, as well as means of recognizing and collecting the return beam. The need for small computers to process the return data has created the demand for compact computer processors and data storage.

Major Companies. The number of companies conducting research and development in sonar systems are few and often specialized. Because sonar systems are expensive and are typically used by large organizations, corporations, and national governments, they are not produced in great numbers and tend to be very costly. Sonar production is especially reliant on military sales, so countries prefer to reserve contracts for sonar sets to their own domestic producers. The vital military nature of sonar also leads to the classification of some designs as national secrets, which further promotes the protection of domestic sonar companies. Consequently, major navies tend to operate domestically produced sonar sets. In the United States, Bendix and Raytheon dominated the sonar industry after World War II, but Raytheon has emerged as the primary provider for the U.S. Navy. Canadian ships feature either American or British-produced sonars.

Limited Availability of Sonar Operators. As the number, uses, and potential of radar continue to develop, the biggest immediate obstacle to sonar operations is the limited number of trained and experienced operators. As the largest user of sonar technology, the major navies of the world are the most significant training establishments for sonar operators, leaving relatively few operators in the civilian sector. Sailors trained in sonar operations can enter the civilian workplace on their discharge from the service, but their skills are so highly valued that most navies provide significant incentives for sonar operators to remain in the military. Also, because military sonars are usually at the cutting edge of technology, sonar systems change and become obsolete relatively quickly. Without constant retraining and remaining familiar with new equipment, transducers, and processors by operators, sonars cannot be used to their greatest potential.

Fascinating Facts About Sonar Technologies

- In the 1970's, a Soviet test of a new Alfa class nuclear submarine generated enough noise that an American sonar array in the Bahamas, 5,000 miles away, detected the sound.
- Modern sonars are so sensitive that operators can distinguish the sound pattern of individual ships and submarines, and the resulting "fingerprint" allows them to identify friendly and potential enemy vessels.
- A World War II-era sonar found on a warship weighed several tons, but the sonar component of an expendable sonobuoy weighs less than ten pounds.
- To facilitate the transmission of sound into the water, shipboard sonars are mounted in large plastic bilges near the bow, requiring the relocation of the ship's anchors to avoid hitting them.
- Because of the distinctive sound, the return echo used on an active sonar beam to find a submerged object was known as "pinging" the target.

Careers and Course Work

Because of their complex mechanical and electronic nature, sonars require a variety of skills to construct, operate, and maintain them. As a sound-based means of detecting hidden objects, a course of study should emphasize a general knowledge of acoustics and specific knowledge of hydroacoustics. Because of its particular operational environment, how to design and construct sonar systems requires a recognition and understanding of the aquatic environment. Sonar design requires extensive knowledge of materials engineering to determine the proper components of the system and marine engineering to allow integration of the radar system into an ocean environment and ensure its successful operation. The clear reception and presentation of the radar signal is also vital, and as this is primarily a function of modern electronics, anyone interested in sonar operations needs to have competency in electronics, computer science, or signal processing. Anyone with these skills should find ready employment in any number of industrial or scientific concerns, especially as the use of nonaquatic sound waves in industry increases.

Social Context and Future Prospects

The use of sonar will only expand as the need for more and more resources from the ocean increases. Growing global populations and demand for fossil fuels will place great emphasis on the exploration for consumable resources. As only a fraction of the ocean floor has been accurately mapped and sonar is the most effective means of doing so, sonar will be an important tool in future economic development. As with any use of technology in an unspoiled environment, the use of sonar will face scrutiny. Environmentalists claim that sonar use can confuse or injure aquatic life, especially dolphins and whales that use biosonar for navigation and hunting. Environmental groups have tried for years, with some success, to limit the use of low-frequency naval sonars on the grounds of its harm to ocean life. At the same time, however, oceanographic research also relies heavily on sonar to study and understand aquatic lifeforms, and finding a happy medium between artificial and natural sonar systems will have to emerge as exploration of the ocean continues.

Steven J. Ramold, Ph.D.

Further Reading

Cox, Albert W. *Sonar and Underwater Sound.* Lexington, Mass.: Lexington Books, 1974. Provides a wide overview of the commercial applications of sonar.

Creasey, D.J., ed. *Advanced Signal Processing.* London: Peregrinus, 1985. Explains the evolution of sound processing and display, among other means of electronic surveillance.

Denny, Mark. *Blip, Ping, and Buzz: Making Sense of Radar and Sonar.* Baltimore: Johns Hopkins University Press, 2007. A good starting point for the theory, application, and use of emanated energy.

Lurton, Xavier. *An Introduction to Underwater Acoustics: Principles and Applications.* 2d ed. Chichester, England: Praxis, 2002. Understanding sonar means understanding how sound works in various aquatic scenarios, and Lurton provides a thorough and understandable discussion of hydroacoustics.

Medwin, Herman. *Sounds in the Sea: From Ocean Acoustics to Acoustical Oceanography.* New York: Cambridge University Press, 2005. The best history of the use of sonar exploration.

Vego, Milan N. *Naval Strategy and Operations in Narrow Seas.* 2d ed. Portland, Oreg.: Frank Cass, 2003. This book contains a detailed history of military sonar development and how navies employ sonar in offensive and defensive operations.

Web Sites

American Society of Naval Engineers
http://www.navalengineers.org/Pages/default.aspx

Raytheon
http://www.raytheon.com

U.S. Geological Survey
Office of Surface Water Hydroacoustics
http://il.water.usgs.gov/adcp

STEAM ENERGY TECHNOLOGY

FIELDS OF STUDY

Mechanical engineering; thermodynamics; heat transfer; fluid mechanics; chemistry; machine design; ship propulsion; electrical power generation; reduction gear design; boiler design; turbine design.

SUMMARY

Steam energy technology is concerned with the conversion of the chemical energy in fuels into the mechanical energy of a rotating shaft. Steam energy is used to propel ships, drive electric generators, and power pumps. Components such as boilers, turbines, pumps, heat exchangers, and piping systems are involved. The majority of electric power in the United States is generated with steam. Steam is less popular for ship propulsion than in the past, but it is still used for nuclear-powered ships and for ships that transport liquefied natural gas. In combined cycle technology, the hot exhaust gas from a gas turbine is used to produce steam that is used to provide additional mechanical energy or to heat buildings.

KEY TERMS AND CONCEPTS

- **Boiler:** Device that uses heat from burning fuel to produce steam.
- **Condenser:** Specialized heat exchanger that uses cooling water from a river or the sea to convert steam leaving a turbine back to water.
- **Fluid Mechanics:** Study of the flow of liquids and gases.
- **Heat Transfer:** Process of transmitting heat. Heat travels through a solid by conduction, is transmitted from a solid surface to a liquid or gas by convection, and is transferred by radiation, which consists of electromagnetic waves.
- **Pump:** Machine that adds mechanical energy to a liquid. The increased energy is indicated by an increase in pressure.
- **Reciprocating Steam Engine:** Machine that uses steam to push pistons up and down in a cylinder. This motion can be used to drive a pump or a ship's propeller.
- **Reduction Gear:** Device that converts the high rotating speed of a turbine to the much lower rotating speed of an electric generator or ship's propeller.
- **Saturated Steam:** Steam that is just at its boiling temperature. While water boils at 212 degrees Fahrenheit at atmospheric pressure, the boiling temperature rises at higher pressures.
- **Superheated Steam:** Steam that has been heated above its boiling temperature. Steam at atmospheric pressure and a temperature above 212 degrees Fahrenheit is superheated.
- **Superheater:** Part of a boiler that superheats steam. Steam flows inside of superheater tubes while hot combustion gases flow around the outside of these tubes. Heat is transferred from the combustion gas to the steam.
- **Thermodynamics:** Study of energy conversion processes.
- **Turbine:** Machine that converts the energy of high-pressure, high-temperature steam into mechanical energy to drive an electric generator or ship's propeller.

Definition and Basic Principles

Modern steam energy technology may involve the production of superheated steam at relatively high pressure and the use of that steam to drive a mechanical device. It may also involve the production of saturated steam at relatively low pressure and the use of that steam to heat buildings or industrial processes.

Most of the electric power in the United States is produced by means of high-pressure, superheated steam driving a turbine. In turn, the turbine drives the electric generator that produces electricity for use in homes, businesses, and factories. This type of steam is also used to drive turbines aboard ships, and those turbines drive the ships' propellers. Because turbines operate best at several thousand revolutions per minute, and propellers operate best at a few hundred revolutions per minute, a speed-reducing gear is used between turbine and propeller.

Many industrial processes in the chemical industry and elsewhere require large amounts of low-pressure saturated steam to heat the materials being processed. For instance, crude oil is made to boil in a distillation column as a way of separating volatile components such as gasoline from nonvolatile ones such as tar. Large buildings in cold climates may be heated by steam produced by a boiler in the basement.

In ancient times, energy was provided by human or animal strength. Steam energy technology was one of the first technologies that humankind used to produce greater force, higher speed, and greater endurance than living things could produce.

Background and History

Hero of Alexandria is credited with the invention of the first steam engine in about 200 B.C.E., although the device does not seem to have been put to practical use. Records indicate that in 1543, Spanish naval officer Blasco de Garay attempted to propel a ship with paddle wheels driven by a steam engine. In the late 1600's, groundwater needed to be pumped out of English mines, and Thomas Savery patented a steam-powered pump on July 25, 1698. By 1767, fifty-seven steam engines were in use in mines near Newcastle, England, with a combined power of about 1,200 horsepower. James Watt was granted a patent for a much-improved engine in 1769.

Robert Fulton launched his Hudson River steamboat in 1807. This vessel traveled from New York City to Albany, New York, a distance of 150 miles, in thirty-two hours. The Brush Electric Light Company in Philadelphia built the first electric generating station in the United States in 1881.

Sir Charles Algernon Parsons of England is regarded as the inventor of the modern steam turbine. He first built a small turbine and used it to drive an electric generator. In 1894, he built the first steam-turbine-powered watercraft. This vessel, the *Turbinia*, achieved the astounding speed of 34.5 knots (just under 40 miles per hour). George Westinghouse acquired the American rights to Parsons's invention in 1895. During the twentieth century, applications of steam energy technology expanded rapidly, and it became the dominant source of energy around the world.

How It Works

Steam is produced in a boiler, where fuel is burned and the heat released is transmitted to water, which boils to form saturated steam. The fuel can be almost anything that burns. Solid fuels include peat, wood, and coal, while liquid fuels range from residual fuel, which is so thick that it must be heated to about 200 degrees Fahrenheit to make it flow easily, to kerosene and gasoline. Natural gas is also used as boiler fuel. Heat is produced as these fuels react with oxygen in the air. Boilers may operate at pressures from slightly above atmospheric pressure to 3,500 pounds per square inch. Boilers at electric generating plants may produce 10 million pounds of steam per hour. There are two basic boiler types: fire tube boilers, in which hot gases produced by combustion pass through tubes surrounded by large quantities of water, and water tube boilers, where water-filled tubes are exposed to hot combustion gases. Fire tube boilers are suitable for low-pressure boilers, and water tube boilers are used for pressures above about 300 pounds per square inch.

Superheating Steam. After steam is produced, it may pass through additional tubes that are exposed on the outside to hot combustion gases. This process, which is called superheating, raises the temperature of the steam above its boiling temperature. Superheating is done to increase the energy content of the steam without changing its pressure. The steam used in turbines is usually superheated to a temperature of 950 to 1,000 degrees Fahrenheit. Special steel alloys must be used in superheater tubes so that they can endure such high temperatures.

Combustion. Liquid fuels are sprayed into a cavity in the boiler called a furnace. Fuel is mixed with air and burned. In a water tube boiler, the hot gases produced by combustion flow first over the superheater tubes and then over tubes containing liquid water, where steam is produced.

Turbines. Superheated steam leaves the boiler and flows to the steam turbine. Here the steam is directed against blades mounted on disks that are attached to the rotating shaft. The shape of the blades deflects the steam, and the steam causes the blades to move in the opposite direction. The process is similar to what happens when someone blows on a pinwheel. In a typical steam turbine, there may be twenty or more disks attached to the shaft. Each disk has many blades arrayed around its rim.

Reduction Gears. Often the desirable speed of a steam turbine is much greater than the speed of the device to which it is connected. For instance, a ship's turbine may rotate at several thousand revolutions per minute, while the propeller should rotate at about one hundred revolutions per minute. A reduction gear is used to convert the high speed of the turbine shaft to the low speed of the propeller. In a reduction gear, a small gear turning at high speed meshes with a large gear that turns at lower speed. Because there is a limit to the ratio of gear sizes, it is often necessary to perform the speed reduction in two steps.

Condensers. Leaving the turbine, the steam enters a condenser. This is a specialized heat exchanger that has cooling water flowing through thousands of tubes. The cooling water may come from a river or from the sea. As the steam comes into contact with the outsides of these tubes, it condenses back to liquid water and drops to the bottom of the condenser. Condensers often operate at 12 or 13 pounds per square inch below atmospheric pressure. This pressure is determined by the cooling water temperature. The lower the condenser pressure, the more energy the turbine is able to extract from the steam.

Pumping and Air Removal. Once the steam has been condensed, a series of pumps transfers the water back to the boiler where the process begins again. These pumps may be driven by electric motors or by small steam turbines. Because the condenser operates below atmospheric pressure, small amounts of air may leak into the water. Most of this air is removed by a vacuum pump, but some of the leaking air may dissolve in the condensing steam. This air is removed either by chemicals or by heating the water to its boiling point without boiling it after a pump has raised its pressure above atmospheric pressure.

Reciprocating Engines. Early steam engines had pistons that moved up and down much as the pistons in an automobile engine do. These engines are known as reciprocating engines. The famous Liberty ships of World War II were powered by reciprocating engines. Turbines are much more efficient than reciprocating engines, so relatively few reciprocating engines remain in use.

Applications and Products

Electric Power Generation. Most of the electricity used in the United States is generated in steam plants, powered by coal, oil-based fuels, natural gas, and nuclear power. Regardless of the type of fuel, these plants tend to be very large. In 2008, the United States had 29 coal-fired plants with capacities of more than 2,000 megawatts. In 2005, there were a total of 605 coal-fired plants in the United States with a combined generating capacity of 336,000 megawatts. In 2004, these plants produced more than 2 billion tons of carbon dioxide, about 8 percent of the world's total. In 2005, the states of Texas, Ohio, Indiana, and Pennsylvania had coal-fired generating capacities of more than 20,000 megawatts each, while Rhode Island and Vermont had no coal-fired plants.

In 2008, the United States had an oil-fired generating capacity of more than 57,000 megawatts, a natural-gas-fired generating capacity of almost 400,000 megawatts, and a nuclear generating capacity of just over 100,000 megawatts. All coal-fired plants and almost all nuclear plants use steam engineering technology. Most oil-fired plants and natural gas plants also use this technology.

A relatively new technology for electric power generation is a combined cycle plant. These use gas turbines driving electric generators for the majority of their power production. However, gas turbines alone have relatively low efficiencies (about 35 percent), because the exhaust gas contains a lot of unused heat energy. In a combined cycle plant, this hot exhaust gas is used to produce steam, and the steam drives a turbine that powers another electric generator. The efficiency of combined cycle plants is very attractive (50 percent or more).

Ship Propulsion. For most of the twentieth century, steam was the dominant source of energy for ship propulsion. Early in the century, reciprocating engines were used, but turbine engines soon took over. The famous liner SS *United States* had four turbine engines producing a total of about 250,000 horsepower. This ship set and held the transatlantic speed record for nearly fifty years. In 1970, most of the ships in the U.S. Navy were steam powered, and the world's merchant fleet was also primarily steam. Later, diesel and gas turbine propulsion became more popular than steam, but even in the early twenty-first century, most of the ships being built to carry liquefied natural gas to Europe and the United States were steam powered.

On ships propelled by steam, the electric generators are also driven by steam turbines. Some of the pumps on these ships are driven by small steam turbines, and steam is used as a heat source for boiling seawater to produce freshwater for human use and for replenishing the water in the steam system.

All nuclear-powered ships are powered by steam

turbines. The nuclear reactor produces very hot water. This water exchanges heat with water at lower pressure in a steam generator, and the lower-pressure water turns to steam. This steam is then used to drive a turbine.

Proposals for the propulsion of liquefied natural gas ships have called for combined cycle systems much like those used for electric power generation. Such systems are much more efficient than conventional steam systems.

Nuclear Power. Nuclear power is used extensively for electric power production, with more than 400 plants worldwide producing about 17 percent of the world's electricity. In France, about 75 percent of the electricity comes from nuclear plants. In 2008, about 104 nuclear power plants were operating in the United States. These plants produce nearly 20 percent of the country's electric energy. Because nuclear plants produce no carbon dioxide and no other airborne pollutants, the United States and other nations are looking to nuclear power to fulfill their future electric needs. There are several different types of nuclear power plants, but nearly all of them make use of steam energy technology. In these plants, the nuclear reactor replaces the boiler furnace as a heat source. Most of the rest of the plant closely resembles a conventional steam plant. Unlike conventional plants, nuclear plants use saturated steam rather than superheated steam because it is not feasible to superheat steam by means of a nuclear reactor.

Industrial Processes. Many industrial processes, such as oil refineries, steel mills, chemical plants, and paper mills, require steam. In oil refineries, steam is used to heat various liquids and is also used in the cracking process that breaks large molecules into smaller ones. A by-product of the cracking process is carbon monoxide. The blast furnace in a steel mill also produces large amounts of carbon monoxide. The carbon monoxide produced is used as fuel in a boiler to create steam. A paper mill uses roughly 10,000 pounds of steam per ton of paper produced. Much of this steam is generated using waste products such as tree bark as fuel. In the chemical industry, steam is used in the making of ethylene and the plastic styrene. It is also used to heat chemicals that would turn solid in their tanks at room temperature.

Steam Heating. Consolidated Edison of New York (Con Edison), New York City's electric power utility, sells steam through 105 miles of pipes that run under Manhattan streets. At three of its electric generating plants, steam leaves the turbines at 350 degrees Fahrenheit and at a pressure of about 115 pounds per square inch higher than atmospheric pressure. Instead of being condensed in the plant, steam flows through large pipes throughout Manhattan. About 100,000 buildings, including the United Nations building, Rockefeller Center, and the Metropolitan Museum of Art buy this steam and use it for heating. Altogether, Con Edison sells about 30 billion pounds of steam for heating each year. A church in lower Manhattan became one of the first steam customers in 1882, and it has been buying steam ever since. Other cities in the United States and Europe have similar systems, but New York's system is the largest.

Impact on Industry

The components that make up a steam energy system—including boilers, turbines, reduction gears, pumps, and heat exchangers—are produced by large industrial corporations. Major boiler manufacturers include Babcock & Wilcox, Foster Wheeler, and Combustion Engineering. Steam turbines are manufactured by General Electric and Westinghouse, among others. Large reduction gears are a very specialized product with relatively few manufacturers. General Electric is one of the major sources.

On land, power plants are usually designed by architecture and engineering companies. KBR (Kellogg, Brown & Root), a company that resulted from the merger of Brown and Root with M. W. Kellogg, is a major company in this field. Fluor Corporation and Stone & Webster, part of the Shaw Group, are also large architecture and engineering companies. These companies perform design services and manage and supervise the construction of plants. Once built, the power plants are operated by electric utility companies such as Con Edison.

Ships are designed by naval architecture and marine engineering companies such as Gibbs & Cox, Alion, and Herbert Engineering. The final product of these companies is a set of drawings and specifications. The shipowner uses these to obtain bids from various shipyards. As the ship is built, the steam machinery is installed, and the shipyard runs the piping that connects everything together.

In the late 1800's, many boilers were poorly designed and constructed. Boiler explosions were common, and many people were killed every year in these explosions. The American Society of Mechanical Engineers (ASME) was originally founded to correct this problem by developing standards for safe boiler design and construction. This organization does not have legal authority to require boilers to be designed to its standards, but many government agencies that do have such authority have specified that boilers must comply with the ASME code. Insurance companies usually require ASME approval before writing insurance on a boiler. The U.S. Coast Guard publishes rules regarding ships, and these rules invoke the ASME code. Individual state governments issue rules for boilers that operate within their jurisdictions, and these state rules also invoke the ASME code. Organizations called classification societies serve as arms of the ship insurance business. Their job is to verify that each insured ship is designed, built, and maintained in such a way that it represents a reasonable insurance risk. These organizations have rules that govern the design, construction, and maintenance of machinery on merchant ships, and these rules draw heavily on the ASME code.

Careers and Course Work

Electric power plants are designed by architecture and engineering companies. Many of the people involved in designing the plants are mechanical engineers with either bachelor's or master's degrees in mechanical engineering. Students in mechanical engineering study advanced mathematics, thermodynamics, fluid mechanics, heat transfer, machine design, and other technical subjects. The design process also requires people with expertise in civil engineering and electrical engineering. The actual equipment—the boilers, turbines, pumps, and so on—are designed by companies that specialize in particular components, while the designers at architecture and engineering firms design the systems that connect these components and make them work together.

The people who manage these plants are usually employees of large public utilities that own the plants. They often have mechanical engineering degrees. In addition to an engineering degree, a manager may have a business degree. A college degree is not required for the people who actually operate and maintain these plants.

Designers of shipboard steam energy systems may have degrees in mechanical engineering or marine engineering. These two fields are closely related. The designers are typically employed by firms specializing in naval architecture and marine engineering.

Operating personnel on U.S. Navy ships are, of course, naval officers and enlisted personnel. Officers on merchant ships must be licensed by the U.S. Coast Guard as operating marine engineers. Many of the officers in the U.S. Merchant Marine are graduates of state maritime academies or the U.S. Merchant Marine Academy, but a college degree is not required to obtain a license. Some people

Fascinating Facts About Steam Energy Technology

- The ancient Greek inventor Archimedes may have used steam cannons that converted about one-tenth of a cup of water into steam to propel hollow clay balls out of cannons during the siege of Syracuse.
- James Watt, developer of an efficient steam engine, coined the word "horsepower" to explain how much power his machine could generate to a potential customer.
- The Marion Steam Shovel Company, founded in 1884 in Marion, Ohio, built the steam shovels that were used to construct the Panama Canal in the early 1900's. The city of Marion became known as the city that built the Panama Canal.
- On April 27, 1865, a boiler on the steamboat *Sultana* exploded as it neared Mound City, Oklahoma, killing between 2,000 and 2,300 people. The steamboat, designed for 300 passengers, was loaded with Union soldiers who had just been released from Confederate prisoner-of-war camps.
- Molten rock, or magma, heats rainwater to create a superheated fluid used in geothermal power plants. The geothermal superheated fluid is converted to steam by crystallizer-reactor clarifier technology.
- According to a West Virginia folktale, John Henry, the greatest steel driver for the C&O Railroad, entered into a contest with a steam-powered drill. John, using two 20-pound hammers, drilled two 7-foot holes in thirty-five minutes, while the steam-powered drill had made only one 9-foot hole. John raised his hammers in victory, then collapsed and died.

obtain the knowledge required for a license by experience and specialized training. They gain this experience by serving in unlicensed positions such as oilers or qualified members of the engineering department on merchant ships.

Social Context and Future Prospects

The demand for electricity continues to grow around the world. At the start of the twenty-first century, China's economy was growing very rapidly, and China's demand for electricity grows along with its economy. The United States has large reserves of coal. Increased use of this coal could reduce its dependence on foreign oil, but coal produces more emissions than any other fuel. In particular, coal produces large amounts of carbon dioxide. Research is ongoing on ways to recapture this carbon dioxide and prevent it from contributing to global warming. If this research is successful, there may be growth in the use of coal.

Supplies of oil and natural gas are finite, and renewable fuels are being sought. A significant advantage of steam energy technology is that boilers can burn all manner of fuels: solid, liquid, and gas. Increased use of solid fuels produced from agriculture may also cause growth in steam engineering technology. Because the combustion of fuels produces carbon dioxide, a major greenhouse gas, efforts are being made to reduce fuel usage. Wind and water power have a role here, but nuclear power, which produces no airborne emissions, appears poised for major growth. Since nearly all nuclear plants use steam, growth in nuclear power may cause growth in use of steam energy technology.

Edwin G. Wiggins, B.S., M.S., Ph.D.

Further Reading

Babcock & Wilcox Co. *Steam: Its Generation and Use.* 39th ed. New York: Babcock and Wilcox, 1978. This classic work includes thorough coverage of marine and on-shore applications.

Bloch, Heinz P., and Murari P. Singh. *Steam Turbines: Design, Applications, and Rerating.* 2d ed. New York: McGraw-Hill, 2009. Features clear explanations of turbine structure and principles of operation.

Gardner, Raymond, et al. *Introduction to Practical Marine Engineering.* New York: Society of Naval Architects and Marine Engineers, 2002. A clear overview of marine steam equipment with excellent drawings.

Kehlhofer, Rolf, et al. *Combined-cycle Gas and Steam Turbine Power Plants.* 3d ed. Tulsa, Okla.: Penn Well, 2009. Examines the operation of combined-cycle power plants.

McBirnie, S. C., and W. J. Fox. *Marine Steam Engines and Turbines.* 4th ed. Boston: Butterworth, 1980. Covers turbines and reciprocating steam engines. Somewhat technical but contains many excellent drawings.

Milton, James H., and Roy M. Leach. *Marine Steam Boilers.* Boston: Butterworth, 1980. Good coverage of the history of boilers with many excellent drawings.

Web Sites

American Society of Mechanical Engineers
http://www.asme.org

Con Edison
Steam Operations
http://www.coned.com/steam/default.asp

Society of Naval Architects and Marine Engineers
http://www.sname.org

U.S. Department of Energy
http://www.energy.gov/index.htm

See also: Gas Turbine Technology; Solar Energy.

STEELMAKING TECHNOLOGIES

FIELDS OF STUDY

Physics; chemistry; mechanical engineering; chemical engineering; materials science; strength of materials; metallurgy; metallography; crystallography; X-ray diffraction analysis; physical chemistry; metallurgy; quantitative analysis; qualitative analysis, instrumental analysis; alloys; ferromagnetism; advanced mathematics; computer science; thermodynamics; phase diagrams; environmental science.

SUMMARY

Steelmaking technologies are processes, such as the Bessemer, open-hearth, and basic oxygen furnace, designed to make an alloy of iron and a small amount of carbon. Depending on the properties desired in the steel, other alloys can be made with such elements as chromium, cobalt, copper, molybdenum, nickel, silicon, and tungsten. Steel is commonly produced as rods, bars, sheets, and wires, which are then employed in constructing automobiles, railroads, ships, and machinery. Steel has also been used to make numerous other things, from delicate surgical instruments to skyscrapers.

KEY TERMS AND CONCEPTS

- **Bessemer Process:** Steelmaking method consisting of forcing compressed air through molten iron contained in a pear-shaped "converter" to remove excess carbon and impurities.
- **Blast Furnace:** Often large structure enclosing raw materials that are smelted into pig iron by high temperatures achieved by injections of air.
- **Coke:** Solid residue produced by heating bituminous coal and other carbonaceous materials, with the concomitant removal of volatile substances by destructive distillation; coke is commonly used in making steel.
- **Flux:** Substance such as calcium carbonate used in smelting that combines with impurities in the ore to produce a vitreous mass (slag) separable from the molten steel.
- **Minimill:** Steelmaking facility making use of pig iron and scrap to manufacture a small range of products.
- **Open-Hearth Process:** Method of steelmaking in which pig iron and scrap in a hearth are alternately heated by waste gases and then by incoming air and fuel gas; also known as the Siemens-Martin process.
- **Oxygen-Furnace Process:** Method of steelmaking in which an oxygen stream is directed downward onto a melted mass of pig iron and scrap, thereby producing high-quality carbon steels; also known as LD process, after the Austrian towns of Linz and Donawitz, where it was developed.
- **Pig Iron:** Basic raw material for steelmaking; pig iron is the product of blast furnaces, and it is made by the reaction of iron ore and coke in the presence of limestone (it averages about 5 percent carbon, 2.5 percent phosphorus, and less than 1 percent of other impurities); also known as cast iron.
- **Puddling:** Process of purifying such impure metals as pig iron by stirring molten materials in an oxidizing environment.
- **Rolling Mill:** Factory in which steel (or other metals) are processed through rollers into sheets, bars, and other forms.
- **Smelting:** Process that uses heat and chemical reactions to change an ore into a metal, for example, iron oxide heated with carbon produces iron and carbon dioxide.
- **Stainless Steel:** Important steel alloy containing iron, chromium, nickel, and carbon; its resistance to rusting has led to wide use.

DEFINITION AND BASIC PRINCIPLES

Steel is an alloy of pure iron, such as wrought iron, and a minute but crucial amount of carbon, which transforms the iron into a material with such desirable qualities as the ability to hold a sharp edge and to tolerate tension, friction, and bending. Even before the chemical nature of steel was understood, artisans had developed steelmaking technologies that resulted in such sophisticated products as samurai swords and fine cutlery. After physicists and chemists discovered and developed phase diagrams, they then possessed the principles that allowed them to predict the properties of multicomponent systems, including steel alloys. Carbon steels could then be described as iron with less than .8 percent carbon, though the carbon content could vary from .2 to 2.1 percent by weight. Low-alloy steels, generally made in the same way as ordinary carbon steels, have less than 5 percent of such elements as chromium, molybdenum, and nickel. High-alloy steels, which require special processing, have large percentages of alloying elements. For example, some stainless steels contain 18 percent chromium and 8 percent nickel. The evolution of steelmaking technologies has led to a diversification of methods and a multiplication of products, with the manufacturing techniques increasingly well adapted to the wide variety of complex steels demanded by many modern industries.

Background and History

Although historians of science and technology often combine their treatment of iron and steel, the two substances had a different—though related—evolution, especially in the modern period. Iron was first smelted from its ore more than 4,000 years ago, and archaeologists have found steel artifacts in several locales during the second millennium B.C.E. Some groups in Africa and Asia made steel weapons, and the Greek philosopher Aristotle discussed steel in the fourth century B.C.E. In the early centuries of the Common Era, certain steels made in the Middle East, such as Damascus steel, became famous for their beauty and ability to maintain a sharp edge. During the European medieval period, iron makers developed improved methods for producing pig and wrought iron as well as steel for such tools as scythes, sickles, spades, and hoes. Ironworks multiplied in eastern Europe from the late medieval period into the sixteenth century, and from the sixteenth to the eighteenth centuries blast furnaces transformed iron and steel manufacturing in central and western Europe. In colonial America, steel was made by baking wrought-iron strips combined with charcoal in sealed clay containers (the so-called cementation process).

Modern steelmaking began in the nineteenth century with the introduction of the Bessemer process in England and the open-hearth process in Germany. These methods of making massive amounts of high-quality steel economically spread to other countries and especially to the United States, whose adoption and improvements of these techniques, along with the discovery of rich iron-ore deposits in the Mesabi Range of Minnesota, led to its becoming, by 1900, the world's largest steel producer, a position it maintained well into the second half of the twentieth century. However, mismanagement, domestic and international competition, and tardiness in relinquishing traditional technologies for such new ones as electric furnaces and minimills resulted in the decline of American steel and the ascendance of such superior new steelmakers as China, Japan, India, and Brazil.

How It Works

Conceptually, the creation of steel, an alloy of iron and carbon, is readily understood since it involves either the removal of impurities and adjusting the carbon content in iron or the addition of the proper amount of carbon to very pure iron. In the long history of steelmaking before the modern era, these subtractive and additive techniques were discovered empirically, by trial and error. As chemists and physicists developed a deeper understanding of elemental metals and alloys, they were able to facilitate the production of traditional steels and create many new steel alloys, which stimulated the progress of the industry in the twentieth century.

Bessemer Process. In the middle of the nineteenth century, the British inventor Henry Bessemer discovered a method of converting pig iron into steel by eliminating its impurities and lowering its carbon content. He accomplished this through a converter, a pear-shaped container with nozzles at its base, through which compressed air was bubbled into the molten iron to oxidize impurities and excess carbon. The American mechanical engineer Alexander Lyman Holley was chiefly responsible for adapting the Bessemer process for U.S. mills, where it became part of a system for manufacturing steel rails for the booming railroad business.

Open-Hearth Process. During the first half of the twentieth century, the open-hearth steelmaking method replaced the Bessemer process in Europe and America. German inventor Karl Siemens and French engineer Pierre-Émile Martin developed this process (also called the Siemens-Martin process), in which the primary agent for transforming iron into steel was a large open-hearth furnace in which a natural-gas flame was directed downward onto the predominantly metal mix, which, in time, was transformed into steel. Although this process took much longer than the Bessemer process, it resulted in large amounts of steel and also allowed technicians to monitor the molten steel, permitting precision steels to be made. Americans increased the sizes of open-hearth furnaces, replaced natural gas by heavy fuel oil, improved the circulatory system, and created efficient instrumentation that controlled the entire process. By the 1930's more than 80 percent of steel made in the United States was done so using the open-hearth method.

Oxygen Steelmaking. During the decades after World War II, oxygen steelmaking replaced the open-hearth process in many countries. The device for transforming a mixture of pig iron and scrap into steel was the basic oxygen furnace, a large, closed barrel-shaped container in which oxygen was directed downward into the heated metals via a water-cooled lance. Within thirty years of its introduction in 1958, 97 percent of U.S. steel was being made by oxygen furnaces and the remaining 3 percent by open-hearth furnaces.

Minimills. During the final decades of the twentieth and the first decade of the twenty-first century, a "minimill revolution" transformed the world steel industry. The central component in a minimill is the electric arc furnace, where scrap metal is melted, purified, and processed into various steel products. Because minimills do not have to make such intermediate products as pig iron, they can minimize expenses for raw materials. New technologies in minimills allow the manufacture of a wide variety of products, from low- and high-carbon steels to specialized alloy steels. In 1970, minimills accounted for about 10 percent of U.S. steel production, but by 2010 they were responsible for more than 60 percent. These gains were due not so much to expansions in scale but to the development of more and better steel products.

Applications and Products

Throughout the history of steelmaking, changing markets have played a pivotal role in determining which applications are viable and which products succeed or fail. Initially, markets existed for military applications such as swords and bayonets as well as such agricultural artifacts as scythes and plows. With the development of the railroad industry the original demand for iron was replaced by increasing orders for steel, since Bessemer steel rails were superior to iron ones in durability and overall cost-effectiveness. However, the railroad industry was eventually replaced by the automotive industry as the leading buyer of steel in all advanced industrialized societies. As plastics substituted for steel in many automobile parts in the second half of the twentieth century, and as aluminum wrested more and more of the market for metal containers from steel, the demand for steel declined. Some scholars classify steelmaking applications in terms of the products sold to other businesses for processing—for example, sheets, strips, bars, rods, pipes, wires, plates, and other structural shapes. Others classify applications in terms of the chief industries making use of steel, such as the railroad, automotive, construction, and machinery industries. The following are representative industrial applications of steel.

Railroad and Shipbuilding Industries. In the United States and other countries, the early history of the railroad industry was dominated by iron, but as thousands of miles of iron rails were laid down and heavily used, they needed to be constantly replaced. In the second half of the nineteenth century and beyond, steel rails proved their superiority, leading to this massive application along with steel wheels and others. Similarly, shipbuilding passed from an iron to a steel age. During the American Civil War, the famous battle of the ironclads, the USS *Monitor* and the CSS *Virginia*, proved that a wooden ship was no match for an ironclad, causing navies around the world to abandon wooden ships in favor of their iron successors. In the late nineteenth century, steel began to replace iron in military and commercial ships because it created lighter and stronger hulls. In the twentieth century, ships constructed mainly of steel became the norm, though steel compositions later changed to increase durability in harsh environments with large temperature fluctuations. In the late twentieth and early twenty-first centuries, aluminum became a serious competitor of steel for high-speed ships.

Automobile Industry. Throughout most of the twentieth century a major consumer of steel has been the automotive industry, particularly in the United States. Steel has been, and continues to be, the chief constituent of car bodies, and hot-rolled sheet steel has become a staple of the auto industry. Steel use in this industry grew after World War II, but a decline developed during the latter decades of the century. Because of the oil and environmental crises, cars were, on the whole, downsized, and steel began to be replaced by plastics and aluminum to reduce weight

Fascinating Facts About Steelmaking Technologies

- Throughout most of the history of steelmaking, the production by individuals and firms was so limited that it was measured in pounds rather than tons.
- When the United States Steel Corporation was founded in 1901, it was the largest industrial company in the world.
- Although 83,000 tons of steel were needed for the Golden Gate Bridge (completed in 1937), only half that amount would be needed in the 2010's, because of new steels and technologies.
- Before World War II, the U.S. steel industry was so dominant that it often produced more than half the world's supply in certain years.
- Steels are generally classified by a four-number system in which the first digit represents the steel type (molybdenum or chromium-nickel), the second indicates the percentage of the chief alloying element, and the last two digits show the average carbon content.
- Since 1970 more than two-thirds of all steel in North America is recycled each year—more than aluminum, glass, paper, and plastic combined.
- Recycling the steel in a single automobile saves 2,500 pounds of iron ore, 1,400 pounds of coal, and 120 pounds of limestone.
- More than 50 percent of the different kinds of steel present in 2010 automobiles did not exist in their 2000 counterparts.

and increase energy efficiency. The steel industry responded by developing better and lighter steels. Furthermore, other materials failed to replace steel in the framework, body panels, and several auto parts under the hood.

Construction Industries. From the end of the nineteenth to the start of the twenty-first century construction has constituted one of the largest markets for steel. It has often rivaled and at times surpassed the auto industry in terms of tons of steel purchased, which is used in constructing high-rise office buildings, bridges, and, increasingly, private homes. Structural steel played an essential role in the building of American skyscrapers. A good example is the Empire State Building, completed in 1931. With 102 stories and height of 1,250 feet it was the tallest building in the world—until the completion of the north tower of the World Trade Center in 1972. Civil engineer John Augustus Roebling and his son, Washington Roebling, were the designer and chief engineer of the Brooklyn Bridge, which opened in 1883. The bridge proved the value of steel wire in the suspension cables and suspenders of this massive construction, then the longest bridge in the world. In the 1930's, the Bethlehem Steel Corporation supplied the structural steel for the George Washington Bridge in New York City and the Golden Gate Bridge in San Francisco.

Special-Purpose Steel-Alloy Applications. Although carbon steels are involved in major applications, the ever-growing variety of alloy steels has resulted in successful products because of their desirable properties and low cost. A good example is stainless steel, sometimes called corrosion-resistant steel, which is conventionally defined as having 11 percent chromium, but austenitic stainless steel is a complex alloy made with iron, chromium, nickel, manganese, silicon, carbon, phosphorus, and sulfur. Elemental components in alloy steels can be varied to suit certain industrial and domestic requirements. Some steels alloyed with cobalt or silicon have applications as permanent magnets. Several heat-resistant steel alloys have been made and successfully marketed, some of which actually have greater strength and dimensional stability at high temperatures.

Impact on Industry

During the first seven decades of the twentieth century, world production of steel generally increased, with the exception of brief declines after World Wars I and II. However, the situation became complex during the twentieth century's concluding decades. For example, industrialized countries, such as the United States, Japan, and Germany, experienced significant declines in steel production from 1974 to 1992, whereas China, Brazil, South Korea, and India had sizable increases for the same period. These trends continued in the twenty-first century. In 2006, when world production of raw steel was 1.23 billion metric tons, U.S. production had declined to 98.2 million metric tons (about 8 percent of the world total). In 2009, China, the world's largest steelmaker, produced more than 567 million tons of crude steel, which was ten times U.S. production.

Since the 1970's developing countries such as Brazil, South Korea, and India have experienced impressive growth in the amount of steel made, which contrasts sharply with the declining output in advanced industrialized countries.

Government and University Research. In the face of daunting competition internationally and domestically, steelmaking in the developed world has responded by pressuring lawmakers and government agencies to reign in unfair trade practices by such developing countries as China, where weak environmental, labor, and safety laws, as well as an undervalued currency, are the norm. Government agencies and university programs have supported advances in steelmaking technologies that will help make developed-world steel better and less expensive than developing-world counterparts. These advances include the expanded employment of computer-based systems to coordinate the various stages of production in minimills. Other research has as its goal the increased automation of all kinds and stages of steelmaking, including the application of robotics.

Industrial and Corporate Sectors. Authors of many books and articles on the decline of the American steel industry offer reasons why the world's principal producer of steel for nearly a century lost its position as manufacturing and technological leader. These include international and domestic competition, mismanagement, reluctance to replace traditional steelmaking technologies with new ones, overly restrictive government regulations, and the overoptimistic belief that accessible and abundant Mesabi Range iron ore and inexpensive Pennsylvania coking coal would last indefinitely. The once mighty Bethlehem Steel Corporation ended its steelmaking in 1995, and United States Steel saw its steel production decline precipitously from 1960 to 2000. With the introduction of the minimill and other advanced steelmaking technologies, some analysts write about a "renaissance of American steel," but others point to how little growth has occurred in world steel production since the 1980's, though steel has continued its impressive rise in developing countries.

Careers and Course Work

In the early history of steelmaking, workers were trained on the job, first through the apprenticeship system and later through participation in what came to be called "shop culture." As steelmaking became professionalized in the late nineteenth century, jobs in production were increasingly staffed with individuals having associate's or bachelor's degrees. Besides general courses in mathematics, physics, chemistry, and materials science, these students would also learn such practical crafts as welding and blueprint reading. As steelmaking technologies became more complex, with the introduction of minimills, electric arc furnaces, continuous casting, and computerized control of processing and product fabrication, master's and doctoral degrees have become essential for students hoping for good jobs in government, industrial administration, and academic research. With employment declining in the developed world and increasing in the developing world, job opportunities for American and European students with advanced degrees and the requisite language skills are becoming available in countries such as China, India, and Brazil.

In 2000, there were more than 200,000 steel-industry wage and salary jobs in the United States, and the rate of decline slowed in the first decade of the twenty-first century. Because ore and coal deposits were located largely in eastern and midwestern states, these were the places where steelmaking careers were and often still are forged. Compensation varies by education, experience, and job level. About 80 percent of American steelmakers are involved in production, and some companies still prefer hiring vocational-school graduates for processing jobs. Those desiring to become machinists, millwrights, or pipefitters generally have to serve an apprenticeship. Nevertheless, executives tend to favor those who have had courses in physics, chemistry, and metallurgy, and engineers have to possess graduate degrees.

Social Context and Future Prospects

Analysts studying the social, political, and economic context of the steelmaking industry, as well as its future prospects, usually make two points. First, steelmaking has been and will continue to be necessary for technological and industrial progress in the developed and developing worlds. Second, steelmaking is facing serious challenges, such as rising costs of energy and raw materials, the expense of adopting new technologies, and the burden of stricter regulations concerning environmental pollution and job safety. For some, especially those who see deleterious competition and unfair trade practices as harming steelmaking's progress, the answer will come in reducing barriers between countries and encouraging technological, economic, and political cooperation. Others believe that the American steel industry can be reinvigorated by strong and effective leaders who are willing to invest in the research and development of new technologies, particularly those that are energy-efficient. In this, they emphasize, the United States must become a leader, not a follower. Steelmaking has been and will continue to be a capital-intensive industry, and to be profitable it is necessary to make wise use of its most precious resources—technological, economic, and human.

Robert J. Paradowski, M.S., Ph.D.

Further Reading

Ahlbrandt, Roger S., Richard J. Fruehan, and Frank Giarratani. *The Renaissance of American Steel: Lessons for Managers in Competitive Industries.* New York: Oxford University Press, 1996. The authors argue that U.S. steelmaking's rebirth is due to downsizing bloated companies and the rise of efficient minimills. Selected bibliography and index.

Diamond, Jared. *Guns, Germs, and Steel: The Fates of Human Societies.* New York: Norton, 1997. This book, which won many awards including the Pulitzer Prize, contends that intersocietal differences in technology, including steelmaking, originate in environmental rather than cultural disparities. Further readings and index.

Hoerr, John P. *And the Wolf Finally Came: The Decline of the American Steel Industry.* Pittsburgh: University of Pittsburgh Press, 1988. Hoerr integrates the collapse of the American steel industry with the decline of the American labor movement. His work is based on primary and secondary sources, including many interviews. Extensive section of notes and an index.

Hogan, William T. *Steel in the Twenty-first Century: Competition Forges a New World Order.* New York: Lexington Books, 1994. Intended for general audiences, this economic study of the steel industry is based on primary and secondary sources as well as interviews with top steel management from all over the world. Bibliography and index.

Reutter, Mark. *Making Steel: Sparrows Point and the Rise and Ruin of American Industrial Might.* Urbana: University of Illinois Press, 2004. Based on more than one hundred interviews, this work emphasizes the human side of the story of American steel.

Rogers, Robert P. *An Economic History of the American Steel Industry.* New York: Routledge, 2009. Intended for undergraduates, this well-documented history of the U.S. steel industry from 1830 to the present uses statistics, changes in technologies and labor unions, and formative events such as the September 11, 2001, tragedy to tell the story of the rise, fall, and resurgence of American steel.

Warren, Kenneth. *Big Steel: The First Century of the United States Steel Corporation, 1901-2001.* Pittsburgh: University of Pittsburgh Press, 2001. Based on the company's archives, this largely chronological account details the origin, development, decline, and resurrection of what had once been a steelmaking giant. Bibliography and index.

Web Sites

American Iron and Steel Institute
http://www.steel.org

National Slag Association
http://www.nationalslag.org

World Steel Association
http://www.worldsteel.org

T

TELECOMMUNICATIONS

FIELDS OF STUDY

Electrical engineering; information technology; computer programming; computer science; physics; chemistry; mathematics.

SUMMARY

Telecommunications literally means "communications from afar." In the modern world, this means primarily electrical and electronic communications, including the technologies that make telecommunications possible, the applications that derive from this technology, and the organizations developing and using it.

KEY TERMS AND CONCEPTS

- **Analogue Signal:** Continuously variable signal that can vary in frequency of amplitude in response to physical phenomenon.
- **Bit:** Either of the digits 1 or 0 in a binary numeration system; also known as binary digit.
- **Cellular Radio Network:** Network of base stations that forms large pockets (cells) of radio signal coverage for mobile phones.
- **Circuit Switching:** Method of establishing a total end-to-end circuit between two parties.
- **Digital Signal:** Signal transmitted in binary code (1's and 0's) indicating pulses of varying levels of magnetic energy.
- **Fiber Optics:** Technology of sending pulses of light down very slender, very pure strands of glass.
- **Internet:** Vast network of communications channels used for data transmission.
- **Long Term Evolution (LTE):** Fourth generation (4G) wireless broadband technology.
- **1G, 2G, 3G, and 4G:** Four generations of technology in the cell phone field; "G" stands for "generation."
- **Packet Switching:** Method for transmitting data by means of small bundles of 1's and 0's that make up a portion of a data message.
- **Wi-Fi:** Trademark of the Wi-Fi Alliance used to describe a group of wireless local area network devices based on the IEEE 802.11 standard.
- **WiMax:** Telecommunications technology that provides both fixed and mobile Internet access; short for Worldwide Interoperability for Microwave Access.

DEFINITION AND BASIC PRINCIPLES

In the information age, data, sound, and video are gathered, stored, disseminated, and manipulated. Senders, receivers, and manipulators are often separated by miles, even continents. The meaning of telecommunications is to transmit data over a distance.

Telecommunications is usually envisioned as taking place between two people, but it is just as likely that information will be transmitted from people to machines, machines to people, or even machines to machines. The first telecommunications were electrical (such as the telegraph) and not electronic. Later, almost all telecommunications came to involve electronics (such as cellular radio, the Internet, and satellite communications systems).

Because telecommunications involves distance, it is important to consider the transmission medium. Early transmissions were carried over copper wire. Then came radio, and still later came coaxial cable and fiber optics. None of these has, or will, become obsolete. Consider, for instance, a video conversation with a colleague in a foreign country. The transmitter is most likely purely electrical. The voice signal is fed to a copper wire; electronics converts the message to an electronic, digital form and feeds it into a fiber-optic cable. The video is probably transmitted through a satellite link. All this lends credence to the idea that every place on Earth is equidistant from every other place, at least from a communications standpoint.

Background and History

Certainly, the progress in communication technology from smoke signals and semaphores was not without effort; it took the work of hundreds of men and women. In 1835, American Samuel F. B. Morse developed a rudimentary telegraph system, which he improved and demonstrated in 1837. In 1864, Scottish physicist and mathematician James Clerk Maxwell predicted that waves of oscillating electric and magnetic fields could travel through empty space at the speed of light. In 1887, German physicist Heinrich Hertz tested Maxwell's theory and successfully demonstrated the existence of electromagnetic waves. His work demonstrated that radio waves could be generated at one location and transmitted to a nearby spot.

American scientist Alexander Graham Bell received a patent for the first telephone system in 1876. Inventor Thomas Edison developed an electrical telegraph known as the quadruplex telegraph, which allowed four signals to be sent and received on a single wire at the same time. In 1877-1878, he developed the carbon transmitter, a telephone microphone used in telephones until the 1980's. Almon Brown Strowger, an undertaker, invented the first automatic switching system, the Strowger switch, which would allow telephone users to direct their own calls. He received a patent in 1891. In 1901, Guglielmo Marconi transmitted the first trans-Atlantic radio signal, from Poldhu, Cornwall, to St. John's, Newfoundland.

In 1904, English electrical engineer and physicist Sir John Ambrose Fleming developed the first diode vacuum tube, and American inventor Lee De Forest inserted a grid in the diode vacuum tube and created the first triode vacuum tube in 1906. In 1948, Bell Telephone Laboratories employees Walter Brattain and John Bardeen applied for a patent for the point-contact transistor, and William Shockley, who had earlier worked with Brattain, applied for a patent for the junction transistor. The three are credited with the invention of the transistor, an electronic device that essentially made the vacuum tube obsolete. In 1958, Jack Kilby, an engineer at Texas Instruments, invented the integrated circuit; six months later, Robert Noyce independently developed the integrated circuit; both are credited with its invention. Without these scientific discoveries and technological adaptations, modern telecommunications would not be possible.

How It Works

Communications involves the conveying of information; telecommunications is communication through a transmission media. For decades, the prevalent means of transmitting information was copper wire. Two wires formed a circuit, and millions of miles of copper wire were strung across the country. Unfortunately, this copper wire had limitations; the desired signal could leak out, and undesirable noise could leak in.

One method of correcting these problems was to use coaxial cable. This type of cable has a single strand of copper at its center, surrounded by insulation and a sheath of conductor such as aluminum or copper. The capacity of coaxial cable is extremely high, and its most well-known application is cable television.

The latest land-based means of transmitting information is fiber optics. A hair-thin strand of glass (very pure and very flexible) does not carry electricity—it carries light. That light, however, is a huge extension of the electromagnetic spectrum and can be triggered on and off at gigabit rates.

Wireless communications (such as radio) have become a significant means of transmitting information. The primary tool of wireless communications, the transistor, was small, used little power, lasted a long time, and was inexpensive. Of equal importance was the speed at which it could operate—it could turn on and off at gigabit rates.

The Transmission Scheme. Equally important is the structure of the signal being sent. Broadly speaking, a signal can be either analogue or digital. An analogue signal is a continuously varying electrical signal that replicates the amplitude or frequency of the originating signal. Thus, the signal being transmitted over a telephone channel is a sine wave that varies in both frequency and amplitude. Unfortunately, such a transmission scheme has inherent problems. A burst of noise (for instance electromagnetic static caused by a nearby motor starting) will be picked up by the channel. Also high-frequency signals (as might be used for data) leak out.

The alternative is a digital transmission scheme. An analogue signal is sampled, a process that changes a continuous signal into a discrete one. The magnitude of each sample is converted into a binary code—a series of ones and zeros—and transmitted down the line. It is possible to apply error-checking techniques to a group of ones and zeros and make corrections for any noise bursts that have entered the system. At the distant end, the digital signal is converted back to an analogue signal. Usually, the communications channel is capable of extremely high speed operation. If so, many individual voice or data channels can be multiplexed at the transmitting end and separated at the receiving end.

The Network. Digitization and the use of fiber optics had a tremendous impact on telecommunications and enabled the Internet and packet switching. The Internet is a vast worldwide network of computers, and packet switching is the method used for transmitting data.

The transmission of a message from point A to point B can be done by establishing a path, or circuit, through this network. Establishing a circuit and transmitting a message is known as circuit switching. This method, however, is quite inefficient, so scientists developed a faster and more versatile method, packet switching.

In packet switching, a complete data message is broken up into thousands of packets. Each packet, in addition to carrying the requisite intelligence, carries information regarding its importance and its destination. A selected packet is sent toward its destination and is received by the node at the end of this particular link and analyzed. It is then sent over a new link to the next node along the way, and the previously used link is released.

Thus, each packet, in its trek across the network, can take a different path. It is left to the intelligence of the computers at each node, and especially to the computer at the receiving end of the overall circuit, to reassemble the packets and deliver them to the recipient. Because of the high speed of this transmission, the time the packet spends at each node (called latency) is less than 1 millisecond.

Applications and Products

The telecommunications industry embraces a number of transmission media (twisted-pair copper, coaxial cable, fiber optics, wireless), a highly efficient transmission scheme (digitization), and a huge, evolving network (the Internet). Uses for these technologies has expanded rapidly and is likely to continue this trend.

Public Switched Telephone Network. The telephone network that serves much of the world is hierarchical in nature. A circuit from a calling party to a receiving party is extended through a series of telephone central offices—both up and down—until a complete circuit has been established. Transmission media most likely will consist of both copper wires and fiber optics. The information communicated starts out as analogue and is transmitted over the majority of its path as a digital signal.

The Internet. Data in the public network use the Internet—that vast array of nodes consisting of computerlike devices, interconnected by wire and fiber optics, and employing packet switching. There is no set path for an entire message to traverse, and transmission speed is so high that the cost of transmitting a single message (no matter where) approaches zero.

E-mail. To receive e-mail, a computer must be connected to the Internet. Connection can be made by coaxial cable or a satellite, as would be the case from a cruise ship. Most often, however, the computer is connected to a telephone line, and digital subscriber line (DSL) technology is used. So that the telephone can be used at the same time as the computer, a DSL circuit using high frequencies is used for the computer and low frequencies for the telephone conversation. A filter is inserted in telephone lines to keep the two separated. A modulation scheme is used to combine the voice and data signals at the user's premises and to separate them at the central office.

Another means of connection that is gaining popularity if the fiber-to-the-home (FTTH) broadband connection, which uses fiber optics to transmit data to the home. Its advantages are faster speeds and greater capacity for data.

Cell Phones. The telecommunications tool used the world over is the cell phone. A cell phone system employs a series of cells, about ten square miles in size. Each cell site, or base station, is equipped with a tower and an antenna and serves subscribers only within its bounds. A cell phone moving from one cell to another will change operating frequencies automatically to conform with the frequencies assigned to the adjacent cell. This scheme makes it possible to reuse frequencies over and over again as the user moves across the country.

The wireless network consists of the wireless (radio) channels connecting the cell phone to the base station. Once a cell phone signal reaches the base station, it travels through copper or fiber to the central office, where it enters the public switched telephone network, or the Internet. The frequencies used for this transmission are extremely valuable, and electronic auctions conducted by the Federal Communications Commission bring in millions of dollars to the U.S. Treasury.

Technologies have been created to improve wireless networks. Wi-Fi (802.11 standard networking) allows connectivity between computers and printers. WiMax (based on the 802.15 standard) is a wireless network that tries to capture the benefits of wireless and broadband technologies. Long Term Evolution (LTE), a fourth generation (4G) wireless broadband technology for wireless networks, was introduced in 2010.

Television. For the most part, cable television employs coaxial cable to connect the head end to the subscribers' premises. The system gets its signal from a satellite 22,300 miles above the Earth's surface, in geosynchronous orbit. Television is also received through satellites, DSL, and fiber optics.

Global Positioning System. Global Positioning Systems (GPSs) are built into many cell phones and smartphones, come with many new cars, and are available as handheld devices. They simply identify the user's longitude and latitude. Associated components—all computer based—translate the longitude and latitude to a location on an electronic map of the city or the area and generate a route to get from one place to another.

Perhaps the most sophisticated part of this is how the GPS determines the user's location. The GPS unit does not transmit a signal to one of the twenty-four satellites that form a part of this program, nor does a satellite transmit a unique signal to any GPS unit. Instead, each of the very accurate clocks on board the satellites produce a pulse that will not miss a beat in thousands of years. A somewhat less accurate clock in each GPS unit will produce an equivalent pulse. The Earth-bound unit measures the time between the beep received from the satellite and the beep generated by the unit. This time difference is used to determine the distance separating the satellite and the Earth-bound unit. Thus, the Earth-bound unit, with its internal maps, can determine its distance from the satellite (although its location could be anywhere on the surface of a sphere with the satellite at its center). A second satellite repeats the process, narrowing the location to the circle that represents the intersection of the spheres around the two satellites, and a third satellite narrows the location of the GPS unit to two points, the intersection of a sphere around the third satellite and the circle formed by the intersection of the first two spheres. A fourth satellite would pinpoint the user's location, but usually one of the two possibilities given by the intersection of the spheres of the three satellites makes sense and the other is nonsensical.

Fascinating Facts About Telecommunications

- In 1948, John and Margaret Walson, who sold televisions and other appliances in the mountains of Pennsylvania, used cables to connect a mountain antenna to their shop and some customers' homes, creating the first community antenna television, or cable television, as it is now called.
- The first weather satellite, the TIROS-1, was launched on April 1, 1960. The data from such satellites, combined with computer programs that use mathematics to help predict weather, have vastly improved weather forecasting.
- DSL (digital subscriber line) was developed to deliver video over copper lines, for video-on-demand services designed to compete with cable television, but it proved to be more popular for providing fast-speed Internet services.
- In July, 2004, the mayor of Grand Haven, Michigan, announced the completion of a citywide WiFi broadband network. It was the first such network in the United States.
- Where, Inc., used Global Positioning System technology to start a family-tracking service called uLocate in 2004. Three years later, it became a location service, offering an app for mobile phones. According to its Web site, it has about 50 million users.
- In 2009, the Federal Trade Commission began enforcing a ban on robocalls made by telemarketers. This ban, however, does not include banks, politicians, charities, or telephone companies, or calls of an informational nature (such as appointment reminders).
- In 2009, in an attempt to curb theft, Saguaro National Park in Arizona began injecting radio frequency identification (RFID) tags into its saguaro cacti, which can sell for $1,000 on the black market.

Radio Frequency Identification. Identification of automobiles, pets, and other items can be made by equipping them with a device called a tag and using radio frequency identification (RFID). The tags are tiny (hardly larger than a piece of glitter) and cost as little as seven cents. A common use is to enable people to enter toll roads without stopping to pay each time. As an automobile passes through the toll gate, a transmitter mounted on the gate sends out a radio frequency signal that is received by a device in the automobile. The device captures a bit of the electrical power being transmitted and triggers a small transmitter. This transmitter reflects the signal it has received, adding identification.

This technology is also used for asset tracking, retailing, and security and access. The U.S. Department of Defense requires all equipment being shipped overseas to carry a tag. The equipment can be tracked as if it had a bar code, although visual contact is not required. The technology permits up to fifty readings per second. Wal-Mart, Target, and Best Buy are among the retailers tagging their stock in an effort to better track inventory, and some retailers are requiring their vendors to provide the tags. U.S. passports incorporate tags that contain a photograph and all other passport information. They can also track travel information. Identification chips are often implanted under the skin of dogs, cats, and horses, and ear tags are used to identify cattle and other farm animals.

Impact on Industry

A competent telecommunications infrastructure is an absolute must for a growing country. The most obvious application for telecommunications—person-to-person communication—is a priority, but telecommunications goes beyond that. People talk to machines, machines talk to people, and increasingly, machines talk to machines. For example, often a person calls someone on the telephone but instead gets an answering machine (or voice messaging). The person records a message, and the machine plays the message for the recipient.

Machines can facilitate communication, as when they are used to telephone or e-mail college students. For example, colleges in Florida, where hurricanes are common, are able to evacuate the campus and conduct classes by telephone and computer.

Just as machines receive information from people, they also deliver information to people. Robocalls, or automated calls, are used by political candidates before elections, by doctor's offices to remind patients of appointments, and by pharmacies to notify patients that their prescriptions are ready.

Communications from machine to machine—although less obvious—are very prevalent. Automobiles use many computers to control parts of the vehicle, such as to engage antilock brakes and to deploy air bags. Large jetliners are programmed to land themselves in the event of an emergency.

The Internet had its origin in the Defense Advanced Research Projects Agency (DARPA), and it has been embraced by the military and industry. Information can be sent quickly and cheaply to anywhere in the world. In fact, it is impossible to tell whether an e-mail originated in a neighboring town or a distant country. People use the Internet to communicate through e-mail, blogs, Facebook, support groups, and other special-interest forums. They also search for information, purchase products, and bank online. The world of finance is highly dependent on telecommunications. Speed—measured in fractions of seconds—is paramount. An individual or broker wishing to buy a few shares of stock can place an order over the Internet. In less than one minute, the stock has been purchased, and a confirmation has been transmitted.

Satellites, originally developed by the government for military purposes, are used to map the world. They collect geographical information and atmospheric conditions and are used for weather prediction, environmental and geological studies, map making, and by individuals and industries. For example, a potential house buyer can often view the property and neighborhood using a computer. Information gathered by satellite is the basis of weather predictions that are disseminated through radio, television, and the Internet, allowing an event planner to monitor the weather and make adequate preparations. Similarly, Global Positioning Systems, originally developed for the military, have many consumer applications.

All of these applications involve a plurality of companies, disciplines, and countries. Transmission equipment and its components, both hardware and software, as well as satellites, are essential in modern telecommunications. Equipment ranges from fiber optics, networking gear, routers, servers, switches, and hubs to transmission equipment such as WiMax,

Wi-Fi, and Long Term Evolution and even spacecraft and drone airplanes.

Careers and Course Work

The telecommunications field encompasses many disciplines, including electrical and computer engineering. A knowledge of programming and software is extremely important, as it is software that drives the hardware. Certainly physics, and increasingly chemistry, play a role in telecommunications. For example, Global Positioning Systems provided inaccurate results until Einstein's special and general theories of relativity were incorporated. Some segments of the telecommunications industry—such as statistical signal processing and antenna design—are so specialized that only a few will understand them.

Anyone interested in the field must begin with science and mathematics courses in high school. Some positions are available for high school graduates or those with associate degrees, often in installing, maintaining, and repairing telecommunications equipment. The more complex the technology, the more likely higher education will be required. At the college level, depending on the area of interest, a major in electrical or computer engineering is a good choice. Because the field is so broad, supplementary course work should be tailored to areas in which the student wants to specialize. To engage in research or teach at the college level, a master's degree or doctorate is required.

Social Context and Future Prospects

People who live by the crystal ball must learn to eat crushed glass. This quaint saying, although unprofessional, is really quite accurate. Predicting the future—especially when the present is moving so quickly—is an impossible task.

In 1965, Gordon Moore, then director of engineering at Fairchild Semiconductor and later cofounder of Intel, observed that the power of the computer chip was doubling every eighteen months. Industry experts welcomed the rapid advances but thought the pace was unsustainable; however, that doubling of capability continues and has been dubbed Moore's law.

Telecommunications has been, and will continue to be, a driving force in the growth of nations. The treatment and transmission of intelligence is faster, cheaper, more comprehensive, and more convenient than ever before. People are aware of what is happening as it happens, and as a result, people are better able to deal with it. People are able to stay connected to work, home, and the world twenty-four hours a day.

There are negative aspects to this trend, and one of the most insidious is privacy. The information being gathered and manipulated is often personal and private, and its availability to virtually any enterprising person is of great concern. In spite of this, data of every sort are flooding the world.

Robert E. Stoffels, B.S.E.E., M.B.A.

Further Reading

Brooks, John. *Telephone: The First Hundred Years.* New York: Harper & Row, 1976. Possibly the most inclusive and important book dealing with the first hundred years of the telephone industry.

Crandall, Robert W. *Competition and Chaos: U.S. Telecommunications Since the 1996 Telecom Act.* Washington, D.C.: Brookings Institution Press, 2005. Examines the effects of deregulation on the telecommunications industry, including the death of the long-distance carriers and the emergence of wireless and broadband technologies.

Dodd, Annabel Z. *The Essential Guide to Telecommunications.* 4th ed. Indianapolis, Ind.: Prentice Hall Professional Technical Reference, 2005. A detailed source of information on the various types of telecommunications, including Wi-Fi and cable.

Fulle, Ronald G. *Telecommunications History and Policy into the Twenty-first Century.* Rochester, N.J.: RIT Press, 2010. Examines the history of telecommunications, analyzing how policy has affected development.

Goggin, Gerard. *Cell Phone Culture: Mobile Technology in Everyday Life.* New York: Routledge, 2006. Looks at the history of cell phones and examines how they are used and their effects on society.

Goleniewski, Lillian. *Telecommunications Essentials: The Complete Global Source.* Upper Saddle River, N.J.: Addison-Wesley, 2007. Begins with the fundamentals of telecommunications and details different types of networks, old, new, and developing.

Hills, Jill. *Telecommunications and Empire.* Urbana: University of Illinois Press, 2007. Discusses the politics involving international communications and the disputes over standards.

WEB SITES

IEEE Communications Society
http://www.comsoc.org

International Telecommunications Education and Research Association
http://www.itera.org

International Telecommunications Society
http://www.itsworld.org

Telecommunications Industry Association
http://www.tiaonline.org/index.cfm

See also: Telephone Technology and Networks; Wireless Technologies and Communication.

TELEMEDICINE

FIELDS OF STUDY

Medicine; human biology; computer science; human-computer interaction; computer programming; communications; information technology; cultural and ethnic studies; applied psychology; educational psychology; sociology; Internet; computer security and reliability; audiovisual.

SUMMARY

Telemedicine is an application of clinical medicine that focuses on medical diagnosis and patient care when the provider and client are separated by distance. Telemedicine can be performed over the telephone, or through e-mail, the Internet, or other communications technology. The development of telemedicine has revolutionized the care of individuals who are separated from medical care by distance or other barriers.

KEY TERMS AND CONCEPTS

- **Information Technology:** Branch of engineering that deals with the use of computers and telecommunications to retrieve, store, and transmit information.
- **Informed Consent:** Consent by a patient to undergo a medical or surgical treatment or to participate in an experiment after the patient understands the risks involved.
- **Internet:** Worldwide system of interconnected networks allowing for data transmission between millions of computers.
- **Real Time:** In telemedicine, when information is received and processed so quickly by a computer that the interaction seems instantaneous.
- **Satellite:** Instrument in orbit around the Earth that is used to amplify, receive, or transmit electromagnetic signals over a wide geographic area.
- **Telecommunication:** Transmitting or receiving information using radio, wire, satellite, fiber-optic, or other media for voice, data, or video communications.
- **Teleconsultation:** Medical interaction between a health care provider and a patient through telemedicine.
- **Telediagnosis:** Diagnosis or detection of a health condition in a remote patient by evaluating electronically transmitted data and information or using remote monitoring of instruments.
- **Telemonitoring:** Use of audio, video, and other peripheral devices to monitor the health of a patient from a distance.
- **Teleradiology:** Transmission of radiographic images, such as X rays, between two locations.
- **Telesurgery:** Surgery performed by a surgeon at a site removed from the patient.
- **Video Conferencing:** Two-way transmission of video images in real time to bring people at different locations together for telemedicine.

Definition and Basic Principles

Telemedicine is an innovative solution to geographic, time, social, and cultural barriers that prevent patients from receiving medical care. Although telemedicine is viewed as experimental and impersonal as compared with a face-to-face medical appointment, strides in teleconferencing and information technology have decreased the detached feeling of distance medicine. Telemedicine can be performed through a variety of communications devices that range from simple telephone calls or e-mails to complicated videoconferencing or satellite options that provide real-time, two-way interaction between the health care practitioner and the patient.

Background and History

The word "telemedicine" comes from the Greek word *tele*, which means "distance" and the Latin word *mederi*, which means "to heal." Although most people think of telemedicine as involving computers, telemedicine began long before there were telephones or the Internet. The early forms of telemedicine were practiced through letters between patients and doctors. These communications could take days or months, depending on the distance. As communications technology progressed to the telegraph, radio, and the telephone, telemedicine became more efficient and more effective as discussions between the patient and health care practitioner approached real time.

In the 1960's, the National Aeronautics and Space Administration (NASA) took telemedicine to a new level as it developed ways to monitor the health of astronauts in space. The equipment created for this purpose, which relayed medical information from the astronauts' specially designed space suits to distant physicians, is among the first peripheral monitoring devices to be developed.

During the 1970's, NASA's Space Technology Applied to Rural Papago Advanced Health Care (STARPAHC) program created one of the first telemedicine clinics on Earth. The clinic provided medical care to the Papago Indians on their remote reservation in Arizona. Remote diagnosis took place through two-way video, audio, and data communication via microwave between a mobile van containing medical equipment and paramedics on the reservation and two remotely located hospitals. The project was considered a success but was terminated in 1977.

In the late 1970's and 1980's, other telemedicine projects funded through the National Institute of Mental Health, the National Library of Medicine, the Memorial University of Newfoundland, and the Australian government took place in Nebraska, Alaska, and Australia. These projects focused on the provision of primary care, specialty care, and psychiatric services to distant or underserved communities.

As access to satellites and other modern telecommunications technology became widespread in the 1990's and 2000's, use of telemedicine grew significantly and has come to be used in many medical facilities. Modern telemedicine uses the Internet, computers, satellites, and related communications technology to provide remote health care services.

How It Works

The most common applications of telemedicine are interactive services that are very similar to ordinary visits with a health care practitioner except that they take place through specialized videoconferencing equipment or an interactive two-way computer-based system. In these telemedicine systems, the patient and health care practitioner discuss symptoms and issues in real time and with the ability to see each other. The patient's location may be equipped with a heart-monitoring electrocardiogram, a stethoscope to listen to the heart and breath, an otoscope to examine the ears, and other special examination cameras and devices that submit information to the remote practitioner. The telemedicine equipment in the patient and health care practitioner sites convey data through a high-speed Internet connection, dedicated high-speed line, or satellite link. Most often these telemedicine services are provided as networked programs between major or tertiary hospitals and outlying clinics and community health centers.

Although interactive sessions involve the patient and health care provider directly in real time, another application of telemedicine is remote monitoring. In remote monitoring, a patient may collect and transmit basic data related to weight, blood pressure, and blood sugar levels to the provider for review. Alternatively, peripheral monitoring devices may be attached to the subject and transmit information directly to the health care provider through telephone lines, wireless connections, satellite links, or other technology. Centralized monitoring centers can be used to provide heart, respiratory, or fetal monitoring in addition to home care services.

Telemedicine is also used to transmit images and data to a health care practitioner for analysis and consultation in a process called store-and-forward or point-to-point connections. This variety of telemedicine is used in nonemergency situations in which an image can be taken with a digital camera and sent to a practitioner for review through the Internet or private high-speed networks. These consultations may be on an informal basis or part of a formal, contracted, centralized reading service. This method of telemedicine can be useful for the review of radiology tests such as X rays, cardiac testing such as echocardiograms, and skin findings such as rashes.

Applications and Products

Routine Remote Diagnosis. At its most basic level, telemedicine technology allows health care practitioners to examine patients remotely and determine a diagnosis from afar. These applications can help individuals affected by geographic, temporal, social, and cultural barriers that prevent them from receiving standard medical care. An example of routine remote diagnosis would be the ability of a pediatrician in Bangor, Maine, to diagnose an ear infection in a sick child on an isolated island off the coast.

Disease Surveillance. Another important application of telemedicine is disease surveillance or monitoring. The course and outcome of many common chronic diseases such as diabetes and hypertension are improved with frequent monitoring of basic health functions. However, often physical, mental, chronological, and financial barriers prevent patients from frequently visiting physicians to receive quick checks of these functions. Telemedicine allows patients to monitor basic health signs at home and transmit them to their physician for review. For example, a patient may be able to use a cell phone to automatically upload data on vital signs and send it to a remote monitoring center. This increased monitoring should increase patient compliance and improve quality of life and disease outcome.

Consultations and Second Opinions. In some areas, primary care physicians are available, but there are few or no specialists who can be consulted to assist in confirming a diagnosis. In these areas, consultations may take the form of live interactive video between the patient and the specialist or between the primary care practitioner and the specialist. In nonemergency situations, specialists may be consulted by transmission of stored diagnostic images, vital signs, video clips, and patient data. For example, a patient with a heart murmur may undergo yearly echocardiograms to monitor the severity of his or her heart issue. As opposed to having multiple cardiologists reviewing echocardiogram images at a specific facility, the hospital or clinic may contract with a central reading facility for the echocardiogram to be reviewed off site. This decreases the number of cardiologists needed at the main site and increases the proficiency of the interpretation as the centralized readers are specially trained and able to compare new images with past studies.

Emergency and Disaster Situations. In disaster and emergency situations when normal medical facilities have been damaged or the need for care is overwhelming, telemedicine can be used through a diverse array of technologies. For example, in late 1992, the U.S. military used telemedicine to treat a variety of infectious diseases, including malaria and dengue fever, in Somalia. Telemedicine enabled military physicians to consult specialists who were unavailable in Somalia or within the military medical community. It was also used in Haiti following the 2010 earthquake to supplement meager medical care and provide specialists who could help with infectious diseases and other conditions not treatable locally.

Training and Provision of Expertise. Telemedicine technology allows experienced medical practitioners to train and observe medical professionals in a variety of specialties. The implementation of two-way communication equipment allows learners to observe surgeries or medical examinations and ask questions in real time. In addition, an observing specialist can watch a trainee perform basic tasks, surgery, or other medical care while offering suggestions and commentary. Specially trained physicians can collaborate and mentor other practitioners through these same procedures. This is particularly useful when conducting rare or specialized medical procedures, in countries without local specialists, or with distance learning.

Telesurgery. Telesurgery, also called remote surgery, is a specialized area of telemedicine. In telesurgery, a distant surgeon can use robotic systems to perform surgery in other locations. Most often the surgeries performed are minimally invasive laparoscopic surgeries that use robotic technology even when the surgeon is in the same room as the patient. Although telesurgery is not yet a common undertaking, the concept has been validated through projects such as Operation Lindbergh, which proved that systems were fast and responsive enough for a surgeon in New York to remove the gallbladder of a patient in France.

Internet Medicine. The ability of data to be easily exchanged over the Internet has furthered the use of the Internet for medical purposes. The Internet provides access to medical articles and numerous Web sites offering medical information, creating an excellent forum for patients and other consumers to obtain specialized health information. For example, on-line discussion groups and support sites provide peer-to-peer support, even for patients with rare diseases.

Impact on Industry

According to market research, the amount spent on telecommunications services in the United States is likely to increase from $7.1 billion in 2009 to $11.6 billion in 2014. Telemedicine is thought to hold the key to reducing health care costs related to routine monitoring and increasing service to underserved communities. Although European countries such as Denmark lead the international medical community in national telemedicine initiatives, the United States has also made substantial strides with its telemedicine programs. Some developing countries, such as India, have had some success with telemedicine projects, but more work is needed.

Government and University Research. The U.S. government and its various agencies have been crucial in the development of increasingly effective telemedicine equipment and protocols. Agencies such as NASA, the United States military, and the National Institutes of Health have spearheaded and funded telemedicine projects since the 1970's. As of 2010, more than twenty government agencies were providing funding in the form of grants and contracting opportunities for telemedicine, telehealth, and health information technology. These and other government agencies have also been among the biggest users of telemedicine. The research being conducted has significantly contributed to validating this effectiveness of telemedicine. For example, after a pilot study funded by the National Institute of Justice, the Bureau of Prisons, the U.S. Department of Justice, and the U.S. Department of Defense, telemedicine began to be used in many government correctional facilities to supplement on- site medical care.

In 2009, members of the U.S. House of Representatives introduced a bill that would allow Medicare to reimburse providers for telehealth services, clarify the credentialing needs of telemedicine practitioners, and create new grant programs to help the medically underserved via telemedicine. This reflects a belief, shared by many, that telemedicine is a cost-effective alternative approach to present-day business practices.

Industry and Business Sectors. The health care sector is a significant user of telemedicine technologies. Telemedicine appeals to health care administrators as a way to centralize the reading of radiology images and thus decrease the need for costly on-site specialists. For example, a community hospital with a twenty-four-hour emergency room might acquire and archive a magnetic resonance image of a brain and then transfer the data to a professional service to read the scan and dictate, transcribe, edit, and communicate the results during the late night hours when the local radiologist is not working.

Telemedicine appeals to health professionals as a way to reach distant patients, improve patient care through more frequent monitoring, and decrease medical care costs.

Medical companies have emerged to provide telemedicine services to consumers, physicians, or major medical centers. These centers may be located in the United States or in other countries such as India.

Fascinating Facts About Telemedicine

- One of the earliest uses of telemedicine was by villagers who used smoke signals to warn people to stay away from their village when serious disease struck.
- More than 200 telehealth networks connect some 2,000 institutions across the United States.
- Telemedicine has provided medical assistance and support to Mount Everest climbers since 1988.
- The top three uses of telemedicine are radiology, dermatology, and psychiatry.
- The first successful transmission of heart rhythms over telephone lines was in the Netherlands in the early 1900's.
- The first state to pass legislation requiring some insurance coverage of telemedicine services was Louisiana in 1995.

Companies based on telemedicine have a variety of functions. For example, several companies serve as centralized reading facilities, which decrease the need for specialized staff physicians to interpret routine cardiac and radiological studies. Other companies focus on monitoring peripheral medical devices for patients in facilities or at home.

Careers and Course Work

Courses in medicine, human biology, computer science, human-computer interaction, communications, information technology, and audiovisual technology make up the foundational requirements for students interested in pursuing careers in telemedicine. Depending on the desired focus—provision of medical care or construction and implementation of the equipment used for telemedicine—a series of studies focused on telecommunications, computer networking, or medicine also are required. Earning a bachelor's degree in biology, computer science, or information technology would serve as an appropriate preparation for graduate work in telemedicine. For provision of care, in most circumstances a master's degree, nursing degree, or medical degree is necessary. However, on the technology side, those without advanced degrees can play roles in program development, networking, and information technology.

Social Context and Future Prospects

The future prospects for the use of telemedicine depend on its ability to increase the quality and accessibility of medical care, to decrease health care costs, and to efficiently use available health care providers. The ability to conduct telemedicine through the Internet has decreased startup costs by reducing equipment costs and eliminating the need for dedicated communication lines.

One of the next important steps in telemedicine is wider-scale reimbursement from insurance companies. Although the American Medical Association supports reimbursement to physicians for any telemedicine services they provide, as of 2010, the level of reimbursement varied significantly from company to company, state to state, and type of telemedicine. For example, although California and eleven other states have passed legislation requiring that insurance companies reimburse for at least some telemedicine consultations or visits, such legislation is lacking in most states.

Although some opponents of telemedicine focus on the impersonal nature of telemedicine, as people are increasingly interacting through computers and other information technologies, they are becoming more comfortable with telemedicine as a concept. Additionally, computer-savvy patients have already begun using the Internet to obtain additional information about their conditions, find support among others similarly afflicted, and gauge whether symptoms they are experiencing are serious enough to require treatment or further analysis.

One controversy brewing in telemedicine is the outsourcing of medicine to other countries. Radiologists have seen an increasing amount of reading and interpretation being transferred to companies based in India, which hire United States board-certified radiologists.

Another on-going issue is concern over privacy when medical data are transmitted through communication technologies. Although information can be encrypted in a manner similar to that used for banking data, standards in telemedicine for privacy have not yet been finalized.

Dawn A. Laney, M.S., C.G.C., C.C.R.C.

Further Reading

Bashshur, Rashid, and Gary Shannon. *History of Telemedicine.* New Rochelle, N.Y.: Mary Ann Liebert, 2009. An interesting review of the history, evolution, context, and transformation of telemedicine.

Maheu, Marlene, Pamela Whitten, and Ace Allen. *E-Health, Telehealth, and Telemedicine: A Guide to Startup and Success.* New York: Jossey-Bass, 2001. A practical and insightful overview of telecommunications and information technologies. Also includes information on telemedicine and the social, legal, economic, and quality issues that affect the U.S. health care system.

Reed, Philip A. "Telemedicine: The Practice of Medicine at a Distance." *The Technology Teacher* (2003): 17-20. A succinct yet thorough review of the history and applications of telemedicine. Particular attention is paid to telemedicine and particular areas of the medical community.

West, Darrell, and Edward Miller. *Digital Medicine: Health Care in the Internet Era.* 2d ed. Washington, D.C.: Brookings Institution Press, 2010. A comprehensive and thoughtful review of costs, concerns, and possible benefits of digital medicine, including telemedicine.

Web Sites

American Telemedicine Association
http://www.americantelemed.org

Association of Telehealth Service Providers
http://tie.telemed.org

International Society for Telemedicine and Health
http://www.isft.net

U.S. Department of Agriculture, Rural Development
Distance Learning and Telemedicine Program
http://www.rurdev.usda.gov/UTP_DLT.html

See also: Telecommunications; Telephone Technology and Networks; Wireless Technologies and Communication.

TELEPHONE TECHNOLOGY AND NETWORKS

FIELDS OF STUDY

Electrical engineering; information technology; computer programming; computer science.

SUMMARY

A country without a solid telecommunications infrastructure faces insurmountable obstacles. Telephones and their networks are an essential part of telecommunications. The ability to connect individuals and government agencies by telephone is necessary to allow nations and economies to flourish. Continuing advances in technology have advanced telephones from large wooden boxes fixed to walls to small handheld portable devices that can do much more than simply enable people to talk at a distance.

KEY TERMS AND CONCEPTS

- **Analogue Signal:** Continuously variable signal that can vary in frequency of amplitude in response to physical phenomenon.
- **Bit:** Either of the digits 1 or 0 in a binary numeration system; also known as binary digit.
- **Broadband:** Circuit permitting high-speed data to be transmitted across a network.
- **Cellular Radio Network:** Network of base stations that form large pockets (cells) of radio signal coverage for mobile phones.
- **Circuit Switching:** Method of establishing a total end-to-end circuit between two parties.
- **Digital Signal:** Signal transmitted in binary code (1's and 0's) indicating pulses of varying levels of magnetic energy.
- **Fiber Optics:** Technology of sending pulses of light down very slender, very pure strands of glass.
- **Hierarchy Of Switching Systems:** Pyramid-like structure of switching centers used to connect two parties in the public switched telephone network.
- **Internet:** Vast network of communications channels used for data transmission.
- **1G, 2G, 3G, And 4G:** Four generations of technology in the cell phone field; "G" stands for "generation."
- **Packet Switching:** Method for transmitting data by means of small bundles of 1's and 0's that make up a portion of a data message.

Definition and Basic Principles

Any communications between two parties—be they people or machines—requires an originator and a receiver, as well as a network connecting the two. The simplest network tying these two parties together is a simple wire (in electrical terms) or a pair of wires. However, as the number of parties increases to three, three hundred, or three thousand, the required network becomes impossibly large. In the legacy telephone network, a hierarchy of switching systems has been developed to allow communications from any user to any other user.

As technology has advanced, other means of communication have been developed. In many cases, the pair of wires connecting the transmitting party and the receiving party has been replaced with radio transmission systems (such as satellite communications, cell phones, microwave systems), coaxial cables (a physical structure that permits high-speed transmission), or fiber optics (which converts electrical signals to light).

Just as the technology used for telecommunications has advanced, so has the structure of the telephone network. The Internet is rapidly replacing the circuit-switched network (the public switched telephone network) that has been in use for more than a century for transmission of not only data but also sound.

Background and History

The telephone industry began on February 14, 1876, when Alexander Graham Bell filed his first patent for a telephone. One month later, on March 10, 1876, he and his assistant Thomas A. Watson produced a model that worked, and the immortal words "Mr. Watson, come here, I want you," were spoken over the device.

In the years that followed, numerous technological advances were made, and the telephone became a practical device. Telephone companies sprang up by the thousands, frequently two or more firms in the same city. Connections among these companies were the exception rather than the rule, and competition became ruthless.

By the middle of the twentieth century, the industry had reached some degree of stability. The Bell System, the largest of the country's telephone companies, had managed to capture about 80 percent of the telephone service in the country.

The Bell System survived a number of government-initiated lawsuits over the years, but in 1982, it agreed to split itself up into two major entities, the telephone operating companies and the parent company, AT&T, which brought with it the long-distance arm, manufacturing facility, and research and development units.

Once again, the industry became highly competitive, with new companies appearing and disappearing each year.

The new technologies embraced by the many startups caused the industry to change dramatically. Advances in radio technology permitted satellite communications, microwave transmissions, and cellular radio. The development of fiber optics allowed data to be transmitted almost at the speed of light and increased the capacity of the country's network. Advances in electronics changed the way information was transmitted. Digital signals replaced analogue. The appearance of the Internet in 1969 and the transmission technique entitled packet switching changed things dramatically.

How It Works

The modern telephone has four major components: the transmitter (which converts sound waves into electrical signals), the receiver (which converts received electrical signals into audio signals), an alerting device (generally called a ringer, although the alerting signal is usually some sort of tone), and a signaling device (either a rotary dial or a set of push-buttons).

The ringer rides the telephone line whenever the telephone is on-hook and is removed from the line when the telephone is in use. The ringing signal for most telephones is 90 volts, alternating current.

A major part of the circuitry of the telephone deals with sidetone. Because the telephone usually operates on a two-wire circuit, sounds spoken into a telephone tend to be heard not only at the distant end of the circuit but also in the local receiver. The circuitry used to combat this reduces the volume of the conversation that is fed back to the local receiver. It is interesting to note that this fed-back conversation is not electronically eliminated entirely, lest the user think the telephone is not operating.

A call is initiated by removing the handset from the cradle. This action removes the ringer from the circuit and connects the telephone's electronic apparatus. A circuit is established through this apparatus, which is detected by the central office. Equipment at the central office sends a dial tone to the subscriber. Hearing this dial tone, the subscriber dials a number—either with a true dial, which interrupts the two-wire circuit the indicated number of times, or a keypad, which transmits two tones to the central office.

The Transmission Medium. The modern transmission medium is a far cry from the two-wire connections of yesteryear. Copper still predominates, but coaxial cable, fiber optics, and radio communications are increasingly used.

Coaxial cable has a greater capacity than does twisted-pair copper and is still the prime transmission medium for cable television. The cable is, in fact, coaxial. A single strand of copper is surrounded by insulation, which is encased in metallic sheathing. This configuration keeps desired signals in and undesirable signals out.

Fiber optics has become the transmission medium of choice. A hair-thin strand of ultra-pure glass carries pulses of light at gigabit rates. By using electronic techniques, thousands of conversations can be multiplexed onto a single strand of fiber.

Several forms of radio communications—satellites, cellular radio, and Wi-Fi—are commonly in use. Satellite communications can be used for voice, television, or data. When used for voice, satellites often result in an undesirable confusion because the time required to bounce a signal off a satellite (parked at an altitude of 22,300 miles, in geostationary orbit) is 0.125 second each way. Therefore, there is a built-in 0.25-second delay in any conversation.

A more talked-about use of radio is the wireless communication provided by cellular radio. The wireless segment of a cell phone conversation is remarkably short; simply the two- or three-mile segment from the cell phone to the cell phone tower. At this point, the signal travels down the tower through copper or fiber optics and goes to the local telephone central office.

A third use of radio transmissions is in local area networks. A radio transmission connects a device attached to a telephone line and a computer. Frequently, this is identified as Wi-Fi. Although the technology is different, a cordless telephone operates in much the same way.

Transmission Path Selection. Some sort of a selection device—for instance, a telephone dial—is required to select the destination of whatever message is being sent. The method of transmitting this address to the network and the network's ability to establish a connection are equally important.

Commonly used transmission devices are telephones, which increasingly are cell phones, smartphones, or handheld computers. The dial is usually a set of pushbuttons (the keypad). The system converts the analogue signals of the unit to a digital signal (a series of ones and zeros).

Since its establishment, the hierarchy of switching systems has served its function well. A transmission path is established from a local switching office to a higher class office and then (if required) to a still higher class office. Then the path is extended downward to finally reach the called party. Thus, a total circuit is established, and the overall process is logically called circuit switching.

This circuit switching network is being replaced with a packet switching network. Instead of establishing and holding a total circuit for the duration of a call (a wasteful process), a message (voice or data) is digitized and then broken into small chunks (called packets). Each packet of this digitized data is preceded with supervisory material that indicates the importance of the packet, its destination, and other information. Then the switch (which of course bears a remarkable resemblance to a computer) establishes a connection to a distant switch and transmits its packet. Each packet, therefore, can travel a different path to the ultimate destination. It is left for the final switch to reassemble the packets in the appropriate sequence.

Applications and Products

The technologies used in the first hundred years of the telephone industry were primarily mechanical and electromechanical. Massive amounts of equipment were necessary, a great deal of space was required for central offices, and the speed of transmission was limited. The advent of electronics—primarily the transistor—changed all this, and new technologies continue to be developed. Some of the technologies that have made major differences in most people's lives are the transistor, the integrated circuit, and central office switching.

The Transistor. In 1947, three scientists from Bell Telephone Laboratories, Walter Brattain, John Bardeen, and William Shockley, invented the transistor, a solid-state device that would replace the vacuum tube. The tiny device was a solid and did not employ a vacuum or special gas. It consumed very little power and could perform amplification or switching just as the vacuum tube could. Furthermore, its life was almost unlimited. Over the years, it also became very inexpensive.

Clearly, the transistor could be used to replace the vacuum tubes used in amplifiers in the telephone network. Also, a rethinking of the switching function in the telephone central offices made possible a new device, the digital switch, and, in later years, the Internet and packet switching.

Although the invention of the transistor was important (it has been called the invention of the century), it by no means ended technological advances in electronics. Transistor technology advanced, and devices were made both smaller and faster, and the result was called the integrated circuit. These advances have not stopped, and there is no evidence that they will. In 1965, Gordon Moore, director of engineering at Fairchild Semiconductor, analyzed the advances being made in transistor technology and determined that the capability of the solid-state device would most likely double each eighteen months. This doubling of capacity is called Moore's law.

Central Office Switching. In the conventional telephone switching system, many telephone lines enter the central office. The switch, based on input instructions (the dialed number), connects one of those lines to another, thus establishing a circuit, a process called circuit switching. Such a circuit is maintained between the two parties (calling and called) for the duration of the telephone call.

Digital switching takes a different approach. The analogue voice signal is sampled very rapidly and converted into a series of ones and zeros. These ones and zeros trigger electronics within the switch, and the identical signal is transmitted from the central office to the next office. Such a switch is more efficient than the legacy analogue switch.

Certainly the transistor, the integrated circuit, and the technology of digital switching were of immense value to the telecommunications industry. Of equal importance, however, was the change that took place in the use of the network itself.

Packet Switching. Circuit switching is inherently inefficient because the dedicated circuit is idle about half the time. A better method, although orders of magnitude more complicated, is packet switching. With packet switching, little bundles, or packets, of ones and zeros are transmitted from computer to computer across the country. Supervisory information preceding each packet of data identifies its priority, its destination, and other pertinent information.

Each packet of information follows its own path to the destination. Because of traffic on the network and the number of stops along the way, a packet sent earlier than another packet may arrive later. If the packets are chunks of digitized voice, the order matters, and it is left to the computer at the receiving point to put the packets in the proper order before delivering them to the recipient.

Two points are extremely important in such a process. The first has to do with the speed of transmission through the network and the time that each packet takes to pass through a node (this is called latency). In actual use, latency has been reduced to about one millisecond. The second point demonstrates the improvement in network efficiency. Each time a packet passes through a node, headed for a transmission link to the next node, the previous link is released. In telephone terms, the computer hangs up. So except for the supervisory information attached to each packet, the efficiency of the system approaches 100 percent.

Wireless Communications. One technology, wireless communications, has become supremely important for telephones. The process of operating telephones over radio is not new, nor is it complicated. Radio transmitting stations, whether public or private, simply mount a very tall antenna on the highest piece of land available and apply enough power to reach all potential users (such as the fire department, police department, radio stations, and taxicab companies). As long as these users stay within range of the antenna, all is well. As many channels as allowed are provided with the system (this number is easily determined with radio and television stations). For a telephone system, however, this poses problems. Just how many channels are required? More important, how many channels are available? All too frequently, too few channels are available to satisfy an entire city. The result would be a lengthy wait for dial tone.

What is the solution? A number of organizations and companies, in the United States and abroad, thought they had an answer. Instead of one large antenna and one set of frequency channels, they would provide a plurality of frequency channels, each assigned to a subset of the geographic community heretofore covered. These geographic subsets would be called cells, and the antenna covering each would be reduced in strength to the point where it would not impinge on nearby cells. This would make it possible to reuse those particular frequency channels in cells somewhat removed from the original cell.

A major problem with this approach, of course, was one of mobility. Would a taxicab driver in one cell, talking on one frequency, after moving to a different cell, have to redial? Certainly, this was not acceptable. By using very sophisticated electronics, the cell phone in use could transfer the call to the cell site antenna it was approaching and drop it from the cell it was leaving. This handoff capability was one of the most important characteristics of the proposed system.

The cell phone has become very popular. About two-thirds of the people in the world have access to a cell phone. In the United States, more and more people with cell phones no longer have a conventional land line. The number of subscribers to conventional phone service is dropping by about 10 percent per year.

Spectrum. This expansion of wireless communications is making it more difficult to find sufficient electromagnetic spectrum to support it. In the United States, toward the end of the twentieth century, the Federal Communications Commission began auctioning off huge chunks of spectrum. They were to be used "to provide a wide array of innovative wireless services and technologies, including voice, data, video and other wireless broadband services. . . ." The value of such spectrum was not lost on the telecommunications service providers. In one such auction, begun August 9, 2006, there were 161 rounds of bidding. The Federal Government took in more than $13 billion from this auction.

1G, 2G, 3G, and 4G. Wireless networks are placed into generations known as 1G, 2G, 3G, and 4G. 1G describes the original, analogue cell phone systems. From 2G on, the systems were digital. 3G improved on 2G, and 4G, which began to appear in 2010, is even more powerful.

IMPACT ON INDUSTRY

The impact that the telephone industry has had on government, industry, commerce, and on individual lives cannot be overemphasized. Instead of going to a library, many people do research on the Internet. Rather than going to a theater, many people watch films at home (or wherever they and their laptops or smartphones are). Rather than staying near a telephone to avoid missing a call, people can take the telephone with them. The cost of telephone communications have continued to drop. It is estimated that a telephone call to the other side of the world will cost less than 1 cent per minute.

These advances are so significant that lawsuits have been filed that challenge the right of individual companies to control the telecommunications network. Also, laws are being passed regarding the use of cell phones in moving vehicles. The marketplace is filled with handheld devices that act as both telephone and computer. Tabletlike electronic devices allow a person to download an entire book, a newspaper, and, in some cases, films and songs. Cell phones are so common that before many public events, the audience is reminded to turn off their cell phones.

The fallout from these advances goes well beyond the telephone itself, or even the devices that serve as telephones. Print newspapers are experiencing problems because so many people have stopped subscribing and receive their news on line. Movie theater attendance is decreasing, as people can get virtually any film they want in their homes with a few key strokes. Advertisers are declining to take out space in magazines because they have developed their own Web sites or Web advertisements. The availability of e-mail has decreased the value and volume of the U.S. mail. Communications around the country and indeed around the world are taking place at the speed of light. Everything is moving fast. Perhaps the greatest effects of the advances in telecommunications are the social implications. Many young people communicate less face to face and more through social networking sites such as Facebook.

Without question, advances in telephony will continue. As more and more people move from wired to wireless communications, transmission techniques will be developed to allow higher speed communications. Applications for handheld devices, or apps, proliferate and are even being developed by high school students. Smart devices are likely to be installed in the home and even in the human body. By dialing a preprogrammed number, people will be able to change the temperature in the house, the voltage applied by an implanted medical device, or the way money is transferred from one bank to another. The possibilities are endless.

Fascinating Facts About Telephone Technology and Networks

- The telephone was invented by people from several countries. In the United States, papers from Alexander Graham Bell and Elisha Gray were filed on the same day in 1876. Bell received the patent, and this patent has withstood the test of time.
- Undertaker Almon Brown Strowger developed the first automatic switching device (patented in 1891), allowing a telephone user to dial directly, because he thought that telephone operators were sending potential customers to his competitors.
- A group of phreakers (telephone system "hackers") found that a whistle from a box of Cap'n Crunch cereal would trigger the electronics in a telephone central office and allow free access to the long-distance network. John T. Draper, self-labeled "Captain Crunch," was arrested for toll fraud after an article was published in *Esquire* in 1971.
- In 1973, Motorola researcher Morton Cooper made the first analogue mobile phone call, using a prototype.
- In 1981, 150 users in Baltimore, Maryland, and Washington, D.C., began a two-year trial of a mobile phone network, with seven towers.
- In 1984, customers lined up to purchase the first mobile phone, Motorola's DynaTAC 8000X, which weighed 2 pounds, was 13 inches long, and cost $3,995.
- In May, 2001, NTT's DoCoMo launched a trial of the third generation (3G) mobile phone network in the Tokyo area. Five months later, it followed with a commercial network.

Careers and Course Work

The services provided by telephone companies and their suppliers no longer amount to what is called POTS (plain old telephone service) but rather are very sophisticated. Anyone who chooses to work in telecommunications, at least on the technological side, must be very conversant with electrical engineering, computer engineering, and information technology.

New terms for new concepts and advancements are constantly being generated. For example, a "Webmaster" is a person who structures, maintains, and expands a Web site. Also, many companies have created a chief information officer (CIO) position. These positions are evidence of the expanding telephone industry.

Social Context and Future Prospects

Moore's law predicts continuing development in computer technology, which is likely to translate to technical advances in the field of telecommunications. In terms of telecommunications, each site on Earth is the same distance from any other site. The cost of transmitting information—voice, data, or video—is extremely low. Communication among people, especially the young, is very frequently through electronic means.

All of this has changed the shape of many companies and institutions. One large computer manufacturer does not have a home office. Dozens of major corporations have offices far removed from their manufacturing facilities. Major universities are offering classes, lesson plans, and other student resources online. Some universities in Florida are set up to allow students to "attend" class even during emergencies, such as hurricanes, when the campus is closed.

Of course, the reliance on online communications does come with some danger. Supposedly secure information frequently turns out to be not secure, and important information (such as credit card numbers) ends up in the hands of unscrupulous people. Systems to avoid problems of this nature are being developed every day but so are means to bypass these new security measures. However, the conveniences outweigh the dangers for most people, who will continue to use a mixture of landlines, cell phones, smart phones, laptops, and desktops to communicate.

Robert E. Stoffels, B.S.E.E., M.B.A.

Further Reading

Bigelow, Stephen J., Joseph J. Carr, and Steve Winder. *Understanding Telephone Electronics.* 4th ed. Boston: Newnes, 2001. Describes how the electronics that make up telephone systems work.

Brooks, John. *Telephone: The First Hundred Years.* New York: Harper & Row, 1976. One of the most inclusive and important histories of the field, dealing with the first hundred years of the telephone industry.

Crystal, David. *Txtng: The Gr8 Db8.* New York: Oxford University Press, 2008. Looks at texting and how and why it is used. The author examines texting in several languages and shows how to decipher it. He concludes that it may help rather than hinder literacy.

Mercer, David. *The Telephone: The Life Story of a Technology.* Westport, Conn.: Greenwood Press, 2006. Discusses how telephone technology has changed the way people live, work, and socialize.

Murphy, John. *The Telephone: Wiring America.* New York: Chelsea House, 2009. A history of the telephone system, including its invention, its spread and usage, and technological advances.

Poole, Ian. *Cellular Communications Explained: From Basics to 3G.* Oxford, England: Newnes, 2006. Examines the technologies and applications of cell phone equipment and networks.

Web Sites

Federal Communications Commission
http://www.fcc.gov

IEEE Communications Society
http://www.comsoc.org

National Exchange Carrier Association
https://www.neca.org

United States Telecom Association
http://ustelecom.org

See also: Telecommunications; Wireless Technologies and Communication.

TELEVISION TECHNOLOGIES

FIELDS OF STUDY

Broadcast engineering; communications; computer programming; computer networking; design and media management; electrical engineering; electromechanical engineering; electronics; film and digital production; materials science; optics; physics; sound engineering; video production.

SUMMARY

Television technologies are concerned with the design, development, operation, and production of televisions and television programming. The field includes further aspects of television production including content creation, signal broadcast, and set up and maintenance of electrical equipment. Television technologies have widespread influence in almost every area of modern life beyond their stated applications in entertainment and education. The television industry continues to evolve rapidly as new broadcast and television technologies emerge.

KEY TERMS AND CONCEPTS

- **Airwaves:** Radio or television broadcasts transmitted as radio-frequency electromagnetic waves.
- **Analogue Television:** Television on which the picture and sound information is transmitted via an analogue signal that conveys the information via deliberate variations in the amplitude or frequency of the signal.
- **Cathode-Ray Tube (CRT):** Glass video-display component of an electronic device (usually a computer or television monitor) composed of a specialized vacuum tube in which images are produced when an electron beam strikes a phosphorescent surface.
- **Digital Television (DTV):** Television on which the picture information is transmitted in digital form rather than via an analogue signal.
- **High-Definition Television (HDTV):** Digital television system that has twice as many scan lines per frame, a wider format, and sharper image quality than conventional television.
- **Nielsen Ratings:** Audience-measurement system developed in an effort to determine the audience size and television viewing habits in the United States.
- **Pixel:** Smallest unit of a displayed digital image that conveys color and intensity. Pixels can code into an electrical signal for transmission.

Definition and Basic Principles

Television technologies are concerned with the design, development, operation, and production of televisions and television programming. The field includes further aspects of television production including content creation, signal broadcast, and set up and maintenance of electrical equipment. The core components of a television are: broadcast audio and video signals, a receiver, a speaker, and a display device.

Background and History

Television is a twentieth-century technological development. The term television was coined in France in 1900 from the Greek *tele* and Latin *vision*, which translates to "far sight." The beginnings of television can be traced back to three major discoveries: the photoconductivity of the element selenium in 1873, the invention of a scanning disk in 1884, and the demonstration of televised moving images in 1926. In 1923, the first television tube was patented by Russian engineer Vladimir Zworykin, although he never built a working model of the system. By 1925, the first moving picture was transmitted in Washington, D.C., and soon after, the Federal Radio Commission (FRC) was created to regulate television and radio broadcasting. In 1926, the National Broadcasting Company (NBC) was founded by the Radio Corporation of America (RCA) as the first major broadcast network. By then the first United States-to-Europe broadcasts were sent and scheduled television broadcasts began. At this time, most commercially made television sets were electromechanical with a radio enhanced by a neon tube and a mechanically spinning disk to produce a small image. In 1934, the FRC was replaced by the Federal Communications Commission (FCC), the government agency that regulates radio, television, wire, satellite, and cable broadcasts. By this time, the first commercially made electronic television sets with cathode-ray tubes became available. On these televisions, all broadcast images were live and in black and white. From 1940 to 1941, broadcasts included the first basketball game, the first political convention, and the first television commercial. By 1945, there were nine commercial television stations in operation in the United States, and televisions were available for purchase at department stores. In 1947, Kinescopes (picture tubes) were first used, and President Harry S. Truman used television to address the public. By 1950, there were eight million television sets in the United States, and color television broadcasts were prevalent by 1953. In 1956, videotaping was first used in a television broadcast. By 1966, all of NBC's news broadcasts were in color. In 1975, cable television expanded to cover a wider area and offer more channels, and in 1976 satellite-television transmission began with Atlanta's WTCG. By 1977, television was present in 71 million homes (97 percent) in the United States. Another milestone in television occurred in 2009 when all television stations were required by FCC regulation to switch from analogue to digital signals for their content broadcasts.

How It Works

The core components of a television are: broadcast audio and video signals, a receiver, a speaker, and a display device. The video signal is composed of the program or television content to be viewed. Television content is most commonly prerecorded. Some visual content is filmed live, such as a news program. Whether prerecorded or live, current images are predominantly broadcast in color. Until recently, analogue signals created color by broadcasting red, green, and blue images in fast succession or combined, which appeared to the viewer to be a single full-color image. Color is conveyed via digital signals as pixels representing specific levels of red, blue, and green.

Also integral to television production is the audio signal, which is the sound broadcast sent with the video signals. The video and audio signals are sent via transmitter and can be broadcast over several distribution systems: cable, satellite, or the airwave transmissions. All of these signals are gathered by the receiver found in each television and transformed into images and sound for viewing. Depending on the television set, the images viewed are projected onto a cathode-ray tube, LCD (liquid crystal display), plasma (gas-charged display), or other display device.

The development of improved television technologies has allowed televisions to become thinner and lighter with sharper images. In addition to improvements in the physical televisions, there have also been improvements in broadcast technology. The current standard for broadcast of television content is digital or high-definition. Digital signals transmit high-quality sound and high-resolution images via electronic code and use less power than the traditional analogue. In fact, in 2009, the FCC mandated that all terrestrial stations move from analogue to digital broadcast signals. This change required the addition of an improved external receiver to standard televisions or purchase of a television able to receive digital signals. In addition to improved images and television sounds, television speakers have also been updated such that improved digital audio can be funneled through home-theater systems.

Applications and Products

Journalism. The use of television as a means of conveying current events and news in the 1950's revolutionized the way the general public obtained information. Through television, the average person could watch extended interviews and speeches by world leaders, creating an almost personal connection between the viewer and the person speaking. It was impossible to achieve the same effect through radio or the newspaper. The exposure to images of local, national, and international events in progress added impact to news stories and documentaries. One of first documentaries to use television images in this fashion was *Harvest of Shame*, a searing indictment of migrant labor practices that was broadcast in 1960. Further examples of the impact of television in journalism include coverage of the first Moon landing in 1969, the wars in Vietnam, Iraq, and Afghanistan, and the devastation wrought by the Indian Ocean tsunami, Hurricane Katrina, and the earthquakes in the early twenty-first century in Haiti, New Zealand, and Japan.

Sporting Events. Television has made sporting events easily accessible to almost anyone. The application of high-resolution television technology has built up an industry of sports announcers and commentators, sports bars for viewing games, and an incredible amount of marketing via television commercials. Outside of the games themselves, television has changed the sports culture, as it spurred on the growth of sponsorship and increased the impact of the sports agent. Most sponsorships and endorsements are made possible by television commercials featuring athletes in addition to closeups of endorsed gear during games. Such athletes supplement their salaries with sponsorships and endorsements ranging from sports drinks and shoes to personal-care products. The players themselves are marketed by their agents to attract sponsors and increase their income and visibility.

Marketing. One of the most effective modes of advertising and marketing is a mass-media campaign, which is often initiated on television. The object of a mass-media campaign is to reach the largest audience possible with a specific message. Beyond mass marketing, an industry of market-research companies that concentrate on television and Nielsen ratings has emerged to help deliver ads to target audiences during particular television shows or time slots.

Nielsen ratings are sets of television-viewer data based on the careful monitoring of the television-viewing habits of a chosen cross-section of households. This system began with requesting chosen families to keep a television log or diary of minutes of each show watched by family members and has moved to electronic monitoring devices that can determine which family member is watching a particular show and when he or she changes the channel. Through data such as that produced by Nielsen ratings, a particular advertising rate can be set for a television commercial based on viewer composition and numbers. An extreme example of this is the incredibly high cost of advertising time during the Super Bowl—around $3 million for a thirty-second spot. Since so many people watch the Super Bowl, most companies are willing to spend the exorbitant fee to advertise their products.

Entertainment. The development of television technologies over time has provided many entertainment options, from the early moving pictures to specialized gaming systems, in the comfort of the home. The use of advances in television can provide a more realistic or personalized experience. The expansion of television channel options through cable and satellite television allows viewers to watch shows focused on their hobby or interest to a degree not previously available. A sports enthusiast can watch a channel such as ESPN for sports news and games, while someone interested in cooking can spend time watching the Food Network or Cooking Channel. Personalization of the television-viewing experience further expanded with the development of the digital video recorder (DVR), which allows multiple programs to be recorded automatically for viewing at a later time. The movement away from watching real-time television programs has expanded with the viewing of television shows, movies, and other programming on the computer or via devices such as the Roku or AppleTV, which allow streaming of movies via Amazon or Netflix and music via Pandora, as well as device-specific programming for children and home-fitness enthusiasts.

Surveillance and Security. The ability to use television technologies to track and identify the movements of individuals or intruders often relies heavily upon images. Closed-circuit television (CCTV) refers to a particular application of television using cameras to transmit images to a selected number of televisions for viewing and recording. CCTV is employed by banks to monitor its automatic teller machines or by other industries to monitor the function of a industrial process during the night shift. In many convenience stores, malls, and airports CCTV is used to watch for suspicious activity and deter crime. On a smaller scale, digital video monitors (nanny cams) can be placed in a baby's bedroom to provide information to parents on how their child is being cared for when they are not there.

Education. Television can be used at home or in the classroom to provide additional visual depictions of particular topics. As students become more media savvy, use of streaming video, games, and interactive features can increase interest in subjects such as history and science. A classic example of educational television is PBS's *Sesame Street*, which was first broadcast in 1969. Television programming also educates through the use of documentaries about world events of historic significance such as World War II. As TV viewers expanded their options from network television into cable and satellite TV, channels entirely devoted to children's television programming, such as Nickelodeon, the Disney Channel, and Sprout, have increased. Programming on public television and educational cable shows are targeted at preschool-age and elementary-school-age children using cartoons, such as *Sid the Science Kid*, which focuses on science; *WordGirl*, which is meant to address vocabulary; and *Liberty's Kids*, an animated history show about the Revolutionary War. As child-focused channels dominate the educational market, other channels have worked on expanding educational television further into the teen and adult market with shows such as the Discovery Channel's *Mythbusters*, which provides a view into scientific method and engineering challenges while the format and topics keep viewers interested. Beyond educational television programming, technology has also allowed students who are ill or live in remote areas to use their televisions to attend school virtually, via teleconferencing. Schools and colleges have also increased their use of distance learning by providing instructional-television classrooms that allow interactive learning with video and voice of instructors in real time.

Impact on Industry

The total value of the television technologies industry is difficult to estimate as television technologies are often joined to the radio industry under the category of broadcasting. According to the U.S. Bureau of Labor Statistics, broadcasting provided about 316,000 wage and salary jobs in 2008 and is expected to experience slow growth, with an increase of 7 percent from 2008 to 2018. Television-broadcasting-industry revenue in 2010 was $35 billion with 1.7 percent growth, according to market research firm IBISWorld.

Advertising and Marketing. The marketing and advertising industry continues to be altered by changes in television technologies. Television continues to allow companies to put a visual image for their product to the most appropriate target audience. However, with the advent of the DVR, streaming Internet videos that allow viewers to bypass or fast-forward through commercials, and other new ways of obtaining television programming, this type of marketing and advertising is now prevalent in a variety of outlets. Marketing and advertising has moved into podcasts, media player ads, pop-up ads related to specific online searches, ads that cannot be skipped before streaming videos on sites such as Hulu, and iPhone Web applications. Product placement and sponsorships are also being used in a broader context online with viral ads that are viewed outside of normal television programming.

Sporting Events. The influence of television on sporting events cannot be overestimated. The vast market of public viewers can be reached at one time only by radio and television, and television is the preferred medium. The broadcast of sports beyond baseball, football, and boxing into others such as bowling, soccer, and extreme sports has increased the interest in these sports and introduced the sports to others.

As an example, the X Games, a sports event focused on skateboarding, moto racing, BMX bike races, and related sports, began in 1995 and quickly brought these relatively unknown sports to the television-viewing public and in particular the teenage demographic. The X Games spawned a new set of sports figures or extreme athletes, such as Tony Hawk and Shaun White, who are now recognizable figures in video games and clothing lines. In addition, the viewing rights for networks for events such as the Super Bowl, World Cup soccer, and the Olympics have a large impact on the viewership and finances of a particular network. The Super Bowl is frequently the most-watched American television program of the year, with Super Bowl XLV in 2011 drawing an audience of 111 million viewers. The network that owns the broadcast rights for a particular Super Bowl not only obtains that volume of viewers for Super Bowl Sunday but also reaps the advertising revenue from the premium-price advertising time slots.

Careers and Course Work

Given the wide-ranging scope of television technologies there are many industry careers that directly correlate to the creation and distribution of television content. Entry-level requirements for these positions will vary significantly by position. Examples of positions in television technologies are television program production personnel, news-related positions such as journalists and TV anchors, on-air talent, sales and marketing jobs, management, and technical roles. Depending on the location and size of market, job categories are more fluid and employees can wear several hats. For example, at a small television station, a manager might also be in charge of marketing and production. Generally, those with a bachelor's degree have the competitive edge to obtain and advance in a position; however, many of the specialized technical positions do not require a specific post-high school degree, as hiring is based on internships, knowledge, and experience. Frequently, an entry-level position may be easier to obtain in a small market, such as a local television news station in a small town or city, although promotion often requires a move to a larger city or network. Given the wide spectrum of careers in television technologies, a sampling of jobs and course-work requirements follows.

Television-production jobs range from supervisor-level positions, such as a director, who manages the overall production, to more targeted technical positions, such as a postproduction assistant or sound designer. Given the specialized needs of scripts, lighting, sound, direction, and editing there are many jobs in television production. Supervisory occupations usually require a bachelor's degree with a major in communications, film and video, or media arts, while more technical jobs do not.

Televised news makes up a significant chunk of local and national television programming, and there are several cable channels entirely devoted to news and current events shows. Some news programs are live and require mobile teams of on-site reporters, camera operators, and associated technical positions to travel and report news. Other news shows are filmed in a studio and require the standard news team. Anchorpersons and the individuals who present the news from studios often have bachelor's degrees in journalism or mass communications. Reporters, who may also present the news, most often have bachelor's degrees in journalism, English, communications, or political science. Technical positions range from supervisory positions, which require a bachelor's degree, to lower-level positions that do not. Many of the technical positions, such as broadcast technician or sound engineering technician, are acquired through formal internships or course work at a technical school. They may not require a formal bachelor's degree, but specialized education in broadcast technology focused on engineering or electronics is helpful in obtaining a position.

Sales and marketing positions in television are varied and may involve developing large-scale marketing plans for the media company, working with advertisers, or direct sales of television services or technologies. Most of these positions require a bachelor's degree in marketing, communications, or a related field although entry-level sales positions may not.

Television writers create the content for television including settings, situations, dialogue, plot lines, and characters. Writers create and develop the idea for a particular television show or movie and work to obtain network support for the production.

In some cases, writers already work within the television company and work in-house producing ideas for shows. In other cases, writers are external to the television networks and must present their ideas for possible production. In either case, the process of creating a television show begins with a writer's idea, which he or she develops into a detailed concept for a show and includes information such as a summary of the show's premise, a breakdown of main characters on the show, ideas for the episodes for one season, and several completed episodes. Then the writer must take the idea or concept and "pitch" it to a producer or other television-company executive. If the producer chooses to purchase the show from an external writer, than a contract or deal will be negotiated for the show and the writer's employment. An agent may assist in this process. Writers often work as part of a team on television shows to create each episode and perform edits as needed during production. Some writers also wear the hat of producer, which entails working on multiple aspects of the television show from development of the show's concept to hiring actors and supervising the show's filming. There are a variety of jobs from the supervisory producer down to staff writers. Writers most often have a bachelor's degree or even a master's in some sort of English discipline. Some universities, such as the University of Southern California, offer bachelor's and

Fascinating Facts About Television Technologies

- The first television commercial was a ten-second advertisement for Bulova watches in 1941.
- In 1960, producers experimented with television systems that released odor during programs called Smell-O-Vision.
- In 1982, Seiko succesfully released a watch that had a 1¼-inch LCD screen on a wristwatch.
- The Brooklyn Dodgers was the first major league baseball team to be televised.
- In 2009, the average American home had 2.5 people and 2.86 television sets.
- The average American spends 20 percent of his or her day watching television.

master's degrees in writing for screen and television or television writing and producing. Other universities offer courses on creative writing for entertainment. Internships and a job as a writer's or producer's assistant in television may also help a writer learn about the television industry and make important contacts as he or she moves into a full-time writer's role.

Social Context and Future Prospects

The ability to record and view people and events will continue to change as new television technologies allow individuals and companies to explore the world further. Future technological innovations will change privacy options as the industry progresses and more information is gathered about television-viewing habits.

The increasing amount of time spent by adults and children watching television is associated with a decrease in exercise and an increase in obesity. In children, it is also associated with development delays, physical and mental milestones in particular. The American Academy of Pediatrics has suggested that televisions should not be placed in children's bedrooms and they should limit television watching from one to two hours per day. The nonprofit Campaign for a Commercial-Free Childhood Nonprofit sponsors events such as Screen-Free Week as a springboard for lifestyle changes that decrease reliance on screens for entertainment.

Television technologies are changing from delivery of television shows and movies on standard television sets to non-television devices such as computers and cellular phones. In addition, television is quickly moving from the standard, linear format of television programming to a nonlinear, on-demand mechanism such as the DVR, which allows viewers to record and watch shows, fast-forwarding through commercials, at a time different from the time when the program was originally broadcast. Television and the Internet have also combined to allow individuals to watch many recent TV programs shown on channels not included in one's cable or satellite subscription. In addition, consumers are watching television programming on computers via the Internet or through a variety of image-enabled devices, such as the iPhone. All of these changes have increased the number of individuals watching a wide variety of television programming.

Dawn A. Laney, B.A., M.S., C.G.C., C.C.R.C.

Further Reading

Blumenthal, Howard J., and Oliver R. Goodenough. *This Business of Television: The Standard Guide to the Television Industry.* 3d ed. New York: Billboard Books, 2006. An extensive guide to television industry production, law, and economics used as a useful resource for television writers, producers, and related professionals in broadcasting. The guide includes useful directories of television-related government agencies, distributors, producers, and associations.

Chamberlain, P. "From Screen to Monitor." *Engineering & Technology* 3, issue 15 (2008): 18-21. A discussion of movie technology, images, and changes in viewing options.

Edgerton, Gary R. *The Columbia History of American Television.* New York: Columbia University Press, 2009. A comprehensive treatment of key events in television history in the United States.

Engelman, Ralph. *Friendlyvision: Fred Friendly and the Rise and Fall of Television Journalism.* New York: Columbia University Press, 2011. A nuanced and thoughtful portrayal of Fred Friendly, one of the foremost figures in the history of American broadcast journalism.

Goldberg, Lee, and William Rabkin. *Successful Television Writing.* Hoboken, N.J.: John Wiley & Sons, 2003. A practical guide for aspiring television writers that breaks down the process of writing a successful proposal (spec script), developing a practiced, confident, and concise pitch, and handling all of the job-related pieces that come next.

Morrow, Robert W. *"Sesame Street" and the Reform of Children's Television.* Baltimore: The Johns Hopkins University Press, 2006. An insightful look at American children's television via the story of *Sesame Street*'s creation.

Web Sites

Directors Guild of America
http://www.dga.org

Society of Motion Picture and Television Engineers
http://www.smpte.org/home

Writers Guild of America, West
http://www.wga.org

See also: Photography

TEMPERATURE MEASUREMENT

FIELDS OF STUDY

Chemistry; electricity; electronics; heat transfer; optics; physics; thermodynamics.

SUMMARY

Temperature is a measure of energy. Objects are said to be in thermal equilibrium when they have the same temperature. Accurate temperature measurements are essential to determine the rate of chemical reactions, the weather, and the health of a living creature, to cite a few examples. Measurements are said to be made in absolute units when they are referred to absolute zero or in relative units when referred to some other standard state. Methods of measuring temperature include those using mechanical probes, those using chemical paints, and nonintrusive optical and sonic methods. The temperature range of interest depends on the application. Temperatures in the Earth's atmosphere range from around 200 to 320 Kelvin (K), while that on Mercury varies from around 70 K on the dark side to more than 670 K on the side facing the Sun. The Sun's surface temperature is nearly 6,000 K.

KEY TERMS AND CONCEPTS

- **Absolute Zero:** Lowest imaginable temperature, a state at which all motion ceases even at the subatomic level.
- **Blackbody Temperature:** Equilibrium temperature corresponding to the spectral distribution of radiation from a perfectly absorbing body.
- **Kinetic Temperature:** Temperature inferred from the amount and distribution of kinetic energy of random motion of the centers of mass of molecules.
- **Kinetic Theory:** Analytical framework relating gas properties to the statistical effects of individual molecules moving in random directions and colliding with each other.
- **Seebeck Effect:** Generation of an electromotive force between two junctions of wires made of certain types of metals, when the junctions are at different temperatures.
- **Stefan-Boltzmann Constant:** Constant of proportionality between the energy radiated per unit time per unit area of a blackbody and the fourth power of absolute temperature. The value of the constant in International System of Units is $5.6704 \times 10^{-8}\ W/m^2 \cdot K^4$.
- **Thermal Equilibrium:** State at which the net rate of heat transfer in any direction is zero.
- **Vibrational Temperature:** Value of temperature inferred from the amount of energy in the vibrational levels of a gas. This is equal to the kinetic temperature at equilibrium.

Definition and Basic Principles

Temperature measurement, or thermometry, is the process of determining the quantitative value of the degree of heat, or the amount of sensible thermal energy contained in matter. The temperature of matter is a manifestation of the motion of atoms and molecules, either relative to each other or because of motions within themselves. Absolute zero is a temperature that creates a state in which molecules come to complete rest.

Temperature can be measured from heat transfer by conduction, convection, or radiation. Household thermometers use either the expansion of metals or other substances or the increase in resistance with temperature. Thermocouples measure the electromotive force generated by temperature difference. Pyrometers measure infrared radiation from a heat source. Spectroscopic thermometry compares the spectrum of radiation against a blackbody spectrum. Temperature-sensitive paints and liquid crystals change intensity of radiation in certain wavelengths with temperature.

Temperature is measured in degrees on either an absolute or a relative scale, with the value of a degree differing from one system of units to another. Two major systems of units in use are degrees Fahrenheit (F) and degrees Celsius (C). The absolute scales of temperature using these units are called the Rankine and the Kelvin scales respectively, both of which refer to the absolute zero of temperature. A change of 5 Kelvin or 5 degrees Celsius corresponds to a change of 9 degrees Rankine or 9 degrees Fahrenheit. Since the freezing point of water at normal pressures is set at 32 degrees Fahrenheit and 0 degrees Celsius, temperatures in degrees Celsius are converted to degrees Fahrenheit by multiplying by 1.8 and adding 32.

Background and History

Danish astronomer Ole Christensen Rømer used the expansion of red wine as the temperature indicator in a thermometer he created in the seventeenth century. He set the zero as the temperature of a salt-ice mixture, 7.5 as the freezing point of water, and 60 as its boiling point. German physicist Daniel Gabriel Fahrenheit invented a mercury thermometer and the Fahrenheit scale, where 0 and 100 were roughly the coldest and hottest temperatures encountered in the European winter and summer. Swedish astronomer Anders Celsius invented the inverted centigrade scale, which was converted to the current Celsius scale by Swedish botanist Carl Linnaeus in the nineteenth century so that 0 represents water's freezing point and 100 its boiling point. British physicist William Thomson, also known as Lord Kelvin, devised the absolute scale in which absolute zero (0 K) corresponds to −273.15 C. Thus, 273.16 K is the triple point of water, defined as the temperature where all three states—ice, liquid water, and water vapor—can coexist. Scottish engineer William John Macquorn Rankine devised an absolute scale measured in units of degrees F, with absolute zero being -459.67 F.

How It Works

Probe Thermometers. Volume expansion thermometers use the expansion of liquids with rising temperature through a narrow tube. The expansion coefficient, defined as the increase in volume per unit volume per unit rise in temperature, is 0.00018 per Kelvin for mercury and 0.00109 per Kelvin for ethyl alcohol colored with dye. Calculating temperature from the actual random thermal motion velocity of every molecule, or the energy contained in a vibrational excitation of every molecule, is impractical. So temperature is measured indirectly in most applications. Different metals expand to different extents when their temperature rises. This difference is used to measure the bending of two strips of metal attached to one another in outdoor thermometers. Thermocouples use the Seebeck or thermoelectric effect discovered by German physicist Thomas Johann Seebeck, in which a voltage difference is produced between two junctions between wires of different metals, when the two junctions are at different temperatures. Some metal combinations produce a voltage that is linear with temperature, and these are used to produce thermocouple sensors. Resistance temperature detectors (RTDs) use the temperature sensitivity of the resistance of certain materials for measurement by including them in an electrical "bridge" circuit of interconnected resistances, such as a Wheatstone bridge, and measuring the voltage drop in a part of the circuit when the resistances are equalized. Probes using carbon resistors are the most stable at low temperatures.

Radiation Thermometers. Infrared thermometers use the different rates of emission in the infrared part of the electromagnetic spectrum from objects, compared to the emissivity of a perfectly absorbing blackbody. The device is calibrated by comparing the radiation coming from the object against that from a reference object of known emissivity at a known temperature. The dependence on emissivity is eliminated using the two-color ratio infrared thermometer, which compares the emission at two different parts of the spectrum against the blackbody spectrum.

Ultrasonic Thermometer. Ultrasonic temperature measurement considers the change in the frequency of sound of a given wavelength traveling through a medium. The frequency increases as the speed of sound increases proportional to the square root of temperature. This technique is difficult to implement with good spatial resolution since the changes in temperature over the entire path of the beam affect the frequency. However, it is suitable for techniques based on mapping multidimensional temperature fields.

Relation Between Temperature and Internal Energy. Nonintrusive laser-based methods generally use the fundamental properties of atoms and molecules. Temperature is a measure of the kinetic energy of the random motion of molecules and is hence proportional to the square of the speed of their random motion. Some energy also goes into the rotation of a molecule about axes passing through its center of mass and into the potential and kinetic energy of vibration about its center of mass. Other energy goes into the excitation of electrons to different energy levels. Quantum theory holds that each of these modes of energy storage occurs only in discrete steps called energy levels. Absorption and emission of energy, which may occur during or after collision with another atom, molecule, subatomic particle, or photon of energy, occurs with a transition by the molecule or atom from one energy level to another. The quantum of energy released or absorbed in such a transition is equal to the difference between the values of energy of the two levels involved in the transition. If the energy is released as radiation, the energy of the transition can be gauged from the frequency of the photon, using Planck's constant.

Applications and Products

Temperature measurement is used in a multitude of applications. Volume expansion thermometers work in the range from about 250 to 475 K, but each thermometer is usually designed for a much narrower range for specific purposes. Examples include the measurement of human body temperature, the atmospheric temperature, and the temperature in ovens used for cooking.

Mercury thermometers were used widely to measure room temperature as well as the temperature of the human body. Human body temperature is usually 98.6 to 99 degrees Fahrenheit, and variations of a few degrees either way may indicate illness or hypothermia. Alcohol thermometers are used in weather sensing and in homes. The atmospheric temperature varies between roughly 200 K in the cold air above the polar regions and more than 330 K above the deserts. With regard to food preparation, temperatures vary from the 256 K of a refrigerator freezer to the 515 K of an oven.

Since the 1990's, expansion thermometers for the lower temperature ranges have mostly been replaced with solid-state devices generally using RTDs with liquid crystal digital (LCD) displays. Platinum film RTDs offer fast and stable response with a constant temperature coefficient of resistance (0.024 ohms per K), whereas coil sensors offer higher sensitivity. Pt100 (platinum-coiled) RTDs have a standard 100 ohms resistance at 273 K and a 0.385 ohms per degree Celsius sensitivity. RTDs are generally limited to temperatures well below 600 K. Thermocouples are made of various metal alloy pairs. Type K (chromel-alumel) thermocouples provide a sensitivity of 41 microvolts per Kelvin in the range −200 C to 1350 C. Type E (chromel-constantan) generates 68 microvolts per Kelvin and is suited to cryogenic temperatures. Types B, R, and S are platinum or platinum-rhodium thermocouples offering around 10 microvolts per Kelvin with high stability and temperature ranges up to 2,030 Kelvin. In using thermocouples, the chemical reaction or catalytic effect of the thermocouple with the environment is important. Some metals, such as tungsten, may get oxidized, while others, such as platinum, catalyze reactions containing hydrogen.

Light waves scattered from different parts of an object interfere with each other. In the case of Rayleigh scattering, where the object such as a molecule is much smaller than the wavelength, the scattering intensity is independent of the direction in which it is scattered. It is much lower in intensity than the Mie scattering that occurs when the particle size is comparable to wavelength. Rayleigh scattering measurement can work only where the Mie scattering from relatively huge dust particles does not drown it out. Rayleigh scattering thermometry uses the fact that in a flow where the static pressure is mostly constant, such as a subsonic jet flame, the amount of light scattered by molecules from an incident laser beam is proportional to the density and inversely proportional to temperature if the composition of the gas does not vary. This simple technique is used to obtain high frequency response; however, there is often a substantial error in assuming that the gas composition is constant. Rayleigh scattering occurs at the same wavelength as the incident radiation, but the frequency band of the scattered spectrum is broadened by the Doppler shift of molecules because of their speed relative to the observer. Where the Doppler shift is due to the random motion, such as in the stagnant or slow-moving air of the atmosphere, the Doppler broadening gives the temperature. This fact is used in high spectral resolution lidar (HSRL) measurements of atmospheric conditions used from ground-based weather stations as well as from satellites.

More sophisticated nonlinear laser-based diagnostic techniques are used to measure temperature in several research applications. The technique of Raman spectroscopy uses the phenomenon of Raman scattering that occurs when molecules change from one state of vibration to another after being excited into a higher vibrational level by laser energy. A variant of this technique is called coherent anti-Stokes Raman spectroscopy (CARS), where the signature of the emission from molecules in given narrow spectral bands is compared against databases of the emission signatures of various gases to determine the composition and temperature of a given gas mixture at high temperatures. This technique is used in developing burners for jet engines and rocket engines. In laser-induced fluorescence, a strong laser pulse excites the molecules in a given small "interrogation volume." Within a few microseconds, the excited molecules release energy as photons at a different frequency than that of the laser pulse energy. A highly sensitive camera relates the spectral distribution to the temperature.

Infrared thermometers are used where nonintrusive sensors are needed or where other electromagnetic fields might interfere with thermocouples. Infrared thermometry is used to capture the change in temperature due to the change in skin friction between laminar and turbulent regions of the flow in the boundary layer over the skin of the space shuttle to determine if there are regions where the flow has separated because of missing tiles or protuberances. Since the space shuttle returned to flight after the *Columbia* disaster in 2003, such diagnostics are performed using ground-based telescope cameras or from chase planes during the upper portions of the liftoff and during orbital passes in order to alert mission controllers if repairs are needed in orbit.

Thermal equilibrium occurs very rapidly when molecules are allowed to collide with each other for some time. At equilibrium, the energy tied up in each of the various modes of energy storage has a specific value. In this case, the temperature measured from the instantaneous value of vibrational energy in the gas will give the same answer as that measured from the rotational or translational energy. However, in some situations, where the pressure and temperature change extremely rapidly or energy is added to a specific mode of storage, the temperature measured from the value of one type of energy storage may be very different from that measured from the translational energy. Examples are strong shocks in supersonic flows and rapid expansion through the nozzle of a gas dynamic laser. A common example is that of a fluorescent light bulb, where energy is added to the electronic excitation of molecules without much exciting of the translational, vibrational, or rotational levels. The gas glows as if it were at a temperature of 10,000 K (the electronic temperature is 10,000 K) whereas the translational temperature may be only 400 K.

Temperature-sensitive paints of various kinds are used to capture the temperature distribution over a surface using either the changes in color (wavelength of emitted radiation) or the intensity at a particular wavelength. Typically these use fluorescent materials. Devices range from simple stick-on tapes to expensive paints that are used on models in high-speed wind tunnel tests.

Fascinating Facts About Temperature Measurement

- The color of stars depends on the shape of their emission spectra and the temperature near their surface. Blue stars, which emit more high-energy radiation, are hotter than white stars, which are hotter than yellow or red ones.
- Glowworms emit visible radiation through a phenomenon called phosphorescence, which occurs with very efficient conversion from thermal energy to emitted radiation, keeping the temperature low.
- The reaction rate in a flammable vapor rises exponentially with temperature per the Arrhenius rate expression. A 3 percent change in temperature can lead to a doubling of reaction rate, leading to a sharp increase in temperature and heat release, causing a flame or explosion.
- When very high frequency sound passes through a liquid, tiny bubbles are formed by the expansion regions of the sound pressure wave, which then collapse suddenly when the compression regions arrive. This collapse generates supersonic shock waves where the local temperature rises to very high levels, accompanied by the emission of visible light. This is called sonoluminescence. Some researchers believe that sonoluminescence can be used to generate conditions suitable for nuclear fusion reactions.
- Researchers have reached very close to the absolute zero of temperature by forcing molecules to come to a stop using the pressure exerted by laser beams.

Impact on Industry

The ability to measure and monitor temperature accurately is crucial in many industries. Many exothermic (heat-releasing) chemical reactions depend on temperature in an exponential fashion, so that beyond a certain temperature the rate increases very rapidly, leading to a flame or explosion. The turbine inlet temperature in a jet engine is the highest temperature in the engine, and its value limits the efficiency and the thrust per unit mass flow rate of the engine. Temperature measurement is used to monitor and control various steps in most manufacturing processes, ranging from cooking in home kitchens to the most sophisticated chemical processing in industry.

Careers and Course Work

A wide variety of physical phenomena can be used to measure temperature. The basic principles come from college-level physics, including optics, and from chemistry. Heat transfer is important, as are material science and electronics, with digital signal processing and image processing used to analyze data and control theory used to turn the measurements into feedback systems to control the temperature of a process or to compensate for the imperfections of an inexpensive measuring system and obtain results as good as those from a much more complicated system. Although few careers can be imagined where temperature measurement is the primary job, several careers involve skills in temperature measurement. Medical doctors, nurses, mothers, and cooks use temperature measurement, as do heating, ventilation, and air-conditioning technicians, nuclear and chemical plant technicians, weather forecasters, astronomers, combustion engineers, and researchers. Preparation for temperature measurement includes courses in physics, chemistry, electrical engineering, and heat transfer.

Social Context and Future Prospects

Temperature measurement will continue to be an area that demands curiosity and scientific thinking across many disciplines. Probes, gauges, and non-intrusive instruments, as well as paints, are all still in heavy use, and each technique appears to be suitable to some particular problem. This diversity appears to be increasing rather than decreasing, so that the field of temperature measurement may be expected to expand and broaden. As energy technologies move away from fossil fuel combustion, where temperatures rarely exceed 2,200 K, into systems based on hydrogen combustion and nuclear fission and fusion, techniques that can measure much higher temperatures may be expected to become much more common.

Narayanan M. Komerath, Ph.D.

Further Reading

Baker, Dean H., E. A. Ryder, and N. H. Baker. *Temperature Measurement in Engineering, Vol. 1.* New York: John Wiley & Sons, 1963. Essential reading and an excellent reference, guiding the user through the various issues.

Benedict, Robert P. *Fundamentals of Temperature, Pressure, and Flow Measurements.* 3d ed. New York: John Wiley & Sons, 1984. Used all over the world in chemical and mechanical engineering courses, this book is a valuable resource for engineers and technicians preparing for professional certification. Contains practical information on instrument interfacing as well as measurement techniques.

Chang, Hasok. *Inventing Temperature: Measurement and Scientific Progress.* New York: Oxford University Press, 2007. Discusses the history of temperature measurement and how researchers managed to establish the reliability and accuracy of the various methods.

Childs, P. R. N., J. R. Greenwood, and C. A. Long. "Review of Temperature Measurement." *Review of Scientific Instruments* 71, no. 8 (2000): 2959-2978. Succinct discussion of the various techniques used in temperature measurement in various media. Discusses measurement criteria and calibration techniques and provides a guide to select techniques for specific applications.

Michalski, L., et al. *Temperature Measurement.* 2d ed. New York: John Wiley & Sons, 2001. Covers basic temperature-measurement techniques at a level suitable for high school students. A section on fuzzy-logic techniques used in thermostats is a novel addition.

Richmond, J. C. "Relation of Emittance to Other Optical Properties." *Journal of Research of the National Bureau of Standards* 67C, no. 3 (1963): 217-226. Regarded as a pioneering piece of work in developing optical measurement techniques for temperature based on the emittance of various materials and media.

Web Sites

American Engineering Association
http://www.aea.org

American Institute of Physics
http://www.aip.org

National Society of Professional Engineers
http://www.nspe.org/index.html

See also: Nuclear Technology

TRANSISTOR TECHNOLOGIES

FIELDS OF STUDY

Physics; chemistry; solid-state physics; semiconductor physics; semiconductor devices; microelectronics; materials science; engineering; electronics.

SUMMARY

Transistor technologies require the expertise of an interdisciplinary group that includes physicists, chemists, and engineers. The semiconductor transistor was invented by physicists, and may be the single most important invention of the twentieth century. As an electronic amplifier and on/off electronic switch, the transistor has revolutionized electronics and by so doing has revolutionized many fields of industry including banking, manufacturing, automobiles, aircraft, military systems, space exploration, medical instrumentation, household appliances, and communication of all sorts. Transistors have been so greatly miniaturized in size that semiconductor chips can contain up to one billion transistors on a silicon chip no larger than a thumbnail.

KEY TERMS AND CONCEPTS

- **Bipolar Transistor:** Transistor that has both electron and positive holes.
- **Chip:** Integrated electronic circuit (for example, a microprocessor) containing very large numbers of transistors and other circuit elements interconnected with very thin, very narrow, metallic interconnections on a piece of silicon.
- **Field Effect Transistor:** Transistor with a source, gate, and drain with an electric field from the gate controlling the flow of electrons from the source to drain in a semiconductor; or the flow of holes from source to drain in a different transistor.
- **Hole:** A defect in a crystal caused by an electron leaving its position in one of the crystal's bonds; equivalent to a positively charged particle.
- **Junction Transistor:** Has two p-n junctions electrically producing potential barriers between emitter and base and between base and collector. Abbreviations include NPN and PNP.
- **N-Type:** Electric conduction using electrons, which have a negative charge.
- **P-Type:** Electric conduction with positive holes.
- **Single Crystal Silicon:** Nearly perfect silicon starting material from which transistors are manufactured.
- **Transistor:** Three-terminal amplifier and on/off switch made from a semiconductor, most frequently silicon but sometimes germanium, germanium-silicon, gallium arsenide, or silicon carbide.
- **Unipolar Transistor:** Uses electrons only or holes only; field effect transistor.

Definition and Basic Principles

NPN and PNP are symbols for the two most important junction transistors. There are three legs or wires leading from each of these transistors, one each from the three separate regions. NPN means that a p-type region with positive holes separates two n-type regions with electrons carrying the electric current. Thus, there are two junctions. They both have natural potential barriers that prevent electrons in the n region from entering the p region and prevent holes from entering the n region until external voltage is applied. When external voltage is applied the transistor functions. The p region is called the base of the transistor. When negative voltage is applied between the first n region and the base, the first n region becomes an emitter of electrons into the base region. Positive voltage is applied to the other n region, between it and the base. That n region and its junction become the collector of the transistor, because it collects the electrons emitted from the emitter. Its potential barrier becomes even larger and an excellent collector of electrons. An electrical signal applied to the emitter modulates the current from emitter to the collector. The input signal is amplified in the collector circuit, which is a high-resistance circuit.

A common current is flowing from a low-voltage input to a high-voltage output. There is both voltage and power gain. There could be current gain if the input signal source were placed in the base leg and the emitter were grounded. The p region has mobile holes, and electrons can be captured by the positive holes in what is called electron-hole recombination. This is detrimental. Recombination is minimized by creating very thin regions and by using semiconductor material of very high quality. The

PNP transistor is identical to the NPN, save for the changing places of the electrons and holes.

Background and History

Walter H. Brattain and John Bardeen, physicists at Bell Laboratories, discovered the transistor effect in December, 1947. They had attempted to produce a semiconductor surface field effect amplifier but failed. In the process of studying surfaces to find the cause of failure, they probed the germanium surfaces with a sharp metal point. They discovered that an electrical signal into one point contact produced an amplified signal in the circuit of a second point contact close to the first. The point contact transistor was born. The name "transistor" was chosen by John R. Pierce, a fellow physicist at Bell Labs. Pierce reasoned that because the new device had current input and voltage output, it should be viewed as a transresistance. Since other devices (conductor, resistor, varistor, thermistor) ended in "-or" the name of the new device should end in "-or," hence the name transistor. Brattain's and Bardeen's supervisor was physicist William B. Shockley, who was not included among the inventors. Shockley went on to invent what would become an even better transistor: a p-n junction transistor. The properties of this transistor could be much more readily designed and controlled than those of a surface-point contact device because Shockley's transistor depended on the "inside" bulk properties of the semiconductor, which were much more easily controlled than the surface device. Some point-contact transistors were manufactured by the early industry, but Shockley's junction transistor dominated the growing industry into the 1960's. As a result, all three, Bardeen, Brattain, and Shockley, received the Nobel Prize in Physics in 1956 for research in semiconductors and the discovery of the transistor effect.

How It Works

Producing Junction Transistors. Junction transistors can be produced during the original crystal growing process for the silicon (or germanium) crystal by adding known n-type and p-type impurities to the molten semiconductor as the solid crystal is slowly pulled from the melt.

An improved method for producing junction transistors uses metal-alloying techniques. For example, aluminum as a dopant (desired impurity) makes silicon p-type. With a small n-type chip of silicon, a small ball of aluminum can be placed onto each of the two surfaces of the silicon chip and the temperature raised to a level high enough to melt the aluminum but not melt the silicon. However, the molten aluminum does alloy with a small surface region of the silicon, and a PNP transistor is produced.

One of the very best ways for producing NPN and PNP junction transistors is by diffusing dopants into the silicon. Diffusion of impurity atoms into silicon is a slow process and must be done at high temperatures without melting the silicon. Double diffusions and triple diffusions have been found useful. One double diffusion method might start with n-type silicon into which a p-type impurity is diffused (such as boron). This produces one junction. An n-type impurity (such as phosphorus) can be diffused on top of the p-type diffusion.

The n diffusion is made to go less deep than the p diffusion, but it must be of higher concentration to overcome and change part of the p region back into n, thereby producing an NPN transistor.

Another way to produce NPN and PNP junction transistors is by means of ion implantation. In this case, atoms of a dopant are ionized and then shot into the surface region of the silicon with very high energy driving the impurity into the silicon to whatever depth is desired in the design. Because this process is very energetic it damages the crystal structure of the silicon. The silicon must then be annealed at some higher temperature in order to remove the damage but at the same time leave the impurity in place. All p-n junction transistors are called bipolar transistors because both n-type and p-type regions are used in each transistor.

Surface Field Effect Transistors. Surface field effect transistors are called unipolar transistors because in any particular transistor only n-type regions and electrons are used, or only p-type regions and holes are used. An external electric field is applied to the surface of the silicon and modulates the electrical conductivity. In place of an emitter-base-collector of the junction transistors, the field effect transistor has a source, gate, and drain. The gate is a thin metal region separated from the silicon by a thin layer of insulating dielectric silicon dioxide. The gate applies the electric field. The source and drain are both n-type if the current flowing through the device is electrons. The source and drain are both p-type if the current flowing is holes. Surface field effect transistors have become the dominant type of transistor used in integrated circuits, which can contain up to one billion transistors plus resistors, capacitors, and the very thinnest of deposited connection wires made from aluminum, copper, or gold. The field effect transistors are simpler to produce than junction transistors and have many favorable electrical characteristics. The names of various field effect transistors go by the abbreviations MOS (metal-oxide semiconductor), PMOS (p-type metal-oxide semiconductor), NMOS (n-type metal-oxide semiconductor), CMOS (complementary metal-oxide semiconductor—uses both p-type unipolar and n-type unipolar).

Applications and Products

One of the great advantages of the transistor, either in the form of a single transistor or many transistors on a chip, is that it is generic. The transistor or chip can be used in any electronic product requiring amplifiers and on-off switches with small size, ruggedness, low power loss, desirable frequency characteristics, and high reliability. The transistor can be made larger for higher power usage, smaller for high-frequency usage, extremely high reliability for military and space usage, and inexpensive for household and consumer products. Transistors are used in all modern electronics. Many different forms of the junction transistor were created at many companies during the first fifteen years of the industry. Engineers at Fairchild Semiconductor learned how to produce reliable surface field effect transistors, and the entire industry moved in that direction in the late 1960's and continues to do so. The industry had also started working with the semiconductor silicon, a more desirable choice than germanium. Many companies used both germanium and silicon producing both diodes and transistors. National Semiconductor was formed in 1959 as the first company to use only silicon and to produce only transistors and its own early integrated circuits. Integrated circuits were originally invented by Robert Noyce at Fairchild, and later Intel, and by Jack Kilby at Texas Instruments in the late 1950's. The integrated circuit (the so-called chip) sealed the future success of the electronics revolution. Intel has become a major force in the semiconductor industry.

It would not be an exaggeration to say that every manufacturing company, every service organization, most homes in the United States, all advanced defense and military equipment, all aircraft, all research, all modern automobiles, and all modern communications use electronic equipment of some kind. In each of thousands upon thousands of applications, the electronic circuitry is different, but the generic transistor in single units or in massive numbers of transistors on single chips are common to all. These transistors and transistor chips are "hidden" in small packages, which in turn are hidden in the larger containers that house the particular electronic equipment. Intel has tried to change the hidden perspective by having computer manufacturers place a label displaying "Intel inside" on the outside covers of personal computers (PCs). Perhaps that helped PC users to realize that what was going on inside the personal computer was not magic but the result of one of the greatest inventions of the twentieth century—the transistor.

Here are a few examples of where transistors and transistor-packed chips are used.

- Aircraft: The entire control instrumentation and "fly by wire" flight systems.
- Automobiles: Engine controls and accessory controls
- Business and personal communication: Communications equipment, cell phones, all of the electronics of the Internet, smart phones.
- Financial institutions: Large computers.
- Household appliances: Microwave oven controls, washing machines, timer controls, televisions.
- Manufacturing: Automatic controls of many products.
- Medical instrumentation: All types of body scanners.
- Military systems: Intercontinental ballistic missiles' computers, inertial guidance systems, and telemetering systems and unmanned drone aircraft.
- Personal computing: Personal computers: desktops, laptops, notepads.
- Research and meteorology: Supercomputers.
- Space exploration: Everything for both manned and unmanned flights.

Software is needed to tie all of the electronics together and to tell the electronics what to do, and the software will be different from one application to another. The software and the electronics hardware are married. The transistors are operating at lightning speed as on/off switches and as amplifiers of smaller electric signals, but switches must be told what to do.

All forms of computers have central processing units (semiconductor chips), semiconductor memories, and analogue-to-digital and digital-to-analogue semiconductor chips for interfacing with the real world.

Impact on Industry

The impact on industry had its early start during World War II. The British had invented radar for the detection of German aircraft but needed U.S. participation to improve and complete radar's development. The Radiation Laboratory at Massachusetts Institute of Technology played a major role. However, it was discovered that semiconductor diodes were needed to detect and rectify the feeble radar signal returning from aircraft or ship targets. That work fell to Bell Laboratories, its manufacturing arm Western Electric, and the New England vacuum tube and lighting manufacturer Sylvania Electric. Suitable point contact silicon and germanium diodes were manufactured by Western Electric and Sylvania, and the chemical company DuPont was discovered to have a useful silicon purification process available. Bell Laboratories' wartime work on semiconductor diodes placed it into the lead position in semiconductors in 1945. Bell Lab physicists invented the point contact transistor shortly thereafter, in late 1947.

Universities played a very small role in the early days of semiconductors. There was a graduate program in semiconductors at the University of Pennsylvania and a graduate program at Purdue University. Semiconductor research and development, however, was almost totally an industry effort.

During the first twenty years of the transistor industry, there was little, if any, financial support from government agencies other than the U.S. Department of Defense and later the U.S. Space Agency now known as the National Aeronautics and Space Administration (NASA). More than ninety percent of the manufactured semiconductor transistors were for military and space program use and only a small percentage of semiconductor devices were used for consumer and industrial products. Texas Instruments did introduce the first transistorized radio, and others introduced hearing aids. As of 2011, most transistors and transistor-packed chips go to industrial and especially consumer products. Perhaps 1 or 2 percent of manufactured transistors go to the military and space programs. During the early years, the U.S. Air Force played a major role in funding the development of integrated circuit chips and other devices.

American industry was very quick to take on the task of developing the semiconductor industry and quickly surpassed its inventors at Bell Labs. Japan was

Fascinating Facts About Transistor Technologies

- The January 2011 issue of *IEEE Spectrum* listed its choices of the top eleven technologies of the decade. All of these eleven applications, except for light-emitting diodes (LEDs), require the use of advanced semiconductor transistors, most in the form of chips with huge numbers of very small transistors. LEDs, however, are produced from semiconductors (of a different kind than silicon or germanium).
- Flexible AC transmission allows more efficient use of huge quantities of electric power to different destinations and allows better incorporation of electricity from renewable—and unreliable—resources such as solar and wind.
- Smart phones contain an abundance of chips and are able to perform an abundance of functions—in addition to talking to another person, from composing e-mail and watching videos to reading the morning paper.
- The telephones and videophones that make up Voice-over Internet Protocol (VoIP) owe their success to chips.
- LEDs use chips of two special semiconductors that give off light and are the future of lighting.
- Multi-core central processing units (CPUs) use several transistor chips in a single package to give more powerful performance.
- Semiconductor chips factor heavily into the success of cloud computing, which allows many computers to use the power of servers located elsewhere.
- The Mars Rovers, Spirit and Opportunity, contain semiconductor chips that receive commands from Earth's Jet Propulsion Lab in Pasadena, California.
- The 2009 Nobel Prize in Physics went to the inventors of the semiconductor charge-coupled device (CCD). The CCD is used in digital cameras and is responsible for recording the image to be photographed.
- Class-D Audio, the very best audio for car stereos, television sets, and personal computers, uses the most advanced transistor chips.

able to obtain considerable knowledge and information from Bell Labs and became a major producer in the industry. However, when Intel decided to concentrate on microprocessor development, Japan decided to concentrate on semiconductor memories. The cutting edge of semiconductor technology turned out to be microprocessors, with Intel dominating the industry worldwide. Taiwan is a major factor in semiconductor production. Most American semiconductor manufacturers have manufacturing facilities in other countries (as well as in the United States), including Asia. Intel has a plant in Ireland. Intel dominates the digital market. Texas Instruments, National Semiconductor, and Analog Devices collectively dominate the analogue market of cell phones and applications of mixed-signal devices. (Texas Instruments is the largest of the three.) Companies in Europe and Asia also manufacture semiconductor devices.

In the new fields of nanoscience and nanotechnology, universities are now playing an important role in nanotransistor research. Both Intel and IBM are active in nanotransistor research as well. This university activity is very different from the early history of the semiconductor industry when universities played a very small role. Future nanotransistor success is uncertain because of the extremely small dimensions required: It is difficult to produce transistor properties of atomic and molecular size.

Careers and Course Work

The semiconductor transistor chip industry probably has the largest percentage of employees with doctoral degrees in all of industry. Highly educated and trained people are also needed in the related industries that supply equipment and materials to the semiconductor industry. A semiconductor fabrication facility (fab plant) is so advanced with environmental requirements and manufacturing equipment that final capital costs can total several billion dollars for just one facility. If nanoscience and nanotechnology are successful in developing mass production processes for nanotransistors on chips, the requirements for extremely clean facilities and for very advanced manufacturing equipment will be huge.

The undergraduate will find many colleges and universities with courses in solid-state physics, microelectronics, and materials science and engineering. These courses are necessary for those expecting to enter the modern field of semiconductors. Following that preparation, it is important to gain experience in one of the nation's many nanoscience and nanotechnology centers, each of which is usually part of a university. University of California, Berkeley, Los Angeles, and Santa Barbara; The Johns Hopkins University; State University of New York at Albany; Purdue University; Carnegie Mellon University; Rice University; Illinois Institute of Technology; Cornell University; Harvard University; University of Massachusetts at Amherst; Columbia University; and the University of Pennsylvania are a few of the many schools that have nanoscience centers.

Social Context and Future Prospects

It would not be an overstatement to claim that virtually every manufacturing company, every service organization, and most homes in the United States use electronic equipment containing transistor-loaded semiconductor chips. Research in nanoscience and nanotechnology will continue to create a culture and a society even more heavily dependent on the availability of advanced electronics.

Edward N. Clarke, B.S., M.S., Ph.D.

Further Reading

Bondyopadhyay, Probir K., Pallab K. Chatterjee, and Utpal K. Chakrabarti, eds. *Proceedings of the IEEE: Special Issue on the Fiftieth Anniversary of the Transistor* 86, no. 1 (January, 1998). Comprehensive collection of papers concerning the invention of the transistor; Moore's law; patents, letters, and notes by William Shockley, John Bardeen, and Walter Brattain; and life in the early days of Silicon Valley.

Callister, William D., Jr., and David G. Rethwisch. *Materials Science and Engineering: An Introduction.* 8th ed. Hoboken, N.J.: John Wiley & Sons, 2010. A popular book for undergraduate engineering students, this text covers many aspects of materials, and has well written sections on semiconductors and semiconductor devices.

Muller, Richard S., and Theodore I. Kamins, with Mansun Chan. *Device Electronics for Integrated Circuits.* 3d ed. New York: John Wiley & Sons, 2003. Very clear description of materials, manufacture, and many types of transistors.

Reid, T. R. *The Chip: How Two Americans Invented the Microchip and Launched a Revolution.* Rev. ed. New York: Simon & Schuster, 2001. The very readable story about Jack Kilby of Texas Instruments and

Robert Noyce of Fairchild Semiconductor and later Intel, who independently came up with different approaches to the invention of the microchip.

Suplee, Curt. *Physics in the Twentieth Century*. Edited by Judy R. Franz and John S. Rigden. New York: Harry N. Abrams, in association with the American Physical Society and the American Institute of Physics, 1999. A well-written and illustrated book prepared on the occasion of the centennial of the American Physical Society; the story of the transistor is found in Chapter 4.

Sze, S. M., and Kwok K. Ng. *Physics of Semiconductor Devices*. 3d ed. Hoboken, N.J.: John Wiley & Sons, 2007. Written by physicists at Bell Labs, this textbook is often used in college courses.

Web Sites

Consumer Electronics Association
http://www.ce.org

Institute of Electrical and Electronics Engineers
http://www.ieee.org

Sematech
http://www.sematech.org

Semi
http://www.semi.org

V

VEHICULAR ACCIDENT RECONSTRUCTION

FIELDS OF STUDY

Mechanical engineering; physics; computer-aided design; forensic investigation; mathematics; photography; vehicle materials; road design; impact injuries; human perception and reaction.

SUMMARY

Vehicle accident reconstruction is the forensic science of determining the factors that contribute to individual motor vehicle accidents. By collecting, documenting, and analyzing evidence recovered from and relating to the scene of a motor vehicle accident, experts can reconstruct the event and determine the primary cause as well as any malfunction, negligence, or criminal intent attached to it. Typically performed in cases involving death or injury, this type of investigation applies knowledge from various sciences to discern the roles of drivers, vehicles, and the environment in the collision. Analysis of the data determines how and why a collision occurred and, in some cases, how similar collisions might be prevented.

KEY TERMS AND CONCEPTS

- **Accident:** Unplanned mishap, yet still with underlying causes.
- **Collision:** Impact of a moving vehicle with another moving vehicle, stationary object, or pedestrian.
- **Forensic Investigation:** Collecting and analyzing data to answer questions of importance to the legal system.
- **Gouge Marks:** Marks made in the earth by a vehicle where it has left the pavement.
- **Scrub Marks:** Curved and irregular black lines left on pavement when a tire is damaged in a collision.
- **Skid Marks:** Straight black stripes left on pavement by tires when sudden braking is applied, and the tires slide instead of spin.
- **Vehicle:** Means of land transportation such as a car, truck, motorcycle, or bicycle.
- **Vehicular Homicide:** Causing the death of a person as the result of negligent driving.
- **Yaw Marks:** Curved black lines left on pavement by tires when the vehicle speed exceeds the capability of the tires, and the tires slide sideways.

DEFINITION AND BASIC PRINCIPLES

Vehicle accident reconstruction is the forensic science of determining the factors that contribute to motor vehicle accidents. The collision of a moving vehicle, such as a car or truck, with another moving vehicle, stationary object, or pedestrian typically results in injury or death. Such an event warrants an investigation into the causative factors; analyses of the evidence are often woven into a reconstruction of the event for presentation in criminal and civil court cases. Physical evidence is collected at the scene, such as the wreckage of all vehicles involved and the scattered debris, alcohol containers, illicit drug paraphernalia, and possible distractions such as cell phones. Photographs are taken of the environmental conditions of the scene, such as landscaping, snow, skid marks, road design, and traffic signs. Relative distances between points of interest are measured and recorded. Any witness statements are taken and survivors are interviewed. The physical states of the drivers, any passengers, and any pedestrians are noted, including accident-related injuries and reaction impairment due to alcohol, drugs, or sleep deprivation. All of this information is analyzed in context to reconstruct the conditions that led to the collision and indicate its primary cause as well as any malfunction, negligence, or criminal intent attached to it.

BACKGROUND AND HISTORY

In 1959, the Insurance Institute for Highway Safety (IIHS), a nonprofit organization, was founded by three major insurance companies that represented

80 percent of the American auto insurance market. Its purpose was to support highway safety research conducted by others to prevent motor vehicle collisions and to minimize injuries in the accidents that still occurred, thus saving lives and reducing financial burdens. Ten years later, IIHS became an independent research organization that studies human behaviors that contribute to collisions, vehicle design and integrity to protect occupants in collisions, and environmental conditions to eliminate obstacles and hazards.

Hugh H. Hurt, Jr., a safety engineer, was the first to study motorcycle accidents. His research, initiated in 1976, was published in 1981 as "The Hurt Report," also titled "Motorcycle Accident Cause Factors and Identification of Countermeasures." His methods of data collection and the remarkable findings that came from keen analysis led to the emergence of the field of vehicle accident reconstruction.

National standardization for the field of vehicle accident reconstruction was first funded by the National Highway Traffic Safety Administration (NHTSA) in 1985. The resulting report was titled "Minimum Training Criteria for Police Traffic Accident Reconstructionists." One recommendation in the report was the formation of a certification board. In 1991, the Accreditation Commission for Traffic Accident Reconstruction was incorporated to promote the recognition of minimum standards outlined in the NHTSA study within the scientific and legal communities.

How It Works

Investigators must approach each vehicle accident reconstruction as a realistic and unbiased narrative of scientific facts in context. Reconstruction involves three phases: investigation, in which evidence is collected; analysis, in which evidence is interpreted in context; and presentation, in which interpretations and conclusions are conveyed to persons who intend to use the information.

Investigation. In the investigation phase, evidence is collected in three areas: the vehicles involved in the collision, the environment in which the collision occurred, and the people involved in the collision. All vehicles are examined to determine the type of collision from the point of impact: head-on, rear-end, right-angle (also called T-bone), oblique (side-swiping), or rollover. Forensic mapping of all surfaces on each vehicle is performed to document the amount and location of damage. The direction of travel before and after the collision and the final resting position is noted for each of the involved vehicles and any collateral vehicles that evaded the collision. Information from event data recorders (EDRs) associated with the air bags are downloaded. Vehicles involved in the collision are thoroughly examined away from the scene. The weight of each vehicle and the engine capability are determined. Mechanical components, including the brakes, tires, and air bags, are analyzed for failure.

Evidence is collected from and about the scene of the collision. In addition to photographs, forensic mapping of the environment using electronic surveying equipment may be performed to produce a computer-generated scale diagram in either two or three dimensions. The precise location of the impact is identified. The length and location of tire marks are documented. The weather conditions, including temperature, visibility, and precipitation on the road surface, are noted. The road design, surface material, and the presence of significant imperfections are documented. The presence and location of traffic control devices are determined. Similarly, investigators document the presence and location of driver information such as warning signs, signs of speed limit changes, and notice of other vehicles entering.

Evidence is also collected about the people involved in the collision. Blood samples may be taken to determine the physical condition of the drivers. Driver qualifications and experience are noted. Logbooks of commercial drivers are obtained. Photographs of impact injuries are taken. Statements are obtained from any witnesses to the collision as well as occupants of the vehicles and responding police officers. Video evidence from speed cameras or security cameras on nearby buildings is also obtained.

Analysis. Evidence is then analyzed to place information in the context of before, during, and after the collision. Measurements of tire marks, the degree of vehicle deformation, and the distance that the vehicles traveled after the collision are used to determine speed and the time of braking. The angle and speed of impact are calculated to determine the timing of the accident and any evasive maneuvers taken to avoid the collision. Two physical laws are considered: the conservation of energy and the conservation of linear momentum. The conservation of energy states that because energy can neither be created nor destroyed, it simply changes forms; in driving, the kinetic energy of motion is normally converted into heat from friction between the tires and the brakes and also the road. In a collision, kinetic energy is dissipated via a skid and via deformation of the vehicle and its occupants. The conservation of linear momentum relates the weights of the vehicles, the angle at which they collided, and the places where they came to rest. Momentum is the product of the vehicle's mass and velocity. Upon impact, the energy of momentum from one vehicle may be transferred to another vehicle, stationary object, or person. Thus, the colliding vehicles may be pushed off the road, a stationary object may be uprooted and displaced, and a pedestrian may become airborne, landing some distance away. One fundamental outcome of this analysis is to determine the speeds and relative positions of the vehicles at various points during the collision sequence.

To analyze the environmental evidence, investigators may use planar photogrammetry, a method that accurately measures three-dimensional objects in photographs and transforms points on multiple photographs to points in scale on a single, comprehensive two- or three-dimensional diagram. In addition, the characteristics of the road design, surface, and maintenance are analyzed to determine their effects on the momentum of the vehicle and the friction value for the dissipation of energy. One fundamental outcome of this analysis is to represent accurately the external conditions under which the collision occurred and the relative location of the collision in time and space. Of particular interest is the perspective of each driver before, during, and after the impact.

To analyze human behavior, blood samples are analyzed to determine alcohol or drug use or other medical conditions that would affect a driver's reaction time. Logbooks are analyzed to determine whether a commercial driver was sleep deprived. Impact injuries are examined to determine each occupant's position in the car, whether a seat belt was worn, and whether an air bag deployed. Evidence is examined to determine whether drivers accelerated, braked, or steered to avoid the accident. One fundamental outcome of this analysis is to calculate time-distance relationships and determine the point of perception for unimpaired and impaired drivers.

Presentation. The interpretations of evidence and the conclusions drawn regarding primary and contributory causes, criminal action, and civil liability may be presented in several forms. Written expert reports propose the most likely scenario of the collision to police and private clients and may include recommendations for the prevention of similar events. Oral trial testimony and depositions present educated opinions and hypotheses supported by evidence and science to attorneys, judges, and juries. Computer-generated scale diagrams provide easily understood visual information and animation produced from the time-distance correlations of the evidence and can present evidence in real time, so observers can appreciate the brevity of the driver reaction time and the length of time that the vehicles continued to travel after the collision. Animation can also present the same scenario from different vantage points, such as those seen by the drivers of the two vehicles.

Applications and Products

Vehicle accident reconstruction is performed in cases of collisions that result in injuries, deaths, and significant property damage. It may be performed to determine criminal action or negligence, including which, if any, traffic laws were violated. Forensic evidence may also indicate intentional action such as homicide, suicide, deliberate property damage, or attempted police evasion. Collision reconstruction may also be performed to determine liability and financial responsibility, particularly in cases in which the parties involved dispute the circumstances of the accident and deny responsibility.

Collision reconstruction may be performed for various kinds of clients: prosecuting attorneys, attorneys for both plaintiffs and defendants, insurance company claims adjusters, trucking companies and other fleets of commercial vehicles, manufacturers contemplating a product recall, and automobile manufacturers and consumer advocates trying to determine crash test ratings.

Collisions may involve single vehicles meeting barriers such as guardrails, trees, overpasses, and walls. They may occur between multiple passenger or commercial vehicles. Vehicles may also collide with pedestrians, animals, and other vehicles such as motorcycles, bicycles, golf carts, snowmobiles, all-terrain vehicles, construction equipment, farm equipment, and coupled trailers.

In the determination of mechanical failure, environmental hazards, and operator error, analysis is based on precise scientific formulas, accurate detailed measurements, and certain knowledge of the multiple and variable conditions present. Environmental measurements are taken with electronic survey equipment such as total station, which uses trigonometry and triangulation to determine absolute locations, elevations, and road camber. It measures distances, horizontal angles, and vertical angles. However, it must be planted upon a known point and have a clear line of sight to landmarks. Vehicle data may be processed by finite element analysis, a mathematical method used in the analysis of elasticity and structure in engineering. It is used in the design and development of vehicles to determine strength while minimizing weight and cost. It is also used in determining the distribution of stresses and displacements in a vehicle collision.

Anthropomorphic test devices, commonly called crash test dummies, come in a variety of sizes corresponding to males and females from infants to adults in the fifth, fiftieth, and ninety-fifth percentile of the American population. They are used to determine how bodies are affected in a crash—what they hit within the interior of the vehicle, how they are ejected, and what injuries may result. The height, weight, and body measurements of the test dummies can be modified to model the actual participants in a specific accident. Injuries that commonly result from a vehicle accident are found in the spine, shoulder, and knee.

Conclusions learned from vehicle accident reconstruction may be applied to amusement ride safety, especially as roller coasters become more complex and exert greater stress on the bodies of the riders, including potential brain injuries. In addition, the impact testing of motorcycle helmets has shown that while a significant blow can reduce the protective capability of the helmet, even after ten more blows, a helmet still retains some capacity to protect the skull, suggesting that wearing a helmet that has already received a forceful impact is still better than not wearing a helmet.

Studies of accident reconstruction data have led to improvements in roof strength requirements to protect occupants in the event of a rollover. Fuel tank integrity requirements have also been upgraded to reduce the likelihood of impacted vehicles bursting into flames, possibly trapping the occupants. Along with air bags, energy-absorbing steering columns have become standard equipment to protect drivers in a collision by preserving the inside space around them. Advanced side-view mirrors and camera-assisted backup systems reduce blind spots, thus reducing the risk of collision.

Impact on Industry

International Organizations. The International Board of Forensic Engineering Sciences is a nonprofit corporation created to certify the professional competence and ethical integrity of forensic engineers, including professional engineers who specialize in vehicular accident reconstruction. Diplomate status from this board confers credibility to forensic engineers testifying in court cases around the world. The Society of Automotive Engineers (SAE), organized in 1905 in New York City, expanded its membership worldwide in 1960, becoming SAE International with affiliate chapters in Japan, Germany, the United Kingdom, India, and elsewhere. Among its values, this organization promotes worldwide automotive safety standards.

Government Agencies. The National Highway Traffic Safety Administration (NHTSA) is an agency of the federal government as part of the Department of Transportation. One of its responsibilities is creating and enforcing safety standards for motor vehicles. Another is creating and maintaining data libraries for the National Center for Statistics and Analysis. In 1985, NHTSA proposed the formation of an accreditation board to standardize the training of police and independent vehicle collision reconstruction personnel. In 1991, the Accreditation Commission for Traffic Accident Reconstruction was formed to promote minimum professional standards within the scientific and legal communities.

Professional Organizations. The National Society of Professional Engineers serves a membership of more than 45,000 licensed professional engineers regardless of discipline. Those members who are forensic engineers, including vehicular accident reconstructionists, may also join the National Academy of Forensic Engineers, an organization to promote the specialty, uphold its professional standards, and support advances in forensic engineering. The National Association of Professional Accident Reconstruction Specialists, founded in 1985, serves more than 800 members who are law enforcement officers, forensic engineers, independent consultants, and government safety personnel in North America. This professional organization strives to disseminate the latest scientific and technical information among its members, educate laypersons about traffic safety concerns, and maintain high standards of ethics and integrity.

Nonprofit Agencies. The Insurance Institute for Highway Safety is a nonprofit technical and promotional organization founded in 1959 and devoted to reducing injuries, deaths, and property losses that result from automobile accidents. Another nonprofit organization, Mothers Against Drunk Driving, was founded by Candy Lightner in 1980 after her daughter, Cari, was killed by a repeat offender drunk driver. Its mission is to raise awareness of the dangers of driving under the influence of alcohol and to encourage protective public policies and personal responsibility.

Careers and Course Work

Vehicular accident reconstructionists typically come from one of two backgrounds: criminal justice or engineering. While the educational paths are different for these two fields, the specialized field of vehicular accident reconstruction has basic principles and methods and requires proficiency in the studies of mathematics, physics, forensic investigation, photography, computer-aided design software, mechanical engineering, and human anatomy and physiology.

Police officers reconstruct collisions to determine if any criminal action contributed to the event. They investigate for evidence of speeding, mechanical violations (such as use of seat belts and headlights or faulty tires, windshield wipers, or brake lights), alcohol use, drug use, and hours-of-service violations, which may cause commercial truckers to become sleep deprived. They earn a degree in criminal justice and then specialize in vehicle accident reconstruction by taking certification courses at accredited universities such as the Northwestern University Traffic Institute in Evanston, Illinois, and the Institute of Police Technology and Management at the University of North Florida in Jacksonville.

Independent vehicular accident investigators are typically hired to determine liability and financial responsibility. While they may be retired or off-duty police officers or automotive technologists, they are more commonly licensed professional engineers in the fields of automotive, materials, mechanical, biomechanical, safety, or structural dynamics engineering. They first earn a bachelor's degree in civil or mechanical engineering and then pursue master's and doctoral degrees to gain depth of specialized knowledge as well as credentials for giving expert testimony in civil and criminal court cases. Such investigators may be hired by attorneys, insurance companies, corporations, or private individuals. Some choose to become professors and conduct research

Fascinating Facts About Vehicular Accident Reconstruction

- In 1965, activist Ralph Nader wrote the book *Unsafe at Any Speed: The Designed-In Dangers of the American Automobile.* This exposé led to the formation of the National Highway Traffic Safety Administration under the Highway Safety Act of 1970.
- The Click It or Ticket program enforcing seat belt use began in North Carolina in 1993. In the first year of the campaign, that state's rate of drivers wearing their seat belts increased from 65 percent to 81 percent. Today, all states conduct this program. In the year 2008, seat belts saved 13,250 lives nationwide.
- The number of pedestrian fatalities has decreased, simply because people are walking less between locations. However, in 2009, nearly 4,100 pedestrians were killed in the United States; this represents 12 percent of all national vehicle accident fatalities.
- Automobile accidents are the leading cause of death for children between the ages of three and fourteen years. Age- and size-appropriate car seats and booster seats have saved the lives of more than 9,000 children since 1975.
- A motorcyclist who does not ride with a helmet is 40 percent more likely to die in a collision than one who does. In 1975, forty-seven states required that helmets be worn by all motorcycle drivers and their passengers; these laws were challenged as contrary to individual rights, and today only twenty states have such a requirement.
- If 10 percent of American intersections with traffic signals were converted to roundabouts (rotaries), which improve traffic flow and control speed, more than 70,000 collisions would be prevented annually, including 450 fatalities.
- When cameras are used to enforce compliance with traffic signals, fatalities caused by vehicles running red lights are reduced by 24 percent.

into subspecialties such as tire marks, impact injuries, and motorcycle safety.

Social Context and Future Prospects

Computer software is becoming more sophisticated. Digital photographs may be entered immediately into software programs for instantaneous measurements. An electronic surveying instrument called a total station can also be used to determine horizontal and vertical angles, slopes, and distances to efficiently map the collision scene. Computer-aided modeling videos offer three-dimensional perspectives of vehicle accident reconstructions that are easily understood, especially by juries. Videos can demonstrate real-time conditions such as how sight lines were obstructed as the vehicle traveled and how little time was available for drivers to react.

Vehicle safety equipment is becoming more complex. Air bag modules receive data from numerous sensors to determine if the air bag should be deployed; some store this information in an event data recorder (EDR) and that data can be downloaded for reconstruction purposes, although it was initially intended for analyzing air bag failures. These EDRs vary with the model of vehicle in which they are installed; those in passenger vehicles generally demonstrate the vehicle deceleration pattern as a result of the collision. Others can record information from the five seconds preceding the accident, including vehicle speed, engine performance, brake light activation, and seat belt use.

As the body of vehicle accident reconstruction experience is growing, so are libraries of information on such things as the characteristics of vehicle materials, tires, driving surfaces, and visibility factors. Data on causative factors are being amassed and used to make recommendations to improve transportation safety at both the manufacturing and the legislative levels.

Bethany Thivierge, B.S., M.P.H.

Further Reading

Brach, Raymond M., and R. Matthew Brach. *Vehicle Accident Analysis and Reconstruction Methods.* Warrendale, Pa.: SAE International, 2011. Covers objective methods to apply and demonstrate the science, mathematic, and engineering principles underlying accident reconstruction.

Brown, John Fiske, Kenneth S. Obenski, and Thomas R. Osborn. *Forensic Engineering Reconstruction of Accidents.* Springfield, Ill.: Charles C. Thomas, 2002. Important concepts written in an easily understood manner for professionals other than engineers.

Franck, Harold, and Darren Franck. *Mathematical Methods for Accident Reconstruction: A Forensic Engineering Perspective.* Boca Raton, Fla.: CRC Press, 2010. The principles of mathematics and physics are applied to the models demonstrating accident reconstruction.

Noon, Randall K. *Forensic Engineering Investigation.* Boca Raton, Fla.: CRC Press, 2001. Covers in detail the engineering principles that are used in concert with forensic investigative methodologies to analyze a variety of accidents and catastrophic events.

Rivers, R. W. *Evidence in Traffic Crash Investigation and Reconstruction: Identification, Interpretation, and Analysis of Evidence and the Traffic Crash Investigation and Reconstruction Process.* Springfield, Ill.: Charles C. Thomas, 2006. This book describes the investigation process, the inspection of evidence, and the consultation of specialists to form a complete understanding of any given collision. It also includes checklists to determine the diligence of an investigation.

Web Sites

Accreditation Commission for Traffic Accident Reconstruction
http://www.actar.org

International Board of Forensic Engineering Sciences
http://www.iifes.org

National Academy of Forensic Engineers
http://www.nafe.org

National Association of Professional Accident Reconstruction Specialists
http://napars.org

Society of Automotive Engineers (SAE)
http://www.sae.org

See also: Computer-Aided Design and Manufacturing; Forensic Science; Photography.

VERTICAL TAKEOFF AND LANDING AIRCRAFT

FIELDS OF STUDY

Aeronautics; aviation; fixed-wing aircraft; rotor-wing aircraft; visual flight rules (VFR) navigation; instrument flight rules (IFR) navigation; terrain flight or contour flight navigation; nap-of-the-earth (NOE) navigation; aircraft structural engineering; propulsion engineering; avionics; air traffic control system.

SUMMARY

Vertical takeoff and landing (VTOL) and short takeoff and landing (STOL) aircraft have emerged as integral segments of the aviation industry, particularly in terms of their military and civilian-transport applications. The helicopter became vital to combat during the Vietnam War, as it was able to deploy troops quickly to a variety of different areas that previously had been accessible only by foot. The most important benefit was the development of medevac (medical evacuation) technology, a practice brought back to the United States by pilots and medical personnel.

KEY TERMS AND CONCEPTS

- **Air Traffic Control System:** Variety of ground-based control stations that a pilot uses to file flight plans, control takeoff and landing, leave or arrive in an airfield, or while flying en route.
- **Autorotation:** Engine-off emergency procedure. In case of a power breakdown in flight, the pilot adjusts the controls for maximum glide with minimum altitude loss. This disconnects the rotors on the helicopter and allows them to free wheel. The air flow reverses and the helicopter lands as an autogiro but without power.
- **Fixed-Wing Aircraft:** Aircraft body usually housing a cabin for crew and passengers; a cargo-storage area; lifting wings that can be located above, below, or in the middle of the body and house the ailerons (used to turn the plane left, right, up, or down); a smaller set of wings usually at the rear of the plane for yaw control; and a tail or rudder assembly.
- **Hybrid Aircraft:** Aircraft that is able to rise and land like a helicopter but convert in flight to the speed of a fixed-wing aircraft.
- **Instrument Flight Rules (IFR):** Navigation using only instruments inside the aircraft to fly the plane.
- **Nap-of-the-Earth (NOE) Navigation:** Flying at low speed with high engine torque near the tree line or as close to the ground as possible to avoid detection.
- **Rotary-Wing Aircraft:** Usually consisting of single or multiple rotor blades above the main cabin with a boom assembly at the rear of the aircraft with a smaller torque rotor at the end of the tail.
- **Short Takeoff And Landing (STOL):** Conventional aircraft designed to use the minimum amount of space necessary for translation from air to ground or ground to air; usually requires greater engine power or outside boosters for translation.
- **Torque:** Tendency of two attached yet freely rotating objects to move in opposite (counterclockwise) directions.
- **Translational Lift:** Amount of power or speed necessary for an aircraft rise into the air.
- **Vertical Takeoff And Landing (VTOL):** Aircraft with the ability to lift itself from the ground to a determined height in a stable transition through the application of power and directional control; most often a helicopter.
- **Visual Flight Rules:** Navigation that uses visual cues with only occasional reference to instruments inside the cockpit.

Definition and Basic Principles

Vertical takeoff and landing (VTOL) aircraft have the capability to take off from a standstill, rise straight up, fly from one place to another, and then set down vertically again.

Short takeoff and landing (STOL) aircraft use a shorter runway for takeoffs and landings, often from unprepared or abnormal surfaces.

Hybrid aircraft can perform both functions to a considerable extent and appear to be the way of the future in the development of new aircraft systems. The existing hybrid craft are used solely by the military. The V-22 Osprey is a tilt-wing system developed by Bell Boeing, and the Harrier Jump Jet is a British creation that uses a variable nozzle system to redirect power.

The autogiro, popular in Europe during the 1920's and 1930's, was a rotary-wing aircraft that had a very short roll distance. The United States even gave the autogiro a trial as an addition to the airmail fleet before determining that the cost far outweighed the benefits.

Background and History

In 1907, French engineer Paul Cornu's "flying bicycle" was the first machine to take off vertically with a pilot and achieve a stabilized, controlled free flight. He is not well known within aviation circles because he was basically overshadowed by the Wright brothers and other aircraft pioneers.

Igor Sikorsky is most often associated with the development of the helicopter and is called the father of the modern helicopter. Sikorsky's VS-300, the first operational helicopter, made its first free flight in 1940. It introduced the practical use of the tail rotor to the helicopter airframe. Its purpose was to counteract the action of torque on the body of the aircraft and keep it from spinning wildly out of control in the opposite direction of the blades overhead.

The Sikorsky R-4 was the first helicopter to fly in combat. It was used for medical evacuation and resupply missions in the China-Burma-India theater of World War II beginning in 1943. The first widespread use of helicopters was during the Korean War. The Bell 47, the bubble cockpit helicopter used for transporting medical cases, is probably the best known. Less well-known models include the OH-23 Raven, Sikorsky's H-19 Chickasaw, and H-5, which was used mostly by the Navy for recovery of pilots downed at sea.

The Vietnam War became known as the "helicopter war." The use of these machines in a variety of roles changed military tactics as well as the public perception for future non-military use.

How It Works

Fixed-Wing Aircraft. The fixed-wing aircraft has wings extending from both sides of the aircraft with one, two, or more engines used to propel the aircraft forward. To accommodate a variety of tasks, both military and civilian, some are being adapted to serve fast transportation needs into and out of small environments. The problems are being approached from two directions: variable wing and ducted variable thrust.

The concept of variable wing thrust is to vary the orientation of the wing relative to ground and the direction of flight to achieve lift or thrust. Ducted variable thrust flight is achieved by manually changing the thrust direction of the engine in a fixed-wing jet aircraft from the rear to directly below the aircraft. The aircraft rises and can be controlled as a helicopter would until the thrust direction is changed for vertical flight or reversed for landing. The Harrier Jump Jet is the most common example of this.

Hybrid Aircraft. The V-22 Osprey has the ability to tilt its wings 90 degrees to surface. The overly large propellers act as rotor blades that turn in opposite directions to cancel out the torque effect. The craft can then act like a helicopter to rise into the air. After clearing ground obstacles and with some forward speed, the pilot engages a lever that causes the wing to tilt into a normal fixed-wing position. The large propellers and standard configuration then allow the plane to fly at speeds of about 300 miles per hour. Upon arrival, the pilot reverses the procedures and again lands as a helicopter would.

Autogiro Aircraft. Autogiros typically have a small body, a pulling or pushing engine, and either short, stubby wings or no wings except for a conventional tail surface and structure. The main distinguishing characteristic is the free-wheeling rotor, usually of three to five blades, that is attached to the body above what would be the wing center of gravity for the plane. This rotor generates the lift by reversing the operation of a rotor on a helicopter. The pushing or pulling engine will move the body along the ground or through the air. Instead of pushing air down as a rotor or airplane wing does, wind rises through the rotor on the autogiro, which is a free-wheeling device, and causes lift.

Rotary-Wing Aircraft. Helicopters use the same principles, with two exceptions. The engine in the helicopter is usually hooked directly to the propeller or rotor. It is also hooked in sync with the shaft that turns the rear or tail rotor. The tail rotor is necessary for the directional control of the aircraft. When the collective, the control system that governs lift, is pulled up, it sends directions for the rotor blades to change their pitch and take a larger bite out of the air to lift the machine. Without the tail rotor, torque takes effect, and this causes the aircraft body to rotate in the opposite direction of the rotor blades. The pitch (or amount of bite into the air of the blades) of the tail rotor is controlled by the foot pedals.

The other exception is in the control. To make an aircraft go in one direction or another, the pilot inputs the movements directly, and the aircraft follows those movements. In a helicopter, the movements are put into the control stick. The control stick is pushed forward and the craft follows suit. Although the pilot is facing forward, the command actually goes to the rotor 90 degrees to his right. As the blade continues to turn, the pilot's command is fully completed and the aircraft moves in the direction the pilot has intended.

Applications and Products

Aviation and aviation technology constitute a multibillion-dollar business. Areas include conceptualization, design, manufacture, and sales of aircraft and aircraft-related components; commercial and private flight; and movement of freight and mail.

Fixed-Wing Aircraft. By far, the most prevalent area of aviation is the fixed wing. However, the variable-duct technology has not yet been applied to the market economy and probably will not be in the near future.

Helicopters. Since their general rise during the Vietnam War, helicopters have become indispensable in a number of fields. Air ambulance and mercy flights allow accident victims and the ill to reach emergency-care facilities within minutes. Air ambulances can be outfitted with the same trauma equipment as a ground ambulance, and they have the added benefit of conveying trained trauma personnel directly to the scene of the accident or injury allowing for quicker diagnosis and treatment.

Originally, this trauma-care element was assigned to police and sheriff's departments, military guard, and reserve units in the United States. But as the use of the helicopters expanded, law-enforcement air units could not keep up with the specialized requirements of trauma and continue to respond to calls for assistance from ground units. Police helicopters shed some of the lifesaving equipment but gained other specialized equipment for their duties. Forward-looking infrared radar (FLIR) allows the helicopter to track subjects at night in a variety of chase and surveillance scenarios. The Nightsun searchlight provides light equal in brightness to daylight, and when it is used at night, it frees those law enforcement officers on the ground to conduct their searches without using flashlights.

Fire departments in many of the larger cities have added helicopters to their inventory of rescue equipment. In some cases, where law enforcement were shedding the medical jobs and the hospitals could not bear the loss of the service, the fire department stepped in to do the job. Fire departments have also found that in some cases they can use the helicopter as a rescue vehicle for individuals who have become trapped or to carry hoses up above the flames where the truck's ladder could not reach.

Heavy-lift helicopters are used to carry things to and from the tops of buildings, where they can hover in one position long enough for the item to be attached or removed from the top of the building without having to make multiple trips via crane. They are also used in logging to carry felled trees from one remote location to a central assembly point for collection or transportation to a mill.

Helicopters are still the best way for damage assessment to be done after, or in some cases, during an emergency. By placing evaluators near the scene or flying dignitaries over it, they can decide how best to respond to the emergency or even speed up the money flow from the government to the disaster site.

Helicopters have also impacted such diverse fields as fishing, crop dusting, intercity and short-range transportation, electricity, and broadcast journalism.

Autogiros. Autogiros have become more of an amusement on film and at air shows than a viable part of the transportation arena. However, there are still enthusiasts who fly these small aircraft for pleasure.

Hybrids. Primarily used by the military, the hybrids are being considered as a possible replacement of the helicopter and fixed wing particularly for civilian transportation. The hybrids are able to land in small areas, like a helicopter, but they also have the speed of a fixed wing for level-distance flying.

Impact on Industry

Aircraft use can be broken down into two major markets: military and civilian. The military has three major categories of use: fixed-wing for combat and transport, rotary-wing for combat and transport, and the emerging hybrid aircraft. The civilian market has two areas for use: transport of cargo and people.

Since the helicopter emerged as a major military component during the Vietnam War, Bell Helicopter has been the prime provider to the Army and Air Force. The UH-1, probably the most famous example, is being phased out by the UH-60 Black Hawk from Sikorsky Aircraft. These are the Army's personnel and cargo workhorses.

The scout field is still being dominated by the

aging yet continuously updated OH-58 Kiowa from Bell and the Hughes 500C Little Bird.

An attack helicopter for all services continues to be the updated versions of the Bell AH-1 Cobra. Created for tactical-battlefield and front-line support during the Vietnam War, these are being replaced by the Apache attack helicopter in the Army aviation units. The Marines continue to use the upgraded versions of the Sea Cobra for their attack helicopter.

The Boeing CH-47 Chinook, a medium-lift helicopter, is the largest-capacity aircraft used by the Army and Air Force for personnel and cargo transport. The Sikorsky CH-54 Sky Crane was the only heavy-lift helicopter in the inventory but was phased out at the end of the Vietnam War because it was too unwieldy. Variants of the Sky Crane are still used in the construction and logging industries.

While the Navy and Marines also use many of the same helicopters as the Army and Air Force, the primary workhorses have been the SH-3 Sea King and the CH-46 Sea Knight. These medium-lift helicopters are used to transport personnel and cargo.

The major deviation is the increasing use of the V-22 Osprey by the Marines. This hybrid aircraft has the variable wing design, which allows it to hover like a helicopter as well as travel and maneuver like a fixed wing between sites.

The civilian market has embraced the helicopter primarily as a means of passenger transport in major cities. It is also used to transport personnel to and from oil rigs based in coastal waters. Bell initially dominated this market with its Bell 222. Cost and maintenance times have contributed to the decline in this market. Trans World Airlines tried using medium-lift helicopters to transport passengers from a heliport in New York City to the major New York airports. Some air-taxi work is still being done on an independent basis, but the commercial airlines no longer provide this service.

Since ground transportation of cargo is still more cost effective, construction, logging, and flights of vital materials (organs, for instance) still tend to dominate this civilian sector. One interesting use is the checking and repair of high-voltage cables. The helicopter hovers next to the cable while a technician does the extremely delicate work of attaching diversion cables to the voltage cable.

Fascinating Facts About Vertical Takeoff and Landing Aircraft

- One of the first successful helicopter flights was made in 1907 by Frenchman Paul Cornu. His flying bicycle rose to a height of about one foot and was able to hover for twenty or thirty seconds.
- In 1936, Germans Heinrich Focke and Gerd Achgelis developed the Focke-Achegelis Fa 61, which was made famous by twenty-five-year-old female test pilot Hanna Reitsch. Reitsch made several test flights in the Fa 61 and flew the aircraft with precision inside the Deutschlandhalle stadium in Berlin in front of awed crowds.
- The most successful helicopter built by Argentinean inventor Raul Pateras de Pescara had sixteen lifting surfaces on two separate counter-rotating rotors and was powered by a 40-horsepower engine with a pusher propeller. The design had one flaw: The pilot sitting under the rotors faced backward.

Careers and Course Work

The most difficult skills to develop in flying are the coordination of movements between the feet, hands, eyes, and the controls. Being able to go seamlessly back and forth between all of these are the basics of a smooth, coordinated flight. In flying helicopters, it is called "finding the hover button." At first, developing these skills is difficult, but with time and experience, they become second nature.

Pilot is the first job that comes to most people's minds. Many helicopter pilots have military backgrounds. The services are usually looking for people with strong math, science, computer, or engineering backgrounds. The Army has less stringent academic requirements because of the extensive use of warrant officers. Warrant officers become experts in a variety of areas within the aviation field including airframe repair, mechanical and engine maintenance, and airfield operations. Upon graduation from a military flight school, one can take the written exam given by the Federal Aviation Administration (FAA) to qualify as a commercial pilot in fixed, rotary wing, or both classifications.

To become a pilot without military training, flying either fixed- or rotary-wing aircraft, one needs the following: a private pilot's license, commercial pilot's license, instrument rating, airline transport pilot certificate (ATP), and a rating for the type of aircraft one will be flying. Many pilots who will be flying for small charter services, product transport services, news agencies, or police agencies may need no more than a commercial pilot's license. For training in rotary-wing aircraft, individuals usually have to attend a private specialty company's ground school then follow up with the flight portion in a separate instruction.

Vocational schools and some two-year colleges train students with specialized skills in airframe construction, power-plant development and testing, maintenance, avionics, and other assorted disciplines applicable to both fixed- and rotary-wing aircraft. Four-year colleges and universities incorporate some aspects of these programs but concentrate on the management side of the aviation industry.

Social Context and Future Prospects

Statistical studies from the FAA and National Transportation Safety Board (NTSB) show that flying is usually safer than driving an automobile. Society has accepted flight as a standard way of commuting from place to place and will put up with the stress and hassles associated with the use of fixed- and rotary-wing aircraft.

Though not yet a reality, a number of visionary inventors and companies are planning for the replacement of existing automobiles with sky cars: hybrids that run on the ground and in the sky. Moller International SE has a mock-up of a sky car with four ducted engines (one on each fender) that it claims will be able to hover, translate to forward flight, and land to a hover again, just like the V-22 Osprey. Other companies working on these kinds of vehicles, such as Terrafugia, are testing long-held design concepts of the automobile as a fixed wing. Alfa Romeo is departing from even that concept by showing the Spix, its concept car, as an air-cushion vehicle.

Traditional helicopter companies are trying to develop single-design helicopters that can be versatile in a number of different roles and configurations. The latest designs in small helicopters have varying interiors and are meant to be used in a variety of different situations.

Michael L. Qualls, M.A., LTC, USAR, Ret., CHS-V

Further Reading

American Institute of Aeronautics and Astronautics. *A Photo History of Experimental VSTOL Aircraft and Their Contributions.* Reston, Va.: Rob Ransone. Available at http://www.aiaa.org/pdf/industry/presentations/xvehicles02ransone.pdf. A short history of this type of aircraft.

Chant, Christopher. *Aviation: An Illustrated History.* London: Orbis, 1983. Covers all areas of aviation in a very comprehensive manner.

Dorr, Robert F. *Chopper: Firsthand Accounts of Helicopter Warfare, World War II to Iraq.* New York: Berkley Books, 2005. An excellent reference for the earliest use of helicopters by the U.S. military in combat situations.

Gunston, Bill. *An Illustrated Guide to Military Helicopters.* Englewood Cliffs, N.J.: Prentice Hall, 1986. Depicts and describes in detail fifty-one military helicopters up through the 1980's.

Jackson, Robert, ed. *Helicopters: Military, Civilian, and Rescue Rotorcraft.* San Diego, Calif.: Thunder Bay Press, 2005. A well-illustrated book that breaks down the individual information of each helicopter that has been used up to 2005.

Web Sites

AHS International: The Vertical Flight Society
http://www.vtol.org

Globalsecurity.org
Vertical and Short Takeoff and Landing Aircraft V/STOL
http://www.globalsecurity.org/military/systems/aircraft/vstol.htm

Helis.com
http://www.helis.com/introduction/prin.php

U.S. Centennial of Flight Commission
http://www.centennialofflight.gov/index.cfm

See also: Air Traffic Control

VITICULTURE AND ENOLOGY

FIELDS OF STUDY

Culinary science; chemistry; bacteriology; botany; horticulture; agriculture; genomics; geology; microbiology.

SUMMARY

Viticulture and enology are fields involved in the cultivation of grapes and the process of winemaking. Specifically, viticulture is the science of growing and cultivating grapes, and enology is the study of wine and winemaking (viniculture refers to the science and process of growing grapes for winemaking). The international wine industry uses techniques developed in antiquity in conjunction with modern research and technology in the effort to maximize the quality and flavor of the finished product. Although many winemakers strive to maintain the traditions of their ancestors, others strive for innovation, creating blends of grapes and hybrid grape varieties with new flavors. The single botanical species used to make wine, *Vitis vinifera* (wine grape), has led to the cultivation of more than 15,000 wine varieties. In the twenty-first century, winemaking occurs on every continent except Antarctica. Worldwide, grape cultivation and winemaking are a multibillion dollar industry, and many universities offer graduate level courses in viticulture and enology.

KEY TERMS AND CONCEPTS

- **Brandy:** Spirit liquor made by distilling grape wine or fermented fruit juice.
- **Cultivar:** Plant created through artificial selection that has been given a unique name because of its distinct characteristics.
- **Distillation:** Process used to purify a liquid by boiling and condensing vapors.
- **Ethyl Alcohol:** Alcohol produced during the fermentation of wine that is responsible for the wine's intoxicating effect.
- **Fermentation:** Process of converting sugar to alcohol, generally involving an enzymatic process occurring in response to the growth of yeast.
- **Finish:** Final phase in wine tasting, it describes the last sensory impression of a wine's flavor when swallowed. A wine's finish will take into account the wine's weight, how long the impression lasts, and the aftertaste.
- **Fortified Wine:** Wine blended with alcohol, typically brandy, to increase the sweetness and alcohol content.
- **Hybrid:** Grape variety created by blending two or more existing varieties of grapes to create unique characteristics.
- **Sommelier:** Specialist in all aspects of procuring, storing, and serving wine, generally employed in the fine dining industry.
- **Varietal:** Wine principally made from one type of grape and given the name of that grape variety.

Definition and Basic Principles

Viticulture is the science of growing and cultivating grapes and grape vines, which includes planting grape seeds, training and maintaining grape vines, and harvesting grapes for the production of grape juice, wine, and other products. Enology is the study of wines and winemaking, which involves the fermentation of grapes and other methods used to create wine and impart specific flavors to the finished product. Viticulture and enology are the backbone of the wine industry, which also includes the commercial production, sale, and evaluation of wine products. Research in enology is used extensively in the culinary sciences, especially in regard to learning how to pair specific wines with certain foods and ingredients, and in teaching wine appreciation to enthusiasts. In addition, enologists and viticulturalists study the factors that contribute to wine flavor and the diseases that affect grapes and grape products.

Background and History

Viticulture and enology have been part of human culture for nearly 8,000 years, when the first wines were made in the ancient Near East (modern Middle East region). The Romans laid down the foundations for grape classification and the aging process of wine. They also pioneered the use of oak fermentation chambers (wooden wine barrels) and cork for bottling. (The earliest extant prose written in Latin is, in fact, a survey on Roman viticulture.) Winemaking was nearly exclusive to Europe until the mid-nineteenth century, when it spread to the Americas.

In 1863, Emperor Napoleon III of France asked chemist Louis Pasteur to investigate the factors that led to deteriorating wine quality. Pasteur's work was a landmark in understanding fermentation and led to the development of enology. The results of his research quickly spread around the world, and in the 1880's, the University of Bordeaux in France and the University of California established the world's first institutions for the study of enology.

During the twentieth and early twenty-first centuries, most of the innovations in wine came from the development of industrial winemaking procedures. Advancements in sanitation during this time also led to the cultivation of higher-quality, longer-lasting wines.

How It Works

The production of wine during a single year, called a vintage, involves three basic components: cultivation, processing and fermentation, and finishing.

Cultivation. Photosynthesis starts the process of converting carbon and chlorophyll to sugar, which is stored in the grapes. There is a target ripeness level for each grape variety, depending on the flavor desired. Viticulturists use tools such as spectrometers to measure the sugar level in the grapes to decide when to harvest them. Variations in environmental factors such as climate and soil quality affect the flavor of the grapes. The ideal condition for grape cultivation consists of warm, relatively dry days alternating with cool nights. Places where these conditions occur naturally, such as Northern California's Russian River Valley, produce the highest-quality wines. Deciding when to harvest is the next variable. Harvesting directly following rain may lead to watery grapes, while harvesting following a dry spell leads to dry grapes with overly intense flavor. Although many industrial winemakers harvest grapes with combines that can pluck thousands of plants in a short time, the highest-quality wines are produced from hand-picked grapes.

Processing and Fermentation. After being harvested, the grapes are juiced. They are placed in a large container with a spout connecting to the bottom. When the container is filled to capacity, gravity will begin to crush the grapes, producing what is called free-run juice. The highest-quality wines are developed from this free-run product. Next, winemakers use air bladders or plungers to press the grapes. With each round of pressing, as liquid from stems, seeds, and skin is filtered into the juice, the quality of the juice declines.

To achieve fermentation, yeast is added, which metabolizes the sugars in the juice and creates alcohol as a by-product. Traditionally, the juice is placed into oak containers, which impart a unique flavor, for fermentation. Modern wineries may also use stainless steel or polished concrete fermentation chambers, although most conduct a portion of their fermentation in the traditional oak containers.

At this point in the process, all wine has a pale color. To make the wine red, the juice is soaked with the grape skins. This imparts a red color to the juice, a process called cold soaking. This process may also increase the complexity, or aroma, and flavors of the wine. To prevent fermentation during cold soaking, the soaking chambers must be kept at cold temperatures—typically between 4 and 15 degrees Celsius (39-50 degrees Fahrenheit). To achieve this, some winemakers cold soak in underground chambers, and others use dry ice or industrial cooling units. After soaking, the temperature is raised to start fermentation.

Finishing the Wine. Once fermented, the wine is filtered to remove sediment and other materials. High-quality wine is filtered through a relatively loose filtration system, and some sediment is left with the wine so as to continue imparting flavor and characteristics. After filtration, wine may be bottled or transferred to oak storage vessels for aging. Many fine wines are stored in oak for periods ranging from a few months to a few years, allowing the flavor to change and develop. Even if the wine is immediately bottled, the wine must rest for several weeks or months within the bottle to achieve a palatable or desired flavor.

Applications and Products

Cultivars. The vast majority of wine is made from grapes that are varieties of a single species, *Vitis vinifera*, the noble or Eurasian grape. Although there are many other grape species, few make satisfactory wines. Most wines are made from varieties of *V. vinifera*, including Merlot, Cabernet Sauvignon, Pinot Noir, Syrah and Shiraz, and Sangiovese, which are among the most important varieties of grapes for red wine; Chardonnay, Pinot Gris, Riesling, and

Sauvignon Blanc are among the most popular varieties for white wines.

Grape Liquors. In addition to wine, grapes can be used to produce a variety of other liquors. Champagne and sparkling wine are made either by injecting carbon dioxide into already-fermented wine or by adding more yeast and sugar to the wine to induce a second fermentation process. With the latter method, the wine is then capped in a special container that traps the carbon dioxide that is released as a by-product of fermentation.

Brandy is another liquor produced by distilling grapes. To make brandy, the fermented juice is boiled until alcohol vapors rise from the liquid, at which point the vapors are captured and recondensed into a spirit. This process, called distillation, can also be used with other types of fermented fruits or vegetables to make spirits.

Fortified wines such as port and sherry have high alcohol content and are typically sweet. During fermentation, winemakers add distilled brandy to the wine. The alcohol kills the yeast in the wine and stops fermentation, thereby preserving more of the sugar in the grapes and simultaneously adding to the alcohol content.

Tartaric Acid. Tartaric acid is an organic crystalline acid found in grapes and several other fruits. During wine fermentation, deposits of tartaric acid collect in the fermentation vessels. Winemakers collect and sell tartaric acid as a food additive. It is used as a solidifying agent in cheese making and added to some foods to increase bitterness. Because of its antioxidant properties, tartaric acid has also become a popular additive for health food and drinks.

Pumice. Pumice is a mat made of the grape leaves, skins, seeds, and stems that collect during the creation of red wine. Wine pumice is collected, dried, pressed into bricks, and sold to health and wellness companies for a variety of products. Because pumice is high in vitamins, minerals, and antioxidants, it has become a popular food additive and is used in a variety of healthful beverages and supplements. Wine pumice has also become exceedingly popular in the cosmetic industry. In vinotherapy, winemaking residue is used to treat the skin. Pumice, for example, may be ground and used in the creation of skin products and makeup. The French winery Château Smith Haut Lafitte is known for its line of beauty products created with its wine pumice.

Fascinating Facts About Viticulture and Enology

- Wine cultivation in the United States first occurred in the state of Ohio, but because of its climate, California became the capital of wine production and still produces 90 percent of the wines made in America.
- From the Middle Ages to the Renaissance, most European wine was produced by Christian monks who developed many of the techniques still used by winemakers.
- Many types of wine and liquor take their name from the area in which they are produced, such as Cognac, a type of brandy first produced in the Cognac region of France.
- It takes about 2.5 pounds of grapes to make an average-sized bottle of wine.
- Of the more than four hundred species of oak, only twenty are considered acceptable for making wine barrels. Winemakers often use oak barrels that have been used for more than a century to ferment wine.
- Ancient Egyptians used not only a variety of grapes but also pomegranates, figs, and dates to produce wine. Pharaohs were entombed with wine, ensuring its consumption in the afterlife.
- The Greeks were largely responsible for spreading wine to much of the ancient world, establishing a wine industry in Western Europe. They even dedicated a god to wine–Dionysus.

Impact on Industry

The wine industry is in a state of growth. Emerging superpowers India and China have begun producing higher-quality wine in the twenty-first century, and wine is produced in all fifty states in the United States. For winegrowers, the market is generally always difficult—enologists and commercial winemakers, however, enjoyed growth in the first decade of the new millennium. Although the 2008-2009 global economic crisis negatively affected the fine wine industry, the production of affordable table wines has increased rapidly.

Wine is subject to regulations that govern alcoholic beverages and other intoxicating substances. In the United States, the Food and Drug Administration (FDA) is responsible for regulating wine, but many states have their own laws governing the production and sale of wine and liquor. In the state of Pennsylvania, for instance, wine and spirits may be distributed only by government-licensed facilities, while other states allow grocery stores to sell wine and liquor.

Careers and Course Work

There are many roads to achieve careers in enology and viticulture. Enology involves knowledge of chemistry, microbiology, bacteriology, and botany. Viticulture involves extensive knowledge of horticulture, and experience in geology and climatology is also beneficial. There are many university programs available in horticulture and botany, and some schools offer courses in enology. Individuals with backgrounds in microbiology, botany, or agricultural studies may choose to focus on viticulture or enology during postgraduate studies. Some wine retailers and culinary education programs also offer classes in viticulture and enology.

Professional enologists may be employed by wine manufacturers or may conduct research for a university. The University of California, Davis, for example, provides training in viticulture and enology at the undergraduate and graduate levels and also employs researchers and research technicians. Funding for wine research can be obtained through a variety of organizations, private sources, and sometimes government and public granting agencies. Several medical agencies have also promoted enological studies, such as the medical benefits of wine and other grape products.

Social Context and Future Prospects

The impact of wine on culture differs widely from country to country. In France and Italy, wine is considered a staple of everyday life, and many Europeans are daily wine consumers. In 2008, Italy overtook France as the world's largest winemaker. In the United States and most of Asia, wine is considered more of a delicacy than a part of a daily routine. Though traditions differ, wine is collected and consumed in nearly every country around the world.

A major debate in the wine industry in the early twenty-first century concerns the use of cork for bottling. Cork has been used as a wine stopper for centuries because of its light and elastic nature—its compressibility allows it to form tight seals—and because it is considered impenetrable to gases and liquids. The problem with cork, however, is that it can be the source of bacteria within the wine or allow leakage. Some wine experts feel that glass or plastic corks may soon overtake natural corks as the most common type of sealant for bottled wine. Other closures gaining ground in the industry include screwcaps.

Micah L. Issitt, B.S.

Further Reading

Goode, Jamie. *The Science of Wine: From Vine to Glass.* Berkeley: University of California Press, 2007. A basic instruction to winemaking and the scientific aspects of enology written for the general reader. Includes information about wine industry debates and controversial issues in the twenty-first century.

Johnson, Hugh. *The Story of Wine.* London: Mitchell Beazley, 2004. An introductory history of wine with information about wine varieties, the development of the industry, and the development of enology as a scientific discipline.

Paul, Harry W. *Science, Wine, and Vine in Modern France.* New York: Cambridge University Press, 2002. Provides information on the history, industry, and development of winemaking and scientific enology in France and around the world, as well as information about wine varieties and careers in the winemaking industry.

Pellechia, Thomas. *Wine: The Eight-Thousand-Year-Old Story of the Wine Trade.* New York: Running Press, 2007. Presents a history of the wine industry from its earliest manifestations to twenty-first-century industrial wine production, along with information on the scientific study of wine and the state of wine research.

Ribereau-Gayon, Pascal, et al., eds. *The Enology Handbook.* Vol. 1. Hoboken, N.J.: John Wiley & Sons, 2006. An introduction to the science and study of wines and winemaking, which requires some knowledge of microbiology and chemistry for full comprehension. Includes information on the history of the wine industry.

Sommers, Brian J. *The Geography of Wine.* New York: Plume, 2008. A basic introduction to the science and art of viticulture. Includes information on microclimate, soil evaluation, harvesting, and the history of grape cultivation.

Web Sites

American Society for Enology and Viticulture
http://asev.org

Wine America: The National Association of American Wineries
http://www.wineamerica.org

See also: Food Science; Horticulture; Zymology and Zymurgy.

WATER-POLLUTION CONTROL

FIELDS OF STUDY

Chemistry; biology; microbiology; economics; hydrology; geology; botany; urban planning; agronomy; animal husbandry; business management; physiology; law; accounting; computer programming; engineering.

SUMMARY

Water-pollution control is a multidisciplinary field that endeavors to reduce or eliminate discharge of chemical and biological compounds into natural waterways. Legal standards set a level playing field, so that the economic burdens and opportunities fall with approximate equality on all enterprises, industries, and municipalities.

KEY TERMS AND CONCEPTS

- **Acid Rain:** Atmospheric emissions of sulfur dioxide and nitrogen oxides, mostly generated by burning fossil fuels, that fall in wet or dry form into bodies of water. When these gases react with water and oxygen, a solution of sulfuric acid and nitric acid is the result. The increased acidity can make certain waters uninhabitable for a variety of plant life and fish.
- **Adsorption:** Surface phenomenon based on electrostatic attraction and repulsion between molecules, useful in separating pollutants from water prior to discharge into a stream, lake, or sea.
- **Best Available Technology:** Most effective technology available to control emissions, which polluters are required by the Federal Water Pollution Control Act Amendments of 1972 to install, unless its use is shown to be infeasible for energy, environmental, or economic reasons.
- **Biological Oxygen Demand (BOD):** Quantity of oxygen consumed by aerobic microorganisms during metabolism of organic wastes.
- **Coliform:** Family of bacteria that live in the digestive tract of animals. In pollution-control monitoring, the count of coliform bacteria in a body of water is used as a measure of organic waste, particularly fecal matter, present in the water. Coliform bacteria themselves are generally harmless, except for the specific strain, *E. coli O157:H7,* and serve as reliable indicators for the presence of harmful pathogens.
- **Dissolve:** Process of one substance (the solute) becoming equally distributed (homogenous) throughout another (the solvent). In water pollution, the relevant applications are that many solids dissolve in water, as do some gases, particularly oxygen. Dissolving is a purely physical process, not one that forms new chemical bonds.
- **Eutrophication:** Process in any body of water, particularly a lake, in which growth of algae or other aquatic plants builds up decomposed plant material at the bottom, eventually converting the area to dry land. A slow natural process, which can be accelerated by phosphate and nitrate pollution.
- **Heavy Metals:** Any metal with an atomic number greater than iron, number 26 on the periodic table. The United States Geological Survey Trace Elements National Synthesis Project monitors groundwater for the presence of heavy metals.
- **Phosphates:** Any of several salts of phosphoric acid, used in detergents and fertilizers, and a critical component for energy storage in living cells. When excess phosphates enter natural water systems, they promote overgrowth of aquatic plants, speeding up eutrophication, and sometimes creating "dead zones" where there is insufficient oxygen to support fish and other animal life.
- **Precipitate:** Solid that forms out of a solution, the chemical process opposite to dissolve, when the concentration of a solute is greater than the saturation point for that substance in a solvent.
- **Sludge:** Any semisolid, semiliquid material, but in water pollution the organic portion of sewage, which is separated out for treatment. Sludge may, however, retain suspended solids such as heavy metals that

remain toxic even after organic compounds have decomposed into harmless substances.

- **Suspend:** When small solids are not actually dissolved in water but retain their own distinct integrity and are carried by the water and have sufficiently low specific gravity that they do not immediately sink to the bottom.

Definition and Basic Principles

Water-pollution control involves preventing excessive contaminants from reaching any usable watercourse. Understanding all the various contaminants that may enter a body of water is an essential part of this subject. It is also necessary to understand the extent of the reactions of receiving waters to wastewater discharges.

Water pollution itself has such a commonly understood everyday meaning that definition can appear almost unnecessary. However, scientific measurement and legal enforcement require precise definition. The Clean Water Act defines "pollutant" as "dredged spoil, solid waste, incinerator residue, sewage, garbage, sewage sludge, munitions, chemical wastes, biological materials, radioactive materials, heat, wrecked or discarded equipment, rock, sand, cellar dirt and industrial, municipal, and agricultural waste discharged into water" (33 U.S.C. Sec. 1362). Toxic pollutant is separately defined to mean those that "after discharge and upon exposure, ingestion, inhalation or assimilation into any organism, either directly from the environment or indirectly by ingestion through food chains, will . . . cause death, disease, behavioral abnormalities, cancer, genetic mutations, physiological malfunctions" in organisms or their offspring. The act then references a table of toxic pollutants assembled by the Committee on Public Works of the House of Representatives and authorizes the administrator of the Environmental Protection Agency (EPA) to add additional pollutants to the list.

Organic wastes include fecal material from humans and animals, either from municipal sewage systems or concentrated livestock agriculture. Most water naturally has some minerals dissolved in it, which vary by locality. When higher concentrations of iron, copper, chromium, platinum, mercury, nickel, zinc, and tin are discharged into waterways because of mining or industrial operations, the concentration can be toxic to aquatic life and cause metabolic diseases or cancer in humans who drink or bathe in the water. Discharges of petroleum and a tremendous number of oil-derived hydrocarbon compounds are toxic or carcinogenic. Common fertilizers, based on phosphorous and nitrogen compounds, cause overgrowth of some aquatic plants, particularly algae.

Background and History

Cities have had sewers and drains since ancient times, but only to flush raw sewage into nearby rivers to be carried away by the current. This certainly contributed to many plagues and epidemics, including the spread of cholera in the nineteenth century. From the dawn of the industrial revolution, manufacturing and mining have routinely been located near sources of water, because water is used in industrial processes and water sources are convenient places to dump waste products. In the United States, concern that refuse and pollutants were creating obstacles to safe navigation inspired the federal Rivers and Harbors Appropriation Act of 1899. Not until the Federal Water Pollution Control Act of 1948 did the first federal legislation tentatively address water pollution as a hazard to human life and safety. The Federal Water Pollution Control Act Amendments of 1972 established the first comprehensive regulatory framework for water-pollution control. Similar measures emerged in North America and in Europe, Australia, and Japan, during the 1960's and 1970's. There are some nineteenth century precedents, including attempts in Britain to regulate major manufacturing sources of pollution and secure clean water for household use, German legislation allowing local authorities to place conditions on methods of manufacturing, and Dutch laws aimed at controlling hazards and nuisances from industrial sources. Still, the German Civil Code of 1873 was typical of the era: It set forth that individuals should tolerate pollution as it was essential to economic development.

How It Works

Water-pollution laws in the United States, since adoption of the Clean Water Act of 1972, begin with the baseline standard that the discharge of any pollutant by any person into any navigable waterway, or tributary flowing into a navigable waterway, is prohibited. Since immediate adherence to this standard could prove physically and economically infeasible, the law then provides that any exception must be authorized by a permit, issued under the National

Pollutant Discharge Elimination System (NPDES), either by the EPA or an authorized state agency. The type of pollutants allowed, concentration, and total amount, are specified in the license. The standard for issuing permits requires use of the best available technology to reduce or eliminate water pollution. The ruling agencies have also weighed economically practical cost.

Europe, Japan, and Developing Nations. Similar standards and enforcement processes are used in Western Europe and are mandated by the European Union–however, many Eastern and central European countries have been slow to adopt best available technology standards, for fear that their economies cannot afford the cost. British pollution-control laws mandate best available technologies not entailing excessive cost (BATNEEC). Japan intensified pollution control from 1965, when industry was investing less than ¥50 billion per year, to 1976, when investment per year rose to ¥1 trillion, about 0.9 percent of the increase in gross domestic product (GDP) during this period. In Italy, at the close of the twentieth century, half of municipal sewage was discharged without treatment into receiving waters. In Latin America as a whole, only about 11 percent of wastewater is properly treated. Developing nations generally lack comprehensive regulatory laws, and up until the turn of the twenty-first century developing economies such as Brazil, China, India, and South Africa had not placed great emphasis on pollution control. These countries are beginning to make significant new investments.

Treatment Methods. Three common types of treatment are known as primary treatment, secondary treatment, and tertiary treatment. Primary treatment extracts pollutants that will settle out by grit removal, screening, grinding, and sedimentation. Secondary treatment utilizes the natural degradation of organic wastes by biological oxidation, using activated sludge, trickling filters, and oxidation ponds. Tertiary treatment involves a wide range of chemicals to remove substances that remain in water after primary and secondary treatments. Sludge removed in either of the first two stages must be disposed of, sometimes by incineration, sometimes by landfill. Tertiary treatment uses chemical and electrical processes to remove more elusive nonorganic pollutants.

Point and Nonpoint Source Pollution. Point source pollution comes from a discrete, easily identifiable source, such as a pipe. Responsibility can be established, chemical content and volume measured, and controls put in place. Nonpoint source pollution enters a watershed through runoff from large areas, such as agricultural land, paved urban surfaces draining into storm drains, even lawns and golf courses, or in precipitation, as in the case of acid rain. While most federal law addresses point source pollution, Section 319 of the Clean Water Act authorized the EPA's Nonpoint Source Management Program (NSMP), which provides grants to states, territories, and tribes, and calls for state-level plans to implement measures for nonpoint source pollution control.

APPLICATIONS AND PRODUCTS

The two sources of water pollutants most effectively targeted by NPDES are manufacturing and domestic waste. Most of the latter enter municipal sewage systems, but so does some manufacturing waste. Industries are sometimes required to pretreat their effluent before discharging it into municipal sewers. This is more difficult to enforce than a specific factory with its own discharge system into a river or lake.

Physical Technology. One of the first steps in primary treatment is a screening system to remove large objects. Removing grit often requires a conveyer system, which allows sand and dirt to settle out of the water stream. Activated sludge process is one of the most visually familiar to anyone who has viewed a common wastewater-treatment plant: large concrete tanks with walkways between them, continually agitated by rotary aerators to mix oxygen into the sludge, which hastens biological processes to break down organic compounds. Thermal conditioning units, a kind of pressure cooker, digest sludge and separate the majority of water from the sludge. Sludge is further dried by vacuum-filtration equipment–a rotating roll spread with sludge with a vacuum applied through the roll to extract more water, leaving a semidried cake that falls off. This can be disposed of in a landfill but is often incinerated.

Chemical Processes. Tertiary treatment uses many methods to remove pollutants that do not settle out in primary or secondary treatment. These include coagulation, flocculation, nitrification, denitrification, adsorption, ion exchange, and electrodialysis. These have high capital investment and operation costs and require highly skilled personnel for operation and maintenance. Flocculation removes solids suspended in water–but it is not effective in removing dissolved compounds. The name derives from "floc," a kind of clumping of different precipitated materials. Generally, one or more chemicals are added to the water being treated, which form a gel or other precipitate that attracts other suspended materials. One common combination is potassium aluminum sulfate and ammonia, which react to form aluminum hydroxide. Adsorption introduces powdered activated carbon to remove even small traces of pollutants. Since there is no chemical bonding, recovered pollutants can be collected and cycled back into industrial processes, offsetting costs or even generating a net profit. Adsorption technology is appropriate for removal of heavy metals and complex organic compounds, including pesticides, silicates, aluminum compounds, wood- and petroleum-based carbon compounds, and resins.

Impact on Industry

Industrial managers often view pollution controls as a burden on commerce. New pollution controls require a substantial up-front capital investment. Complying with requirements for NPDES licensing involves substantial paperwork and record keeping. There are also long-term advantages. The supply of water available for industrial use is improved. The paper industry has long been condemned as a major polluter because its effluent includes suspended solids, dissolved organic and inorganic compounds, sulfur compounds, chlorine, and resin. Waste fiber and wood sugars dissolved in sulfite liquor and discharged into rivers forms a sludge. Paper mills have generally been situated on rivers because a large volume of clean water is essential for turning wood pulp into high-quality paper products–as many as 40,000 gallons are used for one ton of paper.

China. The Academy of Sciences estimated that while China's economic growth in 2005 increased GDP by 11.3 percent, exploitation of natural resources and environmental pollution that year cost 13.9 percent of GDP. China had sixteen of the twenty most polluted rivers in the world at the dawn of the twenty-first century, which has contributed to making cancer the leading cause of death in China. This is in part the product of rapid privatization, combined with continued state supervision of economic development. Pollution emissions control was highlighted in the Eleventh Five-Year Plan (2006-2010) with 1.35 percent of GDP allocated to improved environmental protection. The economic stimulus measures announced in November 2008 allocated $41 billion (almost 270 billion yuan) out of a total of $586 billion for sewage treatment, which is projected to reach 90 percent of China's counties in the subsequent three

Fascinating Facts About Water-Pollution Control

- The Godavari River in southern India has a permanent layer of aluminum hydroxide on the surface. Aluminum hydroxide, one of the toxic wastes from the paper industry in Andhra Pradesh, has completely destroyed aquatic life in the Godavari.
- Emission of organic water pollutants in China in 2004 amounted to more than 6 million kilograms per day. The United States was the only other country in the world with emissions of more than 2 million kilograms.
- In 2000, 11,671 community water systems in the United States, serving 204.1 million people, relied on surface water as their source of supply.
- The Great Lakes contain 6 quadrillion gallons of fresh water, one-fifth of the world's freshwater supply, and 84 percent of the surface water in North America.
- In 2007, the Environmental Protection Agency (EPA) estimated that more than $1 trillion in treatment systems for public water and wastewater were needed in the United States.
- Ammonium perchlorate, a toxic ingredient in rocket fuel, can be found in drinking water, groundwater, or soil in forty-two states. In 2008, nearly 75 percent of common foods and beverages were found to be contaminated with ammonium perchlorate.
- In 22,083 tests performed between 2004 and 2007, drinking water in Houston, Texas, was found to carry forty-six pollutants, eighteen of them at concentrations that exceeded health guidelines.
- In 2008, industries in the United States released a total of 3.8 billion pounds of the 492 toxic wastes on the EPA's list of regulated chemicals. This included 404 million pounds of hydrochloric acid, 214 million pounds of copper compounds, 280 million pounds of nitrate compounds, and 718 million pounds of zinc compounds.

years. The Chinese government candidly admitted that this was the first nationwide sewage-treatment program.

India. Comprehensive regulation is lacking in India. The country has initiated projects such as the Central Ganga Authority in 1985 with Prime Minister Rajiv Gandhi as chair, with a onetime plan to achieve demonstrable improvement in water quality administered through the Ganga Project Directorate within the National Ministry of Environment and Forest. The plan included research on the sources and nature of pollution in the Ganges River, a more rational plan for agriculture, animal husbandry, fisheries, and forest, and financing of treatment and economical use of waste. Sewage-treatment plants, the first ever built in India, were subject to repeated construction delays. However, most planned treatment plants, in 261 sub-projects, were in place by 1998.

Careers and Course Work

A wide variety of careers are available in the public, private for-profit, and private not-for-profit sectors. These include design, manufacture, and operation of wastewater-treatment facilities, both by municipal government and industrial sources. Most government jobs relate to enforcement of water-pollution control laws, preparation of long-term strategies, or the review and funding of research projects. Water is one of four program divisions within the EPA, which also has a research and development program concerned in part with water pollution. EPA includes ten regional offices and laboratories and a network of laboratories and research centers. A large portion of careers in this field involve measuring existing pollution, establishing what is and is not a hazardous substance and what concentrations are tolerable, advocating for new measures, and litigation.

Undergraduate degrees in environmental science typically require courses in biology, natural history, soil science, geology, chemistry, physics, calculus, instrumental methods of analysis, scientific presentations, and interdisciplinary courses such as principles of life sciences, understanding the earth, and ecology. Some colleges offer specific courses on pollution-control technology, toxicology, water-quality analysis and management, solid-waste management, industrial hygiene, and water-pollution biology. In addition to graduate programs in environmental science, many law schools offer degree programs in environmental law, including courses on water-pollution control, natural resources law, regulation of toxic-substance risk, and environmental issues in business transactions.

Environmental law is a considerable specialty, which requires a law degree for attorneys, and for paralegals, course work with specific emphasis on environmental law, preferably with some relevant science courses. Understanding the facts to be litigated requires at least a basic undergraduate familiarity with chemistry and biology. Legal practice options include working for government enforcement agencies at the state or federal level, defending alleged polluters against administrative or judicial actions, advising businesses on compliance, and representing advocacy organizations in civil actions.

Monitoring pollutants requires processing and analyzing huge quantities of data, which requires complex computer software. The EPA uses software called BASINS (Better Assessment Science Integrating Point and Nonpoint Sources) to perform assessments of water quality and of specific watersheds. Compliance with a variety of detailed regulations requires software both for enforcement and helping businesses integrate compliance into their daily operation.

Social Context and Future Prospects

There are literally hundreds of companies in the business of supplying instrumentation for water-quality monitoring, sanitation systems, water-storage systems, filtration systems, chemicals to neutralize pollutants (adding a basic compound to neutralize an acid or vice versa), piping and pumping equipment, and a variety of components and wastewater-treatment products. The market is not going to decline in North America, Europe, or Japan. It is likely to expand exponentially in developing nations such as China, India, Brazil, Nigeria, Indonesia, and South Africa. These countries, particularly China, will develop their own research and production. Finding new methods that would make reclaiming and recycling into profitable operations is an open area of research that has barely been touched. More efficient methods of recovery, which would bring the cost of reclaiming metals in particular below the cost of purchasing newly mined quantities, has yet to be thoroughly explored.

Charles Rosenberg

Further Reading

Abel, P. D. *Water Pollution Biology.* 2d ed. Philadelphia: Taylor & Francis, 1996. A good general reference on the impact of pollution on humans and ecological systems.

Dodson, Roy D. *Storm Water Pollution Control: Municipal, Industrial and Construction NPDES Compliance.* New York: McGraw-Hill, 1999. This text offers perspective on one of the most closely regulated areas of nonpoint-source water pollution.

Helmer, Richard, and Ivanildo Hespanhol, eds. *Water Pollution: A Guide to the Use of Water Quality Management Principles.* London: Spon, 1997. This book offers a global perspective with a summary of primary, secondary, and tertiary treatment methods.

McBeath, Jenifer Huang, and Jerry McBeath. *Environmental Change and Food Security in China.* New York: Springer, 2010. An up-to-date overview of environmental challenges in the world's fastest-growing and second-largest economy.

Sell, Nancy J. *Industrial Pollution Control: Issues and Techniques.* 2d ed. Hoboken, N.J.: John Wiley & Sons, 1992. This book is a more technical compilation of specific processes used in the ferrous and nonferrous metal industries, foundries, cement and glass manufacture, paper and pulp processing, and food industries.

Web Sites

Association for Environmental Studies and Sciences
http://aess.info

Environmental Protection Agency
http://www.epa.gov/waterscience

Society of Environmental Toxicology and Chemistry
http://www.setac.org

Water Environment Federation
http://www.wef.org

WATER SUPPLY SYSTEMS

FIELDS OF STUDY

Hydrology; meteorology; physical geography; civil engineering; climatology; hydrogeology; geology; water resources; water quality; environmental planning; environmental law.

SUMMARY

Water supply systems are designed to provide adequate amounts of water to the public for drinking, bathing, washing, and waste disposal. The water comes from surface water contained in aquifers, rivers, lakes, and reservoirs, groundwater from subsurface sources, and desalinated seawater. Industrial use and water for irrigation forms another large component of the overall water demand. The operators of these systems can be either public or private. In order to meet government standards for potability as drinking water, the raw water must be treated by a mix of mechanical, chemical, and biological means prior to delivery to the customer through a large system of pipes.

KEY TERMS AND CONCEPTS

- **Aquifer:** Water-bearing geologic formation that can yield large amounts of groundwater.
- **Desalination:** Making seawater potable by removing dissolved salts.
- **Exotic Stream:** Water that originates from a humid area and flows through an arid one, such as the Nile.
- **Groundwater:** Subsurface water that is found in the saturated zone in the earth at varying depths.
- **Low-Flush Toilet:** Toilet that uses only 1.6 gallons per flush. All residential toilets installed in the United States after 1994 were required to meet this standard.
- **Per Capita Consumption:** Amount of water used by each person per day, measured in gallons or liters.
- **Permeability:** Capacity of a rock formation or soil to move fluids.
- **Potable:** Suitable for drinking.
- **Reservoir:** Lake, basin, tank, or excavated space that is either natural, partly artificial, or wholly artificial, that is used to store water for drinking, flood control, power, or recreation.
- **Reverse Osmosis:** Removal of contaminants from raw water by forcing it through a semipermeable membrane.
- **Xeriscaping:** Landscaping method used in dry climates to conserve water by utilizing drought-tolerant plants, mulch, and efficient irrigation.

Definition and Basic Principles

Water supply systems comprise water sources, treatment plants, storage systems, and delivery systems to both commercial and residential consumers. These systems involve such apparatus as pumps, chemical feeds, testing laboratories, and networks of both surface and underground pipes. A water-supply facility must be large enough to meet both existing and anticipated future water demands within its service area.

A “community water system” (CWS), as defined by the Environmental Protection Agency (EPA), is a facility that serves water to a community, from small mobile-home parks that have twenty-five year-round residents to large, publicly owned systems, such as the one in New York City, which serves more than 9 million people. The United States has about 167,000 public water systems that serve about 278 million people. The smaller systems generally rely on groundwater from local well fields, while larger systems tend to use surface water from lakes, rivers, and reservoirs. Hydrologically, it should be noted that surface water is partially fed from groundwater as base flow that is part of the hydrologic cycle.

Water quality is a major issue addressed by water supply systems. Raw water sources vary in their quality over time, and operators of water-treatment plants must constantly monitor the biochemical changes that can occur in their plant’s water sources. Heavy precipitation can easily increase the suspended material in the stream, thereby requiring aluminum to make flocs (clot-like masses) that can form and settle these materials to the bottom of a holding basin. State and federal laws govern the quality of potable water for systems of all sizes and purposes. Water-quality standards for industrial uses and irrigation, however, may differ from those for CWSs.

Background and History

The Minoan civilization on the Mediterranean island of Crete about five thousand years ago is believed to have been the first major builder of water-supply aqueducts. In China, archaeologists have discovered ancient wells that were dug to a depth of 1,500 feet. A special form of subsurface tunnel (called a *qanat*) was used in the Middle East (especially Iran) and in North Africa about 1000 B.C.E. Aboveground stone aqueducts were started in Rome in 312 B.C.E., and by 300 B.C.E. there were fourteen major aqueducts bringing in about 40 million gallons per day. The estimated daily per capita use in ancient Rome was an astonishing thirty-eight gallons, as compared with about two gallons for Paris in medieval times.

In the ninth century, a significant water supply system was built in Cordova, Spain. An early pump was installed by Dutch engineer Peter Morice at the London Bridge in 1582, and other sixteenth-century systems were located in Hanover, Germany, and in Mexico. Pumps and piping systems were installed in London and Paris during the early seventeenth century–the earliest residential service probably occurred in London in 1619. Nevertheless, these systems were limited to large urban locales, and many water sources, from the Middle Ages continuing even into the nineteenth century in some cases, were usually surface sources (rivers), where it was quite common for waste to be dumped. The results were deadly: 53,000 people died in London alone during the cholera epidemic of 1848-1849.

Some cities were more advanced, such as Richmond, Virginia, which built slow sand filters to treat its water in 1834, and it was generally during this period that the ability of water to deliver diseases caused by bacteria and other pathogens was recognized. As a result, methods of treating water to rid it of such contaminants were developed, from sand-filtration systems starting in the mid-nineteenth century to chlorination, beginning in England in 1897. It was also at this time that water supply systems had extended beyond the cities to most towns. Jersey City, New Jersey, for example, tried chlorinating raw river water in 1908 and found that the typhoid rate dropped sharply. Consequently, chlorination became a common disinfectant in water-treatment plants in the United States and elsewhere.

How It Works

Water Sources. The first part of a water supply system is its source, and generally there are two: surface sources (rivers and lakes) and groundwater (from underground aquifers). Many cities in the world are favorably situated on rivers or lakes, such as Cairo on the Nile, Chicago on Lake Michigan, London on the Thames, New Orleans on the Mississippi, Paris on the Seine, and Washington, D.C., on the Potomac. Other cities in coastal areas could not use saline water and therefore had to build expensive inland reservoirs. New York City had to build seven reservoirs in the Croton, Catskill, and Delaware watersheds in New York State to serve 9 million people with an average daily consumption of 1.2 billion gallons.

Cities in the southwestern United States are faced with the major problem of population growth without sufficient precipitation for their water supply. The average annual rainfall for Las Vegas and Phoenix is 4.49 and 8.29 inches, respectively. The population for these cities has grown 120 and 63 percent, respectively, from 1990 to a July, 2009, estimate. Consequently, controversy has developed over the large water transfers to California from the other states in the Colorado Basin (Colorado, Wyoming, Utah, New Mexico, Nevada, and Arizona).

Many areas in the United States sit atop extensive permeable formations that contain large amounts of groundwater. The coastal plain province that extends from Cape Cod, Massachusetts, to Florida and continues on to the Gulf of Mexico forms a huge repository of groundwater. New York's Nassau and Suffolk counties have a combined population of 2.85 million people who get all of their groundwater from local sources. Florida is noted for extensive amounts of groundwater in its Biscayne and Floridian aquifers. The glaciated areas in the northern portion of the continental United States, approximately above the Ohio and Missouri rivers, also have favorable supplies of groundwater.

Water Collection and Storage. As part of the hydrologic cycle, a portion of the precipitation that lands on the earth's surface flows slowly into streams that eventually empty, for the most part, into the ocean. Another portion of the precipitation infiltrates permeable layers of the earth and is stored as groundwater in subsurface geologic formations called aquifers. This underground storage also flows slowly back into streams as base flow and can account for a good portion of the total flow of a stream. Groundwater can be extracted from underground sources by pumping from one or more wells. In contrast, surface water from one or more intake pipes can be extracted from a stream, lake, canal, or a reservoir that was built to store water. The raw water from surface intake or a well is transported by pipe to the water plant for physical, chemical, and biological treatment to meet mandated standards for potability prior to distribution.

Water Treatment. The function of all water-supply plants is to treat the incoming raw water so that the finished product is safe to drink. Although there are variations in treatment techniques, most facilities start with screens that block large items that may be floating in the raw water (generally surface), which could include debris, dead animals, and fish. The next step is to move the raw water to holding basins that are designed to settle out suspended sediments such as silt and clay. If the incoming water has large amounts of suspended materials, an additional step called "flocculation/coagulation" is used, whereby chemicals such as alum and soda ash are added to the water to facilitate fine-particle suspension. This coagulation procedure generates floc, which over time will get heavy enough to settle in holding tanks as sludge. About 90 to 99 percent of the viruses that may be in the water are removed by this procedure.

The next step is to let the partially treated water filter through layers of sand and gravel to remove finer particles that may remain. These filters eventually get clogged with sediments, which necessitate removal by backwashing or flushing, leaving a residue that eventually ends up in a sewage plant for additional treatment.

Fluoridation and chlorination generally form the final stages of the water- treatment process. A commonly used compound, which is recommended for maintaining healthy teeth, is sodium fluoride. Approximately half of the U.S. population uses water that is fluoridated at an acceptable concentration level. Chlorination is a very common disinfection method that has been used for nearly a century to kill any residual bacteria and some viruses by adding chlorine gas to the finished water. There are some people who do not like the taste or smell of chlorinated water; adding activated carbon into the treatment process can make the water more palatable. One of the benefits of chlorination is that residual amounts of chlorine can remain in the treated water as it travels many miles in the system. The disinfected water travels with the flow until it reaches all the customers.

Some water utilities use ozone-gas and ultraviolet-light systems as alternatives for killing any residual bacteria and viruses that may still be present in this last stage of water treatment. Ozone is noted as being a strong oxidant of taste and odor compounds in addition to being a very effective disinfectant. Chlorine not only is less expensive than ozone but also has the ability to continue as a disinfectant as it leaves the treatment plant to a household that may be miles away at the end of the distribution system. Ultraviolet (UV) light is very effective in killing almost all of the microbiological organisms that may have slipped through earlier steps in treatment, as the light energy is absorbed in the DNA of the microbes, therefore ending reproduction at the cellular level. UV usage is both expensive and slow, but in select situations, it can be very useful.

Desalination. There are many places in the world that border the oceans. This might be healthy for shipping and tourism but not for drinking water. Seawater has total dissolved solids (TDS) of about 35,000 parts per million (ppm), which is vastly greater than the U.S. safe drinking water standard for TDS of 500 ppm. Two-thirds of the seventy-five hundred desalination plants in the world are found in the Middle East, particularly those countries on the Arabian Peninsula with a mean annual precipitation of only about four to twelve inches. Saudi Arabia itself accounts for more than two thousand of these plants, with a total capacity of about 3.1 billion gallons per day.

One immediate drawback of desalination is that the process requires huge amounts of energy. Desalination expenses are about five times that of filtering freshwater. The techniques that have been tried include heat distillation, chemical (ion exchange) or electrical (ion removal) procedures, freezing, solar humidification, and reverse osmosis (forcing seawater through a semipermeable membrane).

Water Distribution Systems. In industrialized developed countries, the infrastructure for the delivery of potable water to the public, commercial users, and industries includes groundwater wells, intakes for surface water, reservoirs (if available), treatment plants, storage tanks, and an extensive network of pipes. The plants, storage tanks, pipes, and pumps will vary in size based on the expected demand from customers plus a safety factor for fires and emergencies. Treated water can be stored underground or in aboveground tanks. Water storage is necessary to allow sufficient time for disinfection (chlorine contact time) and to handle peak demand, which occurs typically in the morning as many people shower then. The storage facilities within a typical distribution system are designed to hold enough water to meet peak demands that are greater than the maximum daily demand, which is based on 50 percent of the storage capacity of the particular system. In addition, they must have sufficient capacity to handle demand of certain emergencies, such as a fire or water-main break. Most municipal water systems have one or two days' worth of potable water in storage for any kind of emergency that may develop.

It is regrettably apparent that the aforementioned characteristics of a functioning system of treated water and proper distribution to large segments of the population in the developing and undeveloped nations is not occurring at the desired and necessary rate. Some estimates suggest that by 2020, 32 million people, if not more, could die from water-related diseases that could be prevented by adequate treatment and functioning delivery systems. Some estimates suggest that 2 million to 5 million deaths per year can be attributed to water-related diseases that particularly affect small children, such as cholera, typhoid, dysentery, and other diseases associated with diarrhea.

Wastewater Treatment. This field is concerned with the proper treatment of sewage. Its relevance and connection to water supply is immediately obvious if one notes that sewage-treatment plants commonly discharge their treated wastes into the same rivers that water-supply treatment plants use as their source of water. Potable-water plants can quite easily be found downstream of sewage plants. New Orleans is at the downstream end of the Mississippi, and there are many wastewater plants upstream, so the problem is a real one.

Applications and Products

Water Demand. Although only about 5 percent of the finished product is used for drinking, the overwhelming majority of the remaining treated water in the United States and Canada is used for toilet flushing (47 percent) and bathing (37 percent). All the water must meet the same standards, regardless of how it is used. The requirement that all new toilets installed after 1994 must be low-flush bodes well for a reduction in overall water demand.

The U.S. Geological Survey (USGS) has been estimating the use of water in the United States at five-year intervals since 1950. It depends on water-use data from each state as part of its compilation. The report for water use in 2005 was issued in 2009. Water that is distributed to public and private purveyors is used for a variety of domestic, commercial, and industrial purposes. Water distribution also includes public services that may not be billed, such as firefighting, municipal pools and parks, pipe leakage and flushing, and water-tower repairs. About 86 percent (about 258 million people) of the total U.S. population rely on this supply of water for household use. About two-thirds of this water is obtained from surface sources (rivers, lakes, and reservoirs), and the remaining one-third is from groundwater sources.

The U.S. population has increased from 175 million in 1950 to about 300 million in 2005. One would presume that freshwater diversions would have increased at a commensurate rate, but the total diversions actually peaked in 1980. This effect is most probably associated with decreased water diversions for thermoelectric power that require large amounts of cooling water.

Impact on Industry

Government and University Research. The earliest legislation to provide some regulation for drinking-water quality in the U.S. was assigned in 1914 to the U.S. Public Health Service. However, it was voluntary for public systems. Growing concerns about potable water led to the passage of the Safe Drinking Water Act (SDWA) in 1974. This law specified that the EPA has the authority to set standards as a means of protection against the mix of natural or manmade contaminants that could get into public water supplies. Water-quality standards for specified contaminants were established. Public water systems had to follow these standards and treat the finished water to the maximum level specified for each contaminant. The first array of standards included turbidity, microbes (bacteria), inorganic contaminants (such as nitrates and metals), and traces of radioactive emitters (radon, uranium, and radium).

The SDWA was revised in 1986 to include selected contaminant standards. In addition, all surface water systems were required to use established disinfection techniques and had to filter their water source(s) for microbes such as bacteria, viruses, and protozoans. All states had to institute protection procedures for public wells. Water-quality regulations governing wells that service only private homeowners vary from state to state. However, each state can establish distance standards between adjoining homeowners who have their own wells. Depending on the permeability of the underlying geologic formations, states, counties, or cities can specify distances between adjoining wells to minimize well interference.

The EPA established National Primary Drinking Water Regulations (NPDWRs) and National Secondary Drinking Water Regulations (NSDWRs). The NPDWRs are divided into a Maximum Contaminant Level Goal (MCLG) and a Maximum Contaminant Level (MCL). The MCLG refers to the contaminant level in drinking water below which there is no known or expected health risk. These are not enforceable standards and contain a safety margin. In contrast, the MCL denotes the highest contaminant level that is permitted in drinking water and is an enforceable standard. MCLs are set as close to MCLGs as possible, using the best treatment technology available while taking cost into consideration. The list includes 7 microorganisms, 4 disinfection byproducts, 3 disinfectants, 8 inorganic chemicals, 23 organic chemicals, and 4 radionuclides for a total of 49.

The NSDWRs are neither mandatory nor enforceable. The 15 contaminants on the list may have cosmetic effects (discoloration of the skin or teeth) or aesthetic effects (color, taste, or odor).

The SDWA was revised again in 1996 to include a revolving-loan program so that each state could provide money for those communities that needed either to upgrade their systems or protect their source of water. In addition, brief annual water-quality reports had to be provided to each customer served by a community water system so that the customer would be informed if contaminant levels exceeded standards.

Many of the states have Water Resource Research Institutes (WRRI) that are located in leading research universities and are funded by the USGS. These institutes, in turn, fund statewide investigations in the water-resources field that include research in areas related to water supplies and treatment issues. Proposals are solicited from local educational institutions in the state and are then decided on by the WRRI for funding.

Fascinating Facts About Water Supply Systems

- The ancient Romans built a sophisticated system of aqueducts, starting with the Aqua Appia in 312 B.C.E. Six hundred years later, this system delivered water to Rome via 359 miles of aqueducts, 50 miles of which were borne by stone arches.
- The water drunk by Caesar and Cleopatra in ancient Roman times has been circulated innumerable times as part of the hydrologic cycle.
- Water accounts for 66 percent of the human body and 75 percent of the brain.
- Water is the only substance found on Earth that can exist in three forms: solid, liquid, and gas.
- The nutritive value of water is zero; yet it is the dominant ingredient of all flora and fauna.
- The contradiction with water is apparent in the statement made many years ago by Leonardo da Vinci that, "in time and with water, everything changes."
- To maintain a healthy body, humans need to consume about 64 ounces of water per day.
- Public water supply withdrawals in 2005 for the United States indicate that the percentage of the population that was served varied from a low of 56 percent in Maine to a high of 100 percent for Washington, D.C. The national average was 86 percent.

Industry and Business Sectors. An important trend that has been growing in the United States has been the acquisition of public water systems by private companies. Although some of these companies are domestic and have been operating for many decades, many are foreign. Given the increasing complexity and sheer cost of operating a water-treatment plant with the expanding variety of state and federal guidelines that must be followed, this is a new development in the water-supply field. Although as of 2011, the privately held share of the water-treatment industry is only about 9 percent of the total market, the privatization of public water systems is a future possibility.

Careers and Course Work

Careers in water supply occur in a number of fields. The overriding direction is a technical specialty. There are numerous occupations that fall within the purview of water-supply system careers. Some entry-level semi-technical jobs at a water-treatment plant may be available to those with an associate's degree. For those who want to start at a more professional level, a B.A. or B.S. degree in a field such as civil engineering (there are numerous subspecialties and courses in the water-resources field), hydrogeology and geology (groundwater and related geologic subjects), water quality (includes biology and chemistry courses), physical geography (includes courses in geographic information systems and water resources), and the related fields of meteorology and climatology would be required.

Given the increasingly complicated institutional and regulatory aspects of water supply systems, those individuals who might be interested in these aspects should consider environmental planning and law. Simply reading some of the detailed state and federal legislative documents that pertain to water supply issues will provide more information in this area. It is clearly a specialized field that requires someone not only to take appropriate water-related classes but also to take those that deal with environmental planning, law, and regulatory issues.

Social Context and Future Prospects

Water, like food and air, is an absolute necessity for all humanity. One can only sense a continuing problem with global increases in population, especially in countries such as Africa, where inadequate precipitation has resulted in numerous deaths. The Sahelian zone south of the Sahara Desert in northern Africa experienced a drought that began in 1968 and continued through the mid-1980's that killed about 100,000 people and left agriculture and livestock (cattle and goats) heavily affected.

The coming years, regrettably, have a degree of uncertainty regarding the sufficiency of world water supplies. In some places, as in the Sahel, precipitation is very erratic and many people suffer and die. In other areas, such as the United States, problems occur with existing water supplies becoming depleted, as evidenced by the growing "bathtub ring" in Lake Mead behind Hoover Dam on the Colorado River. The bathtub ring is the high water mark that shows where the water level once was. The new problem in the better-watered portions of the world pertains to water-quality issues such as exotic new contaminants (prescription-drug residues) that as of yet are not treated by standard techniques in most water supply plants.

India is a good example of a large and highly populated country that has become increasingly industrialized with a regrettable list of major water-distribution issues. The population is more than 1.1 billion people (it is the second most populous nation in the world) and is estimated to increase 48 percent by 2050. The water-distribution systems in practically all of the cities have fallen into a state of disrepair to the extent that tap water is available for only a few hours a day. About 45 percent of the population does not have access to a workable public wastewater system. If this is the situation in an industrialized nation such as India, it gives one pause to think about all of the other countries that are in worse shape economically.

The future of maintaining the highest level of treatment for drinking water continues to be uncertain, given humanity's apparent ability to generate new and even more exotic materials that manage to become part of the water supply.

Robert M. Hordon, B.A., M.A., Ph.D.

Further Reading

Cech, Thomas V. *Principles of Water Resources: History, Development, Management, and Policy.* 3d ed. Hoboken, N.J.: John Wiley & Sons, 2010. A highly readable and very informative book on a wide variety of useful water resources.

Chin, David A. *Water-Resources Engineering.* 2d ed. Upper Saddle River, N.J.: Pearson Prentice Hall, 2006. Contains a thorough and detailed treatment of the numerous technical aspects of water resources.

Gleick, Peter H., et al. *The World's Water, 2008-2009: The Biennial Report on Freshwater Resources.* Washington, D.C.: Island Press, 2009. A fact-filled book on many aspects of global water issues accompanied by many detailed tables.

Glennon, Robert. *Water Follies: Groundwater Pumping and the Fate of America's Fresh Waters.* Washington, D.C.: Island Press, 2002. A very readable account of the numerous problems that have occurred in various parts of the United States regarding groundwater issues.

Gray, N. F. *Drinking Water Quality: Problems and Solutions.* 2d ed. New York: Cambridge University Press, 2008. A detailed textbook suitable for readers interested in the technical aspects of the water-quality industry.

Mays, Larry W. *Water Resources Engineering.* Hoboken, N.J.: John Wiley & Sons, 2005. A text that covers hydrology, hydraulics, and sustainability of water resources with numerous examples.

Powell, James Lawrence. *Dead Pool: Lake Powell, Global Warming, and the Future of Water in the West.* Berkeley: University of California Press, 2011. An excellent, fact-filled account of the myriad issues surrounding the huge growth of population and the competition for water among those in the Colorado River watershed.

Web Sites

American Geological Institute
http://www.agiweb.org

American Geophysical Union
http://www.agu.org

American Institute of Hydrology
http://www.aih.engr.siu.edu

American Water Resources Association
http://www.awra.org

American Water Works Association
http://www.awwa.org

Geological Society of America
http://www.geosociety.org

National Association of Water Companies
http://www.nawc.org

U.S. Environmental Protection Agency
Drinking Water Contaminants
http://water.epa.gov/drink/contaminants/index.cfm

U.S. Geological Survey
http://www.usgs.gov

Water Environment Federation
http://www.wef.org

WEIGHT AND MASS MEASUREMENT

FIELDS OF STUDY

Physics; mathematics; chemistry; materials science; transportation; economics; aeronautics; logistics; electronics

SUMMARY

Mass is an intrinsic property of matter and remains constant, regardless of the force of gravity, whereas an object's weight varies according to variation in the force of gravity acting upon it. Both mass and weight are relative quantities, rather than absolute. They are fundamental to quantifying human economic and social transactions. The development of standardized systems of weights and measures has been necessary for advancements in science, technology, and commerce. The recognized standard of weights and measures in use today is the metric system.

KEY TERMS AND CONCEPTS

- **Austenitic Steel:** A form of stainless steel containing a high degree of austenite, noted for its resistance to corrosion.
- **Drift:** A slow, steady movement of a scale reading from its initial determination, caused by equilibration of the scale components.
- **Force Restoration Transducer:** A device that uses electronic feedback to adjust the current in an electromagnet until it balances the force put against it by a mass, permitting a highly accurate determination of weight.
- **Hysteresis:** Variation of readings in a scale according to whether the weight being measured is increasing or decreasing.

Definition and Basic Principles

It is not possible to state the absolute mass or weight of any material object. At the most fundamental level, one can only state that a hydrogen atom has the mass of a single proton and a single electron; it is impossible to know the absolute masses of those subatomic particles. This basic relationship, however, provides the means whereby other atoms can be ascribed their corresponding masses and the effects of accumulated mass become observable.

Weight is described as the product of an object's mass and the gravitational force acting on that mass. The hydrogen atom that on Earth has the mass of one proton plus one electron must by definition have the same mass wherever it is located. The gravitational force experienced by that atom will be different in other places, and at each location it will have a somewhat different weight while maintaining the same relative mass to other atoms. The same logic applies to larger quantities of matter because of this atomic relationship.

When the gravitational force experienced by different masses is common to their weights, it is expedient, though not absolutely correct, to use the terms "mass" and "weight" interchangeably in common usage. To quantify these properties in a meaningful and useful way, it also is necessary to define some standard quantity of each to use as a reference to which other masses and weights can be compared. Devices calibrated to correspond to the standard value can be used to measure other material quantities. International standardization of weights and measures works to assist trade fairness and facilitates the comprehension of scientific, technological, and theoretical work in disparate locations.

Background and History

In prehistoric times, measured amounts of materials most likely consisted of handfuls and other amounts, with little or no consideration beyond the equivalency of perceived value. There is an intrinsic conflict in this sort of measurement, as weight versus volume. At some point it would be realized that a container of large seeds had a sensibly different weight than the same container filled with small seeds, arrowheads, fish, or whatever might be traded at the time. The concept of fair exchange on an equivalent weight or equivalent volume basis, or the development of some means of relating weight and volume, would be the beginning of standardization of weights and measures.

History does not record the beginning of standardization, although ancient records reveal that certain standard measurements were used in ancient times, such as cubits for distance and talents and carats for weights. Each measurement related to some basic definition, which was often merely the distance between two body parts and varied from person to person, place to place. Through the sixteenth, seventeenth, and eighteenth centuries, the Imperial or British system of measurement was the accepted standard around the world; it continues to be used today.

In 1790, a commission of the French Academy of Science developed a standardized system of weights and measures that defined unitary weights, volumes, and distances on quantities that were deemed unchangeable and that related to each other in some basic manner. This metric system has since developed into the Système international d'unités (International System of Units; abbreviated SI) and is the universal standard recognized and used around the world.

How It Works

All weight and mass measurements are relative and cannot be known as absolute values. The definition of standard weights and measures provides a reference framework in which units can be treated as though they are absolute measurements.

For the measurement of mass, the metric standard unit is the kilogram (kg), corresponding to 1,000 grams (g). The gram, in turn, is defined as the mass of 1 cubic centimeter (cm^3) of pure water at its temperature of maximum density. Under those conditions, therefore, 1 kg of water occupies a volume of 1,000 cm^3, called 1 cubic decimeter (dm^3) or 1liter (l). This is a readily understandable definition, but in order to have an unchanging standard to which other weights may be readily compared a solid object is required.

The international prototype standard kilogram is housed at the International Bureau of Weights and Measures in Sèvres, France. It is a small cylinder exactly 39 millimeters (mm) in diameter and length, and the alloy is of 90 percent platinum and 10 percent iridium. To protect the constant value of its mass and weight, it is stored under inert gas in a specially constructed triple bell jar to protect it from exposure that could affect the mass of the object. A minimum of sixty countries around the world hold registered precise copies of the prototype standard kilogram, similarly protected. These have traditionally been taken by special secure courier to France, through a series of special precautionary procedures, to be compared with the true prototype kilogram for calibration. Following this procedure the copies are returned to the appropriate national government, where they then serve as the legal standard of measurement until their next calibration.

Given the security measures in place for air transportation and international entry, the method has associated with it a very high degree of fallibility. For all but the most legally sensitive of applications, less stringent standards are normally used. It is sufficient for essentially all purposes that a less precise comparison be used. This permits devices used to determine mass and weight to be designed with preset comparisons and to be arbitrarily adjustable to accepted weight standards. This is typically achieved through the use of standard weight sets that—though neither as stable nor as stringently controlled as the prototype standard—are nevertheless uniformly consistent and conform to the tolerances stated by governing regulations. Typically, these are standardized stainless steel objects or specific electromagnetic conditions known to exert a specific force.

Applications and Products

Because mass is an intrinsic property of matter inextricably tied to weight, its proper measurement is essential in many different fields. The applications and products that have been developed to accommodate the various fields all use the same basic principle of mass comparison, as indeed they must.

Scales and Balances. The measurement of mass or weight (these terms, though often used interchangeably, are technically different) is carried out using scales and balances. A balance functions by directly comparing the mass to be determined with masses that are known. The two are deemed equal when a properly constructed balancing scale indicates equality.

The simplest device, called a two-pan beam balance, consists of a crossbeam from which two pans are suspended to hold the masses to be compared. The crossbeam is balanced on a knife-edge fulcrum and attached to a device that indicates deflection from the neutrality point. The unknown mass is placed on one of the pans and known masses are placed on the other pan until the scale balances at the neutrality point again. At that point, the unknown mass on one pan is equal to the sum of the known masses on the other pan.

Variations of the pan balance have been constructed and used in mass capacities ranging from grams to several tonnes (1 tonne [t], or metric ton = 1,000 kg). The smallest of these have been those used primarily in scientific research that involves small quantities of matter, while the largest variations have been used for weighing large quantities of goods for transport.

With the application of electronics, beam balances are replaced by scale balances (commonly just called scales) that do not directly compare masses. Instead, a scale uses the beam balance principle to quantify an unknown mass by comparing it with a calibrated scale against a known force. The amount of deflection of the scale applied by the unknown mass against the known force determines the quantity of the unknown mass. This tends to simplify the procedure by eliminating the requirement to maintain a set of known comparative masses (a set of standard weights) that are always subject to physical damage, chemical attack, and loss. However, it also requires that the functional force and electromagnetic feedback systems be routinely tested and calibrated.

Science and Engineering. Mass and weight are central concepts of science and engineering, especially the disciplines of chemistry and physics. In chemistry, mass is central to all chemical reactions and processes. All masses are relative, beginning with the mass of the hydrogen atom. This provides the basis for assigning a corresponding mass to every other type of atom that can be identified.

Because of this relative relationship, it is possible to relate the number of atoms or molecules in a sample to a weight that can be measured. The quantity of hydrogen atoms required to make up a bulk mass equal to 1 g must be exactly the same as the number of atoms of carbon (assigned atomic mass = 12) required to make up a bulk mass equal to 12 g, and so on. This equivalency is called the gram-equivalent weight, and the number of atoms or molecules required to make up a gram-equivalent weight is termed 1 mole. Every properly balanced chemical equation specifies the number of moles of each different material in the process, corresponding exactly to the number of atoms or molecules of each that is involved in each single instance of the reaction. Without this concept, the science of modern chemistry would not exist. It is worth noting that this concept would not have developed had medieval alchemists not applied the measurement of material weights in their studies.

Similarly, physics has developed to its present state through the concepts of mass and weight and the application of them as a means to measure forces. In any practical sense a force can only be quantified by measuring its effect relative to a mass that has itself been quantified. In the basic principles of Newtonian mechanics, a force is equal to the product of mass and acceleration ($F = ma$). Newtonian mechanics, however, is not sufficient to describe the small structure of atoms and molecules. This is the realm of quantum mechanics, of which Newtonian mechanics is a subset.

Mass at the atomic and molecular level is especially relevant in the analytical procedure called mass spectrometry. In this methodology, the masses of molecular fragments carrying a positive electrical charge are determined by the magnitude of the interaction of that charge with an applied electromagnetic field. Another procedure, called isotopic labeling, is used to measure the effect of atomic mass on vibrational frequency of specific bonds in spectroscopic analyses and on the rates of reactions in chemical kinetics. These methods, which depend on highly specialized techniques of measuring mass, provide a great deal of essential information about molecular structures and behaviors.

Impact on Industry

National governments around the world maintain a branch with the sole responsibility of ensuring that weighing and measuring devices meet or exceed specified limitations and tolerances, otherwise the device is declared invalid and commercially unusable. In the United States, this regulating body is the National Institute on Standards and Technology (NIST), part of the Department of Commerce. The regulations governing weighing and measuring devices make up most of the 450 pages of NIST publication HB-44 (2010).

Mass and weight affect every aspect of industry and commerce involving material objects. Material transportation—by land, sea, and air—is measured by tonnage. Each tonne or portion thereof has an associated economic cost. The vehicles used for transporting loads —such as cars, trucks, railway cars, ships, and airplanes—must be designed and constructed with sufficient strength to carry their respective loads. The infrastructure upon which they travel must be designed and constructed to accommodate the weights of materials and vehicles moving on them, and they must be maintained against breakdown and failure under those loads. The foundations of buildings must be designed and constructed to bear the weight of their corresponding superstructures and the natural forces that are applied against them. Failure to meet either of these requirements has catastrophic results.

The distinction between mass and weight has become more apparent since the middle of the twentieth century and the advent of space travel. As modern industry increasingly seeks to utilize space and space travel as commercial and industrial resources, the distinction between mass and weight becomes even more essential. The vast majority of measuring devices in use on the planet depend on the constant force of gravity, which does not apply in the state of free fall. Space is an unforgiving environment, and any minor discrepancies in calculations based on mass or weight will be multiplied over the large distances involved, even in Earth orbit. The economic cost of failures in such cases is measured in billions of U.S. dollars. In 1999, for example, failure to properly brake the mass of the Mars Climate Orbiter as it approached that planet resulted in the orbiter's disintegration. Also, it has been observed that variations in the Earth's gravitational field cause the weight of any specific mass to change by as much as 0.8 percent worldwide, and by 0.2 percent just over the contiguous United States. Precise and accurate measurement of weight or mass is therefore essential, not only for space-bound payloads but also for transcontinental transportation in general.

Mundane commercial practices such as bringing agricultural products to market and trade in commodities are based in the exchange of mass for value. A cash-crop farmer, for example, may harvest a grain crop that is then delivered to market through a broker who provides the farmer with a monetary exchange value according to the weight of the amount of grain that was delivered. Similarly, coffee, as the most valuable commodity in the world, is bought and sold by brokers who market the beans to roasters. The roasted beans are ground for brewing, and the brewed beverage is sold by the cupful to billions of retail customers each day.

Precise mass or weight measurement becomes more essential as the value attached to a material increases. Common metals such as copper and aluminum are generally weighed by the kilogram, while such precious metals as gold, silver, and platinum are measured by the gram. Scarcer metals and materials are often measured in milligram (mg) quantities. In past times, when gold and silver were actually used as coins of the realm, the coinage was typically weighed to ensure that the exchange of money for goods was fair. Methods were commonly employed to acquire minute shavings from several coins, to be subsequently combined, melted into bullion form, and traded for actual coins. Today, precious metal coinage is not used in circulation, but it is used for investment purposes. Each coin so struck is subject to stringent quality control and security measures to ensure that it contains the exact amount of precious metal specified by weight.

Careers and Coursework

Mass and weight measurement is not a career field in itself. It is, however, an intrinsic and fundamental practice within a great many career fields. At minimum, a career in such a field will require the individual to have a good comprehension of the sometimes subtle difference between the two and an understanding of their use in the context of the specific career. This understanding is acquired through basic secondary school education in science and mathematics. As career paths become more focused, a greater emphasis is placed on understanding the concepts and applications of mass and weight. These will be central to careers in chemistry, physics, and mechanical, chemical, and civil engineering, for which a college or university degree (bachelor's) is the primary qualification. To a lesser extent, they are also important components in the study of geology and biomechanics.

A career in the aerospace and transportation industries will place a heavy emphasis on the evaluation and manipulation of gross weights and weight changes in logistics. Military transport especially values efficiency in mass handling methods. Electronics and electromagnet technologies are fields necessary to the design and maintenance of measuring devices, and are of particular relevance to regulatory organizations such as NIST. As industry based on space travel grows in importance, its most

Fascinating Facts About Weight and Mass Measurement

- The weight of an object can vary by as much as 0.8 percent according to its location on the planet, but its mass remains absolutely constant.
- Weighing an object determines the force of gravity acting on that object, rather than the mass of the object.
- Mass and weight are used interchangeably within the same gravitational field, though they technically have different meanings.
- Mass is an intrinsic property of matter and remains constant. Its weight is not, and it varies according to the force of gravity acting on that matter.
- The vibrational effect of substituting a single atom of deuterium for a single atom of hydrogen in a molecule is the most readily detected isotope effect because the deuterium atom has twice the mass of a hydrogen atom.
- Beam-balance scales used to weigh large trucks work on the same principle as the smallest pan balances.
- The international standard prototype kilogram is a cylinder made of 90 percent platinum and 10 percent iridium. The cylinder is precisely 39 mm high and 39 mm in diameter.
- Copies of the standard kilogram are kept in more than sixty countries, with each copy housed in a specially designed triple bell jar to prevent changes in its mass from oxidation or the loss of even a single atom.
- The masses of electrically charged pieces of molecules and atoms are directly measured by mass spectrometry.

essential component will be the proper determination of mass quantities. Students who seek a career in this field should expect to undertake studies in mathematics and physics at the most advanced academic levels (master's or doctorate).

Social Context and Future Prospects

The world is a material one, and the concepts of mass and weight are inseparable from matter. All aspects of society are impacted by mass and weight in some manner. The single greatest economic expenditure of commerce and society is the energy required for the movement and transportation of mass. This is closely related to other physical aspects, such as tribology, the study of friction and lubrication, and applied physics, the study of the properties and utilization of condensed matter. Tribological effects, all of which are intimately associated with the movement of mass and the corresponding weight, are estimated to carry an immediate economic cost of about 4 percent of a nation's gross national product, with one-third of all energy expenditures made to overcome friction. If the frictional effects involved in transporting mass could be reduced or eliminated, the economic and social benefits would be immense. Even the movement of heating materials, such as natural gas, exacts an economic cost because of that material's mass and weight.

The green movement, which includes energy-efficient building methods, works toward minimizing the economic impact of mass movement, but at the same time it engenders its own costs in that area. Trade and commerce will always demand mass and weight measurement practices and practitioners to maintain stability, productivity, efficiency, and fairness.

Richard M. Renneboog, M.Sc.

Further Reading

Davidson, S., M. Perkin, and M. Buckley. *The Measurement of Mass and Weight.* Measurement Good Practice Guide 71. Teddington, England: National Physical Laboratory, 2004. Describes the best practices for weighing technique in the context of scientific and technological application.

_______. *Specifications, Tolerances, and Other Technical Requirements for Weighing and Measuring Devices.* http://ts.nist.gov/WeightsAndMeasures/Publications/upload/HB44-10-Complete.pdf. Accessed October, 2011. The official statement of regulations governing the function and operating requirements of weighing and measuring devices in the United States.

Web Sites

National Institute of Standards and Technology
https://standards.nasa.gov/history_metric.pdf

WILDLIFE CONSERVATION

FIELDS OF STUDY

Animal behavior; biology; biotelemetry; botany; computer science; ecology; environmental law; forestry; herpetology; mammalogy; ornithology; photography; biogeography; oceanography; statistics; veterinary medicine; politics.

SUMMARY

Wildlife conservation is a broad discipline that focuses on assuring a future for wildlife in this rapidly changing world. It encompasses a diversity of careers. Some are highly specialized, such as wildlife veterinarian, wildlife researcher, university professor, and environmental lawyer. Others, such as wildlife-refuge manager, wildlife-research technician, wildlife educator, and wildlife law enforcement officer, incorporate the tools and skills of many disciplines coordinated and directed to accomplish wildlife-conservation tasks. Wildlife-conservation careers all involve understanding relationships between wildlife and their environment, the behavioral interactions among species, and the impacts of human-caused and natural changes on wildlife populations.

KEY TERMS AND CONCEPTS

- **Biodiversity:** Diversity of life in a given area.
- **Biogeography:** Study of plant and animal distribution.
- **Conservation Banking:** Market-based entrepreneurial approach to conservation that makes use of privately or publicly owned land managed specifically to provide habitat for endangered or threatened species. A developer or other individual or organization that wishes to alter habitat elsewhere can satisfy legal-mitigation requirements by purchasing mitigation credits from the owner of an approved conservation bank.
- **Ecosystem:** Biological environment with all its diversity of life, interacting with the nonliving features associated with it.
- **Ecotourism:** Leisure travel for the purpose of observing wildlife. Ecotourism provides an economic incentive for protecting wildlife.
- **Extinction:** Loss of every individual of a species. With extinction, a species ceases to exist.
- **Geographic Information System (GIS):** Computer software that allows the user to integrate geographically referenced information to locate and map distribution and movements of animals, the extent of habitats, and the influences of humans on those animals and habitats.
- **Global Positioning System (GPS):** Technology using an array of signal-sending orbiting satellites in conjunction with Earthbound signal receivers. GPS units have become the modern compass and sextant allowing real-time location of researchers and individual animals with great precision anywhere in the world.
- **Habitat Conservation Plan:** Formal plan designed to maintain or restore a healthy population of an endangered or threatened species in a specific area.
- **Mitigation:** In wildlife conservation, refers to money paid or actions taken to compensate for, minimize, or eliminate harm or potential harm to wildlife and their habitat.
- **Mitigation Banking:** Similar to conservation banking, but mitigation credits from an approved wetland mitigation bank are purchased to create, restore, or improve wetlands in order to offset harm done to wetlands elsewhere.
- **Translocation:** Movement of individuals of a species from one area to another, often in an effort to establish a new population or to protect individuals from the impending loss of their habitat.
- **Wildlife Telemetry:** Use of radio transmitters attached to wild animals to monitor their movements and, often, their basic physiological characteristics.

Definition and Basic Principles

Wildlife conservation focuses on the protection, enhancement, and management of wildlife populations in their natural habitats. Understanding ecosystems, habitats, food habits, migration, home ranges, territoriality, reproductive biology, predator-prey relationships, the nature and impact of wildlife diseases, and human impacts on all of these are essential to wildlife conservation. The development and publication of field guides and availability of quality binoculars has increased public interest in wildlife and has greatly facilitated species identification. The development of wildlife telemetry techniques and equipment, enhanced photography and sound recording capabilities, computers that can be used in the field, software such as Geographic Information Systems (GIS), Global Positioning Systems (GPS), and molecular techniques that help identify genetic differences among populations are among the many technological advances of the late twentieth century that have enhanced the quality and scope of information that wildlife conservationists can use. Such technological advances have led to more efficient and more effective wildlife conservation and, along with need associated with declining habitat availability, have driven specialization and growth in wildlife conservation career opportunities. GIS, molecular ecology, wildlife telemetry, fire ecology, translocation, and captive breeding are among many opportunities for specialization within wildlife research and management careers. Wildlife educators, communicators, photographers, and ecotourism operators function at many levels to promote understanding and support of wildlife conservation. Zookeepers, veterinarians, and wildlife rehabilitators contribute greatly to understanding and promotion of wildlife health and to wildlife-reintroduction programs. Environmental law enforcement officers and lawyers are on the front line of wildlife conservation, promoting sound conservation laws and appropriate mitigation efforts and assuring compliance with legal mandates to protect species and their habitat.

Background and History

As human populations grew in the eighteenth century, declines were seen in some wildlife populations, and local laws were sometimes passed to limit hunting, but wildlife conservation in its infancy was largely a voluntary effort, not yet capable of supporting careers. Employment in the field of wildlife conservation is a development of the late nineteenth and twentieth centuries. Early wildlife conservation efforts focused on human needs such as depleted game populations or fisheries resources.

Careers in wildlife conservation began in earnest with the establishment by Congress of the U.S. Fish Commission in 1871 to find ways to conserve declining fish populations. Concern for the welfare of birds and the scientific study of birds grew in the latter half of the nineteenth century, resulting in the founding of the American Ornithologists' Union in 1883 and other professional scientific societies. These had some conservation focus and appointed conservation committees to discuss conservation issues and support efforts to solve conservation problems. Such organizations raised public understanding and awareness of problems and were often led by government employees.

As American agriculture grew, blackbirds and other species were increasingly labeled as pest species. In 1885, the Office of Economic Ornithology was created within the United States Department of Agriculture. This supported wildlife careers in the study of bird food habitats, migration, and habitat use. Although initiated from a negative and economic perspective, the knowledge gained became foundations for later conservation efforts.

In the 1890's, feathers had become fashionable adornments of ladies' hats and other clothing and killing of birds for the millinery industry grew. Hundreds of thousands of egrets, terns, and other birds were shot at for breeding plumes that were worth their weight in gold. Chicks were left to die in the sun. As bird populations plummeted and stories of the slaughter became known, public outrage led to formation of clubs, boycotts of clothing with feathers, and public pressure for legal action to protect birds. Such actions led to formation of the Massachusetts Audubon Society in 1896, other state Audubon societies, and in 1905, to the formation of the National Association of Audubon Societies for the Protection of Wild Birds and Animals. Such nonprofit, conservation-oriented organizations established sanctuaries and hired wardens to protect wildlife. These organizations continue to be major employers in the field of wildlife conservation.

In 1905, the federal government responded to growing conservation sentiments by changing the name of the Office of Economic Ornithology to the Bureau of Biological Survey. The Biological Survey later became the United States Fish and Wildlife Service. Other federal agencies (U.S. Forest Service, Bureau of Land Management, Department of Defense) also developed wildlife-conservation programs associated with lands they managed.

During the twentieth century, habitat losses, species endangerment and extinctions, and pollution led to national (the Endangered Species Act of 1973) and international (International Convention on Trade in Endangered Species of Wild Fauna and Flora) legal protection for wildlife. Government responsibilities for wildlife and government career opportunities in wildlife conservation continue to grow. Nonprofit conservation organizations have also grown in number and in conservation roles, providing further wildlife-conservation career opportunities. Legal protections for wildlife and its habitats also have fostered a large service sector that provides career opportunities in wildlife conservation to assist industry in complying with conservation legal requirements. Finally, public interest in wildlife has grown to include a large, multifaceted ecotourism industry that provides many career opportunities in wildlife conservation.

How It Works

Wildlife conservation began with the protection of animals from overhunting but quickly expanded to include the recognition that destruction and fragmentation of wildlife habitats also reduced wildlife populations. By the mid-twentieth century it had also become obvious that industrial and agricultural chemicals could harm wildlife. The latter realization led to the understanding that humans share these ecosystems with wildlife and that in a very real sense, wildlife was the barometer to the health of the world at large and an indicator of problems that humans could face. Such understanding gave new currency to careers in wildlife conservation. It was also learned that the interconnections between humans and wildlife include shared diseases. The phrase "ecosystem health" took on new meaning. Lyme disease, West Nile virus, rabies, and other diseases created wildlife-conservation career niches for veterinarians and other health professionals. The diversity of careers within the broad field of wildlife conservation continues to grow. Many wildlife conservationists work outdoors; others work at zoos, museums, universities, industries, and in the offices of government agencies and nonprofit organizations.

Refuge, Preserve, or Park Management. Local, state, regional, and national parks, preserves, and wildlife refuges provide significant habitat for wildlife and significant recreational activities for humans. Managers of these lands must deal with wildlife-conservation issues ranging from maintaining and managing habitats to managing hunting and fishing, providing for endangered and threatened species, and conservation education. Land managers must have an understanding of wildlife-habitat needs and the tools for satisfying those needs. It is the land manager's responsibility to make decisions regarding such things as the periodic use of prescribed burning, control of exotic-invasive species, closing of some areas to the public, regulation of hunting, and other recreational or research activities. All of these responsibilities must be carried out with a clear understanding of the natural landscape and the ecosystem within which it is located, the laws that govern the landscape and the wildlife found there, and the human communities of the region. Land managers always work at the human-wildlife interface, and sometimes the work on the human side of things is the most challenging.

Wildlife Telemetry. One of the most useful tools in wildlife conservation is the use of wildlife telemetry. On the local scale this can involve the attachment of a transmitter to an animal and following its signal using a handheld receiver and antenna or using an automated receiver linked to multiple large, fixed antennae to learn patterns of daily or seasonal habitat use. For larger animals, it can also involve monitoring long-distance migrations by using transmitters that communicate with the National Oceanic and Atmospheric Administration's (NOAA) Argos weather satellite. The Argos satellite system was established in 1978 has been used for wildlife telemetry as well as other applications since the late 1980's. As the satellite passes over the animal with a transmitter, it establishes uplink and stores information from the transmitter. Later, as the satellite passes over a receiving station on earth, that information, including location coordinates for the animal, is transmitted by downlink to the receiving station and from there to the researcher. The quality of telemetry equipment and software has steadily increased while at the same time the size of transmitters has decreased, allowing them to be used effectively on insects. Methods of attachment have also changed over the years. Some transmitters are affixed to an animal with a harness or collar, some are glued to a bird's feathers or to the back of a turtle, or attached to a leg or fin, and some are now implanted in or fed to the animal. Use of telemetry can help scientists understand such things as the amount of space a species needs, the habitats it uses, daily and seasonal patterns in behavior, and how the needs of males, females, and juveniles may differ. Limitations on the use of transmitters include size, battery duration, signal strength, and tolerance of the species or individual animal to the transmitter. Capabilities of all of these parameters are continually improving.

Wildlife Law Enforcement. Increasing legal protection for wildlife, and growing and increasingly mobile human populations, have created a diversity of career opportunities. Wildlife law enforcement agents are hired at city, county, state, national, and international levels. In general, this hierarchy reflects increasing pay, increasing responsibility, and increasing specialization. At the local level, responsibilities might be associated with enforcing hunting and anticruelty laws. At the state level they can involve enforcing fish and game laws, helping to manage hunting, enforcing or administering efforts associated with state laws protecting threatened species and their habitats, enforcing laws regulating the keeping of wild animals in captivity, inspecting pet stores, dealing with nuisance animals, and working with veterinarians and others associated with wildlife-health issues. At the national level, wildlife law enforcement agents are often associated with managing and enforcing hunting regulations on federal lands, interstate transport of wildlife, importation of wildlife, and protection of federally protected species. Wildlife law enforcement agents are often associated with U.S. border crossings, where they assist with controlling the importation of wildlife from around the world. At any level, a wildlife law enforcement agent might have to work alone, on weekends, and at all times of day and night. At other times, wildlife law enforcement agents work with other professionals as part of a team dealing with issues that include, but may go far beyond, wildlife conservation issues.

Wildlife Rehabilitation. The care of sick, injured, or orphaned wildlife provides career opportunities at wildlife-rehabilitation centers, zoos, veterinary clinics, theme parks, animal shelters, and research centers. Sometimes this requires problem- or species-specific training and can require advanced education as a veterinarian or radiologist, or other medical specialty. In addition to career opportunities, wildlife rehabilitation provides a great deal of opportunity for volunteer efforts and internships that can provide training for individuals potentially interested in a wildlife-conservation career.

Zoological Medicine. Veterinarians who specialize in working with zoo animals or free-roaming wildlife are certified as diplomates in zoological medicine. To receive this distinction they must have at least three years of professional training beyond that required for a degree in veterinary medicine and fulfill other requirements of the American College of Zoological Medicine. Although sometimes referred to as "wildlife veterinarians," that designation is not an official title but merely a descriptor of what they do. Veterinarians with this expertise are critical to captive breeding and rehabilitation of wildlife.

Applications and Products

Wildlife conservation initially needed little in the way of specialized equipment and produced little in the way of products other than the game that was being protected and managed. The growth of wildlife conservation during the twentieth century, however, has produced a number of things that have been of use outside of wildlife conservation–not the least of which is a land ethic that was fostered by Aldo Leopold in his 1949 book, *A Sand County Almanac.* Not all land use should be determined by economics. Leopold noted that ecosystem "health is the capacity of the land for self-renewal." He defined conservation as "our effort to understand and preserve this capacity." That's what the field of wildlife conservation is all about.

Conservation and Mitigation Banking. Within a few years of the passage of the Endangered Species Act of 1973, the legal burden of providing mitigation for environmental harm done was taking up a good deal of court and industry time. In 1995, the federal and state governments approved mitigation banking as a way to provide predetermined mitigation for environmental harm to wetlands. It was a free-market process whereby an entrepreneur would restore or develop wetlands that would be approved as a mitigation bank. Then, when a court decided that mitigation needed to be done to offset damages to a wetland, the party causing the harm could buy mitigation credits from an approved mitigation bank to satisfy the obligation. The success of mitigation banking for wetlands led to the development of a similar program, called conservation banking, to provide predetermined conservation credits to mitigate harm done to an endangered species.

Bird Banding. Being able to mark a bird so that it can be identified as an individual at a later date was essential to understanding bird longevity, migration and other movements, and bird social behavior. Banding involves placing a serially numbered metal band on a bird's leg, recording details of its sex, age, and where it was banded, and having those data available in a central repository to retrieve if the bird was found by someone else, somewhere else, at a later time. Bird banding began as a hobby in the early 1900's but was quickly realized as a useful scientific tool. In 1920, the Bureau of Biological Survey took over all responsibilities for managing bird banding and banding data to ensure quality of data and so that the technique could be used in wildlife management. Later techniques were developed for adding colored plastic bands along with the serially numbered band, which allowed researchers to identify individuals by their color-band combinations without having to recapture them. These techniques continue to be vital tools in bird conservation.

Translocation Efforts. One of the major conservation problems facing wildlife is fragmentation of habitat and isolation of species in small pockets of habitat. The small number of individuals in such populations results in the population being highly vulnerable to extinction as a result of random events (such as wildfire, tornado, or predation) and also subjects the population to inbreeding, something that can result in genetic harm to the population. What was needed was a way to move individuals from one population to another and to have some assurance that they would stay in the new area and not attempt to return to their original site. The endangered red-cockaded woodpecker is an example of a species facing such problems. Studies of individually color-banded birds revealed that breeding males removed from their home would quickly return from great distances, but that young birds, especially young females, could be moved and would remain at a new site. Translocation efforts using young birds and careful timing have been improved upon through research and are now a common tool for this and other bird species. Similar efforts have also been developed and used for species such as the black-footed ferret. Radio telemetry is often used in conjunction with translocated animals to monitor their movements closely.

Photography. Whereas photography has long been used in many disciplines, it has often been as a supplement or simply for illustration rather than filling a specific need. With the advent of high-quality digital cameras, photography has taken on important roles in documenting wildlife and wildlife-conservation efforts. Researchers take photos for specific purposes and in specific ways rather than merely capturing snapshots. Wildlife photographers now play very important roles in both research and in wildlife-conservation education and communication.

Impact on Industry

During the 1960's and 1970's, there was a global surge in scientific and public recognition of the negative impacts of human population growth and uncontrolled development and exploitation of resources on the natural world. This surge was a result of an increasing rate of plant and animal extinctions, fragmentation and destruction of natural ecosystems, and global pollution problems that threatened not only the natural world but also human health. These events spurred a global interest in wildlife conservation, and that interest led to passage of conservation-related laws, regulations, and conventions around the world. These included the Convention on International Trade in Endangered Species of Wild Fauna and Flora (CITES), which the United States signed onto in 1973. The U.S. Congress also passed the Endangered Species Act of 1973 (ESA), which was quickly signed into law. The ESA stipulated that the federal government would provide money to the individual states if they too would pass endangered-species legislation. This was done. These laws and other environmental laws of the era established regulations that would protect wildlife and to some extent limit habitat destruction and pollution. The regulations provided legal pathways to stop environmental harm, thus providing tools for nonprofit conservation organizations to challenge harmful actions. Many new wildlife conservationists were needed in federal and state government to implement the new laws, and industry became more environmentally aware, hiring environmental consulting firms to monitor and evaluate their impacts on the environment and wildlife and to prepare and often implement habitat-conservation plans associated with mitigation efforts. New wildlife-conservation career paths emerged, and funds became available for the development of techniques and technology needed for wildlife conservation. Although the ESA and state laws have been amended, wildlife conservation continues to grow and diversify.

Careers and Course Work

Minimum wage, entry-level positions are available in the field of wildlife conservation for high school graduates, but these are typically unskilled-labor positions associated with management of wildlife habitat, cleaning cages and feeding animals, and clerical positions for government or nonprofit organizations. These are great positions from which to learn about the long hours and hard work associated with wildlife conservation. They are also positions from which to gain experience to add to a resume and contacts who can serve as references as one moves upward in a career. Internships, paid or unpaid, will provide experience and put one into the arena of wildlife conservation. As with other fields, opportunity, responsibility, and job flexibility come with post-high school education. An associate's degree from a community college with courses in English, speech, biology, chemistry, and basic math is a good stepping-stone to a bachelor's degree. English composition and communication skills are important: Being able to write and speak well are essential throughout a career. A bachelor's degree in biology, environmental studies, environmental science, or wildlife ecology is a good background for many entry-level positions in government, industry, or nonprofit organizations. For most, however, a master's degree with specialization and work, or volunteer internship, experience in a specific field of interest is key to a brighter future. Research positions generally require a Ph.D. or a veterinary medicine degree. As graduate students work under the mentorship of a major professor, when it is time to select a graduate program, one will be best served by choosing a graduate program based on the professor with whom he or she would most like to study. It is in the student's best interest to find an individual who has taken the path that he or she would like to follow.

Social Context and Future Prospects

Wildlife conservation was barely known as a career option two hundred years ago. It has since grown in importance and diversified into areas of specialization as diverse as education, law enforcement, satellite telemetry, zoological medicine, land management, wildlife photography, and the rescue of endangered species. It remains a vibrant, actively growing field that offers opportunity for travel to distant jungles and ocean depths, zoos and research facilities, parks and preserves, backyards and city streets. Any career in wildlife conservation is one from which to learn and grow continually. The challenges are great and change and innovation will be the norm for the future as well. More sophisticated transmitters, greater use of satellite telemetry, and more use of molecular

Fascinating Facts About Wildlife Conservation

- Management for the endangered red-cockaded woodpecker requires fire to control the understory in the old-growth pine forests in which it lives. Lightning-started fires managed the habitat naturally, but because roads act as firebreaks, managers have to introduce fire to their ecosystem.
- Wildlife telemetry has demonstrated that a male Florida panther wanders over about 200 square miles, while a female stays within an area of about 75 square miles.
- Naturalist Paul Bartsch, the first person to use bird banding to monitor bird movements in North America, was not doing it to study the herons he banded. He was interested in how pond snails got from one pond to another and suspected they did so by crawling onto the legs of herons and egrets, in effect, hitching a ride.
- Between 1903 and 1909, President Theodore Roosevelt created 150 national forests, conserving land, wildlife, and habitat for the future.
- President Roosevelt also created fifty-one federal bird reservations. These later became part of the United States' National Wildlife Refuge System.
- The California condor, the largest North American vulture, was at the brink of extinction in 1987 as a result of lead poisoning, habitat loss, and shooting. Only twenty-two individuals remained. These were taken into captivity and through captive breeding, their numbers had increased to 381 by late 2010, 192 of which had been released into the wild.
- Federal law requires that shrimpers use turtle excluder devices (TEDs) on their nets in order to protect sea turtles. When a turtle gets caught, the net opens up and allows the turtle to swim free.
- Scientists tracking killer whales have learned that some groups stay within a local area, feeding on salmon and other fish, while others wander the oceans hunting for seals.

biology are on the horizon. Oil spills, invasive species, habitat losses, as well as human factors such as discarded medicines and diseases, including cancers perhaps caused by factors in the environment, threaten wildlife. Finding causes and cures for them may provide hope for mankind. Challenges for wildlife conservationists in the future will be to connect isolated populations through development of habitat corridors and to find better ways of translocating individuals to preserve genetic diversity, to unlock the secrets of captive breeding of difficult species, and to find the will and the means to restore the habitats they need.

Jerome A. Jackson, Ph.D.

Further Reading

Bonar, Scott A. *The Conservation Professional's Guide to Working with People.* Washington, D.C.: Island Press, 2007. While the focus of wildlife conservation may be wild animals, a wildlife-conservation professional must be able to convince others of needs for actions. This guide provides key advice for success.

Cleva, Sandra. "Working a Beat for Wildlife." *The Wildlife Professional* (Spring, 2010): 46-48. A well-written account of what it is like to be a federal wildlife law enforcement officer.

Fanning, Odom. *Opportunities in Environmental Careers.* New York: McGraw-Hill, 2002. A clearly written, useful book for those interested in environmental careers in many fields. Materials on wildlife management, range management, soil conservation, fisheries, forestry, and others are relevant to wildlife conservation; includes considerable history, educational requirements, and discussion of schools with good programs.

Jackson, Jerome A., William E. Davis, Jr., and John Tautin, eds. *Bird Banding in North America: The First Hundred Years.* Cambridge, Mass.: Harvard University Nuttall Ornithological. Provides a detailed history of the origins, use, and conservation value of bird banding.

Janovy, John, Jr. *On Becoming a Biologist.* 2d ed. Lincoln: University of Nebraska Press, 2004. This enjoyable, inspiring, autobiographical account provides insight into the requirements, challenges, opportunities, and sheer joy associated with working in wildlife conservation.

Leopold, Aldo. *A Sand County Almanac.* New York: Oxford University Press, 1949. A classic, very readable, inspiring series of essays on humakind's relationships with land, wilderness, conservation, and wildlife.

Mazet, J. A., G. E. Hamilton, and L. A. Dierauf. "Educating Veterinarians for Careers in Free-Ranging Wildlife Medicine and Ecosystem Health." *Journal of Veterinary Medical Education* 33, no. 3 (Fall, 2006): 352-360. A technical article that discusses needs, opportunities, and the future of veterinary medical roles in wildlife conservation.

Mittermeier, Cristina. "The Power of Conservation Photography." *The Wildlife Professional* (Fall, 2007): 30-31, 43. Discusses how and why photography is used in conservation and the changing importance of photography in conservation.

Van Riper, C., III, et al. "Career Opportunities in Ornithology." *American Birds* 43, no. 1 (Spring, 1989): 29-37. An assessment of a diversity of bird-focused careers and their prospects for the future.

Wilson, E. O. *Naturalist.* Washington, D.C.: Island Press, 2006. A wonderful autobiographical account of one of the most accomplished ecologists of the past century. Provides examples of conservation issues from around the world and how they are dealt with or should be dealt with.

Web Sites

National Fish and Wildlife Foundation
http://www.nfwf.org

National Wildlife Federation
http://www.nwf.org

National Wildlife Refuge Association
http://www.refugeassociation.org

Space Today Online
http://www.spacetoday.org/Satellites/Tracking/SatTracking.html

World Wide Fund for Nature
http://wwf.panda.org

See also: Fisheries Science; Forestry; Photography.

WIND POWER TECHNOLOGIES

FIELDS OF STUDY

Aerodynamic engineering; applied physics; atmospheric sciences; computer technology and networking; electrical engineering; hydraulics; mechanical engineering; meteorology.

SUMMARY

By 2010, wind power was the fastest-growing source of new electrical energy in the world. Advances in wind-turbine technology adapted from the aerospace industry have reduced the cost of wind power from 38 cents per kilowatt hour (during the early 1980's) to as little as 3 cents per kilowatt hour. This rate is competitive with costs of power generation from fossil fuels, but costs vary according to site.

KEY TERMS AND CONCEPTS

- **Acoustic Noise:** Audible and subaudible sound that is of no use in the operation of a turbine; excess acoustic noise is considered undesirable.
- **Adjusted Availability:** Amount of time that a wind turbine is usable, eliminating time when the turbine is out of operation for various reasons.
- **Aeolus:** Greek god of wind.
- **Air Brake:** Aerodynamic device used to slow a wind-turbine rotor in strong winds, important to prevent damage.
- **Capacity Factor:** Measure of wind-turbine productivity, based on reliability, performance, average wind speed in a given area, and power of the turbine's blades.
- **Cut-Out Wind Speed:** Speed at which a given wind turbine ceases to generate electricity, usually about 65 miles per hour. Some turbines have no cut-out speed but control rotor motion.
- **Fuel-Saver Mode:** Use of a wind generator to replace conventional (usually fossil) fuels in a power plant.
- **Installed Capacity:** Power-generating capacity of a wind turbine.
- **Main Shaft:** Axle or spindle that connects a wind-turbine rotor (with blades) to the transmission apparatus and generator.
- **Multimegawatt Wind Turbine:** Largest turbines, usually with blade diameter of 200 feet or more. The generating capacity is based mainly on the diameter that a blade sweeps.
- **Parallel Generation:** Electric power produced in conjunction with electric utilities, on the electrical grid, the infrastructure by which power is shared between producers and consumers.
- **Setback:** Required area between a wind turbine and the property line meant to keep the facility at a safe distance from other property owners or users of adjacent land.
- **Wind Farm:** Area devoted to generating electricity from an array of wind turbines.

Definition and Basic Principles

Wind power uses kinetic energy in the form of moving air to turn rotors (and attached blades) that in turn power a generator to produce electricity for local use or to be fed into a grid. Wind farms comprise groups of wind turbines. By the early twenty-first century, wind power was becoming competitive in cost with electricity generated by fossil fuels. After 2004, it became less expensive to generate electricity from wind than from fossil fuels in many areas, as its use exploded by 20 to 25 percent annually. As the infrastructure and use of wind power have grown, the unit price has fallen.

Background and History

The use of wind turbines to generate energy is not a new technology. The Spanish author Miguel de Cervantes had his character Don Quixote "tilting at windmills" in the early seventeenth century. The Dutch used their famous windmills to generate modest amounts of energy (as well as to run pumps to empty flooded lowland areas) for hundreds of years; some immigrants to the western United States erected windmills on their farmsteads to generate power before electricity reached them.

What has changed is the size of turbines, their number, efficiency, and organized use to compete with fossil fuels as an energy source in a world experiencing climate change. The use of wind as a serious source of energy in an organized manner associated with electrical grids began with the oil crisis of the

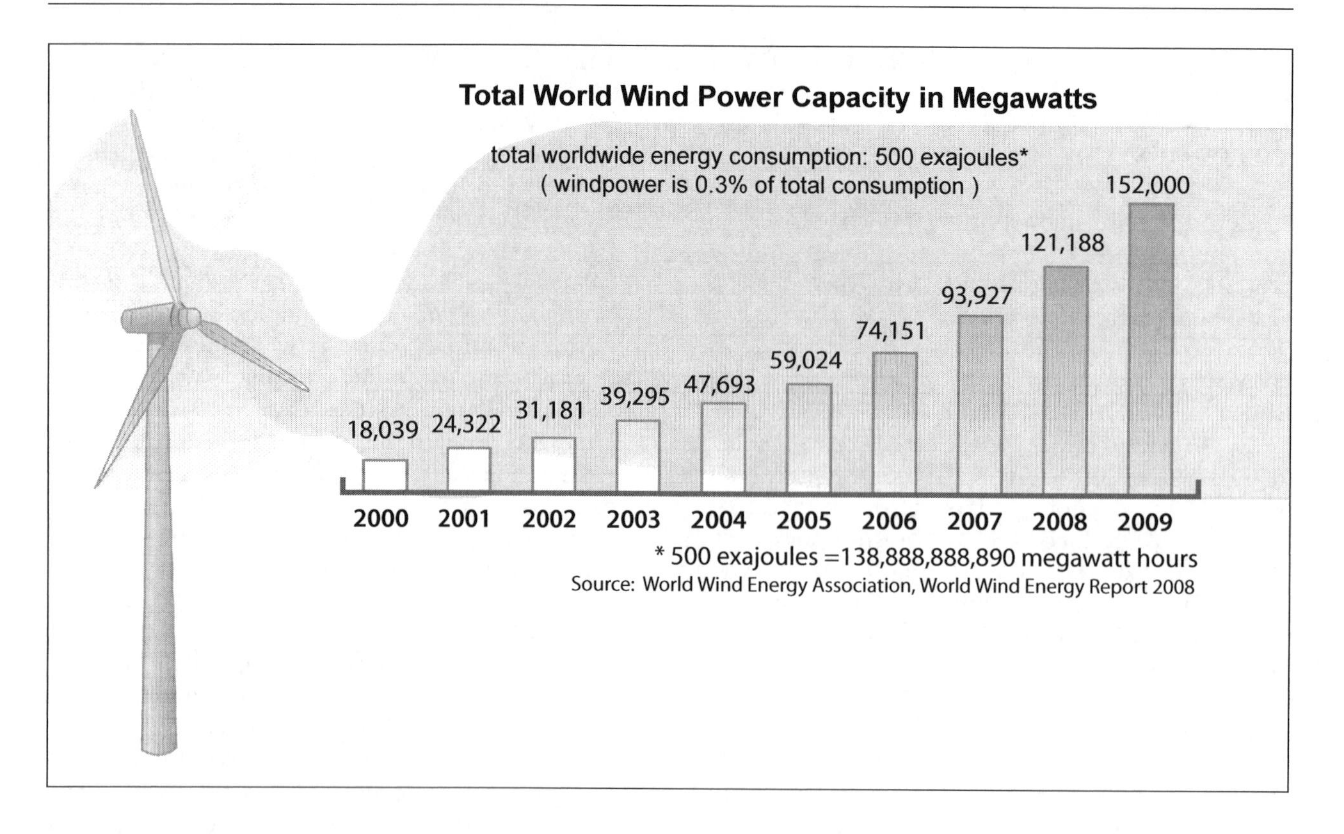

1970's, in which the entire world saw prices of fossil fuels, the major source of power, increase rapidly as a result of disruptions in supply. By the 1980's, wind arose as a serious alternative to fossil fuels in power planning, as concentrations of greenhouse gases in the atmosphere increased and were recognized as a source of climate change.

Denmark, which was dependent on imported oil during the 1970's, made an enduring commitment to achieve energy independence when supplies were embargoed and its economy was, as a result, devastated. Wind power became an important part of this strategy and Denmark has become a world leader in wind-turbine technology. Work on Danish turbines is a major reason that the technology has come to generate electricity that competes in price with oil, coal, and nuclear power. In the meantime, Denmark has built infrastructure that provides several thousand jobs. In matters of advanced technology, Denmark dominates the worldwide wind-power industry. Danish companies have supplied more than half the wind turbines in use worldwide, making wind-energy technology one of the country's largest exports. Some Danish wind turbines have blades almost 300 feet wide–the length of a football field. During January, 2007, a very stormy month, Denmark harvested 36 percent of its electricity from wind, almost double the normal proportion.

Europe's largest onshore wind farm, able to generate enough power for 200,000 homes, was approved in 2008 by the Scottish government. Announcing the approval of the new wind farm before the World Renewable Energy Congress in Glasgow, First Minister Alex Salmond said the 152-turbine Clyde Wind Farm near Abington in South Lanarkshire would make Scotland the green energy capital of Europe.

In 2002, Spain's tiny industrial state of Navarra, generated 25 percent of its electricity through wind power. By 2007, 60 percent of its electricity was from renewable sources (mainly wind, with some solar), and the state had plans to raise that proportion to 75 percent by 2012. The World Wind Energy Report of 2009, published by the World Wind Energy Association, listed these countries as the top five in terms of total wind-generation capacity: United States 35,159 megawatts; Germany 25,777 megawatts; China 26,010 megawatts; Spain 19,149 megawatts; and India 10,925 megawatts.

Electricity generation accounts for about 40 percent of U.S. greenhouse gases. In 2008, wind turbines in the United States generated about 48 billion kilowatt hours of electricity, enough to power about 4.5 million homes. That is a significant rise from previous years but still meets only 1.2 percent of the country's electrical demand. Wind has become a major energy source in some small areas. Following nine years of review and copious controversy, on April 28, 2010, the U.S. federal government approved the United States' first offshore wind farm, Cape Wind in Nantucket Sound, a 130-turbine array off Nantucket, Massachusetts.

Some U.S. state laws require utilities to work toward generating 15 to 25 percent of their electricity from sources that do not produce carbon dioxide. Wind capacity in the Pacific Northwest, where it is often used with hydroelectric, soared from only 25 megawatts in 1998 to about 3,800 megawatts by 2009. During 2006, Washington State added 428 megawatts of wind power, trailing only Texas in new installations. In Seattle by 2009, 90 percent of electricity was from nonfossil sources: traditional hydropower plus an aggressive emphasis on wind by the Bonneville Power Administration. Under best wind conditions, by 2009, 18 percent of the power generated by the Bonneville Power Administration was coming from wind; some projections raised that figure to 30 percent in 2010.

According to Randall Swisher, former executive director of the American Wind Energy Association, the electrical grid in the northwest is especially inviting to wind-power developers because of its hydroelectric distribution network, relatively reliable wind, progressive utility companies, and new state laws that establish preferences for renewable energy. Hydroelectric dams also have been built in river gorges and other valleys that are natural conduits for wind. The transmission lines connecting the dams with urban areas have been created with considerable surplus capacity, a built-in network for wind power. Farmers in the area have been earning $2,000 to $4,000 per site each year by allowing wind turbines on their property.

How It Works

Wind power uses kinetic energy, forcing moving air to turn a rotor attached to a set of blades that powers a generator to produce electric power for local use or to be fed into a grid. Wind farms comprise groups of wind turbines. The power capacity of wind turbines may be restricted only by the size of a blade that can be hauled to a site, mounted, and maintained with some degree of structural integrity. A turbine at 250 feet takes advantage of winds that are 20 percent stronger at that elevation than at 150 feet.

Wind power in its infancy also has been plagued by a shortage of transmission capacity that will allow its transport from turbine sites to users, many of whom live in urban areas. Texas has been a leader in the United States in building transmission capacity. In 2007, the Public Utility Commission of Texas approved transmission lines across the state that eventually will carry up to 25,000 megawatts of wind energy by 2012, a 500 percent increase from 2008. Shell and TXU Energy are planning a 3,000-megawatt wind farm in the Texas Panhandle. In 2007, Texas had installed wind power of 4,356 megawatts. Land values in some counties with wind farms have nearly doubled.

A major problem for wind energy is that the hottest days, when power demand is highest, are usually the least windy. Wind power cannot be used to replace fossil fuels all the time.

Applications and Products

Wind power has developed as a large-scale industry and supplementary source of electricity to which the science and engineering of rotors and supporting assembly (including airline technology) has been applied. To date, the promise of wind energy pertains mainly to large-scale farms that can contribute energy to the electric grid and limit emissions of greenhouse gases from energy produced by fossil fuels.

Wind energy on a smaller scale, useful to individual homeowners or small businesses, also has come into use. The United States invented household-scale wind turbines in about 1920 and is a world leader in that technology. More than 90 percent of the world's household wind turbines have been installed in the U.S. Growth in the number of household wind turbines averaged 14 to 25 percent a year between 2000 and 2009 worldwide, according to Randall Swisher. At least one-half acre of land is required for a residential wind turbine, and regulators must permit a tower of at least 35 feet (the higher the tower, the more efficient the turbine). Technological problems (such as vibration that may damage brick chimneys) have limited small-scale applications to date, but when these problems are solved, small turbines could come into use (possibly supplemented by other sources, such as solar power) to allow small users to achieve independence from fossil fuels.

Impact on Industry

Major corporations, including Shell and BP, have been moving into wind power. At one point, wind power became so popular that a shortage of parts was causing installations to fall behind demand. A new phrase in the English language, "wind rich," describes an area with relatively steady, unimpeded access to turbine-ready breezes. Some pitchmen for wind–including business tycoon T. Boone Pickens and former president George W. Bush–believe that the United States may derive as much as 20 percent of its electricity from wind power by the year 2020.

The distribution system has become a major challenge for wind energy. By 2008, wind-power capacity in the United States was increasing very quickly, but the electrical grid has not kept up with the additional supply in many areas. A $320 million, 200-turbine array at Maple Ridge Wind farm in upstate New York was forced to shut down for lack of transmission capacity–even when the wind blows. Former governor of New Mexico Bill Richardson, who was federal energy secretary under president Bill Clinton, has said, "We still have a third-world grid." In places, the electric grid is almost a century old, with 200,000 miles of lines and hundreds of owners. Some planners are calling for a major, long-term project to modernize the grid that would be akin to the interstate highway system. Only then will the sites with the best wind conditions, which often are sparsely populated, be able to supply major urban areas, where the need for new energy sources is greatest.

Despite the recession that began in 2007, lack of efficient distribution, and tight credit, wind-power capacity in the United States grew 39 percent in 2009, rising to about 2 percent of electrical capacity, or enough to supply the equivalent of 9.7 million homes. More than one-quarter of that power was being generated in Texas. During 2009, about 80 percent of new electricity-generating capacity came from wind, and wind-energy manufacturing employed 85,000 people.

Careers and Course Work

Wind Powering America provides state-by-state lists of wind-energy educational programs, public and private, at two- and four-year colleges. The list also contains helpful annotations. Vestas, the world's largest producer of wind-power technology, has entered into a long-term partnership with the University of Wisconsin's College of Engineering. Plans exist to establish a research-and-development facility that focuses on technology transfer. This partnership will also support professorships that concentrate on wind-energy education-related research.

Careers in wind power range from technicians to electrical engineers, electrical design engineers, environmental architects, and turbine installers. Repair technicians and wind-farm managers will also be in demand as wind power becomes more widespread. Some developmental engineers are employed by large companies, such as General Electric. Education varies from a few months for installers to college degrees (undergraduate and graduate) for engineers.

Wind-power education usually includes courses in advanced mathematics, physics, computer science, electrical engineering, and mechanical engineering.

The educational landscape is new and evolving rapidly. Texas Tech University has a Wind Science and Engineering Research Center. In 2008, Professor Ernest Smith began to teach the first-ever wind law course at the University of Texas School of Law. Cerro Coso Community College in California offers a typical associate's degree. California Wind Tech, in Rancho Cucamonga, offers intensive entry-level training for wind technicians and other professionals in courses as brief as one month that may lead to jobs in circuit troubleshooting, wind-turbine construction, and schematic reading. Iowa State University offers wind-energy science, meteorology, and engineering and manufacturing courses that focus on wind-power generation.

Social Context and Future Prospects

Wind turbines have not been universally welcomed. Some of their neighbors complain that they comprise "visual pollution." The power capacity of wind turbines may be restricted only by the size of a blade that can be hauled to a site, mounted, and maintained with some degree of structural integrity. A turbine at 250 feet takes advantage of winds that are 20 percent stronger than those at 150 feet, said Dr. Mark Z. Jacobson, an associate professor at Stanford

Fascinating Facts About Wind Power Technologies

- Wind power in the United States in 2007 grew by 45 percent, adding 5,244 megawatts of capacity—one-third of all the new electrical capacity in the country. In one year, wind energy employment in the United States doubled to about 20,000.
- The 11,000 residents of Hull, Massachusetts, embraced wind power in a big way, with two turbines in 2008 and more planned. Hull's power plant previously had no generating capacity of its own and was buying energy from another source at an average of about 8 cents per kilowatt-hour. The wind turbines supply it for about half as much.
- While wind power still was a tiny fraction of energy generated in the United States, some areas of Europe (Denmark, as well as parts of Germany and Spain) were using it as a major source. Germany's northernmost state of Schleswig-Holstein was using wind for nearly a third of its electricity by 2009.
- Wind turbines were popping up in some unexpected places. Parts of the United States' Rust Belt, along the south shores of the Great Lakes, are re-tooling as wind-energy centers. In Lackawanna, New York, a suburb of Buffalo on a former Superfund (toxic waste) site, the 2.2 miles of Lake Erie shoreline above the former Bethlehem Steel plant hosts eight white turbines spinning 150-foot blades turning the wind off the lake into power.

University's department of civil and environmental engineering.

The bigger the turbine, the greater the chance it may kill birds, about 40,000 of them a year in the United States by 2006. That amounts to one bird per thirty turbines each year; however, that is a small fraction of the millions killed each year by domestic cats.

Without major advances in ways to store large quantities of electricity or large changes in the way regional power grids are organized, wind may run up against its practical limits sooner than expected. In many places, wind tends to blow best on winter nights, when demand is low. Robert E. Gramlich, policy director with the American Wind Energy Association, said that wind energy could be integrated into an electrical grid on a large enough area that ebbing wind in one part could be balanced by continuing breezes in another.

Noise and vibrations from large wind turbines, which some people compare with a jet taking off, have caused headaches and insomnia in some people who live nearby. Nina Pierpont, a pediatrician who lives in Malone, New York, has given the condition a name, Wind Turbine Syndrome, and plans to write a book about it. Amanda Harry, a physician who lives in the United Kingdom, says the turbines also can cause anxiety, vertigo, and depression. Acoustical engineer Mariana Alves-Pereira, who lives in Portugal, asserts that wind-turbine noise and vibration may contribute to strokes and epilepsy. In 2008, the European Union exonerated wind turbines of culpability in any medical condition except occasional loss of sleep. Curing the syndrome is usually simple: establishing proper setbacks, distances of 1,000 to 2,500 feet, with some people insisting on setbacks of a mile or more.

Bruce E. Johansen, B.A., M.A., Ph.D.

Further Reading

Brown, Lester R. *Plan B 4.0: Mobilizing to Save Civilization.* New York: W. W. Norton, 2009. An excellent text about the role of alternative energy, including wind, in combating global warming.

Burton, Tony, et al. *Wind Energy Handbook.* Hoboken, N.J.: John Wiley & Sons, 2001. A comprehensive, textbook-style compendium of wind-energy technology.

Flannery, Tim. *The Weather Makers: How Man Is Changing the Climate and What It Means for Life on Earth.* New York: Atlantic Monthly Press, 2005. Expounds on why wind energy will play a major role in twenty-first-century energy policy.

Gipe, Paul. *Wind Power: Renewable Energy for Home, Farm, and Business.* White River Junction, Vt.: Chelsea Green, 2004. A handbook that emphasizes how individuals can use wind energy.

McGuire, Bill. *Surviving Armageddon: Strategies for a Threatened Planet.* New York: Oxford University Press, 2005. Chronicles role of wind as a major player in a new energy paradigm.

Musgrove, Peter. *Wind Power.* New York: Cambridge University Press, 2010. Details developments in wind power, including worldwide increases in capacity, changes in technology, politics, and problems.

Web Sites

American Wind Energy Association
http://www.awea.org

Cape Wind
http://www.capewind.org

European Wind Energy Association
http://www.ewea.org

Utility Wind Integration Group
http://www.uwig.org

Wind Powering America
http://www.windpoweringamerica.gov

World Wind Energy Association
http://www.wwindea.org.

WIRELESS TECHNOLOGIES AND COMMUNICATION

FIELDS OF STUDY

Communications system engineering; computer science; computer security; information assurance; mobile computing; networking; systems engineering; usability engineering; telephone technology.

SUMMARY

Wireless technology comprises the hardware, software, and systems that support the transfer of signals, over long or short distances, without the use of electrical conductors or wires. Communications is the transfer of information between a sender and receiver; and wireless communications use wireless technology. Telecommunications, the transfer of messages between a sender and receiver; and data communications, the transfer of data between a sender and receiver, are the two most popular forms of wireless communications.

KEY TERMS AND CONCEPTS

- **Cellular Phones:** Technology that supports wireless telephones, one of the most popular wireless applications.
- **Code-Division Multiplexing:** Technique that creates multiple channels from one physical channel by using code keys to create the channels.
- **Cordless Appliance:** Appliance that operates without electrical cords, including a large number of devices, such as television controllers, cordless computer mice, and cordless phones.
- **Encoding:** Process of changing one representation of data into another; most wireless technologies incorporate a form of digital to analogue encoding, while others, such as code-division multiple access (CDMA), use a digital-to-digital encoding.
- **Modulation:** Process that changes the amplitude, phase, or frequency of a wave; many wireless protocols use modulation.
- **Time-Division Multiplexing:** Popular way to create multiple logical channels from one physical channel by using equally spaced cells.
- **Two-Way Radio:** Communications technology that uses two channels, one to send and one to receive. Two-way radio is used by companies, emergency services, and the military.
- **Wireless Networking:** Connection of computers using wireless technology—a very active area of wireless technology, with many new protocols being developed.

Definition and Basic Principles

Wireless technology includes wireless networking, wireless telecommunications, and other wireless devices. Wireless networking started in the 1980's with work supported by Bell Labs and culminating in the Institute of Electrical and Electronics Engineers (IEEE) 802.11 standard, released in 1997. The IEEE 802.11 standard has improved over the years to become a robust wireless data network. In 1998, Bluetooth was introduced as a peer-to-peer home network and now dominates the peer-to-peer market, even being given an IEEE 802 designation, the IEEE 802.15, in 2002. Recently, the IEEE 802 committee has released a wide group of wireless network standards, including a high-speed metropolitan area standard (the 802.16), several new mobile phone standards (the 802.16e/802.20), and a new standard for remote area access, the 802.22.

Wireless telecommunications not only includes the popular code-division multiple access (CDMA) and Global System for Mobile communications (GSM) cellular telephone technologies but a vast array of other wireless phone systems as well. These include the two-way phone systems used by businesses, such as a power company; by emergency services, such as a fire department; by the military, such as battlefield communications systems; by public service systems, such as the marine VHF radio; and by individual users, such as ham radio operators. There are many other wireless devices in use as well, including infrared devices such as television controllers, cordless mice, garage-door openers, model-car controllers, and several satellite-type devices, such as the Global Positioning Systems installed in cars.

Background and History

In 1880, the first wireless communication device, the photophone, was developed by Alexander Graham Bell. Guglielmo Marconi, an Italian inventor, proved the feasibility of radio communications by sending and receiving the first radio signal in 1895. By the early 1900's, Marconi and Serbian inventor and engineer Nikola Tesla both claimed to have invented the basics of radio transmissions, and both started companies to support radio transmission. A number of radio phones were developed during the early 1900's, but the first modern analogue mobile phone was developed at Motorola by Dr. Martin Cooper in 1973. By 1979, the first cellular phone network was implemented in Japan, based in part on research at Bell Labs, and since then cellular phone networks have seen explosive growth.

In 1929, the first commercial radio transmission was made from KDKA in Pittsburgh. As of 2011, there are many radio, television, and satellite broadcasting stations transmitting a wide variety of news and entertainment.

Infrared and microwave technology operate at a higher frequency than radio but have nearly as many technology applications. Infrared waves were discovered by British-German astronomer Sir William Herschel in 1800. In the late 1990's, many uses for infrared technology were discovered, including remote controls, wireless mice, connecting printers to computers, and even heating saunas. The development of radar during World War II led to many advances in microwave technology, including the accidental discovery of the microwave oven in 1940. The first high-speed microwave network was begun in 1949, and microwave communications is an important component of network infrastructure. In 1965, Bell Labs astronomers Arno Penzias and Robert Wilson discovered cosmic microwave background (CMB) radiation by accident while using a large horn antenna. CMB is significant in that it is considered "noise leftover from the creation of the universe" and points to strong evidence supporting the big bang theory.

How It Works

Wavelength and Antennae. Wireless communications transmit data using sinusoidal waves. The different types of waves can be characterized by their amplitude (the height of half a sine wave), their frequency, how many complete sine waves are in a fixed length, and their phase (the beginning zero point of a sine wave). A popular measure of wireless waves is their hertz (Hz), the number of cycles per second of the wireless wave. For example, radio and television waves are between 10^7 and 10^9 Hz, microwaves are between 10^{10} and 10^{11} Hz, and infrared waves are between 10^{13} and 10^{14} Hz.

Most wireless transmissions require the use of antennae, which are transducers that convert electrical energy into wave energy, to send and receive electromagnetic waves. There are many types of antennae in use. The simplest radio antenna is the dipole antenna, which consists of two wires running in opposite directions connected to a central feed element. Some antennae, such as parabolic or horn antennae, are designed to collect multiple signals into one stronger signal; while others, such as HRS curtain antennae, use an array of simple elements to produce a stronger signal. Antennae can be large, such as the proposed Square Kilometre Array, or small, such as the antenna contained in a smart credit card. Antennae can be directional or nondirectional, fixed or mobile, and designed for sending, receiving, or both. In the design of antennae, many factors are considered, including gain, efficiency, impedance, polarization, and bandwidth.

Wireless technology and communications fundamentals require an understanding of how analogue and digital information are prepared for transmission, how the transmission actually takes place, and how the receiving equipment converts the delivered data into usable information.

Digitizing and Encoding. Some information that is transmitted by wireless devices starts with a digital representation, such as data stored on a computer, and some starts with analogue representation, such as sound. Preparing digital data for transmission is relatively easy to do. In some cases, nothing is done to the data. In others, a minor transformation is done, such as performing Manchester encoding, while in others a fairly complex encoding scheme is applied, such as performing CDMA encoding. If the data to be sent is analogue, then several options are available. Some analogue data, such as a phone conversation, can simply be modulated onto a carrier wave and sent, as with the old analogue telephone system, but most analogue data needs to be digitized before it is sent. There are several approaches to digitizing data, but most of them, such as pulse-code modulation for voice, involve sampling the analogue data, representing the sample data digitally, and normalizing the data.

Modulation and Multiplexing. When data is

transferred from a sender to a receiver over the air, the last step in this process is to modulate the data onto a carrier wave. To modulate analogue data, simply combine the two waves. For digital data, there are a number of modulation techniques in use. The simplest of these are: amplitude-shift keying, using different amplitudes to represent 0/1; frequency-shift keying, using different frequencies to represent 0/1; and phase-shift keying, using different phases to represent 0/1. Other digital-encoding techniques, such as Gaussian minimum-shift keying, are fairly complex.

A single carrier wave often supports multiple channels so that more than one data stream can be sent over a carrier at the same time. This process is called multiplexing. The old analogue phone system used frequency-division multiplexing to carry multiple calls simultaneously. Many digital phone systems use time-division multiplexing, where the digital path is divided into cells, and evenly divided cells create a channel to support multiple paths on one carrier signal. Another popular digital multiplexing technique is code-division multiplexing, in which multiple digital data streams are modified by a code word and then combined into a broader data stream that can be sent over the carrier wave.

Applications and Products

Most of the applications of wireless technology are wireless communications, and the majority of these are mobile phones and wireless networks.

Cellular Phones. Although quite primitive, the first mobile phone was demonstrated by Alexander Graham Bell in 1876. Mobile phones were being used by industry, the military, and government by the 1980's, but they were restricted to a single central antenna. In 1947, Bell Labs introduced the first cellular network architecture, and this has matured into modern cellular phone networks. Cell phones are electronic devices that include a processing unit, a graphics processor, memory, one or more communications chips, and one or more antennae. When operating as a phone, these devices transmit a signal to the nearest cell phone tower, which then forwards it to a controller, which manages multiple cell phone towers. Local calls may be immediately forwarded to the receiver, but for calls at a greater distance, the signal may be transmitted to a satellite or microwave tower and then forwarded to the receiver. Cellular networks derive their name from the fact that each primary cell phone tower creates a cell for phones in range of the tower. Cellular telephone architectures include a process to handle phones being in the range of two towers, and they hand off as a cellular user moves from one tower to another. One of the unique features of cellular phone networks, compared with other wireless communications networks, is its ability to authenticate users and bill them for services.

As of 2011, there are two competing cellular phone architectures, with many more on the horizon. In the United States, the main cellular architecture is code-division multiple access (CDMA). The key characteristic of this architecture is its use of code-division multiplexing to send multiple calls on one signal. The second most popular architecture in the United States is Global System for Mobil (GSM). The most prominent feature of this architecture is that it is an international standard. In fact, worldwide GSM is more popular that CDMA, and it uses time division multiple access (TDMA) to support multiple calls on one signal. Some of the new architectures being proposed to replace CDMA and GSM, such as EDGE and LTE, are upgrades to current standards designed to improve Internet speed, while others, such the IEEE 802.20 standard are completely new standards that have been designed from the ground up to support high-speed Internet.

Wireless Local Area Networks. Wireless networks go back to the Aloha data network and microwave connectivity of data centers in the 1970's. By 1980, a number of scientists were investigating how wireless technology could be used for local area networks (LANs). Two types of connectivity were developed: peer-to-peer, where pairs of stations make a direct connection; and shared access, where all stations share the media. In 1997, the IEEE 802.11 wireless LAN standard was released, and most shared-media LANs developed after that conformed to this standard. In the IEEE 802.11 shared-access mode, each computer in the network uses an access point to connect to the corporate network and the Internet. Handoffs and connection problems for the IEEE 802.11 are handled much like cellular phones. While the IEEE 802.11 supports peer-to-peer connections, Bluetooth, released in 1998, has become the principle standard for these networks. Bluetooth is a standard often used to connect cameras to computers, thermostats to access points, headsets to stereos, and keys to central locking systems. In recognition of Bluetooth's success as a peer-to-peer network the IEEE introduced the IEEE 802.15 as its version of Bluetooth in 2002.

Remote Controls. Infrared waves have a higher frequency than radio waves, and this requires line-of-sight connectivity for applications. Infrared signals can be generated by a number of sources, including the Sun, but a common way for remote controls is with a diode (which is similar in operation to a light-emitting diode or LED). Many devices are used to collect infrared waves at the receiver, but most of them operate like an antenna. One of the most common applications of infrared technology is for remote controls–the most popular example is the one used with the television. A remote control has a number of buttons so users can send different instructions to a receiving device, which has an infrared antenna (sometimes called an infrared demodulator) and decoding mechanism. For TVs, digital video recorders (DVRs), and the like, a standard protocol, developed by Phillips called RC-5, is used for sending and receiving infrared signals.

In addition to remote controls, infrared-technology applications include peripheral connectivity for personal computers, night-vision systems, medical-imaging applications, military tracking systems, interactive game controllers, and saunas.

Satellite Communications and GPS. Satellite communications developed in both the United States and Russia shortly after the launch of Sputnik in 1957. Most satellite transmissions are in the microwave bands, but some use other electromagnetic wavelengths as well. There are a variety of devices for satellite transmissions, but many of them use an enhanced vacuum tube technology with names such as magnetron and gyrotron. Early satellite antennae were very large but have been greatly reduced in size.

From the beginning of satellite communication, scientists realized that measuring the difference in signals, over time, could be used to detect terrestrial positions. This basic observation has been greatly enhanced over the years to become the GPS that can be used by mobile phones and cars to determine their position on the Earth at any time. GPS uses microwave technology in the 1 to 2 megahertz (MHz) range, and one of its greatest successes has been the development of a number of small antennae for use in cell phones and cars.

IMPACT ON INDUSTRY

In the United States, most of the wireless technology and communications development has taken place in private industry. To a smaller extent, government, especially the military and the National Aeronautics and Space Administration (NASA), and universities have contributed to wireless technology and communications development. In other countries, especially Europe, there has been support for wireless technology and communications development in industry, universities, and the government.

Government, Industry, and University Research. All of the early developers of wireless technology (Bell, Marconi, and Tesla) started one, or more, companies to produce wireless devices, and all of their companies did considerable research on wireless technology. AT&T traces its roots back to Alexander Graham Bell, who spawned Bell Labs, which was a leader in the early wireless phone systems, although AT&T later divested Bell Labs. Marconi founded the Wireless Telegraph and Signal Company in England in 1897, and, among other projects, tried to develop a wireless network to connect the British Empire. Tesla started several companies that were active in wireless research and was reported to be working on a death ray at the time of his death in 1943.

Research and development in wireless technology and communications has not slowed down since these early days, with a long list of contributors, including Motorola, Raytheon, Verizon, Cisco, Intel, and many others. For example, in 1964, Amar Bose, an electrical engineering professor at the Massachusetts Institute of Technology, founded Bose Corporation and has since become a leader in loudspeaker research.

The military has always supported wireless communications research, such as the development of radar during World War II. Military research led to the development of the modern infrared night-vision systems and most recently research has been focused on and has made great progress in developing wireless control systems for planes and vehicles. NASA has also been a major contributor to wireless research. In 1961, it launched Telstar, the first communications satellite, and has been a leader in satellite and GPS research ever since. Universities have contributed to wireless research, especially in the area of wireless networks: Much research on the IEEE 802 standard for wireless networking was done in universities.

Fascinating Facts About Wireless Technologies and Communication

- In 1880, Alexander Graham Bell and American inventor Charles Tainter completed the first telephone conversation, using a wireless phone called the photophone.
- In 1947, Bell Labs developed an early cell phone architecture. In the 1960's, AT&T and Motorola competed to develop the first cell phone network in the United States. In1973, Martin Cooper of Motorola made a phone call to AT&T on the DynaTAC, winning the contest.
- In 1898, Nikola Tesla introduced the first remote control, the "teleautomaton," at the Electrical Exhibition in New York City by remotely controlling a model boat. As of 2011, remote controls are indispensable and are used to control TVs, models, drones, and much more.
- The idea for the microwave oven was discovered by engineer Percy LeBaron Spencer of the Raytheon Corporation in 1946 while working on a radar system. He later developed the first microwave oven. Infrared heating elements were used in many appliances, such as hair dryers by the 1990's, demonstrating another novel use of electromagnetic energy.
- Sputnik was launched by the Soviet Union in 1957, and in 1962, NASA launched Telstar, the first U.S. communications satellite. Communication satellites from many countries currently support a wide variety of communications.
- The Telecommunications Act of 1996, the first major overhaul of federal telecommunication regulations in sixty-one years, standardized high-definition television and several new high-speed mobile services.

Industry and Business. The wireless technology and communications industry is extremely large and covers a number of areas. One of the most prominent is the cell phone industry, dominated by Verizon and AT&T in the United States and China Mobile worldwide. A vast number of companies actually build cell phones and other wireless devices, including Apple, LG, Motorola, Samsung, Nokia, and HTC. Companies developing Bluetooth and the IEEE 802 wireless networking components are just as important as the cellular phone companies and include IBM, Cisco, Intel, Microsoft, NETGEAR, and Linksys.

There are thousands of remote-control companies producing many devices to control TVs, appliances, and models, and there are an equally large number of satellite communications companies, including DirecTV, Hughes, and EchoStar. A number of companies are now involved in the smart card industry, with Philips's MIFARE being one of the most important, and many are involved in radio-frequency identification (RFID) systems, used in all stores to track inventory. Wireless appliance companies, such as Whirlpool, manufacture microwave ovens, and a number of companies manufacture infrared heating elements.

Careers and Course Work

Earning a bachelor's degree with a major in electrical engineering, computer engineering, computer science, mathematics, or physics is the way most often selected to prepare for a career in wireless technology and communications. One needs substantial course work in mathematics and physics as a background for this degree. For some positions, such as antenna and infrared heaters design, an engineering background is advisable, while for others, such as developing cell phone applications, a programming background is necessary. For a position involving development of new products, one generally needs a master's or doctorate degree. Those parts of wireless technology related to the construction of devices are generally taught in engineering and physics, while those involved in using and managing wireless communications are usually taught in computer science and mathematics.

There are a wide variety of positions for those seeking careers in wireless technology and communications. Some go into hardware design of mobile wireless devices for companies such as Apple, Intel, and Cisco, while others go to work as wireless network managers for companies such as Verizon and AT&T. Some develop wireless software, such as those used in Microsoft's Windows Mobile 7, Apple's iOS, and Google's Android; still others take jobs in manufacturing to build microwaves and infrared saunas.

Social Context and Future Prospects

While wireless connectivity has improved the ability to communicate, it has also introduced some serious privacy and security issues. Wireless data is easily intercepted and unless encrypted, it is easy for hackers to read and misuse the data. Mobile computing also supports storing data in the cloud on remote servers, and even when the data is encrypted, issues are raised with the owners as to its safety.

Marconi had a dream of a world connected by wireless devices, and as of 2011 that dream is coming true. The IEEE 802 wireless connectivity standards as of 2009 include a personal area network (Bluetooth) that will connect all the devices in a house, car, or office. It has an expanded wireless LAN standard that will support high-speed wireless access of home and office computers into an access point. Both the wireless home network and the wireless LAN can be tied into a high-speed wireless regional network, replacing the current fiber-optic networks. There will be several high-speed cell phone networks that can also tie into the regional wireless system, and when this is added to the current satellite-based wide area networks, the result could be the fully connected wireless world Marconi envisioned.

George M. Whitson III, B.S., M.S., Ph.D.

Further Reading

Forouzan, Behrouz A. *Data Communications and Networking.* 4th ed. New York: McGraw-Hill, 2007. A complete, readable text with excellent coverage of cellular communications.

Liberti, Joseph C., Jr., and Theodore S. Rappaport. *Smart Antennas for Wireless Communications: IS-95 and Third Generation CDMA Applications.* Upper Saddle River, N.J.: Prentice Hall, 1999. Provides complete coverage of cellular phones with an emphasis on CDMA.

Roddy, Dennis. *Satellite Communications.* 4th ed. New York: McGraw-Hill, 2006. Contains thorough coverage of satellites and much material on wireless communications.

Rohde, Ulrich, and Jerry Whitaker. *Communications Receivers: DSP, Software Radios, and Design.* 3d ed. New York: McGraw-Hill, 2001. Provides excellent theoretical coverage of wireless technology.

Stallings, William. *Data and Computer Communications.* 9th ed. Upper Saddle River, N.J.: Prentice Hall, 2011. This text gives an excellent introduction to digital signaling, encoding, and multiplexing.

Web Sites

CTIA: The Wireless Association
http://www.ctia.org

Institute of Electrical and Electronics Engineers
http://www.ieee.org

International Telecommunication Union
http://www.itu.int

Telecommunications Industry Association
http://www.tiaonline.org

See also: Telecommunications; Telephone Technology and Networks.

Z

ZONE REFINING

FIELDS OF STUDY

Chemistry; mathematics; metallurgy; materials science; electronic materials engineering

SUMMARY

Zone refining is a method for producing ultrapure materials that relies on the same chemical principles as the purification of compounds by recrystallization from the melt. The method utilizes a basic process in which a zone of melted material moves relative to the solid phase. This can be accomplished either by slowly translating the melted zone along the length of a solid bar, as is done for most metals, or by slowly withdrawing a solid phase from the liquid melt, as is done with silicon to produce the ultrapure silicon single crystals for integrated circuits. Zone refining can produce materials containing impurities at levels of only a few parts per billion.

KEY TERMS AND CONCEPTS

- **Allotrope:** One of a material's possible forms in a particular phase, characterized by the organization of atoms or molecules within that form.
- **Eutectic Mixture:** The specific composition of a combined material, such as a metallic alloy, at the corresponding eutectic point in the phase diagram of the material.
- **Eutectic Point:** A minimum melting point in the phase diagram of a combined material, typically lower than the melting point of any of the pure components.
- **Phase:** A condition of matter that is characterized by a specific allotropic microstructure or physical state.
- **Phase Diagram:** A chart relating the melting point of combined materials to the relative proportions of the components of the material.

Definition and Basic Principles

Zone refining is a procedure for the purification of materials. When a material containing impurities is melted and allowed to resolidify, the first material to become solid again, or recrystallize, does so without incorporating the impurities. Instead, the impurities remain preferentially dissolved in the liquid state. Zone refining makes use of this principle.

If a relatively small liquid state is maintained and made to travel through the solid mass of the material, like a band of liquid between two regions of solid material, it will accumulate and transport impurities. In practice, the material to be purified is placed in a controlled environment such as a tube. Beginning at one end of the tube and progressing toward the other, a relatively small segment of the material is heated to melting. The heat source is moved along so that the melted region continuously moves along the length of the sample, which freezes again behind the molten region. A cooling system may be employed to assist refreezing the material.

By repeating the process, impurities are collected in the molten region and transported to the terminal end of the sample. Each iteration of the process increases the purity of the sample in a mathematically predictable way.

Background and History

The original process of zone melting was developed at Bell Telephone Laboratories; the basic principles of the methodology were published by W. G. Pfann in July, 1952. The basic method was further developed to employ temperature gradients rather than the original hot-cold melt-freeze zones. The details of this methodology were published by Pfann in September, 1955.

The basic principle is a modification of the long-used practice of producing crystalline compounds of high purity by melting and cooling a suitable material, rather than allowing the crystals to reform from a solvent-based solution. Both solvent and melt crystallization can be thought of as basic methods to isolate pure compounds from impurities.

In one case, the separate solvent represents and contains the entrained impurities, while in the other case the molten material acts as the solvent to contain the entrained impurities. The method was developed to provide high-purity germanium for semiconductor research following the invention of the semiconductor junction transistor in 1947. It has since been developed into different variations, most notably the floating zone method used for the production of high-purity silicon single crystals.

How It Works

Purification by crystallization is one of the first practical laboratory methods learned by beginning chemists, and it has remained essentially unchanged since it was used by alchemists in the Middle Ages. In the process, an impure material is dissolved in the minimum amount of a heated solvent that will dissolve the material. The key to the success of recrystallization is that the material should have a fairly steep solubility-temperature gradient in the particular solvent being used. That is, it should dissolve well in the heated solvent, but only poorly as the solvent cools. Allowing the resulting solution to cool slowly permits highly pure crystals to form while the original impurities remain dissolved in the solvent.

A variation on this method calls for melting the material and allowing the liquid to cool slowly. This method has the same effect as solvent-based crystallization in that the first crystals to form in the molten mass will have the highest purity, while the impurities that were present will be entrained in the last material to solidify. In essence, the molten material becomes the recrystallization solvent. Separating the pure crystals from the impure melt presents a technical difficulty, however, as strict temperature control is required to maintain the molten material without re-melting the newly formed crystals or allowing the impure melt to solidify. The melt process is thus typically used to produce high quality crystals from material that is already pure, rather than as a means of separating pure from impure.

Purification by zone refining relies on the same principles of crystal formation as the solvent and melt methods of crystallization. Zone refining was developed as crystallization from the melt with the restriction that only a portion of the material is molten at any given time. In its original form, the method called for placing a material in a relatively long tube. Beginning at one end of the tube, the material at that location is heated to melting in a narrow band. The heating source is moved along the tube, followed by a cooling mechanism. In this way the molten band travels the length of the solid material in the tube.

Within the molten band, crystallization of higher purity material takes place while the impurities remain dissolved and accumulate in the liquid phase. After the process has been repeated the required number of times to produce the desired material purity, the impurities will have been transported by the molten phase to one end of the material in the tube. The methodology is capable of achieving purities of just a few parts per billion, or greater than 99.9999995 percent purity.

The success of the method depends on the relative motion of the liquid and solid phases. From this point of view, there is no distinction between a molten phase traveling through a solid phase or a solid phase being drawn out of a liquid phase. Both cases have the same effect on the composition of the material. Floating zone refining uses the method in which a solid phase of extreme purity is drawn out of a liquid phase that retains the impurities that were present in the material. This is the method now used for the production of silicon single crystals.

High purity silicon can be produced in two ways. One is to distill volatile silicon chlorides to high purity and then react them with hydrogen gas, producing hydrogen chloride and pure silicon. The other method is to melt a relatively high-purity form of silicon called polysilicon, then slowly draw out a uniform single crystal that forms when a seed crystal is introduced and withdrawn.

The behavior of the material is described by the phase rule—a physicochemical principle that relates the composition of material solutions to their melting points. The purity of a material is characterized by a sharp, distinct melting point. At that specific temperature, the material changes phases from solid to liquid, or from liquid to solid, almost as one. The higher the purity of the material, the sharper and more distinct is its melting/freezing temperature. Materials that contain impurities, however, melt in a range of temperatures, the span of which increases according to the amount of impurities that are entrained within the material. Some solutions have compositions that are eutectic in nature. That is, at a specific composition the solution exhibits a sharp, distinct melting/freezing point, as though it were a single pure compound. Phase theory is fundamental to understanding the process of zone refining.

Applications and Products

Zone refining can be utilized with any material that exhibits suitable melting behavior. The material must melt without thermal decomposition and must not have high volatility in the liquid phase, ideally remaining in the liquid phase (without boiling) over a broad temperature range. If one of these conditions exists, the material will decompose rather than purify or will become subject to significant loss of mass to the vapor phase. This would, in turn, tend to decrease the purity of the remaining material by effectively increasing the relative amounts of impurities in the remaining liquid phase. In addition, if the process is being carried out in a closed system, the generation of a significant vapor phase could have catastrophic results.

The various methods of zone refining are used for one purpose only: to provide materials of extreme purity for other applications. Zone refining is an energy-intensive methodology, however, and tends to limit its general applicability to materials that are required only in small quantities or that have a high associated value. The primary uses of the methods are therefore limited to the production of research-grade compounds or the production of highly desirable materials such as silicon.

Research-Grade Materials. It is important to understand that the methodology of zone refinement is applicable at all temperatures. All that is required is that a small liquid phase be generated and made to travel in an opposite relative direction to the remaining solid phase. For the vast majority of research-grade materials, zone refinement procedures would be a method of last resort.

Chemical compounds that are routinely used in laboratory procedures are simply not required to have the level of purity that can be achieved through zone refinement. In most cases, the desired purity is readily achieved through other means. High-purity materials such as spectroscopic grade solvents at 99.999 percent purity are readily obtained through repeated fractional distillation and other processes, while reagent grade materials are acceptable at 99.9 percent purity. The requirement for purities obtained through zone refinement really only applies to materials of specialist interest and for reference data. For reference spectral data, it is desirable to have spectra that reflect only the compound of interest, with as little background interference from contaminants as possible. Similarly, the characterization data for the material should be influenced as little as possible by entrained impurities.

The physical property values of materials provide empirical data for the verification and refinement of theory. The greater the purity of a material, the more precisely its corresponding characteristic properties will refine foundational theory. For many materials, particularly elemental materials, zone refinement provides an entirely new avenue of study and research, one that focuses on the innate nature of the material.

Ironically, zone refinement is a method of choice for use with thermoplastic polymers. In the polymerization process, many separate polymerization chain reactions take place at once. Each proceeds until the reactive end randomly encounters something that causes the chain reaction to terminate, usually an impurity or contaminant. This randomness produces polymer molecules whose chain lengths and molecular weights are randomly distributed around some average value. Thermoplastics have the property of melting when heated; thermosetting polymers harden and solidify when heated. This makes thermoplastics amenable to zone refinement methods of isolating components that have a narrower distribution of molecular weights and chain lengths.

High-Value Materials. Apart from certain chemical compounds that are required in high purity form, the main application of zone refinement is the production of silicon single crystals for use in integrated-circuit chips. The process requires that a great deal of care be taken so that the resulting material meets the desired specifications.

The material known as polysilicon is typically stacked by hand into the furnace of a device called a crystal puller, along with a small, specified amount of a dopant material (to provide the desired electrical properties to the silicon). The unit is then sealed and purged to eliminate oxygen and other contaminant gases. The polysilicon mass is melted as the crucible is heated to the desired temperature. As the crucible rotates in one direction, a seed crystal attached to a counter-rotating armature is introduced into the melt to begin formation of the single crystal. The growing crystal is slowly withdrawn from the melt, building into a uniform cylindrical shape.

When completed, the crystal is allowed to cool, then checked by Fourier-transform infrared spectroscopy for quality and passed on for slicing into thin wafers, which will become the substrate of the integrated circuits etched onto their highly polished surfaces. Different manufacturers use minor variations of this procedure, such as the floating zone method, in which the crystal grows downward according to the flow of the melt rather than being pulled up from it. It should be noted that the purpose of the method in this application is not to purify the silicon, but to produce a single crystal from a mass of silicon that has been purified by other means.

Impact on Industry

High-purity materials obtained through the process of zone refining are not common industrial assets. The market for such materials is very limited, despite carrying a high value and having had (indirectly) a tremendous impact on the industry through digital electronic technology.

In terms of the purity of metals, particularly coinage metals, it is generally of greater importance just to know the level of purity rather than to ensure the metals be as pure as possible. Silver circulating coinage, minted until 1968 in North America, was minted from a 0.925 silver alloy stock, being 92.5 percent silver with the other 7.5 percent being primarily copper, to provide durability to the metal. In 1968 the silver content was dropped to 50 percent and then eliminated entirely, to be replaced with nickel-steel coinage. Silver and gold coins are now minted for the collector market only, as investment properties. Because these are not circulating coins, their durability is not a consideration. Accordingly, their metal content is restricted to investment grade content, typically 0.9999 silver, gold, or platinum. The small numbers of such coins that are produced annually, primarily to fill subscription orders, only require amounts of material that can be economically produced by zone refining, with that cost being recovered in the purchase of the coinage.

Evaluating the impact of digital electronic technology on industry is difficult. Since the commercialization of transistor technology in the twentieth century, digital electronic technology built upon the high-purity silicon used in the manufacture of integrated circuits and of computer CPU (central processing unit) chips has become the primary means of controlling the function of everything from farm tractors to interplanetary spacecraft and information systems. No aspect of modern existence is untouched by the technology, and potential applications continue to be developed. At the same time, however, researchers are exploring new materials and devices that would outperform silicon-based electronics. If and when such an advance occurs, the value of zone-refined silicon to society can be expected to drop dramatically.

Silicon-based technology is the foundation of multibillion dollar enterprises. The modern personal computer industry exists only because of the ability to economically produce silicon wafers from

Fascinating Facts About Zone Refining

- Zone refining methods can produce materials that are about 99.999999 percent pure, containing impurities of only a few parts per billion.
- At a purity of 5 parts per billion, water would contain the equivalent of approximately 1.5 teaspoons of salt in 200,000 liters of water.
- Crystals can be formed either by precipitation from solution in a solvent or by allowing a melted material to cool slowly.
- The first crystals to form generally have a higher degree of purity than crystals that form afterward.
- Zone refining uses repeated melting and freezing cycles to purify materials through crystallization.
- Zone refining is also known as zone melting, temperature gradient melting, and floating zone refining.
- The theory and process of zone melting was first published in 1952, and it was developed to obtain high purity germanium for semiconductor research.
- A zone refining method is used to produce single crystals of silicon that are typically 30 centimeters in diameter and more than 1 meter in length.
- The major uses of zone refining are the production of silicon crystals for making integrated-circuit chips and for developing ultrapure research materials and precious metals.

large single crystals, which, in turn, is only possible because of zone refining. Each year, several hundred thousand new computers are produced and sold, representing an economic value measured in billions of U.S. dollars. This value does not include ancillary equipment, such as printers and monitors, which are needed for computers to be useful; it also does not include the millions of cellular telephones, calculating devices, gaming consoles, digital cameras, machine control systems, automobiles and other vehicles, global positioning systems (GPSs), satellites, and other devices that use silicon-based electronics. The economic effect that this diffusion of technology has had is beyond measure.

Careers and Coursework

The theoretical principles of zone refining, and therefore the basis for further developments in that field, are defined by advanced mathematics. Students interested in pursuing a career in this field will require a sound footing in advanced mathematics and calculus to understand and use the appropriate concepts of physical chemistry. At the postsecondary level, chemistry, physics, and physical chemistry will be central subjects in a course of study. Students must understand the interactions and states of matter within the theoretical framework of the phase rule. While purification of materials by zone refining methods is not a common practice of organic chemistry, it is more widely used in practical inorganic chemistry and metallurgy and may be part of the procedures to be learned in analytical chemistry. A bachelor's or associate degree from a university or college is the minimum qualification for a career in these fields, and likewise for fields in which zone refining is a central technology, such as electronic materials engineering and precious or specialty metals refining.

If a more advanced, research-oriented career is the goal, students should acquire a graduate degree (master's or doctorate). For these qualifications, coursework will include advanced physical and inorganic chemistry, calculus relevant to equilibrium processes and phase equilibria, fluid dynamics, and crystallography.

Social Context and Future Prospects

Silicon-based electronics are expected to continue their reign as the benchmark of modern technology. Zone refining of silicon to provide the basic material for that electronic technology will therefore continue to be the essential process for many years. It will continue to be so with little change to its present form other than relatively minor improvements to existing techniques.

At the cutting edge of research, however, are new materials that promise the development of new types of transistors that will displace the silicon-based transistor as the workhorse of modern electronics. With that change, the need for the production of high-purity silicon and single crystals will decrease dramatically. Paradoxically, traders and investors in precious metals have become more demanding of high purity precious metals. Zone refinement is the most expedient way of obtaining metals in greater than 99.99 percent purity, and as the demand for precious metals as investment vehicles increases, it also is likely that the desired purity level will increase also.

Richard M. Renneboog, M.Sc.

Further Reading

Munirathnam, N. R., et al. "Zone Refining of Cadmium and Related Characterization." *Bulletin of Materials Science* 28, no. 3 (June, 2005): 209-212. This paper demonstrates the use of zone refining in the advanced study and characterization of specific materials.

Pfann, W. G. "Principles of Zone Melting." *Journal of Metals* (July, 1952): 747-753. The original publication describing the theory and process of zone refining. Clearly lays out the mathematical description of how the process works.

_______. "Temperature Gradient Zone Melting." *Journal of Metals* (September, 1955): 961-964. This paper provides the theory and process of a more technologically sophisticated method of zone refining.

ZYMOLOGY AND ZYMURGY

FIELDS OF STUDY

Brewing and fermentation; wine and liquor production; biochemistry; pharmaceuticals; biogas energy; biotechnology; alternative energy research and development; biosynthetic chemistry; proteomics; laboratory and specialized equipment design and manufacture.

SUMMARY

Zymology is the study and science of fermentation, particularly in the production of wine and beer and in baking. Zymurgy refers to the applied chemistry aspects of alcoholic fermentation in the making of beers, wines, and liquors. (The term "zymurgy" is often used interchangeably with "zymology.") In the process of fermentation, sugar molecules are converted to ethanol and carbon dioxide. Although the most well-known fermentation agent is perhaps *Saccharomyces cerevisiae,* a budding yeast popularly used by brewers and often referred to as brewer's yeast, numerous other strains of yeast and other microbes have been cultured to impart specific properties to the resulting brew, including enhanced alcohol content, altered carbonation, and specific flavorings. In addition, the fermentation process is used in biosynthetic methods to produce specific modifications in chemical substrates that are difficult to achieve by standard in vitro (controlled-environment) chemical methods. Fermentation processes using algae and other microbes have been developed to create biochemicals and specialty pharmaceuticals.

KEY TERMS AND CONCEPTS

- **Biosynthesis:** Synthesis or construction of various chemicals through biological processes instead of traditional laboratory in vitro techniques.
- **Digestion:** Chemical process whereby complex chemical compounds such as proteins, starches, and complex carbohydrates are broken down into smaller, simpler molecules such as amino acids and simple sugars.
- **Enzyme:** Protein molecule that mediates a specific biochemical conversion.
- **Fermentation:** Biochemical conversion process that converts sugars from carbohydrate sources into carbon dioxide, water, ethanol, and other chemical compounds.
- **Fermenter:** Closed apparatus in which a batch fermentation process is carried out.
- **Fermentogen:** Organism or other active agent that initiates and carries out the process of fermentation.
- **In Vitro:** Descriptive term that indicates a process carried out in laboratory glassware; in Latin, it means "within the glass."
- **In Vivo:** Descriptive term that indicates a process carried out within a living system; from the Latin word "vivum," meaning alive.
- **Leavening:** Use of yeast in the baking process to produce a risen dough before baking; the dough has been expanded in volume through the generation of carbon dioxide gas bubbles from yeast-mediated fermentation within the dough.
- **Lyophilization:** Physical process whereby a mixture is quickly frozen and then dehydrated under strong vacuum, also known as freeze-drying.
- **Must:** Raw mixture of grape pulp, water, and sugar in which initial fermentation of wine takes place; analogous to "wort" in the brewing of beer and "mash" in the distillation of liquors.
- **Substrate:** Substance or material on which an enzyme acts.
- **Yeast:** Single-celled plant species that consumes sugars from carbohydrate sources and derives energy through their conversion into ethanol and carbon dioxide.

Definition and Basic Principles

Zymology is a traditional process that predates the sciences of biology and chemistry by several thousand years. The basic process of fermentation has not changed in principle over that entire span of time, however. At its heart, the fermentation process is nothing more than the normal growth of yeast cells or other microbes in a suitable environment. Yeast cells are found in abundance as airborne material. Such airborne yeast is the main reason that many foodstuffs go bad when left exposed to the air, even briefly. In a mixture that contains free sugars in a water-based medium, such as an open pitcher of fruit juice, yeast cells may settle from the air and begin to reproduce in the mixture. The liquid will eventually become carbonated and effervescent because of carbon dioxide produced during the fermentation process, and it will acquire a decidedly acidic or sour taste. The solution may also become cloudy because of the population of microbes.

Zymology and zymurgy have developed as the practical and purposeful application of the fermentation process for the traditional production of alcoholic beverages for human consumption. The scope of their application has been broadened considerably, and they have come to be considered part of the foundation of the biotechnology and bioengineering industries. In practice, zymologists examine ways to apply known fermentation processes to new purposes beyond the traditional baking and brewing applications. This includes the development of new genetic strains of yeasts and other microbes and the development of effective substrate systems to maximize the production of specific materials.

In any fermentation processs, a specific microbe is provided with a controlled environment in which to grow in an uncontrolled manner. This may be a sealed or isolated container of some sort that has been fitted with appropriate equipment to ensure thorough mixing of the contents. Those contents consist of the microbial culture and the nutrient medium in which the culture will grow. The environment may also be an open but sequestered collection of some appropriate substrate material on which the microbial fermentation process will act. The microbial culture is typically grown in a batch process, but continuous production methods can also be employed, depending on the desired output. The output of the process is then captured and recovered by separation methods that return the microbial culture to service.

In some cases, the purpose of the process is to produce more of the microbe, as when yeast cultures are grown to supply needs for particular strains in specific applications. In other cases, the purpose of the process is to use the biochemical processes of the particular microbe to convert one material into other materials, as when biomass is fermented for the production of industrial ethanol, butanol (butyl alcohol), and other compounds. Highly specific and bioengineered microbial cultures are often used to produce proteins and other biochemicals for pharmaceutical and medical research.

Background and History

It is impossible to say with any certainty when the use of microbial fermentation began. It is certain, however, that the effects of yeast activity in producing wine and other alcoholic brews and in the baking of bread have been appreciated for thousands of years. The historical evidence available indicates that beer and bread were staples of the ancient Egyptian diet and often served as wages. In fact, bread loaves have been recovered from Egyptian tombs that are more than 5,000 years old, and the residues of beer found in Egyptian funerary vessels are equally old. Extant records of ancient Egyptian life show that beer and bread were part of every meal, for young and old alike. It is therefore likely that the use of leavening agents (yeast) in baking and brewing activities was well known before that time.

Ironically, knowledge of microbes themselves had to wait for the invention of the compound microscope in the end of the sixteenth century and the astute observations of Dutch scientist Antoni van Leeuwenhoek some seventy years afterward. The relationship between microbes and their observed effects, such as disease and fermentation, was not specifically identified for a further hundred years, when French microbiologist Louis Pasteur provided the first scientific explanations of the basis of fermentation and the brewing of beer. His work is believed to have set the stage for the modern science of biochemistry.

How It Works

Fermentation Processes. Zymology and zymurgy are the study and practice of fermentation and other microbial processes. Microbes are single-celled organisms such as yeast and bacteria. As living organisms, microbes have an active metabolism built on biochemistry. The various biochemical processes and reaction cycles that occur within living organisms function to extract energy for their continued existence through the conversion of materials that serve as foodstuffs. At the most fundamental level, simple sugars such as glucose and fructose are converted to carbon dioxide, water, and ethanol using yeast. The process is, in a sense, the opposite of photosynthesis, in which green plants convert carbon dioxide and water into glucose, and then into various other sugars and carbohydrates. The very same biochemical processes can be used to act on a variety of other substrate systems to produce different outcomes.

Bioengineering. On the other end of the scale are the procedures by which microbes with specific properties are produced. Because of their naturally high reproduction rates, microbes with different properties can be fairly readily identified, isolated, and grown in quantity. In addition, the development of genetic modification techniques allows them to be manipulated in ways that permit the enhancement of specific properties. In this way, yeast strains can be developed that are most suitable for specific varieties of wines and beers and for many other applications. Other microbes have been developed to have specific applications in the production of certain chemical compounds such as enzymes, proteins, and modified biochemicals. The use of microbial methods in the production of isolable chemicals is known as biosynthetic methodology.

Aerobic and Anaerobic Processes. Fermentation and other microbial processes can take place either in the presence of air (aerobic) or isolated from air (anaerobic). The products of the two processes are generally very different. Whereas anaerobic fermentation of a sugar solution produces carbon dioxide and ethanol, the same process carried out in the presence of oxygen will produce acetic acid rather than ethanol. This is what happens to fruit juice that is left out in the open for too long. Free yeast cells from the air settle into the liquid and begin aerobic fermentation, resulting in what is essentially a carbonated solution of flavored vinegar. The same fruit juice, fermented by the same yeast under anaerobic conditions, produces an alcoholic solution that becomes a fine-flavored wine.

Applications and Products

The main processes that rely on microbial action for their success are the brewing of beer, the fermentation of wine, and the baking of bread and other leavened baked goods. The brewing of beer and the fermentation of wine are, in principle, the same process; the differences between them consist primarily of the materials used to provide flavors and the order in which the resulting solutions are handled. At the heart of both is the same simple yeast-mediated fermentation of sugar and water to produce carbon dioxide and ethanol.

Wine. In the fermentation of wine, a mixture of fruit and fruit juice is prepared and added to an appropriate amount of water and allowed to stand for a short period of time to allow the various flavor components and colorings of the fruit to be extracted into the water. This stage is known as the must and is perhaps the single most important step in determining the final flavor and quality of the finished wine. In the next stage, the liquid is separated from the must and combined with the required amount of sugar and a mixture of nutrients needed for the proper growth of the yeast during fermentation–made up to volume with additional water–and then set to ferment in the airfree environment of a fermenter. The isolation of the solution during fermentation is usually achieved by the use of an air-trapping or bubbler system. This system prevents the influx of fresh air while allowing carbon dioxide produced during fermentation to escape from the fermenter.

This initial fermentation stage is followed by a series of decantations, in which the developing wine is separated from the yeast residues and reset to continue the fermentation process. Depending on the type of wine being produced, this may involve the addition of more sugar, other flavoring agents, and yeast of a different strain. The final stages of wine production involve clarification of the liquid, bottling, and storage (or cellaring) of the product so that it ages properly. During aging, a certain amount of fermentation may be desirable to produce a sparkling vintage, which is the procedure for the production of champagne-type wines. During the aging process, the various flavors of the wine meld and blend through various slow chemical reactions to produce the final distinctive flavor of the wine. The alcohol content of the wine produced in this way is typically about 12 to 15 percent.

Beer. The production of beer follows a similar procedure. A solution of sugar and water is prepared and blended with hops and malt, to which is added a robust yeast strain known as *Saccharomyces cerevisiae*, which reproduces quickly. Fermentation of beer takes place at a warmer temperature and at a more rapid rate than that of wine. This eliminates the need for an air-trapping system, as the production of carbon dioxide is so rapid that it continually displaces air that would enter the system. For small-batch brewing such as would be carried out by an individual at home, it is usually sufficient at this stage simply to cover the open vat with a sheet of cloth. The initial fermentation is typically complete in a very few days, after which the rate of fermentation slows considerably as the sugar content of the solution has been converted mainly to carbon dioxide. At this stage, a precisely controlled amount of sugar is added, and the solution is bottled and sealed. The minor fermentation that occurs subsequently forces the carbon dioxide into solution, producing a beverage with an alcohol content of about 5 to 7 percent.

Sugars and Flavors. The type of sugar used in the production of beer and wine is not particularly relevant, since the sugar that is fermented does not contribute to the flavor of the finished product, having been converted completely to carbon dioxide and ethanol. It is therefore of little consequence whether fermentation is carried out using refined white sugar, liquid invert syrup, corn syrup, fructose, or glucose, sometimes referred to as corn sugar. On the other hand, sugars that are part of a flavored mixture, such as brown sugar or honey, will impart distinctive flavors to the final product. Mead, for example, is a type of wine made with the use of honey in the fermentation and has a decidedly honeylike flavor. Similarly, the use of brown sugar imparts a molasses flavor to the product because of the molasses content of the sugar.

Different varieties of grapes are used to produce the vast majority of wines. However, any vegetable substance can be subjected to fermentation to produce a corresponding wine, and wines made from different fruits are becoming ever more popular as consumers are more willing to try and accept other types of wines and wine flavors. Similarly, flavors that derive from materials other than the traditional hops and malt used in brewing beer are also becoming more acceptable, although the range of alternative substrates is more limited, being restricted primarily to wheat, barley, and other grains.

Bread. Bread and other leavened baked goods use yeast-mediated fermentation to produce carbon dioxide gas from sugar in the blend. The strain of yeast used in baking does not produce alcohol in the fermentation process. As the fermentation progresses within the dough, the gas collects and forms bubbles, but its escape is prevented by the elasticity of the dough. The bubbles of carbon dioxide gas act to expand the bulk of the dough, producing a light, fluffy baked good. The relatively hard outer crust forms as the carbon dioxide is driven out of the dough at the surface, which subsequently becomes toasted and caramelized, sealing in the remainder of the gases.

Other Foods. A number of traditional foods rely on bacteria and microbial processes for their food value. Examples of those foods include such well-known concoctions as sauerkraut, kimchi, soy sauce, miso, black bean sauce, lutefisk, and all manner of yogurt. Yogurt in particular is marketed as a probiotic. The digestive tract, or gut, of the human body contains a wide but very specific variety of microbes whose presence is necessary for proper digestion

Fascinating Facts About Zymology and Zymurgy

- Egyptian tombs more than 5,000 years old have contained samples of yeast bread, indicating that people have made use of microbial processes for at least that long.
- The Japanese company Kirin has created beer like that made in ancient Egypt, based on interpretations of hieroglyphs describing the ancient Egyptians' beer-making process and analysis of beer residues found in funerary cups recovered from tombs.
- Yeast cell walls are broken down by treating them with snail gut enzymes so that various biochemicals and enzymes can be recovered from within the yeast cells.
- Almost all the beer brewed in the world is produced through fermentation by the yeast *Saccharomyces cerevisiae*, while every type of wine produced in the world uses a different type of yeast in the fermentation process.
- The bacterium *Escherichia coli* can be fatal to humans, but it is also the source of 6-aminopenicillinic acid used to make penicillin, cephalosporin, and other potent drugs.
- Bacteria such as *Rhodococcus* can be used in the production of polymeric materials such as polyacrylamide.
- Photosynthesis converts carbon dioxide and water into glucose, and fermentation converts glucose into carbon dioxide and ethanol.
- Microbes can convert plant matter into acetone, butyl alcohol, and ethanol.

and to maintain the inherent chemical balance of the digestive system. Probiotic foodstuffs are made to work with and promote the correctly balanced population of gut microbes, thus helping to maintain the optimum health of the host body.

Impact on Industry

The world market for chemicals produced by fermentation methods is predicted to be about $44 billion by 2012, with more than half of that amount involving ethanol fermentation. A short list of economically important materials other than foodstuffs produced by fermentation methodologies includes bulk antibiotics, alcohols (ethanol, butanol, and so on), enzymes, organic acids (amino acids, citric acid, lactic acid, and others), vitamins, polymers, and biogums.

The application of fermentation processes in the production of foodstuffs continues to be a field of constant growth and economic value. For example, the development of a bacterial strain that produces rennin, a compound that is used as a coagulant and ripening agent in the production of cheese, is reported to represent a potential cost reduction in cheese manufacturing in the United States of at least $500,000 annually.

Pharmaceuticals. Yeast fermentation and fermentation processes employing algae and other microbes have become valuable tools in the production of specialty pharmaceuticals and other biochemicals. Using methods similar in concept to those used in the production of wine and beer, the microbial agent is grown in a medium that contains substrate materials. These compounds become involved in the normal metabolic pathways of the microbes and are correspondingly altered by the enzymatic processes of the system. Because these materials are not the correct substrates for the particular enzyme action, they are not consumed but instead are only chemically modified. The modified substrate materials are subsequently harvested from the medium and purified for use. In other approaches, the biochemicals within the microbes themselves are the desired product. The microbial colony is grown to its full extent, and the organisms are harvested. They are subsequently destroyed in a manner that releases their cellular biochemical components to be harvested and purified. This procedure became feasible when a mixture of enzymes obtained from the digestive tract of the snail–called snail gut enzyme–was determined to be capable of digesting and breaking down the cell walls of yeast.

Other Applications. Fermentation-based processes have also been developed for the mass production of certain materials. Most notably, these processes have focused on the bulk production of alternative fuels, particularly ethanol and other solvents, and biogas, which is primarily methane. Processes employing *Clostridia* bacteria, for example, are useful in the production of acetone and 1-butanol, two extremely important organic solvents. In addition, as the applied science and technology behind fermentation becomes more developed, other areas of science will use the ability to tailor microbial processes for the purposes of analytical and test procedures and production methods.

Supporting Technology. The opposite side of the coin in zymology and zymurgy is made up of the businesses and industries that provide the equipment and supplies used in various bioprocesses. A significant industrial base has developed around the manufacture and provision of fermentation vessels, growth media and substrate supply, monitoring systems, isolation and purification equipment, and other ancillary requirements.

Careers and Course Work

Career opportunities in the fields of zymology and zymurgy are numerous and becoming increasingly important as the economic value of zymology increases. The minimum requirement for pursuing a career in these fields is a solid grounding in biological and chemical sciences. Particular areas of study and specialization include microbiology, biochemistry, organic chemistry, physical chemistry, analytical chemistry, and genetics. One may also specialize in the brewing industry through a career in process technology, process operations, oenology (the study of winemaking), viticulture, distillery operations, mechanical engineering, and chemical engineering directed to the application of bioprocessing and biosynthetic techniques.

Social Context and Future Prospects

Several aspects of zymology and zymurgy have become controversial in nature, especially those that have led to the development of bioengineering, gene modification, and the use of mammalian cells as fermentogens for the production of specific proteins and enzymes. However, support is growing for of the use of bioprocesses in solving many of the problems that have developed because of modern industry. For example, bioengineers are looking for microbes that can be used to alleviate pollution in different environmental situations. Particular interest is shown in microbes that can consume crude oil and other contaminants that have been placed in the environment either by accident or design. Others seek a microbial source of cheap but highly nutritious and acceptable food for human consumption. The potential for development of specific microbes is essentially unlimited and holds the possibility of genetically engineering microbes for specific purposes and functions or to produce specific chemical and biochemical compounds. This potential is in itself a controversial issue and promises to be an area in which the ethics and morals of both the scientific community and of society in general will be tested.

Richard M. J. Renneboog, M.Sc.

Further Reading

Bamforth, Charles M. *Food, Fermentation, and Microorganisms.* Ames, Iowa: Blackwell Science, 2005. Examines the science of fermentation and the organisms used in fermenting food. Chapters are devoted to beer, wine, various alcoholic beverages, vinegar, cheese, meat, vegetables, and soy-based products such as miso, soy sauce, and natto.

Hui, Y. H., et al., eds. *Handbook of Food and Beverage Fermentation Technology.* New York: Marcel Dekker, 2004. A good overview of the basic principles of fermentation processes as they are applied to the production of various food and beverage products. Also examines various traditional ethnic foodstuffs from around the world, such as sauerkraut, breads, yogurt, cheese, and kimchi.

McNeil, B., and L. M. Harvey, eds. *Practical Fermentation Technology.* Hoboken, N.J.: John Wiley & Sons, 2008. Hands-on work that defines fermentation and examines working with fermenters.

Panchal, Chandra J., ed. *Yeast Strain Selection.* New York: Marcel Dekker, 1990. Describes collections and studies of various yeast cultures, the characteristics of various strains of yeast, and the selection of yeast strains for specific purposes such as baking and brewing.

Shetty, Kalidas, et al., eds. *Food Biotechnology.* 2d ed. Boca Raton, Fla.: CRC Press, 2006. Covers technical details of many aspects of food biotechnology. An introductory segment describes the relationship between food microbiology and the principles of biochemistry and molecular biology. Other subjects include bioreactor design, gene modification, and bioengineering methods.

Waites, Michael J., Neil L. Morgan, and John S. Rockey. *Industrial Microbiology: An Introduction.* Chichester, England: John Wiley & Sons, 2009. An introductory overview of the physiology of microbes, the methods and processes by which they are employed, and several specific fields of application.

Web Sites

American Society of Brewing Chemists
http://www.asbcnet.org

American Society of Microbiology
Fermentation and Biotechnology Division
http://www.asm.org/division/O/index.html

See also: Biochemical Engineering; Food Science; Genetically Modified Food Production; Industrial Fermentation; Viticulture and Enology.

Appendixes

BIOGRAPHICAL DICTIONARY OF SCIENTISTS

Alvarez, Luis W. (1911-1988): A physicist and inventor born in San Francisco, Alvarez was associated with the University of California, Berkeley, for many years. He explored cosmic rays, fusion, and other aspects of nuclear reaction. He invented time-of-flight techniques and conducted research into nuclear magnetic resonance for which he was awarded the 1968 Nobel Prize in Physics. He contributed to radar research and particle accelerators, worked on the Manhattan Project, developed the ground-controlled approach for landing airplanes, and proposed the theory that dinosaurs were rendered extinct by a massive meteor impacting Earth.

Archimedes (c. 287-c. 212 B.C.E.): A Greek born at Syracuse, Sicily, Archimedes is considered a genius of antiquity, with interests in astronomy, physics, engineering, and mathematics. He is credited with the discovery of fluid displacement (Archimedes' principle) and a number of mathematical advancements. He also developed numerous inventions, including the Archimedes screw to lift water for irrigation (still in use), the block-and-tackle pulley system, a practical odometer, a planetarium using differential gearing, and several weapons of war. He was killed during the Roman siege of Syracuse.

Babbage, Charles (1791-1871): An English-born mathematician and mechanical engineer, Babbage designed several machines that were precursors to the modern computer. He developed a difference engine to carry out polynomial functions and calculate astronomical tables mechanically(which was not completed) as well as an analytical engine using punched cards, sequential control, branching and looping, all of which contributed to computer science. He also made advancements in cryptography, devised the cowcatcher to clear obstacles from railway locomotives, and invented an ophthalmoscope.

Bacon, Sir Francis (1561-1626): A philosopher, statesman, author, and scientist born in England, Bacon was a precocious youth who at the age of thirteen began attending Trinity College, Cambridge. Later a member of Parliament, a lawyer, and attorney general, he rejected Aristotelian logic and advocated for inductive reasoning—collecting data, interpreting information, and carrying out experiments—in his major work, *Novum Organum* (*New Instrument*), published in 1620, which greatly influenced science from the seventeenth century onward. A victim of his own research, he experimented with snow as a way to preserve meat, caught a cold that became bronchitis, and died.

Baird, John Logie (1888-1946): A Scottish electrical engineer and inventor, Baird successfully transmitted black-and-white (in 1925) and color (in 1928) moving television images, and the BBC used his transmitters to broadcast television from 1929 to 1937. He had more than 175 patents for such far-ranging and forward-thinking concepts as big-screen and stereo TV sets, pay television, fiber optics, radar, video recording, and thermal socks. Plagued with ill health and a chronic lack of financial backing, Baird was unable to develop his innovative ideas, which others later perfected and profited from.

Bardeen, John (1908-1991): A Wisconsin-born electrical engineer and physicist, Bardeen worked for Gulf Oil, researching magnetism and gravity, and later studied mathematics and physics at Princeton University, where he earned a doctoral degree. While working at Bell Laboratories after World War II he, Walter Brattain (1902-1987), and William Shockley (1910-1989) invented the transistor, for which they shared the 1956 Nobel Prize in Physics. In 1972, Bardeen shared a second Nobel Prize in Physics for a jointly developed theory of superconductivity; he is the only person to win the same award twice.

Barnard, Christiaan (1922-2001): A heart-transplant pioneer born in South Africa, Barnard was a cardiac surgeon and university professor. He performed the first successful human heart transplant in 1967, extending a patient's life by eighteen days, and subsequent transplants—using innovative operational techniques he devised—allowed new heart recipients to survive for more than twenty years. He was one of the first surgeons to employ living tissues and organs from other species to prolong human life and was a contributor to the effective design of artificial heart valves.

Bates, Henry Walter (1825-1892): A self-taught

naturalist and explorer born in England, Bates accompanied anthropologist-biologist Alfred Russel Wallace (1823-1913) on a scientific expedition to South America between 1848 and 1852, which he described in his 1864 work, *The Naturalist on the River Amazons.* He collected thousands of plant and animal species, most of them unknown to science, and was the first to study the survival phenomenon of insect mimicry. For nearly thirty years he was secretary of the Royal Geographical Society and also served as president of the Entomological Society of London.

Becquerel, Antoine-Henri (1852-1908): A French physicist and engineer born into a family boasting several generations of scientists, Becquerel taught applied physics at the National Museum of Natural History and at the Polytechnic University, both in Paris, and also served as primary engineer overseeing French bridges and highways. He served as president of the French Academy of Sciences and received numerous awards for his work investigating polarization of light, magnetism, and the properties of radioactivity, including the 1903 Nobel Prize in Physics, which he shared with Pierre and Marie Curie.

Bell, Alexander Graham (1847-1922): A Scottish engineer and inventor whose mother and wife were deaf, Bell researched hearing and speech throughout his life. He began inventing practical solutions to problems as a child. His experiments with acoustics led to his creation of the harmonic telegraph, which eventually resulted in the first practical telephone in 1876 and spawned Bell Telephone Company. Bell became a naturalized American citizen and also invented prototypes of flying vehicles, hydrofoils, air conditioners, metal detectors, and magnetic sound and video recording devices.

Benz, Carl (1844-1929): A German engineer and designer born illegitimately as Karl Vaillant, Benz designed bridges before setting up his own foundry and mechanical workshop. In 1888, he invented, built and patented a gas-powered, engine-driven, three-wheeled horseless carriage named the Benz Motorwagen, which was the first automobile available for purchase. In 1895, he built the first trucks and buses and introduced many technical innovations still found in modern automobiles. The Benz Company merged with Daimler in the 1920's and introduced the famous Mercedes-Benz in 1926.

Berzelius, Jons Jakob (1779-1848): A physician and chemist born in Sweden, Berzelius was secretary of the Royal Swedish Academy of Sciences for thirty years. He is credited with discovering the law of constant proportions for inorganic substances and was the first to distinguish organic from inorganic compounds. He developed a system of chemical symbols and a table of relative atomic weights that are still in use. In addition to coining such chemical terms as "protein, "catalysis," "polymer," and "isomer," he identified the elements cerium, selenium, silicon, and thorium.

Bessemer, Henry (1813-1898): The English engineer and inventor is chiefly known for development of the Bessemer process, which eliminated impurities from molten pig iron and lowered costs in the production of steel. Holder of more than one hundred patents, Bessemer also invented items to improve the manufacture of glass, sugar, military ordnance, and postage stamps, and built a test model of a gimballed, hydraulic-controlled steamship to eliminate seasickness. His steel-industry creations led to the development of the modern continuous casting process of metals.

Birdseye, Clarence (1886-1956): The naturalist, inventor, and entrepreneur was born in Brooklyn, New York. He began experimenting in the early 1920's with flash-freezing fish. Using a patented process, he was eventually successful in freezing meats, poultry, vegetables, and fruits, and in so doing changed consumers' eating habits. Birdseye sold his process to the company that later became General Foods Corporation, for whom he continued to work in developing frozen-food technology. His surname—split in two for easy recognition—became a major brand name that is still familiar.

Bohr, Niels (1885-1962): A Danish theoretical physicist, Bohr introduced the concept of atomic structure, in which electrons orbit the nucleus of an atom, and laid the foundations of quantum theory, for which he was awarded the 1922 Nobel Prize in Physics. He later identified U-235, an isotope of uranium that produces slow fission. During World War II, after escaping from Nazi-occupied Denmark, he worked as consultant to the Manhattan Project. Following the war, he returned to Denmark and became a staunch advocate for the nondestructive uses of atomic energy.

Bosch, Carl (1874-1940): The German-born chemist,

metallurgist, and engineer devised a high-pressure chemical technique (the Haber-Bosch process) to fix nitrogen, used in mass-producing ammonia for fertilizers, explosives, and synthetic fuels. He was awarded (along with Friedrich Bergius, 1884-1949) the 1931 Nobel Prize in Chemistry for his work. He was a founder and chairman of the board of IG Farben, for a time the largest chemical company in the world, but was ousted in the late 1930's for criticizing the Nazis.

Brahe, Tycho (1546-1601): A nobleman born of Danish heritage in what is modern-day Sweden, Brahe became interested in astronomy while studying at the University of Copenhagen. He made improvements to the primitive observational instruments of the day but never had access to the telescope. Nonetheless, he was able to study the positions of stars and planets accurately and produced useful catalogs of celestial bodies, particularly for the planet Mars, which helped Johannes Kepler (1571-1630) to formulate the laws of planetary motion. Craters on the Moon and on Mars are named in Brahe's memory.

Brunel, Isambard Kingdom (1806-1859): A British-born civil engineer and inventor, Brunel designed and built tunnels, bridges, and docks—many still in use—often devising ingenious solutions to problems in the process. He is best remembered for developing the SS *Great Britain,* the largest and most modern ship of its time and the first ocean-going iron ship driven by a propeller. Brunel was also a railroad pioneer, serving as chief engineer for Great Western Railway, for which he specified a broad-gauge track to allow higher speeds, improved freight capacity, and greater passenger comfort.

Burbank, Luther (1849-1926): Despite having only an elementary-school education, the Massachusetts-born botanist and horticulturist was a pioneer in the field of agricultural science. Working from a greenhouse and experimental fields in Santa Rosa, California, Burbank developed more than 800 varieties of plants, including new strains of flowers, peaches, plums, nectarines, cherries, peaches, berries, nuts, and vegetables, as well as new crossbred products such as the plumcot. One of his most useful creations, the Russet Burbank, became the potato of choice in food processing, particularly for French fries.

Calvin, Melvin (1911-1997): A Minnesota-born chemist of Russian heritage, Calvin taught molecular biology for nearly fifty years at the University of California, Berkeley, where he founded and directed the Laboratory of Chemical Biodynamics (later the Structural Biology Division) and served as associate director of the Lawrence Berkeley National Laboratory. He and his research team traced the path of carbon-14 through plants during photosynthesis, greatly enhancing understanding of how sunlight stimulates chlorophyll to create organic compounds. He was awarded the 1961 Nobel Prize in Chemistry for his work.

Carnot, Sadi (1796-1832): A French physicist and military engineer, Carnot was an army officer before becoming a scientific researcher, specializing in the theory of heat as produced by the steam engine. His *Reflections on the Motive Power of Fire* focused on the relationship between heat and mechanical energy and provided the foundation for the second law of thermodynamics. His work greatly influenced scientists such as James Prescott Joule (1818-1889), William Thomson (Lord Kelvin, 1824-1907), and Rudolf Diesel (1858-1913) and made possible more practically and efficiently designed engines later in the nineteenth century. Carnot's career was cut short by his death from cholera.

Carson, Rachel (1907-1964): A marine biologist and author born in Springdale, Pennsylvania, Carson worked for the U.S. Bureau of Fisheries before turning full-time to writing about nature. Her popular and highly influential articles, radio scripts and books, including *The Sea Around Us, The Edge of the Sea, Under the Sea-Wind,* and *Silent Spring* enlightened the public about the wonders of nature and the dangers of pesticides such as DDT, which was eventually banned in the United States. Carson is credited with spurring the modern environmental movement.

Celsius, Anders (1701-1774): A Swedish astronomer, Celsius studied the aurora borealis and was the first to link the phenomena to the Earth's magnetic field. He also participated in several expeditions designed to measure the size and shape of the Earth. Founder of the Uppsala Astronomical Observatory, he explored star magnitude, observed eclipses, and compiled star catalogs. He is perhaps best known for the Celsius international temperature scale, which accounts for atmospheric pressure in measuring the boiling and freezing points of water.

Clausewitz, Carl von (1780-1831): As a Prussian-born soldier and military scientist, Clausewitz participated in numerous campaigns, beginning in the early 1790's, and fought in the Napoleonic Wars. After his appointment in 1818 to major general, he taught at the Prussian military academy and helped reform the state army. His principal written work, *On War*, unfinished at the time of his death from cholera, is still considered relevant and continues to influence military thinking via its practical approach to command policies, instruction for soldiers, and methods of planning for strategists.

Colt, Samuel (1814-1862): An inventor born in Hartford, Connecticut, Colt designed a workable multishot pistol while working in his father's textile factory. In the mid-1830's he patented a revolver and set up an assembly line to produce machine-made weapons featuring interchangeable parts. The perfected product, the Colt Peacemaker, was used in the Seminole and Mexican-American wars and became popular during America's western expansion, and Colt became a millionaire. Colt's Manufacturing Company continues to produce a wide variety of firearms for civilian, military, and law-enforcement purposes.

Copernicus, Nicolaus (1473-1543): The Polish mathematician, physician, statesman, artist, linguist, and astronomer is credited with beginning the scientific revolution. His major work, published the year of his death, *De revolutionibus orbium coelestium* (*On the Revolutions of the Heavenly Spheres*), was the first to propose a heliocentric model of the solar system. The book inspired further research by Tycho Brahe (1546-1601), Galileo Galilei (1564-1642), and Johannes Kepler (1571-1630) and stimulated the birth of modern astronomy.

Cori, Gerty Radnitz (1896-1957): A biochemist born in Prague (now the Czech Republic), Cori came to the United States in 1922 and became a naturalized American citizen in 1928. She worked with her husband Carl at what is now Roswell Park Cancer Institute in Buffalo, New York, researching carbohydrate metabolism and discovered how glycogen is broken down into lactic acid to be stored as energy, a process now called the Cori cycle. She was awarded the 1947 Nobel Prize in Physiology or Medicine, the first American woman so honored.

Cousteau, Jacques (1910-1997): A French oceanographer, explorer, filmmaker, ecologist, and author, Cousteau began underwater diving in the 1930's, and it became a lifelong obsession. He coinvented the Aqua-Lung in the 1940's—the precursor to modern scuba gear—and began making nature films during the same decade. He founded the French Oceanographic Campaigns in 1950 and aboard his ship *Calypso* explored and researched the world's oceans for forty years. In the 1970's he created the Cousteau Society, which remains a strong ecological advocacy organization.

Crick, Francis (1916-2004): An English molecular biologist and physicist, Crick designed magnetic and acoustic mines during World War II. He was later part of a biological research team at the Cavendish Laboratory. Focusing on the X-ray crystallography of proteins, he identified the structure of deoxyribonucleic acid (DNA) as a double helix, a discovery that greatly advanced the study of genetics. He and his colleagues, American James D. Watson (b. 1928) and New Zealander Maurice Wilkins (1916-2004), shared the 1962 Nobel Prize in Physiology or Medicine for their groundbreaking work.

Curie, Marie Sklodowska (1867-1934) and Pierre Curie (1859-1906): Polish-born chemist-physicist Marie was the first woman to teach at the University of Paris. She married French physicist-chemist Pierre Curie in 1895, and the couple collaborated on research into radioactivity, discovering the elements polonium and radium. She and her husband shared the 1903 Nobel Prize in Physics for their work; she was the first woman so honored. After her husband died, she continued her research and received the 1911 Nobel Prize in Chemistry, the first person to receive the award in two different disciplines. She founded the Radium (later Curie) Institute.

Daimler, Gottlieb (1834-1900): The German-born mechanical engineer, designer, inventor, and industrial magnate was an early developer of the gasoline-powered internal combustion engine and the automobile. He and fellow industrial designer Wilhelm Maybach (1846-1929) began a partnership in the 1880's to build small, high-speed engines incorporating numerous devices they patented—flywheels, carburetors, and cylinders—still found in modern engines. After creating the first motorcycle they founded Daimler Motors and began selling automobiles in the early 1890's. Their Phoenix model won history's first auto race.

Darwin, Charles Robert (1809-1882): An English naturalist and geologist, Darwin participated in the five-year-long worldwide surveying expedition of the HMS *Beagle* during the 1830's, observing and collecting specimens of animals, plants, and minerals. The voyage inspired numerous written works, particularly *On Natural Selection, On the Origin of the Species,* and *The Descent of Man,* which collectively supported his theory that all species have evolved from common ancestors. Though modern science has virtually unanimously accepted Darwin's findings, his theory of evolution remains a controversial topic among various political, cultural, and religious groups.

Davy, Sir Humphry (1778-1829): A chemist, teacher, and inventor born in England, Davy began conducting scientific experiments as a child. As a teen he worked as a surgeon's apprentice and became addicted to nitrous oxide. He later researched galvanism and electrolysis, discovered the elements sodium, chlorine, and potassium, and contributed to the discovery of iodine. He invented the Davy safety lamp for use in coal mines, was a founder of the Zoological Society of London, and served as president of the Royal Society.

Diesel, Rudolf (1858-1913): A French-born mechanical engineer and inventor of German heritage, Diesel designed an innovative refrigeration system for an ice plant in Paris and improved the efficiency of steam engines. His self-named, patented diesel engine introduced the concept of fuel injection. The efficient diesel engine later became commonplace in trucks, locomotives, ships, and submarines and, after redesign to reduce weight, in modern automobiles. Diesel disappeared while on a ship. His body was later discovered floating in the sea, but it is still unknown whether he fell overboard, committed suicide, or was murdered.

Edison, Thomas Alva (1847-1931): A scientist, inventor and entrepreneur born in Milan, Ohio, Edison worked out of New Jersey in the fields of electricity and communication and profoundly influenced the world. Credited with more than 1,000 patents, he is best known for creating the first practical incandescent light bulb, which has illuminated the lives of humans since 1879. Other inventions include the stock ticker, a telephone transmitter, electricity meters, the mimeograph, an efficient storage battery, and the phonograph and the kinetoscope, which he combined to produce the first talking moving picture in 1913.

Einstein, Albert (1879-1955): A German-born theoretical physicist and author of Jewish heritage, Einstein came to the United States before World War II. Regarded as a genius, and one of the world's most recognized scientists, he was awarded the 1921 Nobel Prize in Physics. He developed general and special theories of relativity, particle and quantum theories, and proposed ideas that continue to influence numerous fields of study, including energy, nuclear power, heat, light, electronics, celestial mechanics, astronomy, and cosmology. Late in life, he was offered the presidency of Israel but declined the honor.

Euclid (c. 330-c. 270 B.C.E.): A Greek mathematician who taught in Alexandria, Egypt, Euclid is considered the father of plane and solid geometry. He is remembered principally for his major extant work, *The Elements*—a treatise containing definitions, postulates, geometric proofs, number theories, discussions of prime numbers, arithmetic theorems, algebra, and algorithms—which has served as the basis for the teaching of mathematics for two thousand years. He also explored astronomy, mechanics, gravity, moving bodies, and music and was one of the first scientists to write about optics and perspective.

Everest, Sir George (1790-1866): A geographer born in Wales, Everest participated for 25 years in the Great Trigonometrical Survey of the Indian subcontinent—which surveyed an area encompassing millions of square miles while locating, measuring, and naming the Himalayan Mountains—and served as superintendent of the project from 1823 to 1843. Later knighted, he served as vice president of the Royal Geographical Society. The world's tallest peak, Mount Everest in Nepal (known locally as Chomolungma), was named in his honor.

Faraday, Michael (1791-1867): A self-educated British chemist and physicist, Faraday served an apprenticeship with chemist Sir Humphry Davy (1778-1829), during which time he experimented with liquefied gases, alloys, and optical glasses. He invented a prototype of what became the Bunsen burner, discovered benzene, and performed experiments that led to his discovery of electromagnetic induction. He built the first electric dynamo, the precursor to the power generator, researched the relationship between magnetism and light, and made numerous other contributions to the studies of electromagnetism and electrochemistry.

Farnsworth, Philo (1906-1971): An inventor born in Utah, Farnsworth became interested in electronics and mechanics as a child. He experimented with television during the 1920's, and late in the decade he demonstrated an electronic, nonmechanical scanning system for image transmissions. During the early 1930's, he worked for Philco but left to carry out his own research. In addition to significant contributions to television, Farnsworth held more than 300 patents and devised a milk-sterilizing process, developed fog lights, an infrared telescope, a prototype of an air traffic control system, and a fusion reaction tube.

Fermi, Enrico (1901-1954): An Italian-born experimental and theoretical physicist and teacher who became an American citizen in 1944, Fermi studied mechanics and was instrumental in the advancement of thermodynamics and quantum, nuclear, and particle physics. Awarded the 1938 Nobel Prize in Physics for his research into radioactivity, he was a member of the team that developed the first nuclear reactor in Chicago in the early 1940's and served as a consultant to the Manhattan Project, which produced the first atomic bomb. He died of cancer from sustained exposure to radioactivity.

Feynman, Richard (1918-1988): A physicist, author, and teacher born in New York, Feynman participated in the Manhattan Project and made numerous contributions to a diverse field of specialized scientific disciplines including quantum mechanics, supercooling, genetics, and nanotechnology. He shared the 1965 Nobel Prize in Physics—with Julian Schwinger (1918-1994) and Sin-Itiro Tomonaga (1906-1979)—for work in quantum electrodynamics, particularly for his lucid explanation of the behavior of subatomic particles. He was a popular and influential professor for many years at California Institute of Technology.

Fleming, Alexander (1881-1955): A Scottish-born biologist, Fleming served in the Royal Army Medical Corps during World War I and witnessed the deaths of many wounded soldiers from infection. He was a professor of bacteriology at a teaching hospital, and he specialized in immunology and chemotherapy research. In 1928 he discovered an antibacterial mold, which over the next decade was purified and mass-produced as the drug penicillin, which played a large part in suppressing infections during World War II. A major contributor to the development of antibiotics, he shared the 1945 Nobel Prize in Physiology or Medicine.

Forrester, Jay Wright (b. 1918): An engineer, teacher, and computer scientist born in Nebraska, Forrester built a wind-powered electrical system while in his teens. Associated with the Massachusetts Institute of Technology as a researcher and professor for many years, he developed servomechanisms for military use, designed aircraft flight simulators, and air defense systems. He founded the field of system dynamics to produce computer-generated mathematical models for such tasks as determining water flow, fluid turbulence, and a variety of mechanical movements.

Fourier, Joseph (1768-1830): A French-born physicist, mathematician, and teacher Fourier accompanied Napoleon's expedition to Egypt, where he served as secretary of the Egyptian Institute and was a major contributor to *Description of Egypt,* a massive work describing the scientific findings that resulted from the French military campaign. After returning to France, Fourier explored numerous scientific fields but is best known for his extensive research on the conductive properties of heat and for his theories of equations, which influenced later physicists and mathematicians.

Franklin, Benjamin (1706-1790): A Boston-born author, statesman, scientist, and inventor, Franklin worked as a printer in his youth and from 1733 to 1758 published *Poor Richard's Almanack.* A key figure during the American Revolution and a founding father of the United States, he established America's first lending library and Pennsylvania's first fire department, served as first U.S. postmaster general and as minister to France and Sweden. He experimented with electricity and is credited with inventing the lightning rod, bifocal glasses, an odometer, the Franklin stove, and a musical instrument made of glass.

Freud, Sigmund (1856-1939): An Austrian of Jewish

heritage, Freud studied neurology before specializing in psychopathology and conducted extensive research into hypnosis and dream analysis to treat hysteria. Considered the father of psychoanalysis and a powerful influence on the field, he originated such psychological concepts as repression, psychosomatic illness, the unconscious mind, and the division of the human psyche into the id, ego, and superego. He fled from the Nazis and went to London. Riddled with cancer from years of cigar smoking, he took morphine to relieve his suffering and hasten his death.

Frisch, Karl von (1886-1982): The Vienna-born son of a surgeon-university professor, von Frisch initially studied medicine before switching to zoology and comparative anatomy. Working as a teacher and researcher out of Munich, Rostock, Breslau, and Graz universities, he focused his research on the European honeybee. He made many discoveries about the insect's sense of smell, optical perception, flight patterns, and methods of communication that have since proved invaluable in the fields of apiology and botany. He was awarded the 1973 Nobel Prize in Physiology of Medicine in recognition of his pioneering work.

Fuller, R. Buckminster (1895-1983): The Massachusetts-born architect, philosopher, engineer, author, and inventor developed systems for lightweight, weatherproof, and fireproof housing while in his twenties. Teaching at Black Mountain College in the late 1940's, Fuller perfected the geodesic dome, built of aluminum tubing and plastic skin, and afterward developed numerous designs for inventions aimed at providing practical and affordable shelter and transportation. He coined the term "synergy" and advocated exploiting renewable sources of energy such as solar power and wind-generated electricity. He was awarded the Presidential Medal of Freedom in 1983.

Galen (129-c. 199): An ancient Roman surgeon, scientist, and philosopher, Galen traveled and studied widely before serving as physician to Roman emperors Marcus Aurelius (121-180), Commodus (161-192), Septimus Severus (146-211), and Caracalla (188-217). In the course of his education he explored human and animal anatomy, became an advocate of proper diet and hygiene, and advanced the practice of surgery by treating the wounds of gladiators and ministering to plague victims. His medical discoveries and healing methods, detailed in numerous written works, influenced medicine for more than 1,500 years.

Galilei, Galileo (1564-1642): A physicist, astronomer, mathematician, and philosopher born in Pisa, Italy, Galileo is known as the father of astronomy and the father of modern science. A keen astronomical observer, he made significant improvements to the telescope, through which he studied the phases of Venus, sunspots, and the Milky Way, and he discovered Jupiter's four largest moons. He risked excommunication and death championing the heretical Copernican heliocentric view of the solar system. He also invented a military compass and a practical thermometer and experimented with pendulums and falling bodies.

Galton, Francis (1822-1911): An anthropologist, geographer, meteorologist, and inventor born in England, Galton was a child prodigy. He traveled widely, exploring the Middle East and Africa, and wrote about his expeditions. Fascinated by numbers, he devised the first practical weather maps for use in newspapers. He also contributed to the science of statistics, studied heredity—coining the term "eugenics" and the phrase "nature or nurture"—and was an early advocate of using fingerprints in criminology. Galton is responsible for inventing a high-frequency whistle used in training dogs and cats.

Gates, Bill (b. 1955): An entrepreneur, philanthropist, and author born in Seattle, Gates became interested in computers as a teenager. He left Harvard in 1975 to cofound and to serve as chairman (until 2006) of Microsoft, which developed software for IBM and other systems before launching its own system in 1985. The result, Microsoft Windows, became the dominant software product in the worldwide personal computer market. Profits from his enterprise made Gates one of the world's richest people, and he has used his vast wealth to assist a wide variety of charitable causes.

Goddard, Robert H. (1892-1945): A physicist, engineer, teacher, and inventor born in Massachusetts, Goddard became interested in science as a child and experimented with kites, balloons, and rockets. He received the first of more than 200 patents in 1914 for multistage and liquid-fuel rockets, and during the 1920's he conducted successful test flights using liquid fuel. Goddard experimented with solid fuels and ion thrusters and is credited with developing tail fins, gyroscopic guidance systems, and many other basics of rocketry that greatly influenced the designs of rocket scientists who came after him.

Grandin, Temple (b. 1947): An animal scientist born in Massachusetts, Grandin was diagnosed with autism as a child. As an adult she earned advanced degrees before receiving a doctorate from the University of Illinois in 1989. A professor at Colorado State University, an author, and an autism advocate, she has made numerous humane improvements to the design of livestock-handling facilities that have been incorporated into meat-processing plants worldwide to reduce or eliminate animal stress, pain, and fear.

Haber, Fritz (1868-1934): A chemist and teacher of Jewish heritage (later a convert to Christianity) born in Germany, Haber developed the Haber process to produce ammonia used in fertilizers, animal feed, and explosives, for which he was awarded the 1918 Nobel Prize in Chemistry. At Berlin's Kaiser Wilhelm Institute (later the Haber Institute) between 1911 and 1933, he developed chlorine gas used in World War I, experimented with the extraction of gold from seawater, and oversaw production of Zyklon B, the cyanide-based pesticide that was employed at extermination camps during World War II.

Halley, Edmond (1656-1742): An English astronomer, mathematician, physicist, and meteorologist, Halley wrote about sunspots and the solar system while a student at Oxford. In the 1670's, he cataloged the stars of the Southern Hemisphere and charted winds and monsoons. Inventor of an early diving bell and a liquid-damped magnetic compass, a colleague of Sir Isaac Newton (1642-1727), and leader of the first English scientific expedition, Halley is best remembered for predicting the regular return of the comet that bears his name.

Heisenberg, Werner (1901-1976): A theoretical physicist and teacher born in Germany, Heisenberg conducted research in quantum mechanics with Niels Bohr (1885-1962) at the University of Copenhagen. There he developed the uncertainty principle, which proves it is impossible to determine the position and momentum of subatomic particles at the same time. Awarded the 1932 Nobel Prize in Physics for his work, he also contributed research on positrons, cosmic rays, spectral frequencies, matrix mechanics, nuclear fission, superconductivity, and plasma physics to the continuing study of atomic theory.

Herschel, William (1738-1822) and Caroline Herschel (1750-1848): A German-born astronomer, composer, and telescope maker who moved to England in his teens, William spent the early part of his career as a musician, playing cello, oboe, harpsichord, and organ and wrote numerous symphonies and concerti. In the 1770's he began building his own large reflecting telescopes and with his diminutive (4 feet, 3 inches) but devoted sister Caroline spent countless hours observing the sky while cataloging nebulae and binary stars. The Herschels are credited with discovering two of Saturn's moons, the planet Uranus and two of its moons, and coining the word "asteroid."

Hersey, Mayo D. (1886-1978): A mechanical engineer born in Rhode Island, Hersey was a preeminent expert on tribology, the study of the relationship between interacting solid surfaces in motion, the adverse effects of wear, and the ameliorating effects of lubrication. He worked as a physicist at the National Institute of Standards and Technology (1910-1920) and the U.S. Bureau of Mines (1922-1926) and taught at the Massachusetts Institute of Technology (1910-1922). He was a consultant to the Manhattan Project and won numerous awards for his contributions to lubrication science.

Hippocrates (c. 460-c. 377 B.C.E.): An ancient Greek physician born on the island of Kos, Hippocrates was the first of his time to separate the art of healing from philosophy and magical ritual. Called the father of Western medicine, he originated the belief that diseases were not the result of superstition but of natural causes, such as environment and diet. Though his concept that illness was the result of an imbalance in the body's fluids (called humors) was later discredited, he pioneered such common modern clinical practices as observation and documentation of patient care. He originated the Hippocratic Oath, which for many centuries served as the guiding principle governing the behavior of doctors.

Hooke, Robert (1635-1703): A brilliant, multitalented British experimental scientist with interests in physics, astronomy, chemistry, biology, geology, paleontology, mechanics, and architecture, Hooke was instrumental as chief surveyor in rebuilding the city of London following the Great Fire of 1666. Among many accomplishments in diverse fields he is credited with inventing the compound microscope—via which he discovered the cells of plants and formulated a theory of fossilization—devised a balance spring to improve the accuracy of timepieces, and either created or refined such instruments as the barometer, anemometer, and hygrometer.

Howlett, Freeman S. (1900-1970): A horticulturist born in New York, Howlett was associated with the Ohio State University as teacher, administrator, and researcher for more than forty-five years and was considered an expert on the history of horticulture. His investigations focused on plant hormones, embryology, fruit setting, reproductive physiology, and foliation for a variety of crops, including fruits, vegetables, and nuts. He created five new varieties of apples popular among consumers. A horticulture and food science building at Ohio State is named in his honor.

Hubble, Edwin Powell (1889-1953): An astronomer born in Missouri, Hubble was associated with the Mount Wilson Observatory in California for more than thirty years. Using what was then the world's largest telescope, he was the first to discover galaxies beyond the Milky Way, which greatly expanded science's concept of the universe. He studied red shifts in formulating Hubble's law, which confirmed the big bang or expanding universe theory. The American space telescope launched in 1990 was named for him, and he was honored in 2008 with a commemorative postage stamp.

Huygens, Christiaan (1629-1695): Born in the Netherlands, Huygens was an astronomer, physicist, mathematician, and prolific author. He made early telescopic observations of Saturn and its moons and was the first to suggest that light is made up of waves. He discovered centrifugal force, proposed a formula for centripetal force, and developed laws governing the collision of celestial bodies. An inveterate inventor, he patented the pendulum clock and the pocket watch and designed an early internal combustion engine. He is also considered a pioneer of science fiction for writing about the possibility of extraterrestrial life.

Jacquard, Joseph Marie (1752-1834): A French inventor, Jacquard created a series of mechanical looms in the early nineteenth century. His experiments culminated in the Jacquard loom attachment, which could be programmed, via punch cards, to weave silk in various patterns, colors, and textures automatically. The labor-saving device became highly popular in the silk-weaving industry, and its inventor received royalties on each unit sold and became wealthy in the process. The loom inspired scientists to incorporate the concept of punch cards for computer information storage.

Jenner, Edward (1749-1823): A surgeon and anatomist born in England, Jenner experimented with cowpox inoculations in an attempt to prevent smallpox, a virulent infectious disease of ancient origin with a high rate of mortality that killed millions of people. In the early nineteenth century, Jenner successfully developed a method of vaccination that provided immunity from smallpox and late in life became personal physician to King George IV. The smallpox vaccination was made compulsory in England and elsewhere, and the disease was declared eradicated worldwide in 1979.

Jobs, Steven (b. 1955-2011): An inventor and entrepreneur of Syrian and American heritage born in San Francisco, Jobs worked at Hewlett-Packard as a teenager and was later employed at Atari designing circuit boards. In 1976, he and coworker Steve Wozniak (b. 1950) and others founded Apple, which designed, built, and sold a popular and highly successful line of personal computers. A multibillionaire and holder of more than 200 patents, Jobs continued to make innovations in interfacing, speakers, keyboards, power adaptation, and myriad other components related to modern computer science until his death in late 2011

Kepler, Johannes (1571-1630): A German mathematician, author and astronomer, he became interested in the cosmos after witnessing the Great Comet of 1577. He worked for Tycho Brahe (1546-1601) for a time and after Brahe's death became imperial mathematician in Prague. A major contributor to the scientific revolution, Kepler studied optics, observed many celestial phenomena, provided the foundation for Sir Issac Newton's theory of gravitation, and developed a set of laws governing planetary motion around the Sun—including the discovery that the orbits of planets are elliptical—that were confirmed by later astronomers.

Krebs, Sir Hans Adolf (1900-1981): A biochemist and

physician born the son of a Jewish surgeon in Germany, Krebs was a clinician and researcher before moving to England after the rise of the Nazis. As a professor at Cambridge University, he explored metabolism, discovering the urea and citric acid cycles—biochemical reactions that promote understanding of organ functions in the body and explain the cellular production of energy—for which he shared the 1953 Nobel Prize in Medicine or Physiology. He was also knighted in 1958 for his work.

Lawrence, Ernest O. (1901-1958): A South Dakota-born physicist and teacher, Lawrence researched the photoelectric effect of electrons at Yale. In 1928, he became a professor at the University of California, Berkeley, where he invented the cyclotron particle accelerator, for which he was awarded the 1939 Nobel Prize in Physics. During World War II, he was involved in the Manhattan Project. Lawrence popularized science and was a staunch advocate for government funding of significant scientific projects. After his death, laboratories at the University of California and the chemical element lawrencium were named in his honor.

Leakey, Louis B. (1903-1972) and Mary Nicol Leakey (1913-1996): Louis, born in Kenya, was an archaeologist, paleontologist, and naturalist who married London-born anthropologist and archaeologist Mary Nicol. Together and often with their sons, Jonathan, Richard, and Philip, they excavated at Olduvai Gorge in East Africa, where they unearthed the tools and fossils of ancient hominids. Their discoveries of the remains of Proconsul africanus, Australopithecus boisei, Homo habilis, Homo erectus, and other large-brained, bipedal primates effectively proved Darwin's theory of evolution and extended human history by several million years.

Leonardo da Vinci (1452-1519): An Italian genius considered the epitome of the Renaissance man, da Vinci was a superb artist, architect, engineer, mathematician, geologist, musician, mapmaker, inventor, and writer. Creator of such famous paintings as the *Mona Lisa* and *The Last Supper,* he is credited with imagining the helicopter, solar power, and the calculator centuries before their invention. His far-ranging mind explored such subjects as anatomy, optics, vegetarianism, and hydraulics, and his journals, written in mirror-image script, are filled with drawings, ideas, and scientific observations that are still closely studied.

Linnaeus, Carolus (1707-1778): A Swedish botanist, zoologist, physician, and teacher, Linnaeus began studying plants as a child. As an adult, he embarked on expeditions throughout Europe observing and collecting specimens of plants and animals and wrote numerous works about his findings. He devised the binomial nomenclature system of classification for living and fossil organisms—called taxonomy—still used in modern science, which provides concise Latin names of genus and species for each example. Linnaeus also cofounded the Royal Swedish Academy of Science.

Lippershey, Hans (c. 1570-c. 1619): A master lens grinder and spectacle maker born in Germany who later became a citizen of the Netherlands, Lippershey is credited with designing the first practical refracting telescope (which he called "perspective glass"). After fruitlessly attempting to patent the device, he built several prototypes for sale to the Dutch government, which distributed information about the telescope across Europe. Other scientists, such as Galileo, soon duplicated and improved upon Lippershey's invention, which became a primary instrument in the science of astronomy.

Lumière, Auguste (1862-1954) and Louis Lumière (1864-1948): The French-born brothers worked at their father's photographic business and devised the dry-plate process for still photographs. From the early 1890's, they patented several techniques—including perforations to guide film through a camera and a color photography process—that greatly advanced the development of moving pictures. From 1895 to 1896, they publicly screened a series of short films to enthusiastic audiences in Asia, Europe, and North and South America, demonstrating the commercial potential of the new medium and launching what would become the multibillion-dollar film industry.

Maathai, Wangari Muta (b. 1940-2011): An environmental and political activist of Kikuyu heritage born in Kenya, Maathai studied biology in the United States before becoming a research assistant and anatomy teacher at the University of Nairobi, where she was the first East African woman to earn a Ph.D. She founded the Green Belt Movement, an organization that plants trees, supports environmental conservation, and advocates for women's rights. A former member of the Kenyan Parliament and former Minister of Environment, she was awarded the 2004 Nobel Peace Prize for her work and is the first African woman to receive the award.

McAdam, John Loudon (1756-1836): A Scottish engineer, McAdam became a surveyor in Great Britain and specialized in road building. He devised an effective method—called "macadam" after its inventor—of creating long-lasting roads using gravel on a foundation of larger stones, with a camber to drain away rainwater, which was adopted around the world. He also introduced hot tar as a binding agent (dubbed "tarmac," an abbreviation of tar-macadam) to produce smoother road surfaces. Modern road builders still use many of the techniques he innovated.

Mantell, Gideon (1790-1852): A British surgeon, geologist, and paleontologist, Mantell began collecting fossil specimens from quarries as a child. As an adult, he was a practicing physician and pursued geology in his spare time. He discovered fossils that were eventually identified as belonging to the Iguanodon and Hylaeosaurus—which he named Megalosaurus and Pelorosaurus—and he became a recognized authority on dinosaurs. His major works were *The Fossils of South Downs: Or, Illustrations of the Geology of Sussex* (1822) and *Notice on the Iguanodon: A Newly Discovered Fossil Reptile* (1825).

Marconi, Guglielmo (1874-1937): An Italian-born electrical engineer and inventor, Marconi experimented with electricity and electromagnetic radiation. He developed a system for transmitting telegraphic messages without the use of connecting wires and by the early twentieth century was sending transmissions across the Atlantic Ocean. His devices eventually evolved into radio, and the transmitter at his factory in England was the first in 1920 to broadcast entertainment to the United Kingdom; he shared the 1909 Nobel Prize in Physics with German physicist Ferdinand Braun (1850-1918).

Maxwell, James Clerk (1831-1879): A Scottish-born mathematician, theoretical physicist, and teacher, Maxwell had an insatiable curiosity from an early age and as a teenager began presenting papers to the Royal Society of Edinburgh. He experimented with color, examined hydrostatics and optics, and wrote about Saturn's rings. His most significant work, however, was performed in the field of electromagnetism, in which he showed that electricity, magnetism, and light are all results of the electromagnetic field, a concept that profoundly affected modern physics.

Mendel, Gregor Johann (1822-1884): Born in Silesia (now part of the Czech Republic), Mendel became interested in plants as a child. In the 1840's he entered an Augustinian monastery, where he studied astronomy, meteorology, apiology, and botany. Called the father of modern genetics, he is best known for his experiments in hybridizing pea plants, which evolved into what later were called Mendel's laws of inheritance. Though his work exerted little influence during his lifetime, his concepts were rediscovered early in the twentieth century and have since proven invaluable to the study of heredity.

Mendeleyev, Dmitri Ivanovich (1834-1907): A Russian chemist, teacher, and inventor, Mendeleyev studied the properties of liquids and the spectroscope before becoming a professor in Saint Petersburg and later serving as director of weights and measures. He created a periodic table of the sixty-three elements then known arranged by atomic mass and the similarity of properties (a revised form of which is still employed in modern science) and used the table to correctly predict the characteristics of elements and isotopes not yet found. Element 101, mendelevium, discovered in 1955, was named in his honor.

Meng Tian (259-210 B.C.E.): A general serving under Qin Shi Huang, first emperor of the Qin Dynasty (221-207 B.C.E.), Meng Tian led an army of 100,000 to drive warlike nomadic tribes north out of China. Descended from architects, he oversaw building of the Great Wall to prevent invasions, cleverly incorporating topographical features and natural barriers into the defensive barricade, which he extended for more than 2,000 miles along the Yellow River. After a coup following Emperor Qin's death, Meng Tian was forced to commit suicide. The Qin Dynasty fell just three years later.

Montgolfier, Joseph Michel (1740-1810) and Jacques-Etienne Montgolfier (1745-1799): Born in France to a prosperous paper manufacturer, the Mongolfier brothers designed and built a hot-air balloon, and in 1783 Jacques-Etienne piloted the first manned ascent in a lighter-than-air craft. The French Academy of Science honored the brothers for their exploits, which inspired further developments in ballooning. The Montgolfier brothers subsequently wrote books on aeronautics and continued experimenting. Joseph is credited with designing a calorimeter and a hydraulic ram, and Jacques-Etienne invented a method for the manufacture of vellum.

Morse, Samuel F. B. (1791-1872): An artist and inventor born in Massachusetts, Morse painted portraits and taught art at the City University of New York before experimenting with electricity. In the mid-1830's, he designed the components of a practical telegraph—a sender, receiver, and a code to translate signals into numbers and words—and in 1844 sent the first message via wire. Within a decade, the telegraph had spread across America and subsequently around the world. The invention would inspire such later advancements in communication as radio, the Teletype, and the fax machine.

Nernst, Walther (1864-1941): A German physical chemist, physicist, and inventor, Nernst discovered the Third Law of Thermodynamics—defining the chemical reactions affecting matter as temperatures drop toward absolute zero—for which he was awarded the 1920 Nobel Prize in Chemistry. He also invented an electric lamp, and developed an electric piano and a device using rare-earth filaments that significantly advanced infrared spectroscopy. He made numerous contributions to the specialized fields of electrochemistry, solid-state chemistry, and photochemistry.

Newton, Sir Isaac (1642-1727): The English physicist, mathematician, astronomer, and philosopher is considered one of the most gifted and scientifically influential individuals of all time. He developed theories of color and light from studying prisms, was instrumental in creating differential and integral calculus, and formulated still-valid laws of celestial motion and gravitation. He was knighted in 1705, the first British scientist so honored. From 1699 until his death he served as master of the Royal Mint and during his tenure devised anticounterfeiting measures and moved England from the silver to the gold standard.

Nobel, Alfred (1833-1896): A Swedish chemist and chemical engineer, Nobel invented dynamite while studying how to manufacture and use nitroglycerin safely. In the course of building a manufacturing empire based on the production of cannons and other armaments, he experimented with combinations of explosive components, also producing gelignite and a form of smokeless powder, which led to the development of rocket propellants. Late in his life, he earmarked the bulk of his vast estate for the establishment of the Nobel Prizes, annual monetary awards given in recognition of outstanding achievements in science, literature, and peace.

Oppenheimer, J. Robert (1904-1967): A brilliant theoretical physicist, researcher, and teacher born to German immigrants in New York City, Oppenheimer was the scientific director of the Manhattan Project, which developed the atomic bombs dropped on Japan during World War II. Following the war, he was primary adviser to the U.S. Atomic Energy Commission and director of the Institute for Advanced Study in Princeton, New Jersey. He contributed widely to the study of electrons and positrons, neutron stars, relativity, gravitation, black holes, quantum mechanics, and cosmic rays.

Owen, Richard (1804-1892): An English biologist, taxonomist, anti-Darwinist, and comparative anatomist, Owen founded and directed the natural history department at the British Museum. He originated the concept of homology, a similarity of structures in different species that have the same function, such as the human hand, the wing of a bat, and the paw of an animal. He also cataloged many living and fossil specimens, contributed numerous discoveries to zoology, and coined the term "dinosaur." Owen advanced the theory that giant flightless birds once inhabited New Zealand long before their remains were found there.

Paré, Ambroise (c. 1510-1590): A French royal surgeon, Paré revolutionized battlefield medicine, developing techniques and instruments for the treatment of gunshot wounds and for performing amputations. He greatly advanced knowledge of human anatomy by studying the effects of violent death on internal organs. He pioneered the lifesaving practices of vascular ligating and herniotomies, designed prosthetics to replace amputated limbs, and was the first to create realistic artificial eyes from such substances as glass, porcelain, silver, and gold.

Pasteur, Louis (1822-1895): A chemist, microbiologist, and teacher born in France, Pasteur focused on researching the causes of diseases and methods for preventing them after three of his children died from typhoid. He proposed a germ theory, demonstrating that microorganisms affect foodstuffs. This ultimately led to his invention of pasteurization—a method of killing bacteria in milk, which was later applied to other substances. A pioneer in immunology, he also developed vaccines to combat anthrax, rabies, and puerperal fever.

Pauli, Wolfgang (1900-1958): An Austrian theoretical physicist of Jewish heritage who converted to Catholicism, Pauli earned a Ph.D. at the age of twenty-one. While lecturing at the Niels Bohr Institute for Theoretical Physics, he researched relativity and quantum physics. He discovered a new law governing the behavior of atomic particles and the characteristics of matter, called the Pauli exclusion principle, for which he was awarded the 1945 Nobel Prize in Physics. During World War II, he moved to the United States and became an American citizen but later relocated to Zurich.

Pauling, Linus (1901-1994): Born in Portland, Oregon, Pauling earned advanced degrees in chemical engineering, physical chemistry, and mathematical physics. A Guggenheim Fellow, he studied quantum mechanics in Munich, Copenhagen, and Zurich before teaching at the California Institute of Technology. He specialized in theoretical chemistry and molecular biology and greatly advanced understanding of the nature of chemical bonds. A political activist who warned of the dangers of nuclear weapons, he became one of a handful of scientists to receive Nobel Prizes in two fields: the 1954 prize in chemistry and the 1982 peace prize.

Pavlov, Ivan (1849-1936): A Russian physiologist and psychologist, Pavlov began investigating the digestive system, which led to experiments with the effects of behavior on the nervous system and the body's automatic functions. He used animals in researching conditioned reflex actions to a variety of visual, tactile, and sound stimuli—including bells, whistles, and electric shocks—to discover the relationship between salivation and digestion and was able to make dogs drool in anticipation of receiving food. He was awarded the 1904 Nobel Prize in Physiology or Medicine for his work.

Planck, Max (1858-1947): A German theoretical physicist credited with founding quantum theory—which affects all matter in the universe—Planck earned a doctoral degree at the age of twenty-one before becoming a professor at the universities of Kiel and Berlin. He explored electromagnetic radiation, quantum mechanics, thermodynamics, blackbodies, and entropy. He formulated the Planck constant, which describes the proportions between the energy and frequency of a photon and provides understanding of atomic structure. He was awarded the 1918 Nobel Prize in Physics for his discoveries.

Ptolemy (c. 100-c. 178): A mathematician, astronomer, and geographer of Greek heritage who worked in Roman-ruled Alexandria, Egypt, Ptolemy wrote several treatises that influenced science for centuries afterward. His *Almagest*, written in about 150, contains star catalogs, constellation lists, Sun and Moon eclipse data, and planetary tables. Ptolemy's eight-volume *Geographia* (*Geography*) followed and incorporates all known information about the geography of the Earth at the time and helped introduce the concept of latitudes and longitudes. His work on astrology influenced Islamic and medieval Latin worlds, and his writings on music theory and optics pioneered study in those fields.

Pythagoras (c. 580-c. 500 B.C.E.): An ancient Greek philosopher and mathematician from Samos, Pythagoras traveled widely seeking wisdom and established a religious-scientific ascetic community in Italy around 530 B.C.E. He had interests in music, astronomy, medicine, and mathematics, and though none of his writings survived, he is credited with the discovery of the Pythagorean theorem governing right triangles (the square of the hypotenuse is equal to the sum of the squares of the other two sides). His life and philosophy exerted considerable influence on Plato (c. 427-347 B.C.E.) and through Plato greatly affected Western thought.

Reiss, Archibald Rodolphe(1875-1929): A chemist, photographer, teacher, and natural scientist born in Germany, Reiss founded the world's first school of forensic science at the University of Lausanne, Switzerland, in 1909. He published numerous works that greatly influenced the new discipline, including *La photographie judiciaire* (*Forensic photography*, 1903) and *Manuel de police scientifique. I Vols et homicides* (*Handbook of Forensic Science: Thefts and Homicides*, 1911). During World War I he investigated alleged atrocities in Serbia and lived there for the rest of his life. The institute he founded more than a century ago has become a major school offering numerous courses in various forensic sciences, criminology, and criminal law.

Röntgen, Wilhelm Conrad (1845-1923): A German physicist, Röntgen studied mechanical engineering before teaching physics at the universities of Strassburg, Giessen, Würzburg, and Munich. He experimented with fluorescence and electrostatic charges. In the process of his work he discovered X rays—and also discovered that lead could effectively block the rays—meanwhile laying the foundations of what would become radiology: the medical specialty that uses radioactive imaging to diagnose disease. He was awarded the first Nobel Prize in Physics in 1901. Element 111, roentgenium, was named in his honor in 2004.

Rutherford, Ernest (1871-1937): A chemist and physicist born in New Zealand, Rutherford studied at the University of Cambridge before teaching physics at McGill University in Montreal and at the University of Manchester. He made some of the most significant discoveries in the field of atomic science, including the relative penetrating power of alpha, beta, and gamma rays, the transmutation of elements via radioactivity, and the concept of radioactive half-life. His work, for which he received the 1908 Nobel Prize in Chemistry, was instrumental in the development of nuclear energy and carbon dating.

Sabin, Albert Bruce (1906-1993): A microbiologist born of Jewish heritage as Albert Saperstein in Russia, Sabin later became an American citizen and changed his name. Trained in internal medicine, he conducted research into infectious diseases and assisted in the development of a vaccine to combat encephalitis. His major contribution to medicine was an effective oral polio vaccine, which was administered in mass immunizations during the 1950's and 1960's and eventually led to the eradication of the disease worldwide. Among other honors, he received the Presidential Medal of Freedom in 1986.

Sachs, Julius von (1832-1897): A German botanist, writer, and teacher, Sachs made great strides in the investigation of plant physiology, morphology, heliotropism, and germination while professor of botany at the University of Würzburg. In addition to numerous written works on photosynthesis, water absorption, and chloroplasts that significantly advanced the science of botany, he also invented a number of devices useful to research, including an auxanometer to measure growth rates, and the clinostat, a device that rotates plants to compensate for the effects of gravitation on botanical growth.

Sakharov, Andrei (1921-1989): A Russian nuclear physicist and human rights activist, Sakharov researched cosmic rays, particle physics, and cosmology. He was a major contributor to the development of the hydrogen bomb but later campaigned against nuclear proliferation and for the peaceful use of nuclear power. He received the 1975 Nobel Peace Prize, and though he received several international honors in recognition of his humanitarian efforts, he spent most of the last decade of his life in exile within the Soviet Union. A human rights center and a scientific prize are named in his honor.

Scheele, Carl Wilhelm (1742-1786): A chemist born in a Swedish-controlled area of Germany, Scheele became a pharmacist at an early age. Though he discovered oxygen through experimentation, he did not publish his findings immediately, and the discovery was credited to Antoine-Laurent Lavoisier (1743-1794) and Joseph Priestly (1733-1804), though science later gave the Scheele recognition he deserved. Scheele also discovered the elements barium, manganese, and tungsten, identified such chemical compounds as citric acid, glycerol, and hydrogen cyanide, experimented with heavy metals, and devised a method of producing phosphorus in quantity for the manufacture of matches.

Shockley, William (1910-1989): A physicist and inventor born to American parents in England, Shockley was raised in California. After earning a doctoral degree, he conducted solid-state physics research at Bell Laboratories. During World War II, he researched radar and anti-submarine devices. Following the war, he was part of the team that invented the first practical solid-state transistor, for which he shared the 1956 Nobel Prize in Physics with John Bardeen (1908-1991) and Walter Brattain (1902-1987). He later set up a semiconductor business that was a precursor to Silicon Valley. His major work, *Electrons and Holes in Semiconductors* (1950), greatly influenced many scientists.

Sikorsky, Igor (1889-1972): A Ukrainian engineer and test pilot who immigrated to the United States and became a naturalized American citizen, Sikorsky was a groundbreaking designer of both airplanes and helicopters. Inspired as a child by the drawings of Leonardo da Vinci (1452-1519), he created and flew the first multi-engine fixed-wing aircraft and the first airliner in the 1910's. He built the first flying boats in the 1930's—the famous Pan Am Clippers—and in 1939 designed the first practical helicopter, which introduced the system of rotors still used in modern helicopters.

Spilsbury, Sir Bernard Henry (1877-1947): The first British forensic pathologist, Spilsbury began performing postmortems in 1905. He investigated cause of death in many spectacular homicide cases—including those of Dr. Crippen and the Brighton trunk murders—that resulted in convictions and enhanced the science of forensics. He was a consultant to Operation Mincemeat, a successful World War II ruse (dramatized in the 1956 film *The Man Who Never Was*) involving the corpse of an alleged Allied courier, which deceived the Axis powers about the invasion of Sicily. Spilsbury was found dead in his laboratory—a victim of suicide.

Stephenson, George (1781-1848): A British mechanical and civil engineer, Stephenson invented a safety lamp for coal mines that provided illumination without the risk of explosions from firedamp. He designed a steam-powered locomotive for hauling coal, which evolved into the first public railway line in the mid-1820's, running on his specified track width of 4 feet, 8.5 inches. This measurement became the worldwide standard railroad gauge. He worked on numerous rail lines, in the process making many innovations in the design and construction of locomotives, tracks, viaducts, and bridges that greatly advanced railroad transport.

Teller, Edward (1908-2003): An outspoken theoretical physicist born in Hungary, Teller came to the United States in the 1930's and taught at George Washington University while researching quantum, molecular, and nuclear physics. A naturalized American citizen, he was a member of the atomic-bomb-building Manhattan Project. A strong supporter for nuclear energy development and testing for both wartime and peacetime purposes, he cofounded and directed Lawrence Livermore National Laboratory and founded the department of applied science at the University of California, Davis.

Tesla, Nikola (1856-1943): Born in modern-day Croatia, the brilliant if eccentric Tesla came to the United States in 1884 to work for Thomas Edison's company and later for Edison's rival George Westinghouse (1846-1914). In 1891, Tesla became a naturalized American citizen. A physicist, mechanical and electrical engineer, and an inventor specializing in electromagnetism, he created fluorescent lighting, pioneered wireless communication, built an alternating- current induction motor, and developed the Tesla coil, variations of which have provided the basis for many modern electrical and electronic devices.

Vavilov, Nikolai Ivanovich (1887-1943): A Russian botanist and plant geneticist, Vavilov served for two decades as director of the Institute of Agricultural Sciences (now the N. I. Vavilov Research Institute of Plant Industry) in Leningrad (now Saint Petersburg). During his tenure, he collected seeds from around the world, establishing the world's largest seed bank—with more than 200,000 samples—and conducted extensive research on genetically improving grain, cereal, and other food crops to produce greater yields to better feed the world. Arrested during World War II for disagreeing with Soviet methods of agronomy, he died of complications from starvation and malnutrition.

Vesalius, Andreas (1514-1564): A physician and anatomist born as Andries van Wesel in the Habsburg Netherlands (now Belgium), Vesalius taught surgery and anatomy at the universities of Padua, Bologna, and Pisa in Italy. Dissatisfied at the inaccuracies in the standard texts of the day—based solely on the 1,400-year-old work of ancient physician Galen, since Rome had long discouraged performing autopsies—he dissected a human corpse in the presence of artists from Titian's studio. This resulted in the seven-volume illustrated work, *De humani corporis fabrica libri septem* (*On the Fabric of the Human Body*, 1543), which served as the foundation for modern anatomy.

Vitruvius (c. 80-c. 15 B.C.E.): A Roman architect and engineer, Vitruvius served in many campaigns under Julius Caesar (100-44 B.C.E.), for whom he designed and built mechanical military weapons, such as the ballista (a projectile launcher) and siege machines. His major written work, *De architectura* (*On Architecture*, c. 27 B.C.E.), set the standard for building structures solidly, usefully, and attractively. The book covers the construction of machines—including cranes, pulleys, sundials, and water clocks. It discusses construction materials and describes ancient Roman building innovations that greatly influenced later architects, particularly during the Renaissance.

Watt, James (1736-1819): A Scottish mechanical and civil engineer, Watt designed a steam engine to pump water out of mines. Refinements of his engine were used in grinding, milling, and weaving, and further improvements—including gauges, throttles, gears, and governors—enhanced the engine's efficiency and safety, making it the prime mover of the Industrial Revolution and the power source of choice for early trains and ships. Watt also devised an early copying machine and discovered a method for producing chlorine for bleaching. The unit of electrical power is named for him.

Wegener, Alfred (1880-1930): A German meteorologist, climatologist, and geophysicist, Wegener was one of the first to employ weather balloons. He was first to advance the theory of continental drift, proposing that the Earth's continents were once a single mass that he called Pangaea; his ideas, however, were not accepted until long after his death. From 1912, he worked in remote areas of Greenland examining polar airflows and drilling into the ice to study past weather patterns. He died in Greenland during his last ill-fated expedition.

Westinghouse, George (1846-1914): Born in Central Bridge, New York, Westinghouse was an engineer, inventor, entrepreneur, and a rival of Thomas Edison (1847-1931). He built a rotary steam engine while still a teenager and in his youth patented several devices—including a fail-safe compressed-air braking system—to improve railway safety. He developed an alternating-current power distribution network that proved superior to Edison's direct-current scheme, invented a power meter still in use, built several successful hydroelectric generating plants, and devised shock absorbers for automobiles.

Whittle, Sir Frank (1907-1996): Born the son of an engineer in England, Whittle joined the Royal Air Force as an aircraft mechanic and advanced to flying officer and test pilot before eventually rising to group captain. While in the Royal Air Force, he began designing aircraft engines that used turbines rather than pistons. In the mid-1930's, he formed a partnership, Power Jets, which produced the first effective turbojet design before the company was nationalized. He later developed a self-powered drill for Shell Oil and wrote a text on gas turbine engines.

Wiener, Norbert (1894-1964): A mathematician born in Missouri, Wiener was a child prodigy. He began college at the age of eleven, earned a bachelor's degree in math at fourteen, and a doctorate in philosophy from Harvard at the age of eighteen. During World War I, he researched ballistics at Aberdeen Proving Ground and afterward spent his career teaching mathematics at Massachusetts Institute of Technology. A pioneer of communication theory, he is credited with the development of theories of cybernetics, robotics, automation, and computer systems, and his work greatly influenced later scientists.

Woodward, John (1665-1728): An English naturalist, physician, paleontologist, and geologist, Woodward was an early collector of fossils, which served as the basis for his *Classification of English Minerals and Fossils* (1729), a work that influenced geology for many years. He also conducted pioneering research into the science of hydroponics. His collection of specimens formed the foundation of Cambridge University's Sedgwick Museum, and his estate was sold to provide a post in natural history, now the Woodwardian Chair of Geology at Cambridge.

Wright, Orville (1871-1948) and Wilbur Wright (1867-1912): The Wright brothers were American aviation pioneers who began experimenting with flight in their teens. In the early 1890's they opened a bicycle sales and repair shop, which financed their research into manned gliders. They soon progressed to designing powered aircraft. They eventually invented and built the first practical fixed-wing aircraft and piloted the world's first sustained powered flight—a distance of more than 850 feet over nearly a minute—in 1903 at Kitty Hawk, North Carolina. The Wright Company later became part of Curtiss-Wright Corporation, a modern high-tech aerospace component manufacturer.

Zeppelin, Ferdinand von (1838-1917): Born in Germany, Zeppelin served in the Prussian army and made a balloon flight while serving as a military observer in the American Civil War. After returning to Europe, he designed and constructed airships and devised a transportation system using lighter-than-air craft. He created a rigid, streamlined, engine-powered dirigible in 1900 and was instrumental in the creation of duralumin, which later led to lightweight all-metal airframes. By 1908, he was providing commercial air service to passengers and mail, which had an enviable record for safety until the *Hindenburg* disaster in 1937.

Zworykin, Vladimir (1889-1982): A Russian who emigrated to the United States after World War I, Zworykin worked at the Westinghouse laboratories in Pittsburgh. An engineer and inventor who patented a cathode ray tube television transmitting and receiving system in 1923, he later worked in development for the Radio Corporation of America (RCA) in New Jersey, where his inventions were perfected in time to be used to telecast the 1936 Olympic Games in Berlin. He also contributed to the development of the electron microscope.

Jack Ewing

GLOSSARY

absolute zero: The complete absence of thermal energy, resulting in a temperature of -273.15 degrees Celsius. This temperature is the basis for the Kelvin scale (starting at 0 Kelvin) developed by the British physicist, Lord Kelvin, in 1848. What living organisms feel as heat or warmth is a difference in temperature between two objects, which results in a transfer of thermal energy. Molecules at absolute zero have no thermal energy to transfer but can receive thermal energy from contact with a warmer object. *See also* cold, heat, temperature.

acid: A compound containing hydrogen ions (with a positive charge) in its molecules, which are released when the acid is dissolved in water. Acids include such familiar hazardous substances as sulphuric, nitric, and hydrochloric acid, essential nutrients such as ascorbic acid (vitamin C), and common flavorings or preservatives such as acetic or ethanoic acid (vinegar). Acids react chemically with substances known as bases. The balance of acids and bases in a solution is measured by the pH scale, from 0 (strongly acidic) to 7 (neutral) to 14 (strongly alkaline). *See also* alkali, basic chemical.

alkali: A base that is dissolved in water. Alkaline substances are identified by a measurement from 8 to 14 on the pH scale. *See also* basic chemical.

alpha particle: One of three common forms of radiation from the nuclei of unstable radioactive elements, consisting of two protons and two neutrons, identical to the nucleus of a helium atom, without its electron shell. It has a velocity in air of one-twentieth the speed of light. *See also* beta particle, gamma ray.

amino acids: Biological molecules that serve as the building blocks of proteins and enzymes. Amino acids are incorporated into proteins by transfer RNA, according to the genetic code contained in DNA. The majority of amino acids have names ending with -ine, and are complex arrangements of atoms of carbon, nitrogen, hydrogen, and oxygen. *See also* enzyme, protein.

animal husbandry: The art and science of breeding, raising, and caring for domesticated animals, primarily in small- or large-scale agriculture, as sources of food, leather, wool, and other products useful to humans. Husbandry skills are not only required for many jobs in agriculture but for zookeepers, maintaining rodent and amphibian populations in laboratories, and for large-scale veterinary and animal-vaccination practices.

antiseptic: Any chemical substance that kills or inhibits the growth of microorganisms causing sepsis—putrefaction, decay, or other infection—generally applied to surface tissues of human or other living organisms or to nonliving surfaces that may harbor microorganisms.

atmosphere: The layers of gas surrounding the solid or liquid surfaces of a planet. The atmosphere of the Earth is 78.08 percent nitrogen, 20.95 percent oxygen, less than one percent argon, and hundredths or thousandths of a percent neon, helium, and hydrogen. The amounts of water vapor, carbon dioxide, methane, nitrous oxide, and ozone vary with biological (and more recently industrial) processes. Water can rise to as high as four percent. The atmosphere has been divided by different studies into five to six distinct layers: the troposphere, tropopause, stratosphere, mesosphere, and ionosphere (or thermosphere), plus the very thin exosphere fading into interplanetary space. The ozone layer is in the upper level of the stratosphere.

atom: The smallest particle of matter that has the characteristics of an element, such as oxygen, iron, calcium, or uranium. Three subatomic particles are common to all atoms: protons, neutrons, and electrons. The characteristics of any atom are determined by the number of these particles, particularly the negatively charged electrons in the outer shell. *See also* compound, element, molecule, periodic table of the elements.

atomic number: The number of protons (positively charged particles) in the nucleus of an atom, also the number of electrons (negative charge) in the atom in its standard form. Ions of an atom have larger or smaller levels of electron charge. *See also* electron, ion, periodic table of the elements, proton.

atomic weight: The total mass of the protons and neutrons in an atomic nucleus, with a tiny addition for the weight of electrons. Uranium has the atomic number 92, for 92 protons and 92 electrons, but different isotopes such as U-235 (atomic weight 235, adding 143 neutrons to 92 protons) or U-238 (atomic weight 238, adding 146 neutrons to 92 protons).

ballistics: The science of propelling objects, from rocks and spears to spacecraft. Mastery of this field requires mathematical precision in determining the energy required to put a stationary object into motion in a desired direction, and adjust its course, considering friction from wind or water, or the absence of friction in relatively empty vacuum, and the effect of any body powerful enough to exert gravitational pull, such as the Earth, Moon, or Sun.

basic chemical: Any substance that reacts with an acid. Bases include some metals, such as sodium, calcium, zinc, and aluminum when not protected by an aluminum oxide coating. Other bases include carbonates, hydroxides, and metal oxides (compounds formed by burning metals in oxygen). When a base reacts with an acid, the result is a metal salt and water. *See also* acid, alkali.

battery: In electricity, any device for storing an electrical charge so that it can be used later to power a machine, heater, or light source. Common types of batteries include lead-acid batteries, used for internal combustion automobiles and backup power for industries and military bases; solid alkaline and carbon-zinc batteries, used for flashlights and portable radios; mercury oxide batteries, used in small electronic equipment such as hearing aids, rechargeable nickel-cadmium and nickel hydride batteries; and lithium-ion batteries, an advanced rechargeable type used in portable computers, iPods, and hybrid or electric motor vehicles. Every battery relies on an oxidation-reduction chemical reaction induced by passing a current through its component materials and a reverse reaction that gives off an electric current when plugged into a circuit.

beta particle: One of three common forms of radiation from the nuclei of unstable radioactive elements, carrying a negative charge, similar to an electron, but moving at a high rate of speed, formed when a neutron (neutral electrical charge) transforms into a proton (positive electrical charge) by ejecting a negative charge. *See also* alpha particle, gamma ray.

binary number system: A mathematical system having only two numerals, 0 and 1, most commonly used in computer hardware and software, because at its most basic, a computer can turn a series of switches on (value = 1) or off (value = 0). *See also* computer.

biosphere: First defined by Austrian geologist Edward Seuss in 1875 as "the place on Earth where life dwells." The concept has been expanded to include all living organisms on Earth, dead organic matter, and the biological component of dynamic processes such as the cycling of carbon, nitrogen, oxygen, phosphorous, and other elements.

British thermal unit (Btu): The heat required to raise one pound of water one degree Fahrenheit, equal to 252 calories. Also, the heat required to produce 779.9 foot-pounds of energy in a mechanical system. This is a common measure of the energy potential in fuels. *See also* heat.

calorie: The heat required to raise the temperature of one gram of water one degree Celsius. This is a common measure of the energy potential in food but can be applied to fuels and mechanical processes. *See also* heat.

carboniferous fuels: Any source of energy obtained from carbon-based compounds, particularly coal and oil. These two common fuel sources accumulated during the Carboniferous period of the Paleozoic era.

catalyst: A substance that makes a chemical reaction between two other substances proceed at a significantly faster rate, without being consumed in the reaction. (See also enzyme.)

ceramics: Inorganic, nonmetallic solids, particularly made from clay, that are processed at high levels of heat, then cooled. In addition to common use for pottery and tableware, ceramics have many industrial applications, such as ceramic-based thermocouples to measure high temperatures, ceramic insulators, laser components, heat storage and diffusion, and capacitors.

chromosome: A basic unit of heredity in living cells. Each chromosome is composed of proteins and DNA, which carry thousands of genes. In a healthy, normal, human cell, there are twenty-three pairs of chromosomes. In sexual reproduction, one chromosome in each pair comes from the father, the other from the mother. *See also* DNA, gene.

climate: Prevailing weather conditions in a specific region or area over the course of a year, characterized by a typical range of temperatures, seasonal or year-round humidity, precipitation, prevailing winds, and extremes of seasonal variation.

cold: The sensation felt when a living organism is in contact with a substance of a lower temperature. Thermal energy naturally flows from a warmer object to a colder object when they are in physical contact. *See also* absolute zero, heat, temperature.

compound: A chemical substance composed of molecules, containing atoms of two or more different elements, forming a single particle. For example, water is a compound, in which each water molecule contains two atoms of the element hydrogen and one atom of the element oxygen. *See also* atom, molecule, element, periodic table of the elements.

computer: Any device that can be programmed to perform mechanical or electrical computation, processing numbers. Since the middle of the twentieth century, the term is commonly used for equipment that can be programmed using a binary number system to perform a variety of work. For a computer to process letters, words, graphic images, or maps, human programmers have to encode nonnumerical data in a numeric form. *See also* binary number system.

cryptology: The science of encrypting or decrypting information, including creating codes for privacy or security and finding means to break a code by working from a message in an unknown code to learn the pattern. Any code in which a symbol is substituted for each letter of the alphabet is particularly vulnerable to decryption, because in a large sample, there is a probability for how often each letter will appear, with "e" being used more often than any other. *See also* encoding.

demography: Variations in human population that can be studied statistically, in defined groups, rather than individual behavior. Almost any characteristic can form the basis for demographic research: ethnicity, diet, religion, wealth, language, education, urbanization, occupation, marriage customs, class or caste distinctions.

desalination: Removing salt from water for use in drinking or agriculture. Ninety-seven percent of the water on earth is salt water, mostly in the oceans. The dissolved salt content is harmful to freshwater land plants and land animals. Removing the salt is energy intensive and therefore has a high cost, but in places where freshwater is in short supply, it is sometimes considered worth the expense.

detergent: A type of surfactant that has the property of removing stains or particles from a surface, keeping it suspended in water, and allowing the suspended solids to be rinsed away. *See also* surfactant.

distillation: Isolating and purifying a liquid substance by heating a solution to the exact boiling point of the desired end product, producing a vapor, then capturing and condensing the vapor in a separate container. Water, perfumes, and alcohols are all common examples of liquids that can be distilled.

DNA (deoxyribonucleic acid): The complex molecule making up genes, encoding inheritance in all living species. It is known for a unique double-helix structure, with the code in an "alphabet" of four types of molecule: cytosine pairs with guanine, while adenine pairs with thymine. DNA is increasingly used for identification, particularly in crime scenes, to determine paternity of a child, and to study inheritance of both individuals and demographic groups. Study of DNA is also leading to new treatments for genetically inherited diseases. *See also* chromosome, gene.

ecology: A branch of biology that studies the complex relationships between living organisms, their environment, the manner in which a variety of plant and animal life forms a mutually interdependent ecosystem, and the competition between life forms within and between ecosystems.

electrical storage: See battery.

electrolysis: Applying an electric current to take apart (decompose) the molecules of a substance in solution. If an electric current is run through a container of water, the hydrogen and oxygen atoms in the water molecule will separate and form diatomic molecules (two atoms in each molecule) of oxygen and hydrogen gas. Electrolysis can be applied to many compounds. Industrial uses include separating chlorine and caustic soda from brine and refining metals such as sodium, calcium, magnesium, and aluminum from common ore compounds.

electromagnetism: A fundamental force of nature acting on all electrically charged particles. The previously separate studies of electricity, magnetism, and optics (initially visible light, which is only one spectrum of electromagnetic radiation) were unified by the work of English scientists Michael

Faraday and Robert Maxwell in the mid-nineteenth century.

electron: A negative charge in the outer shell of every atom. Electrons are sometimes described as particles, but within an atom they act more as electrical charges in shells, rather than particles in orbit. In a stable atom, the number of electrons exactly equals the number of protons in the nucleus. Ions have fewer or additional electric charges. *See also* atom, ion, neutron, proton.

element: A substance formed by atoms of a single type, found in the periodic table of the elements. Hydrogen, helium, and oxygen are all examples of elements. *See also* atom, compound, molecule, periodic table of the elements.

encoding: Representing information by a system of symbols or characters. Encoding processes exist in human memory, heredity, computer programming, and in written or verbal communication, including military or business communications intended to be secret. The most common use is to take a message in a plain language and convert it into a sequence of characters that can be read only by a person instructed in the code—or by a cryptologist who can break the code by mathematically analyzing the pattern. Natural encoding processes include the genetic code, which stores in long molecules of DNA the structure, physiology, and metabolism of a complete living organism. *See also* binary number system, computer, cryptology, DNA.

engineering: Practical application of the knowledge of pure science, and sometimes of art as well, not only to construct buildings, bridges, infrastructure, and engines, but to plan and organize industrial and community processes. There are many branches of engineering, including electrical, industrial, mechanical, civil, aeronautical, geotechnical, transportation, water management, disaster preparedness and management, and telecommunications.

entropy: The spontaneous direction of any natural process, tending to lose energy or to become more chaotic. All things naturally tend toward equilibrium: Two objects of different temperatures, placed in contact, will equalize to a common temperature as heat is transferred from the warmer object to the colder object. A solid dissolved in a solution spreads evenly throughout the solution, gas at two different pressures will equalize, but in none of these examples will objects spontaneously develop uneven temperatures, pressures, or concentrations in solution.

enzyme: A biological catalyst, any protein molecule within a living organism that speeds up biochemical reactions to a rate that will sustain life. The effect may speed up metabolic reactions by a factor of one million, compared with what would occur chemically outside the body. Names and classification of enzymes are regulated by the International Commission on Enzymes. Most enzymes are named by adding -ase to the root of a corresponding substrate, the molecule an enzyme acts upon. Sucrase catalyzes the hydrolysis of sucrose into glucose and fructose. A living cell has a unique set of 3,000 enzymes, each defined by the cell's DNA.

erosion: The process of wind or water wearing away soil or rock. The existence of soil is due in part to erosion of stone surfaces from the time solid rock first formed on Earth. Whole mountain chains have been worn down by erosion—the Appalachian Mountains were once as high as the Rockies. Sandstone is formed by compression of eroded rock under the oceans. Soil erosion can destroy cultivated land. Sheet erosion removes soil in a uniform layer, while rill erosion cuts small channels into the soil, and gully erosion forms deeper channels carrying away a large volume of soil.

eukaryotic cell: A complex cell on which all life more complex than a bacterium or yeast is based, characterized by a nucleus containing the cell's DNA and a number of specialized organelles, supplied with energy by mitochondria, which individually resemble the more primitive prokaryotic cell. *See also* prokaryotic cell.

exosphere: A layer of Earth's atmosphere from 500 kilometers above the surface to between 10,000 and 190,000 kilometers—or halfway to the Moon. At 190,000 kilometers, the force of solar radiation is more powerful than the force of Earth's gravity on the thinly distributed atmospheric molecules, but many scientists consider 10,000 kilometers to be the boundary with interplanetary space. Within the exosphere, a gas molecule can travel hundreds of kilometers before bumping into another gas molecule.

fermentation: A biological process for breaking down complex organic compounds into simpler compounds. One of the most familiar in human history is the conversion by yeast of sugar to carbon dioxide, alcohol, and water. Fermentation also occurs in cells, including animal muscle cells, breaking down glucose to produce lactic acid, lactate, carbon dioxide, and water, as well as adenosine triphosphate, a source of energy. It is less efficient than cellular respiration but occurs when muscles are short of oxygen. Many anaerobic bacteria ferment sugars: Lactobacillus ferment milk to produce yogurt. Fermentation also produces lactic acid in a variety of foods, such as sauerkraut and sourdough bread.

forensic: The application of science to legal concerns. The analysis of crime scenes, firearms, DNA, and the pathology of dead bodies are common subjects of forensic investigation, but dentists, toxicologists, psychiatrists, engineers, and practitioners in many other fields can also be called upon.

fuel cell: A source of electric current that operates in a manner similar to a battery, generating electricity as a by-product of a chemical reaction. The difference is that a fuel cell continues generating power as long as it is supplied with fuel. The hydrogen fuel cell, one of the most commonly known, generates electricity, heat, and water.

gamma ray: One of three common forms of radiation from the nuclei of unstable radioactive elements, which has no mass and no electrical charge. It is made up of photons, the fundamental particle of light, moving at the speed of light, emitted at a high energy level. X rays are similar but originate in electron fields rather than in the nucleus of atoms. *See also* alpha particle, beta particle.

gene: A unit of hereditary information found within a chromosome that determines the characteristics of an organism. Each gene is an ordered series of nucleotides, which are subunits of DNA, composed of a base molecule containing nitrogen, a phosphate molecule, and a pentose sugar molecule. *See also* chromosome, DNA.

Global Positioning System (GPS): A system owned by the United States government and operated by the Air Force for determining the exact position on the surface of the Earth of any user or defined landmark. By receiving signals from any of 24 orbiting satellites that are in direct line of site to a user's position, GPS devices can calculate latitude, longitude, altitude, and time.

gravity: As defined by Sir Isaac Newton's law of universal gravitation, a force that attracts any two objects in the universe. The strength of gravitational attraction depends on the mass of the two objects and decreases according to the distance between them, squared.

halogens: Elements fluorine, chlorine, bromine, iodine, and astatine in the periodic table of the elements. Being one electron short of filling the outermost shell of each atom, halogens form chemical bonds easily with other elements. They often form salts, including common table salt (sodium chloride), and calcium chloride—the salt applied to roads in winter to melt ice.

heat: The transfer of thermal energy, the kinetic energy of molecules, between two objects as the result of a temperature difference. Thermal energy is a vibration at the molecular level, which increases the volume of a substance, or increases the pressure of a gas or liquid in a closed space. Heat can be used to accomplish work mechanically; it is also produced by friction of moving parts, which wastes energy applied to accomplish work in a mechanical system.

husbandry: The cultivation of land to raise edible plant crops (and textile crops) or breed and raise domestic animals for food. *See also* animal husbandry.

hydrocarbons: Among the simplest of organic molecules, made up of a number of carbon and hydrogen atoms. Compounds with a benzene ring in the molecular structure are called aromatic hydrocarbons. Those without a benzene ring are called aliphatic hydrocarbons, which include alkanes (single carbon bond), alkenes (one double bond), and alkynes (one triple bond). There are a nearly unlimited number of derivative carbon compounds that can be formed by adding oxygen atoms to hydrocarbons.

hydrology: The study of water and, more specifically, of the way water cycles through lakes, streams, ponds, rivers, oceans, the atmosphere, and underground water flows and reservoirs, including evaporation, rain-, and snowfall. This includes study of contamination in water, as well as movement and distribution.

hydroponics: Growing plants in a solution of water and selected nutrients, without need for soil. Hydroponics can rely on sunlight or can be established in a closed, indoor environment using grow lights.

inflammable: An object or substance that catches fire easily, from the Latin *inflammare,* meaning to kindle a flame. Often confused, because "flammable" has the same meaning, and for English words, the prefix in- often means opposite, such as invisible. Safety officials have encouraged use of "flammable" to avoid misunderstanding and "nonflammable" to mean a substance that will not burn easily. Inflammability of any substance increases with higher concentration of oxygen in the surrounding atmosphere and decreases with lower oxygen concentration. Burning is an oxidation process—combining oxygen with the molecules of the burning substance—which can begin at lower temperatures with a higher concentration of oxygen. Once started, sufficient heat is given off to make the fire self-sustaining.

ion: An atom that has more or fewer electrons in its outermost shell than protons in its nucleus. *See also* atom, compound, element, molecule, periodic table of the elements.

ionosphere: The outermost layer of Earth's atmosphere, so named because solar radiation causes many atoms at this level to ionize, gaining or losing an electron. *See also* ion.

irradiation: Exposing any substance, or living organism, to radiation most commonly used to destroy bacteria, viruses, fungi (mold), and insects in food or on surfaces used for food preparation. The level of radiation used is not strong enough to disintegrate the nucleus of any atom making up the substance of any food item, so the food itself does not become radioactive. Irradiation is also used in treating cancer, checking luggage at airports, and sterilizing many items.

joule: A measure of heat or energy, used more commonly in scientific research than in industry; the energy required to accelerate a body with a mass of one kilogram using one newton of force over a distance of one meter. Equal to 0.2390 calories or 0.738 foot-pounds.

laser: A light source that emanates from a well-defined wavelength; originally an acronym for light amplified by stimulated emission of radiation.

light-emitting diode (LED): A diode constructed to provide illumination from the movement of electrons through a semiconductor, which is housed in a bulb that concentrates the light in a desired direction. Because there is no filament to warm up, as in a conventional electric light bulb, LEDs use ten percent or less electrical current to provide the same brightness of light and last up to twenty times longer. LEDs can be constructed to provide almost any desired color or hue of light.

lithography: A printing process that is the most common method of printing and publishing. The earliest process, invented by Alois Senefelder in Germany in 1798, is still used by hand-applying a greasy ink to a specially prepared block of limestone, which is moistened with water. The water is repelled by the ink, and in turn repels an oil-based ink, applied with a roller, from moistened areas, creating the desired image when paper is applied against the plate. Modern offset lithography burns a photosensitive metal plate through a negative film image to create a pattern of roughened areas—to which oil- or rubber-based inks will adhere—while a thin layer of water repels the ink from unexposed smooth metal surfaces.

magnetism: One aspect of electromagnetism, involving fields either generated by a current moving through a wire or a magnetized object, in which the molecules are aligned with magnetic north in one uniform direction and magnetic south in the opposite direction.

mean: Sometimes called the arithmetic mean, or average, it is the sum of a series of figures divided by the total number of figures.

median: The middle number in a group of numbers arranged in order from lowest to highest. Comparing the mean and the median can help to correct for the distortion of extreme highs or lows.

mesosphere: A layer of Earth's atmosphere, between the stratosphere and the ionosphere (or thermosphere), variously defined as beginning at an altitude from 30 to 55 kilometers above sea level and continuing to an altitude of about 80 to 90 kilometers.

metabolism: Physical processes and chemical reactions within a living body that convert or use energy, including those associated with digestion, excretion, breathing, blood circulation, growing and using muscle tissues, communication through the nervous system, and body temperature. *See also* enzyme, protein.

metals: A majority of known elements, generally

shiny in appearance and good conductors of heat and electricity. In ionic compounds, metals usually provide the positive ion. Many metals react with acids and therefore act as a base.

metric system: Scales of measurement in which each unit is one tenth of the next largest and ten times the next smallest, simplifying conversion, recording, and mathematical operations. The system is built around the unit of the meter (equivalent to 39.39 inches), which the system's inventors defined as one-ten millionth the distance from the North Pole to the equator.

microbe: Any microscopic form of life, also called a microorganism, particularly bacteria, protozoa, fungi, or virus. Most commonly, this term refers to pathogenic microscopic life—those that cause infection, disease, decay, sepsis, or gangrene. However, biologists are identifying an increasing number of microbes that are beneficial, even essential to life, including a variety of those found in the human intestine.

mitochondria: A type of organelle within eukaryotic cells, where oxygen and nutrients are converted into adenosine triphosphate, the molecule that stores chemical energy for the cell. This process, called aerobic respiration, is possible only in the presence of oxygen. Mitochondria are rod-shaped, have their own DNA, and reproduce independently within the cell, resembling some primitive prokaryotic cells. This suggests that prokaryotic cells were absorbed within the cell walls of evolving eukaryotic cells in a symbiotic relationship. Mitochondria enable cells to produce adenosine triphosphate fifteen times more efficiently than is possible by anaerobic respiration.

mode: The most frequent value in a series of numbers. For example, if the students in a class are all age sixteen, seventeen, or eighteen, and there are more sixteen-year-olds than either seventeen- or eighteen-year-olds, then sixteen would be the mode.

molecule: A particle formed by two or more atoms of the same element or of different elements. A water molecule is formed by one oxygen atom and two hydrogen atoms. Bonds holding molecules together are formed by the electrons in each atom's outer shell. *See also* atom, compound, element, periodic table of the elements.

navigation: The art and science of plotting a course from a starting point to a desired destination, most commonly piloting a boat on water or travel in outer space. The term is sometimes used for travel on land as well, particularly in a desert or flatland without significant landmarks, or in recent years, using a GPS navigation device for ordinary driving. Historically, navigation has been accomplished by using certain stars as reference points for latitude. Longitude was often guesswork, based on the time of day at the starting point compared with the current location, until invention of the seagoing chronometer in 1764. Radio beacons, radar, the gyroscopic compass, accurate maps, and the satellite-based GPS have all provided increased precision in navigation.

neuron: A nerve cell, the basic unit of the spinal cord, nervous system, and brain of human beings and other mammals. Neurons communicate information in chemical and electrical forms throughout an organism. Sensory neurons provide information to the brain from receptors in every part of the body. Motor neurons transmit direction from the brain to muscles. Shortly after birth, neurons stop reproducing, while other cell types continue to do so. Neurons have specialized structures called axons and dendrites to send and receive signals at connections known as synapses.

neutron: A neutral particle within the nucleus of an atom, having neither a positive or negative electrical charge and a weight slightly more than that of a proton. *See also* atom, electron, proton.

noble gases: Elements helium, neon, argon, krypton, xenon, and radon in the periodic table of the elements. They are called "noble" because having their outer electron shell filled to capacity, they do not react with other elements or form compounds. Radon, however, is a radioactive element. All of these elements exist in the form of gases, forming liquids or solids only at extremely low temperatures, approaching absolute zero.

nucleus (atomic): The tightly packed protons and neutrons at the core of every atom, which account for nearly all of an atom's mass. The total size of an atom is defined by its electron shells.

nucleus (cell): An organelle within each living eukaryotic cell that acts as a control center, storing genes on chromosomes, producing messenger RNA molecules (which transfer code for essential proteins from genes in the chromosomes), producing ribosomes, and organizing replication of DNA, including complete copies for cell division.

optics: The study of light, including all systems for gathering, concentrating, and manipulating light, such as mirrors, spectacles, telescopes, microscopes, cameras, spectroscopes, lasers, fiber optic communications, and optical data storage and retrieval.

organic compounds: Compounds that are created in or by living cells, rather than as a result of spontaneous physical processes. The simplest organic compounds are hydrocarbons, composed of carbon and hydrogen. Sugars and some other compounds are composed of carbon, hydrogen, and oxygen. The most complex organic compounds are composed of carbon, hydrogen, oxygen, nitrogen, occasionally sulfur, and sometimes small traces of metals.

oxidation: The process of any other element forming chemical bonds with atoms of oxygen. Common examples include the formation of rust on metal, and the burning of wood or any other inflammable substance. Oxidation occurs at many points in human and animal metabolism, including oxygen binding to hemoglobin in the blood stream and many chemical reactions within each cell in the body.

oxygenation: Infusion of oxygen gas into a solution or organic process, such as the transfer of oxygen through the membranes in the lung to enter the bloodstream.

ozone layer: An outer layer of Earth's atmosphere made up of ozone, a molecule of oxygen containing three atoms, instead of the two atoms of the oxygen breathed by living things. While ozone is toxic to life, causing chemical burns at relatively low concentrations, the ozone layer absorbs much of the ultraviolet radiation in sunlight, protecting life at the planet's surface.

pasteurization: A process discovered by French microbiologist Louis Pasteur for rapidly heating milk to destroy disease-causing bacteria, while leaving the nutritional content of the milk unaffected. It is also used to destroy bacteria in wine and beer manufacturing.

periodic table of the elements: The table of chemical elements arranged according to their atomic number. *See also* atom, atomic number, element, valence.

photon: The particle of light. Since light is a wave of electromagnetic radiation, it is something of a paradox that it also travels in particles. Quantum mechanics is based on the observation that electromagnetic radiation travels in discrete quantities of energy, called quanta. The photon is also the particle that, moving at a high volume of energy, is called a gamma ray.

photosynthesis: A chemical process in which carbon dioxide and water are converted into carbohydrates and oxygen by the energy from a light source, generally the Sun. This reaction, which all plants and many bacteria rely on, is the source of the unnatural presence of oxygen in Earth's atmosphere and supplies the entire food chain upon which animal life depends for existence.

photovoltaic: Electricity from light—any process using the energy of light to generate electricity directly. The most common practical application is the photovoltaic cell, made of a thin semiconductor wafer, treated to form an electric field, with electrical conductors attached to its positive and negative sides. When these poles are connected by a circuit, an electrical current is generated by photons from incoming sunlight, knocking loose electrons in the semiconductor.

physiology: Study of the entire system of physical functions in a living body: its mechanical, biochemical, and bioelectrical processes, the purpose and operation of each organ, and the interaction of different organs and parts.

polarity: The existence of two opposite characteristics, such as the north and south poles of a magnet or the positive and negative poles of a battery.

prokaryotic cell: The earliest and most primitive type of cell, probably the first life form on Earth, lacking a cell nucleus. Most bacteria are prokaryotic. Most one-celled animals, such as paramecium, and simple plants such as algae have the more complex eukaryotic cell.

protein: A long chain of amino acids. There are thousands of different proteins in each cell of the human body, and since each species has slightly different proteins in its cells, there are millions of different proteins in the biosphere. A balance of all necessary proteins is essential to the continued life of any organism. Food consumption must either supply each complete protein that the human body cannot manufacture for itself or a wide variety of incomplete proteins that can be assembled into complete proteins.

proton: The positively charged particles in the nucleus of every atom. Along with neutrons, protons account for most of the weight in an atom, because electrons have very little weight. *See also* atom, electron, neutron.

protoplasm: The living substance of a cell, including the content of the cell membrane and the substance within the cell—a transparent gelatinous material composed of inorganic substances (90 percent water with mineral salts and gases such as oxygen and carbon dioxide), and organic substances (proteins, carbohydrates, lipids, nucleic acids, and enzymes). Protoplasm outside the cell nucleus is called cytoplasm.

radiation: Emission of subatomic particles or of photons at high energy levels from radioactive atoms. *See also* alpha particles, beta particles, gamma rays.

radio waves: A band of wavelengths in the electromagnetic spectrum that can be generated by a spark gap in an electrical circuit and are commonly used for broadcast communication. *See also* electromagnetism, ionosphere.

refrigerant: A compound that transfers thermal energy in cooling systems, including air conditioners, freezers, refrigerators, and low-temperature manufacturing processes. Releasing the refrigerant at low pressure, in tubes that are in physical contact with the area to be cooled, transfers heat to the refrigerant, which is then transferred to radiator coils outside the area to be cooled, by compressing the refrigerant to a liquid state. Common refrigerants include ammonia, dichlorodifluoromethane, propane, hydrochlorofluorocarbon (HCFC), and hydrofluorocarbon (HFC).

RNA (ribonucleic acid): A complex molecule similar to, but less complex than, DNA. RNA molecules transfer genetic information from genes in longer DNA molecules forming the chromosomes of living cells to the active metabolic proteins in a living cell. *See also* DNA.

scientific method: A process for investigating nature by observation and experiment, creating hypotheses to make sense of observations.

seismology: The study of earth movement, particularly the mechanism and causes of earthquakes.

semiconductor: A material that conducts electricity more efficiently than an insulator, but less efficiently than a conductor, useful in constructing diodes, which conduct electricity in only one direction. Common semiconductor materials include silicon, germanium, and selenium. Semiconductors are essential to construction and design of computers and most electronic equipment.

sewage: Waste, usually carried in water, which is either chemical industrial waste or organic waste. Industrial sewage often includes metals, other toxins, and complex molecules that are not naturally metabolized. In most industrialized economies, sewage goes through several stages of treatment to remove solid waste for disposal, and the water is returned to lakes and rivers in relatively clean condition.

spectroscopy: Study of the spectrum of wavelengths of electromagnetic radiation, particularly the wavelengths of light visible to the human eye. When light reflected off any substance is viewed or projected through a spectroscope, each element casts a unique pattern of lines on the visible spectrum. This is useful in astronomy for identifying the chemical composition of stars and planets, since physical samples cannot be obtained, as well as in analytical chemistry.

statistics: Methods of obtaining, organizing, analyzing, interpreting, and presenting numerical data, used in many areas of science and industry, as well as in demographic studies of human populations. *See also* mean, mode, median.

sterilization (of microbes and pathogens): Killing all or most microbes on a working surface, or on the surface of instruments to be used in medical care or food preparation, by means of heat (including pasteurization), irradiation, or chemical antiseptics.

sterilization (pertaining to reproduction): Surgically or chemically preventing an organism from producing offspring. This includes neutering or spaying of pets and domesticated farm or service animals, castration of male animals, irradiating male or female insects as a pest-control measure, and, in humans, severing and tying the Fallopian tubes to prevent pregnancy in a woman or vasectomy to prevent a man inseminating a woman.

stratosphere: A layer of Earth's atmosphere from 18 kilometers to 50 kilometers. Unlike the troposphere, airflow is mostly horizontal with no weather patterns. The ozone layer is in the upper level of the stratosphere.

surfactant: A chemical substance that reduces the surface tension of water. A common use of surfactants is in manufacture of detergents and soaps. Surfactants are complex molecules with one component attracted to water molecules (hydrophilic) and the other component repelled by water molecules (hydrophobic).

temperature: In theory, a measure of the average kinetic energy of molecules: The larger or denser an object is, the more heat is required to raise its temperature by one degree. The most common scales for measuring temperature select arbitrary fixed points and assign arbitrary numerical values. The Fahrenheit scale assigns a value of 32 degrees to the temperature at which water freezes and 212 degrees to the temperature at which water boils. The Celsius scale assigns zero to the freezing point and 100 degrees to the boiling point. The Kelvin scale begins with absolute zero, the temperature at which a substance lacks any thermal energy. *See also* absolute zero, cold, heat.

thermosphere: *See* ionosphere.

toxin: A poison; any substance that will have a toxic effect on organic life, particularly proteins produced by bacteria, plants, or in specialized glands of certain animals. Toxins may damage or paralyze without killing or have a caustic effect, but in high enough concentrations, exposure to most toxins will cause death. Some toxins, in low doses, can be used in medical treatment: Two of the seven types of botulinum toxin are used to inhibit muscle spasms, smooth wrinkles of the upper face (Botox), and treat cervical dystonia.

troposphere: The layer of Earth's atmosphere closest to the surface, up to 14 kilometers, where all weather takes place, with rising and falling air currents. Air pressure at the upper limit of the troposphere is about 10 percent of the pressure at sea level.

ultrasound: Vibrations or sound waves at a higher frequency than the human ear can detect.

vacuum: The absence of matter, including air or other gases, inside an enclosed space or in remote areas of outer space. No perfect vacuum is known, since interstellar space is estimated to contain at least one hydrogen atom per cubic meter.

valence: The capacity of the atoms of an element to combine with other atoms of the same element or another element. *See also* atom, molecule, periodic table of the elements.

Charles Rosenberg

TIMELINE

The Time Line below lists milestones in the history of applied science: major inventions and their approximate dates of emergence, along with key events in the history of science. The developments appear in boldface, followed by the name or names of the person(s) responsible in parentheses. A brief description of the milestone follows.

2,500,000 B.C.E.	**Stone tools:** Stone tools, used by Homo habilis and perhaps other hominids, first appear in the Lower Paleolithic age (Old Stone Age).
400,000 B.C.E.	**Controlled use of fire:** The earliest controlled use of fire by humans may have been about this time.
200,000 B.C.E.	**Stone tools using the prepared-core technique:** Stone tools made by chipping away flakes from the stones from which they were made appear in the Middle Paleolithic age.
100,000-50,000 B.C.E.	**Widespread use of fire by humans:** Fire is used for heat, light, food preparation, and driving off nocturnal predators. It is later used to fire pottery and smelt metals.
100,000-50,000 B.C.E.	**Language:** At some point, language became abstract, enabling the speaker to discuss intangible concepts such as the future.
16,000 B.C.E.	**Earliest pottery:** The earliest pottery was fired by putting it in a bonfire. Later it was placed in a trench kiln. The earliest ceramic is a female figure from about 29,000 to 25,000 B.C.E., fired in a bonfire.
10,000 B.C.E.	**Domesticated dogs:** Dogs seem to have been domesticated first in East Asia.
10,000 B.C.E.	**Agriculture:** Agriculture allows people to produce more food than is needed by their families, freeing humans from the need to lead nomadic lives and giving them free time to develop astronomy, art, philosophy, and other pursuits.
10,000 B.C.E.	**Archery:** Archery allows human hunters to strike a target from a distance while remaining relatively safe.
10,000 B.C.E.	**Domesticated sheep:** Sheep seem to have been domesticated first in Southwest Asia.
9000 B.C.E.	**Domesticated pigs:** Pigs seem to have been domesticated first in the Near East and in China.
8000 B.C.E.	**Domesticated cows:** Cows seem to have been domesticated first in India, the Middle East, and sub-Saharan Africa.
7500 B.C.E.	**Mud bricks:** Mud-brick buildings appear in desert regions, offering durable shelter. The citadel in Bam, Iran, the largest mud-brick building in the world, was built before 500 B.C.E. and was largely destroyed by an earthquake in 2003.
7500 B.C.E.	**Domesticated cats:** Cats seem to have been domesticated first in the Near East.
6000 B.C.E.	**Domesticated chickens:** Chickens seem to have been domesticated first in India and Southeast Asia.
6000 B.C.E.	**Scratch plow:** The earliest plow, a stick held upright by a frame and pulled through the topsoil by oxen, is in use.
6000 B.C.E.	**Electrum:** The substance is a natural blend of gold and silver and is pale yellow in color like amber. The name "electrum" comes from the Greek word for amber.

6000 B.C.E.	**Gold:** Gold is discovered—possibly the first metal to be recognized as such.
6000-4000 B.C.E.	**Potter's wheel:** The potter's wheel is developed, allowing for the relatively rapid formation of radially symmetric items, such as pots and plates, from clay.
5000 B.C.E.	**Wheel:** The chariot wheel and the wagon wheel evolve—possibly from the potter's wheel. One of humankind's oldest and most important inventions, the wheel leads to the invention of the axle and a bearing surface.
4200 B.C.E.	**Copper:** Egyptians mine and smelt copper.
4000 B.C.E.	**Moldboard plow:** The moldboard plow cut a furrow and simultaneously lifted the soil and turned it over, bringing new nutrients to the surface.
4000 B.C.E.	**Domesticated horses:** Horses seem to have been domesticated first on the Eurasian steppes.
4000 B.C.E.	**Silver:** Silver can be found as a metal in nature, but this is rare. It is harder than gold but softer than copper.
4000 B.C.E.	**Domesticated honeybees:** The keeping of bee hives for honey arises in many different regions.
4000 B.C.E.	**Glue:** Ancient Egyptian burial sites contain clay pots that have been glued together with tree sap.
3500 B.C.E.	**Lead:** Lead is first extruded from galena (lead sulfide), which can be made to release its lead simply by placing it in a hot campfire.
c. 3100 B.C.E.	**Numerals:** Numerals appeared in Sumerian, Proto-Elamite, and Egyptian hieroglyphics.
3000 B.C.E.	**Bronze:** Bronze, an alloy of copper and tin, is developed. Harder than copper and stronger than wrought iron, it resists corrosion better than iron.
3000 B.C.E.	**Cuneiform:** The method of writing now known as cuneiform began as pictographs but evolved into more abstract patterns of wedge-shaped (cuneiform) marks, usually impressed into wet clay. This system of marks made complex civilization possible, since it allowed record keeping to develop.
3000 B.C.E.	**Fired bricks:** Humans begin to fire bricks, creating more durable building materials that (because of their regular size and shape) are easier to lay than stones.
3000 B.C.E.	**Pewter:** The alloy pewter is developed. It is 85 to 99 percent tin, with the remainder being copper, antimony, and lead; copper and antimony make the pewter harder. Pewter's low melting point, around 200 degrees Celsius, makes it a valuable material for crafting vessels that hold hot substances.
2700 B.C.E.	**Plumbing:** Earthenware pipes sealed together with asphalt first appear in the Indus Valley civilization. Greeks, Romans, and others provided cities with fresh water and a way to carry off sewage.
2650 B.C.E.	**Horse-drawn chariot (Huangdi):** Huangdi—a legendary patriarch of China—is possibly a combination of many men. He is said to have invented—in addition to the chariot—military armor, ceramics, boats, and crop rotation.

2600 B.C.E.	**Inclined plane:** Inclined planes are simple machines and were used in building Egypt's pyramids. Pushing an object up a ramp requires less force than lifting it directly, although the use of a ramp requires that the load be pushed a longer distance.
c. 2575-c. 2465 B.C.E.	**Pyramids:** Pyramids of Giza are built in Egypt.
1750 B.C.E.	**Tin:** Tin is alloyed with copper to form bronze.
1730 B.C.E.	**Glass beads:** Red-brown glass beads found in South Asia are the oldest known human-formed glass objects.
1600 B.C.E.	**Mercury:** Mercury can easily be released from its ore (such as cinnabar) by simply heating it.
1500 B.C.E.	**Iron:** Iron, stronger and more plentiful than bronze, is first worked in West Asia, probably by the Hittites. It could hold a sharper edge, but it had to be smelted at higher temperatures, making it more difficult to produce than bronze.
1500 B.C.E.	**Zinc:** Zinc is alloyed with copper to form brass, but it will not be recognized as a separate metal until 1746.
1000 B.C.E.	**Concrete:** The ancient Romans build arches, vaults, and walls out of concrete.
1000 B.C.E.	**Crossbow:** The crossbow seems to come from ancient China. Crossbows can be made to be much more powerful than a normal bow.
1000 B.C.E.	**Iron Age:** Iron Age begins. Iron is used for making tools and weapons
700 B.C.E.	**Magnifying glass:** An Egyptian hieroglyph seems to show a magnifying glass.
350 B.C.E.	**Compass:** Ancient Chinese used lodestones and later magnetized needles mostly to harmonize their environments with the principles of feng shui. Not until the eleventh century are these devices used primarily for navigation.
350-100 B.C.E.	**Scientific method (Aristotle):** Aristotle develops the first useful set of rules attempting to explain how scientists practice science.
300 B.C.E.	**Screw:** Described by Archimedes, the screw is a simple machine that appears to be a ramp wound around a shaft. It converts a smaller turning force to a larger vertical force, as in a screw jack.
300 B.C.E.	**Lever:** Described by Archimedes, the lever is a simple machine that allows one to deliver a larger force to a load than the force with which one pushes on the lever.
300 B.C.E.	**Pulley:** Described by Archimedes, the pulley is a simple machine that allows one to change the direction of the force delivered to the load.
221-206 B.C.E.	**Compass:** The magnetic compass is invented in China using lodestones, a mineral containing iron oxide.
215 B.C.E.	**Archimedes' principle (Archimedes of Syracuse):** Archimedes describes his law of displacement: A floating body displaces an amount of fluid the weight of which is equal to the weight of the body.
200 B.C.E.	**Astrolabe:** A set of engraved disks and indicators becomes known as the astrolabe. When aligned with the stars, the astrolabe can be used to determine the rising and setting times of the Sun and certain stars, establish compass directions, and determine local latitude.

40 C.E.	**Ptolemy's geocentric system (Ptolemy):** A world system with the Earth in the center, and the Moon, Venus, Mercury, Sun, Mars, Jupiter, Saturn, and fixed stars surrounding it. The geocentric Ptolemaic system would remain the most widely accepted cosmology for the next fifteen hundred years.
90 C.E.	**Aeolipile (Hero of Alexandria):** The aeolipile—a steam engine that escaping steam causes to rotate like a lawn sprinkler—is developed.
105 C.E.	**Paper and papermaking (Cai Lun):** Although papyrus paper already existed, Cai Lun creates paper from a mixture of fibrous materials softened into a wet pulp that is spread flat and dried. The material is strong and can be cheaply mass-produced.
250 C.E.	**Force pump (Ctesibius of Alexandria):** Ctesibius develops a device that shoots a jet of water, like a fire extinguisher.
815 C.E.	**Algebra (al-Khwārizmī):** al-Khw{amacr}rizm{imacr} develops the mathematics that solves problems by using letters for unknowns (variables) and expressing their relationships with equations.
877 C.E.	**Maneuverable glider (Abbas ibn Firnas):** A ten-minute controlled glider flight is first achieved.
9th century	**Gunpowder:** Gunpowder is invented in China.
1034	**Movable type:** Movable type made of baked clay is invented in China.
1170	**Water-raising machines (al-Jazari):** In addition to developing machines that can transport water to higher levels, al-Jazari invents water clocks and automatons.
1260	**Scientific method (Roger Bacon):** Bacon develops rules for explaining how scientists practice science that emphasize empiricism and experimentation over accepted authority.
1284	**Eyeglasses for presbyopia (Salvino d'Armate):** D'Armate is credited with making the first wearable eyeglasses in Italy with convex lenses. These spectacles assist those with farsightedness, such as the elderly.
1439	**Printing press (Johann Gutenberg):** Gutenberg combined a press, oil-based ink, and movable type made from an alloy of lead, zinc, and antimony to create a revolution in printing, allowing mass-produced publications that could be made relatively cheaply and disseminated to people other than the wealthy.
1450	**Eyeglasses for the nearsighted (Nicholas of Cusa):** Correcting nearsightedness requires diverging lenses, which are more difficult to make than convex lenses.
1485	**Dream of flight (Leonardo da Vinci):** On paper, Leonardo designed a parachute, great wings flapped by levers, and also a person-carrying machine with wings to be flapped by the person. Although these flying devices were never successfully realized, the designs introduced the modern quest for aeronautical engineering.
1543	**Copernican (heliocentric) universe:** Copernicus publishes *De revolutionibus* (*On the Revolutions of the Heavenly Spheres*), in which he refutes geocentric Ptolemaic cosmology and proposes that the Sun, not Earth, lies at the center of the then-known universe (the solar system).

1569	**Mercator projection (Gerardus Mercator):** The Mercator projection maps the Earth's surface onto a series of north/south cylinders.
1594	**Logarithms (John Napier):** Napier's logarithms allow the simplification of complex multiplication and division problems.
1595	**Parachute (Faust Veranzio):** Veranzio publishes a book describing sixty new machines, one of which is a design for a parachute that might have worked.
1596	**Flush toilet (Sir John Harington):** Harington's invention is a great boon to those previously assigned to empty the chamber pots.
1604	**Compound microscope (Zacharias Janssen):** Janssen, a lens crafter, experiments with lenses, leading to both the microscope and the telescope.
1607	**Air and clinical thermometers (Santorio Santorio):** Santorio develops a small glass bulb that can be placed in a person's mouth, with a long, thin neck that is placed in a beaker of water. The water rises or falls as the person's temperature changes.
1608	**Refracting telescope (Hans Lippershey):** Lippershey is one of several who can lay claim on developing the early telescope.
1609	**Improved telescope (Galileo Galilei):** Galileo grinds and polishes his own lenses to make a superior telescope. Galileo will come to be known as the father of modern science.
1622	**Slide rule (William Oughtred):** English mathematician and Anglican minister Oughtred invents the slide rule.
1629	**Steam turbine (Giovanni Branca):** Branca publishes a design for a steam turbine, but it requires machining that is too advanced to be built in his day.
1642	**Mechanical calculator (Blaise Pascal):** Eighteen-year-old Pascal invents the first mechanical calculator, which helps his father, a tax collector, count taxes.
1644	**Barometer (Evangelista Torricelli):** Torricelli develops a mercury-filled barometer, in which the height of the mercury in the tube is a measure of atmospheric pressure.
1650	**Vacuum pump (Otto von Guericke):** After demonstrating the existence of a vacuum, von Guericke explores its properties with other experiments.
1651	**Hydraulic press (Blaise Pascal):** Pascal determines that hydraulics can multiply force. For example, a 50-pound force applied to the hydraulic press might exert 500 pounds of force on an object in the press.
1656	**Pendulum clock (Christiaan Huygens):** Huygens discovers that, for small oscillations, a pendulum's period is independent of the size of the pendulum's swing, so it can be used to regulate the speed of a clock.
1662	**Demography (John Graunt):** Englishman Graunt develops the first system of demography and publishes *Natural and Political Observations Mentioned in the Following Index and Made Upon the Bills of Mortality*, which laid the groundwork for census taking.
1663	**Gregorian telescope (James Gregory):** The Gregorian telescope produces upright images and therefore becomes useful as a terrestrial telescope.

1666	**The calculus (Sir Isaac Newton):** Newton (and independently Gottfried Wilhelm Leibniz) develop the calculus in order to calculate the gravitational effect of all of the particles of the Earth on another object such as a person.
1670	**Spiral spring balance watch (Robert Hooke):** Hooke is also credited as the author of the principle that describes the general behavior of springs, known as Hooke's law.
1672	**Leibniz's calculator (Gottfried Wilhelm Leibniz):** Leibniz develops a calculator that can add, subtract, multiply, and divide, as well as the binary system of numbers used by computers today.
1674	**Improvements to the simple microscope (Antoni van Leeuwenhoek):** Leeuwenhoek, a lens grinder, applies his lenses to the simple microscope and uses his microscope to observe tiny protozoa in pond water.
1681	**Canal du Midi opens:** The 150-mile Canal du Midi links Toulouse, France, with the Mediterranean Sea.
1698	**Savery pump (Thomas Savery):** Savery's pump was impractical to build, but it served as a prototype for Thomas Newcomen's steam engine.
1699	**Eddystone Lighthouse (Henry Winstanley):** English merchant Winstanley designs the first lighthouse in England, located in the English Channel fourteen miles off the Plymouth coast. Winstanley is moved to create the lighthouse after two of his ships are wrecked on the Eddystone rocks.
1700	**Piano (Bartolomeo Cristofori):** Cristofori, a harpsichord maker, constructs an instrument with keys that can be used to control the force with which hammers strike the instrument's strings, producing sound that ranges from piano (soft) to forte (loud)—hence the name "pianoforte," later shortened to "piano."
1701	**Tull seed drill (Jethro Tull):** Before the seed drill, seeds were still broadcast by hand.
1709	**Iron ore smelting with coke (Abraham Darby):** Darby develops a method of smelting iron ore by using coke, rather than charcoal, which at the time was becoming scarce. Coke is made by heating coal and driving off the volatiles (which can be captured and used).
1712	**Atmospheric steam engine (Thomas Newcomen):** Newcomen's engine is developed to pump water out of coal mines.
1714	**Mercury thermometer, Fahrenheit temperature scale (Daniel Gabriel Fahrenheit):** Fahrenheit uses mercury in a glass thermometer to measure temperature over the entire range for liquid water.
1718	**Silk preparation:** John Lombe, owner of the Derby Silk Mill in England, patents the machinery that prepared raw silk for the loom.
1729	**Flying shuttle (John Kay):** On a loom, the shuttle carries the horizontal thread (weft or woof) and weaves it between the vertical threads (warp). Kay develops a shuttle that is named "flying" because it is so much faster than previous shuttles.
1738	**Flute Player and Digesting Duck automatons (Jacques de Vaucanson):** De Vaucanson builds cunning, self-operating devices, or automatons (robots) to charm viewers.

1740	**Steelmaking:** Benjamin Huntsman invents the crucible process of making steel.
1742	**Celsius scale (Anders Celsius):** Celsius creates a new scale for his thermometer.
1745-1746	**Leiden jar (Pieter van Musschenbroek and Ewald Georg von Kleist):** Von Kleist (1745) and Musschenbroek (1746) independently develop the Leiden jar, an early type of capacitor used for storing electric charge.
1746	**Clinical trials prove that citrus fruit cures scurvy (James Lind):** Others had suggested citrus fruit as a cure for scurvy, but Lind gives scientific proof. It still will be another fifty years before preventive doses of foods containing vitamin C are routinely provided for British sailors.
1752	**Franklin stove (Benjamin Franklin):** Franklin develops a stove that allows more heat to radiate into a room than go up the chimney.
1752	**Lightning rod (Benjamin Franklin):** Franklin devises a iron-rod apparatus to attach to houses and other structures in order to ground them, preventing damage during lightning storms.
1756	**Wooden striking clock (Benjamin Banneker):** Banneker's all-wood striking clock operates for the next fifty years. Banneker also prints a series of successful scientific almanacs during 1790's.
1757	**Nautical sextant (John Campbell):** When used with celestial tables, Campbell's sextant allows ships to navigate to within sight of their destinations.
1762	**Marine chronometers (John Harrison):** An accurate chronometer was necessary to determine a ship's position at sea, solving the pressing quest for longitude.
1764	**Spinning jenny (James Hargreaves):** Hargreaves develops a machine for spinning several threads at a time, transforming the textile industry and laying a foundation for the Industrial Revolution.
1765	**Improved steam engine (James Watt):** A steam condenser separate from the working pistons make Watt's engine significantly more efficient than Newcomen's engine of 1712.
1767	**Spinning machine (Sir Richard Arkwright):** Arkwright develops a device to spin fibers quickly into consistent, uniform thread.
1767	**Dividing engine (Jesse Ramsden):** Ramsden develops a machine that automatically and accurately marks calibrated scales.
1770	**Steam dray (Nicolas-Joseph Cugnot):** Cugnot builds his three-wheeled fardier à vapeur to move artillery; the prototype pulls 2.5 metric tons at 2 kilometers per hour.
1772	**Soda water (Joseph Priestley):** Priestley creates the first soda water, water charged with carbon dioxide gas. The following year he develops an apparatus for collecting gases by mercury displacement that would otherwise dissolve in water.
1775	**Boring machine (John Wilkinson):** Wilkinson builds the first modern boring machine used for boring holes into cannon, which made cannon manufacture safer. It was later adapted to bore cylinders in steam engines.

1776	**Bushnell's submarine (David Bushnell):** Bushnell builds the first attack submarine; used unsuccessfully against British ships in the Revolutionary War, it nevertheless advances submarine technology.
1779	**Cast-iron bridge:** Abraham Darby III and John Wilkinson build the first cast-iron bridge in England.
1779	**Spinning mule (Samuel Crompton):** Crompton devises the spinning mule, which allows the textile industry to manufacture high-quality thread on a large scale.
1781	**Uranus discovered (Sir William Herschel):** Herschel observes what he first believes to be a comet; further observation establishes it as a planet eighteen times farther from the Sun than the Earth is.
1782	**Hot-air balloon (Étienne-Jacques and Joseph-Michel Montgolfier):** Shaped like an onion dome and carrying people aloft, the Montgolfiers' hot-air balloon fulfills the fantasy of human flight.
1782	**Oil lamp (Aimé Argand):** Argand's oil lamp revolutionizes lighthouse illumination.
1783	**Parachutes (Louis-Sébastien Lenormand):** Lenormand jumps from an observatory tower using his parachute and lands safely.
1783	**Wrought iron (Henry Cort):** Cort converts crude iron into tough malleable wrought iron.
1784	**Improved steam engine (William Murdock):** In an age when much focus was on steam technology, Murdock works to improve steam pumps that remove water from mines. He will go on to invent coal-gas lighting in 1794.
1784	**Bifocals (Benjamin Franklin):** Tired of changing his spectacles to see things at close range as opposed to objects farther away, Franklin designs eyeglasses that incorporate both myopia-correcting and presbyopia-correcting lenses.
1784	**Power loom (Edmund Cartwright):** Cartwright's power loom forms a major advance in the Industrial Revolution.
1785	**Automated flour mill (Oliver Evans):** Evans's flour mill lays the foundation for continuous production lines. In 1801, he will also invent a high-pressure steam engine.
1790	**Steamboat (John Fitch):** Fitch not only invents the steamboat but also proves its practicality by running a steamboat service along the Delaware River.
1792	**Great clock (Thomas Jefferson):** Jefferson's clock, visible and audible both within Monticello and outside, across his plantation, is designed to maintain efficiency. He also invented an improved portable copying press (1785) and will go on to invent an improved ox plow (1794).
1792	**Coal gas (William Murdock):** Murdock develops methods for manufacturing, storing, and purifying coal gas and using it for lighting.
1793	**Cotton gin (Eli Whitney):** Whitney's engine to separate cotton seed from the fiber transformed the American South, both bolstering the institution of slavery and growing the "cotton is king" economy of the Southern states. Five years later, Whitney develops an assembly line for muskets using interchangeable parts.

1793	**Semaphore (Claude Chappe):** Chappe invents the semaphore.
1796	**Smallpox vaccination (Edward Jenner):** Jenner's vaccine will save millions from death, culminating in the eradication of smallpox in 1979.
1796	**Rumford stove (Benjamin Thompson):** The Rumford stove—a large, institutional stove—uses several small fires to heat the stove top uniformly.
1796	**Hydraulic press (Joseph Bramah):** Bramah builds a practical hydraulic press that operates by a high-pressure plunger pump.
1796	**Lithography (Aloys Senefelder):** Senefelder invents lithography and a process for color lithography in 1826.
1799	**Voltaic pile/electric battery (Alessandro Volta):** Volta creates a pile—a stack of alternating copper and zinc disks separated by brine-soaked felt—that supplies a continuous current and sets the stage for the modern electric battery.
1800	**Iron printing press (Charles Stanhope):** Stanhope invents the first printing press made of iron.
1801	**Pattern-weaving loom (Joseph M. Jacquard):** Jacquard invents a loom for pattern weaving.
1804	**Monoplane glider (George Cayley):** Cayley develops a heavier-than-air fixed-wing glider that inaugurates the modern field of aeronautics. Later models carry a man and lead directly to the Wright brothers' airplane.
1804	**Amphibious vehicle (Oliver Evans):** Evans builds the first amphibious vehicle, which is used in Philadelphia to dredge and clean the city's dockyards.
1805	**Electroplating (Luigi Brugnatelli):** Brugnatelli develops the method of electroplating by connecting something to be plated to one pole of a battery (voltaic pile) and a bit of the plating metal to the other pole of the battery, placing both in a suitable solution.
1805	**Morphine (Friedrich Setürner):** Setürner, a German pharmacist, isolates morphine from opium, but it is not widely used for another ten years.
1806	**Steam locomotive (Richard Trevithick):** After James Watt's patent for the steam engine expires in 1800, Trevithick develops a working steam locomotive. By 1806 he has developed his improved steam engine, named the Cornish engine, which sees worldwide dissemination.
1807	**Internal combustion engine (François Isaac de Rivaz):** De Rivaz builds the first vehicle powered by an internal combustion engine.
1807	**Paddle-wheel steamer (Robert Fulton):** Fulton's steamboat becomes far more commercially successful than those of his competitors.
1808	**Law of combining volumes for gases (Joseph-Louis Gay-Lussac):** Gay-Lussac discovers that, when gaseous elements combine to make a compound, the volumes involved are always simple whole-number ratios.
1810	**Preserving food in sealed glass bottles (Nicolas Appert):** Appert answers Napoleon's call to preserve food in a way that allows his soldiers to carry it with them: He processes food in sealed, air-tight glass bottles.

1810	**Preserving food in tin cans (Peter Durand):** Durand follows Nicolas Appert in preserving food for the French army, but he uses tin-coated steel cans in place of breakable bottles.
1815	**Miner's safety lamp (Sir Humphry Davy):** Davy devises a miner's safety lamp in which the flame is surrounded by wire gauze to cool combustion gases so that the mine's methane-air mixture will not be ignited.
1816	**Macadamization (John Loudon McAdam):** McAdam designs a method of paving roads with crushed stone bound with gravel on a base of large stones. The roadway is slightly convex, to shed water.
1816	**Kaleidoscope (Sir David Brewster):** The name for Brewster's kaleidoscope comes from the Greek words *kalos* (beautiful), *eidos* (form), and *scopos* (watcher). "Kaleidoscope," therefore, literally means "beautiful form watcher."
1816	**Stirling engine (Robert Stirling):** The Stirling engine proves to be an efficient engine that uses hot air as a working fluid.
1818	**First photographic images (Joseph Nicéphore Niépce):** Niépce creates the first lasting photographic images.
1819	**Stethoscope (René-Théophile-Hyacinthe Laënnec):** Laënnec invents the stethoscope to avoid the impropriety of placing his ear to the chest of a female heart patient.
1820	**Dry "scouring" (Thomas L. Jennings):** Jennings discovered that turpentine would remove most stains from clothes without the wear associated with washing them in hot water. His method becomes the basis for modern dry cleaning.
1821	**Diffraction grating (Joseph von Fraunhofer):** Von Fraunhofer's diffraction grating separates incident light by color into a rainbow pattern. The various discrete patterns reveal the structure of specific atomic nuclei, making it possible to identify the chemical compositions of various substances.
1821	**Braille alphabet (Louis Braille):** Braille develops a tactile alphabet—a system of raised dots on a surface—that allows the blind to read by touch.
1821	**Electromagnetic rotation (Michael Faraday):** Faraday publishes his work on electromagnetic rotation, which is the principle behind the electric motor.
1822	**Difference engine (Charles Babbage):** Babbage's "engine" was a programmable mechanical device used to calculate the value of a polynomial—a precursor to today's modern computers.
1823	**Waterproof fabric is used in raincoats (Charles Macintosh):** Macintosh patents a waterproof fabric consisting of soluble rubber between two pieces of cloth. Raincoats made of the fabric are still often called mackintoshes (macs), especially in England.
1824	**Astigmatism-correcting lenses (George Biddell Airy):** Airy develops cylindrical lenses that correct astigmatism. An astronomer, Airy will go on to design a method of correcting compasses used in ship navigation and the altazimuth telescope. He becomes England's astronomer royal in 1835.

1825	**Electromagnet (William Sturgeon):** Sturgeon builds a U-shaped, soft iron bar with a coil of varnished copper wire wrapped around it. When a voltaic current is passed through wire, the bar becomes magnetic—the world's first electromagnet.
1825	**Bivalve vaginal speculum (Marie Anne Victoire Boivin):** Boivin develops the tool now widely used by gynecologists in the examination of the vagina and cervix.
1825	**"Steam waggon" (John Stevens):** Stevens builds the first steam locomotive to be manufactured in the United States.
1826	**Color lithography (Aloys Senefelder):** Senefelder invents color lithography.
1827	**Matches (John Walker):** Walker coats the ends of sticks with a mixture of antimony sulfide, potassium chlorate, gum, and starch to produce "strike anywhere" matches.
1827	**Water turbine (Benoît Fourneyron):** Fourneyron builds the first water turbine; it has six horsepower. His larger, more efficient turbines powered many factories during the Industrial Revolution.
1828	**Combine harvester (Samuel Lane):** Patent is granted to Lane for the combine harvester, which combines cutting and threshing.
1829	**Rocket steam locomotive (George Stephenson):** Stephenson builds the world's first railway line to use a steam locomotive.
1829	**Boiler (Marc Seguin):** Seguin improves the steam engine with a multiple fire-tube boiler.
1829	**Polarizing microscope (William Nicol):** Nicol invents the polarizing microscope, an important forensic tool.
1830	**Steam locomotive (Peter Cooper):** Cooper's four-wheel locomotive with a vertical steam boiler, the *Tom Thumb,* demonstrates the possibilities of steam locomotives and brings Cooper national fame. His other inventions and good management enable Cooper to become a leading industrialist and philanthropist.
1830	**Lawn mower (Edwin B. Budding):** Budding, an English engineer, invents the lawn mower.
1830	**Paraffin (Karl von Reichenbach):** Von Reichenbach, a German chemist, discovers paraffin.
1830	**Creosote (Karl von Reichenbach):** Von Reichenbach distills creosote from beachwood tar. It is used as an insecticide, germicide, and disinfectant.
1831	**Alternating current (AC) generator (Michael Faraday):** Faraday constructs the world's first electric generator.
1831	**Mechanical reaper (Cyrus Hall McCormick):** McCormick's reaper can harvest a field five times faster than earlier methods.
1831	**Staple Bend Tunnel:** The first railroad tunnel in the United States is built in Mineral Point, Pennsylvania.
1832	**Electromagnetic induction (Joseph Henry):** Henry discovers that changing magnetic fields induce voltages in nearby conductors.

1832	**Codeine (Pierre-Jean Robiquet):** French chemist Robiquet isolates codeine from opium. Because of the small amount found in nature, most codeine is synthesized from morphine.
1834	**Hansom cab (Joseph Aloysius Hansom):** English architect Hansom builds the carriage bearing his name.
1835	**Colt revolver (Samuel Colt):** The Colt revolver becomes known as "one of the greatest advances of self-defense in all of human history."
1835	**Photography (Joseph Nicéphore Niépce):** Niépce codevelops photography with Louis-Jacques-Mandé Daguerre.
1836	**Daniell cell (John Frederic Daniell):** Daniell invents the electric battery bearing his name, which is much improved over the voltaic pile.
1836	**Acetylene (Edmund Davy):** Davy creates acetylene by heating potassium carbonate to high temperatures and letting it react to water.
1837	**Electric telegraph (William Fothergill Cooke and Charles Wheatstone):** Wheatstone and Cooke devise a system that uses five pointing needles to indicate alphabetic letters.
1837	**Steam hammer (James Hall Nasmyth):** Nasmyth develops the steam hammer, which he will use to build a pile driver in 1843.
1837	**Steel plow (John Deere):** Previously, plows were made of cast iron and required frequent cleaning. Deere's machine is effective in reducing the amount of clogging farmers experienced when plowing the rich prairie soil.
1837	**Threshing machine (Hiram A. and John A. Pitts):** The Pitts, brothers, develop the first efficient threshing machine.
1838	**Fuel cell (Christian Friedrich Schönbein):** Schönbein's fuel cell might use hydrogen and oxygen and allow them to react, producing water and electricity. There are no moving parts, but the reactants must be continuously supplied.
1838	**Propelling steam vessel (John Ericsson):** Swedish engineer Ericsson invents the double screw propeller for ships allowing them to move much faster than those relying on sails.
1839	**Nitric acid battery (Sir William Robert Grove):** The Grove cell delivered twice the voltage of its more expensive rival, the Daniell cell.
1839	**Daguerreotype (Jacques Daguerre):** Improving on the discoveries of Joseph Nicéphore Niépce, Daguerre develops the first practical photographic process, the Daguerreotype.
1839	**Vulcanized rubber (Charles Goodyear):** Adding sulfur and lead monoxide to rubber, Goodyear processes the batch at a high temperature. The process, later called vulcanization, yields a stable material that does not melt in hot weather or crack in cold.
1840	**Electrical telegraph (Samuel F. B. Morse):** Others had already built telegraph systems, but Morse's system was superior and soon replaced all others.

1841	**Improved electric clock (Alexander Bain):** With John Barwise, Bain develops an electric clock with a pendulum driven by electric impulses to regulate the clock's accuracy.
1841	**First negatives in photography (William Henry Fox Talbot):** Talbot, an English polymath, invents the calotype process, which produces the first photographic negative.
1842	**Commercial fertilizer (John B. Lawes):** Lawes develops superphosphate, the first commercial fertilizer.
1843	**Rotary printing press (Richard March Hoe):** Patented in 1847, the steam-powered rotary press is far faster than the flatbed press.
1843	**Multiple-effect vacuum evaporator (Norbert Rillieux):** Rillieux develops an efficient method for refining sugar using a stack of several pans of sugar syrup in a vacuum chamber, which allows boiling at a lower temperature.
1845	**Suspension bridges (John Augustus Roebling):** A manufacturer of wire cable, Roebling wins a competition for an aqueduct over the Allegheny River and goes on to design other aqueducts and suspension bridges, culminating in the Brooklyn Bridge, which his son, Washington Augustus Roebling, completes in 1883.
1845	**Sewing machine (Elias Howe):** Howe develops a machine that can stitch straight, strong seams faster than those sewn by hand.
1846	**Neptune discovered (John Galle):** German astronomer Galle observes a new planet, based on irregularities in the orbit of Uranus calculated the previous year by England's John Couch Adams and France's Urbain Le Verrier.
1847	**Nitroglycerin (Ascanio Sobrero):** Italian chemist Sobrero creates nitroglycerin.
1847	**Telegraphy applications (Werner Siemens):** Siemens refines a telegraph in which a needle points to the alphabetic letter being sent.
1849	**Laryngoscope (Manuel P. R. Garcia):** Spanish singer and voice teacher, Garcia, known as the father of laryngology, devises the first laryngoscope.
1851	**Foucault's pendulum (Léon Foucault):** Foucault's pendulum proves that Earth rotates.
1851	**Sewing machine (Isaac Merritt Singer):** Singer improves the sewing machine and successfully markets it to women for home use.
1851	**Ophthalmoscope (Hermann von Helmholtz):** Helmholtz invents a device that can be used to examine the retina and the vitreous humor. In 1855, he will invent an ophthalmometer, an instrument that measures the curvature of the eye's lens.
1854	**Kerosene (Abraham Gesner):** Canadian geologist Gesner distills kerosene from petroleum.
1852	**Hypodermic needle (Charles G. Pravaz):** French surgeon Pravaz devises the hypodermic syringe.

1855	**Bunsen burner (Robert Wilhelm Bunsen):** Bunsen—along with Peter Desaga, an instrument maker, and Henry Roscoe, a student—develops a high-temperature laboratory burner, which he and Gustav Kirchhoff use to develop the spectroscope (1859).
1855	**Bessemer process (Sir Henry Bessemer):** Bessemer creates a converter that leads to a process for inexpensively mass-producing steel.
1856	**Synthetic dye (William H. Perkin):** British chemist Perkin produces the first synthetic dye. The color is mauve, which triggers a mauve fashion revolution.
1857	**Safety elevator (Elisha Graves Otis):** Otis's safety elevator automatically stops if the supporting cable breaks.
1858	**Internal combustion engine (Étienne Lenoir):** Lenoir's engine, along with his invention of the spark plug, sets the stage for the modern automobile.
1858	**Transatlantic cable (Lord Kelvin):** Kelvin helps design and install the under-ocean cables for telegraphy between North America and Europe, serving as a chief motivating force in getting the cable completed.
1859	**Signal flares (Martha J. Coston):** Coston's brilliant and long-lasting white, red, and green flares will be adopted by the navies of several nations.
1859	**Lead-acid battery (Gaston Planté):** French physicist Planté invents the lead-acid battery, which led to the invention of the first electric, rechargeable battery.
1860	**Refrigerant (Ferdinand Carré):** French inventor Carré introduces a refrigerator that uses ammonia as a refrigerant.
1860	**Electric incandescent lamp (Joseph Wilson Swan):** Swan produces and patents an incandescent electric bulb; in 1880, two years after Edison's light bulb, Swan will produce a more practical bulb.
1860	**Web rotary printing press (William Bullock):** Bullock's press has an automatic paper feeder, can print on both sides of the paper, cut the paper into sheets, and fold them.
1860	**Henry rifle (Tyler Henry):** American gunsmith Henry designs the Henry rifle, a repeating rifle, the year before the Civil War begins.
1860	**First mail service:** Pony Express opens overland mail service. The service eventually expands to include more than 100 stations, 80 riders, and more than 400 horses.
1861	**Machine gun (Richard Gatling):** Gatling develops the first machine gun, called the Gatling gun. It has six barrels that rotate into place as the operator turns a hand crank; the shells were automatically chambered and fired.
1861	**First color photograph:** Thomas Sutton develops the first color photo based on Scottish physicist James Clerk Maxwell's three-color process.
1861-1862	**USS Monitor (John Ericsson):** Ericsson develops the first practical ironclad ship, which will be use during the Civil War. He goes on to develop a torpedo boat that can fire a cannon from an underwater port.

1862	**Pasteurization (Louis Pasteur):** Pasteur's germ theory of disease leads him to develop a method of applying heat to milk products in order to kill harmful bacteria. He goes on to develop vaccines for rabies, anthrax, and chicken cholera (1867-1885).
1863	**Subway:** The first subway opens in London; it uses steam locomotives. It does not go electric until 1890.
1865	Pioneer **(Pullman) sleeping car (George Mortimer Pullman):** Pullman began working on sleeping cars in 1858, but the *Pioneer* is a luxury car with an innovative folding upper birth to allow the passenger to sleep while traveling.
1866	**Self-propelled torpedo (Robert Whitehead):** English engineer Whitehead develops the modern torpedo.
1866	**Transatlantic telegraph cable:** The first successful transatlantic telegraph cable is laid; it spans 1,686 nautical miles.
1867	**Dynamite (Alfred Nobel):** Nobel mixes clay with nitroglycerin in a one-to-three ratio to create dynamite (Nobel's Safety Powder), an explosive the ignition of which can be controlled using Nobel's own blasting cap. He goes on to patent more than three hundred other inventions and devotes part of the fortune he gained from dynamite to establish and fund the Nobel Prizes.
1867	**Baby formula (Henri Nestlé):** Nestlé combines cow's milk with wheat flour and sugar to produce a substitute for infants whose mothers cannot breast-feed.
1867	**Steam velocipede motorcycle (Sylvester Roper):** Roper spent his lifetime making steam engines lighter and more powerful in order to make his motorized bicycles faster. His velocipede eventually reaches 60 miles per hour.
1867	**Flat-bottom paper bag machine (Margaret E. Knight):** Knight designs a machine that can manufacture flat-bottom paper bags, which can stand open for easy loading.
1867	**Dry-cell battery (Georges Leclanché):** French engineer Lelanche invents the dry-cell battery.
1868	**Typewriter (Christopher Latham Sholes):** American printer Sholes produces the first commercially successful typewriter.
1869	**Periodic table of elements (Dmitry Ivanovich Mendeleyev):** The periodic table, which links chemical properties to atomic structure, will prove to be one of the great achievements of the human race.
1869	**Air brakes for trains (George Westinghouse):** In 1867, Westinghouse developed a signaling system for trains. The air brake makes it easier and safer to stop large, heavy, high-speed trains.
1869	**Transcontinental railroad:** The United States transcontinental railroad is completed.
1869	**Celluloid (John Wesley Hyatt):** American inventor Hyatt produces celluloid, the first commercially successful plastic, by mixing solid pyroxylin and camphor.
1869	**Suez Canal opens:** The canal, 101 miles long, took a decade to build and connects the Red Sea with the eastern Mediterranean Sea.

1871	**Fireman's respirator (John Tyndall):** The respirator grows from Tyndall's studies of air pollution.
1871	**Commercial generator (Zénobe T. Gramme):** Belgian electrical engineer Gramme builds the Gramme machine, the first practical commercial generator for producing alternating current.
1872	**Blue jeans (Levi Strauss):** Miners tore their pockets when they stuffed too many ore samples in them. Strauss makes pants using heavy-duty material with riveted pocket corners so they will not tear out.
1872	**Burbank russet potato (Luther Burbank):** Burbank breeds all types of plants, using natural selection and grafting techniques to achieve new varieties. His Burbank potato, developed from a rare russet potato seed pod, grows better than other varieties.
1872	**Automatic lubricator (Elijah McCoy):** McCoy uses steam pressure to force oil to lubricate the pistons of steam engines.
1872	**Vaseline (Robert A. Chesebrough):** Chesebrough, an American chemist, patents his process for making petroleum jelly and calls it Vaseline.
1873	**QWERTY keyboard (Christopher Latham Sholes):** After patenting the first practical typewriter, Sholes develops the QWERTY keyboard, designed to slow the fastest typists, who otherwise jammed the keys. The basic QWERTY design remains the standard on most computer keyboards.
1874	**Barbed wire (Joseph Farwell Glidden):** An American farmer, Glidden invents and patents barbed wire. Barbed-wire fences make farming and ranching of the Great Plains practical. Without effective fences, animals wandered off and crops were destroyed. At the time of his death in 1906, Glidden is one of the richest men in the country.
1874	**Medical nuclear magnetic resonance imaging (Raymond Damadian):** Damadian and others develop magnetic resonance imaging (MRI) for use in medicine.
1876	**Four-stroke internal combustion engine (Nikolaus August Otto):** In order to deliver more horsepower, Otto's engine compresses the air-fuel mixture. His previous engines operated near atmospheric pressure.
1876	**Ammonia-compressor refrigeration machine (Carl von Linde):** Breweries need refrigeration so they can brew year-round. Linde refines his ammonia-cycle refrigerator to make this possible.
1876	**Telephone (Elisha Gray):** Gray files for a patent for the telephone the same day that Alexander Graham Bell does so. While the case is not clear-cut, and Gray fought with Bell for years over the patent rights, Bell is generally credited with the telephone's invention.
1877	**Phonograph (Thomas Alva Edison):** Edison invents the phonograph—an unexpected outcome of his telephone research.
1878	**First practical lightbulb (Thomas Alva Edison):** Twenty-two people have invented lightbulbs before Edison and Joseph Swan, but they are impractical. Edison's is the first to be commercially viable. Eventually, Swan's company merges with Edison's.

1878	**Loose-contact carbon microphone (David Edward Hughes):** Hughes's carbon microphone advances telephone technology. In 1879, he will invent the induction balance, which will be used in metal detectors.
1878	**Color photography (Frederic Eugene Ives):** American inventor Ives develops the halftone process for printing photographs.
1879	**Saccharin (Ira Remsen):** Remsen synthesizes a compound that is up to three hundred times sweeter than sugar; he also establishes the important *American Chemical Journal*, serving as its editor until 1915.
1880	**Milne seismograph (John Milne):** Milne invents the first modern seismograph for measuring earth tremors. He will come to be called the father of modern seismology.
1881	**Improved incandescent lightbulb (Lewis Howard Latimer):** Latimer develops an improved way to manufacture and to attach carbon filaments in lightbulbs.
1881	**Sphygmomanometer (Karl Samuel Ritter von Basch):** Von Basch invents the first blood pressure gauge.
1882	**Induction motor (Nikola Tesla):** Tesla's theories and inventions make alternating current (AC) practical.
1882	**Two-cycle gasoline engine (Gottlieb Daimler):** Daimler builds a small, high-speed two-cycle gasoline engine. He will also build a successful motorcycle in 1885 and (with Wilhelm Maybach) an automobile in 1889.
1883	**Solar cell (Charles Fritts):** American scientist Fritts designs the first solar cell.
1883	**Shoe-lasting machine (Jan Ernst Matzeliger):** The machine sews the upper part of the shoe to the sole and reduces the cost of shoes by 50 percent.
1884	**Fountain pen (Lewis Waterman):** The commonly told story is that Waterman was selling insurance and lost a large contract when his pen leaked all over it, prompting him to invent the leak-proof fountain pen.
1884	**Vector calculus (Oliver Heaviside):** Heaviside develops vector calculus to represent James Clerk Maxwell's electromagnetic theory with only four equations instead of the usual twenty.
1884	**Roll film (George Eastman):** Roll film will replace heavy plates, making photography both more accessible and more convenient. In 1888, Eastman and William Hall invent the Kodak camera. These developments open photography to the masses.
1884	**Roll film (George Eastman):** Roll film will replace heavy plates, making photography both more accessible and more convenient. In 1888, Eastman and William Hall invent the Kodak camera. These developments open photography to the masses.
1884	**Steam turbine (Charles Parsons):** Designed for ships, Parsons's steam turbine is smaller, more efficient, and more durable than the steam engines in use.
1884	**Census tabulating machine (Herman Hollerith):** Hollerith's machine uses punch cards to tabulate 1890 census data. He goes on to found the company that later becomes International Business Machines (IBM).

1885	**Machine gun (Hiram Stevens Maxim):** Maxim patents a machine gun that can fire up to six hundred bullets per minute.
1885	**Bicycle (John Kemp Starley):** English inventor Starley is responsible for producing the first modern bicycle, called the Rover.
1885	**First gasoline-powered automobile (Carl Benz):** Benz not only manufactures the first gas-powered car but also is first to mass-produce automobiles.
1885	**Incandescent gas mantle (Carl Auer von Welsbach):** The Austrian scientist invents the incandescent gas mantle.
1886	**Dictaphone (Charles Sumner Tainter):** Tainter, an American engineer who frequently worked with Alexander Graham Bell, designs the Dictaphone.
1886	**Dishwasher (Josephine Garis Cochran):** Like modern washers, Cochran's dishwasher cleans dishes with sprays of hot, soapy water and then air-dries them.
1886	**Electric transformer (William Stanley):** Stanley, working at Westinghouse, builds the first practical electric transformer.
1886	**Gramophone (Emile Berliner):** A major contribution to the music recording industry, Berliner's gramophone uses flat record discs for recording sound. Berliner goes on to produce a helicopter prototype (1906-1923).
1886	**Linotype machine (Ottmar Mergenthaler):** Pressing keys on the machine's keyboard releases letter molds that drop into the current line. The lines are assembled into a page and then filled with molten lead.
1886	**Electric-traction system (Frank J. Sprague):** Sprague's motor can propel a tram up a steep hill without its slipping.
1886	**Hall-Héroult electrolytic process (Charles Martin Hall and Paul Héroult):** The industrial production of aluminum from bauxite ore made aluminum widely available. Prior to the electrolytic process, aluminum was a precious metal with a value about equal to that of silver.
1886	**Coca-Cola (John Stith Pemberton):** Developed as pain reliever less addictive than available opiates, the original Coca-Cola contains cocaine from cola leaves and caffeine from kola nuts. It achieves greater success as a beverage marketed where alcohol is prohibited.
1886	**Yellow pages (Reuben H. Donnelly):** Yellow paper was used in 1883 when the printer ran out of white paper. Donnelly now purposely uses yellow paper for business listings.
1886	**Fluorine (Henri Moissan):** French chemist Moissan isolates fluorine and is awarded the Nobel Prize in Chemistry in 1906. Compounds of fluorine are used in toothpaste and in public water supplies to help prevent tooth decay.
1887	**Radio transmitter and receiver (Heinrich Hertz):** Hertz will use these devices to discover radio waves and confirm that they are electromagnetic waves that travel at the speed of light; he also discovers the photoelectric effect.
1887	**Distortionless transmission lines (Oliver Heaviside):** Heaviside recommends that induction coils be added to telephone and telegraph lines to correct for distortion.

1887	**Olds horseless carriage (Ransom Eli Olds):** Olds develops a three-wheel horseless carriage using a steam engine powered by a gasoline burner.
1887	**Synchronous multiplex railway telegraph (Granville T. Woods):** Woods patents a variation of the induction telegraph that allows messages to be sent between moving trains and between trains and railway stations. He will eventually obtain sixty patents on electrical and electromechanical devices, most of them related to railroads and communications.
1888	**Cordite (Sir James Dewar):** Dewar, with Sir Frederick Abel, invents cordite, a smokeless gunpowder that is widely adopted for munitions.
1888	**Pneumatic rubber tire (John Boyd Dunlop):** Dunlop's pneumatic tires revolutionize the ride for cyclists and motorists.
1888	**Kodak camera:** George Eastman, founder of Eastman Kodak, introduces the first Kodak camera.
1889	**Electric drill (Arthur James Arnot):** Arnot's drill is used to cut holes in rock and coal.
1889	**Bromine extraction (Herbert Henry Dow):** Dow's method for extracting bromine from brine enables bromine to be widely used in medicines and in photography.
1889	**Rayon (Louis-Marie-Hilaire Bernigaud de Chardonnet):** Bernigaud de Chardonnet, a French chemist, invents rayon, the first artificial fiber, as an alternative to silk.
1889	**Celluloid film:** George Eastman replaces paper film with celluloid.
1890	**Improved carbon electric arc (Hertha Marks Ayrton):** The carbon arc produces an intense light that is used in streetlights.
1890	**Pneumatic (air) hammer (Charles B. King):** A worker with a pneumatic hammer can break up a concrete slab many times faster than can a worker armed with only a sledgehammer.
1890	**Smokeless gunpowder (Hudson Maxim):** Maxim (perhaps with brother Hiram) develops a version of smokeless gunpowder that is adopted for modern firearms; he goes on to develop a smokeless cannon powder that will be used during World War I.
1890	**Rubber gloves in the operating room:** American surgeon William Stewart Halsted introduces the use of sterile rubber gloves in the operating room.
1891	**Rubber automobile tires (André and Édouard Michelin):** The Michelin brothers manufacture air-inflated tires for bicycles and later automobiles, which leads to a successful ad campaign, featuring the Michelin Man (Bibendum).
1891	**Carborundum (Edward Goodrich Acheson):** Attempting to create artificial diamonds, Acheson instead synthesizes silicon carbide, the second hardest substance known. He will develop an improved graphite-making process in 1896.
1892	**Kinetoscope (Thomas Alva Edison):** Edison completes Kinetoscope; the first demonstration is held a year later.
1892	**Calculator (William Seward Burroughs):** Burroughs builds the first practical key-operated calculator; it prints entries and results.

1892	**Dewar flask (Sir James Dewar):** Dewar invents the vacuum bottle, a vacuum-jacketed vessel for storing and maintaining the temperature of hot or cold liquids.
1892	**Artificial silk (Charles F. Cross and Edward J. Bevan):** British chemists Cross and Bevan create viscose artificial silk (cellulose acetate).
1893	**Color photography plate (Gabriel Jonas Lippmann):** Also known as the Lippmann plate for its inventor, the color photography plate uses interference patterns, rather than various colored dyes, to reproduce authentic color.
1893	**Alternating current calculations (Charles Proteus Steinmetz):** Steinmetz's calculations make it possible for engineers to determine alternating current reliably, without depending on trial and error, when designing a new motor.
1894	**Cereal flakes (John Harvey Kellogg):** Kellogg, a health reformer who advocates a diet of fruit, nuts, and whole grains, invents flaked breakfast cereal with the help of his brother, Will Keith Kellogg. In 1906 Kellogg established a company in Battle Creek, Michigan, to manufacture his breakfast cereal.
1894	**Automatic loom (James Henry Northrop):** Northrop builds the first automatic loom.
1895	**Streamline Aerocycle bicycle (Ignaz Schwinn):** Through hard work and dedication, Schwinn develops a bicycle that eventually makes his name synonymous with best of bicycles.
1895	**Victrola phonographs (Eldridge R. Johnson):** Johnson develops a spring-driven motor for phonographs that provides the constant record speed necessary for good sound reproduction.
1895	**Cinématographe (Auguste and Louis Lumière):** The Lumière brothers' combined motion-picture camera, printer, and projector helps establish the movie business. Using a very fine-grained silver-halide gelatin emulsion, they cut photographic exposure time down to about one minute.
1895	**Antenna:** Aleksandr Stepanovich Popov demonstrated radio reception with a coherer, which he also used as a lightning detector.
1896	**Wireless telegraph system (Guglielmo Marconi):** Marconi is the first to send wireless signals across the Atlantic Ocean, inaugurating a new era of telecommunications.
1896	**Aerodromes (Samuel Pierpont Langley):** Langley's "Aerodrome number 6," using a small gasoline engine, makes an unmanned flight of forty-eight hundred feet.
1896	**Four-wheel horseless carriage (Ransom Eli Olds):** Oldsmobile patents Olds's internal combustion engine and applies it to his four-wheel horseless carriage, naming it the "automobile."
1896	**High-frequency generator and transformer (Elihu Thomson):** Thomson produces an electric air drill, which advances welding to improve the construction of new appliances and vehicles. He will also invent other electrical devices, including an improved X-ray tube.

1896	**X-ray tube (Wilhelm Conrad Röntgen):** After discovering X radiation, Röntgen mails an X-ray image of a hand wearing a ring and paving the way for the medical use of X-ray imaging—one of the most important discoveries ever made for medical science.
1896	**Better sphygmomanometer (Scipione Riva-Rocci):** Italian pediatrician Riva-Rocci develops the most successful and easy-to-use blood-pressure gauge.
1897	**Modern submarine (John Philip Holland):** Holland's submarine is the first to use a gasoline engine on the surface and an electric engine when submerged.
1897	**Oscilloscope (Karl Ferdinand Braun):** The oscilloscope is an invaluable device used to measure and display electronic waveforms.
1897	**Jenny coupler (Andrew Jackson Beard):** Beard's automatic coupler connects the cars in a train without risking human life. The introduction of automatic couplings reduces coupling-related injuries by a factor of five.
1897	**Escalator (Charles Seeberger):** Before Seeberger built the escalator in its now-familiar form, it was a novelty ride at the Coney Island amusement park.
1897	**Automobile components (Alexander Winton):** The Winton Motor Carriage Company is incorporated, and Winton begins manufacturing automobiles. His popular "reliability runs" helps advertise automobiles to the American market. He will produce the first American diesel engine in 1913.
1897	**Diesel engine (Rudolf Diesel):** Diesel's internal combustion engine rivals the efficiency of the steam engine.
1897	**Electron discovered (J. J. Thomson):** Thomson uses an evacuated tube with a high voltage across electrodes sealed in the ends. Invisible particles (later named electrons) stream from one of the electrodes, and Thomson establishes the particles' properties.
1898	**Flashlight (Conrad Hubert):** Hubert combines three parts—a battery, a lightbulb, and a metal tube—to produce a flashlight.
1898	**Mercury vapor lamp (Peter Cooper Hewitt):** Hewitt's mercury vapor lamp proves to be more efficient than incandescent lamps.
1899	**Alpha particle discovered (Ernest Rutherford):** Rutherford detects the emission of helium 4 nuclei (alpha particles) in the natural radiation from uranium.
1900	**Aspirin:** Aspirin is patented by Bayer and sold as a powder. In 1915 it is sold in tablets.
1900	**Dirigibles (Ferdinand von Zeppelin):** Von Zeppelin flies his airship three years before the Wright brothers' airplane.
1900	**Gamma ray discovered (Paul Villard):** Villard discovers gamma rays in the natural radiation from uranium. They resemble very high-energy X rays.
1900	**Brownie camera:** George Eastman introduces the Kodak Brownie camera. It is sold for $1 and the film it uses costs 15 cents. The Brownie made photography an accessible hobby to almost everyone.
1901	**Acousticon hearing aid (Miller Reese Hutchison):** Hutchison invents a battery-powered hearing aid in the hopes of helping a mute friend speak.

1901	**Vacuum cleaner (H. Cecil Booth):** Booth patents his vacuum cleaner, a machine that sucks in and traps dirt. Previous devices, less effective, had attempted to blow the dirt away.
1901	**String galvanometer (electrocardiograph) (Willem Einthoven):** Einthoven's device passes tiny currents from the heart through a silver-coated silicon fiber, causing the fiber to move. Recordings of this movement can show the heart's condition.
1901	**Silicone (Frederick Stanley Kipping):** English chemist Kipping studies the organic compounds of silicon and coins the term "silicone."
1902	**Airplane engine (Charles E. Taylor):** Taylor begins building engines for the Wright brothers' airplanes.
1902	**Lionel electric toy trains (Joshua Lionel Cowen):** Cowen publishes the first Lionel toy train catalog. Lionel miniature trains and train sets become favorite toys for many years and are prized by collectors to this day.
1902	**Air conditioner (Willis Carrier):** Whole-house air-conditioning becomes possible.
1903	**Windshield wipers (Mary Anderson):** At first, the driver operated the wiper with a lever from inside the car.
1903	**Wright Flyer (Wilbur and Orville Wright):** The Wright Flyer is the first heavier-than-air machine to solve the problems of lift, propulsion, and steering for controlled flight.
1903	**Safety razor with disposable blade (King Camp Gillette):** Gillette's razor used a disposable and relatively cheap blade, so there was no need to sharpen it.
1903	**Space-traveling projectiles (Konstantin Tsiolkovsky):** Tsiolkovsky publishes "The Exploration of Cosmic Space by Means of Reaction-Propelled Apparatus," in which he includes an equation for calculating escape velocity (the speed required to propel an object beyond Earth's field of gravity). He is also recognized for the concept of rocket propulsion and for the wind tunnel.
1903	**Ultramicroscope (Richard Zsigmondy):** Zsigmondy builds the ultramicroscope to study colloids, mixtures in which particles of a substance are dispersed throughout another substance.
1903	**Spinthariscope (Sir William Crookes):** Crookes invents a device that sparkles when it detects radiation. He also develops and experiments with the vacuum tube, allowing later physicists to identify alpha and beta particles and X rays in the radiation from uranium.
1903	**Crayola crayons (Edwin Binney):** With his cousin C. Harold Smith, Binney invents dustless chalk and crayons marketed under the trade name Crayolas.
1903	**Motorcycle:** Harley-Davidson produces the first motorcycle, built to be a racer.
1903	**Electric iron:** Earl Richardson introduces the lightweight electric iron.
1904	**Glass bottle machine:** American inventor Michael Joseph Owens designs a machine that produces glass bottles automatically.

1905	**Novocaine (Alfred Einkorn):** While researching a safe local anesthetic to use on soldiers, German chemist Einkorn develops novocaine, which becomes a popular dental anesthetic.
1905	**Special relativity (Albert Einstein):** At the age of twenty-six, Einstein uses the constancy of the speed of light to explain motion, time, and space beyond Newtonian principles. During the same year, he publishes papers describing the photoelectric effect and Brownian motion.
1905	**Intelligence testing:** French psychologist Alfred Binet devises the first of a series of tests to measure an individual's innate ability to think and reason.
1906	**Hair-care products (Madam C. J. Walker):** Walker trains a successful sales force to go door-to-door and sell directly to women. Her saleswomen, beautifully dressed and coiffed, are instructed to pamper their clients.
1906	**Broadcast radio (Reginald Aubrey Fessenden):** In broadcast radio, sound wave forms are added to a carrier wave and then broadcast. The carrier wave is subtracted at the receiver leaving only the sound.
1906	**Klaxon horn (Miller Reese Hutchison):** Hutchison files a patent application for the electric automobile horn.
1906	**Chromatography (Mikhail Semenovich Tswett):** Tswett, a Russian botanist, invents chromatography.
1906	**Freeze-drying (Jacques Arsène d'Arsonval and George Bordas):** D'Arsonval and Bordas invent freeze-drying, but the practice is not commercially developed until after World War II.
1907	**Sun valve (Nils Gustaf Dalén):** Dalén's device uses sunlight to activate a lighthouse beacon. His other inventions make automated acetylene beacons in lighthouses possible.
1907	**Mantoux tuberculin skin test (Charles Mantoux):** French physician Mantoux develops a skin-reaction test to diagnose tuberculosis. He builds on the work of Robert Koch and Clemens von Pirquet.
1908	**Helium liquefaction (Heike Kamerlingh Onnes):** Kamerlingh Onnes produces liquid helium at a temperature of about 4 kelvins. He will also discover superconductivity in several materials cooled to liquid helium temperature.
1908	**"Tin Lizzie" (Model T) automobile (Henry Ford):** Ford's development of an affordable automobile, manufactured using his assembly-line production methods, revolutionize the U.S. car industry.
1908	**Electrostatic precipitator (Frederick Gardner Cottrell):** The electrostatic precipitator is invaluable for cleaning stack emissions.
1908	**Geiger-Müller tube (Hans Geiger):** Geiger invents a device, popularly called the Geiger counter, that is a reliable, portable radiation detector. Later his student Walther Müller helps improve the instrument.
1908	**Vacuum cleaner (James Murray Spangler):** Spangler receives a patent on his electric sweeper, and his Electric Suction Sweeper Company eventually becomes the Hoover Company, the largest such company in the world.

1908	**Cellophane (Jacques Edwin Brandenberger):** Brandenberger builds a machine to mass-produce cellophane, which he has earlier synthesized while unsuccessfully attempting to develop a stain-resistant cloth.
1908	**Water treatment:** Chlorine is used to purify water for the first time in the United States, in New Jersey, helping to reduce waterborne illnesses such as cholera, typhoid, and dysentery.
1908	**Audion (Lee De Forest):** De Forest invents a vacuum tube used in sound amplification. In 1922, he will develop talking motion pictures, in which the sound track is imprinted on the film with the pictures, instead of on a record to be played with the film, leading to exact synchronization of sound and image.
1909	**Synthetic fertilizers (Fritz Haber):** Haber also invents the Haber process to synthesize ammonia on a small scale.
1909	**Maxim silencer (Hiram Percy Maxim):** The silencer reduces the noise from firing the Maxim machine gun.
1909	**pH scale:** Danish chemist Søren Sørensen introduces the pH scale as a standard measure of alkalinity and acidity.
1910	**Chlorinator (Carl Rogers Darnall):** Major Darnall builds a machine to add liquid chlorine to water to purify it for his troops. His method is still widely used today.
1910	**Bakelite (Leo Hendrik Baekeland):** Bakelite is the first tough, durable plastic.
1910	**Neon lighting (Georges Claude):** Brightly glowing neon tubes revolutionize advertising displays.
1910	**Syphilis treatment:** German physician Paul Ehrlich and Japanese physician Hata Sahachirō discover the effective treatment of arsphenamine (named Salvarsan by Ehrlich) for syphilis.
1911	**Colt .45 automatic pistol (John Moses Browning):** Commonly called the Colt Model 1911, an improved version of the Colt Model 1900, the Colt .45 is the first autoloading pistol produced in America. Among Browning's other inventions are the Winchester 94 lever-action rifle and the gas-operated Colt-Browning machine gun.
1911	**Gyrocompass (Elmer Ambrose Sperry):** Sperry receives a patent for a nonmagnetic compass that indicates true north.
1911	**Atomic nucleus identified (Ernest Rutherford):** Rutherford discovered the nucleus by bombarding a thin gold foil with alpha particles. Some were deflected through large angles showing that something small and hard was present.
1911	**Ductile tungsten (William David Coolidge):** Coolidge also invented the Coolidge tube, an improved X-ray producing tube.
1911	**Ochoaplane (Victor Leaton Ochoa):** In addition to inventing this plane with collapsible wings, Ochoa also developed an electricity-generating windmill.
1911	**Automobile electric ignition system (Charles F. Kettering):** Kettering invents the first electric ignition system for cars.
1912	**Automatic traffic signal system (Garrett Augustus Morgan):** Morgan also invents a safety hood that served as a rudimentary gas mask.

1913	**Gyrostabilizer (Elmer Ambrose Sperry):** Sperry develops the gyrostabilizer, a device to control the roll, pitch, and yaw of a moving ship. He will go on to invent the flying bomb, which is guided by a gyrostabilizer and by radio control.
1913	**Erector set (Alfred C. Gilbert):** Erector sets provide hands-on engineering experience for countless children.
1913	**Zipper (Gideon Sundback):** While others had made zipper-like devices but had never successfully marketed them, Sundback designs a zipper in approximately its present form. He also invents a machine to make zippers.
1913	**Improved electric lightbulb (Irving Langmuir):** Langmuir fills his lightbulb with a low-pressure inert gas to retard evaporation from the tungsten filament.
1913	**Industrialization of the Haber process (Carl Bosch):** Bosch scales up Haber's process for making ammonia to an industrial capacity. The process comes to be known as the Haber-Bosch process.
1913	**Bergius process (Friedrich Bergius):** Bergius develops high-pressure, high-temperature process to produce liquid fuel from coal.
1913	**Electric dishwasher:** The Walker brothers of Philadelphia produce the first electric dishwasher.
1913	**Stainless steel (Harry Brearley):** Brearley invents stainless steel.
1913	**Thermal cracking (William Burton and Robert Humphreys):** Standard Oil chemical engineers Burton and Humphreys discover thermal cracking, a method of oil refining that significantly increases gasoline yields.
1914	**Backless brassiere (Caresse Crosby):** The design of a new women's undergarment leads to the expansion of the U.S. brassiere industry. Caresse was originally a marketing name that Mary Phelps Jacob eventually adopted as her own.
1915	**Panama Canal opens:** The passageway between the Atlantic and Pacific oceans creates a boon for the shipping industry.
1915	**General relativity (Albert Einstein):** Einstein refines his 1905 theory of relativity (now called special relativity) to describe the theory that states that uniform accelerations are almost indistinguishable from gravity. Einstein's theory provides the basis for physicists' best understanding of gravity and of the framework of the universe.
1915	**Jenny (Glenn H. Curtiss):** The Jenny becomes a widely used World War I biplane, and Curtis becomes a general manufacturer of airplanes and airplane engines.
1915	**Pyrex:** Corning's brand name for glassware is introduced.
1915	**Warfare:** Depth-charge bombs are first used by the Allies against German submarines.
1916	**By-products of sweet potatoes and peanuts (George Washington Carver):** Carver publishes his famous bulletin on 105 ways to prepare peanuts.
1919	**Proton discovered (Ernest Rutherford):** After bombarding nitrogen gas with alpha particles (helium 4 nuclei), Rutherford observes that positive particles with a single charge are knocked loose. They are protons.
1919	**Toaster (Charles Strite):** Strite invents the first pop-up toaster.

1920	**Microelectrode (Ida H. Hyde):** Hyde's electrode is small enough to pierce a single cell. Chemicals can also be very accurately deposited by the microprobe.
1921	**Ready-made bandages:** Johnson & Johnson puts Band-Aids on the market.
1921	**Antiknock solution (Thomas Midgley, Jr.):** While working at a General Motors subsidiary, American mechanical engineer Midgley develops an antiknock solution for gasoline.
1921	**Insulin:** University of Toronto researchers Frederick Banting, J. J. R. Macleod, and Charles Best first extract insulin from a dog, and the first diabetic patient is treated with purified insulin the following year. Banting and Macleod win the 1923 Nobel Prize in Physiology or Medicine for their discovery of insulin.
1923	**Improved telephone speaker (Georg von Békésy):** Békésy's studies of the human ear lead to an improved telephone earpiece. He will also construct a working model of the inner ear.
1923	**Quick freezing (Clarence Birdseye):** Birdseye's quick-freezing process preserves food's flavor and texture better than previously used processes.
1924	**Coincidence method of particle detection (Walther Bothe):** Bothe's method proves invaluable in the use of gamma rays to discover nuclear energy levels.
1924	**Ultracentrifuge (Theodor Svedberg):** Svedberg's ultracentrifuge can separate isotopes, such as uranium 235 from uranium 238, from each other—a critical step in building the simplest kind of atomic bomb.
1924	**EEG:** German scientist Hans Berger records the first human electroencephalogram (EEG), which shows electrical patterns in the brain.
1925	**Leica I camera:** Leitz introduces the first 35-millimeter Leica camera at the Leipzig Spring Fair.
1925	**First U.S. television broadcast:** Charles Francis Jenkins transmits the silhouette image of a toy windmill.
1926	**Automatic power loom (Sakichi Toyoda):** Toyoda's loom helps Japan catch up with the western Industrial Revolution.
1926	**Liquid-fueled rocket (Robert H. Goddard):** A solid-fueled rocket is either on or off, but a liquid-fueled rocket can be throttled up or back and can be shut off before all the fuel is expended.
1927	**Aerosol can (Erik Rotheim):** Norwegian engineer Rotheim patents the aerosol can and valve.
1927	**Adiabatic demagnetization (William Francis Giauque):** Adiabatic demagnetization is part of a refrigeration cycle that, when used enough times, can chill a small sample to within a fraction of a kelvin above absolute zero.
1927	**All-electronic television (Philo T. Farnsworth):** Farnsworth transmits the first all-electronic television image using his newly developed camera vacuum tube, known as the image dissector. Previous systems combined electronics with mechanical scanners.
1927	**First flight across the Atlantic:** Charles Lindbergh flies the Spirit of St. Louis across the Atlantic. He is the first to make a solo, nonstop flight across the ocean.

1927	**Iron lung (Philip Drinker):** Drinker, a Harvard medical researcher, assisted by Louis Agassiz Shaw, devises the first modern practical respirator using an iron box and two vacuum cleaners. Drinker calls the device the iron lung.
1927	**Garbage disposal (John W. Hammes):** American architect Hammes develops the first garbage disposal to make cleaning up the kitchen easier for his wife. It is nicknamed the "electric pig" when it first goes on the market.
1927	**Adjustable-temperature iron:** The Silex Company begins to sell the first iron with an adjustable temperature control.
1927	**Analogue computer (Vannevar Bush):** Bush builds the first analogue computer. He is also the first person to describe the idea of hypertext.
1928	**Sliced bread (Otto F. Rohweddeer):** Bread that came presliced was advertised as "the greatest forward step in the baking industry since bread was wrapped." Today the phrase "the greatest thing since sliced bread" is used to describe any innovation that has a broad, positive impact on daily life.
1928	**First television programs:** First regularly scheduled television programs in the United States air. They are produced out of a small, experimental station in Wheaton, Maryland.
1928	**Link Trainer (Edwin Albert Link):** Link's flight simulator created realistic conditions in which to train pilots without the expense or risk of an actual air flight. Link also developed a submersible decompression chamber.
1928	**New punch card:** IBM introduces a new punch card that has rectangular holes and eight columns.
1928	**Radio network:** NBC establishes the first coast-to-coast radio network in the United States.
1928	**Pap smear (George N. Panpanicolaou):** Greek cytopathologist Panpanicolaou patents the pap smear, a test that helps detect uterine cancer.
1928	**Portable offshore drilling (Louis Giliasso):** Giliasso creates an efficient portable method of offshore drilling by mounting a derrick and drilling outfit onto a submersible barge.
1929	**Iconoscope (Vladimir Zworykin):** Zworykin claims that he, not Philo T. Farnsworth, should be credited with the invention of television.
1929	**Strobe light (Harold E. Edgerton):** Edgerton's strobe is used as a flash bulb. He pioneers the development of high-speed photography.
1929	**Dymaxion products (R. Buckminster Fuller):** Fuller's "Dymaxion" products feature an energy-efficient house using prefabricated, easily shipped parts.
1929	**Van de Graaff generator (Robert Jemison van de Graaff):** Van de Graaff invents the Van de Graaff generator, which accumulates electric charge on a moving belt and deposits it in a hollow glass sphere at the top.

1929-1936	**Cyclotron (Ernest Orlando Lawrence and M. Stanley Livingston):** Lawrence and Livingston are studying particle accelerators and develop the cyclotron, which consists of a vacuum tank between the poles of a large magnet. Alternating electric fields inside the tank can accelerate charged particles to high speeds. The cyclotron is used to probe the atomic nucleus or to make new isotopes of an element, including those used in medicine.
1930	**Schmidt telescope (Bernhard Voldemar Schmidt):** Schmidt's telescope uses a spherical main mirror and a correcting lens at the front of the scope. It can photograph large fields with little distortion.
1930	**Pluto discovered (Clyde Tombaugh):** Tombaugh observes a body one-fifth the mass of Earth's moon. Pluto comes to be regarded as the ninth planet of the solar system, but in 2006 it is reclassified as one of the largest-known Kuiper Belt objects, a dwarf planet.
1930	**Freon refrigeration and air-conditioning (Charles F. Kettering):** After inventing an electric starter in 1912 and the Kettering Aerial Torpedo in 1918 (the world's first cruise missile), Kettering and Thomas Midgley, Jr., use Freon gas in their cooling technology. (Freon will later be banned because of the effects of chlorofluorocarbons on Earth's ozone layer.)
1930	**Synthetic rubber (Wallace Hume Carothers):** Carothers synthesizes rubber and goes on to develop nylon in 1935. His work professionalizes polymer chemistry as a scientific field.
1930	**Scotch tape (Richard G. Drew):** After inventing masking tape, Drew invents the first waterproof, see-through, pressure-sensitive tape that also acted as a barrier to moisture.
1930	**Military and commercial aircraft (Andrei Nikolayevich Tupolev):** Tupolev emerges as one of the world's leading designers of military and civilian aircraft. His aircraft set nearly eighty world records.
1930's	**Washing machine (John W. Chamberlain):** Chamberlain invents a washing machine that enables clothes to be washed, rinsed, and have the water extracted from them in a single operation.
1931	**Electric razor (Jacob Schick):** Schick introduces his first electric razor, which allows dry shaving. It has a magazine of blades held in the handle.
1931	**Radio astronomy (Karl G. Jansky):** One of the founders of the field of radio astronomy, Janksy detects radio static coming from the Milky Way's center.
1932	**Positron discovered (Carl D. Anderson):** Anderson discovers the positron, a positive electron and an element of antimatter.
1932	**Neoprene (Julius Nieuwland):** The first synthetic rubber is marketed.
1932	**Neutron discovered (James Chadwick):** Chadwick detects the neutron, an atomic particle with no charge and a mass only slightly greater than that of a proton. Except for hydrogen 1, the atomic nuclei of all elements consist of neutrons and protons.

1932	**Phillips-head screw (Henry M. Phillips):** The Phillips-head screw has an X-shaped slot in the head and can withstand the torque of a machine-driven screwdriver, which is greater than the torque that can be withstood by the conventional screw.
1932	**Duplicating device for typewriters (Beulah Louise Henry):** Henry's invention uses three sheets of paper and three ribbons to produce copies of a document as it is typewritten. Henry also develops children's toys—for example, a doll the eye color of which can be changed.
1932	**Cockroft-Walton accelerator (John Douglas Cockcroft and Ernest Thomas Sinton Walton):** The Cockcroft-Walton accelerator is used to fling charged particles at atomic nuclei in order to investigate their properties.
1932	**Richter scale (Charles Francis Richter):** Richter develops a scale to describe the magnitude of earthquakes; it is still used today.
1932	**Neutron (Sir James Chadwick):** Chadwick proves the existence of neutrons; he is awarded the 1935 Nobel Prize in Physics for his work.
1933	**Nuclear chain reaction (Leo Szilard):** Szilard conceives the idea of a nuclear chain reaction. He becomes a key figure in the Manhattan Project, which eventually builds the atomic bomb.
1933	**Magnetic tape recorder (Semi Joseph Begun):** Begun builds the first tape recorder, a dictating machine using wire for magnetic recording. He also develops the first steel tape recorder for mobile radio broadcasting and leads research into telecommunications and underwater acoustics.
1933	**Electron microscope (Ernst Ruska):** Ruska makes use of the wavelengths of electrons—shorter than those of visible light—to build a microscope that can image details at the subatomic level.
1933	**Recording:** Alan Dower Blumlein's patent for stereophonic recording is granted.
1933	**Polyethylene (Eric Fawcett and Reginald Gibson):** Fawcett and Gibson of Imperial Chemical Industries in London accidentally discover polyethylene. Hula hoops and Tupperware are just two of the products made with the substance.
1933	**Modern airliner:** Boeing 247 becomes the first modern airliner.
1933	**Solo flight:** Wiley Post makes the first around-the-world solo flight.
1934	**First bathysphere dive:** Charles William Beebe and Otis Barton make the first deep-sea dive in the Beebe-designed bathysphere off the Bermuda coast.
1934	**Langmuir-Blodgett films (Katharine Burr Blodgett):** A thin Langmuir-Blodgett film deposited on glass can make it nearly nonreflective.
1934	**Passenger train:** The Burlington Zephyr, America's first diesel-powered streamlined passenger train, is revealed at the World's Fair in Chicago.
1935	**Frequency modulation (Edwin H. Armstrong):** Armstrong exploits the fact that, since there are no natural sources of frequency modulation (FM), FM broadcasts are static-free.

1935	**Diatometer (Ruth Patrick):** Patrick's diatometer is a device placed in the water to collect diatoms and allow them to grow. The number of diatoms is sensitive to water pollution.
1935	**Kodachrome color film (Leopold Mannes and Leopold Godowsky, Jr.):** Mannes and Godowsky invent Kodachrome, a color film that is easy to use and produces vibrant colors. (With the digital revolution of the late twentieth century, production of Kodachrome is finally retired in 2009.)
1935	**Physostigmine and cortisone (Percy Lavon Julian):** Julian synthesizes physostigmine, used to treat glaucoma, and cortisone, used for arthritis. He will hold more than 130 patents and will become the first African American chemist inducted into the National Academy of Sciences.
1935	**Mobile refrigeration (Frederick McKinley Jones):** Mobile refrigeration enables the shipping of heat-sensitive products and compounds, from blood to frozen food.
1935	**Radar-based air defense system (Sir Robert Alexander Watson-Watt):** Watson-Watt's technical developments and his efforts as an administrator will be so important to the development of radar that he will be called the "father of radar."
1935	**Fallingwater (Frank Lloyd Wright):** Wright designs and builds a showcase house blending its form with its surroundings. One of the greatest architects of the twentieth century, he will produce many architectural innovations in structure, materials, and design.
1936	**Field-emission microscope (Erwin Wilhelm Müller):** Müller completes his dissertation, "The Dependence of Field Electron Emission on Work Function," and goes on to develop the field-emission microscope, which can resolve surface features as small as 2 nanometers.
1936	**Pentothal (Ernest Volwiler and Donalee Tabern):** Pentothal is a fast-acting intravenous anesthetic.
1937	**Muon discovered (Seth Neddermeyer):** Neddermeyer, working with Carl Anderson, J. C. Street, and E. C. Stevenson discover the muon (a particle similar to a heavy electron) while examining cosmic-ray tracks in a cloud chamber.
1937	**Concepts of digital circuits and information theory (Claude Elwood Shannon):** Shannon's most important contributions were electronic switching and using information theory to discover the basic requirements for data transmission.
1937	**X-ray crystallography (Dorothy Crowfoot Hodgkin):** Hodgkin uses X-ray crystallography to reveal the structure of molecules. She goes on to win the 1964 Nobel Prize in Chemistry.
1937	**Model K computer (George Stibitz):** The model K, an early electronic computer, employs Boolean logic.
1937	**Artificial sweetener:** American chemist Michael Sveda invents cyclamates, which is used as a noncaloric artificial sweetener until it is banned by the U.S. government in 1970 because of possible carcinogenic effects.
1937	**First pressurized airplane cabin:** The first pressurized airplane cabin is achieved in the United States with Lockheed's XC-35.

1937	**Antihistamines (Daniel Bovet) :** Swiss-born Italian pharmacologist Bovet discovers antihistamines. He is awarded the 1957 Nobel Prize in Physiology or Medicine for his work.
1938	**Teflon (Roy J. Plunkett):** Plunkett accidentally synthesizes polytetrafluoroethylene (PTFE), now commonly known as Teflon, while researching chlorofluorocarbon refrigerants.
1938	**Electron microscope (James Hillier and Albert Prebus):** Adapting the work of German physicists, Hillier and Prebus develop a prototype of the electron microscope; and in 1940 Hillier produces the first commercial electron microscope available in the United States.
1938	**Xerography (Chester F. Carlson):** Xerography uses electrostatic charges to attract toner particles to make an image on plain paper. A hot wire then fuses the toner in place.
1938	**Walkie-talkie (Alfred J. Gross):** Gross's portable, two-way radio allows the user to move around while sending messages without remaining tied to a bulky transmitter. Gross invents a pager in 1949 and a radio tuner in 1950 that automatically follows the drift in carrier frequency due to movement of a sender or receiver.
1937-1938	**Analogue computer (George Philbrick):** Philbrick builds the Automatic Control Analyzer, which is an electronic analogue computer.
1939	**Helicopter (Igor Sikorsky):** Sikorsky, formerly the chief construction engineer and test pilot for the first four-engine aircraft, tests his helicopter, the Vought-Sikorsky 300, which after improvements will emerge as the world's first working helicopter.
1939	**Jet engine (Hans Joachim Pabst von Ohain):** The first jet-powered aircraft flies in 1939, while the first jet fighter will fly in 1941.
1939	**Atanasoff-Berry Computer (John Vincent Atanasoff and Clifford Berry):** The ABC, the world's first electronic digital computer, uses binary numbers and electronic switching, but it is not programmable.
1939	**DDT (Paul Hermann Müller):** Müller discovers the insect-repelling properties of DDT. He is awarded the 1948 Nobel Prize in Physiology or Medicine.
1940's	**Solar technology (Maria Telkes):** Telkes develops the solar oven and solar stills to produce drinking water from ocean water.
1940	**Cavity magnetron (Henry Boot and John Randall):** Boot and Randall develop the cavity magnetron, which advances radar technology.
1940	**Penicillin:** Sir Howard Walter Florey and Ernst Boris Chain isolate and purify penicillin. They are awarded, with Sir Alexander Fleming, the 1945 Nobel Prize in Physiology or Medicine.
1940	**Blood bank (Charles Richard Drew):** Drew establishes blood banks for World War II soldiers.
1940	**Color television (Peter Carl Goldmark):** Goldmark produces a system for transmitting and receiving color-television images using synchronized rotating filter wheels on the camera and on the receiver set.

1940	**Paintball gun (Charles and Evan Nelson):** The gun and paint capsules, invented to mark hard-to-reach trees in the forest, are eventually used for the game of paintball (1981), in which people shoot each other with paint.
1940	**Audio oscillator (William Redington Hewlett):** Hewlett invents the audio oscillator, a device that creates one frequency (pure tone) at a time. It is the first successful product of his Hewlett-Packard Company.
1940	**Antibiotics (Selman Abraham Waksman):** Waksman, through study of soil organisms, finds sources for the world's first antibiotics, including streptomycin and actinomycin.
1940	**Plutonium (Glenn Theodore Seaborg):** Seaborg synthesizes one of the first transuranium elements, plutonium. He becomes one of the leading figures on the Manhattan Project, which will build the atomic bomb. While he and others urged the demonstration of the bomb as a deterrent, rather than its use on the Japanese civilian population, the latter course was taken.
1940	**Thompson submachine gun (John T. Thompson):** Thompson works with Theodore Eickhoff and Oscar Payne to invent the American version of the submachine gun.
1940	**Automatic auto transmission:** General Motors offers the first modern automatic automobile transmission.
1941	**Jet engine (Sir Frank Whittle):** Whittle develops the jet engine independent of Hans Joachim Pabst von Ohain in Germany. After World War II, they meet and become good friends.
1941	**Solid-body electric guitar (Les Paul):** Paul's guitar lays the foundation for rock music. He also develops multitrack recording in 1948.
1941	**Z3 programmable computer (Konrad Zuse):** Zuse and his colleagues complete the first general-purpose, programmable computer, the Z3, in December. In 1950, Zuse will sell a Z4 computer—the only working computer in Europe.
1941	**Velcro (Georges de Mestral):** Burrs sticking to his dog's fur give de Mestral the idea for Velcro, which he perfects in 1948.
1941	**Dicoumarol:** The anticoagulant drug dicoumarol is identified and synthesized.
1941	**RDAs:** The first Recommended Dietary Allowances (RDAs), nutritional guidelines, are accepted.
1942	**Superglue (Harry Coover and Fred Joyner):** After developing superglue (cyanoacrylate), Coover rejects it as too sticky for a 1942 project. Coover and Joyner rediscover superglue in 1951, when Coover recognizes it as a marketable product.
1942	**Aqua-Lung (Jacques-Yves Cousteau and Émile Gagnon):** The Aqua-Lung delivers air at ambient pressure and vents used air to the surroundings.
1942	**Controlled nuclear chain reaction (Enrico Fermi):** In 1926 Fermi helped develop Fermi-Dirac statistics, which describe the quantum behavior of groups of electrons, protons, or neutrons. He now produces the first sustained nuclear chain reaction.

1942	**Synthetic vitamins (Max Tishler):** After synthesizing several vitamins during the 1930's, Tishler and his team develop the antibiotic sulfaquinoxaline to treat coccidiosis. He also develops fermentation processes to produce streptomycin and penicillin.
1942	**Bazooka:** The United States military first uses the bazooka during the North African campaign in World War II.
1943	**Meteorology:** Radar is first used to detect storms.
1944	**Electromechanical computer (Howard Aiken and Grace Hopper):** The Mark series of computers is built, designed by Aiken and Hopper. The U.S. Navy uses it to calculate trajectories for projectiles.
1944	**Colossus:** Colossus, the world's first vacuum-tube programmable logic calculator, is built in Britain for the purpose of breaking Nazi codes.
1944	**Phased array radar antennas (Luis W. Alvarez):** Alvarez's phased array sweeps a beam across the sky by turning hundreds of small antennas on and off and not by moving a radar dish.
1944	**V-2 rocket (Wernher von Braun):** Working for the German government during World War II, von Braun and other rocket scientists develop the V-2 rocket, the first long-range military missile and first suborbital missile. Arrested for making anti-Nazi comments, he later emigrates to the United States, where he leads the team that produces the Jupiter-C missile and launches vehicles such as the Saturn V, which help make the U.S. space program possible.
1944	**Quinine:** Robert B. Woodward and William von Eggers Doering synthesize quinine, which is used as an antimalarial.
1945	**Automatic Computing Engine (Alan Mathison Turing):** While the Automatic Computing Engine (ACE) was never fully built, it was one of the first stored-program computers.
1945	**Atomic bomb (J. Robert Oppenheimer):** Oppenheimer, the scientific leader of the Manhattan Project, heads the team that builds the atomic bomb. On the side of military use of the bomb to end World War II quickly, Oppenheimer saw this come to pass on August 6, 1945, when the bomb was dropped over Hiroshima, Japan, killing and maiming 150,000 people; a similar number of casualties ensued in Nagasaki on August 9, when the second bomb was dropped. Japan surrendered on August 14.
1945	**Dialysis machine (Willem Johan Kolff):** Kolff designs the first artificial kidney, a machine that cleans the blood of patients in renal failure, and refuses to patent it. He will construct the artificial lung in 1955.
1945	**Radioimmunoassay (RIA) (Rosalyn Yalow):** RIA required only a drop of blood (rather than the tens of milliliters previously required) to find trace amounts of substances.
1945	**Electronic Sackbut (Hugh Le Caine):** Le Caine builds the first music synthesizer, joined by the Special Purpose Tape Recorder in 1954, which could simultaneously change the playback speed of several recording tracks.

1945	**ENIAC computer (John William Mauchly and John Presper Eckert):** The Electronic Numerical Integrator and Computer, ENIAC, is the first general-purpose, programmable, electronic computer. (The Z3, developed independently by Konrad Zuse from 1939 to 1941 in Nazi Germany, did not fully exploit electronic components.) Built to calculate artillery firing tables, ENIAC is used in calculations for the hydrogen bomb.
1945	**Microwave oven (Percy L. Spencer):** The microwave oven grew out of the microwave generator, the magnetron tube, becoming more affordable.
1946	**Tupperware (Earl S. Tupper):** Tupper exploits plastics technology to develop a line of plastic containers that he markets at home parties starting in 1948.
1946	**Carbon-14 dating (Willard F. Libby):** Libby uses the half-life of carbon 14 to develop a reliable means of dating ancient remains. Radiocarbon dating has proven to be invaluable to archaeologists.
1946	**Magnetic tape recording (Marvin Camras):** Camras develops a magnetic tape recording process that will be adapted for use in electronic media, including music and motion-picture sound recording, audio and videocassettes, floppy disks, and credit card magnetic strips. For many years his method is the primary way to record and store sound, video, and digital data.
1946-1947	**Audiometer (Georg von Békésy):** Békésy invents a pure-tone audiometer that patients themselves can control to measure the sensitivity of their own hearing.
1946	**Radioisotopes for cancer treatment:** The first nuclear-reactor-produced radioisotopes for civilian use are sent from the U.S. Army's Oak Ridge facility in Tennessee to Brainard Cancer Hospital in St. Louis.
1947	**Transistor (John Bardeen, Walter H. Brattain, and William Shockley):** Hoping to build a solid-state amplifier, the team of Bardeen, Brattain, and Shockley discover the transistor, which replaces the vacuum tube in electronics. Bardeen is later part of the group that develops theory of superconductivity.
1947	**Platforming (Vladimir Haensel):** American chemical engineer Haensel invents platforming, a process that uses a platinum catalyst to produce cleaner-burning high-octane fuels.
1947	**Tubeless tire:** B.F. Goodrich announces development of the tubeless tire.
1948	**Holography (Dennis Gabor):** Gabor publishes his initial results working with holograms in Nature. Holograms became much more spectacular after the invention of the laser.
1948	**Long-playing record (LP) (Peter Carl Goldmark):** Goldmark demonstrates the LP playing the cello with CBS musicians. The musical South Pacific is recorded in LP format and boosts sales, making the LP the dominant form of recorded sound for the next four decades.
1948	**Gamma-ray pinhole camera (Roscoe Koontz):** Working to make nuclear reactors safer, Koontz invents the gamma-ray pinhole camera. The pinhole should act like a lens and form an image of the gamma source.
1948	**Instant photography (Edwin Herbert Land):** Land develops the simple process to make sheets of polarizing material. He perfects the Polaroid camera in 1972.

1948	**Synthetic penicillin (John C. Sheehan):** Sheehan develops the first total synthesis of penicillin, making this important antibiotic widely available.
1949	**First peacetime nuclear reactor:** Construction on the Brookhaven Graphite Research Reactor at Brookhaven Laboratory on Long Island, New York, is completed.
1949	**Magnetic core memory (Jay Wright Forrester):** Core memory is used from the early 1950's to the early 1970's.
1950's	**Fortran (John Warner Backus):** Backus develops the computer language Fortran, which is an acronym for "formula translation." Fortran allows direct entry of commands into computers with Englishlike words and algebraic symbols.
1950	**Planotron (Pyotr Leonidovich Kapitsa):** Kapitsa invents a magnetron tube for generating microwaves. He becomes a corecipient of the Nobel Prize for Physics in 1978 for discovering superfluidity in liquid helium.
1950	**Purinethol (Gertrude Belle Elion):** Elion develops the first effective treatment for childhood leukemia, 6-mercaptopurine (Purinethol). Elion later discovers azathioprine (Imuran), an immunosuppressive agent used for organ transplants.
1950	**Artificial pacemaker (John Alexander Hopps):** Hopps develops a device to regulate the beating of the heart to treat patients with erratic heartbeats. By 1957, the device is small enough to be implanted.
1950	**Contact lenses (George Butterfield):** Oregon optometrist Butterfield develops a lens that is molded to fit the contours of the cornea.
1951	**Fiber-optic endoscope (fibroscope) (Harold Hopkins):** Hopkins fastened together a flexible bundle of optical fibers that could convey an image. One end of the bundle could be inserted into a patient's throat, and the physician could inspect the esophagus.
1951	**The Pill (Carl Djerassi):** The birth-control pill, which becomes the world's most popular and is possibly most widely used contraceptive, revolutionizes not only medicine but also gender relations and women's status in society. Its prolonged use is later revealed to have health consequences.
1951	**Field-emission microscope (Erwin Wilhelm Müller):** Müller develops the field-ion microscope, followed by an atom-probe field-ion microscope in 1963, which can detect individual atoms.
1951	**Maser (Charles Hard Townes):** The maser (microwave amplification by stimulated emission of radiation) is a "laser" for microwaves. Discovered later, the "laser" patterned its name the acronym "maser."
1951	**Artificial heart valve (Charles Hufnagel):** Hufnagel develops an artificial heart valve and performs the first heart-valve implantation surgery in a human patient the following year.
1951	**UNIVAC (John Mauchly and John Presper Eckert):** Mauchly and Eckert invent the Universal Automatic Computer (UNIVAC). UNIVAC is competitor of IBM's products.
1952	**Bubble chamber (Donald A. Glaser):** In a bubble chamber, bubbles form along paths taken by subatomic particles as they interact, and the bubble trails allow scientists to deduce what happened.

1952	**Photovoltaic cell (Gerald Pearson):** The photovoltaic cell converts sunlight into electricity.
1952	**Improved electrical resistor (Otis Boykin):** Boykin's resistor had improved precision, and its high-frequency characteristics were better than those of previous resistors.
1952	**Language compiler (Grace Murray Hopper):** Hopper invents the compiler, an intermediate program that translates English-language instructions into computer language, followed in 1959 by Common Business Oriented Language (COBOL), the first computer programming language to translate commands used by programmers into the machine language the computer understands.
1952	**Amniocentesis (Douglas Bevis):** British physician Bevis develops amniocentesis.
1952	**Gamma camera (Hal Anger):** Nuclear medicine pioneer Anger creates the first prototype for the gamma camera. This leads to the inventions of other medical imaging devices, which detect and diagnose disease.
1953	**Medical ultrasonography (Inge Edler and Carl H. Hertz):** Edler and Hertz adapt an ultrasound probe used in materials testing in a shipyard for use on a patient. Their technology makes possible echograms of the heart and brain.
1953	**Inertial navigation systems (Charles Stark Draper):** Draper's inertial navigation system (INS) is designed to determine the current position of a ship or plane based on the initial location and acceleration.
1953	**Heart-lung machine (John H. Gibbon, Jr.):** American surgeon Gibbon conducts the first successful heart surgery using a heart-lung machine that he constructed with the help of his wife, Mary.
1953	**First frozen meals:** Swanson develops individual prepackaged frozen meals. The first-ever meal consists of turkey, cornbread stuffing, peas, and sweet potatoes.
1954	**Geodesic dome ® (Buckminster Fuller):** After developing the geodesic dome, Fuller patents the structure, an energy-efficient house using prefabricated, easily shipped parts.
1954	**Atomic absorption spectroscopy (Sir Alan Walsh):** Atomic absorption spectroscopy is used to identify and quantify the presence of elements in a sample.
1954	**Synthetic diamond (H. Tracy Hall):** Hall synthesizes diamonds using a high-pressure, high-temperature belt apparatus that can generate 120,000 atmospheres of pressure and sustain a temperature of 1,800 degrees Celsius in a working volume of about 0.1 cubic centimeter.
1954	**Machine vision (Jerome H. Lemelson):** Machine vision allows a computer to move and measure products and to inspect them for quality control.
1954	**Hydrogen bomb (Edward Teller):** The first hydrogen bomb, designed by Teller, is tested at the Bikini Atoll in the Pacific Ocean.
1954	**Silicon solar cells (Calvin Fuller):** Silicon solar cells have proven to be among the most efficient and least expensive solar cells.

1954	**First successful kidney transplant (Joseph Edward Murray):** American surgeon Murray performs the first successful kidney transplant, inserting one of Ronald Herrick's kidneys into his twin brother, Richard. Murray shares the 1990 Nobel Prize for Physiology or Medicine with E. Donnall Thomas, who developed bone marrow transplantation.
1954	**Transistor radio:** The first transistor radio is introduced by Texas Instruments.
1954	**IBM 650:** The IBM 650 computer becomes available. It is considered by IBM to be its first business computer, and it is the first computer installed at Columbia University in New York.
1954	**First nuclear submarine:** The United States launches the first nuclear- powered submarine, the USS *Nautilus.*
1955	**Color television's RGB system (Ernst Alexanderson):** The RGB system uses three image tubes to scan scenes through colored filters and three electron guns in the picture tube to reconstruct scenes.
1955	**Floppy disk and floppy disk drive (Alan Shugart):** Working at the San Jose, California, offices of International Business Machines (IBM), Shugart develops the disk drive, followed by floppy disks to provide a relatively fast way to store programs and data permanently.
1955	**Hovercraft (Sir Christopher Cockerell):** Cockerell files a patent for his hovercraft, an amphibious vehicle. He earlier invented several important electronic devices, including a radio direction finder for bombers in World War II.
1955	**Pulse transfer controlling device (An Wang):** The device allows magnetic core memory to be written or read without mechanical motion and is therefore very rapid.
1955	**Polio vaccine (Jonas Salk):** Salk's polio vaccine, which uses the killed virus, saves lives and improves the quality of life for millions afflicted by polio.
1956	**Fiber optics (Narinder S. Kapany):** Kapany, known as the father of fiber optics, coins the term "fiber optics." In high school, he was told by a teacher that light moves only in a straight line; he wanted to prove the teacher wrong and wound up inventing fiber optics.
1956	**Scotchgard (Patsy O'Connell Sherman):** Sherman develops a stain repellent for fabrics that is trademarked as Scotchgard.
1956	**Ovonic switch (Stanford Ovshinsky):** Ovshinsky invents a solid-state, thin film switch meant to mimic the actions of neurons.
1956	**Videotape recorder (Charles P. Ginsburg):** The video recorder allows programs to be shown later, to provide instant replays in sports, and to make a permanent record of a program.
1956	**Liquid Paper (Bette Nesmith Graham):** Graham markets her "Mistake Out" fluid for concealing typographical errors.
1956	**Dipstick blood sugar test (Helen M. Free):** Free and her husband Alfred co-invent a self-administered urinalysis test that allows diabetics to monitor their sugar levels and to adjust their medications accordingly.

1956	**350 RAMAC:** IBM produces the first computer disk storage system, the 350 RAMAC, which retrieves data from any of fifty spinning disks.
1957	**Wankel rotary engine (Felix Wankel):** Having fewer moving parts, the Wankel rotary engine ought to be sturdier and perhaps more efficient than the common reciprocating engine.
1957	**Laser (Gordon Gould, Charles Hard Townes, Arthur L. Schawlow, Theodore Harold Maiman):** Having conducted research on using light to excite thallium atoms, Gould tries to get funds and approval to build the first laser, but he fails. Townes (inventor of the maser) and Schawlow of Bell Laboratories will first describe the laser, and Maiman will first succeed in building a small optical maser. Gould coins the term "laser," which stands for light amplification by stimulated emission of radiation.
1957	**Intercontinental ballistic missile (ICBM):** The Soviet Union develops the ICBM.
1957	**First satellite:** The Soviet Union launches Sputnik, the first man-made satellite.
1958	**CorningWare:** CorningWare cookware is introduced. It is based on S. Donald Stookey's 1953 discovery that a heat-treatment process can transform glass into fine-grained ceramics.
1958	**Integrated circuit (Robert Norton Noyce and Jack St. Clair Kilby):** The microchip, independently discovered by Noyce and Kilby, proves to be the breakthrough that allows the miniaturization of electronic circuits and paves the way for the digital revolution.
1958	**Ultrasound:** Ultrasound becomes the most common method for examining a fetus.
1958	**Planar process (Jean Hoerni):** Hoerni develops the first planar process, which improves the integrated circuit.
1960's	**Lithography:** Optical lithography, a process that places intricate patterns onto silicon chips, is used in semiconductor manufacturing.
1960	**Measles vaccine (John F. Enders):** Enders, an American physician, develops the first measles vaccine. It is tested the following year and is hailed a success.
1960	**Echo satellite (John R. Pierce):** The first passive-relay telecommunications satellite, Echo, reflected signals. The signals, received from one point on Earth, "bounce" off the spherical satellite and are reflected back down to another, far distant, point on Earth.
1960	**Automatic letter-sorting machine (Jacob Rabinow):** Rabinow's machine greatly increased the speed and efficiency of mail delivery in the United States. He also invented an optical character recognition (OCR) scanner.
1960	**Ruby laser (Theodore Harold Maiman):** Maiman produces a ruby laser, the world's first visible light laser.
1960	**Helium-neon gas laser (Ali Javan):** Javan produces the world's second visible light laser.
1960	**Chardack-Greatbatch pacemaker (Wilson Greatbatch and William Chardack):** Greatbatch and Chardack create the first implantable pacemaker.

1960	**Radionuclide generator:** Powell Richards and Walter Tucker and their colleagues at Brookhaven Laboratory in New York invent a short half-life radionuclide generator for use in nuclear medicine diagnostic imaging procedures.
1961	**Audio-animatronics (Walt Disney):** Disney established WED, a research and development unit that developed the inventions he needed for his various enterprises. WED produced the audio-animatronic robotic figures that populated Disneyland, the 1964-1965 New York World's Fair, films, and other attractions. Audio-animatronics enabled robotic characters to speak or sing as well as move.
1961	**Ruby laser:** The ruby laser is first used medically by Charles Campbell and Charles Koester to excise a patient's retinal tumor.
1961	**First person in space:** Soviet astronaut Yuri Gagarin becomes the first person in space when he orbits the Earth on April 12.
1962	**Soft contact lenses (Otto Wichterle):** Wichterle's soft contacts can be worn longer with less discomfort than can hard contact lenses.
1962	**Continuously operating ruby laser (Willard S. Boyle and Don Nelson):** The invention relies on an arc lamp shining continuously (rather than the flash lamp used by Theodore Maiman in 1960).
1962	**Light-emitting diode (Nick Holonyak, Jr.):** Holonyak makes the first visible-spectrum diode laser, which produces red laser light but also stops lasing yet remains a useful light source. Holonyak has invented the red light-emitting diode (LED), the first operating alloy device—the "ultimate lamp."
1962	**Telstar satellite (John R. Pierce):** The first satellite to rebroadcast signals goes into operation, revolutionizing telecommunications.
1962	**Quasar 3C 273 (Maarten Schmidt):** Schmidt shows that this quasar is very distant and hence very bright. Further research shows quasars to be young galaxies with active, supermassive black holes at their centers.
1962	**First audiocassette:** The Philips company of the Netherlands releases the audiocassette tape.
1962	**Artificial hip (Sir John Charnley):** British surgeon Charnley invents the low-friction artificial hip and develops the surgical techniques for emplacing it.
1963	**Learjet (Bill Lear):** The Learjet, a small eight-passenger jet with a top speed of 560 miles (900 kilometers) per hour, can shuttle VIPs to meetings and other engagements.
1963	**Self-cleaning oven:** General Electric introduces the self-cleaning electric oven.
1963	**Artificial heart (Paul Winchell):** Winchell receives a patent (later donated to the University of Utah's Institute for Biomedical Engineering) for an artificial heart that purportedly became the model for the successful Jarvick-7.
1963	**6600 computer (Seymour Cray):** The 6600 was the first of a long line of Cray supercomputers.
1963	**Carbon fiber (Leslie Philips):** British engineer Philips develops carbon fiber, which is much stronger than steel.

1964	**Three-dimensional holography (Emmett Leith):** Leith and Juris Upatnieks present the first three-dimensional hologram at the Optical Society of America conference. The hologram must be viewed with a reference laser. The hologram of an object can then be viewed from different angles, as if the object were really present.
1964	**Moog synthesizer (Robert Moog):** The Moog synthesizer uses electronics to create and combine musical sounds.
1964	**Cosmic background radiation (Arno Penzias and Robert Wilson):** Penzias and Wilson detect the cosmic background radiation, which corresponds to that which would be radiated by a body at 2.725 kelvins. It is thought to be greatly redshifted primordial fireball radiation left over from the big bang.
1964	**BASIC programming language (John Kemeny and Thomas Kurtz):** Kemeny and Kurtz develop the BASIC computer programming language. BASIC is an acronym for Beginner's All-purpose Symbolic Instruction Code.
1965	**Minicomputer (Ken Olsen):** Perhaps the first true minicomputer, the PDP-8 is released by Digital Equipment Corporation. Founder Olsen makes computers affordable for small businesses.
1965	**Aspartame (James M. Schlatter):** Schlatter discovers aspartame, an artificial sweetener, while trying to come up with an antiulcer medication.
1965	**First space walk:** Soviet astronaut Aleksei Leonov is the first person to walk in space.
1966	**Gamma-electric cell (Henry Thomas Sampson):** Sampson works with George H. Miley to produce the gamma-electric cell, which converts the energy of gamma rays into electrical energy.
1966	**Handheld calculator (Jack St. Clair Kilby):** While working for Texas Instruments, Kilby does for the adding machine what the transistor had done for the radio, inventing a handheld calculator that retails at $150 and becomes an instant commercial success.
1966	**First unmanned moon landing:** Soviet spacecraft Luna 9 lands on the moon.
1967	**Electrogasdynamic method and apparatus (Meredith C. Gourdine):** Gourdine develops electrogasdynamics, which involves the production of electricity from the conversion of kinetic energy in a moving, ionized gas.
1967	**Pulsars (Jocelyn Bell and Antony Hewish):** Pulsars, rapidly rotating neutron stars, are discovered.
1968	**Practical liquid crystal displays (James Fergason):** Fergason develops an liquid crystal display (LCD) screen that has good visual contrast, is durable, and uses little electricity.
1968	**Lasers in medicine:** Francis L'Esperance begins using the argon-ion laser to treat patients with diabetic retinopathy.
1968	**Computer mouse (Douglas Engelbart):** Engelbart presents the computer mouse, which he had been working on since 1964.
1968	**Apollo 7:** Astronauts on Apollo 7, the first piloted Apollo mission, take photographs and transmit them to the American public on television.

1968	**Interface message processors:** Bolt Beranek and Newman Incorporated win a Defense Advanced Research Projects Agency (DARPA) contract to develop the packet switches called interface message processors (IMPs).
1969	**Rubella vaccine:** The rubella vaccine is available.
1969	**First person walks on the moon:** Neil Armstrong, a member of the U.S. Apollo 11 spacecraft, is the first person to walk on the moon.
1969	**Boeing 747:** The Boeing 747 makes its first flight, piloted by Jack Waddell.
1969	**Concorde:** The Concorde makes its first flight, piloted by André Turcat.
1969	**Charge-coupled device (Willard S. Boyle and George E. Smith):** Boyle and Smith develop the charge-coupled device, the basis for digital imaging.
1969	**ARPANET launches:** The Advanced Research Projects Agency starts ARPANET, which is the precursor to the Internet. UCLA and Stanford University are the first institutions to become networked.
1970's	**Digital seismology:** Digital seismology is used in oil exploration and increases accuracy in finding underground pools.
1970's	**Mud pulse telemetry:** Mud pulse telemetry becomes an oil-industry standard; pressure pulses are relayed through drilling mud to convey the location of the drill bit.
1970	**Optical fiber (Robert Maurer and others):** Maurer, joined by Donald Keck, Peter Schultz, and Frank Zimar, produces an optical fiber that can be used for communication.
1970	**Compact disc (James Russell):** The compact disc (CD) revolutionizes the way digital media is stored.
1970	**UNIX (Dennis Ritchie and Kenneth Thompson):** Bell Laboratories employees Ritchie and Thompson complete the UNIX operating system, which becomes popular among scientists.
1970	**Network Control Protocol:** The Network Working Group deploys the initial ARPANET host-to-host protocol, called the Network Control Protocol (NCP), establishing connections, break connections, switch connections, and control flow over the ARPANET.
1971	**Computerized axial tomography (Godfrey Newbold Hounsfield):** In London, doctors performed the first CAT scan of a living patient and detected a brain tumor. In a CAT (or CT) scan, X rays are taken of a body like slices in a loaf of bread. A computer then assembles these slices into a detail-laden three-dimensional image.
1971	**First videocassette recorder:** Sony begins selling the first videocassette recorder (VCR) to the public.
1971	**Microprocessor (Ted Hoff):** The computer's central processing unit (CPU) is reduced to the size of a postage stamp.
1971	**Electronic switching system for telecommunications (Erna Schneider Hoover):** Hoover's system prioritizes telephone calls and fixes an efficient order to answer them.
1971	**Intel microprocessors:** Intel builds the world's first microprocessor chip.

1971	**Touch screen (Sam Hurst):** Hurst's touch screen can detect if it has been touched and where it was touched.
1972	**First recombinant DNA organism (Stanley Norman Cohen, Paul Berg, and Herbert Boyer):** The methods to combine and transplant genes are discovered when this team successfully clones and expresses the human insulin gene in the Escherichia coli.
1972	**Far-Ultraviolet Camera (George R. Carruthers):** The Carruthers-designed camera is used on the Apollo 16 mission.
1972	**Cell encapsulation (Taylor Gunjin Wang):** Wang develops ways to encapsulate beneficial cells and introduce them into a body without triggering the immune system.
1972	**Pioneer 10:** The U.S. probe Pioneer 10 is launched to get information about the outer solar system.
1972	**Networking goes public:** ARPANET system designer Robert Kahn organizes the first public demonstration of the new network technology at the International Conference on Computer Communications in Washington, D.C.
1972	**Pong video game (Nolan K. Bushnell and Ted Dabney):** Bushnell and Dabney register the name of their new computer company, Atari, and issue Pong shortly thereafter, marking the rise of the video game industry.
1973	**Automatic computerized transverse axial (ACTA) whole-body CT scanner (Robert Steven Ledley):** The first whole-body CT scanner is operational. Ledley goes on to spend much of his career promoting the use of electronics and computers in biomedical research.
1973	**Packet network interconnection protocols TCP/IP (Vinton Gray Cerf and Robert Kahn):** Cerf and Kahn develop transmission control protocol/Internet protocol (TCP/IP), protocols that enable computers to communicate with one another.
1973	**Automated teller machine (Don Wetzel):** Wetzel receives a patent for his ATM. To make it a success, he shows banks how to generate a group of clients who would use the ATM.
1973	**Food processor:** The Cuisinart food processor is introduced in the United States.
1973	**Air bags in automobiles:** The Oldsmobile Tornado is the first American car sold equipped with air bags.
1973	**Space photography:** Astronauts aboard Skylab, the first U.S. space station, take high-resolution photographs of Earth using photographic remote-sensing systems. The astronauts also take photographs with handheld cameras.
1974	**Kevlar (Stephanie Kwolek):** Kwolek receives a patent for the fiber Kevlar. Bullet-resistant Kevlar vests go on sale only one year later.
1975	**Ethernet (Robert Metcalfe and David Boggs):** Metcalfe and Boggs invent the Ethernet, a system of software, protocols, and hardware allowing instantaneous communication between computer terminals in a local area.

1975	**Semiconductor laser:** Scientists working at Diode Labs develop the first commercial semiconductor laser that will operate continuously at room temperature.
1976	**First laser printer:** IBM's 3800 Printing System is the first laser printer. The ink jet is invented in the same year, but it is not prevalent in homes until 1988.
1976	**Apple computer (Steve Jobs):** Jobs cofounds Apple Computer with Steve Wozniak.
1976	**Jarvik-7 artificial heart (Robert Jarvik):** The Jarvik-7 allows a calf to live 268 days with the artificial heart. Jarvik combined ideas from several other workers to produce the Jarvik-7.
1976	**Apple II (Steve Wozniak):** Wozniak develops the Apple II, the best-selling personal computer of the 1970's and early 1980's.
1976	**First Mars probes:** The National Aeronautics and Space Administration (NASA) launches Viking 1 and Viking 2, which land on obtain images of Mars.
1976	**Kurzweil Reading Machine (Ray Kurzweil):** Kurzweil develops an optical character reader (OCR) able to read most fonts.
1976	**Microsoft Corporation (Bill Gates):** Gates, along with Paul Allen, found Microsoft, a software company. Gates will remain head of Microsoft for twenty-five years.
1976	**Conductive polymers:** Hideki Shirakawa, Alan G. MacDiarmid, and Alan J. Heeger discover conductive polymers. They are awarded the 2000 Nobel Prize in Chemistry.
1977	**Global Positioning System (GPS) (Ivan A. Getting):** The first GPS satellite is launched, designed to support a navigational system that uses satellites to pinpoint the location of a radio receiver on Earth's surface.
1977	**Fiber-optic telephone cable:** The first fiber-optic telephone cables are tested.
1977	**Echo-planar imaging (Peter Mansfield):** British physicist Mansfield first develops the echo-planar imaging (EPI).
1977	**Gossamer Condor (Paul MacCready):** MacCready designs the Gossamer Albatross, which enables human-powered flight.
1978	**Smart gels (Toyoichi Tanaka):** Tanaka discovers and works with "smart gels," polymer gels that can expand a thousandfold, change color, or contract when stimulated by minor changes in temperature, magnetism, light, or electricity. This capacity makes them useful in a broad range of applications.
1978	**Charon discovered (James Christy):** Charon is discovered as an apparent bulge on a fuzzy picture of Pluto. Its mass is about 12 percent that of Pluto.
1978	**First cochlear implant surgery:** Graeme Clark performs the first cochlear implant surgery in Australia.
1978	**The first test-tube baby:** Louise Brown is born in England.
1978	**First MRI:** The first magnetic resonance image (MRI) of the human head is taken in England.
1979	**First laptop (William Moggridge):** Moggridge, of Grid Systems in England, designs the first laptop computer.

1979	**First commercially successful application:** The VisiCalc spreadsheet for Apple II, designed by Daniel Bricklin and Bob Frankston, helps drive sales of the personal computer and becomes its first successful business application.
1979	**USENET (Tom Truscott, Jim Ellis and Steve Belovin):** Truscott, Ellis, and Belovin create USENET, a "poor man's ARPANET," to share information via e-mail and message boards between Duke University and the University of North Carolina, using dial-up telephone lines.
1979	**In-line roller skates (Scott Olson and Brennan Olson):** After finding some antique in-line skates, the Olson brothers begin experimenting with modern materials, creating Rollerblades.
1980's	**Controlled drug delivery (Robert S. Langer):** Langer develops the foundation of controlled drug delivery technology used in cancer treatment.
1980	**Alkaline battery (Lewis Urry):** Eveready markets alkaline batteries under the trade name Energizer. Urry's alkaline battery lasts longer than its predecessor, the carbon-zinc battery.
1980	**Interferon (Charles Weissmann):** Weissmann produces the first genetically engineered human interferon, which is used in cancer treatment.
1980	**TCP/IP:** The U.S. Department of Defense adopts the TCP/IP suite as a standard.
1981	**Ablative photodecomposition (Rangaswamy Srinivasan):** Srinivasan's research on ablative photodecomposition leads to multiple applications, including laser-assisted in situ keratomileusis (LASIK) surgery, which shapes the cornea to correct vision problems.
1981	**Scanning tunneling microscope (Heinrich Rohrer and Gerd Binnig):** The scanning tunneling microscope shows surfaces at the atomic level.
1981	**Improvements in laser spectroscopy (Arthur L. Schawlow and Nicolaas Bloembergen):** Schawlow shares the Nobel Prize in Physics with Nicolaas Bloembergen for their work on laser spectroscopy. While most of Schawlow's inventions involved lasers, he also did research in superconductivity and nuclear resonance.
1981	**First IBM personal computer:** The first IBM PC, the IBM 5100, goes on the market with a $1,565 price tag.
1982	**Compact discs appear:** Compact discs are now sold and will start replacing vinyl records.
1982	**First artificial heart:** Seattle dentist Barney Clark receives the first permanent artificial heart, and he survives for 112 days.
1983	**Cell phone (Martin Cooper):** The first mobile (wireless) phone, the DynaTAC 8000X, receives approval by the Federal Communications Commission (FCC), heralding an age of wireless communication.
1983	**Internet:** ARPANET, and networks attached to it, adopt the TCP/IP networking protocol. All networks that use the protocol are known as the Internet.
1983	**Cyclosporine:** Immunosuppressant cyclosporine is approved for use in transplant operations in the United States.

1983	**Polymerase chain reaction (Kary B. Mullis):** While driving to his cottage in Mendocino, California, Mullis develops the idea for the polymerase chain reaction (PCR). PCR will be used to amplify a DNA segment many times, leading to a revolution in recombinant DNA technology and a 1993 Nobel Prize in Chemistry for Mullis.
1984	**Domain name service is created:** Paul Mockapetris and Craig Partridge develop domain name service, which links unique Internet protocol (IP) numerical addresses to names with suffixes such as .mil, .com, .org, and .edu.
1984	**Mac is released:** Apple introduces the Macintosh, a low-cost, plug-and-play personal computer with a user-friendly graphic interface.
1984	**CD-ROM:** Philips and Sony introduce the CD-ROM (compact disc read-only memory), which has the capacity to store data of more than 450 floppy disks.
1984	**Surgery in utero:** William A. Clewall performs the first successful surgery on a fetus.
1984	**Cloning:** Danish veterinarian Steen M. Willadsen clones a lamb from a developing sheep embryo cell.
1984	**AIDS blood test (Robert Charles Gallo):** Gallo and his colleagues identify the virus HTLV-3/LAV (later renamed human immunodeficiency virus, or HIV) as the cause of acquired immunodeficiency syndrome, or AIDS. Gallo creates a blood test that can identify antibodies specific to HIV. This blood test is essential to keeping the supply in blood banks pure.
1984	**Imaging X-ray spectrometer (George Edward Alcorn):** Alcorn patents his device, which makes images of the source using X rays of specific energies, similar to making images with a specific wavelength (color) of light. It is used in acquiring data on the composition of distant planets and stars.
1984	**DNA profiling (Alec Jeffreys):** Noticing similarities and differences in DNA samples from his lab technician's family, Jeffreys discovers the principles that lead to DNA profiling, which has become an essential tool in forensics and the prosecution of criminal cases.
1985	**Windows operating system (Bill Gates):** The first version of Windows is released.
1985	**Implantable cardioverter defibrillator:** The U.S. Food and Drug Administration (FDA) approves Polish physician Michel Mirowski's implantable cardioverter defibrillator (ICD), which monitors and corrects abnormal heart rhythms.
1985	**Industry Standard Architecture (ISA) bus (Mark Dean and Dennis Moeller):** Dean and Moeller design the standard way of organizing the central part of a computer and its peripherals, the ISA bus, which is patented in this year.
1985	**Atomic force microscope:** Calvin Quate, Christoph Gerber, and Gerd Binnig invent the atomic force microscope, which becomes one of the foremost tools for imaging, measuring, and manipulating matter at the nano scale.
1986	**Mir:** The Soviet Union launches the Mir space station, the first permanent space station.
1986	**Burt Rutan's Voyager:** Dick Rutan (Burt's brother) and Jeana Yeager make the first around-the-world, nonstop flight without refueling in the Burt Rutan-designed Voyager. The Voyager is the first aircraft to accomplish this feat.

1986	**High-temperature superconductor (J. Georg Bednorz and Karl Alexander Müller):** Bednorz and Müller show that a ceramic compound of lanthanum, barium, copper, and oxygen becomes superconducting at 35 kelvins, a new high- temperature record.
1987	**Azidothymidine:** The FDA approves azidothymidine (AZT), a potent antiviral, for AIDS patients.
1987	**Echo-planar imaging:** Echo-planar imaging is used to perform real-time movie imaging of a single cardiac cycle.
1987	**Parkinson's treatment:** French neurosurgeon Alim-Louis Benabid implants a deep-brain electrical-stimulation system into a patient with advanced Parkinson's disease.
1987	**First corneal laser surgery:** New York ophthalmologist Steven Trokel performs the first laser surgery on a human cornea. He had refined his technique on a cow's eye. Trokel was granted a patent for the Excimer laser to be used for vision correction.
1987	**UUNET and PSINet:** Rick Adams forms UUNET and Bill Schrader forms PSINet to provide commercial Internet access.
1988	**Transatlantic fiber-optic cable:** The first transatlantic fiber-optic cable is installed, linking North America and France.
1988	**Laserphaco probe (Patricia Bath):** Bath's probe is used to break up and remove cataracts.
1989	**Method for tracking oil flow underground using a supercomputer (Philip Emeagwali):** Emeagwali receives the Gordon Bell Prize, considered the Nobel Prize for computing, for his method, which demonstrates the possibilities of computer networking.
1989	**World Wide Web (Tim Berners-Lee and Robert Cailau):** Berners-Lee finds a way to join the idea of hypertext and the young Internet, leading to the Web, coinvented with Cailau.
1989	**First dial-up access:** The World debuts as the first provider of dial-up Internet access for consumers.
1990's	**Environmentally friendly appliances:** Water-saving and energy-conserving washing machines and dryers are introduced.
1990	**Hubble Space Telescope:** The Hubble Space Telescope is launched and changes the way scientists look at the universe.
1990	**Human Genome Project begins:** The U.S. Department of Energy and the National Institutes of Health coordinate the Human Genome Project with the goal of identifying all 30,000 genes in human DNA and determining the sequences of the three billion chemical base pairs that make up human DNA.
1990	**BRCA1 gene discovered (Mary-Claire King):** King finds the cancer-associated gene on chromosome 17. She demonstrates that humans and chimpanzees are 99 percent genetically identical.

1991	**Nakao Snare (Naomi L. Nakao):** The Snare is a device that captures polyps that have been cut from the walls of the intestine, solving the problem of "lost polyp syndrome."
1991	**America Online (AOL):** Quantum Computer Services changes its name to America Online; Steve Case is named president. AOL offers e-mail, electronic bulletin boards, news, and other information.
1991	**Carbon nanotubes (Sumio Iijima):** Although carbon nanotubes have been seen before, Iijima's 1991 paper establishes some basic properties and prompts other scientists' interest in studying them.
1991	**The first hot-air balloon crosses the Pacific (Richard Branson and Per Lindstrad):** Branson and Lindstrad, who teamed up in 1987 to cross the Atlantic, make the 6,700-mile flight in 47 hours and break the world distance record.
1992	**Newton:** Apple introduces Newton, one of the first handheld computers, or personal digital assistants, which has a liquid crystal display operated with a stylus.
1993	**Mosaic (Marc Andreessen):** Andreessen launches Mosaic, followed by Netscape Navigator in 1995—the first Internet browsers. Both Mosaic and Netscape allow novices to browse the World Wide Web.
1993	**Flexible tailored elastic airfoil section (Sheila Widnall):** Widnall applies for a patent for this device, which addresses the problem of being able to measure fluctuations in pressure under unsteady conditions. She serves as secretary of the Air Force (the first woman to lead a branch of the military) and also serves on the board investigating the space shuttle Columbia accident of 2003.
1993	**Light-emitting diode (LED) blue and UV (Shuji Nakamura):** Nakamura's blue LED makes white LED light possible (a combination of red, blue, and green).
1994	**Genetically modified (GM) food:** The Flavr Savr tomato, the first GM food, is approved by the FDA.
1994	**Channel Tunnel:** Channel Tunnel, or Chunnel, opens, connecting France and Britain by a railway constructed beneath the English Channel.
1995	**51 Pegasi (Michel Mayor and Didier Queloz):** Mayor and Queloz detect a planet orbiting another normal star, the first extrasolar planet (exoplanet) to be found. As of June, 2009, 353 exoplanets were known.
1995	**Saquinavir:** The FDA approves Saquinavir for the treatment of AIDS. It is the first protease inhibitor, which reduces the ability of the AIDS virus to spread to new cells.
1995	**iBot (Dean Kamen):** Kamen invents iBOT, a super wheelchair that climbs stairs and helps its passenger to stand.
1995	**Global Positioning System (Ivan A. Getting):** The GPS becomes fully operational.
1995	**Illusion transmitter (Valerie L. Thomas):** A concave mirror can produce a real image that appears to be three-dimensional. Thomas's system uses a concave mirror at the camera and another one at the television receiver.
1996	**LASIK:** The first computerized excimer laser (LASIK), designed to correct the refractive error myopia, is approved for use in the United States.

1996	**First sheep is cloned:** Scottish scientist Ian Wilmut clones the first mammal, a Finn Dorset ewe named Dolly, from differentiated adult mammary cells.
1997	**Robotic vacuum:** Swedish appliance company Electrolux is the first to create a prototype of a robotic vacuum cleaner.
1998	**PageRank (Larry Page):** The cofounder of Google with Sergey Brin, Page devises PageRank, the count of Web pages linked to a given page and a measure how valuable people find that page.
1998	**UV Waterworks (Ashok Gadgil):** The device uses UV from a mercury lamp to kill waterborne pathogens.
1998	**Napster:** College dropout Shawn Fanning creates Napster, an extremely popular peer-to-peer file-sharing platform that allowed users to download music for free. In 2001 the free site was shut down because it encouraged illegal sharing of copyrighted properties. The site then became available by paid subscription.
1999	**Palm VII:** The Palm VII organizer is on the market. It is a handheld computer with 2 megabytes of RAM and a port for a wireless phone.
1999	**BlackBerry (Research in Motion of Canada):** A wireless handheld device that began as a two-way pager, the BlackBerry is also a cell phone that supports Web browsing, e-mail, text messaging, and faxing—it is the first smart phone.
2000	**Hoover-Diana production platform:** A joint venture by Exxon and British Petroleum (BP), the Hoover-Diana production platform goes into operation in the Gulf of Mexico. Within six months it is producing 20,000 barrels of oil a day.
2000	**Clone of a clone:** Japanese scientists clone a bull from a cloned bull.
2000	**Minerva:** The Library of Congress initiates a prototype system called Minerva (Mapping the Internet Electronic Resources Virtual Archives) to collect and preserve open-access Web resources.
2000	**Supercomputer:** The ASCI White supercomputer at the Lawrence Livermore National Laboratory in California is operational. It can hold six times the information stored in the 29 million books in the Library of Congress.
2001	**XM Radio:** XM Radio initiates the first U.S. digital satellite radio service in Dallas-Ft. Worth and San Diego.
2001	**Human cloning:** Scientists at Advanced Cell Technology in Massachusetts clone human embryos for the first time.
2001	**iPod (Tony Fadell):** Fadell introduces the iPod, a portable hard drive-based MP3 player with an Internet-based electronic music catalog, for Apple.
2001	**Segway PT (Dean Kamen):** Kamen introduces his personal transport device, a self-balancing, electric-powered pedestrian scooter.
2003	**First digital books:** Lofti Belkhir introduces the Kirtas BookScan 1200, the first automatic, page-turning scanner for the conversion of bound volumes to digital files.
2003	**Aqwon (Josef Zeitler):** The hydrogen-powered scooter Aqwon can reach 30 miles (50 kilometers) per hour. Its combustion product is water.

2003	**Human Genome Project is completed:** After thirteen years, the 25,000 genes of the human genome are identified and the sequences of the 3 million chemical base pairs that make up human DNA are determined.
2004	**Stem cell bank:** The world's first embryonic stem cell bank opens in England.
2004	**SpaceShipOne and SpaceShipTwo (Burt Rutan):** Rutan receives the U.S. Department of Transportation's first license issued for suborbital flight for SpaceShipOne, which shortly thereafter reaches an altitude of 328,491 feet. Rutan's rockets are the first privately funded manned rockets to reach space (higher than 100 kilometers above Earth's surface).
2004	**Columbia supercomputer:** The NASA supercomputer Columbia, built by Silicon Graphics and Intel, achieves sustained performance of 42.7 trillion calculations per second and is named the fastest supercomputer in the world. It is named for those who lost their lives in the explosion of the space shuttle Columbia in 2003. Because technology evolves so quickly, the Columbia will not be the fastest for very long.
2005	**Blue Gene/L supercomputer:** The National Nuclear Security Administration's BlueGene/L supercomputer, built by IBM, performs at 280.6 trillion operations per second and is now the world's fastest supercomputer.
2005	**Eris (Mike Brown):** Working with C. A. Trujillo and D. L. Rabinowitz, Brown discovers Eris, the largest known dwarf planet and a Kuiper Belt object. It is 27 percent more massive than Pluto, another large Kuiper Belt object.
2005	**Nix and Hydra discovered (Pluto companion team):** The Hubble research team—composed of Hal Weaver, S. Alan Stern, Max Mutchler, Andrew Steffl, Marc Buie, William Merline, John Spencer, Eliot Young, and Leslie Young—finds these small moons of Pluto.
2006	**Digital versus film:** Digital cameras have almost wholly replaced film cameras. *The New York Times* reports that 92 percent of cameras sold are digital.
2007	**First terabyte drive:** Hitachi Global Storage Technologies announces that it has created the first one-terabyte (TB) hard disk drive.
2007	**iPhone (Apple):** Apple introduces its smart phone, a combined cell phone, portable media player (equal to a video iPod), camera phone, Internet client (supporting e-mail and Web browsing), and text messaging device, to an enthusiastic market.
2008	**Roadrunner:** The Roadrunner supercomputer, built by IBM and Los Alamos National Laboratory, can process more than 1.026 quadrillion calculations per second. It works more than twice as fast as the Blue Gene/L supercomputer and is housed at Los Alamos in New Mexico.
2008	**Mammoth Genome Project:** Scientists sequence woolly mammoth genome, the first of an extinct animal.
2008	**Columbus lands:** The space shuttle Atlantis delivers the Columbus science laboratory to the International Space Station. The twenty-three-foot long laboratory is able to conduct experiments both inside and outside the space station.

2008	**Retail DNA test (Anne Wojcicki):** Wojcicki (wife of Google founder Sergey Brin) offers an affordable DNA saliva test, 23andMe, to determine one's genetic markers for ninety traits. The product heralds what *Time* magazine dubs a "personal-genomics revolution."
2009	**Large Hadron Collider:** The Large Hadron Collider (LHC) becomes the world's highest energy particle accelerator.
2009	**Hubble Space Telescope repairs (NASA):** STS-125 astronauts conducted five space walks from the space shuttle Atlantis to upgrade the Hubble Space Telescope, extending its life to at least 2014.
2009	**AIDS vaccine:** Scientists in Thailand create a vaccine that seems to reduce the risk of contracting the AIDS virus by more than 31 percent.
2010	**Jaguar supercomputer:** The Oak Ridge National Laboratory in Tennessee is home to Jaguar, the world's fastest supercomputer, the peak speed of which is 2.33 quadrillion floating point operations per second.

Charles W. Rogers, Southwestern Oklahoma State University, Department of Physics; updated by the editors of Salem Press

GENERAL BIBLIOGRAPHY

Aaboe, Asger. *Episodes from the Early History of Astronomy.* New York: Springer-Verlag, 2001.

Abbate, Janet. *Inventing the Internet.* Cambridge, Mass.: MIT Press, 2000.

Abell, George O., David Morrison, and Sidney C. Wolff. *Exploration of the Universe.* 5th ed. Philadelphia: Saunders College Publishing, 1987.

Achilladelis, Basil, and Mary Ellen Bowden. *Structures of Life.* Philadelphia: The Center, 1989.

Ackerknecht, Erwin H. *A Short History of Medicine.* Rev. ed. Baltimore: The Johns Hopkins University Press, 1982.

Aczel, Amir D. *Fermat's Last Theorem: Unlocking the Secret of an Ancient Mathematical Problem.* Reprint. New York: Four Walls Eight Windows, 1996.

Adler, Robert E. *Science Firsts: From the Creation of Science to the Science of Creation.* Hoboken, N.J.: John Wiley & Sons, 2002.

Alberts, Bruce, et al. *Molecular Biology of the Cell.* 2d ed. New York: Garland, 1989.

Alcamo, I. Edward. *AIDS: The Biological Basis.* 3d ed. Boston: Jones and Bartlett, 2003.

Aldersey-Willliams, Hugh. *The Most Beautiful Molecule: An Adventure in Chemistry.* London: Aurum Press, 1995.

Alexander, Arthur F. O'Donel. *The Planet Saturn: A History of Observation, Theory, and Discovery.* 1962. Reprint. New York: Dover, 1980.

Alioto, Anthony M. *A History of Western Science.* 2d ed. Upper Saddle River, N.J.: Prentice Hall, 1993.

Allen, Oliver E., and the editors of Time-Life Books. *Atmosphere.* Alexandria, Va.: Time-Life Books, 1983.

Ames, W. F., and C. Rogers, eds. *Nonlinear Equations in the Applied Sciences.* San Diego: Academic Press, 1992.

Andriesse, Cornelis D. *Christian Huygens.* Paris: Albin Michel, 2000.

Angier, Natalie. *Natural Obsessions: Striving to Unlock the Deepest Secrets of the Cancer Cell.* Boston: Mariner Books/Houghton Mifflin, 1999.

Annaratone, Donnatello. *Transient Heat Transfer.* New York: Springer, 2011.

Anstey, Peter R. *The Philosophy of Robert Boyle.* London: Routledge, 2000.

Anton, Sebastian. *A Dictionary of the History of Science.* Pearl River, N.Y.: Parthenon Publishing, 2001.

Archimedes. *The Works of Archimedes.* Translated by Sir Thomas Heath. 1897. Reprint. New York: Dover, 2002.

Arms, Karen, and Pamela S. Camp. *Biology: A Journey into Life.* 3d ed. Philadelphia: Saunders College Publishing, 1987.

Armstrong, Neil, Michael Collins, and Edwin E. Aldrin. *First on the Moon.* New York: Williams Konecky Associates, 2002.

Arrizabalaga, Jon, John Henderson, and Roger French. *The Great Pox: The French Disease in Renaissance Europe.* New Haven, Conn.: Yale University Press, 1997.

Arsuaga, Juan Luis. *The Neanderthal's Necklace: In Search of the First Thinkers.* Translated by Andy Klatt. New York: Four Walls Eight Windows, 2002.

Artmann, Benno. *Euclid: The Creation of Mathematics.* New York: Springer- Verlag, 1999.

Asimov, Isaac. *Exploring the Earth and the Cosmos.* New York: Crown, 1982.

_______. *The History of Physics.* New York: Walker, 1984.

_______. *Jupiter, the Largest Planet.* New York: Ace, 1980.

Aspray, William. *John von Neumann and the Origins of Modern Computing.* Boston: MIT Press, 1990.

Astronomical Society of the Pacific. *The Discovery of Pulsars.* San Francisco: Author, 1989.

Audesirk, Gerald J., and Teresa E. Audesirk. *Biology: Life on Earth.* 2d ed. New York: Macmillan, 1989.

Aughton, Peter. *Newton's Apple: Isaac Newton and the English Scientific Revolution.* London: Weidenfeld & Nicolson, 2003.

Aujoulat, Norbert. *Lascaux: Movement, Space, and Time.* New York: Harry N. Abrams, 2005.

Aveni, Anthony F., ed. *Skywatchers.* Rev. ed. Austin: University of Texas Press, 2001.

Baggott, Jim. *Perfect Symmetry: The Accidental Discovery of Buckminsterfullerene.* New York: Oxford University Press, 1994.

Baine, Celeste. *Is There an Engineer Inside You? A Comprehensive Guide to Career Decisions in Engineering.*

2d ed. Belmont, Calif.: Professional Publications, 2004.

Baker, John. *The Cell Theory: A Restatement, History and Critique.* New York: Garland, 1988.

Baldwin, Joyce. *To Heal the Heart of a Child: Helen Taussig, M.D.* New York: Walker, 1992.

Barbieri, Cesare, et al., eds. *The Three Galileos: The Man, the Spacecraft, the Telescope: Proceedings of the Conference Held in Padova, Italy on January 7-10, 1997.* Boston: Kluwer Academic, 1997.

Barkan, Diana Kormos. *Walther Nernst and the Transition to Modern Physical Science.* New York: Cambridge University Press, 1999.

Barrett, Peter. *Science and Theology Since Copernicus: The Search for Understanding.* Reprint. Dorset, England: T&T Clark, 2003.

Bartusiak, Marcia. *Thursday's Universe.* New York: Times Books, 1986.

Basta, Nicholas. *Opportunities in Engineering Careers.* New York: McGraw-Hill, 2003.

Bates, Charles C., and John F. Fuller. *America's Weather Warriors, 1814-1985.* College Station: Texas A&M Press, 1986.

Bazin, Hervé. *The Eradication of Smallpox: Edward Jenner and the First and Only Eradication of a Human Infectious Disease.* Translated by Andrew Morgan and Glenise Morgan. San Diego: Academic Press, 2000.

Beatty, J. Kelly, and Andrew Chaikin, eds. *The New Solar System.* 3d rev. ed. New York: Cambridge University Press, 1990.

Becker, Wayne, Lewis Kleinsmith, and Jeff Hardin. *The World of the Cell.* New York: Pearson/Benjamin Cummings, 2006.

Berlinski, David. *A Tour of the Calculus.* New York: Vintage Books, 1997.

Bernstein, Jeremy. *Three Degrees Above Zero: Bell Labs in the Information Age.* New York: Charles Scribner's Sons, 1984.

Bernstein, Peter L. *Against the Gods: The Remarkable Story of Risk.* New York: John Wiley & Sons, 1996.

Bertolotti, M. *Masers and Lasers: An Historical Approach.* Bristol, England: Adam Hilger, 1983.

Bickel, Lennard. *Florey: The Man Who Made Penicillin.* Carlton South, Victoria, Australia: Melbourne University Press, 1995.

Bizony, Piers. *Island in the Sky: Building the International Space Station.* London: Aurum Press Limited, 1996.

Blackwell, Richard J. *Galileo, Bellarmine, and the Bible.* London: University of Notre Dame Press, 1991.

Bliss, Michael. *The Discovery of Insulin.* Chicago: University of Chicago Press, 1987.

Blumenberg, Hans. *The Genesis of the Copernican World.* Translated by Robert M. Wallace. Cambridge, Mass.: MIT Press, 1987.

Blunt, Wilfrid. *Linnaeus: The Compleat Naturalist.* Princeton, N.J.: Princeton University Press, 2001.

Bodanis, David. *Electric Universe: The Shocking True Story of Electricity.* New York: Crown Publishers, 2005.

Bohm, David. *Causality and Chance in Modern Physics.* London: Routledge & Kegan Paul, 1984.

Bohren, Craig F. *Clouds in a Glass of Beer: Simple Experiments in Atmospheric Physics.* New York: John Wiley & Sons.

Boljanovic, Vukota. *Applied Mathematics and Physical Formulas: A Pocket Reference Guide for Students, Mechanical Engineers, Electrical Engineers, Manufacturing Engineers, Maintenance Technicians, Toolmakers, and Machinists.* New York: Industrial Press, 2007.

Bolt, Bruce A. *Inside the Earth: Evidence from Earthquakes.* New York: W. H. Freeman, 1982.

Bond, Peter. *The Continuing Story of the International Space Station.* Chichester, England: Springer-Praxis, 2002.

Boorstin, Daniel J. *The Discoverers.* New York: Random House, 1983.

Bottazzini, Umberto. *The Higher Calculus: A History of Real and Complex Analysis from Euler to Weierstrass.* New York: Springer-Verlag, 1986.

Bourbaki, Nicolas. *Elements of the History of Mathematics.* Translated by John Meldrum. New York: Springer, 1994.

Bowler, Peter J. *Charles Darwin: The Man and His Influence.* Cambridge, England: Cambridge University Press, 1996.

_______. *Evolution: The History of an Idea.* Rev. ed. Berkeley: University of California Press, 1989.

_______. *The Mendelian Revolution: The Emergence of Hereditarian Concepts in Modern Science and Society.* Baltimore: The Johns Hopkins University Press, 1989.

Boyer, Carl B. *A History of Mathematics.* 2d ed., revised by Uta C. Merzbach. New York: John Wiley & Sons, 1991.

Bracewell, Ronald N. *The Fourier Transform and Its Applications.* 3d rev. ed. New York: McGraw-Hill, 1987.

Brachman, Arnold. *A Delicate Arrangement: The Strange Case of Charles Darwin and Alfred Russel Wallace.* New York: Times Books, 1980.

Bredeson, Carmen. *John Glenn Returns to Orbit: Life on the Space Shuttle.* Berkeley Heights, N.J.: Enslow, 2000.

Brock, Thomas, ed. *Milestones in Microbiology, 1546-1940.* Washington, D.C.: American Society for Microbiology, 1999.

Brock, William H. *The Chemical Tree: A History of Chemistry.* New York: W. W. Norton, 2000.

Brooks, Paul. *The House of Life: Rachel Carson at Work.* 2d ed. Boston: Houghton Mifflin, 1989.

Browne, Janet. *Charles Darwin: The Power of Place.* New York: Knopf, 2002.

Brush, Stephen G. *Cautious Revolutionaries: Maxwell, Planck, Hubble.* College Park, Md.: American Association of Physics Teachers, 2002.

Brush, Stephen G., and Nancy S. Hall. *Kinetic Theory of Gases: An Anthology of Classic Papers With Historical Commentary.* London: Imperial College Press, 2003.

Bryant, Stephen. *The Story of the Internet.* London: Pearson Education, 2000.

Buffon, Georges-Louis Leclerc. *Natural History: General and Particular.* Translated by William Smellie. Avon, England: Thoemmes Press, 2001.

Burger, Edward B., and Michael Starbird. *Coincidences, Chaos, and All That Math Jazz: Making Light of Weighty Ideas.* New York: W. W. Norton, 2005.

Burke, Terry, et al., eds. *DNA Fingerprinting: Approaches and Applications.* Boston: Birkhauser, 2001.

Byrne, Patrick H. *Analysis and Science in Aristotle.* Albany: State University of New York Press, 1997.

Calder, William M., III, and David A. Traill, eds. *Myth, Scandal, and History: The Heinrich Schliemann Controversy.* Detroit: Wayne State University Press, 1986.

Calinger, Ronald. *A Contextual History of Mathematics.* Upper Saddle River, N.J.: Prentice Hall, 1999.

Canning, Thomas N. *Galileo Probe Parachute Test Program: Wake Properties of the Galileo.* Washington, D.C.: National Aeronautics and Space Administration, Scientific and Technical Information Division, 1988.

Cantor, Geoffrey. *Michael Faraday: Sandemanian and Scientist: A Study of Science and Religion in the Nineteenth Century.* New York: St. Martin's Press, 1991.

Carlisle, Rodney. *Inventions and Discoveries: All the Milestones in Ingenuity—from the Discovery of Fire to the Invention of the Microwave.* Hoboken, N.J.: John Wiley & Sons, 2004.

Carlson, Elof Axel. *Mendel's Legacy: The Origin of Classical Genetics.* Woodbury, N.Y.: Cold Spring Harbor Laboratory Press, 2004.

Carola, Robert, John P. Harley, and Charles R. Noback. *Human Anatomy and Physiology.* New York: McGraw-Hill, 1990.

Carpenter, B. S., and R. W. Doran, eds. *A. M. Turing's ACE Report of 1946 and Other Papers.* Cambridge, Mass.: MIT Press, 1986.

Carpenter, Kenneth J. *The History of Scurvy and Vitamin C.* Cambridge England: Cambridge University Press, 1986.

Carrigan, Richard A., and W. Peter Trower, eds. *Particle Physics in the Cosmos.* New York: W. H. Freeman, 1989.

_______, eds. *Particles and Forces: At the Heart of the Matter.* New York: W. H. Freeman, 1990.

Cassanelli, Roberto, et al. *Houses and Monuments of Pompeii: The Works of Fausto and Felice Niccolini.* Los Angeles: J. Paul Getty Museum, 2002.

Caton, Jerald A. *A Review of Investigations Using the Second Law of Thermodynamics to Study Internal-Combustion Engines.* London: Society of Automotive Engineers, 2000.

Chaikin, Andrew. *A Man on the Moon: The Voyages of the Apollo Astronauts.* New York: Penguin Group, 1998.

Chaisson, Eric J. *The Hubble Wars.* New York: HarperCollins, 1994.

Chaisson, Eric J., and Steve McMillan. *Astronomy Today.* 5th ed. Upper Saddle River, N.J.: Pearson Prentice Hall, 2004.

Chandrasekhar, Subrahmanyan. *Eddington: The Most Distinguished Astrophysicist of His Time.* Cambridge, England: Cambridge University Press, 1983.

Chang, Hasok. *Inventing Temperature: Measurement and Scientific Progress.* Oxford, England: Oxford University Press, 2004.

Chang, Laura, ed. *Scientists at Work: Profiles of Today's Groundbreaking Scientists from "Science Times."* New York: McGraw-Hill, 2000.

Chant, Christopher. *Space Shuttle.* New York: Exeter Books, 1984.

Chapman, Allan. *Astronomical Instruments and Their Users: Tycho Brahe to William Lassell.* Brookfield, Vt.: Variorum, 1996.

Chase, Allan. *Magic Shots.* New York: William Morrow, 1982.

Check, William A. *AIDS.* New York: Chelsea House, 1988.

Cheng, K. S., and G. V. Romero. *Cosmic Gamma-Ray Sources.* New York: Springer-Verlag, 2004.

Christianson, John Robert. *On Tycho's Island: Tycho Brahe and His Assistants, 1570-1601.* New York: Cambridge University Press, 2000.

Chung, Deborah D. L. *Applied Materials Science: Applications of Engineering Materials in Structural, Electronics, Thermal and Other Industries.* Boca Raton, Fla.: CRC Press, 2001.

Clark, Ronald W. *The Life of Ernst Chain: Penicillin and Beyond.* New York: St. Martin's Press, 1985.

_______. *The Survival of Charles Darwin: A Biography of a Man and an Idea.* New York: Random House, 1984.

Cline, Barbara Lovett. *Men Who Made a New Physics.* Chicago: University of Chicago Press, 1987.

Clos, Lynne. *Field Adventures in Paleontology.* Boulder, Colo.: Fossil News, 2003.

Clugston, M. J., ed. *The New Penguin Dictionary of Science.* 2d ed. New York: Penguin Books, 2004.

Coffey, Patrick. *Cathedrals of Science: The Personalities and Rivalries That Made Modern Chemistry.* New York: Oxford University Press, 2008.

Cohen, I. Bernard. *Benjamin Franklin's Science.* Cambridge, Mass.: Harvard University Press, 1990.

_______. *The Newtonian Revolution.* New York: Cambridge University Press, 1980.

Cohen, I. Bernard, and George E. Smith, eds. *The Cambridge Companion to Newton.* New York: Cambridge University Press, 2002.

Cole, K. C. *The Universe and the Teacup: The Mathematics of Truth and Beauty.* Fort Washington, Pa.: Harvest Books, 1999.

Cole, Michael D. *Galileo Spacecraft: Mission to Jupiter: Countdown to Space.* New York: Enslow, 1999.

Collin, S. M. H. *Dictionary of Science and Technology.* London: Bloomsbury Publishing, 2003.

Connor, James A. *Kepler's Witch: An Astronomer's Discovery of Cosmic Order Amid Religious War, Political Intrigue, and the Heresy Trial of His Mother.* San Francisco: HarperSanFrancisco, 2004.

Conrad, Lawrence, et al., eds. *The Western Medical Tradition, 800 B.C. to A.D. 1800.* New York: Cambridge University Press, 1995.

Cook, Alan. *Edmond Halley: Charting the Heavens and the Seas.* New York: Oxford University Press, 1998.

Cooke, Donald A. *The Life and Death of Stars.* New York: Crown, 1985.

Cooper, Geoffrey M. *Oncogenes.* 2d ed. Boston: Jones and Bartlett, 1995.

Cooper, Henry S. F., Jr. *Imaging Saturn: The Voyager Flights to Saturn.* New York: H. Holt, 1985.

Corsi, Pietro. *The Age of Lamarck: Evolutionary Theories in France, 1790-1830.* Berkeley: University of California Press, 1988.

Coulthard, Malcolm, and Alison Johnson, eds. *The Routledge Handbook of Forensic Linguistics.* New York: Routledge, 2010.

Craven, B. O. *The Lebesgue Measure and Integral.* Boston: Pitman Press, 1981.

Crawford, Deborah. *King's Astronomer William Herschel.* New York: Julian Messner, 2000.

Crease, Robert P., and Charles C. Mann. *The Second Creation: Makers of the Revolution in Twentieth Century Physics.* New York: Macmillan, 1985.

Crewdson, John. *Science Fictions: A Scientific Mystery, A Massive Cover-Up, and the Dark Legacy of Robert Gallo.* Boston: Little, Brown, 2002.

Crick, Francis. *What Mad Pursuit: A Personal View of Scientific Discovery.* New York: Basic Books, 1988.

Crump, Thomas. *A Brief History of Science as Seen Through the Development of Scientific Instruments.* New York: Carroll & Graf, 2001.

Cunningham, Andrew. *The Anatomical Renaissance: The Resurrection of the Anatomical Projects of the Ancients.* Brookfield, Vt.: Ashgate, 1997.

Cutler, Alan. *The Seashell on the Mountaintop: A Story of Science, Sainthood, and the Humble Genius Who Discovered a New History of the Earth.* New York: Dutton/Penguin, 2003.

Dalrymple, G. Brent. *The Age of the Earth.* Stanford, Calif.: Stanford University Press, 1991.

Darrigol, Oliver. *Electrodynamics from Ampère to Einstein.* Oxford, England: Oxford University Press, 2000.

Dash, Joan. *The Longitude Prize.* New York: Farrar, Straus and Giroux, 2000.

Daston, Lorraine. *Classical Probability in the Enlightenment.* Princeton, N.J.: Princeton University Press, 1988.

Davies, John K. *Astronomy from Space: The Design and Operation of Orbiting Observatories.* New York: John Wiley & Sons, 1997.

Davies, Paul. *The Edge of Infinity: Where the Universe Came from and How It Will End.* New York: Simon & Schuster, 1981.

Davis, Joel. *Flyby: The Interplanetary Odyssey of Voyager 2.* New York: Atheneum, 1987.

Davis, Martin. *Engines of Logic: Mathematicians and the Origin of the Computer.* New York: W. W. Norton, 2000.

Davis, Morton D. *Game Theory: A Nontechnical Introduction.* New York: Dover, 1997.

Davis, William Morris. *Elementary Meteorology.* Boston: Ginn, 1894.

Dawkins, Richard. *The Ancestor's Tale: A Pilgrimage to the Dawn of Evolution.* New York: Houghton Mifflin, 2004.

_______. *River Out of Eden: A Darwinian View of Life.* New York: Basic Books, 1995.

Day, Michael H. *Guide to Fossil Man.* 4th ed. Chicago: University of Chicago Press, 1986.

Day, William. *Genesis on Planet Earth.* 2d ed. New Haven, Conn.: Yale University Press, 1984.

Dean, Dennis R. *James Hutton and the History of Geology.* Ithaca, N.Y.: Cornell University Press, 1992.

Debré, Patrice. *Louis Pasteur.* Translated by Elborg Forster. Baltimore: The Johns Hopkins University Press, 1998.

DeJauregui, Ruth. *100 Medical Milestones That Shaped World History.* San Mateo, Calif.: Bluewood Books, 1998.

De Jonge, Christopher J., and Christopher L. R. Barratt, eds. *Assisted Reproductive Technologies: Current Accomplishments and New Horizons.* New York: Cambridge University Press, 2002.

Delaporte, François. *The History of Yellow Fever: An Essay on the Birth of Tropical Medicine.* Cambridge, Mass.: MIT Press, 1991.

Dennett, Daniel C. *Darwin's Dangerous Idea: Evolution and the Meanings of Life.* New York: Simon & Schuster, 1995.

Dennis, Carina, and Richard Gallagher. *The Human Genome.* London: Palgrave Macmillan, 2002.

DeVorkin, David H. *Race to the Stratosphere: Manned Scientific Ballooning in America.* New York: Springer-Verlag, 1989.

Dewdney, A. K. *The Turing Omnibus.* Rockville, Md.: Computer Science Press, 1989.

Diamond, Jared. *The Third Chimpanzee: The Evolution and Future of the Human Animal.* New York: HarperCollins, 1992.

DiCanzio, Albert. *Galileo: His Science and His Significance for the Future of Man.* Portsmouth, N.H.: ADASI, 1996.

Dijksterhuis, Eduard Jan. *Archimedes.* Translated by C. Dikshoorn, with a new bibliographic essay by Wilbur R. Knorr. Princeton, N.J.: Princeton University Press, 1987.

Dijksterhuis, Fokko Jan. *Lenses and Waves: Christiaan Huygens and the Mathematical Science of Optics in the Seventeenth Century.* Dordrecht, the Netherlands: Kluwer Academic, 2004.

Dimmock, N. J., A. J. Easton, and K. N. Leppard. *Introduction to Modern Virology.* 5th ed. Malden, Mass.: Blackwell Science, 2001.

Dore, Mohammed, Sukhamoy Chakravarty, and Richard Goodwin, eds. *John Von Neumann and Modern Economics.* New York: Oxford University Press, 1989.

Drake, Stillman. *Galileo: Pioneer Scientist.* Toronto: University of Toronto Press, 1990.

_______. *Galileo: A Very Short Introduction.* New York: Oxford University Press, 2001.

Dreyer, John Louis Emil, ed. *The Scientific Papers of Sir William Herschel.* Dorset, England: Thoemmes Continuum, 2003.

Duck, Ian. *One Hundred Years of Planck's Quantum.* River Edge, N.J.: World Scientific, 2000.

Dudgeon, Dan E., and Russell M. Mersereau. *Multidimensional Digital Signal Processing.* Englewood Cliffs, N.J.: Prentice Hall, 1984.

Dunham, William. *The Calculus Gallery: Masterpieces from Newton to Lebesgue.* Princeton, N.J.: Princeton University Press, 2005.

_______. *Euler: The Master of Us All.* Washington, D.C.: Mathematical Association of America, 1999.

_______. *Journey Through Genius.* New York: John Wiley & Sons, 1990.

Durham, Frank, and Robert D. Purrington. *Frame of the Universe.* New York: Cambridge University Press, 1983.

Easton, Thomas A. *Careers in Science.* 4th ed. Chicago: VGM Career Books, 2004.

Edelson, Edward. *Gregor Mendel: And the Roots of Genetics.* New York: Oxford University Press, 2001.

Edey, Maitland A., and Donald C. Johanson. *Blueprints: Solving the Mystery of Evolution.* Boston: Little, Brown, 1989.

Edwards, Robert G., and Patrick Steptoe. *A Matter of Life.* New York: William Morrow, 1980.

Ehrenfest, Paul, and Tatiana Ehrenfest. *The Conceptual Foundations of the Statistical Approach in Mechanics.* Mineola, N.Y.: Dover, 2002.

Ehrlich, Melanie, ed. *DNA Alterations in Cancer: Genetic and Epigenetic Changes.* Natick, Mass.: Eaton, 2000.

Eisen, Herman N. *Immunology: An Introduction to Molecular and Cellular Principles of the Immune Responses.* 2d ed. Philadelphia: J. B. Lippincott, 1980.

Espejo, Roman, ed. *Biomedical Ethics: Opposing Viewpoints.* San Diego: Greenhaven Press, 2003.

Evans, James. *The History and Practice of Ancient Astronomy.* New York: Oxford University Press, 1998.

Fabian, A. C., K. A. Pounds, and R. D. Blandford. *Frontiers of X-Ray Astronomy.* London: Cambridge University Press, 2004.

Fara, Patricia. *An Entertainment for Angels: Electricity in the Enlightenment.* New York: Columbia University Press, 2002.

_______. *Newton: The Making of a Genius.* New York: Columbia University Press, 2002.

_______. *Sex, Botany, and the Empire: The Story of Carl Linnaeus and Joseph Banks.* New York: Columbia University Press, 2003.

Farber, Paul Lawrence. *Finding Order in Nature: The Naturalist Tradition from Linnaeus to E. O. Wilson.* Baltimore: The Johns Hopkins University Press, 2000.

Fauvel, John, and Jeremy Grey, eds. *The History of Mathematics: A Reader.* 1987. Reprint. Washington, D.C.: The Mathematical Association of America, 1997.

Feferman, S., J. W. Dawson, and S. C. Kleene, eds. *Kurt Gödel: Collected Works.* 2 vols. New York: Oxford University Press, 1986-1990.

Feldman, David. *How Does Aspirin Find a Headache?* New York: HarperCollins, 2005.

Ferejohn, Michael. *The Origins of Aristotelian Science.* New Haven, Conn.: Yale University Press, 1991.

Ferguson, Kitty. *The Nobleman and His Housedog: Tycho Brahe and Johannes Kepler—The Strange Partnership That Revolutionized Science.* London: Headline, 2002.

Ferris, T. *Coming of Age in the Milky Way.* New York: Doubleday, 1989.

Ferris, Timothy. *Galaxies.* New York: Harrison House, 1987.

Field, George, and Donald Goldsmith. *The Space Telescope.* Chicago: Contemporary Books, 1989.

Field, J. V. *The Invention of Infinity: Mathematics and Art in the Renaissance.* New York: Oxford University Press, 1997.

Fincher, Jack. *The Brain: Mystery of Matter and Mind.* Washington, D.C.: U.S. News Books, 1981.

Finlayson, Clive. *Neanderthals and Modern Humans: An Ecological and Evolutionary Perspective.* New York: Cambridge University Press, 2004.

Finocchiaro, Maurice A., ed. *The Galileo Affair: A Documentary History.* Berkeley: University of California Press, 1989.

Fischer, Daniel. *Mission Jupiter: The Spectacular Journey of the Galileo Spacecraft.* New York: Copernicus Books, 2001.

Fischer, Daniel, and Hilmar W. Duerbeck. *Hubble Revisited: New Images from the Discovery Machine.* New York: Copernicus Books, 1998.

Fisher, Richard B. *Edward Jenner, 1741-1823.* London: Andre Deutsch, 1991.

Flowers, Lawrence O., ed. *Science Careers: Personal Accounts from the Experts.* Lanham, Md.: Scarecrow Press, 2003.

Ford, Brian J. *The Leeuwenhoek Legacy.* London: Farrand, 1991.

_______. *Single Lens: The Story of the Simple Microscope.* New York: Harper & Row, 1985.

Fournier, Marian. *The Fabric of Life: Microscopy in the Seventeenth Century.* Baltimore: The Johns Hopkins University Press, 1996.

Fowler, A. C. *Mathematical Models in the Applied Sciences.* New York: Cambridge University Press, 1997.

Foyer, Christine H. *Photosynthesis.* New York: Wiley-Interscience, 1984.

Frängsmyr, Tore, ed. *Linnaeus: The Man and His Work.* Canton, Mass.: Science History Publications, 1994.

Franklin, Benjamin. *Autobiography of Benjamin Franklin.* New York: Buccaneer Books, 1984.

French, A. P., and P. J. Kennedy, eds. *Niels Bohr: A Centenary Volume.* Cambridge, Mass.: Harvard University Press, 1985.

French, Roger. *William Harvey's Natural Philosophy.* New York: Cambridge University Press, 1994.

Fridell, Ron. *DNA Fingerprinting: The Ultimate Identity.* New York: Scholastic, 2001.

Friedlander, Michael W. *Cosmic Rays.* Cambridge, Mass.: Harvard University Press, 1989.

Friedman, Meyer, and Gerald W. Friedland. *Medicine's Ten Greatest Discoveries.* New Haven, Conn.: Yale University Press, 2000.

Friedman, Robert Marc. *Appropriating the Weather: Vilhelm Bjerknes and the Construction of a Modern Meteorology.* Ithaca, N.Y.: Cornell University Press, 1989.

Friedrich, Wilhelm. *Vitamins.* New York: Walter de Gruyter, 1988.

Friedrichs, Günter, and Adam Schaff. *Microelectronics and Society: For Better or for Worse, a Report to the Club of Rome.* New York: Pergamon Press, 1982.

Frist, William. *Transplant.* New York: Atlantic Monthly Press, 1989.

Fuchs, Thomas. *The Mechanization of the Heart: Harvey and Descartes.* Rochester, N.Y.: University of Rochester Press, 2001.

Gallo, Robert C. *Virus Hunting: AIDS, Cancer, and the Human Retrovirus: A Story of Scientific Discovery.* New York: Basic Books, 1993.

Galston, Arthur W. *Life Processes of Plants.* New York: Scientific American Library, 1994.

Gamow, George. *The New World of Mr. Tompkins.* Cambridge, England: Cambridge University Press, 1999.

Gani, Joseph M., ed. *The Craft of Probabilistic Modeling.* New York: Springer-Verlag, 1986.

García-Ballester, Luis. *Galen and Galenism: Theory and Medical Practice from Antiquity to the European Renaissance.* Burlington, Vt.: Ashgate, 2002.

Gardner, Eldon J., and D. Peter Snustad. *Principles of Genetics.* 7th ed. New York: John Wiley & Sons, 1984.

Gardner, Robert, and Eric Kemer. *Science Projects About Temperature and Heat.* Berkeley Heights, N.J.: Enslow Publishers, 1994.

Garner, Geraldine. *Careers in Engineering.* 3d ed. New York: McGraw-Hill, 2009.

Gartner, Carol B. *Rachel Carson.* New York: Frederick Ungar, 1983.

Gasser, James, ed. *A Boole Anthology: Recent and Classical Studies in the Logic of George Boole.* Dordrecht, the Netherlands: Kluwer, 2000.

Gay, Peter. *The Enlightenment: The Science of Freedom.* New York: W. W. Norton, 1996.

_______. *Freud: A Life for Our Time.* New York: W. W. Norton, 1988.

Gazzaniga, Michael S. *The Social Brain: Discovering the Networks of the Mind.* New York: Basic Books, 1985.

Geison, Gerald. *The Private Science of Louis Pasteur.* Princeton, N.J.: Princeton University Press, 1995.

Gell-Mann, Murray. *The Quark and the Jaguar: Adventures in the Simple and the Complex.* New York: W. H. Freeman, 1994.

Georgotas, Anastasios, and Robert Cancro, eds. *Depression and Mania.* New York: Elsevier, 1988.

Gerock, Robert. *Mathematical Physics.* Chicago: University of Chicago Press, 1985.

Gest, Howard. *Microbes: An Invisible Universe.* Washington, D.C.: ASM Press, 2003.

Gesteland, Raymond F., Thomas R. Cech, and John F. Atkins, eds. *The RNA World: The Nature of Modern RNA Suggests a Prebiotic RNA.* 2d ed. Cold Spring Harbor, N.Y.: Cold Spring Harbor Laboratory Press, 1999.

Gigerenzer, Gerd, et al. *The Empire of Chance: How Probability Theory Changed Science and Everyday Life.* New York: Cambridge University Press, 1989.

Gilder, Joshua, and Anne-Lee Gilder. *Heavenly Intrigue: Johannes Kepler, Tycho Brahe, and the Murder Behind One of History's Greatest Scientific Discoveries.* New York: Doubleday, 2004.

Gillispie, Charles Coulston, Robert Fox, and Ivor Grattan-Guinness. *Pierre-Simon Laplace, 1749-1827: A Life in Exact Science.* Princeton, N.J.: Princeton University Press, 2000.

Gingerich, Owen. *The Book Nobody Read: Chasing the Revolutions of Nicolaus Copernicus.* New York: Walker, 2004.

_______. *The Eye of Heaven: Ptolemy, Copernicus, Kepler.* New York: Springer-Verlag, 1993.

Glashow, Sheldon, with Ben Bova. *Interactions: A Journey Through the Mind of a Particle Physicist and the Matter of This World.* New York: Warner Books, 1988.

Glass, Billy. *Introduction to Planetary Geology.* New York: Cambridge University Press, 1982.

Gleick, James. *Chaos: Making a New Science.* New York: Penguin Books, 1987.

_______. *Isaac Newton.* New York: Pantheon Books, 2003.

Glen, William. *The Road to Jaramillo: Critical Years of the Revolution in Earth Science.* Stanford, Calif.: Stanford University Press, 1982.

Glickman, Todd S., ed. *Glossary of Meteorology.* 2d ed. Boston: American Meteorological Society, 2000.

Goddard, Jolyon, ed. *National Geographic Concise History of Science and Invention: An Illustrated Time Line.* Washington, D.C.: National Geographic, 2010.

Goding, James W. *Monoclonal Antibodies: Principles and Practice.* New York: Academic Press, 1986.

Godwin, Robert, ed. *Mars: The NASA Mission Reports.* Burlington, Ont.: Apogee Books, 2000.

_______. *Mars: The NASA Mission Reports.* Vol. 2. Burlington, Ont.: Apogee Books, 2004.

_______. *Space Shuttle STS Flights 1-5: The NASA Mission Reports*. Burlington, Ont.: Apogee Books, 2001.

Goetsch, David L. *Building a Winning Career in Engineering: 20 Strategies for Success After College*. Upper Saddle River, N.J.: Pearson/Prentice Hall, 2007.

Gohlke, Mary, with Max Jennings. *I'll Take Tomorrow*. New York: M. Evans, 1985.

Gold, Rebecca. *Steve Wozniak: A Wizard Called Woz*. Minneapolis: Lerner, 1994.

Goldsmith, Donald. *Nemesis: The Death Star and Other Theories of Mass Extinction*. New York: Berkley Publishing Group, 1985.

Goldsmith, Maurice, Alan Mackay, and James Woudhuysen, eds. *Einstein: The First Hundred Years*. Elmsford, N.Y.: Pergamon Press, 1980.

Golinski, Jan. *Science as Public Culture: Chemistry and Enlightenment in Britain, 1760-1820*. Cambridge, England: Cambridge University Press, 1992.

Golthelf, Allan, and James G. Lennox, eds. *Philosophical Issues in Aristotle's Biology*. Cambridge, England: Cambridge University Press, 1987.

Gooding, David, and Frank A. J. L. James, eds. *Faraday Rediscovered: Essays on the Life and Work of Michael Faraday, 1791-1867*. New York: Macmillan, 1985.

Gordin, Michael D. *A Well-Ordered Thing: Dmitrii Mendeleev and the Shadow of the Periodic Table*. New York: Basic Books, 2004.

Gornick, Vivian. *Women in Science: Then and Now*. New York: Feminist Press at the City University of New York, 2009.

Gould, James L., and Carol Grant Gould. *The Honey Bee*. New York: Scientific American Library, 1988.

Gould, Stephen Jay. *Time's Arrow, Time's Cycle: Myth and Metaphor in the Discovery of Geological Time*. Cambridge, Mass.: Harvard University Press, 1987.

_______. *Wonderful Life: The Burgess Shale and the Nature of History*. New York: W. W. Norton, 1989.

Govindjee, J. T. Beatty, H. Gest, and J.F. Allen, eds. *Discoveries in Photosynthesis*. Berlin: Springer, 2005.

Gow, Mary. *Tycho Brahe: Astronomer*. Berkeley Heights, N.J.: Enslow, 2002.

Graham, Loren R. *Science, Philosophy, and Human Behavior in the Soviet Union*. New York: Columbia University Press, 1987.

Grattan-Guinness, Ivor. *The Norton History of the Mathematical Sciences*. New York: W. W. Norton, 1999.

Gray, Robert M., and Lee D. Davisson. *Random Processes: A Mathematical Approach for Engineers*. Englewood Cliffs, N.J.: Prentice-Hall, 1986.

Greene, Mott T. *Geology in the Nineteenth Century: Changing Views of a Changing World*. Ithaca, N.Y.: Cornell University Press, 1982.

Gregory, Andrew. *Harvey's Heart: The Discovery of Blood Circulation*. London: Totem Books, 2001.

Gribbin, John. *Deep Simplicity: Bringing Order to Chaos and Complexity*. New York: Random House, 2005.

_______. *Future Weather and the Greenhouse Effect*. New York: Delacorte Press/Eleanor Friede, 1982.

_______. *The Hole in the Sky: Man's Threat to the Ozone Layer*. New York: Bantam Books, 1988.

_______. *In Search of Schrödinger's Cat: Quantum Physics and Reality*. New York: Bantam Books, 1984.

_______. *In Search of the Big Bang*. New York: Bantam Books, 1986.

_______. *The Omega Point: The Search for the Missing Mass and the Ultimate Fate of the Universe*. New York: Bantam Books, 1988.

_______. *The Scientists: A History of Science Told Through the Lives of Its Greatest Inventors*. New York: Random House, 2002.

Gribbin, John, ed. *The Breathing Planet*. New York: Basil Blackwell, 1986.

Gutkind, Lee. *Many Sleepless Nights: The World of Organ Transplantation*. New York: W. W. Norton, 1988.

Hackett, Edward J., et al., eds. *The Handbook of Science and Technology Studies*. 3d ed. Cambridge, Mass.: MIT Press, 2008.

Hald, Anders. *A History of Mathematical Statistics from 1750 to 1950*. New York: John Wiley & Sons, 1998.

_______. *A History of Probability and Statistics and Their Applications Before 1750*. New York: John Wiley & Sons, 1990.

Hall, A. Rupert. *The Scientific Revolution, 1500-1750*. 3d ed. New York: Longman, 1983.

Halliday, David, and Robert Resnick. *Fundamentals of Physics: Extended Version*. New York: John Wiley & Sons, 1988.

Halliday, David, Robert Resnick, and Jearl Walker. *Fundamentals of Physics*. 7th ed. New York: John Wiley & Sons, 2004.

Hankins, Thomas L. *Science and the Enlightenment*. Reprint. New York: Cambridge University Press, 1991.

Hanlon, Michael, and Arthur C. Clarke. *The Worlds of Galileo: The Inside Story of NASA's Mission to Jupiter*. New York: St. Martin's Press, 2001.

Hanson, Earl D. *Understanding Evolution*. New York: Oxford University Press, 1981.

Hargittai, István. *Martians of Science: Five Physicists Who Changed the Twentieth Century.* New York: Oxford University Press, 2006.

Harland, David M. *Jupiter Odyssey: The Story of NASA's Galileo Mission.* London: Springer-Praxis, 2000.

_______. *Mission to Saturn: Cassini and the Huygens Probe.* London: Springer-Praxis, 2002.

_______. *The Space Shuttle: Roles, Missions, and Accomplishments.* New York: John Wiley & Sons, 1998.

Harland, David M., and John E. Catchpole. *Creating the International Space Station.* London: Springer-Verlag, 2002.

Harrington, J. W. *Dance of the Continents:* New York: V. P. Tarher, 1983.

Harrington, Philip S. *The Space Shuttle: A Photographic History.* San Francisco: Brown Trout, 2003.

Harris, Henry. *The Birth of the Cell.* New Haven, Conn.: Yale University Press, 1999.

Harrison, Edward R. *Cosmology: The Science of the Universe.* Cambridge England: Cambridge University Press, 1981.

Hart, Michael H. *The 100: A Ranking of the Most Influential Persons in History.* New York: Galahad Books, 1982.

Hart-Davis, Adam. *Chain Reactions: Pioneers of British Science and Technology and the Stories That Link Them.* London: National Portrait Gallery, 2000.

Hartmann, William K. *The Cosmic Voyage: Through Time and Space.* Belmont, Calif: Wadsworth, 1990.

_______. *Moons and Planets.* 5th ed. Belmont, Calif.: Brooks-Cole Publishing, 2005.

Hartwell, L. H., et al. *Genetics: From Genes to Genomes.* 2d ed. New York: McGraw-Hill, 2004.

Harvey, William. *The Circulation of the Blood and Other Writings.* New York: Everyman's Library, 1990.

Haskell, G., and Michael Rycroft. *International Space Station: The Next Space Marketplace.* Boston: Kluwer Academic, 2000.

Hathaway, N. *The Friendly Guide to the Universe.* New York: Penguin Books, 1994.

Havil, Julian. *Gamma: Exploring Euler's Constant.* Princeton, N.J.: Princeton University Press, 2003.

Hawking, Stephen W. *A Brief History of Time.* New York: Bantam Books, 1988.

Haycock, David. *William Stukeley: Science, Religion, and Archeology in Eighteenth-Century England.* Woodbridge, England: Boydell Press, 2002.

Hazen, Robert. *The Breakthrough: The Race for the Superconductor.* New York: Summit Books, 1988.

Headrick, Daniel R. *Technology: A World History.* New York: Oxford University Press, 2009.

Heath, Sir Thomas L. *A History of Greek Mathematics: From Thales to Euclid.* 1921. Reprint. New York: Dover Publications, 1981.

Heilbron, J. L. *The Dilemmas of an Upright Man: Max Planck As a Spokesman for German Science.* Berkeley: University of California Press, 1986.

_______. *Electricity in the Seventeenth and Eighteenth Centuries: A Study in Early Modern Physics.* Mineola, N.Y.: Dover Publications, 1999.

_______. *Elements of Early Modern Physics.* Berkeley: University of California Press, 1982.

_______. *Geometry Civilized: History, Culture, and Technique.* Oxford, England: Clarendon Press, 1998.

Heilbron, J. L., and Robert W. Seidel. *Lawrence and His Laboratory: A History of the Lawrence Berkeley Laboratory.* Berkeley: University of California Press, 1989.

Heisenberg, Elisabeth. *Inner Exile: Recollections of a Life with Werner Heisenberg.* Translated by S. Cappelari and C. Morris. Boston: Birkhäuser, 1984.

Hellegouarch, Yves. *Invitation to the Mathematics of Fermat-Wiles.* San Diego: Academic Press, 2001.

Henig, Robin Marantz. *The Monk in the Garden: The Lost and Found Genius of Gregor Mendel, the Father of Genetics.* New York: Mariner Books, 2001.

Henry, Helen L., and Anthony W. Norman, eds. *Encyclopedia of Hormones.* 3 vols. San Diego: Academic Press, 2003.

Henry, John. *Moving Heaven and Earth: Copernicus and the Solar System.* Cambridge, England: Icon, 2001.

Herrmann, Bernd, and Susanne Hummel, eds. *Ancient DNA: Recovery and Analysis of Genetic Material from Paleographic, Archaeological, Museum, Medical, and Forensic Specimens.* New York: Springer-Verlag, 1994.

Hershel, Sir John Frederic William, and Pierre-Simon Laplace. *Essays in Astronomy.* University Press of the Pacific, 2002.

Hillar, Marian, and Claire S. Allen. *Michael Servetus: Intellectual Giant, Humanist, and Martyr.* New York: University Press of America, 2002.

Hobson, J. Allan. *The Dreaming Brain.* New York: Basic Books, 1988.

Hodge, Paul. *Galaxies.* Cambridge, Mass.: Harvard University Press, 1986.

Hodges, Andrew. *Alan Turing: The Enigma.* 1983. Reprint. New York: Walker, 2000.

Hofmann, James R., David Knight, and Sally Gregory Kohlstedt, eds. *André-Marie Ampère: Enlightenment and Electrodynamics.* Cambridge, England: Cambridge University Press, 1996.

Holland, Suzanne, Karen Lebacqz, and Laurie Zoloth, eds. *The Human Embryonic Stem Cell Debate: Science, Ethics, and Public Policy.* Cambridge, Mass.: MIT Press, 2001.

Holmes, Frederic Lawrence. *Antoine Lavoisier, the Next Crucial Year: Or, the Sources of His Quantitative Method in Chemistry.* Princeton, N.J.: Princeton University Press, 1998.

Horne, James. *Why We Sleep.* New York: Oxford University Press, 1988.

Hoskin, Michael A. *The Herschel Partnership: As Viewed by Caroline.* Cambridge, England: Science History, 2003.

_______. *William Herschel and the Construction of the Heavens.* New York: Norton; 1964.

Howse, Derek. *Greenwich Time and the Discovery of Longitude.* New York: Oxford University Press, 1980.

Hoyt, William G. *Planet X and Pluto.* Tucson: University of Arizona Press, 1980.

Hsü, Kenneth J. *The Great Dying.* San Diego: Harcourt Brace Jovanovich, 1986.

Huerta, Robert D. *Giants of Delft, Johannes Vermeer and the Natural Philosophers: The Parallel Search for Knowledge During the Age of Discovery.* Lewisburg, Pa.: Bucknell University Press, 2003.

Hummel, Susanne. *Fingerprinting the Past: Research on Highly Degraded DNA and Its Applications.* New York: Springer-Verlag, 2002.

Hunter, Michael, ed. *Robert Boyle Reconsidered.* New York: Cambridge University Press, 1994.

Hynes, H. Patricia. *The Recurring Silent Spring.* New York: Pergamon Press, 1989.

Ihde, Aaron J. *The Development of Modern Chemistry.* New York: Dover, 1984.

Irwin, Patrick G. J. *Giant Planets of Our Solar System: Atmospheres, Composition, and Structure.* London: Springer-Praxis, 2003.

Isaacson, Walter. *Benjamin Franklin: An American Life.* New York: Simon & Schuster, 2003.

Jackson, Myles. *Spectrum of Belief: Joseph Fraunhofer and the Craft of Precision Optics.* Cambridge, Mass.: MIT Press, 2000.

Jacobsen, Theodor S. *Planetary Systems from the Ancient Greeks to Kepler.* Seattle: University of Washington Press, 1999.

Jacquette, Dale. *On Boole.* Belmont, Calif.: Wadsworth, 2002.

Jaffe, Bernard. *Crucibles: The Story of Chemistry.* New York: Dover, 1998.

James, Ioan. *Remarkable Mathematicians: From Euler to Von Neumann.* Cambridge, England: Cambridge University Press, 2002.

Janowsky, David S., Dominick Addario, and S. Craig Risch. *Psychopharmacology Case Studies.* 2d ed. New York: Guilford Press, 1987.

Jeffreys, Diarmuid. *Aspirin: The Remarkable Story of a Wonder Drug.* London: Bloomsbury Publishing, 2004.

Jenkins, Dennis R. *Space Shuttle: The History of the National Space Transportation System: The First 100 Missions.* Stillwater, Minn.: Voyageur Press, 2001.

Johanson, Donald, and B. Edgar. *From Lucy to Language.* New York: Simon and Schuster, 1996.

Johanson, Donald C., and Maitland A. Edey. *Lucy: The Beginnings of Humankind.* New York: Simon & Schuster, 1981.

Johanson, Donald C., and James Shreeve. *Lucy's Child: The Discovery of a Human Ancestor.* New York: William Morrow, 1989.

Johnson, George. *Strange Beauty: Murray Gell-Mann and the Revolution in Twentieth-Century Physics.* New York: Alfred A. Knopf, 1999.

Jones, Henry Bence. *Life and Letters of Faraday.* 2 vols. London: Longmans, Green and Co., 1870.

Jones, Meredith L., ed. *Hydrothermal Vents of the Eastern Pacific: An Overview.* Vienna, Va.: INFAX, 1985.

Jones, Sheilla. *The Quantum Ten: A Story of Passion, Tragedy, and Science.* Toronto: Thomas Allen, 2008.

Jones, W. H. S., trans. *Hippocrates.* 4 vols. 1923-1931. Reprint. New York: Putnam, 1995.

Jordan, Paul. *Neanderthal: Neanderthal Man and the Story of Human Origins.* Gloucestershire, England: Sutton, 2001.

Joseph, George Gheverghese. *The Crest of the Peacock: The Non-European Roots of Mathematics.* London: Tauris, 1991.

Jungnickel, Christa, and Russell McCormmach. *Cavendish: The Experimental Life.* Lewisburg, Pa.: Bucknell University Press, 1999.

Kaplan, Robert. *The Nothing That Is: A Natural History of Zero.* New York: Oxford University Press, 2000.

Kargon, Robert H. *The Rise of Robert Millikan: Portrait of a Life in American Science.* Ithaca, N.Y.: Cornell University Press, 1982.

Katz, Jonathan. *The Biggest Bang: The Mystery of Gamma-Ray Bursts.* London: Oxford University Press, 2002.

Kellogg, William W., and Robert Schware. *Climate Change and Society: Consequences of Increasing Atmospheric Carbon Dioxide.* Boulder, Colo.: Westview Press, 1981.

Kelly, Thomas J. *Moon Lander: How We Developed the Apollo Lunar Module.* Washington, D.C.: Smithsonian Books, 2001.

Kemper, John D., and Billy R. Sanders. *Engineers and Their Profession.* 5th ed. New York: Oxford University Press, 2001.

Kepler, Johannes. *New Astronomy.* Translated by William H. Donahue. New York: Cambridge University Press, 1992.

Kermit, Hans. *Niels Stensen: The Scientist Who Was Beatified.* Translated by Michael Drake. Herefordshire, England: Gracewing 2003.

Kerns, Thomas A. *Jenner on Trial: An Ethical Examination of Vaccine Research in the Age of Smallpox and the Age of AIDS.* Lanham, Md.: University Press of America, 1997.

Kerrod, Robin. *Hubble: The Mirror on the Universe.* Richmond Hill, Ont.: Firefly Books, 2003.

_______. *Space Shuttle.* New York: Gallery Books, 1984.

Kevles, Bettyann. *Naked to the Bones: Medical Imaging in the Twentieth Century.* Reading, Mass.: Addison Wesley, 1998.

Kiessling, Ann, and Scott C. Anderson. *Human Embryonic Stem Cells: An Introduction to the Science and Therapeutic Potential.* Boston: Jones and Bartlett, 2003.

King, Helen. *Greek and Roman Medicine.* London: Bristol Classical, 2001.

_______. *Hippocrates' Woman: Reading the Female Body in Ancient Greece.* New York: Routledge, 1998.

Kirkham, M. B. *Principles of Soil and Plant Water Relations.* St. Louis: Elsevier, 2005.

Kline, Morris. *Mathematical Thought from Ancient to Modern Times.* New York: Oxford University Press, 1990.

Klotzko, Arlene Judith, ed. *The Cloning Sourcebook.* New York: Oxford University Press, 2001.

Knipe, David, Peter Howley, and Diane Griffin. *Field's Virology.* 2 vols. New York: Lippincott Williams and Wilkens, 2001.

Knowles, Richard V. *Genetics, Society, and Decisions.* Columbus, Ohio: Charles E. Merrill, 1985.

Koestler, Arthur. *The Sleepwalkers.* New York: Penguin Books, 1989.

Kolata, Gina Bari. *Clone: The Road to Dolly, and the Path Ahead.* New York: William Morrow, 1998.

Komszik, Louis. *Applied Calculus of Variations for Engineers.* Boca Raton, Fla.: CRC Press, 2009.

Kramer, Barbara. *Neil Armstrong: The First Man on the Moon.* Springfield, N.J.: Enslow, 1997.

Krane, Kenneth S. *Modern Physics.* New York: John Wiley & Sons, 1983.

Lagerkvist, Ulf. *Pioneers of Microbiology and the Nobel Prize.* River Edge, N.J.: World Scientific Publishing, 2003.

Lamarck, Jean-Baptiste. *Lamarck's Open Mind: The Lectures.* Gold Beach, Ore.: High Sierra Books, 2004.

_______. *Zoological Philosophy: An Exposition with Regard to the Natural History of Animals.* Translated by Hugh Elliot with introductory essay by David L. Hull and Richard W. Burckhardt, Jr. Chicago: University of Chicago Press, 1984.

Landes, Davis S. *Revolution in Time: Clocks and the Making of the Modern World.* Rev. ed. Cambridge, Mass.: Belknap Press, 2000.

Langone, John. *Superconductivity: The New Alchemy.* Chicago: Contemporary Books, 1989.

Lappé, Marc. *Broken Code: The Exploitation of DNA.* San Francisco: Sierra Club Books, 1984.

_______. *Germs That Won't Die.* Garden City, N.Y.: Doubleday, 1982.

La Thangue, Nicholas B., and Lasantha R. Bandara, eds. *Targets for Cancer Chemotherapy: Transcription Factors and Other Nuclear Proteins.* Totowa, N.J.: Humana Press, 2002.

Laudan, R. *From Mineralogy to Geology: The Foundations of a Science, 1650-1830.* Chicago: University of Chicago Press, 1987.

Lauritzen, Paul, ed. *Cloning and the Future of Human Embryo Research.* New York: Oxford University Press, 2001.

Leakey, Mary D. *Disclosing the Past.* New York: Doubleday, 1984.

Le Grand, Homer E. *Drifting Continents and Shifting Theories.* New York: Cambridge University Press, 1988.

Levy, David H. *Clyde Tombaugh: Discoverer of Planet Pluto.* Tucson: University of Arizona Press, 1991.

Lewin, Benjamin. *Genes III.* 3d ed. New York: John Wiley & Sons, 1987.

_______. *Genes IV.* New York: Oxford University Press, 1990.

Lewin, Roger. *Bones of Contention: Controversies in the Search for Human Origins.* New York: Simon & Schuster, 1987.

Lewis, Richard S. *The Voyages of Columbia: The First True Spaceship.* New York: Columbia University Press, 1984.

Lindberg, David C. *The Beginnings of Western Science.* Chicago: University of Chicago Press, 1992.

Lindley, David. *Degrees Kelvin: A Tale of Genius, Invention, and Tragedy.* Washington, D.C.: Joseph Henry Press, 2004.

Linzmayer, Owen W. *Apple Confidential: The Real Story of Apple Computer, Inc.* San Francisco: No Starch Press, 1999.

Lloyd, G. E. R., and Nathan Sivin. *The Way and the Word: Science and Medicine in Early China and Greece.* New Haven, Conn.: Yale University Press, 2002.

Logan, J. David. *Applied Mathematics.* 3d ed. Hoboken, N.J.: John Wiley & Sons, 2006.

Logsdon, John M. *Together in Orbit: The Origins of International Participation in the Space Station.* Washington, D.C.: National Aeronautics and Space Administration, 1998.

Longrigg, James. *Greek Medicine: From the Heroic to the Hellenistic Age.* New York: Routledge, 1998.

_______. *Greek Rational Medicine: Philosophy and Medicine from Alcmaeon to the Alexandrians.* New York: Routledge, 1993.

Lorenz, Edward. *The Essence of Chaos.* Reprint. St. Louis: University of Washington Press, 1996.

Lorenz, Ralph, and Jacqueline Mitton. *Lifting Titan's Veil: Exploring the Giant Moon of Saturn.* London: Cambridge University Press, 2002.

Loudon, Irvine. *The Tragedy of Childbed Fever.* New York: Oxford University Press, 2000.

Luck, Steve, ed. *International Encyclopedia of Science and Technology.* New York: Oxford University Press, 1999.

Lutgens, Frederick K., and Edward J. Tarbuck. *The Atmosphere: An Introduction to Meteorology.* 2d ed. Englewood Cliffs, N.J.: Prentice-Hall, 1982.

Lyell, Charles. *Elements of Geology.* London: John Murray, 1838.

_______. *The Geological Evidences of the Antiquity of Man with Remarks on Theories of the Origin of Species by Variation.* London: John Murray, 1863.

_______. *Principles of Geology, Being an Attempt to Explain the Former Changes of the Earth's Surface by Reference to Causes Now in Operation.* 3 vols. London: John Murray, 1830-1833.

Lynch, William T. *Solomon's Child: Method in the Early Royal Society of London.* Stanford, Calif.: Stanford University Press, 2000.

Ma, Pearl, and Donald Armstrong, eds. *AIDS and Infections of Homosexual Men.* Stoneham, Mass: Butterworths, 1989.

MacDonald, Allan H., ed. *Quantum Hall Effect: A Perspective.* Boston: Kluwer Academic Publishers, 1989.

MacHale, Desmond. *George Boole: His Life and Work.* Dublin: Boole Press, 1985.

Machamer, Peter, ed. *The Cambridge Companion to Galileo.* New York: Cambridge University Press, 1998.

Mactavish, Douglas. *Joseph Lister.* New York: Franklin Watts, 1992.

Mader, Sylvia S. *Biology.* 3d ed. Dubuque, Iowa: Win. C. Brown, 1990.

Magner, Lois. *A History of Medicine.* New York: Marcel Dekker, 1992.

Mahoney, Michael Sean. *The Mathematical Career of Pierre de Fermat, 1601-1665.* 2d rev. ed. Princeton, N.J.: Princeton University Press, 1994.

Mammana, Dennis L., and Donald W. McCarthy, Jr. *Other Suns, Other Worlds? The Search for Extrasolar Planetary Systems.* New York: St. Martin's Press, 1996.

Mandelbrot, B. B. *Fractals and Multifractals: Noise, Turbulence, and Galaxies.* New York: Springer-Verlag, 1990.

Mann, Charles, and Mark Plummer. *The Aspirin Wars: Money, Medicine and 100 Years of Rampant Competition.* New York: Knopf, 1991.

Marco, Gino J., Robert M. Hollingworth, and William Durham, eds. *Silent Spring Revisited.* Washington, D.C.: American Chemical Society, 1987.

Margolis, Howard. *It Started with Copernicus.* New York: McGraw-Hill, 2002.

Marshak, Daniel R., Richard L. Gardner, and David Gottlieb, eds. *Stem Cell Biology.* Woodbury, N.Y.: Cold Spring Harbor Laboratory Press, 2002.

Martzloff, Jean-Claude. *History of Chinese Mathematics.* Translated by Stephen S. Wilson. Berlin: Springer, 1987.

Massey, Harrie Stewart Wilson. *The Middle Atmosphere as Observed by Balloons, Rockets, and Satellites.* London: Royal Society, 1980.

Masson, Jeffrey M. *The Assault on Truth: Freud's Suppression of the Seduction Theory*. New York: Farrar, Straus and Giroux, 1984.

Mateles, Richard I. *Penicillin: A Paradigm for Biotechnology*. Chicago: Canadida Corporation, 1998.

Mayo, Jonathan L. *Superconductivity: The Threshold of a New Technology*. Blue Ridge Summit, Pa.: TAB Books, 1988.

McCarthy, Shawn P. *Engineer Your Way to Success: America's Top Engineers Share Their Personal Advice on What They Look for in Hiring and Promoting*. Alexandria, Va.: National Society of Professional Engineers, 2002.

McCay, Mary A. *Rachel Carson*. New York: Twayne, 1993.

McDonnell, John James. *The Concept of an Atom from Democritus to John Dalton*. Lewiston, N.Y.: Edwin Mellen Press, 1991.

McEliece, Robert. *Finite Fields for Computer Scientists and Engineers*. Boston: Kluwer Academic, 1987.

McGraw-Hill Concise Encyclopedia of Science and Technology. 6th ed. New York: McGraw-Hill, 2009.

McGraw-Hill Dictionary of Scientific and Technical Terms. 6th ed. New York: McGraw-Hill, 2002.

McGrayne, Sharon Bertsch. *Nobel Prize Women in Science: Their Lives, Struggles and Momentous Discoveries*. 2d ed. Washington, D.C.: Joseph Henry Press, 1998.

McIntyre, Donald B., and Alan McKirdy. *James Hutton: The Founder of Modern Geology*. Edinburgh: Stationery Office, 1997.

McLester, John, and Peter St. Pierre. *Applied Biomechanics: Concepts and Connections*. Belmont, Calif.: Thomson Wadsworth, 2008.

McMullen, Emerson Thomas. *William Harvey and the Use of Purpose in the Scientific Revolution: Cosmos by Chance or Universe by Design?* Lanhan, Md.: University Press of America, 1998.

McQuarrie, Donald A. *Quantum Chemistry*. Mill Valley, Calif: University Science Books, 1983.

Menard, H. W. *The Ocean of Truth: A Personal History of Global Tectonics*. Princeton, N.J.: Princeton University Press, 1986.

Menzel, Donald H., and Jay M. Pasachoff. *Stars and Planets*. Boston: Houghton Mifflin Company, 1983.

Merrell, David J. *Ecological Genetics*. Minneapolis: University of Minnesota Press, 1981.

Mettler, Lawrence E., Thomas G. Gregg, and Henry E. Schaffer. *Population Genetics and Evolution*. 2d ed. Englewood Cliffs, N.J.: Prentice-Hall, 1988.

Meyers, Robert A., ed. *Encyclopedia of Physical Science and Technology*. 18 vols. San Diego: Academic Press, 2005.

Meyerson, Daniel. *The Linguist and the Emperor: Napoleon and Champollion's Quest to Decipher the Rosetta Stone*. New York: Ballantine Books, 2004.

Middleton, W. E. Knowles. *A History of the Thermometer and Its Use in Meteorology*. Ann Arbor, Mich.: UMI Books on Demand, 1996.

Miller, Ron. *Extrasolar Planets*. Brookfield, Conn.: Twenty-First Century Books, 2002.

Miller, Stanley L. *From the Primitive Atmosphere to the Prebiotic Soup to the Pre-RNA World*. Washington, D.C.: National Aeronautics and Space Administration, 1996.

Mishkin, Andrew. *Sojourner: An Insider's View of the Mars Pathfinder Mission*. New York: Berkeley Books, 2003.

Mlodinow, Leonard. *Euclid's Window: A History of Geometry from Parallel Lines to Hyperspace*. New York: Touchstone, 2002.

Monmonier, Mark. *Air Apparent: How Meteorologists Learned to Map, Predict, and Dramatize Weather*. Chicago: University of Chicago Press, 2000.

Moore, Keith L. *The Developing Human*. Philadelphia: W. B. Saunders, 1988.

Moore, Patrick. *Eyes on the University: The Story of the Telescope*. New York: Springer-Verlag, 1997.

_______. *Patrick Moore's History of Astronomy*. 6th rev. ed. London: Macdonald, 1983.

Morell, V. *Ancestral Passions: The Leakey Family and the Quest for Humankind's Beginnings*. New York: Simon & Schuster, 1995.

Morgan, Kathryn A. *Myth and Philosophy from the Presocratics to Plato*. New York: Cambridge University Press, 2000.

Moritz, Michael. *The Little Kingdom: The Private Story of Apple Computer*. New York: Morrow, 1984.

Morris, Peter, ed. *Making the Modern World: Milestones of Science and Technology*. 2d ed. Chicago: KWS Publishers, 2011.

Morris, Richard. *The Last Sorcerers: The Path from Alchemy to the Periodic Table*. Washington, D.C.: Joseph Henry Press, 2003.

Morrison, David. *Voyages to Saturn*. NASA SP-451. Washington, D.C.: National Aeronautics and Space Administration, 1982.

Morrison, David, and Tobias Owen. *The Planetary System*. 3d ed. San Francisco: Addison Wesley, 2003.

Morrison, David, and Jane Samz. *Voyage to Jupiter.* NASA SP-439. Washington, D.C.: Government Printing Office, 1980.

Moss, Ralph W. *Free Radical: Albert Szent-Györgyi and the Battle over Vitamin C.* New York: Paragon House, 1988.

Muirden, James. *The Amateur Astronomer's Handbook.* 3d ed. New York: Harper & Row, 1987.

Mullis, Kary. *Dancing Naked in the Mind Field.* New York: Pantheon Books, 1998.

Mulvihill, John J. *Catalog of Human Cancer Genes: McKusick's Mendelian Inheritance in Man for Clinical and Research Oncologists.* Foreword by Victor A. McKusick. Baltimore: The Johns Hopkins University Press, 1999.

Nahin, Paul J. *Oliver Heaviside: Sage in Solitude.* New York: IEEE Press, 1987.

Ne'eman, Yuval, and Yoram Kirsh. *The Particle Hunters.* New York: Cambridge University Press, 1986.

Netz, Reviel. *The Shaping of Deduction in Greek Mathematics: A Study in Cognitive History.* New York: Cambridge University Press, 2003.

Neu, Jerome, ed. *The Cambridge Companion to Freud.* New York: Cambridge University Press, 1991.

North, John. *The Norton History of Astronomy and Cosmology.* New York: W. W. Norton, 1995.

Nutton, Vivian, ed. *The Unknown Galen.* London: Institute of Classical Studies, University of London, 2002.

Nye, Mary Jo. *Before Big Science: The Pursuit of Modern Chemistry and Physics, 1800-1940.* New York: Twayne, 1996.

Nye, Robert D. *Three Psychologies: Perspectives from Freud, Skinner, and Rogers.* Pacific Grove, Calif.: Brooks-Cole, 1992.

Oakes, Elizabeth H. *Encyclopedia of World Scientists.* Rev ed. New York: Facts on File, 2007.

Olson, James S., and Robert L. Shadle, ed. *Encyclopedia of the Industrial Revolution in America.* Westport, Conn.: Greenwood Press, 2002.

Olson, Steve. *Mapping Human History: Genes, Races and Our Common Origins.* New York: Houghton Mifflin, 2002.

Ozima, Minoru. *The Earth: Its Birth and Growth.* Translated by Judy Wakabayashi. Cambridge, England: Cambridge University Press, 1981.

Pagels, Heinz R. *The Cosmic Code.* New York: Simon & Schuster, 1982.

_______. *The Cosmic Code: Quantum Physics As the Law of Nature.* New York: Bantam Books, 1984.

_______. *Perfect Symmetry: The Search for the Beginning of Time.* New York: Simon & Schuster, 1985.

Pai, Anna C. *Foundations for Genetics: A Science for Society.* 2d ed. New York: McGraw-Hill, 1984.

Pais, Abraham. *The Genius of Science: A Portrait Gallery of Twentieth-Century Physicists.* New York: Oxford University Press, 2000.

Palmer, Douglas. *Neanderthal.* London: Channel 4 Books, 2000.

Parker, Barry R. *The Vindication of the Big Bang: Breakthroughs and Barriers.* New York: Plenum Press, 1993.

Parslow, Christopher Charles. *Rediscovering Antiquity: Karl Weber and the Excavation of Herculaneum, Pompeii, and Stabiae.* New York: Cambridge University Press, 1998.

Parson, Ann B. *The Proteus Effect: Stem Cells and Their Promise.* Washington, D.C.: National Academies Press, 2004.

Pedrotti, L., and F. Pedrotti. *Optics and Vision.* Upper Saddle River, N.J.: Prentice Hall, 1998.

Peitgen, Heinz-Otto, and Dietmar Saupe, eds. *The Science of Fractal Images.* New York: Springer-Verlag, 1988.

Peltonen, Markku, ed. *The Cambridge Companion to Bacon.* New York: Cambridge University Press, 1996.

Penrose, Roger. *The Emperor's New Mind: Concerning Computers, Minds, and the Laws of Physics.* New York: Oxford University Press, 1989.

Persaud, T. V. N. *A History of Anatomy: The Post-Vesalian Era.* Springfield, Ill.: Charles C. Thomas, 1997.

Peterson, Carolyn C., and John C. Brant. *Hubble Vision: Astronomy with the Hubble Space Telescope.* London: Cambridge University Press, 1995.

_______. *Hubble Vision: Further Adventures with the Hubble Space Telescope.* 2d ed. New York: Cambridge University Press, 1998.

Pfeiffer, John E. *The Emergence of Humankind.* 4th ed. New York: Harper & Row, 1985.

Piggott, Stuart. *William Stukeley: An Eighteenth-Century Antiquary.* New York: Thames and Hudson, 1985.

Pike, J. Wesley, Francis H. Glorieux, David Feldman. *Vitamin D.* 2d ed. Academic Press, 2004.

Plionis, Manolis, ed. *Multiwavelength Cosmology.* New York: Springer, 2004.

Plotkin, Stanley A., and Edward A. Mortimer. *Vaccines.* 2d ed. Philadelphia: W. B. Saunders, 1994.

Polter, Paul. *Hippocrates.* Cambridge, Mass.: Harvard University Press, 1995.

Popper, Karl R. *The World of Parmenides: Essays on the Presocratic Enlightenment.* Edited by Arne F. Petersen and Jørgen Mejer. New York: Routledge, 1998.

Porter, Roy. *The Greatest Benefit to Mankind: A Medical History of Humanity, from Antiquity to the Present.* New York: W. W. Norton, 1997.

Porter, Roy, ed. *Eighteenth Century Science.* Vol. 4 in *The Cambridge History of Science.* New York: Cambridge University Press, 2003.

Poundstone, William. *Prisoner's Dilemma.* New York: Doubleday, 1992.

Poynter, Margaret, and Arthur L. Lane. *Voyager: The Story of a Space Mission.* New York: Macmillan, 1981.

Principe, Lawrence. *The Aspiring Adept: Robert Boyle and His Alchemical Quest.* Princeton, N.J.: Princeton University Press, 1998.

Prochnow, Dave. *Superconductivity: Experimenting in a New Technology.* Blue Ridge Summit, Pa.: TAB Books, 1989.

Pullman, Bernard. *The Atom in the History of Human Thought.* New York: Oxford University Press, 1998.

Pycior, Helena M. *Symbols, Impossible Numbers, and Geometric Entanglements: British Algebra Through the Commentaries on Newton's Universal Arithmetick.* New York: Cambridge University Press, 1997.

The Rand McNally New Concise Atlas of the Universe. New York: Rand McNally, 1989.

Rao, Mahendra S., ed. *Stem Cells and CNS Development.* Totowa, N.J.: Humana Press, 2001.

Raup, David M. *The Nemesis Affair: A Story of the Death of Dinosaurs and the Ways of Science.* New York: W. W. Norton, 1986.

Raven, Peter H., and George B. Johnson. *Biology.* 2d ed. St. Louis: Times-Mirror/Mosby, 1989.

Raven, Peter H., Ray F. Evert, and Susan E. Eichhorn. *Biology of Plants.* 6th ed. New York: W. H. Freeman, 1999.

Reader, John. *Missing Links: The Hunt for Earliest Man.* Boston: Little, Brown, 1981.

Reichhardt, Tony. *Proving the Space Transportation System: The Orbital Flight Test Program.* NASA NF-137-83. Washington, D.C.: Government Printing Office, 1983.

Remick, Pat, and Frank Cook. *21 Things Every Future Engineer Should Know: A Practical Guide for Students and Parents.* Chicago: Kaplan AEC Education, 2007.

Repcheck, Jack. *The Man Who Found Time: James Hutton and the Discovery of the Earth's Antiquity.* Reading, Mass.: Perseus Books, 2003.

Rescher, Nicholas. *On Leibniz.* Pittsburgh, Pa.: University of Pittsburgh Press, 2003.

Reston, James. *Galileo: A Life.* New York: HarperCollins, 1994.

Rhodes, Richard. *The Making of the Atomic Bomb.* New York: Simon & Schuster, 1986.

Rigutti, Mario. *A Hundred Billion Stars.* Translated by Mirella Giacconi. Cambridge, Mass.: MIT Press, 1984.

Ring, Merrill. *Beginning with the Presocratics.* 2d ed. New York: McGraw-Hill, 1999.

Riordan, Michael. *The Hunting of the Quark.* New York: Simon & Schuster, 1987.

Roan, Sharon. *Ozone Crisis: The Fifteen-Year Evolution of a Sudden Global Emergency.* New York: John Wiley & Sons, 1989.

Robinson, Daniel N. *An Intellectual History of Psychology.* 3d ed. Madison: University of Wisconsin Press, 1995.

Rogers, J. H. *The Giant Planet Jupiter.* New York: Cambridge University Press, 1995.

Rose, Frank. *West of Eden: The End of Innocence at Apple Computer.* New York: Viking, 1989.

Rosenthal-Schneider, Ilse. *Reality and Scientific Truth: Discussions with Einstein, von Laue, and Planck.* Detroit: Wayne State University Press, 1980.

Rossi, Paoli. *The Birth of Modern Science.* Translated by Cynthia De Nardi Ipsen. Oxford, England: Blackwell, 2001.

Rowan-Robinson, Michael. *Cosmology.* London: Oxford University Press, 2003.

Rowland, Wade. *Galileo's Mistake: A New Look at the Epic Confrontation Between Galileo and the Church.* New York: Arcade, 2003.

Rudin, Norah, and Keith Inman. *An Introduction to Forensic DNA Analysis.* Boca Raton, Fla.: CRC Press, 2002.

Rudwick, Martin, J. S. *The Great Devonian Controversy: The Shaping of Scientific Knowledge Among Gentlemanly Specialists.* Chicago: University of Chicago Press, 1985.

Rudwick, M. J. S. *The Meaning of Fossils: Episodes in the History of Paleontology.* Chicago: University of Chicago Press, 1985.

Ruestow, Edward Grant. *The Microscope in the Dutch Republic: The Shaping of Discovery.* New York: Cambridge University Press, 1996.

Ruse, Michael. *The Darwinian Revolution: Science Red in Tooth and Claw.* 2d ed. Chicago: University of Chicago Press, 1999.

Ruspoli, Mario. *Cave of Lascaux.* New York: Harry N. Abrams, 1987.

Sagan, Carl. *Cosmos.* New York: Random House, 1980.

Sagan, Carl, and Ann Druyan. *Comet.* New York: Random House, 1985.

Sandler, Stanley I. *Chemical and Engineering Thermodynamics.* New York: John Wiley & Sons, 1998.

Sang, James H. *Genetics and Development.* London: Longman, 1984.

Sargent, Frederick. *Hippocratic Heritage: A History of Ideas About Weather and Human Health.* New York: Pergamon Press, 1982.

Sargent, Rose-Mary. *The Diffident Naturalist: Robert Boyle and the Philosophy of Experiment.* Chicago: University of Chicago Press, 1995.

Sauer, Mark V. *Principles of Oocyte and Embryo Donation.* New York: Springer, 1998.

Schaaf, Fred. *Comet of the Century: From Halley to Hale-Bopp.* New York: Springer-Verlag, 1997.

Schatzkin, Paul. *The Boy Who Invented Television: A Story of Inspiration, Persistence, and Quiet Passion.* Silver Spring, Md.: TeamCon Books, 2002.

Schiffer, Michael Brian. *Draw the Lightning Down: Benjamin Franklin and Electrical Technology in the Age of the Enlightenment.* Berkeley: University of California Press, 2003.

Schlagel, Richard H. *From Myth to Modern Mind: A Study of the Origins and Growth of Scientific Thought.* New York: Peter Lang Publishing, 1996.

Schlegel, Eric M. *The Restless Universe: Understanding X-Ray Astronomy in the Age of Chandra and Newton.* London: Oxford University Press, 2002.

Schliemann, Heinrich. *Troy and Its Remains: A Narrative of Researches and Discoveries Made on the Site of Ilium and in the Trojan Plain.* London: J. Murray, 1875.

Schneider, Stephen H. *Global Warming: Are We Entering the Greenhouse Century?* San Francisco: Sierra Club Books, 1989.

Schofield, Robert E. *The Enlightened Joseph Priestley: A Study of His Life and Work from 1773 to 1804.* University Park: Pennsylvania State University Pres, 2004.

Scholz, Christopher, and Benoit B. Mandelbrot. *Fractals in Geophysics.* Boston: Kirkäuser, 1989.

Schonfelder, V. *The Universe in Gamma Rays.* New York: Springer-Verlag, 2001.

Schopf, J. William, ed. *The Earth's Earliest Biosphere.* Princeton, N.J.: Princeton University Press, 1983.

Schorn, Ronald A. *Planetary Astronomy: From Ancient Times to the Third Millennium.* College Station: Texas A&M University Press, 1999.

Schwinger, Julian. *Einstein's Legacy.* New York: W. H. Freeman, 1986.

Sears, M., and D. Merriman, eds. *Oceanography: The Past.* New York: Springer-Verlag, 1980.

Seavey, Nina Gilden, Jane S. Smith, and Paul Wagner. *A Paralyzing Fear: The Triumph over Polio in America.* New York: TV Books, 1998.

Segalowitz, Sid J. *Two Sides of the Brain: Brain Lateralization Explored.* Englewood Cliffs, N.J.: Prentice-Hall, 1983.

Segrè, Emilio. *From X-Rays to Quarks.* San Francisco: W. H. Freeman, 1980.

Sekido, Yataro, and Harry Elliot. *Early History of Cosmic Ray Studies: Personal Reminiscences with Old Photographs.* Boston: D. Reidel, 1985.

Sfendoni-Mentzou, Demetra, et al., eds. *Aristotle and Contemporary Science.* 2 vols. New York: P. Lang, 2000-2001.

Shank, Michael H. *The Scientific Enterprise in Antiquity and the Middle Ages.* Chicago: University of Chicago Press, 2000.

Sharratt, Michael. *Galileo: Decisive Innovator.* Cambridge, Mass.: Blackwell, 1994.

Shectman, Jonathan. *Groundbreaking Scientific Experiments, Investigations, and Discoveries of the Eighteenth Century.* Westport, Conn.: Greenwood Press, 2003.

Sheehan, John. *The Enchanted Ring: The Untold Story of Penicillin.* Cambridge, Mass.: MIT Press, 1982.

Shilts, Randy. *And the Band Played On: Politics, People, and the AIDS Epidemic.* New York: St. Martin's Press, 1987.

Silk, Joseph. *The Big Bang.* Rev. ed. New York: W. H. Freeman, 1989.

Silverstein, Arthur M. *A History of Immunology.* San Diego: Academic Press, 1989.

Simmons, John. *The Scientific Hundred: A Ranking of the Most Influential Scientists, Past and Present.* Secaucus, N.J.: Carol, 1996.

Simon, Randy, and Andrew Smith. *Superconductors: Conquering Technology's New Frontier.* New York: Plenum Press, 1988.

Simpson, A. D. C., ed. *Joseph Black, 1728-1799: A Commemorative Symposium.* Edinburgh: Royal Scottish Museum, 1982.

Singh, Simon. *Fermat's Enigma: The Epic Quest to Solve the World's Greatest Mathematical Problem.* New York: Anchor, 1998.

Slayton, Donald K., with Michael Cassutt. *Deke! U.S. Manned Space: From Mercury to the Shuttle.* New York: Forge, 1995.

Smith, A. Mark. *Ptolemy and the Foundations of Ancient Mathematical Optics.* Philadelphia: American Philosophical Society, 1999.

Smith, G. C. *The Boole-De Morgan Correspondence, 1842-1864.* New York: Oxford University Press, 1982.

Smith, Jane S. *Patenting the Sun: Polio and the Salk Vaccine.* New York: Anchor/Doubleday, 1991.

Smith, Robert W. *The Space Telescope: A Study of NASA, Science, Technology and Politics.* New York: Cambridge University Press, 1989.

Smyth, Albert Leslie. *John Dalton, 1766-1844.* Aldershot, England: Ashgate, 1998.

Snider, Alvin. *Origin and Authority in Seventeenth-Century England: Bacon, Milton, Butler.* Toronto: University of Toronto Press, 1994.

Sobel, Dava. *Galileo's Daughter: A Historical Memoir of Science, Faith, and Love.* New York: Penguin Books, 2000.

_______. *Longitude: The True Story of a Lone Genius Who Solved the Greatest Scientific Problem of His Time.* New York: Penguin Books, 1995.

Spangenburg, Ray, and Diane Kit Moser. *Modern Science: 1896-1945.* Rev ed. New York: Facts on File, 2004.

Spilker, Linda J., ed. *Passage to a Ringed World: The Cassini-Huygens Mission to Saturn and Titan.* Washington, D.C.: National Aeronautics and Space Administration, 1997.

Stanley, H. Eugue, and Nicole Ostrowsky, eds. *On Growth and Form: Fractal and Non-Fractal Patterns in Physics.* Dordrecht, the Netherlands: Martinus Nijhoff, 1986.

Starr, Cecie, and Ralph Taggart. *Biology.* 5th ed. Belmont, Calif.: Wadsworth, 1989.

Stefik, Mark J., and Vinton Cerf. *Internet Dreams: Archetypes, Myths, and Metaphors.* Cambridge, Mass.: MIT Press, 1997.

Stein, Sherman. *Archimedes: What Did He Do Besides Cry Eureka?* Washington, D.C.: Mathematical Association of America, 1999.

Steiner, Robert F., and Seymour Pomerantz. *The Chemistry of Living Systems.* New York: D. Van Nostrand, 1981.

Stewart, Ian, and David Tall. *Algebraic Number Theory and Fermat's Last Theorem.* 3d ed. Natick, Mass.: AK Peters, 2002.

Stigler, Stephen M. *The History of Statistics.* Cambridge, Mass.: Harvard University Press, 1986.

Stine, Gerald. *AIDS 2005 Update.* New York: Benjamin Cummings, 2005.

Strathern, Paul. *Mendeleyev's Dream: The Quest for the Elements.* New York: Berkeley Books, 2000.

Streissguth, Thomas. *John Glenn.* Minneapolis, Minn.: Lerner, 1999.

Strick, James. *Sparks of Life: Darwinism and the Victorian Debates over Spontaneous Generation.* Cambridge, Mass.: Harvard University Press, 2000.

Strogatz, Steven H. *Nonlinear Dynamics and Chaos: With Applications to Physics, Biology, Chemistry and Engineering.* Reading, Mass.: Perseus, 2001.

Struik, Dirk J. *The Land of Stevin and Huygens: A Sketch of Science and Technology in the Dutch Republic During the Golden Century.* Boston: Kluwer, 1981.

Stryer, Lubert. *Biochemistry.* 2d ed. San Francisco: W. H. Freeman, 1981.

Stukeley, William. *The Commentarys, Diary, & Common-Place Book & Selected Letters of William Stukeley.* London: Doppler Press, 1980.

Sturtevant, A. H. *A History of Genetics.* 1965. Reprint. Woodbury, N.Y.: Cold Spring Harbor Laboratory Press, 2001.

Sullivan, Woodruff T., ed. *Classics in Radio Astronomy.* Boston: D. Reidel, 1982.

_______. *The Early Years of Radio Astronomy. Reflections Fifty Years After Jansky's Discovery.* New York: Cambridge University Press, 1984.

Sulston, John, and Georgina Ferry. *The Common Thread: A Story of Science, Politics, Ethics, and the Human Genome.* Washington, D.C.: Joseph Henry Press, 2002.

Sutton, Christine. *The Particle Connection.* New York: Simon & Schuster, 1984.

Suzuki, David T., and Peter Knudtson. *Genethics.* Cambridge, Mass.: Harvard University Press, 1989.

Swanson, Carl P., Timothy Merz, and William J. Young. *Cytogenetics: The Chromosome in Division, Inheritance, and Evolution.* 2d ed. Englewood Cliffs, N.J.: Prentice-Hall, 1980.

Swetz, Frank, et al., eds. *Learn from the Masters.* Washington, D.C.: Mathematical Association of America, 1995.

Tanford, Charles. *Franklin Stilled the Waves.* Durham, N.C.: Duke University Press, 1989.

Tarbuck, Edward J., and Frederick K. Lutgens. *The Earth: An Introduction to Physical Geology.* Columbus, Ohio: Charles E. Merrill, 1984.

Tattersall, Ian. *The Last Neanderthal: The Rise, Success, and Mysterious Extinction of Our Closest Human Relatives.* New York: Macmillan, 1995.

Taub, Liba Chaia. *Ptolemy's Universe: The Natural Philosophical and Ethical Foundations of Ptolemy's Astronomy.* Chicago: Open Court, 1993.

Tauber, Alfred I. *Metchnikoff and the Origins of Immunology: From Metaphor to Theory.* New York: Oxford University Press, 1991.

Taubes, Gary. *Nobel Dreams: Power, Deceit and the Ultimate Experiment.* New York: Random House, 1986.

Taylor, Michael E. *Partial Differential Equations I: Basic Theory.* 2d ed. New York: Springer, 2011.

Taylor, Peter Lane. *Science at the Extreme: Scientists on the Cutting Edge of Discovery.* New York: McGraw-Hill, 2001.

Thomas, John M. *Michael Faraday and the Royal Institution: The Genius of Man and Place.* New York: A. Hilger, 1991.

Thompson, A. R., James M. Moran, and George W. Swenson, Jr. *Interferometry and Synthesis in Radio Astronomy.* New York: John Wiley & Sons, 1986.

Thompson, D'Arcy Wentworth. *On Growth and Form.* Mineola, N.Y.: Dover, 1992.

Thoren, Victor E., with John R. Christianson. *The Lord of Uraniborg: A Biography of Tycho Brahe.* New York: Cambridge University Press, 1990.

Thrower, Norman J. W., ed. *Standing on the Shoulders of Giants: A Longer View of Newton and Halley.* Berkeley: University of California Press, 1990.

Thurman, Harold V. *Introductory Oceanography.* 4th ed. Westerville, Ohio: Charles E. Merrill, 1985.

Tietjen, Jill S., et al. *Keys to Engineering Success.* Upper Saddle River, N.J.: Prentice-Hall, 2001.

Tillery, Bill W., Eldon D. Enger, and Frederick C. Ross. *Integrated Science.* New York: McGraw-Hill, 2001.

Tiner, John Hudson. *Louis Pasteur: Founder of Modern Medicine.* Milford, Mich.: Mott Media, 1990.

Todhunter, Isaac. *A History of the Mathematical Theory of Probability: From the Time of Pascal to that of Laplace.* Sterling, Va.: Thoemmes Press, 2001.

Tombaugh, Clyde W., and Patrick Moore. *Out of Darkness: The Planet Pluto.* Harrisburg, Pa.: Stackpole Books, 1980.

Toulmin, Stephen, and June Goodfield. *The Fabric of the Heavens: The Development of Astronomy and Dynamics.* Chicago: University of Chicago Press, 1999.

Townes, Charles H. *How the Laser Happened: Adventures of a Scientist.* New York: Oxford University Press, 1999.

Traill, David A. *Schliemann of Troy: Treasure and Deceit.* London: J. Murray, 1995.

Trefil, James S. *The Dark Side of the Universe. Searching for the Outer Limits of the Cosmos.* New York: Charles Scribner's Sons, 1988.

_______. *From Atoms to Quarks: An Introduction to the Strange World of Particle Physics.* New York: Charles Scribner's Sons, 1980.

_______. *Space, Time, Infinity: The Smithsonian Views the Universe.* New York: Pantheon Books, 1985.

_______. *The Unexpected Vista.* New York: Charles Scribner's Sons, 1983.

Trefil, James, and Robert M. Hazen. *The Sciences: An Integrated Approach.* New York: John Wiley & Sons, 2003.

Trefil, James, ed. *The Encyclopedia of Science and Technology.* New York: Routledge, 2001.

Trento, Joseph J. *Prescription for Disaster: From the Glory of Apollo to the Betrayal of the Shuttle.* New York: Crown, 1987.

Trinkhaus, Eric, ed. *The Emergence of Modern Humans: Biocultural Adaptations in the Later Pleistocene.* Cambridge, England: Cambridge University Press, 1989.

Trounson, Alan O., and David K. Gardner, eds. *Handbook of In Vitro Fertilization.* 2d ed. Boca Raton, Fla.: CRC Press, 1999.

Tucker, Tom. *Bolt of Fire: Benjamin Franklin and His Electrical Kite Hoax.* New York: Public Affairs Press, 2003.

Tucker, Wallace H., and Karen Tucker. *Revealing the Universe: The Making of the Chandra X-Ray Observatory.* Cambridge, Mass.: Harvard University Press, 2001.

Tunbridge, Paul. *Lord Kelvin: His Influence on Electrical Measurements and Units.* London, U.K.: P. Peregrinus, 1992.

Tuplin, C. J., and T. E. Rihll, eds. *Science and Mathematics in Ancient Greek Culture.* New York: Oxford University Press, 2002.

Turnill, Reginald. *The Moonlandings: An Eyewitness Account.* New York: Cambridge University Press, 2003.

United States Office of the Assistant Secretary for Nuclear Energy. *The First Reactor.* Springfield, Va.: National Technical Information Service, 1982.

University of Chicago Press. *Science and Technology Encyclopedia.* Chicago: Author, 2000.

Van Allen, James A. *Origins of Magnetospheric Physics.* Expanded ed. 1983. Reprint. Washington, D.C.: Smithsonian Institution Press, 2004.

Van Dulken, Stephen. *Inventing the Nineteenth Century: One Hundred Inventions That Shaped the Victorian Age.* New York: New York University Press, 2001.

Van Heijenoort, Jean. *From Frege to Gödel: A Source Book in Mathematical Logic, 1879-1931.* Cambridge, Mass.: Harvard University Press, 2002.

Verschuur, Gerrit L. *Hidden Attraction: The History and Mystery of Magnetism.* New York: Oxford University Press, 1993.

_______. *The Invisible Universe Revealed: The Story of Radio Astronomy.* New York: Springer-Verlag, 1987.

Villard, Ray, and Lynette R. Cook. *Infinite Worlds: An Illustrated Voyage to Planets Beyond Our Sun.* Foreword by Geoffrey W. Marcy and afterword by Frank Drake. Berkeley: University of California Press, 2005.

Viney, Wayne. *A History of Psychology: Ideas and Context.* Boston: Allyn & Bacon, 1993.

Vogt, Gregory L. *John Glenn's Return to Space.* Brookfield, Conn.: Millbrook Press, 2000.

Von Bencke, Matthew J. *The Politics of Space: A History of U.S.-Soviet/Russian Competition and Cooperation in Space.* Boulder, Colo.: Westview Press, 1996.

Wagener, Leon. *One Giant Leap: Neil Armstrong's Stellar American Journey.* New York: Forge Books, 2004.

Wakefield, Robin, ed. *The First Philosophers: The Presocratics and the Sophists.* New York: Oxford University Press, 2000.

Waldman, G. *Introduction to Light.* Englewood Cliffs, N.J.: Prentice Hall, 1983.

Walker, James S. *Physics.* 2d ed. Upper Saddle River, N.J.: Pearson Prentice Hall, 2004.

Wallace, Robert A., Jack L. King, and Gerald P. Sanders. *Biosphere: The Realm of Life.* 2d ed. Glenview, Ill.: Scott, Foresman, 1988.

Waller, John. *Einstein's Luck: The Truth Behind Some of the Greatest Scientific Discoveries.* New York: Oxford University Press, 2002.

_______. *Fabulous Science: Fact and Fiction in the History of Science Discovery.* Oxford, England: Oxford University Press, 2004.

Walt, Martin. *Introduction to Geomagnetically Trapped Radiation.* New York: Cambridge University Press, 1994.

Wambaugh, Joseph. *The Blooding.* New York: Bantam Books, 1989.

Wang, Hao. *Reflections on Kurt Gödel.* Cambridge, Mass.: MIT Press, 1985.

Watson, James D. *The Double Helix: A Personal Account of the Discovery of the Structure of DNA.* Reprint. New York: W. W. Horton, 1980.

Watson, James D., and John Tooze. *The DNA Story.* San Francisco: W. H. Freeman, 1981.

Watson, James D., et al. *Molecular Biology of the Gene.* 4th ed. Menlo Park, Calif.: Benjamin/Cummings, 1987.

Weber, Robert L. *Pioneers of Science: Nobel Prize Winners in Physics.* 2d ed. Philadelphia: A. Hilger, 1988.

Weedman, Daniel W. *Quasar Astrophysics.* Cambridge, England: Cambridge University Press, 1986.

Wells, Spencer. *The Journey of Man: A Genetic Odyssey.* Princeton, N.J.: Princeton University Press, 2002.

Westfall, Richard S. *Never at Rest: A Biography of Isaac Newton.* New York: Cambridge University Press, 1980.

Wheeler, J. Craig. *Cosmic Catastrophe: Supernovae and Gamma-Ray Bursts.* London: Cambridge University Press, 2000.

Whiting, Jim, and Marylou Morano Kjelle. *John Dalton and the Atomic Theory.* Hockessin, Del.: Mitchell Lane, 2004.

Whitney, Charles. *Francis Bacon and Modernity.* New Haven, Conn.: Yale University Press, 1986.

Whyte, A. J. *The Planet Pluto.* New York: Pergamon Press, 1980.

Wilford, John Noble. *The Mapmakers.* New York: Alfred A. Knopf, 1981.

_______. *The Riddle of the Dinosaur.* New York: Alfred A. Knopf, 1986.

Wilkie, Tom, and Mark Rosselli. *Visions of Heaven: The Mysteries of the Universe Revealed by the Hubble Space Telescope.* London: Hodder & Stoughton, 1999.

Will, Clifford M. *Was Einstein Right?* New York: Basic Books, 1986.

Williams, F. Mary, and Carolyn J. Emerson. *Becoming Leaders: A Practical Handbook for Women in Engineering, Science, and Technology.* Reston, Va.: American Society of Civil Engineers, 2008.

Williams, Garnett P. *Chaos Theory Tamed.* Washington, D.C.: National Academies Press, 1997.

Williams, James Thaxter. *The History of Weather.* Commack, N.Y.: Nova Science, 1999.

Williams, Trevor I. *Howard Florey: Penicillin and After.* London: Oxford University Press, 1984.

Wilmut, Ian, Keith Campbell, and Colin Tudge. *The Second Creation: The Age of Biological Control by the Scientists That Cloned Dolly.* London: Headline, 2000.

Wilson, Andrew. *Space Shuttle Story.* New York: Crescent Books, 1986.

Wilson, Colin. *Starseekers.* Garden City, N.Y.: Doubleday, 1980.

Wilson, David B. *Kelvin and Stokes: A Comparative Study in Victorian Physics.* Bristol, England: Adam Hilger, 1987.

Wilson, Jean D. *Wilson's Textbook of Endocrinology.* 10th ed. New York: Elsevier, 2003.

Windley, Brian F. *The Evolving Continents.* 2d ed. New York: John Wiley & Sons.

Wojcik, Jan W. *Robert Boyle and the Limits of Reason.* New York: Cambridge University Press, 1997.

Wolf, Fred Alan. *Taking the Quantum Leap.* San Francisco: Harper & Row, 1981.

Wollinsky, Art. *The History of the Internet and the World Wide Web.* Berkeley Heights, N.J.: Enslow, 1999.

Wolpoff, M. *Paleoanthropology.* 2d ed. Boston: McGraw-Hill, 1999.

Wood, Michael. *In Search of the Trojan War.* Berkeley: University of California Press, 1988.

Wormald, B. H. G. *Francis Bacon: History, Politics, and Science, 1561-1626.* New York: Cambridge University Press, 1993.

Yen, W. M., Marc D. Levenson, and Arthur L. Schawlow. *Lasers, Spectroscopy, and New Ideas: A Tribute to Arthur L. Schawlow.* New York: Springer-Verlag, 1987.

Yoder, Joella G. *Unrolling Time: Huygens and the Mathematization of Nature.* New York: Cambridge University Press, 2004.

Yolton, John W. ed. *Philosophy, Religion, and Science in the Seventeenth and Eighteenth Centuries.* Rochester, N.Y.: University of Rochester Press, 1990.

Zeilik, Michael. *Astronomy: The Evolving Universe.* 4th ed. New York: Harper & Row, 1985.

Index

SUBJECT INDEX

Note: Page numbers in **bold** indicate main discussion

B

N

Q

R